PRODUCT AND PROCESS DESIGN PRINCIPLES

International Student Version

PRODUCT AND PROCESS DESIGN PRINCIPLES

Synthesis, Analysis, and Evaluation

Third Edition

International Student Version

Warren D. Seider

Department of Chemical and Biomolecular Engineering
University of Pennsylvania

J.D. Seader

Department of Chemical Engineering
University of Utah

Daniel R. Lewin

Department of Chemical Engineering
Technion—Israel Institute of Technology

Soemantri Widagdo

3M Company
Display and Graphics Business Laboratory

WILEY

Product and Process Design Principles
Synthesis, Analysis, and Evaluation
Third Edition, ISV

Authorized reprint by Wiley India Pvt. Ltd., 4435-36/7, Ansari Road, Daryaganj, New Delhi - 110002.

Cover Image: naqiewei/Getty Images

Reprint: 2015

Printed at: Chaudhary Press, Delhi.

ISBN: 978-81-265-5748-6

Dedication

To the memory of my parents, to Diane, and to Benjamin, Deborah, Gabriel, Joe, Jesse, and Idana.

To the memory of my parents, to Sylvia, and to my children.

To my parents, Harry and Rebeca Lewin, to Ruti, and to Noa and Yonatan.

To the memory of my father, Thodorus Oetojo Widagdo, to my mother, and to Richard.

To the memory of Richard R. Hughes, a pioneer in computer-aided simulation and optimization, with whom two of the authors developed many concepts for carrying out and teaching process design.

About the Authors

Warren D. Seider is Professor of Chemical and Biomolecular Engineering at the University of Pennsylvania. He received a B.S. degree from the Polytechnic Institute of Brooklyn and M.S. and Ph.D. degrees from the University of Michigan. Seider has contributed to the fields of process analysis, simulation, design, and control. He co-authored *FLOWTRAN Simulation—An Introduction* in 1974 and has coordinated the design course at Penn for nearly 30 years, involving projects provided by many practicing engineers in the Philadelphia area. He has authored or co-authored over 100 journal articles and authored or edited seven books. Seider was the recipient of the AIChE Computing in Chemical Engineering Award in 1992 and co-recipient of the AIChE Warren K. Lewis Award in 2004 with J. D. Seader. He served as a Director of AIChE from 1984 to 1986 and has served as chairman of the CAST Division and the Publication Committee. He helped to organize the CACHE (Computer Aids for Chemical Engineering Education) Committee in 1969 and served as its chairman. Seider is a member of the Editorial Advisory Board of *Computers and Chemical Engineering*.

J. D. Seader is Professor Emeritus of Chemical Engineering at the University of Utah. He received B.S. and M.S. degrees from the University of California at Berkeley and a Ph.D. from the University of Wisconsin. From 1952 to 1959, he designed chemical and petroleum processes for Chevron Research, directed the development of one of the first computer-aided process design programs, and co-developed the first widely used computerized vapor–liquid equilibrium correlation. From 1959 to 1965, he conducted rocket engine research for Rocketdyne on all of the engines that took man to the moon. Before joining the faculty at the University of Utah in 1966, Seader was a professor at the University of Idaho. He is the author or co-author of 111 technical articles, eight books, and four patents. Seader is co-author of the section on distillation in the sixth and seventh editions of *Perry's Chemical Engineer's' Handbook*. He is co-author of *Separation Process Principles*, published in 1998, with a second edition in 2006. Seader was Associate Editor of *Industrial and Engineering Chemistry Research* for 12 years, starting in 1987. He was a founding member and trustee of CACHE for 33 years, serving as Executive Officer from 1980 to 1984. For 20 years, he directed the use by and distribution of Monsanto's FLOWTRAN process simulation computer program to 190 chemical engineering departments worldwide. Seader served as Chairman of the Chemical Engineering Department at the University of Utah from 1975 to 1978, and as a Director of AIChE from 1983 to 1985. In 1983, he presented the 35th Annual Institute Lecture of AIChE. In 1988, he received the Computing in Chemical Engineering Award of the CAST Division of AIChE. In 2004, he received the CACHE Award for Excellence in Computing in Chemical Engineering Education from the ASEE. In 2004, he was co-recipient, with Professor Warren D. Seider, of the AIChE Warren K. Lewis Award for Chemical Engineering Education.

Daniel R. Lewin is Professor of Chemical Engineering, the Churchill Family Chair, and the Director of the Process Systems Engineering (PSE) research group at the Technion, the Israel Institute of Technology. He received his B.Sc. from the University of Edinburgh and his D.Sc. from the Technion. His research focuses on the interaction of process design and process control and operations, with emphasis on model-based methods. He has authored or co-authored over 100 technical publications in the area of process systems engineering, as well as the first and second editions of this textbook and the multimedia CD-ROM that accompanies them. Professor Lewin has been awarded a number of prizes for research excellence, and twice received the Jacknow Award and the Alfred and Yehuda Weissman

Award in recognition of teaching excellence at the Technion. He served as Associate Editor of the *Journal of Process Control* and is a member of the International Federation of Automatic Control (IFAC) Committee on Process Control.

Soemantri Widagdo is Manager of Multifunctional Surfaces and Adhesives, in the Display and Graphics Business at 3M. He received his B.S. degree in chemical engineering from Bandung Institute of Technology, Indonesia, and his M.Ch.E. and Ph.D. degrees from Stevens Institute of Technology. Early in his career, he developed the first electric generator in Indonesia that used biomass gasification technology. After the completion of his graduate studies, he began his career in the United States with the Polymer Processing Institute (PPI), Hoboken, New Jersey. As the head of its computation group, he led the development of an analysis software package for twin-screw compounding. During his tenure at PPI, he was also Research Professor of Chemical Engineering at Stevens Institute of Technology. He joined 3M in 1998 and has served as the technology leader for polymer compounding, as a Six-Sigma Black Belt, and in a number of technology management positions. He has been involved in a variety of technology and product-development programs involving renewable energy, industrial and transportation applications, consumer office products, electrical and electronics applications, health care and dentistry, and display and graphics applications. He has authored and co-authored over 20 technical publications.

Preface

OBJECTIVES

A principal objective of this textbook and accompanying Web site, referred to here as courseware, is to describe modern strategies for the design of chemical products and processes, with an emphasis on a systematic approach. Since the early 1960s, undergraduate education has focused mainly on the engineering sciences. In recent years, however, more scientific approaches to product and process design have been developed, and the need to teach students these approaches has become widely recognized. Consequently, this courseware has been developed to help students and practitioners better utilize the modern approaches to product and process design. Like workers in thermodynamics; momentum, heat, and mass transfer; and chemical reaction kinetics, product and process designers apply the principles of mathematics, chemistry, physics, and biology. Designers, however, utilize these principles, and those established by engineering scientists, to create chemical products and processes that satisfy societal needs while returning a profit. In so doing, designers emphasize the methods of synthesis and optimization in the face of uncertainties—often utilizing the results of analysis and experimentation prepared in cooperation with engineering scientists—while working closely with their business colleagues.

In this courseware, the latest design strategies are described, most of which have been improved significantly with the advent of computers, numerical mathematical programming methods, and artificial intelligence. Since most curricula place little emphasis on design strategies prior to design courses, this courseware is intended to provide a smooth transition for students and engineers who are called upon to design innovative new products and processes.

The first edition of this textbook focused on the design of commodity chemical processes. While this material was updated and augmented to include new developments, the second edition broadened this focus to include the design of chemical products, with emphasis on specialty chemicals involving batch, rather than continuous, processing. It also introduced design techniques for industrial and configured consumer products. This third edition expands upon the strategies for product design beginning with the need for a project charter, followed by the creation of an innovation map in which potential new technologies are linked to consumer needs. Then, it focuses on the Stage-Gate™ Product-Development Process (SGPDP) for the design of basic, industrial, and configured consumer chemical products. Eight new case studies have been added to illustrate these product design strategies.

This courseware is intended for seniors and graduate students, most of whom have solved a few open-ended problems but have not received instruction in a systematic approach to product and process design. To guide this instruction, the subject matter is presented in five parts. The introductions to Parts One, Two, and Three show how these parts relate to the entire design process and to each other. Part One focuses on the design of basic chemical products, Part Two on industrial chemical products, and Part Three on configured consumer chemical products. All of the materials are presented at the senior level.

After Chapter 1 introduces chemical product design, Chapter 2 covers the product-development process. In so doing, the latter introduces many steps in product design that are business oriented, for example, creating a pipeline for new product development, carrying out a market assessment, determining customer needs, and carrying out an opportunity assessment. Chapter 2 is, in effect, the transition chapter between Chapter 1 and Parts One, Two, and Three, in which the technical methods of product and process design are covered, concentrating on each of the three kinds of chemical products (basic, industrial, and configured consumer). Then, within each of the three parts, in Chapters 13, 15, and 17, the new case studies are presented for eight chemical products.

More specifically, in Part One, which deals with basic chemicals, consumer needs for chemical products are usually satisfied by meeting well-defined physical and thermophysical properties. Usually, a search for the appropriate molecules or mixtures of molecules is followed by process design. The *concept* stage of the SGPDP then focuses on process synthesis, for which the process design procedures were established in our second edition. Hence, Part One of our third edition contains all of the process synthesis coverage in the second edition, updated to include additional subjects and/or improved discussions, when appropriate.

Parts Two and Three of this third edition are new. These parts begin by discussing the new technologies upon which industrial and configured consumer chemical products are based. Then, they present case studies involving the design of specific chemical products. While various process/manufacturing technologies are presented, they are in connection with the specific chemical products. Unlike for basic chemicals, whose physical and thermophysical properties are usually well defined, the unit operations for industrial and configured consumer chemical products usually depend on the technology platforms upon which the new products are based; for example, extrusion, forming, and packaging devices for thin polymer films, and mixing and homogenization devices to generate stable emulsions in pastes and creams. Consequently, no attempt is made in our third edition to discuss general process synthesis techniques for industrial and configured consumer chemical products. Rather, the focus is on case studies involving specific technologies. Examples and homework exercises are provided that enable students to master the approaches to product design—permitting them to apply these approaches to the design of new products that involve other technologies.

Stated differently, for *process* design, the coverage is similar to that in our second edition. The emphasis throughout Part One, especially, is on process invention and detailed process synthesis; that is, process creation and the development of a base-case design(s). For the former, methods of generating the tree of alternative process flowsheets are covered. Then, for the most promising flowsheets, a base-case design(s) is developed, including a detailed process flow diagram, with material and energy balances. The base-case design(s) then enters the detailed design stage, in which the equipment is sized, cost estimates are obtained, a profitability analysis is completed, and optimization is carried out, as discussed in Part Four of this third edition.

LIMITED TIME—PROCESS OR PRODUCT DESIGN?

When limited time is available, some faculty and students may prefer to focus on *process* design rather than *product* design. This can be accomplished, using the materials that have been updated from our second edition, by skipping Chapter 2 and studying Parts One, Four, and Five. In Part One, Chapters 4–12 emphasize process synthesis, simulation, and optimization. Then, in Part Four, Chapters 18–24 cover strategies for detailed design, equipment sizing, and optimization. Finally, Chapter 26 in Part Five covers design reports, both written and oral.

Courses that focus on *product* design rather than *process* design could begin with Chapters 1 (Sections 1.0–1.3) and 2. For basic chemical products, emphasis could be placed on Chapter 3, *Materials Technology: Molecular-Structure Design for Basic Chemicals*; Chapter 11, *Batch Process Design*; and Chapter 13, *Product Design Case Studies for Basic Chemicals*. Then, emphasis might shift to the innovation maps and case studies for the industrial and configured consumer chemical products in Chapters 14–17, as well as Chapter 25, *Design for Six-Sigma*, and Chapter 26, *Written and Oral Reports*. Further recommendations for *product* design courses are provided under Feature 2 below.

ONE OR TWO DESIGN COURSES?

In a recent survey conducted by John Wiley, with responses from 50 departments of chemical engineering in the United States, half of the departments teach one design course while the other half teach two design courses. With two courses available, it is possible to build a lecture course that emphasizes both product and process design, covering selected subjects from Chapters 1 and 2 and Parts One through Five, depending on the subjects covered in prior courses. Students would solve homework exercises and take midsemester and final exams but would not work on a comprehensive design project, the latter being reserved for a design project course in the second semester.

Alternatively, one of the two courses might focus on process design with the other focusing on product design. For such a sequence, this textbook provides instruction in most of the topics covered in both courses.

For departments with just one design course, a comprehensive process design project would be included. For such a course, instructors must be more careful in their selection of lecture materials, which should be presented in time for their use in solving the design project. Note that single design courses are often offered by departments that cover design-related topics in other courses. For example, many departments teach economic analysis before students take a design course. Other departments teach the details of equipment design in courses on transport phenomena and unit operations. This textbook and its Web site are well suited for these courses because they provide much reference material that can be covered as needed.

PROCESS SIMULATORS

Throughout this courseware, various methods are utilized to perform extensive process design calculations and provide graphical results that are visualized easily, including the use of computer programs for simulation and design optimization. The use of these programs is an important attribute of this courseware. We believe that our approach is an improvement over an alternative approach that introduces the strategies of process synthesis *without computer methods*, emphasizing heuristics and back-of-the-envelope calculations. We favor a blend of heuristics and analysis using the computer. Since the 1970s, many faculty have begun to augment the heuristic approach with an introduction to the analysis of prospective flowsheets using simulators such as ASPEN PLUS, ASPEN HYSYS, UNISIM, PRO-II, CHEMCAD, FLOWTRAN, BATCH PLUS, and SUPERPRO DESIGNER. Today, most schools use one of these simulators, but often without adequate teaching materials. Consequently, the challenge for us, in the preparation of this courseware, has been to find the proper blend of modern computational approaches and simple heuristics.

PLANTWIDE CONTROL

As processes become more integrated to achieve more economical operation, their responses to disturbances and setpoint changes become more closely related to the design integration; consequently, the need to assess their controllability gains importance. Chapter 12, *Plantwide Control*, teaches students a simple strategy for qualitatively configuring plantwide control systems in the *concept* stage of process design. It is recommended that this strategy be used during the *concept* stage to screen potential plants for ease of control, noting that the reliability of the screening is significantly enhanced by employing the quantitative methods provided in the file, Supplement_to_Chapter_12.pdf, in the PDF Files folder, which can be downloaded from the Wiley Web site associated with this book.

FORMAT OF COURSEWARE

This courseware takes the form of a conventional textbook accompanied by computer programs to be utilized by the reader in various aspects of his or her design studies. As the design strategies have been elucidated during the development of this courseware, fewer specifics have been provided in the chapters concerning the software packages involved. Instead, multimedia modules have been developed to give many examples of the simulator input and output, with frame-by-frame instructions, to discuss the nature of the models provided for the processing units, with several example calculations presented as well. These modules, that can be obtained by writing to acadmktg@wiley.com use voice, video, and animation to introduce new users of steady-state simulators to the specifics of two of the most widely used process simulation programs, ASPEN PLUS and HYSYS (either ASPEN HYSYS or UNISIM). These include several tutorials that provide instruction on the solution of problems for courses in material and energy balances, thermodynamics, heat transfer, separations, and reactor design. In many cases, students will have been introduced to process simulators in these courses. Also, video segments show portions of a petrochemical complex in operation, including distillation towers, heat exchangers, pumps and compressors, and chemical reactors. The Web site also includes files, in the Program and Simulation Files folder, that contain the solutions for more than 60 examples using either ASPEN PLUS or HYSYS, as well as problems solved using GAMS, an optimization package, and the MATLAB scripts in Chapter 12. The files are feferred to in each example and can easily be used to vary parameters and explore alternative solutions.

As indicated in the Table of Contents for the textbook, supplemental sections of several chapters are provided in PDF files on the Web site, in the PDF Files folder, with only a brief summary of the material presented in the textbook. Furthermore, Appendix II provides a list of design projects whose detailed statements are provided in the file, Supplement_to_Appendix_II.pdf, in the PDF File folder on the Web site. These involve the design of chemical prodjcts and processes in several industries. Many are derived from the petrochemical industry, with much emphasis on environmental and safety considerations, including the reduction of sources of pollutants and hazardous wastes and purification before streams are released into the environment. Several orginate in the biochemical industry, including fermentations to produce pharmaceuticals, foods, and chemicals. Others are involved in the manufacture of polymers and electronic materials. Each design problem has been solved by groups of two, three, or four students at the University of Pennsylvania, with copies of their design reports available through Interlibrary Loan from the Engineering Library at the university.

INSTRUCTOR RESOURCES

- Solutions Manual
- PDF Files
- Lecture Slides
- Recitation Slides
- Program and Simulation Files
- Module Instruction Sequence

Please write to acadmktg@wiley.com to request

ADVICE TO STUDENTS AND INSTRUCTORS

In using this textbook and additional resources, students and instructors are advised to take advantage of the following five features:

Feature 1: Key Steps in Product and Process Design

The textbook is organized around the key steps in product and process design shown in Figures PI.1 (p. 56), PII.1 (p. 372), and PIII.1 (p. 408). These steps reflect current practice and provide a sound sequence of instruction, yet with much flexibility in permitting the student and instructor to place emphasis on preferred subjects. Instructors may wish to refer to these figures often while teaching process and/or product design.

Students can study Chapters 4, 5, and 6 in sequence. Although these chapters provide many examples and exercises, the multimedia modules which can be downloaded from the Web site can be referred to for details of the process simulators. Chapters in the remainder of Part One and in Part Four can be studied as needed. There are many cross-references throughout the text—especially to reference materials needed when carrying out designs. For example, students can begin to learn heuristics for heat integration in Chapters 4 and 6, learn algorithmic methods in Chapter 9, learn the strategies for designing heat exchangers and estimating their costs in Part Four (Chapters 18 and 22), and learn the importance of examining the controllability of heat exchanger networks in Chapter 12.

Instructors can begin with Chapters 4, 5, and 6 and design their courses to cover the other chapters as desired. Because each group of students has a somewhat different background depending on the subjects covered in prior courses, the textbook is organized to give instructors much flexibility in their choice of subject matter and the sequence in which it is covered. Furthermore, design instructors often have difficulty deciding on a subset of the many subjects to be covered. This book provides sufficiently broad coverage to permit the instructor to emphasize certain subjects in lectures and homework assignments, leaving others as reference materials students can use when carrying out their design projects. In a typical situation of teaching students to (1) generate design alternatives, (2) select a base-case design, and (3) carry out its analysis, the textbook enables the instructor to emphasize one or more of the following subjects: synthesis of chemical reactor networks (Chapter 7), separation train synthesis (Chapter 8), heat exchanger network heat (Chapter 9), process unit design (e.g., heat exchangers—Chapter 18), and plantwide control (Chapter 12).

Feature 2: Numerous Product Design Examples

This textbook introduces the key steps in *product* design with numerous examples. These steps have been developed with the assistance and recommendations of successful practitioners of product design in industry.

Students can begin in Sections 1.1, 1.2, and 1.3 to learn, when developing new products, about: (1) the infrastructure of an operating business unit in a large manufacturing operation; (2) product- and technology-development frameworks; (3) the distinctions between basic, industrial, and configured consumer chemical products; and (4) innovation maps that show the links between new technologies and customer needs. Then, in Chapter 2, they can learn the steps in the product-development process, including creating a project charter, carrying out a market assessment, determining customer needs, and carrying out an opportunity assessment, among many others. In Part One, on Product Design for Basic Chemicals, in Chapter 3, they can learn to find chemicals and chemical mixtures having desired properties and performance; that is, to carry out molecular-structure design. Chapter 4 shows how to synthesize a batch process for the manufacture of tissue plasminogen activator (tPA)—a protein that helps dissolve clots to reduce the chances of a stroke or heart attack—and Chapter 5 introduces the methods of batch process simulation as applied to the tPA process. Then, students can turn to Chapter 12 to learn how to optimize the design and scheduling of batch processes. Both Parts Two and Three concentrate on the design of more complex chemical products—industrial chemicals and configured consumer chemical products. Chapters 14 and 16 show how to create innovation maps that link new technologies to customer needs for five different products. The use of these innovation maps alerts students to the importance of patents in the development of new products. Chapters 15 and 17 present case studies of product designs.

Instructors can create a course in *product* design using the materials and exercises referred to in the preceding paragraph. The product designs in Chapters 13, 15, and 17 can be expanded upon and/or used as the basis of design projects for student design teams. In our experience, students can frequently formulate their own product design projects based on their own experience and awareness of consumer needs.

Feature 3: Process Synthesis—Heuristic to Algorithmic Methods

Process synthesis is introduced using mostly heuristics in Chapters 4 and 6, whereas Chapters 7–11 provide more detailed algorithmic methods for chemical reactor network synthesis, separation train synthesis, heat and power integration, mass integration, and the optimal design and sequencing of batch processes.

This feature enables the student to begin carrying out process designs using easy-to-understand rules of thumb when studying Chapters 4–6. As these ideas are mastered, the student can learn algorithmic approaches that enable him or her to produce better designs. For example, Chapter 4 introduces two alternative sequences for the separation of a three-component mixture (in the vinyl-chloride process), whereas Chapter 8 shows how to generate and evaluate many alternative sequences for the separation of multicomponent mixtures, both ideal and nonideai.

This organization provides the instructor the flexibility to emphasize those subjects most useful to his or her students. Chapters 4–6 can be covered fairly quickly, giving the students enough background to begin work on a process design-oriented project. This can be important at schools where only one semester is allotted for the design course. Then, while the students are working on their design projects, the instructor can cover more systematic, algorithmic methods, as well as optimization methods, which students can apply to improve their designs. In a typical situation when covering Chapters 4–6, the instructor would not cover nonideal separations such as azeotropic, extractive, or reactive distillations. Consequently, most students would begin to create simple designs involving reactors followed by separation trains. After the instructor covers the subject matter in Chapter 8, the students would begin to take advantage of more advanced separation methods.

Feature 4: Process Simulators

Process simulators (steady state, dynamic, and batch) are referred to throughout the textbook (e.g., ASPEN PLUS, ASPEN HYSYS, UNISIM, CHEMCAD, PRO/II, ASPEN DYNAMICS, BATCH PLUS, and SUPERPRO DESIGNER). These simulators permit access to large physical property, equipment, and cost databases, and the examination of aspects of numerous chemical processes. Emphasis is placed on the usage of simulators to obtain data and perform routine calculations.

Through the use of the process simulators, which are widely used in industry, students learn how easy it is to obtain data and perform routine calculations. They learn effective approaches to building up knowledge about a process through simulation. The multimedia modules, which can be downloaded from the Web site, provide students with the details of the methods used for property estimation and equipment modeling. They learn to use simulators intelligently and to check their results. For example, in Chapter 4, examples show how to use simulators to assemble a preliminary database and perform routine calculations when computing heat loads, heats of reaction, and vapor/liquid equilibria. Then, in Chapter 5, two examples show how to use the simulators to assist in the synthesis of toluene hydrodealkylation and monochlorobenzene separation processes. Most of the remaining chapters include examples of the use of simulators to obtain additional information, including equipment sizes, costs, profitability analyses, and the performance of control systems.

Because the book and Web site contain so many routine self-study examples of how the simulators are useful in building up a process design, the instructor has time to emphasize other aspects of process design. Through the examples and multimedia instruction on the Web site, with emphasis on ASPEN PLUS and HYSYS (ASPEN HYSYS and UNISIM) students obtain the details they need to use the simulators effectively, saving the instructor class time, as well as time in answering detailed questions as the students prepare their designs. Consequently, students obtain a better understanding of the design process and are exposed to a broader array of concepts in process design. In a typical situation when creating a base-case design, students use the examples in the text and the encyclopedic modules and tutorials on the Web site to learn how to obtain physical property estimates, heats of reaction, flame temperatures, and phase distributions. Then, students learn how to create a reactor section, using the simulators to perform routine material and energy balances. Next, they create a separation section and may eventually add recycle streams. Thanks to the coverage of the process simulators in Chapters 4–6 and the Web site, instructors need review only the highlights in class.

Feature 5: Detailed Design Techniques

Part Four includes chapters that provide instruction and examples of the design of heat exchangers; multistage and packed towers; pumps, compressors, and expanders; and polymer compounding devices (extruders). In addition, Chapter 22 provides guidelines for selecting processing equipment and equations for estimating the purchase costs of a broad array of equipment items. Furthermore, this chapter shows how to use the Icarus Process Evaluator (IPE) of Aspen Tech, along with the process simulators, to estimate purchase costs and the total permanent investment for a chemical plant.

Students can use the chapters in Part Four when carrying out their design projects. In this book, most of the information they need for estimating equipment sizes, purchase costs, and operating costs and for carrying out profitability analyses is provided.

Instructors can use the chapters on equipment design to supplement or provide review of the subjects covered in earlier courses, selecting topics most appropriate for their students.

Acknowledgments

In the preparation of this courseware, several graduate and post-doctoral students made significant contributions, including Charles W. White III, George J. Prokopakis, Joseph W. Kovach III, Tulio R. Colmenares, Miriam L. Cygnarowicz, Alden N. Provost, David D. Brengel, Soemantri Widagdo, Amy C. Sun, Roberto Irrizary-Rivera, Leighton B. Wilson, James R. Phimister, Pramit Sarma, and Thomas A. Adams II, at the University of Pennsylvania; and Oren Weitz, Boris Solovyev, Eyal Dassau, Joshua Golbert, Eytan Filiba, and Eran Nahari at the Technion. Colmenares developed the lecture notes, with many examples, that are the basis for Chapter 9, *Heat Exchanger Networks*. Widagdo co-authored the review article on azeotropic distillation in which many of the concepts in Section 8.5, Sequencing Azeotropic Distillation Columns, are presented. Phimister was the teaching assistant during the semester in which many new concepts were introduced. Subsequently, he wrote introductory material on the use of GAMS, which appears on the Web site. The case study on penicillin manufacture in Chapter 25 is based on Dassau's Ph.D. research. Adams assisted in revising the Aspen IPE Course Notes. The successes in our product and process design courses are closely related to the many contributions of these graduate and post-doctoral students. Their help is very much appreciated.

Technion students Eyal Dassau, Joshua Golbert, Garry Zaiats, Daniel Schweitzer, and Eytan Filiba; and University of Pennsylvania students Murtaza Ali, Scott Winters, Diane M. Miller, Michael DiTillio, Christopher S. Tanzi, Robert C. Chang, Daniel N. Goldberg, Matthew J. Fucci, Robyn B. Nathanson, Arthur Chan, Richard Baliban, Joshua Levin, Larry Dooling, and Thomas Dursch implemented the multimedia Web site and/or assisted in many other ways. Their efforts are also appreciated. In this regard, seed money for the initial development of the Web site was provided by Dean Gregory Farrington, University of Pennsylvania, and is acknowledged gratefully.

Several colleagues at the University of Pennsylvania and industrial consultants from local industry in the Philadelphia area were very helpful as these materials evolved, especially Arnold Kivnick, Leonard A. Fabiano, Scott L. Diamond, John C. Crocker, Talid Sinno, Adam Brostow (Air Products and Chemicals), Robert M. Busche (Bio-en-gene-er Associates, Wilmington, DE), F. Miles Julian, Robert F. Hoffman, Robert Nedwick (Penn State University—formerly ARCO Chemical Company), Robert A. Knudsen (Lyondell), and David Kolesar (Rohm & Haas).

Four faculty—Michael E. Hanyak, Jr. (Bucknell), Daniel W. Tedder (Georgia Tech), Dale E. Briggs (Michigan), and Colin S. Howat (Kansas)—reviewed a preliminary version of the first edition. Three additional faculty—John T. Baldwin (Texas A&M), William L. Luyben (Lehigh), and Daniel A. Crowl (Michigan Tech)—reviewed a preliminary version of the second edition. In addition, Professor Ka Ng (Hong Kong University of Science and Technology), Dr. Soemantri Widgado (3M), Professor Costas Maranas (Penn State), and Professor Luke Achenie (Connecticut) reviewed selected chapters for the second edition. Three additional faculty—Barry Barkel (Michigan), Miguel Bagajewicz (Oklahoma), and Dimitrios V. Papavassiliou (Oklahoma)—reviewed a preliminary version of this third edition. Their suggestions and critiques were particularly helpful. Also, for this third edition, Arunprakesh Karunanithi assisted us in preparing Section 3S.1 on the crystallization of organic solids. Finally, Lorenz T. Biegler (CMU) provided helpful suggestions concerning the organization of the initial version of Chapter 24, *Design Optimization*. Some of the material for this chapter was adopted from Chapter 13 by L.T. Biegler in *FLOWTRAN Simulation—An Introduction* (Seader et al., 1987).

The cooperation of Jila Mahalec, Vladimir Mahalec, Herbert I. Britt, Atilla Forouchi, Sanjay Patnaik, Lawrence Fry, Lorie Roth, Robert Steinberger, Siva Natarajan, and Lawrence B. Evans at Aspen Technology; and Bill Svrcek, Rich Thomas, and James Holoboff at Hyprotech (now part of Aspen Technology) has been especially valuable, and is acknowledged with appreciation.

It is of special note that, during the preparation of the first edition, Professors Christodoulos A. Floudas (Princeton) and William L. Luyben (Lehigh) provided W.D. Seider an opportunity to lecture in their classes and utilize some of these materials as they were being developed. Their interactions and insights have been very helpful. This cooperation, and some of the work on this project, was facilitated in part by NSF Grant No. EEC-9527441 from the Combined Research and Curriculum Development Program.

The support of 3M during the preparation of this manuscript is gratefully acknowledged, especially that of Jay Ihlenfeld, Fred Palenski, Robert Finochiarro, Alan Hulme-Lowe, Terry Potts, and Ann Meitz, who supported the sabbatical leave of W.D. Seider during the spring of 2007. Also, S. Widagdo recognizes his 3M colleagues John Huizinga, Hua Chan, Paul Driscol, Mike Martin, Steve Lenius, and Bridget Bentz, for helping him grow professionally and for their dedication during his tenure as Technology Leader at the Engineering System Technology Center. He is also indebted to David Todd (PPI), Costas Gogos (PPI), and Zehev Tadmor (Technion) for guiding his work on polymer processing and compounding during his tenure at the Polymer Processing Institute (PPI). Thanks are extended to Eduardo Canedo, who provided the twin-screw simulation software, TXS, for solution of the examples in Chapter 21. At 3M, S. Widagdo thanks numerous mentees who taught him more than they realized, especially Dan Scott, Mary Boon, Patrick Fischer, and Tony Hollobaugh. Finally, he thanks Professor Saswinadi Sasmojo, who provided his first experience in product design involving an electrical generator using biomass gasification.

Throughout the development of the first and second editions of this textbook and its Web site, A. Wayne Anderson, the Editor for College Publishing at John Wiley & Sons, was extremely helpful. Wayne's excellent advice and guidance is very much appreciated. Since our second edition appeared, Wayne was replaced by Jennifer Welter, who provided much advice and guidance in formulating our plans for and preparing this third edition.

It is important to acknowledge the secretarial support provided in the most efficient and effective manner by John Linscheid, who made it possible to prepare the first edition, and the helpful assistance of Meghan Godfrey during the production phase of this third edition. Finally, W.D. Seider received two Lady Davis Visiting Professorships at the Technion during the spring of 1996 and 2002, and D.R. Lewin was a Visiting Professor at the University of Pennsylvania during the summer of 1997. In addition, W.D. Seider spent a sabbatical leave at 3M Company in the spring of 2007. Financial support in connection with these sabbatical leaves enabled them to work on the manuscript and is very much appreciated.

August 2008
W.D. Seider,
J.D. Seader,
D.R. Lewin,
S. Widagdo

Brief Contents

Contents

APPENDIXES

INDICES

Chapter 1

Product Design Introduction

1.0 OBJECTIVES

This chapter introduces a broad array of considerations that confront chemical engineers in developing new chemical products and processes. It also introduces a framework typically used for developing new products, together with *innovation maps* to help identify the connections between technological inventions and the voice of the customer. Special emphasis is placed on the growing importance of protecting the environment and ensuring safe and reliable chemical products, as well as manufacturing facilities, considerations that are prominent in the minds of product and process design teams. In addition, this chapter has a section that introduces the crucial role of engineering ethics in the work of product and process designers.

After studying this chapter, the reader should:

1. Be knowledgeable about the organizational structures involved in product and process design, and their interactions, at chemical companies.
2. Have an appreciation of the key steps in carrying out a product and/or process design, including the Stage-Gate™ product- and technology-development framework.
3. Be aware of the many kinds of environmental issues and safety considerations prevalent in the design of new chemical products and processes.
4. Appreciate the importance of maintaining high ethical principles in product and process design.

Although you will not solve any design problems in this chapter, you will obtain the background information that will be expanded upon and referred to throughout the remaining chapters of this text.

1.1 INTRODUCTION

In the last few decades of the 20th century, chemical engineering graduates became increasingly involved in the development of new consumer products. The shift away from traditional process design activities, which dominate the manufacture of basic chemicals, coupled with the lack of education in product design, have left new chemical engineering graduates inadequately prepared for the challenge. To counter this, their employers have included them on new product-development teams, providing on-the-job training, through exposure to the art of developing new products as practiced by experienced engineers. Because this strategy has had just limited success, the third edition of this textbook is intended to define the principles of the new product-development and design process, while continuing to cover the basic principles of process design.

Large manufacturing corporations, such as GE (General Electric Company), P&G (Procter & Gamble), 3M (formerly Minnesota Mining and Manufacturing Company), Motorola, Nortel (formerly Northern Telecom Limited), Apple (formerly Apple Computer), and many others, have created, over many years, a working structure that supports the new product-development process, as illustrated in Figure 1.1. The origin of a new product idea often derives from the interactions among these organizations and their interactions with their customers. In the remainder of this section, the roles of organizations involved in product design are discussed, one-by-one, focusing on: (1) building external and internal relationships, (2) anticipating changes in the markets, and (3) providing intellectual property protection.

The *business-development* role in a typical company includes corporate alliances, new business development, sales and marketing, major account management, and mergers and acquisitions. To achieve this, business professionals are active in building and managing relationships, both external and internal, and in anticipating business changes. The management of external relationships includes networking to establish and maintain relationships with business leaders; understanding business issues, needs, and trouble areas; learning the customer's corporate structures; and working with the decision makers in their major accounts. The management of internal relationships includes gaining a

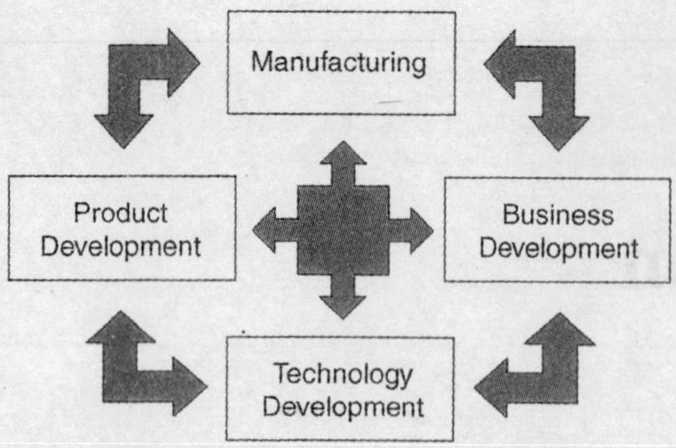

Figure 1.1 Infrastructure of an operating business unit in a large manufacturing corporation.

thorough understanding of the business; building relationships with the key decision makers within their organizations; and educating the company about the market conditions.

Another key role of business professionals is in anticipating changes in the marketplace, including close observation of market trends; understanding emerging technologies and their implications in the marketplace; and critically evaluating current solutions and anticipating future dominant market solutions. The organization is staffed with *business-development* professionals, career sales and marketers, and in some cases, seasoned technical professionals with technical service or product-application backgrounds.

In parallel, the *technology-development* role in a typical company includes strategic technology planning, new technology development, technology commercialization, and technology protection. To achieve this, technical professionals are active in managing technology, both external and internal, and anticipating technological changes. The management of external relationships includes networking to establish and maintain relationships with technology leaders at leading research universities, independent research institutions, national laboratories, and venture capitalists and technology incubators; and understanding technological issues and needs in the market segments of interest. The management of internal relationships includes gaining a thorough understanding of the strategic plan of various businesses within the company and aligning the technology-development efforts with these strategic plans; building relationships with key decision makers (that is, new-technology *early* adopters); and tracking and educating the company about technology trends and emerging future dominant technologies in the marketplace.

Another key role is in managing technology protection for the company, including management of the submission of inventions, patent filing and prosecution, patent portfolio management, and tracking the patent portfolios of their major competitors. This normally involves a group of intellectual-property lawyers and professionals that works closely with the technology-development organization.

Depending on the size of the company, each business unit or subsidiary may have its own laboratory to support its business, in addition to a corporate-wide R&D (research and development) organization. In this situation, the latter focuses on longer-term technology development, while the former concentrates on shorter-term development. Naturally, these efforts must be synchronized to the extent possible. Most R&D organizations are comprised of chemical engineering, chemistry, biology, physics, materials science, mechanical engineering, electrical engineering, and computer science graduates, especially those with advanced degrees, who have been trained to carry out research and develop new technologies.

The *product-development* role in a typical company includes strategic product planning, new product development, product launch, customer service, and product protection. To achieve this, product-development professionals focus on the *products*. While the business-development and technical professionals are active in managing business relationships and technologies, the product-development professionals manage *products*, both external and internal, and anticipate product offering changes. The management of external relationships includes networking to establish and maintain relationships with *early* product adopters, such as product design centers in their market segments; and understanding product issues, needs, and trouble areas in the market segments of interest. The management of internal relationships includes gaining a thorough understanding of the strategic product-offering plans of various segments of the business and aligning the product-development efforts with these strategic plans; building relationships with the key decision makers; and educating the company about product trends and changes in the marketplace.

When anticipating changes in the marketplace, the product-development professionals focus on the emerging products and their trends; understanding the emerging technologies and their implications for potential product offerings in the marketplace; critically evaluating current product solutions and product offering changes by their major competitors; and extrapolating potential market needs. These professionals have a similar role to the technical professionals in the protection of intellectual property, where the focus is on managing *product* protection, including the management of the submission of *product* inventions, patent filing and prosecution, and *product* patent portfolio management. They, too, work closely with a group of intellectual-property lawyers and professionals. The product-development organization is somewhat broader than the R&D organization, being staffed with multidisciplinary engineering professionals, including chemical engineering graduates, often with bachelor's and graduate degrees.

Finally, the *manufacturing* role in a typical company includes strategic manufacturing planning, new manufacturing process development and qualification, manufacturing supply-chain development, raw-material sourcing, and product-quality assurance. To achieve this, manufacturing professionals are active in managing the manufacturing of various products, both external and internal, and anticipating manufacturing-technology changes. The management of the external relationships includes networking to establish and maintain relationships with toll manufacturing facilities and engineering companies; and understanding product-manufacturing issues, needs, and trouble areas. The management of internal relationships includes gaining a thorough understanding of the strategic product-manufacturing plan of various product platforms and aligning the manufacturing-technology development effort with these strategic plans; building relationships with key decision makers; and educating the company about product-manufacturing trends and changes in the marketplace, including outsourcing and multiple-sourcing practices.

When anticipating changes in the marketplace, like the product-development professionals, the manufacturing professionals focus on the emerging manufacturing technologies and practices and their impact on manufacturing yield, cost, and product-quality assurance, as well as extrapolating future manufacturing needs. The manufacturing organization is normally staffed with engineering professionals, including chemical engineering graduates with a broad background.

To run a large company smoothly, and achieve a company's goals, the four organizations in Figure 1.1 must communicate, collaborate, and work together closely and effectively. As companies grow, the management of these interactions becomes more difficult, but remains crucial to the success of the company. Design teams have become increasingly interdisciplinary and more distributed around the world—as businesses have become more globally oriented, providing products for local markets as well as for customers with differing needs worldwide. Beginning in the next section, and extending into Chapter 2, a framework to facilitate these interactions during the early stages of the product-development process is discussed. It arranges for the concerted effort of these four organizations in the successful introduction of new products.

The remainder of this book, however, focuses mainly on just the two principal building blocks involving engineers and scientists: technology development and product development, together with their interactions. While the roles of business development and manufacturing development are discussed, they are not the foci of the product design discussions. In recent years, many major companies have adopted the Stage-Gate™ Product-Development Process (SGPDP) to manage product design. The success of the commercialization of a new technology often hinges on connecting the new technology to the market, with the connections between new technologies and the market needs being new products. In the next two sections, the concept of the *innovation map* is introduced to help locate the connections between various technologies (materials, process/manufacturing, and product technologies) and the market needs.

1.2 PRODUCT- AND TECHNOLOGY-DEVELOPMENT FRAMEWORK

Consider, in Figure 1.1, the two building blocks: technology development and product development, and their interactions. In recent years, as discussed above, many companies have adopted the Stage-Gate™ Product-Development Process (SGPDP) to manage the latter (Cooper, 2005). For the former, very recently Cooper (2001, 2002) introduced the Stage-Gate™ Technology-Development Process (SGTDP). Together, these provide a combined framework for the technology- and product-development processes, as shown in Figure 1.2.

The Stage-Gate™ Technology-Development Process (SGTDP™) is comprised of three stages: *technology scoping*, *technology assessment*, and *technology transfer. Technology scoping* is a relatively inexpensive stage of brief duration involving idea generation, literature searching, and evaluating alternative ways for conducting the technology-development project. The *technology assessment* stage is more extensive, as it is designed to demonstrate technical feasibility; that is, to show that a new technology functions properly and is worthy of further attention. The *technology transfer* stage usually involves a full experimental and/or modeling effort to advance the technology and justify the identification of potential

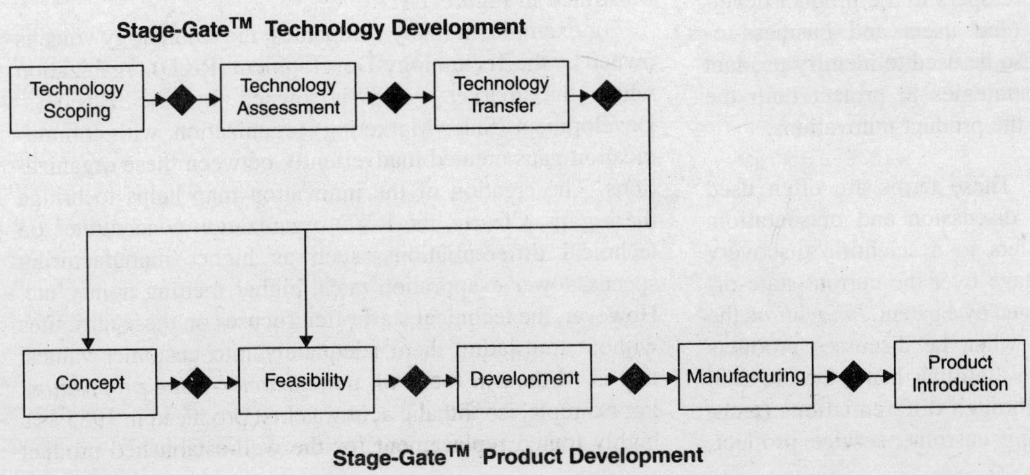

Figure 1.2 Stage-Gate™ technology- and product-development framework (Cooper, 2001, 2002).

applications, products, and/or new manufacturing processes. During this latter stage, potential new products may be identified and defined, with preliminary market and business analyses conducted. Often, connections to potential business partners are identified and explored to learn their potential interests in technology transfer. When technology transfer takes place, ideally it is connected to a product-development effort, and is most likely to be transferred to the *concept* or *feasibility* stages of the Stage-Gate™ Product-Development Process (SGPDP) shown in Figure 1.2.

While the technology-development process (SGTDP) is not discussed in detail in this textbook due to space and scope limitations, it is important to note that not all technology-development efforts yield new products immediately. Often the new technologies, while successfully demonstrated, are ahead of their time, not finding a place in the market until long after they are developed. In fact, some new technologies never reach the market, or quickly fail in the marketplace, for reasons including a lack of customer acceptance, business infeasibility, and obsolescence relative to other new technologies. The network PC falls in the category of failed customer acceptance; no one wants a terminal when he or she can have a full-blown PC. Some new technologies result in a single product or a family of products having a wide range of applications. In the latter case, the terminology *technology platform* is often used. Examples include the Internet, polymers, wireless, lighting, and display technologies, and many others.

As shown in Figure 1.2, the SGPDP consists of five stages: concept, feasibility, development, manufacturing, and product introduction. These are discussed in Section 2.3.

1.3 INNOVATION MAP AND CLASSES OF CHEMICAL PRODUCTS

This section begins with an introduction to the *innovation map* (Widagdo, 2006), which was developed to help identify the connections between various technological inventions (materials, process/manufacturing, and product technologies) and the *voice of the customer* (market or customer needs). It is these connections, or intersections in space, that present new product opportunities. The *innovation map* helps to connect the technology developers to the product developers and their customers (end users and business-to-business customers). It can also be used to identify product platforms or families, and strategies to protect both the technological inventions and the product innovations.

Invention and Innovation. These terms are often used interchangeably in a casual discussion and presentation. More formally, *invention* refers to a scientific discovery with a clear technical advantage over the current state-of-the-art, and is generally protected by a patent. *Innovation*, the favorite word of Bill Gates when he discusses products produced by Microsoft, is, on the other hand, the creation of business/economic value through differentiations (technical, business, sales, marketing, customer service, product, etc.). The transformation of these differentiations into customer values, satisfying a *customer-value proposition*, discussed below, is the key for successful new product development. Product differentiation is important to differentiate internally a family of product offerings and to differentiate their products from those of the competition.

Innovation Map

A product-development effort can be *technology* or *market driven*. A *market-driven*, new product development starts with a known market or customer need for a solution. A good example of a market-driven technology is Scotch® Magic™ Tape of the 3M Company. In 1959, when the Xerox plain-paper photocopier using xerography began replacing the Photostat copier, which produced a negative print, a new era began. Consumers could now paste up a sheet to be copied, using pieces cut from various sources. The pieces could be pasted onto the sheet or fastened to the sheet with cellophane tape. However, the Xerox copy would clearly show the outline of the piece if pasted or a smudged surface where cellophane tape was used. Scotch® Magic™ Tape, with its frosty matte finish and absence of acetate that reflects light, came to the rescue. This tape was invisible on the Xerox copy.

On the other hand, a *technology-driven*, new product development begins with a technological invention, involving materials, process/manufacturing, and/or product technologies. Such an invention was the PC (personal computer), whose future was predicted in 1962 by John W. Mauchly, who in 1944 at the University of Pennsylvania, with J. Presper Eckert, designed the ENIAC, the first general-purpose, electronic, digital computer. At that time, the future need for PCs was predicted by very few prognosticators.

In either case (market driven or technology driven), it is necessary to match the market/customer needs with the technological invention. The successful match creates product innovation, with the *innovation map* providing the linkages between the technology and market voices. These linkages, while straightforward to create, are often overlooked outside of the business world, as the technology and market voices are often owned by the different organizations illustrated in Figure 1.1.

For example, in many companies, the technology voice is owned by the Technology-Development (R&D) organization while the market voice is owned by the Business-Development (Sales/Marketing) organization, with communication gaps created inadvertently between these organizations. The creation of the innovation map helps to bridge these gaps. Clearly, the R&D organization concentrates on technical differentiations, such as higher manufacturing speeds, lower evaporation rates, higher melting points, etc. However, the technical staff often focuses on these attributes without translating them adequately into customer values; that is, without a focus on the *customer-value proposition*. For example, isophthalic acid was first produced in 1955 as a highly touted replacement for the well-established product phthalic anhydride, because the acid was clearly superior to the anhydride for applications in synthetic coatings for wood and metal, home paint, automobile and appliance finishes, and such plastic articles as luggage and lightweight boats. However, consumers were unwilling to pay the 15 percent higher cost for the acid and the plant was eventually shut down.

From the other perspective, as the Sales/Marketing organization focuses on the *customer-value proposition*, it may undervalue the advantages of the technical differentiation. In many cases, the creation of these linkages in the early design stages has been crucial to successful new-product commercialization. For success, the necessary and sufficient technological components for a new product to deliver the intended solution to market or customer needs must be identified.

The *innovation map*, many examples of which are shown later, relates the technological components of product developments to the technical advantages—that is, showing the technical differentiation—and ultimately to the satisfaction of the *customer-value proposition*. The construction of an innovation map begins with the identification of its six layers and the various elements associated with each layer:

Materials Technology: materials that enable the new product.

Process/Manufacturing Technology: processes that enable the manufacturing of the new product or its components.

Product Technology: product components or precursors, usually intended for business-to-business customers and not for end users.

Technical-Value Proposition: technical differentiations or advantages.

Products: a single product, product family, or product platform.

Customer-Value Proposition: product attributes, advantages, and differentiations, expressed from the customer point of view.

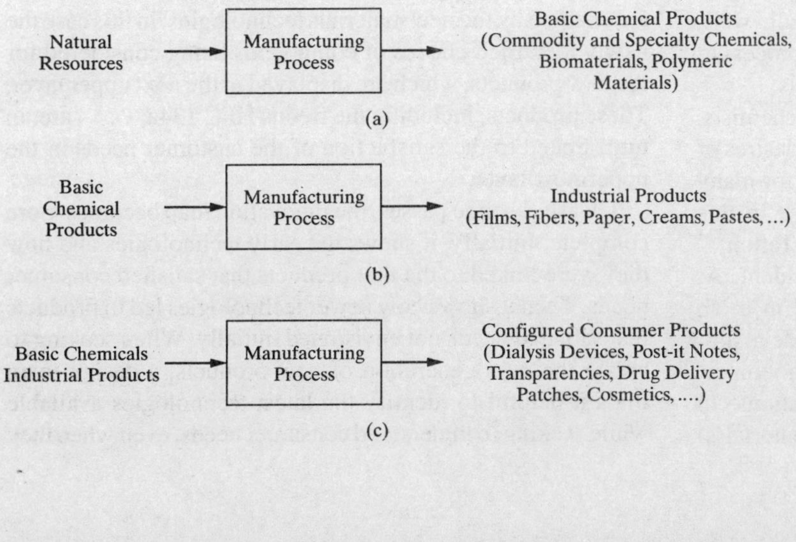

Once the elements are identified and placed on the appropriate layers in the map, the connectivity among them is drawn to show the interplay between the technological elements, the technical-value proposition, and ultimately the customer-value proposition.

The *innovation map* evolves during product design, being updated periodically during the new product-development and commercialization processes. Often, product-development leaders use innovation maps to manage new product-development efforts, and to spot unmet customer needs as targets for next-generation products.

Here, innovation maps are introduced for three kinds of chemical products: *basic chemicals*, *industrial chemicals*, and *configured consumer products*. But, first, this introduction is preceded by a definition and discussion of these three classes of products.

Classes of Chemical Products

Thousands of chemical products are manufactured, with companies like 3M having developed over 60,000 chemical products since being founded in 1904. Who has not used 3M's Magic Tape^{TM}? The scope of chemical products is extremely broad. They can be roughly classified as: (1) basic chemical products, (2) industrial products, and (3) configured consumer products.

As shown in Figure 1.3a, *basic chemical* products are manufactured from natural resources. They include commodity and specialty chemicals (e.g., commodity chemicals—ethylene, acetone, vinyl chloride; and specialty chemicals—difluoroethylene, ethylene glycol monomethyl ether, diethyl ketone), biomaterials (e.g., pharmaceuticals, tissue implants), and polymeric materials (e.g., ethylene copolymers, polyvinyl chloride [PVC], polystyrene). They normally involve well-defined molecules and mixtures of molecules, and are normally not sold directly to the consumer.

The manufacture of *industrial chemical* products begins with the *basic chemical* products, as shown in Figure 1.3b. *Industrial chemical* products include films, fibers (woven and

Figure 1.3 Manufacture of chemical products.

nonwoven), paper, creams, and pastes. While they are characterized by thermophysical and transport properties (like *basic chemicals*), other properties are normally dominant in satisfying customer needs, including microstructure, particle-size distribution, and functional, sensorial, rheological, and physical properties. See the introduction to Part Two for more specifics. Like *basic chemicals*, few *industrial chemicals* are purchased by the consumer.

Finally, as shown in Figure 1.3c, *configured consumer chemical* products are manufactured from *basic chemical* and *industrial chemical* products. These include dialysis devices, hand warmers, Post-it notes™, ink-jet cartridges, detachable wall hangers, solar desalination devices, transparencies for overhead projectors, drug-delivery patches, fuel cells, cosmetics, detergents, pharmaceuticals, etc. Unlike *basic* and *industrial chemical* products, *configured consumer chemical* products are normally sold to the consumer. In most cases, they are characterized by properties similar to those of *industrial chemicals* and, in some cases, their three-dimensional configurations are crucial in satisfying consumer needs. For more specifics, see the introduction to Part Three.

Many chemical products, especially specialty products, are manufactured in small quantities, and the design of a product focuses on identifying the chemicals or mixture of chemicals that have the desired properties, such as strength, stickiness, porosity, permeability, and therapeutic effectiveness, to satisfy specific consumer needs. For these, the challenge is to create a product that can be protected by patents and has sufficiently high market demand to command an attractive selling price. After the chemical mixture is identified, it is often necessary to design a manufacturing process, often involving small-batch operations.

Other chemical products, often referred to as *commodity* chemicals, are required in large quantities. These are often intermediates in the manufacture of specialty chemicals and industrial and configured consumer products. These include ethylene, propylene, butadiene, methanol, ethanol, ethylene oxide, ethylene glycol, ammonia, nylon, and caprolactam (for carpets); together with solvents like benzene, toluene, phenol, methyl chloride, and tetrahydrofuran; and fuels like gasoline, kerosene, and diesel. These are manufactured in large-scale processes that produce billions of pounds annually in continuous operation. Since they usually involve small, well-defined molecules, the focus of the design is on the process to produce these chemicals from various raw materials.

Often chemicals originate in the research labs of chemists, biochemists, and engineers who seek to satisfy the desires of customers for chemicals with improved properties for many applications (e.g., textiles, carpets, plastic tubing). In this respect, several well-known products, such as Teflon™ (polytetrafluoroethylene), were discovered by accident. At DuPont, a polymer residue that had accumulated in a lab cylinder of tetrafluoroethylene was found to provide a slippery surface for cookware, capable of withstanding temperatures up to 250°C, among many similar applications. In other cases, an inexpensive source of a raw material(s) becomes available and process engineers are called on to design processes that use this chemical, often with new reaction paths and methods of separation.

Basic Chemicals Innovation Maps

Because *basic chemicals* are usually well-defined molecules and mixtures of molecules, without complicating functional, sensorial, rheological, and physical properties that normally characterize *industrial chemicals* and *configured consumer chemical* products, technological inventions are normally associated with new materials, and less often with new process/manufacturing and product technologies. Hence, their innovation maps are usually the simplest, as illustrated next for a new environmentally friendly refrigerant.

As discussed in Section 3.2, beginning in the 1930s, Thomas Midgely, Jr., sought to develop a refrigerant product for a broad range of household, automotive, and industrial applications. Over the next 50 years, this led to several inventions that, while successful in many respects, led to serious ozone-depletion problems in the earth's stratosphere. Consequently, in the 1980s, it became necessary to find alternative, environmentally safe refrigerants.

The progression of materials inventions through the 20th century, together with the customer-value proposition (that is, customer needs), is shown in the innovation map of Figure 1.4, and discussed in detail in Section 3.2. Suffice it to observe, at this point, that both proceed in parallel, with the customer initially seeking low-cost refrigeration and air conditioning, involving nontoxic chemicals that are safe (that is, nonflammable), and subsequently adding the requirements to avoid ozone depletion and smog production. Meanwhile, in response, Midgely began by restricting his search to compounds involving C, N, O, S, and H atoms with the halogens F and Cl, having high latent heats of vaporization, low viscosity, and low melting points. In the 1980s, when ozone and smog problems were identified, the search was restricted to compounds involving C, H, and F, eventually with the addition of O and S, but with the omission of Cl.

Of special note in Figure 1.4 are the two intermediate layers of the innovation map. The elements above the materials technology layers show technical differentiations that are enabled by the new materials technologies, in this case the newly identified classes of compounds being considered for the new products, which are displayed in the next upper layer. These products, including the freons HFC 134a, . . . , are, in turn, linked to the satisfaction of the customer needs in the uppermost layer.

Clearly, as time passes, the innovation map becomes more complete. Initially, it shows the early technologies and how they were linked to the first products that satisfied consumer needs. Then it shows how newer technologies led to products that satisfied needs not envisioned initially. When seeking to create the next generation of new products, a design team finds it helpful to identify the latest technologies available while seeking to understand consumer needs, even when they

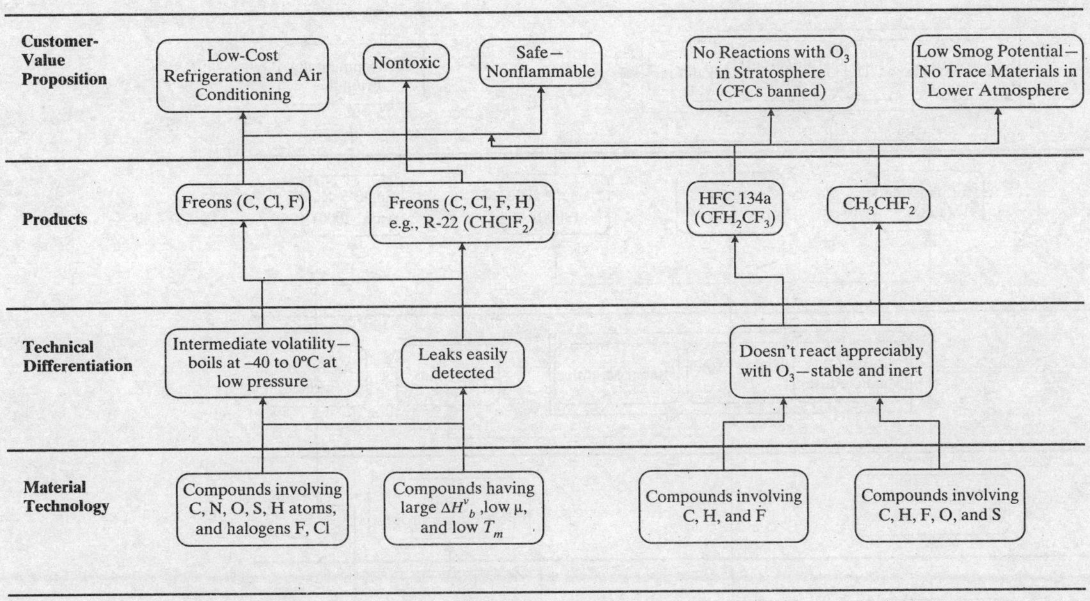

Figure 1.4 Environmentally friendly refrigerant innovation map (also Figure 3.1).

are dormant (possibly because consumers don't recognize the potential for the new technologies). Gradually, as the team begins its product design, it seeks to fill in the new technologies and the customer needs, and to create outstanding new products that link them together. In short, when beginning to develop a new product, an innovation map can help to suggest new products that provide these linkages. Gradually, during the product-development process, the technologies and customer needs can be refined, leading to improved products.

The innovation map in Figure 1.4 is further developed and discussed in detail in Section 3.2, where optimization algorithms are presented to search for the new refrigerant products.

Industrial Chemicals Innovation Maps

Industrial chemicals are normally characterized by properties beyond those of molecular structure and composition. For example, pastes and creams are colloids that are characterized by microstructures and particle-size distributions, among other properties. Often, to achieve stable emulsions or the like, new products involve the design of new process/manufacturing techniques, such as microscale mixing and extrusion devices. Similarly, fibers, both woven and nonwoven, often involve high-throughput extrusion devices that provide uniform diameters and the like. For these reasons, the innovation maps for industrial chemicals usually involve an additional new-technology layer, that is, the process/manufacturing layer. This is illustrated next for thin glass substrates in LCD displays.

Active-matrix, liquid-crystal displays (AM-LCD) consist of two glass layers with liquid-crystal materials sandwiched between these two substrates. The front glass layer, known as the front panel, has a color filter embedded on it, and the rear glass layer, known as the back panel, has embedded electronic switches (electrodes and transistors). The back panel, on which amorphous silicon is deposited, must: (1) be able to withstand processing temperatures up to 400°C without significant deformation; (2) be transparent at 350 nm for photolithography on the amorphous silicon; (3) have a coefficient of thermal expansion (CTE) similar to that of amorphous silicon; and (4) contain no elements that contaminate silicon, especially alkali cations. In summary, panel makers require that the LCD glass substrates have high thermal and dimensional stability, high durability (when exposed to etching chemicals), be alkali free, and be compatible with downstream processes that manufacture the front and back panels.

As discussed in detail in Section 14.2, specialty glass has been the leading technology for LCD substrates, with the market leader since the 1980s being Corning Incorporated. Based upon their technologies, Figure 1.5 introduces an innovation map that traces the development of specialty glasses (at the lowest level) to satisfy the needs of the manufacturers of LCD front and back panels (at the highest level). At this point, it should be sufficient to become familiar with the development of the new materials and process/manufacturing technologies that satisfy the increasingly demanding customer needs. A more thorough discussion is given in Section 14.2.

At the lowest level of the innovation map, Corning, in 1987, introduced the Corning-7059 glass substrate, that is, a barium boron silicate (BaO-B_2O_3-Al_2O_3-SiO_2) glass. Unfortunately, although it used an effective high-viscosity glass-fusion process, its CTE was too high, causing deformation at low temperatures. At the highest level of the innovation map, it satisfied customer needs for transparent and alkali-free substrates, but didn't provide a sufficiently low CTE and durability when exposed to etching chemicals.

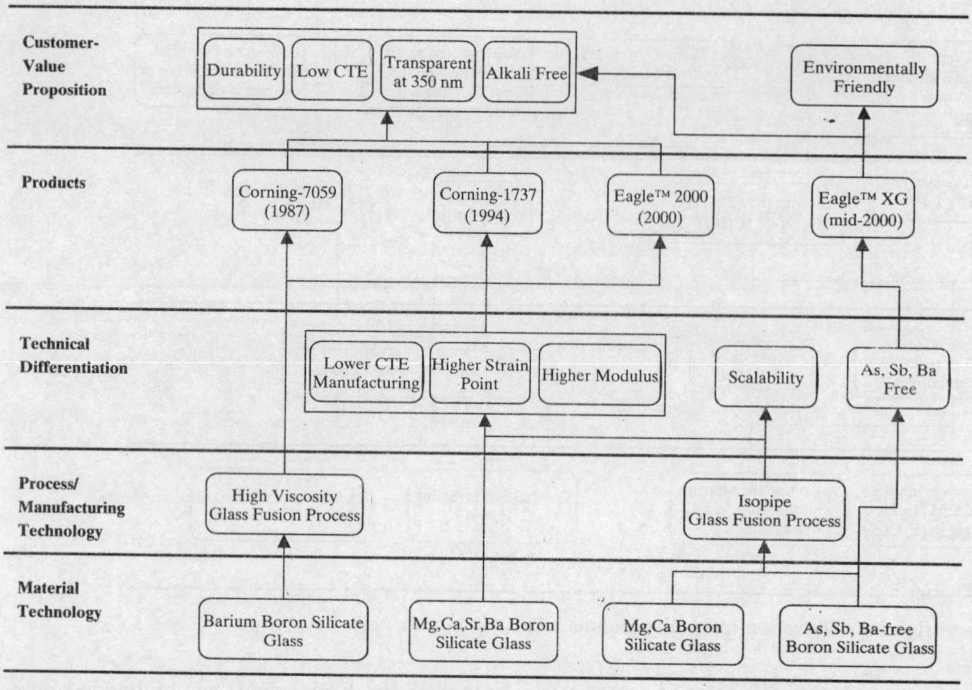

Figure 1.5 Innovation map for thin glass substrates in liquid-crystal displays (LCDs).

To satisfy these customer needs, Corning, beginning in 1994, introduced three new products based upon new materials technologies. The Corning-1737 glass substrate replaced BaO with an optimized mix of four alkaline-earth compounds: MgO, CaO, SrO and BaO, which provided improved technical differentiations, that is, a lower CTE (3.8 compared with 4.6 ppm/°C, providing a closer match to the CTE of α-Si), a higher strain point, a lower density (2.55 compared with 2.74 g/cc), a higher modulus (with less sagging), higher durability, and a very high liquidus viscosity that enables precision sheet forming.

Subsequently, in 2000, Corning introduced Eagle™ 2000 LCD substrates, which replaced all of the BaO and most of the SrO with CaO, further lowering the CTE and density, and further increasing the strain-point temperature. In addition, the amount of boron oxide was increased to minimize the liquidus temperature and lower the melting point.

Also in this period, Corning filed patents on a novel glass-fusion process (that is, a new process/manufacturing technology), in which the molten glass mixture is fed to a trough, called an isopipe (Helfinstein et al., U.S. Patent 6,974,786). The molten glass evenly overflows the two longitudinal sides of the isopipe, fusing just below the bottom tip of the isopipe. Because the glass forms at a free boundary, exposed only to air, its surface remains flawless and smooth, requiring no polishing or grinding. Another major advantage of the isopipe glass-fusion process is its scalability with respect to size (width) and thickness to accommodate the market needs for larger and thinner displays (as thin as 100 μm). This is a key technical differentiation, which together with a new boron silicate glass that is free of arsenic, antimony, and barium, led to the new Corning product, Eagle™XG, which satisfied the new customer need to be more environmentally friendly.

As discussed for basic chemicals, the innovation map gains both new technologies and customer needs in time—with generations of new products linking the new technologies and customer needs. From the perspective of a product-design team, moving to the right on the innovation map, the latest new technologies and customer needs are entered. As the team proceeds, following the SGPDP, for example, the linkages are improved through better products with better definition of the new technologies and customer needs.

Configured Consumer Chemical Product Innovation Maps

Many *configured consumer chemical* products add the three-dimensional configuration, which introduces yet another opportunity for technical invention, often referred to as product technology. For example, when designing halogen light bulbs to provide longer life and softer colors, the high operating temperatures often require a secondary casing, to protect against burns, in addition to the normal primary casing (for example, a quartz bulb). As discussed in Section 17.2, this introduces a new product technology, which is represented in the product-technology layer of the innovation map, to be introduced next.

For this discussion, to provide a brief introduction to innovation maps for configured consumer products, an innovation map is presented that traces back to the initial invention associated with light bulbs. Note that a more detailed history is presented in Section 16.2.

Beginning in the early 1800s, Humphrey Davy created light by passing electrical current through a platinum filament, with the heat-generated radiating light in the visible range. Then, about 80 years later, low-wattage light bulbs were manufactured using carbon filaments, with the disadvantage that their combustion products turned the bulbs dark black. This was overcome in 1903, when William Coolidge invented an improved method of making tungsten filaments, which outlasted all other types of filaments, enabling Coolidge to manufacture light bulbs at practical costs.

Shortly thereafter, in 1906, the General Electric Company patented a method of making tungsten filaments for use in incandescent light bulbs. Tungsten filaments offer a high melting temperature and low vapor pressures, which translate to a lower evaporation rate of tungsten vapor and reduced *blackening*. Subsequently, another GE researcher, Irving Langmuir, suppressed the tungsten evaporation by filling the light bulb with an inert gas that wouldn't burn the filament. However, the inert gas circulated in the bulb, carrying away too much heat, which, in turn, significantly reduced the brightness of the bulb. To reduce heat losses, Langmuir invented the tight-coil filament, the basis for modern incandescent light bulbs.

The innovation map in Figure 1.6 begins with the dark grey elements to the left that show the progression of materials and process/technology inventions, together with the customer-value proposition (that is, the customer needs), through the early 1900s. These are discussed in detail in Section 16.2. At this point, it is sufficient to recognize that these proceed in parallel, with the customer initially seeking light bulbs lasting for 750 hr, versatile in shape, having various light qualities, and at low cost. These needs were eventually met by a progression of inventions involving the use of tungsten, inert gases, and the Coolidge process for the manufacture of ductile tungsten rods.

Consider the dark grey entries to the left in Figure 1.6. Of the six layers in that figure, the fourth shows the technical differentiations enabled by the materials and process/manufacturing technologies, in this case low-cost manufacturing, a high tungsten melting point, and a low tungsten evaporation rate. In the third layer are the new product technologies enabled by the technical differentiations (that is, tightly coiled filaments, gas-filled bulbs, and high-wattage bulbs). These, in turn, lead to the principal product in the second layer, the incandescent light bulb, which in the early 1900s satisfied the four customer needs in the first layer. Conveniently, the innovation map shows all of these linkages very clearly.

The next linkages, in dotted boxes in the innovation map, trace the development of the halogen light bulb, which is discussed in detail in Sections 16.2 and 17.2. By the 1980s, the customer needs had been extended to include longer-life bulbs, on the order of 2,000 hr, with improved light quality, including warm, cool, and daylight qualities. To fulfill these, the discovery of Frederick Mosby that halogen gases react with tungsten at high temperature (~3,100 K) in chemical equilibrium, permitting tungsten vapor to redeposit on the tungsten filament, added a key materials technology. This, coupled with a quartz primary casing to contain the hot gases, provided the technical differentiations, that is, the high-temperature reaction and equilibrium deposition that led to a new product technology (a secondary casing, to prevent burns). This, in turn, led to the small halogen light bulb products that satisfied the five customer needs (four in the dark grey boxes, one in a dotted box).

Throughout the second half of the 20th century, a competing technology, fluorescent light, was developed. This involves a gas-discharge lamp that uses electricity flowing between electrodes at both ends of a fluorescent tube, which excites mercury vapor and produces shortwave light of ultraviolet photons. These photons collide with the phosphor coating on the inside of the fluorescent tube, creating light in the visible region. A magnetic ballast is required to turn on the fluorescent lamp.

To illustrate these advances, the innovation map in Figure 1.6 is extended with entries having a cross-hatched background.

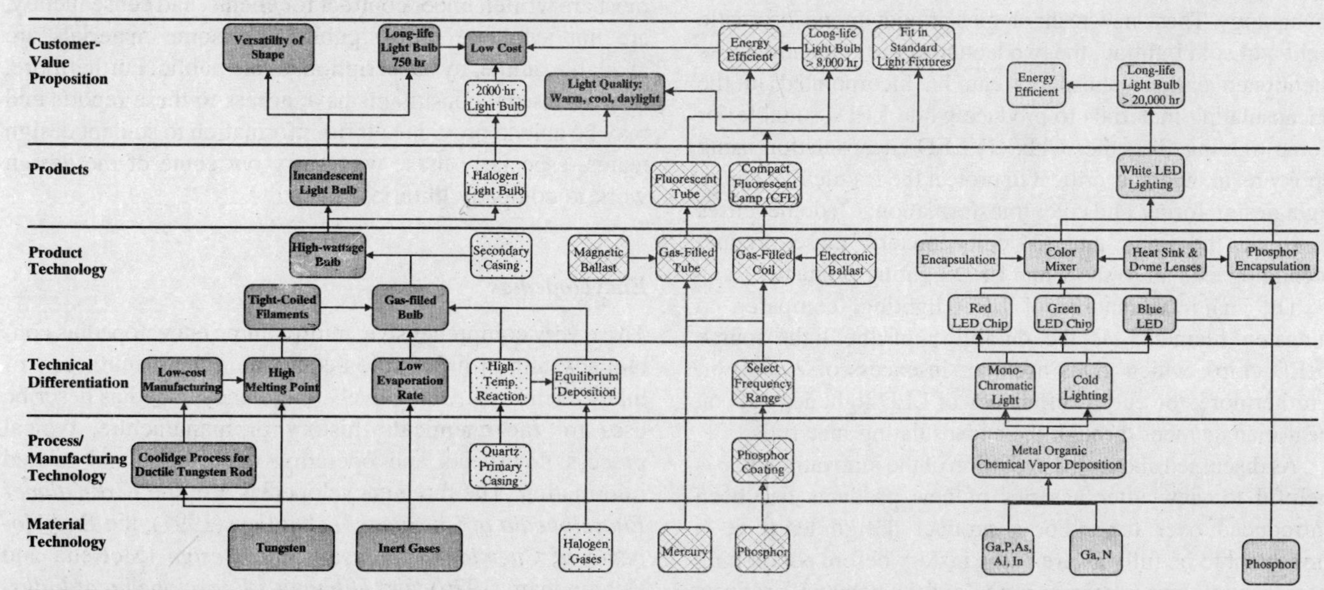

Figure 1.6 Innovation map for light bulbs.

Here, the customer needs were longer-life bulbs, exceeding 8,000 hr, especially for display and industrial lighting, as well as increased energy efficiency. The new materials technologies involved mercury, in small nontoxic quantities, and phosphors, which, coupled with phosphor coatings, were deposited using a new process/manufacturing technology, to give a select frequency range, the desired technical differentiation. Initially, new product technologies—that is, a magnetic ballast and gas-filled tubes—provided the fluorescent tube product, which satisfied the two additional customer needs.

However, more recently, in the 1990s, when the consumer needs expanded to include compact bulbs that would fit into standard light fixtures, a new electronic ballast was invented, as a new product technology. This eliminated the slow starting and flickering of fluorescent tubes, and permitted the introduction of the compact fluorescent lamp (CFL).

Even more recently, in the early 2000s, an emerging technology for home lighting is LED-based lighting, a solid-state technology (with no moving or loose parts) formed using Group III-V semiconductor materials. As a current passes through the p-n junctions created in these materials, light is emitted. Depending on the selection of the materials (GaP, GaAs, AlGaAs, AlInGaP), various monochromatic lights (red and yellow) are produced. The introduction of GaN offered the ability to produce blue and green LEDs. For home lighting, white LEDs are produced by incorporating phosphors into the encapsulating materials (epoxies) and using light management lenses.

LED technologies are introduced into the innovation map in Figure 1.6 using elements in light grey, toward the right boundary. Here, the consumer needs are expanded to even longer-life bulbs exceeding 20,000 hr, also at high-energy efficiencies. The new materials technologies include compounds of Ga, P, As, Al, In, and N. The manufacturing process for the LED wafer—that is, metal-organic, chemical-vapor deposition (MOCVD)—is a key new process/manufacturing technology. These new technologies provide monochromatic light and cold lighting, the two technical differentiations. As mentioned earlier, phosphors can be incorporated in the encapsulating materials to produce white LEDs suitable for home lighting. Together with: (1) LED encapsulation using epoxy resin, which is critical to protect the fragile LED chip, light design forms, and color transformation; (2) dome lenses to direct the light; and (3) color mixers, these product technologies lead to the white LED lighting product.

The major advantage of LED lighting, compared to tungsten filaments, is the durability of the light source (LED chip), which gives lifetimes in excess of 20,000 hr. Furthermore, the energy efficiency of LED light depends on heat management through the encapsulating materials.

As discussed above, the evolution of the innovation map is helpful to view after a series of new products has been introduced over time. For a product design team, it is important to be fully aware of the history before positioning new technologies, customer needs, and the potential products that link them together.

Literature Survey

When creating innovation maps, design teams in industry have access to company employees, company files, and the open literature, including patents. These resources provide helpful leads to specific problems, as well as information about related products, thermophysical property and transport data, possible flowsheets, equipment descriptions, and process models. If the company has been manufacturing the principal products, or related chemicals, information available to the design team provides an excellent starting point, enabling the team to consider variations to current practice very early in the design cycle. In spite of this, even when designing a next-generation product or plant to expand the production of a chemical product, or retrofitting a plant to eliminate bottlenecks and expand its production, the team may find that many opportunities exist to improve the processing technologies. Several years normally separate products, plant startups, and retrofits, during which technological changes are often substantial. For example, consider the recent shift in distillation, particularly under vacuum conditions, from trays to high-performance packings. For this reason, it is important to make a thorough search of the literature to uncover the latest data, flowsheets, equipment, and models that can lead to improved products and more profitable designs. Several literature resources are widely used by design teams. These include the Stanford Research Institute (SRI) Design Reports, encyclopedias, handbooks, indexes, patents (most of which are available electronically), and the Google™ search engine.

SRI Design Reports

SRI, a consortium of several hundred chemical companies, publishes detailed documentation for many chemical processes. While their reports provide a wealth of information, most are written under contract for clients, and consequently, are not available to the public. Yet some materials are available online, by subscription, to the public. Furthermore, most industrial consultants have access to these reports and may be able to provide helpful information to student design teams, especially those who carry out some of the design work in company libraries.

Encyclopedias

Three very comprehensive, multivolume encyclopedias contain a wealth of information concerning the manufacture of many chemicals. Collectively, these encyclopedias describe uses for the chemicals, history of manufacture, typical process flowsheets and operating conditions, and related information. The three encyclopedias are: the *Kirk-Othmer Encyclopedia of Chemical Technology* (1991), the *Encyclopedia of Chemical Processing and Design* (McKetta and Cunningham, 1976), and *Ullmann's Encyclopedia of Industrial Chemistry* (1988). For a specific chemical or substance, it is not uncommon for one or more of these encyclopedias to have 5 to 10 pages of pertinent information, together with literature references for more detail and background. Although the encyclopedias are updated too infrequently to always provide the latest technology, the information they contain is normally very helpful to a design team when beginning to assess a design problem. Other encyclopedias may also be helpful, including the *McGraw-Hill Encyclopedia of Science and Technology* (1987), *Van Nostrand's Scientific Encyclopedia* (Considine, 1995), the *Encyclopedia of Fluid Mechanics* (Cheremisinoff, 1986), the *International Encyclopedia of Heat and Mass Transfer* (Hewitt et al., 1997), and the *Encyclopedia of Material Science and Engineering* (Bever, 1986).

Another encyclopedia of rapidly growing importance and usefulness is Wikipedia, a multilingual, Web-based, free-content encyclopedia project written collaboratively for the Internet by volunteers. The vast majority of Wikipedia articles can be edited by anyone with access to the Internet. Its primary servers are in Tampa, Florida, with additional servers in Amsterdam and Seoul. Because Wikipedia is Web-based, articles can be created within minutes or hours of an announced event or development, and can be constantly updated. This makes Wikipedia the encyclopedia of choice for the latest technology, with convenient access through the Google™ search engine. For example, if up-to-date information is desired on the new compact fluorescent light bulb, one need only open www.google.com on the Internet and enter the search keywords "fluorescent light bulb wiki." At the top of the search results is the Web address for a Wikipedia article entitled "Compact Fluorescent Lamp."

Wikipedia was created in 2001, and as of mid-2008, contained more than 2.5 million articles. Because Wikipedia articles can be created and edited by anyone with Internet access, critics claim that it is susceptible to errors and unchecked information. While this is true, recent studies suggest that Wikipedia is broadly as reliable as the *Encyclopaedia Britannica*.

Handbooks and Reference Books

Several key handbooks and reference books are well known to chemical engineers. These include *Perry's Chemical Engineer's Handbook* (Green and Perry, 2008), the CRC *Handbook of Chemistry and Physics* (the so-called *Rubber Handbook*, published annually by CRC Press, Boca Raton, FL) (Lide, 1997), *JANAF Thermochemical Tables* (Chase, 1985), *Riegel's Handbook of Industrial Chemistry* (Kent, 1992), the *Chemical Processing Handbook* (McKetta, 1993a), the *Unit Operations Handbook* (McKetta, 1993b), *Process Design and Engineering Practice* (Woods, 1995b), *Data for Process Design and Engineering Practice* (Woods, 1995a), the *Handbook of Reactive Chemical Hazards* (Bretherick, 1990), and the *Standard Handbook of Hazardous Waste Treatment and Disposal* (Freeman, 1989), among many other sources. A useful Internet site for many important chemical engineering topics is www.cheresources.com.

Indexes

To search the current literature, especially the research and technology journals, several indexes are extremely helpful. These provide access to a broad spectrum of journals, including electronic access to issues since the late 1970s, with rapidly improving search engines. These indexes provide links to most of the articles published during this period, including kinetics data, thermophysical property data, and much related information for many chemicals. Of primary interest to a design team are the *Applied Science and Technology Index* (with electronic access to 350 journals since 1983), the *Engineering Index* (with access to 4,500 journals, technical reports, and books, electronically since 1985), *Chemical Abstracts* (one of the most comprehensive scientific indexing and abstracting services in biochemistry, organic chemistry, macromolecular chemistry, physical and analytical chemistry, and applied chemistry and chemical engineering—available electronically with entries since 1907), and the *Science Citation Index* (with access to 3,300 journals since 1955, available electronically since 1980, with searches that indicate where the author's work has been cited). The Google™ search engine also provides a convenient method for searching for scholarly literature, called Google Scholar. Searches are made from keywords such as topics and authors. For example, if a Google Scholar search is made on the keyword "stage-gate," more than 1,000 articles are cited, with a 1990 article entitled "Stage-gate systems: A new tool for managing new products" by Robert G. Cooper at the top of the list. Frequently, the full-text article can be downloaded or viewed for a fee or through a service such as Science Direct, to which many libraries subscribe.

Patents

These are important sources with which the design team must be aware to avoid the duplication of designs protected by patents. Perhaps more significantly, patent searches are often indispensable in tracing the development of new technologies when creating innovation maps. After the 17 years that protect patented products and processes in the United States are over, patents are often helpful in the design of next-generation processes to produce the principal chemicals, or improved chemical products that have preferable properties, chemical reactions, and so on. However, many patents withhold important know-how that may be vital to success. Since a patent is legal property, like a house or a car, it, and perhaps the know-how, can be owned, bought, and sold. Often patents are licensed for fees on the order of 3–6 percent of gross sales. This can be important when a design team decides to incorporate a patented chemical product in its design. Patents from the United States, Great Britain, Germany, Japan, and other countries are available on the Internet, with the details of carrying out a patent search discussed in Section 2.4.

Google™ Search Engine

A key aspect of the accelerated development of the Internet in the 2000s has been the emergence and remarkable performance of the Google search engine. Thanks to advanced search algorithms and massive parallelism (over 1,000,000 computers working in parallel), millions of people regularly search the entire Internet using search terms often known only to a few individuals. Yet, Google regularly returns extensive lists of pertinent information in the form of Web site addresses in the order of popularity, which is normally very helpful. The Google search engine is so effective and easy to use that its use is becoming ubiquitous. While its use as a first step in information retrieval is common, the quality of the information it retrieves, especially technical information for the design of new products, must be critically assessed. Over the years, the investments of technical publishers in journals and books, edited by outstanding technical persons and subject to peer review, have produced an authoritative literature that is normally more reliable than the results of an open-ended search using a general search engine. The Google search engine continues to evolve and add features. Recent additions to Google search products include: Book Search, Catalogs, Directory, Images, Maps, News, Patent Search, Product Search, Scholar, and Translate.

Stimulating Invention and Innovation

The evolution from farming in the 19th century to industrialization in the 20th century and to information technology and globalization in the 21st century has dramatically altered humankind. In the latest era especially, with time and space contracting and fast-growing global competition, speed to the market has become significantly more critical to business success.

In the 1990s, innovation was expressed in new technology, quality control, and cost efficiency. In the 2000s, information technology and globalization have extended innovation beyond new products to involve the reinvention of business models, internal processes, networking with existing and potential customers and partners, and branding. This has led company executives to increase their focus on innovation, placing it among their top three initiatives. Yet, nearly half are dissatisfied with their outcomes, as reported in an annual survey of the most innovative companies (*Business Week*, 2006). Therein, four major obstacles to innovation are cited:

1. Slow development time, it being recognized that time-to-market increasingly makes or breaks a new product launch.
2. Lack of coordination of product-development efforts. Here, early decision-making processes that establish priorities for new product developments are critical, it being important to identify early winning product-development efforts in the face of payoff uncertainties.
3. Problems in selecting the proper metrics to drive innovation engines. While successful new product launches are achieved, it is important to create environments that stimulate new innovations. Striking a balance between these two can be very difficult.
4. Obtaining good customer insights, recognizing that the "unmet needs" of customers are difficult to detect.

The discussion that follows is limited to technology innovations by chemists, physicists, material scientists, and chemical engineers. However, it is recognized that, to win in the 21st century marketplace, other kinds of innovations are equally important.

Technological inventions and their commercialization into new products (chemical products, in particular) are stimulated by corporations that encourage interactions among their working groups (researchers, marketers, salespeople, manufacturing engineers, and others), as well as interactions with existing and potential partners and customers. These have been discussed in Section 1.1 and illustrated in Figure 1.1. In this regard, *Business Week* and the Boston Consulting Group have listed many of the top innovative companies, including Apple, Google, 3M, Toyota, Microsoft, GE, Procter & Gamble, Nokia, BMW, IKEA, Samsung, Sony, Starbucks, Virgin, and IBM (*Business Week*, 2006). Note the inclusion of non-technology companies (i.e., Virgin, Starbucks, and IKEA), which were selected based on their unique business innovations. In this regard, their leaders are recognized for corporate policies that seek to maintain a climate in which invention and innovation flourish. Several examples of the initiatives in practice at the aforementioned companies are discussed next. Note that while many of these companies do not concentrate on chemical products/processes, most of their products involve significant materials and process inventions and innovations. The following approaches, developed by several different companies to stimulate innovation, are applicable throughout the basic chemical, industrial, and configured consumer product industries.

Fifteen Percent Rule, Tech Forums, Stretch Goals, Process Innovation Tech Centers—3M Company

Fifteen Percent Rule. At 3M, managers are expected to allow employees 15 percent of their time to work on projects of their own choosing. This rule, which has become a fundamental part of the 3M culture, is assessed nicely by Bill Coyne, a research and development manager: "The 15 percent part of the Fifteen Percent Rule is essentially meaningless. Some of our technical people use much more than 15 percent of their time on projects of their own choosing. Some use less than that; some use none at all. The number is not so important as the message, which is this: the system has some slack in it. If you have a good idea, and the commitment to squirrel away time to work on it, and the raw nerve to skirt your lab manager's expressed desires, then go for it" (Gundling, 2000).

Tech Forum. This terminology, which is used at 3M, is typical of organizational structures designed to encourage technical exchange and a cross-fertilization of ideas between persons working in many corporate divisions at widely disparate locations. At 3M, the Tech Forum is organized into chapters and committees, with the chapters focused on technology, including the Physics Chapter, the Life Sciences Chapter, and the Product Design Chapter. Chapters hold seminars related to their own areas of technology, presented by outside speakers or 3M employees. Some chapters do not have a technical focus. For example, the Intellectual Property Chapter is primarily targeted at patent attorneys. Also at 3M, the Tech Forum hosts a two-day "Annual Event" at the St. Paul headquarters, with each of the 3M labs invited to assemble a booth. Since the company rewards labs when other divisions use their technology, employees have an incentive to participate in this internal trade show.

Stretch Goals. Another 3M policy, intended to *stretch* the pace of innovation, is the rule that at least 30 percent of annual sales should come from products introduced in the past four years. The policy has recently been refined to establish an even greater sense of urgency, such that "10 percent of sales should come from products that have been in the market for just one year" (Gundling, 2000).

Process Innovation Technology Centers. Since 75 percent of manufacturing at 3M is done internally, two technology centers are provided. One is staffed with chemical engineers and material scientists to help researchers scale-up a new idea for a product from the bench to production, with a focus on core technologies. The other center handles the development and scale-up for key manufacturing process technologies such as coating, drying, and inspection and measurement. The latter is staffed primarily with chemical and mechanical engineers and software development personnel. These centers work closely with researchers and engineers involved in product development and equipment design.

Synergistic Innovations—Apple

When launching its iPod^{TM}, Apple is cited for its success in using no fewer than seven types of innovations, including a novel agreement among music companies to sell their songs online and a novel business model to sell individual songs for 99 cents. These are in addition to the key design innovation that created a new vehicle for users to listen to their music. Clearly, Apple has created an environment for the synergistic combination of innovations in their new products.

Value Innovation Strategy—Toyota

Yet another successful company, Toyota, a well-known product innovator, launched its *value innovation* strategy when producing its top-selling Prius^{TM}. Rather than squeeze its suppliers to reduce the cost of single parts, Toyota seeks to identify savings that span entire vehicle systems, beginning earlier in the product-design process.

Open Innovation Concept—P&G and IBM

At P&G and IBM, an open innovation concept has been embraced, with P&G having transformed its R&D process into an open innovation environment called "connect and develop," tapping collective knowledge and technologies from around the world. The executives of these companies have set targets seeking 50 percent of their new products from external sources. To speed up the introduction of new products, they seek to outsource some of their technology-development effort, while maintaining their core technology competencies. They are leveraging the worldwide pool of inventors, scientists, and suppliers to develop new products internally. In related approaches, service companies like NineSigma establish links between companies and researchers at university, government, and private labs; YourEncore connects retired scientists and engineers with businesses; and Yet2.com offers an online marketplace for intellectual properties. IBM even owns a company that purchases patents in open markets and provides free licenses to software developers to promote their products.

Central Innovation Coordination—BMW

At BMW, the focus has been on the coordination of innovation from the center. For the development of each new car, they relocate all of their project team members—consisting of 200 to 300 personnel from various disciplines including engineering, design, production, marketing, purchasing, and finance—to the BMW Research and Innovation center, called FIZ, for up to three years. This helps to improve communications and ensure early collaborations in the design stage among members of the various disciplines.

Cross-Functional Collaboration—Southwest Airlines and GE

To achieve cross-functional collaboration, Southwest Airlines secluded its in-flight, ground, maintenance, and dispatch personnel for short periods to brainstorm ideas addressing broad issues such as high-impact changes to make their operations more efficient. Over 100 ideas were presented to senior management, three of which involved sweeping operational changes. Similarly, GE addresses major issues by assembling teams of top performers from various disciplines for one to two weeks. Their recommendations are presented to senior executives, with many implemented and involving corporate-wide changes.

Internet Surfing—Samsung

To spark their innovative spirits, Samsung mandates that its engineers allocate a portion of their time to surfing the Internet for technical and business news. Their observations of worldwide trends often lead to new foci in their product-development processes.

Encouraging Entrepreneurial Behavior—GE

To encourage entrepreneurial actions, including external foci and risk taking, GE has implemented innovative leadership measures for their management teams. These are in addition to their traditionally rigid performance rankings.

Learning Journeys—Starbucks, 3M, and Nokia

To obtain a better appreciation of local cultures, behavior patterns, and fashions, Starbucks has formulated *learning journeys* in which groups of product developers visit several countries for a few months. This immersion into foreign cultures has led to new product ideas.

Similarly, in the current information age, with people increasingly using personal data assistants (PDAs) and smart phones, 3M perceived this as a threat to its Post-it note™ business. Consequently, by observing how users share digital photo collections, a 3M Office Supply Business team developed Post-it Picture™ paper.

In another example, after a long immersion into Chinese, Indian, and Nepalese cultures, a team of Nokia developers created a low-cost phone for illiterate customers unable to comprehend combinations of numbers and letters. Their phone has an "iconic" menu that permits illiterate customers to navigate lists of contact images.

Keystone Innovation—Corning

Finally, the Corning *keystone* innovation strategy (Graham and Shuldiner, 2001) recognizes that many of their "top-hit" innovations combine three important ingredients for market success: they (1) address mega-trend market needs, (2) address performance barriers, and (3) provide differentiated and sustainable solutions. Beginning with their successful light bulb in the late 1870s, which involved a glass envelope surrounding a filament, Corning addressed the mega-trend application for electricity promoted by Edison's incandescent light bulb. In addition, they addressed the performance barrier caused by air and other gases, which destroy the light-emitting filaments, by inventing advanced manufacturing processes that were difficult to replicate by their closest competitors. The combination of these three elements sustained the light bulb envelope business throughout the 20th century. In their most recent major innovation (early 2000s)—ultra-thin, super-flat glass substrates for liquid-crystal displays (LCDs)—Corning took advantage of the mega-trend in the growing demand for high-quality displays having small footprints. As discussed in Sections 14.3 and 15.3, they addressed a different performance barrier, that is, the non-uniform thickness of glass substrates for high-resolution images, by inventing a novel glass formulation and the Isopipe™ manufacturing process, which are also difficult to replicate. And consequently, their *keystone* innovation strategy has created a glass substrate business poised to withstand growing competition from OLED display technologies.

Summary

Innovative cultures require environments with greater degrees-of-freedom, which collide directly with environments having few, if any, degrees-of-freedom, typical of the later product-development stages. Successful companies maintain this delicate balance, enabling them to launch more successful products into the market in record times.

Pharmaceutical Products

While many firms seek to stimulate invention and innovation in these ways, special considerations are needed for the design of pharmaceutical products. As the design team creates its innovation map and carries out the Stage-Gate™ Product-Development Process (SGPDP), it is important to be aware of the typical *development cycle* or time line for the discovery and development of new pharmaceutical molecules, as discussed thoroughly by Pisano in *The Development Factory* (1997). The four key steps are examined next.

Discovery

Exploratory research is intended to identify molecules that will prove safe and effective in the treatment of disease. This step involves working backward to isolate classes of compounds or specific molecular structures that are likely to have the desired therapeutic effect, such as blocking a particular enzyme that may cause elevated blood pressure. Most of the work involves literature and patent searches, isolation or synthesis of test tube quantities, and testing on laboratory animals. This is an iterative process that usually involves the exploration of thousands of compounds to locate a handful that are sufficiently promising for further development. Increasingly, it involves methods of genomic analysis, with laboratory testing using microfluidic devices, and the application of data-mining techniques to locate the most promising proteins (and cells within which they can be grown), from numerous laboratory databases.

Preclinical Development

During this phase, a company seeks to obtain sufficient data on a drug to justify the more expensive and risky step of testing in humans. First, the drug is injected into animal species to determine its toxicity. In addition, pharmacological and pharmacokinetic studies are undertaken to quantify the main and side effects and the speeds of absorption and metabolism. In parallel, formulations are devised for administering the drug (e.g., in tablets, capsules, microcapsules, injections, or cream). This phase ends with the preparation and filing of an Investigational New Drug (IND) application that is filed with the FDA in the United States. This application seeks approval to begin testing the drug on humans. Note that about 50 percent of the potential drugs are eliminated for some reason during this phase. While the preclinical development is underway, process research is initiated in which alternative synthetic routes are considered and evaluated on a laboratory scale.

Clinical Trials

These trials are administered over three phases, each of which has a duration of one to two years. In Phase 1 trials, the drug is tested in multiple doses on healthy volunteers to determine whether there are significant side effects and to identify maximum tolerable doses. When Phase 1 is successful, Phase 2 trials begin, involving afflicted patients. Both drug and *placebo* treatments are administered to a control group, where the patients are unaware of whether they have received the drug, a placebo, or a substitute drug. During Phase 1 and especially during Phase 2, the development of a pilot plant facility is accelerated, as the demand for test quantities increases, leading into Phase 3 trials. During the latter phase, the drug is administered to thousands of patients at many locations over several years. The intent is to confirm the safety and efficacy of the drug over long-term use, as compared with existing drugs. When successful, the data are submitted to the FDA. Note that when approval is granted, only the expanded pilot plant, constructed for Phase 3, is permitted to produce the drug for commercial distribution.

Approval

Together with the data from the clinical trials, an application is prepared for the FDA, requesting permission to sell the drug. The FDA evaluates the application during a period that can last up to two years.

Summary

In summary, extensive work to create new molecules, usually proteins, that have the appropriate therapeutic properties begins in the concept stage of the Stage-Gate^{TM} Product-Development Process (SGPDP). This work is discussed further in Section 3.3, "Searching for New Materials—Basic Chemical Products." Then, as Phases 1 and 2 of the clinical trials proceed, process design is undertaken to produce large quantities of the drug, first for Phase 3 testing and then for commercial operation, as discussed in Chapter 4, "Process Synthesis—Basic Chemicals," in Section 4.4 for a plant to produce tissue plasminogen activator (tPA), a drug that can dissolve blood clots, which cause most heart attacks and strokes.

Socio-Technical Aspects of Product Design

While the technical contributions of engineers and scientists toward the design of new products were well recognized throughout the 20th century, the growing influences of social issues on product designs have been recognized only during the past few decades. As mentioned in connection with Figure 1.1, the roles of engineers and scientists in corporate infrastructures have become more broadly defined, ranging from technology development, to product development, to manufacturing, and to business development. Clearly, technically oriented professionals are attracted to these fields by their curiosities and desires to master complex phenomena, in the hope of creating new products for the benefit of their fellow human beings.

Initially technical education concentrates on the basic principles, such as the first and second laws of thermodynamics, chemical and biochemical kinetics, and momentum, heat, and mass transfer. This is followed by design, which introduces extensions that teach students to apply the basic principles to the creation of new products and processes. In its best mode, design involves the creation of alternative approaches to satisfying societal needs from which the best approach can be determined, making use of expertise from a wide variety of disciplines.

Historically, the focus of engineers and scientists was often limited, concentrating on the "what" and "how" dimensions of engineering design; that is, the *technical* dimensions. It was common to begin with new technologies, like the Coolidge process for drawing thin tungsten rods, and focus on "what" to manufacture (for example, light bulbs having longer life), and "how" to manufacture them (for example, providing tungsten filaments that evaporate slowly, prolonging the time between filament failures).

Recently, however, engineers and scientists have come to recognize the existence of two *social* dimensions, "who" and "why," that are crucial to the success of new products. Using these coordinates, designers better appreciate for whom the products are being designed and why the products are potentially useful. For example, while a new refrigerated pharmaceutical is attractive to consumers in the Western world, its utility in developing countries, without refrigerators, is nonexistent. As emphasized above, when creating *innovation maps* for new products, it is important to obtain the *voice of the customers*; that is, to fully understand customer needs, likes, and dislikes. Normally, the standard of living and commercial infrastructure are key to the *customer-value proposition*, with the consumer often not ready to take advantage of technical inventions, no matter how exciting and difficult they are to uncover.

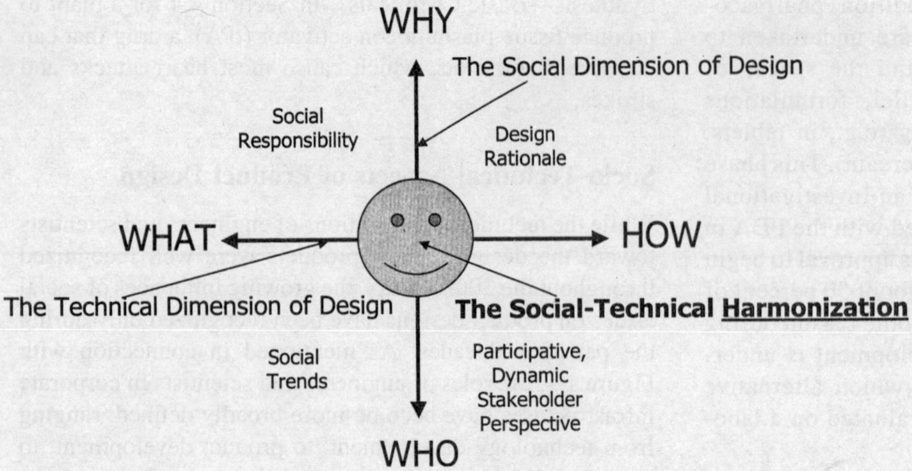

Figure 1.7 Socio-technical engineering design.

The interaction of the technical and social dimensions is shown schematically in Figure 1.7, which was introduced by Lu (2004), who summarized the results of a workshop on engineering design. In this diagram, the two technical axes ("what," "how") are orthogonal to the two social axes ("why," "who"). At the origin, the need for *social-technical harmonization* is depicted.

Each of the quadrants shows different aspects of the socio-technical interactions. Arbitrarily beginning with the first quadrant, the relationship of the "why" to the "how" dimensions represents the *design rationale*; that is, the recognition that *why* a product is being designed has a significant impact on *how* the design is accomplished. For example, to provide a product primarily used for travel, light materials having structural integrity are important.

Turning to the second quadrant, the impact of the "how" on the "who" dimensions represents the *stakeholder dynamics*; that is, it shows *how* the technical invention often limits the consumer base capable of benefiting from it. For example, the ability to reversibly generate heat by freezing at room temperatures permits the design of reversible hand warmers for use at sporting events in cold climates.

The third quadrant shows the impact of the "who" on the "what" dimensions, representing *social trends*; that is, it captures the impact of a group of consumers on the selection of products to be produced. Here an example is the influence of beachgoers on the colloidal microstructure of sunscreens for easy application at high temperatures in the presence of salt water.

Finally, the fourth quadrant shows the impact of the "what" on the "why" dimensions, representing *social responsibility*; that is, capturing the impact of the choice of products on the reasons consumers justify their usage. For example, given the political difficulties in dealing with oil-producing nations, the choice of fuels, like ethanol generated from biomass, is an important justification to customers concerned about sustainability and energy efficiency.

When carrying out product design using the Stage-Gate™ Product-Development Process (SGPDP), to be discussed in Chapter 2, design teams are increasingly cognizant of these socio-technical interactions. As will be seen, these are particularly important during the *concept* stage when superior concepts are being generated and selected.

1.4 ENVIRONMENTAL PROTECTION

One of the most significant changes that has occurred since the late1970s throughout the manufacturing and transportation sectors within the United States, and those of many other industrialized nations, is the transformation of environmental protection from a secondary to a primary issue. Through the Environmental Protection Agency (EPA), with the cooperation of the U.S. Congress, tighter environmental regulations have been legislated and enforced over this period. This has resulted in a noticeable improvement in air quality (especially in urban areas), a reduction in water pollution, and considerable progress in the remediation of many waste dumps containing toxic chemicals. In short, the United States and many other industrialized nations are rapidly increasing their emphases on maintaining a clean environment. To bring this about, in recent years large investments have been made by the chemical process industries to eliminate sources of pollution. These have increased the costs of manufacturing, which, in turn, have been transmitted to consumers through increased costs of end products. Because most producers are required to satisfy the same regulations, the effect has been to translate the costs to most competitors in an evenhanded manner. Problems have arisen, however, when chemicals are produced in countries that do not have strict environmental standards and are subsequently imported into the United States at considerably lower prices. Issues such as this are discussed regularly at international conferences on the environment, which convene every two or three years with the objective of increasing the environmental standards of all countries.

In the 21st century, the desire to achieve *sustainability* (that is, to meet the needs of society today while respecting the ability of future generations to meet their needs) in the selection of chemical products and raw materials has gained widespread attention. As discussed below, this involves the selection of carbon sources for raw materials and fuels, and their subsequent oxidation to carbon dioxide, with its influence on global warming due to the *greenhouse effect.* In addition, there are growing concerns about the politics of oil-producing nations, which provide fuels as well as the raw materials for carbon-based chemical products.

In this section, several of the more pressing environmental issues are reviewed, followed by a discussion of many environmental factors in process design. Then, a few primitive problem statements are reviewed, with reference to the more complete statements provided in the file, Supplement_to_Appendix_II.pdf, in the PDF Files folder, which can be downloaded from the Wiley Web site associated with this book. For more comprehensive coverage in these areas, the reader is referred to the discussions of "Environmental Protection, Process Safety, and Hazardous Waste Management" in *Frontiers in Chemical Engineering* (Amundson, 1988); *Environmental Considerations in Process Design and Simulation* (Eisenhauer and McQueen, 1993); *Pollution Prevention for Chemical Processes* (Allen and Rosselot, 1997); and *Green Engineering: Environmentally Conscious Design of Chemical Processes* (Allen and Shonnard, 2002). Early efforts to protect the environment focused on the removal of pollutants from waste gas, liquid, and solid streams. Effort has now shifted to *waste minimization* (e.g., waste reduction, pollution prevention) and *sustainability* in the selection of raw materials and chemical products.

www.wiley.com/go/global/seider

Environmental Issues

At the risk of excluding many key environmental issues, the following are singled out as being closely related to the design of chemical products and processes.

Burning of Fossil Fuels for Power Generation and Transportation

Because fossil fuels are the predominant sources of power worldwide, their combustion products are a primary source of several pollutants, especially in the urban centers of industrialized nations. More specifically, effluent gases from burners and fires contain sizable concentrations of SO_2, the nitrogen oxides (NO_x), CO, CO_2, soot, ash, and unburned hydrocarbons. These, in turn, result in many environmental problems, including acid rain (principally concentrated in H_2SO_4), smog and hazes (concentrated in NO_x), the accumulation of the so-called *greenhouse gas* (CO_2), volatile toxic compounds (e.g., formaldehyde, phenol), and organic gases (e.g., CO), which react with NO_x, especially on hot summer days, altering the O_3 level. As the adverse impacts of pollutants on animals, plant life, and humans are being discovered by scientists and engineers, methods are sought to reduce their levels significantly. In some cases, this is accomplished by one of several methods, such as separating the sources (e.g., sulfur compounds) from fuels; adjusting the combustion process (e.g., by reducing the temperature and residence time of the flame to produce less NO_x); separating soot, ash, and noxious compounds from effluent gases; reacting the effluent gases in catalytic converters; or through the use of algae to consume (through photosynthesis) large quantities of CO_2 in flue gases (a recently proposed technique now under study). As a rule of thumb, it should be noted that the cost of cleaning combustion products is approximately an order of magnitude less than the cost of removing contaminants from fuel. This is an important heuristic, especially when designing processes that are energy intensive, requiring large quantities of fuel.

Sustainability and Life-Cycle Design

By selecting sustainable raw materials and producing sustainable products, designers attempt to meet the needs of society today while respecting the anticipated needs of future generations. Such choices are also intended to avoid harming the environment and limiting the choices of future generations. In some cases, this translates to the use of so-called green raw materials and the production of so-called green products. These often help to resolve health problems, provide environmental protection, preserve natural resources, and prevent climate change.

As mentioned above, the growing emphasis on *sustainability* is closely related to the increasing recognition of global warming due to the *greenhouse effect* as well as political problems associated with the traditional suppliers of oil and natural gas. Historically, most chemical products have been derived from methane, ethane, propane, and aromatics, normally obtained from oil and natural gas. Furthermore, a large percentage of energy for manufacturing (on the order of 80 percent) and wastes produced in manufacturing (also on the order of 80 percent) are associated with the chemical industries, including petroleum refining, chemicals production, forest products, steel, aluminum, glass, and cement. To achieve sustainability while producing high-quality products, it is desirable to use small amounts of raw materials and energy, and to produce small amounts of waste.

When planning for sustainability in the 21st century, the rapid growth of the large developing nations, especially China and India, is important. Some estimates project that the world population will stabilize at 9–10 billion people, with the consumption of commodities (steel, chemicals, lumber, . . .) increasing by factors of 5–6 and energy by a factor of 3.5. Furthermore, the choices of resources are complicated by sustainability considerations. Decisions to take advantage of today's cheap prices and easy accessibility may result in expensive or inaccessible raw materials for future generations.

With the price of oil having quadrupled in just three years, there has been a move toward the usage of renewable "green" resources. Following the lead in Brazil over the past few decades, biomass (e.g., sugars, corn, and cellulosic wastes) is being converted to ethanol, principally as a gasoline substitute. In addition, biomass has been used to produce chemicals (e.g., 1,3-propanediol and tetrahydrofuran). However, such carbon sources, when burned as fuel or incinerated as waste, produce carbon dioxide, and consequently, it has become increasingly important to find practical ways to sequester carbon dioxide rather than release it into the atmosphere. For these reasons, alternate energy sources, such as hydrogen, nuclear, wind, solar, and geothermal, are gaining increased attention.

It is also becoming common to consider the full *life cycle* when designing chemical products. A growing class of products, formed from biomass, are biodegradable. For example, biodegradable microcapsules carrying pharmaceuticals are injected into the bloodstream for delayed drug delivery over extended periods on the order of one month. In these cases, the raw materials are *renewable* and there are no waste-disposal issues.

Handling of Toxic Wastes

In the chemical and nuclear power industries, large quantities of toxic wastes are produced annually, largely in wastewater streams, which in 1988 amounted to 97 percent of the wastes produced, as shown in the pie chart of Figure 1.8. While a small portion is incinerated (on the order of 3 percent in the late 1980s), the bulk is disposed of in or on the land, with a variety of methods having been introduced over the past century to bury these wastes. Since the late 1960s, many of the burial sites (e.g., Love Canal, Times Beach) have threatened the health of nearby residents and, more broadly, have threatened to contaminate the underground water supply throughout entire states and countries. In this regard, studies by the state of California have shown that aqueous waste streams from the processing of electronic materials are posing widespread threats to the groundwater in California's Silicon Valley. In fact, this area has a leading number of sites on the U.S. National Priority List of toxic waste dumps (which is comprised of approximately 10,000 sites throughout the United States). In process design, it is essential that facilities be included to remove pollutants from wastewater streams. The design of mass-exchange networks (MENs) for this and other purposes is the subject of Chapter 10.

Bioaccumulated Chemicals

Probably the most well-known cases of chemicals that have been discovered to *bioaccumulate* in the soil and plant life are the insecticide DDT (1,1-bis(4-chlorophenyl)-2,2,2-trichloroethane; $C_{14}H_9Cl_5$) and the solvent PCBs (polychlorinated biphenyls). DDT was sprayed in large quantities by low-flying airplanes to kill insects and pests throughout the 1950s. Unfortunately, although effective for protecting crops, forests, and plant life, toxic effects in birds, animals, and humans were strongly suspected, as discussed in Section 1.6. Consequently, DDT was banned by the U.S. EPA in 1972. Its effect, however, will remain for some time due to its having bioaccumulated in the soil and plant life.

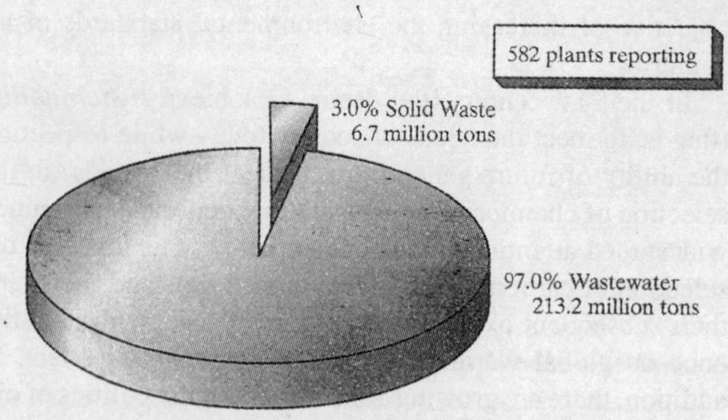

Figure 1.8 Hazardous waste generation in the United States in 1988 (Eisenhauer and McQueen, 1993).

Toxic Metals and Minerals

In this category, major changes have taken place since the late 1960s in response to the discoveries of the toxic effects of lead, mercury, cadmium, and asbestos on animals and humans. After lead poisoning (accompanied by brain damage, disfigurement, and paralysis) was related to the ingestion of lead-based paints by children (especially in older buildings that are not well maintained), the EPA banned lead from paints as well as from fuels. In fuel, tetraelhyl lead had been used as an octane enhancer throughout the world. It was subsequently replaced by methyl *tertiary*-butyl ether (MTBE), which is also being replaced due to reports that it can contaminate ground water. Mercury, which has been the mainstay of manometers in chemistry laboratories, has similarly been found to be extremely toxic, with disastrous effects of accidental exposure and ingestion reported periodically. In the case of asbestos, its toxic effects have been known since the late 1940s, yet it remains a concern in all buildings built before then. Gradually, as these buildings are being renovated, sheets of asbestos insulation and asbestos ceiling tiles are being removed and replaced by nontoxic materials. Here, also, the incidents of asbestos poisoning are associated most often with older buildings that have not been well maintained.

Summary

As the adverse effects of these and other chemicals becomes better understood, chemical engineers are being called on to satisfy far stricter environmental regulations. In many cases, these regulations are imposed to be safe even before sufficient data are available to confirm toxic effects. For these reasons, chemical companies are carefully reexamining their existing products and processes, and evaluating all proposed plants to confirm that they are environmentally sound, at least insofar as meeting the regulations imposed, or anticipated to be imposed, by the environmental regulation agencies.

Environmental Factors in Product and Process Design

The need to retrofit existing plants and to design new, environmentally sound plants has required chemical engineers to become far more proficient in accounting for environmentally related factors. In this section, a few of the better-recognized factors are discussed. Additional coverage is included related to purges in Section 6.3, energy conservation in Chapter 9, and wastewater treatment in Chapter 10. More complete coverage may be found in the comprehensive textbook *Green Engineering: Environmentally Conscious Design of Chemical Processes* (Allen and Shonnard, 2002).

Reaction Pathways to Reduce Byproduct Toxicity

The selection of reaction pathways to reduce byproduct toxicity is a key consideration during preliminary process synthesis, when the reaction operations are positioned. As the reaction operations are determined by chemists and biochemists in the laboratory, the toxicity of all of the chemicals, especially chemicals recovered as byproducts, needs to be evaluated. For this purpose, companies have toxicity laboratories and, in many cases, large repositories of toxicity data. One useful source, especially for students at universities, is the *Pocket Guide to Chemical Hazards* of the National Institute for Occupational Safety and Health (NIOSH, 1987). Clearly, when large quantities of toxic chemicals are anticipated, other reaction pathways must be sought; when these cannot be found, design concepts are rejected, except under unusual circumstances.

Reducing and Reusing Wastes

Environmental concerns have caused chemical engineers to place even greater emphasis on recycling, not only unreacted chemicals but also product and byproduct chemicals. In so doing, design teams commonly anticipate the *life cycles* of their products and byproducts, paying special attention to the waste markets, so as to select the appropriate waste quality. Stated differently, the team views the proposed plant as a producer of engineering scrap and attempts to ensure that there will be a market for the chemicals produced after their useful life is over. Clearly, this is a principal consideration in the production of composite materials and polymers. In this connection, it is important to plan on producing segregated wastes when they are desired by the waste market, and in so doing, to avoid overmixing the waste streams.

Avoiding Nonroutine Events

To reduce the possibilities for accidents and spills, with their adverse environmental consequences, processes are often designed to reduce the number of transient operations, cleanup periods, and catalyst regeneration cycles. In other words, emphasis is on the design of a process that is easily controlled at or near a nominal steady state, with reliable controllers and effective fault-detection sensors.

Materials Characterization

Often, waste chemicals are present in small amounts in gaseous or liquid effluents. To maintain low concentrations of such chemicals below the limits of environmental regulations, it is important to use effective and rapid methods for measuring or deducing their concentrations from other measurements. In this regard, the design team needs to understand the effect of concentration on toxicity, which can vary significantly in the dilute concentration range. Yet another consideration is to design the plant to use recycled chemicals—that is, someone else's waste. When this is accomplished, it is necessary to know the range of compositions within which the waste chemicals are available.

Design Objectives, Constraints, and Optimization

Environmental objectives are normally not well defined because economic objective functions normally involve profitability measures, whereas the value of reduced pollution is not easily quantified by economic measures. As a consequence, design teams often formulate mixed objective functions that attempt to express environmental improvements in financial terms. In other cases, the team may settle for the optimization of an economic objective function, subject to bounds on the concentrations of the solutes in the waste streams. It is important to assess whether the constraints are *hard* (not allowed to be violated) or *soft* (capable of being violated under unusual circumstances). Emphasis must be placed on the formulation of each constraint and the extent to which it must be honored.

Regulations

As mentioned previously, some environmental regulations can be treated as constraints to be satisfied during operation of the process being designed. When a mathematical model of the proposed process is created, the design team can check that these constraints are satisfied for the operating conditions being considered. When an objective function is formulated, the design variables can be adjusted to obtain the maximum or minimum while satisfying the constraints. Other regulations, however, are more difficult to quantify. These involve the expectations of the public and the possible backlash should the plant be perceived as a source of pollution. In a similar vein, constraints may be placed on the plant location, principally because the local government may impose zoning regulations that require chemical plants to be located in commercial areas, beyond a certain distance from residential neighborhoods. To keep these regulations from becoming too prohibitive, chemical companies have a great incentive to gain public confidence by satisfying environmental regulations and maintaining excellent safety records, as discussed in Section 1.5.

Intangible Costs

Like the regulations imposed by local governments, some of the economic effects of design decisions related to the environment are very difficult to quantify. These include the cost of liability when a plant is found to be delinquent in satisfying regulations, and in this connection, the cost of legal fees, public relations losses, and delays incurred when environmental groups stage protests. Normally, because these costs cannot be estimated reliably by a design team, mixed objectives are not formulated and no attempts are made to account for them in an optimization study. Rather, the design team concentrates on ensuring that the regulations will be satisfied, thereby avoiding legal fees, public relations losses, and the complications associated with public demonstrations.

Properties of Dilute Streams

Most pollutants in the effluent and purge streams from chemical plants are present in dilute concentrations. Furthermore, since the regulations often require that their concentrations be kept below parts per million or parts per billion, reliable and fast analysis methods are needed to ensure that the regulations are satisfied. Beyond that, it is often important to understand the impact of the concentration on the kinetics of these species in the environment—for example, the rates of chemical reaction of organic species, such as CO, with NO_x in the atmosphere to produce O_3, and the rate at which other reaction byproducts are formed. With this knowledge, a company can help regulatory agencies arrive at concentration limits more scientifically and, in some cases, at limits that are less restrictive, and cost companies, and the consumers of their products, less in the long run. Note that, in urban smog, high concentrations of ozone often create problems for people with respiratory ailments.

Properties of Electrolytes

Many aqueous streams contain inorganic compounds that dissociate into ionic species, including acids, bases, and salts, often in dilute concentrations. These electrolytic solutions commonly occur in the manufacture of inorganic chemicals (e.g., soda ash, Na_2CO_3), in the strong solvents used in the pulp and paper industry, in the aqueous wastes associated with the manufacture of electronic materials (e.g., silicon wafers, integrated circuits, photovoltaic films), and in many other industries. Strong electrolytes dissociate into ionic species whose interactions with water and organic molecules are crucial to understanding the state of a mixture—that is, the phases present (vapor, water, organic liquid, solid precipitates, etc.) at a given temperature and pressure. Hence, when designing processes that involve electrolytes, a design team needs to include the properties of ionic species in its thermophysical properties database. Fortunately, to provide assistance for designers, databases and facilities for estimating the thermophysical properties of a broad base of ionic species over an increasing range of temperatures and pressures, are available in process simulators.

Environmental Design Problems

Since the late 1970s, the number of design projects focusing on the solution of environmental problems has increased significantly. These, in turn, are closely related to environmental regulations, which have become increasingly strict. Although it is beyond the scope of this book to provide a comprehensive treatment of the many kinds of designs that have been completed, it is important that the reader gain a brief introduction to typical design problems. This is accomplished through the design projects listed in Table 1.1. As can be seen, a large fraction of the design projects are concerned with air quality; others involve water treatment; two involve soil treatment; one involves the conversion of waste fuel to chemicals; one proposes the use of a biochemical conversion to consume solid waste and produce ethanol fuel; and several involve the production of fuels and chemicals from renewable resources. The problem statements for these design projects, as they were presented to student groups, are reproduced in the file, Supplement_to_Appendix_II.pdf, in the PDF Files folder, which can be downloaded from the Wiley Web site associated with this book. Keep in mind that, as the designs proceeded, the design teams often upgraded the information provided, and in some cases created variations that were not anticipated by the originator of the problem statement.

A closer look at Table 1.1 shows that the projects address many aspects of air-quality control. Two alternative approaches to sulfur removal from fuels are proposed, one involving desulfurization of the fuel, the other the recovery of sulfur from its combustion products. One is concerned with NO_x removal from combustion products, and three involve the recovery of hydrocarbons from effluent gases. One explores the interesting possibility of growing algae by the photosynthesis of CO_2 from combustion gases as a vehicle for reducing the rate at which CO_2 is introduced into the atmosphere. Under water treatment, the projects involve the recovery of organic and inorganic chemicals from aqueous waste streams. Two alternative approaches to soil treatment are proposed, including the use of phytoremediation; that is, using plants to absorb lead and other heavy metals. All of the projects involve chemical reactions, and consequently, the design teams are comprised of chemical engineers, chemists, and biochemists. In this respect, it seems clear that chemistry and biology are the key ingredients that qualify chemical engineers to tackle these more challenging environmental problems.

Table 1.1 Environmental Design Projects

Project	Location in Book†
Environmental—Air Quality	
R134a Refrigerant (2001)	App. IIS—Design Problem A-IIS.9.1
Biocatalytic Desulfurization of Diesel Oil (1994)	App. IIS—Design Problem A-IIS.9.2
Sulfur Recovery Using Oxygen-Enriched Air (1993)	App. IIS—Design Problem A-IIS.9.3
California Smog Control (1995)	App. IIS—Design Problem A-IIS.9.4
Zero Emissions (1991)	App. IIS—Design Problem A-IIS.9.5
Volatile Organic Compound Abatement (1994)	App. IIS—Design Problem A-IIS.9.6
Recovery and Purification of HFC by Distillation (1997)	App. IIS—Design Problem A-IIS.9.7
Carbon Dioxide Fixation by Microalgae for Mitigating the Greenhouse Effect (1993)	App. IIS—Design Problem A-IIS.9.8
Hydrogen Generation for Reformulated Gasoline (1994)	App. IIS—Design Problem A-IIS.9.9
R125 Refrigerant Manufacture (2004)	App. IIS—Design Problem A-IIS.9.10
Zero-Emissions Solar Power Plant (2008)	App. IIS—Design Problem A-IIS.9.11
Removing CO_2 from Stack Gas and Sequestration Technologies (2008)	App. IIS—Design Problem A-IIS.9.12
Environmental–Water Treatment	
Effluent Remediation from Wafer Fabrication (1993)	App. IIS—Design Problem A-IIS.10.1
Recovery of Germanium from Optical Fiber Manufacturing Effluents (1991)	App. IIS—Design Problem A-IIS.10.2
Solvent Waste Recovery (1997)	App. IIS—Design Problem A-IIS.10.3
Environmental–Soil Treatment	
Phytoremediation of Lead-Contaminated Sites (1995)	App. IIS—Design Problem A-IIS.11.1
Soil Remediation and Reclamation (1993)	App. IIS—Design Problem A-IIS.11.2
Environmental–Renewable Fuels and Chemicals	
Fuel Processor for 5 KW PEM Fuel Cell Unit (2002)	App. IIS—Design Problem A-IIS.12.1
Production of Low-Sulfur Diesel Fuel (2000)	App. IIS—Design Problem A-IIS.12.2
Waste Fuel Upgrading to Acetone and Isopropanol (1997)	App. IIS—Design Problem A-IIS.12.3
Conversion of Cheese Whey (Solid Waste) to Lactic Acid (1993)	App. IIS—Design Problem A-IIS.12.4
Ethanol for Gasoline from Corn Syrup (1990)	App. IIS—Design Problem A-IIS.12.5
Furfural and Methyl-tetrahydrofuran-based Biorefinery (2008)	App. IIS—Design Problem A-IIS.12.6
Furfural and THF in China – Corn to Clothes (2008)	App. IIS—Design Problem A-IIS.12.7
Diethyl Succinate Manufacture within a Biorefinery (2008)	App. IIS—Design Problem A-IIS.12.8
1-3 Propanediol from Corn Syrup (2008)	App. IIS—Design Problem A-IIS.12.9
Biobutanol as Fuel (2008)	App. IIS—Design Problem A-IIS.12.10
Green Diesel Fuel – A Biofuel Process (2008)	App. IIS—Design Problem A-IIS.12.11
Environmental–Miscellaneous	
Combined Cycle Power Generation (2001)	App. IIS—Design Problem A-IIS.13.1

†In the file, Supplement_to_Appendix_II.pdf, in the PDF Files folder, which can be downloaded from the Wiley Web site associated with this book.

1.5 SAFETY CONSIDERATIONS

A principal objective in the design and operation of chemical processes is to maintain safe conditions for operating personnel and inhabitants who live in the vicinity of the plants. Unfortunately, the importance of meeting this objective is driven home periodically by accidents, especially accidents in which lives are lost and extensive damage occurs. To avoid this, all companies have extensive safety policies and procedures to administer them. In recent years, these have been augmented through cooperative efforts coordinated by technical societies, for example, the Center for Chemical Plant Safety of the American Institute of Chemical Engineers, which was formed in 1985, shortly after the accident in Bhopal, India, on December 3, 1984. In this accident, which took place in a plant partially owned by Union Carbide and partially owned locally, water (or some other substance—the cause is still uncertain) accidentally flowed into a tank in which the highly reactive intermediate, methyl isocyanate (MIC) was stored, leading to a rapid increase in temperature accompanied by boiling, which caused toxic MIC vapors to escape from the tank. The vapors passed through a pressure-relief system and into a scrubber and flare system that had been installed to consume the MIC in the event of an accidental release. Unfortunately, these systems were not operating, and approximately 25 tons of toxic MIC vapor were released, causing a dense vapor cloud that escaped and drifted over the surrounding community, killing more than 3,800 civilians and seriously injuring an estimated 30,000 more.

Like Section 1.4 on environmental issues, this section begins with a review of two safety issues that are considered by many design teams, followed by an introduction to many of the design approaches for dealing with these issues. For more comprehensive coverage of these areas, the reader is referred to *Chemical Process Safety: Fundamentals with Applications* (Crowl and Louvar, 1990); *Plant Design for Safety—A User-Friendly Approach* (Kletz, 1991); a collection of student problems, *Safety, Health, and Loss Prevention in Chemical Processes: Problems for Undergraduate Engineering Curricula—Student Problems* (American Institute of Chemical Engineers, 1990); and *Guidelines for Engineering Design for Process Safety, CCPS*, AIChE (1993).

The U.S. Chemical Safety and Hazard Investigation Board (CSB), established by the Clean Air Act Amendments of 1990, is an independent federal agency with the mission of ensuring the safety of workers and the public by preventing or minimizing the effects of chemical incidents. They attempt to determine the root and contributing causes of chemical accidents. Their Web site at http://www.csb.gov is a very useful source of brief and detailed accident reports.

Safety Issues

Of the many potential safety issues, two are singled out for coverage here because they must be confronted often in the design of chemical, petroleum, and petrochemical plants and in other plants in which exothermic reactions and operations occur at elevated pressures.

Fires and Explosions

In organic chemical processes, it is not uncommon for sizable concentrations of flammable chemicals to accumulate in air or pure oxygen with the possibilities of ignition or even explosion. For this reason, laboratory studies have been carried out to determine the flammability limits for many of the common organic chemical vapors. These limits at 25°C and 1 atm are listed for many chemicals in Table 1.2, where the LFL is the lower flammability limit (that is, the volume percent of the species in air below which flammability does not occur) and the UFL is the upper flammability limit (above which flammability does not occur). Within these limits, flames and explosions can occur and, consequently, design teams must be careful to keep the concentrations outside the flammability range. In addition, Table 1.2 provides auto-ignition temperatures, above which a flammable mixture is capable of extracting enough energy from the environment to self-ignite. At lower temperatures, an ignition source must be present. The flash point, given in the second column of Table 1.2, is the lowest temperature at which sufficient vapor exists in air to form an ignitable mixture. At the flash point, the vapor burns, but only briefly, as insufficient vapor is formed to sustain combustion.

Table 1.2 pertains to pure chemicals. For mixtures, the flammability limits are often estimated using the Le Chatelier equation, an empirical equation that must be used with caution:

$$\mathrm{LFL_{mix}} = \frac{1}{\sum_{i-1}^{c}(y_i/\mathrm{LFL}_i)} \quad \mathrm{UFL_{mix}} = \frac{1}{\sum_{i-1}^{c}(y_i/\mathrm{UFL}_i)} \tag{1.1}$$

where LFL_i and UFL_i are the flammability limits of species i, y_i is the mole fraction of species i in the vapor, and C is the number of chemical species in the mixture, excluding air.

To extend the flammability limits to elevated temperatures and pressures, the following equations have been developed:

$$\mathrm{LFL}_T = \mathrm{LFL}_{25}\left[1 - \frac{0.75(T-25)}{\Delta H_c}\right] \tag{1.2a}$$

$$\mathrm{UFL}_T = \mathrm{UFL}_{25}\left[1 + \frac{0.75(T-25)}{\Delta H_c}\right] \tag{1.2b}$$

and

$$\mathrm{UFL}_p = \mathrm{UFL} + 20.6(\log P + 1) \tag{1.3}$$

where T is the temperature (in °C), ΔH_c is the net heat of combustion (in kcal/mol at 25°C), P is the pressure (in MPa absolute), and UFL is the upper flammability limit at 101.3 kPa (1 atm). The lower flammability limit is not observed to vary significantly with the pressure. These equations, plus others to estimate the flammability limits for species not listed in Table 1.2, are presented by Crowl and Louvar (1990), with a more complete discussion and references to their sources. With this kind of information, the process designer makes sure that flammable mixtures do not exist in the process during startup, steady-state operation, or shutdown.

Toxic Releases and Dispersion Models

In chemical processing, it is desirable to avoid working with chemicals that are toxic to animals, humans, and plant life. This is an important consideration as design teams select from among the possible raw materials and consider alternate reaction paths, involving intermediate chemicals and byproducts. In some cases, decisions can be made to work with nontoxic chemicals. However, toxicity problems are difficult to avoid, especially at the high concentrations of chemicals in many process streams and vessels. Consequently, the potential for a release in toxic concentrations during an accident must be considered carefully by design teams. In so doing, a team must identify the ways in which releases can occur; for example, due to the buildup of pressure in an explosion, the rupture of a pipeline due to surges at high pressure, or the collision of a tank car on a truck or train. It is also important for the team to select protective devices and processing units, to assess their potential for failure, and, in the worse case, to model the spread of a dense, toxic vapor. Given the potential for the rapid spreading of a toxic cloud, it is often necessary to find an alternative design,

Table 1.2 Flammability Limits of Liquids and Gases

Compound	Flash Point (°F)	LFL (%) in air	UFL (%) in air	Autoignition temperature (°F)
Acetone	0.0[a]	2.5	13	1,000
Acetylene	Gas	2.5	100	
Acrolein	−14.8	2.8	31	
Acrylonitrile	32	3.0	17	
Aniline	158	1.3	11	
Benzene	12.0[b]	1.3	7.9	1,044
n-Butane	−76	1.6	8.4	761
Carbon monoxide	Gas	12.5	74	
Chlorobenzene	85[b]	1.3	9.6	1,180
Cyclohexane	−1[b]	1.3	8	473
Diborane	Gas	0.8	88	
Dioxane	53.6	2.0	22	
Ethane	−211	3.0	12.5	959
Ethyl alcohol	55	3.3	19	793
Ethylene	Gas	2.7	36.0	914
Ethylene oxide	−20[a]	3.0	100	800
Ethyl ether	−49.0[b]	1.9	36.0	180
Formaldehyde		7.0	73	
Gasoline	−45.4	1.4	7.6	
n-Heptane	24.8	1.1	6.7	
n-Hexane	−15	1.1	7.5	500
Hydrogen	Gas	4.0	75	1,075
Isopropyl alcohol	53[a]	2.0	12	850
Isopropyl ether	0	1.4	7.9	830
Methane	−306	.5	15	1,000
Methyl acetate	15	3.1	16	935
Methyl alcohol	54[a]	6	36	867
Methyl chloride	32	8.1	17.4	1,170
Methyl ethyl ketone	24[a]	1.4	11.4	960
Methyl isobutyl ketone	73	1.2	8.0	860
Methyl methacrylate	50[a]	1.7	8.2	790
Methyl propyl ketone	45	1.5	8.2	941
Naphtha	−57	1.2	6.0	550
n-Octane	55.4	1.0	6.5	
n-Pentane	−40	1.51	7.8	588
Phenol	174	1.8	8.6	
Propane	Gas	2.1	9.5	
Propylene	−162	2.0	11.1	927
Propylene dichloride	61	3.4	14.5	1,035
Propylene oxide	−35	2.3	36	869
Styrene	87[b]	1.1	7.0	914
Toluene	40	1.2	7.1	997

[a]Open-cup flash point

[b]Closed-cup flash point

Source: Martha W. Windholtz, Ed., *The Merck Index: An Encyclopedia of Chemicals, Drugs, and Biologicals*, 10th ed. (Merck, Rahway, NJ, 1983). p. 1124; Gressner G. Hawley, Ed., *The Condensed Chemical Dictionary*, 10th ed. (Van Nostrand Reinhold, New York, 1981), pp. 860–861: Richard A. Wadden and Peter A. Scheff, *Engineering Design for the Control of Workplace Hazards* (McGraw-Hill, New York, 1987), pp. 146–156.

not involving this chemical, rather than lake the chance of exposing the surrounding community to a serious health hazard. Although it is beyond the scope of this discussion, it should be noted that dispersion models are developed by chemical engineers to predict the movement of vapor clouds under various conditions—for example, a continuous point release, at steady state, with no wind; a puff with no wind; a transient, continuous point release with no wind; as well as all of the previously mentioned factors with wind. These and other models are described by Crowl and Louvar (1990) and de Nevers (1995), accompanied by example calculations.

Design Approaches Toward Safe Chemical Plants

In the previous discussion of two important safety issues, design approaches to avoid accidents have been introduced. This section provides a more complete enumeration without discussing implementational details, which are covered by Crowl and Louvar (1990).

Techniques to Prevent Fires and Explosions

One method of preventing tires and explosions is *inerting*—that is, the addition of an inert gas to reduce the oxygen concentration below the minimum oxygen concentration (MOC), which can be estimated using the LFL and the stoichiometry of the combustion reaction. Another method involves avoiding the buildup of static electricity and its release in a spark that can serve as an ignition source. Clearly, the installation of grounding devices, and the use of antistatic additives that increase conductivity, reducing the buildup of static charges, can help to reduce the incidence of sparking. In addition, explosion-proof equipment and instruments are often installed; for example, explosion-proof housings that do not prevent an explosion, but that do absorb the shock and prevent the combustion from spreading beyond the enclosure. Yet another approach is to ensure that the plant is well ventilated, in most cases constructed in the open air, to reduce the possibilities of creating flammable mixtures that could ignite. Finally, sprinkler systems are often installed to provide a rapid response to fires and a means to contain them effectively.

Relief Devices

In processes where pressures can build rapidly, especially during an accident, it is crucial that the design team provide a method for relieving the pressure. This is accomplished using a variety of relief devices, depending on whether the mixtures are vapors, liquids, solids, or combinations of these phases. In some cases the vessels can be vented to the atmosphere; in other cases they are vented to containment systems, such as scrubbers, flares, and condensers. The devices include relief and safety valves, knock-out drums, rupture disks, and the like. Relief system design methodologies are presented in detail in the AIChE publication *Emergency Relief System Design Using DIERS Technology* (1992).

Hazards Identification and Risk Assessment

Hazards identification and risk assessment are key steps in process design. As shown in Figures PI.1, PII.1, and PIII.1, and discussed in the introductions to Parts One, Two, and Three, they are normally carried out in connection with the preparation of the final design. In these steps, the plant is carefully scrutinized to identify all sources of accidents or hazards. This implies that the design team must consider the propagation of small faults into catastrophic accidents, an activity that is complicated by the possibility of two or more faults occurring either simultaneously or in some coordinated fashion. At some point, especially after the economics satisfy the feasibility test, the design team normally prepares a HAZOP study in which all of the possible paths to an accident are identified. Then when sufficient probability data are available, a fault tree is created and the probability of the occurrence for each potential accident is computed. Clearly, this requires substantial experience in operating comparable facilities, which is generally available in the large chemical companies. Note that an introduction to HAZOP analysis is presented in the supplement to this chapter. See the file, Supplement_to_Chapter_1.pdf, in the PDF Files folder, which can be downloaded from the Wiley Web site associated with this book.

Material Safety Data Sheets

A process design should be accompanied by a Material Safety Data Sheet (MSDS) for every chemical appearing in the process. These sheets, which are developed by chemical manufacturers and kept up to date under OSHA (Occupational Safety and Health Agency of the federal government) regulations, contain safety and hazard information, physical and chemical characteristics, and precautions on safe handling and use of the chemical. The MSDSs, which usually involve several pages of information, are available on the Internet at:

http://hazard.com/msds/

http://www.ilpi.com/msds/

http://www.msdssearch.com

1.6 ENGINEERING ETHICS

In 1954, the National Society of Professional Engineers (NSPE) adopted the following statement, known as the Engineers' Creed:

As a Professional Engineer, I dedicate my professional knowledge and skill to the advancement and betterment of human welfare.

I pledge:

To give the utmost of performance;
To participate in none but honest enterprise;
To live and work according to the laws of man and the highest standards of professional conduct;
To place service before profit, the honor and standing of the profession before personal advantage, and the public welfare above all other considerations.
In humility and with need for Divine Guidance, I make this pledge.

In 1977, a similar statement was approved by the Accreditation Board for Engineering and Technology (ABET), as follows:

Engineers uphold and advance the integrity, honor, and dignity of the engineering profession by:

I. Using their knowledge and skill for the enhancement of human welfare;

II. Being honest and impartial, and serving with fidelity the publics, their employees;

III. Striving to increase the competence and prestige of the engineering profession; and

IV. Supporting the professional and technical societies of their disciplines.

These two statements have to do with ethics, also called *moral philosophy*, which is derived from the Greek *ethika*, meaning character. Thus, ethics deals with standards of conduct or morals. Unfortunately, there are no universal standards; only the ethics of Western civilization are considered in detail here. There is a movement toward the development of global ethics, which is described briefly at the end of this section.

Engineering ethics is concerned with the personal conduct of engineers as they uphold and advance the integrity, honor, and dignity of engineering while practicing their profession. This conduct of behavior has obligations to (1) self, (2) employer and/or client, (3) colleagues and co-workers, (4) public, and (5) environment.

Specific examples of these obligations are given in individual codes of ethics adopted by the various engineering societies (e.g., AIChE, ASCE, ASME, and IEEE) and by the NSPE. The following is the Code of Ethics adopted by the American Institute of Chemical Engineers (AIChE):

Members of the American Institute of Chemical Engineers shall uphold and advance the integrity, honor, and dignity of the engineering profession by: being honest and impartial and serving with fidelity their employers, their clients, and the public; striving to increase the competence and prestige of the engineering profession; and using their knowledge and skill for the enhancement of human welfare. To achieve these goals, members shall:

1. *Hold paramount the safety, health, and welfare of the public in performance of their professional duties.*
2. *Formally advise their employers or clients (and consider further disclosure, if warranted) if they perceive that a consequence of their duties will adversely affect the present or future health or safety of their colleagues or the public.*
3. *Accept responsibility for their actions and recognize the contributions of others; seek critical review of their work and offer objective criticism of the work of others.*
4. *Issue statements and present information only in an objective and truthful manner.*
5. *Act in professional matters for each employer or client as faithful agents or trustees, and avoid conflicts of interest.*
6. *Treat fairly all colleagues and co-workers, recognizing their unique contributions and capabilities.*
7. *Perform professional services only in areas of their competence.*
8. *Build their professional reputations on the merits of their services.*
9. *Continue their professional development throughout their careers, and provide opportunities for the professional development of those under their supervision.*

A more detailed code of ethics for engineers was adopted initially by the NSPE in July 1964. Since then, it has been updated 24 times and will probably continue to receive updates. The January 2006 version is shown in Figure 1.9. Some idea of the direction in which engineering ethics is moving may be gleaned from the following changes made since 1996, as taken from the NSPE Web site at: http://www.nspe.org.

February 2001—The NSPE Board approved the following change to the Code of Ethics: Deletion of Section III. I.e. "Engineers shall not actively participate in strikes, picket lines, or other collective coercive action."

July 2002—The NSPE Board approved the following changes to the Code of Ethics: New Section II. I.e. "Engineers shall not aid or abet the unlawful practice of engineering by a person or firm." Old Section II.I.e. was renumbered as new Section II.I.f.

January 2003—The NSPE Board approved a new section (III.9.e.) to the Code of Ethics that reads: "Engineers shall continue their professional development throughout their careers and should keep current in their specialty fields by engaging in professional practice, participating in continuing education courses, reading in the technical literature, and attending professional meetings and seminars."

January 2006—The NSPE Board approved a new section (III.2.d.) to the Code of Ethics that reads: "Engineers shall strive to adhere to the principles of sustainable development in order to protect the environment for future generations." Footnote 1. "Sustainable development" is the challenge of meeting human needs for natural resources, industrial products, energy, food, transportation, shelter, and effective waste management while conserving and protecting environmental quality and the natural resource base essential for future development.

It is important for an engineer, or one preparing for entry into the profession, to develop the ability to address, in an

Preamble

Engineering is an important and learned profession. As members of this profession, engineers are expected to exhibit the highest standards of honesty and integrity. Engineering has a direct and vital impact on the quality of life for all people. Accordingly, the services provided by engineers require honesty, impartiality, fairness, and equity, and must be dedicated to the protection of the public health, safety, and welfare. Engineers must perform under a standard of professional behavior that requires adherence to the highest principles of ethical conduct.

I. Fundamental Canons

Engineers, in the fulfillment of their professional duties, shall:

1. Hold paramount the safety, health, and welfare of the public.
2. Perform services only in areas of their competence.
3. Issue public statements only in an objective and truthful manner.
4. Act for each employer or client as faithful agents or trustees.
5. Avoid deceptive acts.
6. Conduct themselves honorably, responsibly, ethically, and lawfully so as to enhance the honor, reputation, and usefulness of the profession.

II. Rules of Practice

1. Engineers shall hold paramount the safety, health, and welfare of the public.

   a. If engineers' judgment is overruled under circumstances that endanger life or property, they shall notify their employer or client and such other authority as may be appropriate.
   b. Engineers shall approve only those engineering documents that are in conformity with applicable standards.
   c. Engineers shall not reveal facts, data, or information without the prior consent of the client or employer except as authorized or required by law or this Code.
   d. Engineers shall not permit the use of their name or associate in business ventures with any person or firm that they believe is engaged in fraudulent or dishonest enterprise.
   e. Engineers shall not aid or abet the unlawful practice of engineering by a person or firm.
   f. Engineers having knowledge of any alleged violation of this Code shall report thereon to appropriate professional bodies and, when relevant, also to public authorities, and cooperate with the proper authorities in furnishing such information or assistance as may be required.

2. Engineers shall perform services only in the areas of their competence.

   a. Engineers shall undertake assignments only when qualified by education or experience in the specific technical fields involved.
   b. Engineers shall not affix their signatures to any plans or documents dealing with subject matter in which they lack competence, nor to any plan or document not prepared under their direction and control.
   c. Engineers may accept assignments and assume responsibility for coordination of an entire project and sign and seal the engineering documents for the entire project, provided that each technical segment is signed and sealed only by the qualified engineers who prepared the segment.

3. Engineers shall issue public statements only in an objective and truthful manner.

   a. Engineers shall be objective and truthful in professional reports, statements, or testimony. They shall include all relevant and pertinent information in such reports, statements, or testimony, which should bear the date indicating when it was current.
   b. Engineers may express publicly technical opinions that are founded upon knowledge of the facts and competence in the subject matter.
   c. Engineers shall issue no statements, criticisms, or arguments on technical matters that are inspired or paid for by interested parties, unless they have prefaced their comments by explicitly identifying the interested parties on whose behalf they are speaking, and by revealing the existence of any interest the engineers may have in the matters.

4. Engineers shall act for each employer or client as faithful agents or trustees.

   a. Engineers shall disclose all known or potential conflicts of interest that could influence or appear to influence their judgment or the quality of their services.
   b. Engineers shall not accept compensation, financial or otherwise, from more than one party for services on the same project, or for services pertaining to the same project, unless the circumstances are fully disclosed and agreed to by all interested parties.

Figure 1.9 NSPE code of ethics for engineers—January 2006 version.

c. Engineers shall not solicit or accept financial or other valuable consideration, directly or indirectly, from outside agents in connection with the work for which they are responsible.
d. Engineers in public service as members, advisors, or employees of a governmental or quasi-governmental body or department shall not participate in decisions with respect to services solicited or provided by them or their organizations in private or public engineering practice.
e. Engineers shall not solicit or accept a contract from a governmental body on which a principal or officer of their organization serves as a member.

5. Engineers shall avoid deceptive acts.

a. Engineers shall not falsify their qualifications or permit misrepresentation of their or their associates' qualifications. They shall not misrepresent or exaggerate their responsibility in or for the subject matter of prior assignments. Brochures or other presentations incident to the solicitation of employment shall not misrepresent pertinent facts concerning employers, employees, associates, joint venturers, or past accomplishments.
b. Engineers shall not offer, give, solicit, or receive, either directly or indirectly, any contribution to influence the award of a contract by public authority, or which may be reasonably construed by the public as having the effect or intent of influencing the awarding of a contract. They shall not offer any gift or other valuable consideration in order to secure work. They shall not pay a commission, percentage, or brokerage fee in order to secure work, except to a bona fide employee or bona fide established commercial or marketing agencies retained by them.

III. Professional Obligations

1. Engineers shall be guided in all their relations by the highest standards of honesty and integrity.

a. Engineers shall acknowledge their errors and shall not distort or alter the facts.
b. Engineers shall advise their clients or employers when they believe a project will not be successful.
c. Engineers shall not accept outside employment to the detriment of their regular work or interest. Before accepting any outside engineering employment, they will notify their employers.
d. Engineers shall not attempt to attract an engineer from another employer by false or misleading pretenses.
e. Engineers shall not promote their own interest at the expense of the dignity and integrity of the profession.

2. Engineers shall at all times strive to serve the public interest.

a. Engineers shall seek opportunities to participate in civic affairs; career guidance for youths; and work for the advancement of the safety, health, and well-being of their community.
b. Engineers shall not complete, sign, or seal plans and/or specifications that are not in conformity with applicable engineering standards. If the client or employer insists on such unprofessional conduct, they shall notify the proper authorities and withdraw from further service on the project.
c. Engineers shall endeavor to extend public knowledge and appreciation of engineering and its achievements.
d. Engineers shall strive to adhere to the principles of sustainable development in order to protect the environment for future generations.

3. Engineers shall avoid all conduct or practice that deceives the public.

a. Engineers shall avoid the use of statements containing a material misrepresentation of fact or omitting a material fact.
b. Consistent with the foregoing, engineers may advertise for recruitment of personnel.
c. Consistent with the foregoing, engineers may prepare articles for the lay or technical press, but such articles shall not imply credit to the author for work performed by others.

4. Engineers shall not disclose, without consent, confidential information concerning the business affairs or technical processes of any present or former client or employer, or public body on which they serve.

a. Engineers shall not, without the consent of all interested parties, promote or arrange for new employment or practice in connection with a specific project for which the engineer has gained particular and specialized knowledge.
b. Engineers shall not, without the consent of all interested parties, participate in or represent an adversary interest in connection with a specific project or proceeding in which the engineer has gained particular specialized knowledge on behalf of a former client or employer.

5. Engineers shall not be influenced in their professional duties by conflicting interests.

a. Engineers shall not accept financial or other considerations, including free engineering designs, from material or equipment suppliers for specifying their product.
b. Engineers shall not accept commissions or allowances, directly or indirectly, from contractors or other parties dealing with clients or employers of the engineer in connection with work for which the engineer is responsible.

6. Engineers shall not attempt to obtain employment or advancement or professional engagements by untruthfully criticizing other engineers, or by other improper or questionable methods.

Figure 1.9 *(Continued)*

a. Engineers shall not request, propose, or accept a commission on a contingent basis under circumstances in which their judgment may be compromised.
b. Engineers in salaried positions shall accept part-time engineering work only to the extent consistent with policies of the employer and in accordance with ethical considerations.
c. Engineers shall not, without consent, use equipment, supplies, laboratory, or office facilities of an employer to carry on outside private practice.

7. Engineers shall not attempt to injure, maliciously or falsely, directly or indirectly, the professional reputation, prospects, practice, or employment of other engineers. Engineers who believe others are guilty of unethical or illegal practice shall present such information to the proper authority for action.

a. Engineers in private practice shall not review the work of another engineer for the same client, except with the knowledge of such engineer, or unless the connection of such engineer with the work has been terminated.
b. Engineers in governmental, industrial, or educational employ are entitled to review and evaluate the work of other engineers when so required by their employment duties.
c. Engineers in sales or industrial employ are entitled to make engineering comparisons of represented products with products of other suppliers.

8. Engineers shall accept personal responsibility for their professional activities, provided, however, that engineers may seek indemnification for services arising out of their practice for other than gross negligence, where the engineer's interests cannot otherwise be protected.

a. Engineers shall conform with state registration laws in the practice of engineering.
b. Engineers shall not use association with a nonengineer, a corporation, or partnership as a "cloak" for unethical acts.

9. Engineers shall give credit for engineering work to those to whom credit is due, and will recognize the proprietary interests of others.

a. Engineers shall, whenever possible, name the person or persons who may be individually responsible for designs, inventions, writings, or other accomplishments.
b. Engineers using designs supplied by a client recognize that the designs remain the property of the client and may not be duplicated by the engineer for others without express permission.
c. Engineers, before undertaking work for others in connection with which the engineer may make improvements, plans, designs, inventions, or other records that may justify copyrights or patents, should enter into a positive agreement regarding ownership.
d. Engineers' designs, data, records, and notes referring exclusively to an employer's work are the employer's property. The employer should indemnify the engineer for use of the information for any purpose other than the original purpose.
e. Engineers shall continue their professional development throughout their careers and should keep current in their specialty fields by engaging in professional practice, participating in continuing education courses, reading in the technical literature, and attending professional meetings and seminars.

Footnote 1 "Sustainable development" is the challenge of meeting human needs for natural resources, industrial products, energy, food, transportation, shelter, and effective waste management while conserving and protecting environmental quality and the natural resource base essential for future development.

As Revised January 2006

"By order of the United States District Court for the District of Columbia, former Section 11(c) of the NSPE Code of Ethics prohibiting competitive bidding, and all policy statements, opinions, rulings or other guidelines interpreting its scope, have been rescinded as unlawfully interfering with the legal right of engineers, protected under the antitrust laws, to provide price information to prospective clients; accordingly, nothing contained in the NSPE Code of Ethics, policy statements, opinions, rulings or other guidelines prohibits the submission of price quotations or competitive bids for engineering services at any time or in any amount."

Statement by NSPE Executive Committee

In order to correct misunderstandings which have been indicated in some instances since the issuance of the Supreme Court decision and the entry of the Final Judgment, it is noted that in its decision of April 25, 1978, the Supreme Court of the United States declared: "The Sherman Act does not require competitive bidding."
It is further noted that as made clear in the Supreme Court decision:

1. Engineers and firms may individually refuse to bid for engineering services.
2. Clients are not required to seek bids for engineering services.
3. Federal, state, and local laws governing procedures to procure engineering services are not affected, and remain in full force and effect.

Figure 1.9 *(Continued)*

4. State societies and local chapters are free to actively and aggressively seek legislation for professional selection and negotiation procedures by public agencies.

5. State registration board rules of professional conduct, including rules prohibiting competitive bidding for engineering services, are not affected and remain in full force and effect. State registration boards with authority to adopt rules of professional conduct may adopt rules governing procedures to obtain engineering services.

6. As noted by the Supreme Court, "nothing in the judgment prevents NSPE and its members from attempting to influence governmental action . . ."

Figure 1.9 *(Continued)*

ethical fashion, significant workplace problems that may involve difficult choices. For this purpose, the Online Ethics Center (OEC) for Engineering and Science (formerly the World Wide Web Ethics Center for Engineering and Science), was established in 1995 under a grant from the National Science Foundation (NSF). The Center, located at Case-Western Reserve University, provides very extensive educational resources, including more than 100 case studies, at the Web site:

http://onlineethics.org/

Figure 1.10 provides just a sample of Center case studies dealing with public safety and welfare. The Center also sponsors conferences, addresses the ABET Readiness Committee call for a *Guide to Ethics for Dummies*, and provides sample student assignments, from the freshman to senior level, on practical ethics for use by instructors.

A breach of ethics or a courageous show of ethics is frequently newsworthy. An example of a breach of ethics is that of an MIT student who used university hardware to distribute commercial software over the Internet. Recent examples of a more courageous show of ethics, which are presented as case studies by the WWW Ethics Center for Engineering and Science, include:

1. The attempts by Roger Boisjoly to avert the *Challenger* space disaster.
2. The emergency repair by William LeMessurier of structural supports for the Citicorp Tower in New York City.
3. The campaign of Rachel Carson for control of the use of pesticides.

The work of Rachel Carson has had a significant impact on stirring the world to action on environmental protection,

- **Suspected Hazardous Waste**
  A supervisor instructs a student engineer to withhold information from a client about the suspected nature of waste on the client's property, to protect what the supervisor takes to be the client's interest.
- **Clean Air Standards and a Government Engineer**
  An engineer defies immediate supervisor because he believes supervisor's instruction would pose an environmental health hazard.
- **The Responsibility for Safety and the Obligation to Preserve Client Confidentiality**
  Tenants sue their building's owner, and the owner employs an engineer who finds structural defects not mentioned in the tenant's lawsuit. Issues of public safety versus client confidentiality.
- **Code Violations with Safety Implications**
  Engineer discovers deficiencies in a building's structural integrity, and it would breach client confidentiality to report them to a third party.
- **Whistleblowing City Engineer**
  An engineer privately informs other city officials of an environmental threat, a problem her supervisor has ordered her not to disclose.
- **Safety Considerations and Request for Additional Engineering Personnel**
  An engineer is concerned for worker safety during construction but yields to his client's objections at the cost of an on-site representative.
- **Engineer's Dispute with Client Over Design**
  A client believes an engineer's designs are too costly, but the engineer fears that anything less may endanger the public.
- **Do Engineers Have a Right to Protest Shoddy Work and Cost Overruns?**
  An engineer who is employed by a government contractor objects to a subcontractor's poor performance and is ignored and silenced by management.
- **Change of Statement of Qualifications for a Public Project**
  An engineering firm takes measures to remedy a deficit in a particular area of expertise needed to successfully compete for and carry out a public project.
- **Knowledge of Damaging Information**
  An engineer has a conflict between honoring an agreement to an employer and reporting a hazard to protect the public interest.

Figure 1.10 Engineering ethics cases in the Ethics Center for Engineering & Science (http://www.cwru.edu/affil/wwwethics).

beginning with concerted efforts on college campuses. Carson was a U.S. Fish and Wildlife Service biologist who, in 1951, published *The Sea Around Us*, which won the National Book Award. In 1962, Carson's book *Silent Spring* was published. In that book, she criticized the widespread use of chemical pesticides, fertilizers, and weed killers, citing case histories of damage to the environment. In particular, she cited the disappearance of songbirds (thus, the title of the book) due to the use of the synthetic, chlorine-containing pesticide DDT (previously discussed in Section 1.4), which kills insects by acting as a nerve poison. Synthetic pesticides were developed during a period of great economic development after World War II in an attempt to reduce insect-caused diseases in humans and to increase food production. More specifically, the use of DDT practically eliminated the anopheles mosquito that had caused malaria in many countries in Asia, Africa, and South and Central America. Carson claimed that the problems created by DDT were worse than the problems it solved. DDT disrupted reproductive processes and caused bird eggs to be infertile or deformed. Because DDT breaks down very slowly in the soil, its concentration builds up in the food chain as larger organisms eat smaller ones. Even though no adverse effects of DDT on humans have been found, its use in the United States was banned in 1972. However, it is still manufactured in the United States, and it is still used in parts of the world for malaria control.

Concern over the environment has led to much interest in the development of global ethics. Considerable information on this subject is available on the Web site of The Institute for Global Ethics, which is located at Camden, Maine:

http://www.globalethics.org

The Institute exists because many believe in their statements:

1. *Because we will not survive the twenty-first century with the twentieth century's ethics.*
2. *The immense power of modern technology extends globally. Many hands guide the controls and many decisions move those hands. A good decision can benefit millions, while an unethical one can cripple our future.*

The Institute strongly believes that education in ethics must begin at the middle- and high-school level. Accordingly, they provide instructional materials suitable for that level. They also stress the concept of moral courage, which they are in the process of defining. In a recent white paper, the Institute makes the following statements:

Moral courage is different from physical courage. Physical courage is the willingness to face serious risk to life or limb instead of fleeing from it.

Moral courage is not about facing physical challenges that could harm the body. It's about facing mental challenges that could harm one's reputation, emotional well-being, self-esteem, or other characteristics. These challenges, as the term implies, are deeply connected with our moral sense—our core moral values.

Moral courage, . . . has four salient characteristics:

- *It is the courage to be moral—to act with fairness, respect, responsibility, honesty, and compassion even when the risks of doing so are substantial.*
- *It requires a conscious awareness of those risks. The sleepwalker on the ridgepole is not courageous unless, waking up, he or she perceives the danger and goes forward anyway.*
- *It is never formulaic or automatic, but requires constant vigilance against its opposite (moral timidity) and its counterfeit (moral foolhardiness).*
- *It can be promoted, encouraged, and taught through precept, example, and practice.*

The teaching of engineering ethics to senior engineering students can be difficult, especially when students raise questions from their personal experiences. Years ago, a student in a senior design class, at an appointment in the instructor's office, asked the following question: "Two weeks ago, I accepted an offer of employment from a company that had set a deadline for accepting the offer. Yesterday, I received a better offer from another company for a better job opportunity at a higher starting salary. What should I do?" At that time, the instructor was inclined to tell the student to stand by the commitment to the first company. Several years later the tables were turned. A senior student told the instructor that he had accepted an offer with an excellent company and then rejected two other offers that he had received. One month later, the student informed the instructor that the company to which he had committed reneged on the offer because of a downturn in the economy. Furthermore, job offers from the two other companies that had made offers were no longer available. From then on, when asked, the instructor recited these two episodes and told students to look out for their own best interests. If they got a better offer after accepting an earlier offer, renege on the first and take the second. Was the instructor giving ethical advice?

1.7 SUMMARY

Having studied this chapter, the reader should

1. Be acquainted with the organizational structures in product and process design and have an appreciation of the key steps in the Stage-Gate™ product and technology development framework.
2. Be familiar with the distinctions between basic chemicals, industrial chemicals, and configured consumer chemical products.
3. Have examined the innovation maps presented herein for a basic chemical (environmentally friendly

refrigerant), an industrial chemical (thin-glass substrate for an LCD display), and a configured consumer product (light bulb). The reader should also be familiar with the approach to preparing innovation maps when designing new chemical products.

4. Be knowledgeable about the principal issues in environmental protection and the many safety considerations that are foremost in the minds of product and process designers. The reader should also have some familiarity with the many design methods used to protect our environment and provide safe chemical processes.

5. Understand that it is crucial for engineers to maintain high ethical principles, especially as they relate to protecting the public against environmental and safety problems. At the minimum, the reader should be familiar with the codes of ethics presented herein and recognize that engineers are often confronted with difficult choices that must be resolved using high ethical standards.

REFERENCES

1. ALLEN, D. T., and K. S. ROSSELOT, *Pollution Prevention for Chemical Processes*, John Wiley & Sons, New York (1997).

2. ALLEN, D. T., and D. R. SHONNARD, *Green Engineering: Environmentally Conscious Design of Chemical Processes*, Prentice-Hall, Englewood Cliffs, New Jersey (2002).

3. American Institute of Chemical Engineers, *Safety, Health, and Loss Prevention in Chemical Processes: Problems for Undergraduate Engineering Curricula—Student Problems*, AIChE, New York (1990).

4. American Institute of Chemical Engineers, *Emergency Relief System Design Using DIERS Technology*, AIChE, New York (1992).

5. AMUNDSON, N. R., Ed., *Frontiers in Chemical Engineering: Research Needs and Opportunities*, National Research Council, National Academy Press, Washington, DC (1988).

6. BEVER, M. B., Ed., *Encyclopedia of Materials Science and Engineering*, Pergamon Press, Oxford (1986).

7. BRETHERICK, L., *Handbook of Reactive Chemical Hazards*, Butterworth, London (1990).

8. *Business Week*, "The World's Most Innovative Companies," April 24, 2006.

9. CHASE, M. W., Ed., *JANAF Thermochemical Tables*, 3rd ed., Parts 1 and 2, in *J. Phys. Chem. Ref. Data*, 14 (Suppl. 1) (1985).

10. CHEREMISINOFF, N. P., Ed., *Encyclopedia of Fluid Mechanics*, Gulf Publishing Co., Houston (1986).

11. CONSIDINE, D. M., Ed., *Van Nostrand's Scientific Encyclopedia*, 8th ed., Van Nostrand, New York (1995).

12. COOPER, R. G., *Winning at New Products: Accelerating the Process from Idea to Finish*, 3rd Ed., Perseus Publ., Cambridge, Mass., 2001.

13. COOPER, R. G., *Product Leadership: Creating and Launching Superior New Products*, Perseus Publ., Cambridge, Mass., 2002.

14. COOPER, R. G., *Product Leadership: Creating and Launching Superior New Products*, 2nd ed., Basic Books Cambridge, Mass., 2005.

15. CROWL, D. A., and J. F. LOUVAR, *Chemical Process Safety: Fundamentals with Applications*, Prentice-Hall, Englewood Cliffs, New Jersey (1990).

16. de NEVERS, N., *Air Pollution Control Engineering*, McGraw-Hill, New York (1995).

17. EISENHAUER, J., and S. MCQUEEN, *Environmental Considerations in Process Design and Simulation*, Energetics, Inc., Columbia, Maryland (1993).

18. FREEMAN, H. M., Ed., *Standard Handbook of Hazardous Waste Treatment and Disposal*, McGraw-Hill, New York (1989).

19. GRAHAM, M. B. W., and A. T. SHULDINER, *Corning and the Craft of Innovation*, Oxford University Press, 2001.

20. GREEN, D. W., and R. H. PERRY, Ed., *Perry's Chemical Engineer's Handbook*, 8th ed. McGraw-Hill, New York (2008).

21. *Guidelines for Engineering Design for Process Safety*, CCPS, AIChE, 1993.

22. GUNDLING, E., *The 3M Way to Innovation: Balancing People and Profit*, Kodansha International, Tokyo (2000).

23. HEWITT, G. F., *International Encyclopedia of Heat and Mass Transfer*, CRC Press, Boca Raton, FL, 1997.

24. KENT, J. A., Ed., *Riegel's Handbook of Industrial Chemistry*, 9th ed., Van Nostrand Reinhold, New York (1992).

25. *Kirk-Othmer Encyclopedia of Chemical Technology*, 4th ed. Wiley-Interscience, New York (1991).

26. KLETZ, T., *Plant Design for Safety—A User-Friendly Approach*, Hemisphere, Washington, DC (1991).

27. LIDE, D. R., Ed., *Handbook of Chemistry and Physics*, 78th ed. CRC Press, Boca Raton, Florida (1997).

28. LU, S., *Social Aspects of Engineering Design*, Report of Focus Group 3, NSF Strategic Workshop on Engineering Design in Year 2030, 2004. Report available from Prof. S. Lu, S. Cal. Univ.

29. *McGraw-Hill Encyclopedia of Science and Technology*, 6th ed. McGraw-Hill, New York (1987).

30. MCKETTA, J. J., Ed., *Chemical Processing Handbook*, Marcel Dekker, New York (1993a).

31. MCKETTA, J. J., Ed., *Unit Operations Handbook*, Marcel Dekker, New York (1993b).

32. MCKETTA, J. J., and W. A. CUNNINGHAM, Eds.,*Encyclopedia of Chemical Processing and Design*, Marcel Dekker, New York (1976).

33. National Institute for Occupational Safety and Health, *Pocket Guide to Chemical Hazards*, NIOSH, Cincinnati, Ohio (1987).

34. PISANO, G. P., *The Development Factory: Unlocking the Potential of Process Innovation*, Harvard Business School Press, Cambridge (1997).

35. *Ullmann's Encyclopedia of Industrial Chemistry*, 5th ed., VCH, Deerfield Beach, Florida (1988).

36. WIDAGDO, S,"Incandescent Light Bulb: Product Design and Innovation,"*Ind. Eng. Chem. Res.*, **45**, 8231–8233 (2006).

37. WOODS, D. R., *Data for Process Design and Engineering Practice*, Prentice-Hall, Englewood Cliffs, New Jersey (1995a).

38. WOODS, D. R., *Process Design and Engineering Practice*, Prentice-Hall, Englewood Cliffs, New Jersey (1995b).

Patents

39. U.S. Patent 6,974,786.Helfinstine, J. D., D. J. Liebner, J. L. Martin, D. V. Neubauer, and W. R. Powell, *Sag Control of Isopipes Used in Making Sheet Glass by the Fusion Process* (2005).

Chapter 2

Steps in Product-Development

2.0 OBJECTIVES

This chapter introduces approaches for developing new chemical products. It discusses the strengths and pitfalls of the Stage-Gate™ Product-Development Process (SGPDP). In addition, the application of the SGPDP for different classes of projects, ranging from *product extensions* to the development of *new-to-the-world* products, is discussed. Emphasis is on selecting the key questions to be answered, the key deliverables, and the clusters of tools for finding answers to these questions.

After studying this chapter, the reader will be prepared to read Part One of this book, in which the SGPDP is applied to the design of *basic* chemical products, concluding with three case studies in Chapter 13. Then, in Parts Two and Three, the SGPDP is applied to the design of *industrial* chemical products and *configured consumer* products using selected case studies.

Alternatively, because many steps in product design are business oriented—for example, creating a pipeline for new product development, carrying out a market assessment, determining customer needs, and carrying out an opportunity assessment—some readers may prefer to skip the details of Chapter 2 and proceed directly to Part One. Then, in Chapters 13, 15, and 17, when these subjects are discussed in connection with the product design case studies, the reader can refer back to Chapter 2 for the details as necessary.

After studying this chapter, the reader should:

1. Be able to use the elements of the Stage-Gate™ Product-Development Process (SGPDP) for the design of new products.
2. Be aware of the need to create a pipeline for new product development.
3. Be able to create a project charter to begin a product-development effort.
4. Know the role of new technologies in developing new products—seeking to relate them to customer needs in an *innovation map* before product development begins.
5. Be able to carry out a market assessment, that is, determine the value proposition, and carry out market segmentation and value-chain analysis in the *concept* stage.
6. Be able to determine customer requirements, that is, the *voice of the customer*, and to translate customer voices into customer requirements.
7. Be able to identify the technical requirements for a new product, beginning with customer requirements, and to formulate the House of Quality.
8. Be able to evaluate new solution concepts using the Pugh matrix.
9. Know how to use Porter five-force and intellectual-property analysis (to search the patent literature) when carrying out an opportunity assessment to determine the risks associated with a new product.
10. Appreciate the key steps in the *feasibility, development, manufacturing*, and *product-introduction* stages of the SGPDP.

2.1 INTRODUCTION

To generate new products effectively over many years and decades, it is important to create a pipeline for new product development, the health of which depends on four key factors (Cooper, 2005). These are illustrated in Figure 2.1 and listed next with brief explanations:

1. ***Environment for Innovation.*** To achieve innovation-friendly environments involving people at many levels, companies have introduced several approaches, as discussed in the section on *Stimulating Invention and Innovation* in Section 1.3. In many cases, successful companies are known for empowering their employees to take calculated risks.
2. ***Product and Technology Strategy.*** From the business management perspective, the clear definition of business goals and strategies is similarly effective in achieving successful product-development environments. As the

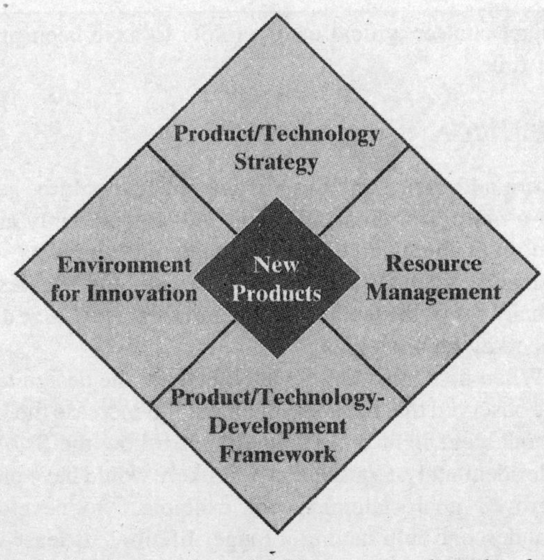

Figure 2.1 Key factors for a healthy new-product-development pipeline.

former are defined, their communication and translation into new technology- and product-development strategies is important.

3. ***Resource Management.*** Also from the business management perspective, the adherence to the disciplined management of a company's product and technology portfolio is important. This involves the careful deployment of resources to top-priority programs to achieve business objectives.
4. ***Technology- and Product-Development Framework.*** This fourth block recognizes the need to adopt company-wide processes for executing, monitoring, and evaluating technology and product-development efforts.

Regarding the fourth block, a key factor has been the impact of globalization, which has significantly increased worldwide competition. This, in turn, has led to increased pressures to reduce the time-to-market while increasing the success rates of new-product introductions. And, consequently, the need for universal product- and technology-development frameworks, to manage, monitor, and control the execution of the product- and technology-development strategies, is more widely recognized. In this chapter, only the new product-/technology-development framework is discussed, with emphasis on the framework developed by Cooper (2001, 2002, 2005).

The business success of the commercialization of a new product depends on two factors: (1) the selection of the new-product ideas to be developed, and (2) the execution of the transformation of the new-product ideas into products to be launched. In this chapter, it is assumed that the strategic selection of these ideas has been accomplished, being beyond the scope of discussion. Rather, focus is on the execution of the Stage-Gate™ Product-Development Process (SGPDP). It is also assumed that the technological inventions are mostly in place. The alternative, to develop the required technologies during product development, normally increases the cycle times and delays times-to-market significantly, and consequently, is preferably avoided. For this reason, the foci of technology-development organizations are usually shifted toward the *next* technologies upon which *future* products will be based.

There are two major incentives for adopting a systematic product-development process such as the SGPDP. These involve:

1. ***Reduction in the new product-development time.*** Depending on the types of new products, ranging from minor modifications to existing products to new products for new markets, the development times range from a few months to a few years. Because totally new products are often complicated by the concurrent development of underlying technologies, these are often avoided. Ideally, the developments of new technologies precede their inclusions in new products.
2. ***Increased success rate of new products launched into the market.*** During the product-development process, before large investments are committed, it is important to verify the feasibility of the product, its critical parameters for manufacture, and the size of the market opportunity. The SGPDP is designed to reduce the risks during development; that is, to increase the probabilities of successes. Furthermore, as products are developed, it is not uncommon for the size of the market opportunity to decrease. Often, projections of the *true values* become better defined as feasible market segments are identified. The product of the probability of success and the total market opportunity is often a good measure of the health of a new product-development effort.

Before beginning a discussion of the SGPDP in Section 2.3, consider next the first steps in initiating a product design. These include the formulation of a design team and the preparation of its project charter, followed by the search for new technologies upon which the product design is based—that is, the creation of an *innovation map*, as discussed in Section 1.3.

2.2 PROJECT CHARTER AND NEW TECHNOLOGIES

With a pipeline for new product development in place, as promising ideas are generated, design teams are formulated, which begin their work by creating a project charter.

Project Charter

A project charter is the starting point for a product-development effort. Its key elements are specific goals, a project scope, deliverables, and a time line, which are discussed next in this subsection.

First the reader should be aware that, when developing a project charter, design teams often seek to follow the

SMART principle; that is, to focus on *specific, measurable, agreed-upon, realistic,* and *time-based* aspects of the product design. By *specific*, it is implied that the charter is well defined and clear to persons with a basic knowledge of the project. The term *measurable* implies that well-recognized, clear indicators are available to denote when the project objectives have been achieved. By *agreed-upon*, it is implied that all of the stakeholders are in agreement with the goals of the project. The term *realistic* implies that these goals can be achieved using available knowledge, time, and resources. And, finally, the term *time-based* implies that the datelines for completion of the overall project and its stages are sound.

The identification of specific goals, that is, specific targets, is very important to prevent so-called *project scope creep*. Often, these goals are expressed by stating the objectives of the project or, alternatively, by posing several questions to be answered during the course of the project.

The project scope defines the boundary of the project. It is helpful to classify its associated elements as *in-scope* or *out-of-scope* using an *in-scope and out-of-scope* exercise, as illustrated in Example 2.1, recognizing that the boundaries and scope often change as the project progresses.

The deliverables are items to be completed during the project. They help to define the desired outcomes of the project, and are often subdivided into groups to be presented at the gate review following each stage, as illustrated in Figure 2.3(b) in Section 2.3. Normally, the gate reviewers focus their evaluations on these deliverables.

The time line presents projections of the project completion date and the dates for the gate reviews. Of course, these dates are frequently renegotiated at the completion of each stage.

Next, Example 2.1 illustrates the creation of a project charter for the development of a new incandescent light bulb product in the mid-1980s. Note that a complete case study for this product design is presented in Section 17.2.

EXAMPLE 2.1

The time frame for this example is taken to be the mid-1980s, when a market leader in general lighting had just announced a new incandescent light bulb with an improved lifetime from 750 to 1,000 hr. To provide competition, hopefully in the near future, a development team of the leading competitor was created to carry out a design project to increase the lifetime of incandescent light bulbs, with the goal of at least doubling their lifetimes to 2,000 hr while maintaining their costs. Assume you were a member of the team charged with first developing a charter for the project, and:

1. Write a goal statement for the lifetime-improvement project.
2. Carry out the *in-scope and out-of-scope* exercise.
3. Determine the deliverables and timeline.

Based upon your knowledge of incandescent light bulbs in the mid-1980s (see the history in Section 16.2), prepare a project charter typical of one likely to have been prepared at that time.

SOLUTION

Before addressing the project charter, it should be recognized that a multidisciplinary design team was probably assembled, involving chemical engineers, material scientists, electrical engineers, product-development persons, and businesspersons. In many cases, the team would include representative developers of the key technologies.

When first considering the objectives, the design team probably observed that the goal statement "to increase the lifetime of incandescent light bulbs" would not follow the SMART principle adequately. Consequently, it likely would have prepared an improved goal statement, for example, "to develop a new incandescent bulb having a longer lifetime, at least twice that of existing products (namely, 2,000 hr), while not significantly altering the price." Subsequently, adhering more closely to the SMART principle, it likely created an initial project charter typical of that in Table 2.1.

As is common in initial project charters, the initial scope likely defined only the major boundaries of the project, and, consequently, it was probably decided to refine the scope using the *in-scope and out-of-scope* exercise. This exercise would not only have further defined the product specifications, but also would have likely clarified the objectives of the development project in terms of the technical requirements, manufacturing requirements, etc. As shown in the case studies in Chapters 13, 15, and 17, the initial scope is often altered as the product design proceeds.

In the *in-scope and out-of-scope* exercise, given the goal statement and the initial project scope, it is recommended that

Table 2.1 Initial Project Charter

Project Name	Longer-lifetime incandescent light bulb
Project Champions	Business Director of the Home Lighting Business
Project Leader	John Doe
Specific Goals	Incandescent light bulbs with a lifetime of at least 2,000 hr, or twice that of the current product line, at the same cost
Project Scope	In-scope: • Light bulbs for household lighting • Minimal changes to the current manufacturing capability Out-of-scope: • Light bulbs for non-household applications
Deliverables	• Business opportunity assessment • Technical feasibility assessment • Manufacturing capability assessment • Product life-cycle assessment
Time Line	Product prototypes for market testing within 12 months

each team member list the items he/she believes to be in-scope (included) or out-of-scope (excluded) in the project. With the views of the individual team members represented, the entire team discusses the controversial items and achieves a consensus list. Note that in those rare incidents when an item cannot be resolved, it is common to resolve the ambiguities with the assistance of the project sponsors.

In this example, the initial scope in Table 2.1 would likely have been expanded into:

In-Scope
• Lifetime should be at least 2,000 hr
• Light bulb for general lighting for home usage
• Compact and fits standard fittings
• No significant capital increase in manufacturing
• Based on incandescent technology or its variants
• Maintain profit margin
Out-of-Scope
• Display lighting
• Automobile headlights
• Wide spectrum (daylight) lighting
• Emerging technology

Next, the product-development team probably agreed upon its *list of deliverables* to be completed during the project. Note that often the deliverables are best defined as the answers to a list of critical questions, providing data and/or recommendations that influence the *business decision makers* on the project. Because this light-bulb project was likely based upon a limited *voice of the customer*, that is, typical customer opinions, it would have been prudent to ask if the needs for longer-life light bulbs were *real*. Stated differently, key questions were likely: How large is the business opportunity for longer-life light bulbs? Is it technically feasible? Is it possible to save energy (although energy savings were not a priority in the mid-1980s, due to the availability of low-cost fuel)? In obtaining answers, it is likely the design team contacted the proponents, or *champions*, among its customers, seeking answers to these questions in the form of: (1) a business opportunity assessment, (2) a technical feasibility assessment, and (3) a manufacturing capability assessment. In addition, with companies required by law to comply with EHS (environmental, health, and safety) regulations and standards, the team needed to plan for product life-cycle management.

For this project, given that a competitive, new light bulb had just been announced, the time-to-market would have been a critical element. For this reason, the project *champions/sponsors* had likely asked the team to provide a product prototype within one year, with the *concept* stage (to be discussed) completed in three months and the *feasibility* stage completed in 12 months.

Having completed the *in-scope and out-of-scope* exercise, the list of deliverables, and the time line, the design team likely prepared its final project charter similar to that in Table 2.2. After receiving approval by the project sponsors, the team likely began the *concept* stage of the SGPDP, or equivalent, as discussed in the next subsection.

Table 2.2 Final Project Charter

Project Name	Longer-lifetime incandescent light bulb
Project Champions	Business Director of the Home Lighting Business
Project Leader	John Doe
Specific Goals	Incandescent light bulbs with a lifetime of at least 2,000 hr, or twice that of the current product line, at the same cost
Project Scope	In-scope: • Lifetime is at least 2,000 hours • Light bulbs for household lighting o General lighting o Compact and fits standard fittings o Based on incandescent light technology • Minimal changes to the current manufacturing capability o No significant capital increase • Maintain profit margin
	Out-of-scope: • Light bulbs for non-household applications o Display lighting o Automotive headlights o Wide spectrum (daylight) bulbs • Emerging technology
Deliverables	• Business opportunity assessment: o How large is the business opportunity for longer-life light bulbs? • Technical feasibility assessment: o Is it technically feasible? • Manufacturing capability assessment o Can it be manufactured without significant capital investment? • Product life-cycle assessment o Would it satisfy the current regulatory requirements?
Time Line	Product prototypes for market testing within 12 months

New Technologies

As mentioned above, as the design team begins its work, it is assumed that most of the technological inventions are in place. Before deciding to proceed with the SGPDP, the Stage-Gate™ Technology-Development Process (SGTDP), discussed in Section 1.2, must be completed. In short, the design team identifies the new materials, process/manufacturing, and product technologies upon which the new product is to be based. Then, it matches its perception of the *voice of the customer* with the new technologies in an *innovation map*, as discussed and illustrated in Section 1.3.

Figure 2.2 summarizes the steps in creating a project charter and identifying the new technologies upon which the new product is to be based. Having created its project

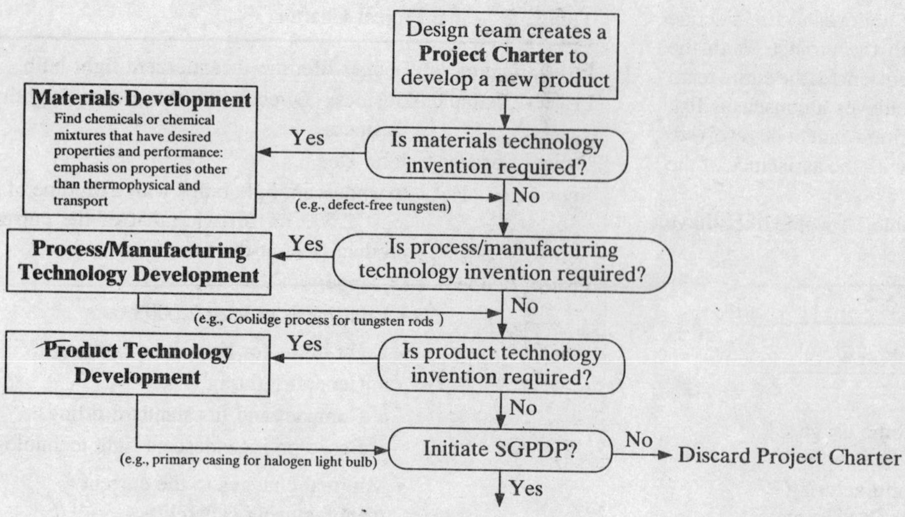

Figure 2.2 Steps in creating a project charter and identifying new technologies.

charter, when designing a configured consumer product the design team identifies the new materials, process/manufacturing, and product technologies. When these are judged to be sufficiently promising, it may create an initial *innovation map* and proceed to begin the SGPDP, as discussed in the next section. Note that as the design team proceeds through the SGPDP, it normally refines the new technologies and gains a better understanding of the customer needs and, consequently, modifies its initial *innovation map*.

2.3 STAGE-GATE™ PRODUCT-DEVELOPMENT PROCESS (SGPDP)

As introduced in Section 1.2, the Stage-Gate™ Product-Development Process (SGPDP) was developed by Robert Cooper in the 1980s (Cooper, 2001, 2002, 2005). It provides a roadmap for transforming new ideas into products that satisfy customer needs, ready to be launched. The SGPDP has been used successfully by companies around the world, often adapted by practitioners to specific needs and environments. It is essentially a gating process. The SGPDP consists of several stages, between which gate reviews are conducted involving key stakeholders and decision makers from business, technical, manufacturing, supply-chain, and environmental, health, and safety organizations. At each gate review, a decision is made to either: (1) advance the design project to the next stage, (2) retain the design project at the current stage until pending critical issues are resolved, or (3) cancel the design project when a need is no longer recognized, or when roadblocks have been encountered that render the project infeasible.

To advance from stage to stage, a product-development project must pass the gate reviews; that is, receive *passing* grades. Each gate is intended to reduce the risk by verifying the manufacturability and matching the size of the opportunity and product features and performance to the customer needs. To facilitate the reviews, several tasks are designed to be accomplished in each stage. The gate reviews serve to check and verify that key questions have been answered satisfactorily before the design team is permitted to proceed to the next stage. Since its introduction, there have been many variations and adaptations of the original Stage-Gate™ process. Typically, however, a SGPDP consists of five stages, as shown in Figure 2.3. In Figure 2.3(a), the stages are represented by rectangular blocks, and the gates are represented by diamonds. Brief labels below the rectangles indicate the principal steps to be accomplished in that stage. Above the diamonds are brief summaries of the items to be screened or evaluated during each gate review. In addition, Figure 2.3(b) provides brief statements of the goals to be achieved in each stage, as well as the deliverables to be presented at the gate reviews. Note that the specific items listed in Figures 2.3(a) and 2.3(b) depend on the kind of chemical product. The introduction to Part One shows the items for *basic* chemicals in Figure PI.1; the introduction to Part Two shows the items for industrial chemicals in Figure PII.1; and the introduction to Part Three shows the items for *configured consumer* chemical products in Figure PIII.1.

The major drawback of the SGPDP occurs when new development teams overemphasize the tasks and deliverables, losing sight of the critical issues in ensuring the success of their projects. To discourage this, in recent years many companies, including Procter & Gamble, Caterpillar, and Rohm & Haas, have modified the SGPDP to focus more on the critical issues in an outer layer, with the tasks and deliverables used to resolve the critical issues. This approach is introduced in the discussion that follows.

2.4 CONCEPT STAGE

The *concept* stage, being first in the Stage Gate™ Product-Development Process, primarily serves as the business and product definition step. During this stage, an extensive investigation is carried out to define the product and to verify the attractiveness of the product prior to committing significant funds. When building an entirely new product family, this stage involves an extensive market study to define, not only the market opportunity, but also the market segment(s) and customer needs. And consequently, the latter becomes

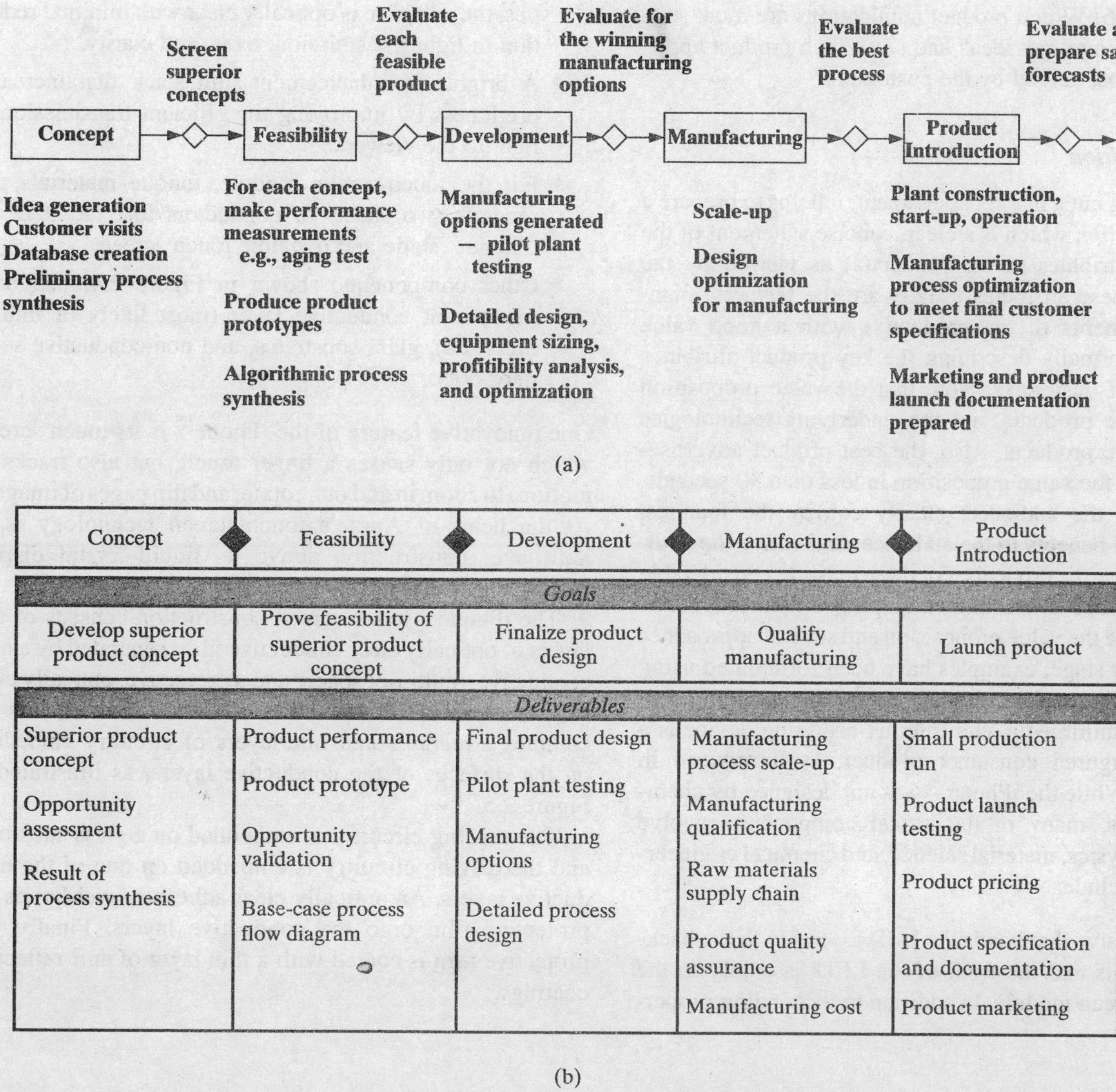

Figure 2.3 Schematic of the Stage-Gate™ Product-Development Process (SGPDP).

the basis for developing the new product concepts. For new product extensions in an existing market, although the market segment(s) and customer needs are generally known, a careful verification of market viability and product strategy, as they relate to customer needs, is still required. This often applies for new *basic* chemical products when the variations in chemical structure, properties, and raw materials are small.

In general, the goals of the *concept* stage are twofold: to (1) define the product and (2) build the business case. Note that in building the business case, a market opportunity assessment needs to be conducted. Because resources to obtain these data are usually not readily available in the public domain, it is recommended that students work with their librarians to obtain market data using the resources listed in the reference section and other sources. To accomplish the two goals, several tasks are recommended, including: carrying out a market assessment, determining customer requirements, determining product requirements, creating product concepts, and carrying out an opportunity assessment. These are discussed next.

Market Assessment

A new product-development program often begins with a product or technology idea for which the new product-development team seeks to identify and analyze its *value creation* and *value capture* (Murray, 2007). For value creation, it aims to determine whether the idea would add significant value for customers, while for value capture, it aims to determine whether its inherent economic value can be captured in the face of competition.

More specifically, when analyzing the value creation of a new idea, many questions are typically answered, including: (1) Who are the customers? (2) Who is likely to buy? (3) Who should be approached for sales? (4) Which markets should be considered? (5) Which customers are most likely to purchase? (6) Which product applications are most suitable for the technology idea? and (7) Which product applications are most valued by the customers?

Value Proposition

When carrying out a market assessment, it helps to prepare a *value proposition*, which is a clear, concise statement of the compelling attributes of the product(s) as viewed by the customers. These attributes usually are the features, functions, and benefits of the product(s), with a good value proposition normally describing the key product attributes for a group of customers. Note that the value proposition belongs to the products, not the underlying technologies that enable the products. Also, the best product advertisements convey the value proposition in less than 30 seconds. They capture the audience; clearly convey the features, functions, and benefits to the audience; and entice the audience to purchase the products. For these reasons, considerable effort is expended in creating value-proposition statements.

To illustrate the value proposition and several approaches in the *concept* stage, examples have been formulated using the 2007 Apple product, the iPhone™. This modern communication, multimedia, and Internet-browsing device is a type of configured consumer product, as introduced in Section 1.3. While the iPhone™ was not designed by chemical engineers, many of its critical components involve chemistry, physics, material science, and chemical engineering. These include:

- An adhesive that bonds the LCD assembly to the backlight units and the front of the LCD assembly to the touchscreen module. In addition to its bonding properties, the adhesive is optically clear with minimal reduction in light transmission, haze, and clarity.
- A brightness-enhancement film stack that increases brightness by improving the efficient transmission of light to the viewers.
- For the touch-screen module, unique materials that separate two sheets of conductors that facilitate the iPhone™ static and dynamic touch screen.
- Other components, shown in Figure 2.4, include a transparent conductive layer (most likely of indium tin oxide), glass substrates, and non-conductive separator dots.

One innovative feature of the iPhone™ is its touch screen, which not only senses a finger touch, but also tracks its motion (to zoom in and out, rotate, and flip pages of images). At the heart of Apple's touch-screen technology is its multilayer construction above its liquid-crystal display (LCD), as illustrated in Figure 2.4.

The iPhone™ touch-screen construction consists of two layers of optically clear conductive films separated by a non-conductive, multi-dot spacer and a protective, optically clear surface. A closer examination of these layers reveals the complex structures and interlayers of circuitry embedded on the surfaces of the conductive layers, as illustrated in Figure 2.5.

The sensing circuitry is embedded on a glass substrate, and the driving circuitry is embedded on one of the conductive layers. An optically clear adhesive layer bonds the protective film onto the conductive layers. Finally, the protective film is coated with a thin layer of anti-reflective coatings.

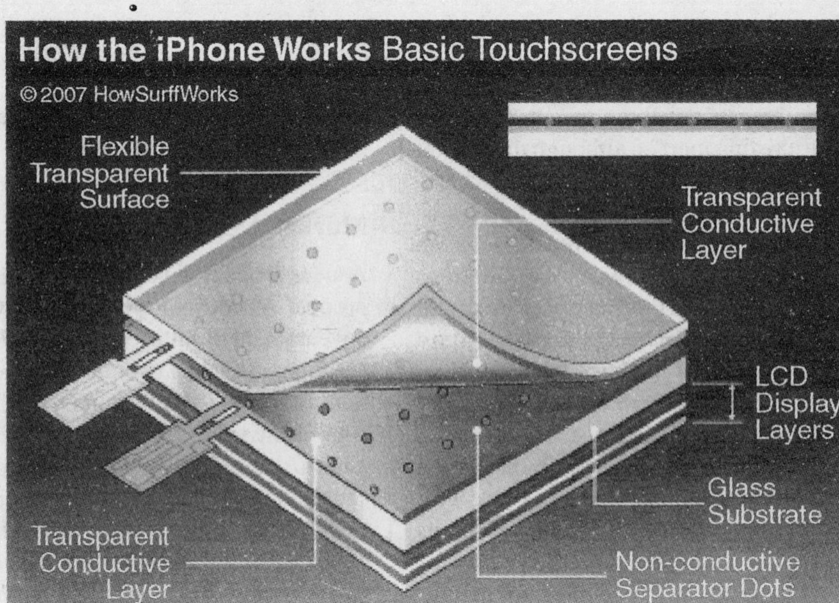

Figure 2.4 iPhone™ touch-screen construction (*Source*: Howstuffworks Web site). Reprinted with permission.

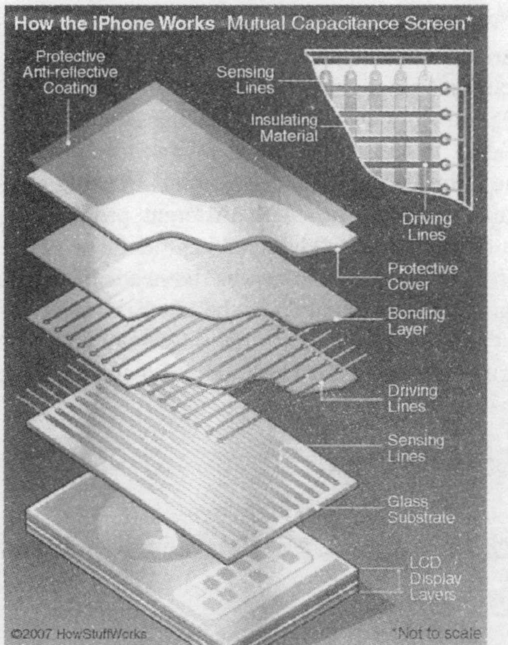

Figure 2.5 Details of iPhone™ construction (*Source*: Howstuffworks Web site). Reprinted with permission.

EXAMPLE 2.2

You are a member of the product-development team for the Apple iPhone™. The iPhone™ is a *smart* phone, a multimedia and Internet-enabled mobile phone announced by the Apple CEO, Steve Jobs, during his keynote address at the Macworld Conference & Expo in January 2007, and launched on June 29, 2007.

(a) Identify the features, functions, and benefits of the iPhone™.

(b) Formulate a compelling value proposition for the iPhone™.

SOLUTION

Note that much of the information in this solution was obtained from the iPhone™ article in the Wikipedia encyclopedia, the Howstuffworks Web site, and advertisements on the Apple Web site.

The iPhone™ functions include those of a camera phone, a multimedia player, and a mobile phone, and Internet services such as e-mail, text messaging, Web browsing, visual voice mail, and wireless connectivity. Inputs are accomplished using a touch screen with a virtual keyboard and buttons.

In more detail, the iPhone™ has the following features, functions, and benefits:

1. As a mobile phone, iPhone™ allows phone calls to be initiated by touching a finger to the name or number in its address book, favorites list, or call log. It also synchronizes all of a user's contacts on a PC, Macintosh computer, or Internet service. In addition, users may resequence voice mail messages, like e-mail, to be accessed in a preferred sequence.
2. The iPhone™ is a widescreen Apple iPod™ with touch controls that permit music, audiobooks, videos, TV shows, and movies to be accessed using a high-resolution, 3.5-inch widescreen display. It also synchronizes the content of Apple iTunes™ libraries on a PC and Macintosh computer. And, subsequently, these items are accessed by touching the screen.
3. For Internet communications, the iPhone™ features an HTML e-mail client and Safari, a Web browser for a portable device that synchronizes with bookmarks on a PC and Macintosh computer. Safari also includes facilities for Google and Yahoo searches. The iPhone™ multitasks, permitting Web pages to be read while e-mail is downloaded in the background using Wi-Fi or EDGE.
4. The iPhone™ is a small, lightweight, handheld device.

Based on these features, functions, and benefits, Apple wrote the following advertisement: "iPhone™ combines three amazing products—a revolutionary mobile phone, a widescreen iPod™ with touch controls, and a breakthrough Internet communications device with desktop-class e-mail, web browsing, maps, and searching—into one small and lightweight handheld device. iPhone™ also introduces an entirely new user interface based on a large multi-touch display and pioneering new software, letting you control everything with just your fingers. So it ushers in an era of software power and sophistication never before seen in a mobile device, completely redefining what you can do on a mobile phone." Note that the use of advertising terms like "amazing," "revolutionary," "breakthrough," "pioneering," "ushers in an era," "never before seen," and "completely redefining" is designed to attract customers, and possibly to generate enthusiasm among members of the product-development team, and eventually the business decision makers. A clearer, more concise statement can be prepared, devoid of these advertising terms: "iPhone™ is a lightweight, widescreen smart phone with visual touch control that also performs as a fully functioning media player (iPod) and an Internet communication device with desktop-class e-mail, Web browsing, maps, and searching." This value-proposition statement effectively focuses on the value the iPhone™ brings to its targeted customers, and, while more technical, lacks the advertising punch provided by the terms in quotes. Advertising is best left to those trained in advertising.

Market Segmentation

Often a product can be sold in different markets or for different applications. For example, the Apple iPhone™ can be marketed as a smart phone, a portable media player, or a personal digital assistant (PDA). While the iPhone™ was designed to provide a combination of these three functions, the new product-development team should consider how to best capture the value of its product. To answer this question, several related questions should be addressed:

- Which customers are willing and capable of paying the most?
- Which customers would benefit most from the product features and functions?
- Which applications are on the path to significant progress, that is, being developed most rapidly?
- Which applications are served best by the product?

Note that the last two questions address the suitability of the technology that enabled the products, rather than customer needs for a specific product(s). Regardless, when addressing these questions, the success of a market segmentation depends on the customers' behavior in the selected segment; more specifically, on the customers' ability to adopt a new product, where their needs may have been fully or partially fulfilled by an existing product.

A market is defined as a group of people (potential customers) who, as individuals or in collective organizations, have needs for products in a product category and who have the ability and willingness to purchase such products. While so-called end-use customers (consumers) benefit from the products, but not as profiteers, many companies and businesses buy products for resale, for direct use in production of other products, or for consumption in daily operation. As an example, office supplies are consumed daily by their producers.

Traditionally, markets are categorized by geography and demography. In demographic segmentation, groups of consumers are segmented by age, sex, ethnicity, income, education, occupation, family life cycle, family size, religion, or social class. Geographic segmentation classifies groups of consumers by subculture values, population size, population growth, population density, natural resources, natural terrain, etc.

Also, markets can be characterized by demand clusters having similar needs or patterns. Three typical demand clusters are:

- *Homogeneous demand*, where the customers have uniform demands for products for similar reason(s). This commonly applies to products that satisfy basic needs such as food staples.
- *Clustered demand*, where customer demands can be segmented into two or more distinct clusters. For example, cars are often segmented into luxury, basic, sport, or spacious (family-oriented) clusters.
- *Diffused demand*, where customer preferences are varied. Here, product differentiations are more costly to establish and more difficult to communicate. For example, in the cosmetics market, companies are compelled to offer hundreds of shades of lipstick. Often, given such diffused demands, companies try to structure customer demands by developing market segments of moderate size.

Individuals or organizations with diverse product needs have heterogeneous needs or demand patterns. For these situations, market segmentation is the process of dividing a total market into market groups consisting of people having similar product needs; that is, forming clusters of needs. Its purpose is to create a market mix that more closely matches the needs of the individuals in the selected market segment(s). In summary, the resulting market segment(s) consist of individuals, groups, or organizations that share one or more characteristics leading to similar product needs.

When developing new products, having identified diverse market segments, the product-development team can either focus on a single market segment (finding the best match for its product with the needs of that segment) or design a product to satisfy the needs of multiple market segments. The latter is often achieved by: (1) offering a new product family or portfolio, with different products designed to match the needs of selected segments, or (2) designing a *universal* product that serves the needs of multiple market segments. One example of the latter is universal duct tapes that are effective in dry or humid climates. Note that market segmentation can also be used to organize the supply chain for this product.

EXAMPLE 2.3

You are a member of the product-development team for the Apple iPhone™. Identify its market segments and their characteristics.

SOLUTION

Because of the multiple product features within this product, it seems likely the iPhone™ development team tried to capture different market segments with a single product. Historically, Apple introduced its iPod™ in the early 1990s, establishing the media-player market segment. Shortly thereafter, in the late 1990s and early 2000s, the mobile-phone market segment was developed in the United States. Given the advanced and smart features offered by the iPhone™, it seems clear that the development team intended to capture the smart phone and Internet segments, very likely attempting to create a confluence of the media-player, smart phone, and Internet markets. The results of its strategy to segment the market remain to be seen. Its success will be measured by sales revenues over the next few years.

Value-Chain Analysis

The capturing or realization of the potential economic value of a new product, in the face of competition, is crucial to the success of the new-product-development team. During the *concept* stage, the team needs answers to the following related questions:

- Should we sell products and/or services? Note that Apple does both. New customers can subscribe to AT&T phone service through the iTunes™ application.
- How should we protect the competitive advantage expressed in the value-proposition statement?
- How far along the value chain should we go to capture the maximum value?

In this subsection, value-chain analysis is discussed, beginning with a definition of the *value chain*. Then, an activity in the value chain is identified with which the most value is realized (from both the customer and the manufacturer perspective), and with which the most protection from competitors (of technology, product, and customers) can be achieved.

Wireless Value Chain

	Chipsets/ Infrastructure/ Platforms	Handset Makers	Application Developers	Content Providers	Operators	Retailers
Categories	Chipsets/ Infrastructure/ Platforms	Handset Makers	Application Developers	Content Providers	Operators	Retailers
Examples (from N.America)	. Intel . Oualcomm . Ericsson . Openwave . Symbian . Sun	. Nokia . Samsung . LGE . NEC . Sharp	. Lightsurf . Webraska	. CNN . Disney . ESPN . Sony	. ATTWS . Cingular . Nextel . Sprint . T-Mobile . Verizon	. Best Buy . Circuit City . Radio Shack

Figure 2.6 Mobile-phone value chain.

A *value chain* in a business is comprised of activities or functions in the creation and delivery of a product(s) to its end users. These are classified (Porter, 1998a,b) as *primary*, that is, *directly* needed in the production and delivery of the product(s), and *secondary*, that is, supporting activities *not directly* involved.

In Figure 2.6, an example of a mobile-phone value chain for the wireless market is shown, which consists of the entities chipset/infrastructure/platforms, handset makers, application developers, content providers, wireless operators, and retailers. For the chipset/infrastructure/platforms, the key players include Qualcomm, Intel, and Ericsson, who provide the chipset and the digital wireless technologies (e.g., Bluetooth™). The handset makers serve as integrators of the many elements of wireless devices, including the telecommunication processors and chipsets, housing designs, displays, keyboards, and user interfaces. Typical companies associated with this part of the value chain are Nokia, Samsung, LG Electronics, and NEC. The application developers and content providers provide application accessories and contents; for example, news application developers may work with news agencies, such as CNN or ESPN, to deliver content to the end users. Next in the value chain are the so-called network operators such as AT&T, Sprint, T-Mobile, and Verizon, which provide network wireless services. Finally, the last entity in the wireless value chain is the retailers, such as Best Buy and Circuit City, which sell the mobile phones to the end users—often serving as agents for the network operators included in the services they provide to their customers.

EXAMPLE 2.4

As a member of the product-development team for the Apple iPhone™, identify the portions of the value chain in which Apple chose to participate.

SOLUTION

When Apple partnered with AT&T as the service provider, it chose not to be a wireless service provider; and, consequently, it seems likely that Apple did not design and produce its own telecommunication processors. Apple sells iPhone™ directly to customers and manufactures iPhone™. Furthermore, it developed applications such as the touch-screen user interface, the media player, Internet communications, and iTunes™. Apple does not provide contents such as news, music, and movies. In summary, Apple chose to participate as a handset maker, application developer, and retailer, choosing to increase its profits by participating in three entities of the value chain for wireless devices.

Customer Requirements

Until the last quarter of the 20th century, new-product development was generally viewed as the exclusive domain of scientists and engineers in the R&D organization. They sought to create technological innovations that excite customers. These were transferred to other organizations, such as manufacturing, product development, and business development (see Figure 1.1), which carried them into the marketplace.

By the early 1980s, Japan had emerged as a world force in new-product development, having captured significant world market shares in industries as diverse as automotive, consumer electronics, and heavy manufacturing. Japanese companies and manufacturers placed heavy emphasis on quality, going beyond the simple methods for manufacturing-defect reduction taught by W. Edwards Deming, extending their quest for quality to the initial product design phase; that is, the *concept* stage (Deming, 1950). High atop their priority lists were studies of customer needs in every new product-development effort. To accomplish this, the Japanese developed the Quality Functional Deployment (QFD) methodology (Katz, 2004), where QFD begins with comprehensive lists of customer needs. Over time, the lists of needs and the process of obtaining them became known as the *voice of the customer* (VOC).

It has been traditional for the R&D organization to have little, if any, contact with customers. Rather, the business-development organization, with its marketing, sales, and technical-services groups, has maintained contacts and close working relationships with customers. As a result, marketing, sales, and technical service personnel have been added to product-development teams today to capture the VOC.

The process of obtaining the VOC often involves primary and secondary research. In secondary research, no customers are contacted and interviewed, but rather, general market requirements are collected through market studies and analyses. To distinguish this from primary research, it is said that secondary research provides the *voice of the market* (VOM), a more generic assessment of customer needs.

To a novice in product development, collecting the VOC might imply visiting and interviewing a few key customers. However, over the years, the science of listening, observing, interviewing, processing and analyzing customer needs, and converting them into useful product requirements has matured significantly. The first studies of the VOC, which continue to be highly regarded, were conducted and reported by Griffin and Hauser (1993). The principal steps of their process to obtain the VOC involve: (a) selecting customers, (b) preparing questionnaires, (c) conducting customer interviews, (d) processing and analyzing customer needs, and (e) translating the customer needs into product requirements. Next, these steps are introduced briefly, with readers referred to the chapter by Katz (2004) and the paper by Griffin and Hauser (1993) for further details.

Selection of the customers is a critical step, often on par with determination of the sample size and population in a survey. In this selection, the contacts with current customers, as well as non-customers, are equally important, with the latter often providing insights regarding their decisions not to purchase products. Similarly, contacts with major accounts, as well as small accounts, have comparable importance. Often, the latter are the least satisfied customers. It is also important to select customers associated with the various entities in the value chain for a product(s).

EXAMPLE 2.5

As a member of the product-development team for the Apple iPhone™, develop a list of groups of customers for collecting the VOC.

SOLUTION

Two obvious groups of potential customers are mobile-phone users and non-users, with the latter group involving many persons resistant to owning mobile phones. Both groups can be subdivided by income level, education level, rural and metropolitan dwellings, age, race, etc. As the iPhone™ development team may have initially envisioned a universal wireless device, users of media players, personal digital assistants (PDAs), and handheld computers may have been sought. Based on market studies, business travelers may have been a targeted group of customers.

Having selected the groups of customers, it remains to prepare questions for them, which are often best assembled systematically. Initially, it is recommended that the design team address the purpose of obtaining the VOC, followed by the *specific* issues for which customer views are sought. Given a consensus, the design team would be prepared to generate a list of questions for its potential customers.

EXAMPLE 2.6

As a member of the product-development team for the Apple iPhone™, develop a list of questions that address its preferred user interface.

SOLUTION

The purpose of obtaining the VOC is to determine the preferred user interface for a wireless device. The desired outcomes might be the:

- Types of wireless devices used by customers
- Kinds of user interfaces commonly used in their wireless devices
- Usages for their wireless devices
- Features of their user interfaces used most and least frequently
- Features of their user interfaces most liked or disliked
- Features of the proposed user interface not available in their current devices

Note that these outcomes focus on understanding the needs of the customers, not on solutions that address their needs. Also, this example is intended to determine the preferred features for the user interface, not for the entire product. (For the latter, see Exercise 2.10.)

Next, questions are assembled for customer interviews that address these outcomes. The questions should induce open-ended responses, not *yes* or *no* answers. For example, the following questions address the types of their wireless devices:

- How many wireless devices do you own? Please identify them.
- Of your wireless devices, which one do you use the most?
- How do you use this device?
- What do you like the most about this device? Ditto for its user interface?
- What other brands did you consider before you decided to purchase this device?
- How do you compare the user interfaces of the devices you considered with the user interface of the device you purchased?
- Which features of the user interface were compromised in your selection?

Clearly, these lists can become lengthy. To obtain the desired outcome, it is important that the product-development team agree on the top three questions, positioning them early in the interview. Similar lists of questions can be generated to determine customer preferences regarding the desired features and functionalities of a smart phone; for example, its use as a phone, a multimedia player, an Internet browser, a camera, and a PDA. (For the latter, see Exercise 2.10.)

The interviews are normally conducted by a three-person team comprised of an interviewer, a note taker, and an observer. The interviewer conducts the interview process posing questions and follow-up questions, guided by the prepared list of questions. To be most effective, it is normally best to adapt to the discussion thread, rather than follow a sequence of prepared questions. In parallel, the note taker records a transcript of the interview process, and the observer seeks to observe the images (unexpressed voices), reactions, body language, and other non-verbal responses offered during the interview, including samples, processes, charts, and diagrams, etc.

During and after the interview, it is important to define the VOC, which is comprised of a comprehensive set of customer *wants* and *needs*, with the needs perceived to be more important. These are usually best:

- Expressed in the customer's language or images (unexpressed voices).
- Organized to follow the customer's thought process, and his or her potential uses and interactions with the proposed products or services.
- Prioritized by the customer in terms of importance and performance, as well as current satisfaction with existing alternatives.

A sample of a verbatim customer interview concerning a new office-seating product is given in Table 2.3. While most of the specifics are noteworthy to few readers, this interview demonstrates how significant user experiences are commonly extracted by interview teams. The capturing of these experiences is crucial to the creation of successful products.

Clearly, the interviewer adjusted his/her questions according to the responses offered. The use of follow-up, or clarifying, questions to explore wants, likes, or dislikes is nicely illustrated by the last question in Table 2.3.

In some cases, customers offer solutions that address their needs. Consider, for instance, the following response to an interviewer's question: "The exterior wall should be an alloy of aluminum and titanium," which led the interviewer to ask,

Table 2.3 Verbatim Customer Interview for a New Office-Seating Product (Adapted from Hallowell, 2005)

Question: Please tell me about your current office chair. What does it look like? What features does it have?
Answer: It's fairly old, and is beige with a low back. It has *small tufts that sometimes attach to my slacks or skirt*. But, it's quite comfortable.

Question: What makes it comfortable?
Answer: *There's much room to move about,* especially when I work at my computer for many hours. I tend *to switch from position-to-position* quite often. You might say, it *moves with me.*

Question: Is it leather or fabric?
Answer: It's fabric, and unfortunately, *its arms are becoming frayed.* That's in part due to my use, but our cleaning staff is also responsible. After vacuuming, the custodians rapidly push the chairs under the desks and tables, often *causing the arms to be scraped.*

Question: Does this happen to leather chairs?
Answer: I had a leather chair at my previous company. While the arms were not frayed, it was very uncomfortable.

Question: Why?
Answer: Because on warm days, *due to perspiration, my skin often adhered to the chair, making it difficult to move freely.*

Question: Should I assume you prefer fabric?
Answer: Definitely.

Question: Why?
Answer: Because *it breathes* more. *I perspire less and can move about more easily.*

Question: Does your current chair have other advantages?
Answer: Its *elevation is easy to adjust*. Also, after a year, I learned that the stiffness of the tilt can be adjusted when leaning backwards.

Question: Wasn't that obvious?
Answer: It was described in its manual, but *I misplaced the manual shortly after receiving the chair.*

Question: What makes the elevation easy to adjust?
Answer: The chair moves upward when a paddle underneath, on the side, is lifted, and moves downward when sitting on the chair with the paddle depressed. Other chairs, with the paddle underneath in the front, are *more awkward to depress when wearing a skirt.*

Question: Does your current chair have disadvantages?
Answer: It doesn't move easily; for example, from my desk to my computer table.

Question: Can you be more specific?
Answer: *The wheels are very stiff*, and consequently, I must push aggressively to move just four feet. I wish it were *easier to move about my workspace.*

"Why is this a good solution?" The resulting response stated the underlying need: "It needs to be both lightweight and strong." Note that some customers even offer a desired specification, for example: "The thickness of the film should be less than 2 mils." Here, a helpful follow-up question would ask why 2 mils is the desired thickness. The most effective interviewers ask follow-up questions that probe the actual needs, wants, and reasons for customer likes and dislikes.

To resolve differences in interpreting customer voices, interview teams are advised to review their notes shortly after each interview, with follow-up interviews scheduled to clarify and resolve the differences. Having reached a consensus, the next step is to extract the customer voices—that is, the *wants* and *needs* useful in defining the product requirements—from the verbatim interview notes.

EXAMPLE 2.7

Extract the customer voices from the interview transcript associated with the office-seating product in Table 2.3.

SOLUTION

To obtain the customer voices, note that the *wants* and *needs* useful in defining the product requirements are closely related to the expressions in *italics* in Table 2.3. These are:

- Small tufts that sometimes attach to my slacks or skirt.
- There's much room to move about.
- To switch from position-to-position.
- Moves with me.
- Its arms are becoming frayed.
- Causing the arms to be scraped.
- Due to perspiration, my skin often adhered to the chair, making it difficult to move freely.
- It breathes.
- I perspire less and can move about more easily.
- Its elevation is easy to adjust.
- I misplaced the manual shortly after receiving the chair.
- More awkward to depress when wearing a skirt.
- The wheels are very stiff.
- Easier to move about my workspace.

Example 2.7 shows that in just one interview, 10–15 customer voices are typically obtained. With 20–50 customers interviewed, it is not uncommon to accumulate hundreds, possibly thousands, of customer voices. Of course, many are redundant or similar, especially those associated with common needs, wants, or likes and dislikes, and consequently, their processing can be a challenging task.

The KJ process, also known as the *Affinity Diagram* process, is commonly used for the processing and analysis of the VOC, where KJ refers to the Japanese anthropologist Jiro Kawakita, who studied the science of the intellect (Kawakita, 1977). The KJ process, which is used for defining, clarifying, and organizing qualitative data based upon language, involves: (1) grouping the voices into groups of similar voices based upon content or relevance, (2) providing a representative title for each group, and (3) rank-ordering the groups according to importance. These steps are repeated with increasing levels of abstraction; that is, including increasingly abstract voices, leading to the addition of more abstract subgroups having more abstract titles, as illustrated in Figure 2.7. See also Hallowell (2008). Finally, the relationships (that is, *affinities*) among these higher-level groups of voices are established.

In the end, not only are the voices grouped, but their structural relationships are well defined. This is illustrated by Anderson and Sanchez (1993), who present the KJ process conducted by the Bose® Corporation when defining customer requirements for their first high-fidelity speakers. For a more complete discussion of the implementation and mechanics of the KJ process for developing comprehensive sets of customer requirements, readers are referred to Creveling et al. (2003).

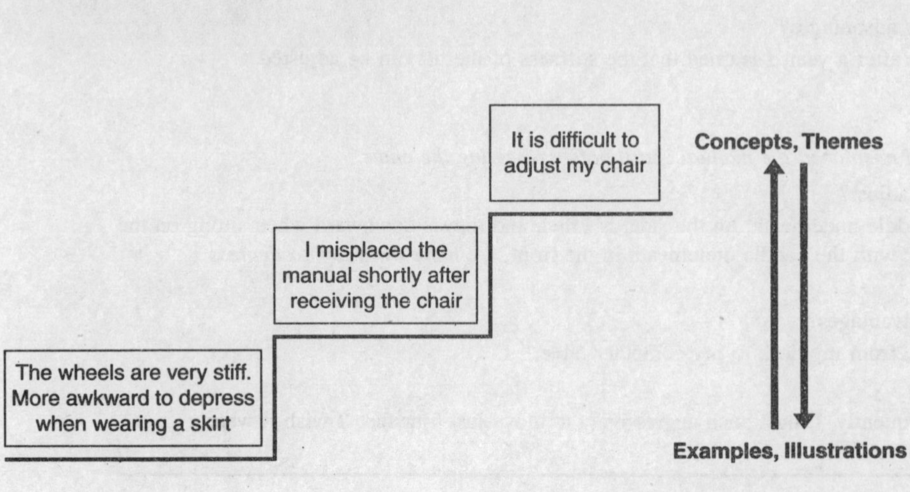

Figure 2.7 Ladder of abstraction of customer voices.

Translating Customer Voices into Customer Requirements

Because the VOCs differ significantly from the product developers' definition of new products, a translation into customer requirements is necessary. The need for this translation is illustrated by automotive customers who *want* a "roomy front seat," while product developers need to assign dimensions that provide leg-, shoulder-, hip-, and head-room, among many others. Other examples include customers who seek shampoos that provide "healthy-looking hair," and mobile phones that have "hassle-free registration." The translation and prioritization of VOCs into customer requirements is the first step in defining product requirements. Note that a KJ process is often applied to the customer requirements as well.

EXAMPLE 2.8

Translate the customer voices in Example 2.7 from the interview transcript associated with the office-seating product in Table 2.3 into a set of customer requirements.

SOLUTION

A customer voice, or a group of customer voices, may be translated into one or more customer requirements. Below is a list of customer requirements obtained from the voices in Example 2.7:

- Nothing on the surface that can attach to customer's clothing.
- Much room to move about within the chair.
- Easy to change positions within the chair: leaning forward, backward, or to either side.
- Arms that do not fray, even when moved aggressively into and out of tight spaces.
- A chair that "breathes": keeping customers from perspiring and adhering to the surfaces.
- Easy and intuitive to adjust the elevation and back-tilt of the chair.
- Accessible instructions for adjusting the chair.
- Customer should not assume embarrassing positions to adjust the chair.
- Easy movement about workspaces.

Note that the customer requirements should not contain solutions, such as "customers prefer fabric," or problems, such as "the wheels are very stiff," because these may bias the generation of potential solutions later in the *concept* stage. Furthermore, customer requirements should be sufficiently specific, at the right level of abstraction. Very abstract requirements suggest a broad range of solutions, while requirements that are defined too specifically may unduly limit the range of potential solutions. As an example of the latter, the customer requirement that "the wheels must rotate freely" limits the solution concepts to chairs with wheels.

Once a set of customer requirements has been prepared, a questionnaire may be sent to the interviewees to validate the requirements and establish the relative importance of each requirement. Typically, to rank their importance, customers are asked to distribute 100 points among the requirements listed.

Product Requirements

Given a set of customer requirements, the product-development team seeks to express them as product requirements using a more technical language involving quantitative and measurable variables.

The House of Quality (HOQ), also known as the Quality Function Deployment (QFD), relates the various requirements (customer, product, manufacturing) to one another. When first formulated, in the *concept* stage, the HOQ relates the customer requirements to the overall product requirements.

In general, an HOQ consists of six blocks, as shown in Figure 2.8. Block A is associated with the customer requirements, and Block B is associated with the quantitative and measurable technical requirements that correspond to at least one of the customer requirements. Then the relationships between the customer and technical requirements are described by the correlation matrix in Block C. Here, the entries represent a qualitative relationship (*yes* or *no*) or, more quantitatively, represent its strength (0 = none, 1 = weak, 3 = moderate, and 9 = strong). Turning to the roof of the HOQ, Block D, the synergies and conflicts among the technical requirements are represented. Note that synergistic requirements change in the same direction, as they either more or less fulfill the requirements. On the other hand, conflicting requirements change in opposite directions, signaling the need for compromises. Finally, Block E gives weighting factors for the customer requirements, and Block F represents the capabilities of the competitors in fulfilling the customer requirements.

Often the customer requirements are categorized as *fitness-to-standard* (FTS) or *new-unique-and-difficult* (NUD). The former are basic requirements that must be satisfied for customers to purchase a product(s)—requirements that are related to satisfaction levels well known to the market, while for *new-to-the-world* products based on *disruptive* technologies, customers (early adopters) are able to knowingly and willingly accept substandard FTS performance. For instance, for the first digital cameras, early adopters were willing to accept poorer picture quality in exchange for the ability to transmit pictures over the Internet or manipulate images before printing.

Similarly, customers are rarely willing to pay more for the FTS features, while NUD requirements often attract customers, as well as higher prices, and are often available in limited supplies. Because the NUD requirements provide competitive advantages for companies, their fulfillment has priority in the *concept* stage, as illustrated by the Apple iPhone™.

Note that some practioners include only the NUD requirements in Block A. Also, in Block D, the direction of

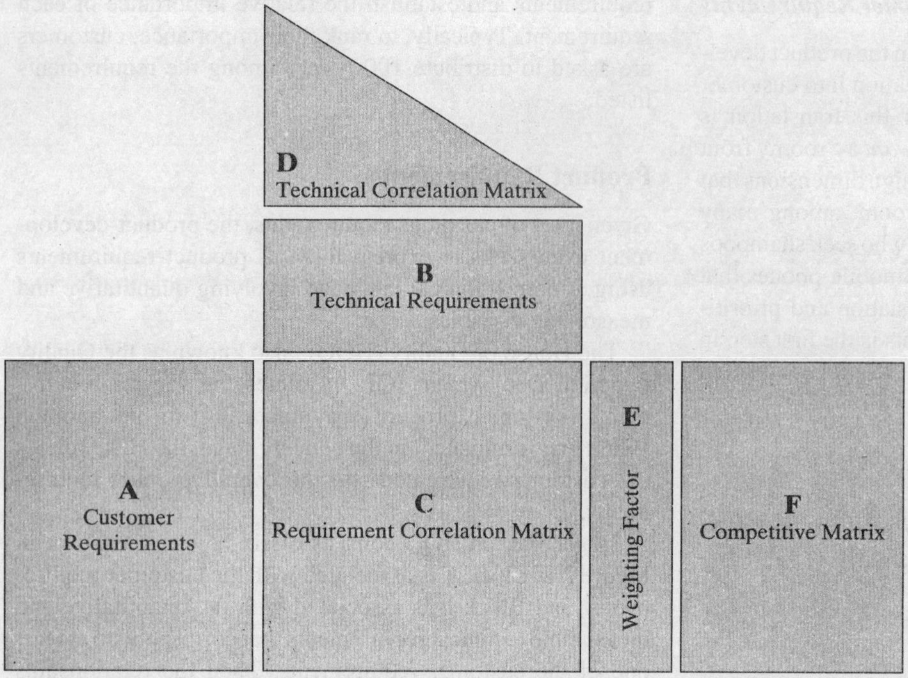

Figure 2.8 Elements of a House of Quality.

fulfillment of the NUD requirements is shown with a (+) rating indicating that as the technical requirement increases, it better fulfills an associated technical requirement. Similarly, a (−) rating indicates the opposite effect.

HOQs are created to focus on different aspects of the Stage-Gate™ Product-Development Process. For example, some HOQs are for overall systems, while others are for product subsystems or components. Yet others are for manufacturing subsystems and components. Hence, HOQs are often created, or extended, at different stages in the SGPDP.

To illustrate the creation of a first HOQ in the *concept* stage, consider the product design of a longer-lifetime incandescent light bulb, which is discussed in detail in Section 17.2. In Table 2.4, the lower rectangular matrix pairs the customer requirements in the first column with at least one quantitative technical requirement or parameter in the adjacent columns. For example, the customer requirement that the bulbs fit in lamps with shades, recess lamps, or tracking light fixtures imposes a maximum operating temperature; that is, a technical variable with an upper bound to prevent fires. Similarly, the quality of light, *warm* or *cool*, imposes an operating temperature known as the *color* temperature. In other cases, the customer requirement translates directly into the technical requirement, for example, the lifetime.

At the top of the house, the *interaction matrix* shows the synergistic technical requirements or parameters—for example, the lifetime and color temperature. As introduced, (+) indicates that both variables increase or decrease; (−) indicates that when one variable increases, the other decreases, and vice versa; and a blank entry indicates no significant relationship between the variables. For the incandescent light bulb, the higher the color temperature, the higher the bulb temperature, and consequently, the shorter the lifetime. Thus, the (−) signals the need for a compromise between the color temperature and the lifetime.

The selection of the FTS and NUD requirements is followed by identification of the technical requirements to fulfill the NUD requirements. These steps, together with the creation of the first HOQ, are illustrated in detail for three products in Chapters 15 and 17—that is, for thin-glass substrates for LCDs (Section 15.2); for halogen light bulbs (Section 17.2); and for a home hemodialysis product (Section 17.3). For further details on creating the HOQ, readers are referred to Creveling et al. (2003).

Product Concepts

Having decided upon the technical requirements for the product(s), the product-development team begins to develop its new product concepts, that is, potential solutions that satisfy the NUD and FTS requirements. For complex products with many components, parallel development efforts, component by component, often lead to a collection of concepts for the combined product. Clearly, this approach is risky, as unanticipated interactions may be overlooked in the *concept* stage. For this reason, it is often necessary to defer judgment until a prototype can be created in the *feasibility* stage. In this respect, flexibility is crucial when deciding to carry a concept through to the *feasibility* stage.

Ideally, a multifunctional product-development team creates the solution concepts, with the team comprised of multidisciplinary personnel (Creveling et al., 2003), including:

- Scientists and engineers who have developed the underlying technologies likely to be used in the new product.

Table 2.4 First House of Quality for Longer-Lifetime Incandescent Light Bulbs

Interaction Matrix

+ synergistic

	Lifetime (hours)	Cost $/Watt	Max Op. Temp	Color Temp	Energy Eff.	Lum. Ef
Lum. Eff						1
Energy Eff.					1	+
Color Temp				1	+	–
Max Op. Temp			1			
Cost $/Watt		1	–	–	–	
Lifetime (hours)	1	+		–		

Customer Requirement	Lifetime (hours)	Cost $/Watt	Max Op. Temp	Color Temp	Energy Eff.	Lum. Ef	Weight
Lifetime (NUD)	X						0.3
No Cost Premium (FTS)		X					0.2
Fit Various Fixtures (FTS)							0.1
Lamps with Shade			X				
Recess Lamps			X				
Tracking Lights			X				
Colors of Light (FTS)							0.1
Warm Light				X			
Cool Light				X			
Energy Efficient (NUD)					X	X	0.3

- Development engineers in related fields.
- Senior manufacturing engineers, such as those who developed previous generations of the product.
- Technical- and customer-service personnel who worked on previous generations of the product, or who have extensive experience in handling technical problems and customer concerns.
- Marketing and sales personnel.
- Supply-chain specialists.
- Health, safety, environmental, and regulatory specialists.

This internally focused team is often complemented with technology-development partners from industry, academia, and/or the government. It may also include selected customers, with their levels of involvement dependent on the needs of the product-development projects. Usually, they serve as consultants rather than core members of the product-development team.

When generating new solution concepts, the so-called Pugh matrix (Pugh, 1996), in which each solution concept (partial and complete) is judged against a reference solution, is useful for screening purposes. Given the reference solution, which is usually the best known in the market, each concept is evaluated against the reference solution and assigned a qualitative valuation of inferior (–), superior (+), or equal (0). Note that in refined variations of the valuation method, ratings such as (– – –) or (+++) indicate different levels of superiority or inferiority.

In most cases, solution concepts don't satisfy all of the requirements. When the violating requirements differ, combinations of the solution concepts are often attempted, with the combined concepts having a reduced number of violated requirements. Usually, the resulting concept having the

Table 2.5 Pugh Matrix and Superior Concept

Requirement	Reference Concept	Concept A	Concept B	Concept A + B
NUD-1	Best Solution	+	−	+
NUD-2	on the Market	−	+	+
NUD-3		+	0	+
NUD-4		−	0	0
NUD-5		−	−	−
NUD-6		0	0	0

minimum number of unmet requirements is selected to be the *superior* concept. Rarely, however, is a superior concept located that satisfies all of the NUD requirements.

Table 2.5 shows a typical Pugh matrix in which two solution concepts are rated against the best solution on the market for six NUD requirements. While concept A is rated superior to the best solution for NUD-1 and NUD-3, it is inferior to the best solution for NUD-2, 4, and 5, and equal to the best solution for NUD-6. Alternatively, concept B is rated superior to the best solution for NUD-2, inferior to the best solution for NUD-1 and NUD-5, and equal to the best solution for NUD-3, 4, and 6. As anticipated, the combined concept A+B is superior compared to the best solution for NUD-1, 2, and 3, and equal to the best solution for NUD-4 and 6. Still, it remains inferior to the best solution for NUD-5. Clearly, concept A+B is the *superior* concept, as it has the minimum number of inferior (−) ratings. Note that the weighting factors are often ignored when examining the new concepts. They gain importance when selecting the superior concept.

The usage of the Pugh matrix to select superior concepts is illustrated for three products in Chapters 15 and 17—that is, for thin-glass substrates for LCDs (Section 15.2); for halogen light bulbs (Section 17.2); and for a home hemodialysis product (Section 17.3). For further details on the Pugh matrix, readers are referred to Creveling et al. (2003).

Opportunity Assessments

In the *concept* stage, product-development teams normally focus on assessing their product opportunities by carrying out preliminary product cost estimates and risk analyses. To proceed to the *feasibility* stage, their approximate cost estimates must be promising, with more accurate estimates compiled as the team proceeds from stage to stage through the SGPDP. For the design of basic chemical products, as shown in Chapter 13, the focus is usually on the profitability of the manufacturing processes, while for industrial and configured consumer products, as shown in Chapters 15 and 17, the estimates are more closely associated with the product selling prices and their mass production. As mentioned earlier, before recommending a sizable investment, it is crucial to assess the risks of not capturing the potential economic value, which involve estimating the economic value and the associated competition. Due to space limitations, just two assessment approaches, the Porter five-forces analysis and the intellectual-property (IP) analysis, are covered in this section.

Porter Five-Forces Analysis

Competitive analysis commonly focuses on the competition forces within markets or industries. In 1979, Porter argued that in addition to the market competition force, there are four other forces to be considered: (1) the bargaining power of the suppliers, (2) the bargaining power of the customers, (3) the threat of new entrants, and (4) the threat of substitute products. To these, a sixth force was added by critics of the Porter approach; that is, the bargaining power of other stakeholders such as government, shareholders, environmental agencies, etc.

The competition forces within the market influence and are influenced by the market growth rate, the number and diversity of competitors, the brand equity, the advertisement expense budget, and exit barriers. The bargaining power of the suppliers includes differentiation of the supplied materials; the relative cost of the input materials, as compared with the total product cost and the selling price; the cost involved in switching suppliers; the presence of substitute input materials; etc. In contrast, the bargaining power of the customers includes the purchase volume, buyer price sensitivity, presence of substitute or alternate products, etc. The threat of the new entrants is related to the capital cost of changing products, the barrier-to-entry, the brand equity, the switching cost, supply-chain access, etc. Finally, the threat of product substitutes includes buyer switching costs, the propensity to substitute, the relative price performance of substitutes, and the perceived product differentiation, which indicates the value of the product features beyond the competition offering.

The product-development team knows most of these forces, or is familiar with most of them, for routine product extensions. However, for *new-to-the-world* products, and even products intended for an adjacent market, it is critical to assess these forces.

EXAMPLE 2.9

Assess the barrier-to-entry forces probably considered by Apple when introducing the iPhone™, its first wireless device.

SOLUTION

Apple is one of the most innovative technology companies worldwide. Its revolutionary products have included the Macintosh computer, with its friendly user interface. In the early 2000s, Apple introduced another *disruptive* product, iPod™, a record-breaking media player.

Recently, Apple decided to expand its market to include the mobile-phone and PDA markets by introducing the iPhone™. Clearly, its major barrier to entry was its prior absence from these markets. Second, to develop its product differentiation, Apple reverted to its success in creating world-class user interfaces, having introduced the first active touch-screen user interface. Yet another barrier was the channel-to-market, which Apple strategically overcame by creating an exclusive alliance with the AT&T mobile-phone service carrier. In this regard, AT&T made a strategic move to work with Apple to revolutionize its customer usage of mobile phones. In one major example, they introduced a nonsequential, voice mail retrieval method.

Intellectual-Property Analysis

Intellectual-property, or patent, analysis is used to assess technical competitiveness, to forecast technological trends, and to plan for potential competition from disruptions and displacements by new technologies, all of which are important when developing a new product. As an example, the introduction of a new product to the market can make obsolete a new-product-development effort. For these reasons, an early awareness of the new technologies that may displace a product concept is crucial in realizing the return on the investment for a new product.

A patent for an invention is defined as the grant of a property right to the inventor that excludes others from practicing the invention. It is used to protect a product, giving the inventor an exclusivity to produce the product for a limited time, 17 years in the United States. Patentable inventions may include: (1) operating methods or processes, (2) physical structures such as the composition of matter, and (3) product features or articles. Recently, these have been expanded to include: (4) algorithms and (5) business processes.

Using the *innovation map* described in Section 1.3, the product-development team can identify key inventions in material, process/manufacturing, and product technology that enable product differentiation in the marketplace. These inventions are often protected by patents or defensive publications, or are kept as trade secrets. When *picket-fencing* a product, alternate technologies or pathways to produce a similar product are often protected as well.

The primary strength of patent analysis is as a leading indicator of technological change, offering an efficient way of identifying technological discontinuity. Patent analysis usually begins with a patent search, followed by analysis of the patents located. Until recently, patent searches were carried out by experts using specialized tools and databases. Today, patent searches are easily achieved using the Google advanced patent search on the Internet, as shown in Figure 2.9.

A Google advanced patent search locates patents using user-supplied keywords, patent numbers, patent titles, inventor names, assignee names, patent classifications, issue dates, and filing dates. Normally, searches are first carried out using keywords. After appropriate keywords have been found,

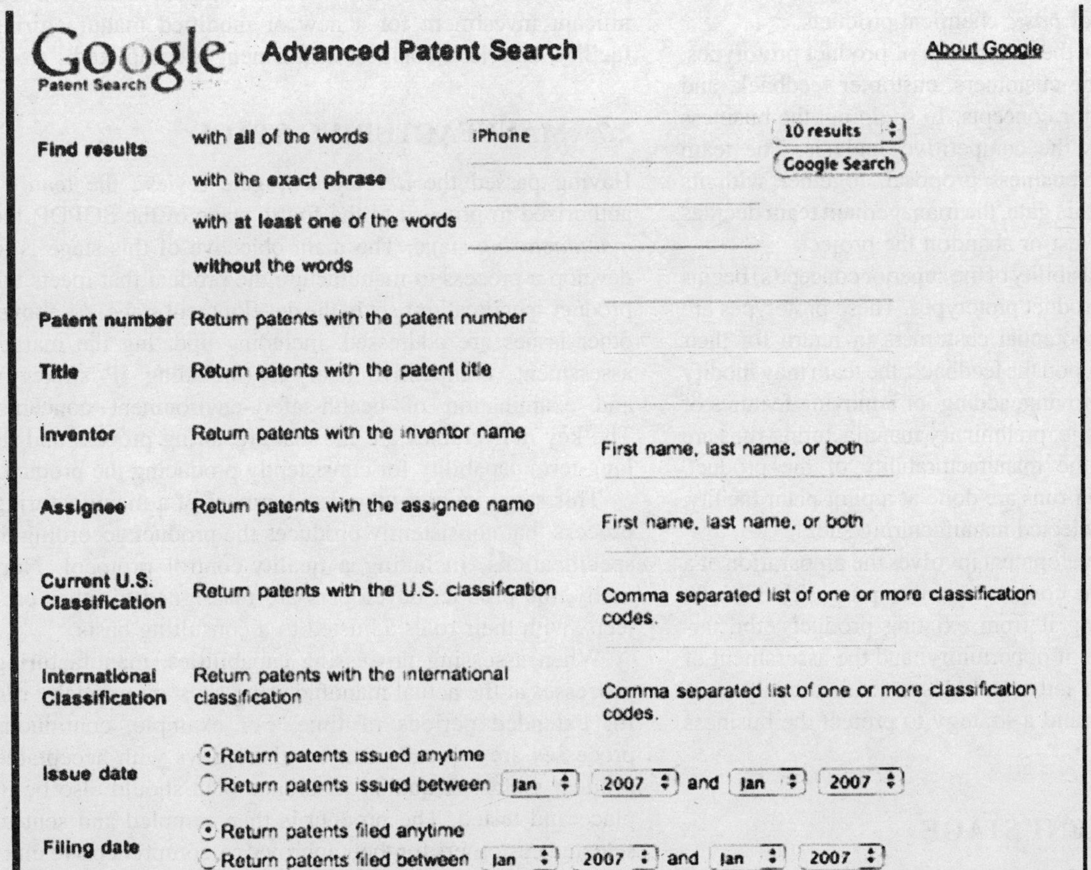

Figure 2.9 Google advanced patent search.

searches can be limited by other attributes such as the inventor, assignee (company name), and filing or issue dates. Because issue dates are often many years after filing dates, searches often use the filing date to obtain a historical perspective of competitive filings.

Patent searches are illustrated in detail for four products in Chapters 14–17—that is, for thin-glass substrates for LCDs (Section 15.2); for washable crayon mixtures (Section 14.3); for halogen light bulbs (Section 16.2); for a home hemodialysis product (Section 16.3); and for a lab-on-a-chip for the high-throughput screening of kinase inhibitors (Section 17.4).

2.5 FEASIBILITY STAGE

The second stage of the Stage-Gate™ Product-Development Process is the *feasibility* stage. The main objectives of this stage are to validate the superior concept(s) generated during the *concept* stage against the customer requirements, and to build the business case for the project. In addition, other issues are addressed including updating the market assessment, competitive analysis (including IP strategy), and examination of health-safety-environment concerns.

The key deliverables are: (1) an assessment of the extent to which the superior concepts fulfill the customer requirements, (2) a business case for capturing the potential economic value of the product in the face of competition, and (3) a base-case process flow diagram, when applicable, especially for the design of *basic* chemical products.

This stage involves the generation of product prototypes, their evaluation by the customers, customer feedback, and the redesign of superior concepts. In so doing, the business case is revised, as is the competitive analysis. The team prepares a complete business proposal, together with its recommendation. At this gate, the management team decides whether to further invest or abandon the project.

Validation of the feasibility of the superior concept(s) begins with the building of product prototypes. These prototypes are shared with selected potential customers in return for their feedback. Depending upon the feedback, the team may modify the concept(s) by improving, adding, or removing features of the product. At this stage, preliminary manufacturing runs are conducted to verify the manufacturability of the product. Usually, manufacturing runs are done at a pilot-plant facility, with a few runs at a selected manufacturing site.

A business case development involves the preparation of a business proposal that covers the value proposition of the product, differentiating it from existing products; the targeted market; the size of opportunity; and the assessment of the business risk. The latter includes a detailed analysis of competitive offerings and a strategy to protect the business (IP strategy).

2.6 DEVELOPMENT STAGE

Having passed the feasibility gate, the team is authorized to proceed to the third stage of the SGPDP, the *development* stage. The main objective of this stage is to fully develop the product, ensuring that it is manufacturable and delivers the promised value proposition to its customers. As necessary, detailed design, equipment-sizing, profitability analysis, and optimization are carried out. In addition, other issues are addressed including updating the market assessment, competitive analysis (including IP strategy), and examination of health-safety-environment concerns. The key deliverables are the product specifications, the manufacturing feasibility assessment, and the detailed process design—especially for the *basic* chemical products.

This stage involves the development of the product construction specifications. In this stage, the construction, features, and complete specifications are developed. Also, customers are contacted more frequently in an attempt to align their FTS and NUD requirements more closely with the final product specifications. The team prepares a complete manufacturing assessment focusing on the manufacturing feasibility evaluation of the new product. Of particular interest is the capital investment required to manufacture the product.

Manufacturing assessment often involves several manufacturing runs at existing manufacturing sites to evaluate their suitability for producing the product. In this regard, the team normally prepares a risk analysis of using the existing equipment. When necessary, a capital investment estimate is prepared for a new or modified manufacturing facility. Potential manufacturing sites are also identified and their performance is evaluated. When the project involves a significant investment for a new or modified manufacturing facility, a revised business assessment is completed.

2.7 MANUFACTURING STAGE

Having passed the *development* gate review, the team is authorized to proceed to the fourth stage of the SGPDP, the *manufacturing* stage. The main objective of this stage is to develop a process to manufacture the product that meets the product specifications set in the development stage. As above, other issues are addressed, including updating the market assessment, competitive analysis (including IP strategy), and examination of health-safety-environment concerns. The key deliverables are the manufacturing process and its long-term capability for consistently producing the product.

This stage involves the development of a manufacturing process that consistently produces the product according to specifications, including a quality-control protocol. Normally, the product developers are reassigned to other projects, with their roles adjusted to a consulting basis.

When assessing processing capabilities, manufacturing processes at the actual manufacturing sites are normally run for extended periods of time. For example, continuous processes are typically run for three days with acceptable product yields. A quality-assurance plan should also be in place and tested. The product is then sampled and sent to selected customers for their approval. A comprehensive unit-cost analysis is prepared and the business plan is updated to account for the new unit costs, as necessary.

2.8 PRODUCT-INTRODUCTION STAGE

Having passed the *manufacturing* gate review, the team is authorized to proceed to the last stage of the SGPDP, the *product-introduction* stage. The main objective of this stage is to prepare a product-launch plan that includes product literature containing the final product specifications, pricing strategy, branding strategy, advertisements, and new-product-announcements. In addition at this stage, the product inventory is normally built for about two months of sales. The key deliverables are the product-introduction plan and the product inventory.

The product-launch plan is developed by the sales and marketing team members, as well as the technical-services forces. Product literature is developed, as well as training materials to be used by the sales and technical-services personnel. Pricing strategy is also developed, as well as the channel for releasing the products. For example, for the iPhone™, the product release was to Apple and AT&T stores. Apple decided not to utilize common retailers such as Best Buy or Circuit City. Yet another key activity is branding. For instance, branding commonly involves combining the strengths of two or more brands to maximize the sales and acceptance of the product.

Pricing strategy is a key consideration in launching a new product(s), as it is a major factor in positioning the product(s) in the market. While there is no simple recipe to set prices, several steps are involved:

1. Develop a marketing strategy by performing a market analysis, which involves market segmentation, targeting, and positioning.
2. Estimate the demand curve, that is, the relationship between the sales volume and the product price. Clearly, lower prices usually lead to higher sales volumes. For existing products, estimates of sales volumes at prices below or above the current price indicate the price elasticity. With inelastic demand, price increases are feasible.
3. Calculate the unit cost, including the fixed and variable costs, in manufacturing the new product. The unit cost sets the lower bound on the product price and determines the profit margin at higher prices.
4. Understand the environmental conditions to evaluate competitive responses and legal constraints. Prices set too low may invite an undesirable response such as a price war, while prices set too high may induce new competitors to enter the market. Legally, firms are not free to price products at any level. At the low extreme, firms may be accused of predatory pricing for a global product. Also, prices should be uniform to prevent allegation of price discrimination.
5. Set pricing objectives, for example, profit maximization, revenue maximization, sales volume maximization, profit-margin maximization, or price stabilization. As noted in Chapter 23, each pricing objective has different economic impacts. For example, sales volume maximization usually increases the market share by decreasing long-term costs due to the economy of scale.

Clearly, all of the above factors must be considered when setting prices. For new products, the pricing objective is usually to maximize either the profit margin or the sales volume (market share). To achieve these objectives, so-called *skim* or *penetration* pricing strategies are often employed (Dean, 1976), as discussed next.

Skim pricing aims to sell the new product to less price-sensitive customers near the top of the market pyramid. This strategy is appropriate when large cost savings are not anticipated or are difficult to attain at high volumes. It is also appropriate when company resources are inadequate for the large capital expenditure required for high-volume production.

For penetration pricing, the new product is sold at a low price to gain market share rapidly, soon after production begins. This strategy is appropriate when: (1) a large cost reduction is anticipated at high volumes, (2) the product is anticipated to gain mass adoption quickly, or (3) a sizable threat of impending competition is on the horizon.

To set prices that achieve the various pricing objectives, several strategies are commonly used. These are:

- Cost plus pricing, where the price is set at the unit cost plus a desired profit margin.
- Target return pricing, where the price is set to achieve a specified return on investment (ROI).
- Value-based pricing, where the price is set at the effective value to the customer relative to alternative products.
- Psychological pricing, where the price is set based on product quality at an acceptable price point, which the consumer perceives to be a fair price.

In addition to the *list* price for the end user, a discounted price should be set for distributors (wholesalers), dealers, and select end users.

As the product life cycle progresses, the demand curve commonly changes; that is, the relationships between the demand and costs change. Consequently, pricing policies are normally reevaluated over time.

EXAMPLE 2.10

Assess the pricing strategies for highly innovative products such as the iPhone™ or Post-it™.

SOLUTION

Most highly innovative products involving complex manufacturing processes and high R&D costs usually adopt the skim pricing strategy early in the product life cycle to maximize profit margins. For the iPhone™, the high price targets the product toward customers at the top of the market pyramid. Furthermore, Apple used value-based and psychological pricing techniques, given that no similar products were available in the market. In addition, early marketing stimulated much excitement, and the new touch-screen technology induced high anticipation among many consumers.

For Post-its™, 3M effectively protected the adhesives inventions used, successfully preventing significant competition. It is likely that a value-based pricing strategy was used.

Henderson's Law

When estimating the demand curve, companies are unable to project reliably the settling price for a product. Instead, asymptotic prices can be predicted fairly well using the so-called experience curve. The latter is a log-log curve whose ordinate is the product price, or cost, and whose abscissa is the cumulative number of units produced. As the number of manufactured units increases, so does the experience of producing them, and consequently, the price drops. This relationship, often attributed to Henderson (1974), can be expressed:

$$p_n = p_1 n^{-\alpha} \tag{2.1}$$

where p_1 is the price (or cost) of the first unit of production, n is number of production units, p_n is price (or cost) of the nth unit of production, and α is the slope of the log-log curve, or the so-called price *elasticity*. Hence, when projecting the asymptote at high n, a value of α must be estimated, with typical values within 0.7–0.9. This relationship applies to the prices of numerous products, including LCD panels, ballpoint pens, VCRs, flat-panel televisions, and calculators. Based upon values of α for existing products, companies often estimate the price elasticity for new products.

Eq. (2.1) can be justified as follows: Initially, when the number of manufactured units is low, the price (and cost) per unit is high because the volume is low, primarily because the manufacturing process has not been scaled-up to a more economical scale. In addition, often the manufacturing technology is immature initially. As an example, initially, a new chemical compound is produced in a pilot-plant at a high price per pound. At anticipated high throughputs, using continuous processing, the price (and cost) per pound will be sharply reduced. As another example, Figure 2.10 illustrates the projected cost of electricity (from 2010 to 2050) using photovoltaic technology as a function of the installed capacity, prepared by the Energy Research Center of the Netherlands (www.ecn.nl/fileadmin/ecn/units/bs/PHOTEX/photex.xls).

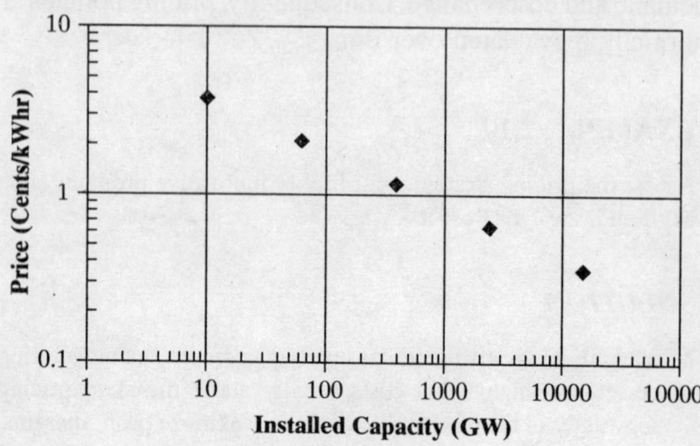

Figure 2.10 Experience curve of electricity production using photovoltaic technology.

EXAMPLE 2.11

The price of electricity production using photovoltaic technology for the next decades as a function of the installed capacity is given in Table 2.6 (and graphed in Figure 2.10).

Table 2.6 Price of Electricity Production

Installed Capacity (GW)	Price (Cents/kWhr)
10	3.75
63	2.09
387	1.16
2,399	0.64
14,851	0.36

For expanded capacities projected beyond 2050, graph the price per kWhr for various values of price elasticity: 0.75, 0.8, and 0.9.

SOLUTION

Each experience curve can be constructed easily using the following rule: as the cumulative production is doubled, the price decreases by the price elasticity. For example, when $\alpha = 0.75$, as the cumulative installed capacity doubles from 14,851 to 29,702 GW, the price of electricity decreases by 25% $= 0.75 \times 0.36$ cents $= 0.27$ cents. Repeating this for size expansions:

Installed Capacity (GW)	Price (Cents/kWhr)
14,851	0.36
29,703	0.27
59,406	0.20
118,811	0.15
273,623	0.11
475,245	0.09
950,490	0.06
1,900,980	0.05
3,801,961	0.04

Also, repeating this with $\alpha = 0.8$ and 0.9, the experience curves in Figure 2.11 are graphed.

Clearly, the choice of the elasticity, α, has a significant impact on the price projections. Note that $\alpha \sim 0.78$ for the data in Table 2.6 and Figure 2.10. As indicated above, for price projections into the future, it is common to set α within 0.7–0.9.

For many products beyond the economy-of-scale, initially the profit margin is high, as there are no competitors. However, as the market grows, the profit margin decreases.

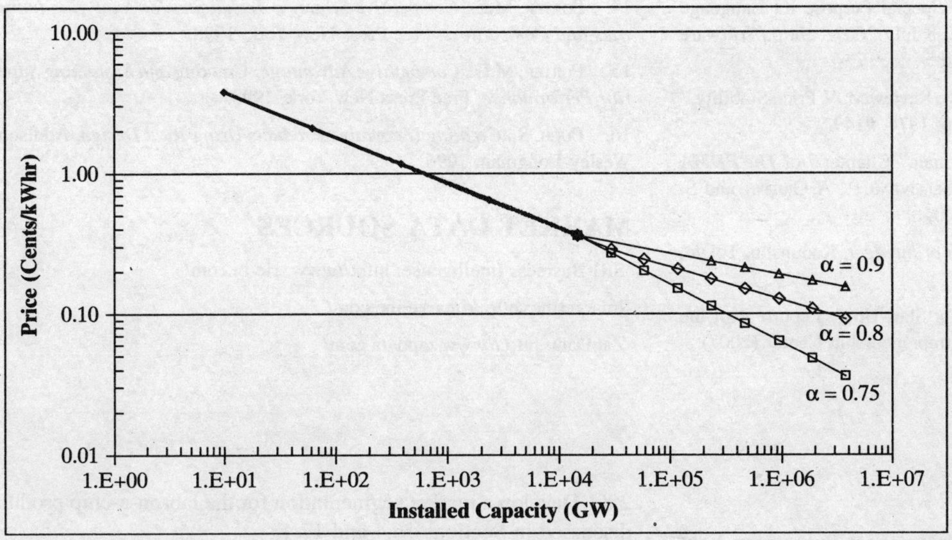

Figure 2.11 Projected experience curve of electricity production using photovoltaic technology at $\alpha = 0.75$, 0.8, and 0.9.

For initial price (and cost) estimates, Eq. (2.1) is often applied. Note that these estimates can be made throughout the SGPDP, not just in the *product-introduction* stage. Normally, however, costs rather than prices are estimated in the early stages (*concept,* etc.). Serious price projections usually wait until the *product-introduction* stage.

Summary

In summary, the pricing of a new product is very important, as it can determine not only the successful commercialization of an innovative technology into the new product, but also the realization of the return on investment for the R&D required to produce the new product. A balance must be struck between product acceptance in the marketplace and the maximization of profit. For a *new-to-the-world* product with significant perceived value to the customers, prices commonly seem exorbitant. For example, the price of an iPhone™ is close to the price of a low-end computer, and the price of Post-it™ notes is much higher than that of colored paper. In many of these cases, consumers accept the price premium to obtain perceived valuable functionalities and features.

2.9 SUMMARY

Having studied this chapter, the reader should:

1. Be acquainted with the need to develop a pipeline for new-product development and the steps in beginning a new-product-development effort, that is, the creation of a project charter and an *innovation map* for the new product.
2. Understand the five stages in the Stage-Gate™ Product-Development Process (SGPDP), especially the *concept* stage.
3. Be able to carry out the *concept* stage, involving a market assessment, determination of customer requirements (*voice of the customer*) and product requirements, evaluation of new product concepts, and carrying out an opportunity assessment. While the techniques are introduced using general terms, the reader should be familiar with their application in the case studies of Chapters 13, 15, and 17, involving ammonia, environmentally friendly refrigerants, water-dispersible β-carotene, thin-glass substrates for LCDs, washable crayons, halogen light bulbs, home hemodialysis, and lab-on-a-chip products.
4. Be acquainted with the steps in the *feasibility, development, manufacturing*, and *product-introduction* stages of the SGPDP.

REFERENCES

1. Anderson, E., and J. Sanchez, "Application of Concept Engineering on the Bose Enchilada Project," *Center for Quality of Management Journal*, **2**(3), 23–32 (1993).

2. Cooper, R. G., *Winning at New Products: Accelerating the Process from Idea to Finish*, 3rd ed., Perseus Publ., Cambridge, Mass., 2001.

3. Cooper, R. G., *Product Leadership: Creating and Launching Superior New Products*, Perseus Publ., Cambridge, Mass., 2002.

4. Cooper, R. G., *Product Leadership: Creating and Launching Superior New Products*, 2nd ed., Basic Books, Cambridge, Mass., 2005.

5. Creveling, C. M., J. L. Slutsky, and D. Antis, Jr., *Design for Six Sigma in Technology and Product Development*, Pearson Education, 2003.

6. Dean, J., "Pricing Policies for New Products," *Harvard Business Review*, Reprint 76604 (1976).

7. Deming, W. E., *Elementary Principles of the Statistical Control of Quality*, Nippon Kagaku Gijutsu Renmei, Tokyo, 1950.

8. Griffin, A., and J. Hauser, "The Voice of the Customer," *Marketing Science*, **12**(1), 1–27 (1993).

9. HALLOWELL, D. L.,"Effective Use of Special Purpose KJ Language Processing,"22 June 2005; iSixSigma.com, 8 July 2007. <http://software.isixsigma.com/library/content/c050622b.asp>.

10. HENDERSON, B., "The Experience Curve Reviewed: V. Price Stability," *Perspectives*, The Boston Consulting Group, 1974, #149.

11. KATZ, G. M., "The Voice of the Customer," Chapter 7 of *The PDMA ToolBook 2 for New Product Development*, BELLIVEAU, P., A. GRIFFIN, and S. SOMERMEYER (eds.), John Wiley & Sons, 2004.

12. KAWAKITA, J., *A Scientific Exploration of Intellect*, Kodunshu, Tokyo, 1977.

13. MURRAY, F., "Technology Strategy for Start-Ups," Lecture 4 of the Nuts and Bolts of Business Plans, MIT Entrepreneurship Center (2007).

14. PORTER, M.E., *Competitive Strategy: Techniques for Analyzing Industries and Competitors*, Free Press New York, 1998.

15. PORTER, M.E., *Competitive Advantage: Creating and Sustaining Superior Performance*, Free Press New York, 1998.

16. PUGH, S., *Creating Innovative Products Using Total Design*, Addison-Wesley-Longman, 1996.

MARKET DATA SOURCES

SRI Business Intelligence: http://www.sric-bi.com/

ImarketInc: http://imarketinc.com/

ZapData: http://www.zapdata.com/

EXERCISES

Project Charter

2.1 Develop a project charter for the compact fluorescent light bulb discussed in Section 1.3.

2.2 Develop a project charter for the lab-on-a-chip product discussed in Sections 16.4 and 17.4.

2.3 Develop a project charter for the home hemodialysis product discussed in Sections 16.3 and 17.3.

Value Propositon

2.4 Write a value-proposition statement for the compact fluorescent light bulb discussed in Section 1.3.

2.5 Write a value-proposition statement for the lab-on-a-chip product discussed in Sections 16.4 and 17.4.

2.6 Write a value-proposition statement for the home hemodialysis product discussed in Sections 16.3 and 17.3.

Market Segmentation

2.7 Develop a market segmentation for the home hemodialysis product discussed in Sections 16.3 and 17.3.

2.8 Develop a market segmentation for the lab-on-a-chip product discussed in Sections 16.4 and 17.4.

2.9 Develop a market segmentation for the iPhone™ discussed in Example 2.3 using a demographic approach. Use the Internet to obtain the list price for this product.

Voice of the Customer

2.10 Generate lists of questions for determining the desired product features and functionalities for a smart phone such as the iPhone™. Consider features and functionalities in several areas, for example, for use as a phone, a multimedia player, a camera, an Internet browser, and a PDA.

2.11 Carry out a KJ analysis for the customer voices in Table 2.3:

(a) Group the customer voices into groups with similar voices. Each group should have three to five voices.

(b) Assign a title representing the customer needs to each group.

Part One

Product Design for Basic Chemicals

Part One presents, in 11 chapters, strategies for the design of *basic chemical* products and the processes to produce them. It follows the Stage-Gate™ Product-Development Process (SGPDP), which is introduced in Chapters 1 and 2, and presented for the design of *basic chemical* products in Figure PI.1. As discussed in Section 1.3, *basic chemicals* are normally well-defined molecules and mixtures of molecules characterized by thermophysical and transport properties, but are not normally described by other properties, including microstructure; particle-size distribution; and functional (e.g., cleansing, adhesion, shape), sensorial (e.g., feel, smell), rheological (non-Newtonian viscosity), and physical (e.g., stability) properties. The latter are normally the focus of so-called *industrial chemicals* (e.g., fibers, films, creams, and pastes), which are covered in Part Two, and *configured consumer products*, which are covered in Part Three. As indicated in Figure 1.3, few *basic chemicals* are purchased by the consumer. Rather, they are the ingredients for *industrial chemicals* and *configured consumer products*.

MATERIALS TECHNOLOGY DEVELOPMENT: MOLECULAR STRUCTURE DESIGN

After the design team creates its product charter, as discussed in Section 2.2, it seeks to identify appropriate materials technologies to achieve its objectives when they are needed. This is the step at the top left of Figure PI.1, which, for *basic chemicals*, usually involves a search for the appropriate molecules or mixtures of molecules to satisfy property specifications that align closely with customer needs. Examples include:

1. Thin polymer films to protect electronic devices having a high glass-transition temperature and low water solubility;
2. Refrigerants that boil and condense at desired temperatures and low pressures, while not reacting with ozone in the earth's stratosphere;
3. Environmentally friendly solvents for cleaning, for example, to remove ink pigments, and for separations, as in liquid-liquid extraction;
4. Low-viscosity lubricants;
5. Proteins for pharmaceuticals that have the desired therapeutic effects;
6. Solutes for hand warmers that remain supersaturated at normal temperatures, solidifying at low temperatures when activated; and
7. Ceramics having high tensile strength and low viscosity for processing.

Often design problems are formulated in which the molecular structure is manipulated, using optimization methods, to achieve the desired properties. For this purpose, methods of property estimation are needed, which often include group contribution methods, and, increasingly, molecular simulations (using molecular dynamics and Monte-Carlo methods). The search for molecular structure is often iterative, involving heuristics, experimentation, and the need to evaluate numerous alternatives in parallel, especially in the discovery of pharmaceutical proteins, as discussed in Chapter 3.

When specifying desired properties and selecting potential chemicals and chemical mixtures, design teams must be acutely aware of environmental and safety issues and regulations. These issues are so important that they are discussed separately in Sections 1.4 and 1.5 and throughout the book. See, for example, the selection of environmentally friendly refrigerants and solvents in Chapter 3 and the need to avoid producing and storing hazardous intermediates in Section 6.2.

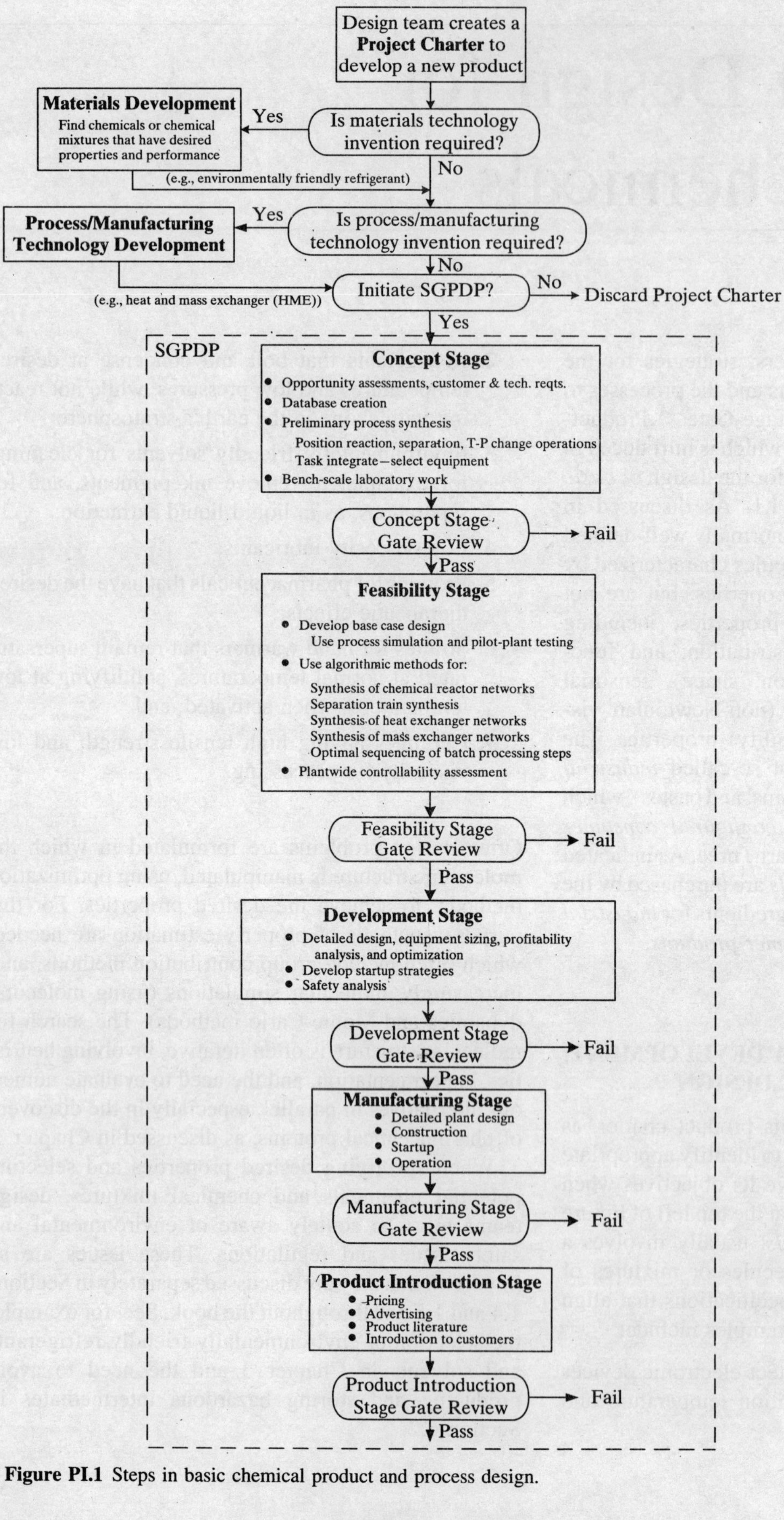

Figure PI.1 Steps in basic chemical product and process design.

PROCESS/MANUFACTURING TECHNOLOGY DEVELOPMENT

In the development of some basic chemical products, new process/manufacturing technologies play an important role. For example, as discussed in Section 13.2, heat and mass exchanger technologies help to improve the profitability in the production of ammonia. Consequently, before beginning the SGPDP, the design team normally checks whether new process/manufacturing technologies can be helpful in producing the product.

After the new technologies are assessed, as shown in Figure PI.1, a decision is made regarding whether to initiate the Stage-Gate™ Product-Development Process (SGPDP), which is shown in the dashed box.

CONCEPT STAGE

The *concept* stage for basic chemical products begins with opportunity assessments and the determination of customer and technical requirements, as discussed in Section 2.4. Then, it normally focuses on the process creation (or synthesis) step, usually beginning with the assembly of a preliminary database that is comprised of thermophysical property data, including vapor-liquid equilibrium data, flammability data, toxicity data, chemical prices, and related information needed for preliminary process synthesis. In some cases, experiments are initiated to obtain important missing data that cannot be accurately estimated, especially when the primitive problem does not originate from a laboratory study. In this regard, experimental reaction data are always required, as is experimental separation data when mixtures to be separated are moderately to highly nonideal. Then, preliminary process synthesis begins with the design team creating flowsheets involving just the reaction, separation, and temperature- and pressure-change operations. Process equipment is selected in a so-called *task-integration* step. This latter step involves the selection of the operating mode; that is, continuous, batch, or semicontinuous. Only those flowsheets, that show a favorable gross profit are explored further; the others are rejected. In this way, detailed work on the process is avoided when the projected cost of the raw materials exceeds that of the products. These steps are described in detail in Chapter 4, in which typical flowsheets of operations are synthesized to address the problem of increasing the production of vinyl chloride, and the problem of producing the pharmaceutical tissue plasminogen activator (tPA). Finally, as promising flowsheets are assembled, bench-scale experiments are often undertaken.

FEASIBILITY STAGE

Development of Base-Case Process

To address the most promising flowsheet alternatives for the manufacture of basic chemicals, the design team is usually expanded or assisted by specialized engineers, to develop *base-case designs*. This usually involves the development of just one flow diagram for each favorable process. As described in Section 4.5, the design team begins by creating a detailed process flow diagram, accompanied by material and energy balances, and a list of the major equipment items. A material balance table shows the state of each stream; that is, the temperature, pressure, phase, flow rate, and composition, plus other properties as appropriate. In many cases, the material and energy balances are performed, at least in part, by a computer-aided process simulator, such as ASPEN PLUS, ASPEN HYSYS, UNISIM, CHEMCAD, and PRO/II for commodity chemicals, and BATCH PLUS and SUPERPRO DESIGNER for specially chemicals, especially pharmaceuticals. Then, the design team seeks opportunities to improve the designs of the process units and to achieve more efficient process integrations for the production of commodity chemicals, applying the methods of *heat and power integration*, for example, by exchanging heat between hot and cold streams, and *mass integration* to minimize raw materials and wastes.

For each base-case design, three additional activities usually take place in parallel. Given the detailed process flow diagram, the design team refines the preliminary database to include additional data such as transport properties and reaction kinetics, feasibilities of the separations, matches to be avoided in heat exchange (i.e., forbidden matches), heuristic parameters, equipment sizes and costs as a function of throughput, and so on. This is usually accompanied by pilot-plant testing to confirm that the various equipment items will operate properly and to refine the database. If unanticipated data are obtained, the design team may need to revise the flow diagram. In some cases, equipment vendors run tests, as well as generate detailed equipment specifications. To complement these activities, a simulation model is prepared for the base-case design. Process simulators are often useful in generating databases because of their extensive data banks of pure-component properties and physical property correlations for ideal and nonideal mixtures. When not available, simulation programs can regress experimental data taken in the laboratory or pilot plant for empirical or theoretical curve fitting.

In developing a base-case design, the design team checks regularly to confirm that the process remains promising. When this is not the case, the team often returns to one of the steps in process creation or redevelops the base-case design.

Finally, before leaving this topic, the reader should note that *process creation* and *development of a base-case process* are the subjects of Chapters 4–6 in Part One of this book.

Detailed Process Synthesis Using Algorithmic Methods

While the design team develops one or more base-case designs, detailed process synthesis may be undertaken using algorithmic methods as described in Chapters 7–11. For continuous processes, these methods: (1) create and evaluate chemical reactor networks for conversion of feed to product chemicals (Chapter 7) and separation trains for recovering species in multicomponent mixtures (Chapter 8), and (2) create and evaluate efficient networks of heat exchangers with turbines for power recovery and high thermodynamic efficiency (Chapter 9), and networks of mass exchangers (Chapter 10) to reduce waste. For batch processes, these methods create and evaluate optimal sequences and schedules for batch operation (Chapter 11). With the results of these methods, the design team compares the base case with other promising alternatives, and in many cases identifies flowsheets that deserve to be developed along with, or in place of, the base-case design. More specifically, the supplement to Chapter 9 discusses second-law analysis, which provides an excellent vehicle for screening the base-case design or alternatives for energy efficiency. In this analysis, the lost work is computed for each process unit in the flowsheet. When large losses are encountered, the design team seeks methods to reduce them. Chapter 9 also presents algorithmic methods to synthesize networks of heat exchangers, turbines, and compressors that satisfy the heating, cooling, and power requirements of the process. These methods, which place emphasis on the minimization of utilities such as steam and cooling water, are used by the design team to provide a high degree of heat and power integration in the most promising processes.

Plantwide Controllability Assessment

An assessment of the controllability of the process is initiated after the detailed process flow diagram has been completed, beginning with the qualitative synthesis of control structures for the entire flow diagram, as discussed in Chapter 12. Then measures are utilized that can be applied before the equipment is sized in the manufacturing stage, where detailed process design is carried out, to assess the ease of controlling the process and the degree to which it is inherently resilient to disturbances. These measures permit alternative processes to be screened for controllability and resiliency with little effort and, for the most promising processes, they identify suitable control structures. Subsequently, control systems are added and rigorous dynamic simulations are carried out to confirm the projections using the approximate measures discussed previously. This is also covered in Chapter 12, which addresses the subject of plantwide controllability assessment.

DEVELOPMENT STAGE

Detailed Design, Equipment Sizing, Profitability Analysis, and Optimization

For a new process to produce commodity or specialty basic chemicals, after completing the base-case design the design team usually receives additional assistance in carrying out the detailed process design, equipment sizing and capital-cost estimation, profitability analysis, and optimization of the process. These topics are covered in separate chapters in Part Four, which begins the *development* stage of the SGPDP. Although these chapters describe several methods, having a range of accuracy, for computing equipment sizes, cost estimates, and profitability analyses, it is important to recognize that the more approximate methods are often sufficient to distinguish between alternatives during product conception and process creation (the *concept* stage of the SGPDP—Chapters 3–6), and detailed process synthesis (the *feasibility* stage of the SGPDP—Chapters 7–11). Throughout these chapters, references are made to the approximate costing methods included in Chapters 22 and 23. Selected optimization techniques are presented in Chapter 24.

When the detailed process design is completed, the economic feasibility of the process is checked to confirm that the company's profitability requirements have been met. If this proves unsatisfactory, the design team determines whether the process is still promising. If so, the team returns to an earlier step to make changes that it hopes will improve the profitability. Otherwise, this process design is rejected.

Develop Startup Strategies

While carrying out these steps, the design team formulates startup strategies to help identify the additional equipment that is usually required. In some cases, using dynamic simulators, the team extends the model for the dynamic simulation of the control system and tests startup strategies, modifying them when they are not implemented easily. In addition, the team often prepares its recommendations for the initial operating strategies after startup has been completed.

Safety Analysis

Another crucial activity involves a *formal* analysis of the reliability and safety of the proposed process. Note that, as discussed in Section 1.5 and throughout the book, these considerations must be foremost throughout the design process. It is common practice to carry out formal safety analysis, that is, systematic analysis of the Piping and Instrumentation Diagram (P&ID), to reduce or eliminate the risks that typical faults (valve and pump failures, leaks, etc.) propagate through plants—creating accidents, such as explosions, toxic vapor clouds, fires, etc. This analysis is generally referred to as a HAZOP (*Haz*ard and *Op*erability) study. Methods for and examples of HAZOP analysis, together with risk assessment, are presented in the supplement to Chapter 1. Also, the reader is referred to the texts by Crowl and Louvar (1990) and Kletz (1992) and the following books developed by the Center for Chemical Process Safety of the American Institute of Chemical Engineers:

1. *Safety, Health, and Loss Prevention in Chemical Processes: Problems for Undergraduate Engineering Curricula—Student Problems* (1990).
2. *Guidelines for Hazard Evaluation Procedures, Second Edition with Worked Examples* (1992).
3. *Self-Study Course: Risk Assessment* (2002).

The latter reference is particularly noteworthy for instructors because it provides a Microsoft PowerPoint file that can be integrated into a safety lecture.

MANUFACTURING STAGE

Plant Design, Construction, Startup, and Operation

Detailed plant design, construction, startup, and operation are carried out in the *manufacturing* stage of the SGPDP, as shown in Figure PI.1. In creating the plant design for a basic chemical process, much detailed work is done, often by contractors, using many mechanical, civil, and electrical engineers. For processes that produce commodity and specialty basic chemicals, they complete equipment drawings, piping diagrams, instrumentation diagrams, the equipment layout, the construction of a scale model, and the preparation of bids. Then the construction phase is entered, in which engineers and project managers play a leading role. The design team often returns to assist in plant startup and operation. Note that chemical engineers do not usually play leading roles in the final design and construction activities.

PRODUCT-INTRODUCTION STAGE

As the plant comes online, product-launch strategies are normally implemented. These include setting the product price, advertising to perspective customers, preparing and distributing product literature, and introducing the product to selected customers. These are normally the responsibilities of sales and marketing personnel, many of whom have been trained as chemical engineers.

SUMMARY

This brief introduction to Figure PI.1 should give the reader a good appreciation of the subjects to be learned in the design of basic chemical products and processes, and how this text is organized to describe the design methodologies.

REFERENCES

1. American Institute of Chemical Engineers, *Safety, Health, and Loss Prevention in Chemical Processes: Problems for Undergraduate Engineering Curricula—Student Problems*, AIChE, New York 1990.

2. American Institute of Chemical Engineers, *Guidelines for Hazard Evaluation Procedures, Second Edition with Worked Examples*, AIChE, New York (1992).

3. American Institute of Chemical Engineers, *Self-Study Course: Risk Assessment*, AIChE, New York (2002).

4. Crowl, D. A., and J. F. Louvar, *Chemical Process Safety: Fundamentals with Applications*, Prentice-Hall, Englewood Cliffs, New Jersey (1990).

5. Kletz, T., *HAZOP and HAZAN*, Third Edition, IChemE (1992).

Chapter 3

Materials Technology: Molecular-Structure Design for Basic Chemicals

3.0 OBJECTIVES

As discussed in Section 1.3, *basic chemical* products are nearly pure molecular species, or mixtures defined by composition only—as compared with *industrial chemical* products, which are more complex materials characterized by additional properties, such as particle-size distribution, emulsion viscosity, micropore structure, etc. When designing new *basic chemical* products, the design team often begins with customer needs that can be translated into desired physical properties, such as liquid density, glass-transition temperature, and solubility of a solute.

To facilitate this, the creation of *innovation maps* for typical *basic chemicals* is discussed first. Then, several methods for *molecular-structure design* are introduced. These methods select from among numerous permutations of atoms and molecular groups to identify molecules, and mixtures of molecules, that satisfy specifications for properties and performance. For this purpose, methods of property estimation, including semi-empirical and group-contribution methods, are introduced. Methods for Monte-Carlo and molecular-dynamics estimation of properties are also introduced, with emphasis on their promising role in product design, but are not used in the examples presented. To satisfy property specifications, where appropriate, optimization algorithms are covered. These adjust the number and position of preselected atoms and molecular groups to minimize the difference between the property estimates and specifications. Where property estimation techniques are not effective, the role of experimental methods, with emphasis on the discovery of pharmaceuticals, is covered. Examples and discussions show how to use the methods of molecular-structure design to locate:

- **a.** polymers that have desired properties, such as density, glass-transition temperature, and water absorption,
- **b.** refrigerants that boil and condense at desired temperatures and low pressures, while not reacting with ozone,
- **c.** solvents that suspend solids, such as ink pigments and paints, drying rapidly while being nontoxic and environmentally friendly,
- **d.** solvents for liquid-liquid extraction, extractive distillation, or azeotropic distillation,
- **e.** macromolecules as pharmaceuticals, such as proteins that function as antibodies, which are Y-shaped multidomain proteins that bind specific antigens or receptors with exquisite selectivity,
- **f.** solutes for hand warmers that remain supersaturated at low temperatures, crystallizing exothermically, only when activated,
- **g.** solvents for the crystallization of organic solids, like ibuprofen, with desired properties and morphology (that is, crystal structure).

The design of other chemical products, not covered in this chapter, includes lubricants having low viscosity that withstand high engine temperatures (Dare-Edwards; 1991), and ceramics having high tensile strength and low density (Giannelis, 1989).

In this chapter, the iterative nature of *molecular-structure design* is emphasized, often involving heuristics, experimentation, and the need to evaluate numerous alternatives in parallel, as in the discovery of pharmaceuticals. Typical work processes in industry are covered.

After studying this chapter, the reader should:

1. Be able to construct an *innovation map* for a basic chemical.
2. Be able to identify critical inventions and innovations involving materials technologies for basic chemical products.
3. Be aware of typical considerations in specifying the physical properties and performance of potential chemical products.

4. Know how to set up a search for chemicals and chemical mixtures that satisfy specifications for physical properties.
5. Understand the role of group-contribution methods, and other molecular modeling techniques, in estimating properties during molecular-structure design.
6. Know how to apply optimization methods to locate molecular structures having the desired properties.
7. Appreciate the role of parallel experimentation in searching for pharmaceuticals.
8. Be aware of the many kinds of chemical products discovered using molecular-structure design.

3.1 INTRODUCTION

Often, chemical engineers in industry are challenged by the need to develop new products that satisfy consumer needs. This chapter focuses on the development of appropriate materials technologies; that is, the search for appropriate molecules or mixtures of molecules to satisfy property specifications that align closely with customer needs—as briefly discussed in the introduction to Part One that describes Figure PI.1. It introduces the discovery of new chemicals, that is, new materials technology, for the design of new products. These new materials are intended to create a product that satisfies customer needs, while offering a competitive advantage. To achieve these objectives, the concept of an *innovation map* is developed showing the connections between the technological components and customer satisfaction, that is, the *customer-value proposition*. The success of a new product often relies on careful attention to this interplay. In Section 3.2, before an *innovation map* of the type introduced in Section 1.3 is created, the history of the development of environmentally friendly refrigerants is covered, which, in hindsight, is the basis for the creation of the *innovation map*. In an actual product design activity, the innovation map is revised multiple times to reflect the progress of the design.

3.2 INNOVATION MAP FOR ENVIRONMENTALLY FRIENDLY REFRIGERANTS

To create an *innovation map*, which guides the development of new technologies, it is important to examine the key technological inventions that have accompanied chemical products that are closely related to the product being designed. These inventions in materials technology for basic chemicals are critical to attracting customers and satisfying their perceived needs. For each technology element (invention), its critical parameters must be understood. In addition, the *innovation map* helps to protect new technologies; that is, to identify the key invention(s) and a strategy for their *intellectual-property* (IP) protection. To illustrate the creation of an *innovation map* for a basic chemical product, the materials technologies related to the design of environmentally friendly refrigerants are reviewed next.

Environmentally Friendly Refrigerant Inventions

To provide cooling for foods, pharmaceuticals, and the like, in refrigerators and freezers, as well as enclosed air spaces, refrigeration cycles are required. As discussed in Section 9S.6, the latter involve evaporators, compressors, condensers, and valves (or turbines). Hence, when searching for a refrigerant (i.e., a working fluid in a refrigeration cycle), it is important to locate stable, volatile compounds that boil at typical refrigeration temperatures.

This was the challenge faced by Thomas Midgely, Jr., who sought to develop a refrigerant product for a broad range of household, automotive, and industrial applications. In 1937, Midgely, working for General Motors, published the first comprehensive study of the design of small molecules for refrigerants. Through examination of the periodic table, he concluded that inert gases are too light (having very low boiling points) for most applications, and that the metals are impractical due to their potential for freezing when the refrigeration system is shut down. Consequently, he concentrated on compounds involving C, N, O, S, and H atoms, and the halogens, F, Cl, Br, and I. Although compounds containing F tend to be more flammable, he considered their relatively low toxicity to be an overriding advantage. While compounds containing Cl are less flammable, they are more toxic, but he considered them not sufficiently toxic to be excluded. Compounds containing Br and I were considered to be far too toxic. Midgely also recognized the desirability of refrigerants having: (1) a large latent heat of vaporization, to reduce their throughput when removing a specified heat duty; (2) a low viscosity, to reduce the recirculation power; and (3) a low freezing point, to reduce the possibility of freezing. His work led to the development of a number of refrigerants containing C, Cl, and F atoms, called Freons®. Further work by DuPont led to additional Freon® refrigerants containing H atoms as well.

In the years that followed Midgely's research, with the increasing usage of refrigerants for home refrigerators and air-conditioning systems, especially automobile air conditioners, the production of Freon 21®, $CHCl_2F$, which has a normal boiling point of 8.9°C, as well as other CHClF forms, grew rapidly. However, because the concentration of these compounds increased to parts-per-billion in the stratosphere, and chlorine atoms were found to react with ozone, decreasing significantly the earth's ozone layer, CFCs (compounds containing Cl) were banned in the Montreal protocol of 1987. This led to a search for new refrigerant products having comparable properties, but excluding chlorine. One possibility, Freon 11®, CF_3H, was rejected because it boils at too low a temperature, −82°F. Another, HFC-134a, CFH_2CF_3, with a normal boiling point of −26.6°C, has become a popular alternative.

Innovation Map and Product Design for Environmentally Friendly Refrigerants

The preceding subsection provides an historical account of the invention of refrigerants for household, automotive, and commercial purposes. Through hindsight, the relationships among the inventions that led to the successful production of environmentally friendly refrigerants have been constructed. Note, however, that during the design of a new product, often the components of the discoveries are incomplete, and consequently, these relationships are not clear, complicating the path to the successful commercialization of the new product. Gaining clarity here, in turning technical advantage into competitive advantage, is the key to commercializing a successful new product.

In the next subsection, an *innovation map* is created to provide this clarity and guide the new product-development efforts. But, first, the reader is referred to Section 1.3, which discusses the relationship of a scientific discovery, that is, an *invention*, and its transformation into an *innovation*, followed by a discussion of the interplay between the technical developments and customer satisfaction, that is, the *customer-value proposition*.

Innovation Map

As introduced in Section 1.3, the *innovation map* relates the technological components of product developments to the technical advantages, that is, the map shows the technical differentiation, ultimately to the satisfaction of the *customer-value proposition*. The *innovation map* evolves during product design, being updated periodically during the new-product-development and commercialization processes.

For the product development of environmentally friendly refrigerants, consider the *innovation map* in Figure 3.1. Note that while this map is readily generated in hindsight, the generation of such a map should be the objective of product design strategies.

To construct the *innovation map* for environmentally friendly refrigerants in hindsight, one must first identify the elements in its four levels, moving from the bottom to the top of the map:

1. *Materials Technology:* compounds involving C, N, O, S, H atoms, and halogens F and Cl; compounds having large ΔT_b^v, low μ, and low T_m; compounds involving C, H, and F; compounds involving C, H, F, O, and S.
2. *Technical Differentiation (Technical-Value Proposition):* intermediate volatility—boils at −40 to 0°C at low pressure; leaks easily detected; doesn't react appreciably with O_3—stable and inert.
3. *Products:* Freons® (C, Cl, F); Freons® (C, Cl, F, H), e.g., R-22—$CHClF_2$; HFC 134a (CFH_2CF_3); CH_3CHF_2.
4. *Customer-Value Proposition:* low-cost refrigeration and air conditioning; nontoxic; safe—nonflammable; no reactions with O_3 in stratosphere (CFCs banned); low smog potential—no trace materials in lower atmosphere.

After identifying the elements at all four levels of the innovation map, their connectivity in the map is added to show the interplay between the technological elements, the *technical-value proposition*, and ultimately the *customer-value proposition*. An unmet customer need, such as lower smog potential, can be an objective for the next generation of products.

The choice of new materials, in this case, small molecules for refrigerants, is the technological challenge in satisfying the customers' perceived needs for low-cost, nontoxic,

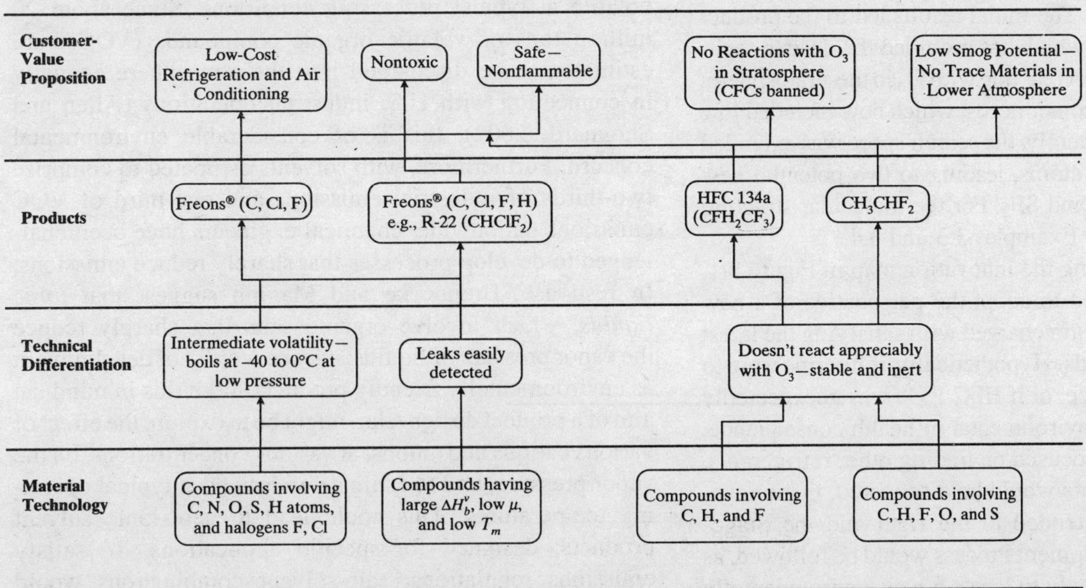

Figure 3.1 Environmentally friendly refrigerant innovation map.

nonflammable refrigerants. Initially, as viewed by Midgely in 1937, thousands of compounds involving all of the elements in the periodic table were potential refrigerants. To make intelligent selections, Midgely reasoned that only compounds involving C, N, O, S, and H atoms could be sufficiently volatile, but not too volatile (like inert gases), to serve as working fluids (i.e., ones that vaporize and condense at boiling points in the range of −40 to 0°C at low pressure). The addition of the halogen atoms, F and Cl, made possible a number of compounds having the proper volatility while increasing their latent heat of vaporization, and providing sufficiently low viscosities and melting-point temperatures. The selection of these seven atoms sharply reduced the search space of the *molecular-structure design* problem. Still, thousands of potential compounds remained to be considered. In the end, fewer than 30 Freons®, involving C, Cl, and F, evolved as successful products—and eventually, when the H atom was added, products like Freon 21®, $CHCl_2F$, became very popular. In summary, in the 1940s, the innovation map in Figure 3.1 shows that these new materials, involving just seven atoms, provided the technical differentiations, that is, intermediate volatilities and easy detection of leaks. In turn, these technical differentiations led to the Freon® products that satisfied the customer-value propositions; that is, providing low-cost refrigeration and air conditioning with nontoxic and nonflammable refrigerants.

These products, while successful in satisfying these original customer needs, unfortunately created a significant environmental problem. By the mid-1980s, measurements confirmed that parts-per-billion concentrations of these compounds had accumulated in the stratosphere and had reacted with O_3 to create ozone holes at the South and North Poles. Consequently, protocols were adopted in Montreal that banned the use of chlorine-containing Freon® compounds.

This re-opened the refrigerant design problem, with the search space for new materials further restricted; that is, with the Cl atom eliminated. The initial results led to the product HFC 134a (CFH_2CF_3), which has provided the desired technical differentiation of not reacting with O_3 in the stratosphere, while satisfying the customer needs, which now included low smog potential. Subsequently, the search space was extended to include the O and S atoms, leading to two potential new refrigerants, CH_3CHF_2 and SF_2. For the molecular-structure design calculations, see Examples 3.3 and 3.4.

Finally, when studying the innovation map in Figure 3.1, the reader is encouraged to adopt the perspective of a new product-development team charged with satisfying the latest perceived customer needs. Hypothetically, if fluorine were to become a scarce resource, or if HFC 134a was unexpectedly tied to some negative environmental or health consequence, R&D efforts might be focused on finding other refrigerants. As these new refrigerants would be discovered, the innovation map would be extended to the right and the Stage-Gate™ Product-Development Process would be followed, as illustrated in the case study to locate a new environmentally friendly refrigerant in Section 13.3.

3.3 SEARCHING FOR NEW MATERIALS—BASIC CHEMICAL PRODUCTS

Often, the search for new chemical products is motivated by a desire to improve the capabilities and performance of existing products. Increasingly, new products are sought that are lighter, stronger, biodegradable, safer to manufacture, less toxic, and more environmentally friendly. Just prior to the *concept* stage of product design, when creating a project charter (see Section 2.2), these kinds of objectives are identified, and as the designer(s) focuses on the most promising ideas, more quantitative specifications are established for the properties of a chemical or a chemical mixture.

From one perspective, the search for new basic chemicals to achieve desired properties is referred to as "reverse property prediction." When molecules are known, their properties are estimated using many property estimation methods (see Section 3.4) or, when estimation methods are not available or sufficiently accurate, experimental measurements are obtained. Clearly, to obtain desired properties of unknown chemicals or chemical mixtures, the latter can be selected iteratively until the desired properties are achieved. Alternatively, an optimization strategy can be implemented to adjust the molecular structure automatically, as discussed in Section 3.5. Stated differently, instead of determining the properties of specific chemicals or chemical mixtures, the chemical structure is determined, given the specified properties; that is, the "reverse property prediction" problem is solved, as discussed by Gani (2004).

Often, product development is closely related to the discoveries of a research and development group. As an example, consider the search for improved liquid solvents, which are ubiquitous in the processing and conveyance of chemicals. A key concern, noted by Brennecke and Maginn (2001), is that, due to a narrow liquidus range (difference between boiling and freezing points), most solvents are quite volatile at typical processing conditions. Since about 20 million tons of volatile organic compounds (VOCs) are estimated to be discharged into the atmosphere annually in connection with U.S. industrial operations (Allen and Shonnard, 2002), this is of considerable environmental concern. Furthermore, with solvents estimated to comprise two-thirds of industrial emissions and one-third of VOC emissions nationwide, chemical engineers have been challenged to develop processes that sharply reduce emissions. In response, Brennecke and Maginn suggest that *ionic liquids*, which involve organic salts that sharply reduce the vapor pressure of liquids, may be worthy of development as environmentally friendly products. With this in mind, an aim of a product design team might be to explore the effect of various cations and anions, at various concentrations, on the vapor pressure, and to estimate emissions at typical operating temperatures. This could lead to new ionic solvent products, designed for specific applications. To satisfy emissions regulations, salt–solvent combinations would be sought that reduce emissions at low cost, involving small salt concentrations. Promising possibilities include quaternary ammonium salts, commonly used for phase-transfer catalysis and for gas separations, for example, to recover water vapor or carbon dioxide. In these salts, the organic "R" group can be adjusted to affect such properties as the water solubility of the ionic solvent.

When designing these products, either measurements of the vapor pressure as a function of salt concentration, or property estimation methods, are needed. As discussed in the next section, the latter are often available for the mixtures involved. For example, the calculation of vapor–liquid equilibria for aqueous electrolytes is carried out commonly using property estimation systems, such as those provided by the OLI electrolyte engine (Aspen, 1999; Zemaitis et al., 1986) and the ASPEN PLUS simulator (Chen et al., 2001). Recently, methods have been developed to calculate equilibria for organic electrolyte solutions, which should be applicable for many ionic solvents, and are implemented in ASPEN PLUS (*Getting Started*, 1999).

Like ionic liquids, many chemical products are designed using property estimation methods, as discussed in the next section. These include polymer membranes and refrigerants. However, for pharmaceuticals, the properties of proteins are normally not estimated. Rather, they are determined experimentally in the laboratory, as discussed next.

Pharmaceuticals Product Design

For the discovery of new drugs having the proper therapeutic properties, as discussed by Pisano (1997), two distinct strategies have evolved: either chemical synthesis or synthesis through recombinant biotechnology methods.

Synthetic Chemical Drugs

These drugs are synthesized through a sequence of chemical reactions that either add or subtract atoms. Consequently, synthesis of potential molecules begins with identification of the starting materials and selection of the reactions, which comprise the *synthetic route*. Initially, much work is done conceptually, outside of the laboratory, where chemists explore routes using journal articles and computer simulations, to the extent possible. Gradually, a tree of alternative routes leading to potentially attractive drugs evolves. As the tree is pruned, several alternatives are modeled more carefully to locate the most promising routes and reject the least attractive. Then, heuristics are applied to select those routes that have desirable characteristics; for example, the fewest reaction steps, with high selectivity and yields, involving nonhazardous byproducts, easy separations to recover the desired products, and safe implementation.

Normally, the limitations of theory, such as the inability to predict kinetic rates and conversions, require the chemist to carry out small-scale experiments for the most promising routes. Mostly, experiments are performed one reaction at a time, rather than several in parallel, to better understand the kinetics, energy requirements, solvents needed, etc.

Gradually, an attractive route is developed, with a pilot plant designed to manufacture sufficient quantities of product for preclinical trials, while the search continues for more attractive routes. The scale-up from laboratory to pilot-plant quantities involves many of the considerations introduced in the next chapter on process creation; that is, *process synthesis*. As illustrated in Section 4.4 for the manufacture of tissue plasminogen activator (tPA), the flowsheet of process operations is synthesized and the equipment is selected, together with operating strategies that involve batch processing almost entirely. Then, as described in Section 5.5 and subsequent chapters, chemical engineers on the design team implement the methods of process simulation, equipment sizing, and cost estimation.

The search for attractive routes is best characterized as iterative, with new leads continually appearing due to successes and failures in the laboratory and pilot plant. While some theory leads to the initial chemical routes, automated methods of route synthesis are not yet practical, especially for the synthesis of proteins having on the order of 500–600 amino acids. In fact, chemists have been severely limited in the choice of smaller target molecules, which can be synthesized with far fewer chemical reaction steps. To generate proteins, which have become key therapeutic drugs, chemists and biochemists have turned to cell cultures.

Genetically Engineered Drugs

As discussed by Pisano (1997), the key breakthrough came when Herbert Cohen and Stanley Boyer, of the University of California at San Francisco, invented a means of inserting genes into bacterial cells. By *expressing* the gene for a protein into the DNA of a bacterial or mammalian cell, the latter becomes capable of producing that protein. For example, as discussed in Section 4.4, the tPA gene can be isolated from human melanoma cells inserted into Chinese hamster ovary (CHO) cells, which then generate the tPA protein.

To genetically engineer a drug, chemists and biochemists begin by identifying target proteins; that is, proteins that have the desired therapeutic properties such as a monoclonal antibody or insulin. Then, the gene sequence that *codes* for the protein must be identified, together with a host cell to be used for growing the protein. Identification of the desired therapeutic properties usually begins with knowledge of a disease, leading the chemist or biochemist to work backward to find a protein that inhibits a chemical reaction involved in that disease. This was the approach used at Eli Lilly to find Prozac®, a serotonin inhibitor, for treating depression.

At the heart of process development is the need to precisely measure the quantity and purity of protein expression. These measurements are crucial for determining the rate and conversion of protein growth and the purification yield, which are the basis for the design of cultivation, fermentation, and separation equipment. Initially, process researchers focus on identifying the most promising cell lines—cells that can produce the desired protein—that have a high rate of production and require the least expensive nutrients for growth. Many iterations are usually required in this search, involving many prospective cell lines and operating conditions. For this purpose, automated labs-on-a-chip are often employed, permitting hundreds and thousands of cell clones to be evaluated experimentally in parallel. As an example, see the discussions of product designs for labs-on-a-chip to high-throughput screen potential kinase inhibitors in Sections 16.4 and 17.4.

3.4 PROPERTY ESTIMATION METHODS

Theoretical approaches to molecular-structure design require accurate estimates of physical and transport properties. These are derived commonly from the principles of thermodynamics and transport phenomena, often using molecular simulations. Since the literature abounds with estimation methods, reference books and handbooks are particularly useful sources. One of the most widely used, *Properties of Gases and Liquids* (Poling et al., 2001), provides an excellent collection of estimation methods and data for chemical mixtures in the vapor and liquid phases. For polymers, *Properties of Polymers* (van Krevelen, 1990) provides a collection of group-contribution methods and data for a host of polymer properties.

In recent years, property information systems have become widely available in computer packages. Some are available on a stand-alone basis, such as PPDS Version 4.1 (1997; www.ppds.co.uk) and ASPEN PROPERTIES (2006), to be used in custom stand-alone programs and specific process simulators, such as ASPEN PLUS, ASPEN HYSYS (formerly HYSYS.Plant), and BATCH PLUS. Others are available for use within specific process simulators such as UNISIM, PRO/II, CHEMCAD, and SUPERPRO DESIGNER. Note that the property information systems often can be accessed through the CAPE-OPEN standard interface (McGough and Halloran, 2005; Pons, 2005). Commonly, constants and parameters are stored for a few thousand chemical species, with programs provided to estimate the property values of mixtures and determine the constants and parameters for species that are not in the data bank, using estimation methods or the regression of experimental data. Virtually all of the property systems estimate the properties of mixtures of organic chemicals in the vapor and liquid phases. Methods are also provided for electrolytes and some solids, but these are less predictive and less accurate.

Computer Data Banks

Data banks for the pure species may be viewed as a collection of *data records*, each containing the constants and parameters for a single chemical [e.g., the critical properties (T_c, P_c, v_c), the normal boiling point (T_{nbp}), vapor pressure coefficients, heat capacity coefficients, acentric factor, etc.]. One such data bank, which is utilized in ASPEN PROPERTIES, is that compiled by Poling et al. (2001) (also Reid et al., 1977, 1987) in Appendix A of the *Properties of Gases and Liquids*. This data bank, which is referred to as the ASPENPCD (ASPEN PLUS Pure Component Data Bank), contains data for 472 chemicals, using data solely from Reid et al. (1977). Another data bank, known as PURECOMP, originates from the DIPPR® (Design Institute for Physical Property Data, sponsored by the AIChE) data bank, with information supplemented by Aspen Technology, Inc. (e.g., the UNIFAC group contributions) and parameters from the ASPENPCD data bank. It is an updated DIPPRPCD data bank and has superseded the ASPENPCD data base as the main source of parameters for pure components. For Version 2006 of ASPEN PLUS, the PURECOMP data base was renamed PURE20. The PURECOMP data bank contains data for over 1,973 chemicals (mostly organic), not including the ionic species in electrolytes. In addition, ASPEN PLUS provides access to the AQUEOUS data bank for over 1,676 ionic species, to be used for electrolytes. Recently, the NIST-TRC database was included in ASPEN PROPERTIES 2006. This includes property parameters and experimental data for approximately 13,000 chemicals (mostly organic), increasing the total number of chemicals to over 15,000. Note that these entries were collected and evaluated by the Thermodynamics Research Center (TRC) using the NIST ThermoData Engine (TDE) and the NIST-TRC Source Data Archival System for experimental thermophysical and thermochemical property data.

When the constants and parameters are not stored for a chemical species, most of the property information systems have programs for the regression of experimental data (e.g., tables of vapor pressures, liquid densities, and heat capacities as a function of temperature). Finally, when insufficient experimental data are available, programs are often provided to estimate the properties based upon the molecular structure, using group- and bond-contribution methods, often utilizing limited data (e.g., the normal boiling point). These are particularly useful in the early stages of product and process design before a laboratory or pilot-plant study is initiated.

Property Estimation

Each of the property information systems has an extensive set of subroutines to determine the parameters for vapor pressure equations (e.g., the extended Antoine equation), heat capacity equations, etc., by regression and to estimate the thermophysical and transport properties. The latter subroutines are called to determine the state of a chemical mixture (phases at equilibrium) and its properties (density, enthalpy, entropy, etc.). When calculating phase equilibria, the fugacities of the species are needed for each of the phases. A review of the phase equilibrium equations, as well as the facilities provided by the process simulators for the calculation of phase eqilibria, is provided on the multimedia models, which can be downloaded from the Wiley Web site associated with this book—follow the path *ASPEN → Physical Property Estimation* and *HYSYS → Physical Property Estimation*.

www.wiley.com/go/global/seider

As mentioned above, when a data record for a pure species cannot be located in one of its data banks, each of the property information systems permits the designer to enter the missing constants and parameters. Furthermore, methods are provided to estimate the constants and parameters when the designer cannot provide these. This is especially important when laboratory and pilot-plant data are not available.

Usually, bond- or group-contribution methods are used to estimate the constants and parameters for pure species, with the designer providing the molecular structure of the chemical species, as shown, for example, for trifluoropropylene:

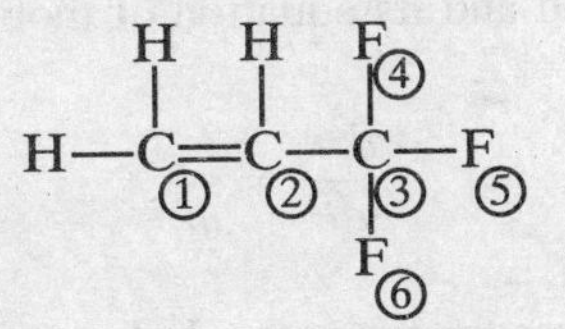

Here, all atoms, with the exception of hydrogen, are numbered, and the bonds associated with each carbon atom and its adjacent numbered atoms can be specified as follows:

Atom 1		Atom 2		
Number	Type	Number	Type	Bond Type
1	C	2	C	Double Bond
2	C	3	C	Single Bond
3	C	4	F	Single Bond
3	C	5	F	Single Bond
3	C	6	F	Single Bond

Using bond- and group-contribution techniques, as discussed by Poling et al. (2001) and Joback and Reid (1987), many properties can be estimated, including the critical volume, normal boiling point, liquid density and heat of vaporization at the normal boiling point, and ideal-gas heat capacity coefficients. Similarly, using the UNIFAC group-contribution method, the activity coefficients of trifluoropropylene in solution with other chemical species can be estimated for use in computing phase equilibria.

Polymer Property Estimation

As mentioned above, van Krevelen (1990) presents semi-empirical group-contribution methods and data for each group in a polymer "repeating unit." Data are provided to estimate a host of polymer properties, including the density, specific heat, glass-transition temperature, water absorption, and refractive index. For a specific property, these are in one of two forms:

$$p[n] = \frac{\sum_{i=1}^{N} A_i n_i}{\sum_{i=1}^{N} B_i n_i} \tag{3.1}$$

and

$$p[n] = \left(\frac{\sum_{i=1}^{N} A_i n_i}{\sum_{i=1}^{N} B_i n_i} \right)^d \tag{3.2}$$

where n_i is the number of groups of type i in the polymer "repeating unit," N is the number of types of groups in the repeating unit, A_i is the contribution associated with group i, B_i is the molecular weight of group i, and d is an exponent for each property to be estimated. Note that the denominator summation is either the molecular weight or the specific volume of the repeat unit. In most cases, these property estimates lie within 5 to 10 percent of experimental values, which are often sufficiently close to permit the selection of repeating units to meet property specifications.

EXAMPLE 3.1

Estimate the glass-transition temperature of polyvinyl chloride, T_g, with repeating unit—(CH_2CHCl)—, using the following group contributions (van Krevelen, 1990):

Group	A_i	B_i
$—CH_2—$	2,700	14
—CHCl—	20,000	48.5

SOLUTION

$$T_g = \frac{2,700 \times 1 + 20,000 \times 1}{14 \times 1 + 48.5 \times 1} = 363\text{ K}$$

This compares fairly well with the experimental value of 356 K (van Krevelen, 1972, page 114).

Caution. When using property estimation methods, especially group- and bond-contribution methods, care must be taken to avoid large differences from experimental values, especially when the molecules, temperatures, and pressures are substantially different from those used to estimate the parameters of the methods.

Microsimulation

Two methods of a more fundamental nature than group- and bond-contribution methods are being used increasingly to improve estimates of thermophysical and transport properties. These involve molecular dynamics (MD) and Monte-Carlo simulations, with small collections (typically 100–10,000) of interacting molecules, and are commonly referred to as *microsimulations*. In general, Monte-Carlo simulations are numerical statistical methods that utilize sequences of random numbers. The name Monte Carlo was coined during the Manhattan Project of World War II, which resulted in the development of the atomic bomb. Monte Carlo is the capital of Monaco, a world center for games of chance. A simple example of Monte-Carlo simulation is the evaluation of the integral of a complex function, $y = f\{x\}$, over the interval $\{x_1, x_2\}$. A plot of $f\{x\}$ against x over the specified interval is prepared. A range of y is selected from 0 to y_1 where y_1 is greater than the largest value of y in the interval. A random-number generator is then used to select pairs of y–x values within the range of the rectangle bounding the x-interval and the y-range. Suppose that 900 pairs are selected. If 387 pairs fall under the $f\{x\}$ curve of the plot, then the value of the integral is:

$$\frac{387}{900}(y_1 - 0)(x_2 - x_1)$$

Molecular Dynamics

This method involves the numerical integration of the equations of motion ($F = ma$) for each of the molecules, subject to intermolecular forces, in time. The molecules are positioned arbitrarily in a simulation cell, that is, a three-dimensional cube, with initial velocities also specified arbitrarily. Subsequently, the velocities are scaled so that the summation of the kinetic energies of the molecules, $3NkT/2$, gives the specified temperature, T, where N is the number of molecules and k is the Boltzmann constant. Note that after many collisions with the walls and the other molecules, the relative positions and velocities of the molecules are independent of the initial conditions.

During the simulation, the force on each molecule is calculated as the sum of the forces of interaction with all of the surrounding molecules. These are the dispersion forces, also referred to as the London and van der Waals forces, which depend on the separation distance, r, between two molecules, as shown in Figure 3.2. These are represented by the dimensionless form of the commonly used Lennard-Jones pair potential, $U\{r\}/\epsilon$:

$$\frac{U[r]}{\epsilon} = 4\left[\left(\frac{\sigma}{r}\right)^{12} - \left(\frac{\sigma}{r}\right)^{6}\right] \tag{3.3}$$

which expresses the intermolecular potential between two molecules as a function of the distance, r, between them. In

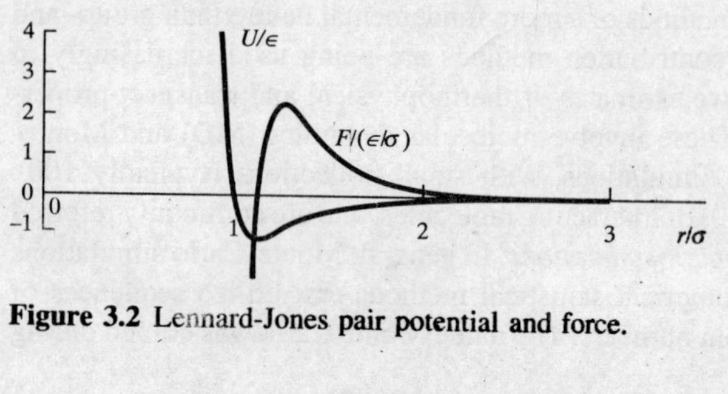

Figure 3.2 Lennard-Jones pair potential and force.

this equation, $U\{r\}$ is the intermolecular potential energy, ϵ is the maximum energy of attraction between a pair of molecules, and σ is the collision diameter of the molecules. Differentiating Eq. (3.3) with respect to r, the negative gradient of the dimensionless intermolecular potential is the dimensionless force, $F\{r\}/(\epsilon/\sigma)$, between them:

$$\frac{F\{r\}}{\epsilon/\sigma} = -24\left[2\left(\frac{\sigma}{r}\right)^{13} - \left(\frac{\sigma}{r}\right)^{7}\right] \tag{3.4}$$

Then, for each molecule i, the summation of all forces acting on it is computed and its equation of motion:

$$\frac{d^2 r_i}{dt^2} = \frac{F_i}{m_i} \tag{3.5}$$

is integrated across the time step. Values of ϵ and σ for many molecules and methods of estimating them for other molecules are given by Bird et al. (2002). To obtain property estimates, the set of differential equations, one for each molecule, is integrated using picosecond time steps over several hundred thousand time steps. Then, time averages are computed to give properties such as the configurational energy, the pressure, and the self-diffusion coefficient.

With the increased availability of software (e.g., DISCOVER in CERIUS^2 by Materials Studio, ETOMICA: Kofke and Mihalick, 2002) and faster computer clusters, these simulations are being carried out more routinely for the estimation of thermophysical and transport properties, as well as for the calculation of phase equilibria. The estimates are tuned to match experimental data by adjusting the energy and size parameters.

Monte-Carlo Methods

In Monte-Carlo simulations, the energy of the molecular system is minimized by randomly moving the molecules in accordance with a desired probability distribution. After each move, the energy of each molecule is computed. When the total energy is reduced, the move is accepted and the molecules are redistributed. Moves are continued until equilibrium is achieved. As for molecular dynamics simulations, potential functions are provided. After convergence, the thermophysical properties, at equilibrium, are computed by averaging. Monte-Carlo methods, which are particularly effective for the calculation of thermophysical properties, including phase equilibria, are considered in detail by Rowley (1994).

3.5 OPTIMIZATION TO LOCATE MOLECULAR STRUCTURE

Molecular-structure design relies on accurate property estimation methods. When sufficiently accurate, the atoms and groups in the molecular structure are adjusted to minimize the sum of the squares of the differences between the property estimates and the specified values:

$$\min_{n} \sum_{j=1}^{P} \left(p_j\{n\} - p_j^{spec}\right)^2 \qquad (3.6)$$

where p_j^{spec} is specified by the designer for property j in an array of P target properties. Often this minimization is carried out subject to specified bounds; for example:

$$p_j^L \leq p_j\{n\} \leq p_j^U \qquad (3.7)$$

and

$$n_i \epsilon \{n_i^L, n_i^U\}, \quad i = 1, \ldots, N \qquad (3.8)$$

where n_i is the number of groups of type i in molecule j, and N is the number of types of molecular groups in molecule j. It is also necessary to ensure that when a new molecular group is added to a molecule that the number of free attachments available for bonding is zero, and when added to the repeating unit of a polymer, the number of free attachments is two. This can be checked by computing the number of free attachments:

$$f = \sum_{i=1}^{N} (\nu_i - 2) n_i + 2 \qquad (3.9)$$

where ν_i is the valence, or number of free bonds, associated with molecular group i. The molecular group can be added to a molecule when $f = 0$ and to a repeating unit when $f = 2$. To locate the molecular structure at the optimum, the objective function (3.6) is combined with the constraints (3.7)–(3.9) into a mixed-integer nonlinear program (MINLP),* which is solved using a mathematical programming solver, such as GAMS. In the next three subsections, these steps are illustrated for polymer, refrigerant, and solvent designs.

Polymer Design

Having discussed the estimation of polymer properties in the previous section, the methods of polymer design are described in connection with the design of a polymer film in Example 3.2.

EXAMPLE 3.2

A polymer film is needed to protect an electronic device. Since the device will operate at temperatures below 60°C and must be protected by a fairly dense layer, which absorbs small concentrations of water, a design team has prepared the following product-quality specifications: (1) density = 1.5 g/cm^3, (2) glass-transition temperature = 383 K (50 degrees above the operating temperature), and (3) water absorption = 0.005 g/g polymer. As stated initially by Derringer and Markham (1985), candidate molecular groups, together with their group contributions, are:

i	Group	Y_i	V_i	H_i	M_i
1	—CH_2—	2,700	15.85	0.000033	14
2	—CO—	27,000	13.40	0.11	28
3	—COO—	8,000	23.00	0.075	44
4	—O—	4,000	10.00	0.02	16
5	—CONH—	12,000	24.90	0.75	43
6	—CHOH—	13,000	19.15	0.75	30
7	—CHCl—	20,000	29.35	0.015	48.5

where M_i, V_i, Y_i, and H_i are the contributions for group i in estimating the molecular weight, M, the molar volume, V, the glass-transition temperature, T_g, and the water absorption, W, according to:

$$M = \sum_{i=1}^{7} M_i n_i \quad \text{g/mol}$$

$$V = \sum_{i=1}^{7} V_i n_i \quad \text{cm}^3/\text{mol}$$

$$Y = \sum_{i=1}^{7} Y_i n_i \quad \text{K(g/mol)}$$

$$H = \sum_{i=1}^{7} H_i n_i \quad \text{mol H}_2\text{O/mol polymer}$$

$$\rho = M/V \quad \text{g/cm}^3$$

$$T_g = Y/M \quad \text{K}$$

$$W = 18H/M \quad \text{g H}_2\text{O/g polymer}$$

Formulate the mixed-integer nonlinear program and use GAMS to obtain the optimal solution. For an introduction to GAMS, see the file GAMS.pdf in the PDF Files folder, which can be downloaded from the Wiley Web site associated with this book.

SOLUTION

Using the objective function in Eq. (3.6), with relative differences, the nonlinear program is:

$$\min_{\substack{\text{w.r.t} \\ n}} \left(\frac{\rho - \rho^{spec}}{\rho^{spec}}\right)^2 + \left(\frac{T_g - T_g^{spec}}{T_g^{spec}}\right)^2 + \left(\frac{W - W^{spec}}{W^{spec}}\right)^2$$

$$\text{s.t.}: \quad M = \sum_{i=1}^{7} M_i n_i$$

$$V = \sum_{i=1}^{7} V_i n_i$$

$$Y = \sum_{i=1}^{7} Y_i n_i$$

$$H = \sum_{i=1}^{7} H_i n_i$$

$$\rho = M/V$$

$$T_g = Y/M$$

$$W = 18H/M$$

$$0 \leq n_i \leq 7 \quad i = 1, \ldots, 7$$

$$1 \leq \rho \leq 1.5$$

$$298 \leq T_g \leq 673$$

$$0 \leq W \leq 0.18$$

Note that Eq. (3.9) is not included because each group has just two attachments.

*For an introduction to MINLPs, see Section 24.3 and Example 9.16

GAMS Program

This mixed-integer nonlinear program is coded in GAMS as follows:

```
VARIABLES
   n1, n2, n3, n4, n5, n6, n7, TG, RHO, W, H, V, Y, M,
   STG, SRHO, SW, Z;
POSITIVE VARIABLES
   STG, SRHO, SW, W, H, V, Y, TG, M, RHO;
INTEGER VARIABLES
   n1, n2, n3, n4, n5, n6, n7;
   n1.LO = 0;     n1.UP = 7;
   n2.LO = 0;     n2.UP = 7;
   n3.LO = 0;     n3.UP = 7;
   n4.LO = 0;     n4.UP = 7;
   n5.LO = 0;     n5.UP = 7;
   n6.LO = 0;     n6.UP = 7;
   n7.LO = 0;     n7.UP = 7;
   M.LO = 14;
   V.LO = 10;
   H.LO = 3.3E-5;
   Y.LO = 2700;
   W.LO = 0;          W.UP = 0.18;
   TG.LO = 298;       TG.UP = 673;
   RHO.LO = 1;        RHO.UP = 1.5;
   STG.L = 383; SRHO.L = 1.50; SW.L = .005;
EQUATIONS
   SPEC1, SPEC2, SPEC3, MOLWEIGHT, GLASSTMP,
   YTOT, HTOT, VTOT, DENSITY, ABSORBANCE, OBJ;
OBJ.. Z=E=((SQR((STG - TG)/STG))
          + (SQR ((SRHO - RHO)/SRHO))
          + (SQR((SW - W)/SW)));
SPEC1.. SRHO =E= 1.5;
SPEC2.. STG  =E= 383;
SPEC3.. SW   =E= .005;
MOLWEIGHT.. M =E= n1*(14) + n2*(28) + n3*(44)
                  +n4*(16) + n5*(43) +n6*(30)
                  +n7*(48.5);
YTOT.. Y =E=n1*(2700) + n2*(27000) + n3*(8000)
            +n4*(4000) + n5*(12000) +n6*(13000)
            +n7*(20000);
HTOT.. H =E= n1*(3.3E-5) + n2*(0.11) + n3*(0.075)
            +n4*(0.02) + n5*(0.75) +n6*(0.75)
            +n7*(0.015);
VTOT.. V =E= n1*(15.85) + n2*(13.40) + n3*(23)
            +n4*(10) + n5*(24.9) +n6*(19.15)
            +n7*(29.35);
GLASSTMP..   TG  =E= (Y/M);
DENSITY..    RHO =E= (M/V);
ABSORBANCE.. W   =E= ((18*H)/M)
MODEL GROUPS /ALL/;
SOLVE GROUPS USING MINLP MINIMIZING Z;
OPTION DECIMALS = 4;
DISPLAY TG.L, RHO.L, W.L, n1.L, n2.L, n3.L, n4.L,
n5.L, n6.L, n7.L, Z.L;
```

All variables in GAMS must be declared. Then, n_i, $i = 1 \cdots \cdot 7$, are declared as integer variables, with lower and upper bounds specified. The remaining variables are real and are declared positive, with bounds specified for ρ, T_g, and W. Note that each equation is assigned a name, including the objective function, which is named Z. The SOLVE statement indicates that the MINLP is to minimize Z using the MINLP solver. Finally, the variables to be displayed in the solution are identified, with the .L suffix indicating the level (final) value computed.

GAMS Solution

```
*    VARIABLE TG.L =     384.6847
*    VARIABLE RHO.L      = 1.4889
*    VARIABLE W.L        = 0.0049
*    VARIABLE n1.L       = 3.0000
*    VARIABLE n2.L       = 0.0000
*    VARIABLE n3.L       = 0.0000
*    VARIABLE n4.L       = 0.0000
*    VARIABLE n5.L       = 0.0000
*    VARIABLE n6.L       = 0.0000
*    VARIABLE n7.L       = 6.0000
*    VARIABLE Z.L        = 0.0007
```

At the minimum, the repeat unit has three —CH_2— groups and six —CHCl— groups; that is, —$[(CH_2)_3, (CHCl)_6]$—. The objective function, Z = 0.0007, and the three properties lie within 2% of the specifications.

For further discussion of the optimal design of polymer repeat units, the reader is referred to Maranas (1996).

Refrigerant Design

In Section 3.2, the history of refrigerant design, beginning with the work of Thomas Midgely, Jr. (1937), is traced to the 1980s where Freon® refrigerants, such as R-21, were replaced by new Freons®, such as R-134a, that do not react with ozone in the stratosphere.

In this section, the problem of designing a new refrigerant product is considered, given the temperatures at which heat is to be absorbed by the evaporator and rejected from the condenser of a refrigerator. Note that the design of a conventional refrigerator is discussed in Sections 9S.6 and 9S.8 (in the file, Supplement_to_Chapter_9.pdf, in the PDF Files folder, which can be downloaded from the Wiley web site associated with this book) and in most books on engineering thermodynamics. Beginning with the selection of k molecular groups, each of which can appear in a candidate refrigerant n times, up to n_{max} times, the number of distinct molecular designs is:

$$\sum_{n=2}^{n_{max}} C\{k, n\} = \sum_{n=2}^{n_{max}} \frac{(k+n-1)!}{n!(k-1)!} \tag{3.10}$$

where $C\{k, n\}$ is the number of combinations of k groups taken n at a time (Joback and Stephanopoulos, 1989). Clearly, this number can become very large, on the order of millions, for as few as 10 molecular groups. To illustrate the problem of selecting from among such a large number of combinations, consider the next example. Note that a less restrictive formulation was solved initially by Joback and Stephanopoulos (1989). Subsequently, Gani and co-workers (1991) excluded oxygen atoms and added restrictions that limited the scope of the search for new molecules. Yet another formulation, which includes oxygen atoms, was provided by Duvedi and Achenie (1996). Note also that chlorine is included in the search to show that molecules containing chlorine are the most desirable when the ozone layer is disregarded. In the example that follows, a measure used by Duvedi and Achenie (1996) to estimate the ozone depletion potential (ODP) is shown to lead to different refrigerants.

EXAMPLE 3.3

It is desired to design a refrigerant that can absorb heat at temperatures as low as 30°F(−1.1°C) and reject heat at temperatures as high as 110°F(43.3°C). For the design, consider 13 molecular groups: CH_3, CH_2, CH, C, OH, O, NH_2, NH, N, SH, S, F, and Cl. When combining these groups, compounds with double or triple bonds, which tend to polymerize, should be avoided. Also, compounds involving both nitrogen and halides should be avoided, as these tend to be explosive. Desirable refrigerants should have: (1) a vapor pressure, $P^s\{-1.1°C\} > 1.4$ bar, to ensure that leaks are from the refrigeration system (rather than from vacuum operation, into which air and water vapor can leak), (2) $P^s\{43.3°C\} < 14$ bar, to keep the compression ratio from exceeding 10, (3) an enthalpy of vaporization, $\Delta H^v\{-1.1°C\} > 18.4$ kJ/mol, to reduce the amount of refrigerant needed (where 18.4 kJ/mol is the latent heat of vaporization of Freon 12®, the refrigerant banned in 1987), and (4) a liquid heat capacity, $C_{pl}\{21.1°C\} < 32.2$ cal/(mol K), to reduce the amount of refrigerant that flashes across the valve (where 32.2 cal/(mol K) is the heat capacity of liquid Freon 12®). Note that 21.1°C is the average of the extreme temperatures.

SOLUTION

The estimation methods by Duvedi and Achenie (1996) are used:

1. Normal boiling point and critical properties (Joback and Reid, 1987).

$$T_b = 198.2 + \sum_{i=1}^{N} T_{b_i} n_i$$

$$T_c = T_b\left[0.584 + 0.965\sum_{i=1}^{N} T_{c_i} n_i - \left(\sum_{i=1}^{N} T_{c_i} n_i\right)^2\right]^{-1}$$

$$P_c = \left(0.113 + 0.0032 n_A - \sum_{i=1}^{N} P_{c_i} n_i\right)^{-2}$$

where the temperatures and pressures are in K and bar, and n_A is the total number of atoms in the molecule.

2. Vapor pressure—Riedel–Plank–Miller method (Reid et al., 1977).

$$\ln P_r^s = \frac{-G[1 - T_r^2 + k(3 + T_r)(1 - t_r)^3]}{T_r}$$

$$G = 0.4835 + 0.4605h$$

$$h = T_{b_r}\frac{\ln P_c}{1 - T_{b_r}}$$

$$k = \frac{\left[\frac{h}{G} - (1 + T_{b_r})\right]}{(3 + T_{b_r})(1 - T_{c_r})^2}$$

where T_r and P_r are the reduced temperature and pressure.

3. Liquid heat capacity—Chueh and Swanson method (Reid et al., 1987).

$$c_{pl} = 0.239\sum_{i=1}^{N} c_{pl_i} n_i$$

where c_{pl} is in cal/mol K.

4. Latent heat of vaporization.

At normal boiling point—Vetere modification of Kistiakowsky eqn. (Duvedi and Achenie, 1996):

$$\Delta H_b^v = S_{vb} T_b$$

$$S_{vb} = 44.367 + 15.33\log T_b + 0.39137 T_b/M + 0.00433 T_b^2/M - 5.627 \times 10^{-6} T_b^3/M$$

At other temperatures (Reid et al., 1987):

$$\Delta H^v\{T\} = \Delta H_b^v\left[\frac{1 - T/T_c}{1 - T_b/T_c}\right]^n$$

$$n = \left[\frac{0.00264(\Delta H_b^v)}{RT_b} + 0.8794\right]^{10}$$

where the latent heat of vaporization is in J/mol and M is the molecular weight.

The group contributions for use in the above equations from Joback and Reid (1987) are:

Group	Valence	T_c	P_c	V_c	T_b	n_i	c_{pl}	M
—CH_3	1	0.0141	−0.0012	65	23.58	4	36.8	15.04
—CH_2—	2	0.0189	0	56	22.88	3	30.4	14.03
—CH=	3	0.0164	0.002	41	21.74	2	21	13.02
=C=	4	0.0067	0.0043	27	18.25	1	7.36	12.01
—OH	1	0.0741	0.0112	28	92.88	2	44.8	17.01
—O—	2	0.0168	0.0015	18	22.42	1	35	16
—NH_2	1	0.0243	0.0109	38	73.23	3	58.6	16.03
—NH—	2	0.0295	0.0077	35	50.17	2	43.9	15.02
—N=	3	0.0169	0.0074	9	11.74	1	31	14.01
—S—	2	0.0119	0.0049	54	68.78	1	33	32.07
—SH	1	0.0031	0.0084	63	63.56	2	44.8	33.08
—F	1	0.0111	−0.0057	27	−0.03	1	17	19
—Cl	1	0.0105	−0.0049	58	38.13	1	36	35.45

Using a mixed-integer nonlinear program with various objective functions, Duvedi and Achenie (1996) found three compounds that satisfy the specified constraints. These are:

Compound	ΔH^v, kJ/mol at −1.1°C	c_{pl} cal/mol-K at 21.1°C	P^s, bar at −1.1°C	P^s, bar at 43.3°C
CCl_2F_2	18.76	27.1	2.94	10.67
CF_3OH	19.77	24.7	2.69	13.57
CH_3Cl	20.37	17.4	2.39	8.72

ΔH^v is reported at the lowest temperature; that is, the temperature at which evaporation occurs in a refrigerator, while c_{pl} is reported at the average temperature. Note that the values of ΔH^v and P^s are computed using experimental T_b(CCl_2F_2 = 244.2K, CF_3OH = 251.48K, CH_3Cl = 249.1K) because the group-contribution method is not sufficiently accurate. Note that the differences in ΔH^v from those reported by Duvedi and Achenie (1996) are due to the differences in T_b and simplifications in the methods for estimating S_{vb}.

Since CH_3Cl contains chlorine, which depletes ozone in the earth's stratosphere, Example 3.4 repeats the search using the ozone depletion potential.

EXAMPLE 3.4

Redesign the refrigerant molecules using the ozone depletion potential (ODP) defined for molecules having one carbon atom:

$$ODP = 0.585602 n_{Cl}^{-0.0035} e^{M/238.563},$$

and having two carbon atoms

$$ODP = 0.0949956 n_{Cl}^{-0.00404477} e^{M/83.7953}$$

Repeat the search in Example 3.3 by minimizing the ODP.

SOLUTION

Using a mixed-integer nonlinear program with various objective functions, Duvedi and Achenie (1996) found two compounds that satisfy the specified constraints. These are:

Compound	ΔH^v, kJ/mol at −1.1°C	C_{pl}, cal/mol at 21.1°C	P^s, bar at −1.1°C	P^s, bar at 43.3°C	ODP
SF_2	18.3	16.0	3.84	13.9	0
CH_3CHF_2	20.6	21.9	2.08	7.91	0

Note that the latent heat of vaporization of SF_2 is sufficiently close to 18.4 kJ/mol to be acceptable, given the approximate estimation methods. Also, neither molecule contains chlorine, and hence, the ODP is zero.

Solvent Design

Organic solvents play a key role in many aspects of chemical processing and in the delivery of chemicals to consumers. In chemical processing, solvents are often used: (1) to mobilize solids, frequently dissolving them; (2) to clean equipment, as in removing grease and grime; and (3) in cleaning clothing, as in dry cleaning. In contrast, in the delivery of chemicals to consumers, solvents often convey particles onto surfaces in coatings, such as in paint and printing ink.

Until the past decade, the solvent market was dominated by a few principal products, solvents known for their ability to "dissolve most anything" (Kirschner, 1994). These included acetone, mixed xylenes, and 1,1,1-trichloroethane, which are manufactured in large-scale processes by the major chemical companies. For environmental and health reasons, over the past decade there has been a gradual shift away from these solvents. The U.S. Environmental Protection Agency maintains a Toxic Release Inventory (TRI), which includes acetone, 1,1,1-trichloroethane, and other common solvents whose emissions into the air have been gradually reduced. Other solvents are included on the hazardous air pollutants (HAP) list of the 1990 Clean Air Act; for example, 1,1,1-trichloroethane, which, like Freon® refrigerants, accumulates in the stratosphere and destroys ozone. Furthermore, other solvents, like monomethylether, monoethylether, and their acetates, have been associated with high miscarriage rates.

Chemical companies are increasingly challenged to reduce the usage of these targeted solvents, and consequently, a host of solutions are being sought, including shifts toward: (1) aqueous solvents, where possible, (2) more concentrated paints and coatings, containing less solvent, and (3) hot-melt, ultraviolet-cured, and waterborne adhesives. For example, in the manufacture of cosmetics and personal care products, solvents are being dropped from some formulations due to their high content of volatile organic compounds (VOCs) and are being replaced, for example, by water-based hair sprays and solid deodorant sticks.

In meeting the challenge, chemical companies are designing a growing number of environmentally friendly solvents—that is, *engineered* or *designer* solvents—that satisfy the specifications for each application; and end users are altering their usage patterns. As a result, new solvents are appearing gradually, designed as specialty chemicals to replace the use of commodity solvents; for example, diacetone alcohol cleaner is a replacement for acetone in the shipbuilding industry. In some cases, the cleaning methods themselves are changing; for example, a one-step, vapor-degreasing process is replaced by a two-step process, involving a dip-tank rinse followed by drying. Another example includes the recycling and reuse of cleaning solvents. In the dry-cleaning business, because the principal solvent, perchloroethylene, is suspected to be a carcinogen and appears on the list of HAPs, used solvent is being filtered and recycled, refrigerated condensers are being installed to recover vapor emissions, and new water and steam-cleaning (wet) processes are being developed.

The search for each new *specialty* solvent is a product design problem. As shown for environmentally friendly refrigerants in Section 3.2, it is helpful to create an *innovation map* that shows the new materials technologies, that is, the new classes of chemicals, and their connections to the customer needs—that is, the *customer-value proposition.* Clearly the success of a new product often relies on the careful attention to this interplay.

In this section, the focus is on strategies to search for the new chemicals. Two examples are presented in which: (1) a new solvent is designed as a replacement for 1,1,1-trichloroethane for cleaning surfaces in the lithographic printing industry (Sinha et al., 1999), and (2) a solvent is selected to remove a solute from a mixture in a liquid–liquid extraction process (Gani et al., 1991; Pretel et al., 1994). Like the previous examples on the design of polymers and refrigerants (Examples 3.2–3.4), initially desired properties are selected, together with a set of candidate molecular groups and target property values. In practice, of course, it is crucial that these properties reflect the needs of potential customers. Then, chemical structures are determined whose property estimates, using group-contribution methods, are closest to the target properties.

Property Estimation

For solvent design, in addition to the normal boiling point, liquid density, and the latent heat of vaporization, solubility and related properties, as well as health and safety properties, must be estimated. Estimation methods for these properties are discussed next.

Solubility and Related Measures

For the design of solvents to clean surfaces, to apply coating resins, and to swell cured elastomers, the Hansen solubility parameter

$$\delta_T = \sqrt{\delta_D^2 + \delta_P^2 + \delta_H^2} \tag{3.11}$$

provides a useful measure of solvent performance. As defined in Eq. (3.11), this parameter is comprised of three solubility parameters: (1) δ_D, to account for nonpolar (dispersive) interactions, (2) δ_P, to account for polar interactions, and (3) δ_H, to account for hydrogen-bonding interactions. These three contributions may be estimated using group-contribution methods:

$$\delta_D = \frac{\sum_{i=1}^{N} n_i F_{D_i}}{V_0 + \sum_{i=1}^{N} n_i V_i} \tag{3.12}$$

$$\delta_P = \frac{\sqrt{\sum_{i=1}^{N} n_i (1{,}000\, F_{P_i})}}{V_0 + \sum_{i=1}^{N} n_i V_i} \tag{3.13}$$

$$\delta_H = \sqrt{\frac{\sum_{i=1}^{N} n_i (-U_{H_i})}{V_0 + \sum_{i=1}^{N} n_i V_i}} \tag{3.14}$$

where n_i is the number of groups of type i in the solvent molecule, N is the number of group types in the solvent molecule, and F_{D_i}, F_{P_i}, U_{H_i}, and V_i are the contributions associated with group i. The latter are tabulated for common groups by van Krevelen and Hoftyzer (1976—F_{D_i}, F_{P_i}), Hansen and Beerbower (1971—U_{H_i}), and Constantinou and Gani (1994—V_i) Note that the constant associated with the molar volume prediction for liquids is $V_0 = 12.11\ \text{cm}^3/\text{mol}$.

Given estimates of these three solubility parameters, a solvent is likely to dissolve a solute when:

$$4(\delta_D - \delta_D^*)^2 + (\delta_p - \delta_p^*)^2 + (\delta_H - \delta_H^*)^2 \leq (R^*)^2 \tag{3.15}$$

where R^*, referred to as the radius of interaction as defined by Hansen (1969), and δ_D^*, δ_P^*, and δ_H^*, are parameters related to the solute. Note that the left-hand side of inequality (3.15) is the distance between the solute and solvent molecules, a measure of the solute-solvent interaction.

For other applications, like the selection of solvents for the liquid–liquid extraction of solutes from mixtures, solubility and related measures are determined on the basis of the liquid-phase activity coefficients, γ_{ij}, for solute-solvent pairs. Usually, for screening purposes, it is sufficient to estimate the liquid-phase activity coefficient at infinite dilution, γ_{ij}^∞, using group-contribution methods.

When considering solvent S for extraction of solute A from species B, Pretel and co-workers (1994) use the UNIFAC group-contribution method to obtain estimates of four solvent properties:

$$\text{Solvent Selectivity} = S_s = \beta = \frac{x_{A,S}}{x_{B,S}} = \frac{\gamma_{B,S}^\infty MW_A}{\gamma_{A,S}^\infty MW_B} \tag{3.16}$$

$$\text{Solvent Power} = S_P = x_{A,S} = \frac{1}{\gamma_{A,S}^\infty}\frac{MW_A}{MW_S} \tag{3.17}$$

$$\text{Solute Distribution Coefficient} = m = K_D = \frac{x_{A,S}}{x_{A,B}} = \frac{\gamma_{A,B}^\infty}{\gamma_{A,S}^\infty}\frac{MW_B}{MW_S} \tag{3.18}$$

$$\text{Solvent Loss} = S_l = x_{S,B} = \frac{1}{\gamma_{S,B}^\infty}\frac{MW_S}{MW_B} \tag{3.19}$$

where x are mass fractions. Clearly, a desirable solvent will have large selectivity, solvent power, and distribution coefficient; and low solvent loss, as discussed in Example 3.6 below.

Health and Safety Measures

Several empirically defined properties are useful in restricting the selection of solvents to those having low impacts on health and safety. These are presented next.

Bioconcentration Factor This factor, which is related to the likelihood of a solvent accumulating in and harming living tissue, was correlated by Veith and Konasewich (1975) as:

$$\log_{10}\text{BCF} = 0.76 \log_{10} K_{ow} - 0.23 \tag{3.20}$$

where K_{ow} is the octanol-water partition coefficient, which is expressed as:

$$\log_{10} K_{ow} = \sum_{i=1}^{N} n_i \chi_i^0 + 0.12 \sum_{i=1}^{N} n_i \chi_i^1 \tag{3.21}$$

where χ_i^0 and χ_i^1 are the *fragment* and *factor* of group i, respectively, as tabulated by Hansch and Leo (1979).

Toxicity Measure The lethal concentration of a solvent. LC_{50}, a useful measure of toxicity, has been correlated by Konemann (1981) as:

$$\log_{10}\text{LC}_{50} = -0.87 \log_{10} K_{ow} - 0.11 \tag{3.22}$$

Flash Point The flash-point temperature is a measure of the explosive potential of vapor mixtures in air. For paraffins, aromatics, and cycloparaffins, it has been correlated as a function of the normal boiling point (Butler et al., 1956; Lyman et al., 1981):

$$T_f = 0.683\, T_b - 119 \tag{3.23}$$

where the temperatures are in kelvin.

EXAMPLE 3.5

In a lithographic printing process, ink is conveyed to an impression plate by means of a train of rubber rollers known as "blankets." These blankets must be cleaned regularly since their cleanliness is crucial for the production of high-quality images. It is desired to replace the current solvent, 1,1,1-trichloroethane, with an environmentally friendly solvent having the ability to dissolve dried ink rapidly and having a short drying time; that is, having a small latent heat of vaporization, and consequently, a short drying time and low utility costs for vaporization. In addition, the solvent should cause negligible swelling of the blanket and be nonflammable. These are the desired product-quality specifications.

SOLUTION

This solution is based upon that presented by Sinha and co-workers (1999). A set of 12 molecular groups is selected upon which the search for solvent molecules is based. These are: CH_3—, —CH_2—, Ar—(C_6H_5—), Ar= (C_6H_4=), —OH, CH_3CO—, —CH_2CO—, —COOH, CH_3COO—, —CH_2COO—, —CH_3O, and —CH_2O—. Note that chlorine is omitted to avoid ozone-depletion problems.

Next, specifications are provided to define the desired properties of the solvent molecules to be designed. The ink residue is assumed to consist of phenolic resin, Super Bakacite 1001, for which the following parameters were estimated: $\delta_D^* = 23.3\ \mathrm{MPa}^{1/2}$, $\delta_P^* = 6.6\ \mathrm{MPa}^{1/2}$, $\delta_H^* = 8.3\ \mathrm{MPa}^{1/2}$, and $R^* = 19.8\ \mathrm{MPa}^{1/2}$. For the lithographic blanket, which is typically polyisoprene rubber, swelling is avoided when $\delta_p > 6.3\ \mathrm{MPa}^{1/2}$. Furthermore, the bioconcentration factor is sufficiently low and the lethal concentration is sufficiently high when $\log_{10} K_{ow} < 4.0$. Finally, to ensure that the solvent is liquid at ambient pressure, it is required that $T_b > 323$ K and $T_m < 223$ K, where T_m is the melting-point temperature. Note that while no bounds are placed upon the standard latent heat of vaporization of 298 K, ΔH^v, it is minimized to reduce the drying time and the cost of heating utilities.

Group Contributions

The following group contributions have been taken from van Krevelen and Hoftyzer (1976), Hansen and Beerbower (1971), and Constantinou and Gani (1994).

To estimate the normal boiling point, the melting point, and the standard latent heat of vaporization at 298 K:

$$T_b = T_{b0} \ln\left(\sum_{i=1}^{N} n_i T_{bi}\right)$$

$$T_m = T_{m0} \ln\left(\sum_{i=1}^{N} n_i T_{mi}\right)$$

$$\Delta H^v = \Delta H_0^v + \sum_{i=1}^{N} n_i H_{vi}$$

where $T_{b0} = 204.2$ K, $T_{m0} = 102.4$ K, and $\Delta H_0^v = 6.829$ KJ/mol.

Sinha and co-workers (1999) formulate a mixed-integer nonlinear program that minimizes ΔH^v to locate three compounds that satisfy the specified constraints. These are:

Compound	ΔH^v, kJ/mol	T_b, K	T_m, K	δ_p, $\mathrm{MPa}^{1/2}$	$\log_{10} K_{ow}$
Methyl ethyl ketone	35.5	354.9	193.2	9.66	1.59
Diethyl ketone	40.1	385.7	206.6	8.21	2.37
Ethylene glycol monomethyl ether	47.6	387.4	200.2	11.5	−0.65

EXAMPLE 3.6

It is desired to locate a solvent for the liquid–liquid extraction of ethanol from its azeotrope with water. This dehydration has been carried out principally by heterogeneous azeotropic distillation using benzene, now known to be a carcinogen, as an entrainer. If such a solvent can be located, liquid–liquid extraction could become the preferred processing technique.

SOLUTION

Potential molecular groups for the solvents are selected from among those in the UNIFAC VLE (vapor–liquid equilibrium) tables (Hansen et al., 1991). Solvents are sought that have the following properties: MW < 300, $T_b - T_{b,\mathrm{furfural}} > 50$ K, $S_s > 7$ wt./wt., m > 1.0 wt%/wt%, and $S_l > 0.1$ wt%.

Group	Valence	T_{bi}	T_{mi}	F_{Di}	F_{Pi}	U_{Hi}	H_{Vi}	V_i	χ_i^0	χ_i^1
CH_3—	1	0.8894	0.464	420	0	0	4.116	26.14	0.89	1
—CH_2—	2	0.9225	0.9246	270	0	0	4.65	16.41	0.66	1
Ar—	1	6.2737	7.5434	1,430	110	0	33.042	70.25	1.9	1
Ar=	2	6.2737	7.5434	1,430	110	0	33.042	70.25	1.67	1
—OH	1	3.2152	3.5979	210	500	−19,500	24.529	5.51	−1.64	1
CH_3CO—	1	3.566	4.8776	210	800	−2,000	18.99	36.55	−0.44	2
—CH_2CO—	2	3.8967	5.6622	560	800	−2,000	20.41	28.16	−0.67	2
—COOH	1	5.8337	11.563	409	450	−11,500	43.046	22.32	−1.11	1
CH_3COO—	1	3.636	4.0823	806	510	−3,300	22.709	45	−0.6	2
—CH_2COO—	2	3.3953	3.5572	609	510	−3,300	17.759	35.67	−0.83	2
—CH_3O	1	2.2536	2.9248	520	410	−4,800	10.919	32.74	−0.93	2
—CH_2O—	2	1.6249	2.0695	370	410	−4,800	7.478	23.11	−1.16	2

Pretel and co-workers (1994) estimate these properties, as well as the solvent power, S_p, and the solvent density, ρ_s, for many candidate solvents. They observe that the constraints are not satisfied for any of the candidates, and consequently conclude that liquid–liquid extraction is not a favorable process for the dehydration of the ethanol–water azeotrope.

Solvent Design for Crystallization of Organic Solids

The selection of solvents is of special significance in determining the morphology of organic crystals, especially in the manufacture of pharmaceuticals. In this section, which appears in the file, Supplement_to_Chapter_3.pdf in the PDF Files folder, which can be downloaded from the Wiley Web site associated with this book, solvent selection for the crystallization of ibuprofen is discussed.

Solutes for Hand Warmers

This section, which appears in the file, Supplement_to_Chapter_3.pdf, briefly introduces hand-warmer products, with emphasis on the selection of solutes to provide heat at comfortable temperatures.

www.wiley.com/go/global/seider

3.6 SUMMARY

This chapter has concentrated on the design of new basic chemical products. Initially, the *innovation map* is introduced to connect the search for new molecules with customer needs. The latter are the basis for thermophysical and transport property specifications being sought by a design team for a new chemical product(s). Emphasis has been placed on the use of group-contribution methods for the property estimates. Molecular simulation methods, which are gaining favor, are introduced briefly, but are not used in the examples presented. Through examples for the design of polymer repeating units, refrigerants, solvents for removal of printing ink, and solvents for liquid–liquid extraction, optimization methods are employed to locate the best molecules.

REFERENCES

1. Allen, D.T., and D.R. Shonnard, *Green Engineering: Environmentally Conscious Design of Chemical Processes*, Prentice-Hall, Englewood Cliffs, New Jersey (2002).

2. *Aspen OLI User Guide,* Version 10.2, Aspen Technology, Inc., Cambridge, MA (1999).

3. *Aspen Properties Toolkit Manual*, Aspen Technology, Inc., Cambridge, MA (2006).

4. Bird, R.B., W.E. Stewart, and E.N. Lightfoot, *Transport Phenomena*, 2nd ed., John Wiley & Sons, New York (2002).

5. Brennecke, J.F., and E.J. Maginn, "Ionic Liquids: Innovative Fluids for Chemical Processing," *AIChE J.*, **47**(11), 2384–2389 (2001).

6. Butler, R.M., G.M. Cooke, G.G. Lukk, and B.G. Jameson, "Prediction of Flash Points for Middle Distillates, *Ind. Eng. Chem.*, **48**, 808–812 (1956).

7. Chen, C.-C., C.P. Bokis, and P. Mathias, "Segment-based Excess Gibbs Energy Model for Aqueous Organic Electrolytes," *AIChE J.*, **47**(11), 2593–2602 (2001).

8. Constantinou, L., and R. Gani, "New Group Contribution Method for Estimating Properties of Pure Compounds," *AIChE J.*, **40**, 1697–1710 (1994).

9. Dare-Edwards, M.P., "Novel Family of Traction Fluids Derived from Molecular Design," *J. Synth. Lubr.*, **8**(3), 197 (1991).

10. Derringer, G.C., and R.L. Markham, "A Computer-Based Methodology for Matching Polymer Structures with Required Properties," *J. Appl. Polymer Sci.*, **30**, 4609–4617 (1985).

11. Duvedi, A.P., and L.E.K. Achenie, "Designing Environmentally Safe Refrigerants Using Mathematical Programming," *Chem. Eng. Sci.*, **51**(15), 3727–3729 (1996).

12. Gani, R., "Computer-aided Methods and Tools for Chemical Product Design," *Trans. IChemE, Part A, Chem. Eng. Res. Design*, **82**(A11), 1494–1504 (2004).

13. Gani, R., B. Nielsen, and A. Fredenslund, "A Group Contribution Approach to Computer-aided Molecular Design," *AIChE J.*, **37**(9), 1318–1332 (1991).

14. *Getting Started Modeling Processes with Electrolytes*, Version 10.2, Aspen Technology, Inc., Cambridge, MA (1999).

15. Giannelis, E.P., "Molecular Engineering of Ceramics. Chemical Approaches to the Design of Materials," *Eng.: Cornell Q.*, **23**(2), 15 (1989).

16. Hansch, C., and A.J. Leo *Substituent Constants for Correlation Analysis in Chemistry and Biology*, Wiley, New York (1979).

17. Hansen, C.M., "The Universality of the Solubility Parameter," *Ind. Eng. Chem. Prod. Res. Dev.*, 2–11 (1969).

18. Hansen, C.M., and A. Beerbower "Solubility Parameters," in A. Standen, Ed., *Kirk-Othmer Encyclopedia of Chemical Technology*, Wiley-Interscience, New York (1971).

19. Hansen, H.K., P. Rasmussen, Aa. Fredenslund, M. Schiller, and J. Gmehling, "Vapor–liquid Equilibria by UNIFAC Group Contribution: 5. Revision and Extension," *Ind. Eng. Chem. Res.*, **30**, 2352 (1991).

20. Joback, K.G., and R.C. Reid, "Estimation of Pure-Component Properties from Group Contributions," *Chem. Eng. Commun.*, **57**, 233–243 (1987).

21. Joback, K.G., and G. Stephanopoulos, "Designing Molecules Possessing Desired Physical Property Values," in J.J. Sirola, I.E. Grossmann, and G. Stephanopoulos, Eds., Proceedings of *Foundations of Computer-aided Process Design, (FOCAPD'89)*, pp. 363–387, AIChE, New York (1989).

22. Kirschner, E.M., "Environment, Health Concerns Force Shift in Use of Organic Solvents," *Chem. Eng. News*, 13–20, June 20 (1994).

23. Kofke, D.A., and B.C. Mihalick, "Web-based Technologies for Teaching and Using Molecular Simulation," *Fluid Phase Equil.*, **194–197**, 327–335 (2002).

24. KONEMANN, H., "Quantitative Structure–Activity Relationships in Fish Toxicity Studies. 1. Relationship for 50 Industrial Chemicals," *Toxicology*, **19**, 209–221 (1981).

25. LYMAN, W.J., W.F. REEHL, and D.H. ROSENBLATT, *Handbook of Chemical Property Estimation Methods*, McGraw-Hill, New York (1981).

26. MARANAS, C.D., "Optimal Computer-Aided Molecular Design: A Polymer Design Case Study," *Ind. Eng. Chem. Res.*, **35**, 3403–3414 (1996).

27. MCGOUGH, A., and M. HALLORAN, "Overview of AspenTech's CAPE-OPEN Support," 2nd Annual U.S. CAPE-OPEN Meeting, Morgantown, WV, May 25–26, 2005. See conference abstracts at: http://www.netl.doe.-gov/publications/proceedings/05/CAPE-OPEN/conf_pro.html#Papers

28. MIDGELY, T., "From Periodic Table to Production," *Ind. Eng. Chem.*, **29**, 241–244 (1937).

29. PISANO, G.P., *The Development Factory: Unlocking the Potential of Process Innovation*, Harvard Business School Press, Cambridge (1997).

30. POLING, B.E., J.M. PRAUSNITZ, and J.P. O'CONNELL, *Properties of Gases and Liquids*, 5th ed., McGraw-Hill, New York (2001).

31. PONS, M. "Cape-Open and Simulis Thermodynamics Enable You to Use Rigorous Thermodynamics in Matlab," 2005AIChE Annual Meeting, Cincinnati, OH, Oct. 30–Nov. 4, 2005. See abstract at: http://aiche.confex.com/aiche/2005/techprogram/P31460.HTM

32. *PPDS2 for Windows: User Manual and Reference Guide*, Nat'l. Eng. Lab., E. Kilbridge, Glasgow, UK (1997).

33. PRETEL, E.J., P. ARAYA LOPEZ, S.B. BOTTINI, and E.A. BRIGNOLE, "Computer-aided Molecular Design of Solvents for Separation Processes," *AIChE J.*, **40**(8), 1349–1360 (1994).

34. REID, R.C., J.M. PRAUSNITZ, and B.E. POLING, *The Properties of Gases & Liquids*, 4th ed., McGraw-Hill, New York (1987).

35. REID, R.C., J.M. PRAUSNITZ, and T.K. SHERWOOD, *The Properties of Gases & Liquids*, 3rd ed., McGraw-Hill, New York (1977).

36. ROWLEY, R.L., *Statistical Mechanics for Thermophysical Property Calculations*, Prentice-Hall, Englewood Cliffs, New Jersey (1994).

37. SINHA, M., L.E.K. ACHENIE and G.M. OSTROVSKY, "Environmentally Benign Solvent Design by Global Optimization," *Comput. Chem. Eng.*, **23**, 1381–1394 (1999).

38. VAN KREVELEN, D.W., *Properties of Polymers*, 3rd ed., Elsevier, Amsterdam (1990).

39. VAN KREVELEN, D.W., *Properties of Polymers: Correlation with Chemical Structure*, Elsevier, Amsterdam (1972).

40. VAN KREVELEN, D.W., and P.J. HOFTYZER, *Properties of Polymers: Their Estimation and Correlation with Chemical Structure*, Elsevier, Amsterdam (1976).

41. VEITH, G.D., and D.E. KONASEWICH "Structure–activity Correlations in Studies of Toxicity and Bioconcentration with Aquatic Organisms," Proceedings of Symposium in Burlington, Ontario—Canada Center for Inland Water (1975).

42. ZEMAITIS, J.F., D.M. CLARKE, M. RAFAL, and N.C. SCRIVNER, *Handbook of Aqueous Electrolyte Thermodynamics*, DIPPR, AIChE, New York, NY (1986).

EXERCISES

3.1 For the polymer film in Example 3.2, use GAMS to locate the repeat units having the second-and third-lowest values of the objective function.

3.2 Many companies and municipalities are reluctant to handle chlorine, either in processing or in incinerating wastes. Resolve Example 3.2 without the —CHCl— group.

3.3 For an electronic device designed to operate at higher temperatures near a furnace, a high glass-transition temperature, 423 K, is required. Resolve Example 3.2 with this constraint.

3.4 It is desired to find a refrigerant that removes heat at −20°C and rejects heat at 32°C. Desirable refrigerants should have $P^s\{-20°C\} > 1.4$ bar, $P^s\{32°C\} < 14$ bar, $\Delta H^v\{-20°C\} > 18.4$ kJ/mol, and $c_{pl}\{6°C\} < 32.2$ cal/mol-K. For the candidate groups CH_3, CH, F, and S, formulate a mixed-integer nonlinear program and use GAMS to solve it. Use the group-contribution method in Section 1 of the solution to Example 3.3 to estimate T_b. *Hint:* Maximize the objective function, $\Delta H^v\{-20°C\}$.

3.5 Using the group contributions in Example 3.5, determine whether methyl propyl ketone, methyl butyl ketone, and methyl isobutyl ketone are suitable solvents. For —CH═, use the group contributions for —CH_2—.

Chapter 4

Process Synthesis—Basic Chemicals

4.0 OBJECTIVES

This chapter covers the steps under the block "Concept Stage" in Figure PI.1, which provides an overview of the steps in designing new basic chemical products and processes. Because the first step, "Opportunity assessment, customer, and technical requirements," has been discussed in Section 2.4, this chapter begins with the next step, "Preliminary Database Creation."

After studying this chapter, the reader should:

1. Understand how to go about assembling design data and creating a preliminary database.
2. Be able to implement the steps in creating flowsheets involving reactions, separations, and *T–P* change operations. In so doing, many alternatives are identified that can be assembled into a synthesis tree containing the most promising alternatives.
3. Know how to select the principal pieces of equipment and to create a detailed process flow diagram, with a material and energy balance table and a list of major equipment items.
4. Understand the importance of building a pilot plant to test major pieces of equipment where some uncertainty exists.
5. Have an initial concept of the role of a process simulator in obtaining data and in carrying out material and energy balances. This subject is expanded upon in Chapter 5, where the use of process simulators to make calculations for continuous and batch processes is presented.

4.1 INTRODUCTION

This chapter focuses on the steps often referred to as *process creation*, which are implemented by a design team when designing a process to manufacture a basic chemical product. It describes the components of the preliminary database and suggests several sources, including the possibility of carrying out laboratory experiments. Then, using the database, it shows how to create a synthesis tree, with its many promising flowsheets, for consideration by the design team. This is accomplished first for the design of a continuous process to produce a commodity chemical, vinyl chloride, and subsequently, for the design of a batch process to produce a pharmaceutical.

For each of the most promising alternatives in the synthesis tree, a base-case design is created. Because this is central to the work of all design teams, the strategy for creating a detailed process flow diagram is covered and the need for pilot-plant testing is discussed.

4.2 PRELIMINARY DATABASE CREATION

Having completed an initial assessment of the need for a process design, and having conducted a literature search, the design team normally seeks to organize the data required for the design into a compact database, one that can be accessed with ease as the team proceeds to create process flowsheets and develop a base-case design. At this stage, several alternatives are being considered, involving several raw materials, the desired products, and several byproducts and reaction intermediates. For these chemicals, basic thermophysical properties are needed, including molecular weight, normal boiling point, freezing point, critical properties, standard enthalpy and Gibbs free energy of formation, and vapor pressures, densities, heat capacities, and latent heats as a function of temperature. If chemical reactions are involved, some rudimentry information concerning the rates of the principal chemical reactions, such as conversion and product distribution as a function of space velocity, temperature, and pressure, is often needed before initiating the process synthesis steps. When necessary, additional data are located, or measured in the laboratory, especially when the design team gains enthusiasm for a specific processing concept. In addition, the team needs environmental and safety data, including information on the toxicity of the chemicals, how they affect animals and humans, and flammability in air. Material Safety Data Sheets (MSDSs) will be available for chemicals already being produced but will have to be developed for new chemicals. Also, for preliminary economic evaluation, chemical prices are needed. Additional information, such as transport properties, detailed chemical kinetics, the corrosivity of the chemicals, heuristic parameters, and data for sizing equipment, is normally not needed during process creation. It is added by the design team, after a detailed process flow diagram has been created, and before work on the detailed design of the equipment commences.

When the data are assembled, graphs are often prepared with curves positioned to provide a good representation, especially for experimental data with scatter. Alternatively, the coefficients of equations, theoretical or empirical, are computed using regression analysis programs. This is especially common for thermophysical property data, such as the vapor pressure, P^s, as a function of the temperature, T, and vapor–liquid equilibrium data, as discussed later in this section.

If molecular-structure design has been carried out previously, as discussed in Chapter 3, most of the pertinent data will have been collected. This is especially the case for protein pharmaceuticals where automated labs-on-a-chip are often employed, permitting hundreds and thousands of cell clones to be evaluated experimentally in parallel, as discussed in Section 3.1. Also, Sections 16.4 and 17.4 discuss the design of labs-on-a-chip for the high-throughput screening of potential kinase inhibitor drugs for treating cancer.

Thermophysical Property Data

For basic properties such as molecular weight, normal boiling point, melting point, and liquid density (often at 20°C), the CRC *Handbook of Chemistry and Physics* (CRC Press, Boca Raton, FL, annual) provides a compilation for a large number of organic and inorganic compounds. In addition, it provides vapor pressure data and enthalpies and free energies of formation for many of these compounds, as well as selected properties, such as the critical temperature, for just a few of these compounds. Similar compilations are provided by *Perry's Chemical Engineers' Handbook* (Green and Perry, 2008), *Properties of Gases and Liquids* (Poling et al., 2001), and *Data for Process Design and Engineering Practice* (Woods, 1995). In addition, extensive databases for as many as 15,000 compounds are provided by process simulators (e.g., ASPEN PLUS, ASPEN HYSYS, UNISIM, CHEMCAD, PRO/II, BATCH PLUS, and SUPERPRO DESIGNER), as discussed in Section 3.4. These are extremely useful as they are accessed by large libraries of programs that carry out material and energy balances, and estimate equipment sizes and costs.

Because phase equilibria are important in most chemical processes, design teams usually spend considerable time assembling data, especially vapor–liquid and liquid–liquid equilibrium data. Over the years, thousands of articles have been published in which phase equilibria data are provided. These can be accessed by a literature search, although the need to search the literature has largely been negated by the extensive compilation provided in *Vapor–Liquid Equilibrium Data Collection* (Gmehling et al., 1980). In this DECHEMA data bank, which is available both in more than 20 volumes and electronically, the data from a large fraction of the articles can be found easily. In addition, each set of data has been regressed to determine interaction coefficients for the binary pairs to be used to estimate liquid-phase activity coefficients for the NRTL, UNIQUAC, Wilson, etc., equations. This database is also accessible by process simulators. For example, with an appropriate license agreement, data for use in ASPEN PLUS can be retrieved from the DECHEMA database over the Internet. For nonideal mixtures, the extensive compilation of Gmehling (1994) of azeotropic data is very useful.

In this section, space is not available to discuss the basics of phase equilibrium; for this material, the reader is referred to many excellent thermodynamics books (e.g., Balzhiser et al., 1972; de Nevers, 2002; Kyle, 1984; Sandler, 2006; Smith et al., 1997; Walas, 1985). Yet process designers usually need to work with phase equilibria data to obtain reasonable predictions for phase conditions and separations of specific mixtures in the temperature and pressure ranges anticipated. This usually requires data regression using models that are best suited for the compositions, temperatures, and pressures under study. Consequently, in this section, two examples are presented in which methods of data regression are needed. To assist the reader, a review of the basics of phase equilibrium is presented in the multimedia modules, which can be downloaded from the Wiley Web sites associated with this textbook—follow the path *ASPEN → Physical Property Estimation,* in which the equations are derived, the data banks are summarized, and many of the phase equilibrium models are tabulated and discussed briefly.

EXAMPLE 4.1

This example involves vapor–liquid equilibrium (VLE) data for the design of a distillation tower to dehydrate ethanol. A portion of the T–x–y data for an ethanol–water mixture, measured at 1.013 bar (1 atm) using a Gillespie still (Rieder and Thompson, 1949), is shown in Figure 4.1a. Here, it is desired to use regression analysis to enable the UNIQUAC equation to represent the data accurately over the entire composition range.

SOLUTION

Using ASPEN PLUS and data from the DECHEMA data bank, with the details described on the multimedia modules that accompany this textbook (*ASPEN → Physical Property Estimation → Equilibrium Diagrams → Property Data Regression*), the x–y diagram in Figure 4.1b is obtained, which compares the data points with a curve based on the following built-in

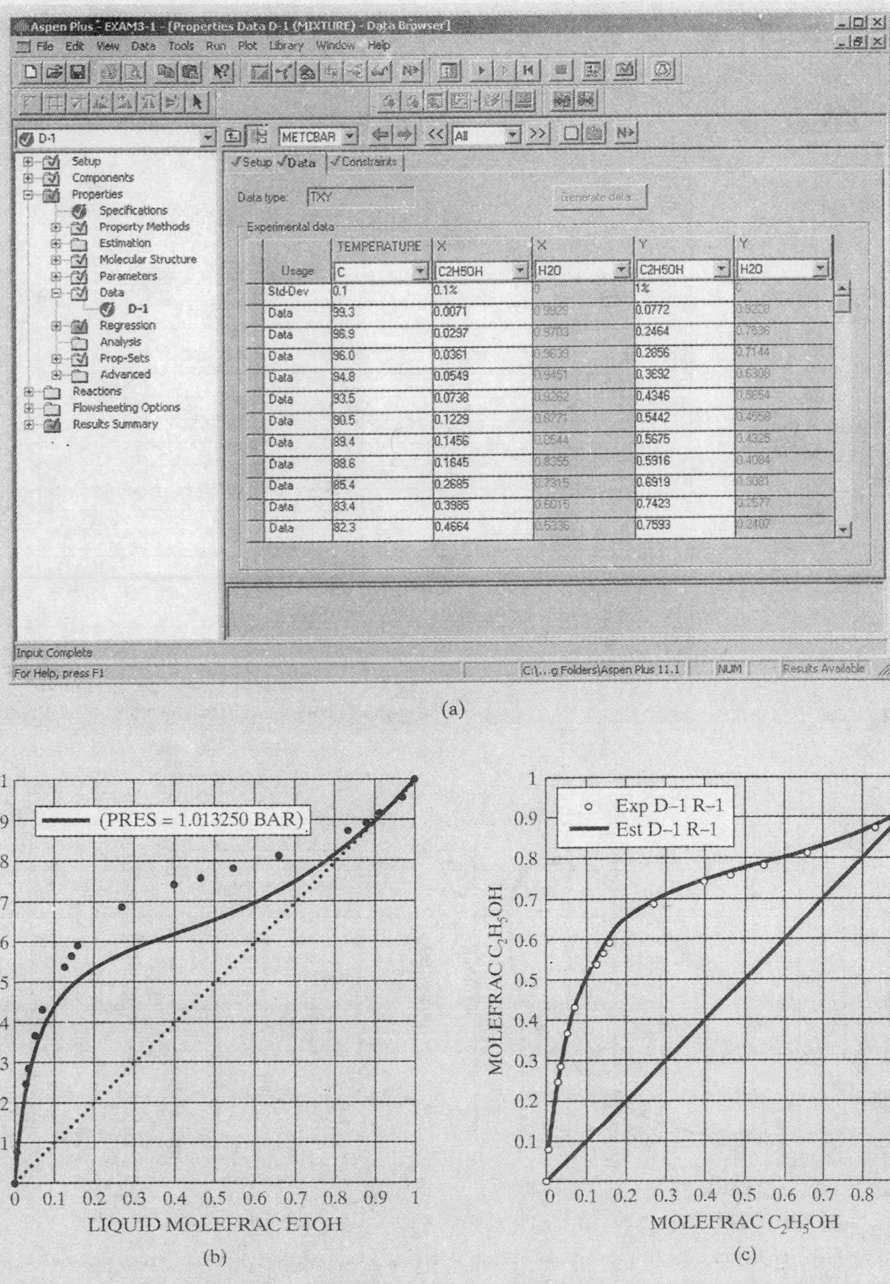

Figure 4.1 Regression of ethanol–water data using UNIQUAC: (a) VLE data (Rieder and Thompson, 1949); (b) x–y diagram before regression; (c) x–y diagram after regression.

interaction coefficients retrieved from the VLE-IG data bank: $a_{E,W} = 2.0046$, $a_{W,E} = -2.4936$, $b_{E,W} = -728.97$, and $b_{W,E} = 756.95$. Then the data regression system is used with the Rieder and Thompson data and much better agreement between the data and the VLE estimates is obtained, as shown in Figure 4.1c. Note that the data regression system adjusts the interaction coefficients to $a_{E,W} = 3.8694$, $a_{W,E} = -3.9468$, $b_{E,W} = -1,457.2$, and $b_{W,E} = 1,346.8$.

Clearly, data regression is needed to obtain a rigorous design for the distillation. Furthermore, in this case, the UNIQUAC equation represents the nonidealities of this polar mixture quite well. When the Peng–Robinson (Reid et al., 1987) equation is used instead, as shown on the multimedia CD-ROM, the data are not represented as well after the data regression is completed.

EXAMPLE 4.2

A second example is provided in which vapor–liquid equilibrium data for a CH_4–H_2S mixture are utilized in connection with the design of a natural gas expander plant. In this case, a portion of the P–x–y data, measured by Reamer et al. (1951), is shown in Figure 4.2a, and regression analysis is used to enable the Soave–Redlich–Kwong (SRK) equation to represent the data better.

SOLUTION

ASPEN PLUS is used with the SRK equation:

$$P = \frac{RT}{V-b} - \frac{a}{V(V+b)}$$

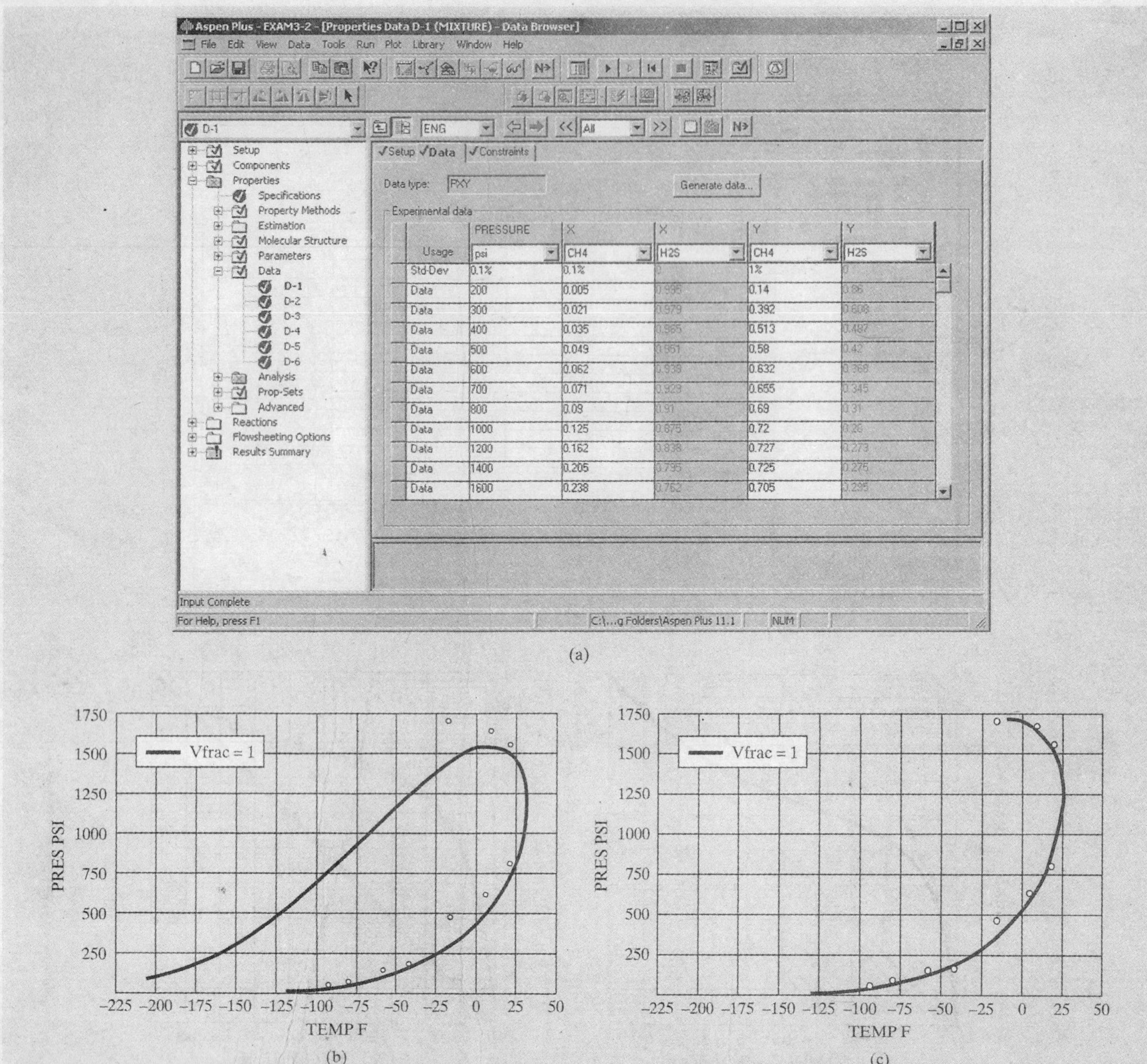

Figure 4.2 Regression of CH_4–H_2S data using the Soave–Redlich–Kwong equation: (a) VLE data (Reamer et al., 1951); (b) phase envelope before regression—75.2 mol% CH_4; (c) phase envelope after regression—75.2 mol% CH_4.

$$a = \sum_{i=1}^{2}\sum_{j=1}^{2} x_i x_j (a_i a_j)^{0.5}(1 - k_{ij})$$

$$b = \sum_{i=1}^{2} x_i b_i$$

$$a_i = f\{T, T_{ci}, P_{ci}, \omega_i\}$$

$$b_i = f\{T_{ci}, P_{ci}\}$$

$$k_{ij} = k_{ji}$$

and P is the pressure, T is the absolute temperature, V is the molar volume of the mixture; a_i and b_i are pure-component constants, T_{ci}, P_{ci} and ω_i are the critical temperature and pressure and the acentric factor for species i, and k_{ij} and k_{ji} are the binary interaction coefficients. Using the built-in parameters retrieved from the EOS-LIT data bank, the P–T phase envelope in Figure 4.2b is obtained, with the details described on the multimedia modules, which can be downloaded from the Wiley Web site associated with this book—follow the path, *ASPEN → Physical Property Estimation → Equilibrium Diagram → Property Data Regression.* Then the data regression system is used to adjust three of the parameters to

$$k_{CH_4,H_2S} = -0.11399,\ \omega_{CH_4} = -0.3344,\ \text{and}\ \omega_{H_2S} = 0.04377.$$

The result is a significant improvement of the phase envelope when compared with the experimental data, as shown in Figure 4.2c. Note, especially, the improvement in the critical region at the elevated pressures.

Environmental and Safety Data

As mentioned in Section 1.4, design teams need toxicity data for raw materials, products, byproducts, and intermediates incorporated in a process design. In toxicology laboratories operated by chemical companies and governmental agencies, such as the U.S. Environmental Protection Agency (EPA) and the U.S. Food and Drug Administration (FDA), tests are run to check the effects of various chemicals on laboratory animals. The chemicals are administered in varying dosages, over differing periods, and in different concentrations, stimulating effects that are measured in many ways, including effects on the respiratory system, the skin, and the onset of cancer. In most cases, the results are provided in extensive reports or journal articles. In some cases, chemicals are difficult to classify as toxic or nontoxic.

Already it is well known that a number of common chemicals are toxic to humans and need to be avoided. One source of information on these chemicals is the Toxic Chemical Release Inventory (TRI), which is maintained by the U.S. EPA, and includes over 600 chemicals. A list of these chemicals is available at the Internet site:

http://www.epa.gov/tri/chemical/index.htm

Another source is provided by the ratings of the National Fire Protection Association (NFPA), which are tabulated for many chemicals in *Data for Process Design and Engineering Practice* (Woods, 1995). The first of three categories is titled "Hazard to Health" and entries are rated from 0 to 4, with 0 meaning harmless and 4 meaning extremely hazardous.

As seen in Table 1.2 and discussed in Section 1.5, data on the flammability of organic compounds are tabulated and, for those compounds not included in the table, methods are available to estimate the data. In addition, tables of flammability data are also available for aerosols and polymers in *Perry's Chemical Engineers' Handbook* (Green and Perry, 2008). The NFPA ratings provide a less quantitative source for many chemicals under "Flammability Hazard," which is the second of the three categories (also rated from 0 to 4).

Chemical Prices

Economics data are often related to supply and demand, and consequently they fluctuate and are much more difficult to estimate. Most companies, however, carry out market studies and have a basis for projecting market size and chemical prices. In view of the uncertainties, to be safe, economic analyses are often conducted using a range of chemical prices to determine the sensitivity of the results to specific prices.

One widely used source of prices of commodity chemicals is from *ICIS Chemical Business* (formerly *Chemical Market Reporter*), a weekly publication. Their Web site, http://www.icis.com/StaticPages/Students.htm, provides information for students in their Knowledge Zone. It should be noted, however, that these prices may not reflect the market situation in a particular location; nevertheless, they provide a good starting point. In addition, commodity chemical prices may be found via ICIS pricing. This service publishes weekly pricing benchmarks to the industry and offers samples of reports that are approximately six months old via the following link: http://www.icispricing.com/il_shared/il_splash/chemicals.asp?llink%. Obviously, to obtain better estimates, at least for the immediate future, the manufacturers of the chemicals should be contacted directly. Lower prices than those listed can often be negotiated. Articles on chemicals of commerce in trade magazines can be searched for on the Web site http://www.findarticles.com. Subscribers to *ICIS Chemical Business* can also obtain more recent market trends and data derived from both their ICIS news and ICIS pricing services.

In some cases, it may be desirable to estimate the prices of utilities, such as steam, cooling water, and electricity, during process creation. Here also, appropriate prices can be obtained from local utility companies. As a start, however, values are often tabulated, as provided in Table 23.1.

Summary

To the extent possible using the literature, company files, computer data banks, and similar sources, the design team assembles a preliminary database for use in preliminary process synthesis, the subject of Section 4.4. Typically, the database contains thermophysical property data, rudimentary reaction-rate data, data concerning toxicity and flammability of the chemicals, and chemical prices. In cases where data cannot be located, estimation methods are often available. However, when the results are sensitive to the estimates, conclusions must be drawn with caution. In most cases, when a process looks promising, an experimental program is initiated, as discussed in the next section. Note that other kinds of data are normally not necessary until the detailed process flow diagram has been developed for the base-case design, and the design team is preparing to complete the detailed design of the equipment items. Note also that when molecular-structure design has been used to select the chemical product, experimental data and/or theoretical estimates are usually available in data banks, especially in drug development.

4.3 EXPERIMENTS

Many design concepts are the result of extensive experiments in the laboratory, which provide valuable data for the design team. Often, however, laboratory experiments are carried out in small vessels, using small quantities of expensive solvents, and under conditions where the conversion and selectivity to the desired product are far from optimal. For this reason, as a design concept becomes more attractive, it is common for the design team to request additional experiments at other conditions of compositions, temperatures, and pressures, and using solvents that are more representative of those suitable for large-scale production. In cases where no previous in-house experimental work has been done, laboratory programs are often initiated at the request of the design team, especially when estimates of the rates of reaction are not very reliable. When chemical reactions involve the use of catalysts, it is essential that experiments be conducted on catalyst life using feedstocks that are representative of those to be used for large-scale production, and that may contain potential catalyst poisons.

Laboratory experiments may also be necessary to aid in the selection and preliminary design of separation operations. The separation of gas mixtures requires consideration of absorption, adsorption, and gas permeation, all of which may require the search for an adequate absorbent, adsorbent, and membrane material, respectively. When nonideal liquid mixtures are to be separated, laboratory distillation experiments should be conducted early because the possibility of azeotrope formation can greatly complicate the selection of adequate separation equipment, which may involve the testing of one or more solvents or entrainers. When solids are involved, early laboratory tests of such operations as crystallization, filtration, and drying are essential.

Clearly, as data are obtained in the laboratory, they are tabulated and usually regressed, to allow addition to the preliminary database for use by the design team in preliminary process synthesis, the subject of the next section.

4.4 PRELIMINARY PROCESS SYNTHESIS[1]

Design teams use many kinds of processing operations to carry out chemical reactions and to separate products and byproducts from each other and from unreacted raw materials. In many respects, one of the greatest challenges in process design involves the synthesis of configurations that produce chemicals in a reliable, safe, and economical manner, and at high yield with little or no waste. Until recently, this part of the design process, often referred to as *process synthesis*, in which many kinds of process operations are configured into flowsheets, was performed from experience gained in similar processing situations, with little formal methodology.

Thanks to research over the past 35 years, coupled with methods of decision-tree analysis and mathematical programming, synthesis strategies have become more quantitative and scientific. In Chapters 7–11 of this text, a primary objective is to cover many of the modern strategies for synthesizing process flowsheets. The objective of this introductory section, however, is simply to show some of the steps and decision processes, mostly by example. After examining two case studies involving the synthesis of a vinyl chloride process and of a process to manufacture tissue plasminogen activator (tPA), the reader should have a good appreciation of the principal issues in process synthesis.

As discussed earlier, preliminary process synthesis occurs after an alternative processing concept has been created. Having defined the concept and assembled the preliminary database, usually with some experimentation, the design team sets out to synthesize a flowsheet of process operations to convert the raw materials to the desired products. First, it decides on the *state* of the raw materials, products, and byproducts, before assembling different configurations of the process operations.

To introduce this approach, this section begins by reviewing the concept of the chemical state, followed by a review of the principal operations, before covering several of the key steps in process synthesis, and utilizing them to create the vinyl chloride and tPA processes. Throughout this development, it should be clear that the synthesis or invention of a chemical process involves the generation and solution of a large combinatorial problem. Here, intuition and experience are as important to the design team as to the composer or artist. The emphasis in this section is on the use of *heuristics*, or rules of thumb, for the synthesis step. However, throughout this text, it will be evident, especially in Chapters 7–11, that many quantitative methods of synthesis, combined with optimization, are available to the design team to generate the most promising process flowsheets.

Chemical State

As the first step in process synthesis, the design team must decide on raw-material and product specifications. These are referred to as *states*. Note that the state selections can be changed later with modifications to the flowsheets. To define the state, values of the following conditions are needed:

1. Mass (flow rate)
2. Composition (mole or mass fraction of each chemical species of a unique molecular type)
3. Phase (solid, liquid, or gas)
4. Form, if solid phase (e.g., particle size distribution and particle shape)
5. Temperature
6. Pressure

In addition, some well-defined properties, such as the intrinsic viscosity, average molecular weight, color, and odor of a polymer, may be required. These are often defined in connection with the research and marketing departments, which work to satisfy the requests and requirements of their customers. It is not uncommon for a range of conditions and properties to be desired, some of which are needed intermittently by various customers as their downstream requirements vary. When this is the case, care must be taken to design a process that is sufficiently flexible to meet changing demands.

For most chemicals, the scale (i.e., production level or flow rate) of the process is a primary consideration early in the design process. Working together with the marketing people, the scale of the process is determined on the basis of the projected demand for the product. Often the demographics of

[1] Adapted from Myers and Seider (1976), Chap. 3.

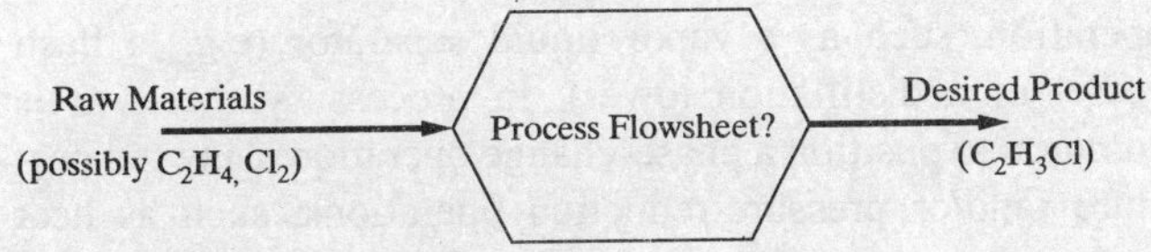

Figure 4.3 Process synthesis problem.

the most promising customers have an important impact on the location of the plant and the choice of its raw materials. As the scale and the location are established, the composition, phase, form, temperature, and pressure of each product and raw-material stream are considered as well. When the desired states of these streams have been identified, the problem of process synthesis becomes better defined. As shown in Figure 4.3, for the production of vinyl chloride, it remains to insert the process operations into the flowsheet.

It is noteworthy that once the state of a substance is fixed by conditions 1–6, all physical properties (except for the form of a solid), including viscosity, thermal conductivity, color, refractive index, and density, take on definite values. Furthermore, the state of a substance is independent of its position in a gravitational field and its velocity. Although there are other conditions (magnetic field strength, surface area) whose values are needed under certain conditions, the six conditions listed above are usually sufficient to fix the state of a substance.

Process Operations

Throughout the chemical engineering literature, many kinds of equipment, so-called *unit operations*, are described, including distillation columns, absorbers, strippers, evaporators, decanters, heat exchangers, filters, and centrifuges, just to mention a few. The members of this large collection, many of which are listed in Tables 5.1 and 5.2 in connection with process simulators, all involve one or more of these basic operations:

1. Chemical reaction
2. Separation of chemical mixtures
3. Phase separation
4. Change of temperature
5. Change of pressure
6. Change of phase
7. Mixing and splitting of streams or batches
8. Operations on solids, such as size reduction and enlargement

Since these are the building blocks of nearly all chemical processes, it is common to create flowsheets involving these basic operations as a first step in process synthesis. Then, in a *task integration* step, operations are combined where feasible. In the remainder of this section, before considering the steps in process synthesis, each of the basic operations is considered in some detail.

Chemical reaction operations are at the heart of many chemical processes. They are inserted into a flowsheet to effect differences in the molecular types between raw-material and product streams. To this end, they involve the chemistry of electron transfers, free-radical exchanges, and other reaction mechanisms, to convert the molecular types of the raw materials into products of other molecular types that have the properties sought by a company's customers. Clearly, the positioning of the reaction operations in the flowsheet involves many considerations, including the degree of conversion, reaction rates, competing side reactions, and the existence of reactions in the reverse direction (which can result in constraints on the conversion at equilibrium). These, in turn, are related closely to the temperature and pressure at which the reactions are carried out, the methods for removing or supplying energy, and the catalysts that provide competitive reaction rates and selectivity to the desired products. In the next subsections, many of these issues are considered in the context of process synthesis. These are revisited throughout the text, especially in Sections 6.2, 6.3, and 6.5 and Chapter 7.

Separation operations appear in almost every process flowsheet. They are needed whenever there is a difference between the desired composition of a product or an intermediate stream and the composition of its source, which is either a feed or an intermediate stream. Separation operations are inserted when the raw materials contain impurities that need to be removed before further processing, such as in reactors, and when products, byproducts, and unreacted raw materials coexist in a reactor effluent stream. The choice of separation operations depends first on the phase of the mixture and second on the differences in the physical properties of the chemical species involved. For liquid mixtures, when differences in volatilities (i.e., vapor pressure) are large, it is common to use vapor–liquid separation operations (e.g., distillation), which are by far the most common. For some liquid mixtures, the melting points differ significantly and solid–liquid separations, involving crystallization, gain favor. When differences in volatilities and melting points are small, it may be possible to find a solvent that is selective for some components and not others, and to use a liquid–liquid separation operation. For other mixtures, particularly gases, differences in absorbability (in an absorbent), adsorbability (on an adsorbent; e.g., activated carbon, molecular sieves, or zeolites), or permeability through a membrane may be exploited with adsorption and membrane separation operations. These and many other separation operations are considered throughout this text, especially in Chapters 6 and 8. The first example of process synthesis that follows introduces some of the considerations in the positioning of distillation operations, and Section 6.4 and Chapter 19 contribute to create the underpinnings for a comprehensive treatment of the synthesis of separation trains in Chapter 8. Many separation operations require phase-separation operations, which may be accomplished by vessels called flash drums for vapor–liquid separation, by decanters for liquid–liquid separation, and by filters and centrifuges for liquid–solid separation.

The need to change temperatures usually occurs throughout a chemical process. In other words, there are often differences in the temperatures of the streams that enter or leave the process or that enter or leave adjacent process operations, such as reaction and separation operations. Often a process stream needs to be heated or cooled from its *source* temperature to its *target* temperature. This is best accomplished through heat exchange with other process streams that have complementary cooling and heating demands. Since the energy crisis in 1973, numerous strategies, some of which are covered in Chapter 9, have been invented to synthesize networks of heat exchangers that minimize the need for heating and cooling utilities, such as steam and cooling water. In the examples of process synthesis, heating and cooling operations are inserted into the flowsheet to satisfy the heating and cooling demands, and a few of the concepts associated with heat integration are introduced. Then, in Section 6.5 and Chapter 18, additional concepts are presented to accompany Chapter 9 on heat and power integration.

The positioning of pressure-change operations such as gas compressors, gas turbines or expanders, liquid pumps, and pressure-reduction valves in a process flowsheet is often ignored in the early stages of process design. As will be seen, it is common to select the pressure levels for reaction and separation operations. When this is done, pressure-change operations will be needed to decrease or increase the pressure of the feed to the particular operation. In fact, for processes that have high power demands, usually for gas compression, there is often an opportunity to obtain much of the power through integration with a source of power, such as turbines or expanders, which are pressure-reduction devices. In process synthesis, however, where alternative process operations are being assembled into flowsheets, it has been common to disregard the pressure drops in pipelines when they are small relative to the pressure level of the process equipment. Liquid pumps to overcome pressure drops in lines and across control valves and to elevate liquid streams to reactor and column entries often have negligible costs. Increasingly, as designers recognize the advantages of considering the controllability of a potential process while developing the base-case design, the estimation of pressure drops gains importance because flow rates are controlled by adjusting the pressure drop across a valve. In the first example of process synthesis, some of the important considerations in positioning the pressure-change operations in a flowsheet are introduced. These are developed further in Sections 6.6, 9S.9 (in the file, Supplement_to_Chapter_9.pdf, in the PDF Files folder, which can be downloaded from the Wiley Web site associated with this book), 9.8, and Chapter 20.

Often there are significant differences in the phases that exit from one process operation and enter another. For example, hot effluent gases from a reactor are condensed, or partially condensed, often before entering a separation operation, such as a vapor–liquid separator (e.g., a flash vessel or a distillation tower). In process synthesis, it is common to position a phase-change operation using temperature- and/or pressure-reduction operations, such as heat exchangers and valves.

The mixing operation is often necessary to combine two or more streams and is inserted when chemicals are recycled and when it is necessary to blend two or more streams to achieve a product specification. In process synthesis, mixing operations are inserted usually during the distribution of chemicals, a key step that is introduced in the first example of process synthesis and expanded upon in Section 6.3. Because the impact of mixing on the thermodynamic efficiency and the utilization of energy is often very negative, as discussed in Section 9S.4, it is usually recommended that mixer operations not be introduced unless they are necessary—for example, to avoid discarding unreacted chemicals. In this regard, it is noteworthy that mixing is the reverse of separation. Although there is an energy requirement in separating a stream into its pure constituents, mixing can be accomplished with no expenditure of energy other than the small amount of energy required when an agitator is used to speed up the mixing process. In cases where the streams are miscible and of low viscosity, mixing is accomplished easily by joining two pipes, avoiding the need for a mixing vessel. Splitting a stream into two or more streams of the same temperature, pressure, and composition is also readily accomplished in the piping.

Synthesis Steps

Given the states of the raw-material and product streams, process synthesis involves the selection of processing operations to convert the raw materials to products. In other words, each operation can be viewed as having a role in eliminating one or more of the property differences between the raw materials and the desired products. As each operation is inserted into a flowsheet, the effluent streams from the new operation are closer to those of the required products. For example, when a reaction operation is inserted, the stream leaving often has the desired molecular types, but not the required composition, temperature, pressure, and phase. To eliminate the remaining differences, additional operations are needed. As separation operations are inserted, followed by operations to change the temperature, pressure, and phase, fewer differences remain. In one parlance, the operations are inserted with the goal of reducing the differences until the streams leaving the last operation are identical in state to the required products. Formal, logic-based strategies, involving the proof of theorems that assert that all of the differences have been eliminated, have been referred to as *means–end analysis*. In process synthesis, these formal strategies have not been developed beyond the synthesis of simple processes. Rather, an informal approach, introduced by Rudd, Powers, and Siirola (1973) in a book entitled *Process Synthesis*, has been adopted widely. It involves positioning the process operations in the following steps to eliminate the differences:

Synthesis Step	**Process Operations**
1. Eliminate differences in molecular types	Chemical reactions
2. Distribute the chemicals by matching *sources* and *sinks*	Mixing
3. Eliminate differences in composition	Separation
4. Eliminate differences in temperature, pressure, and phase	Temperature, pressure, and phase change
5. Integrate tasks; that is, combine operations into *unit processes* and decide between continuous and batch processing	

Rather than discuss these steps in general, it is probably more helpful to describe them as they are applied to the synthesis of two processes, in this case, the vinyl chloride and tPA processes, which are synthesized in the next subsections.

Several general observations, however, are noteworthy before proceeding with the examples. First, like the vinyl chloride and tPA processes to be discussed next, most chemical processes are built about chemical reaction and/or separation operations. Consequently, the steps involved in synthesizing these processes are remarkably similar to those for the manufacture of other chemicals. As the syntheses proceed, note that many alternatives should be considered in the application of each step, many of which cannot be eliminated before proceeding to the next steps. The result is that, at each step, a new set of candidate flowsheets is born. These are organized into *synthesis trees* as the steps are applied to create the vinyl chloride and tPA processes. The synthesis trees are compact representations of the huge combinatorial problem that almost always develops during process synthesis. As will be seen, approaches are needed to eliminate the least promising branches as soon as possible, to simplify the selection of a near-optimal process flowsheet. These approaches are further refined in subsequent chapters. The decision between continuous and batch processing is introduced briefly next, before proceeding with the two examples.

Continuous or Batch Processing

When selecting processing equipment in the task-integration step, the production scale strongly impacts the operating mode. For the production of commodity chemicals, large-scale continuous processing units are selected, whereas for the production of many specialty chemicals as well as industrial and configured consumer chemical products, small-scale batch processing units are preferable. The choice between continuous or batch, or possibly semi-continuous, operation is a key decision. See Section 11.1 for a more complete discussion of this subject.

Example of Process Synthesis: Manufacture of Vinyl Chloride

Consider the need to manufacture vinyl chloride,

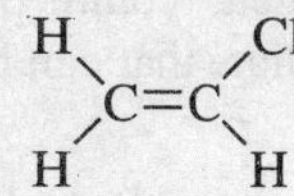

a monomer intermediate for the production of polyvinyl chloride,

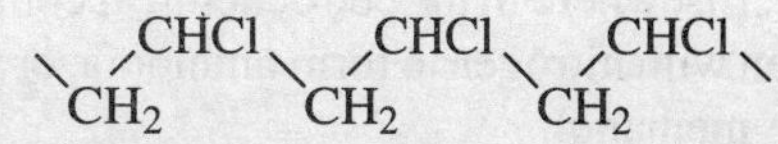

an important polymer (usually referred to as just vinyl) that is widely used for rigid plastic piping, fittings, and similar products. Over the years, large commercial plants have been built, some of which produce over 1 billion lb/yr. Hence, polyvinyl chloride, and the monomer from which it is derived, is referred to commonly as a commodity chemical that is produced continuously, rather than in batch, virtually everywhere. Historically, vinyl chloride was discovered in 1835 in the laboratory of the French chemist Regnault, and the first practical method for polymerizing vinyl chloride was developed in 1917 by the German chemists Klatte and Rollett (Leonard, 1971). Vinyl chloride is an extremely toxic substance and, therefore, industrial plants that manufacture it or process it must be designed carefully to satisfy government health and safety regulations.

Consider an opportunity that has arisen to satisfy a new demand, on the order of 800 million pounds per year, for vinyl-chloride monomer in a petrochemical complex on the Gulf Coast, given that an existing plant owned by the company produces 1 billion pounds per year of this commodity chemical. Consider further that a design team has been formulated, has prepared a project charter, and has begun the *concept* stage of the Stage-Gate^{TM} Product-Development Process. Four potential alternatives have been identified, including:

Alternative 1. A competitor's vinyl-chloride plant, which produces 2 MMM (billion) lb/yr of vinyl chloride and is located about 100 miles away, might be expanded to produce the required amount, which would be shipped by truck or rail in tank car quantities. In this case, the design team projects the purchase price and designs storage facilities. This might be the simplest solution to provide the monomer required to expand the local PVC plant.

Alternative 2. Purchase and ship, by pipeline from a nearby plant, chlorine from the electrolysis of NaCl solution. React the chlorine with in-house ethylene to produce the monomer and HCl as a byproduct.

Alternative 3. Because the existing company petrochemical complex produces HCl as a byproduct in many processes (e.g., in chloroform and carbon tetrachloride manufacture) at a depressed price because large quantities are produced, HCl is normally available at low prices. Reactions of HCl with acetylene, or ethylene and oxygen, could produce 1,2-dichloroethane, an intermediate that can be cracked to produce vinyl chloride.

Alternative 4. Design an electrolysis plant to produce chlorine. One possibility is to electrolyze the HCl, available from within the petrochemical complex, to obtain H_2 and Cl_2. React chlorine, according to alternative 2. Elsewhere in the petrochemical complex, react hydrogen with nitrogen to form ammonia, or with CO to produce methanol.

These are typical of the alternatives that might be selected from a large number of ideas, and which serve as a base on which to begin the engineering of a product or a process.

In this example, only the production of the monomer is considered, with a focus on alternatives 2 and 3. In the *concept* stage, the objective is to create several promising flowsheets, as candidate solutions, to be inserted later into Figure 4.3. In addition to data from the chemistry laboratory, two patents (Benedict, 1960; B.F. Goodrich Co., 1963) play a key role in process synthesis. These were located by the design team during their literature search and entered into the preliminary database. When appropriate, they will be referred to in connection with the synthesis steps that follow.

Step 1 Eliminate Differences in Molecular Type: For the manufacture of vinyl chloride, data from the chemistry laboratory focus on several promising chemical reactions involving the chemicals shown in Table 4.1. Note that since vinyl chloride has been a commodity chemical for many years, these chemicals and the reactions involving them are well known. For newer substances, the design team often begins to carry out process synthesis as the data are emerging from the laboratory. The challenge, in these cases, is to guide the chemists away from those reaction paths that lead to processes that are costly to build and operate, and to arrive at designs as quickly as possible, in time to capture the market before a competitive process or chemical is developed by another company.

Table 4.1 Chemicals That Participate in Reactions to Produce Vinyl Chloride

Chemical	Molecular weight	Chemical formula	Chemical structure
Acetylene	26.04	C_2H_2	H−C≡C−H
Chlorine	70.91	Cl_2	Cl−Cl
1,2-Dichloroethane	98.96	$C_2H_4Cl_2$	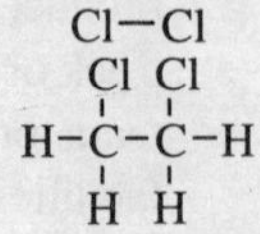
Ethylene	28.05	C_2H_4	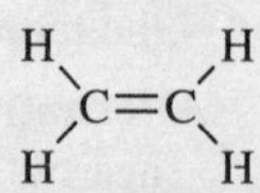
Hydrogen chloride	36.46	HCl	H−Cl
Vinyl chloride	62.50	C_2H_3Cl	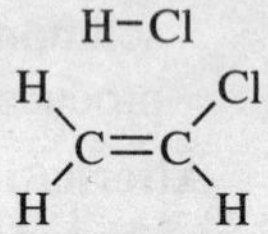

Returning to the manufacture of vinyl chloride, the principal reaction pathways are as follows.

1. ***Direct Chlorination of Ethylene***

$$C_2H_4 + Cl_2 \rightarrow C_2H_3Cl + HCl \qquad (4.1)$$

This reaction appears to be an attractive solution to design alternative 2. It occurs spontaneously at a few hundred degrees Celsius, but unfortunately does not give a high yield of vinyl chloride without simultaneously producing large amounts of byproducts such as dichloroethylene. Another disadvantage is that one of the two atoms of expensive chlorine is consumed to produce the byproduct hydrogen chloride, which may not be sold easily.

2. ***Hydrochlorination of Acetylene***

$$C_2H_2 + HCl \rightarrow C_2H_3Cl \qquad (4.2)$$

This exothermic reaction is a potential solution for the concept denoted as alternative 3. It provides a good conversion (98%) of acetylene to vinyl chloride at 150°C in the presence of mercuric chloride ($HgCl_2$) catalyst impregnated in activated carbon at atmospheric pressure. These are fairly moderate reaction conditions, and hence, this reaction deserves further study.

3. ***Thermal Cracking of Dichloroethane from Chlorination of Ethylene***

$$C_2H_4 + Cl_2 \rightarrow C_2H_4Cl_2 \qquad (4.3)$$
$$C_2H_4Cl_2 \rightarrow C_2H_3Cl + HCl \qquad (4.4)$$
$$\overline{C_2H_4 + Cl_2 \rightarrow C_2H_3Cl + HCl} \quad \text{(overall)} \; (4.1)$$

The sum of reactions (4.3) and (4.4) is equal to reaction (4.1). This two-step reaction path has the advantage that the conversion of ethylene to 1,2-dichloroethane in exothermic reaction (4.3) is about 98% at 90°C and 1 atm with a Friedel–Crafts catalyst such as ferric chloride ($FeCl_3$). Then, the dichloroethane intermediate is converted to vinyl chloride by thermal cracking according to the endothermic reaction (4.4), which occurs spontaneously at 500°C and has conversions as high as 65%. The overall reaction presumes that the unreacted dichloroethane is recovered entirely from the vinyl chloride and hydrogen chloride and recycled. This reaction path has the advantage that it does not produce dichloroethylene in significant quantities, but it shares the disadvantage with reaction path 1 of producing HCl. It deserves further examination as a solution to design alternative 2.

4. *Thermal Cracking of Dichloroethane from Oxychlorination of Ethylene*

$$C_2H_4 + 2HCl + \tfrac{1}{2}O_2 \rightarrow C_2H_4Cl_2 + H_2O \quad (4.5)$$
$$C_2H_4Cl_2 \rightarrow C_2H_3Cl + HCl \quad (4.4)$$
$$C_2H_4 + HCl + \tfrac{1}{2}O_2 \rightarrow C_2H_3Cl + H_2O \quad \text{(overall)} \quad (4.6)$$

In reaction (4.5), which *oxychlorinates* ethylene to produce 1,2-dichloroethane, HCl is the source of chlorine. This highly exothermic reaction achieves a 95% conversion of ethylene to dichloroethane at 250°C in the presence of cupric chloride ($CuCl_2$) catalyst, and is an excellent candidate when the cost of HCl is low. As in reaction path 3, the dichloroethane is cracked to vinyl chloride in a pyrolysis step. This reaction path should be considered also as a solution for design alternative 3.

5. *Balanced Process for Chlorination of Ethylene*

$$C_2H_4 + Cl_2 \rightarrow C_2H_4Cl_2 \quad (4.3)$$
$$C_2H_4 + 2HCl + \tfrac{1}{2}O_2 \rightarrow C_2H_4Cl_2 + H_2O \quad (4.5)$$
$$2C_2H_4Cl_2 \rightarrow 2C_2H_3Cl + 2HCl \quad (4.4)$$
$$2C_2H_4 + Cl_2 + \tfrac{1}{2}O_2 \rightarrow 2C_2H_3Cl + H_2O \quad \text{(overall)} \quad (4.7)$$

This reaction path combines paths 3 and 4. It has the advantage of converting both atoms of the chlorine molecule to vinyl chloride. All of the HCl produced in the pyrolysis reaction is consumed in the oxychlorination reaction. Indeed, it is a fine candidate for the solution of design alternative 2.

Given this information, it seems clear that the design team would reject reaction path 1 on the basis of its low *selectivity* with respect to the competing reactions (not shown) that produce undesirable byproducts. This leaves the other reaction paths as potentially attractive to be screened on the basis of the chemical prices. Although it is too early to estimate the cost of the equipment and its operation, before the remaining process operations are in place, the design team normally computes the *gross profit* (i.e., the profit excluding the costs of equipment and the operating costs) for each reaction path and uses it as a vehicle for screening out those that cannot be profitable. To illustrate this process for the production of vinyl chloride, Table 4.2 provides a representative set of prices for the principal chemicals, obtained from a source such as the ICIS Business Americas (formerly the *Chemical Marketing Reporter*), as discussed earlier. The gross profit is computed as the income derived from the sales of the products and byproducts less the cost of the raw materials. It is computed by first converting to a mass basis, as illustrated for reaction path 3:

Table 4.2 Assumed Cost of Chemicals Purchased or Sold in Bulk Quantities

Chemical	Cost (cents/lb)
Ethylene	30
Acetylene	80
Chlorine	18
Vinyl chloride	35
Hydrogen chloride	25
Water	0
Oxygen (air)	0

	C_2H_4	$+ Cl_2$	$= C_2H_3Cl$	$+ HCl$
lbmol	1	1	1	1
Molecular weight	28.05	70.91	62.50	36.46
lb	28.05	70.91	62.50	36.46
lb/lb of vinyl chloride	0.449	1.134	1	0.583
cents/lb	30	18	35	25

Then, the gross profit is $35(1) + 25(0.583) - 30(0.449) - 18(1.134) = 15.69$ cents/lb of vinyl chloride. Similar estimates are made for the overall reaction in each of the reaction paths, it being assumed that complete conversion can be achieved without any side reactions (not shown), with the results shown in Table 4.3.

Even without the capital costs (for construction of the plant, purchase of land, etc.) and the operating costs (for labor, steam, electricity, etc.), the gross profit for reaction path 2 is negative, whereas

Table 4.3 Gross Profit for Production of Vinyl Chloride (Based on Chemical Prices in Table 4.2)

Reaction path	Overall reaction	Gross profit (cents/lb of vinyl chloride)
2	$C_2H_2 + HCl = C_2H_3Cl$	−16.00
3	$C_2H_4 + Cl_2 = C_2H_3Cl + HCl$	15.69
4	$C_2H_4 + HCl + \frac{1}{2}O_2 = C_2H_3Cl + H_2O$	6.96
5	$2C_2H_4 + Cl_2 + \frac{1}{2}O_2 = 2C_2H_3Cl + H_2O$	11.32

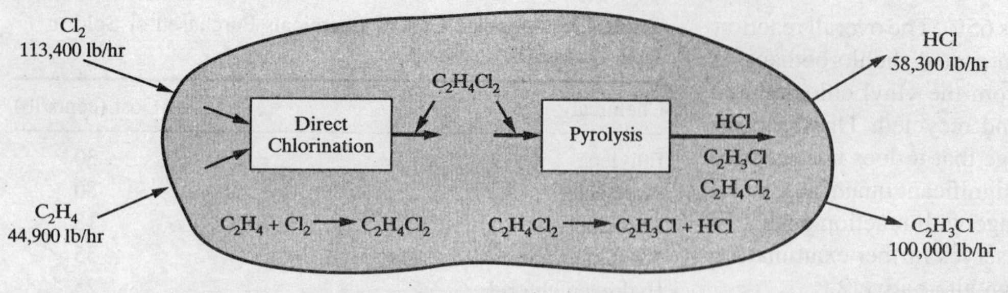

Figure 4.4 Reaction operations for the thermal cracking of dichloroethane from the chlorination of ethylene (reaction path 3).

the gross profits for the other reaction paths are positive. This is principally because acetylene is very expensive relative to ethylene. The fairly high price of HCl also contributes to the inevitable conclusion that vinyl chloride cannot be produced profitably using this reaction path. It should be noted that the price of HCl is often very sensitive to its availability in a petrochemical complex. In some situations, it may be available in large quantities as a byproduct from another process at very low cost. At a much lower price, reaction path 2 would have a positive gross profit, but would not be worthy of further consideration when compared with the three reaction paths involving ethylene. Turning to these paths, all have sufficiently positive gross profits, and hence are worthy of further consideration. It is noted that the price of HCl strongly influences the gross profits of reaction paths 3 and 4, with the gross profit of reaction path 5 midway between the two. Before proceeding with the synthesis, the design team would be advised to examine how the gross profits vary with the price of HCl.

Figure 4.4 shows the first step toward creating a process flowsheet for reaction path 3. Each reaction operation is positioned with arrows representing its feed and product chemicals. The *sources* and *sinks* are not shown because they depend on the *distribution of chemicals*, the next step in process synthesis. The flow rates of the external sources and sinks are computed assuming that the ethylene and chlorine sources are converted completely to the vinyl chloride and hydrogen chloride sinks. Here, a key decision is necessary to set the scale of the process, that is, the production rate at capacity. In this case, a capacity of 100,000 lb/hr (~800 million lb/yr, assuming operation 330 days annually—an operating factor of 0.904) is dictated by the opportunity presented above. Given this flow rate for the product (principal sink for the process), the flow rates of the HCl sink and the raw-material sources can be computed by assuming that the raw materials are converted to the products according to the overall reaction. Any unreacted raw materials are separated from the reaction products and recycled. By material balance, the results in Figure 4.4 are obtained, where each flow rate in lbmol/hr is 1,600.

Similar flowsheets, containing the reaction operations for reaction paths 4 and 5, would be prepared to complete step 1 of the synthesis. These are represented in the synthesis tree in Figure 4.9, which will be discussed after all of the synthesis steps have been completed. Note that their flowsheets are not included here due to space limitations, but are requested in Exercise 4.5 at the end of the chapter. As the next steps in the synthesis are completed for reaction path 3, keep in mind that they would be carried out for the other reaction paths as well. Note, also, that only the most promising flowsheets are developed in detail, usually by an expanded design team or, in some cases, by a competitive design team.

Step 2 **Distribute the Chemicals:** In step 2, where possible, the sources and sinks for each of the chemical species in Figure 4.4 are matched so that the total mass flow into a reactor equals the total mass flow out. This often entails the introduction of mixing operations to eliminate differences in flow rates when a single sink is supplied by two or more sources. In other cases, a single source is divided among several sinks. To achieve the distribution of chemicals in Figure 4.5, the ethylene and chlorine sources are matched with their sinks into the chlorination reactor. It is assumed that ethylene and chlorine enter the reactor in the stoichiometric ratio of 1:1 as in reaction (4.3). Because the raw materials are in this ratio, no differences exist between the flow rates of the sources and sinks, and hence, no mixers are needed. Flow rates of 113,400 lb/hr of chlorine and 44,900 lb/hr of ethylene produce 158,300 lb/hr of dichloroethane. When it is desired to have an excess of one chemical in relation to the other so as to completely consume the other chemical, which may be toxic or very expensive (e.g., Cl_2), the other raw material (e.g., C_2H_4) is mixed with recycle and fed to the reactor in excess. For example, if the reactor effluent contains unreacted C_2H_4, it is separated from the dichloroethane product and recycled to the reaction operation. Note that the recycle is the source of the excess chemical, and the flow rate of the external source of C_2H_4 for a given production rate of dichloroethane is unaffected. This alternative distribution of chemicals is discussed

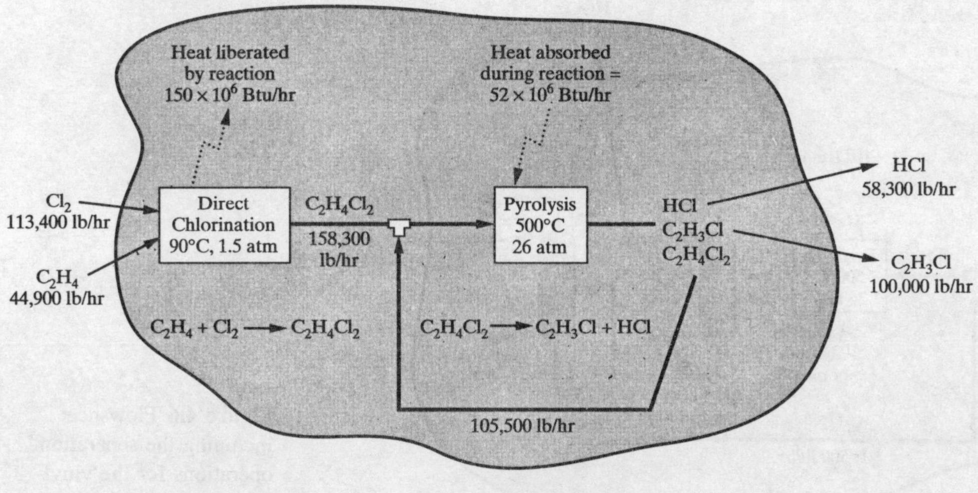

Figure 4.5 Flowsheet showing a distribution of chemicals for thermal cracking of dichloroethane from chlorination of ethylene (reaction path 3).

further in Section 6.3 and illustrated in Figure 6.1. Returning to the distribution of chemicals in Figure 4.5, note that, at reactor conditions of 90°C and 1.5 atm, experimental data indicate that 98% of the ethylene is converted to dichloroethane, with the remainder converted to unwanted byproducts such as trichloroethane. This loss of yield of main product and small fraction of byproduct is neglected at this stage in the synthesis.

Next, the dichloroethane source from the chlorination operation is sent to its sink in the pyrolysis operation, which operates at 500°C. Here only 60% of the dichloroethane is converted to vinyl chloride with a byproduct of HCl, according to reaction (4.4). This conversion is within the 65% conversion claimed in the patent. To satisfy the overall material balance, 158,300 lb/hr of dichloroethane must produce 100,000 lb/hr of vinyl chloride and 58,300 lb/hr of HCl. But a 60% conversion only produces 60,000 lb/hr of vinyl chloride. The additional dichloroethane needed is computed by mass balance to equal $[(1-0.6)/0.6] \times 158,300$ or 105,500 lb/hr. Its source is a recycle stream from the separation of vinyl chloride from unreacted dichloroethane, from a mixing operation inserted to combine the two sources, to give a total of 263,800 lb/hr. The effluent stream from the pyrolysis operation is the source for the vinyl-chloride product, the HCl byproduct, and the dichloroethane recycle. To enable these chemicals to be supplied to their sinks, one or more separation operations are needed and are addressed in the next synthesis step.

Figure 4.5 also shows the heats of reaction for the two reaction steps. These are computed at the temperatures and pressures of the reaction operations from heats of formation and heat capacities as a function of temperature. There are many sources of this data, especially the process simulators that are discussed in Chapter 5. When a simulator, such as ASPEN PLUS, is used, it is convenient to define each of the reaction operations and to perform an energy balance at reactor conditions. The simulators report the rate at which heat must be transferred to or from the reactor to achieve exit conditions from given inlet conditions or, if operated adiabatically, the exit conditions for no heat transfer, as discussed on the multimedia modules, which can be downloaded from the Wiley Web site associated with this book—follow the paths, *ASPEN → Chemical Reactors* and *HYSYS → Chemical Reactors*. For reaction path 3, the chlorination operation provides a large source of energy, 150 million Btu/hr, but at a low temperature, 90°C, whereas the pyrolysis operation requires much less energy, 52 million Btu/hr, at an elevated temperature, 500°C. Since this heat source cannot be used to provide the energy for pyrolysis, other uses for this energy should be sought as the synthesis proceeds. These and other sources and sinks for energy are considered during task integration in step 5.

www.wiley.com/go/global/seider

As for the pressure levels in the reaction operations, 1.5 atm is selected for the chlorination reaction to prevent the leakage of air into the reactor to be installed in the task-integration step. At atmospheric pressure, air might leak into the reactor and build up in sufficiently large concentrations to exceed the flammability limit. For the pyrolysis operation, 26 atm is recommended by the B.F. Goodrich patent (1963) without any justification. Since the reaction is irreversible, the elevated pressure does not adversely affect the conversion. Most likely, the patent recommends this pressure to increase the rate of reaction and, thus, reduce the size of the pyrolysis furnace, although the tube walls must be thick and many precautions are necessary for operation at elevated pressures. The pressure level is also an important consideration in selecting the separation operations, as will be discussed in the next synthesis step.

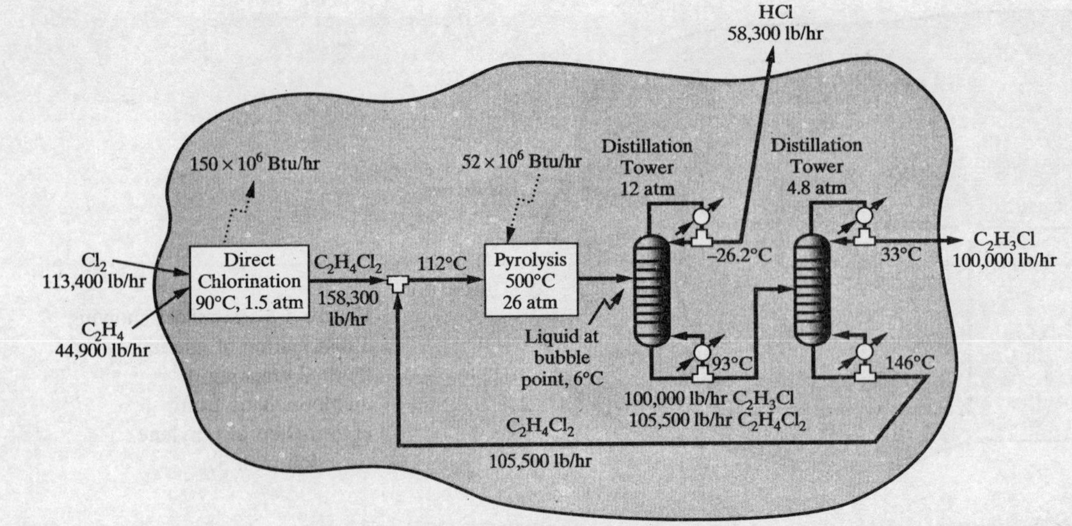

Figure 4.6 Flowsheet including the separation operations for the vinyl-chloride process.

Referring to Figure 4.9, at the "Distribution of Chemicals" level, two branches have been added to the synthesis tree to represent the two distributions in connection with reaction path 3. Each of these branches represents a different partially completed flowsheet, that is, Figures 4.5 and 6.1. Other distributions arise in connection with reaction paths 4 and 5. These are represented using dashed lines in the synthesis tree.

Step 3 Eliminate Differences in Composition: As mentioned earlier, for each distribution of chemicals, the needs for separation become obvious. In Figure 4.5, for example, it is clear that the pure effluent from the chlorination reaction operation needs no separation, but the effluent from the pyrolysis operation is a mixture that needs to be separated into nearly pure species. Here, the source of the three species in the effluent is at a composition far different from the compositions of the three sinks: vinyl chloride product, HCl byproduct, and the dichloroethane for recycle. To eliminate these composition differences, one or more separation operations are needed.

One possibility is shown in Figure 4.6, in which two distillation towers in series are inserted into the flowsheet. Distillation is possible because of the large volatility differences among the three species. This can be seen by examining the boiling points in Table 4.4, which can be obtained from vapor pressure data in the preliminary database, or from a process simulator. In the first column, HCl is separated from the two organic chemicals. In the second column, vinyl chloride is separated from dichloroethane. At 1 atm, the boiling point of HCl is very low, −84.8°C, and hence if HCl were recovered at 1 atm as the distillate of the first tower, very costly refrigeration would be necessary to condense the reflux stream. At 26 atm (the pyrolysis reaction pressure), HCl boils at 0°C, and much less costly refrigeration could be used. The B.F. Goodrich patent recommends operation at 12 atm without any justification. At this pressure, HCl boils at −26.2°C and the bottoms product, comprised of vinyl chloride and dichloroethane with trace quantities of HCl, has a bubble point of 93°C, which can be calculated by a process simulator. The bottoms product at this reduced temperature and pressure is farther away from the critical points of vinyl chloride–dichloroethane mixtures at the bottom of the distillation column. It is likely, therefore, that B.F. Goodrich selected this lower pressure to avoid operation in the critical region where the vapor and liquid phases approach each other and are much more difficult to disengage (i.e., have small flooding velocities and require very large diameters and tray spacings). Furthermore, low-pressure steam is adequate for the reboiler. When this distillation tower is inserted into the flowsheet, the conditions of its feed stream, or sink, need to be identified. If the feed is a saturated liquid, the temperature is 6°C at 12 atm, with a mild refrigerant required for cooling. A preferable feed temperature would be 35°C or higher, which could be achieved by completing the cooling and partial condensation of the pyrolysis reactor effluent with cooling water, but the introduction of vapor into the column would increase the refrigeration load of the condenser at

Table 4.4 Boiling Points and Critical Constants

Chemical	Normal boiling point (1 atm, °C)	Boiling point (°C) 4.8 atm	12 atm	26 atm	Critical constants T_c(°C)	P_c(atm)
HCl	−84.8	−51.7	−26.2	0	51.4	82.1
C_2H_3Cl	−13.8	33.1	70.5	110	159	56
$C_2H_4Cl_2$	83.7	146	193	242	250	50

−26.2°C. Upon making this specification, key differences (temperature, pressure, and phase) appear between the effluent from the pyrolysis operation and the feed to the distillation column. These are eliminated in the next synthesis step by inserting temperature and pressure change operations, with each temperature specification leading to a somewhat different flowsheet.

After the first distillation operation is inserted into the flowsheet, the second follows naturally. The bottoms from the HCl-removal tower is separated into nearly pure species in the second tower, which is specified at 4.8 atm, as recommended by the B.F. Goodrich patent. Under these conditions, the distillate (nearly pure vinyl chloride) boils at 33°C and can be condensed with inexpensive cooling water, which is available at 25°C. The bottoms product boils at 146°C, and hence, the vapor boilup can be generated with medium-pressure steam, which is widely available in petrochemical complexes.

Alternative separation operations can be inserted into Figure 4.5. When distillation is used, it is also possible to recover the least volatile species, dichloroethane, from the first column, and separate HCl from vinyl chloride in the second column. Yet another possibility is to use a single column with a side stream that is concentrated in the vinyl-chloride product. Absorption with water, at atmospheric pressure, can be used to remove HCl. The resulting vapor stream, containing vinyl chloride and dichloroethane, could be dried by adsorption and separated using distillation. With so many alternatives possible, the process designer needs time or help to select the most promising separation operations. As mentioned previously, this topic is considered in detail in Chapter 8.

Furthermore, as before, the synthesis tree in Figure 4.9 is augmented. In this case, the new branches represent the different flowsheets for the alternative separation operations. Clearly, as each step of the synthesis is completed, the tree represents many more possible flowsheets.

Step 4 Eliminate Differences in Temperature, Pressure, and Phase: When the reaction and separation operations are positioned, the states of their feed and product streams are selected. This is accomplished usually by adjusting the temperature and pressure levels to achieve the desired reaction conversions and separation factors. Subsequently, after the flowsheets have been created, these are often adjusted toward the economic optimum, often using the optimizers in the process simulators discussed in Chapter 24. In this synthesis step, however, the states are assumed to be fixed and operations are inserted to eliminate the temperature, pressure, and phase differences between the feed sources, the product sinks, and the reaction and separation operations.

Figure 4.7 shows one possible flowsheet. It can be seen that liquid dichloroethane from the recycle mixer at 112°C and 1.5 atm undergoes the following operations:

1. Its pressure is increased to 26 atm.
2. Its temperature is raised to the boiling point, which is 242°C at 26 atm.

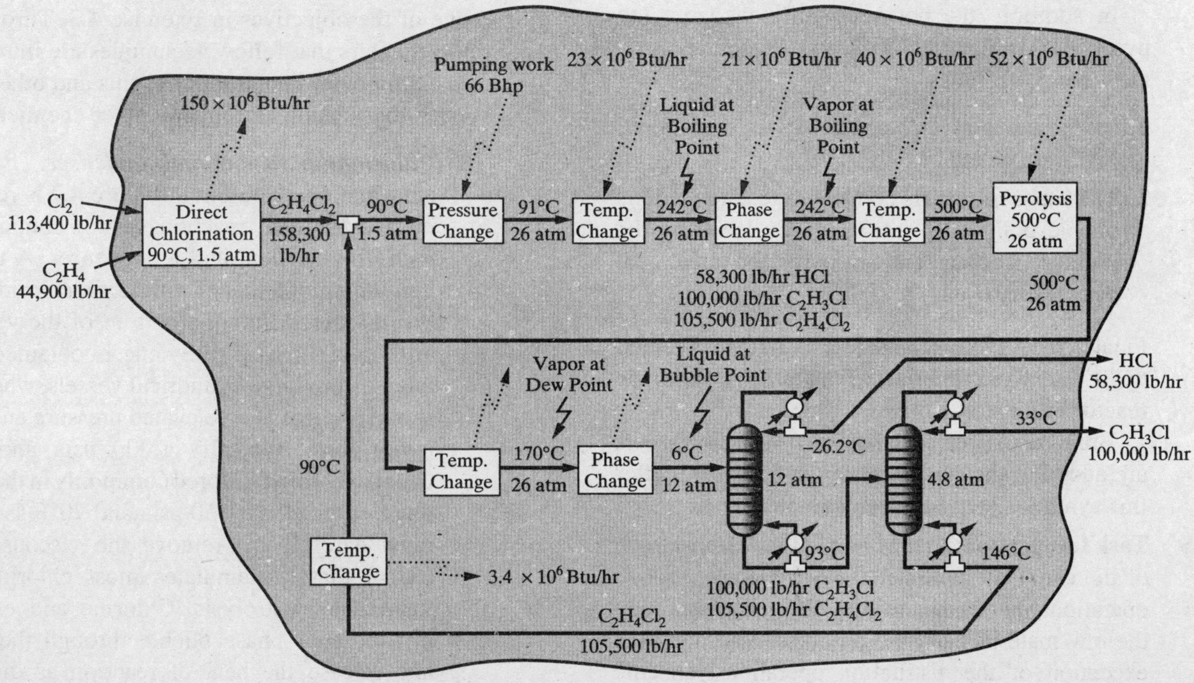

Figure 4.7 Flowsheet with temperature-, pressure-, and phase-change operations in the vinyl-chloride process.

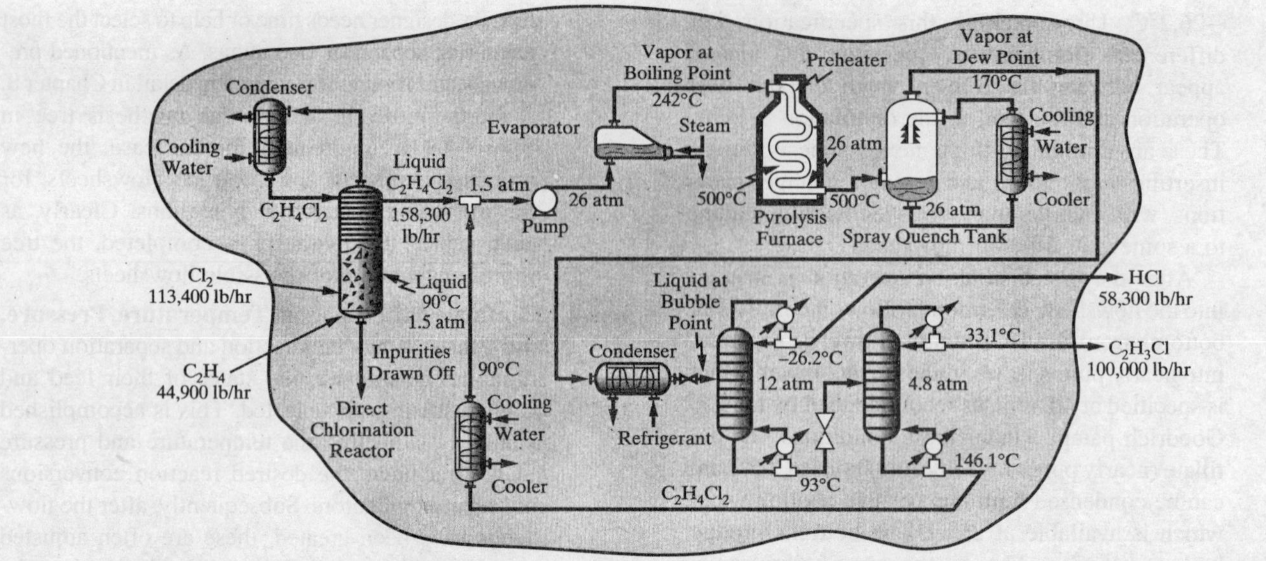

Figure 4.8 Flowsheet showing task integration for the vinyl-chloride process.

3. Dichloroethane liquid is vaporized at 242°C.
4. Its temperature is raised to the pyrolysis temperature, 500°C.

Note that an alternative flowsheet would place operations 1 and 2 after operation 3. However, this is very uneconomical, as the cost of compressing a vapor is far greater than the cost of pumping a liquid because the molar volume of a vapor is so much greater than that of a liquid (typically, a factor of 100 times greater). For a more complete discussion of this observation, which is just one of many design heuristics or rules of thumb, see Section 6.7.

In addition, the hot vapor effluent from the pyrolysis operation (at 500°C and 26 atm) is operated upon as follows:

1. Its temperature is lowered to its dew point, 170°C at 26 atm.
2. The vapor mixture is condensed to a liquid at its bubble point, 6°C at 12 atm, by lowering the pressure, cooling, and removing the latent heat of condensation.

Finally, the dichloroethane recycle stream is cooled to 90°C to avoid vaporization when mixed with the reactor effluent at 1.5 atm.

Branches to represent the two new flowsheets are added to the synthesis tree in Figure 4.9 after this synthesis step has been completed.

Step 5 **Task Integration:** At the completion of step 4, each of the candidate flowsheets has a complete set of operations that eliminates the differences between the raw materials and the products. Still, with the exception of the distillation operations, specific equipment items are not shown. The selection of the processing units, often referred to as unit operations, in which one or more of the basic operations are carried out, is known as *task integration.* To assist in this selection, the reader is referred to *Chemical Process Equipment* (Walas, 1988).

Figure 4.8 shows one example of task integration for the vinyl-chloride process. At this stage in process synthesis, it is common to make the most obvious combinations of operations, leaving many possibilities to be considered when the flowsheet is sufficiently promising to undertake the preparation of a base-case design. As you examine this flowsheet, with the descriptions of the process units that follow, see if you can suggest improvements. This is one of the objectives in Exercise 4.3. Throughout the chapters that follow, techniques are introduced to obtain better integration for this and other processes that manufacture many other chemicals.

1. ***Chlorination reactor and condenser.*** The direct chlorination operation in Figure 4.7 is replaced by a cylindrical reaction vessel, containing a rectifying section, and a condenser. A pool of liquid dichloroethane, with ferric chloride catalyst dissolved, fills the bottom of the vessel at 90°C and 1.5 atm. Ethylene is obtained commonly from large cylindrical vessels, where it is stored as a gas at an elevated pressure and room temperature, typically 1,000 psia and 70°F. Chlorine, which is stored commonly in the liquid phase, typically at 150 psia and 70°F, is evaporated carefully to remove the viscous liquid (taffy) that contaminates most chlorine produced by electrolysis. Chlorine and ethylene in the vapor phase bubble through the liquid and release the heat of reaction as dichloroethane is produced. This heat causes the

dichloroethane to vaporize and rise up the rectifying section into the condenser, where it is condensed with cooling water. Note that heat is needed to drive the reboiler in the first distillation column at 93°C, but the heat of reaction cannot be used for this purpose unless the temperature levels are adjusted. How can this be accomplished?

Most of the condensate is mixed with the effluent from the recycle cooler to be processed in the pyrolysis loop. However, a portion is refluxed to the rectifying section of the column, which has several trays, to recover any of the less volatile species (e.g., trichloroethane) that may have vaporized. These *heavies* accumulate at the bottom of the liquid pool and are removed periodically as impurities.

2. ***Pump.*** Since the pressure-change operation involves a liquid, it is accomplished by a pump, which requires only 66 Bhp, assuming an 80% efficiency. The enthalpy change in the pump is very small and the temperature does not change by more than 1°C.
3. ***Evaporator.*** This unit, in the form of a large kettle, with a tube bundle inserted across the bottom, performs the temperature- and phase-change operations. Saturated steam that passes through the tubes condenses as the dichloroethane liquid is heated to its boiling point and vaporized. The large vapor space is provided to enable liquid droplets, entrained in the vapor, to coalesce and drop back into the liquid pool, that is, to disengage from the vapor that proceeds to the pyrolysis furnace.
4. ***Pyrolysis furnace.*** This unit also performs two operations: It preheats the vapor to its reaction temperature, 500°C, and it carries out the pyrolysis reaction. The unit is constructed of refractory brick, with natural gas-fired heaters, and a large bundle of Nickel, Monel, or Inconel tubes, within which the reaction occurs. The tube bundle enters the coolest part of the furnace, the so-called *economizer* at the top, where the preheating occurs.
5. ***Spray quench tank and cooler.*** The quench tank is designed to rapidly quench the pyrolysis effluent to avoid carbon deposition in a heat exchanger. Cold liquid (principally dichloroethane) is showered over the hot gases, cooling them to their dew point, 170°C. As the gases cool, heat is transferred to the liquid and removed in the adjacent cooler. The warm liquid, from the pool at the base of the quench vessel, is circulated to the cooler, where it is contacted with cooling water. Any carbon that deposits in the quench vessel settles to the bottom and is bled off periodically. Unfortunately, this carbon deposition, as well as the corrosive HCl, is anticipated to prevent the use of the hot effluent gases in the tubes of the evaporator, which would have to be serviced often to remove carbon and replace corroded tubes. Note that coke formation in the pyrolysis products is discussed by Borsa et al. (2001). Consequently, large amounts of heat are transferred to cooling water, and the fuel requirements for the process are high. As noted later in the section on pilot-plant testing, the design team is likely to measure the rate of carbon deposition and, if it is not very high, may decide to implement a design with a feed/product heat exchanger.
6. ***Condenser.*** To produce a saturated liquid at 6°C, the phase-change operation is carried out by a condenser that transfers heat to a mild refrigerant. Then the pressure is lowered to 12 atm across a valve.
7. ***Recycle cooler.*** To prevent vapor from entering the pump when the recycle stream is mixed with effluent from the direct chlorination reactor, the recycle stream is cooled to 90°C (below the boiling point of chloroethane at 1.5 atm) using cooling water.

This completes the task integration in Figure 4.8. Can you suggest ways to reduce the need for fuel and hot utilities such as steam?

Synthesis Tree

Throughout the synthesis of the vinyl-chloride process, branches have been added to the synthesis tree in Figure 4.9 to represent the alternative flowsheets being considered. The bold branches trace the development of just one flowsheet as it evolves in Figures 4.3–4.8. Clearly, there are many alternative flowsheets, and the challenge in process synthesis is to find ways to eliminate whole sections of the tree without doing much analysis. By eliminating reaction paths 1 and 2, as much as 40% of the tree is eliminated in the first synthesis step. Similar screening techniques are applied by the design team in every step, as discussed throughout this book.

To satisfy the objective of generating the most promising flowsheets, care must be taken to include sufficient analysis in each synthesis step to check that each step does not lead to a less profitable flowsheet or exclude the most profitable flowsheet prematurely. For this reason, it is common practice in industry to mix these synthesis steps with analysis using the simulators introduced in the next chapter.

Heuristics

It is important to keep in mind that, when carrying out the steps in preliminary process synthesis, the resulting synthesis

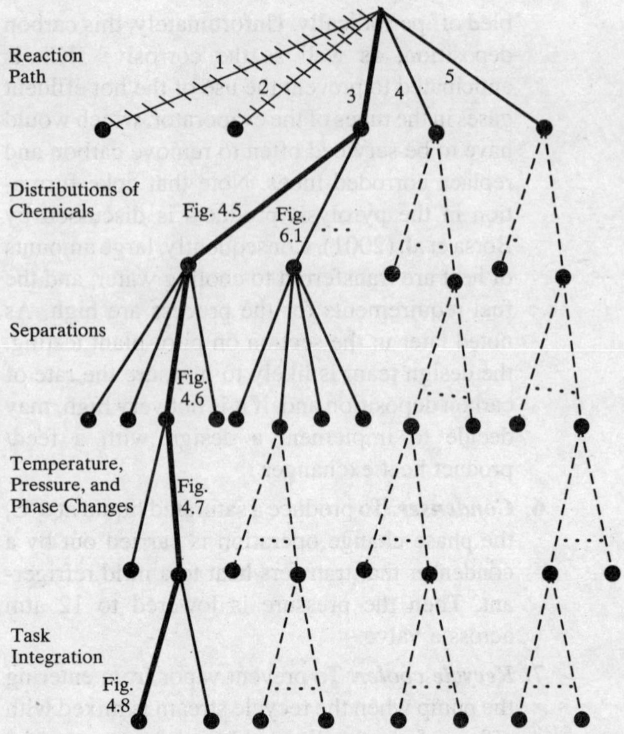

Figure 4.9 Inverted synthesis tree for the production of vinyl chloride.

tree is closely related to any heuristics or rules of thumb used by the design team. In the vinyl-chloride example, emphasis was placed on the synthesis steps, and not on the use of heuristics by the design team. An exception is the heuristic that it is cheaper to pump a liquid than compress a gas. Heuristics are covered more thoroughly in Chapter 6, where it will become clear that the synthesis tree can be improved significantly. See also *Conceptual Design of Chemical Processes* (Douglas, 1988) and Walas (1988), where many heuristics are presented.

Example of Process Synthesis: Manufacture of Tissue Plasminogen Activator (tPA)

In the manufacture of pharmaceuticals, consider the possible production of plasminogen activators, which are powerful enzymes that trigger the proteolytic (breaking down of proteins to form simpler substances) degradation of blood clots that cause strokes and heart attacks. Since the mid-1980s, Genentech, a U.S. company, has manufactured tissue plasminogen activator (tPA), which they sold for $2,000 per 100-mg dose in the early 2000s, with annual sales of $300 MM/yr (MM in American engineering units is thousand-thousand or 1 million). Given that their patent was set to expire in 2003, Genentech developed a next-generation, Food and Drug Administration (FDA)-approved, plasminogen activator called TNK-tPA, which is easier and safer for clinicians to use. With a rapidly growing market, the question arose as to whether an opportunity existed for another company to manufacture a generic (i.e., without a brand name) form of tPA that could compete favorably with TNK-tPA.

To examine this possibility, a design team was formulated. It prepared a project charter and began the *concept* stage of the Stage-Gate™ Product-Development Process. Two potential alternatives were identified, including:

Alternative 1. While a generic form of tPA may not compete well against TNK-tPA in the United States, it may be possible to market a low-cost generic tPA in foreign markets, where urokinase and streptokinase are low-cost alternatives, which sell for only $200/dose, but are associated with increased bleeding risks. Market analysis suggests that a maximum production rate of 80 kg/yr would be appropriate over the next five years.

Alternative 2. Given the possibility that lower health care reimbursements are received by hospitals in the United States, it may be reasonable to develop a similar process that competes favorably with TNK-tPA in the United States.

Other promising alternatives were likely to arise, often initiated by successes in a research laboratory.

Tissue plasminogen activator (tPA) is a recombinant therapeutic protein comprised of 562 amino acids, as shown schematically in Figure 4.10. Note that tPA is produced using a recombinant cell, which results from a recombination of genes. To eliminate blood clots, tPA activates plasminogen to plasmin, an enzyme, which dissolves fibrin formations that hold blood clots in place. In this way, blood flow is reestablished once the clot blockage dissolves, an important effect for patients suffering from a heart attack (microcardial infarction) or stroke. This example shows the steps in synthesizing a process to address the challenges posed by the opportunity posed in alternative 1: that is, to manufacture less expensive forms of tPA that can be sold for $200 per 100-mg dose. Note that it leads to a batch process involving many small process units that must be scheduled for the manufacture of tPA, rather than a large-scale continuous process as for the manufacture of vinyl chloride.

Stated differently, based upon extensive research in the biochemistry laboratory, the tPA gene was isolated from human melanoma cells, and the process synthesis problem in Figure 4.11 created. As shown, tPA is produced using mammalian [e.g., Chinese hamster ovary (CHO)] cells that have tPA-DNA as part of their genetic contents (genome). In an aerobic bioreaction operation, the tPA-CHO cells grow in a nutrient media, HyQ PF-CHO—Hyclone media, a blend of nutrients, salts (including $NaHCO_3$), amino acids, insulin, growth factors, and transferrin, specifically for growth of CHO cells. Other ingredients include sterilized water, air, and CO_2. In addition to tPA, endotoxins may be a contaminant of the product, which must be removed because they elicit a variety of inflammatory responses in animals. Other byproducts include cell debris, wastewater, and gas emissions, especially N_2 from air, unconsumed O_2 from air, and CO_2, which regulates the pH. An important source of data, in

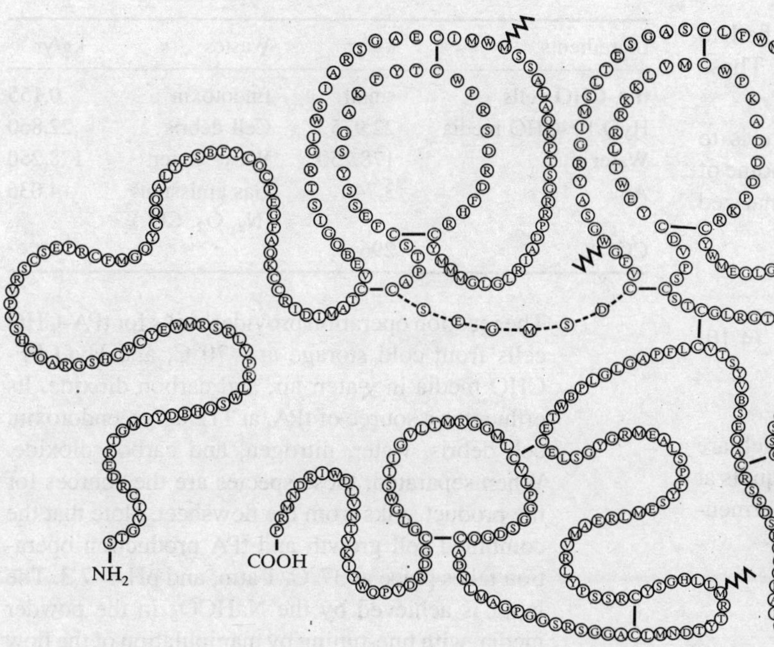

Figure 4.10 Schematic of tissue plasminogen activator (tPA).

addition to that taken in the biochemistry laboratory, is a U.S. patent, filed by Genentech (Goeddel et al., 1988), which provides considerable qualitative and quantitative information. See also the design report by Audette et al. (2000).

Step 1 Eliminate Differences in Molecular Type: In the manufacture of a macromolecule like tPA through cell growth, a complex array of chemical reactions is often approximated by global reactions that are understood far less than the well-defined reactions for the manufacture of a simple monomer, like vinyl chloride. In terms of global reactions to manufacture tPA, two principal reaction paths are provided by the biochemist, as follows.

1. ***Mammalian Cells.*** Into CHO cells, the tPA-DNA sequence must be inserted and expressed.

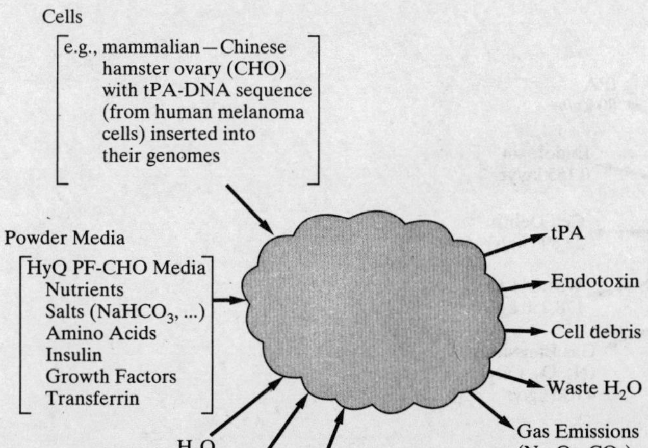

Figure 4.11 Process synthesis problem.

The resulting tPA-CHO cells are specially selected CHO cells with many copies of tPA-DNA inserted into their genomes, and which secrete high levels of tPA. This tPA-DNA insertion step is summarized in the reaction:

$$\begin{aligned} \text{tPA} - \text{DNA sequence} + \text{CHO cells} \\ \rightarrow \text{selected high-expressing tPA} \\ - \text{CHO cells} \end{aligned} \quad (4.8)$$

The product of this "catalyst preparation" is a master stock of tPA-CHO cells, which are prepared in the laboratory and stored in 1-mL aliquots at −70°C to be used as inoculum for the bioreaction:

$$\begin{aligned} \text{tPA} - \text{CHO cells} + \text{HyQ PF} - \text{CHO media} \\ + O_2 \rightarrow \text{Increased cell numbers} \end{aligned} \quad (4.9)$$

As the cells grow in this aerobic cultivation at a rate of 0.39×10^6 cell/(ml-day), oxygen from air is consumed at the rate of 0.2×10^{-12} mol O_2/(cell-hr), and tPA is produced at the rate of 50 pico gram tPA/(cell-day). The latter is secreted gradually into the liquid media solution. Note that reaction (4.8) is carried out once during the research and development phase. Initially, 1–10 mg of tPA-DNA are added to 10^6 cells to produce a few tPA-CHO cells in many unmodified CHO cells. After careful selection, one tPA-CHO cell (the "founder" cell) is selected and amplified to yield about 10^6 cells/mL in 10–100 L. These cells are frozen in aliquots.

2. ***Bacterial Cells.*** A promising alternative is to insert the tPA-DNA sequence into the genome of *Escherichia coli* (*E. coli*) cells, as summarized by the reaction:

$$\begin{aligned} &\text{tPA} - \text{DNA sequence} + E.\,coli\ \text{cells} \\ &\quad\rightarrow \text{selected high-expressing tPA} \\ &\quad - E.\,coli\ \text{cells} \end{aligned} \tag{4.10}$$

Then, the tPA–*E. coli* bacteria cells, which are grown in the laboratory, are frozen in aliquots at −70°C to be used as inoculum for the fermentation reaction:

$$\begin{aligned} &\text{tPA} - E.\,coli\ \text{cells} + \text{powder media} \\ &\quad + O_2 \rightarrow \text{increased cell numbers} \end{aligned} \tag{4.11}$$

A batch fermentation of tPA–*E. coli* can produce 5–50 mg tPA/L-broth at harvest. *Escherichia coli* may require disruption to release tPA, which is then more difficult to separate. Should a process be synthesized based upon this reaction path, reaction rate data from the laboratory will be needed. Unlike CHO cells, *E. coli* cells do not add sugar groups (glycosylation) to tPA. Like CHO cells, tPA–*E. coli* cells are produced and frozen during the research and development phase.

Returning to the reaction path with CHO cells, using laboratory data, the reaction operation is inserted onto the flowsheet, as shown in Figure 4.12. At a production rate of 80 kg/yr of tPA, the lab reports that the following ingredients are consumed and waste products are produced:

Ingredients	kg/yr	Wastes	kg/yr
tPA-CHO cells	small	Endotoxin	0.155
HyQ PF-CHO media	22,975	Cell debris	22,860
Water	178,250	Waste water	178,250
Air	3,740	Gas emissions (N_2, O_2, CO_2)	4,036
CO_2	296		

The reaction operation provides sinks for tPA-CHO cells from cold storage at −70°C, and HyQ PF-CHO media in water, air, and carbon dioxide. Its effluent is a source of tPA, at 112 kg/yr, endotoxin, cell debris, water, nitrogen, and carbon dioxide. When separated, these species are the sources for the product sinks from the flowsheet. Note that the combined cell growth and tPA production operation takes place at 37°C, 1 atm, and pH = 7.3. The latter is achieved by the $NaHCO_3$ in the powder media, with fine-tuning by manipulation of the flow rate of CO_2.

Before accepting a potential reaction path, it is important to examine the gross profit; that is, the difference between the sales revenues and the cost of ingredients. To accomplish this, the sales price of tPA is projected (e.g., \$200 per 100-mg dose), and the costs of ingredients are projected, with estimates often obtained from the suppliers. A typical list of cost estimates is shown in Table 4.5. The cost of water for injection (WFI) is based upon estimates of the cost of sterilizing municipal water (12 cents/kg = 45 cents/gal = \$450/1,000 gal, which is far higher than the typical cost of process water = \$0.50/1.000 gal). The costs of sterilized air and carbon dioxide are for industrial cylinders of compressed gases. The cost of the tPA-CHO cells is not included, as it is associated with the cost of research, which is subsequently estimated as an operating cost.

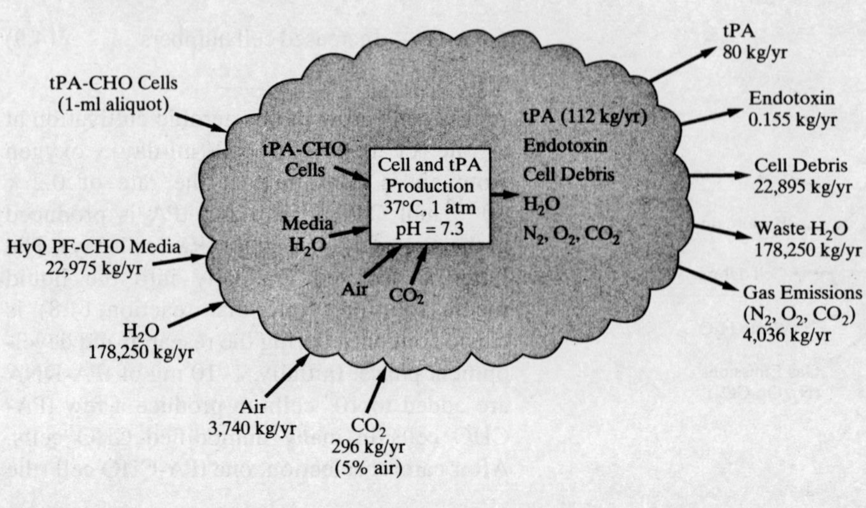

Figure 4.12 Reaction operations using mammalian CHO cells.

Table 4.5 Assumed Cost of Chemicals Produced or Sold

Chemical	kg/kg tPA	Cost ($/kg)
tPA	1	2,000,000
HyQ PF CHO powder media	287.2	233
Water for injection (WFI)	2,228	0.12
Air	46.8	1,742
CO_2	3.7	1,447
tPA-CHO cells	—	[a]

[a]Not included in gross profit estimate—related to cost of research, an operating cost.

Using these costs, the gross profit is estimated:

$$\begin{aligned}\text{Gross Profit} &= 2,000,000 - 287.2 \times 233 \\ &\quad - 2,228 \times 0.12 - 3.7 \times 1,447 \\ &\quad - 46.8 \times 1,742 \\ &= \$1,846,000/\text{kg tPA}\end{aligned}$$

Clearly, this is very high for tPA, a typical pharmaceutical. However, the gross profit does not account for the operating costs, which include the cost of research, the cost of utilities, and the investment cost, and are high for separations that involve expensive mass separating agents. With such a promising gross profit, the process synthesis proceeds at an accelerated pace.

Step 2 Distribute the Chemicals: In this step, the sources and sinks for each species in Figure 4.12 are matched so that the total mass flow rate into the reaction operation equals the mass flow rate out. This often entails the introduction of mixing operations, as illustrated in the previous example for vinyl chloride.

In this case, only one mixing operation is introduced, in which the HyQ PF-CHO powder media is mixed with water, as shown in Figure 4.13. Otherwise the sources and sinks are matched directly. However, the effluent from the cell growth, tPA production reactor must be separated before its species are matched with the product sinks.

Step 3 Eliminate Differences in Composition: For most distributions of chemicals, composition differences exist between streams to be separated and the sinks to which these species are sent. Clearly, in Figure 4.13, the effluent from the cell growth, tPA production reactor must be separated.

Many separation system possibilities exist, with one provided in Figure 4.14. Here, the reactor effluent is sent to a separator for recovery of the gas emissions from the liquid mixture, with the latter sent to a centrifuge to remove wet cell debris from the harvest media or clarified broth. Note that because proteins lose their activity at temperatures above ~0°C, the centrifuge, and all other separation operations, are operated at 4°C, slightly above the freezing point of water. The harvest media is mixed with arginine hydrochloride, an amino acid:

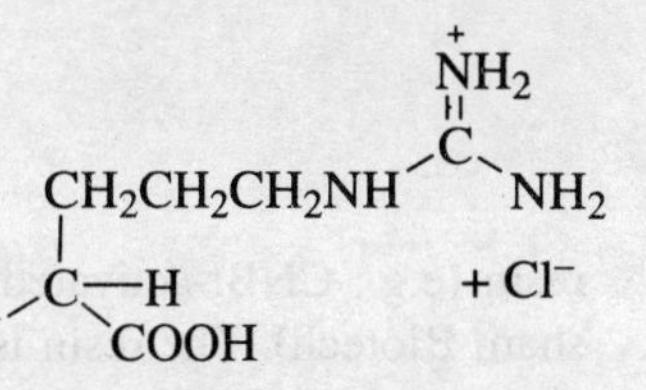

which prevents tPA from self-aggregating. Note that 45,870 kg/yr provides a concentration of 2.0 molar, which is sufficient to prevent aggregation.

The resulting mixture is sent to microfilters to remove large quantities of wastewater, which passes through the filters. For this step, alternate separators, like gel filtration and an Acticlean Etox resin (by Sterogene), should be considered. The retentate from the filter, which contains tPA, other proteins, endotoxin, arginine hydrochloride, and some water, is sent to an affinity chromatography operation. Here, tPA is selectively adsorbed on a

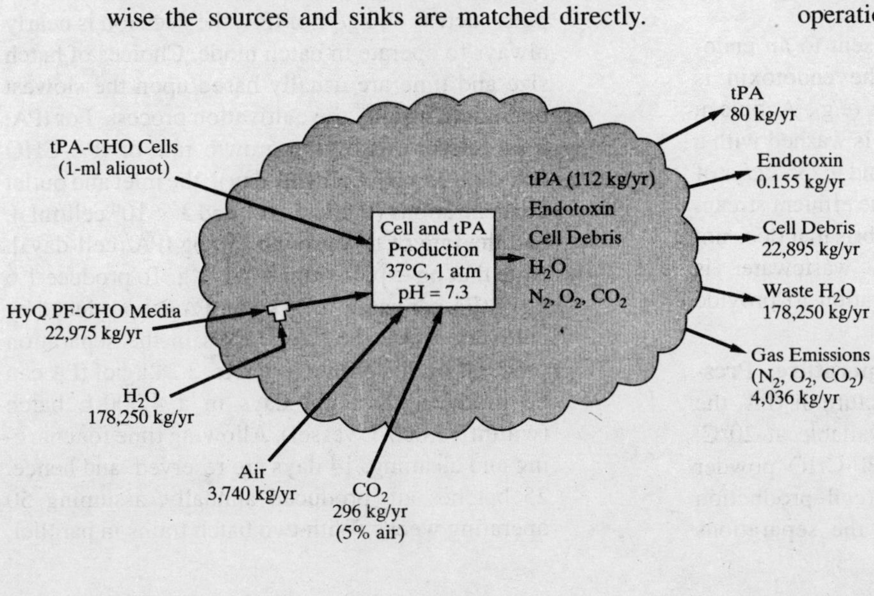

Figure 4.13 Flowsheet showing a distribution of chemicals for the tPA process.

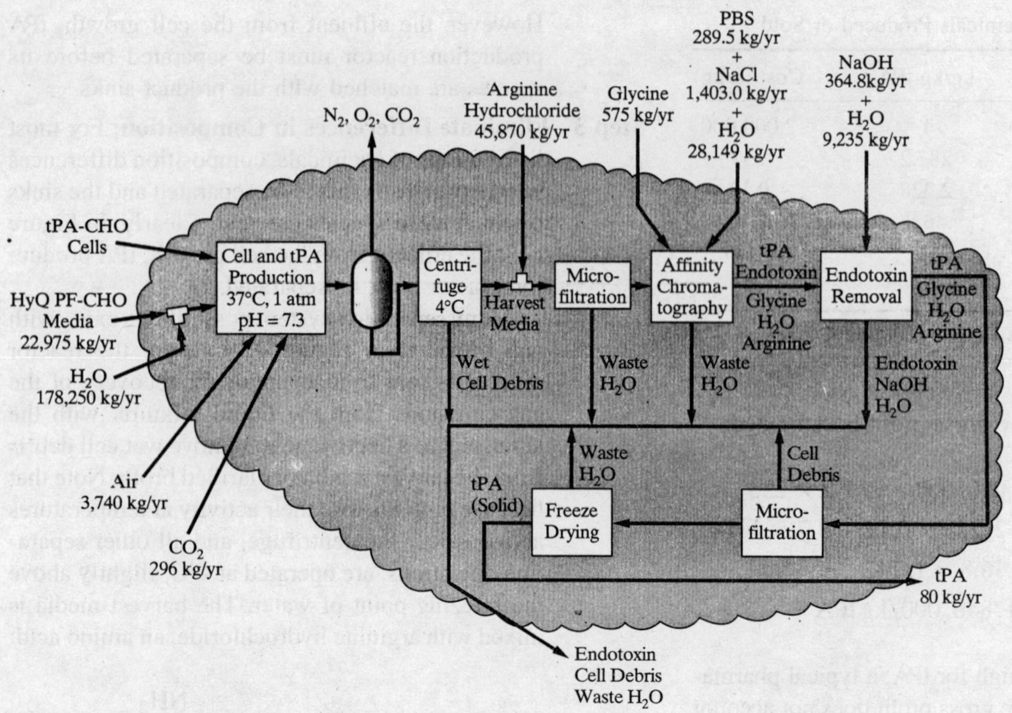

Figure 4.14 Flowsheet including the separation operations for the tPA process.

resin (e.g., CNBr-activated Sepharose, by Amersham Biotech). The resin is then eluted with glycine, an amino acetic acid:

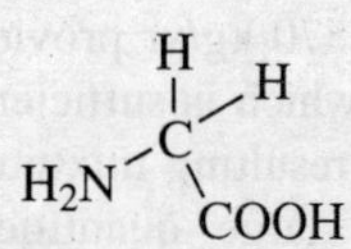

From lab measurements, 575 kg/yr of glycine are sufficient for the elution process. After the column is eluted, it is equilibrated with a mixture of 289.5 kg/yr of phosphate buffer solution (PBS) and 1,403.0 kg/yr of NaCl, with the quantities determined in the lab.

The resulting tPA solution is sent to an endotoxin removal column where the endotoxin is adsorbed selectively onto a resin (e.g., Acticlean Etox by Sterogene). This column is washed with a mixture of 364.8 kg/yr of NaOH and 9,235 kg/yr of water to remove the endotoxin. The effluent stream is microfiltered, to remove cell debris that does not pass through the filter. Then, wastewater is removed in a freeze-drying operation to provide tPA in powder form.

Step 4 Eliminate Differences in Temperature, Pressure, and Phase: In the manufacture of tPA, the ingredients are assumed to be available at 20°C, water is mixed with the HyQ PF-CHO powder media at 4°C, the cultivations (cell production operations) occur at 37°C, and the separations occur at 4°C. The exothermic heat of the cultivation is removed at 37°C. Only small pressure changes occur and can be neglected at this stage of process synthesis. Similarly, no phase-change operations are added to the flowsheet. Hence, only a few temperature-change operations are added to Figure 4.14, with the resulting flowsheet shown in Figure 4.15.

Step 5 Task Integration: At this stage in the synthesis, various items of equipment are selected, often combining two or more adjacent operations into a single equipment item; that is, in *task integration.* The first key decision involves whether to operate in continuous or batch mode. For small throughputs, such as 80 kg/yr of tPA, the decision is nearly always to operate in batch mode. Choices of batch size and time are usually based upon the slowest operation, usually the cultivation process. For tPA, it is determined by the growth rate of tPA-CHO cells [0.39×10^6 cell/(ml-day)], the inlet and outlet concentrations (0.225×10^6 and 3×10^6 cell/mL), and the rate of tPA growth [50 pg tPA/(cell-day)]. Note that pg $\equiv$ picogram $\equiv 10^{-12}$g. To produce 1.6 kg of tPA per batch, 2.24 kg of tPA are produced by cultivation, allowing for losses in the separation process. At this production rate, 2.24 kg of tPA can be produced in eight days in a 4,000-L batch (within a 5,000-L vessel). Allowing time for charging and cleaning, 14 days are reserved, and hence, 25 batches are produced annually, assuming 50 operating weeks. With two batch trains in parallel,

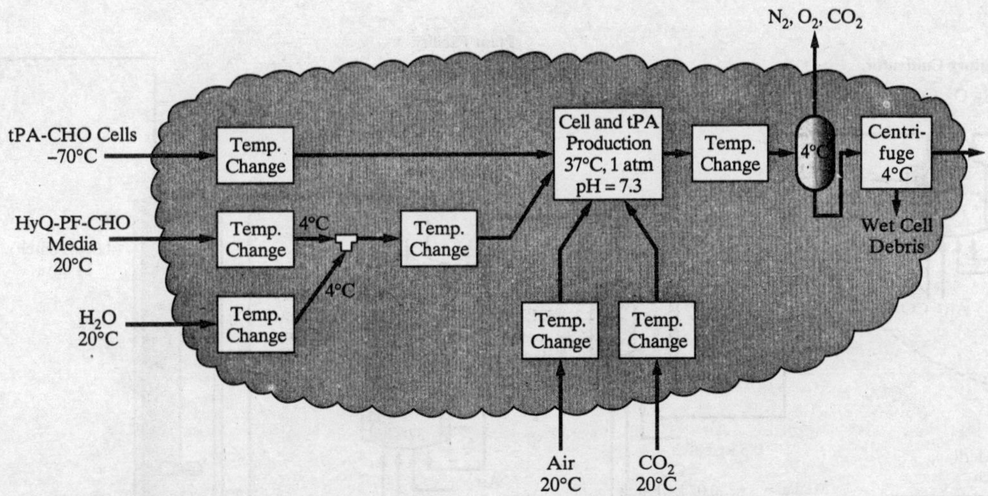

Figure 4.15 Flowsheet with the temperature-change operations in the tPA process.

50 batches are produced annually, that is, 1.6 kg of tPA are produced per batch.

The flowsheet in Figure 4.16a begins in a 1-L laboratory cultivator, into which a 1-mL aliquot of tPA-CHO cells is charged from cold storage at −70°C. To this, HyQ PF-CHO media, water, air, and CO_2 are added. Cultivation takes place over five days to produce 1.2 kg/batch of inoculum, which is emptied from the cultivator and transferred to the plant in one day. This effluent inoculates three cultivators in series, which carry out the cell and tPA production operation. The first is 40 L, with a 30-L batch that grows cells from 1.05×10^6 to 3×10^6 cell/mL in five days, with two additional days for loading and cleaning. The second is 400 L, with a 300-L batch that grows cells from 0.25×10^6 to 3×10^6 cell/mL in seven days, with 2.5 additional days for loading and cleaning. Finally, the third is 5,000 L with a 4,000-L batch that grows cells from 0.225×10^6 to 3×10^6 cell/mL in eight days, with six additional days for loading and cleaning. Note that gas emissions, containing N_2, O_2, and CO_2, are vented continuously from the cultivators. A 5,000-L mixing tank is installed to load and mix the powder media and water in two days. Note the tank jacket through which refrigerant is circulated. This vessel is followed by a microfilter, which sterilizes the mixture by removing bacteria, and a hot water heat exchanger. One last vessel, a 5,000-L holding tank, is provided to hold the contents of one cultivator batch (2.24, 457.17, 0.0031, 3,565 kg/batch of tPA, tPA-CHO cells, endotoxin, and water, respectively), in the event that the centrifuge is taken off-line for servicing. The effluent from the third cultivator is cooled to 4°C in the shell-and-tube heat exchanger, which is cooled by a refrigerant on the shell side.

Turn next to the separation section in Figure 4.16b. The centrifuge is designed to handle small batches, at a rate of 400 L/hr over 10 hr. It rotates at high speed with the wet cell mass (which contains all of the tPA-CHO cells, five wt% of the tPA, 20 wt% of the water, and none of the endotoxin fed to the centrifuge) thrown to the outside collection volume and removed. Note that at this stage in process synthesis, recovery fractions are estimated using heuristics and experimental data when available. Also, since the endotoxin contaminant must be removed entirely, it is assumed to be entirely recovered (100%) in the effluent from the microfilters. The clarified broth (2,854 kg/batch) exits through the central tube overhead. It enters a mixing tank in which arginine hydrochloride is added to form a 1.1 molar solution, which is microfiltered to remove 3,494 kg/batch of wastewater. The concentrated product, at 207 L/batch and containing 98, 5.62, and 5.62 wt% of the tPA, arginine hydrochloride, and water fed to the microfilter, is mixed with 67.4 kg/batch of arginine in a second mixing vessel to give 2.0 molar arginine. This solution is microfiltered to remove particulate matter before being sent to the affinity holding tank. The effluent, which contains 95, 98, 100, and 98 wt% of the tPA, arginine, endotoxin, and water fed to the microfilter, is loaded into a 58-L affinity chromatography column, which adsorbs 100, 100, 2, and 2 wt% of tPA, endotoxin, arginine, and water, as shown in Figure 4.16c. Most of the adsorbed tPA, 1.69 kg/batch, is eluted with a stream containing glycine (523 kg/batch at 2.2, 43.5, and 54.3 wt% of glycine, arginine, and water, respectively) and sent to a 500-L holding tank (405.7 kg/batch containing 1.69, 8.7, 175.6, 0.0026, and 219.7 kg/batch of tPA, glycine, arginine, endotoxin, and water, respectively). Note that the elution buffer recovers 85

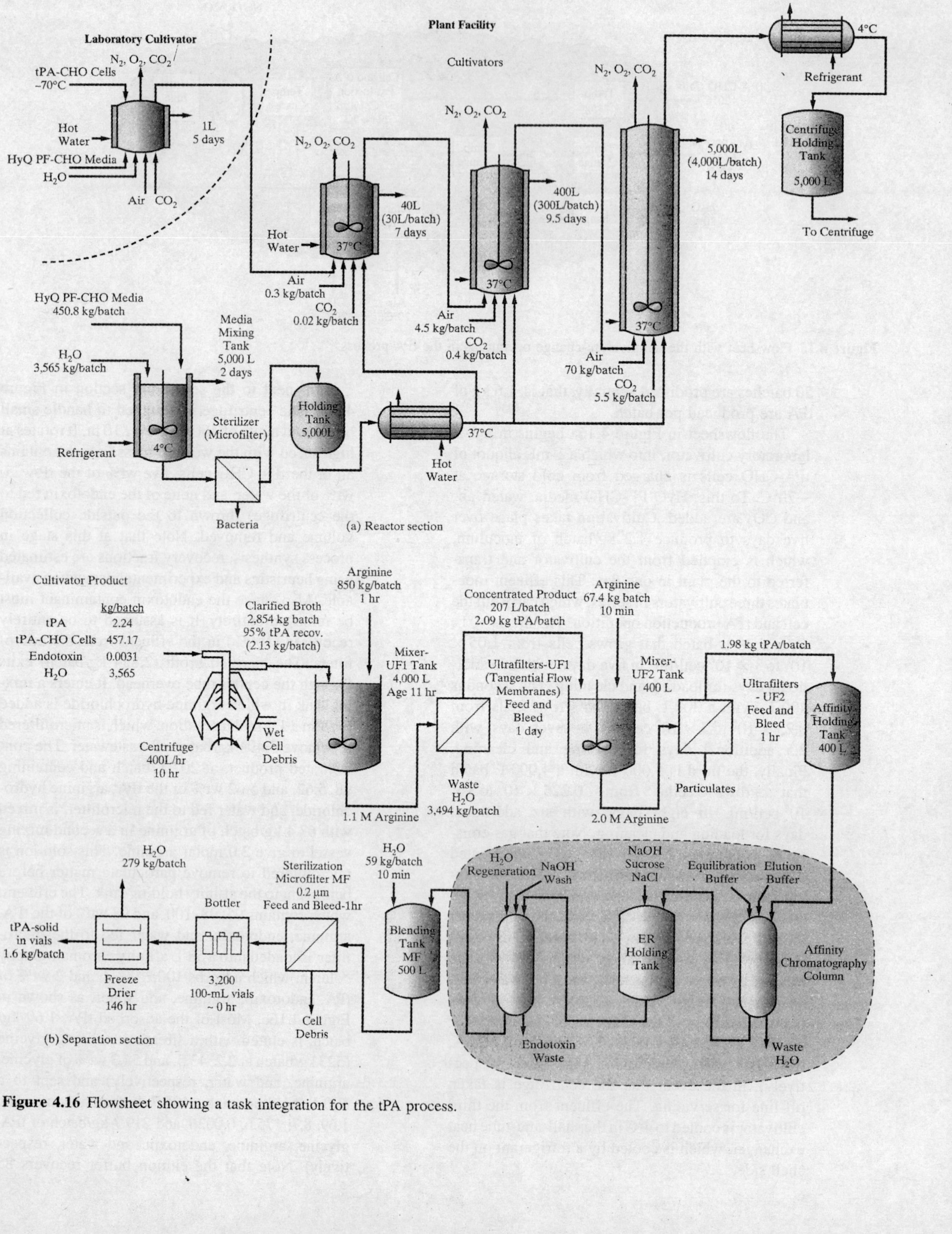

Figure 4.16 Flowsheet showing a task integration for the tPA process.

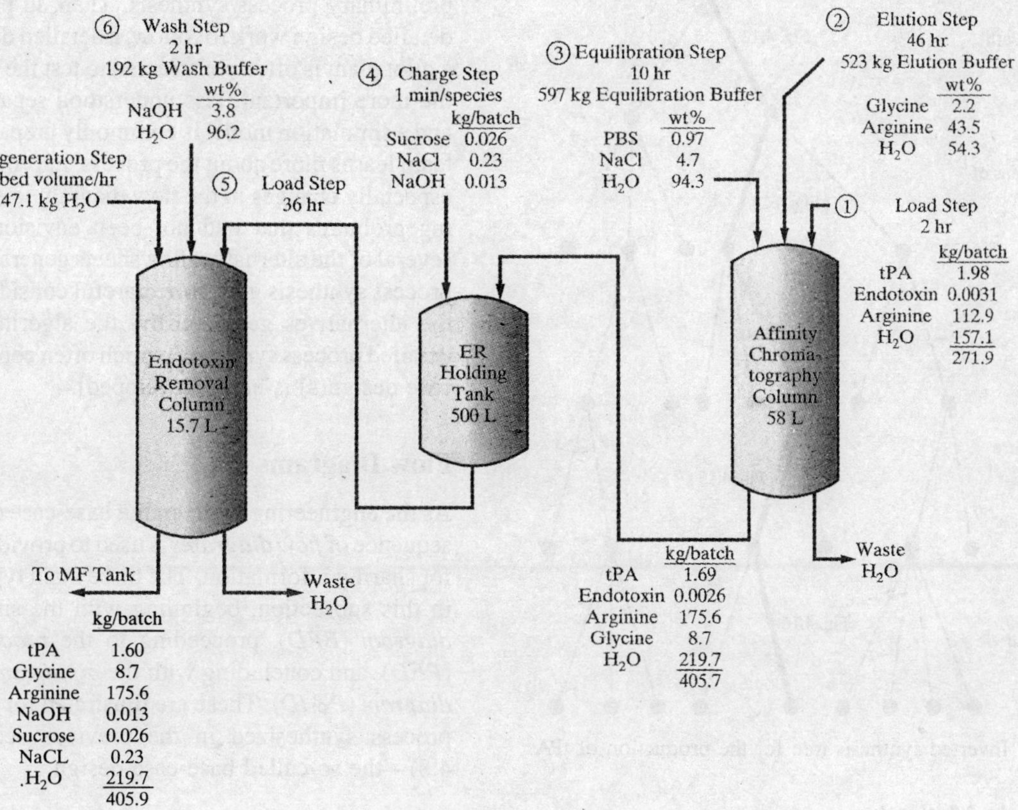

(c) Detailed separation section

Figure 4.16 *(Continued)*

wt% of the tPA and endotoxin from the resin. The affinity chromatography column is equilibrated with an equilibration buffer (597 kg/batch containing 0.97, 4.7, and 94.3 wt% PBS, NaCl, and water, respectively). After a caustic and sucrose mix is added to the holding tank (0.013, 0.026, and 0.33 kg/batch of NaOH, sucrose, and NaCl, respectively), the mixture is loaded into the endotoxin removal column (406.0 kg/batch). In this 15.7-L column, the endotoxins are adsorbed, and removed, by washing with caustic (192 kg/batch containing 3.8 and 96.2 wt% NaOH and water, respectively), which is discarded. The endotoxin removal column is regenerated with 47.1 kg/batch of water, while the endotoxin-free solution (405.9 kg/batch containing 1.6, 8.7, 175.6, 0.013, 0.026, 0.23, and 219.7 kg/batch of tPA, glycine, arginine, NaOH, sucrose, NaCl, and water, respectively) is sent to a holding tank, where 59 kg/batch of water are added. After sterilization with a microfilter to remove cell debris, from which 99.7% of the tPA is recovered, the solution is sent to a bottler and 100-mL vials, each containing 100 mg of tPA, are conveyed to a freeze-drier, where the water is evaporated.

It is important to recognize that the batch sizes in Figure 4.16 are representative. However, as discussed subsequently in Section 5.5 and Chapter 11, the batch times and vessel sizes are key design variables in scheduling and optimizing batch processes.

Synthesis Tree

Clearly, at each step in the synthesis of the process flowsheet, alternatives are generated and the synthesis tree fills in. For the tPA process, a schematic of a synthesis tree is shown in Figure 4.17. Note that the bold branch corresponds to the flowsheets in Figures 4.12–4.16. In design synthesis, the engineer strives to identify the most promising alternatives, eliminating the least promising alternatives by inspection, wherever possible. Initially, heuristic rules help to make selections. Eventually, algorithmic methods involving optimization can be introduced to check the heuristics and identify more promising alternatives, as discussed in Chapter 11. It should be emphasized, however, that the design window, beginning during Phases 1 and 2 of the clinical trials, is small, typically on the order of 12–16 months, before Phase 3 begins (see Section 1.3). Consequently, emphasis is normally placed on the rapid development of a promising design, and less on design optimization. Stated differently, for high-priced pharmaceuticals, it is far more important to be first-to-market rather than to achieve relatively small savings in the capital investment or operating expenses for the plant through design optimization. For further discussion, see Pisano (1997).

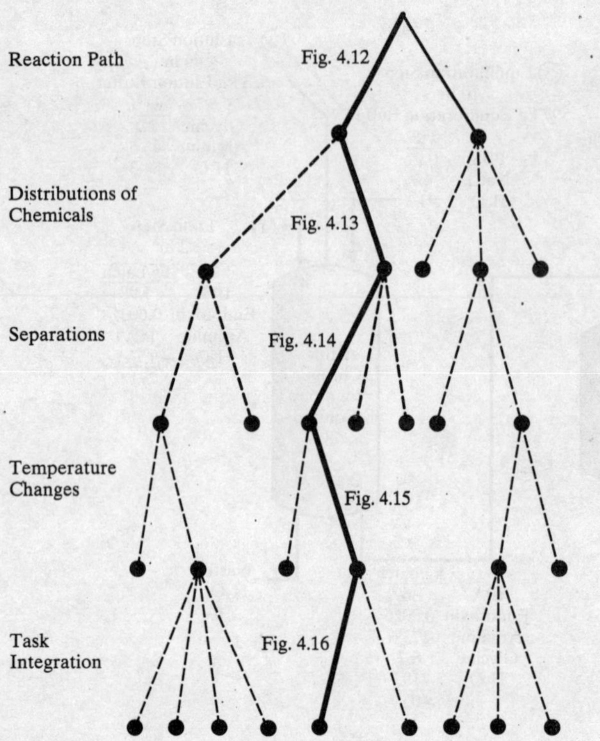

Figure 4.17 Inverted synthesis tree for the production of tPA.

Algorithmic Methods

Finally, before leaving this section on preliminary process synthesis, the limitations of the heuristic approaches should not be overlooked. Many algorithmic methods are very effective for the synthesis of alternative flowsheets, their analysis, and optimization. These methods are usually used by design teams in parallel with their work on the development of the base-case design, which is the subject of the next section. The algorithmic methods are easily implemented and are illustrated with many examples in Chapters 7–11.

4.5 DEVELOPMENT OF THE BASE-CASE DESIGN

At some point in the synthesis of alternative flowsheets, it becomes important to select one or two of the most promising alternatives for further development into the so-called *base-case design*(s). To accomplish this, the design team is usually expanded, mostly with chemical engineers, or assisted by more specialized engineers, as the engineering workload is increased significantly. With expanded engineering involvement, the design team sets out to create a detailed process flow diagram and to improve the task integration begun in preliminary process synthesis. Then, in preparation for the detailed design work to follow, a detailed database is created, a pilot plant is often constructed to test the reaction steps and the more important, less understood separation operations, and a simulation model is commonly prepared. As the design team learns more about the process, improvements are made, especially changes in the flow diagram to eliminate processing problems that had not been envisioned. In so doing, several of the alternative flowsheets generated in preliminary process synthesis gain more careful consideration, as well as the alternatives generated by the algorithmic methods, in detailed process synthesis [which often continues as the base-case design(s) is being developed].

Flow Diagrams

As the engineering work on the base-case design proceeds, a sequence of *flow diagrams* is used to provide a crucial vehicle for sharing information. The three main types are introduced in this subsection, beginning with the simplest *block flow diagram (BFD)*, proceeding to the *process flow diagram (PFD)*, and concluding with the *piping and instrumentation diagram (P&ID)*. These are illustrated for the vinyl-chloride process synthesized in the previous section (see Figure 4.8)—the so-called base-case design.

Block Flow Diagram (BFD)

The block flow diagram represents the main processing sections in terms of functional blocks. As an example, Figure 4.18 shows a block flow diagram for the vinyl-chloride process, in which the three main sections in the process—namely, ethylene chlorination, pyrolysis, and separation—are shown, together with the principal flow topology. Note that the diagram also indicates the overall material balances and the conditions at each stage, where appropriate. This level of detail is helpful to summarize the principal processing sections and is appropriate in the early design stages, where alternative processes are usually under consideration.

Process Flow Diagram (PFD)

Process flow diagrams provide a more detailed view of the process. These diagrams display all of the major processing units in the process (including heat exchangers, pumps, and compressors), provide stream information, and include the main control loops that enable the process to be regulated under normal operating conditions. Often, preliminary PFDs are constructed using the process simulators. Subsequently,

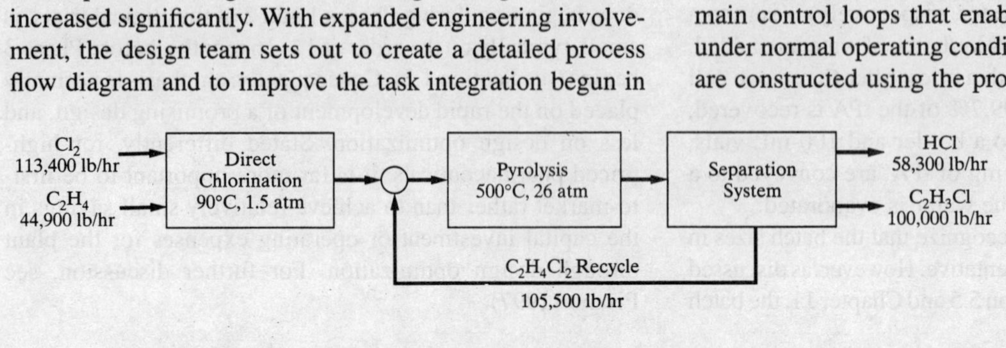

Figure 4.18 Block flow diagram for the vinyl-chloride process.

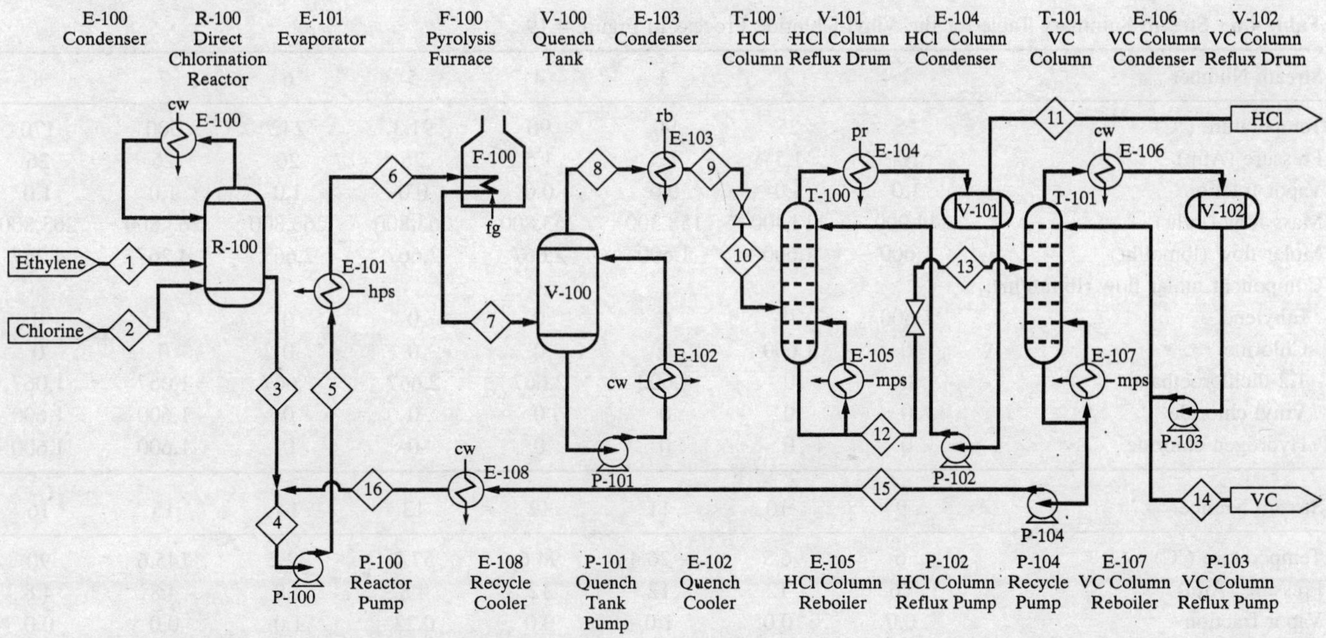

Figure 4.19 Process flow diagram for the vinyl-chloride process.

more detailed PFDs are prepared using software such as AUTOCAD and VISIO, the latter having been used to prepare Figure 4.19 for the vinyl-chloride process. The conventions typically used when preparing PFDs are illustrated using this figure and are described next.

Processing Units Icons that represent the units are linked by arcs (lines) that represent the process streams. The drawing conventions for the unit icons are taken from accepted standards, for example, the ASME (American Society for Mechanical Engineers) standards (ASME, 1961). A partial list of typical icons is presented in Figure 4.20. Note that each unit is labeled according to the convention U-XYY, where U is a single letter identifying the unit type (V for vessel, E for exchanger, R for reactor, T for tower, P for pump, C for compressor, etc.). X is a single digit identifying the process area where the unit is installed, and YY is a two-digit number identifying the unit itself. Thus, for example, E-100 is the identification code for the heat exchanger that condenses the overhead vapors from the chlorination reactor. Its identification code indicates that it is the 00 item installed in plant area 1.

Stream Information Directed arcs that represent the streams, with flow direction from left to right wherever possible, are numbered for reference. By convention, when streamlines cross, the horizontal line is shown as a continuous arc, with the vertical line broken. Each stream is labeled on the PFD by a numbered diamond. Furthermore, the feed and product streams are identified by name. Thus, streams 1 and 2 in Figure 4.19 are labeled as the ethylene and chlorine feed streams, while streams 11 and 14 are labeled as

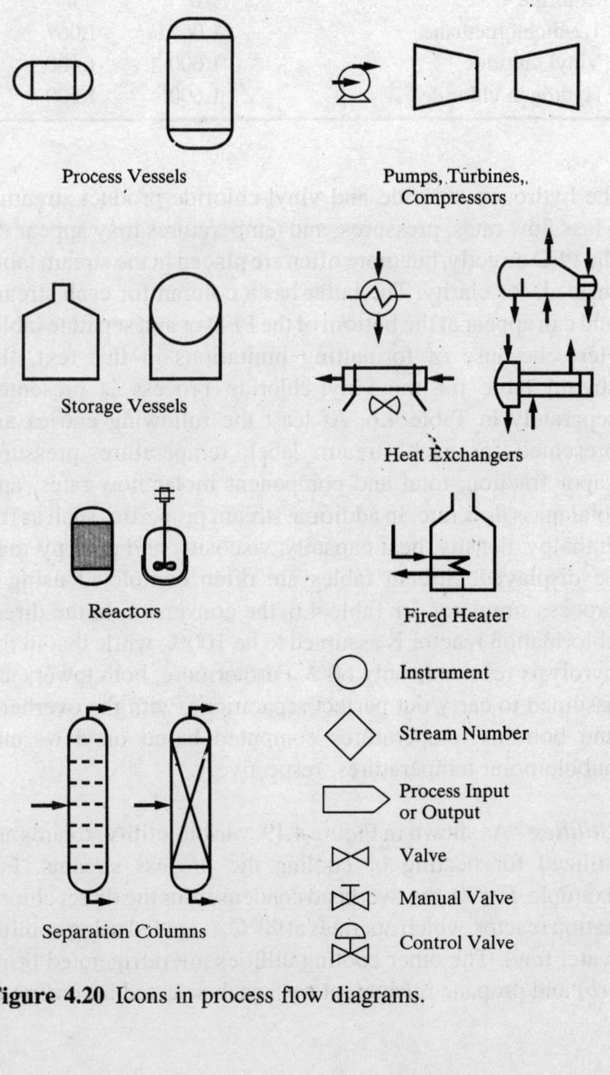

Figure 4.20 Icons in process flow diagrams.

Table 4.6 Stream Summary Table for the Vinyl-Chloride Process in Figure 4.19

Stream Number	1	2	3	4	5	6	7	8
Temperature (°C)	25	25	90	90	91.3	242	500	170
Pressure (Atm)	1.5	1.5	1.5	1.5	26	26	26	26
Vapor fraction	1.0	1.0	0.0	0.0	0.0	1.0	1.0	1.0
Mass flow (lb/hr)	44,900	113,400	158,300	263,800	263,800	263,800	263,800	263,800
Molar flow (lbmol/hr)	1,600	1,600	1,600	2,667	2,667	2,667	4,267	4,267
Component molar flow (lbmol/hr):								
Ethylene	1,600	0	0	0	0	0	0	0
Chlorine	0	1,600	0	0	0	0	0	0
1,2-dichloroethane	0	0	1,600	2,667	2,667	2,667	1,067	1,067
Vinyl chloride	0	0	0	0	0	0	1,600	1,600
Hydrogen chloride	0	0	0	0	0	0	1,600	1,600
Stream Number	**9**	**10**	**11**	**12**	**13**	**14**	**15**	**16**
Temperature (°C)	6	6.5	−26.4	94.6	57.7	32.2	145.6	90
Pressure (Atm)	26	12	12	12	4.8	4.8	4.8	4.8
Vapor fraction	0.0	0.0	1.0	0.0	0.23	1.0	0.0	0.0
Mass flow (lb/hr)	263,800	263,800	58,300	205,500	205,500	100,000	105,500	105,500
Molar flow (lbmol/hr)	4,267	4,267	1,600	2,667	2,667	1,600	1,067	1,067
Component molar flow (lbmol/hr):								
Ethylene	0	0	0	0	0	0	0	0
Chlorine	0	0	0	0	0	0	0	0
1,2-dichloroethane	1,067	1,067	0	1,067	1,067	0	1,067	1,067
Vinyl chloride	1,600	1,600	0	1,600	1,600	1,600	0	0
Hydrogen chloride	1,600	1,600	1,600	0	0	0	0	0

the hydrogen chloride and vinyl-chloride product streams. Mass flow rates, pressures, and temperatures may appear on the PFD directly, but more often are placed in the stream table instead, for clarity. The latter has a column for each stream and can appear at the bottom of the PFD or as a separate table. Here, because of formatting limitations in this text, the stream table for the vinyl-chloride process is presented separately in Table 4.6. At least the following entries are presented for each stream: label, temperature, pressure, vapor fraction, total and component molar flow rates, and total mass flow rate. In addition, stream properties such as the enthalpy, density, heat capacity, viscosity, and entropy may be displayed. Stream tables are often completed using a process simulator. In Table 4.6, the conversion in the direct chlorination reactor is assumed to be 100%, while that in the pyrolysis reactor is only 60%. Furthermore, both towers are assumed to carry out perfect separations, with the overhead and bottoms temperatures computed based on dew- and bubble-point temperatures, respectively.

Utilities As shown in Figure 4.19, various utility streams are utilized for heating or cooling the process streams. For example, E-100, the overhead condenser for the direct chlorination reactor, which operates at 90°C, is cooled using cooling water (*cw*). The other cooling utilities are refrigerated brine (*rb*) and propane refrigerant (*pr*), each selected according to the temperature level of the required utility. Heating utilities are fuel gas (*fg*), high-pressure steam (*hps*), and medium-pressure steam (*mps*). A list of heating and cooling utilities, with temperature ranges, and the abbreviations commonly used on PFDs is presented in Table 4.7 (see also Table 23.1 and the subsection on *utilities* in Section 23.2).

Equipment Summary Table This provides information for each equipment item in the PFD, with the kind of information typically provided for each type of unit shown in Table 4.8. Note that the materials of construction (MOC), and operating temperature and pressure, are required for all units. Also note that suggestions for the materials of construction are provided in Appendix III.

In summary, the PFD is the most definitive process design document, encapsulating much of the commonly referred to design information. As such, it is used and updated throughout much of process design. However, it lacks many details required to begin the construction engineering work for the plant. Many of these details are transmitted in the *Piping and Instrumentation Diagram.*

Piping and Instrumentation Diagram (P&ID)

This is the design document transmitted by the process design engineers to the engineers responsible for plant

Table 4.7 Heating and Cooling Utilities—Identifiers and Temperature Ranges

Identifier	Utility	Typical Operating Range (°F)
Hot Utilities—In increasing cost per BTU:		
lps	Low-pressure steam, 15 to 30 psig	250 to 275°F
mps	Medium-pressure steam, 100 to 150 psig	325 to 366°F
hps	High-pressure steam, 400 to 600 psig	448 to 488°F
fo	Fuel oils	
fg	Fuel gas	Process waste stream
po	Petroleum oils	Below 600°F
dt	Dowtherms	Below 750°F
Cold Utilities—In increasing cost per BTU:		
bfw	Boiler feed water	Used to raise process steam
ac	Air cooling	Supply at 85 to 95°F—temperature approach to process 40°F
rw	River water	Supply at 80 to 90°F (from cooling tower), return at 110°F
cw	Cooling water	Supply at 80 to 90°F (from cooling tower), return at 115 to 125°F
cw	Chilled water	45 to 90°F
rb	Refrigerated brine	0 to 50°F
pr	Propane refrigerant	−40 to 20°F

construction. It is also used to support the startup, operation of the process, and operator training. Consequently, it contains items that do not appear in the PFD, such as the location and type of all measurement and control instruments, the positioning of all valves, including those used for isolation and control, and the size, schedule, and materials of construction of piping. As a result, a number of interconnected P&IDs are prepared for a process that is represented on a single PFD. For more instruction on the preparation of P&IDs, the reader is referred to the book by Sandler and Luckiewicz (1993).

Table 4.8 Equipment Summary Specifications

Equipment type	Required specification
Vessel	Height, diameter, orientation, pressure, temperature, materials of construction (MOC)
Towers	Height, diameter, orientation, pressure, temperature, number of and type of trays, height and type of packing, MOC
Pumps	Driver type, flow, suction and discharge pressures, temperature, shaft power, MOC
Compressors	Driver type, inlet flow, suction and discharge pressures, temperature, shaft power, MOC
Heat exchangers	Type, area, duty, number of shell and tube passes, for both shell and tubes: operating temperature, pressure, pressure drop, and MOC
Fired heaters	Type, tube pressure and temperature, duty, radiant and convective heat transfer area, MOC

Calculations Supporting Flow Diagrams

As indicated when discussing the stream table (Table 4.6), and emphasized when synthesizing the vinyl-chloride process in the previous section, the material balances for the process streams are nearly complete after preliminary process synthesis. These are conducted by means of spreadsheets and by process simulators, as discussed in Chapter 5. At this stage, the design team checks the assumptions. It also completes the material and energy balances associated with heat addition and removal, without attempting to carry out heat and power integration. As indicated in the section that follows on process integration, the design team carries out heat and power integration just prior to the detailed design stage.

It should also be noted that, during the synthesis of the vinyl-chloride process, no attempt was made to complete calculations to determine the number of stages and reflux ratios for the distillation towers, and furthermore, perfect splits may be assumed. Hence, the condenser and reboiler heat duties are not yet known. The vapor stream, S1, is assumed to be saturated, pure dichloroethane, which releases its heat of vaporization, 143.1 Btu/lb, to cooling water, which is heated from 30° to 50°C. Both the direct chlorination reactor and the pyrolysis furnace are assumed to operate adiabatically, and natural gas is assumed to have a lower heating value of 23,860 Btu/lb (heat of combustion at 25°C). The liquid effluent from the quench is assumed to have a composition in vapor–liquid equilibrium at 150°C and 26 atm. The stream is cooled to 50°C with cooling water to release the heat necessary to cool the pyrolysis products from 500 to 170°C (4.66×10^7 Btu/hr). No attempt is made to calculate the amount of propane refrigerant necessary to remove the heat to cool the pyrolysis effuent to its bubble point at 6°C (5.20×10^7 Btu/hr); this calculation is completed during process integration, when the heat and power integration is completed.

These calculations could have been completed using the process simulators, which are commonly used to calculate the heats of reaction, enthalpy changes upon heating and cooling, and vapor–liquid equilibria, as well as to perform material and energy balances using approximate models involving specifications of split fractions in separators and conversions in chemical reactors. Note that a complete simulation is usually not justified until the design team is ready to begin the detailed design. Gradually, additional detail is added to the simulation model; for example, the number of stages and the reflux ratio are selected for the distillation columns, and the material and energy balances are completed with recycle streams that are not assumed to be pure. As mentioned previously, this is the subject matter of Chapter 5, in which the methods of building a simulation model are introduced. After studying Chapter 5, the reader should be able to prepare a simulation for the vinyl-chloride process (see Exercise 5.5) and prepare a more accurate representation of the flow diagram in Figure 4.19 and Table 4.6.

Process Integration

With the detailed process flow diagram completed, the task-integration step, which was initiated in the preliminary process synthesis, is revisited by the design team. The assumptions are checked and opportunities are sought to improve the designs of the processing units, and to achieve a more efficient process integration. In the latter, attempts are made to match cold streams that need to be heated with hot streams that have cooling requirements, so as to reduce the need for external utilities such as steam and cooling water. In addition, where possible, power is extracted from hot streams at elevated pressures, so as to drive compressors and pumps. Also, when solvents, such as water, are used as mass separating agents, opportunities are sought to reduce the amount of solvent used through mass integration. Often, significant improvements can be made in the process design beyond those achievable in the preliminary process synthesis. The algorithmic methods in Chapter 9 for heat and power integration and in Chapter 10 for mass integration are commonly applied by the design team; they provide a systematic approach to minimizing the utilities, matching the hot and cold streams, inserting turbines (as a part of heat engines), minimizing the amount of solvent used, and so on.

Detailed Database

Having completed the process flow diagram (PFD), the design team seeks to check its key assumptions further and to obtain the additional information needed to begin work on the detailed design. As discussed earlier, this usually involves three activities in parallel, the first of which is to create a detailed database by refining and adding to the preliminary database. In the other two activities, a pilot plant is constructed to confirm that the equipment items operate properly and to provide data for the detailed data bank, and a simulation model is prepared to enable the team to project the impact of changes in the design and operation parameters, such as temperatures, pressures, reflux ratios, and the number of stages.

In creation of the detailed database, it is common to add transport and kinetics data, as well as data concerning the feasibility of the separations, the identity of any forbidden matches in heat exchange, heuristic parameters, and data for sizing the equipment. Each process requires somewhat different data, and hence it is inappropriate to generalize. However, it is instructive to examine the mix of data needed by a design team in connection with the vinyl-chloride process in Figure 4.19.

Beginning with the chlorination reactor, data are needed to determine the impact of the concentrations of C_2H_4, Cl_2, and $FeCl_3$ catalyst in the $C_2H_4Cl_2$ pool on the intrinsic rate of the chlorination reaction (in kmol/m^3 hr). With these data, the team can determine the order of the reaction and its rate constant as a function of temperature, and eventually compute the residence time to achieve nearly complete conversion.

Similar data are required for the pyrolysis reactor. In this case, the intrinsic rate of reaction is needed as a function of concentration, temperature, and pressure. Furthermore, since the rate of reaction may be limited by the rate at which heat is transferred to the reacting gases, it is probably desirable to estimate the tube-side heat transfer coefficient, h_i, as a function of the Reynolds and Prandtl numbers in the tubes. The appropriate equations and coefficients, which are described in Chapter 18, would be added to the database.

In the vinyl-chloride process, because of the significant differences in the volatilities of the three principal chemical species, distillation, absorption, and stripping are prime candidates for the separators, especially at the high production rates specified. For other processes, liquid–liquid extraction, enhanced distillation, adsorption, and membrane separators might become more attractive, in which case the design team would need to assemble data that describe the effect of solvents on species phase equilibrium, species adsorption isotherms, and the permeabilities of the species through various membranes.

A key limitation in the flowsheets in Figures 4.8 and 4.19 is that the cold $C_2H_4Cl_2$ stream is not heated by the pyrolysis products because the rate of carbon deposition in such a feed/product heat exchanger is anticipated to be high, and would cause the heat exchanger to foul with carbon. As discussed above, the design team would normally apply the methods of heat and power integration to design a network of heat exchangers that would effect significant economies. Hence, it is important to learn more about the rate of carbon deposition. Before the team proceeds to the detailed design stage, it needs data to confirm the validity of this perception above—that is, to enable it to characterize the intrinsic rate of carbon deposition. If the rate is found to be sufficiently low, the team may decide to cool the hot pyrolysis products through heat exchange with the cold streams. For maintenance, to remove carbon deposits periodically, two heat exchangers could be installed in parallel, one of which would be operated while the other is being cleaned. This would provide substantial savings in fuel and cooling water utilities. On the other hand, if the rate of carbon deposition is high, the design team would avoid the exchange of heat between these two streams; that is, it would continue to consider the exchange of that heat to be a so-called *forbidden match.*

The additional data for sizing the equipment are typically maximum pressure drops, tube lengths, and baffle spacings in heat exchangers, surface tensions and drag coefficients for estimating the flooding velocities (to be used in determining the tower diameters), specifications for tray spacings in multi-staged towers, and residence times in flash vessels and surge tanks. Examples of the use of this type of data for the detailed design of heat exchangers are provided in Chapter 18, and for the detailed design of a distillation tower in Chapter 19.

Pilot-Plant Testing

Clearly, as the detailed database is assembled, the needs for pilot-plant testing become quite evident. For the manufacture of new chemicals, a pilot plant can produce quantities of product suitable for testing and evaluation by potential customers. Very few processes that include reaction steps are constructed without some form of pilot-plant testing prior to doing detailed design calculations. This is an expensive, time-consuming step that needs to be anticipated and planned for by the design team as early as possible, so as to avoid extensive delays. Again, although it is inappropriate to generalize, the vinyl-chloride process provides good examples of the need for pilot-plant testing and the generation of data for detailed design calculations.

As mentioned in the previous subsection, kinetic data are needed for both the chlorination and pyrolysis reactors, as well as to determine the rate of carbon deposition. In all three cases, it is unlikely that adequate data can be located in the open literature. Consequently, unless sufficient data exist in company files, or were taken in the laboratory and are judged to be adequate, pilot-plant testing is needed. Generally, the pilot-plant tests are conducted by a development team working closely with the design team. As the data are recorded, regression analyses are commonly used to compute the coefficients of compact equations to be stored in the database.

As mentioned in connection with the need for laboratory experiments, pilot-plant tests also help to identify potential problems that arise from small quantities of impurities in the feed streams, and when unanticipated byproducts are produced, usually in small quantities, that have adverse effects such as to impart an undesired color or smell to the product. When a catalyst is used, the impact of these species needs to be studied, and, in general, the useful life of the catalyst needs to be characterized. Pilot plants can also verify separation schemes developed during process design.

Process Simulation

As mentioned throughout the discussion of preliminary process synthesis and the creation of the process flow diagram, the process simulator usually plays an important role, even if a simulation model is not prepared for the entire flowsheet. When parts of a simulation model exist, it is common for the design team to assemble a more comprehensive model, one that enables the team to examine the effect of parametric changes on the entire process. In other cases, when the process simulators have not been used for design, a simulation model is often created for comparison with the pilot-plant data and for parametric studies.

High-speed PCs and laptop computers, which have excellent graphical user interfaces (GUIs), have replaced workstations as the preferred vehicle for commercial simulators, and are now finding widespread use throughout the chemical process industries. The use of simulators, which is the subject of the next chapter, has become commonplace in assisting the design team during process creation.

4.6 SUMMARY

Having studied this chapter, the reader should

1. Be able to create a preliminary database for use in preliminary process synthesis—involving the manufactures of basic chemical products.
2. Understand the steps in preliminary process synthesis and be able to use them to develop other flowsheets for the manufacture of vinyl chloride and tPA (corresponding to the other branches of the synthesis trees in Figures 4.9 and 4.17), as well as for the manufacture of other chemicals.
3. Understand the steps taken by the design team in preparing one or more base-case designs. For the manufacture of vinyl chloride, or another chemical, you should be able to create a detailed process flow diagram and understand the need to complete the task-integration step begun during preliminary process synthesis and carry out the process integration step. In addition, you should be able to determine whether continuous or batch operation is more suitable.

4. Know how to prepare for the detailed design step, that is, to expand the database to include important kinetics data and the like, and to seek data from a pilot plant when necessary. You should also recognize the need for a model of the process, usually implemented by process simulators, to be covered in Chapter 5.

REFERENCES

1. ASME. *Graphical Symbols for Process Flow Diagrams.* ASA Y32.11. Amer. Soc. Mech. Eng., New York (1961).

2. AUDETTE, M., C. METALLO, and K. NOOTONG. *Human Tissue Plasminogen Activator*, Towne Library. University of Pennsylvania, Philadelphia, Pennsylvania (2000).

3. BALZHISER, R.E., M.R. SAMUELS, and J.D. ELIASSEN. *Chemical Engineering Thermodynamics*, Prentice-Hall, Englewood Cliffs, New Jersey (1972).

4. Benedict, D.B., Process for the Preparation of Olefin Dichlorides, U.S. Patent 2,.929,.852, March 22 (1960).

5. B.F. Goodrich Co., Preparaton of Vinyl Chloride. British Patent 938,.824,. October 9 (1963) *Instrument Symbols and Identification*, Instrument Society of America Standard ISA-S5–1, Research Triangle Park, North Carolina (1975).

6. BORSA, A.G., A.M. HERRING, J.T. MCKINNON, R.L. MCCORMICK, and G.H. KO. "Coke and Byproduct Formation during 1.2-Dichloroelhane Pyrolysis in a Laboratory Tubular Reactor," *Ind. Eng. Chem. Res.*, **40**, 2428–2436 (2001).

7. deNEVERS, N., *Physical and Chemical Equilibrium for Chemical Engineers*, Wiley-Interscience, New York (2002).

8. DOUGLAS, J.M., *Conceptual Design of Chemical Processes.* McGraw-Hill, New York (1988).

9. GMEHLING, J., *Azeotropic Data.* VCH Publishers, Deerfield Beach, Florida (1994).

10. GMEHLING, J., U. ONKEN, W. ARLT, P. GRENZHEUSER, U. WEIDLICH, and B. KOLBE, *Vapor–Liquid Equilibrium Data Collection*, 13 Parts, DECHEMA, Frankfurt, Germany (1980).

11. GOEDDEL, D.V., W.J. KOHR, D. PENNICA, and G.A. VEHAR, Human Tissue Plasminogen Activator, U.S. Patent 4,.766,075, August 23 (1988).

12. GREEN, D.W., and R.H. PERRY, Ed., *Perry's Chemical Engineers' Handbook*, 8th ed., McGraw-Hill, New York (2008).

13. KOHN, J.P., and F. KURATA, Heterogeneous Phase Equilibria of the Methane–Hydrogen Sulfide System, *AIChE J.*, 4(2), 211 (1958).

14. KYLE, B.J., *Chemical and Process Thermodynamics*, Prentice-Hall, Englewood Cliffs, New Jersey (1984).

15. LEONARD, E.C., Ed., *Vinyl and Diene Monomers, Part 3.* Wiley-Interscience, New York (1971).

16. LIDE, D.R., Ed., *Handbook of Chemistry and Physics*, CRC Press, Boca Raton, Florida, annual.

17. MYERS, A.L., and W.D. SEIDER, *Introduction to Chemical Engineering and Computer Calculations*, Prentice-Hall, Englewood Cliffs, New Jersey (1976).

18. PISANO, G.P., *The Development Factory: Unlocking the Potential of Process Innovation*, Harvard Business School Press, Boston (1997).

19. POLING, B.E., J.M. PRAUSNITZ, and J.P. O'CONNELL, *Properties of Gases and Liquids*, 5th ed., McGraw-Hill, New York (2001).

20. REAMER, H.H., B.H. SAGE, and W.N. LACEY, "Phase Equilibria in Hydrocarbon Systems," *Ind. Eng. Chem.*, **43**, 976 (1951).

21. REID, R.C., J.M. PRAUSNITZ, and B.E. POLING, *The Properties of Gases & Liquids*, 4th ed., McGraw-Hill, New York (1987).

22. RIEDER, R.M., and A.R. THOMPSON, "Vapor–Liquid Equilibrium Measured by a Gillespie Still," *Ing. Eng. Chem.*, **41** (12), 2905 (1949).

23. RUDD, D.F., G.J. POWERS, and J.J. SIIROLA, *Process Synthesis*, Prentice-Hall, Englewood Cliffs, New Jersey (1973).

24. SANDLER, H.J., and E.T. LUCKIEWICZ, *Practical Process Engineering*, XIMIX, Philadelphia, Pennsylvania (1993).

25. SANDLER, S.J., *Chemical, Biochemical, and Engineering Thermodynamics*, 4th ed., Wiley, New York (2006).

26. SMITH, J.M., H.C.Van, NESS, and M.M. ABBOTT, *Chemical Engineering Thermodynamics*, 5th ed., McGraw-Hill, New York (1997).

27. WALAS, S.M., *Chemical Process Equipment*, Butterworth, London (1988).

28. WALAS, S.M., *Phase Equilibria in Chemical Engineering*, Butterworth, London (1985).

29. WOODS, D.R., *Data for Process Design and Engineering Practice*, Prentice-Hall, Englewood Cliffs, New Jersey (1995).

EXERCISES

4.1 For an equimolar solution of *n*-pentane and *n*-hexane, compute:

(a) The dew-point pressure at 120°F

(b) The bubble-point temperature at 1 atm

(c) The vapor fraction, at 120°F and 0.9 atm, and the mole fractions of the vapor and liquid phases

4.2 For the manufacture of vinyl chloride, assemble a preliminary database. This should include thermophysical property data, MSDSs for each chemical giving toxicity and flammability data, and the current prices of the chemicals.

4.3 Consider the flowsheet for the manufacture of vinyl chloride in Figure 4.8.

(a) If the pyrolysis furnace and distillation towers are operated at low pressure (1.5 atm), what are the principal disadvantages? What alternative means of separation could be used?

(b) For the process shown, is it possible to use some of the heat of condensation from the $C_2H_4Cl_2$ to drive the reboiler of the first distillation tower? Explain your response. If not, what process change would make this possible?

(c) Consider the first reaction step to make dichloroethane. Show the distribution of chemicals when ethylene is 20% in excess of the stoichiometric amount and the chlorine is entirely converted. Assume that 100,000 lb/hr of vinyl chloride are produced.

(d) Consider the first distillation tower. What is the advantage of cooling the feed to its bubble point at 12 atm as compared with introducing the feed at its dew point?

(e) Why isn't the feed to the pyrolysis furnace heated with the hot pyrolysis products?

(f) What is the function of the trays in the direct chlorination reactor?

(g) Suggest ways to reduce the need for fuel and hot utilities such as steam.

4.4 **(a)** To generate steam at 60 atm, two processes are proposed:

(1) Vaporize water at 1 atm and compress the steam at 60 atm.

(2) Pump water to 60 atm followed by vaporization.

Which process is preferred? Why?

(b) In a distillation tower, under what circumstances is it desirable to use a partial condenser?

4.5 Synthesize a flowsheet for the manufacture of vinyl chloride that corresponds to one of the other branches in the synthesis tree in Figure 4.9. It should begin with reaction path 4 or 5.

4.6 Using the chemical engineering literature, complete the detailed database for the detailed design of the base-case process in Figure 4.19. When appropriate, indicate the kind of data needed from a pilot plant and how this data should be regressed.

Chapter 5

Process Simulation

5.0 OBJECTIVES

In Chapters 3 and 4, the emphasis was on finding molecules and chemical mixtures that have desired properties and on the creation of alternative process flowsheets that arise from the product design problem. The steps in generating the preliminary database, carrying out experiments, performing preliminary process synthesis, preparing a process flow diagram for the base-case design, and developing a detailed database and carrying out pilot-plant testing, prior to preparing the detailed design, were described in Chapter 4. For the production of vinyl-chloride monomer, a synthesis tree was generated and a base-case design was initiated. Throughout, emphasis was on calculations to obtain bubble- and dew-point temperatures, heats of reaction, and so on, and to satisfy material and energy balances, calculations carried out routinely by process simulators. Similarly, for the production of tissue plasminogen activator (tPA), the laboratory tests to locate the target protein, the gene sequence that codes for the protein, and the host cell to be used for growing the protein, were introduced in Chapter 3. Then, a synthesis tree was generated and a base-case design was initiated in Chapter 4. However, no instruction was provided on the use of the process simulators. This is the objective of the current chapter, which focuses on the basics of steady-state process simulation and describes the key role that process simulators play in assisting the design team in process creation. Four of the major process simulators are introduced for steady-state simulation: ASPEN PLUS by Aspen Technology, Inc.; ASPEN HYSYS by Aspen Technology, Inc., and UNISIM by Honeywell Process Solutions (originally by Hyprotech, Ltd.); CHEMCAD by ChemStations, Inc.; and PRO/II by Simulation Sciences, Inc. On the multimedia modules, which can be downloaded from the Wiley Web site associated with this book, detailed instructions are provided on the use of the process simulators, with current emphasis on ASPEN PLUS and ASPEN HYSYS. Because ASPEN HYSYS and UNISIM are variants of HYSYS by Hyprotech, Ltd., all subsequent references to HYSYS in this book refer both to ASPEN HYSYS and to UNISIM, noting that minor differences exist in their user interfaces. Chapter 5 concludes with an introduction to batch process simulation, placing emphasis on BATCH PLUS by Aspen Technology, Inc., and SUPERPRO DESIGNER by Intelligen, Inc.

www.wiley.com/go/global/seider

After studying this chapter, and the associated CD-ROM, the reader should,

1. Understand the role of process simulators in process creation and be prepared to learn about their roles in equipment sizing and costing, profitability analysis, optimization, and dynamic simulation in the chapters that follow.
2. For steady-state simulation, be able to create a simulation flowsheet, involving the selection of models for the process units and the sequence in which process units associated with recycle loops are solved to obtain converged material and energy balances.
3. Understand degrees of freedom in modeling process units and flowsheets, and be able to make design specifications and follow the iterations implemented to satisfy them. When using HYSYS, the reader will learn that its implementation of *bidirectional information flows* is very efficient in satisfying many specifications.
4. Learn the step-by-step procedures for using ASPEN PLUS and HYSYS. The CD-ROM covers many of these steps. Additional assistance is available by consulting the extensive user manuals distributed with the software.
5. Be able to use the process simulators systematically during process creation, following sequences similar to those illustrated later in this chapter for a toluene hydrodealkylation process. The reader will learn to simulate portions of the process (the reactor section, the distillation section, etc.) before attempting to simulate the entire process with its recycle loops. Many examples and exercises enable the reader to master these techniques.
6. Be able to use the batch process simulators to carry out material and energy balances, and to prepare an operating schedule in the form of a Gantt chart for the process.

5.1 INTRODUCTION

Having concentrated on the generation of process flowsheets in Chapter 4, this chapter focuses on the role of analysis, that is, the solution of the material and energy balances coupled with phase equilibria, and the equations of transport and chemical kinetics. The emphasis in this chapter is on finding suitable operating conditions for processes (temperatures, pressures, etc.). Computing packages that model the process units are introduced and utilized to model the highly integrated flowsheets commonly designed to achieve more profitable operation. As has been mentioned, these packages are referred to as *process simulators*, most of which are used to simulate potential processes in the steady state—that is, to determine the unknown temperatures, pressures, and component and total flow rates at steady state. More recently, these packages have been extended to permit the dynamic simulation of processes and their control systems as they respond to disturbances and changes in operating points. Chapter 5 concentrates on steady-state simulation and the scheduling of batch processes during process creation; Chapter 12 shows how to use the HYSYS dynamic simulator to confirm that a potential process is easily controlled as typical disturbances arise. Both Chapters 5 and 12 are accompanied by extensive coverage on the multimedia modules that accompany this book, which explains how to use ASPEN PLUS and HYSYS. The Aspen Icarus Process Evaluator (Aspen IPE) package, provided by Aspen Technology, Inc., is used for cost estimation and an economics spreadsheet is used for profitability analysis. These topics are covered separately in Sections 22.7 and 23.8, respectively. Finally, the packages have extensive facilities for process optimization, with Chapter 24 and the multimedia modules concentrating on optimization in ASPEN PLUS and HYSYS.

Often during the synthesis steps and the creation of the base-case design, as mentioned when synthesizing the vinyl-chloride process in Sections 4.4 and 4.5, process simulators are utilized by the design team to calculate heat duties, power requirements, phase and chemical equilibria, and the performance of multistaged towers, among many other calculations. For the production of commodity chemicals, as the alternative flowsheets evolve, it is common to perform these calculations assuming operation in the steady state; hence, many *steady-state simulators* have become available to process engineers. For the production of specialty chemicals in batch processes, it is common to perform similar calculations using batch process simulators.

In this chapter, the principles behind the use of several widely used flowsheet simulators are introduced. For processes in the steady state, these include ASPEN PLUS, HYSYS, CHEMCAD, and PRO/II. For batch processes, these include BATCH PLUS and SUPERPRO DESIGNER.

The multimedia modules, which can be downloaded from the Wiley Web site associated with this book also explain how to use the dynamic simulators. Emphasis is placed on HYSYS. Using HYSYS, the design team can complete a steady-state simulation, add controllers, and activate the integrator to carry out a dynamic simulation. Similar facilities are provided in ASPEN DYNAMICS by Aspen Technology, Inc.

A primary objective of this chapter is to show how to use the process simulators during process synthesis to better define the most promising processes. After the basic principles are covered, a case study is presented in which the simulators are used to help synthesize the reactor and separation sections of a toluene hydrodealkylation process. Finally, a case study is presented in which the BATCH PLUS simulator is used to help synthesize the operating schedule for a tissue plasminogen activator (tPA) process. Many of the details concerning the process simulators are presented on the multimedia modules associated with this book. The latter cover the ASPEN PLUS and HYSYS simulators, with step-by-step audio instructions for completing the input forms. Some coverage of BATCH PLUS is also provided. In addition, the multimedia modules provide video segments from a large-scale petrochemical complex to illustrate some of the equipment being modeled, tutorials on the estimation and regression of physical property data. Also, .bkp and .hsc files for the ASPEN PLUS and HYSYS examples throughout the book can be downloaded from the Program and Simulation Files folder on the Wiley Web site associated with this book.

5.2 PRINCIPLES OF STEADY-STATE FLOWSHEET SIMULATION

Given the detailed process flow diagram for a base-case design (e.g., Figure 4.19), or a process flow diagram after the task-integration step in process synthesis, or even an incomplete flow diagram after one of the earlier steps, it is often possible to use a process simulator to solve for many of the unknown temperatures, pressures, and flow rates in the steady state. For an existing process, analysis using a process simulator is performed routinely to study potential changes in the operating conditions or the possibility of a retrofit to improve its profitability.

In this section, the objective is to cover the basics of steady-state simulation, with an introduction to ASPEN PLUS, HYSYS, CHEMCAD, and PRO/II. However, no attempt is made to show how to use these simulators when carrying out the step-by-step strategy in Chapter 4. This is accomplished in Section 5.3, in which a case study is presented involving the synthesis of a process to hydrodealkylate toluene by reaction with hydrogen to produce benzene and methane. Readers who have experience in using steady-state simulators may find it preferable to skim through these materials, to identify those sections that can add to their understanding, and to proceed to Section 5.3. Others with little or no experience are advised to study this section at least through the subsection on "Flash Vessel Control" before proceeding to the next section.

Process and Simulation Flowsheets

Process flowsheets are the *language* of chemical processes. Like a work of art, they describe an existing process or a hypothetical process in sufficient detail to convey the essential features.

Analysis, or simulation, is the tool chemical engineers use to interpret process flowsheets, to locate malfunctions, and to predict the performance of processes. The heart of analysis is the mathematical model, a collection of equations that relate the process variables, such as stream temperature, pressure, flow rate, and composition, to surface area, valve settings, geometrical configuration, and so on. The steady-state simulators solve for the unknown variables, given the values of certain known quantities.

There are several levels of analysis. In order of increasing complexity, they involve: material balances, material and energy balances, equipment sizing, and profitability analysis. Additional equations are added at each level. New variables are introduced, and the equation-solving algorithms become more complicated.

Fortunately, most *basic* chemical processes involve conventional process equipment: heat exchangers, pumps, distillation columns, absorbers, and so on. For these process units, the equations do not differ among chemical processes. The physical and thermodynamic properties and chemical kinetics constants differ, but not the equations. Hence, it is possible to prepare one or more equation-solving algorithms for each process unit to solve the material and energy balances and to compute equipment sizes and costs. A library of subroutines or models, usually written in FORTRAN or C, that automate such equation-solving algorithms is at the heart of process simulators. These subroutines or models are hereafter referred to as *procedures, modules*, or *blocks*. As discussed at the end of this section, in a small but growing class of simulators (e.g., gPROMS, and as options in ASPEN PLUS and HYSYS), equations that model a process unit are stored, rather than embedded in FORTRAN or C subroutines that solve the equations associated with the model for each process unit. Given the interconnecting streams, equations for the units in a process are assembled to be solved simultaneously by an equation solver such as the Newton–Raphson method.

To use a flowsheet simulator effectively, it is very helpful to distinguish between process flowsheets and the so-called *simulation flowsheets* associated with process simulators. These distinctions are drawn in the next subsections.

Process Flowsheets

A process flowsheet is a collection of icons to represent process units and arcs to represent the flow of materials to and from the units. The process flowsheet emphasizes the flow of material and energy in a chemical process, as illustrated in Figure 5.1.

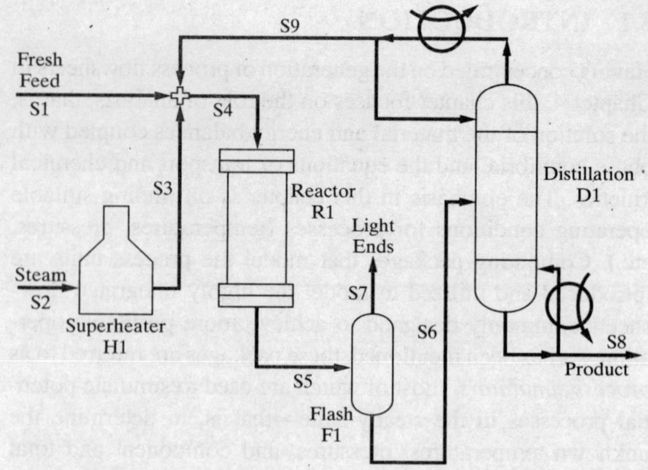

Figure 5.1 Process flowsheet.

Simulation Flowsheets

A simulation flowsheet, on the other hand, is a collection of simulation units to represent computer programs (subroutines or models) that *simulate* the process units and arcs to represent the flow of information among the simulation units. A simulation flowsheet emphasizes information flows. The analogy between the process flowsheet and the simulation flowsheet is illustrated by comparing Figures 5.1 and 5.2a. The latter has been prepared specifically for ASPEN PLUS. The simulation flowsheet may use blocks or icons to represent the process units. For ASPEN PLUS, Figure 5.2a uses blocks whereas Figure 5.2b uses icons. Figures 5.2c, 5.2d, and 5.2e for HYSYS, CHEMCAD, and PRO/II, respectively, use icons.

Several constructs appear in Figure 5.2:

1. The *arcs* represent the transfer of flow rates, temperature, pressure, enthalpy, entropy, and vapor and liquid fractions for each stream. The stream names can be thought of as the names of vectors that store stream variables in a specific order, as illustrated for ASPEN PLUS in the unnumbered table that follows Figure 5.2.

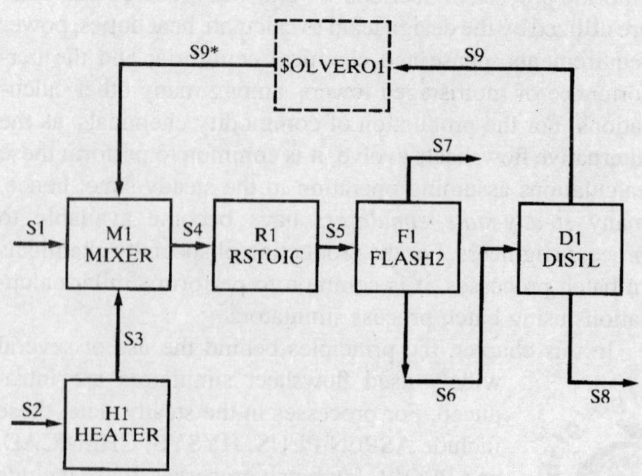

(a)

Figure 5.2 Simulation flowsheet: (a) ASPEN PLUS blocks; (b) ASPEN PLUS icons; (c) HYSYS icons; (d) CHEMCAD icons; (e) PRO/II icons.

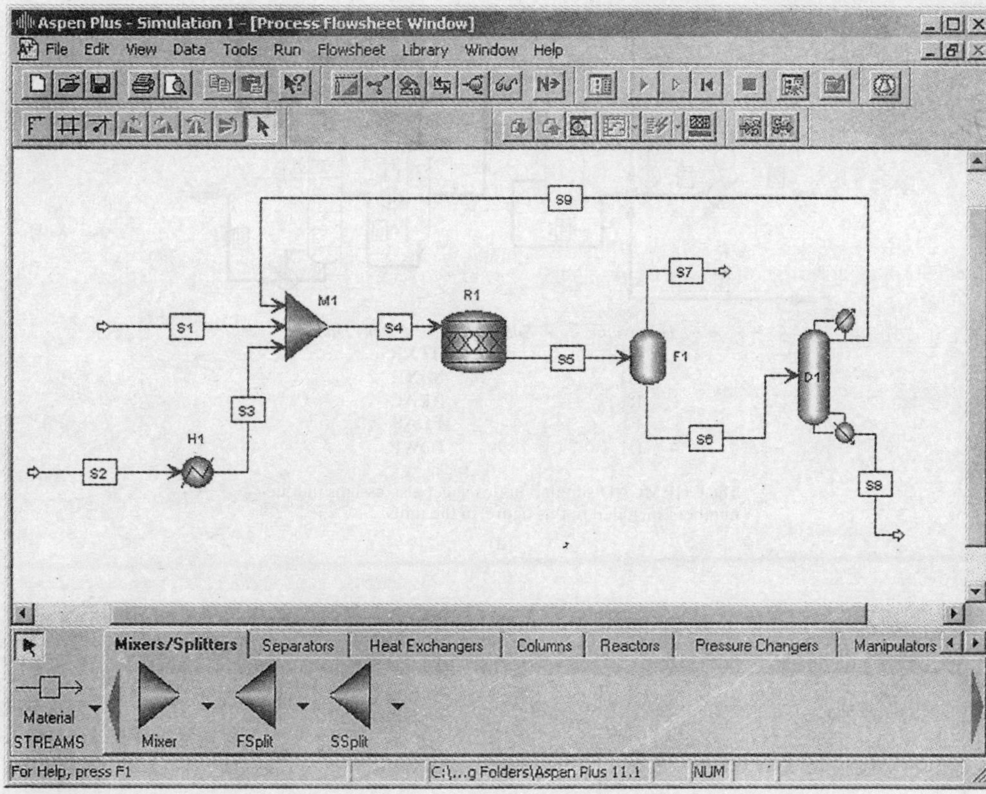

(b)

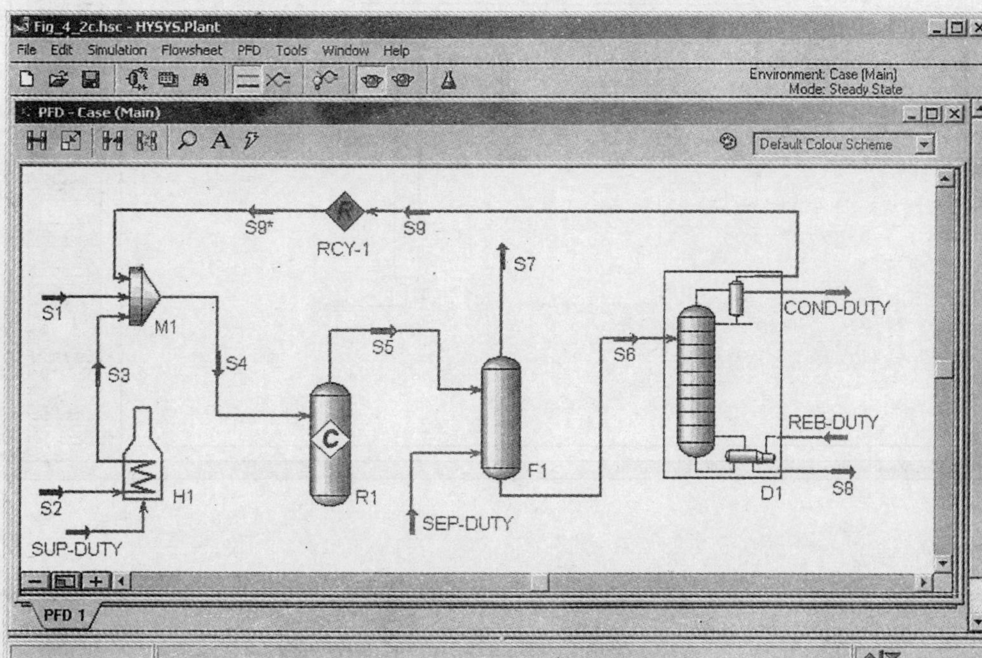

Unit Name	*Model Name*
H1	Heater
M1	Mixer
R1	Reactor
F1	Separator
D1	Column

(c)

Figure 5.2 (*Continued*)

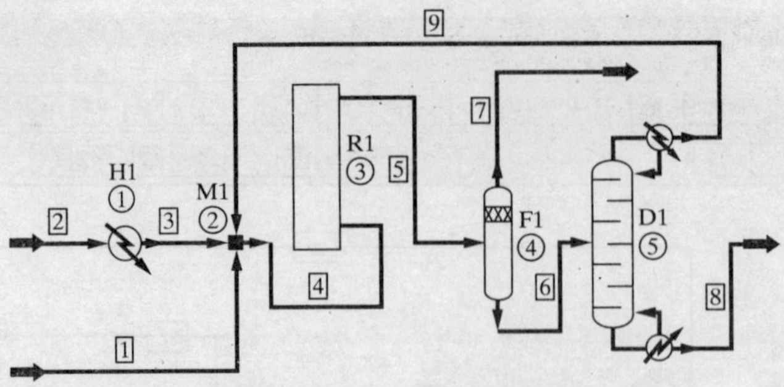

Unit Name	Subroutine (or Block) Name
H1	HTXR
M1	MIXE
R1	REAC
F1	FLAS
D1	TOWR

The CHEMCAD simulation flowsheet also assigns unique numbers, included in this figure, to the units.

(d)

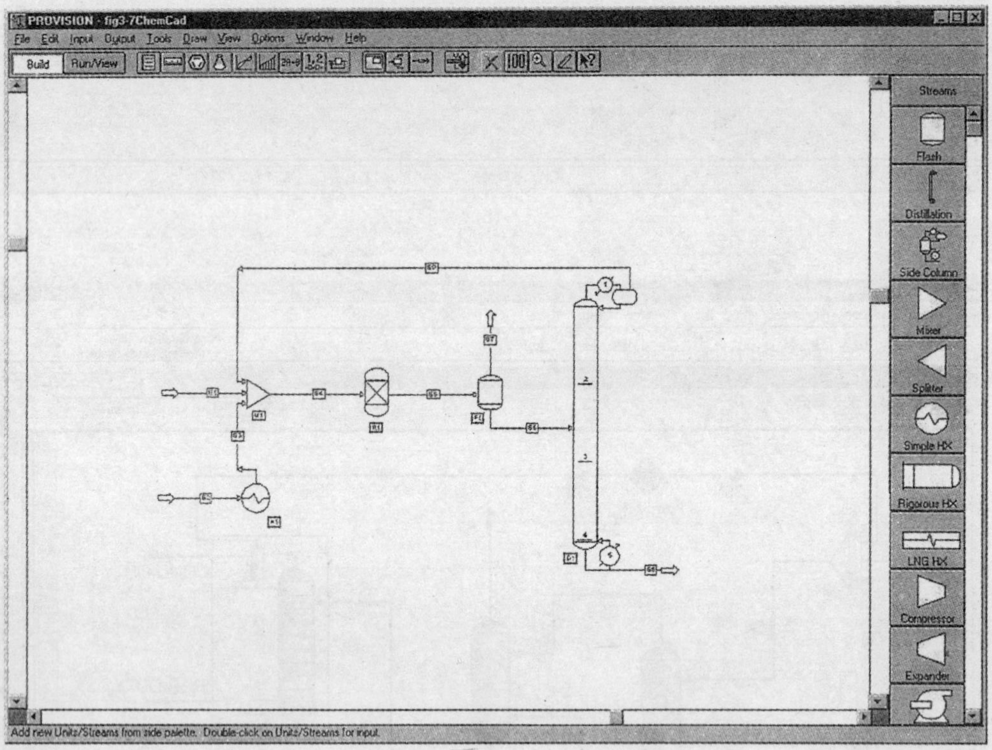

(e)

Figure 5.2 (*Continued*)

Vector Element		Vector Element (Continued)	
1 to C	chemical flow rates, kmol/s	$C+5$	vapor fraction (molar)
$C+1$	total flow rate, kmol/s	$C+6$	liquid fraction (molar)
$C+2$	temperature, K	$C+7$	mass entropy, J/kg K
$C+3$	pressure, MPa	$C+8$	density, kg/m^3
$C+4$	mass enthalpy, J/kg	$C+9$	molecular weight, kg/kmol

where C is the number of chemical species.

2. The *solid-line rectangles* in Figure 5.2a, and the icons in Figures 5.2b–5.2e, represent *simulation units*. In Figure 5.2a, the upper character string provided by the user is a unique name of the simulation unit or *unit name*. The lower character string is the name of the subroutine or model, or so-called *block name* in many of the simulators. Although *model* or *block* are commonly used, the term *subroutine* is used throughout this book to emphasize that the models are computer codes. The equations to model a process unit involve equipment parameters, such as area, number of equilibrium stages, or valve settings. Although different values of the parameters characterize each occurrence of a process unit in the process flowsheet, the same subroutine or model is often used several times in a simulation flowsheet. In Figure 5.2c, for HYSYS, the unit names are in upper case, having been selected to be identical to those in Figure 5.2a, and the model names are tabulated separately in boldface (for emphasis, since HYSYS does not use upper case to represent its model names). In Figure 5.2d, for CHEMCAD, the unit names are the upper character string, and the subroutine (or block) names are tabulated separately. In Figure 5.2e, for PRO/II, the unit names are shown, and the subroutine names are on the menu of icons.

3. The *dashed-line rectangle* in Figure 5.2a represents a mathematical convergence unit that uses a subroutine to adjust the stream variables in the information recycle loop because iterative calculations are necessary. These are discussed under "Recycle." Note that when entering a simulation flowsheet into most of the simulators, the mathematical convergence unit is not specified or shown in the flowsheet. Rather, it is positioned by the simulator unless the user intervenes. HYSYS is an exception, as shown in Figure 5.2c, in which the user positions the recycle convergence unit, RCY-1. In Figures 5.2d and 5.2e, CHEMCAD and PRO/II do not show the convergence unit, but it exists and is transparent to the user.

To convert from a process flowsheet to a simulation flowsheet, one replaces the process units with appropriate simulation units. For each simulation unit, a subroutine (or block, or model) is assigned to solve its equations. Each of the simulators has an extensive list of subroutines (or blocks, or models) to model and solve the process unit equations. In most cases, the models range from approximate to detailed and rigorous, with the most approximate models used during the initial steps of process synthesis and the more rigorous models gradually substituted as fewer flowsheets remain competitive. To make effective usage of the simulators, process engineers need to become familiar with the underlying assumptions in the models provided by each simulator. These are described in user manuals that accompany simulator software. It is an objective of this section, and especially the multimedia modules, which can be downloaded from the Wiley Web site associated with this book, to discuss the principal models available. Partial lists for the four major simulators are provided in Table 5.1.

Table 5.1 Unit Subroutines

(a) ASPEN PLUS [excluding solids-handling equipment—see Seider et al. (1999) Table A-IV.2]		
Mixers and splitters	MIXER	Stream mixer
	FSPLIT	Stream splitter
Separators	SEP	Component separator—multiple outlets
	SEP2	Component separator—two outlets
Flash drums	FLASH2	Two-outlet flash drums
	FLASH3	Three-outlet flash drums
Decanters	DECANTER	Two liquid phase decanter (FLASH3 without vapor)
Approximate distillation	DSTWU	Winn–Underwood–Gilliland design
	DISTL	Edmister simulation
	SCFRAC	Edmister simulation—complex columns
Multistage separation	RADFRAC	Two and three phases, with or without reaction
(Equilibrium-based simulation)	MULTIFRAC	Ditto—with interlinked column sections
	PETROFRAC	Ditto—for petroleum refining
	EXTRACT	Liquid-liquid extractors
(Mass transfer simulation)	RATESEP	Two phases—mass transfer model for staged or packed columns
Heat exchange	HEATER	Heater or cooler
	HEATX	Two-stream heat exchanger
	MHEATX	Multistream heat exchanger
	HETRAN	Interface to B-JAC program for shell-and-tube heat exchangers
	AEROTRAN	Interface to B-JAC program for air-cooled heat exchangers
	HTRIXIST	Interface to HTRI program for shell-and-tube heat exchangers
	HXFLUX	Heat transfer (radiation or convection) calculation module

(Continued)

Table 5.1 Unit Subroutines (*Continued*)

(a) ASPEN PLUS [excluding solids-handling equipment—see Seider et al. (1999) Table A-IV.2] (*Continued*)		
Reactors	RSTOIC	Extent of reaction specified
	RYIELD	Reaction yields specified
	RGIBBS	Multiphase, chemical equilibrium
	REQUIL	Two-phase, chemical equilibrium
	RCSTR	Continuous-stirred tank reactor
	RPLUG	Plug-flow tubular reactor
Pumps, compressors, and turbines	PUMP	Pump or hydraulic turbine
	COMPR	Compressor or turbine
	MCOMPR	Multistage compressor or turbine
	VALVE	Control valves and pressure reducers
Pipeline	PIPE	Pressure drop in a pipe
	PIPELINE	Pressure drop in a multisegment pipeline
Stream manipulators	MULT	Stream multiplier
	DUPL	Stream duplicator
	SELECTOR	Stream selector
	QTVEC	Loads stream manipulator
(b) HYSYS		
Mixers and splitters	**Mixer**	Stream mixer
	Tee	Stream splitter
Separators	**Component Splitter**	Component separator—two outlets
Flash drums	**Separator**	Multiple feeds, one vapor and one liquid product
	3-Phase Separator	Multiple feeds, one vapor and two liquid products
	Tank	Multiple feeds, one liquid product
Approximate distillation	**Shortcut Column**	Fenske–Underwood design
Multistage separation (Equilibrium-based simulation)	**Column**	Generic multiphase separation, including absorber, stripper, rectifier, distillation, liquid-liquid extraction. Additional strippers and pump-arounds can be added. All models support two or three phases and reactions. Physical property models are available for petroleum refining applications.
Heat exchange	**Cooler/Heater**	Cooler or heater
	Heat Exchanger	Two-stream heat exchanger
	Lng	Multistream heat exchanger
Reactors	**Conversion Reactor**	Extent of reaction specified
	Equilibrium Reactor	Equilibrium reaction
	Gibbs Reactor	Multiphase chemical equilibrium (stoichiometry not required)
	CSTR	Continuous-stirred tank reactor
	PFR	Plug-flow tubular reactor
Pumps, compressors, and turbines	**Pump**	Pump or hydraulic turbine
	Compressor	Compressor
	Expander	Turbine
	Valve	Adiabatic valve
Pipeline	**Pipe Segment**	Single/multiphase piping with heat transfer
(c) CHEMCAD		
Mixers and splitters	MIXE	Stream mixer
	DIVI	Stream splitter
Separators	CSEP	Component separator—multiple outlets
	CSEP	Component separator—two outlets
Flash drums	FLAS	Two-outlet flash drums
	LLVF	Three-outlet flash drums
	VALV	Valve

(*Continued*)

Table 5.1 Unit Subroutines (*Continued*)

(c) CHEMCAD (*Continued*)		
Approximate distillation	SHOR	Winn–Underwood–Gilliland design
Multistage separation	SCDS, TOWR	Two and three phases, with or without reaction
(Equilibrium-based and mass	TPLS	Ditto—with interlinked column sections
transfer simulation)	EXTR	Liquid–liquid extractors
Heat exchange	HTXR	Heater or cooler
	HTXR	Two-stream heat exchanger
	LNGH	Multistream heat exchanger
	FIRE	Fired heater
Reactors	REAC	Extent of reaction specified
	EREA	Two-phase, chemical equilibrium
	GIBS	Multiphase, chemical equilibrium
	KREA	Continuous-stirred tank reactor
	KREA	Plug-flow tubular reactor
Pumps, compressors, and turbines	PUMP	Pump or hydraulic turbine
	COMP, EXPN	Compressor or turbine
Pipeline	PIPE	Pressure drop in a pipe
Stream manipulators	SREF	Stream multiplier
	SREF	Stream duplicator
(d) PRO/II		
Mixers and splitters	MIXER	Combines two or more streams
	SPLITTER	Splits a single feed or mixture of feeds into two or more streams
Flash drums	FLASH	Calculates the thermodynamic state of any stream when two variables are given by performing phase equilibrium calculations
Distillation column	COLUMN	Splits feed stream(s) into its components based on temperature and pressure
		By default, a distillation column includes a condenser and reboiler
Heat exchanger	HX	Heats or cools a single process stream, exchanges heat between two process streams or between a process and utility stream
	HXRIG	Rates a TEMA shell-and-tube heat exchanger, rigorously calculating heat transfer and pressure drop
	LNGHX	Exchanges heat between any number of hot and cold streams; identifies zone temperature crossovers and pinch points
Reactors	REACTOR	Models simultaneous reactions defined by fraction converted
	EQUILIBRIUM	Models one reaction defined as an approach to equilibrium temperature or as a fractional approach to chemical equilibrium
	GIBBS	Simulates a single-phase reactor at minimum Gibbs free energy
	CSTR	Simulates a continuously fed, perfectly mixed reactor; adiabatic, isothermal, or constant volume
	PLUG	Simulates a tubular reactor exhibiting plug-flow behavior (no axial mixing or heat transfer)
Pumps, compressors, and turbines	PUMP	Increases P of a stream
	COMPRESSOR	Compresses the feed stream according to specifications
	EXPANDER	Expands stream to the specified conditions and determines the work produced
	VALVE	Simulates the pressure drop
	PIPE	Simulates the pressure drop in a pipe

In most simulators, new subroutines (or blocks, or models) may be programmed by a user and inserted into the library. These can call, in turn, upon the extensive libraries of subroutines and data banks provided by the process simulators for estimation of the thermophysical and transport properties (see the multimedia modules), equipment sizes and costs (see Section 22.7), and so on.

Observe that a mixing unit, modeled using the MIXER subroutine in ASPEN PLUS, the **Mixer** model in HYSYS, and the MIXE and MIXER subroutines in CHEMCAD and PRO/II, is

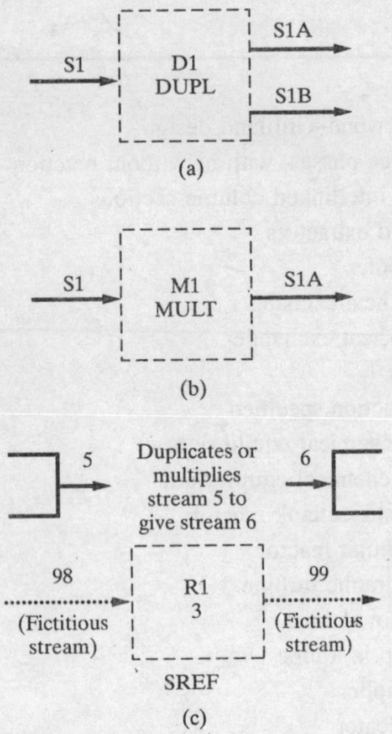

Figure 5.3 Stream manipulators: (a) duplication in ASPEN PLUS; (b) multiplication in ASPEN PLUS; (c) duplication and multiplication in CHEMCAD.

introduced to simulate the mixing of streams, even though mixing is often performed in an actual process in the pipeline. Similarly, a stream dividing unit, modeled using the FSPLIT subroutine in ASPEN PLUS, the **Tee** model in HYSYS, and the DIVI and SPLITTER subroutines in CHEMCAD and PRO/II, is needed to branch a flow to two or more destinations. Note that simulators often have units to duplicate a process stream or multiply the flow rate of a process stream. In ASPEN PLUS, the DUPL subroutine is used to prepare two identical copies (S1A and S1B) of stream S1, as shown in Figure 5.3a. In Figure 5.3b, the MULT subroutine multiplies the flow rate of stream S1 to give stream S1A. Dashed lines are used because these simulation units do not correspond to process units in a chemical plant. Duplication and multiplication are accomplished quite differently and with the same subroutine, SREF, in CHEMCAD. Referring to Figure 5.3c, streams 98 and 99 entering and leaving, respectively, unit R1, which is modeled by the SREF subroutine, are fictitious and are not streams in the actual process. The SREF subroutine requires that the user state the name of the stream to be duplicated or multiplied, for example, stream 5, and the name of the resulting duplicated stream, for example, stream 6.

The steady-state models in simulators do *not* solve time-dependent equations. They simulate the steady-state operation of process units (operation subroutines) and estimate the sizes and costs of process units (cost subroutines). Two other types of subroutines are used to converge recycle computations (convergence subroutines) and to alter the equipment parameters (control subroutines). These subroutines are discussed in this section.

Each of the simulators has a similar syntax for specifying the topology of the simulation flowsheet. In ASPEN PLUS, for the simulation flowsheet in Figure 5.4a, the engineer draws the flowsheet in Figure 5.4b. Similarly, in HYSYS, CHEMCAD, and PRO/II, the simulation flowsheets are shown in Figures 5.4c–5.4e. Because the instructions for a new user of ASPEN PLUS are involved, new users are referred to the module *ASPEN → Principles of Flowsheet Simulation → Creating a Simulation Flowsheet* in the multimedia modules, which can be downloaded from the Wiley Web site associated with this book. The ASPEN PLUS *Getting Started* manual is another good source of these instructions. For a new user of HYSYS, instructions are found in the multimedia modules by referring to the module *HYSYS → Principles of Flowsheet Simulation → Getting Started in HYSYS*.

www.wiley.com/go/global/seider

When using the process simulators, it is important to recognize that, with some exceptions, most streams are comprised of chemical species that distribute within one or more solution phases that are assumed to be in phase equilibrium. The exceptions are streams involving so-called nonconventional components, which are usually solids such as coal, ash, and wood. ASPEN PLUS has facilities for modeling these streams, but new users should not be concerned with these models until they have gained experience with streams in phase equilibrium.

For each stream in vapor–liquid equilibrium, there are $C + 2$ degrees of freedom, where C is the number of chemical species. These degrees of freedom can be satisfied by specifying C species flow rates (or $C - 1$ species mole fractions and the total flow rate) and two intensive variables such as the temperature, pressure, vapor fraction, or enthalpy. For example, when specifying the species flow rates for a stream and its pressure and temperature, all of the intensive properties are computed by solving the vapor–liquid equilibrium equations. These properties include the vapor fraction, enthalpy, and entropy. Alternatively, when the pressure and vapor fraction are specified, the remaining intensive properties are computed. Bubble-point and dew-point temperatures are computed by specifying the vapor fraction to be zero and unity, respectively.

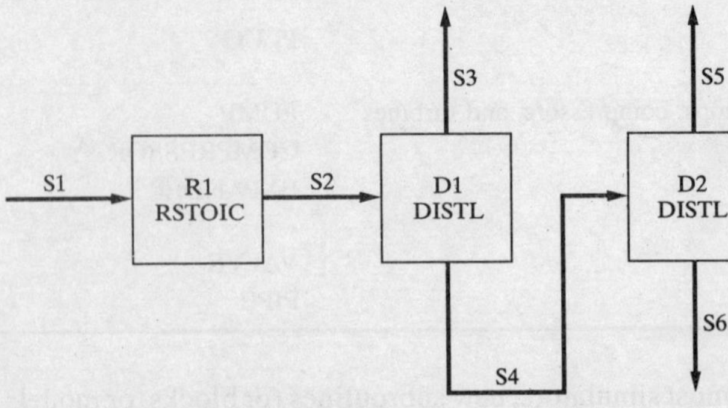

Figure 5.4 Acyclic flowsheet: (a) simulation flowsheet; (b) ASPEN PLUS flowsheet form; (c) HYSYS PFD; (d) CHEMCAD simulation flowsheet; (e) PRO/II simulation flowsheet.

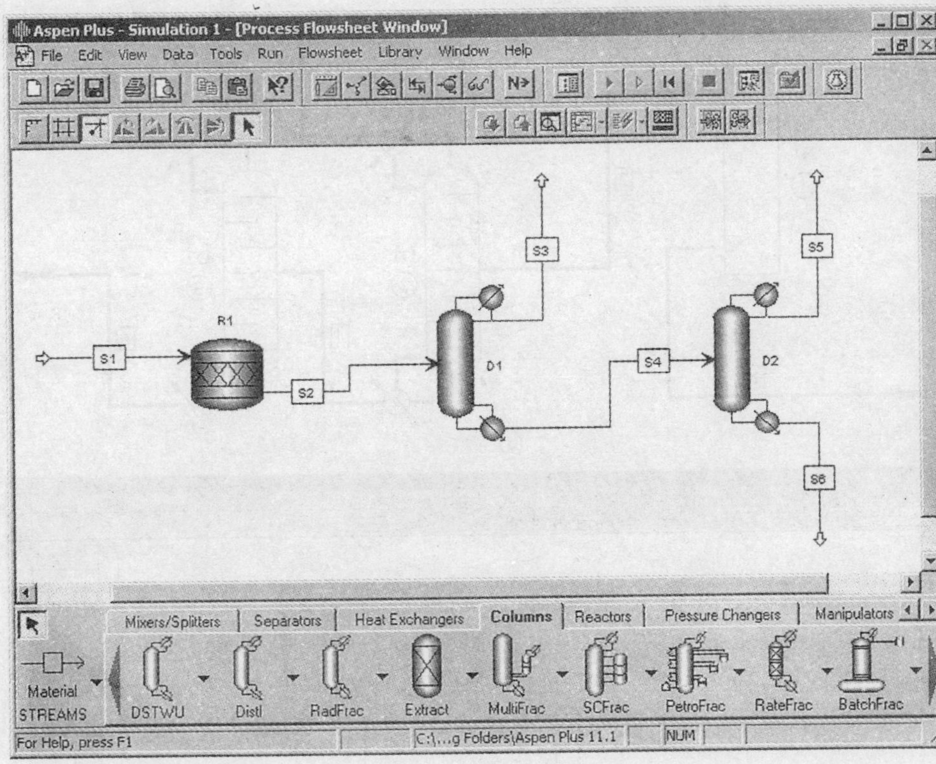

(b)

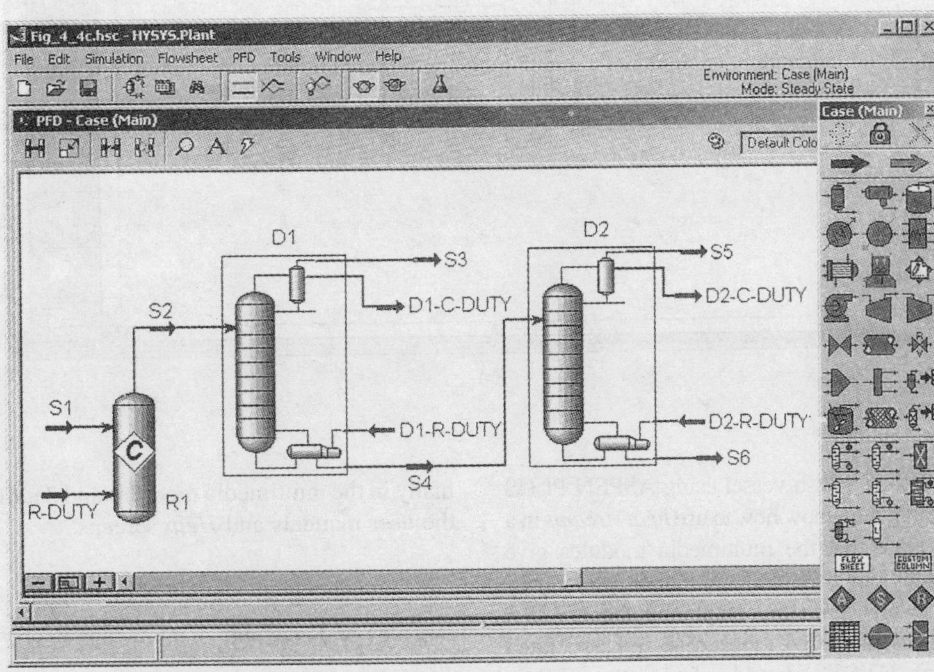

(c)

Figure 5.4 *(Continued)*

Since phase equilibrium is a major segment of courses in thermodynamics, it does not seem appropriate to devote space in a chapter on process simulation to this subject. Yet it is important, when learning to use the process simulators, to understand how they apply the theory of phase equilibrium in modeling streams as well as so-called *flash vessels*, that is, vapor-liquid separators. In the multimedia module *ASPEN → Separators → Phase Equilibria and Flash* and in the module *HYSYS → Separations → Flash*, concepts on phase equilibria and flash separations are reviewed. In addition, and perhaps of more use to many readers, these modules present the

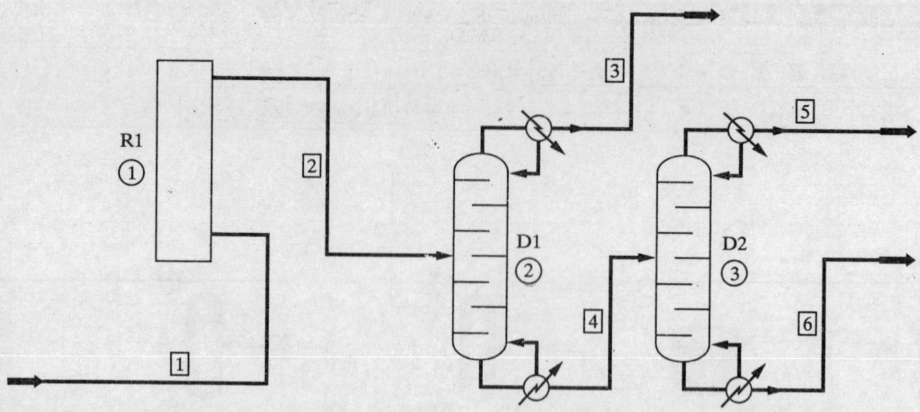

(d)

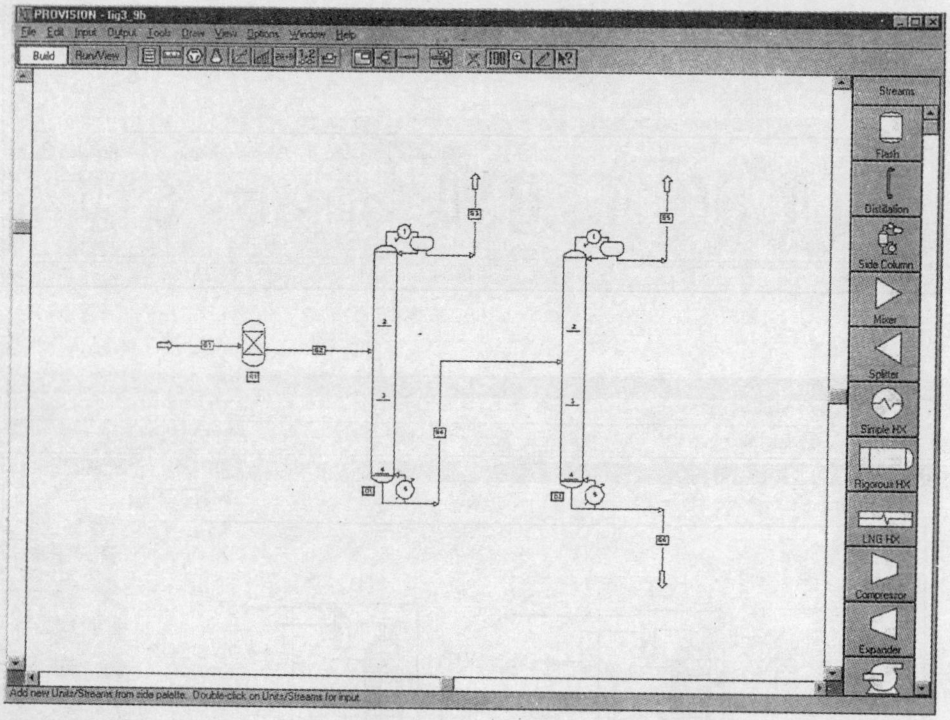

(e)

Figure 5.4 (*Continued*)

solution to a simulation of a flash vessel using ASPEN PLUS and HYSYS. In so doing, they show how to use *heat streams* in a process simulation. Note that the multimedia modules give audio tutorials for completing the ASPEN PLUS input forms and examining the results, and for completing the HYSYS simulation flowsheet, its *Workbook* specsheets, and associated inputs. Readers using HYSYS should refer to the module *HYSYS → Principles of Flowsheet Simulation → Getting Started in HYSYS → Units Catalog* for a list of the unit models, with links to modules that provide detailed information.

Unit Subroutines

Table 5.1 lists the unit subroutines (or blocks, or models) in each of the four simulators. Several of the subroutines are referred to in the sections that follow, with descriptions of many in the multimedia modules and detailed descriptions in the user manuals and *Help* screens.

Degrees of Freedom

A degrees-of-freedom analysis (Smith, 1963; Rudd and Watson, 1968; Myers and Seider, 1976) is incorporated in the development of each subroutine (or block, or model) that simulates a process unit. These subroutines solve sets of $N_{\text{Equations}}$ involving $N_{\text{Variables}}$, where $N_{\text{Equations}} < N_{\text{Variables}}$. Thus, there are $N_D = N_{\text{Variables}} - N_{\text{Equations}}$ degrees of freedom, or input (decision) variables. Most subroutines are written for known values of the input stream variables, although HYSYS permits specification of a blend of input and output stream variables, or output stream variables entirely.

EXAMPLE 5.1

Consider the cooler in Figure 5.5, in which the binary stream S1, containing benzene and toluene at a vapor fraction of $\phi_1 = 0.5$, is condensed by removing heat, Q. Carry out a degrees-of-freedom analysis.

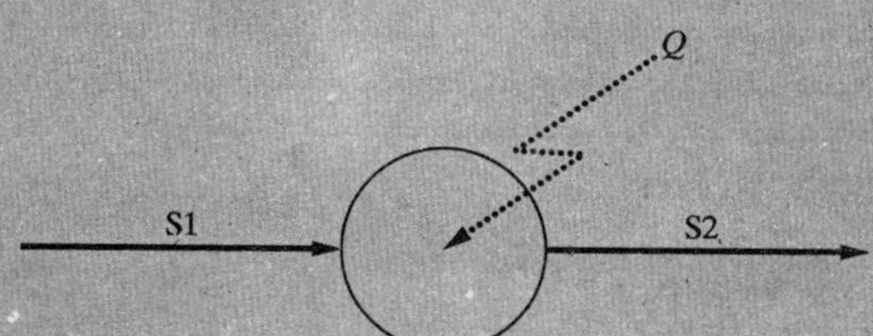

Figure 5.5 Schematic of a cooler.

SOLUTION

At steady state, the material and energy balances are

$$F_1x_{B1} = F_2x_{B2} \quad (5.1)$$

$$F_1x_{T1} = F_2x_{T2} \quad (5.2)$$

$$F_1h_1 + Q = F_2h_2 \quad (5.3)$$

where F_i is the molar flow rate of stream i, x_{ji} is the mole fraction of species j in stream i, and h_i is the enthalpy of stream i, which can be expressed as

$$h_i = h_i\{P_i, \phi_i, \underline{x}_i\} \quad i = 1, 2 \quad (5.4)$$

and $x_{Ti} = 1 - x_{Bi}$, $i = 1, 2$. Note that, in this case, the pressure, P, and vapor fraction, ϕ, accompany the mole fractions as the $C + 2$ intensive variables that provide the enthalpy and other intensive variables of each stream. For this model, $N_{\text{Equations}} = 7$ and $N_{\text{Variables}} = 13(F_i,\ h_i,\ P_i,\phi_i,\ x_{Bi},\ \text{and}\ x_{Ti},\ i = 1, 2, \text{and}\ Q)$. Hence, $N_D = 13 - 7 = 6$, and one set of specifications is comprised of the variables of the feed stream $(F_1, P_1, \phi_1, x_{B1})$ and P_2 and Q. In the process simulators, so-called heater and cooler subroutines are provided to solve the equations for specifications like these.

EXAMPLE 5.2

Consider the mixer in Figure 5.6, in which binary streams S1 and S2, also containing benzene and toluene, are mixed isobarically to form stream S3. Carry out a degrees-of-freedom analysis.

SOLUTION

At steady state, its material and energy balances are

$$F_1x_{B1} + F_2x_{B2} = F_3x_{B3} \quad (5.5)$$

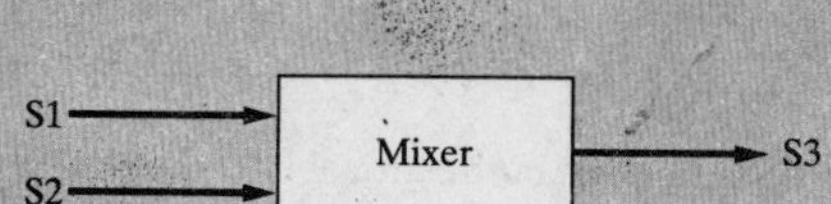

Figure 5.6 Benzene–toluene mixer.

$$F_1x_{T1} + F_2x_{T2} = F_3x_{T3} \quad (5.6)$$

$$F_1h_1 + F_2h_2 = F_3h_3 \quad (5.7)$$

Using temperature and pressure as the intensive variables, Eq. (5.4) becomes

$$h_i = h_i\{T_i, P, \underline{x}_i\} \quad i = 1, 2, 3 \quad (5.8)$$

and $x_{Ti} = 1 - x_{Bi}$, $i = 1, 2, 3$. For this model, $N_{\text{Equations}} = 9$, and $N_{\text{Variables}} = 16$ $(F_i,\ h_i,\ T_i,\ x_{Bi},\ \text{and}\ x_{Ti},\ i = 1, 2, 3, \text{and}\ P)$. Hence, $N_D = 16 - 9 = 7$, and a common set of specifications is comprised of the variables of the feed streams $(F_1, x_{B1}, T_1, \text{and}\ F_2, x_{B2}, T_2)$ and P.

Consider the information flows between a unit subroutine and the stream and equipment vectors in Figure 5.7 for the FLASH2 subroutine of ASPEN PLUS. These are typical of the subroutines (or models, or blocks) in all of the simulators: In ASPEN PLUS, FORTRAN subroutines that model the process units have access to vectors containing the *inlet (feed)* and *outlet (product) stream variables*, and *equipment parameters*, respectively. The equipment parameter vectors are created as the ASPEN PLUS forms described in the multimedia module *ASPEN → Separators → Phase Equilibria and Flash → Flash vessels → FLASH2* are completed by the user; in this case, specifications for the temperature and pressure are entered. Assume that the process consists only of a flash vessel, modeled by the FLASH2 subroutine. Then, the variables for the stream FEED are entered into its vector. Estimates of the enthalpy, vapor and liquid fractions, entropy, and density are computed by the property estimation system. After all of the forms have been completed, an ASPEN PLUS program is generated by ASPEN PLUS. This program is a compact representation of the specifications provided on the forms. It has many paragraphs, two of which are shown in Figure 5.7 (see the multimedia module for the entire program). Next, ASPEN PLUS interprets the program, generates a calculation sequence (providing the order in which the simulation units are computed), and calls the appropriate subroutine (model) for each simulation unit.

During execution of a unit subroutine, the stream vectors and equipment parameters are accessed, from a so-called B vector in ASPEN PLUS, and changes are recorded when new values are computed as the equations are solved. In most of the simulators, the unit subroutines take the variables of the feed streams as input and compute the variables of the product streams; most equipment parameters are specified, but some are computed and stored.

In the schematic of the FLASH2 subroutine in Figure 5.7, on the second line, the B vector, which contains the stream vectors and equipment parameters for all of the streams and simulation units, is referred to in the B common storage.

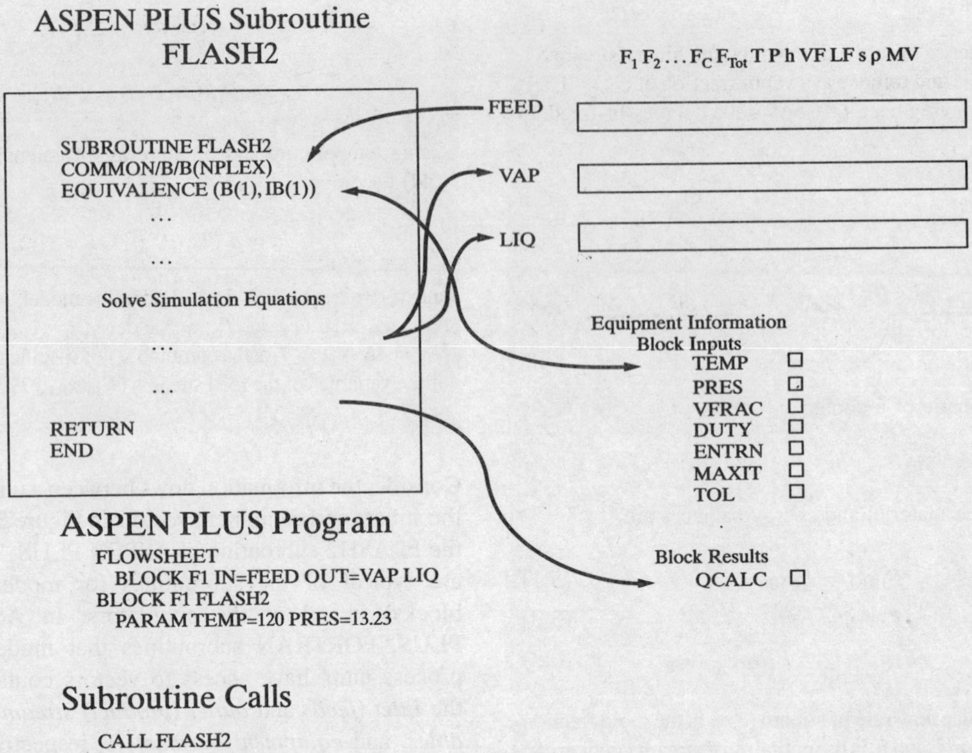

Figure 5.7 ASPEN PLUS unit subroutine—information transfer.

When simulating the flash vessel, F1, the stream variables are taken from the FEED vector and two of the equipment parameters, TEMP, PRES, VFRAC, and DUTY, are taken from the subroutine inputs. As the flash equations are solved, the stream variables are stored in the VAP and LIQ vectors and the heat duty is stored as a parameter, QCALC.

The parameters to be entered by the user for each simulation unit are summarized in connection with its input form, and the associated *Help* information, as well as in Volume 1 of the *ASPEN PLUS Reference Manual → Unit Operation Models Reference Manual*. Note that many default values are provided by ASPEN PLUS that are replaced easily by the user as desired. Upon completion of all of the unit computations, the latest values of all of the stream variables and equipment parameters are displayed on the forms or placed in a report file for printing.

For each unit, the vector of parameters computed by a unit subroutine is saved for display and printing, and to initiate iterative computations for subsequent executions of the subroutine.

In HYSYS, the models are programmed to reverse the information flow when appropriate, that is, to accept values for the variables of the product streams and to compute the variables of the feed streams. HYSYS implements the so-called *bidirectional information flow*, as described next.

Bidirectional Information Flow

In nearly all of the flowsheet simulators, the material and energy balances for the process units are solved given specifications for the inlet streams and the equipment parameters, along with selected variables of the outlet streams (e.g., temperatures and pressures). The unknown variables to be computed are usually those of the outlet streams (typically, the flow rates and compositions). The HYSYS simulator is a notable exception in that most combinations of specifications are permitted for each simulation model. With this flexibility, HYSYS can implement a *reverse information flow*, in which specifications are provided for the product streams and the unknown variables of the inlet streams are computed. More commonly, HYSYS implements a *bidirectional information flow*, involving the calculation of the unknown variables associated with the inlet and outlet streams. Whenever a stream variable is altered, the adjacent process units are recomputed. This causes the information to flow in parallel to the material streams when a unit downstream is recomputed, or opposite to the material streams when a unit upstream is recomputed.

EXAMPLE 5.3 ***(Example 5.1 Revisited)***

For the cooler in Figure 5.5, it is desired to specify the vapor fraction of the effluent stream, ϕ_2, the heat duty, Q, and the pressure drop. Can this be accomplished with bidirectional information flow?

SOLUTION

In the HYSYS simulator, bidirectional information flow is utilized to compute the vapor fraction of the feed stream, ϕ_1. This cannot be accomplished directly in the other process simulators, where, instead, an iteration loop is created in which a guess is provided for

ϕ_1 and iterations are carried out until the specified value of ϕ_2 is obtained, as discussed in the next subsection. Note that for the heater or cooler model in most simulators, the vapor fraction can be specified for both streams and the heat duty computed.

Control Blocks—Design Specifications

Occasionally, the need arises to provide specifications for variables or parameters that are not permitted by a unit subroutine (or block, or model). To accomplish this, all of the simulators provide a facility for iterative adjustment of the variables and parameters that are permitted to be specified so as to achieve the desired specifications. Guesses are made for the so-called manipulated variables. Then, the simulation calculations are performed and a *control* subroutine compares the calculated values with the desired specifications, which may be called *set points*. When significant differences, or errors, are detected, the control subroutine prepares new guesses, using numerical methods, and transfers control to repeat the simulation calculations. Since the procedure is analogous to that performed by feedback controllers in a chemical plant (which are designed to reject disturbances during dynamic operation), it is common to refer to these convergence subroutines as *feedback control* subroutines (Henley and Rosen, 1969).

In the HYSYS simulator, this is accomplished by the **Adjust** operation, in CHEMCAD by the CONT subroutine, and in PRO/II by the CONTROLLER subroutine. In ASPEN PLUS, the equivalent is accomplished with so-called *design specifications*. The latter terminology is intended to draw a distinction between simulation calculations, where the equipment parameters and feed stream variables are specified, and design calculations, where the desired properties of the product stream (e.g., temperature, composition, flow rate) are specified and the equipment parameters (area, reflux ratio, etc.) and feed stream variables are calculated. In HYSYS, the **Adjust** operation is used to adjust the equipment parameters and some feed stream variables to meet the specifications of the stream variables. Furthermore, the **Set** object is used to adjust the value of an attribute of a stream in proportion to that of another stream. For assistance in the use of the **Adjust** and **Set** objects, the reader is referred to the multimedia module *HYSYS → Principles of Flowsheet Simulation → Getting Started in HYSYS → Convergence of Simulation.* As was discussed in the subsection on bidirectional information flow, for *all* of its subroutines, HYSYS provides a *bidirectional information flow,* that is, when product stream variables are specified, the subroutines calculate most of the unknown inlet-stream variables. In CHEMCAD, a control unit, with one inlet stream and one outlet stream (which may be identical to the inlet stream), is placed into the simulation flowsheet using the CONT subroutine. The parameters of the control unit are specified so as to achieve the desired value of a stream variable (or an expression involving stream variables) or an equipment parameter (or an expression involving equipment parameters) by manipulating an equipment parameter or a stream variable. This is the *feed-backward mode,* which requires that the control unit be placed downstream of the units being simulated. The CONT subroutine also has *a feed-forward mode.*

As an example of using a feedback control subroutine in ASPEN PLUS, return to the benzene–toluene mixer in Example 5.2.

EXAMPLE 5.4 *(Example 5.2 Revisited)*

For an equimolar feed stream, S1, at 1,000 lbmol/hr and 100°F, the flow rate of a toluene stream, S2, at 50°F is adjusted to achieve a desired temperature of the mixer effluent (e.g., 85°F), as shown in Figure 5.8a. Convergence units for feedback control (design specifications) are shown on simulation flowsheets as *dotted circles* connected to streams and simulation units by *dotted arcs.* The arcs represent the information flow of stream variables to the control unit and information flow of adjusted equipment parameters to simulation units. Note that the control units of most simulators can adjust the flow rates of the streams. After the calculations by the MIXER subroutine are completed, the control subroutine samples the effluent temperature. It adjusts the flow rate of stream S2 when the specified temperature is not achieved and transfers to the MIXER subroutine to repeat the mixing calculations. This cycle is repeated until the convergence criteria are satisfied or the maximum number of iterations is exceeded.

Instructions to create a design specification using ASPEN PLUS for the mixing unit M1 are provided in the multimedia module *ASPEN → Principles of Flowsheet Simulation → Control Blocks—Design Specifications.*

Based on the input specifications in this module, ASPEN PLUS generates the program in the module, and the simulator reports that

```
SEQUENCE USED WAS: $OLVER01 M1
                   (RETURN $OLVER01)
```

The iterative procedure used by \$OLVER01 is initiated in the manner shown in Figure 5.9a. As indicated above, an initial guess for the manipulated variable (800 lbmol/hr), and the minimum and maximum values of the manipulated variable (0 and 2,000 lbmol/hr), are provided. Then, \$OLVER01 adjusts the manipulated variable, using one of several convergence algorithms, until the convergence tolerances are satisfied with $F_2 = 402.3$ lbmol/hr. When the upper or lower bound is reached, a message is provided

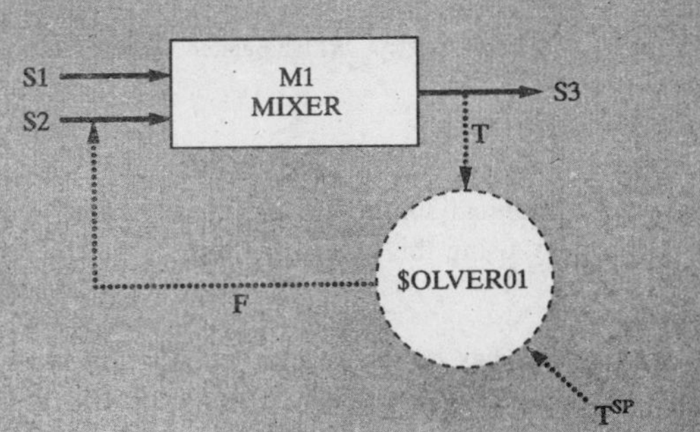

(a)

Figure 5.8 Feedback control—design specifications for the benzene–toluene mixer: (a) ASPEN PLUS blocks; (b) HYSYS icons.

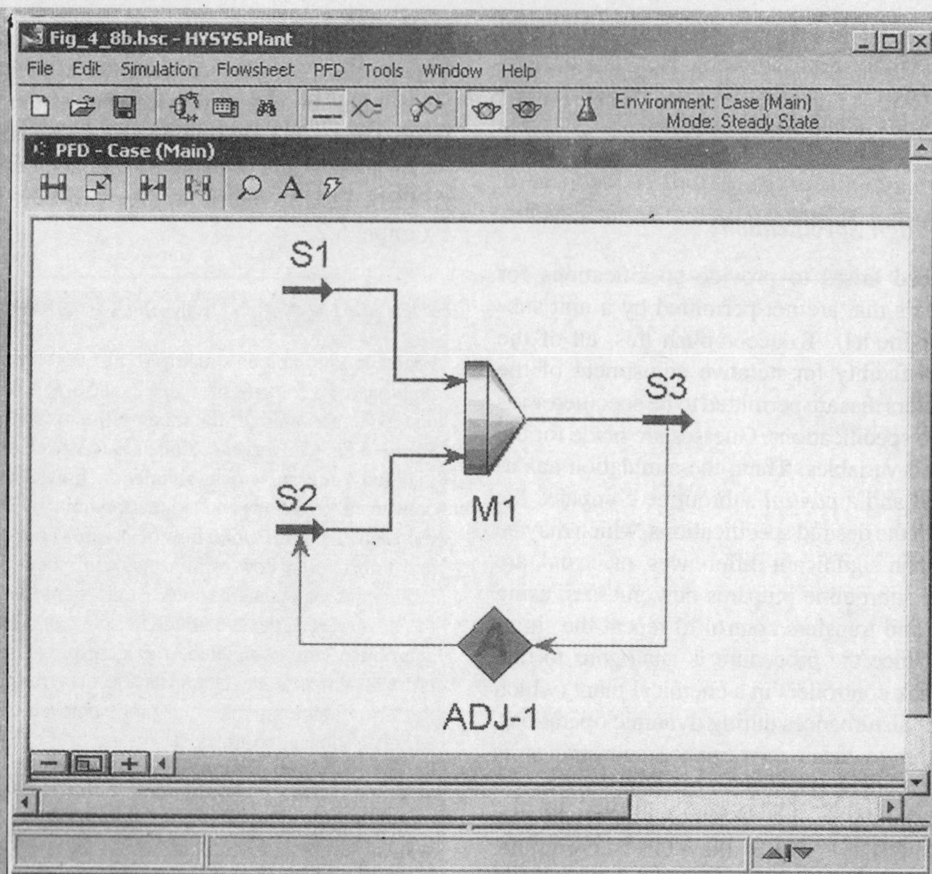

(b)

Figure 5.8 (*Continued*)

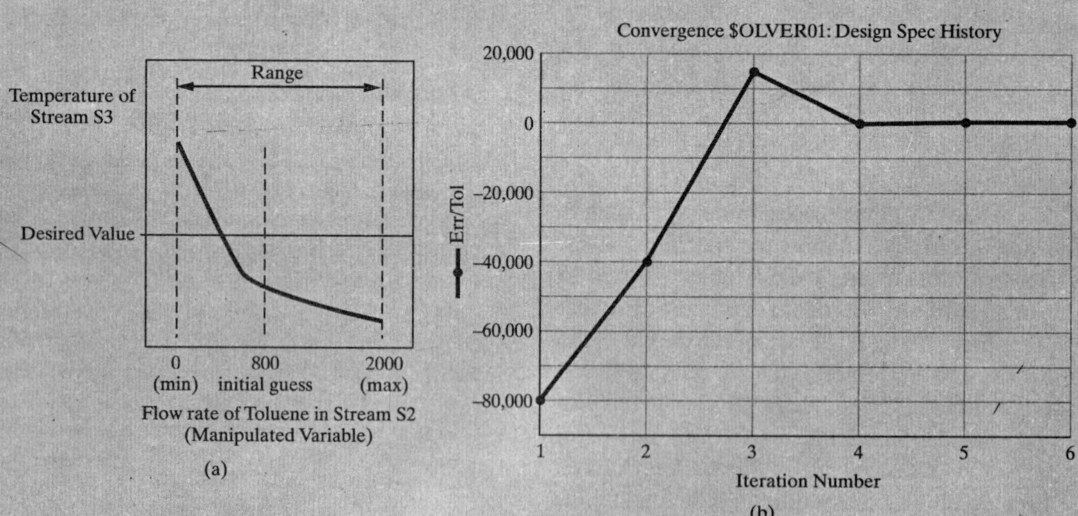

Figure 5.9 Graphical solution of the mixer control problem: (a) specifications for the manipulated variable; (b) ASPEN PLUS iteration history using the secant method.

that convergence has not been achieved. For this example, the secant method was used to achieve convergence, with the iteration history displayed in Figure 5.9b.

For the benzene–toluene mixer, Figure 5.8b shows the HYSYS simulation flowsheet in which the **Adjust** operation manipulates the flow rate of stream S2 to achieve the desired temperature.

Calculation Order

In most process simulators, the units are computed (simulated) one at a time. The calculation order is automatically computed to be consistent with the flow of information in the simulation flowsheet, where the information flow depends on the specifications for the chemical process. Usually, the variables of the process feed streams are specified and information flows parallel to the material flows. In other words, the calculations proceed from unit to unit, beginning with units for which all of the feed streams have been specified. For the flowsheet in Figure 5.4, the units are calculated in the order R1, D1, and D2, that is, starting from the feed end of the process. Before initiating the computations, ASPEN PLUS is provided with data for the variables of the feed stream, S1, and equipment parameters for the three units. The calculation orders for HYSYS, CHEMCAD, and PRO/II are the same. For HYSYS, the simulation flowsheet is shown in Figure 5.4c, using the **Conversion Reactor, Column,** and **Column** models, respectively. Similarly, the CHEMCAD simulation flowsheet is shown in Figure 5.4d, using the REAC, TOWR, and TOWR subroutines. Finally, the PRO/II simulation flowsheet is shown in Figure 5.4e, using the REACTOR, COLUMN, and COLUMN subroutines.

After the subroutine (or model, or block) computations are completed, all of the stream variables and equipment parameters may be displayed or printed, as illustrated in the report files for ASPEN PLUS in the multimedia module *ASPEN → Principles of Flowsheet Simulation → Interpretation of Input and Output: Program Output*.

www.wiley.com/go/global/seider

Recycle

Flowsheets are rarely acyclic, as in Figure 5.4. In process synthesis, most distributions of chemicals involve recycle streams as in Figure 5.1. For the simpler distributions, where the fractional conversions or the extents of reaction are known, the split fractions are specified, and no purge streams exist, as in the vinyl-chloride process (Figures 4.8 and 4.19), the flow rates of the species in the recycle streams can be calculated directly (without iteration).

When the reaction operations involve reversible reactions or competing reactions, the split fractions of the species leaving the separators are complex functions of the operating conditions (such as the temperatures, pressures, and reflux ratios), and purge streams exist, then iterative calculations are necessary. In these cases, the simulation flowsheets usually contain information recycle loops, that is, cycles for which too few stream variables are known to permit the equations for each unit to be solved independently. For these processes, a solution technique is needed to solve the equations for all of the units in an information recycle loop. One solution technique is to *tear* one stream in the recycle loop, that is, to guess the variables of that stream (Henley and Rosen, 1969; Myers and Seider, 1976; Westerberg et al., 1979). Based on *tear stream* guesses, information is passed from unit to unit until new values of the variables of the tear stream are computed. These new values are used to repeat the calculations until the convergence tolerances are satisfied. The variables of the tear streams are often referred to as *tear variables*.

In process simulators, recycle convergence units are inserted into the tear stream. These units can be represented by dashed rectangles, as illustrated in Figures 5.2a and 5.10a. In so doing, an additional stream vector is created. Convergence units use convergence subroutines to compare the newly computed variables (in the feed stream to the convergence unit) with guessed values (in the product stream from the convergence unit) and to compute new guess values when the two streams are not identical to within convergence tolerances.

In most process simulators, the convergence units are positioned automatically. Consider the flowsheet in Figure 5.10a. The process feed is stream S10, which the user would specify. Unit H1 could then be calculated. The set of units M1, R1, D1, and D2 constitutes a recycle loop. A convergence unit must be placed somewhere in this loop. In a recycle loop, calculations begin with the streams leaving the convergence unit. Each of the units in the loop is then computed, returning to the convergence unit,

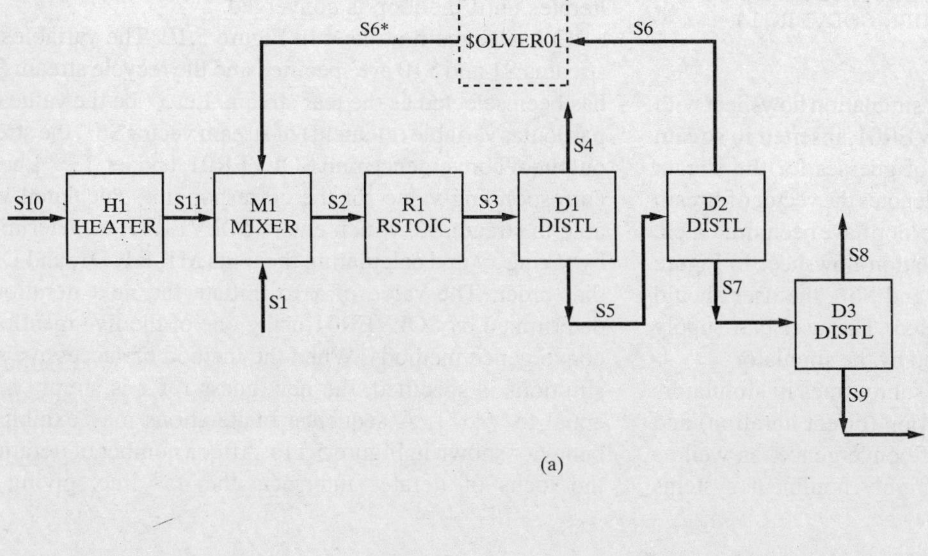

Figure 5.10 Process with recycle: (a) simulation flowsheet; (b) ASPEN PLUS simulation flowsheet.

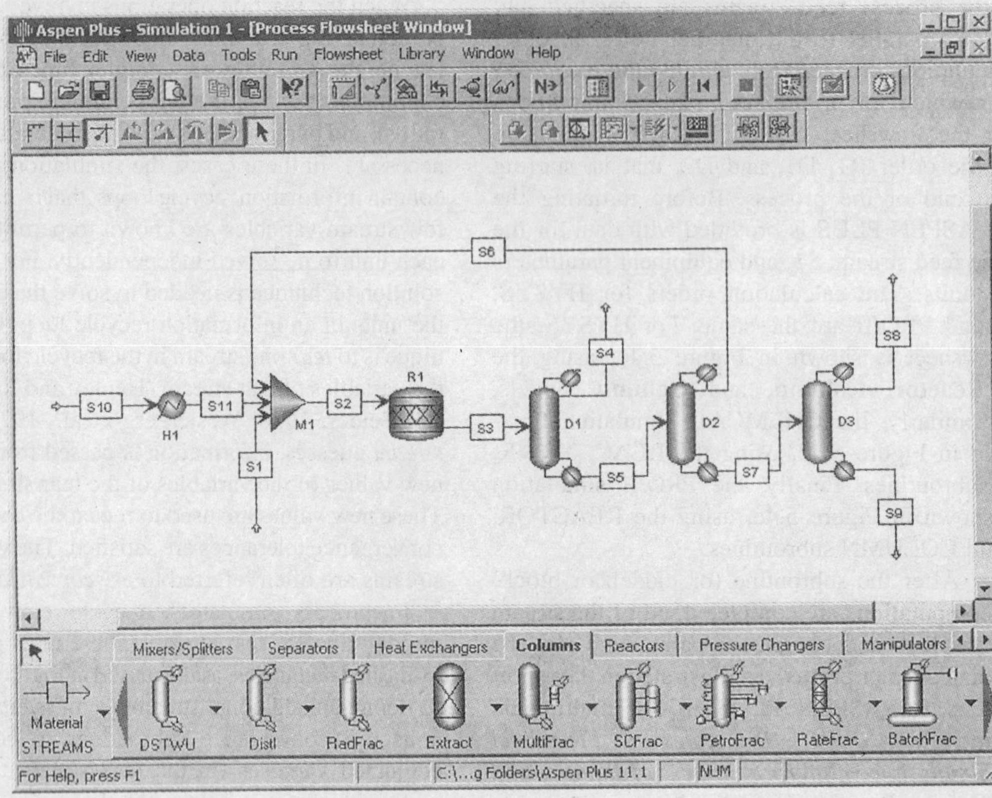

(b)

Figure 5.10 (*Continued*)

where convergence is checked. When convergence is not achieved, the simulator repeats the loop calculations. Upon satisfying the convergence criteria, control is transferred to the unit following the recycle loop in the calculation order. In Figure 5.10a, that unit is D3. ASPEN PLUS names the recycle convergence units \$OLVER01, \$OLVER02, . . . , in sequence. The names of the convergence units are reported in the calculation sequence output, which is illustrated below for the flowsheet in Figure 5.10a:

```
SEQUENCE USED WAS: H1
                   $ OLVER01 M1 R1 D1 D2
                   (RETURN $OLVER01)
                   D3
```

Note that Figure 5.10a shows the simulation flowsheet with the recycle convergence unit, \$OLVER01, inserted in stream S6. Here, S6* denotes the vector of guesses for the stream variables of the tear stream, and S6 denotes the vector of stream variables after the units in the recycle loop have been simulated. Although the ASPEN PLUS simulation flowsheet in Figure 5.10b does not show \$OLVER01 and S6*, the user should recognize that they are implemented. The user can supply guesses for S6*, or they are supplied by the simulator.

All of the recycle convergence subroutines in simulators implement the successive substitution (direct iteration) and the bounded Wegstein methods of convergence, as well as more sophisticated methods for highly nonlinear systems where the successive substitution or Wegstein methods may fail or may be very inefficient. These other methods include the Newton–Raphson method, Broyden's quasi-Newton method, and the dominant-eigenvalue method (Wegstein, 1958; Henley and Rosen, 1969; Myers and Seider, 1976; Westerberg et al., 1979). Each of these five methods determines whether the relative difference between the guessed variables (e.g., for S6* in Figure 5.10a) and calculated variables (e.g., stream S6 in Figure 5.10a) are all less than a prespecified tolerance. If not, the convergence subroutine computes new guesses for its output stream variables and iterates until the loop is converged.

Consider the flowsheet in Figure 5.10. The variables for streams S1 and S10 are specified and the recycle stream (S6) has been selected as the tear stream. Let x^* be the value of a particular variable (element) of stream vector S6*, the stream output of convergence unit \$OLVER01, and let $f\{x^*\}$ be the corresponding value for the corresponding calculated variable in stream S6, which enters \$OLVER01, as determined by taking x^* and calculating the units M1, R1, D1, and D2 in that order. The value of x to initiate the next iteration is determined by \$OLVER01 using one of the five mentioned convergence methods. When the method of successive substitutions is specified, the new guess for x is simply made equal to $f\{x^*\}$. A sequence of iterations may exhibit the behavior shown in Figure 5.11a. After a number of iterations, the locus of iterates intersects the 45° line, giving the

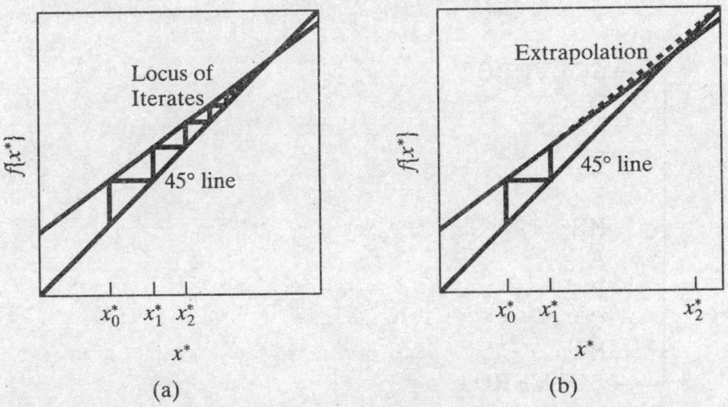

Figure 5.11 Convergence of a recycle loop: (a) successive substitution method; (b) Wegstein's method.

converged value of x in stream S6. When the slope of the locus of iterates $(f\{x\}, x)$ is close to unity (for processes with high recycle ratios), a large number of iterations may be required before convergence occurs.

Wegstein's method can be employed to accelerate convergence when the method of successive substitutions requires a large number of iterations. As shown in Figure 5.11b, the previous two iterates of $f\{x^*\}$ and x^* are extrapolated linearly to obtain the next value of x as the point of intersection with the 45° line. The equation for this straight-line extrapolation is derived easily as

$$x = \left(\frac{s}{s-1}\right)x^* - \left(\frac{1}{s-1}\right)f\{x^*\} \tag{5.9}$$

where s is the slope of the extrapolated line. A more convenient form of Eq. (5.9) uses a weighting function defined by $q = s/(s-1)$, giving

$$x = qx^* + (1-q)f\{x^*\} \tag{5.10}$$

Thus, weights q and $1-q$ are applied, respectively, to x^* and $f\{x^*\}$. Equation (5.10), with q defined by the slope, is usually employed when the slope is less than 1, such that $q < 0$. Typically, q is bounded between -20 and 0 to ensure stability and a reasonable rate of convergence. Wegstein's method reduces to the method of successive substitutions, $x = f\{x^*\}$, when $q = 0$.

When the tear stream is determined automatically by the process simulator, it is possible to override it. For example, ASPEN PLUS selects stream S2, but it can be replaced with stream S6. To do so, select *Convergence* from the *Data* pulldown menu. Then select *Tear*, which produces the *Tear Streams Specifications* form. Enter S6 as the tear stream. Other simulators permit the override in a similar manner.

Figure 5.12a shows a simulation flowsheet with two recycle loops for ASPEN PLUS. Flowsheets for CHEMCAD and PRO/II are identical except for the subroutine (or model) names for the units. Note that no recycle convergence units are shown. This is typical of the simulation flowsheets displayed by most process simulators. The flowsheet for HYSYS is an exception because the recycle convergence unit(s) are positioned by the user and appear explicitly in the flowsheet. For ASPEN PLUS, CHEMCAD, and PRO/II, to complete the simulation flowsheet, either one or two convergence units are inserted, as described below. Note that a single convergence unit suffices because stream S6

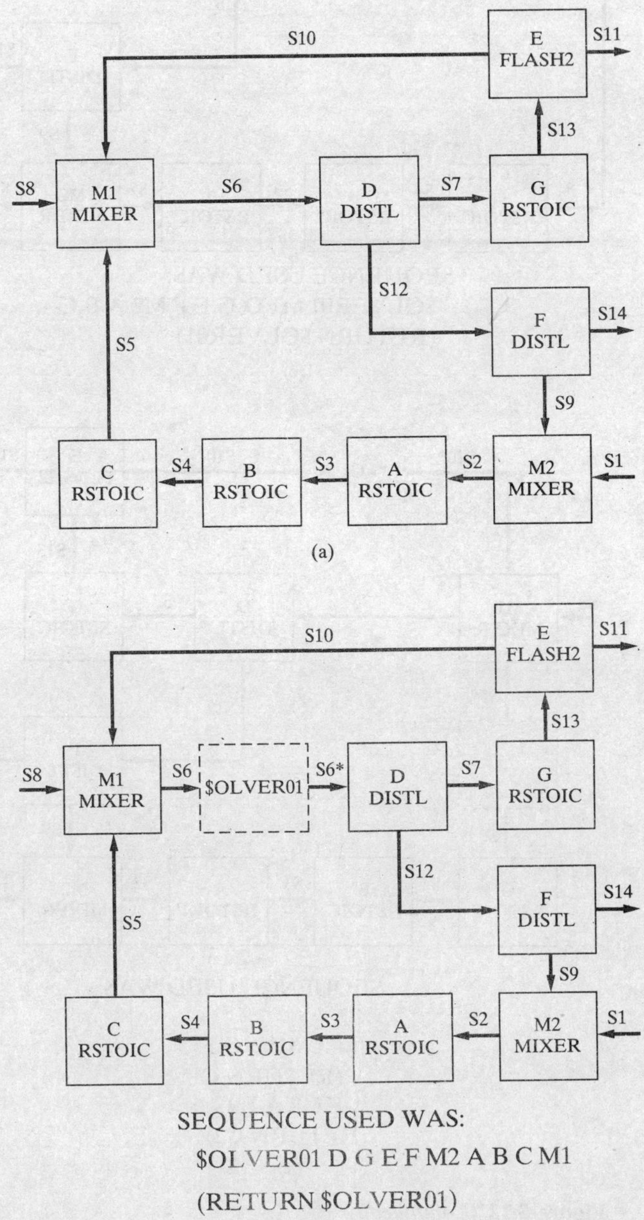

Figure 5.12 Nested recycle loops: (a) Incomplete simulation flowsheet; (b) simulation flowsheet with a single tear stream and a single recycle convergence unit; (c) simulation flowsheet with two tear streams and a single recycle convergence unit; (d) simulation flowsheet with two tear streams and two recycle convergence units.

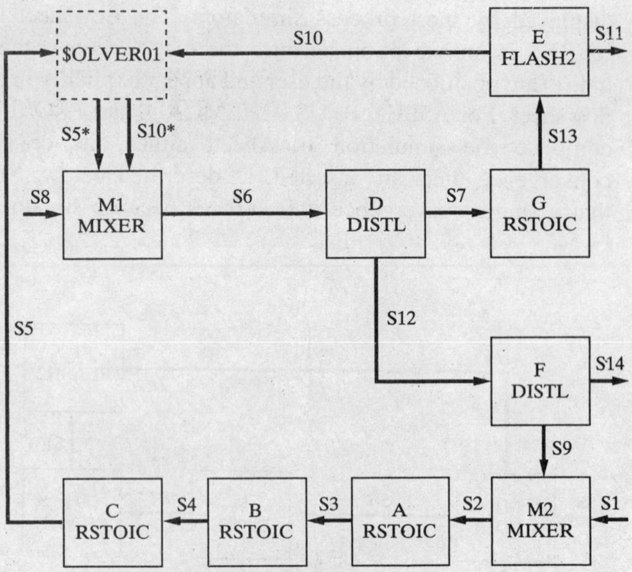

SEQUENCE USED WAS:
$OLVER01 M1 D G E F M2 A B C
(RETURN $OLVER01)

(c)

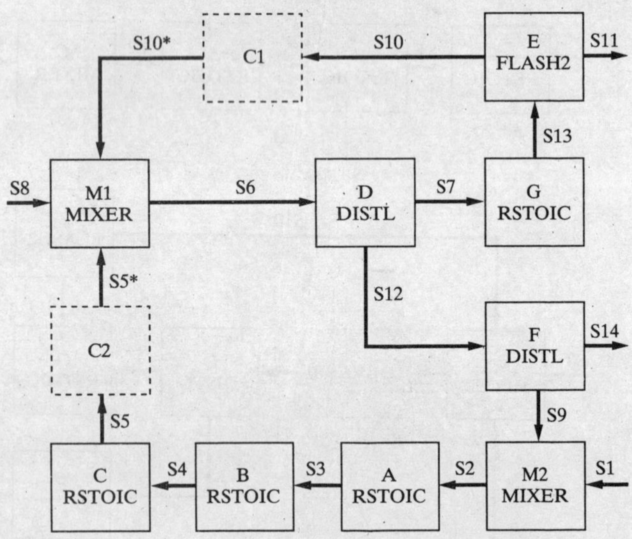

SEQUENCE USED WAS:
C2
C1 M1 D G E
(RETURN C1)
F M2 A B C
(RETURN C2)

(d)

Figure 5.12 (*Continued*)

is common to both loops, as illustrated in Figure 5.12b. Stream S6 is torn into two streams, S6 and S6*, with guesses provided for the variables in S6*. Since no units are outside of the loops, all units are involved in the iterative loop calculations. The calculation sequence is

```
→ $OLVER01
  D
  G
  E
  F
  M2
  A
  B
  C
  M1
└─$OLVER01
```

In ASPEN PLUS, the calculation sequence output is

```
SEQUENCE USED WAS:
 $OLVER01 D G E F M2 A B C M1
 (RETURN $OLVER01)
```

Note that this is the calculation sequence prepared by ASPEN PLUS. Alternatively, when the user prefers to provide guesses for the two recycle streams, S5 and S10, the simulation flowsheet in Figure 5.12c is utilized. To accomplish this in ASPEN PLUS, select *Convergence* from the *Data* pulldown menu. Then, select *Tear,* which produces the *Tear Streams Specifications* form. Enter S5 and S10 as the tear streams. Then the calculation sequence output becomes

```
SEQUENCE USED WAS:
 $OLVER01 M1 D G E F M2 A B C
 (RETURN $OLVER01)
```

In this case, a single convergence unit, $OLVER01, checks for convergence and adjusts the guess values for streams S5 and S10 simultaneously.

Yet another sequence, shown in Figure 5.12d, can be programmed for ASPEN PLUS, with instructions for completing the ASPEN PLUS forms provided in the multimedia module *ASPEN → Principles of Flowsheet Simulation → Recycle → Multiple Recycle Loops.* This results in the calculation sequence output

```
SEQUENCE USED WAS:
 C2
 C1 M1 D G E
 (RETURN C1)
 F M2 A B C
 (RETURN C2)
```

In this sequence, the internal loop, C1, is converged during every iteration of the external loop, C2 (which includes C1). This may be efficient when the units outside C1 require extensive computations.

A more complex flowsheet, which contains three recycle loops, is shown in Figure 5.13a. Two calculation sequences are illustrated in Figure 5.13b and 5.13c. These involve the minimum number of tear streams, S5 and S8, and result in the following output from ASPEN PLUS:

Option 1

```
SEQUENCE USED WAS:
 CONV2 F G
   CONV1 D A B C
   (RETURN CONV1)
 E
 (RETURN CONV2)
```

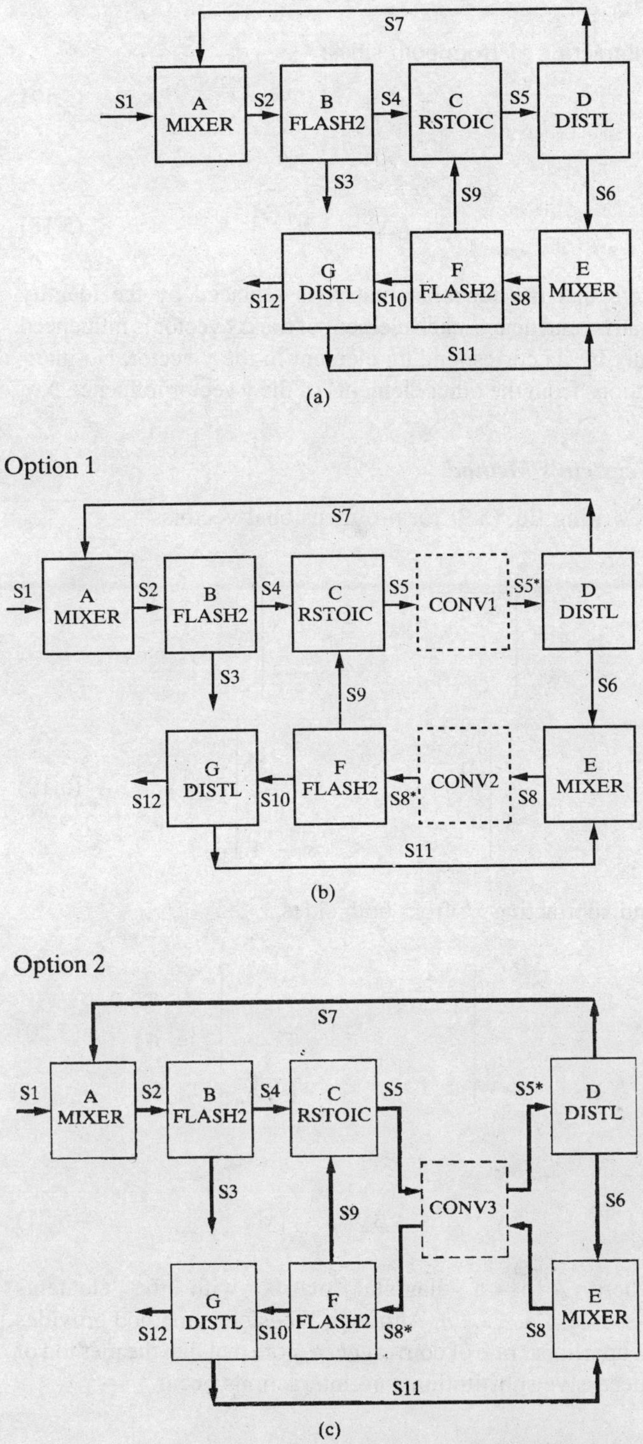

Figure 5.13 Three recycle loops: (a) incomplete simulation flowsheet; (b) simulation flowsheet with two tear streams and two recycle convergence units; (c) simulation flowsheet with two tear streams and one recycle convergence unit.

Option 2

```
SEQUENCE USED WAS:
  CONV3 F G D A B C E
  (RETURN CONV3)
```

In both options, guesses are provided for the variables in streams S5 and S8. In option 1, the internal loop, CONV1, is converged during every iteration of the external loop, CONV2. In option 2, both loops are converged simultaneously. Note that the minimum number of tear streams may not provide for the most rapid convergence. An alternative solution procedure for this flowsheet involves three tear streams, for example, S7, S9, and S11, with one convergence unit.

When using ASPEN PLUS, the details of the convergence forms and the CONVERGENCE paragraph generated can be found under *Help → Using Aspen Plus → Convergence*. See also the multimedia modules in *ASPEN → Principles of Flowsheet Simulation → Recycle*. For HYSYS, the user can consult the modules under *HYSYS → Principles of Flowsheet Simulation → Getting Started in HYSYS → Convergence of Simulation → Recycle* and for CHEMCAD and PRO/II, their user manuals.

Recycle Convergence Methods

In the previous subsection, the successive substitution and Wegstein methods were introduced as the two methods most commonly implemented in recycle convergence units. Other methods, such as the Newton–Raphson method, Broyden's quasi-Newton method, and the dominant-eigenvalue method, are candidates as well, especially when the equations being solved are highly nonlinear and interdependent. In this subsection, the principal features of all five methods are compared.

For the recycle convergence unit in Figure 5.14, let

$$\underline{y} = \underline{f}\{\underline{x}^*\} - \underline{x}^* \tag{5.11}$$

where $\underline{x}^*$ is the vector of guesses for n recycle (tear) variables and $\underline{f}\{\underline{x}^*\}$ is the vector of the recycle variables computed from the guesses after one pass through the simulation units in the recycle loop. Clearly, the objective of the convergence unit is to adjust $\underline{x}^*$ so as to drive y toward zero.

Newton–Raphson Method

The Newton–Raphson second-order method can be written as

$$\underline{\underline{J}}\{\underline{x}^*\}\underline{\Delta x} = -\underline{y}\{\underline{x}^*\} \tag{5.12}$$

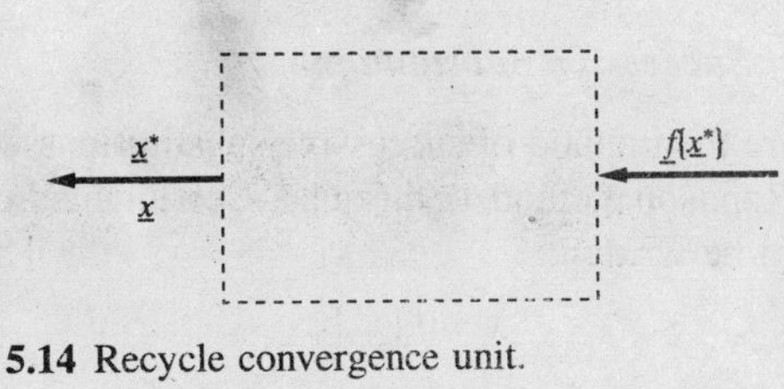

Figure 5.14 Recycle convergence unit.

where $\underline{\Delta x} = \underline{x} - \underline{x}^*$. Substituting and rearranging, the new values of the recycle variables, $\underline{x}$, are

$$\underline{x} = \underline{x}^* - \underline{\underline{J}}^{-1}\{\underline{x}^*\}\underline{y}\{\underline{x}^*\} \tag{5.13}$$

In these equations, the Jacobian matrix:

$$\underline{\underline{J}}\{\underline{x}^*\} = \begin{bmatrix} \frac{\partial y_1}{\partial x_1} & \frac{\partial y_1}{\partial x_2} & \cdots & \frac{\partial y_1}{\partial x_n} \\ \frac{\partial y_2}{\partial x_1} & \frac{\partial y_2}{\partial x_2} & \cdots & \frac{\partial y_2}{\partial x_n} \\ \vdots & \vdots & & \vdots \\ \frac{\partial y_n}{\partial x_1} & \frac{\partial y_n}{\partial x_2} & \cdots & \frac{\partial y_n}{\partial x_n} \end{bmatrix}_{\underline{x}^*} \tag{5.14}$$

is evaluated at $\underline{x}^*$.

In each iteration of the Newton–Raphson method, when the guesses are close to the true values, the length of the error vector, $||\underline{y}||$, is the square of its length after the previous iteration; that is, when the length of the initial error vector is 0.1, the subsequent error vectors are reduced to 0.01, 10^{-4}, 10^{-8}, ... However, this rapid rate of convergence requires that n^2 partial derivatives be evaluated at $\underline{x}^*$. Since most recycle loops involve many process units, each involving many equations, the chain rule for partial differentiation cannot be implemented easily. Consequently, the partial derivatives are evaluated by numerical perturbation; that is, each guess, x_i, $i = 1, \ldots, n$, is perturbed, one at a time. For each perturbation, δx_i, $i = 1, \ldots, n$, a pass through the recycle loop is required to give y_j^p, $j = 1, \ldots, n$. Then the partial derivatives in the ith row are computed by difference:

$$\frac{\partial y_j}{\partial x_i} \simeq \frac{y_j^p - y_j}{\delta x_i} \qquad j = 1, \ldots, n \tag{5.15}$$

This requires $n + 1$ passes through the recycle loop to complete the Jacobian matrix for just one iteration of the Newton–Raphson method; that is, for $n = 10$, 11 passes are necessary, usually involving far too many computations to be competitive.

Alternatively, so-called *secant methods* can be used to approximate the Jacobian matrix with far less effort (Westerberg et al., 1979). These provide a *superlinear* rate of convergence; that is, they reduce the errors less rapidly than the Newton–Raphson method, but more rapidly than the method of successive substitutions, which has a *linear* rate of convergence (i.e., the length of the error vector is reduced from 0.1 to 0.01, 10^{-3}, 10^{-4}, 10^{-5}, ...). These methods are also referred to as *quasi-Newton methods*, with Broyden's method being the most popular.

Method of Successive Substitutions

To compare the method of successive substitutions with the Newton–Raphson method, or the quasi-Newton methods, the former can be written:

$$\underline{x} = \underline{f}\{\underline{x}^*\} \tag{5.16}$$

Subtracting $\underline{x}^*$ from both sides:

$$\underline{x} - \underline{x}^* = \underline{f}\{\underline{x}^*\} - \underline{x}^* \tag{5.17}$$

or

$$-\underline{\underline{I}}\,\underline{\Delta x} = -\underline{y}\{\underline{x}^*\} \tag{5.18}$$

Note that the Jacobian matrix is replaced by the identity matrix, and hence, each element of the $\underline{\Delta x}$ vector is influenced only by its corresponding element in the $\underline{y}$ vector. No interactions from the other elements of the $\underline{y}$ vector influence Δx_i.

Wegstein's Method

Rewriting Eq. (5.9) for n-dimensional vectors,

$$\underline{x} = \begin{bmatrix} \frac{s_1}{s_1 - 1} & & \\ & \ddots & \\ & & \frac{s_n}{s_n - 1} \end{bmatrix} \underline{x}^* - \begin{bmatrix} \frac{1}{s_1 - 1} & & \\ & \ddots & \\ & & \frac{1}{s_n - 1} \end{bmatrix} \underline{f}\{\underline{x}^*\} \tag{5.19}$$

and subtracting $\underline{x}^*$ from both sides,

$$\underline{x} - \underline{x}^* = \begin{bmatrix} \frac{1}{1 - s_1} & & \\ & \ddots & \\ & & \frac{1}{1 - s_n} \end{bmatrix} \underline{y}\{\underline{x}^*\} \tag{5.20}$$

or

$$-\underline{\underline{A}}\,\underline{\Delta x} = -\underline{y}\{\underline{x}^*\} \tag{5.21}$$

where $\underline{\underline{A}}$ is a diagonal matrix with the elements $1 - s_i$, $i = 1, \ldots, n$. Although Wegstein's method provides a superlinear rate of convergence, note that like the method of successive substitutions, no interactions occur.

Dominant-Eigenvalue Method

In the dominant-eigenvalue method, the largest eigenvalue of the Jacobian matrix is estimated every third or fourth iteration and used in place of s_i in Eq. (5.20) to accelerate the method of successive substitutions, which is applied at the other iterations (Orbach and Crowe, 1971; Crowe and Nishio, 1975).

Flash with Recycle Problem

To master the concepts of recycle analysis, it is recommended that the reader solve several of the exercises at the end of the

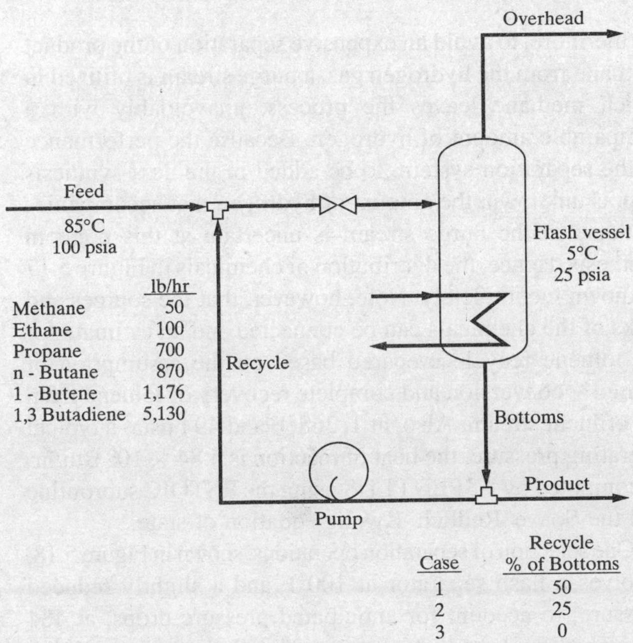

Figure 5.15 Flash with recycle process.

chapter. Of these, the so-called flash with recycle problem (Exercise 5.1a) should be tackled first. Although it involves just one recycle loop, it demonstrates a very important principle. See if you can identify it!

Consider the simple process in Figure 5.15. For the three cases, compare and discuss the flow rates and compositions of the product streams. The model for the flash vessel is presented in the multimedia modules under *HYSYS → Separations → Flash* and *ASPEN → Separators → Phase Equilibria and Flash → Flash Vessels* which include a narrated video of an industrial flash vessel, and the models for the mixer and splitter are rather straightforward. The pump model deserves special attention and is discussed in the multimedia modules under *HYSYS → Pumps, Compressors, & Expanders → Pumps* and *ASPEN → Pumps, Compressors, & Expanders → Pumps* which include a narrated video of an industrial pump.

www.wiley.com/go/global/seider

Note that the flash with recycle process is a good representation of a quench vessel, in which hot gases, typically from an exothermic reactor, are quenched by a cold liquid recycle. Quenches are often needed to provide rapid cooling of a reactor effluent by direct-contact heat transfer. Cold liquid is showered over hot, rising gases. As some of the liquid vaporizes, the latent heat of vaporization is absorbed, and cooling occurs. Quenches are particularly effective for the rapid cooling of organic vapors so as to avoid, or at least reduce, the deposition of solid carbon by chemical reaction. Any solid that is deposited can be bled with the condensate from the vessel bottoms rather easily. The alternative, shell-and-tube heat exchangers, often become fouled with solids and must be shut down periodically for cleaning.

Flash Vessel Control

Next, it is recommended that the reader solve a variation on the flash with recycle problem. In this variation (Exercise 5.1b), case 3 is modified so as to determine the flash temperature to obtain 850 lb/hr of overhead vapor.

Equation-Oriented Architectures

In the discussion thus far, unit subroutines (or blocks, or models) have been utilized to solve the equations that model the process units, given values for the degrees of freedom (i.e., a consistent set of specifications) associated with each process unit. The simulators determine a calculation sequence, which can be altered by the user, for proceeding from equipment subroutine to equipment subroutine in solving the equations associated with the entire process flowsheet. In most simulators, information flows in parallel with the flow of material and energy in the process flowsheet. In HYSYS, bidirectional information flow enables the simulator to return to execute a subroutine when one of its degrees of freedom has been altered, either upstream or downstream. When using subroutines, it is necessary to tear streams in recycle loops and perform iterative calculations. Similarly, when specifying degrees of freedom that require the calculation of equipment parameters, such as the area of a heat exchanger, iterative calculations are necessary to satisfy these so-called *design specifications.*

In contrast, so-called *equation-oriented* simulators have been developed. These include gPROMS (Process Systems Enterprise, Ltd.), and as options in ASPEN PLUS Version 2006 and Version 3.0.1 of ASPEN HYSYS (Aspen Technology, Inc.). In these simulators, libraries of equations are stored to represent the model associated with each process unit. Using the connectivity of the process flowsheet, that is, the streams that connect the process units, a set of equations is assembled for the entire flowsheet. Then, the degrees of freedom are determined by the simulator. The user is required to make enough specifications to satisfy the degrees of freedom. The simulator then solves the independent set of equations. Typically, a variation on the Newton–Raphson method is utilized and convergence is achieved when the residuals of the equations are sufficiently small.

To construct equation-oriented models for an entire process, it becomes important to identify specifications that are consistent, avoiding overspecifying or underspecifying subsets of the equations. When convergence is not achieved, facilities are provided to examine the values of selected variables and the residuals of selected equations. This requires well-designed programs that can display subsets of variables and equation residuals.

Clearly, equation-oriented simulators avoid iterations through subroutines in converging recycle loops and design specifications. Given good initial guesses, this is a major advantage. In ASPEN PLUS, good guesses can be provided by solving the unit subroutines in an initial pass through the flowsheet. Furthermore, ASPEN PLUS permits the creation of a hybrid simulation, in which subroutines are used to solve the equations associated with some process units while the equations associated with the remaining process units are solved simultaneously.

5.3 SYNTHESIS OF THE TOLUENE HYDRODEALKYLATION PROCESS

In this section, process simulators are utilized to assist in carrying out the steps introduced in Sections 4.4 and 4.5 for the synthesis of a process to hydrodealkylate toluene. This process was used actively following World War II, when it became favorable to convert large quantities of toluene, which was no longer needed to make the explosive TNT, to benzene for use in the manufacture of cyclohexane, a precursor of nylon. In this case, a product design alternative involves the conversion of toluene to benzene and, for this purpose, the principal reaction path is well defined. It involves

$$C_7H_8 + H_2 \rightarrow C_6H_6 + CH_4 \qquad \text{(R1)}$$

which is accompanied by the side reaction

$$2C_6H_6 \rightarrow C_{12}H_{10} + H_2 \qquad \text{(R2)}$$

Laboratory data indicate that the reactions proceed irreversibly without a catalyst at temperatures in the range of 1,200–1270°F with approximately 75 mol% of the toluene converted to benzene and approximately 2 mol% of the benzene produced in the hydrodealkylation reaction converted to biphenyl. Since the reactions occur in series in a single processing unit, just a single reaction operation is positioned in the flowsheet, as shown in Figure 5.16. The plant capacity is based on the conversion of 274.2 lbmol/hr of toluene, or approximately 200 MMlb/yr, assuming operation 330 days per year.

One distribution of chemicals involves a large excess of hydrogen gas to prevent carbon deposition and absorb much of the heat of the exothermic hydrodealkylation reaction. Furthermore, to avoid an expensive separation of the product methane from the hydrogen gas, a purge stream is utilized in which methane leaves the process, unavoidably with a comparable amount of hydrogen. Because the performance of the separation system, to be added in the next synthesis step, is unknown, the amount of hydrogen that accompanies methane in the purge stream is uncertain at this point in synthesis. Hence, the distribution of chemicals in Figure 5.17 is known incompletely. Note, however, that the sources and sinks of the chemicals can be connected and an estimate for the toluene recycle prepared based on the assumption of 75 mol% conversion and complete recovery of toluene from the effluent stream. Also, at 1,268°F and 494 psia, a typical operating pressure, the heat of reaction is 5.84×10^6 Btu/hr, as computed by ASPEN PLUS using the RSTOIC subroutine and the Soave–Redlich–Kwong equation of state.

One selection of separation operations, shown in Figure 5.18, involves a flash separator at 100°F and a slightly reduced pressure, to account for anticipated pressure drops, at 484 psia. The liquid product is sent to a distillation train in which H_2 and CH_4 are recovered first, followed by C_6H_6 and then C_7H_8. Note that the pressures of the distillation columns have not yet been entered. These are computed to permit the usage of cooling water in the condensers; that is, the pressures are adjusted to set the bubble- or dew-point temperatures of the vapor streams to be condensed at 130°F or greater. This is accomplished using ASPEN PLUS for simulation of the distillation section, to be discussed shortly.

The next synthesis step involves positioning operations to change the temperatures, pressures, and phases where differences exist between the reaction and separation operations, as well as the sources of the raw materials and sinks for the product chemicals. For this process, the toluene and hydrogen feed streams are assumed to be available at elevated pressure, above that required in the hydrodealkylation reactions. When this is not the case, the appropriate operations to increase the pressure must be inserted. One arrangement of the temperature-, pressure-, and phase-change operations is shown in Figure 5.19 for the reaction section only. Clearly, large quantities of heat are needed to raise the temperature of the feed chemicals to 1,200°F, and similarly large quantities

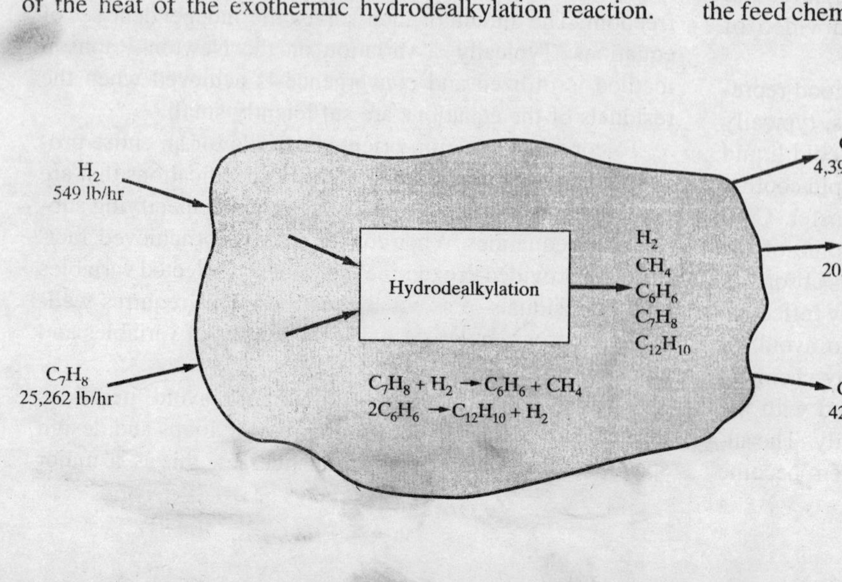

Figure 5.16 Reaction operation for the hydrodealkylation of toluene.

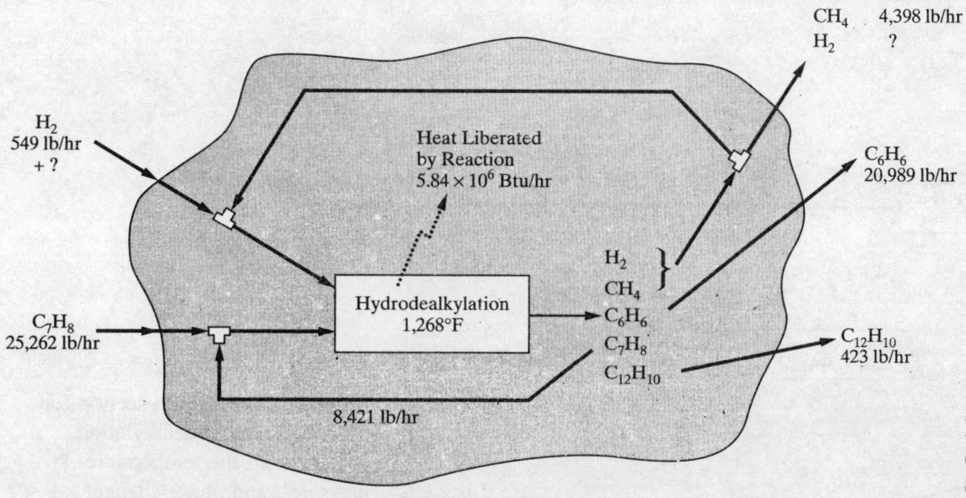

Figure 5.17 Distribution of chemicals for the hydrodealkylation of toluene.

of heat must be removed to partially condense the reactor effluent. These heat loads are calculated by ASPEN PLUS as discussed shortly.

The next synthesis step involves task integration, that is, the combination of operations into process units. In one task integration, shown in Figure 5.20, reactor effluent is quenched rapidly to 1,150°F, primarily to avoid the need for a costly high-temperature heat exchanger, and is sent to a feed/product heat exchanger. There, it is cooled as it heats the mixture of feed and recycle chemicals to 1,000°F. The stream is cooled further to 100°F, the temperature of the flash separator. The liquid from the quench is the product of the reactor section, yet a portion of it is recycled to quench the reactor effluent. The vapor product is recycled after a portion is purged to keep methane from building up in the process. This recycle is compressed to the pressure of the feed chemicals, 569 psia. Returning to the feed/product heat exchanger, the hot feed mixture leaves at 1,000°F and is sent to a gas-fired furnace for further heating to 1,200°F, the temperature of the feed to the reactor. Note that the gases are heated in a tube bank that resides in the furnace, and hence a high pressure drop is estimated (70 psia). On the other hand, the hydrodealkylation reactions take place in a large-diameter vessel that has negligible pressure drop. Clearly, at a later stage in the process design, these pressure drops, along with pressure drops in the connecting pipes, can be estimated. Normally, however, small errors in the pressure drops have only a small impact on the equipment sizes and costs as well as the operating costs.

Process Simulation

As mentioned during the discussion of the synthesis steps, process simulators are very useful. They are used to calculate

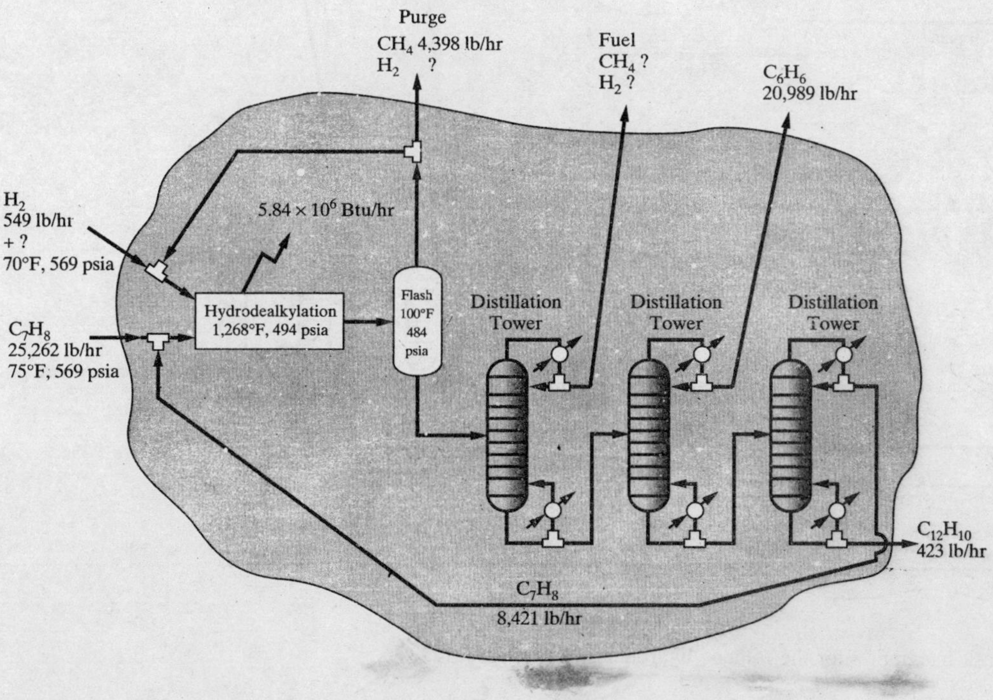

Figure 5.18 Flowsheet including the separation operations for the toluene hydrodealkylation process.

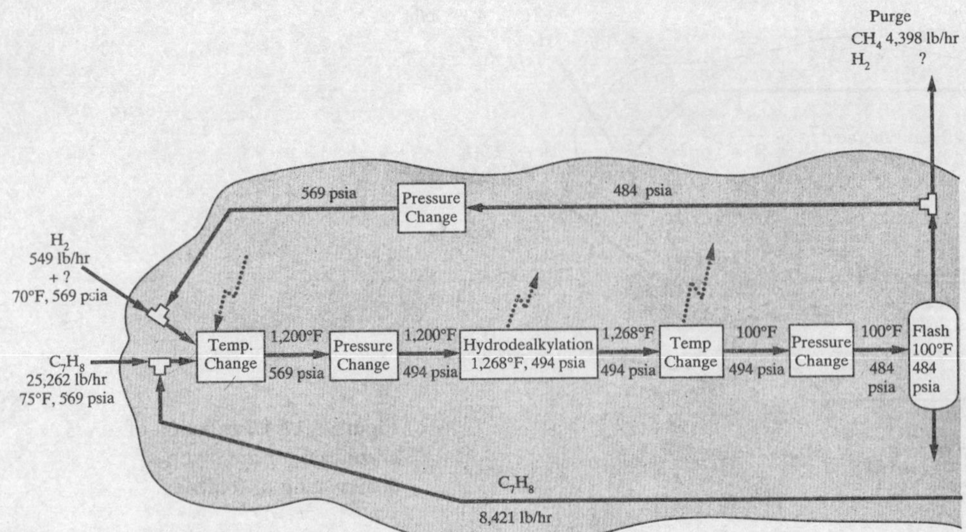

Figure 5.19 Reaction section for the toluene hydrodealkylation process with the temperature-, pressure-, and phase-change operations.

heats of reaction, heat added to or removed from a stream, power requirements for pumps and compressors, performance of a flash separator at various temperatures and pressures, and bubble- and dew-point temperatures associated with distillates and bottoms products, among many other quantities.

In this subsection, three simulations are suggested that, when carried out as exercises, show the more comprehensive role process simulators normally play during process synthesis. The first simulation involves the reactor section of the proposed process. It is intended to provide a better understanding of its performance. Note that several assumptions are made concerning the recycle streams, so as not to complicate the analysis. Then the separation section, involving three distillation towers, is examined, with specifications made for the flow rates and compositions of the product streams. Finally, after obtaining a better understanding of the performance of these two sections, the entire process is simulated. In this simulation, the flow rates and compositions of the recycle and purge streams are computed to satisfy

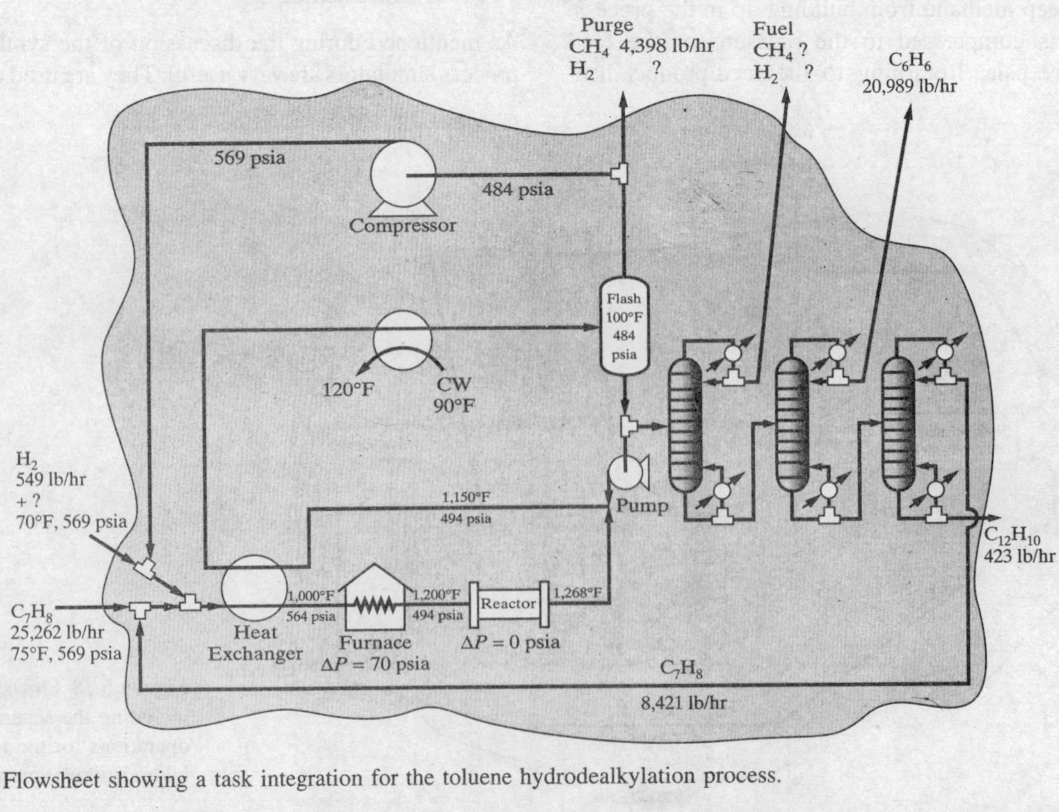

Figure 5.20 Flowsheet showing a task integration for the toluene hydrodealkylation process.

material and energy balances. Of course, during any of these simulations, the specifications can be varied to gain a better understanding of the performance of the process. In Exercise 23.21, you will have an opportunity to use the Aspen Icarus Process Evaluator (Aspen IPE) to size all of the equipment, estimate its installation costs, and perform a profitability analysis.

Simulation of the Reactor Section

The conditions for this simulation are shown in Figure 5.21 and summarized in Exercise 5.2. As mentioned before, representative values are assumed for the flow rates of the species in the gas and toluene recycle streams. Also, typical values are provided for the heat transfer coefficients in both heat exchangers, taking into consideration the phases of the streams involved in heat transfer, as discussed in Section 18.3. Subroutines and models for the heat exchangers and reactor are described in the ASPEN and HYSYS multimedia modules on *Heat Exchangers* and *Chemical Reactors*. In ASPEN PLUS and HYSYS, there are no models for furnaces, and hence it is recommended that you calculate the heat required using the HEATER subroutine and the **Heater** model, respectively. For estimation of the thermophysical properties, it is recommended that the Soave–Redlich–Kwong equation of state be used.

Species Flow Rates (lbmol/hr)	Feed	Recycle	Gas Recycle
H_2	0	0	2,045.9
CH_4	0	0	3,020.8
C_6H_6 (benzene)	0	3.4	42.8
C_7H_8 (toluene)	274.2	82.5	5.3
$C_{12}H_{10}$ (biphenyl)	0	1.0	0

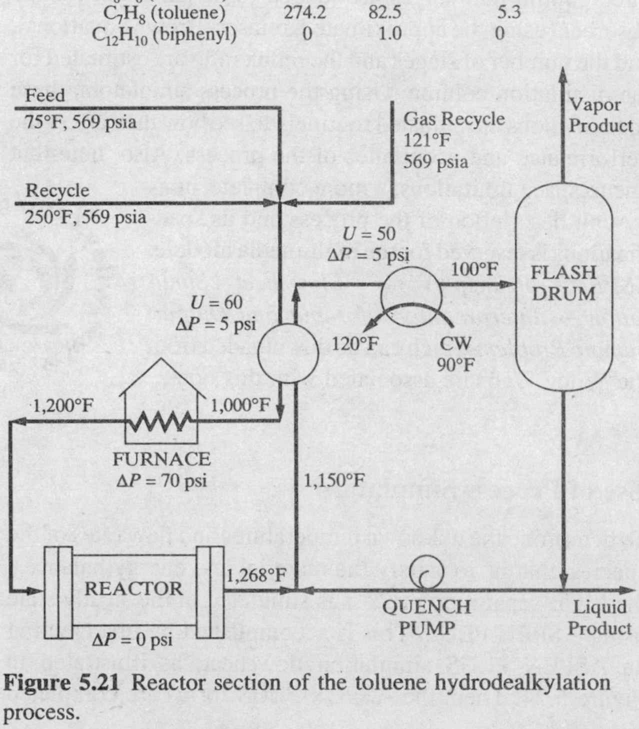

Figure 5.21 Reactor section of the toluene hydrodealkylation process.

Species	Flow Rate (lbmol/hr) Reactor Product Stream	Product 1	Product 2	Product 3
H_2	1.5	1.5		
CH_4	19.3	19.2	0.1	
C_6H_6	262.8	1.3	258.1	3.4
C_7H_8	84.7		0.1	84.6
$C_{12}H_{10}$	5.1			5.1

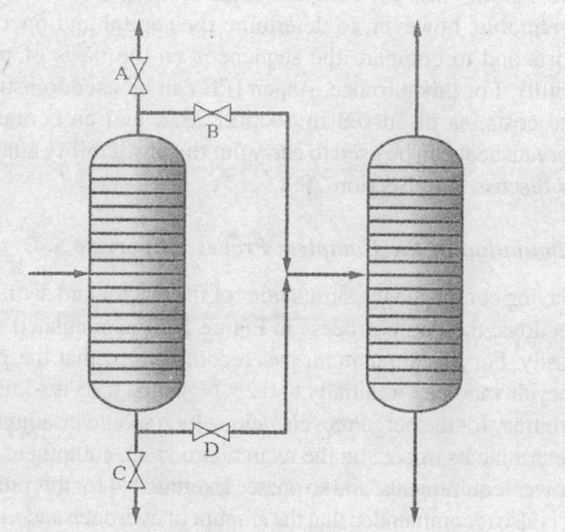

Figure 5.22 Toluene hydrodealkylation process—distillation section.

Simulation of the Distillation Section

The specifications for the distillation section are provided in Figure 5.22, and summarized in Exercise 5.3, in which three product streams are specified. The objective is to determine the tower pressures, number of equilibrium stages, and reflux ratios. In this problem, toluene and biphenyl are lumped together as a single product. Two configurations are examined for separating hydrogen and methane, as a single product, from benzene, and from toluene and biphenyl. Subsequently, the distillation column to separate toluene from biphenyl can be designed.

An objective is to examine the two separation sequences shown. In the *direct sequence*, valves A and D are open, B and C are closed, and product 1 (H_2 and CH_4) is recovered in the distillate of the first tower. Alternatively, in the *indirect sequence*, valves B and C are open, and product 3 (C_7H_8 and $C_{12}H_{10}$) is recovered in the bottoms product of the first tower.

Using the flowsheet simulators, design calculations are needed to estimate the reflux ratio and the theoretical tray requirements for the two towers in each of the sequences. In ASPEN PLUS, this is accomplished with the DSTWU subroutine, which is described in the multimedia module *ASPEN → Separators → Distillation → FUG Shortcut Design*. In HYSYS, the **Shortcut Column** model is used, which is described in the modules under *HYSYS → Separations → Distillation → Shortcut Distillation Column*. The reflux ratio is set, arbitrarily to 1.3 times the minimum and the column pressures are adjusted to

www.wiley.com/go/global/seider

obtain distillate temperatures greater than or equal to 130°F (to permit the use of cooling water for condensation). However, no column pressure is permitted to be less than 20 psia (to avoid vacuum operation). Total condensers are used, except a partial condenser is needed when methane is taken overhead.

The operating pressures, reflux rates, and numbers of trays are a good basis for comparison of the two sequences. It is preferable, however, to determine the capital and operating costs and to compare the sequences on the basis of profitability. For this purpose, Aspen IPE can be used to estimate the costs, as discussed in Section 22.7, and an economics spreadsheet can be used to carry out the profitability analysis, as discussed in Section 23.8.

Simulation of the Complete Process (Exercise 5.4)

Having completed the simulations of the reactor and distillation sections, the entire process in Figure 5.20 is simulated rather easily. For this simulation, it is recommended that the purge/recycle ratio be set initially to 0.25. Note that 0.25 is somewhat arbitrary for the purge/recycle ratio, which should be adjusted to determine its impact on the recirculation rate, equipment sizes, power requirements, and so on; see Exercise 5.4 for this purpose. It is also recommended that the amount of hydrogen added to the process feed stream be adjusted to the amount of hydrogen leaving in the purge stream. This can be accomplished in ASPEN PLUS using a design specification. Also, the initial guesses for the recycle streams can be set equal to the values assumed when simulating the reactor section of the process. For the distillation columns, the RADFRAC subroutine can be used for simulation with the number of stages and the reflux ratio previously calculated by the DSTWU subroutine. See the multimedia module *ASPEN → Separators → Distillation → MESH Equations → RADFRAC* for an example using the RADFRAC subroutine. In HYSYS, the **Column** model is used, as described in the module *HYSYS → Separations → Distillation → Column Setup*.

As mentioned in Section 4.5, "Development of the Base-Case Design," the simulation model prepared for the complete process is often the source of the stream conditions in the PFD (e.g., Figure 4.19). Furthermore, as the design team completes the process integration step, the model can be improved to represent the more complete PFD.

In this section, several subroutines have been recommended for usage with ASPEN PLUS and HYSYS. These recommendations can be extended readily to permit the simulations to be carried out with CHEMCAD or PRO/II.

5.4 STEADY-STATE SIMULATION OF THE MONOCHLOROBENZENE SEPARATION PROCESS

Another process, which is considered throughout this text, involves the separation of a mixture consisting of HCl, benzene, and monochlorobenzene (MCB), the effluent from a reactor to produce MCB by the chlorination of benzene. As discussed in Chapter 8, when separating a light gaseous species, such as HCl, from two heavier species, it is common to vaporize the feed partially, followed by separation of the vapor and liquid phases in a *flash separator*. To obtain nearly pure HCl, the benzene and MCB can be absorbed in an absorber. Then, since benzene and MCB have significantly different boiling points, they can be separated by distillation. The process that results from this synthesis strategy is shown in Figure 5.23. Included on the diagram is the design basis (or specifications). Note that a portion of the MCB product is used as the absorbent.

As shown in the flowsheet, the feed is partially vaporized in the preheater, H1, and separated into two phases in the flash vessel, F1. The vapor from F1 is sent to the absorber, A1, where most of the HCl vapor passes through, but the benzene is largely absorbed using recycled MCB as the absorbent. The liquid effluents from F1 and A1 are combined, treated to remove the remaining HCl with insignificant losses of benzene and MCB, and distilled in D1 to separate benzene from MCB. The distillate rate is set equal to the benzene flow rate in the feed to D1, and the reflux ratio is adjusted to obtain the indicated MCB impurity in the distillate. The bottoms are cooled to 120°F in the heat exchanger, H2, after which one-third of the bottoms is removed as MCB product, with the remaining two-thirds recycled to the absorber. Note that this fraction recycled is specified during the *distribution of chemicals* in process synthesis, along with the temperature of the recycle, in an attempt to absorb benzene without sizable amounts of HCl. Furthermore, the temperature of stream S02 is specified to generate an adequate amount of vapor, three equilibrium stages are judged to be sufficient for the absorber (using the approximate Kremser–Brown equations), and the number of stages and the reflux ratio are estimated for the distillation column. Using the process simulators, these specifications are adjusted routinely to see how they affect the performance and economics of the process. Also, note that due to space limitations, a more complete, unit-by-unit description of the process and its specifications is reserved for the multimedia module, *ASPEN → Principles of Flowsheet Simulation → Interpretation of Input and Output: Sample Problem* which can be downloaded from the Wiley Web site associated with this book.

Use of Process Simulators

To determine the unknown temperatures and flow rates of the species, that is, to satisfy the material and energy balances, the MCB separation process is simulated in the steady state using ASPEN PLUS. This is accomplished by first creating an ASPEN PLUS simulation flowsheet, as illustrated in Figure 5.24. Then, the ASPEN PLUS forms are completed

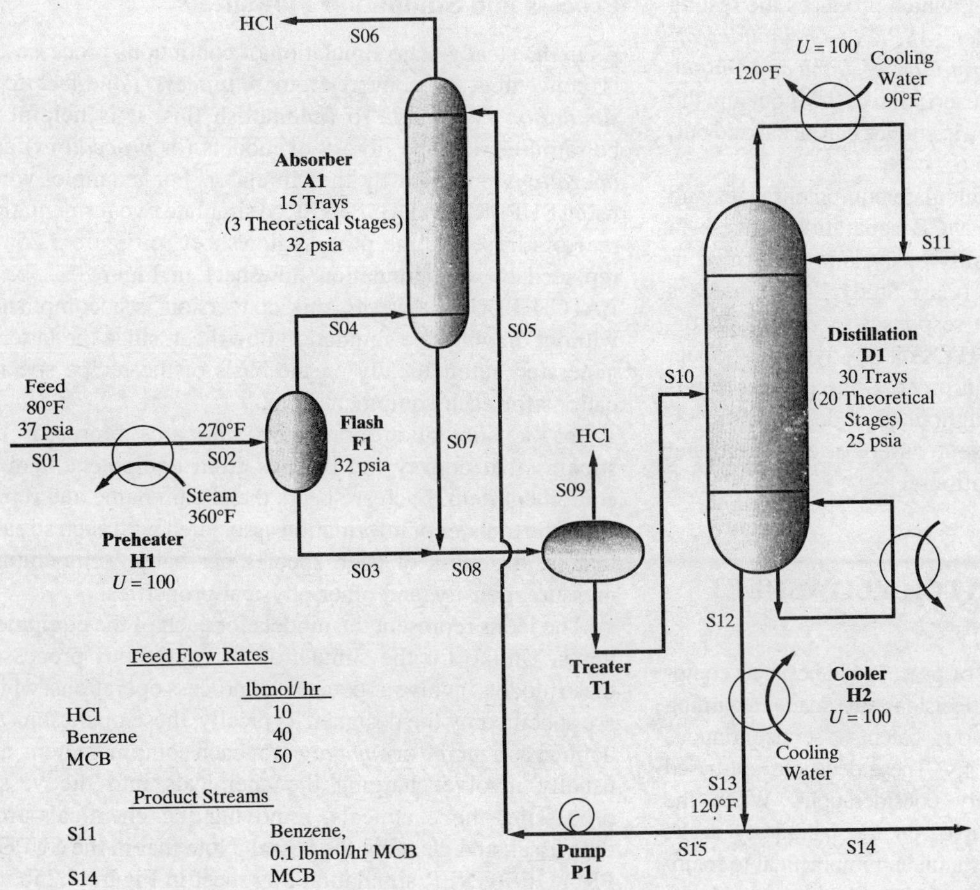

Figure 5.23 Process flowsheet for the MCB separation process.

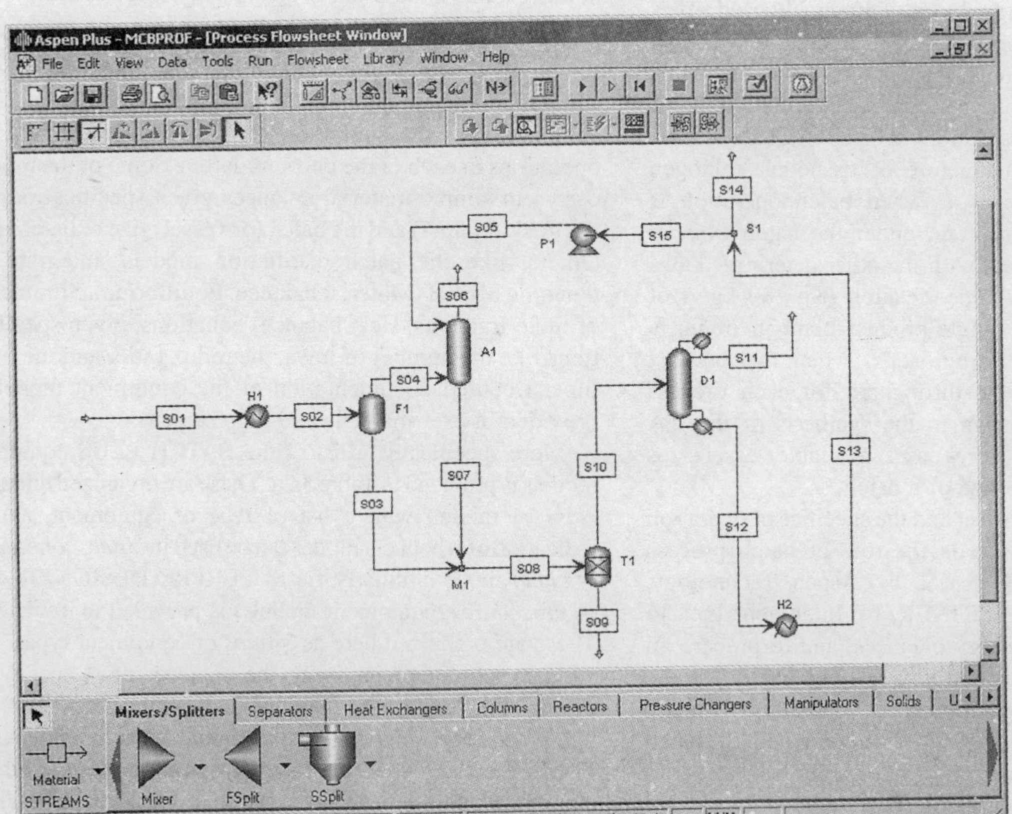

Figure 5.24 ASPEN PLUS simulation flowsheet for the MCB separation process.

and the *Run* button is depressed, which produces the results shown in the modules under *ASPEN → Principles of Flowsheet Simulation → Interpretation of Input and Output* which provides a unit-by-unit description of the input and the computer output. Then, parametric studies can be carried out, as recommended in Exercise 5.6.

Aspen IPE is also used to calculate equipment sizes and estimate capital costs for the MCB separation process in Section 22.7. Then, a profitability analysis is performed in Section 23.8. In Section 12S.5 (Case Study 12S.3), process controllers are added and their responses to various disturbances are computed using HYSYS in dynamic mode. Hence, for the MCB separation process, the process simulators have been used throughout the design process, although most design teams use a variety of computational tools to carry out these calculations.

5.5 PRINCIPLES OF BATCH FLOWSHEET SIMULATION

During the task-integration step of process synthesis, as equipment items are selected, key decisions are made regarding whether they operate in continuous, batch, or semicontinuous modes, as discussed in Section 4.4. These decisions are based upon throughput and flexibility considerations. When the throughput is small, for example, on the laboratory scale, continuous operation is often difficult and impractical to maintain, it usually being simpler and more profitable to complete a batch in hours, days, or weeks. Even for larger throughputs, where multiple products are produced, with variably sized orders received regularly, batch processes offer the ease of switching from the production of one product to another; that is, flexibility, which is more difficult to achieve in continuous operation. These and other issues are discussed in more detail in Chapter 11.

As shown for the manufacture of tissue plasminogen activator (tPA) in Section 4.4, when batch operation is selected for an equipment item, either the batch time or batch size must be selected, with the other determined as a function of the throughput specification (e.g., 80 kg/yr of tPA). Furthermore, for a single-product plant involving a serial sequence of processing steps, when the product throughput is specified, the throughput for each process unit is determined, as shown in the synthesis of the tPA process in Section 4.4. In many cases, available vessel sizes are used to determine the size of a batch.

Given the process flowsheet and the specifics of operation for each equipment item, it is the role of batch process simulators, like BATCH PLUS, by Aspen Technology, Inc., and SUPERPRO DESIGNER, by Intelligen, Inc., to carry out material and energy balances, and to prepare an operating schedule in the form of a Gantt chart for the process. Then, after the equipment and operating costs are estimated and profitability measures are computed, the batch operating parameters and procedures can be varied to increase the profitability of the design.

Process and Simulation Flowsheets

As in the steady-state simulation of continuous processes, it is convenient to convert from a *process flowsheet* to a *simulation flowsheet*. To accomplish this, it is helpful to be familiar with the library of models (or *procedures*) and *operations* provided by the simulator. For example, when using SUPERPRO DESIGNER to simulate two fermentation reactors in series, the process flowsheet in Figure 5.25a is replaced by the simulation flowsheet in Figure 5.25b. In BATCH PLUS, however, this conversion is accomplished without drawing the simulation flowsheet, since the latter is generated automatically on the basis of the recipe specifications for each equipment item.

In the simulation flowsheets, the arcs represent the streams that convey the batches from equipment item to equipment item. Each arc bears the stream name and represents the transfer of information associated with each stream; that is, the mass of each species per batch, temperature, pressure, density, and other physical properties.

The icons represent the models for each of the equipment items. Unlike for the simulation of continuous processes, these models involve a sequence of process operations, which are specified by the designer. Typically, these operations are defined as a *recipe* or *campaign* for each equipment item, and usually involve charging the chemicals into the vessel, processing the chemicals, removing the chemicals from the vessel, and cleaning the vessel. Note that in the SUPERPRO DESIGNER simulation flowsheet in Figure 5.25b, the microfiltration model represents both the microfilter and its holding tank in the process flowsheet, Figure 5.25a.

Equipment Models

Table 5.2 lists the equipment models (or procedures) and operations in each of the two simulators. Some of the models carry out simple material balances given specifications for the feed stream(s) and the batch (or vessel) size or batch time. Others, like the batch distillation models, integrate the dynamic MESH (Material balance, Equilibrium, Summation of mole fractions, Heat balance) equations, given specifications like the number of trays, the reflux ratio, and the batch time. Detailed documentation of the equipment models is provided in user manuals and help screens.

More specifically, a list of the BATCH PLUS equipment models is provided in Table 5.2a. These are organized under the *class* of model, with a list of *type* of equipment, and an indication of whether a model can be used in *batch, continuous*, or *either* mode. Similarly, for SUPERPRO DESIGNER, a list of *procedures* (equipment models) is provided in Table 5.2c. These are organized here as *groups* of equipment types.

For each equipment item, the engineer must specify the details of its operations. These include specifications for charging, processing, emptying, and cleaning. When using BATCH PLUS, these are specified in the steps in a recipe, with the equipment items defined as the steps are specified. A full list of

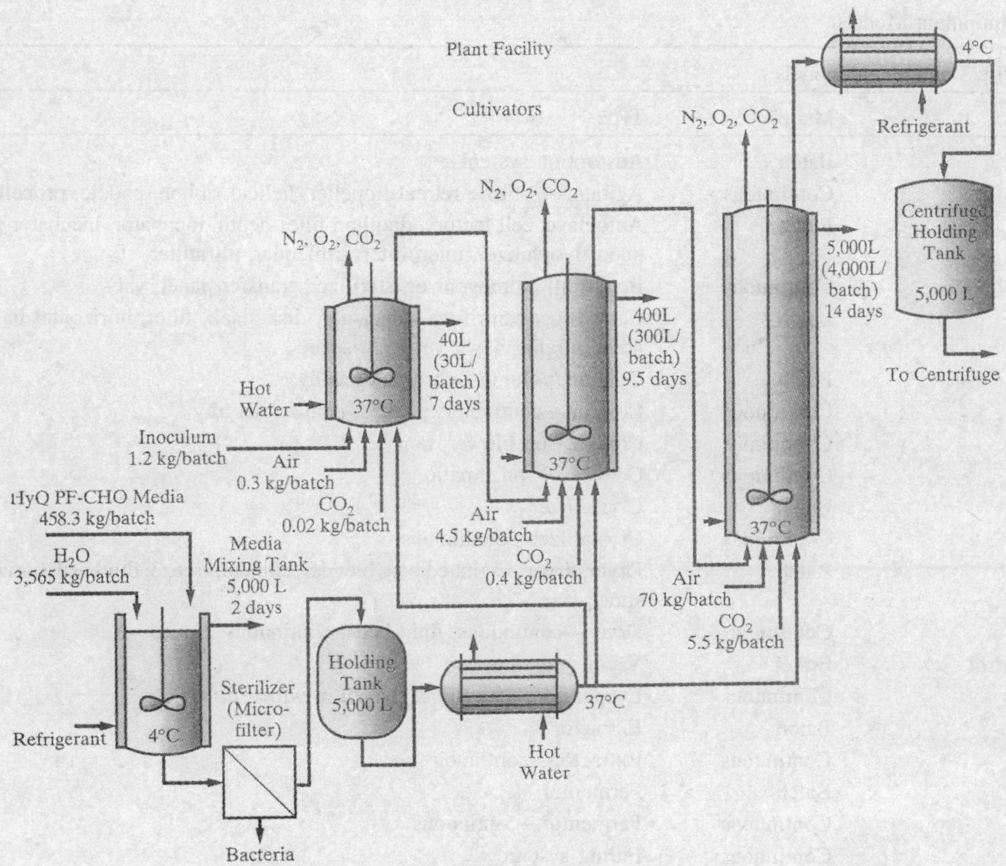

(a) Process flowsheet

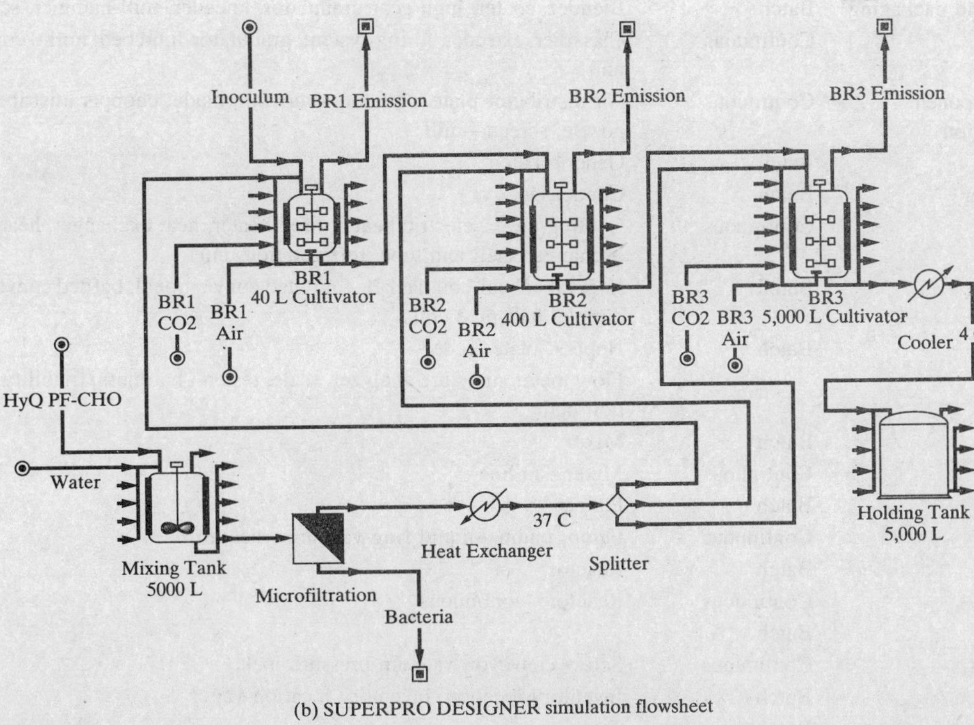

(b) SUPERPRO DESIGNER simulation flowsheet

Figure 5.25 Flowsheets.

Table 5.2 Equipment Models

(a) BATCH PLUS equipment models

Class	Mode	Type
Adsorption	Batch	Adsorption system
Agitator	Continuous	Agitator—3-blade retreat impeller, helical ribbon, paddle, propeller, turbine
Biotech	Batch	Autoclave, cell factory, diafilter, filter-depth, incubator, incubator-shaker, laminar flow hood, lyophilizer, microfilter, triblender, ultrafilter
	Continuous	Bead mill, homogenizer, sterilizer, transfer panel, valve
Centrifuge	Batch	Centrifuge, centrifuge—decanter, disk-stack, filter, horizontal basket, multichamber-bowl, tubular-bowl, vertical basket
Column	Batch	Column, column—chromatography
	Continuous	Column—continuous packed, continuous tray
Compressor	Continuous	Compressor, blower, fan
Conveyor	Continuous	Conveyor—pneumatic
Crystallizer	Batch	Crystallizer
	Continuous	Crystallizer—continuous
Dryer	Batch	Dryer, dryer—agitated pan, blender, conical, freeze, fluid bed, horizontal paddle, rotary, spray, tray
	Continuous	Dryer—continuous, fluid bed—continuous
Emission control	Either	Vapor emission vent
Evaporator	Continuous	Evaporator—long tube, thin film, wiped film
Extractor	Batch	Extractor
	Continuous	Extractor—continuous
Fermenter	Batch	Fermentor
	Continuous	Fermentor—continuous
Filling	Continuous	Filling system
Filter	Batch	Filter—agitated Nutsche, air, bag, belt, cross flow, dryer, in-line, pot, press, sparkler, tank sheet
	Continuous	Filter—continuous
Formulation and packaging	Batch	Blender, coater, high gear granulator, kneader, mill-hammer, screen, sifter
	Continuous	Classifier, extruder, filling system, granulator-fluid bed, mill—continuous, jet; tableting unit
Changing component for formulation	Continuous	Air distributor plate, agitator—impeller, blade; chopper, distributor plate, filter socks, nozzle, screen—mill
Generic	Batch	Generic batch
Heat exchanger	Batch	Condenser
	Continuous	Cooling tower, electric heater, fired heater, heat exchanger, heat exchanger plate, heat exchanger shell and tube, refrigeration unit
Heat transfer	Batch	Internal helical-coil, jacket—agitated conventional, baffled conventional, conventional, dimple, half-pipe coils
Hopper	Batch	Hopper, plate feeder
Instrument		Flow meter, moisture analyzer, scale, tester—hardness, friability, thickness; disintegration bath
Mixer	Batch	Mixer
	Continuous	Mixer—in-line
Piping	Batch	Piping
Pump	Continuous	Pump, pump—liquid ring vacuum, vacuum
Reactor	Batch	Reactor
	Continuous	Reactor—continuous
Scrubber	Batch	
Solid transport	Continuous	Screw conveyor, vacuum-pressure lock
Storage location	Batch	Inventory location, inventory location-vapor
Tank	Batch	Tank
Miscellaneous	Continuous	After burner, cyclone, demister, dust collector, ejector, hydrocyclone, steam jet

(*Continued*)

Table 5.2 Equipment Models (*Continued*)

(b) BATCH PLUS Operations

Batch operations	Age, centrifuge, charge, clean, cool, concentrate, crystallize, decant, distill, dry, evacuate, extract, filter, filter-in-place, heat, heat-to-reflux-and-age, line-blow, line-flush, open/close-vent, pH-adjust, pressurize, purge, QC-test, quench, quench-in-place, react, react-distill, start-sweep, stop-sweep, transfer, transfer-through-heat-exchanger, utilize, vent, wash-cake, yield-react
Chromatography operations	Elute-column, equilibrate-column, load-column, regenerate-column, wash-column
Continuous operations	Crystallize-continuously, Distill-continuously, Dry-continuously, Extract-continuously, filter-continuously, react-continuously
Biotech operations	Cell-disrupt, centrifuge-by-settling, depth-filter, diafilter, ferment, ferment-continuously, microfilter, sterilize, transfer-through-sterilizer, ultrafilter

(c) SUPERPRO DESIGNER Procedures (Equipment Models)

Group	Mode	Type
Vessel	Batch	Reactor, fermentor, seed fermentor, airlift fermentor
Continuous reaction	Continuous	Stoichiometric (CSTR, PFR, fermentor, seed fermentor, airlift fermentor)
	Continuous	Kinetics (CSTR, PFR, fermentor, seed fermentor)
	Continuous	Equilibrium (CSTR)
	Continuous	Environmental (Well-mixed aerobic, biooxidation, . . .)
Filtration	Batch	Microfiltration, ultrafiltration, reverse osmosis, diafiltration, dead end filtration, Nutsche filtration, plate and frame filtration, baghouse filtration, electrostatic precipitation
	Feed and Bleed (continuous)	Microfiltration, ultrafiltration, reverse osmosis
	Either	Rotary, vacuum filtration, air filtration, belt filtration, granular media filtration, baghouse filtration, electrostatic precipitation
Centrifugation	Batch	Basket centrifuge
	Either	Decanter centrifuge, disk-stack centrifuge, bowl centrifuge, Centritech centrifuge, cyclone, hydroclone
Homogenization	Either	High pressure, bead milling
Chromatography/adsorption	Batch	Gel filtration, packed bed adsorption (PBA) chromatography, granular activated carbon (GAC)—liquid and gaseous stream
Drying	Batch	Tray drying, freeze drying
	Either	Spray drying, fluid bed drying, drum drying, rotary drying, sludge drying
Sedimentation	Either	Decanting (2-liquid phases), clarification, inclined plane (IP) clarification, thickener basin, dissolved air flotation tank, oil separator
Distillation	Batch	Shortcut batch distillation
	Either	Flash drum, shortcut distillation
Extraction	Either	Mixer-settler, differential column extractor, centrifugal extractor
Phase change	Either	Condensation for gas streams, multiple-effect evaporation, crystallization
Adsorption/stripping	Either	Absorber, stripper, degasifier
Storage	Batch	Hopper, equalization tank, junction box mixing
	Either	Blending tank, flat bottom tank, receiver, horizontal tank, vertical on legs tank, silo
Heat exchange	Either	Heating, electrical heating, cooling, heat exchanging (2-streams), heat sterilization
Mixing	Batch	Bulk flow (tumble mixer)
	Either	Bulk flow (2–9 streams), discrete flow (2–9 streams)
Splitting	Either	Bulk flow (2–9 streams), discrete flow (2–9 streams), component flow (2–9 streams)
Size reduction	Either	Grinding (bulk or discrete flow), shredding (bulk or discrete flow)

(*Continued*)

Table 5.2 Equipment Models (*Continued*)

Formulation and packaging	Either	Extrusion, blow molding, injection molding, trimming, filling, assembly, printing, labeling, boxing, tableting
Transport (near)	Either	Liquid (pump) Gas (compressor, fan) Solids (belt conveyor—bulk or discrete flow, pneumatic conveyor—bulk or discrete flow, screw conveyor—bulk or discrete flow, bucket elevator—bulk or discrete flow)
Transport (far)	Either	By land (truck—bulk or discrete flow, train—bulk or discrete flow) By sea (ship—bulk or discrete flow) By air (airplane—bulk or discrete flow)

(d) SUPERPRO DESIGNER Operations

Absorb	Adsorb	Agitate	Assemble
Bio-oxidize	Bioreact	Centrifuge	Charge
Clarify	Clean-in-place (CIP)	Compress	Concentrate (batch)
Concentrate (feed & bleed)	Condense	Convert to bulk	Convert to discrete
Convey	Cool	Crystallize	Cyclone
Cycloning	Decant	Degasify	Diafilter
Distill	Dry	Dry cake	Elevate
Elute	Equalize	Equilibrate	Evacuate
Exchange heat	Extract/phase split	Extrude	Ferment (kinetic)
Ferment (stoichiometric)	Fill	Filter	Flash
Flotate	Gas sweep	Grind	Handle solids flow
Heat	Hold	Homogenize	Incinerate
Label	Load	Mix	Mix solids
Mold	Neutralize	Oxidize	Pack
Pass through	Precipitate	Pressurize	Print
Pump	Pump gas	Purge/inlet	Radiate
React (equilibrium)	React (kinetic)	React (stoichiometric)	Regenerate
Separate oil	Shred	Split	Steam-in-place (SIP)
Sterilize	Store	Store solids	Strip
Tablet	Thicken	Transfer in	Transfer out
Transport	Trim	Vaporize/concentrate	Vent
Wash	Wash cake		

the operations available is provided in Table 5.2b. Following this discussion, the results for a BATCH PLUS simulation of the reactor section of the tPA process are provided in Example 5.5, with detailed instructions for specifying the operations and equipment items provided in the multimedia tutorial *ASPEN → Tutorials → Batch Process Simulation → tPA Manufacture*. In SUPERPRO DESIGNER, since the engineer provides a simulation flowsheet, the operations are specified unit-by-unit. Its list of operations is provided in Table 5.2d.

www.wiley.com/go/global/seider

Combined Batch and Continuous Processes

Since it is possible to have adjacent equipment items operating in batch and continuous modes, it is important to understand the conventions used when preparing a mixed simulation with batch and continuous operations. In most cases, it is desirable to install a holding tank to moderate the surges that would otherwise occur.

In SUPERPRO DESIGNER, each flowsheet is defined by the engineer as either *batch* or *continuous*. In batch mode, stream results are reported on a per-batch basis, even for streams associated with continuous processes in a batch flowsheet. Each equipment item is designated as operating in batch/semicontinuous or continuous mode. Scheduling information must be included for all items designated as operating as batch/semicontinuous. Semicontinuous units operate continuously while utilized, but are shut down between uses. Equipment items designated as continuous are assumed to operate at all times, and are excluded from operation schedules (and Gantt charts).

When a SUPERPRO DESIGNER flowsheet is defined to be in continuous mode, streams are reported on a per-hour basis. Scheduling information is not required, and no overall batch time is calculated. Individual batch processes can be inserted into the flowsheet, with their batch and turnaround times specified.

In BATCH PLUS, every simulation is for an overall batch process, with stream values always reported on a per-batch basis. Continuous operations, however, can be inserted. For these units, a feed is loaded, the vessel is filled to its surge volume, and an effluent stream immediately begins to transfer the product downstream. This differs from normal batch operation, which involves loading all of the feed and completing the processing steps before unloading. Specific units in BATCH PLUS, such as the *Fermenter*, can also operate *as fed-batch*. In such operations, a feed is added continuously to the batch while an operation is taking place.

With SUPERPRO DESIGNER and BATCH PLUS, caution must be exercised when introducing continuous operations into batch processes, as no warnings are provided when a continuous process unit is running dry. When a feed to a continuous unit runs dry, the simulator assumes that this unit is shut down and restarted when the feed returns. Clearly, such operation is infeasible for many units, such as distillation columns and chemical reactors. Consequently, when continuous processes are included, it is important to check the results computed by the batch simulators to be sure that unreasonable assumptions have not been made.

An advantage of adding continuous operations arises when the process bottleneck is transferred to the continuous unit. When a schedule is devised such that the continuous unit is always in operation, batch cycling is avoided.

EXAMPLE 5.5 *tPA Cultivators*

As discussed in Section 4.4, tPA-CHO cells are used to produce tPA. These cells are duplicated to a density of 3.0×10^6 cell/mL, after which the culture becomes too dense and the tPA-CHO cells die at a high rate. For this reason, engineers cultivate tPA-CHO cells in a sequence of bioreactors, each building up mass to a density of 3.0×10^6 cell/mL, with the accumulated cell mass used to inoculate the next largest reactor, until the desired cell mass is reached.

In this example, the objective is to determine the effective time between batches; that is, the *cycle time*, which is less than the total occupied time of a sequence of batch operations. The cycle time is smaller because while one batch is moving through the sequence, other batches are being processed simultaneously in other pieces of equipment both upstream and downstream. Therefore, the effective time between batches, or the cycle time, is determined by the equipment unit that requires the most processing time. This equipment unit is known as the *bottleneck*, and consequently, to reduce the cycle time, engineers seek to reduce the processing time of the bottleneck as much as possible. Usually, the bottleneck is associated with the largest process unit, often the main bioreactor, because these reactors involve the largest cultivation times. See Chapter 11 for a more complete discussion of the cycle time and bottleneck.

For this example, the BATCH PLUS simulator is used to determine the cycle time for a portion of the tPA process that involves just two cultivators, as shown in Figure 5.26. Initially, a mixing tank is charged with 3,565 kg of water and 458.3 kg of HyQ PF-CHO media, with a charge time of one hour. The material in the tank is cooled to 4°C for one day and aged for two days to allow for quality-assurance testing. Then, this material is transferred to a 0.2-μm microfilter for sterilization, to remove bacteria over a two-hour period, and sent to a holding tank. Next, the first cultivator is charged with 1.2 kg of tPA-CHO cells in one hour.

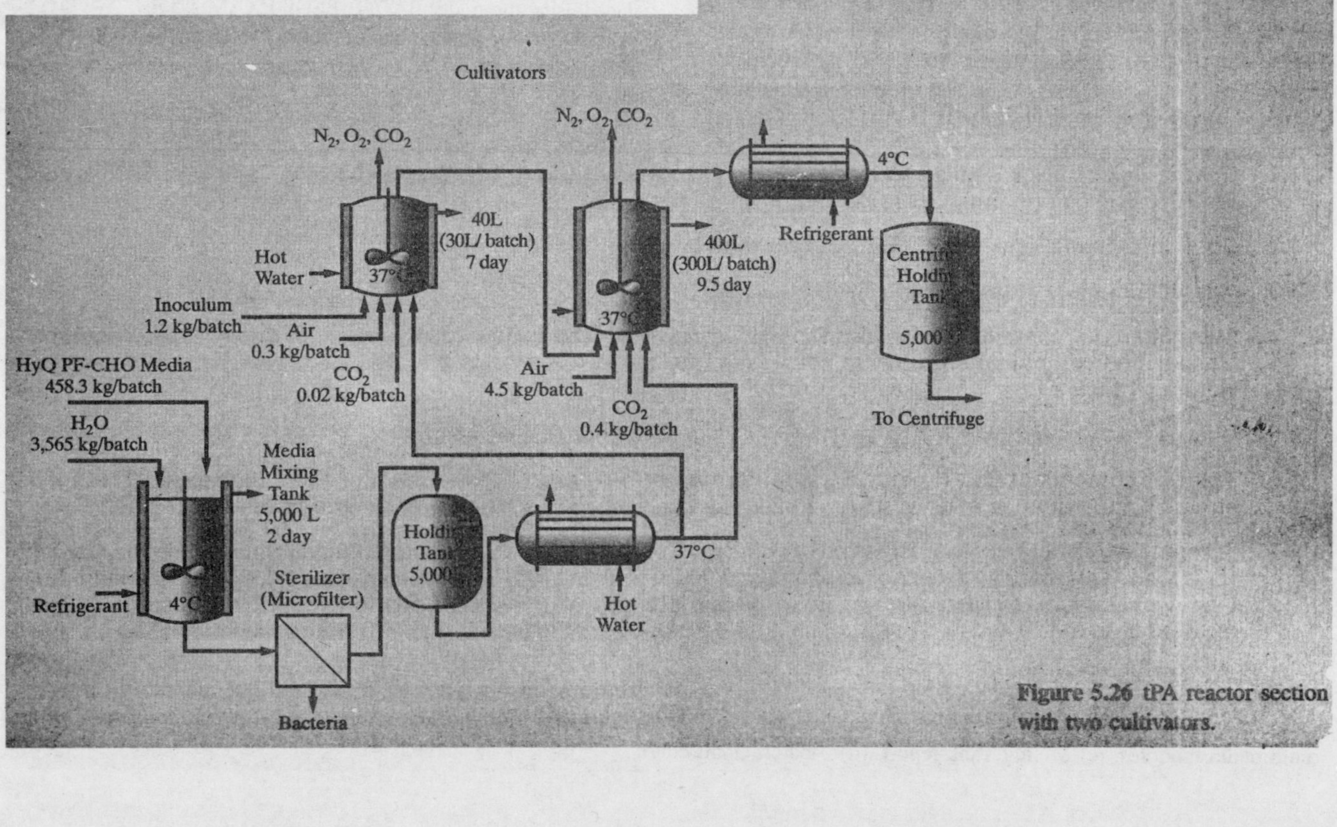

Figure 5.26 tPA reactor section with two cultivators.

Then, 21.2 kg of material from the holding tank are heated in a heat exchanger to 37°C and added to the first cultivator in 0.5 day, after which cultivation takes place over the next five days. The yield from the cultivation is 15.3 wt% tPA-CHO cells, 0.01 wt% endotoxin, 84.7 wt% water, and 0.01 wt% tPA. The products of Cultivator 1 are fed to Cultivator 2 in 0.5 day. Then, 293.5 kg of media from the holding tank are heated to 37°C and fed to Cultivator 2 in 0.5 day, after which the cultivation takes place over seven days. Immediately after Cultivator 1 is emptied, it is cleaned-in-place using 60 kg of water over 20 hours. Note that to override the BATCH PLUS estimate, a charge time of 1 min should be entered. Then, it is sterilized at 130°C for two hours and cooled to 25°C (with one-hour heat-up and cool-down times). The yield of the cultivation in Cultivator 2 is 11.7 wt% tPA-CHO cells, 7.67×10^{-4} wt% endotoxin, 88.3 wt% water, and 0.039 wt% tPA. After this cultivation, the contents of Cultivator 2 are cooled in a heat exchanger to 4°C and transferred to the centrifuge holding tank over 0.5 day. After Cultivator 2 is emptied, it is cleaned-in-place using 600 kg of water over 20 hours, and sterilized using the procedure for Cultivator 1.

To determine the cycle time and the bottleneck unit, create a multiple-batch Gantt chart using BATCH PLUS. Generate equipment content and capacity reports to determine the sizes of the equipment items. Examine the stream table report to monitor the production of tPA-CHO cells and tPA in the process.

SOLUTION

When using BATCH PLUS, as discussed step-by-step on the multimedia modules (see *ASPEN → Tutorials → Batch Process Simulation → tPA Manufacture*), the materials are specified; that is, tPA-CHO cells, tPA, media, water, nitrogen, oxygen, and carbon dioxide. Then, each equipment item is entered with its recipe of operations. Note that there is no cultivator model in BATCH PLUS, and consequently, the *Fermenter* model is used in its place. Given this information, BATCH PLUS generates a recipe of operations for the process, shown in Figure 5.27a, and prepares a simulation flowsheet (using Microsoft VISIO). BATCH PLUS also generates a table including the per-batch flow rates of each stream in the process in a time-dependent manner, a portion of which is shown in Figure 5.27b. Study of this report allows the monitoring of the growth of tPA-CHO cells, and their production of tPA, as they travel from vessel to vessel. The third column from the last in the report indicates that the final stream in the process contains 36.9 kg of tPA-CHO cells, 0.12 kg of tPA, 0.0024 kg of endotoxin, and 278.8 kg of water. For the simulation flowsheet and the entire stream table, see *ASPEN → Tutorials → Batch Process Simulation → tPA Manufacture* in the multimedia modules.

In addition, BATCH PLUS uses Microsoft EXCEL to prepare an *Equipment Contents Report*, which displays, for each vessel in the process, an inventory of the contents of the vessel during each step the vessel is utilized. This includes the mass of components, as well as overall liquid and solid volume and mass. Inspection of these reports allows estimation of required vessel sizes. The report for the Mixing Tank, shown in Figure 5.27c, indicates a maximum liquid and solid volume of 4,050 L after operation 1.1. It can, therefore, be concluded that the Mixing Tank unit must be larger than 4,050 L; for example, 5,000 L. Similarly, the reports for Fermenters 1 and 2, shown in *ASPEN → Tutorials → Batch Process Simulation → tPa Manufacture*, indicate maximum volumes of 22.8 L and 322 L after operations 1.7 and 1.8. On this basis, 40-L and 400-L vessels are selected for Fermenters 1 and 2.

Finally, when the Gantt chart prepared by BATCH PLUS is extended to show three batches, as shown in Figure 5.27d, the bottleneck of the process is determined quite easily. For each vessel, solid blocks show the time period during which it is in operation. Solid blocks are for the first, second, and third batches, respectively. The bottleneck is associated with the equipment unit that is utilized at all times; that is, for which the red, blue, and green blocks touch each other. Clearly, this unit determines the cycle time. Note that these results can be reproduced using the folder BATCH PLUS-EXAM 5-5 in the Program and Simulation Files folder, which can be downloaded from the Wiley Web site associated with this book.

1.1. Charge Mixing Tank with 458.3 kg of Media. The charge time is 1 h. Charge Mixing Tank with 3565 kg of WATER. The charge time is 1 h.

1.2. Cool unit Mixing Tank to 4 C. The cooling time is 1 day.

1.3. Age the contents of unit Mixing Tank for 2 day.

1.4. Microfilter the contents of Mixing Tank in Microfilter. The mode of operation is Batch Concentration. Unspecified components go to the Permeate. The operation time is 2 h. The permeate stream is sent to Holding Tank.

1.5. Charge Fermenter 1 with 1.2 kg of tPA-CHO Cells. The charge time is 1 h.

1.6. Transfer contents of unit Holding Tank to Fermenter 1 through heat exchanger Heat Exchanger. The final stream temperature is 37 C. Transfer 21.2 kg of vessel contents. The transfer time is 0.5 day.

1.7. Ferment in unit Fermenter 1. The yield of tPA-CHO Cells in the Solid phase is 0.153, of Endotoxin in the Liquid phase is 0.0001, of tPA in the Liquid phase is 0.0001, of WATER in the Liquid phase is 0.847, of Media in the Liquid phase is 0, of Media in the Solid phase is 0 and of tPA-CHO Cells in the Liquid phase is 0. The fermentation time is 5 day. Continuously add 0.02 kg of CARBON-DIOXIDE. Continuously add 0.3 kg of AIR.

Figure 5.27 BATCH PLUS simulation for Example 5.5. (a) Operations recipe for the process - BATCH PLUS output. (b) Last three columns of stream table (for the complete table, see *ASPEN → Tutorials → Batch Process Simulation → tPa Manufacture* in the multimedia modules. (c) Mixing Tank report. (d) 3-batch Gantt chart.

Start Parallel

Series

1.8. Transfer contents of unit Fermenter 1 to Fermenter 2. Transfer 100% of vessel contents. The transfer time is 0.5 day.

1.9. Transfer contents of unit Holding Tank to Fermenter 2 through heat exchanger Heat Exchanger. The final stream temperature is 37 C. Transfer 293.5 kg of vessel contents. The transfer time is 0.5 day.

1.10. Ferment in unit Fermenter 2. The yield of tPA-CHO Cells in the Solid phase is 0.117, of Endotoxin in the Liquid phase is 7.67e-6, of tPA in the Liquid phase is 0.00039, of WATER in the Liquid phase is 0.883, of Media in the Solid phase is 0, of Media in the Liquid phase is 0 and of tPA-CHO Cells in the Liquid phase is 0. The fermentation time is 7 day. Continuously add 0.4 kg of CARBON-DIOXIDE. Continuously add 4.5 kg of AIR.

Series

1.11. Clean unit Fermenter 1. Clean with 60 kg of WATER. The feed time is 1 min. Cleaning time is 20 h.

1.12. Sterilize the contents of Fermenter 1. The sterilization temperature is 130 C. The heat-up time is 1 h. Maintain the temperature for 2 h. The cool-down time is 1 h.

End Parallel

1.13. Transfer contents of unit Fermenter 2 to Centrifuge Holding Tank through heat exchanger Cooler. The final stream temperature is 4 C. Transfer 100% of vessel contents. The transfer time is 0.5 day.

1.14. Clean unit Fermenter 2. Clean with 600 kg of WATER. The feed time is 1 min. Cleaning time is 20 h.

1.15. Sterilize the contents of Fermenter 2. The sterilization temperature is 130 C. The heat-up time is 1 h. Maintain the temperature for 2 h. The cool-down time is 1 h.

(a)

Step Stream Table

Process (Version):	Reactor Growth Chain (1.0)	Key Input Intermediate:	Media
Step (Version):	Step1 (1.0)	Key Output Intermediate:	tPA
Simulation Date:	12/16/2001 17:10	Number of Batches:	1
		Plan Quantity:	0.12 kg

BATCH PLUS Stream Label			1.13. Transfer-Through-Heat-Exchanger-107	1.14. Clean-111	1.14. Clean-112
Operation			1.13. Transfer-Through-Heat-Exchanger	1.14. Clean	1.14. Clean
Start Time		(min)	24,000.00	24,720.00	25,921.00
End Time		(min)	24,720.00	24,721.00	25,922.00
Total Time		(min)	720.00	1.00	1.00
From Unit			Cooler		Fermenter 2
To Unit			Centrifuge Holding Tank	Fermenter 2	
Stream Type			Intermediate	Input	Output
Mass - (kg)	Per Batch	Mol Wt			
Total			3159000	600.0000	600.0000
Endotoxin		18.02	0.0024		
tPA		18.02	0.1232		
AIR		28.95			
CARBON-DIOXIDE		44.01			
WATER		18.02	278.8286	600.0000	600.0000
Media		18.02			
tPA-CHO Cells		18.02	36.9456		
Total Mass		(kg)	315.90	600.00	600.00
Total Volume		(liter)	312.09	603.73	603.73
Mass Flowrate		(kg/min)	0.44	37.27	37.27
Volume Flowrate		(liter/h)	26.01	2,250.00	2,250.00
Composite Product Factor			2,565.12	4,872.03	4,872.03
Phase			Liquid1+Solid	Liquid1	Liquid1
Temperature		(C)	4.00	25.00	25.00
Average Density		(kg/Cubic m)	1,012.20	993.81	993.81
Average Viscosity		(cp)	1.35	0.92	0.92
Average Heat Capacity		(kJ/kg-K)	4.21	4.18	4.18
Average Molecular Weight			18.02	18.02	18.02

(b)

Figure 5.27 *(Continued)*

Step Equipment Contents

Process (Version):	Reactor Growth Chain (1.0)	Key Input Intermediate:	Media
Step (Version):	Step1 (1.0)	Key Output Intermediate:	tPA
Simulation Date:	12/16/01 17:10	Number of Batches:	1
		Plan Quantity:	0.12 kg

Mixing Tank

Operation		START	1.1. Charge	1.2. Cool	1.3. Age	1.4. Microfilter
Time	(min)	0.00	60.00	1,500.00	4,380.00	4,500.00
Mass - (kg)	Mol Wt					
Total		5.8957	4,024,4221	4,024,5086	4,024,6086	6,3424
WATER	18.02		3,565.0000	3,565.0000	3,565.0000	
NITROGEN	28.01	4.5223	0.8607	1.0038	1.0038	4.8650
OXYGEN	32.00	1.3734	0.2614	0.3048	0.3048	1.4774
Media	18.02		458.3000	458.3000	458.3000	
Liquid+Solid Mass	(kg)	0.00	4,023.30	4,023.30	4,023.30	0.00
Liquid+Solid Volume	(liter)	0.00	4,048.34	3,968.36	3,968.36	0.00
Phase		Gas	Gas+Liquid1	Gas+Liquid1	Gas+Liquid1	Gas
Temperature	(C)	25.00	25.00	4.00	4.00	4.00
Pressure	(kPa)	101.33	101.33	101.33	101.33	101.33
Average Liq+Sol Density	(kg/Cubic m)	0.00	993.81	1,013.84	1,013.84	0.00
Average Liq+Sol Viscosity	(cp)	0.00	0.92	1.53	1.53	0.00
Average Liq+Sol Heat Capacity	(kJ/kg-K)	0.00	4.18	4.21	4.21	0.00
Average Liq+Sol Molecular Weight		0.00	18.02	18.02	18.02	0.00

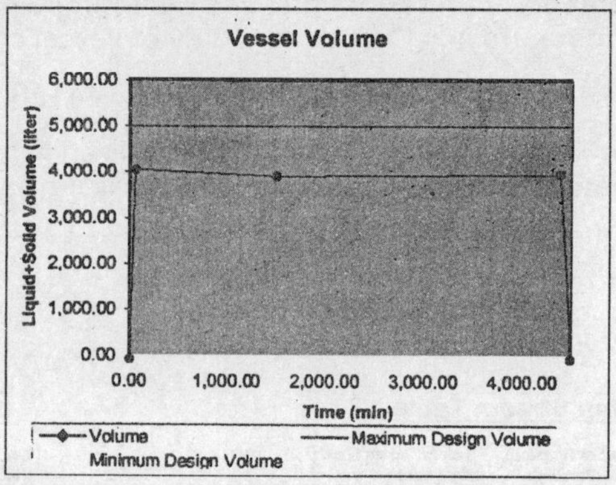

(c)

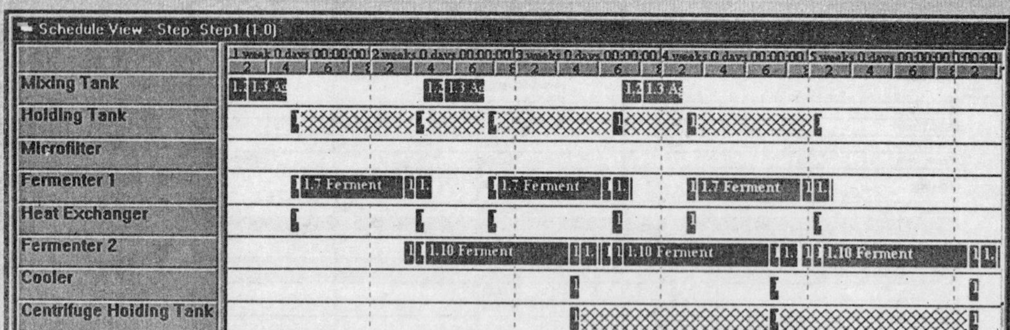

(d) Solid entries from left to right are associated with the first, second, and third batches – red, blue, and green in the Gantt chart prepared by BATCH PLUS.

5.6 SUMMARY

Having studied this chapter and the accompanying material on the multimedia CD-ROM, the reader should

1. Be able to prepare a simulation flowsheet, beginning with a process flowsheet.
2. Be able to prepare a steady-state simulation using ASPEN PLUS and HYSYS, and be familiar with the capabilities of CHEMCAD and PRO/II.
3. Be able to set up simulations involving recycle loops and design specifications, which are usually implemented with iterative control algorithms.
4. Understand how to follow the calculation sequences implemented automatically by ASPEN PLUS, CHEMCAD, and PRO/II for recycle calculations and when satisfying design specifications. For recycle

calculations, the reader should be able to alter the calculation sequence by specifying tear streams.

5. Understand the many places that process simulators are useful during process synthesis and when preparing the base-case design, as discussed in Sections 4.4 and 4.5.

6. Be able to prepare a simulation of a batch process using BATCH PLUS and SUPERPRO DESIGNER.

7. Have completed several exercises involving steady-state simulation using one of the four simulators ASPEN PLUS, HYSYS, CHEMCAD, and PRO/II and involving batch process simulation using one of the two simulators BATCH PLUS and SUPERPRO DESIGNER.

REFERENCES

1. CROWE, C.M., and M. NISHIO, "Convergence Promotion in the Simulation of Chemical Processes—The General Dominant Eigenvalue Method," *AIChE J.*, **21**, 3 (1975).

2. HENLEY, E.J., and E.M. ROSEN, *Material and Energy Balance Computations*, Wiley, New York (1969).

3. MYERS, A.L., and W.D. SEIDER, *Introduction to Chemical Engineering and Computer Calculations*, Prentice-Hall, Englewood Cliffs, New Jersey (1976).

4. ORBACH, O., and C.M. CROWE, "Convergence Promotion in the Simulation of Chemical Processes with Recycle—The Dominant Eigenvalue Method," *Can. J. Chem. Eng.*, **49**, 509 (1971).

5. RUDD, D.F., and C.C. WATSON, *Strategy of Process Engineering*, Wiley, New York (1968).

6. SEIDER, W.D., J.D. SEADER, and D.R. LEWIN, *Process Design Principles: Synthesis, Analysis, and Evaluation*, John Wiley & Sons (1999).

7. SMITH, B.D., *Design of Equilibrium Stage Processes*, McGraw-Hill, New York (1963).

8. WEGSTEIN, J.H., "Accelerating Convergence of Iterative Processes," *Commun. Assoc. Comp. Machinery*, **1**(6), 9 (1958).

9. WESTERBERG, A.W., H.P. HUTCHISON, R.L. MOTARD, and P. WINTER, *Process Flowsheeting*, Cambridge University Press, Cambridge (1979).

EXERCISES

5.1 *Flash with recycle* (a) Consider the flash separation process shown in Figure 5.15. If using ASPEN PLUS, solve all three cases using the MIXER, FLASH2, FSPLIT, and PUMP subroutines and the RK-SOAVE option set for thermophysical properties. Compare and discuss the flow rates and compositions for the overhead stream produced by each of the three cases.

(b) Modify case 3 of Exercise 5.1a so as to determine the flash temperature necessary to obtain 850 lb/hr of overhead vapor. If using ASPEN PLUS, a design specification can be used to adjust the temperature of the flash drum to obtain the desired overhead flow rate.

5.2 *Toluene hydrodealkylation process—reactor section.* As discussed in Section 5.3, toluene (C_7H_8) is to be converted thermally to benzene (C_6H_6) in a hydrodealkylation reactor. The main reaction is $C_7H_8 + H_2 \rightarrow C_6H_6 + CH_4$. An unavoidable side reaction occurs that produces biphenyl: $2C_6H_6 \rightarrow C_{12}H_{10} + H_2$.

The reactor section of the process is shown in Figure 5.21, as are the conditions for the feed and two recycle streams. The flow rate of the quench stream should be such that the reactor effluent is quenched to 1,150°F. Conversion of toluene in the reactor is 75 mol%. Two mole percent of the benzene present after the first reaction occurs is converted to biphenyl. Use a process simulator to perform material and energy balances with the Soave–Redlich–Kwong equation of state (RK-SOAVE option in ASPEN PLUS).

5.3 *Toluene hydrodealkylation process—separation section.* As discussed in Section 5.3, the following stream at 100°F and 484 psia is to be separated by two distillation columns into the indicated products.

	lbmol/hr			
Species	**Feed**	**Product 1**	**Product 2**	**Product 3**
H_2	1.5	1.5		
CH_4	19.3	19.2	0.1	
C_6H_6 (benzene)	262.8	1.3	258.1	3.4
C_7H_8 (toluene)	84.7		0.1	84.6
$C_{12}H_{10}$ (biphenyl)	5.1			5.1

Two different distillation sequences are to be examined, as shown in Figure 5.22. In the first sequence, H_2 and CH_4 are removed in the first column.

If using ASPEN PLUS, the DSTWU subroutine is used to estimate the reflux ratio and theoretical tray requirements for both sequences. In addition, the RK-SOAVE option is used. Specify a reflux ratio equal to 1.3 times the minimum. Use design specifications to adjust the isobaric column pressures so as to obtain distillate temperatures of 130°F; however, no column pressure should be less than 20 psia. Also, specify total condensers, except note that a partial condenser is used when H_2 and CH_4 are taken overhead.

5.4 Complete a simulation of the entire process for the hydrodealkylation of toluene in Figure 5.20. Initially, let the purge/recycle ratio be 0.25; then, vary this ratio and determine its effect on the performance of the process. Use a design specification to determine the unknown amount of hydrogen to be added to the feed stream (equal to that lost in the purge). If using ASPEN PLUS, the distillation columns can be simulated using the RADFRAC subroutine with the number of stages and reflux ratio previously computed by the DSTWU subroutine. (To carry out an economic analysis of the process, see Exercise 23.21.)

5.5 (a) Complete a steady-state simulation of the vinyl-chloride process in Figure 4.19. First, create a simulation flowsheet. Assume that:

Cooling water is heated from 30 to 50°C.
Saturated steam is available at 260°C (48.4 atm).

If using ASPEN PLUS, use the UNIQUAC option set for thermophysical properties.

(b) Carry out process integration and repeat the steady-state simulation.

5.6 For the monochlorobenzene separation process in Figure 5.23, the results of an ASPEN PLUS simulation are provided in the multimedia modules under *ASPEN → Principles of Flowsheet Simulation → Interpretation of Input and Output*. Repeat the simulation with:

(a) 25% of MCB recycled at 130°F
Stream S02 at 250°F
15 theoretical stages in the distillation column
(b) Other specifications of your choice

5.7 *Cavett Problem.* A process having multiple recycle loops formulated by R.H. Cavett [*Proc. Am. Petrol. Inst.,* **43,** 57 (1963)] has been used extensively to test tearing, sequencing, and convergence procedures. Although the process flowsheet requires compressors, valves, and heat exchangers, a simplified ASPEN PLUS flowsheet is shown in Figure 5.28 (excluding the recycle convergence units). In this form, the process is the equivalent of a four-theoretical-stage, near-isothermal distillation (rather than the conventional near-isobaric type), for which a patent by A. Gunther [U.S. Patent 3,575,007 (April 13, 1971)] exists. For the specifications shown on the flowsheet, use a process simulator to determine the component flow rates for all streams in the process.

5.8 Use a process simulator to model a two-stage compression system with an intercooler. The feed stream consists of 95 mol% hydrogen and 5 mol% methane at 100°F and 30 psia; 440 lbmol/hr is compressed to 569 psia. The outlet temperature of the intercooler is 100°F and its pressure drop is 2 psia. The centrifugal compressors have an isentropic efficiency of 0.9 and a mechanical efficiency of 0.98.

Determine the power requirements and heat removed for three intermediate pressures (outlet from the first three stages): 100, 130, 160 psia. If using ASPEN PLUS, use the MCOMPR subroutine and the RK-SOAVE option.

5.9 Consider the ammonia process in which N_2 and H_2 (with impurities Ar and CH_4) are converted to NH_3 at high pressure (Figure 5.29). If using ASPEN PLUS, use the following subroutines:

Compressor	COMPR
Reactor	RSTOIC
Heat Exchanger	HEATER
High-Pressure Separator	FLASH2
Low-Pressure Separator	FLASH2
Recirculating Compressor	COMPR

You are given the feed stream and fraction purged in the splitter. Prepare a simulation flowsheet and, when applicable, show the calculation sequence prepared by the process simulator (if using ASPEN PLUS, complete SEQUENCE USED WAS:).

Component	lbmol/hr
N_2	358.2
CO_2	4,965.6
H_2S	339.4
C_1	2,995.5
C_2	2,395.5
C_3	2,291.0
iC_4	604.1
nC_4	1,539.9
iC_5	790.4
nC_5	1,129.9
nC_6	1,764.7
nC_7	2,606.7
nC_8	1,844.5
nC_9	1,669.0
nC_{10}	831.7
nC_{12}	1,214.5

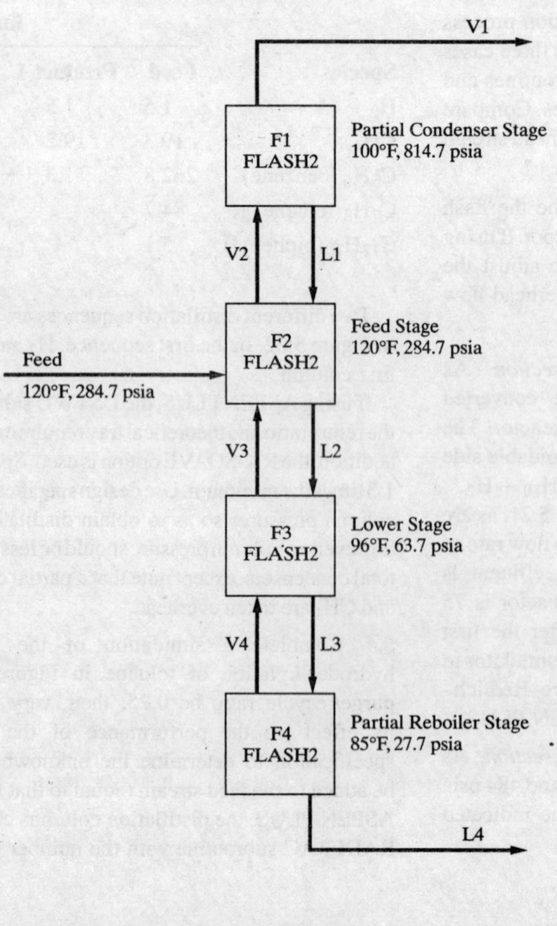

Figure 5.28 Near-isothermal distillation process.

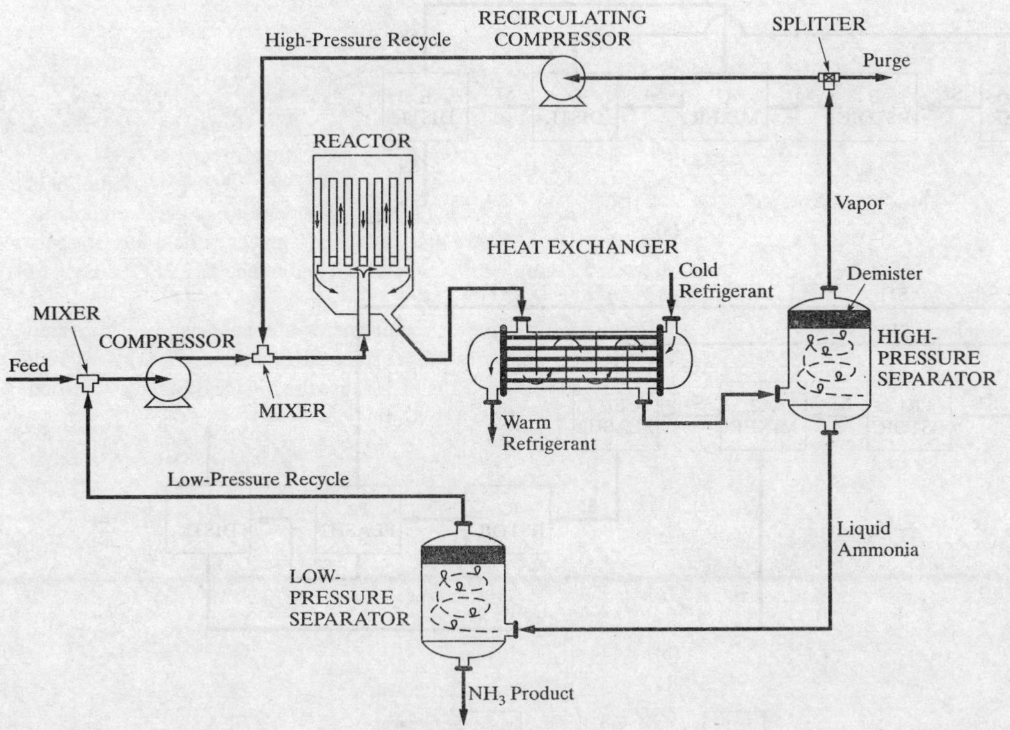

Figure 5.29 Ammonia reaction loop.

5.10 The feed (equimolar A and B) to a reactor is heated from 100°F to 500°F in a 1–2 parallel-counterflow heat exchanger with a mean overall heat transfer coefficient of 75 Btu/hr ft^2°F. It is converted to C by the exothermic reaction A + B → C, in an adiabatic plug-flow tubular reactor (Figure 5.30). For a process simulator, prepare a simulation flowsheet and show the calculation sequence to determine:

(a) Flow rates and unknown temperatures for each stream

(b) Heat duty and area of the countercurrent shell-and-tube heat exchanger

5.11 Consider the simulation flowsheets in Figure 5.31, which were prepared for ASPEN PLUS. The feed stream, S1, is specified, as are the parameters for each process unit.

Complete the simulation flowsheets using sequences acceptable to ASPEN PLUS. If any of the streams are torn, your flowsheets should include the recycle convergence units. In addition, you should indicate the calculation sequences.

This problem is easily modified if you are working with HYSYS, CHEMCAD, or PRO/II.

5.12 Use a process simulator to determine the flow rate of saturated vapor benzene at 176.2°F and 1 atm to be mixed with 100 lbmol/hr of liquid benzene to raise its temperature from 25 to 50°F. Prepare a good initial estimate. *Note:* $\lambda_{NBP} =$ 13,200 Btu/lbmol, $c_p = 0.42$ Btu/lb°F.

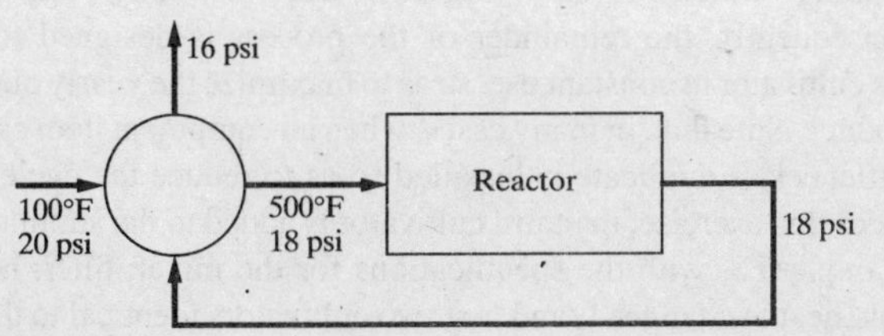

Figure 5.30 Reactor with feed/product heat exchanger.

5.13 A distillation tower is needed to separate an equimolar mixture, at 77°F and 1 atm, of benzene from styrene. The distillate should contain 99 mol% benzene and 95 mol% of the benzene fed to the tower.

Use a process simulator to determine the minimum number of trays at total reflux (N_{min}), the minimum reflux ratio (R_{min}), and the theoretical number of trays at equilibrium when $R = 1.3R_{min}$.

5.14 Use a process simulator to determine the heat required to vaporize 45 mol% of a liquid stream entering an evaporator at 150°F and 202 psia and containing

	lbmol/hr
Propane	250
n-Butane	400
n-Pentane	350

Assume that the evaporator product is at 200 psia. Use the Soave–Redlich–Kwong equation of state.

5.15 For an equimolar mixture of *n*-pentane and *n*-hexane at 10 atm, use a process simulator to compute:

(a) The bubble-point temperature

(b) The temperature when the vapor fraction is 0.5

5.16 Hot gases from the toluene hydrodealkylation reactor are cooled and separated as shown in the flowsheet of Figure 5.32. In a steady-state simulation, can the composition of the recycle stream be determined without iterative recycle calculations? Explain your answer.

5.17 Given the feed streams and the parameters of the process units as shown in Figure 5.33, complete the simulation flowsheet for ASPEN PLUS and show the calculation sequence (i.e., complete the statement SEQUENCE USED WAS:). If any of the streams are torn, your flowsheet should include the stream convergence units.

The simulation flowsheet can be modified for HYSYS, CHEMCAD, or PRO/II and the exercise repeated.

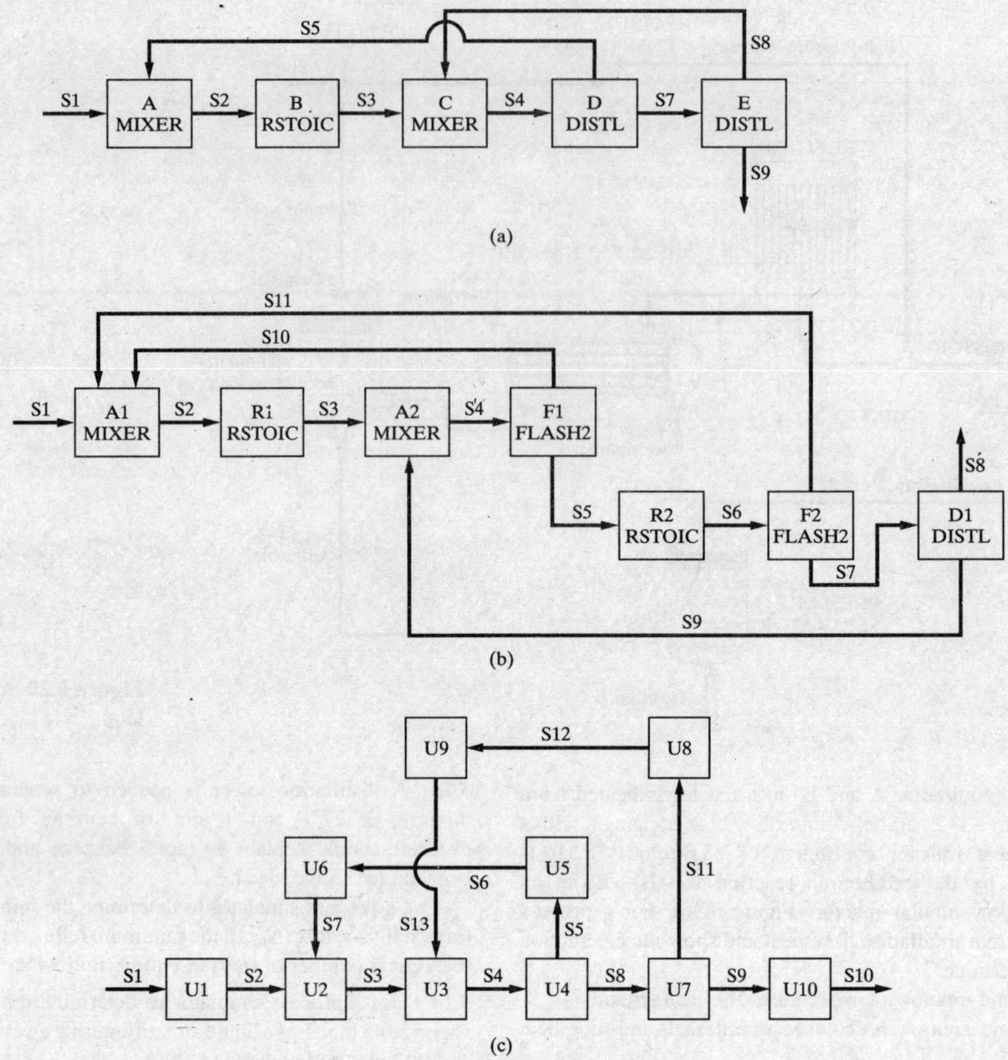

Figure 5.31 Interlinked recycle loops.

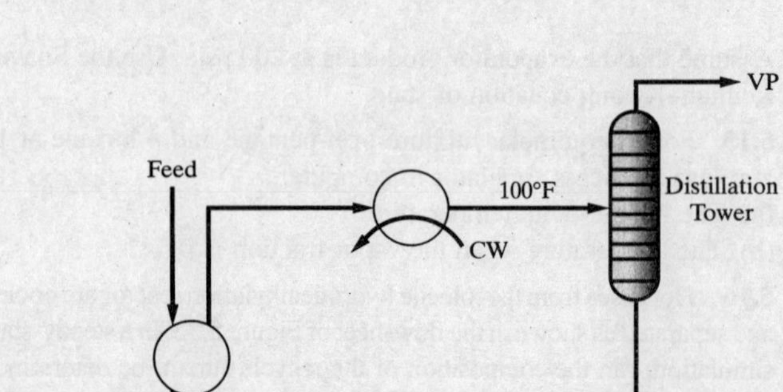

Figure 5.32 Combined quench/distillation process.

5.18 Suppose 100 lbmol/hr of steam (stream STI) at 500°F and 1 atm heats 40 lbmol/hr of cold water (stream CWI) from 70°F to 120°F in a shell-and-tube heat exchanger. For the simulation flowsheet in ASPEN PLUS (Figure 5.34), determine the outlet temperature of the steam.

The problem can be modified for the usage of HYSYS, CHEMCAD, or PRO/II.

5.19 *Debottlenecking reactor train.* When the third tPA cultivator in Section 4.4 is added to the two cultivators in Example 5.5, as shown in Figure 5.25a, a significant time strain is placed on the process because the combined feed, cultivation, harvest, and cleaning time in this largest vessel is long and rigid. Consequently, the remainder of the process is designed to keep this cultivator in constant use, so as to maximize the yearly output of product. Note that, in many cases, when an equipment item causes a bottleneck, a duplicate is installed so as to reduce the cycle time.

For this exercise, the third cultivator is added to the simulation in Example 5.5, with the specifications for the mixer, filter, holding tank, heat exchanger 1, and first two cultivators identical to those in Example 5.5. After the cultivation is completed in Cultivator 2, its

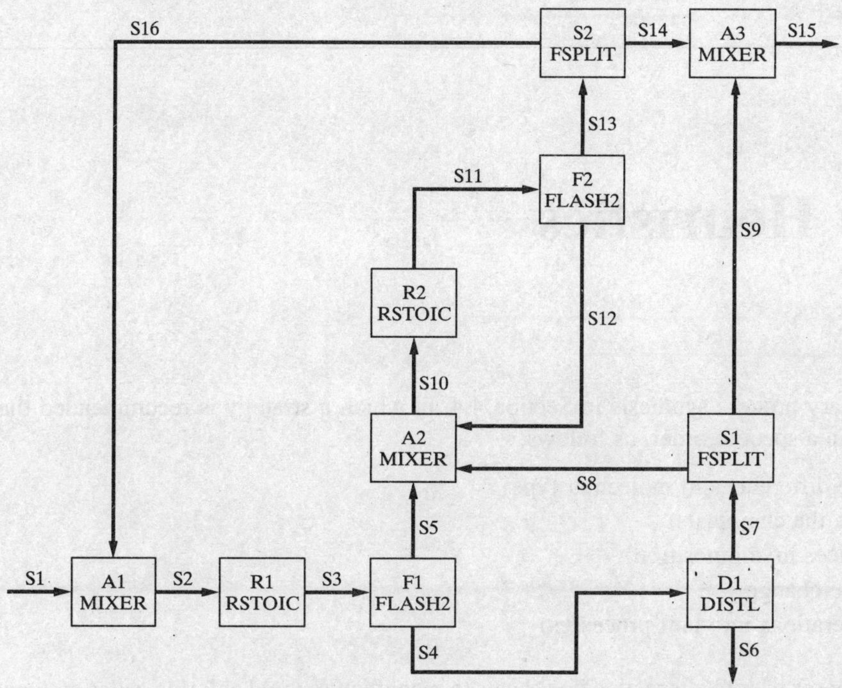

Figure 5.33 Incomplete simulation flowsheet.

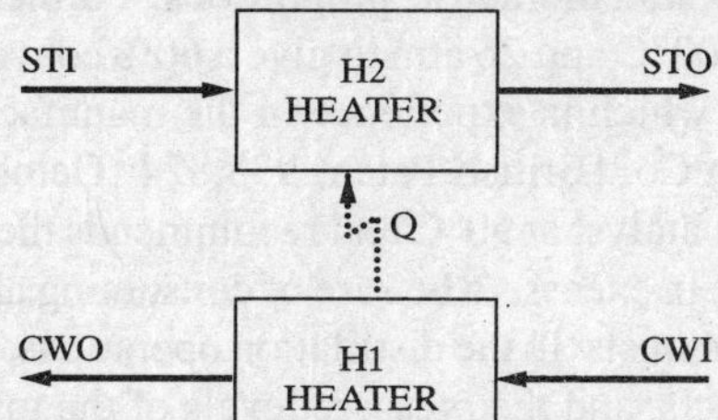

Figure 5.34 Simulation flowsheet with a heat stream.

cell mass is transferred as inoculum to Cultivator 3 over 0.5 day. Then, the remaining media from the mixing tank is heated to 37°C and added over 1.5 days, after which cultivation takes place over eight days. Immediately after the transfer from Cultivator 2 to Cultivator 3, Cultivator 2 is cleaned-in-place using 600 kg of water over 20 hr. The yield of the cultivation in Cultivator 3 is 11.4 wt% tPA-CHO cells, 7.7×10^{-5} wt% endotoxin, 88.9 wt% water, and 0.0559 wt% tPA. When the cultivation is completed in Cultivator 3, its contents are cooled in a heat exchanger to 4°C and transferred to the centrifuge holding tank over one day, and Cultivator 3 is cleaned using 600 kg of water over 67 hr and sterilized using the procedure for Cultivators 1 and 2.

To eliminate an undesirable bottleneck(s), and reduce the cycle time to 14 days (total operation time of Cultivator 3), it may be necessary to add an equipment unit(s).

Print and submit the text recipes and 3-batch schedules for both the original process and the modified process, if debottlenecking is necessary, as prepared by BATCH PLUS.

5.20 *tPA Process simulation.* For the entire process flowsheet in Figure 4.16a, b, c of Section 4.4, complete a BATCH PLUS simulation. Print and submit the text recipes and 3-batch schedules for both the original process and the modified process, as prepared by BATCH PLUS.

Note that the operating times, batch sizes, and recovery percentages are shown in Figure 4.16a, b, c. Unfortunately, BATCH PLUS determines that 15,000 kg/batch of elution buffer are required to elute the affinity chromatography column. To circumvent this, after 523 kg have been fed, specify a "cut" of 404 kg. The latter is collected as the elution effluent while the difference is rejected in the wastewater. Also, BATCH PLUS does not model the selective adsorption of endotoxin without tPA, arginine, sucrose, glycine, NaCl, and NaOH. Consequently, to obtain the desired effluent streams, one approach is to adsorb endotoxin, while recovering the waste effluent (which contains tPA, arginine, sucrose, glycine, NaCl, and NaOH). Then, the elution buffer (500 kg/batch of water) is used to elute the endotoxin, which is rejected as wastewater.

Chapter 6

Process Synthesis Heuristics

6.0 OBJECTIVES

This chapter returns to the steps of preliminary process synthesis in Section 4.4, in which a strategy is recommended that involves assembling the process operations in a specific order, as follows:

1. Chemical reactions (to eliminate differences in molecular type)
2. Mixing and recycle (to distribute the chemicals)
3. Separation (to eliminate differences in composition)
4. Temperature, pressure, and phase change
5. Task integration (to combine operations into unit processes)

In Section 4.4, as the operations are inserted into alternative flowsheets to manufacture vinyl chloride, *rules of thumb* or *heuristics* are utilized. For example, when positioning the direct chlorination operation, it is assumed that because the reaction is nearly complete at 90°C, ethylene and chlorine can be fed in stoichiometric proportions. Furthermore, when positioning the pyrolysis operation, the temperature and pressure are set at 500°C and 26 atm to give a 60% conversion. These assumptions and specifications are based on many factors, not the least of which is experience in the manufacture of vinyl chloride and similar chemicals. In this case, a patent by the B.F. Goodrich Co. [British Patent 938,824 (October 9, 1963)] indicates the high conversion of ethylene and chlorine over a ferric chloride catalyst at 90°C and recommends the temperature and pressure levels of the pyrolysis reaction. The decision not to use ethylene in excess, to be sure of consuming all of the toxic chlorine, is based on the favorable conversions reported experimentally by chemists. In the distillation operations, the choice of the key components, the quality of the feed streams and the distillate products, and the pressure levels of the towers are also based on rules of thumb. In fact, heuristics like these and many others can be organized into an *expert system*, which can be utilized to synthesize sections of this and similar chemical processes.

Normally, design teams use heuristics when generating the alternatives that make up a synthesis tree, such as that shown in Figure 4.9. For the most part, heuristics are easy to apply; that is, they involve the setting of temperatures, pressures, excess amounts of chemicals, and so on. Often, they require little analysis in that simple material balances can be completed without iterations before proceeding to the next synthesis step. Consequently, several promising flowsheets are generated rapidly, with relatively little effort. Then, as described in Section 4.5, the emphasis of the design team shifts to the creation of a *base-case design.* The assumptions are checked, a process flow diagram is assembled (e.g., Figure 4.19), and a complete material and energy balance is carried out, often using the process simulators discussed in Chapter 5.

Clearly, the heuristics used by a design team to generate the synthesis tree are crucial in the design process. Section 4.4 provides just a brief introduction to these heuristics, and hence it is the objective of this chapter to describe the principal heuristics used in process design more thoroughly. A total of 53 heuristics are presented in Sections 6.2 through 6.9. In many cases, the heuristics are accompanied by examples. For quick reference, the heuristics are collected together in Table 6.2 at the end of this chapter. Additional guidance in the selection of equipment is given in Chapters 22 and 23 when determining equipment purchase and operating costs, Chapter 7 when designing reactors, Chapter 18 when sizing heat exchangers, Chapter 19 when sizing distillation towers, and Chapter 20 when sizing pumps, compressors, and gas expanders.

After studying this chapter and the heuristics in Table 6.2, the reader should

1. Understand the importance of selecting reaction paths that do not involve toxic or hazardous chemicals, and when unavoidable, to reduce their presence by shortening residence times in the process units and avoiding their storage in large quantities.
2. Be able to distribute the chemicals, when generating a process flowsheet, to account for the presence of inert species, to purge species that would otherwise build up to unacceptable concentrations, to achieve a high selectivity

to the desired products, and to accomplish, when feasible, reactions and separations in the same vessels (e.g., reactive distillations).

3. Be able to apply heuristics in selecting separation processes to separate liquids, vapors, vapor–liquid mixtures, and other operations involving the processing of solid particles, including the presence of liquid and/or vapor phases.
4. Be able to distribute the chemicals, by using excess reactants, inert diluents, and cold (or hot) shots, to remove the exothermic (supply the endothermic) heats of reaction. These distributions can have a major impact on the resulting process integration.
5. Understand the advantages, when applicable, of pumping a liquid rather than compressing a vapor.

Through several examples and the exercises at the end of the chapter, the reader should be able to apply the heuristics in Table 6.2 when generating a synthesis tree. Also, he or she should obtain an appreciation of the role of heuristics and recognize that the heuristics covered are only a subset of many that are widely applied in chemicals processing.

6.1 INTRODUCTION

In a chapter on heuristics for process synthesis, it is important to emphasize that heuristics are used commonly by design teams to expedite the generation of alternative flowsheets in preliminary process synthesis. Then, as the alternatives are generated, or afterward, it is common to perform material and energy balances, and related forms of analysis, often using a process simulator. Although this chapter is devoted to heuristics, in the remainder of this section, emphasis is placed on the kinds of analyses usually used to improve upon designs suggested by heuristics. It is important to understand the consequences of heuristics and to recognize, at least through one typical design, the way in which process simulators are used to explore and alter heuristics.

Return to the design of the toluene hydrodealkylation process, as it is presented in Section 5.3. In the reactor section, after heuristics are utilized to set (1) the large excess of H_2 in the hydrodealkylation reactor, (2) the temperature level of the quenched gases that enter the feed-product heat exchanger, and (3) the temperature in the flash vessel, the simulator is used to complete the material and energy balances and to examine the effects of these heuristics on the performance of the reactor section. In the distillation section, after heuristics are used to set (1) the quality of the feed, (2) the use of partial or total condensers, (3) the use of cooling water in the condensers, and (4) the ratio of the reflux ratio to the minimum reflux ratio (R/R_{min}), the simulator is used to determine the appropriate pressure levels in the columns, to estimate the number of stages and the position of the feed stage, to estimate the reflux ratio, and, most important, to compute the distillate and bottoms products. The simulator provides an excellent vehicle for studying the effect of departures from the heuristics on the performance of the separation train.

Eventually, the two sections of the plant are combined and heuristics are used to set the purge-to-recycle ratio. Here, the simulator determines the recycle streams, which in the analysis heretofore were specified, once again using heuristic rules. Then the simulator provides an easy-to-use vehicle for studying the effect of the purge-to-recycle ratio on the performance of the process.

Having completed a simulation of the flowsheet, or possibly after working with the reactor and separation sections alone, the designer can estimate the capital and operating costs and can compute and optimize measures of the profitability, as discussed in Chapters 22–24. Furthermore, the simulator is used often by the engineer to study the effect of making small changes in the structure of the flowsheet (e.g., to recover toluene and biphenyl from the bottom of the first distillation tower). By examining their impact on the performance and profitability, the designer implements an *evolutionary* synthesis strategy, often using the process simulator. Some prefer to refer to this approach as process synthesis by *interactive analysis*. The basic approach is to examine sections of a plant one-by-one, generating alternative structures with heuristics and experimenting with them, retaining the most promising as operations are added to complete the flowsheets.

As mentioned above, in this chapter the focus is on the heuristics, because they are crucial in generating quickly the most promising structures. Subsequently, more systematic methods for generating many alternative flowsheets are considered in Chapters 7–11. Throughout this chapter, examples are provided to show how to use simulators to assist in evaluating the effect of the heuristics on the performance of the processes being designed. Even when the so-called *algorithmic* approaches to process synthesis are introduced in Chapters 7–11, the heuristics discussed in this chapter play an important role.

Eventually, the methods of mathematical programming are introduced in Chapters 9 and 24. Through the use of mixed-integer nonlinear programs (MINLPs), these methods are designed to optimize *superstructures* involving all of the potential streams and process units to be considered during the optimization, with many streams and process units turned off because they are associated with suboptimal solutions. When MINLPs can be formulated and solved rigorously, the use of heuristic rules can be sharply reduced or even eliminated. However, MINLPs are difficult to formulate in the early stages of process design because there are substantial uncertainties, and consequently, it is important to place emphasis on the rules of thumb needed to get started in the design process. As will be shown in this and subsequent chapters, these rules often lead to near-optimal designs.

6.2 RAW MATERIALS AND CHEMICAL REACTIONS

Heuristic 1: ***Select raw materials and chemical reactions to avoid, or reduce, the handling and storage of hazardous and toxic chemicals.***

As discussed in Chapter 4, the selection of raw materials and chemical reactions is often suggested by chemists, biologists, biochemists, or other persons knowledgeable about the chemical conversions involved. In recent years, with the tremendous increase in awareness of the need to avoid handling hazardous and toxic chemicals, in connection with environmental and safety regulations (as discussed in Sections 1.4 and 1.5), raw materials and chemical reactions are often selected to protect the environment and avoid the safety problems that are evident in Material Safety Data Sheets (MSDSs). For example, recall that when the vinyl-chloride process was synthesized in Section 4.4, the reaction of acetylene with HCl was rejected because of the high cost of acetylene. Today, in addition, this reaction path would be rejected on the basis of the high reactivity of acetylene and the difficulty of ensuring safe operation in the face of unanticipated disturbances.

In connection with the handling of hazardous chemicals, the 1984 accident in Bhopal, India, in which water was accidentally mixed with the active intermediate methyl isocyanate, focused worldwide attention on the need to reduce the handling of highly reactive intermediates. As discussed in Section 1.5, within an hour of the accident, a huge vapor cloud swept across Bhopal, leading to the death of over 3,800 victims in the vicinity of the Union Carbide plant. This accident, together with the discovery of polluted groundwaters adjacent to chemical plants, especially those that process nuclear fuels, have led safety and environment experts to call for a sharp reduction in the handling of hazardous chemicals.

For these reasons, societal needs are increasingly being formulated that call for new processes to avoid or sharply reduce the handling of hazardous chemicals. As an example, consider the manufacture of ethylene glycol, the principal ingredient of antifreeze. Ethylene glycol is produced commonly by two reactions in series:

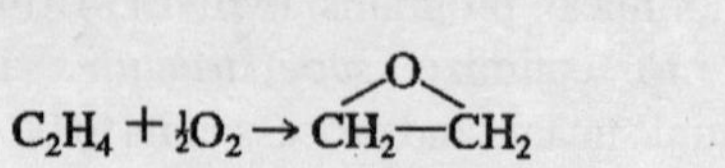
(R1)

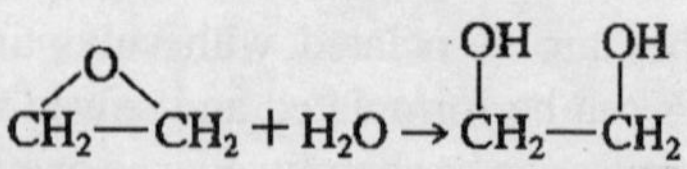
(R2)

The first reaction involves the partial oxidation of ethylene over an Ag-gauze catalyst. Since both reactions are highly exothermic, they need to be controlled carefully. More important from a safety point of view, a water spill into an ethylene oxide storage tank could lead to an accident similar to the Bhopal incident. Yet it is common in processes with two reaction steps to store the intermediates so as to permit the products to be generated continuously, even when maintenance problems shut down the first reaction operation.

Given the societal need to eliminate the storage of large quantities of reactive intermediates such as ethylene oxide, four alternative processing concepts are possible:

1. Eliminate the storage tank(s), causing intermittent interruptions in the production of ethylene glycol when the oxidation reaction shuts down.
2. Use costly chlorine and caustic (compared to oxygen from air) in a single reaction step:

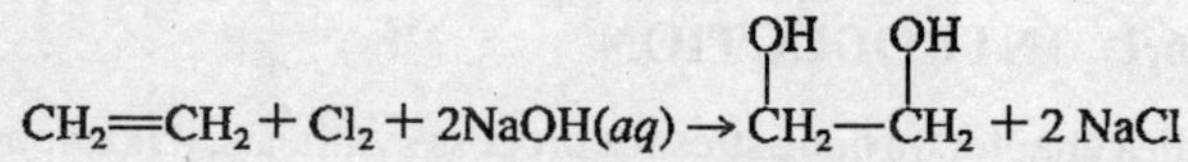
(R3)

   This alternative requires more expensive raw materials, but completely avoids the intermediate.
3. As ethylene oxide is formed, react it with carbon dioxide to form ethylene carbonate, a much less active intermediate. This reaction

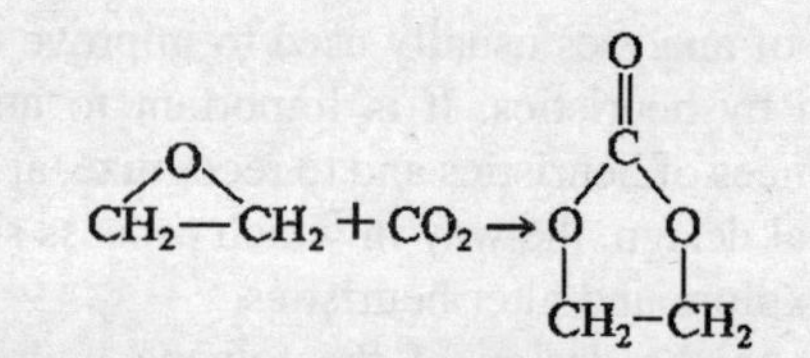
(R4)

   occurs smoothly over a tetraethylammonium bromide catalyst. Ethylene carbonate can be stored safely and hydrolyzed to form the ethylene glycol product as needed.
4. Carry out reactions (R1) and (R4) consecutively over an Ag-gauze catalyst by reacting ethylene in a stream containing oxygen and carbon monoxide. To consider this as an alternative processing concept, laboratory or pilot-plant data on the rates of reaction are necessary.

In summary, there is an increasing emphasis on retrofitting processes to eliminate active intermediates, and in the design of new processes to avoid these chemicals entirely. Furthermore, the designers of new processes are being asked, with increasing frequency, to select raw materials and reactions accordingly. These have become important considerations in the early stages of process design.

6.3 DISTRIBUTION OF CHEMICALS

Heuristic 2: ***Use an excess of one chemical reactant in a reaction operation to consume completely a valuable, toxic, or hazardous chemical reactant. The MSDSs will indicate which chemicals are toxic and hazardous.***

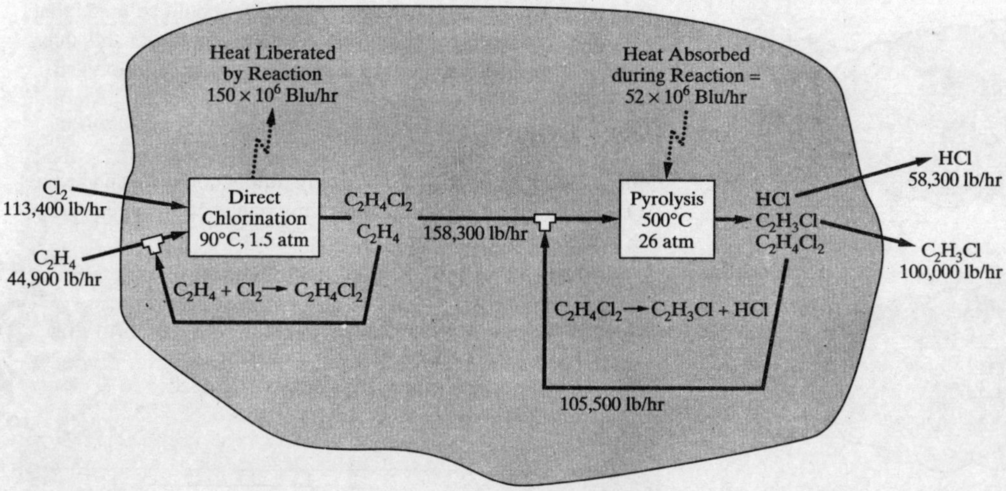

Figure 6.1 Distribution of chemicals for the production of vinyl chloride involving an excess of ethylene.

After the reaction operations are positioned in a process flowsheet, the sources of chemicals (i.e., the feed streams and reactor effluents) are distributed among the sinks for chemicals (i.e., the feed streams to the reaction operations and the products from the process). In this distribution, decisions are made concerning (1) the use of one chemical reactant in excess in a reaction operation, (2) the handling of inert species that enter in the feed streams, and (3) the handling of undesired byproducts generated in side reactions. For example, as we have seen in Figure 4.5, one distribution of chemicals for the vinyl-chloride process involves stoichiometric amounts of ethylene and chlorine fed to the direct-chlorination reactor. Alternatively, an excess of ethylene can be utilized as shown in Figure 6.1. In this distribution, the reactor is designed to consume completely the hazardous and toxic chlorine, but the recovery of unreacted ethylene from the dichloroethane product is required. Clearly, an important consideration is the degree of the excess, that is, the ethylene/chlorine ratio. It governs the costs of separation and recirculation, and often plays a key role in the process economics. In many design strategies, this ratio is set using heuristics, with larger ratios used to ensure consumption of the most hazardous chemicals. Eventually, as a *base-case design* evolves, the ratio is varied systematically, often using a process simulator. In mathematical programming strategies, it is treated as a *design variable*, to be varied during optimization, with a lower bound. Note that for exothermic reactions, the excess chemical often serves the useful function of absorbing the heat of reaction and thereby maintaining more moderate temperatures. This is an important approach to handling large heats of reaction and is considered with several common alternatives in Section 6.5, in the subsection on heat removal from exothermic reactors. An excess of one chemical reactant is also used to increase conversion of the other (limiting) reactant when the extent of reaction is limited by equilibrium. Also, side reactions can be minimized by using an excess of one reactant.

Inert Species

Heuristic 3: ***When nearly pure products are required, eliminate inert species before the reaction operations when the separations are easily accomplished and when the catalyst is adversely affected by the inert, but not when a large exothermic heat of reaction must be removed.***

Often impure feed streams contain significant concentrations of species that are inert in chemical reaction operations. When nearly pure products are required, an important decision concerns whether impurities should be removed before or after reaction operations. As an example, consider the flowsheet in Figure 6.2a, in which two reaction operations have been positioned. An impure feed stream of reactant C contains the inert species D, and hence a decision is required concerning whether to remove D before or after reaction step 2, as shown in Figures 6.2b and 6.2c, respectively. Clearly, the ease and cost of the separations, that is, D from C, and D from E (plus unreacted A and C), must be assessed. This can be accomplished by examining the physical properties on which the separations are based. For example, when considering distillation, estimates of the relative volatilities are used. When the mixtures are ideal, the relative volatility, α_{ij}, is simply a ratio of the vapor pressures ($\alpha_{ij} = P_i^s/P_j^s$). Otherwise, activity coefficients are needed ($\alpha_{ij} = \gamma_i P_i^s/\gamma_j P_j^s$). When the relative volatilities differ significantly from unity, an easy and inexpensive separation is anticipated. Similarly, when considering crystallization, the differences in the freezing points are examined, and for dense membrane separations, the permeabilities of the pure species are estimated as the product of the solubility in the membrane and the molecular diffusivity. Other considerations are the sizes of the reactors and separators, with larger reactors required when the separations are postponed. Also, for exothermic reactions the inert species

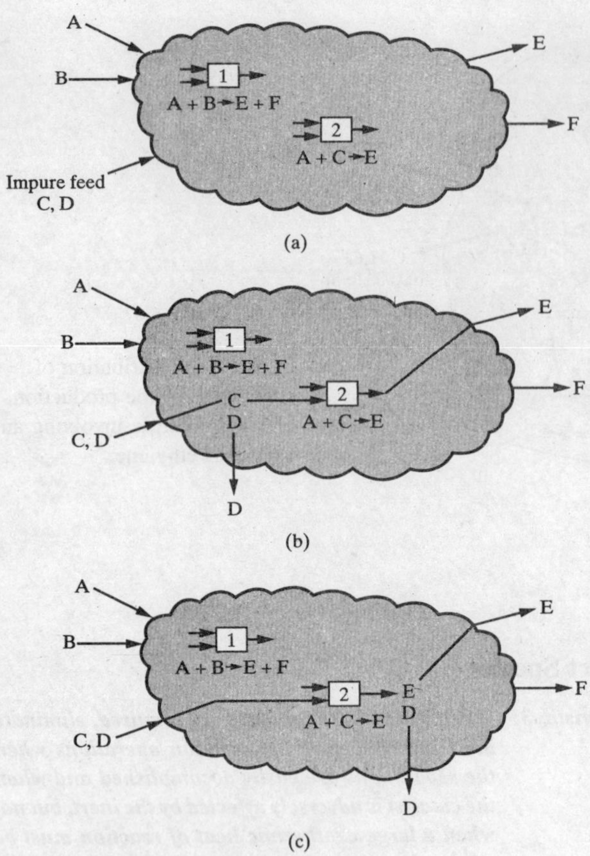

Figure 6.2 Partial distribution of chemicals showing the alternatives for removing inert species D: (a) reaction operations; (b) recovery before reaction;(c) recovery after reaction.

absorb some of the heat generated, thereby lowering the outlet temperatures of the reactors.

EXAMPLE 6.1 *Distribution of Chemicals*

To satisfy the Clean Air Act of 1990, gasoline must have a minimum oxygen atom content of 2.0 mol%. In the 1990s, the most common source of this oxygen was methyl *tertiary-butyl* ether (MTBE), which is manufactured by the reaction

$$CH_3OH + iso\text{-butene} \rightleftharpoons MTBE$$

It is desired to construct an MTBE plant at your refinery, located on the Gulf Coast of Texas. Methanol will be purchased and *iso*-butene is available in a mixed-C_4 stream that contains

	Wt%
1-Butene	27
iso-Butene	26
1,3-Butadiene	47

During process synthesis, in the distribution of chemicals, a key question involves whether it is preferable to remove 1-butene and 1,3-butadiene before or after the reaction operation. In this example, distillation is considered, although other separation methods normally are evaluated as well. It should be noted that recently MTBE was found to contaminate groundwater and, thus, in most locations is no longer the preferred source of oxygen.

SOLUTION

These hydrocarbon mixtures are only mildly nonideal, and hence it is satisfactory to examine the boiling points of the pure species or, better yet, their vapor pressures. These can be tabulated and graphed as a function of temperature using a simulator; for example, the following curves are obtained from ASPEN PLUS (and can be reproduced using the EXAM6-1.bkp file in the Program and Simulation Files folder, which can be downloaded from the Wiley Web site associated with this book.

www.wiley.com/go/global/seider

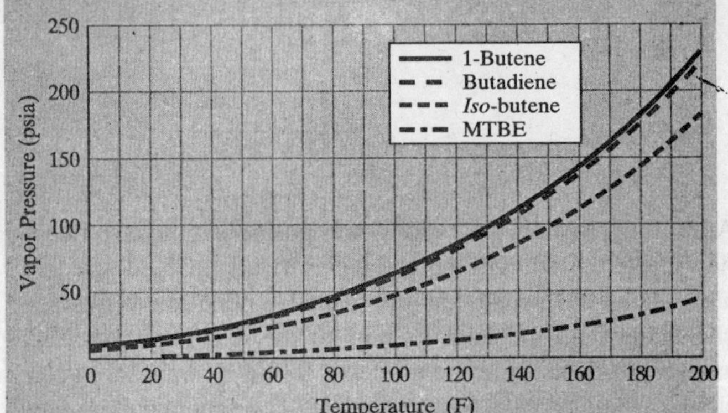

With respect to MTBE, the relative volatilities, $\alpha = P^s/P^s_{MTBE}$, at 200°F, are 5.13 (1-butene), 4.96 (1,3-butadiene), and 4.04 (*iso*-butene). Clearly, the relative volatilites of 1-butene and 1,3-butadiene are very close, but each differs significantly from the value for *iso*-butene. On this basis, the former two compounds can be separated by distillation before or after the reaction operation. Other considerations, such as their impact on the catalyst, the volumes of the reactors and distillation towers, and the temperature levels in the exothermic reactors, should be evaluated in making this decision.

EXAMPLE 6.2 *Positioning an Equilibrium Reaction Operation*

Consider the reaction and distillation operations for the isomerization of *n*-butane to isobutane according to the reaction

$$n\text{-}C_4H_{10} \rightleftharpoons i\text{-}C_4H_{10}$$

The feed to the process is a refinery stream that contains 20 mol% *iso*-butane. Show the alternatives for positioning the reaction and distillation operations.

SOLUTION

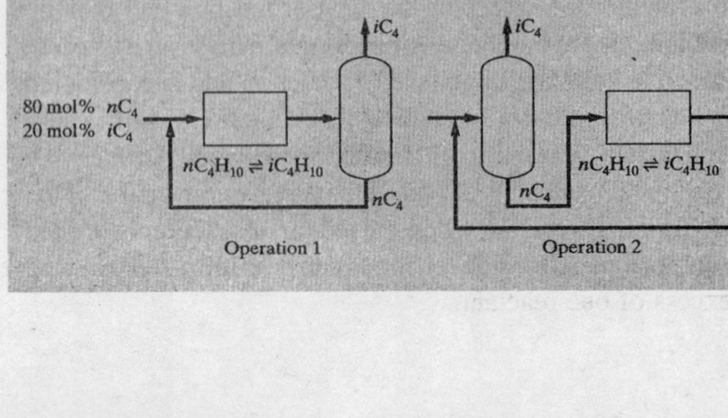

As shown in the diagram, either operation can be placed first. By positioning the distillation column first, a nearly pure feed is sent to the reaction operation, providing a higher conversion to *iso*-butane. The effectiveness of this configuration depends on the relative difficulty of achieving the distillation separation. To determine this, the two configurations should be simulated.

Purge Streams

Heuristic 4: ***Introduce purge streams to provide exits for species that enter the process as impurities in the feed or are formed in irreversible side reactions, when these species are in trace quantities and/or are difficult to separate from the other chemicals. Lighter species leave in vapor purge streams, and heavier species exit in liquid purge streams.***

Trace species, often introduced as impurities in feed streams or formed in side reactions, present special problems when the chemicals are distributed in a flowsheet. In a continuous process, these accumulate continuously unless a means is provided for their removal, either by reaction, separation, or through purge streams. Since the reaction or separation of species in low concentration is usually costly, purge streams are used when the species are nontoxic and have little impact on the environment. Purge streams are also used for removing species present in larger amounts when their separation from the other chemicals in the mixture is difficult. As an example, consider the distribution of chemicals in the ammonia process ($N_2 + 3H_2 \rightleftharpoons 2NH_3$) in Figure 6.3. Trace amounts of argon accompany nitrogen, which is recovered from air, and trace amounts of methane accompany hydrogen, which is produced by steam reforming ($CH_4 + H_2O \rightleftharpoons 3H_2 + CO$). After reforming, the carbon monoxide and unreacted methane and steam are recovered, leaving trace quantities of methane. Although nitrogen and hydrogen react at high pressures, in the range of 200–400 atm depending on the process throughput, the conversion is low (usually in the range of 15–20 mol%), and large quantities of unreacted nitrogen and hydrogen are recirculated. The purge stream provides a sink for the argon and methane, which otherwise would build to unacceptable concentrations in the reactor bed, which is packed with reduced iron oxide catalyst. The light gases from the flash vessel are split into purge and recycle streams, with the purge/recycle ratio being a key decision variable. As the purge/recycle ratio increases, the losses of nitrogen and hydrogen increase, with an accompanying reduction in the production of ammonia. This is counterbalanced by a decrease in the recirculation rate. In the early stages of process synthesis, the purge/recycle ratio is often set using heuristics. Eventually, it can be varied with a process simulator to determine its impact on the recirculation rates and equipment sizes. Then it can be adjusted, also using a process simulator, to optimize the return on investment for the process, as discussed in Chapters 23 and 24. Note that the alternative of separating trace species from the vapor stream, thereby avoiding the purge of valuable nitrogen and hydrogen, may also be considered. These separations—for example, adsorption, absorption, cryogenic distillation, and gas permeation with a membrane—may be more expensive. Finally, it should be recognized that argon and methane are gaseous species that are purged from the vapor recycle stream. Other processes involve heavy impurities that are purged from liquid streams.

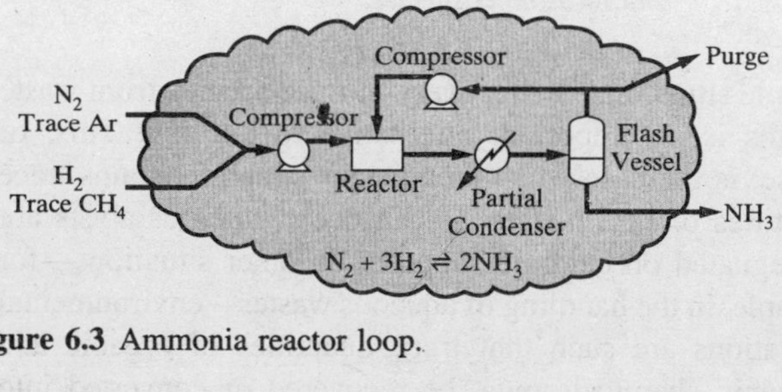

Figure 6.3 Ammonia reactor loop.

EXAMPLE 6.3 ***Ammonia Process Purge***

In this example, the ammonia reactor loop in Figure 6.3 is simulated using ASPEN PLUS to examine the effect of the purge-to-recycle ratio on the compositions and flow rates of the purge and recycle streams. For the ASPEN PLUS flowsheet below, the following specifications are made:

Simulation Unit	Subroutine	T (°F)	P (atm)
R1	REQUIL	932	200
F1	FLASH2	−28	136.3

and the Chao–Seader option set is selected to estimate the thermophysical properties. Note that REQUIL calculates chemical equilibria at the temperature and pressure specified, as shown on the multimedia module *ASPEN → Chemical Reactors → Equilibrium Reactors → REQUIL.*

The combined feed stream, at 77°F and 200 atm, is comprised of

	lbmol/hr	Mole Fraction
N_2	24	0.240
H_2	74.3	0.743
Ar	0.6	0.006
CH_4	1.1	0.011
	100.0	1.000

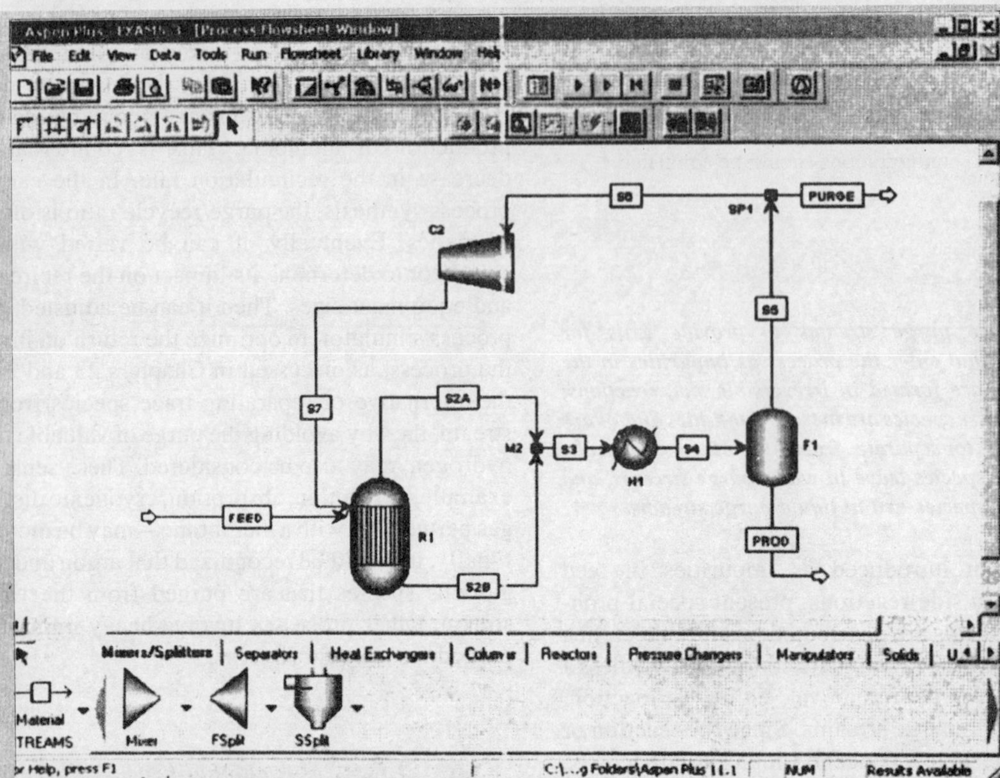

SOLUTION

Several variables are tabulated as a function of the purge/recycle ratio:

Purge/Recycle Ratio	PROD Flow Rate (lbmol/hr)	Recycle Flow Rate (lbmol/hr)	Purge Flow Rate (lbmol/hr)	Purge Mole Fraction Ar	Purge Mole Fraction CH_4
0.1	39.2	191.0	19.1	0.028	0.052
0.08	40.75	209.3	16.7	0.033	0.060
0.06	42.4	233.9	14.0	0.040	0.074
0.04	44.3	273.5	10.9	0.053	0.093
0.02	45.8	405.6	8.1	0.072	0.133

In all cases, the mole fractions of Ar and CH_4 in the purge are significantly greater than in the feed. As the purge/recycle ratio is decreased, the vapor effluent from the flash vessel becomes richer in the inert species and less H_2 and N_2 are lost in the purge stream. However, this is accompanied by a significant increase in the recycle rate and the cost of recirculation, as well as the reactor volume. Note that the EXAM6-3.bkp file (in the Program and Simulation Files folder, which can be downloaded from the Wiley Web site associated with this book) can be used to reproduce these results. Although not implemented in this file, the purge/recycle ratio can be adjusted parametrically by varying the fraction of stream S5 purged in a *sensitivity analysis*, which is one of the *model analysis tools* found in most simulators. The capital and operating costs can be estimated and a profitability measure optimized as a function of the purge/recycle ratio.

Heuristic 5: ***Do not purge valuable species or species that are toxic and hazardous, even in small concentrations (see the MSDSs). Add separators to recover valuable species. Add reactors to eliminate, if possible, toxic and hazardous species.***

In some situations, the recovery of trace species from waste streams is an important alternative to purging. This, of course, is the case when an aqueous stream contains trace quantities of rare metals, as can occur when catalysts are impregnated on ceramic supports. In other situations—for example, in the handling of aqueous wastes—environmental regulations are such that trace quantities of organic and inorganic chemicals must be recovered or converted into

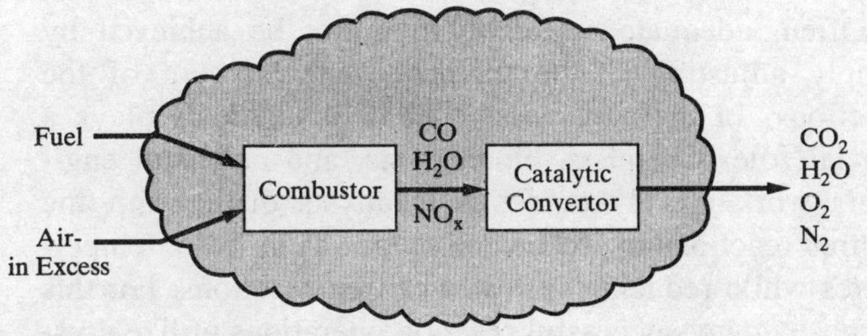

Figure 6.4 Catalytic conversion of combustion effluents.

an environmentally acceptable form. One process to treat aqueous streams in the vicinity of leaking tanks is *supercritical oxidation*, using acoustic waves or lasers to produce plasmas. In this process, the waste species [including chlorinated hydrocarbons, pesticides, phenols (e.g., *p*-nitrophenol), and esters] are oxidized at temperatures and pressures associated with supercritical water (Hua et al., 1995a, b). Yet another example involves the catalytic conversion of hydrocarbons and carbon monoxide in the exhaust gases from internal combustion engines. As illustrated in Figure 6.4, rather than purge the exhaust gases from a combustion engine, catalytic converters commonly convert carbon monoxide and nitrogen oxides to carbon dioxide and nitrogen, respectively. Again, the decision to insert a reaction step, rather than to separate or purge, in the early stages of process design is made often based on the availability of a catalyst and experience; that is, heuristics.

Recycle to Extinction

Heuristic 6: ***Byproducts that are produced in reversible reactions, in small quantities, are usually not recovered in separators or purged. Instead, they are usually recycled to extinction.***

Often small quantities of chemicals are produced in side reactions, such as the reaction of benzene to form biphenyl in the toluene hydrodealkylation process. When the reaction proceeds *irreversibly*, small quantities of byproducts must be separated away, as in Figure 5.20, or purged; otherwise they will build up in the process until the process must be shut down.

When the reaction proceeds *reversibly*, however, it becomes possible to achieve an equilibrium conversion at steady state by recycling product species without removing them from the process. In so doing, it is often said that undesired byproducts are *recycled to extinction*. It is important to recognize this when distributing chemicals in a potential flowsheet so as to avoid the loss of chemicals through purge streams or the insertion of expensive separation operations. Recycle to extinction, which is considered in more detail in Section 7S.4, is most effective when the equilibrium conversion of the side reaction is limited by a small chemical equilibrium constant at the temperature and pressure conditions in the reactor.

EXAMPLE 6.4 *Reversible Production of Biphenyl*

This example follows the simulation of the complete toluene hydrodealkylation process at the end of Section 5.3 and is presented without a solution because it is the basis for Exercise 6.4. To recycle the biphenyl to extinction, the flowsheet in Figure 5.20 is modified to eliminate the last distillation column, and unreacted toluene from the second column is recycled with biphenyl. This is accomplished by the reversible reaction

$$2C_6H_6 \rightleftharpoons C_{12}H_{10} + H_2$$

Note that this eliminates one of the two waste streams from the process. The other, which loses large quantities of H_2, is the gas purge stream. To avoid this loss, the use of membrane or adsorption separators should be considered.

Selectivity

Heuristic 7: ***For competing reactions, both in series and parallel, adjust the temperature, pressure, and catalyst to obtain high yields of the desired products. In the initial distribution of chemicals, assume that these conditions can be satisfied. Before developing a base-case design, obtain kinetics data and check this assumption.***

When chemical reactions compete in the formation of a desired chemical, the reaction conditions must be set carefully to obtain a desirable distribution of chemicals. Consider, for example, the series, parallel, and series-parallel reactions in Figure 6.5, where species B is the desired product. For these and similar reaction systems, it is important to consider the temperature, pressure, ratio of the feed chemicals, and the residence time

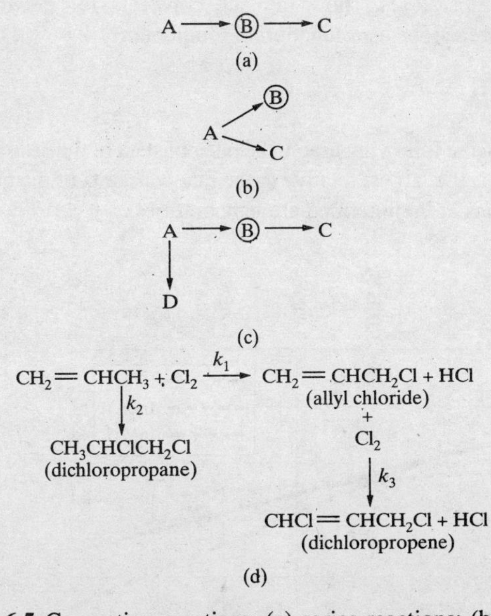

Figure 6.5 Competing reactions: (a) series reactions; (b) parallel reactions; (c) series-parallel reactions; (d) exothermic allyl chloride reactions.

Table 6.1 Heats of Reaction and Kinetics Constants for the Allyl Chloride Process[a]

Reaction	ΔH_R (Btu/lbmol)	k_0 [lbmol/(hr ft^3 atm^2)]	E/R (°R)
1	−4,800	206,000	13,600
2	−79,200	11.7	3,430
3	−91,800	4.6×10^8	21,300

[a]Biegler and Hughes, 1983.

when distributing the chemicals. One example of series-parallel reactions occurs in the manufacture of allyl chloride. This reaction system, which involves three competing second-order exothermic reactions, is shown in Figure 6.5d, with the heats of reaction, ΔH_R, activation energies, E, and preexponential factors, k_0, in Table 6.1. Note that because $E_1/E_2 > 1$ and $E_1/E_3 < 1$, the conversion to allyl chloride is highest at intermediate temperatures. In the early stages of process synthesis, when distributing the chemicals, these considerations are helpful in setting the temperature, pressure, and the ratio of propylene/chlorine in the feed.

When selectivity is the key to the success of a process design, it is not uncommon to carry out an extensive analysis of the reactor alone, before distributing the chemicals, and proceeding with the synthesis of the flowsheet. In other cases, using simulation models, the distribution of chemicals is carried out as the process is optimized to achieve an economic objective.

EXAMPLE 6.5 ***Selectivity of the Allyl Chloride Reactions***

To demonstrate the advantages of running the reactions at intermediate temperatures, show the rate constants for the three competing reactions as a function of temperature.

SOLUTION

As shown in the following graph, the rate constant of the desirable Reaction 1 is the largest relative to the rate constants of the other two reactions at the intermediate temperatures.

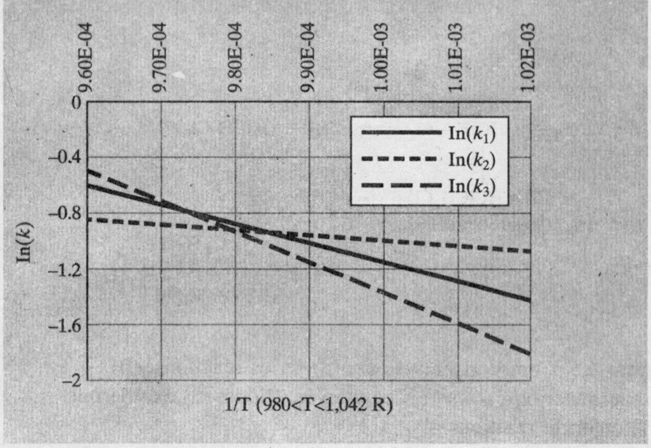

Often, adequate selectivities cannot be achieved by simply adjusting the temperature and pressure of the reactions. In these cases, the field of catalysis plays a crucial role. Chemists, biochemists, and chemical engineers work closely to design catalysts that permit the desired reactions to proceed more rapidly at lower temperatures while reducing the rates of side reactions. For this purpose, many successful reaction operations utilize crystalline zeolites, which are very effective *shape-selective catalysts*. In fact, it is possible to synthesize zeolite structures in which the cavities are just small enough to prevent the reactant molecules for the undesired side reactions from migrating to the reaction sites. Other commonly used catalysts involve the rare metals; for example, platinum, palladium, and rhodium. Clearly, when distributing chemicals in process synthesis, it is crucial to understand the relative rates of the competing reactions. Company laboratories are a key source of this information, as are patents and articles in the scientific journals. Process designers spend considerable time searching the extensive literature to locate appropriate catalysts.

Of course, many fine books have been written on the subject of catalysis. To study how to design reactors to achieve the desired selectivity, several outstanding textbooks are available; consider, for example, *Elements of Chemical Reaction Engineering* (Fogler, 2005) and *The Engineering of Chemical Reactions* (Schmidt, 2004).

Reactive Separations

Heuristic 8: ***For reversible reactions especially, consider conducting them in a separation device capable of removing the products, and hence driving the reactions to the right. Such reaction-separation operations lead to very different distributions of chemicals.***

The last step in process synthesis recommended in Section 4.4 is *task integration,* that is, the combination of operations into process units. In the synthesis steps recommended there, reaction operations are positioned first, chemicals are distributed (as discussed earlier in this section), and separation operations are positioned, followed by temperature-, pressure-, and phase-change operations, before task integration occurs. In some cases, however, this strategy does not lead to effective combinations of reaction and separation operations, for example, reactive distillation towers, reactive absorption towers, and reactive membranes. Alternatively, when the advantages of merging these two operations are examined by a design team, a combined reaction-separation operation is placed in the flowsheet before chemicals are distributed, with a significant improvement in the economics of the design. Although the subject of reactive separations is covered in Section 8.5, a brief introduction to reactive distillation is provided next.

Reactive distillation is used commonly when the chemical reaction is reversible, for example,

$$aA + bB \rightleftharpoons cC + dD$$

and there is a significant difference in the relative volatilities of the chemicals at the conditions of temperature and pressure suitable for the reaction. In conventional processing, when a reversible reaction operation is followed by a distillation column, it is common to use an excess of a feed chemical to drive the reaction forward. Alternatively, when the reaction takes place in the gas phase, the pressure is raised or lowered, depending on whether the summation of the stoichiometric coefficients is negative or positive. An advantage of reactive distillation, as shown for the production of methyl acetate,

$$MeOH + HOAc \rightleftharpoons MeOAc + H_2O$$

in Figure 6.6, is that the product chemicals are withdrawn from the reaction section in vapor and liquid streams, thereby driving the reaction forward without excess reactant or changes in pressure. Since methanol is more volatile than acetic acid, it is fed to the bottom of the reaction zone, where it concentrates in the vapor phase and contacts acetic acid, which is fed at the top of the reaction zone and concentrates in the liquid phase. As methyl acetate is formed, it concentrates in the vapor phase and leaves the tower in the nearly pure distillate. The water product concentrates in the liquid phase and is removed from the tower in a nearly pure bottoms stream.

In summary, when the advantages of a combined operation (involving a reversible reaction and distillation, in this case) are clear to the design team, the operation can be inserted into the flowsheet *before* the chemicals are distributed in process synthesis. This is a heuristic design procedure that can simplify the synthesis steps and lead to a more profitable design.

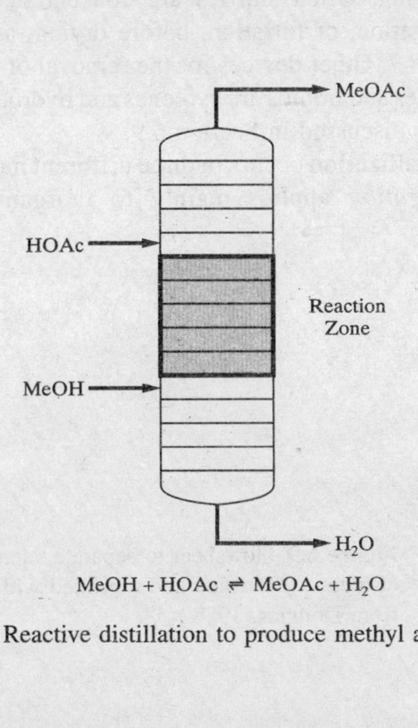

Figure 6.6 Reactive distillation to produce methyl acetate.

Optimal Conversion

Consider the case of a single reaction with a large chemical equilibrium constant such that it is possible to obtain a complete conversion. However, the optimal conversion may not be complete conversion. Instead, an economic balance between a high reactor section cost at high conversion and a high separation section cost at low conversion determines the optimum. Unfortunately, a heuristic for the optimal conversion is not available because it depends on many factors. This subject is considered in more detail in Section 7S.3 on reactor-separator-recycle networks.

6.4 SEPARATIONS

Separations Involving Liquid and Vapor Mixtures

Heuristic 9: ***Separate liquid mixtures using distillation, stripping, enhanced (extractive, azeotropic, reactive) distillation, liquid–liquid extraction, crystallization, and/or adsorption. The selection between these alternatives is considered in Chapter 8.***

Heuristic 10: ***Attempt to condense or partially condense vapor mixtures with cooling water or a refrigerant. Then, use Heuristic 9.***

Heuristic 11: ***Separate vapor mixtures using partial condensation, cryogenic distillation, absorption, adsorption, membrane separation, and/or desublimation. The selection among these alternatives is considered in Chapter 8.***

The selection of separation processes is dependent on the phase of the stream to be separated and the relative physical properties of its chemical species. Liquid and vapor streams are separated often using the strategy recommended by Douglas (1988) in *Conceptual Design of Chemical Processes*. This strategy is reproduced here using the original figures, slightly modified, with the publisher's permission. It is expanded upon in Chapter 8. Note that the choice of type of separator is often influenced by the scale of the process, with distillation often favored by economies-of-scale at large throughputs, and adsorption and membrane separation gaining favor as throughputs decrease.

When the reaction products are in the liquid phase, Douglas recommends that a liquid-separation system be inserted in the flowsheet, as shown in Figure 6.7. The liquid-separation system involves one or more of the following separators: distillation and enhanced distillation, stripping, liquid–liquid extraction, and so on, with the unreacted chemicals recovered in a liquid phase and recycled to the reaction operation.

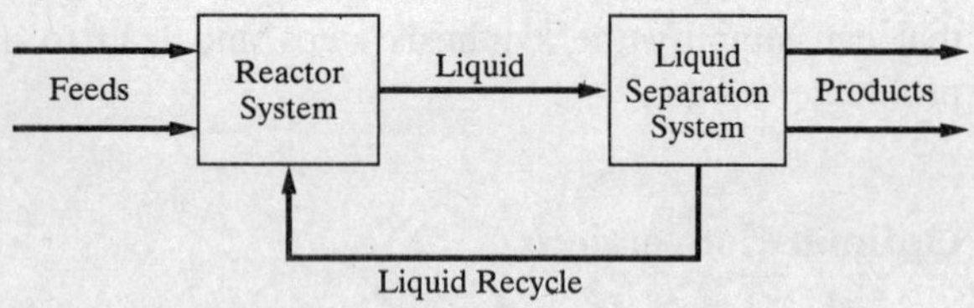

Figure 6.7 Flowsheet to separate liquid reactor effluents. (Reprinted with permission from Douglas, 1988.)

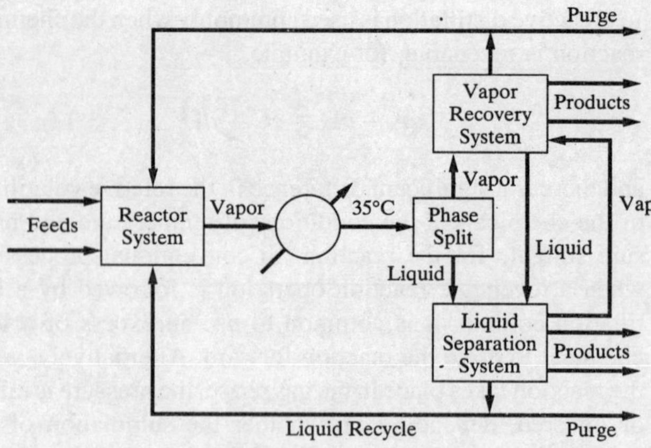

Figure 6.8 Flowsheet to separate vapor reactor effluents. (Modified and reprinted with permission from Douglas, 1988.)

For reaction products in the vapor phase, Douglas recommends that an attempt be made to partially condense them by cooling with cooling water or a refrigerant. Cooling water can cool the reaction products typically to 35°C, as shown in Figure 6.8. However, in warm climates, a higher temperature, e.g., 45°C, is required. The usual objective is to obtain a liquid phase, which is easier to separate, without using refrigeration, which involves an expensive compression step. When partial condensation occurs, a liquid-separation system is inserted, with a liquid purge added when necessary to remove trace inerts that concentrate in the liquid and are not readily separated. The vapor phase is sent to a vapor recovery system, which involves one or more of the following separations: partial condensation (at elevated pressures and cryogenic temperatures), cryogenic distillation, absorption, adsorption, membrane separation, and desublimation. Unreacted chemicals are recycled to the reactor section and vapor products are removed. A vapor purge is added when necessary to remove inerts that concentrate in the vapor and are not readily separated. Any liquid produced in the vapor recovery system is sent to the liquid recovery system for product recovery and the recycle of unreacted chemicals.

When the reactor effluent is already distributed between vapor and liquid phases, Douglas combines the two flowsheets, as shown in Figure 6.9. It should be recognized that the development of the separation systems for all three flowsheets involves several heuristics. First, certain separation devices, such as membrane separators, are not considered for the separation of liquids. Second, to achieve a partial condensation, cooling water is utilized initially, rather than compression and refrigeration. In this regard, it is presumed that liquid separations are preferred. An attempt is made to partially condense the vapor products, but no attempt is made to partially vaporize the liquid products. While these and other heuristics are based on considerable experience and usually lead to profitable designs, the designer needs to recognize their limitations and be watchful for situations in which they lead to suboptimal designs. Furthermore, for the separation of multicomponent streams, formal methods have been developed to synthesize separation trains involving vapors or liquids. These are covered in Chapter 8.

Separations Involving Solid Particles

For streams that involve solid phases or species that crystallize or precipitate, additional considerations are necessary when selecting a separation system because several steps may be necessary due to the impossibility of removing dry solids directly from a liquid. When separating inorganic chemicals, in an aqueous solution especially, the stream is often cooled or partially evaporated to recover solids by crystallization, followed by filtration or centrifugation, and then drying. Often slurries are concentrated by settling, centrifugation, or filtration, before drying, as discussed in Section 8.7. Other devices for the removal of solid particles from gases and liquids are cyclones and hydroclones, respectively, as discussed in Section 6.9.

Crystallization occurs in three different modes. *Solution crystallization* applies mainly to inorganic chemicals,

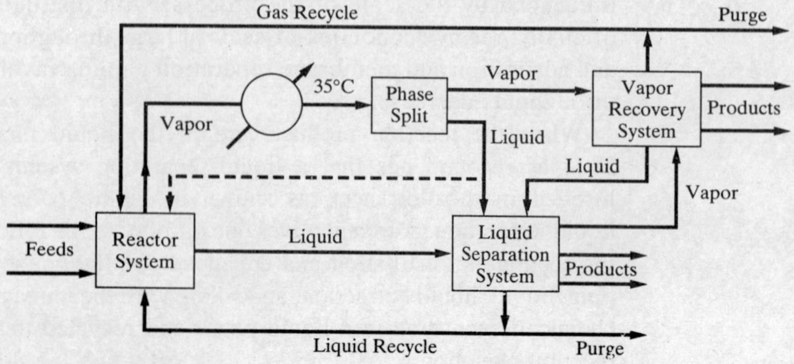

Figure 6.9 Flowsheet to separate vapor/liquid reactor effluents. (Modified and reprinted with permission from Douglas, 1988.)

which are crystallized from a solvent, often water, with an operating temperature far below the melting point of the crystals. *Precipitation* is fast solution crystallization that produces large numbers of very small crystals. It usually refers to the case where one product of two reacting solutions is a solid of low solubility, for example, the precipitation of insoluble silver chloride when aqueous solutions of silver nitrate and sodium chloride are mixed together. In *melt crystallization*, two or more chemicals of comparable melting points are separated at an operating temperature in the range of the melting points. Crystallization is capable of producing very pure chemicals when conducted according to the following heuristics, noting that recovery by any mode of crystallization may be limited by a eutectic composition.

Heuristic 12: ***Crystallize inorganic chemicals from a concentrated aqueous solution by chilling when solubility decreases significantly with decreasing temperature. Keep the solution at most 1 to 2°F below the saturation temperature at the prevailing concentration. Use crystallization by evaporation, rather than chilling, when solubility does not change significantly with temperature.***

Heuristic 13: ***Crystal growth rates are approximately the same in all directions, but crystals are never spheres. Crystal growth rates and sizes are controlled by limiting the extent of supersaturation, $S = C/C_{saturation}$, where C is concentration, usually in the range $1.02 < S < 1.05$. Growth rates are influenced greatly by the presence of impurities and of certain specific additives that vary from case to case.***

Heuristic 14: ***Separate organic chemicals by melt crystallization with cooling, using suspension crystallization, followed by removal of crystals by settling, filtration, or centrifugation. Alternatively, use layer crystallization on a cooled surface, with scraping or melting to remove the crystals. If the melt forms a solid solution, instead of a eutectic, use repeated melting and freezing steps, called fractional melt crystallization, or zone melting to obtain nearly pure crystalline products.***

Prior to crystallization, it is common to employ evaporation to concentrate a solution, particularly an aqueous solution of inorganic chemicals. Because of the relatively high cost of evaporating water, with its very large heat of vaporization, the following heuristics are useful for minimizing the cost.

Heuristic 15: ***Using multiple evaporators (called effects) in series, the latent heat of evaporation of water is recovered and reused. With a single evaporator, the ratio of the amount of water evaporated to the amount of external steam supplied to cause the evaporation is typically 0.8. For two effects, the ratio becomes 1.6; for three effects 2.4, and so forth. The magnitude of the boiling-point elevation caused by the dissolved inorganic compounds is a controlling factor in selecting the optimal number of effects. The elevation is often in the range of 3 to 10°F between solution and pure water boiling points. When the boiling-point rise is small, minimum evaporation cost is obtained with 8 to 10 effects. When the boiling-point rise is appreciable, the optimal number of effects is small, 6 or less. If necessary, boost interstage steam pressures with steam-jet or mechanical compressors.***

Heuristic 16: ***When employing multiple effects, the liquid and vapor flows may be in the same or different directions. Use forward feed, where both liquid and vapor flow in the same direction, for a small number of effects, particularly when the liquid feed is hot. Use backward feed, where liquid flows in a direction opposite to vapor flows, for cold feeds and/or a large number of effects. With forward feed, intermediate liquid pumps are not necessary, whereas they are for backward feed.***

Solution crystallization produces a slurry of crystals and mother liquor, which is partially separated by filtration or centrifugation into a wet cake and a mother liquor. Filtration through a filter medium of porous cloth or metal may be carried out under gravity, vacuum, or pressure. Centrifugation may utilize a solid bowl or a porous bowl with a filter medium. Important factors in the selection of equipment include: (1) moisture content of the cake, (2) solids content of the mother liquor, (3) fragility of the crystals, (4) crystal particle size, (5) need for washing the crystals to replace mother liquor with pure water, and (6) filtration rate. Filtration rate is best determined by measuring the rate of cake thickness buildup using a small-scale laboratory vacuum leaf filter test with the following criteria: Rapid, 0.1 to 10 cm/s; Medium, 0.1 to 10 cm/min; Slow, 0.1 to 10 cm/hr.

Heuristic 17: ***When crystals are fragile, effective washing is required, and clear mother liquor is desired, use: gravity, top-feed horizontal pan filtration for slurries that filter at a rapid rate; vacuum rotary-drum filtration for slurries that filter at a moderate rate; and pressure filtration for slurries that filter at a slow rate.***

Heuristic 18: ***When cakes of low moisture content are required, use: solid-bowl centrifugation if solids are permitted in the mother liquor; centrifugal filtration if effective washing is required.***

Wet cakes from filtration or centrifugation operations are sent to dryers for removal of remaining moisture. A large number of different types of commercial dryers have been developed to handle the many different types of feeds, which include not only wet cakes, but also pastes, slabs, films, slurries, and liquids. The heat for drying may be supplied from a hot gas in direct contact with the wet feed or it may be supplied indirectly through a wall. Depending on the thickness of the feed and the degree of agitation, drying times can range from seconds to hours. The following heuristics are useful in making a preliminary selection of drying equipment:

Heuristic 19: ***For granular material, free flowing or not, of particle sizes from 3 to 15 mm, use continuous tray and belt dryers with direct heat. For free-flowing granular solids that are not heat sensitive, use an inclined rotary cylindrical dryer, where the heat may be supplied directly from a hot gas or indirectly from tubes, carrying steam, that run the length of the dryer and are located in one or two rings concentric to and located just inside the dryer rotating shell. For small, free-flowing particles of 1 to 3 mm in diameter, when rapid drying is possible, use a pneumatic conveying dryer with direct heat. For very small free-flowing particles of less than 1 mm in diameter, use a fluidized-bed dryer with direct heat.***

Heuristic 20: ***For pastes and slurries of fine solids, use a drum dryer with indirect heat. For a liquid or pumpable slurry, use a spray dryer with direct heat.***

6.5 HEAT REMOVAL FROM AND ADDITION TO REACTORS

After positioning the separation operations, the next step in process synthesis, as recommended in Section 4.4, is to insert the operations for temperature, pressure, and phase changes. To accomplish this, many excellent algorithms for heat and power integration have been developed. These are presented in Chapter 9. The objective of this section, which considers several approaches to remove the heat generated in exothermic reaction operations and to add the heat required by endothermic reaction operations, is more limited. The subject is discussed at this point because several of the approaches for heat transfer affect the distribution of chemicals in the flowsheet and are best considered after the reaction operations are positioned. These approaches are discussed next, together with other common approaches to remove or add the heat of reaction. First, the methods for removing the heat generated by exothermic reaction operations are presented. Then, some distinctions are drawn for the addition of heat to endothermic reaction operations. For the details of heat removal from or addition to complex reactor configurations, the reader is referred to Section 7.2.

Heat Removal from Exothermic Reactors

Given an exothermic reaction operation, an important first step is to compute the *adiabatic reaction temperature*, that is, the maximum temperature attainable, in the absence of heat transfer. Note that this can be accomplished readily with any of the process simulators. Furthermore, algorithms have been presented for these iterative calculations by Henley and Rosen (1969) and Myers and Seider (1976), among many sources.

EXAMPLE 6.6 ***Adiabatic Reaction Temperature***

Consider the reaction of carbon monoxide and hydrogen to form methanol:

$$CO + 2H_2 \rightarrow CH_3OH$$

With the reactants fed in stoichiometric amounts at 25°C and 1 atm, calculate the standard heat of reaction and the adiabatic reaction temperature.

SOLUTION

In ASPEN PLUS, the RSTOIC subroutine is used with a feed stream containing 1 lbmol/hr CO and 2 lbmol/hr H_2 and the PSRK method (Soave–Redlich–Kwong equation of state with Holderbaum–Gmehling mixing rules). To obtain the heat of reaction, the fractional conversion of CO is set at unity, with the product stream temperature at 25°C and the vapor fraction at 1.0. The latter keeps the methanol product in the vapor phase at 2.44 psia, and hence both the reactants and product species are vapor. The heat duty computed by RSTOIC is −38, 881 Btu/hr, and hence the heat of reaction is $\Delta H_r = -38,881$ Btu/lbml CO.

To obtain the adiabatic reaction temperature for complete conversion, the heat duty is set at zero and the pressure of the methanol product stream is returned to 1 atm. This produces an effluent temperature of 1, 158°C (2, 116°F), which is far too high for the Cu-based catalyst and the materials of construction in most reactor vessels. Hence, a key question in the synthesis of the methanol process, and similar processes involving highly exothermic reactions, is how to lower the product temperature. In most cases, the designer is given or sets the maximum temperature in the reactor and evaluates one of the heat-removal strategies described in this section.

Note that these results can be reproduced using the EXAM6-6.bkp file in the Program and Simulation Files folder, which can be downloaded from the Wiley Web site associated with this book. Also, RSTOIC-type subroutines are described in Section 7.1.

www.wiley.com/go/global/seider

Heuristic 21: ***To remove a highly exothermic heat of reaction, consider the use of excess reactant, an inert diluent, or cold shots. These affect the distribution of chemicals and should be inserted early in process synthesis.***

Heuristic 22: ***For less exothermic heats of reaction, circulate reactor fluid to an external cooler, or use a jacketed vessel or cooling coils. Also, consider the use of intercoolers between adiabatic reaction stages.***

To achieve lower temperatures, several alternatives are possible, beginning with those that affect the distribution of chemicals.

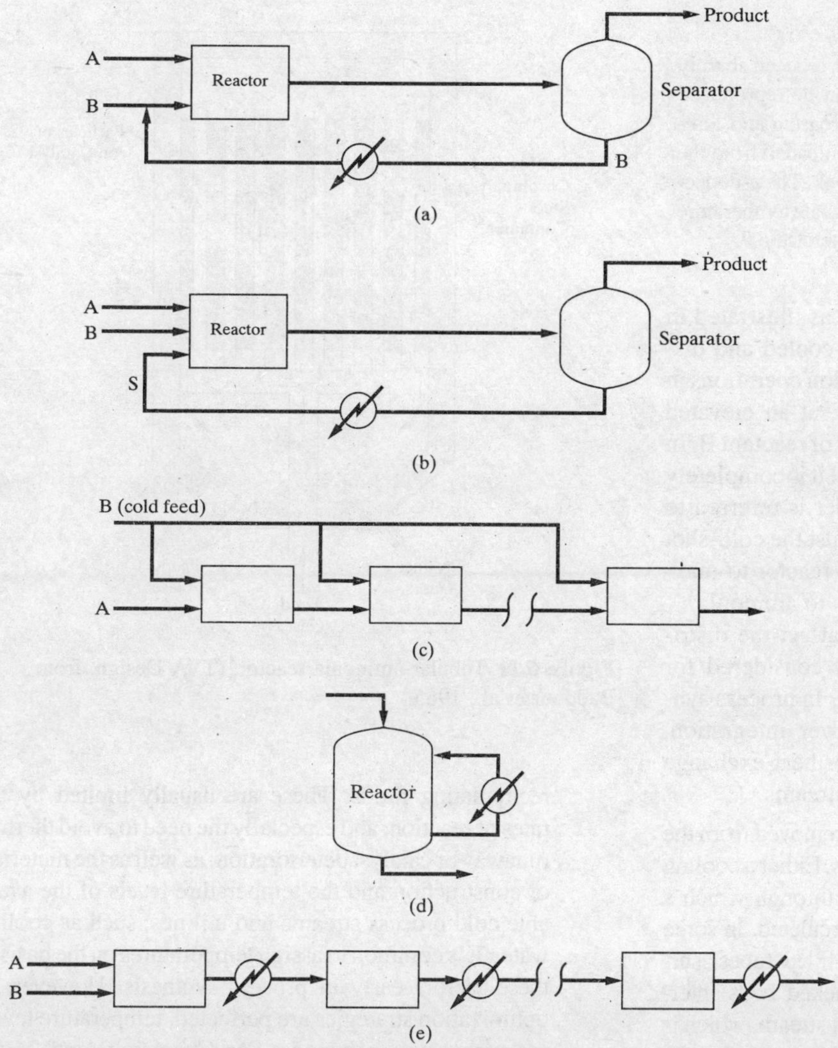

Figure 6.10 Flowsheets to remove the heat of reaction: (a) use of excess reactant; (b) use of inert diluent, S; (c) use of cold shots; (d) diabatic operation; (e) use of intercoolers.

1. Use an excess of one reactant to absorb the heat. This alternative was discussed earlier in Section 6.3 and is illustrated in Figure 6.10a, where excess B is recovered from the separator and recirculated to the reaction operation. Heat is removed in the separator or by exchange with a cold process stream or a cold utility (e.g., cooling water).

EXAMPLE 6.7 *Excess Reactant*

Returning to Example 6.6, an excess of H_2 is specified such that the mole ratio of H_2/CO is arbitrarily 10 and the temperature of the reactor effluent stream is computed.

SOLUTION

Again, using the RSTOIC subroutine in ASPEN PLUS with a complete conversion of CO, the effluent temperature is reduced to 337°C (639°F), a result that can be reproduced using the EXAM6-7.bkp file in the Program and Simulation Files folder, which can be downloaded from the Wiley Web site associated with this book. The Sensitivity command can be used to compute the effluent temperature as a function of the H_2/CO ratio, as in Exercise 6.8.

www.wiley.com/go/global/seider

2. Use of an inert diluent, S. Figure 6.10b illustrates this alternative. One example occurs in the manufacture of methanol, where carbon monoxide and hydrogen are reacted in a fluidized bed containing catalyst. In the Marathon Oil process, a large stream of inert oil is circulated through the reactor, cooled, and recirculated to the reactor. Note that diluents like oil, with large heat capacities, are favored to maintain lower temperatures with smaller recirculation rates. The disadvantage of this approach, of course, is that a new species, S, is introduced, which, when separated from the reaction mix, cannot be removed entirely from the desired product. As in Alternative 1, heat is removed in the separator or by exchange with a cold process stream or a cold utility.

EXAMPLE 6.8 *n-Dodecane Diluent*

Returning to Example 6.6, 5 lbmol/hr of *n*-dodecane is added to the reactor feed (1 lbmol/hr CO and 2 lbmol/hr H_2) and the temperature of the reactor effluent stream is computed.

SOLUTION

In this case, the effluent temperature is reduced sharply to 77.6°C (171.7°F), a result that can be reproduced using the EXAM6-8.bkp file in the Program and Simulation Files folder, which can be downloaded from the Wiley Web site associated with this book. The *n*-dodecane flow rate is adjusted to give an adequate temperature reduction. This is accomplished in Exercise 6.9.

3. Use of cold shots. In this alternative, as illustrated in Figure 6.10c, one of the reactants is cooled and distributed among several adiabatic reaction operations in series. The effluent from each stage, at an elevated temperature, is cooled with a cold shot of reactant B. In each stage, additional A is reacted until it is completely consumed in the last stage. The reader is referred to Example 7.3, which shows how to adjust the cold-shot distribution in an ammonia synthesis reactor to maximize the conversion of synthesis gas to ammonia.

The next two alternatives do not affect the distribution of chemicals and are usually considered for moderately exothermic reactions, later in process synthesis—that is, during heat and power integration, when opportunities are considered for heat exchange between high- and low-temperature streams.

4. Diabatic operation. In this case, heat is removed from the reaction operation in one of several ways. Either a cooling jacket is utilized or coils are installed through which a cold process stream or cold utility is circulated. In some cases, the reaction occurs in catalyst-filled tubes, surrounded by coolant, or in catalyst-packed beds interspersed with tubes that convey a coolant stream, which is often the reactor feed stream, as illustrated for the ammonia reactor (TVA design) in Figure 6.11. Here, heat transfer from the reacting species in the catalyst bed preheats the reactants, N_2 and H_2, flowing in the cooling tubes, with the tube bundle designed to give adequate heat transfer as well as reaction. The design procedure is similar to that for the design of heat exchangers in Chapter 18. It is noted that for large-capacity, ammonia-synthesis reactors, the optimal design usually calls for a series of adiabatic beds packed with catalyst, with cooling achieved using cold shots, as shown in Figure 6.10c. Most commercial ammonia reactors use a combination of these two cooling arrangements. Yet another alternative is to circulate a portion of the reacting mixture through an external heat exchanger in which the heat of reaction is removed, as shown in Figure 6.10d.

5. Use of intercoolers. As shown in Figure 6.10e, the reaction operation is divided into several adiabatic stages, with heat removed by heat exchangers placed between each stage. Here, also, heat is transferred either to cold process streams that require heating or to cold utilities.

In all of these alternatives, the design team selects acceptable temperature levels and flow rates of the recirculating fluids. These are usually limited by the rates of reaction, and especially the need to avoid thermal runaway or catalyst deterioration, as well as the materials of construction and the temperature levels of the available cold process streams and utilities, such as cooling water. It is common to assign temperatures on the basis of these factors early in process synthesis. However, as optimization strategies are perfected, temperature levels are varied within bounds. See Chapters 9 and 24 for discussions of the use of optimization in process synthesis and optimization of process flowsheets, as well as Example 7.3 to see how constrained optimization is applied to design an ammonia cold-shot converter.

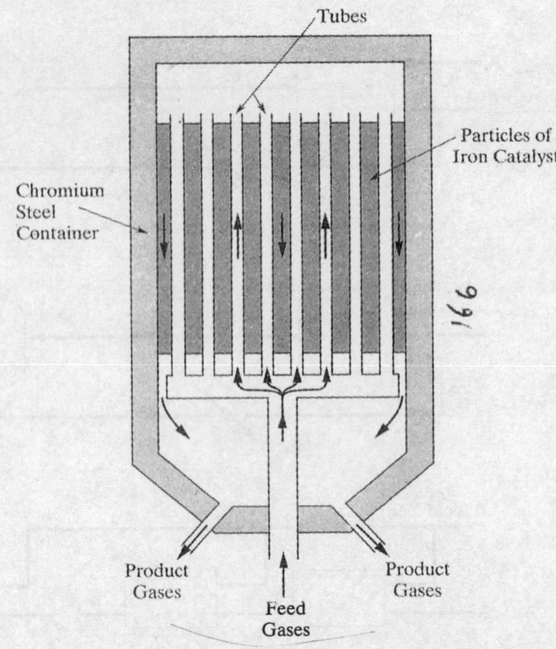

Figure 6.11 Tubular ammonia reactor. (TVA Design, from Baddour et al., 1965.)

Heat Addition to Endothermic Reactors

Heuristic 23: ***To control temperature for a highly endothermic heat of reaction, consider the use of excess reactant, an inert diluent, or hot shots. These affect the distribution of chemicals and should be inserted early in process synthesis.***

Heuristic 24: ***For less endothermic heats of reaction, circulate reactor fluid to an external heater, or use a jacketed vessel or heating coils. Also, consider the use of interheaters between adiabatic reaction stages.***

For endothermic reaction operations, sources of the heat of reaction are needed. As in the case of exothermic reaction operations, the heat of reaction and the adiabatic reaction temperature can be computed using a simulator. The latter provides a lower bound on the temperature of the reactor effluent.

When adding heat to endothermic reaction operations, three approaches for heat addition affect the distribution of chemicals in the flowsheet and are best considered immediately after the reaction operations are positioned. Given that the feed stream is preheated, these approaches closely parallel the first three approaches for heat removal from exothermic reaction operations in Figures 6.10a–6.10c. When an excess of a reactant is utilized, the decrease in effluent temperature varies inversely with the degree of excess. Similarly, when an inert diluent is added, the effluent temperature is decreased inversely with the amount of diluent. For example, when ethylbenzene is pyrolyzed to produce styrene,

$$\text{Ethylbenzene} \rightarrow \text{Styrene} + H_2$$

superheated steam is added to provide the heat of reaction and keep the reaction temperature elevated. Of course, the addition of steam significantly increases the reactor volume and both the operating and installation costs. Finally, it is possible to add hot shots of reactants to a series of reactors, in a way similar to the cold shots added in Figure 6.10c.

The same two alternatives that do not affect the distribution of chemicals apply for the addition of heat to endothermic reaction operations. In this case, the diabatic operation in Figure 6.10d involves the addition of heat in ways similar to the removal of heat. Jackets, coils, and heat exchanger designs are very common. Also, interheaters between stages, as in Figure 6.10e, are used in many situations in place of intercoolers.

6.6 HEAT EXCHANGERS AND FURNACES

As discussed in the previous two sections, heat exchange is commonly used in conjunction with separation and reaction steps to change the temperature and/or phase condition of a process stream. When using a process simulator to perform heat-exchange calculations, it is necessary to select a method of heat exchange from the following six possibilities, where all but the last two involve the separation, by a solid wall, of the two streams exchanging heat.

1. Heat exchange between two process fluids using a double-pipe, shell-and-tube, or compact heat exchanger.
2. Heat exchange between a process fluid and a utility, such as cooling water or steam, using a double-pipe, shell-and-tube, air-cooled, or compact heat exchanger.
3. High-temperature heating of a process fluid using heat from the products of combustion in a furnace (also called a fired heater).
4. Heat exchange within a reactor or separator, rather than in an external heat-exchange device such as a shell-and-tube heat exchanger or furnace, as described in Section 6.5.
5. Direct heat exchange by mixing the two streams that are exchanging heat.
6. Heat exchange involving solid particles.

The following heuristics are useful in selecting an initial basis for the heat-exchange method and the operating conditions. Details of heat exchanger selection and design are presented in Chapter 18.

Heuristic 25: ***Unless required as part of the design of the separator or reactor, provide necessary heat exchange for heating or cooling process fluid streams, with or without utilities, in an external shell-and-tube heat exchanger using countercurrent flow. However, if a process stream requires heating above 750°F, use a furnace unless the process fluid is subject to chemical decomposition.***

Preliminary estimates of exit temperatures of streams flowing through a heat exchanger can be made with the following heuristics.

Heuristic 26: ***Near-optimal minimum temperature approaches in heat exchangers depend on the temperature level as follows:***

10°F or less for temperatures below ambient.

20°F for temperatures at or above ambient up to 300°F.

50°F for high temperatures.

250 to 350°F in a furnace for flue gas temperature above inlet process fluid temperature.

As an example, suppose it is desired to heat 25,000 lb/hr of toluene at 100°F and 90 psia with 25,000 lb/hr of styrene at 300°F and 50 psia. Under these conditions, assume that both streams will be liquid, but this must be verified by flash calculations after the exit temperatures and pressures have been determined. From the previous two heuristics, use a shell-and-tube heat exchanger with countercurrent flow and a minimum approach temperature of 20°F. Let the average specific heats of the two streams be 0.43 Btu/lb-°F for toluene and 0.44 Btu/lb-°F for styrene. Initially it is not known to which end of the heat exchanger the 20°F minimum approach applies. Assume it applies at the toluene inlet end. If so, the styrene exit temperature is 100 + 20 = 120°F. This gives a heat-exchanger duty, based on styrene, of:

$$Q = 25,000(0.44)(300 - 120) = 1,980,000 \text{ Btu/hr}$$

Using this duty, the exit temperature of toluene, $T_{\text{toluene out}}$, can be computed from:

$$Q = 1,980,000 = 25,000(0.43)(T_{\text{toluene out}} - 100)$$

Solving, $T_{\text{toluene out}} = 284.2$°F. But this gives a temperature approach of 300 − 284.2 = 15.8°F at the styrene inlet end, which is less than the minimum approach of 20°F. Therefore, the minimum approach must be applied to the styrene inlet end. Similar calculations give $T_{\text{toluene out}} = 280$°F and $T_{\text{styrene out}} = 124.1$°F. This corresponds to an approach temperature at the toluene inlet end of 24.1°F, which is greater than the minimum approach temperature and, therefore, is acceptable.

Heuristic 27: ***When using cooling water to cool or condense a process stream, assume a water inlet temperature of 90°F (from a cooling tower) and a maximum water outlet temperature of 120°F.***

When cooling and condensing a gas, both sensible and latent heat can be removed in a single heat exchanger. However, because of the many two-phase flow regimes that can occur when boiling a fluid, it is best to provide three separate heat exchangers when changing a subcooled liquid to a superheated gas, especially when the difference between the bubble point and dew point is small. The first exchanger preheats the liquid to the bubble point; the second boils the liquid; the third superheats the vapor.

Heuristic 28: ***Boil a pure liquid or close-boiling liquid mixture in a separate heat exchanger, using a maximum overall temperature driving force of 45°F to ensure nucleate boiling and avoid undesirable film boiling as discussed in Section 18.1.***

As discussed in detail in Section 18.1, the minimum approach temperature in a countercurrent-flow heat exchanger may occur at an intermediate location rather than at one of the two ends when one of the two streams is both cooled and condensed. If the minimum temperature approach is assumed to occur at one of the two ends of the heat exchanger, a smaller approach or a temperature crossover that violates the second law of thermodynamics may occur at an intermediate location. To avoid this situation, the following heuristic should be applied:

Heuristic 29: ***When cooling and condensing a stream in a heat exchanger, a zone analysis, described in Section 18.1, should be made to make sure that the temperature difference between the hot stream and the cold stream is equal to or greater than the minimum approach temperature at all locations in the heat exchanger. The zone analysis is performed by dividing the heat exchanger into a number of segments and applying an energy balance to each segment to determine corresponding stream inlet and outlet temperatures for the segment, taking into account any phase change. A process simulation program conveniently accomplishes the zone analysis.***

When using a furnace to heat and/or vaporize a process fluid, the following heuristic is useful for establishing inlet and outlet heating medium temperature conditions so that fuel and air requirements can be estimated.

Heuristic 30: ***Typically, a hydrocarbon gives an adiabatic flame temperature of approximately 3,500°F when using the stoichiometric amount of air. However, use excess air to achieve complete combustion and give a maximum flue gas temperature of 2,000°F. Set the stack gas temperature at 650 to 950°F to prevent condensation of corrosive components of the flue gas.***

Pressure drops of fluids flowing through heat exchangers and furnaces may be estimated with the following heuristics.

Heuristic 31: ***Estimate heat-exchanger pressure drops as follows:***

1.5 psi for boiling and condensing.

3 psi for a gas.

5 psi for a low-viscosity liquid.

7–9 psi for a high-viscosity liquid.

20 psi for a process fluid passing through a furnace.

Unless exotic materials are used, heat exchangers should not be used for cooling and/or condensing process streams with temperatures above 1,150°F. Instead, use the following heuristic for direct heat exchange.

Heuristic 32: ***Quench a very hot process stream to at least 1,150°F before sending it to a heat exchanger for additional cooling and/or condensation. The quench fluid is best obtained from a downstream separator as in Figure 5.21 for the toluene hydrodealkylation process. Alternatively, if the process stream contains water vapor, liquid water may be an effective quench fluid.***

Streams of solid particles are commonly heated or cooled directly or indirectly. Heat transfer is much more rapid and controllable when using direct heat exchange.

Heuristic 33: ***If possible, heat or cool a stream of solid particles by direct contact with a hot gas or cold gas, respectively, using a rotary kiln, a fluidized bed, a multiple hearth, or a flash/pneumatic conveyor. Otherwise, use a jacketed spiral conveyor.***

6.7 PUMPING, COMPRESSION, PRESSURE REDUCTION, VACUUM, AND CONVEYING OF SOLIDS

As mentioned in the previous section, it is common to consider the integration of all temperature- and pressure-change operations. This is referred to as *heat and power integration* and is covered in Chapter 9 after important thermodynamic considerations are presented first in Sections 9S.0–9S.10. At this point, however, there are several important heuristics that are useful in determining what type of operations to insert into the flowsheet to increase or decrease pressure. Details of the equipment used to perform pressure-change operations are presented in Chapter 20.

Increasing the Pressure

Gases: To increase the pressure, the most important consideration is the phase state (vapor, liquid, or solid) of the stream. If the stream is a gas, the following heuristic applies for determining whether a *fan, blower,* or *compressor* should be used.

Heuristic 34: ***Use a fan to raise the gas pressure from atmospheric pressure to as high as 40 inches water gauge (10.1 kPa gauge or 1.47 psig). Use a blower or compressor to raise the gas pressure to as high as 206 kPa gauge or 30 psig. Use a compressor or a staged compressor system to attain pressures greater than 206 kPa gauge or 30 psig.***

In Figure 5.20 for the toluene hydrodealkylation process, the pressure of the recycle gas leaving the flash drum at 100°F and 484 psia is increased with a compressor to 569 psia, so that, after pressure drops of 5 psia through the heat exchanger and 70 psia through the furnace, it enters the reactor at a required pressure of 494 psia.

The following heuristic is useful for estimating the exit temperature, which can be significantly higher than the entering temperature, and the power requirement when increasing the gas pressure by a single stage of reversible, adiabatic compression.

Heuristic 35: ***Estimate the theoretical adiabatic horsepower (THp) for compressing a gas from:***

$$\text{THp} = \text{SCFM}\left(\frac{T_1}{8{,}130a}\right)\left[\left(\frac{P_2}{P_1}\right)^a - 1\right] \quad (6.1)$$

where SCFM = standard cubic feet of gas per minute at 60°F and 1 atm (379 SCF/lbmol); T_1 = gas inlet temperature in °R; inlet and outlet pressures, P_1 and P_2, are absolute pressures; and $a = (k - 1)/k$, with k = the gas specific heat ratio, C_p/C_v.

Estimate the theoretical exit temperature, T_2, for a gas compressor from:

$$T_2 = T_1(P_2/P_1)^a \quad (6.2)$$

For example, if air at 100°F is compressed from 1 to 3 atm (compression ratio = 3) using $k = 1.4$, the THp is computed to be 128 Hp/standard million ft^3/day, with an outlet temperature = 306°F.

When using a compressor, the gas theoretical exit temperature should not exceed approximately 375°F, the limit imposed by most compressor manufacturers. This corresponds to a compression ratio of 4 for $k = 1.4$ and $T_2 = 375$°F. When the exit gas temperature exceeds the limit, a single gas compression step cannot be used. Instead, a multistage compression system, with intercoolers between each stage, must be employed. Each intercooler cools the gas back down to approximately 100°F. The following heuristic is useful for estimating the number of stages, *N*, required and the interstage pressures.

Heuristic 36: ***Estimate the number of gas compression stages, N, from the following table, which assumes a specific heat ratio of 1.4 and a maximum compression ratio of 4 for each stage.***

Final Pressure/Inlet Pressure	Number of Stages
<4	1
4 to 16	2
16 to 64	3
64 to 256	4

Optimal interstage pressures correspond to equal Hp for each compressor. Therefore, based on the above equation for theoretical compressor Hp, estimate interstage pressures by using approximately the same compression ratio for each stage with an intercooler pressure drop of 2 psi or 15 kPa.

For example, in Exercise 5.8, a feed gas at 100° F and 30 psia is to be compressed to 569 psia. From the above table, with an overall compression ratio of 569/30 = 19, a 3-stage system is indicated. For equal compression ratios, the compression ratio for each stage of a 3-stage system = $19^{1/3} = 2.7$. The estimated stage pressures are as follows, taking into account a 2 psi drop for each intercooler and its associated piping:

Stage	Compressor Inlet Pressure, psia	Compressor Outlet Pressure, psia
1	30	81
2	79	213
3	211	569

When compressing a gas, the entering stream must not contain liquid, and the exiting stream must be above its dew point so as not to damage the compressor. To remove any entrained liquid droplets from the entering gas, a vertical knock-out drum equipped with a demister pad is placed just upstream of the compressor. To prevent condensation in the compressor, especially when the entering gas is near its dew point, a heat exchanger should also be added at the compressor inlet to provide sufficient preheat to ensure that the exiting gas is well above its dew point.

Liquids: If the pressure of a liquid is to be increased, a pump is used. The following heuristic is useful for determining the types of pumps best suited for a given task, where the head in feet is the pressure increase across the pump in psf (pounds force/ft^2) divided by the liquid density in lb/ft^3.

Heuristic 37: ***For heads up to 3,200 ft and flow rates in the range of 10 to 5,000 gpm, use a centrifugal pump. For high heads up to 20,000 ft and flow rates up to 500***

gpm, use a reciprocating pump. Less common are axial pumps for heads up to 40 ft for flow rates in the range of 20 to 100,000 gpm and rotary pumps for heads up to 3,000 ft for flow rates in the range of 1 to 1,500 gpm.

For liquid water, with a density of 62.4 lb/ft^3, heads of 3,000 and 20,000 ft correspond to pressure increases across the pump of 1,300 and 8,680 psi, respectively.

When pumping a liquid from an operation at one pressure, P_1, to a subsequent operation at a higher pressure, P_2, the pressure increase across the pump must be higher than $P_2 - P_1$ in order to overcome pipeline pressure drop, control valve pressure drop, and possible increases in elevation (potential energy). This additional pressure increase may be estimated by the following heuristic.

Heuristic 38: ***For liquid flow, assume a pipeline pressure drop of 2 psi/100 ft of pipe and a control valve pressure drop of at least 10 psi. For each 10-ft rise in elevation, assume a pressure drop of 4 psi.***

For example, in Figure 4.8 the combined chlorination reactor effluent and dichloroethane recycle at 1.5 atm is sent to a pyrolysis reactor operating at 26 atm. Although no pressure drops are shown for the two temperature-change and one phase-change operations, they may be estimated at 10 psi total. The line pressure drop and control valve pressure drop may be estimated at 15 psi. Take the elevation change as 20 ft, giving 8 psi. Therefore, the total additional pressure increase is $10 + 15 + 8 = 33$ psi or 2.3 atm. The required corresponding pressure increase across the pump (pressure-change operation) is, therefore, $(26 - 1.5) + 2.3 = 26.8$ atm. For a liquid density of 78 lb/ft^3 or 10.4 lb/gal, the required pumping head is $26.8(14.7)(144)/78 = 730$ ft. The flow rate through the pump is $(158,300+105,500)/10.4/60 = 422$ gpm. Using Heuristic 37, select a centrifugal pump.

The following heuristic provides an estimate of the theoretical pump Hp. Unlike the case of gas compression, the temperature change across the pump is small and can be neglected.

Heuristic 39: ***Estimate the theoretical horsepower (THp) for pumping a liquid from:***

$$\text{THp} = (\text{gpm})(\text{Pressure increase, psi})/1{,}714 \tag{6.3}$$

For example, the theoretical Hp for pumping the liquid in Figure 4.8, using the above data, is (422)(26.8)(14.7)/1,714 = 97 Hp.

Decreasing the Pressure

The pressure of a gas or liquid stream can be reduced to ambient pressure or higher with a single throttle valve or two or more such valves in series. The adiabatic expansion of a gas across a valve is accompanied by a decrease in the temperature of the gas. The exiting temperature is estimated from Eq. (6.2) above for gas compression. For a liquid, the exit temperature is almost the same as the temperature entering the valve. In neither case is shaft work recovered from the fluid. Alternatively, it is possible to recover energy in the form of shaft work that can be used elsewhere by employing a turbine-like device. For a gas, the device is referred to as an expander, expansion turbine, or turboexpander. For a liquid, the corresponding device is a power-recovery turbine. The following heuristics are useful in determining whether a turbine should be used in place of a valve.

Heuristic 40: ***Consider the use of an expander for reducing the pressure of a gas or a pressure-recovery turbine for reducing the pressure of a liquid when more than 20 Hp and 150 Hp, respectively, can be recovered.***

Heuristic 41: ***Estimate the theoretical adiabatic horsepower (THp) for expanding a gas from:***

$$\text{THp} = \text{SCFM}\left(\frac{T_1}{8,130a}\right)\left[1 - \left(\frac{P_2}{P_1}\right)^a\right] \tag{6.4}$$

Heuristic 42: ***Estimate the theoretical horsepower (THp) for reducing the pressure of a liquid from:***

$$\text{THp} = (\text{gpm})(\text{Pressure decrease, psi})/1,714 \tag{6.5}$$

In Figure 4.7, the pyrolysis effluent gas, following cooling to 170°C and condensation to 6°C at 26 atm, is reduced in pressure to 12 atm before entering the first distillation column. The flowsheet in Figure 4.8 shows the use of a valve, following the condenser, to accomplish the pressure reduction. Should a pressure-recovery turbine be used? Assume a flow rate of 422 gpm. The pressure decrease is $(26 - 12)\ (14.7) = 206$ psi. Using Eq. (6.5), $\text{THp} = (422)\ (206)/1,714 = 51$, which is much less than 150 Hp. Therefore, according to the above heuristic, a valve is preferred. Alternatively, the pressure reduction step could be inserted just prior to the condenser, using an expander on the gas at its dew point of 170°C. The total flow rate is $(58,300 + 100,000 + 105,500)/60 = 4,397$ lb/min. The average molecular weight is computed to be 61.9, giving a molar flow rate of 71 lbmol/min. The corresponding SCFM (standard cubic feet per minute at standard conditions of 60°F and 1 arm) is $(71)(379) = 26,900$. Assume $k = 1.2$, giving $a = (1.2 - 1)/1.2 = 0.167$. With a decompression ratio of $12/26 = 0.462$ and $T_1 = 797$°R, Eq. (6.4) above gives 1,910 THp, which is much more than 20 Hp. Therefore, according to the above heuristic, an expander should be used. The theoretical temperature of the gas exiting the expander, using the above Eq. (6.2) is $797(0.462)^{0.167} = 701$°R = 241°F = 116°C.

Pumping a Liquid or Compressing a Gas

When it is necessary to increase the pressure between process operations, it is almost always far less expensive to pump a

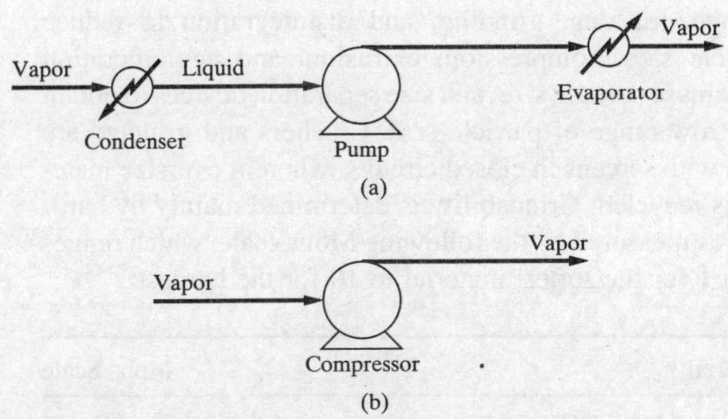

Figure 6.12 Alternatives for raising the pressure of a vapor stream: (a) pump a liquid; (b) compress the vapor.

liquid rather than compress a vapor. This is because the power required to increase the pressure of a flowing stream is

$$W = \int_{P_1}^{P_2} V\,dP \tag{6.6}$$

where V is the volumetric flow rate, which is normally far less for liquid streams—typically two orders of magnitude less. Hence, it is common to install pumps having approximately 10 Hp, whereas comparable compressors require approximately 1,000 Hp and are far more expensive to purchase and install. For these reasons, if the low-pressure stream is a vapor and the phase state is also vapor at the higher pressure, it is almost always preferable to condense the vapor, pump it, and revaporize it, rather than compress it, as illustrated in Figure 6.12. The exception to this heuristic is when refrigeration is required for condensation, which, as discussed in Section 9.2, involves extensive compression of the working fluid. If the low-pressure stream is a liquid and the high-pressure stream is a vapor, it is preferable to increase the pressure first with a pump and then vaporize the liquid, rather than vaporize the liquid and then compress it. This is the subject of Exercise 6.12 and the following example.

Heuristic 43: ***To increase the pressure of a stream, pump a liquid rather than compress a gas, unless refrigeration is needed.***

EXAMPLE 6.9 ***Feed Preparation of Ethylbenzene***

Ethylbenzene is to be taken from storage as a liquid at 25°C and 1 atm and fed to a styrene reactor as a vapor at 400°C and 5 atm at 100,000 lb/hr. In this example, two alternatives are considered for positioning the temperature- and pressure-increase operations.

SOLUTION

1. Pump first. Using the PUMP and HEATER subroutines in ASPEN PLUS discussed on the multimedia modules, which can be downloaded from the Wiley Web site associated with this book, 12.5 brake Hp are required to pump the liquid followed by 4.67×10^7 Btu/hr to vaporize the liquid and heat the vapor to 400°C.
2. Vaporize the liquid first. Using the HEATER and COMPR subroutines, discussed on the multimedia CD-ROM, 4.21×10^7 Btu/hr are required to vaporize the liquid and heat it to 349.6°C, followed by 1,897 brake Hp to compress the vapor to 5 atm at 400°C.

Clearly, the power requirement is substantially less when pumping a liquid. Note that these results can be reproduced using the EXAM6-9.bkp file in the Program and Simulation Files folder, which can be downloaded from the Wiley Web site associated with this book.

Vacuum

When process pressures less than the ambient pressure are required, special devices and considerations are necessary. Vacuum operation is most common in crystallization, distillation, drying, evaporation, and pervaporation operations. A vacuum inside the equipment causes inleakage of ambient-pressure air. A vacuum system is used to remove this air together with any associated vapor in the process stream that passes through the equipment. For continuous processes, vacuums are predominantly in the range of 1 to 760 torr (0.13 to 101.3 kPa). In this range, it is common to use a vacuum pump, which compresses the gas from vacuum (suction pressure) to ambient pressure, or a jet-ejector system, which uses a flow of pressurized water or steam to mix with and remove the gas to create the vacuum. To design the vacuum system, it is necessary to estimate the inleakage of air, determine the total amount of gas (inleakage plus associated vapor) to be removed, and select an appropriate system for the vacuum level required. The following heuristics are useful. Details of vacuum equipment are presented in Section 22.6.

Heuristic 44: ***Estimate inleakage of air by:***

$$w = kV^{0.667} \tag{6.7}$$

where w = lb/hr of air inleakage, $V = ft^3$ of volume of the equipment under vacuum, and k = 0.2 for pressures greater than 90 torr, 0.15 for pressures between 21 and 89 torr, 0.10 for pressures between 3.1 and 20 torr, and 0.051 for pressures between 1 and 3 torr.

Heuristic 45: ***To reduce the amount of gas sent to the vacuum system if its temperature is greater than 100°F, add a condenser using cooling water before the vacuum system. The gas leaving the condenser will be at a dew-point temperature of 100°F at the vacuum pressure.***

Heuristic 46: ***For pressures down to 10 torr and gas flow rates up to 10,000 ft^3/min at the inlet to the vacuum system, use a liquid-ring vacuum pump. For pressures down to 2 torr and gas flow rates up to 1,000,000 ft^3/min at the inlet to the vacuum system, use a steam-jet ejector system (one-stage for 100 to 760 torr, two-stage for 15 to 100 torr, and three-stage***

for 2 to 15 torr). Include a direct-contact condenser between stages.

Heuristic 47: ***For a three-stage steam-jet ejector system used to achieve a vacuum of 2 torr, 100 pounds of 100 psig steam per pound of gas are required.***

Conveying Granular Solids

The movement of streams of granular solids horizontally and/or vertically is achieved with conveyors and elevators. When selecting the type of equipment, important considerations are stickiness, abrasiveness, particle size, and density of the solid particles.

Heuristic 48: ***If the solid particles are small in size, low in particle density, and not sticky or abrasive, use pneumatic conveying with air at 1 to 7 ft^3/ft^3 of solids and 35 to 120 ft/s air velocity for distances up to 400 ft. Otherwise, for sticky and/or abrasive solids of any size and density, use a screw conveyer and/or bucket elevator for distances up to 150 ft. For solid particles of any size and shape, and not sticky, use a belt conveyor, with inclination up to 30° if necessary, for long distances up to a mile or more.***

Changing the Pressure of Granular Solids

Continuous processes frequently involve granular solids, either alone, slurried in a liquid, or fluidized in a gas. In many cases, the streams containing the solids are at ambient pressure and are moved with ease by conveyors and elevators. However, in some cases, elevated pressure or vacuum may be required. If solid particles alone are being processed, they are placed in a closed hopper where the pressure is then adjusted to the required pressure. The solids are then conveyed under pressure or vacuum. For a slurry, the solids are placed in a hopper, from which they are discharged into a liquid at ambient pressure. The resulting slurry is then pumped to the desired pressure. A gas–solid particles mixture is formed by discharging solid particles from a hopper, through a rotary valve (also referred to as a rotary air-lock valve), into a gas stream at elevated pressure. For discharge pressures up to 15 psig, a standard rotary valve is used. For discharge pressures in the range of 15 psig to almost 100 psig, high-performance rotary valves are available. Rotary valves must be carefully designed to minimize or avoid gas leakage, prevent bridging of solids in the valve, and avoid wear of the vanes in the valve.

6.8 CHANGING THE PARTICLE SIZE OF SOLIDS AND SIZE SEPARATION OF PARTICLES

It is frequently necessary to change the particle size of solids to meet product specifications or change reaction or drying rate. Methods to accomplish changes in particle size, discussed in detail in *Chemical Process Equipment—Selection and Design* by S.M. Walas (1988) and in *Perry's Chemical Engineers' Handbook*, 8th Edition (Green and Perry, 2008), include crushing, grinding, and disintegration to reduce particle size; compression, extrusion, and agglomeration to increase particle size; and size separation devices to obtain a narrow range of particle size. Crushers and grinders are used with screens in closed circuits, wherein oversize material is recycled. Grindability is determined mainly by hardness as measured by the following Mohs scale, which ranges from 1 for the softest material to 10 for the hardest:

Material	Mohs Scale
Talc, $Mg_3SiO_{10}(OH)_2$	1
Gypsum, $CaSO_4 \cdot 2H_2O$	2
Calcite, $CaCO_3$	3
Fluorite, CaF_2	4
Apatite, $Ca_5(PO_4)_3(OH, F, Cl)$	5
Feldspar, Na, Ca, K, Al silicates	6
Quartz, SiO_2	7
Topaz, $Al_2SiO_4(F, OH)_3$	8
Corundum, Al_2O_3	9
Diamond, C	10

Materials with a Mohs scale from 1 to 3 are considered soft and include graphite, many plastics, asphalt, sulfur, many inorganic salts, marble, and anthracite coal. Intermediate hardness extends from a Mohs scale of 4 to 6 and includes limestone, asbestos, and glass. Hard materials are characterized by a Mohs scale of 7 to 10 and include sand, granite, and emery. The following heuristics apply to particle-size reduction. Size of small particles is commonly stated in terms of screen mesh size according to the following U.S. standard (ASTM EII), where not all mesh sizes are given:

Mesh (Openings/inch)	Sieve Opening, mm
4	4.75
6	3.35
8	2.36
12	1.70
16	1.18
20	0.841
30	0.600
40	0.420
50	0.300
70	0.212
100	0.149
140	0.106
200	0.074
270	0.053
400	0.037

Heuristic 49: ***Crushing of coarse solids. Use a jaw crusher to reduce lumps of hard, abrasive, and/or sticky materials of 4 inches to 3 feet in diameter to slabby particles of 1 to 4 inches in size. Use a gyratory***

crusher to reduce slabby materials of 8 inches to 6 feet in size to rounded particles of 1 to 10 inches in diameter. Use a cone crusher to reduce less hard and less sticky materials of 2 inches to 1 foot in diameter to particles of 0.2 inch (4 mesh) to 2 inches in diameter.

Heuristic 50: ***Grinding to fine solids. Use a rod mill to take particles of intermediate hardness as large as 20 mm and reduce them to particles in the range of 10 to 35 mesh. Use a ball mill to reduce particles of low to intermediate hardness of 1 to 10 mm in size to very small particles of less than 140 mesh.***

Heuristic 51: ***Particle-size enlargement. Use compression with rotary compression machines to convert powders and granules into tablets of up to 1.5 inches in diameter. Use extruders with cutters to make pellets and wafers from pastes and melts. Use roll compactors to produce sheets from finely divided materials; the sheets are then cut into any desired shape. Use rotating drum granulators and rotary disk granulators with binders to produce particles in the size range of 2 to 25 mm.***

Heuristic 52: ***Size separation of particles. Use a grizzly of spaced, inclined, vibrated parallel bars to remove large particles greater than 2 inches in diameter. Use a revolving cylindrical perforated screen to remove intermediate-size particles in the size range of 0.25 inch to 1.5 inches in diameter. Use flat, inclined woven screens (U.S. standard) that are vibrated, shaken, or impacted with bouncing balls to separate small particles in the size range of 3 to 80 mesh. Use an air classifier to separate fine particles smaller than 80 mesh.***

6.9 REMOVAL OF PARTICLES FROM GASES AND LIQUIDS

Fine particles are most efficiently removed from dilute suspensions in gases and liquids by using centrifugal force in cyclones and hydroclones, respectively.

Heuristic 53: ***Use a cyclone separator to remove, from a gas, droplets or solid particles of diameter down to 10 microns (0.01 mm). Use a hydroclone separator to remove, from a liquid, insoluble liquid droplets or solid particles of diameter down to 5 microns (0.005 mm). However, small amounts of entrained liquid droplets are commonly removed from gases by vertical knock-out drums equipped with mesh pads to help coalesce the smallest droplets.***

6.10 CONSIDERATIONS THAT APPLY TO THE ENTIRE FLOWSHEET

The preceding discussion was directed primarily to the initial development of the process flowsheet, by considering particular sections of it and specific types of equipment within it. However, the flowsheet(s) resulting from the application of the 53 heuristics presented above need(s) additional development by applying some general considerations that may be able to improve the process, particularly with respect to efficiency, simplicity, and economics. The following general considerations, suggested by A. Brostow of Air Products and Chemicals, Inc., are typical of those used in process design by industrial chemical engineers:

a. To increase second-law efficiency and reduce energy consumption, avoid, if possible, the mixing of streams of different temperature, pressure, or composition. This is considered in detail in Sections 9S.0–9S.10 on second-law analysis.

b. For a new process, determine how it differs from a similar conventional process and pinpoint the advantages and disadvantages of the new process, making changes where disadvantages are uncovered.

c. For a new process, determine the maximum production rate and yield, and look for opportunities to increase the production rate and yield. Then, calculate theoretical efficiencies by applying lost-work analysis as presented in Sections 9S.0–9S.10. Look for ways to increase the efficiency.

d. Carefully examine the process flowsheet, looking for ways to eliminate equipment by combining, rearranging, or replacing process steps.

e. Perform preliminary economic evaluations at different production rates and corresponding plant sizes using simple scaling methods, noting that what is not economical at a small size may be economical at a large size and vice versa.

6.11 SUMMARY

Having studied this chapter, the reader should

a. Be able to implement the steps in Section 4.4 for process synthesis more effectively, using the many heuristics presented herein and summarized in Table 6.2. The examples and exercises should enable him or her to gain experience in their application.

b. Recognize the limitations of the heuristics in Table 6.2 and the role of the process simulator in permitting the systematic variation of parameters and the examination of alternative designs. The reader should also recognize that the heuristics listed are a subset of the many rules of thumb that have been applied by design teams in carrying out process synthesis.

Table 6.2 Heuristics in Chapter 6

	Heuristic
Reaction operations	
1	Select raw materials and chemical reactions to avoid, or reduce, the handling and storage of hazardous and toxic chemicals.
Distribution of chemicals	
2	Use an excess of one chemical reactant in a reaction operation to consume completely a valuable, toxic, or hazardous chemical reactant. The MSDSs will indicate which chemicals are toxic and hazardous.
3	When nearly pure products are required, eliminate inert species before the reaction operations when the separations are easily accomplished and when the catalyst is adversely affected by the inert, but not when a large exothermic heat of reaction must be removed.
4	Introduce purge streams to provide exits for species that enter the process as impurities in the feed or are formed in irreversible side reactions, when these species are in trace quantities and/or are difficult to separate from the other chemicals. Lighter species leave in vapor purge streams, and heavier species exit in liquid purge streams.
5	Do not purge valuable species or species that are toxic and hazardous, even in small concentrations (see the MSDSs). Add separators to recover valuable species. Add reactors to eliminate, if possible, toxic and hazardous species.
6	Byproducts that are produced in ***reversible*** reactions, in small quantities, are usually not recovered in separators or purged. Instead, they are usually **recycled to extinction**.
7	For competing reactions, both in series and parallel, adjust the temperature, pressure, and catalyst to obtain high yields of the desired products. In the initial distribution of chemicals, assume that these conditions can be satisfied. Before developing a base-case design, obtain kinetics data and check this assumption.
8	For reversible reactions especially, consider conducting them in a separation device capable of removing the products, and hence driving the reactions to the right. Such reaction-separation operations lead to very different distributions of chemicals.
Separation operations—liquid and vapor mixtures	
9	Separate liquid mixtures using distillation, stripping, enhanced (extractive, azeotropic, reactive) distillation, liquid–liquid extraction, crystallization, and/or adsorption. The selection between these alternatives is considered in Chapter 8.
10	Attempt to condense or partially condense vapor mixtures with cooling water or a refrigerant. Then, use Heuristic 9.
11	Separate vapor mixtures using partial condensation, cryogenic distillation, absorption, adsorption, membrane separation, and/or desublimation. The selection among these alternatives is considered in Chapter 8.
Separation operations—involving solid particles	
12	Crystallize inorganic chemicals from a concentrated aqueous solution by chilling when solubility decreases significantly with decreasing temperature. Keep the solution at most 1 to 2°F below the saturation temperature at the prevailing concentration. Use crystallization by evaporation, rather than chilling, when solubility does not change significantly with temperature.
13	Crystal growth rates are approximately the same in all directions, but crystals are never spheres. Crystal growth rates and sizes are controlled by limiting the extent of supersaturation, $S = C/C_{saturation}$, where C is concentration, usually in the range $1.02 < S < 1.05$. Growth rates are influenced greatly by the presence of impurities and of certain specific additives that vary from case to case.
14	Separate organic chemicals by melt crystallization with cooling, using suspension crystallization, followed by removal of crystals by settling, filtration, or centrifugation. Alternatively, use layer crystallization on a cooled surface, with scraping or melting to remove the crystals. If the melt forms a solid solution instead of a eutectic, use repeated melting and freezing steps, called fractional melt crystallization, or zone melting to obtain nearly pure crystalline products.
15	Using multiple evaporators (called effects) in series, the latent heat of evaporation of water is recovered and reused. With a single evaporator, the ratio of the amount of water evaporated to the amount of external steam supplied to cause the evaporation is typically 0.8. For two effects, the ratio becomes 1.6; for three effects 2.4, and so forth. The magnitude of the boiling-point elevation caused by the dissolved inorganic compounds is a controlling factor in selecting the optimal number of effects. The elevation is often in the range of 3 to 10°F between solution and pure water boiling points. When the boiling-point rise is small, minimum evaporation cost is obtained with 8 to 10 effects. When the boiling-point rise is appreciable, the optimal number of effects is small, 6 or less. If necessary, boost interstage steam pressures with steam-jet or mechanical compressors.

Table 6.2 Heuristics in Chapter 6 (*Continued*)

	Heuristic
16	When employing multiple effects, the liquid and vapor flows may be in the same or different directions. Use forward feed, where both liquid and vapor flow in the same direction, for a small number of effects, particularly when the liquid feed is hot. Use backward feed, where liquid flows in a direction opposite to vapor flows, for cold feeds and/or a large number of effects. With forward feed, intermediate liquid pumps are not necessary, whereas they are for backward feed.
17	When crystals are fragile, effective washing is required, and clear mother liquor is desired, use: gravity, top-feed horizontal pan filtration for slurries that filter at a rapid rate; vacuum rotary-drum filtration for slurries that filter at a moderate rate; and pressure filtration for slurries that filter at a slow rate.
18	When cakes of low moisture content are required, use: solid-bowl centrifugation if solids are permitted in the mother liquor; centrifugal filtration if effective washing is required.
19	For granular material, free flowing or not, of particle sizes from 3 to 15 mm, use continuous tray and belt dryers with direct heat. For free-flowing granular solids that are not heat sensitive, use an inclined rotary cylindrical dryer, where the heat may be supplied directly from a hot gas or indirectly from tubes, carrying steam, that run the length of the dryer and are located in one or two rings concentric to and located just inside the dryer rotating shell. For small, free-flowing particles of 1 to 3 mm in diameter, when rapid drying is possible, use a pneumatic conveying dryer with direct heat. For very small free-flowing particles of less than 1 mm in diameter, use a fluidized-bed dryer with direct heat.
20	For pastes and slurries of fine solids, use a drum dryer with indirect heat. For a liquid or pumpable slurry, use a spray dryer with direct heat.
Heat removal and addition	
21	To remove a highly exothermic heat of reaction, consider the use of excess reactant, an inert diluent, or cold shots. These affect the distribution of chemicals and should be inserted early in process synthesis.
22	For less exothermic heats of reaction, circulate reactor fluid to an external cooler, or use a jacketed vessel or cooling coils. Also, consider the use of intercoolers between adiabatic reaction stages.
23	To control temperature for a highly endothermic heat of reaction, consider the use of excess reactant, an inert diluent, or hot shots. These affect the distribution of chemicals and should be inserted early in process synthesis.
24	For less endothermic heats of reaction, circulate reactor fluid to an external heater, or use a jacketed vessel or heating coils. Also, consider the use of interheaters between adiabatic reaction stages.
Heat exchangers and furnaces	
25	Unless required as part of the design of the separator or reactor, provide necessary heat exchange for heating or cooling process fluid streams, with or without utilities, in an external shell-and-tube heat exchanger using countercurrent flow. However, if a process stream requires heating above 750°F, use a furnace unless the process fluid is subject to chemical decomposition.
26	Near-optimal minimum temperature approaches in heat exchangers depend on the temperature level as follows: 10°F or less for temperatures below ambient. 20°F for temperatures at or above ambient up to 300°F. 50°F for high temperatures. 250 to 350°F in a furnace for flue gas temperature above inlet process fluid temperature.
27	When using cooling water to cool or condense a process stream, assume a water inlet temperature of 90°F (from a cooling tower) and a maximum water outlet temperature of 120°F.
28	Boil a pure liquid or close-boiling liquid mixture in a separate heat exchanger, using a maximum overall temperature driving force of 45°F to ensure nucleate boiling and avoid undesirable film boiling as discussed in Section 18.1.
29	When cooling and condensing a stream in a heat exchanger, a zone analysis, described in Section 18.1, should be made to make sure that the temperature difference between the hot stream and the cold stream is equal to or greater than the minimum approach temperature at all locations in the heat exchanger. The zone analysis is performed by dividing the heat exchanger into a number of segments and applying an energy balance to each segment to determine corresponding stream inlet and outlet temperatures for the segment, taking into account any phase change. A process simulation program conveniently accomplishes the zone analysis.

(*Continued*)

Table 6.2 Heuristics in Chapter 6 (*Continued*)

	Heuristic
30	Typically, a hydrocarbon gives an adiabatic flame temperature of approximately 3,500°F when using the stoichiometric amount of air. However, use excess air to achieve complete combustion and give a maximum flue gas temperature of 2,000°F. Set the stack gas temperature at 650 to 950°F to prevent condensation of the corrosive components of the flue–gas.
31	Estimate heat-exchanger pressure drops as follows: 1.5 psi for boiling and condensing. 3 psi for a gas. 5 psi for a low-viscosity liquid. 7–9 psi for a high-viscosity liquid. 20 psi for a process fluid passing through a furnace.
32	Quench a very hot process stream to at least 1,150°F before sending it to a heat exchanger for additional cooling and/or condensation. The quench fluid is best obtained from a downstream separator as in Figure 5.21 for the toluene hydrodealkylation process. Alternatively, if the process stream contains water vapor, liquid water may be an effective quench fluid.
33	If possible, heat or cool a stream of solid particles by direct contact with a hot gas or cold gas, respectively, using a rotary kiln, a fluidized bed, a multiple hearth, or a flash/pneumatic conveyor. Otherwise, use a jacketed spiral conveyor.
Pressure increase operations	
34	Use a fan to raise the gas pressure from atmospheric pressure to as high as 40 inches water gauge (10.1 kPa gauge or 1.47 psig). Use a blower or compressor to raise the gas pressure to as high as 206 kPa gauge or 30 psig. Use a compressor or a staged compressor system to attain pressures greater than 206 kPa gauge or 30 psig.
35	Estimate the theoretical adiabatic horsepower (THp) for compressing a gas from: $$\text{THp} = \text{SCFM}\left(\frac{T_1}{8{,}130a}\right)\left[\left(\frac{P_2}{P_1}\right)^a - 1\right] \quad (6.1)$$ where SCFM = standard cubic feet of gas per minute at 60°F and 1 atm (379 SCF/lbmol); T_1 = gas inlet temperature in °R; inlet and outlet pressures, P_1 and P_2, are absolute pressures; and $a = (k-1)/k$, with k = the gas specific heat ratio, C_p/C_v. Estimate the theoretical exit temperature, T_2, for a gas compressor from: $$T_2 = T_1(P_2/P_1)^a \quad (6.2)$$
36	Estimate the number of gas compression stages, N, from the following table, which assumes a specific heat ratio of 1.4 and a maximum compression ratio of 4 for each stage. Final Pressure/Inlet Pressure — Number of Stages: <4 — 1 4 to 16 — 2 16 to 64 — 3 64 to 256 — 4 Optimal interstage pressures correspond to equal Hp for each compressor. Therefore, based on the above equation for theoretical compressor Hp, estimate interstage pressures by using approximately the same compression ratio for each stage with an intercooler pressure drop of 2 psi or 15 kPa.
37	For heads up to 3,200 ft and flow rates in the range of 10 to 5,000 gpm, use a centrifugal pump. For high heads up to 20,000 ft and flow rates up to 500 gpm, use a reciprocating pump. Less common are axial pumps for heads up to 40 ft for flow rates in the range of 20 to 100,000 gpm and rotary pumps for heads up to 3,000 ft for flow rates in the range of 1 to 1,500 gpm.

Table 6.2 Heuristics in Chapter 6 (*Continued*)

	Heuristic
38	For liquid flow, assume a pipeline pressure drop of 2 psi/100 ft of pipe and a control valve pressure drop of at least 10 psi. For each 10-ft rise in elevation, assume a pressure drop of 4 psi.
39	Estimate the theoretical horsepower (THp) for pumping a liquid from: $$\text{THp} = (\text{gpm})(\text{Pressure increase, psi})/1,714 \quad (6.3)$$
Pressure decrease operations	
40	Consider the use of an expander for reducing the pressure of a gas or a pressure-recovery turbine for reducing the pressure of a liquid when more than 20 Hp and 150 Hp, respectively, can be recovered.
41	Estimate the theoretical adiabatic horsepower (THp) for expanding a gas from: $$\text{THp} = \text{SCFM}\left(\frac{T_1}{8{,}130a}\right)\left[1-\left(\frac{P_2}{P_1}\right)^a\right] \quad (6.4)$$
42	Estimate the theoretical horsepower (THp) for reducing the pressure of a liquid from: $$\text{THp} = (\text{gpm})(\text{Pressure decrease, psi})/1,714 \quad (6.5)$$
Pumping liquid or compressing gas	
43	To increase the pressure of a stream, pump a liquid rather than compress a gas, unless refrigeration is needed.
Vacuum	
44	Estimate inleakage of air by: $$w = kV^{0.667} \quad (6.6)$$ where w = lb/hr of air inleakage, V = ft^3 of volume of the equipment under vacuum, and $k = 0.2$ for pressures greater than 90 torr, 0.15 for pressures between 21 and 89 torr, 0.10 for pressures between 3.1 and 20 torr, and 0.051 for pressures between 1 and 3 torr.
45	To reduce the amount of gas sent to the vacuum system if its temperature is greater than 100°F, add a condenser using cooling water before the vacuum system. The gas leaving the condenser will be at a dew-point temperature of 100°F at the vacuum pressure.
46	For pressures down to 10 torr and gas flow rates up to 10, 000 ft^3/min at the inlet to the vacuum system, use a liquid-ring vacuum pump. For pressures down to 2 torr and gas flow rates up to 1, 000, 000 ft^3/min at the inlet to the vacuum system, use a steam-jet ejector system (one-stage for 100 to 760 torr, two-stage for 15 to 100 torr, and three-stage for 2 to 15 torr). Include a direct-contact condenser between stages.
47	For a three-stage steam-jet ejector system used to achieve a vacuum of 2 torr, 100 pounds of 100 psig steam per pound of gas are required.
Conveying granular solids	
48	If the solid particles are small in size, low in particle density, and not sticky or abrasive, use pneumatic conveying with air at 1 to 7 ft^3/ft^3 of solids and 35 to 120 ft/s air velocity for distances up to 400 ft. Otherwise, for sticky and/or abrasive solids of any size and density, use a screw conveyer and/or bucket elevator for distances up to 150 ft. For solid particles of any size and shape, and not sticky, use a belt conveyor, with inclination up to 30° if necessary, for long distances up to a mile or more.
Solid particle size change and separation	
49	Crushing of coarse solids. Use a jaw crusher to reduce lumps of hard, abrasive, and/or sticky materials of 4 inches to 3 feet in diameter to slabby particles of 1 to 4 inches in size. Use a gyratory crusher to reduce slabby materials of 8 inches to 6 feet in size to rounded particles of 1 to 10 inches in diameter. Use a cone crusher to reduce less hard and less sticky materials of 2 inches to 1 foot in diameter to particles of 0.2 inch (4 mesh) to 2 inches in diameter.
50	Grinding to fine solids. Use a rod mill to take particles of intermediate hardness as large as 20 mm and reduce them to particles in the range of 10 to 35 mesh. Use a ball mill to reduce particles of low to intermediate hardness of 1 to 10 mm in size to very small particles of less than 140 mesh.

(*Continued*)

Table 6.2 Heuristics in Chapter 6 (*Continued*)

	Heuristic
51	Particle-size enlargement. Use compression with rotary compression machines to convert powders and granules into tablets of up to 1.5 inches in diameter. Use extruders with cutters to make pellets and wafers from pastes and melts. Use roll compactors to produce sheets from finely divided materials; the sheets are then cut into any desired shape. Use rotating drum granulators and rotary disk granulators with binders to produce particles in the size range of 2 to 25 mm.
52	Size separation of particles. Use a grizzly of spaced, inclined, vibrated parallel bars to remove large particles greater than 2 inches in diameter. Use a revolving cylindrical perforated screen to remove intermediate-size particles in the size range of 0.25 inch to 1.5 inches in diameter. Use flat, inclined woven screens (U.S. standard) that are vibrated, shaken, or impacted with bouncing balls to separate small particles in the size range of 3 to 80 mesh. Use an air classifier to separate fine particles smaller than 80 mesh.
53	Use a cyclone separator to remove, from a gas, droplets or solid particles of diameter down to 10 microns (0.01 mm). Use a hydroclone separator to remove, from a liquid, insoluble liquid droplets or solid particles of diameter down to 5 microns (0.005 mm). However, small amounts of entrained liquid droplets are commonly removed from gases by vertical knock-out drums equipped with mesh pads to help coalesce the smallest droplets.

REFERENCES

1. Baddour, R.F., P.L.T. Brian, B.A. Logeais, and J.P. Eymery, "Steady-State Simulation of an Ammonia Synthesis Converter," *Chem. Eng. Sci.*, **20**, 281 (1965).

2. Biegler, L.T., and R.R. Hughes, "Process Optimization: A Comparative Case Study," *Comput. Chem. Eng.*, **7**(5), 645 (1983).

3. Douglas, J.M., *Conceptual Design of Chemical Processes*, McGraw-Hill, New York (1988).

4. Fogler, H.S., *Elements of Chemical Reaction Engineering*, 4th ed., Prentice-Hall, Englewood Cliffs, NJ (2005).

5. B.F. Goodrich Co., "Preparation of Vinyl Chloride," British Patent 938,824 (October 9, 1963).

6. Henley, E.J., and E.M. Rosen, *Material and Energy Balance Computations*, Wiley, New York (1969).

7. Hua, I., R.H. Hochemer, and M.R. Hoffmann, "Sonochemical Degradation of *p*-Nitrophenol in a Parallel-Plate Near-Field Acoustical Processor," *Environ. Sci. Technol.*, **29** (11), 2790 (1995a).

8. Hua, I., R.H. Hochemer, and M.R. Hoffmann, "Sonolytic Hydrolysis of *p*-Nitrophenyl Acetate: The Role of Supercritical Water," *J. Phys. Chem.*, **99**, 2335 (1995b).

9. Myers, A.L., and W.D. Seider, *Introduction to Chemical Engineering and Computer Calculations*, Prentice-Hall, Englewood Cliffs, New Jersey (1976).

10. Green, D.W., and R.H. Perry, Ed., *Perry's Chemical Engineers' Handbook*, 8th ed., McGraw-Hill, New York (2008).

11. Schmidt, L.D., *The Engineering of Chemical Reactions*, 2nd ed., Oxford University Press, Oxford (2004).

12. Walas, S.M., *Chemical Process Equipment—Selection and Design*, Butterworth, Stoneham, Massachusetts (1988).

EXERCISES

6.1 For the production of ethylene glycol, how much is the gross profit per pound of ethylene glycol reduced when chlorine and caustic are used to avoid the production of ethylene oxide?

6.2 Consider ethyl-tertiary-butyl-ether (ETBE) as an alternative gasoline oxygenate to MTBE. While the latter appears to have the best combination of properties such as oxygen content, octane number, energy content, and cost, the former can be manufactured using ethanol according to:

$$C_2H_5OH + \textit{Iso}\text{-butene} \rightleftharpoons \text{ETBE}$$

Since ethanol can be manufactured from biomass, it is potentially more acceptable to the environment.

(a) Rework Example 6.1 for this process.

(b) Is reactive distillation promising for combining the reaction and separation operations? If so, suggest a distribution of chemicals using a reactive distillation operation.

6.3 For the ammonia process in Example 6.3, consider operation of the reactor at 932°F and 400 atm. Use a simulator to show how the product, recycle, and purge flow rates, and the mole fractions of argon and methane, vary with the purge-to-recycle ratio. How do the

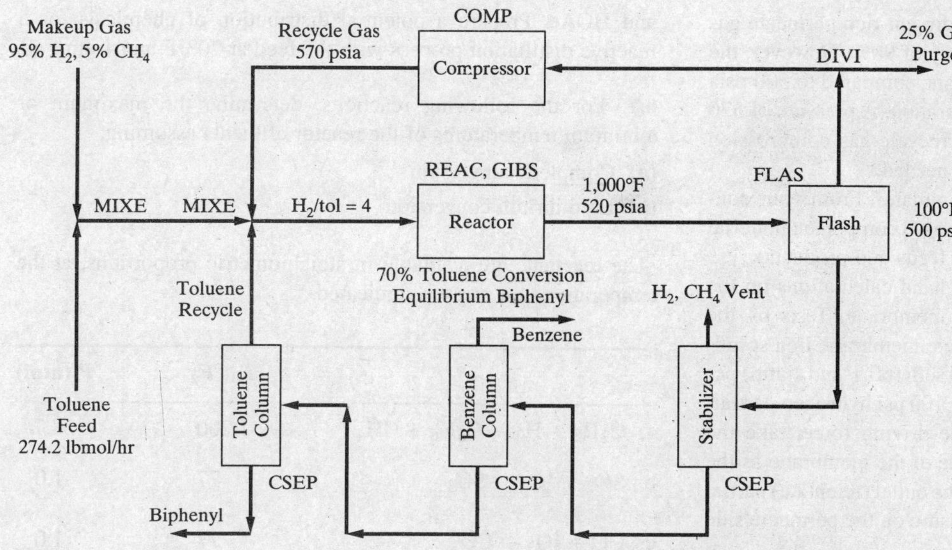

Figure 6.13 Flowsheet for the toluene hydrodealkylation process.

power requirements for compression vary, assuming 3 atm pressure drop in the reactor and 1 atm pressure drop in the heat exchanger?

6.4 *Revamp of a toluene hydrodealkylation process.* This problem considers some waste-minimization concepts. Our operating toluene hydrodealkylation unit, shown in Figure 6.13, involves the hydrogenation of toluene to benzene and methane. An equilibrium side reaction produces a small quantity of biphenyl. To be more competitive, and eliminate waste, the process needs to be studied for a possible revamp. The customer for our small production of biphenyl has informed us that it will not renew its contract with us, and we have no other prospective buyer for biphenyl. Also, a membrane separator company believes that if we install their equipment, we can reduce our makeup hydrogen requirement. Make preliminary process design calculations with a simulator to compare the two alternatives below, and advise me of the technical feasibility of the second alternative and whether we should consider such a revamp further. For your studies, you will have to perform mainly material balance calculations. You will not make detailed distillation calculations, and liquid pumps need not be modeled. For the second alternative, calculate the required area in square feet of the membrane unit and determine if it is reasonable.

Alternative 1. Do no revamp and use the biphenyl for its fuel value.

Alternative 2. Eliminate operation of the toluene column and recycle the biphenyl (with the toluene) to extinction. This should increase the yield of benzene. Also, install a membrane separation unit to reduce hydrogen consumption.

Current plant operation: The current plant operation can be adequately simulated with CHEMCAD, using the equipment models indicated in the flow diagram. Alternatively, any other simulator can be used with appropriate models. Note that the flow diagram of the process includes only the reactor, separators, and recycle-gas compressor. The plant operating factor is 96% (8,410 hr/yr). The feedstock is pure toluene at a flow rate of 274.2 lbmol/hr, which is fixed for both alternatives, because any additional benzene that we can make can be sold. The makeup hydrogen is 95 mol% hydrogen and 5 mol% methane. Our reactor outlet conditions are 1,000°F and 520 psia. The hydrogen-to-toluene molar ratio in the feed to the reactor must be 4 to prevent coke formation. The toluene conversion is 70%. The biphenyl in the reactor effluent is the chemical equilibrium amount. The flash drum conditions are 100°F at 500 psia. The flash vapor is not separated into hydrogen and methane, but is purged to limit methane buildup in the recycle gas. The purge gas, which has fuel value, is 25% of the vapor leaving the flash vessel. Perfect separations can be assumed for the three columns. Based on this information, you can obtain the current plant material balance.

Alternative 1. Simulate the current plant operation. Note that the process has two recycle loops that must be converged. The SRK equation of state is adequate for K-values and enthalpies. From your converged material balance, summarize the overall component material balance in pounds per year (i.e., process feeds and products).

Alternative 2. Eliminate the toluene column and rerun the simulation. Since the biphenyl will be recycled to extinction, the benzene production should increase. Replace the stream divider, which divides the flash vapor into a purge and a gas recycle, with a membrane separation unit that can be modeled with a CSEP (black-box separator) unit.

For Alternative 2, the vendor of the membrane unit has supplied the following information:

Hydrogen will pass through the membrane faster than methane. Vapor benzene, toluene, and biphenyl will not pass through the membrane.

The hydrogen-rich permeate will be the new recycle gas. The retentate gas will be used for fuel.

Tests indicate that the purity of the hydrogen-rich permeate gas will be 95 mol% with a hydrogen recovery of 90%. However, the pressure of the permeate gas will be 50 psia, compared to 500 psia for the recycle gas in the current plant operation. A pressure of 570 psia is required at the discharge of the recycle-gas compression system. Thus, a new compressor will be needed.

Run the revamped process with the simulator. From your converged material balance, summarize the overall component material balance in pounds per year (i.e., process feeds and products).

The membrane unit is to be sized by hand calculations on the basis of the hydrogen flux through the membrane. Tests by the vendor using a nonporous cellulose acetate membrane in a spiral-wound module indicate that this flux is 20 scfh (60°F and 1 atm) per square foot of membrane surface area per 100 psi hydrogen partial-pressure driving force. To determine the driving force, take the hydrogen partial pressure on the feed side of the membrane as the arithmetic average between the inlet and the outlet (retentate) partial pressures. Take the hydrogen partial pressure on the permeate side as that of the final permeate.

Summarize and discuss your results in a report and make recommendations concerning cost studies.

6.5 For the reaction system:

$$A + B \xrightarrow{k_1} C + A \xrightarrow{k_3} E$$
$$\quad\ \ \searrow_{k_2} D + A \nearrow_{k_4}$$

select an operating temperature that favors the production of C. The pre-exponential factor and activation energy for the reactions are tabulated as follows:

Rxtn	k_0 ($m^6/kmol^2 \cdot s$)	E (kJ/kmol)
1	3.7×10^6	65,800
2	3.6×10^6	74,100
3	5.7×10^6	74,500
4	1.1×10^7	74,400

6.6 Propylene glycol mono-methyl-ether acetate (PMA) is produced by the esterification of propylene glycol mono-methyl ether (PM) in acetic acid (HOAc):

$$\underset{\text{PM}}{CH_3CH(OH)CH_2OCH_3} + \underset{\text{HOAc}}{CH_3COOH} \rightleftharpoons$$
$$\underset{\text{PMA}}{CH_3CH(OC(CH_3)O)CH_2OCH_3} + H_2O$$

Conventionally, the reaction takes place in a fixed-bed reactor followed by the recovery of PMA from water, and unreacted PM and HOAc. Prepare a potential distribution of chemicals for a reactive distillation process with the feed at 203°F and 1 atm.

6.7 For the following reactions, determine the maximum or minimum temperatures of the reactor effluents assuming:

(a) Complete conversion

(b) Equilibrium conversion

The reactants are available in stoichiometric proportions, at the temperature and pressure indicated.

	T_0(°F)	P (atm)
a. $C_7H_8 + H_2 \rightarrow C_6H_6 + CH_4$	1,200	38.7
b. $SO_2 + \frac{1}{2}O_2 \rightarrow SO_3$	77	1.0
c. $CO + \frac{1}{2}O_2 \rightarrow CO_2$	77	1.0
d. $C_2H_4Cl_2 \rightarrow C_2H_3Cl + HCl$	932	26.0

Also, find the heats of reaction at the conditions of the reactants.

6.8 For Example 6.7, use a simulator to graph the effluent temperature of the methanol reactor as a function of the H_2/CO ratio.

6.9 For Example 6.8, use a simulator to graph the effluent temperature of the methanol reactor as a function of the dodecane flow rate.

6.10 Divide the methanol reaction operation in Example 6.6 into five consecutive stages in series. Feed the CO reactant entirely into the first operation at 25°C and 1 atm. Divide the H_2 reactant into five cold shots and vary the temperature of H_2 before dividing it into cold shots. Assuming that the reaction operations are adiabatic, determine the maximum temperature in the flowsheet as a function of the temperature of the cold shots. How does this compare with the adiabatic reaction temperature?

6.11 Repeat Exercise 6.10 using intercoolers instead of cold shots and an unknown number of reaction stages. The feed to the first reactor is at 25°C and 1 atm. Throughout the reactors, the temperature must be held below 300°C. What is the conversion of CO in the first reactor? How many reaction stages and intercoolers are necessary to operate between 25°C and 300°C?

6.12 *Alternatives for preparing a feed.* A process under design requires that 100 lbmol/hr of toluene at 70°F and 20 psia be brought to 450°F and 75 psia. Develop at least three flowsheets to accomplish this using combinations of heat exchangers, liquid pumps, and/or gas compressors. Discuss the advantages and disadvantages of each flowsheet, and make a recommendation as to which flowsheet is best.

Chapter 7

Design of Reactors and Reactor Networks

7.0 OBJECTIVES

The design of the reactor section of the process flow diagram addresses the need to eliminate differences in molecular type. More specifically, it is desired to ensure sufficient yield and selectivity of the required product species, by appropriate selection of a single reactor or network of reactors. The presence of at least one chemical reactor and one or more separation sections for the separation of the effluent mixture leaving the reactor(s) characterizes many chemical processes. In almost all cases, one or more of the streams leaving the separation section(s) are recycled to the reactor.

This chapter begins with the design of individual reactors and networks of reactors, without regard for the separation section(s) and possible recycle therefrom. These topics are presented in Sections 7.1 to 7.5. The reader should then refer to Chapter 8, where the synthesis of separation trains is discussed in the absence of consideration of the reactor section. An introduction to the interaction between the reactor and separation sections was presented in Chapter 6 by examining a few examples of this interaction. Here, in the supplement to Chapter 7, namely Sections 7S.1 to 7S.5, that introduction is extended to a detailed treatment of the design of reactor-separator-recycle networks.

After studying this chapter, the reader should

1. Be familiar with the types of reactor models available in simulators and their use in process simulation. Further assistance is provided in the multimedia modules, which can be downloaded from the Wiley Web site associated with this book (*ASPEN → Chemical Reactors and HYSYS → Chemical Reactors*).
2. Be able to design a system for heat transfer in association with the reactor, so as to sustain an exothermic or endothermic reaction at its desired temperature level, and study the design using simulation.
3. Be able to determine if a reactor network should be considered and, if so, design it using the concept of the attainable region.
4. Be aware of the effect of velocity and temperature profiles on conversion in tubular-flow reactors and how to account for these profiles.
5. Be able to determine the best location for the separation section, either before or after the reactor.
6. Understand the tradeoffs between purge-to-recycle ratio, recycle ratio, and raw-material loss when dealing with inert or byproduct chemicals that are difficult to separate from the reactants.
7. Understand the need to determine the optimal reactor conversion, involving the tradeoff between the cost of the reactor section and the cost of the separation section(s) in the presence of recycle, even when chemical equilibrium greatly favors the products of the reaction.
8. Understand the conditions under which the recycle of byproducts to extinction can be employed to reduce waste and increase yield.
9. Be aware of the snowball effect in a reactor-separator-recycle network and the importance of designing an adequate control system, as discussed in Example 12.11.

www.wiley.com/go/global/seider

7.1 INTRODUCTION

As mentioned above, this chapter treats reactors as single entities and in combination with other reactors and/or separators, including the effect of recycle from a separator to a reactor. Section 7.2 describes the types of single reactor models available in flowsheet simulators. Section 7.3 discusses the applicability of the reactor types to model particular process circumstances, which may involve complex configurations, particularly for reactors that require heat exchange so as to sustain a stable operation at the desired temperature level. Here, examples are presented from the processing industry. In Section 7.4, attainable-region analysis is introduced as a tool for the optimal design of reactor networks. Section 7.5 discusses the limitations of the reactor models in flowsheet simulators and introduces advanced tubular-flow models that take into account the effects of velocity and temperature profiles when they are not flat.

Beginning with Section 7S.1, reactors are considered in combination with separators. The feed to a reactor section of a chemical process almost always is a combined feed consisting of a fresh feed mixed with one or more recycle streams, as shown later in Figure 8.1. Although Figure 8.1 shows only one reactor section, multiple reactor sections are sometimes required, with separation sections located between each pair of reactor sections. Of major importance is the fact that fresh reactor feeds rarely contain only the reactants for the desired reaction. Besides the reactants, they may contain inert chemicals, potential reactants for side reactions, catalyst poisons, and products of the desired reaction(s). Recycle streams are intended to contain only unconverted reactants of the desired reaction(s). However, more commonly, recycle streams also contain products of the desired reaction(s), products of undesired side reactions, and inert chemicals. Reactor effluents are almost never products that meet purity specifications. Besides the products, effluents may contain reactants, inerts, products of undesired side reactions, and feed impurities. Thus, almost every chemical process that involves a chemical reaction section also involves one or more separation sections in addition to one or more recycle streams.

A major challenge of process design is devising an optimal scheme for uniting the reaction and separation functions of a process. Section 7S.1 discusses the problem of where best to locate the separation section with respect to the reaction section. Although it might seem that the reaction section should logically precede the separation section, such a sequence is not always optimal. Section 7S.2 extends the treatment in Section 7S.1 by discussing the tradeoffs in processes involving recycle back to the reaction section. Because of recycle from a separation section, an optimal reactor conversion per pass-through the reactor exists, which is examined in Section 7S.3. In some cases, an optimal yield and minimization or elimination of waste products can be achieved by recycling to extinction the products of undesirable side reactions, as discussed in Section 7S.4. Then, Section 7S.5 introduces an important problem in control, namely snowball effects in the control of processes involving recycle from a separator to a reactor. Finally, Section 7S.6 describes computational fluid dynamics (CFD) models for tubular chemical reactors when radial velocity and temperature profiles have a significant effect on conversion.

7.2 REACTOR MODELS

Chemical reactors, particularly for continuous processes, are often custom-designed to involve multiple phases (e.g., vapor, liquid, reacting solid, and solid catalyst), different geometries (e.g., stirred tanks, tubular flows, converging and diverging nozzles, spiral flows, and membrane transport), and various regimes of momentum, heat, and mass transfer (e.g., viscous flow, turbulent flow, conduction, radiation, diffusion, and dispersion). There are so many configurations, involving different combinations of these attributes, that attempts to develop generalized reactor models have met with limited success.

Most of the process simulators provide four kinds of reactor models, including: (1) a stoichiometric model that permits the specification of reactant conversions and extents of reaction for one or more specified reactions; (2) a model for multiple phases (vapor, liquid, and solid) in chemical equilibrium, where the approach to equilibrium for individual reactions can be specified; (3) a kinetic model for a continuous-stirred-tank reactor (CSTR) that assumes perfect mixing of homogeneous phases (liquid or vapor); and (4) a kinetic model for a plug-flow tubular reactor (PFTR or PFR), for homogeneous phases (liquid or vapor) and assuming no backmixing (dispersion). These ideal models are used in the early stages of process synthesis, when the details of the reactor designs are less important but reactor effluents and heat duties are needed.

The ideal reactor models are replaced by custom-made models as the details gain significance. For this purpose, all of the flowsheet simulators provide facilities for the insertion of user-generated models. These are refined often as the design proceeds and as reactor data from the laboratory or pilot plant are regressed, with some of the simulators providing facilities for estimating the parameters of kinetic models by nonlinear regression.

When working with the ideal reactor models, the reader should refer to available textbooks on reactor analysis and design, for example, *Elements of Chemical Reaction Engineering* by H.S. Fogler (2005), *The Engineering of Chemical Reactions* by L.D. Schmidt (1998), *Chemical Reaction Engineering* by O. Levenspiel (1999), and to the user manuals and tutorial segments of the flowsheet simulators. The following discussion of the ideal reactor models used in simulators is preceded by a brief review of reaction stoichiometry and reaction extent, which together provide the basis for calculation of the conversion reaction model in the simulators. Advanced models for tubular reactors that account for non-plug flow are introduced in Section 7.5.

Reaction Stoichiometry

For most of the reactor models in the flowsheet simulators, it is necessary to provide R chemical reactions involving C chemical species:

$$\sum_{j=1}^{C} \nu_{ij} A_j = 0, \; i = 1, \ldots, R \tag{7.1}$$

where A_j is the chemical formula for species j and ν_{ij} is the stoichiometric coefficient for species j in reaction i (negative for reactants, positive for products). As an example, for the manufacture of methanol, let the chemicals be ordered according to decreasing volatility, that is, (1) H_2, (2) CO, and (3) CH_3OH. The reaction can be written as $-2H_2 - CO + CH_3OH = 0$, and the stoichiometric coefficient matrix is

$$\nu_{ij} = \begin{matrix} H_2 \\ CO \\ CH_3OH \end{matrix} \begin{bmatrix} -2 \\ -1 \\ 1 \end{bmatrix}$$

Extent of Reaction

Consider a single reaction. In the stoichiometric reactor models, one specifies the fractional conversion, X_k, of key reactant k,

$$X_k = \frac{n_{k,in} - n_{k,out}}{n_{k,in}}, \qquad (7.2)$$

where $n_{k,\,in}$ and $n_{k,\,out}$ are the moles of species k entering and leaving the reactor and $0 \leq X_k \leq 1$, or the extent (number of moles extent) of reaction i,

$$\xi_i = \frac{\Delta n_{ij}}{\nu_{ij}}, \; j = 1, \ldots, C, \qquad (7.3)$$

is specified. The molar flow rates of the components in the reactor effluent, $n_{j,out}$, are computed from the component molar flow rates in the reactor feed, $n_{j,in}$, by a component material balance equation that is consistent with the reaction stoichiometry. If a specification of the fractional conversion of the key component, k, is made with Eq. (7.2):

$$n_{j,out} = n_{j,in} - n_{k,in} X_k \left(\frac{\nu_j}{\nu_k}\right), \; j = 1, \ldots, C \qquad (7.4)$$

If the extent of the reaction is specified:

$$n_{j,out} = n_{j,in} + \xi \nu_j, \quad j = 1, \ldots, C \qquad (7.5)$$

For example, for the conversion of CO and H_2 to CH_3OH, assuming an initial feed of 100 kmol/hr CO and 600 kmol/hr H_2 and 70% conversion of CO (the key component), using Eq. (7.4), the molar flow rates of the three components in the reactor effluent are:

$$n_{H_2,out} = 600 - 100(0.7)(-2/-1) = 460 \text{ kmo/hr}$$

$$n_{CO,out} = 100 - 100(0.7)(-1/-1) = 30 \text{ kmol/hr}$$

$$n_{CH_3OH,out} = 0 - 100(0.7)(1/-1) = 70 \text{ mol/hr}$$

The mole fraction of methanol in the reactor effluent is $y_{CH_3OH} = 70/(460 + 30 + 70) = 0.125$. If, instead, the extent of reaction is specified as 70 kmol/hr, Eq. (7.5) gives the same results.

For multiple reactions, the reactions must be specified as series or parallel. The former is equivalent to having reactors in series with the feed to each reactor except the first, being the product from the previous reactor. Each reaction can have a different key reactant. For parallel reactions, it is preferable to specify the extent of reaction for each reaction, which results in:

$$n_{j,out} = n_{j,in} + \sum_{i=1}^{R} \xi_i \nu_{ij}, \qquad j = 1, \ldots, C \qquad (7.6)$$

Equilibrium

A chemical reaction can be written as a general stoichiometric equation, in terms of reactants *A*, *B*, etc., and products *R*, *S*, etc.,

$$aA + bB + \cdots = rR + sS + \cdots \qquad (7.7)$$

In writing this equation, it is very important that, unless otherwise stated, each reactant and product is understood to be the pure component in a separate and designated phase: gas, liquid, or solid. A reaction is characterized by two important thermodynamic quantities, namely the heat of reaction and the Gibbs (free) energy of reaction. Furthermore, these two quantities are functions of temperature and pressure. Thermodynamic data are widely available in simulators and elsewhere for more than a thousand components, for the calculation of these two quantities under standard state conditions, for example, at a reference temperature of 25°C and 1 bar with all components in a designated phase, usually as an ideal gas. The effect of temperature on the heat of reaction depends on the heat capacities of the reactants and products and the effect of temperature on those heat capacities. For many reactions, the effect of temperature on the heat of reaction is relatively small. For example, the reaction of CO and H_2 to form gaseous methanol,

$$CO + 2H_2 \rightleftharpoons CH_3OH, \qquad (7.8)$$

has a standard heat of reaction, ΔH°_{rxn}, at 25°C of −90,400 kJ/kmol of methanol, while at 800°C, the heat of reaction, ΔH°_{rxn}, is −103,800 kJ/kmol, a relatively small change for such a large change in temperature. By contrast, the effect of temperature on the Gibbs energy of reaction can be very large. For example, for the same methanol formation reaction, the standard Gibbs energy of reaction, ΔG°_{rxn}, is −25,200 kJ/kmol at 25°C. Already, at 500°C, the Gibbs energy of reaction, ΔG°_{rxn}, has undergone a drastic change to +88,000 kJ/kmol.

Many reactions of industrial importance are limited by chemical equilibrium, with partial conversion of the limiting reactant and, with the rate of the reverse reaction equal to the rate of the forward reaction. For a specified feed composition and final temperature and pressure, the product composition at chemical equilibrium can be computed by either of two methods: (1) chemical equilibrium constants (K-values) computed from the Gibbs energy of reaction combined with material balance equations for a set of independent reactions, or (2) the minimization of the Gibbs energy of the reacting system. The first method is applicable when the stoichiometry can be specified for all reactions being considered. The second method requires only a list of the possible products.

For the first method, a chemical equilibrium constant, K, is computed for each independent stoichiometric reaction in the set, using the equation

$$K = \frac{a_R^r a_S^s \cdots}{a_A^a a_B^b \cdots} = \exp\left(-\frac{\Delta G^\circ_{\text{rxn}}}{RT}\right) \quad (7.9)$$

where a_i is the component activity.

For a gas solution, the activity is given by

$$a_i = \bar{\phi}_i y_i P = \bar{\phi}_i P_i, \quad (7.10)$$

where $\bar{\phi}_i$, is the fugacity coefficient of component i in the gas mixture, equal to 1.0 for an ideal gas solution, and P_i is the partial pressure. In general, $\bar{\phi}_i$ is a function of T, P, and composition. At low to moderate pressures, $\bar{\phi}_i = 1.0$, so that the activity is equal to the partial pressure in bar. It is common to replace the activity in the equation for K with the above equation to give:

$$K = \frac{y_R^r y_S^s \cdots}{y_A^a y_B^b \cdots}\left(\frac{\bar{\phi}_R^r \bar{\phi}_S^s \cdots}{\bar{\phi}_A^a \bar{\phi}_B^b \cdots}\right) P^{r+s+\cdots-a-b-\cdots}$$
$$= \frac{y_R^r y_S^s \cdots}{y_A^a y_B^b \cdots} K_{\bar{\phi}} P^{r+s+\cdots-a-b-\cdots} = \frac{P_R^r P_S^s \cdots}{P_A^a P_B^b \cdots} K_{\bar{\phi}} \quad (7.11)$$

where $K_{\bar{\phi}} = 1.0$ for low to moderate pressures.

For a liquid solution, the activity is given by:

$$a_i = x_i \gamma_i \exp\left[\frac{\bar{V}_i}{RT}(P - P_i^s)\right] \quad (7.12)$$

where γ_i is the activity coefficient of component i in the liquid mixture and is equal to 1.0 for an ideal liquid solution, $\bar{V}_i$ is the partial molar volume of component i, and P_i^s is the vapor pressure of component i. The pressures are in bar. In general, γ_i is a function of T, P, and composition. For ideal liquid solutions at low to moderate pressures, $\gamma_i = 1.0$, so that the activity is equal to the mole fraction. It is common to replace the activities in Eq. (7.9) with Eq. (7.12) to give

$$K = \frac{x_R^r x_S^s \cdots}{x_A^a x_B^b \cdots}\left(\frac{\gamma_R^r \gamma_S^s \cdots}{\gamma_A^a \gamma_B^b \cdots}\right) = \frac{x_R^r x_S^s \cdots}{x_A^a x_B^b \cdots} K_\gamma \quad (7.13)$$

where $K_\gamma = 1.0$ for ideal liquid solutions at low to moderate pressures. Most textbooks on chemical thermodynamics present charts of $\log_{10} K$ as a function of temperature for many chemical reactions. The van't Hoff equation relates K to temperature by

$$\left(\frac{d \ln K}{dT}\right)_P = \frac{\Delta H^\circ_{\text{rxn}}}{RT^2} \quad (7.14)$$

If the heat of reaction is assumed independent of temperature over a particular range of temperature, integration and conversion to $\log_{10}$ form gives the approximate correlating equation:

$$\log_{10} K = A + B/T \quad (7.15)$$

where T is the absolute temperature. Many chemical equilibrium curves are represented with reasonable accuracy by this equation. For example, for the gas-phase reaction of CO and H_2 to form methanol, over a temperature range of 273 to 773 K,

$$\log_{10} K = -12.275 + 4{,}938/T \quad (7.16)$$

Typically, the methanol synthesis reaction is catalyzed by copper-zinc oxide, at a pressure of 100 bar and a temperature of 300°C. A large excess of hydrogen is used to help absorb the relatively high heat of reaction. At these conditions, $K = 0.0002202$ and $K_{\bar{\phi}} = 0.61$. Therefore,

$$\frac{(y_{CH_3OH})}{(y_{CO})(y_{H_2})^2} = \frac{K}{K_{\bar{\phi}}} P^2 = \frac{0.0002202}{0.61}(100)^2 = 3.61 \quad (7.17)$$

If X is the equilibrium fractional conversion of the limiting reactant, CO, then using the same initial feed composition as before and the stoichiometry for the reaction, the equilibrium mole fractions are

$$y_{CH_3OH} = \frac{100\,X}{700 - 200\,X}$$

$$y_{CO} = \frac{100 - 100\,X}{700 - 200\,X}$$

$$Y_{H_2} = \frac{600 - 200\,X}{700 - 200\,X}$$

Combining the above four equations to give a nonlinear equation in X and solving gives $X = 0.7087$.

The second method for computing chemical equilibrium is to apply the criterion that the total Gibbs energy, G, is a minimum at constant temperature and pressure. Alternatively, one could use the entropy, S, as a maximum or the Helmholtz energy, A, as a minimum, but the Gibbs energy is most widely applied. Two advantages of this second method are: (1) the avoidance of having to formulate stoichiometric equations (only the possible products need to be specified), and (2) the ease of formulation for multiple phases and simultaneous phase equilibrium. For a single phase, the total Gibbs energy at a specified T and P is given by

$$G = \sum_{i=1}^{C} N_i \bar{G}_i \quad (7.18)$$

where N_i is the mole number of component i and $\overline{G}_i$, is the partial molar Gibbs energy of component i in the equilibrium mixture. The components are those in the feed plus those that may be produced by chemical reactions. The Gibbs energy is minimized with respect to N_i, which are constrained by atom balances. The method is readily extended to multiple phases.

It would seem that for simplicity or usefulness, the second method would be preferred because when using this model, an independent set of chemical reactions need not be specified. Hence, the designer is not required to identify the reactions that take place. However, since most reactors are designed to emphasize desired reactions and curtail or exclude undesired reactions, the chemical reactions that take place in the reactor are usually known by the time the reactor is to be designed. Thus, the first model may be preferred, with the second model being useful for preliminary exploration of the thermodynamic possibilities. To make the second model more useful, restrictions should be placed on certain improbable reactions. If this is not done, the second method can produce results that are incorrect because they implicitly require reactions that are not kinetically feasible.

For instructions on the use of the equilibrium constant and Gibbs reactor models in the process simulators, see the multimedia modules, which can be downloaded from the Wiley Web site associated with this book (*ASPEN → Chemical Reactors → Equilibrium Reactors → REQUIL* or *RGIBBS* and *HYSYS → Chemical Reactors → Setting Up Reactors → Equilibrium* or *Gibbs*).

Kinetics

Fractional conversion and equilibrium reactor models are useful in the early stages of process design when conducting material and energy balance studies. However, eventually reactor systems must be configured and sized. This requires knowledge of reaction kinetics, which is obtained by conducting laboratory experiments. For homogeneous non-catalytic reactions, power-law expressions are commonly used for regression of laboratory kinetic data. These expressions are not always based on the stoichiometric equation because several elementary reaction steps may be involved, the sum of which is the stoichiometric equation, but one of which may control the overall reaction rate. Elementary reaction steps rarely involve more than two molecules. The general power-law kinetic equation is

$$-r_j = -\frac{dC_j}{dt} = k\prod_{i=1}^{C} C_i^{\alpha_i} \tag{7.19}$$

where $-r_j$ is the rate of disappearance of component j (in mol/time-volume), C_i is the concentration of component i (in mol/volume), t is time, k is the reaction rate coefficient, α_i is the order of reaction with respect to component i, and C is the number of components.

For gas-phase reactions, the partial pressure, P_i, is sometimes used in place of the concentration, C_i, in the kinetic equation. The reaction rate coefficient is a function of temperature as given by the empirical Arrhenius equation:

$$k = k_o \exp\left(-E/RT\right) \tag{7.20}$$

where k_o is the pre-exponential factor, and E is the activation energy.

For reactions that are catalyzed by solid porous catalyst particles, the sequence of elementary steps may include adsorption on the catalyst surface of one or more reactants and/or desorption of one or more products. In that case, a Langmuir–Hinshelwood (LH) kinetic equation is often found to fit the experimental kinetic data more accurately than the power-law expression of Eq. (7.19). The LH formulation is characterized by a denominator term that includes concentrations of certain reactants and/or products that are strongly adsorbed on the catalyst. The LH equation may also include a prefix, η, called an overall effectiveness factor, that accounts for mass and heat transfer resistances, both external and internal, to the catalyst particles. As an example, laboratory kinetic data for the air-oxidation of SO_2 to SO_3 are fitted well by the following LH equation:

$$-r_{SO_2} = \frac{\eta k\left(P_{SO_2}P_{O_2} - \dfrac{P_{SO_3}^2}{K^2 P_{SO_2}}\right)}{\left[1 + K_1 P_{SO_2}^{1/2} + K_2 P_{SO_3}^{1/2}\right]} \tag{7.21}$$

where K is the chemical equilibrium constant and K_1 and K_2 are adsorption equilibrium constants.

Ideal Kinetic Reaction Models—CSTRs and PFRs

CSTR

The simplest kinetic reactor model is the CSTR (continuous-stirred-tank reactor), in which the contents are assumed to be perfectly mixed. Thus, the composition and the temperature are assumed to be uniform throughout the reactor volume and equal to the composition and temperature of the reactor effluent. However, the fluid elements do not all have the same residence time in the reactor. Rather, there is a residence-time distribution. It is not difficult to provide perfect mixing of the fluid contents of a vessel to approximate a CSTR model in a commercial reactor. A perfectly mixed reactor is used often for homogeneous liquid-phase reactions. The CSTR model is adequate for this case, provided that the reaction takes place under adiabatic or isothermal conditions. Although calculations only involve algebraic equations, they may be nonlinear. Accordingly, a possible complication that must be considered is the existence of multiple solutions, two or more of which may be stable, as shown in the next example.

EXAMPLE 7.1 ***Adiabatic CSTR for Hydrolysis of Propylene Oxide***

Propylene glycol (PG) is produced from propylene oxide (PO) by liquid-phase hydrolysis with excess water under adiabatic and near-ambient conditions, in the presence of a small amount of soluble sulfuric acid as a homogeneous catalyst:

$$C_3H_6O + H_2O \rightarrow C_3H_8O_2 \tag{7.22}$$

Because the exothermic heat of reaction is appreciable, excess water is used. Furthermore, because PO is not completely soluble in water, methanol is added to the feed, which enters the reactor at 23.9°C, with the following flow rates:

Propylene oxide:	18.712 kmol/hr
Water, to be determined, within the range:	160 to 500 kmol/hr
Methanol:	32.73 kmol/hr

It is proposed to consider the use of an existing agitated reactor vessel, which can be operated adiabatically at 3 bar (to suppress vaporization), with a liquid volume of 1.1356 m^3. The reaction occurs in a sequence of elementary steps, with the controlling step involving two molecules of PO. The power-law kinetic equation is:

$$-r_{PO} = 9.15 \times 10^{22} \exp(-1.556 \times 10^5/RT)C_{PO}^2 \tag{7.23}$$

where, C_{PO} is in kmol/m^3, $R = 8.314$ kJ/kmol-K, and T is in K. Carry out a sensitivity analysis to investigate the effect of the water feed rate on the operating temperature and the PO conversion.

SOLUTION

www.wiley.com/go/global/seider

As shown on the multimedia modules, which can be downloaded from the Wiley Web site associated with this book (following the links: *HYSYS → Chemical Reactors → Setting Up Reactors → CSTR* or *ASPEN → Chemical Reactors → Kinetic Reactors → CSTRs → RCSTR*), analysis of this process shows the possibility of multiple steady states. For example, at a water flow rate of 400 kmol/hr, the following steady states are obtained: (1) conversion of 83% with an effluent temperature of 62°C, (2) conversion of 45% with an effluent temperature of 44°C, and (3) conversion of 3% with an effluent temperature of 25°C. The intermediate steady state at 45% conversion is unstable, while the other two steady states are stable. Furthermore, a controllability and resiliency (C&R) analysis on this process is carried out in Case Study 12S.1, where a design involving a single CSTR is compared with one utilizing two CSTRs in series.

PFR

More complex is the plug-flow tubular reactor (PFR or PFTR), in which the composition of the fluid, flowing as a plug, gradually changes down the length of the reactor, with no composition or temperature gradients in the radial direction. Furthermore, mass- and heat-transfer rates are negligible in the axial direction. Thus, the PFR is completely unmixed, with all fluid elements having the same residence time in the reactor. If the reactor operates under adiabatic or nonisothermal conditions, the temperature of the flowing fluid changes gradually down the length of the reactor.

All simulators provide one-dimensional, plug-flow models that neglect axial dispersion. Thus, there are no radial gradients of temperature, composition, or pressure; and mass diffusion and heat conduction do not occur in the axial direction. Operation of the reactor can be adiabatic, isothermal, or nonadiabatic, nonisothermal. For the latter, heat transfer to or from the reacting mixture occurs along the length of the reactor.

Consider the case of adiabatic operation with one chemical reaction. A mole balance for the limiting reactant, A, can be written as:

$$F_{A0}\frac{dX}{dV} = -r_A\{X, T\} \tag{7.24}$$

where F_{A0} is the molar flow rate of A entering the reactor, X is the fractional conversion of A, V is the reactor volume, and r_A is the rate of reaction of A written as a function of fractional conversion and temperature. Because the process simulators compute enthalpies referred to the elements, as described by Felder and Rousseau (2000), with values for the standard enthalpy of formation built into the component properties data bank, the heat of reaction is handled automatically, and the energy balance for adiabatic operation becomes simply:

$$H\{X, T\} = H\{X = 0, T = T_0\} \tag{7.25}$$

where H is the enthalpy flow rate of the reacting mixture in energy/unit time, and T_0 is the entering temperature. Equations (7.24) and (7.25) are solved by numerical integration. The following example illustrates the use of the simulator models for a PFR to size a plug-flow adiabatic reactor for the noncatalytic hydrodealkylation of toluene.

EXAMPLE 7.2 ***Adiabatic PFR for Toluene Hydrodealkylation***

A hydrodealkylation reactor feed at 1,200°F and 494 psia consists of

Component	lbmol/hr
Hydrogen	2,049.1
Methane	3,020.8
Benzene	39.8
Toluene	362.0
Biphenyl	4.2
Total	5,475.9

These molar flow rates account for the small extent of reaction for the secondary reaction of benzene to biphenyl (2%), and ignore the negligible rate of the reverse reaction, leaving the main reaction to be considered:

$$H_2 + C_7H_8 \rightarrow CH_4 + C_6H_6 \tag{7.26}$$

Laboratory studies have shown that in the absence of a catalyst, this is a free-radical chain reaction that proceeds in three elementary steps:

$$\begin{gathered} H_2 \rightleftharpoons 2H\bullet \\ H\bullet + C_6H_5CH_3 \rightarrow C_6H_6 + CH_3\bullet \\ CH_3\bullet + H_2 \rightarrow CH_4 + H\bullet \end{gathered} \qquad (7.27)$$

Equilibrium is established rapidly in the first step to provide hydrogen free radicals. The sum of the next two steps is the stoichiometric equation. Step two is the slow or rate-controlling step. Thus, the overall reaction rate is not proportional to the product of the hydrogen and toluene concentrations as given by the law of mass action when applied to the stoichiometric reaction. Instead, the overall reaction rate is proportional to the product of the hydrogen free-radical and toluene concentrations as given in the second elementary step. For the above hydrodealkylation chain reaction, the power-law kinetic equation is derived as follows. Because the first elementary step approaches equilibrium:

$$K_1 = \frac{C_{H\bullet}^2}{C_{H_2}} \qquad (7.28)$$

Rearrangement gives

$$C_{H\bullet} = K_1^{1/2} C_{H_2}^{1/2} \qquad (7.29)$$

The power-law kinetic equation for the second elementary step, which determines the overall reaction rate, is

$$-\frac{dC_{\text{toluene}}}{dt} = k_2 C_{H\bullet} C_{\text{toluene}} \qquad (7.30)$$

Combining the last two equations,

$$-\frac{dC_{\text{toluene}}}{dt} = k_2 K_1^{1/2} C_{H_2}^{1/2} C_{\text{toluene}} \qquad (7.31)$$

This rate expression correlates well with laboratory kinetic data for temperatures in the range of 500–900°C and pressures from 1 to 250 atm, with $k_2 K_1^{1/2} = 6.3 \times 10^{10} \exp(-52{,}000/RT)$, concentrations in kmol/m^3, time in s, T in K, and $R = 1.987$ cal/mol-K. Use the PFR model in a process simulator to determine the length of a cylindrical plug-flow reactor with a length-to-diameter ratio of six that yields a toluene conversion of 75%. Use the Peng–Robinson equation of state to estimate the thermophysical properties for this vapor-phase reaction.

SOLUTION

To see how an adiabatic PFR is designed to provide a 75% conversion of toluene, see the multimedia modules, which can be downloaded from the Wiley Web site associated with this book. Follow the link *HYSYS → Chemical Reactors → Setting Up Reactors → PFR* for a solution obtained with HYSYS, and *ASPEN → Chemical Reactors → Kinetic Reactors → PFTRs → RPLUG* for a solution with ASPEN PLUS.

www.wiley.com/go/global/seider

Note that the results provided by these simulators are almost identical; the HYSYS result calls for a reactor volume of 3,690 ft^3($D = 9.2$ ft, $L = 55.3$ ft), while ASPEN PLUS gives a volume of 3,774 ft^3($D = 9.3$ ft, $L = 55.8$ ft). The main reason for the slight discrepancy is due to the neglected pressure drop in the HYSYS simulation (the ASPEN PLUS calculation assumes a pressure drop of 5 psia).

At high flow rates (high Reynolds numbers) in a long tubular reactor, the PFR model is generally assumed to be valid because turbulent flow may approximate plug flow without appreciable axial mass and heat transfer. However, when radial transport effects are significant, the use of the plug-flow assumption may lead to significant errors, based on the work of Churchill and co-workers, as discussed below in Section 7.5. At Reynolds numbers below 2,100, laminar flow persists and the PFR model is not valid because of the parabolic (nonplug) velocity profile. A partially mixed condition exists with a residence-time distribution for fluid elements. More rigorous models that account for fluid mechanics and transport are discussed in Section 7.5. Computational fluid dynamics (CFD), as discussed in Section 7S.6, may be used to solve such models.

For liquid-phase reactions, a single PFR- or CSTR-type reactor is often used. For a single reaction at isothermal conditions, the volume of a PFR is smaller than that of a CSTR for the same conversion and temperature. However, for (1) autocatalytic reactions, where the reaction rate depends on the concentration of a product, or (2) autothermal reactions, where the feed is cold, but the reaction is highly exothermic, the volume of a CSTR can be smaller than that of a PFR, such that axial dispersion in a tubular reactor may actually be beneficial. In general, a CSTR is not used for a gas-phase reaction because of the difficulty in obtaining perfect mixing in the gas.

For noncatalytic homogeneous reactions, a tubular reactor is widely used because it can handle liquid or vapor feeds, with or without phase change in the reactor. The PFR model is usually adequate for the tubular reactor if the flow is turbulent and if it can be assumed that when a phase change occurs in the reactor, the reaction takes place predominantly in one of the two phases. The simplest thermal modes are isothermal and adiabatic. The nonadiabatic, nonisothermal mode is generally handled by a specified temperature profile or by heat transfer to or from some specified heat source or sink and a corresponding heat-transfer area and overall heat-transfer coefficient. Either a fractional conversion of a limiting reactant or a reactor volume is specified. The calculations require the solution of ordinary differential equations.

For fixed-bed catalytic reactors, a PFR model with a pseudo-homogeneous kinetic equation is usually adequate and is referred to as a 1-D (one-dimensional) model. However, if the reactor is nonadiabatic with heat transfer to or from the wall, the PFR model is not usually adequate and a 2-D model, involving the solution of partial differential equations for variations in temperature and composition in both the axial and radial directions, is necessary. Simulators do not include 2-D models, but they can be generated by the user and inserted into the simulator.

Models for fluidized-bed catalytic reactors are the most complex and cannot be adequately modeled with either the CSTR or PFR models. Because some of the gas passing through the fluidized bed can bypass the suspended catalyst, the conversion in a fluidized bed can be less than that predicted by the CSTR model.

www.wiley.com/go/global/seider

The multimedia modules that accompany this book provide more complete coverage of the modeling of reactions and reactors using the process simulators. For ASPEN PLUS, follow the link: *ASPEN → Chemical Reactors → Overview*. For HYSYS, see: *HYSYS → Chemical Reactors → Overview*.

7.3 REACTOR DESIGN FOR COMPLEX CONFIGURATIONS

As discussed in Section 6.5, temperature control is an important consideration in reactor design. Adiabatic operation is always considered first because it provides the simplest and least-expensive reactor. However, when reactions are highly exothermic or endothermic, it is often desirable to exercise some control over the temperature. Methods for accomplishing this, as shown in Figure 7.1, include: (a) heat transfer to or from the reacting fluid, across a wall, to or from an external cooling or heating agent; (b) an inert or reactive heat carrier or diluent in the reacting fluid; (c) a series of reactor beds with a heat exchanger for cooling or heating between each pair of beds; and (d) cold-shot cooling (also called direct-contact quench) or hot-shot heating, where the combined feed is split into two or more parts, one of which enters at the reactor entrance while the remaining parts enter the reactor at other locations. The following are industrial examples of the application of these four methods. In considering these examples, a useful measure of the degree of exothermicity or endothermicity of a reaction is the adiabatic temperature rise (ATR) for complete reaction with reactants in the stoichiometric ratio.

An example of the industrial use of a heat-exchanger reactor in Figure 7.1a is in the manufacture of phthalic anhydride, produced by the oxidation of orthoxylene with air in the presence of vanadium pentoxide catalyst particles, as discussed by Rase (1977). The reaction, which is carried out at about 375°C and 1.2 atm, is highly exothermic with an ATR of about 1,170°C, even with nitrogen in the air providing some dilution. Adiabatic operation is not feasible. The reactor resembles a vertical shell-and-tube heat exchanger. Hundreds of long tubes of small diameter, inside the shell, are packed with catalyst particles and through which the reacting gas passes downward. A heat-transfer medium consisting of a sodium nitrite-potassium nitrate-fused salt circulates outside the tubes through the shell to remove the heat of reaction. Water is ruled out as a heat-transfer medium in this case because the required water pressure would be very high. The heat-transfer rate distribution is not adequate to maintain isothermal conditions for the reacting fluid, but its temperature changes by less than 40°C. In some applications, the arrangement involves catalyst-packed beds

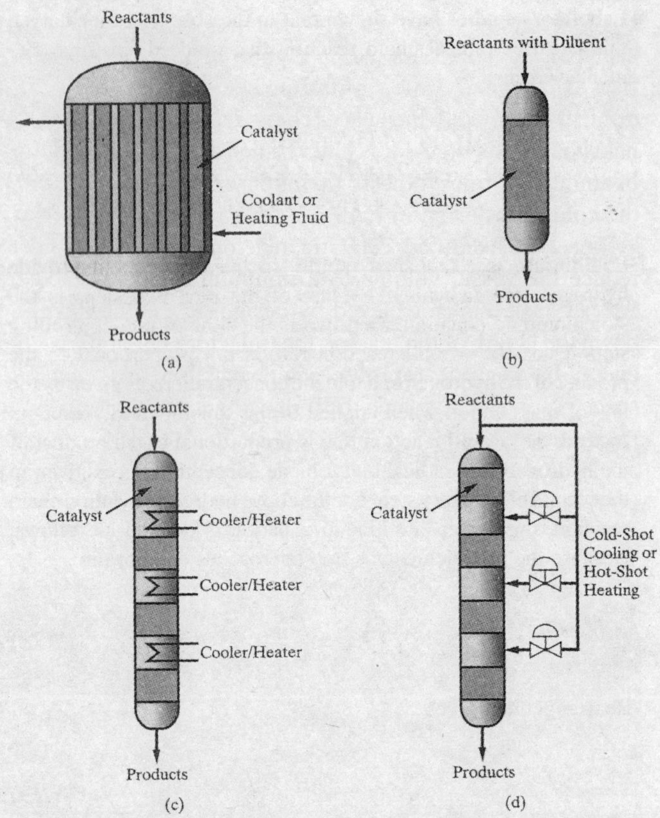

Figure 7.1 Reactors for handling large adiabatic temperature changes: (a) heat-exchanger reactor; (b) use of diluent; (c) external heat exchange; (d) hot/cold shot.

interspersed with tubes conveying a coolant (e.g., the TVA design in Figure 6.11).

Styrene is produced by the catalytic dehydrogenation of ethylbenzene at 1.2 atm and a temperature of about 575°C, as described by Smith (1981). The reaction is sufficiently endothermic, with an ATR of about −460°C, such that if the reactor were operated adiabatically with a feed of pure ethylbenzene, the temperature of the reacting fluid would decrease to such an extent that the reaction rate would be unduly compromised, resulting in a very large reactor volume. To maintain a reasonable temperature, a large amount of steam is added to the feed (typically with a molar ratio of steam to ethylbenzene equal to 20:1), which is preheated to 625°C before entering the reactor (Figure 7.1b). The steam is inert and is easily recovered from the reactor effluent by condensation. The presence of the steam reduces the reaction rate because the styrene concentration is reduced, but the reactor can be operated adiabatically in a simple manner.

Sulfur trioxide, which is used to make sulfuric acid, is produced by catalytic oxidation of sulfur dioxide in air with vanadium pentoxide catalyst particles at 1.2 atm and a temperature of about 450°C, as discussed by Rase (1977). Adiabatic operation is not feasible since the reaction is highly exothermic with an ATR of about 710°C, even with nitrogen in the air providing some dilution. Hence, the reactor system consists of four adiabatic reactor beds, of the same diameter but different height, in series, with a heat exchanger between each pair of beds, as shown in Figure 7.1c. The temperature rises adiabatically in each reactor bed, and the hot reactor effluent is cooled in the heat exchanger positioned before the next bed. When the ATR is higher, such as in the manufacture of ammonia from synthesis gas, as described by Rase (1977), the cold-shot design in Figure 7.1d is recommended.

For 1-D fixed-bed catalytic reactors, it is desirable to reduce the vessel volume to a minimum. As presented by Aris (1960), this is achieved by design. If z is the direction down the length of the reactor, the trajectory of the mass- and energy-balance equations for a single reaction in $X(z)$ and $T(z)$ space is adjusted to match the trajectory corresponding to the maximum reaction rate, (X^*, T^*) [curved line in Figure 7.2], as closely as possible. Thus, tube-cooled (or heated) reactors, cold-shot (or hot-shot) converters, and multiple adiabatic beds with intercoolers (or interheaters) need to be carefully designed in such a way that $(X(z), T(z)) \approx (X^*, T^*)$.

As an example, consider an exothermic reversible reaction in a PFR. For this case, the rate of the reverse reaction increases more rapidly with increasing temperature than does the rate of the forward reaction. Also, the reverse reaction is slow and the forward reaction fast at low temperatures. Thus, for a maximum rate of reaction, the temperature should be high at low conversions and low at high conversions. This is shown in Figure 7.2, taken from Smith (1981), where the reaction rate for a sequence of fractional conversions, X, starting with $X_1 = 0$, is plotted against temperature, T. For each value of X, the reaction rate curve in Figure 7.2 shows a maximum value. A locus of maximum rates is shown, corresponding to the solid and dashed line passing through the points C and B, with the maximum reaction rate decreasing with increasing fractional conversion. At each conversion level, the desired temperature corresponds to the maximum reaction rate. In Figure 7.2, the feed enters at temperature T_A, with a reaction rate at point A. Although the maximum reaction rate for X_1 is not shown, it is clear that T_A is not the temperature corresponding to the maximum rate. If the entering temperature cannot be increased, it is best to operate isothermally at T_A until the conversion at point C is reached, and then follow the optimal profile CB to the desired conversion. Suppose the reactor exit conversion is X_4. Then the desired reactor temperature trajectory is the solid line ACB, with reactor exit temperature T_B. Corresponding to this trajectory, but not shown in Figure 7.2, is a heat-duty profile, which must be matched by heat exchange to achieve the optimal reaction rate trajectory. Alternatively, isothermal operation of the reactor at T_A corresponds to the trajectory ACD in Figure 7.2. In this case, the reaction rates are not at their maximum values except at point C, requiring a larger reactor volume. If instead of a PFR, a CSTR were used, the optimal temperature of operation for achieving conversion X_4 would be T_B, which corresponds to the maximum reaction rate for that conversion. By specifying a temperature profile for a PFR or an exit temperature for a CSTR, the optimal reactor volume can be determined together with the required corresponding heat-duty profile.

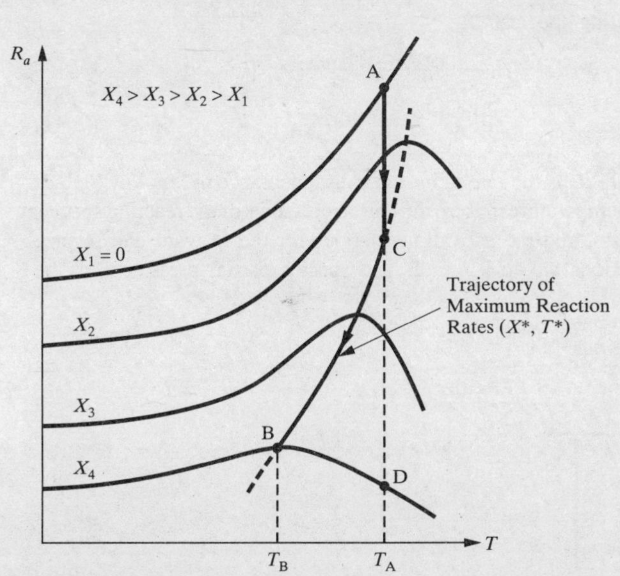

Figure 7.2 Temperature trajectories for an exothermic reversible reaction in a PFR (Smith, 1981).

As discussed by Van Heerden (1953), the reactor feed temperature has an important effect on the stability of an autothermal reactor, that is, a reactor whose feed is preheated by its effluent. For a reversible exothermic reaction, as in ammonia synthesis, the heat generation rate varies nonlinearly with the reaction temperature, as shown by curve (a) in Figure 7.3. At low temperatures, the rate of heat generation is limited by the low rate of the forward reaction to ammonia. At very high temperatures, the rate of reaction is limited by equilibrium, so that again, low heat generation rates are to be expected. However, at some intermediate temperature, the reaction rate exhibits a maximum value. In contrast, because heat transfer by convection is dominant, the rate of heat removal is almost linear with the reaction temperature, with a slope dependent on the degree of heat exchange between the outlet and the inlet. Thus, the intersection of the heat removal line (b) and the heat generation line (a) sometimes leads to three possible operating conditions: (O) the non-reacting state, (I) the ignition point, and (S) the desired operating point. Both the non-reacting and the desired operating points are stable, since a small positive

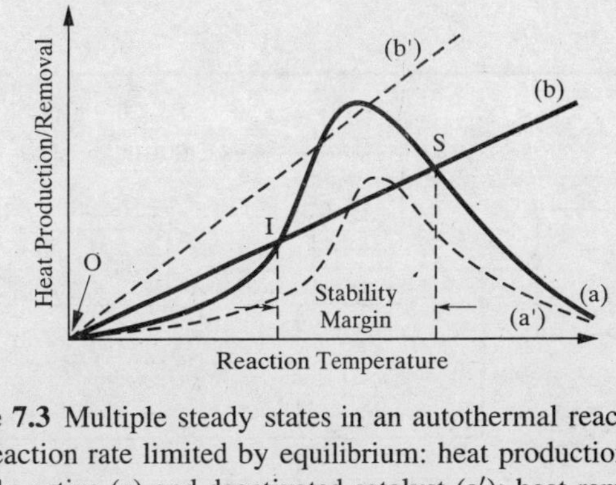

Figure 7.3 Multiple steady states in an autothermal reactor, with reaction rate limited by equilibrium: heat production rates for fully active (a) and deactivated catalyst (a′); heat removal rates for normal (b), and increased heat transfer (b′).

perturbation in the reactor temperature causes the heat removal rate to exceed the heat generation rate, decreasing the reactor temperature. Similarly, a small negative perturbation in the reactor temperature has the opposite effect, leading to a temperature rise. Using the same arguments, the ignition point is *unstable* because a small positive perturbation in temperature leads to a jump to the desired, stable operating point, whereas a negative perturbation leads to a so-called "blow-out," to the stable, non-reacting state. Note that a similar analysis for an adiabatic CSTR in Case Study 12S.1 also detects the possibility of three steady states.

For tube-cooled converters, Van Heerden (1953), and for cold-shot converters, Stephens and Richards (1973), refer to the temperature difference between operating points I and S as the "stability margin.' Clearly, operation at S with larger stability margins would be more robust to disturbances. Thus, a design with increased rate of heat transfer, indicated by the line (b′) in Figure 7.3, would clearly have a lower stability margin. In such cases, the proximity of the stable operating point to the unstable ignition point leads to an increased likelihood of loss of control in the face of process upsets. For example, ammonia synthesis catalyst undergoes deactivation, mainly by poisoning due to feed impurities, or by high temperature sintering, which reduces the catalyst surface area. Lewin and Lavie (1984) studied the effect of catalyst deactivation on the optimal operation of a tube-cooled ammonia converter, which can lead to loss of stability, since decreased catalyst activity leads to lower heat generation rates, as shown by line (a′) in Figure 7.3.

The following example illustrates how a cold-shot reactor is designed to maximize conversion while satisfying stability margins.

EXAMPLE 7.3 *Optimal Bypass Distribution in a Three-Bed, Cold-Shot Ammonia Synthesis Converter*

A reactor for the synthesis of ammonia consists of three adiabatic beds, shown in Figure 7.4. As summarized in Table 7.1, the reactor feed consists of two sources, the first of which is a make-up feed stream of 20,000 kmol/hr at 25°C and 150 atmospheres containing mainly hydrogen and nitrogen in the stoichiometric molar ratio of 3:1. Since ammonia synthesis gas is produced from naphtha and air, it contains small concentrations of methane from naphtha, and argon from air. Both of these species reduce the partial pressures of the reagents, and thus affect the reaction rate. The second feed, which contains larger concentrations of the inert components, is a recycle stream of 40,000 kmol/hr at 25°C and 150 atmospheres consisting of unreacted synthesis gas, recovered after removing the ammonia product. The converter consists of three cylindrical, 2-m-diameter adiabatic beds, packed with catalyst for bed lengths of 1.5 m, 2 m, and 2.5 m, respectively. The reactor feed is split into three branches, with the first branch becoming the main feed entering the first bed after being preheated by the hot reactor effluent from the third bed. The second and third branches, with flow fractions ϕ_1 and ϕ_2, respectively, are controlled by adjusting valves V-1 and V-2, and provide cold-shot cooling at the first and second bed effluents, respectively. It is desired to optimize the allocation of the bypass fractions to maximize the conversion in the converter.

Table 7.1 Ammonia Converter—Make-up Feed and Recycle Streams

		Make-up stream	Recycle stream
Flow rate (kmol/hr)		20,000	40,000
Temperature (°C)		25	
Pressure (atm)		150	
Compositions (mol %):	H_2	72	61
	N_2	24	20
	NH_3	0	1.5
	CH_4	3	13
	Ar	1	4.5

SOLUTION

Ammonia is synthesized in a reversible reaction, whose rate is correlated by the Tempkin equation (Tempkin and Pyzev, 1940), expressed in terms of the partial pressures, in atmospheres, of the reacting species:

$$R_a = 10^4 e^{-91,000/RT}[P_{N_2}]^{0.5}[P_{H_2}]^{1.5} - 1.3 \times 10^{10} e^{-140,000/RT}[P_{NH_3}] \quad (7.32)$$

where R_a is the rate of nitrogen disappearance in kmol/m^3–s, T is the temperature in K, P_i the partial pressures of the reacting species in atm, and the activation energies for the forward and reverse reactions are in kJ/kmol. The species partial pressures can be

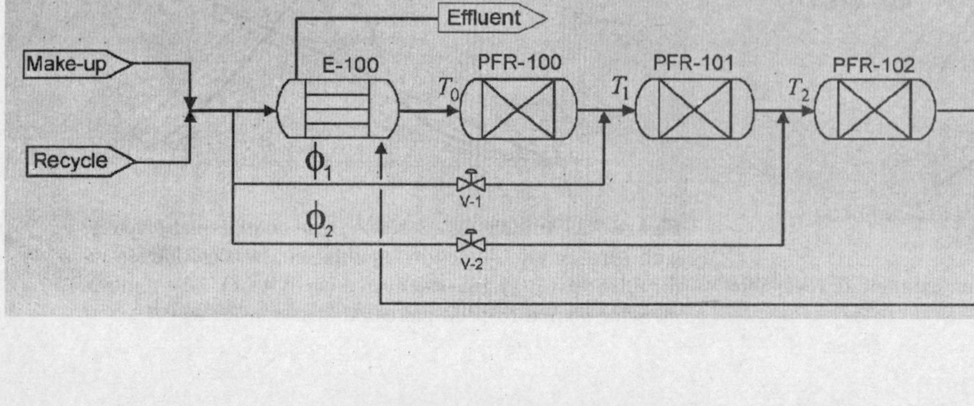

Figure 7.4 Cold-shot ammonia synthesis converter.

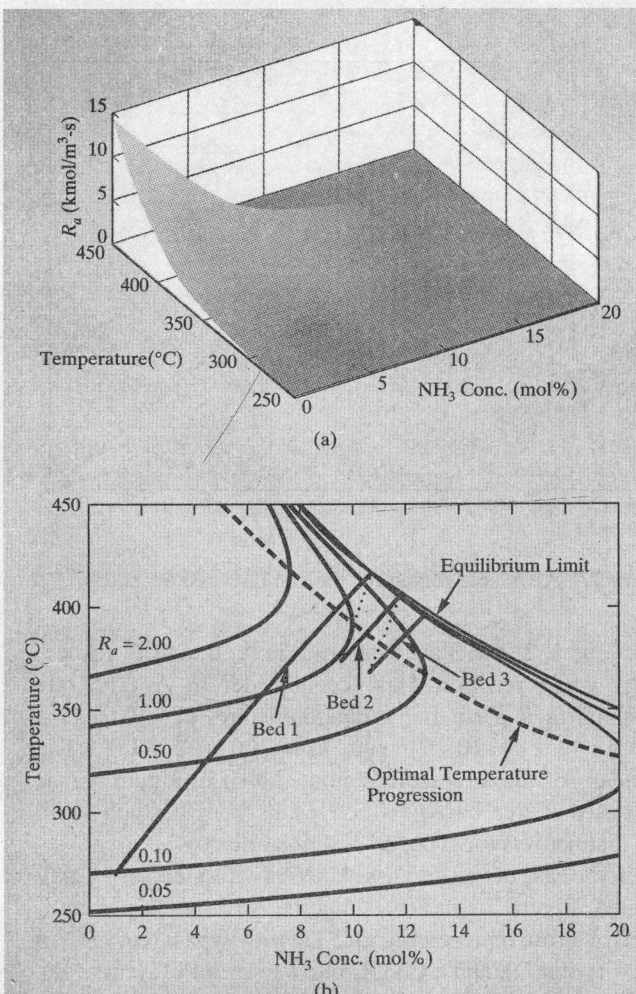

Figure 7.5 Composition–temperature space for ammonia synthesis converter: (a) reaction rate as a function of ammonia mole percent and temperature; (b) suboptimal cold-shot composition–temperature trajectory, plotted over reaction rate contours, with bypasses set to $\underline{\phi} = [0.1,\ 0.1]^T$.

expressed in terms of the ammonia mole fraction, x_{NH_3}, and the original feed composition:

$$P_{H_2} = \frac{F_o x_{f,H_2} - 1.5\xi}{F_o - \xi} P,\ P_{N_2} = \frac{F_o x_{f,N_2} - 0.5\xi}{F_o - \xi} P \text{ and}$$
$$P_{NH_3} = \frac{F_o x_{f,NH_3} + \xi}{F_o - \xi} P = x_{NH_3} P \tag{7.33}$$

where F_o is the total molar flow rate of the combined reactor feed, $\xi = F_o(x_{NH_3} - x_{f,NH_3})/(1 + x_{NH_3})$ is the molar conversion, P is the operating pressure, and $x_{f,i}$ is the feed mole fraction of species i. Consequently, the rate of reaction can be computed as a function of the temperature and x_{NH_3}, as shown in Figure 7.5a for an operating pressure of 150 atm. The ridge of maximum reaction rate in composition-temperature space defines an optimal decreasing temperature progression that is to be approximated by the appropriate design and operation of the converter.

The composition–temperature trajectory in the converter is plotted over contours of reaction rate in Figure 7.5b for suboptimal bypass fractions, $\underline{\phi} = [0.1,\ 0.1]^T$. Note that in the figure, the trajectories in the converter beds are solid lines, while those at the cold-shot mixing junctions are dotted lines. The temperature in Bed 1(PFR-100) rises to 415°C, close to the equilibrium limit. The first cold shot cools the gas to 370°C. In Bed 2 (PFR-101), the temperature rises to 405°C, again close to the equilibrium limit. Before entering Bed 3, the final cold shot cools the gas to 360°C. The ammonia effluent concentration from the last bed is 12.7 mol%. Figure 7.5b also includes a dashed line for the optimal temperature progression when, instead of cold-shot cooling, heat is continuously removed while the reaction proceeds in a single PFR. This line is the locus of maximum ammonia concentrations as a function of the reaction rate.

To maximize the conversion in the reactor, the following nonlinear program (NLP), of the type discussed in Chapter 24, is formulated:

$$\max_{\text{w.r.t. } \phi_1,\ \phi_2} \zeta \tag{7.34}$$

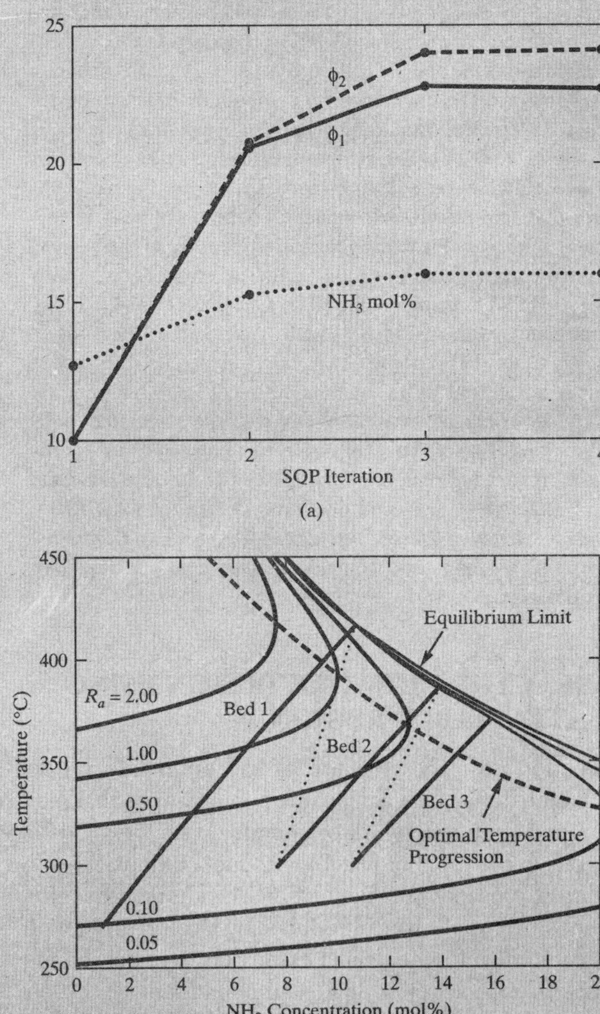

Figure 7.6 Optimal selection of bypass fractions for cold-shot ammonia converter: (a) convergence to the optimal solution; (b) optimal cold-shot profile in composition–temperature space with bypasses set to $\phi_2 = [0.227, 0.240]^T$.

Subject to (s. t.)

$$f(\underline{x}) = 0 \tag{7.35}$$

$$T_1 > 300°C \tag{7.36}$$

$$T_2 > 300°C \tag{7.37}$$

$$\phi_1 + \phi_2 \leq 0.6 \tag{7.38}$$

where Eq. (7.35) refers to the kinetics and material and energy balances for the converter in Figure 7.4, and Eqs. (7.36) and (7.37) define lower limits for the combined feed temperatures to the second and third beds. These minimum values are taken arbitrarily as 300°C, but are representative of minimum ignition temperatures. Note that in the first bed, where the rate of reaction, and with it the heat generation rate, is higher, the feed temperature is maintained constant at the lower value of 270°C, by appropriate design of the heat-exchanger, E-100. Finally, Eq. (7.38) ensures that a total maximum bypass of 60% is not exceeded, noting that this upper limit is arbitrary.

The NLP in Eqs. (7.34)–(7.38) is solved efficiently using successive quadratic programming (SQP), as described in Chapter 24. Figure 7.6a shows the optimal converged solution, which is obtained in four iterations. The final ammonia composition in the converter effluent is 15.9 mol%, obtained with optimal bypass fractions $\phi_1 = 0.227$ and $\phi_2 = 0.240$. The composition–temperature trajectories for the optimal bypass distribution, shown in Figure 7.6b, confirm that the overall performance of the three beds is significantly improved through increased utilization of the second and third beds. These results can be reproduced with HYSYS, using the file NH3_CONVERTOR_OPT.hsc, and with ASPEN PLUS, using the file NH3_CONVERTOR.OPT.bkp. For full details, the reader is referred to the multimedia modules, which can be downloaded from the Wiley web site associated with this book. This example is presented in multimedia tutorials under: *HYSYS → Tutorials → Reactor Design → Ammonia Converter Design*, and *ASPEN → Tutorials → Reactor Design Principles → Ammonia Converter Design*. Using simulators, complex reactor configurations are readily designed.

7.4 REACTOR NETWORK DESIGN USING THE ATTAINABLE REGION

This section describes the use of the attainable region (AR), which defines the achievable compositions that may be obtained from a network of chemical reactors. This is analogous to the topic of feasible product compositions in distillation, presented in Section 8.5. The attainable region in composition space was introduced by Horn (1964), with more recent developments and extensions by Glasser and co-workers (Glasser et al. 1987; Hildebrandt et al., 1990). Figure 7.7 illustrates the attainable region for van de Vusse kinetics (van de Vusse, 1964), based on the reactions:

$$\begin{aligned} &A \underset{k_2}{\overset{k_1}{\rightleftarrows}} B \xrightarrow{k_3} C \\ &2A \xrightarrow{k_4} D \end{aligned} \tag{7.39}$$

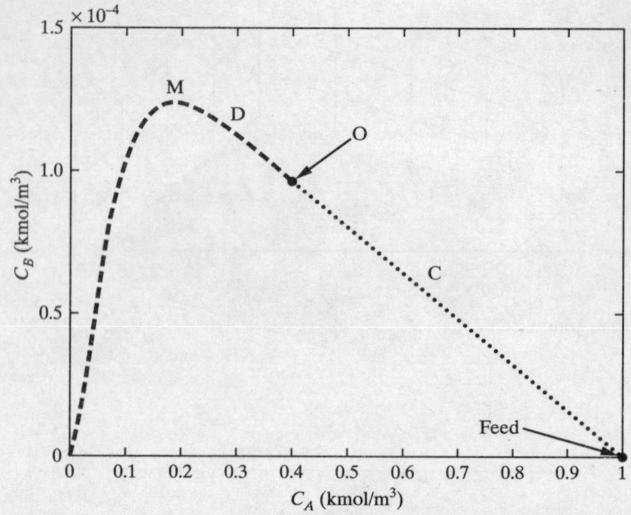

Figure 7.7 Attainable region for the van de Vusse reactions.

Reactions 1, 2, and 3 are first-order in A, B, and B, respectively, while reaction 4 is second-order in A. The rate constants at a particular temperature are: $k_1 = 0.01\,s^{-1}$, $k_2 = 5\,s^{-1}$, $k_3 = 10\,s^{-1}$, and $k_4 = 100\,m^3/kmol \bullet s$. The boundary of the attainable region, shown in Figure 7.7, is composed of arcs, each of which results from the application of a distinct reactor type, as described next.

For the case of van de Vusse kinetics with a feed of 1 kmol/m^3 of A, Figure 7.7 indicates that the AR boundary is composed of an arc representing a CSTR with bypass (curve C), a CSTR (point O), and a CSTR followed by a PFR (curve D). Within the region bounded by the three arcs and the horizontal base line ($C_B = 0$), product compositions can be achieved with some combination of these reactor configurations. The appropriate reactor configuration along the boundary of the attainable region depends on the desired effluent concentration of A. When $1 > C_A > 0.38$ kmol/m^3, a CSTR with bypass (curve C) provides the maximum concentration of B, while when $C_A < 0.38$ kmol/m^3, this is achieved using a CSTR (point O), followed by a PFR (Curve D). Note that the maximum achievable concentration, $C_B = 1.25 \times 10^{-4}$ kmol/m^3, is obtained using a CSTR followed by a PFR (at point M along curve D). Evidently, the attainable region provides helpful assistance in the design of optimal reactor networks. A procedure for the construction of attainable regions is discussed next.

Construction of the Attainable Region

A systematic method for the construction of the attainable region using CSTRs and PFRs, with or without mixing and bypass, for a system of chemical reactions, as presented by Hildebrandt and Biegler (1995), is demonstrated for van de Vusse kinetics:

Step 1: *Begin by constructing a trajectory for a PFR from the feed point, continuing to the complete conversion of A or chemical equilibrium.* In this case, the

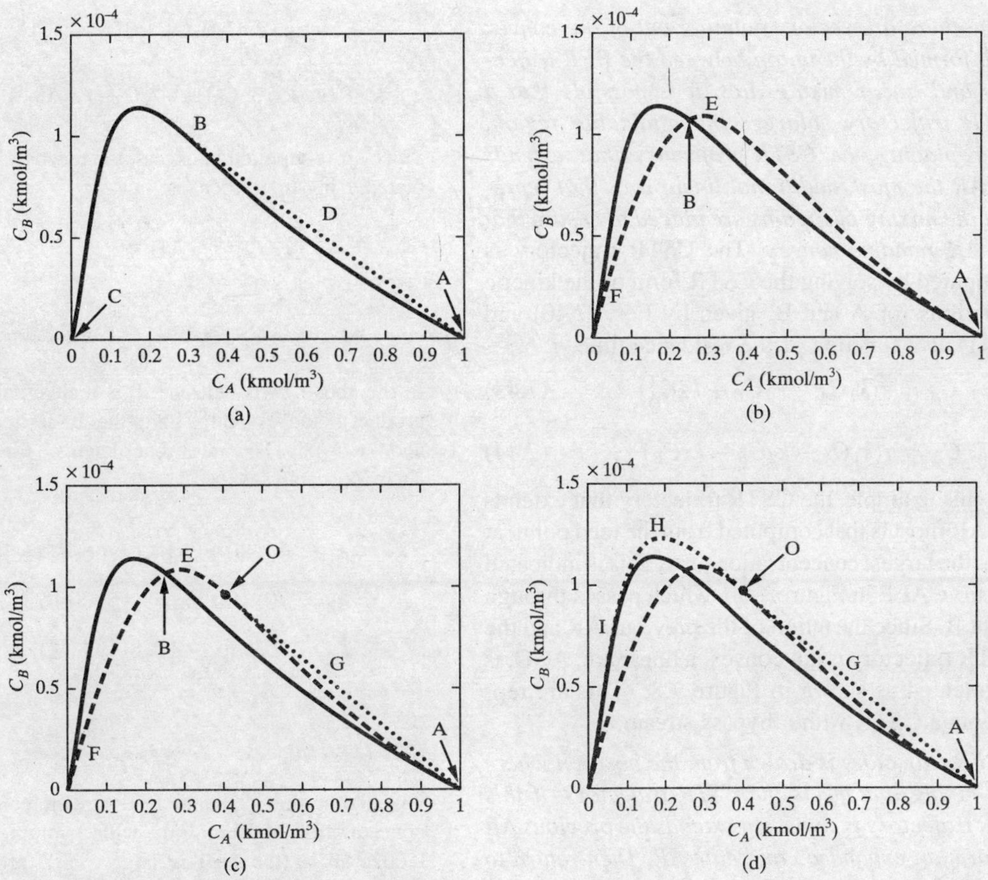

Figure 7.8 Construction of the attainable region for the van de Vusse reaction: (a) PFR trajectory from $\underline{C}(0) = [1, 0]^T$ (solid line), with mixing line (dotted line); (b) CSTR trajectory from $\underline{C}(0) = [1, 0]^T$ (dashed line); (c) addition of bypass to CSTR (dotted line); (d) addition of PFR in series with CSTR (dot-dashed line).

PFR trajectory is computed by solving simultaneously the kinetic equations for A and B:

$$\frac{dC_A}{d\tau} = -k_1 C_A + k_2 C_B - k_4 C_A^2 \tag{7.40}$$

$$\frac{dC_B}{d\tau} = k_1 C_A - k_2 C_B - k_3 C_B \tag{7.41}$$

where τ is the PFR residence time. Note that kinetic equations for C and D are not required for the construction of the attainable region in two-dimensional space because their compositions do not appear in Eqs. (7.40) and (7.41). The trajectory in $C_A - C_B$ space is plotted in Figure 7.8a as curve ABC. In this example, component A is completely converted.

Step 2: *When the PFR trajectory bounds a convex region, this constitutes a candidate attainable region. A convex region is one in which all straight lines drawn from one point on the boundary to any other point on the boundary lie wholly within the region or on the boundary. If not, the region is nonconvex. When the rate vectors,* $[dC_A/d\tau,\ dC_B/d\tau]^T$, *at concentrations outside of the candidate AR do not point back into it, the current limits are the boundary of the AR and the procedure terminates.* In this example, as seen in Figure 7.8a, the PFR trajectory is not convex from A to B, so proceed to the next step to determine if the attainable region can be extended beyond the curve ABC.

Step 3: *The PFR trajectory is expanded by linear arcs, representing mixing between the PFR effluent and the feed stream, extending the candidate attainable region. Note that a linear arc connecting two points on a composition trajectory is expressed by the equation:*

$$\underline{c}^* = \alpha \underline{c}_1 + (1 - \alpha)\underline{c}_2 \tag{7.42}$$

where $\underline{c}_1$ *and* $\underline{c}_2$ *are vectors for two streams in the composition space,* $\underline{c}^*$ *is the composition of the mixed stream, and* α *is the fraction of the stream with composition* $\underline{c}_1$ *in the mixed stream. The linear arcs are then tested to ensure that no rate vectors positioned on them point out of the AR. If there are such vectors, proceed to the next step, or return to step 2.* As shown in Figure 7.8a, a linear arc, ADB, is added, extending the attainable region to ADBC. Since for this example, rate vectors computed along this arc are found to point out of the extended AR, proceed to the next step.

Step 4: *Since there are vectors pointing out of the convex hull, formed by the union between the PFR trajectory and linear mixing arcs, it is possible that a CSTR trajectory enlarges the attainable region. After placing the CSTR trajectory that extends the AR the most, additional linear arcs that represent the mixing of streams are placed to ensure that the AR remains convex.* The CSTR trajectory is computed by solving the CSTR form of the kinetic equations for A and B, given by Eqs. (7.40) and (7.41) as a function of the residence time, τ:

$$C_{A0} - C_A = \tau(k_1 C_A - k_2 C_B + k_4 C_A^2) \tag{7.43}$$

$$C_B = \tau(k_1 C_A - k_2 C_B - k_3 C_B) \tag{7.44}$$

For this example, the CSTR trajectory that extends the AR most is that computed from the feed point, at C_{A0}, the largest concentration of A. This is indicated as curve AEF in Figure 7.8b, which passes through point B. Since the union of the previous AR and the CSTR trajectory is not convex, a linear arc, AGO, is augmented as shown in Figure 7.8c. This arc represents a CSTR with a bypass stream.

Step 5: *A PFR trajectory is drawn from the position where the mixing line meets the CSTR trajectory. If this PFR trajectory is convex, it extends the previous AR to form an expanded candidate AR. Then return to Step 2. Otherwise, repeat the procedure from Step 3.* As shown in Figure 7.8d, the PFR trajectory, OHI, leads to a convex attainable region. The boundaries of the region are: (a) the linear arc, AGO, which represents a CSTR with bypass stream; (b) the point O, which represents a CSTR; and (c) the arc OHI, which represents a CSTR followed by a PFR in series. It is noted that the maximum composition of B is obtained at point H, using a CSTR followed by a PFR.

Clearly, the optimal reactor design minimizes the annualized cost, computed to account for the capital and operating costs, and not simply the design that maximizes the yield or selectivity. Nonetheless, the maximum attainable region identifies the entire space of feasible concentrations. The following example shows how the attainable region is used to select the most appropriate reactor network to maximize the yield of a desired product where a number of competing reactions occur.

EXAMPLE 7.4 ***Reaction Network Synthesis for the Manufacture of Maleic Anhydride***

Maleic anhydride, $C_4H_2O_3$, is manufactured by the oxidation of benzene with excess air over vanadium pentoxide catalyst (Westerlink and Westerterp, 1988). The following reactions occur:

$$\text{Reaction 1: } C_6H_6 + \tfrac{9}{2}O_2 \rightarrow C_4H_2O_3 + 2CO_2 + 2H_2O \tag{7.45}$$

$$\text{Reaction 2: } C_4H_2O_3 + 3O_2 \rightarrow 4CO_2 + H_2O \tag{7.46}$$

$$\text{Reaction 3: } C_6H_6 + \tfrac{15}{2}O_2 \rightarrow 6CO_2 + 3H_2O \tag{7.47}$$

Since air is supplied in excess, the reaction kinetics are approximated using first-order rate laws:

$$A \xrightarrow{r_1} P \xrightarrow{r_2} B, \qquad A \xrightarrow{r_3} C$$

$$r_1 = k_1 C_A,\ r_2 = k_2 C_p,\ \text{and } r_3 = k_3 C_A \tag{7.48}$$

In the above, A is benzene, P is maleic anhydride (the desired product), and B and C are the undesired byproducts (H_2O and CO_2). The rate coefficients for Eqs. (7.48) are (in m^3/kg catalyst • s):

$$\left.\begin{aligned} k_1 &= 4,280 \exp[-12,660/T(\text{K})] \\ k_2 &= 70,100 \exp[-15,000/T(\text{K})] \\ k_3 &= 26 \exp[-10,800/T(\text{K})] \end{aligned}\right\} \tag{7.49}$$

SOLUTION

Given that the available feed stream contains benzene at a concentration of 10 mol/m^3, with a volumetric flow rate, v_0, of 0.0025 m^3/s (the feed is largely air), propose a network of isothermal reactors to maximize the yield of maleic anhydride.

First, an appropriate reaction temperature is selected. Following Heuristic 7 in Chapter 6, Figure 7.9 shows the effect of temperature on the three rate coefficients, and indicates that in the range $366 < T < 850$ K, the rate coefficient, k_1, of the desired reaction to MA is larger than those of the competing reactions. An operating temperature at the upper end of this range is recommended, as the rate of reaction increases with temperature.

Since all of the reaction rate expressions involve only benzene and maleic anhydride, the system can be expressed in a two-dimensional composition space. For this system, the attainable

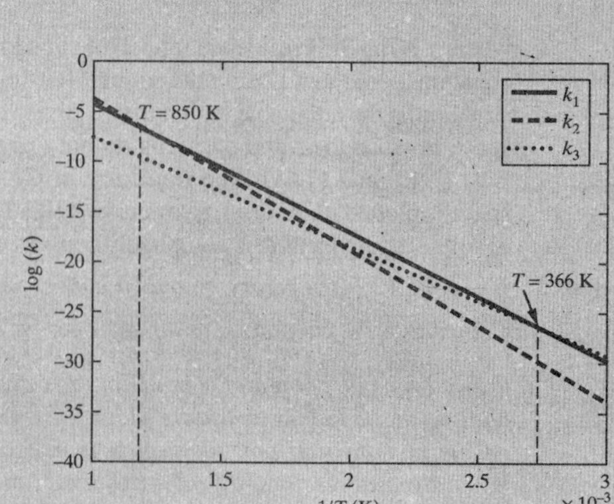

Figure 7.9 Influence of temperature on rate constants for MA manufacture.

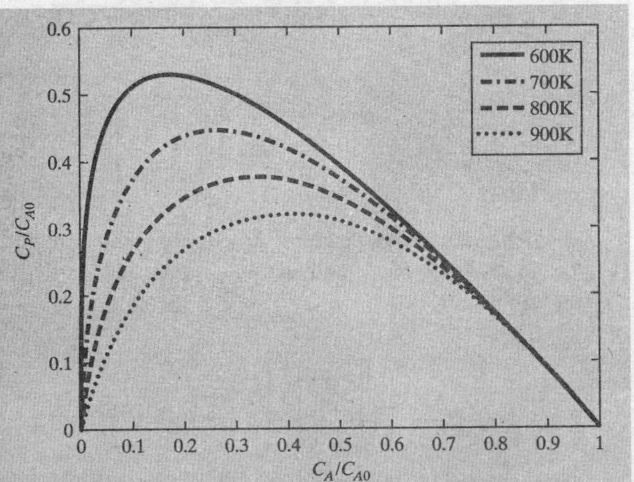

Figure 7.10 Attainable regions for MA manufacture at various temperatures.

region is straight forward to construct. This begins by tracing the composition space trajectory for a packed-bed reactor (PBR), modeled as a PFR, which depends on the solution of the molar balances:

$$v_0 \frac{dC_A}{dW} = -k_1 C_A - k_3 C_A, \quad C_A(0) = C_{A0} = 10\,\text{mol/m}^3 \tag{7.50}$$

$$v_0 \frac{dC_P}{dW} = k_1 C_A - k_2 C_P, \quad C_P(0) = 0\,\text{mol/m}^3 \tag{7.51}$$

where W is the kg of catalyst. In the above equations, the temperature-dependent rate constants are computed using Eq. (7.49). Figure 7.10 presents solutions of these equations as trajectories in $C_A - C_P$ space for several operating temperatures. Since these trajectories are convex, and rate vectors computed along their boundaries are tangent to them, it is concluded that each trajectory bounds the attainable region for its corresponding temperature.

Evidently, a single plug-flow reactor (or packed-bed reactor in this case) provides the maximum production of maleic anhydride, with the required space velocity being that which brings the value of C_P to its maximum in Figure 7.10. At 800 K, it is determined that the maximum concentration of maleic anhydride is 3.8 mol/m^3, requiring 4.5 kg of catalyst. At 600 K it is 5.3 mol/m^3, but at this low temperature, 1,400 kg of catalyst is needed. A good compromise is to operate the PBR at an intermediate temperature, for example, 770 K, with a maximum concentration of maleic anhydride of 4.0 mol/m^3, requiring 8 kg of catalyst.

Figure 7.11a shows composition profiles for all species as a function of bed length (proportional to the weight of catalyst), indicating that the optimal catalyst loading, where the concentration of maleic anhydride is a maximum, is about 8 kg for isothermal operation at 770 K. Figure 7.11b indicates that the yield (the ratio of the desired product rate and feed rate) under these conditions is 61%, while the selectivity is only about 10%. The selectivity (the ratio of the desired product and total products) for this reaction system is poor, due to the large amounts of CO_2 and H_2O produced, with the highest selectivity, achieved by repressing both of the undesired reactions, at 22%.

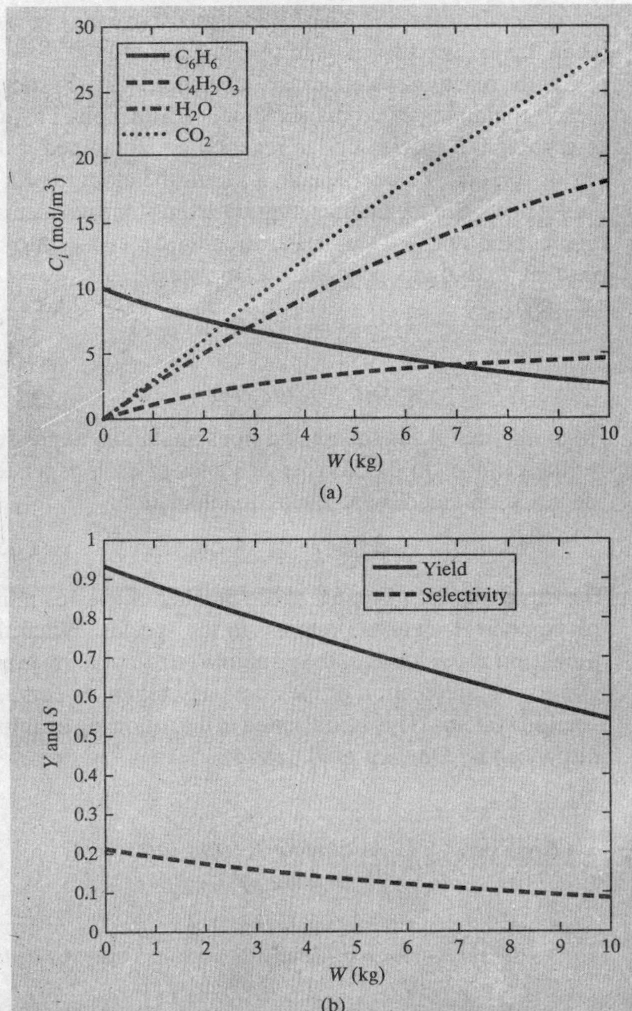

Figure 7.11 Composition profiles for MA manufacture in an isothermal PBR at 770 K: (a) composition profiles; (b) selectivity and yield.

Thus far, the attainable region has been shown for the analysis of systems with two key compositions to be tracked. In the following, the principle of reaction invariants is used to reduce the composition space in systems of larger dimension.

The Principle of Reaction Invariants

Because the attainable region depends on geometric constructions, it is effectively limited to the analysis of systems involving two independent species. However, as shown by Omtveit et al. (1994), systems involving higher dimensions can be analyzed using the two-dimensional AR approach by applying the principle of reaction invariants of Fjeld et al. (1974). The basic idea consists of imposing atom balances on the reacting species. These additional linear constraints impose a relationship between the reacting species, permitting the complete system to be projected onto a reduced space of independent species. The AR analysis above may be used when this reduced space is in two dimensions.

Let the reacting system consist of n_i moles of each species i, each containing a_{ij} atoms of element j. The molar changes in each of the species due to reaction are combined in the vector $\Delta\underline{n}$, and the coefficients a_{ij} form the atom matrix $\underline{\underline{A}}$, noting that since the number of gram-atoms for each element remain constant, $\underline{\underline{A}}\,\Delta\underline{n} = 0$. Partitioning $\Delta\underline{n}$ and $\underline{\underline{A}}$ into dependent, d, and independent, i, components:

$$\underline{\underline{A}} = [\underline{\underline{A}}_d \,|\, \underline{\underline{A}}_i] \tag{7.52}$$

$$\Delta\underline{n}^T = [\Delta\underline{n}_d^T \,|\, \Delta\underline{n}_i^T] \tag{7.53}$$

Assuming that $\underline{\underline{A}}_d$ is square and nonsingular, an expression for the changes in the number of moles of each dependent species is obtained by algebraic manipulation:

$$\Delta\underline{n}_d = -\underline{\underline{A}}_d^{-1}\underline{\underline{A}}_i\Delta\underline{n}_i \tag{7.54}$$

The dimension of i is equal to the number of species minus the number of elements (atoms) in the species. When this dimension is two or less, the principle of reaction invariants permits the application of the attainable region to complex reaction systems. This is illustrated in the following example, introduced by Omtveit et al. (1994).

EXAMPLE 7.5 ***Attainable Region for Steam Reforming of Methane***

Construct the attainable region for the steam reforming of methane at 1,050 K, and use it to identify the networks that provide for the maximum composition and selectivity of CO.

SOLUTION

The following reactions, involving five species and three elements, dominate in the steam reforming of methane:

$$CH_4 + 2H_2O \rightleftharpoons CO_2 + 4H_2 \tag{7.55}$$

$$CH_4 + H_2O \rightleftharpoons CO + 3H_2 \tag{7.56}$$

$$CO + H_2O \rightleftharpoons CO_2 + H_2 \tag{7.57}$$

By evoking the principle of reaction invariants, the number of species that need to be tracked for this system is reduced to two so that the attainable region can be shown in two dimensions. Accordingly, the vector of molar changes is

$$\begin{aligned}\Delta\underline{n}^T &= [\Delta\underline{n}_d^T \,|\, \Delta\underline{n}_i^T] \\ &= [\Delta n_{H_2}, \Delta n_{H_2O}, \Delta n_{CO_2} \,|\, \Delta n_{CH_4}, \Delta n_{CO}]^T \end{aligned} \tag{7.58}$$

where methane and carbon monoxide have been selected as the independent components. The atom balances for the three elements, C, H, and O, are

$$\begin{aligned} \text{C balance}: &\quad \Delta n_{CO_2} + \Delta n_{CH_4} + \Delta n_{CO} = 0 \\ \text{H balance}: &\quad 2\Delta n_{H_2} + 2\Delta n_{H_2O} + 4\Delta n_{CH_4} = 0 \\ \text{O balance}: &\quad \Delta n_{H_2O} + 2\Delta n_{CO_2} + \Delta n_{CO} = 0 \end{aligned}$$

The atom matrix $\underline{\underline{A}}$ with rows corresponding to C, H, and O, respectively, is

$$\underline{\underline{A}} = [\underline{\underline{A}}_d \,|\, \underline{\underline{A}}_i] = \left[\begin{array}{ccc|cc} 0 & 0 & 1 & 1 & 1 \\ 2 & 2 & 0 & 4 & 0 \\ 0 & 1 & 2 & 0 & 1 \end{array}\right] \tag{7.59}$$

The dependent molar changes, $\Delta\underline{n}_d$, are expressed in terms of the molar changes in methane and carbon monoxide, using Eq. (7.54):

$$\begin{aligned}\Delta\underline{n}_d &= \begin{bmatrix} \Delta n_{H_2} \\ \Delta n_{H_2O} \\ \Delta n_{CO_2} \end{bmatrix} = -\underline{\underline{A}}_d^{-1}\underline{\underline{A}}_i\Delta\underline{n}_i \\ &= \begin{bmatrix} -4 & -1 \\ 2 & 1 \\ -1 & -1 \end{bmatrix} \begin{bmatrix} \Delta n_{CH_4} \\ \Delta n_{CO} \end{bmatrix} \end{aligned} \tag{7.60}$$

For example, if $\Delta n_{CH_4} = -5$ mol and $\Delta n_{CO} = 3$ mol, Eq. (7.60) gives $\Delta n_{H_2} = 17$ mol, $\Delta n_{H_2O} = -7$ mol, and $\Delta n_{CO_2} = 2$ mol. The feed would have to contain more than 5 moles of methane and 7 moles of water.

Xu and Froment (1989) provide the kinetic expressions for the reversible reactions in Eqs. (7.55)–(7.57) in terms of the partial pressures of the participating species. Noting that the number of moles increases by two in each of the reactions in which methane is consumed, the total number of moles in the system is given by

$$n_T = [n_{H_2} + n_{H_2O} + n_{CO_2} + n_{CH_4} + n_{CO}]_0 - 2\Delta n_{CH_4} \tag{7.61}$$

The partial pressure of each of the five species is expressed as: $P n_i/n_T$, for i, where the number of moles of the dependent species, H_2, H_2O, and CO_2, are expressed in terms of the number of moles of CH_4 and CO using Eq. (7.60). This allows the construction of the attainable region for the steam reforming reactions at 1,050 K, which was computed by Omtveit et al. (1994) as follows:

Step 1: *Begin by constructing a trajectory for a PFR from the feed point, continuing to the complete conversion of methane or chemical equilibrium.* Here, the PFR trajectory is computed by solving the kinetic equations for the reactions of Eqs. (7.55)–(7.57) to give the mole numbers of CH_4 and CO. This leads to trajectory (1) in Figure 7.12, which tracks the compositions from the feed point, A, to chemical equilibrium at point B.

Step 2: *When the PFR trajectory bounds a convex region, this constitutes a candidate attainable region. When the rate vectors at concentrations outside of the candidate AR do not point back into it, the current limits are the boundary of the AR and the procedure terminates.* In Figure 7.12, the PFR trajectory is not convex, so proceed to the next step.

Step 3: *The PFR trajectory is expanded by linear arcs representing mixing between the PFR effluent and the feed stream, extending the candidate attainable region.*

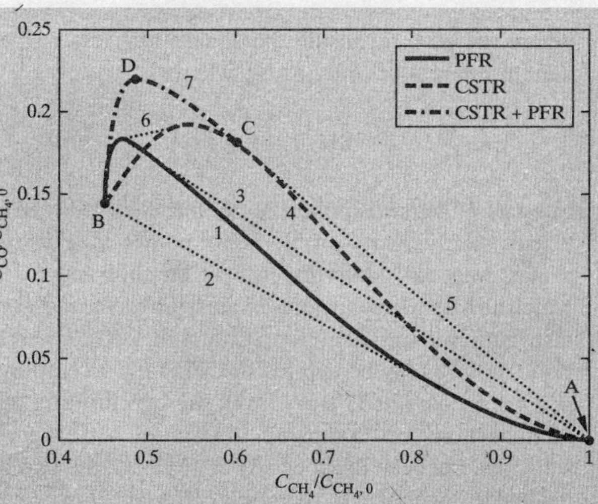

Figure 7.12 Development of the attainable region for steam reforming reactions at $T = 1{,}050$ K.

Here, two linear arcs are introduced to form a convex hull, tangent to the PFR trajectory from below, connecting to the chemical equilibrium point B (line 2), and from the feed point to a point tangent to the PFR trajectory from above (line 3). In this example, line 2 constitutes the lower boundary of the attainable region. It is found that rate trajectories point out of the convex hull, so proceed to the next step.

Step 4: *Since there are vectors pointing out of the convex hull formed by the union between the PFR trajectory and linear mixing arcs, a CSTR trajectory may enlarge the attainable region. After placing the CSTR trajectory that extends the AR the most, additional linear arcs that represent the mixing of streams are placed to ensure that the AR remains convex.* Here, the CSTR trajectory is computed by solving the molar balances for CH_4 and CO as the residence time, τ, varies. This gives trajectory (4), augmented by two linear arcs, connecting the feed point to a point tangent to the CSTR trajectory (line 5) at point C, and an additional line (6) connecting the CSTR to the PFR trajectories at two tangent points. This forms the new candidate attainable region on which trajectories are identified that point outward.

Step 5: *A PFR trajectory is drawn from the position where the mixing line meets the CSTR trajectory. When this PFR trajectory is convex, it extends the previous AR to form an expanded candidate AR. Return to Step 2. Otherwise, repeat the procedure from Step 3.* As shown in Figure 7.12, the PFR trajectory (line 7) leads to a convex attainable region. The boundaries of the region are: (a) the linear arc (line 5) from A to C, which represents a CSTR with a bypass stream; (b) the point C, which represents a CSTR; and line 7 from C to B, which represents a CSTR followed by a PFR in series. Note that the maximum composition of CO is obtained at point D, using a CSTR and PFR in series. The maximum selectivity, defined by the ratio of CO/CH_4, is also achieved at point D, where the ratio is 0.47, as compared to point C, where the ratio is only 0.30.

7.5 RIGOROUS MODELS FOR TUBULAR CHEMICAL REACTORS

In the previous sections of this chapter, ideal reactor models (PFR and CSTR) were developed and applied to simple and complex configurations, and to networks. These ideal models, which represent the two extremes of no axial mixing and complete mixing, are widely available in computer-aided simulation programs, and only require thermodynamic and reaction-kinetics data. Continuous stirred-tank reactors, which are used often for liquid-phase reactions, may be confidently simulated with the ideal CSTR model if adequate agitation of the contents of the vessel is provided. However, for application to tubular-flow reactors, the ideal PFR model—which ignores axial transport of heat and mass and assumes no gradients of velocity, temperature, or composition in the radial direction—can sometimes, as shown in this section, be in serious error. With the advent of easily applied numerical methods and/or the use of CFD (computational fluid dynamics) programs, it is now possible, as discussed in this section, to routinely account for radial gradients in tubular-flow reactors. This section deals with an introduction to rigorous solutions for tubular-flow reactors, particularly those operating under laminar-flow conditions. In the final section of this chapter, Section 7S.6, a detailed reactor design is presented using a CFD model with the COMSOL program.

Isothermal Conditions

The simplest case of an isothermal, tubular-flow reactor that does not conform to the PFR model is a laminar-flow reactor as presented by Cleland and Wilhelm (1956). They assumed a first-order, irreversible chemical reaction occurring under isothermal, continuous, steady-state, constant density (no volume change on reaction), with fully developed laminar-flow conditions in a straight, circular tube of constant diameter. They neglected free (natural)-convection effects and axial diffusion of species, but took into account the parabolic velocity distribution and the radial diffusion of species. The radial velocity distribution causes a radial distribution of residence times, which in turn causes a radial concentration gradient. However, the concentration gradient is diminished by radial molecular diffusion. Thus, if radial diffusion is rapid, the assumption of plug flow is approached; if slow, however, the effect on the length of reactor required, compared to that for plug flow, can be very significant, as will be shown.

Under the assumptions of Cleland and Wilhelm, the governing equation, a parabolic partial differential equation, is a single differential mass balance, at a point in the reactor, for the species controlling the rate of reaction. Thus,

$$2v_{avg}\left[1-\left(\frac{r}{R}\right)^2\right]\frac{\partial c_i}{\partial z} = D_i\left(\frac{\partial^2 c_i}{\partial r^2}+\frac{1}{r}\frac{\partial c_i}{\partial r}\right) - kc_i \quad (7.62)$$

where v_{avg} is the average flow velocity, r is the radial distance from the tube axis, R is the inside radius of the tube, c_i is the molar concentration of the rate-controlling species, z is the axial distance from the reactor inlet, D_i is the molecular diffusivity of the rate-controlling species, and k is the first-order reaction-rate constant. If the entering concentration is c_{i_0}, the boundary conditions for Eq. (7.62) are $c_i = c_{i_0}$ at $z = 0$ and $\partial c_i/\partial r = 0$ at both the wall ($r = R$) and the centerline ($r = 0$). Eq. (7.62) cannot be solved analytically, except at the two limits of the ratio of molecular diffusivity to reaction-rate constant, expressed as $D_i/kR^2 = \alpha$. When α approaches infinity (rapid diffusion compared to reaction rate and/or small-diameter tube), the plug-flow assumption is approached because the concentration is uniform over the cross section of the tube at each axial position. Then, Eq. (7.62) with its boundary conditions can be integrated to give,

$$c_i = c_{i_0} e^{-kz/v_{avg}} = c_{i_0} e^{-k\tau_{avg}} \tag{7.63}$$

where τ_{avg} is the average residence time after traveling the distance z. Note that Eq. (7.63) is identical to that for an isothermal, constant-density batch or plug-flow reactor, with first-order irreversible kinetics, and is applicable to both laminar- and turbulent-flow conditions.

When α approaches zero (slow diffusion compared to reaction rate and/or a large-diameter tube), a radial concentration gradient forms because of the distribution of residence times of the laminar-flow streamlines over the cross section of the tube. Integration of Eq. (7.62) gives a more complicated result, where the concentration depends on r and z:

$$c_i = c_{i_0}\left[2\int_1^{\infty} \frac{e^{-\frac{kz}{2v_{avg}}Y}}{Y^3}\,dY\right] = c_{i_0}\left[2\int_1^{\infty} \frac{e^{-\frac{k\tau_{avg}}{2}Y}}{Y^3}\,dY\right]$$
$$= 2c_{i_0}E_3(k\tau_{avg}/2) \tag{7.64}$$

where the integral inside the square brackets is the exponential integral of order 3, E_3, where

$$Y = 1/\left[1 - \left(\frac{r}{R}\right)^2\right]$$

Eq. (7.64) is not applicable to turbulent-flow conditions because the velocity profile is not parabolic. For that case, a mathematical model given by Churchill and Yu (2006), which utilizes a very accurate expression for the radial velocity distribution in turbulent flow (Churchill, 2001; Churchill and Zajic, 2002), may be applied or CFD may be employed.

Using Eqs. (7.63) and (7.64), the two limiting cases for laminar flow are compared in Figure 7.13, where the i subscript has been dropped. At first glance, the two limiting results do not appear to be widely different. However, in fact, as the fractional conversion $[1 - (c_i/c_{i_0})]$ approaches 1 (i.e., as c/c_0, the fraction unconverted, approaches 0 in Figure 7.13), the difference between the two results becomes very significant when the required average residence time or its equivalence, length of the reactor, is considered. This is shown in Figure 7.14, where the ratio of reactor length for zero radial diffusion-to-reactor length for infinite radial diffusion (plug flow) is plotted against the fraction unreacted on a logarithmic scale. Included are the results of Cleland and Wilhelm for a first-order reaction ($n = 1$), together with results from Bosworth (1948) for zero-order ($n = 0$), and Denbigh (1951) for second-order, reactions ($n = 2$). Consider a fractional conversion of 0.99 (i.e., 0.01 or 1% unreacted). Figure 7.14 gives limiting ratios of reactor lengths of 1.82, 1.53, and 1.34 for zero-, first-, and second-order reactions, respectively. Thus, the effect of a lack of radial diffusion in laminar flow can be substantial.

Actual laminar-flow reactors will operate between the two extremes of Figure 7.13. For particular conditions between the extremes, numerical solutions of Eq. (7.62), which are readily obtained by standard finite-difference, finite element,

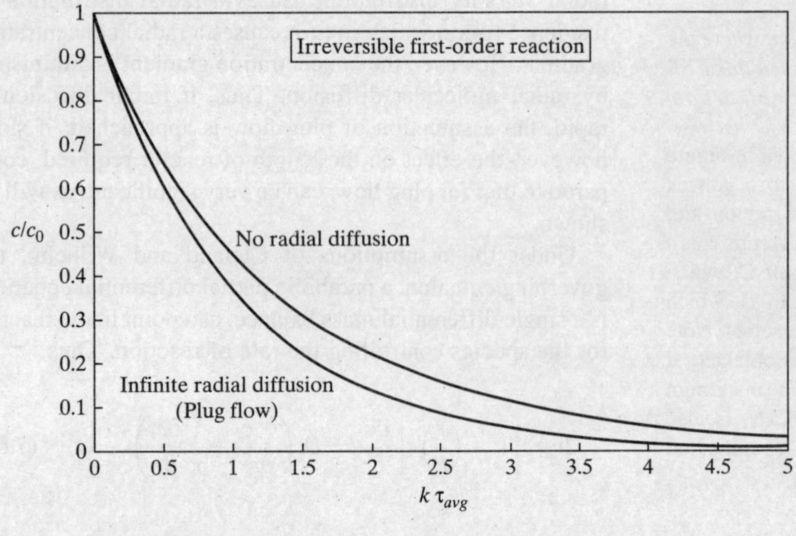

Figure 7.13 Limits of operation of an isothermal, tubular, laminar-flow reactor.

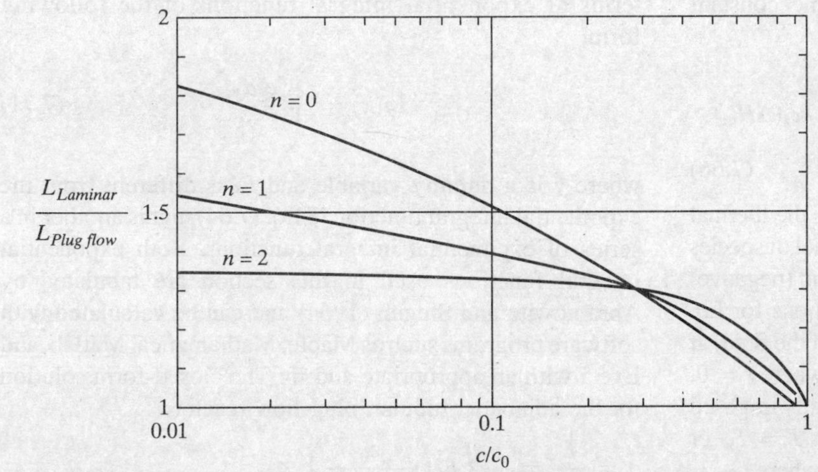

Figure 7.14 Effect of reaction order on the ratio of tubular reactor length for isothermal laminar flow with no radial diffusion to plug flow.

collocation, or CFD procedures for solving partial differential equations, can be employed. Cleland and Wilhelm used an implicit finite-difference method, which is discussed in detail by Lapidus (1962).

EXAMPLE 7.6 ***Suitability of a Laminar-Flow Reactor***

As discussed by Cleland and Wilhelm, liquid-phase reactions are more likely candidates for laminar-flow reactors than gas-phase reactions. Consider the liquid-phase hydrolysis of a dilute aqueous solution of acetic anhydride, a reaction considered by Cleland and Wilhelm (1956). At dilute conditions, the reaction can be conducted almost isothermally. At a temperature of 35°C, the reaction under dilute conditions is effectively first-order in acetic anhydride with a rate constant of 0.273 min^{-1}. The diffusivity of acetic acid in water at 35°C is estimated to be 1.23×10^{-5} cm^2/s. The viscosity and density of the solution can be approximated as that of water, or 0.72 g/cm-s and 1.0 g/cm^3, respectively. Assume that a 98% conversion of acetic anhydride is desired, corresponding to $c_i/c_{i_0} = 0.02$. For a tubular plug-flow reactor, using Eq. (7.63), the average residence time required is:

$$\tau_{avg} = \ln(c_i/c_{i_0})/(-k) = \ln(0.02)/(-0.0273)$$
$$= 14.3 \text{ minutes}$$

Assume a reactor length of 20 ft. Therefore, the average velocity in the reactor tube is low at 20/14.3 = 1.40 ft/min or 0.71 cm/s. Now compute the Reynolds number, $N_{Re} = 2Rv_{avg}\rho/\mu$, for a series of inside tube diameters (2*R*).

Tube Diameter, inches	Reynolds Number
1	250
2	500
4	1,000
6	1,500
8	2,000

Thus, for a reactor tube length of 20 ft, laminar flow would occur for tube diameters ranging from 1 to 8 inches. The number of tubes required would depend on the total flow rate of reactor feed and the tube diameter selected. For this example, a laminar-flow reactor is certainly a possible choice. Suppose a 4-inch inside diameter were chosen, could the plug-flow assumption be made?

SOLUTION

To answer this question, compute the value of $\alpha = D_i/kR^2 = [(1.23 \times 10^{-5})(60)]/(0.273)[(4/2)(2.54)]^2 = 0.0001$. This is a very small number, indicating that radial diffusion will be essentially negligible, invalidating the plug-flow assumption. Thus, Figure 7.14 applies where, for $c_i/c_{i_0} = 0.02$, the residence time for the reaction will have to be 50% greater than for a plug-flow reactor, or 1.5(14.3) = 21.5 minutes. The reactor diameter can be maintained if the reactor tube length is increased to 30 ft. Alternatively, if the reactor length is maintained at 20 ft, the fluid velocity must be reduced to 0.93 ft/min or 0.473 cm/s. Then, laminar flow would persist up to a tube diameter of 12 inches. Experimental chemical-reaction results of Seader and Southwick (1981) show that when radial diffusion is small, as in this example for a laminar-flow tube reactor, the required reactor length can be reduced so as to approach that for plug flow if a figure-eight coiled reactor tube is used. Such a coil causes the crossing of flow from one side of the tube to the other at each intersection of the lobes of the coil, thus flattening the axial velocity profile so that it approximates plug flow.

Non-Isothermal Conditions

When isothermal flow cannot be assumed, the reaction-rate constant in Eq. (7.62) will not be constant, but will depend on temperature, according to the Arrhenius equation

$$k = k_{\infty}e^{-\frac{E}{RT}} \tag{7.65}$$

where k_{∞} is the pre-exponential factor, *E* is the activation energy per mole for the reaction, *R* is the universal gas constant, and *T* is the absolute temperature. Furthermore, a thermal energy balance must be coupled to the differential species mass balance, Eq. (7.62). If we assume constant properties, the differential energy balance is

$$2v_{avg}\left[1-\left(\frac{r}{R}\right)^2\right]\rho C_p\frac{\partial T}{\partial z}=k_T\left(\frac{\partial^2 T}{\partial r^2}+\frac{1}{r}\frac{\partial T}{\partial r}\right)-kc_i(\Delta H_{rx}) \tag{7.66}$$

where ρ is the density, C_p is the specific heat, k_T is the thermal conductivity, and ΔH_{rx} is the heat of reaction per mol of species i reacted, (positive) for an endothermic reaction and (negative) for an exothermic reaction. The boundary conditions for Eq. (7.66) include $T = T_0$ (the entering temperature of the fluid) at $z = 0$ and $\partial T/\partial r = 0$ at the centerline of the tube where $r = 0$. The additional required boundary condition is usually imposed at the inside wall of the tube, $r = R$, with either $T = T_w$ or $Q = Q_w$, with an adiabatic condition of $Q_w = 0$, where $Q_w = -k_T(\partial T/\partial r)_w$ is the heat flux at the inside wall of the tube.

Consider the simplest case of an adiabatic tubular reactor at the extreme condition of rapid radial diffusion compared to the reaction rate and/or a very small tube diameter. This is equivalent to the plug-flow assumption. The radial gradients in Eqs. (7.62) and (7.66) disappear so that these two equations become ordinary differential equations that can be combined by eliminating the differential dz, resulting in the following relationship between the species concentration and temperature, which is applicable to both tubular laminar- and turbulent-flow conditions.

$$T-T_0=(c_i-c_{i_0})\left(\frac{\Delta H_{rx}}{\rho C_p}\right) \tag{7.67}$$

Now, assuming a first-order, irreversible reaction,

$$-\frac{dc_j}{dt}=k_\infty c_i e^{-\frac{E}{RT}}, \tag{7.68}$$

Eq. (7.67) can be combined with (7.68) to eliminate either c_i or T, such that the resulting equation can be integrated. As shown by Churchill (1974, 2005), it is most convenient to eliminate T and introduce new variables:

$$\text{Fractional conversion} = X = 1-\frac{c_i}{c_{i_0}} \tag{7.69}$$

and adiabatic reactor outlet temperature, T^*, for complete conversion ($X = 1$), is

$$T^*=T_0+c_{i_0}\left(\frac{-\Delta H_{rx}}{\rho C_p}\right) \tag{7.70}$$

where the term in parentheses is assumed to be independent of temperature.

The resulting single ordinary differential equation is then integrated by parts to give a closed-form solution in terms of exponential integral functions of the following form,

$$Ei(x)=\int_{-\infty}^{x}\frac{e^y dy}{y} \tag{7.71}$$

where y is a dummy variable and Ei is different from the exponential integral function in Eq. (7.64), but is another in a series of exponential integral functions. Both exponential integral functions used in this section are tabulated by Abramowitz and Stegun (1964) and can be calculated with software programs such as Maple, Mathematica, Matlab, and Excel with an appropriate add-in. The closed-form solution for the adiabatic, tubular, plug-flow reactor is

$$\tau=\frac{L}{v_{avg}}=\frac{1}{k_\infty}\left\{Ei\left(\frac{\varepsilon_0}{1+X/Y}\right)-Ei(\varepsilon_0)+\exp(-\varepsilon^*)\left[Ei\left(\frac{\varepsilon^*}{Y}\right)-Ei\left(\frac{(1-X)\varepsilon^*}{X+Y}\right)\right]\right\} \tag{7.72}$$

where $Y = T_0/(T^* - T_0)$, and $\varepsilon^* = E/RT^*$, and $\varepsilon_0 = E/RT_0$. For a specified outlet fractional conversion, Eq. (7.72) can be used to calculate the reactor length and Eq. (7.67) can give the outlet absolute temperature. Closed-form solutions for irreversible reactions of other orders in adiabatic, plug-flow reactors are given by Douglas and Eagleton (1962). These solutions also involve exponential integral functions. When the assumption of rapid radial diffusion is not valid, the transport Equations (7.62) and (7.66) must be solved simultaneously by appropriate numerical methods or with a CFD program, as illustrated in Section 7S.6.

7.6 SUPPLEMENTAL TOPICS

The following sections, which involve reactor-separator-recycle networks and a computational fluid mechanics model for tubular reactors, are in the supplement to Chapter 7, in the PDF Files folder, which can be downloaded from the Wiley Web site associated with this book. See the file Supplement_for_Chapter 7.pdf.

- 7S.1 Locating the Separation Section with Respect to the Reactor Section
- 7S.2 Tradeoffs in Processes Involving Recycle
- 7S.3 Optimal Reactor Conversion
- 7S.4 Recycle to Extinction
- 7S.5 Snowball Effects in the Control of Processes Involving Recycle
- 7S.6 Computational Fluid Dynamics (CFD) Models for Tubular Chemical Reactors

7.7 SUMMARY

This chapter has introduced the design of chemical reactors and reactor networks. Different methods of reaction temperature control have been presented, with emphasis on the use of multiple adiabatic reactors or beds, using cold shots or heat exchangers between reactors or beds. The attainable region is presented to define the reactor network that maximizes either the yield or the selectivity of a desired product, given the feed to the reactor. However, since reactor yield is often sacrificed in favor of selectivity, conversion is rarely complete, with unreacted species recycled. Thus, the optimal reactor feed conditions depend on the plant economics and the reactor network should be synthesized as part of an overall plant design.

When the potential for the loss of reactants to side reactions is significant and recycle to extinction cannot be employed, a recent study by Ward, Mellichamp, and Doherty (2004) can be employed to determine the optimal tradeoff between the performance of the reactor network and the performance of the separation system. The tradeoff considers the effect of selectivity losses at high reactant conversion as opposed to recycle costs at low reactant conversion. In the optimization, the use of recycle flow rate(s), rather than reactant conversion, reduces the complexity of the calculations. Of paramount interest in determining the optimal policy is the kinetics of the competing reactions. Results differ dramatically depending on whether the competing reaction is of the same or different order with respect to the controlling reactant. Cases involving both CSTR and PFR reactors are considered. The analysis is of particular value when (1) neither capital costs nor operating costs are dominant, (2) the order of the side reaction is greater than that of the main reaction, with respect to the controlling reactant, and (3) the process design has built-in flexibility.

After completing this chapter and reviewing the CD-ROM that accompanies this book, the reader should

1. be able to use effectively ASPEN PLUS and/or HYSYS to model chemical reactors, implementing complex configurations involving tube cooling and cold shots.
2. have an appreciation for the complex configurations that are often used in commercial reactor designs, especially when it is required to handle highly exothermic or endothermic reactions.
3. be able to define the combination of CSTRs and/or PFRs that maximize the yield or selectivity of the desired reactor product for a particular feed composition, given the reaction kinetics, using attainable region analysis.
4. understand when the ideal reactor models can be used and when they should be abandoned in favor of CFD models.
5. understand the considerations in determining the best locations, with respect to the reactor section, of the separation sections.
6. be aware of the many tradeoffs between the reactor section and the separation section(s) when recycle is used, particularly when fast side reactions can occur.
7. know that the optimal fractional conversion of the limiting reactant in the reactor section is usually less than 100% of the equilibrium conversion.
8. be able to apply the concept of recycle to extinction to reduce waste and increase the yield of the main product.
9. be aware that the snowball effect can occur in a reactor-separator-recycle network.

REFERENCES

1. Abramowitz, M., and Stegun, I.A., Eds., *Handbook of Mathematical Functions with Formulas, Graphs, and Mathematical Tables*, NBS Applied Mathematics Series 55, U.S. Government Printing Office, Washington, DC (1964).

2. Aris, R., "The Optimum Design of Adiabatic Reactors with Several Beds," *Chem. Eng. Sci.*, **12**, 243 (1960).

3. Bosworth, R.C.L., *Phil. Mag.*, **39**, 847 (1948).

4. Churchill, S.W., "Interaction of Chemical Reactions and Transport. 1. An Overview," *Ind. Eng. Chem. Res.*, **44**, 5199 (2005).

5. Churchill, S.W., *The Interpretation and Use of Rate Data. The Rate Process Concept*, McGraw-Hill Scripta, Washington, DC (1974).

6. Churchill, S.W., "Turbulent Flow and Convection: The Prediction of Turbulent Flow and Convection in a Round Tube," in J.P. Hartnett and T.F. Irvine, Jr., Ser. Eds., *Advances in Heat Transfer*, Academic Press, New York, **34**, 255–361 (2001).

7. Churchill, S.W., and B. Yu, "Effects of Transport on Reactions in Homogeneous Tubular Flow," *Ind. Eng. Chem. Res.*, **45**, 8583–8593 (2006).

8. Churchill, S.W., and S.C. Zajik, "Prediction of Fully Developed Convection with Minimal Explicit Empiricism," *AIChE Journal*, **48**, 927 (2002).

9. Cleland, F.A., and R.H. Wilhelm, "Diffusion and Reaction in Viscous-flow Tubular Reactor," *AIChE Journal*, **2**, 489 (1956).

10. Denbigh, K.G., *J. Appl. Chem.*, **1**, 227 (1951).

11. Douglas, J.M., and L.C. Eagleton, "Analytical Solutions for Some Adiabatic Reactor Problems," *Ind. Eng. Chem. Fundamentals*, **1**, 116 (1962).

12. Felder, R.M., and R.W. Rousseau, *Elementary Principles of Chemical Processes*, Third Edition, John Wiley & Sons, New York, pp. 451–452 (2000).

13. Fjeld, M., O.A. Asbjornsen, and H.J. Astrom, "Reaction Invariants and Their Importance in the Analysis of Eigenvectors, Stability and Controllability of CSTRs," *Chem. Eng. Sci.*, **30**, 1917 (1974).

14. Fogler, H.S., *Elements of Chemical Reaction Engineering*, Fourth Edition , Prentice-Hall, Englewood Cliffs, New Jersey (2005).

15. Glasser, D., C. Crowe, and D.A. Hildebrandt, "A Geometric Approach to Steady Flow Reactors: The Attainable Region and Optimization in Concentration Space," *Ind. Eng. Chem. Res.*, **26** (9), 1803 (1987).

16. Hildebrandt, D., and L.T. Biegler, "Synthesis of Chemical Reactor Networks," *AIChE Symp. Ser.* No. 304, Vol. **91**, 52 (1995).

17. Hildebrandt, D.A., D. Glasser, and C. Crowe, "The Geometry of the Attainable Region Generated by Reaction and Mixing with and without Constraints," *Ind. Eng. Chem. Res.*, **29**(1), 49 (1990).

18. Horn, F.J.M., "*Attainable Regions in Chemical Reaction Technique,*" in the *Third European Symposium of Chemical Reaction Engineering*, Pergamon Press, London (1964).

19. Lapidus, L., *Digital Computation for Chemical Engineers*, McGraw-Hill Book Co., New York (1962).

20. Levenspiel, O., *Chemical Reaction Engineering*, Third Edition, Wiley, New York (1999).

21. Lewin, D.R., and R. Lavie, "Optimal Operation of a Tube Cooled Ammonia Converter in the Face of Catalyst Bed Deactivation," *I. Chem. Eng. Symp. Ser.*, **87**, 393 (1984).

22. OMTVEIT, T., J. TANSKANEN, and K.M. LIEN, "Graphical Targeting Procedures for Reactor Systems," *Comput.Chem. Eng.*, **18**(S), S113 (1994).

23. RASE, H.F., *Chemical Reactor Design for Process Plants, Volume Two: Case Studies and Design Data*, Wiley-Interscience, New York (1977).

24. SCHMIDT, L.D., *The Engineering of Chemical Reactions*, Oxford University Press, Oxford (1998).

25. SEADER, J.D., and L.M. SOUTHWICK, "Soponification of Ethyl Acetate in Curved-Tube Reactors," *Chem. Eng. Commun.*, **9**, 175 (1981).

26. SMITH, J.M., *Chemical Engineering Kinetics*, Third Edition, McGraw-Hill, New York (1981).

27. STEPHENS, A.D., and R.J. RICHARDS, "Steady State and Dynamic Analysis of an Ammonia Synthesis Plant," *Automatica*, **9**, 65 (1973).

28. TEMKIN, M., and V. PYZHEV, "Kinetics of Ammonia Synthesis on Promoted Iron Catalyst," *Acta Physicochim. U.R.S.S.*, **12**(3), 327 (1940).

29. TRAMBOUZE, P.J., and E.L. PIRET, "Continuous Stirred Tank Reactors: Designs for Maximum Conversions of Raw Materials to Desired Product," *AIChE Journal*, **5**, 384 (1959).

30. VAN DE VUSSE, J. G., "Plug Flow vs. Tank Reactor," *Chem. Eng. Sci.*, **19**, 994 (1964).

31. VAN HEERDEN, C., "Autothermic Processes—Properties and Reactor Design," *Ind. Eng. Chem.*, **45**(6), 1242 (1953).

32. WARD, J.D., D.A. MELLICHAMP, and M.F. DOHERTY, "Importance of Process Chemistry in Selecting the Operating Policy for Plants with Recycle," *Ind. Eng. Chem. Res.*, **43**, 3957 (2004).

33. WESTERLINK, E.J., and K.R. WESTERTERP, "Safe Design of Cooled Tubular Reactors for Exothermic Multiple Reactions: Multiple Reaction Networks," *Chem. Eng. Sci.*, **43**(5), 1051 (1988).

34. XU, J., and G. FROMENT, "Methane Steam Reforming: Diffusional Limitations and Reactor Simulation," *AIChE Journal*, **35**(1), 88 (1989).

EXERCISES

7.1 Carry out a modified design for the ammonia converter in Example 7.3, consisting of three diabatic reactor bed sections, each of 2 m diameter and 2 m length (note that the total bed length is the same as before). Assuming the same reactor inlet temperature of 270°C, compute the optimal heat duties and effluent temperatures for each bed, such that the effluent ammonia mole fraction for the reactor is maximized. Plot the temperature-composition trajectory for the modified converter design and compare it to the three-bed cold-shot design of Example 7.3.

7.2 A system of three parallel reactions (Trambouze and Piret, 1959) is:

$$A \xrightarrow{k_1} B \quad A \xrightarrow{k_2} C \quad A \xrightarrow{k_3} D$$

where the reactions are zero-order, first-order, and second-order, respectively, with $k_1 = 0.025$ mol/L-min, $k_2 = 0.2\ \text{min}^{-1}$, and $k_3 = 0.4$ L/mol-min, and the initial concentration of $C_A = 1$ mol/L. Use the attainable region algorithm to find the reactor network that maximizes the selectivity of C from A.

7.3 Repeat Exercise 7.2, taking the first two reactions as first-order, and the last as second-order, with $k_1 = 0.02\ \text{min}^{-1}$, $k_2 = 0.2\ \text{min}^{-1}$, and $k_3 = 2.0$ L/mol-min, and the initial concentration of $C_A = 1$ mol/L. Use the attainable region method to find the reactor network that maximizes the selectivity of C from A.

7.4 For the reaction system:

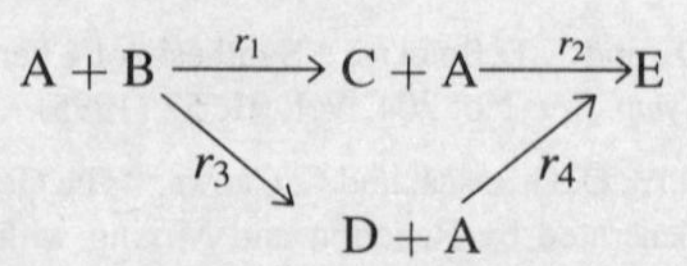

where $r_1 = k_1 C_A^2$, $r_2 = k_2 C_c$, $r_3 = k_3 C_A$, and $r_4 = k_4 C_A$. The rate constants are $k_1 = 3\ \text{m}^3/\text{kmol-min}$, $k_2 = 10\ \text{min}^{-1}$, $k_3 = 0.5\ \text{min}^{-1}$, $k_4 = 1.5\ \text{min}^{-1}$, and the feed concentration of A is 1 kmol/m^3. Use the attainable region method to find the reactor network that maximizes the selectivity of C from A.

7.5 In Example 7.5, choose methane and hydrogen as independent components. Derive relationships for the mole numbers of the remaining components in terms of methane and hydrogen.

7.6 *Cumene process with drag (purge) streams* In Section 7S.1, a process for producing cumene by the alkylation of benzene with propylene is described. The flowsheet for the process is given in Figure 7S.1. However, that flowsheet does not provide for the removal of water, ethane, isobutane, MCP, MCH, toluene, n-propylbenzene, t-BB, and p-cymene. For their removal, it is proposed to add two drag (purge) streams to the flowsheet, one from the distillate of the benzene recovery column, C2, and the other from the bottoms of the cumene recovery column, C3. Also, the flowsheet in Figure 7S.1 does not provide for an exit for the heavies produced in the alkylation and trans-alkylation reactors in the event that their amounts are too large to be included in the allowable impurity in the cumene product. Thus, it may be necessary to add a fourth distillation column, C4, following C3, with the distillate from C4 fed to the trans-alkylation reactor and the bottoms from C4 being a heavies product. If so, the heavies must not contain more than 5% of the DIPBs and lighter entering C4.

Most of the data for the cumene process is given in Section 7S.1. However, missing are the product distributions for the two reactors. These are as follows from laboratory studies:

Component	Alkylation Reactor Change in pounds per 100 pounds of propylene in the combined feed	Trans-Alkylation Reactor Change in pounds per 100 pounds of propylene in the combined feed to the Alkylation Reactor
Propylene	−100.0000	0.0000
1-Butene	−0.0039	
Benzene	−168.1835	−16.3570
Toluene	−0.0214	
Cumene	232.7018	50.7652
n-Propylbenzene	0.0346	0.0087
p-Cymene	0.0306	−0.0025
t-BB	0.0080	−0.0007

m-DIPB	20.3314	−20.2323
p-DIPB	14.7797	−14.4953
Alkylation Heavies	0.3227	
Trans-alkylation Heavies	0.0000	0.3121
Total change	0	0

Note, again, that the conversion of DIPBs in the trans-alkylation reactor is only 50%.

Using the above data and the data in Section 7S.1, revise the flowsheet in Figure 7S.1 and produce a complete material balance with the component flow rates in lbmol/hr for each stream in your flowsheet. Try to maximize the production of cumene. Be sure to add two drag streams for removal of byproducts, and a fourth distillation column, if necessary. Compute the overall percent conversion of benzene to cumene and the annual production of cumene in lb/yr if the operating factor is 0.95. If a heavies product is produced, what could it be used for?

7.7 The feed to a pentane isomerization process consists of 650 kmol/hr of n-pentane and 300 kmol/hr of isopentane. The effluent from the catalytic isomerization reactor will contain 6.5 moles of isopentane for every mole of n-pentane. The catalyst prevents the formation of neopentane. If the isopentane product, produced by separating isopentane from n-pentane by distillation, is to contain only 2 wt% n-pentane and the separation system is to be placed before the reactor, calculate the total flow rate and composition of the reactor effluent, the combined feed to the reactor, and the bottoms product from the distillation column. Design the distillation column. Repeat the material balance calculations and the design of the distillation column if the separation system is placed after the reactor. Based on your results and without determining any capital or operating costs, which separation system placement is preferred?

Chapter 8

Separation Train Synthesis

8.0 OBJECTIVES

Most chemical processes are dominated by the need to separate multicomponent chemical mixtures. In general, a number of separation steps must be employed, where each step separates between two components of the feed to that step. During process design, separation methods must be selected and sequenced for these steps. This chapter discusses some of the techniques for the synthesis of separation trains. More detailed treatments are given by Douglas (1995), Barnicki and Siirola (1997), and Doherty and Malone (2001).

After studying this chapter, the reader should

1. Be familiar with the more widely used industrial separation methods and their basis for separation.
2. Understand the concept of the separation factor and be able to select appropriate separation methods for vapor, liquid, and solid–fluid mixtures.
3. Understand how distillation columns are sequenced and how to apply heuristics to narrow the search for a near-optimal sequence.
4. Be able to apply algorithmic methods to determine an optimal sequence of distillation-type separations.
5. Be familiar with the difficulties in and techniques for determining feasible sequences when azeotropes can form.
6. Be able to determine feasible separation systems for gas mixtures and solid–fluid systems.

8.1 INTRODUCTION

Almost all chemical processes require the separation of mixtures of chemical species (components). In Section 6.4, three flowsheets (Figures 6.7, 6.8, and 6.9) are shown for processes involving a reactor followed by a separation system. A more general flowsheet for a process involving one reactor system is shown in Figure 8.1, where separation systems are shown before as well as after the reactor section. A *feed separation system* may be required to purify the reactor feed(s) by removing catalyst poisons and inert species, especially if they are present as a significant percentage of the feed. An *effluent separation system*, which follows the reactor system and is almost always required, recovers unconverted reactants (in gas, liquid, and/or solid phases) for recycle to the reactor system and separates and purifies products and byproducts. Where separations are too difficult, purge streams are used to prevent buildup of certain species in recycle streams. Processes that do not involve a reactor system also utilize separation operations if the feed is a mixture that requires separation. Frequently, the major investment and operating costs of a process will be those costs associated with the separation equipment, rather than with the chemical reactor(s).

Feed Separation System

As shown in Figure 8.1, the combined feed to a reactor section may consist of one or more feed streams and one or more recycle streams when conversion of reactants is incomplete. When a feed separation system is needed and more than one feed enters the process, it is usually preferable to provide separate separation operations for the individual feed streams before mixing them with each other and with any recycle streams. Some industrial examples of chemical processes that require a feed separation system are:

1. Production of polypropylene from a feed of propylene and propane. Propane, which is not involved in the propylene polymerization reaction, is removed from the propylene by distillation.
2. Production of acetaldehyde by the dehydrogenation of ethanol using a chromium-copper catalyst. If the feed is a dilute solution of ethanol in water, distillation is used to concentrate the ethanol to the near-azeotrope composition (89.4 mol% ethanol at 1 atm) before it enters the reactor.
3. Production of formaldehyde by air-oxidation of methanol using a silver catalyst. The entering air is scrubbed

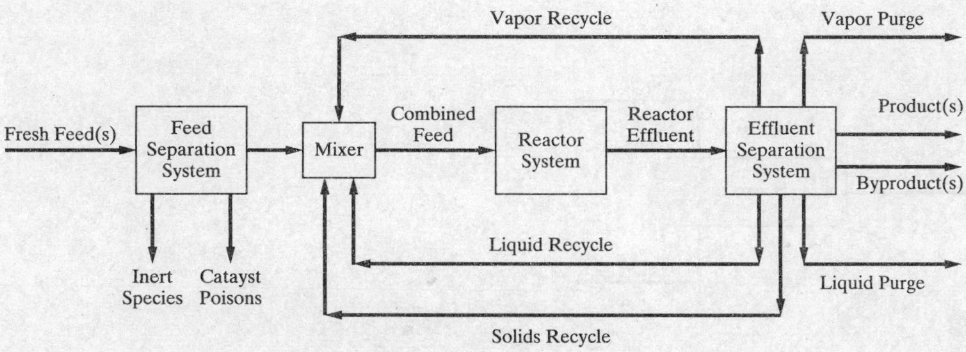

Figure 8.1 General flowsheet for a process with one reactor system.

with aqueous sodium hydroxide to remove any SO_2 and CO_2, which are catalyst poisons.

4. Production of vinyl chloride by the gas-phase reaction of HCl and acetylene with a mercuric chloride catalyst. Small amounts of water are removed from both feed gases by adsorption to prevent corrosion of the reactor vessel and acetaldehyde formation.
5. Production of phosgene by the gas-phase reaction of CO and chlorine using an activated carbon catalyst. Both feed gases are treated to remove oxygen, which poisons the catalyst; sulfur compounds, which form sulfur chlorides; hydrogen, which reacts with both chlorine and phosgene to form HCl; and water and hydrocarbons, which also form HCl.

Phase Separation of Reactor Effluent

In Figure 8.1, the reactor effluent may be a heterogeneous (two or more phases) mixture, but most often is a homogeneous (single-phase) mixture. When the latter, it is often advantageous to change the temperature and/or (but less frequently) the pressure to obtain a partial separation of the components by forming a heterogeneous mixture of two or more phases. Following the change in temperature and/or pressure, phase equilibrium is rapidly attained, resulting in the following possible phase conditions of the reactor effluent, where two liquid phases may form (phase splitting) when both water and hydrocarbons are present in the reactor effluent:

Possible Phase Conditions of Reactor Effluent

Vapor	Liquid
Vapor and Liquid	Liquid 1 and Liquid 2
Vapor, Liquid 1, and Liquid 2	Liquid and Solids
Vapor and Solids	Liquid 1, Liquid 2, and Solids
Vapor, Liquid, and Solids	Solids
Vapor, Liquid 1, Liquid 2, and Solids	

With the reasonable assumption that the phases in a heterogeneous mixture are in phase (physical) equilibrium for a given reactor effluent composition at the temperature and pressure to which the effluent is brought, process simulators can readily estimate the amounts and compositions of the phases in equilibrium by an isothermal (two-phase)-flash calculation, provided that solids are not present. When the possibility of two liquid phases exists, it is necessary to employ a three-phase flash model, rather than the usual two-phase flash model. The three-phase model considers the possibility that a vapor phase may also be present, together with two liquid phases.

In the absence of solids, the resulting phases are separated, often by gravity, in a flash vessel for the V-L case, or in a decanter for the V-L1-L2 or L1-L2 cases. For the latter two cases, centrifugal force may be employed if gravity settling is too slow because of small liquid-density differences or high liquid viscosities. If solids are present with one or two liquid phases, it is not possible to separate completely the solids from the liquid phase(s). Instead, a centrifuge or filter is used to deliver a wet cake of solids that requires further processing to recover the liquid and dry the solids.

Several examples of phase-separation equipment are shown in Figure 8.2. Each exiting phase is either recycled to the reactor, purged from the system, or, most often, sent to separate vapor, liquid, or slurry separation systems, as shown in Figure 8.3. The effluents from these separation systems are products, which are sent to storage; byproducts, which also leave the process; *reactor-system recycle streams*, which are sent back to the reactor; or *separation-system recycle streams*, which are sent to one of the other separation systems. Purges and byproducts are either additional valuable products, which are sent to storage; fuel byproducts, which are sent to a fuel supply or storage system; and/or waste streams, which are sent to waste treatment, incineration, or landfill.

Consider the following examples of phase-equilibria calculations for industrial reactor effluents:

1. ***Vapor-liquid case.*** The reactor effluent for a toluene hydrodealkylation process, of the type discussed in Section 5.3, is a gas at 1,150°F and 520 psia. When brought to 100°F at say 500 psia by a series of heat exchangers, the result is a vapor phase in equilibrium with a single liquid phase. A two-phase flash calculation using the SRK equation of state gives the following results:

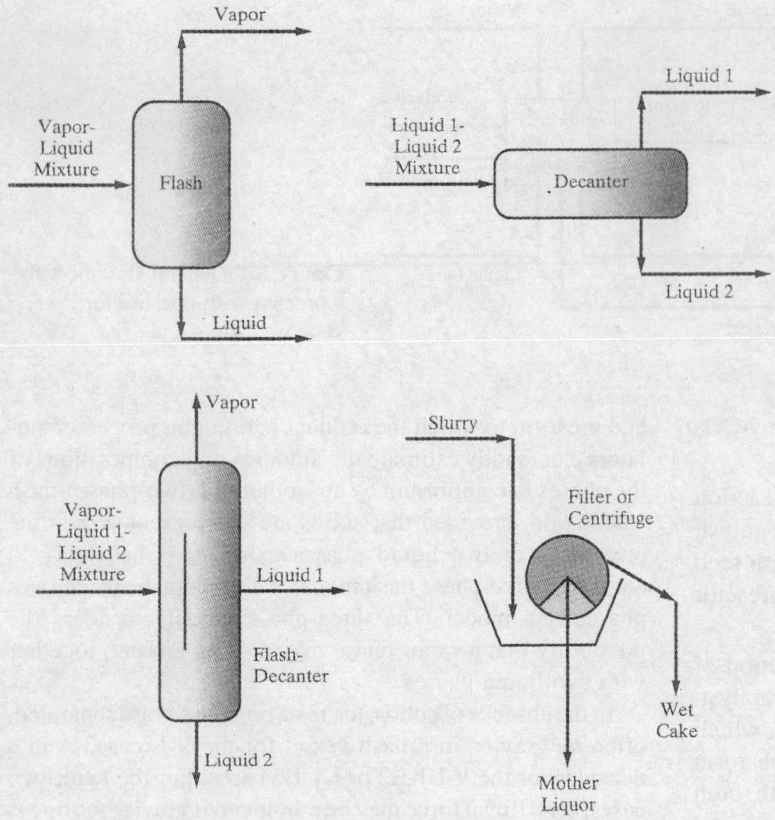

Figure 8.2 Examples of phase-separation devices.

Reactor Effluent Phase Equilibrium for a Toluene Hydrodealkylation Process

Component	Effluent (lbmol/hr)	Vapor (lbmol/hr)	Liquid (lbmol/hr)
H_2	1,292	1,290	2
CH_4	1,167	1,149	18
Benzene	280	16	264
Toluene	117	2	115
Biphenyl	3	0	3
Total	2,859	2,457	402

As seen, a reasonably good separation is made between the light gases, H_2 and CH_4, and the three less-volatile aromatic hydrocarbons. The vapor is sent to a vapor separation system to recover CH_4 as a byproduct and H_2 for recycle. The liquid is sent to a liquid separation system to recover benzene as the main product, toluene for recycle to the reactor, and biphenyl as a fuel byproduct. Alternatively, the vapor can be divided, without component separation, into a reactor recycle stream and a vapor purge stream to prevent buildup of CH_4, while the biphenyl can be separated with the toluene and recycled to extinction. These two alternatives are shown in Figure 8.4.

2. ***Vapor-liquid 1-liquid 2 case.*** The reactor effluent in a styrene production process, involving the reaction

$$\text{methanol} + \text{toluene} = \text{styrene} + \text{hydrogen} + \text{water}$$

with a side reaction, of the same reactants, that produces ethylbenzene and water, is a gas at 425°C and 330 kPa. When brought to 38°C at say 278 kPa by a series of heat exchangers, the result is a vapor phase in equilibrium with an organic-rich liquid phase and a water-rich liquid phase. A three-phase flash calculation using the NRTL method for estimating liquid-phase activity coefficients gives the following results:

Reactor Effluent Phase Equilibrium for a Styrene Process

Component	Effluent (kmol/hr)	Vapor (kmol/hr)	Liquid 1 (kmol/hr)	Liquid 2 (kmol/hr)
H_2	352.2	352.2	0.0	0.0
Methanol	107.3	9.9	31.0	66.4
Water	489.1	8.0	0.5	480.6
Toluene	107.3	1.7	105.5	0.1
Ethylbenzene	140.7	0.5	140.0	0.2
Styrene	352.2	2.0	350.1	0.1
Total	1,548.8	374.3	627.1	547.4

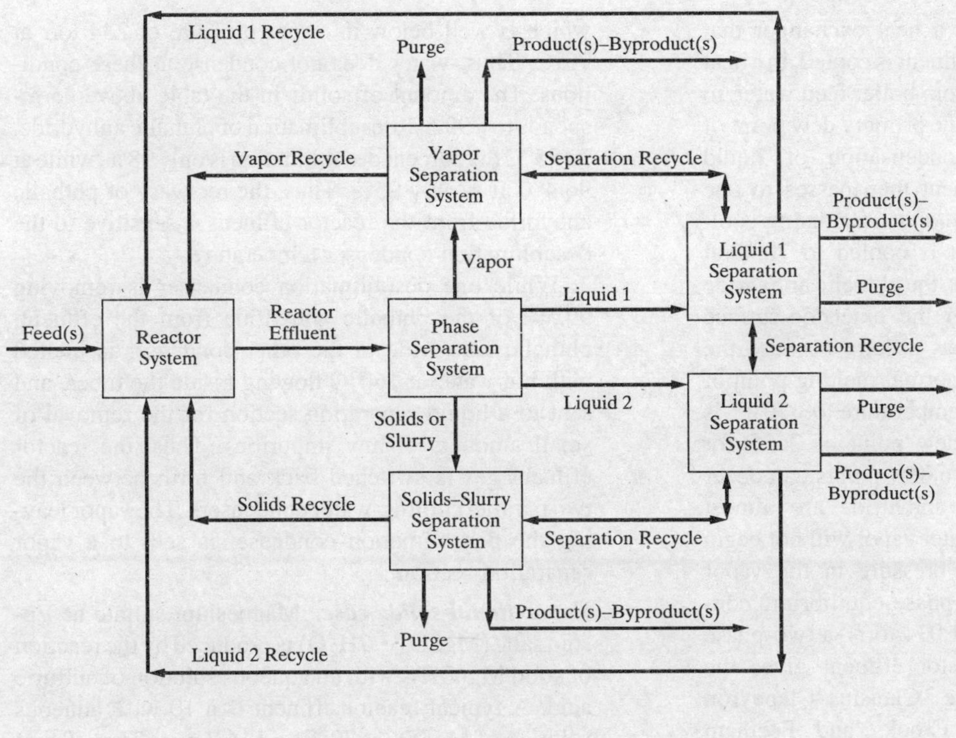

Figure 8.3 Process flowsheet showing separate separation systems with reactor-system and separation-system recycles.

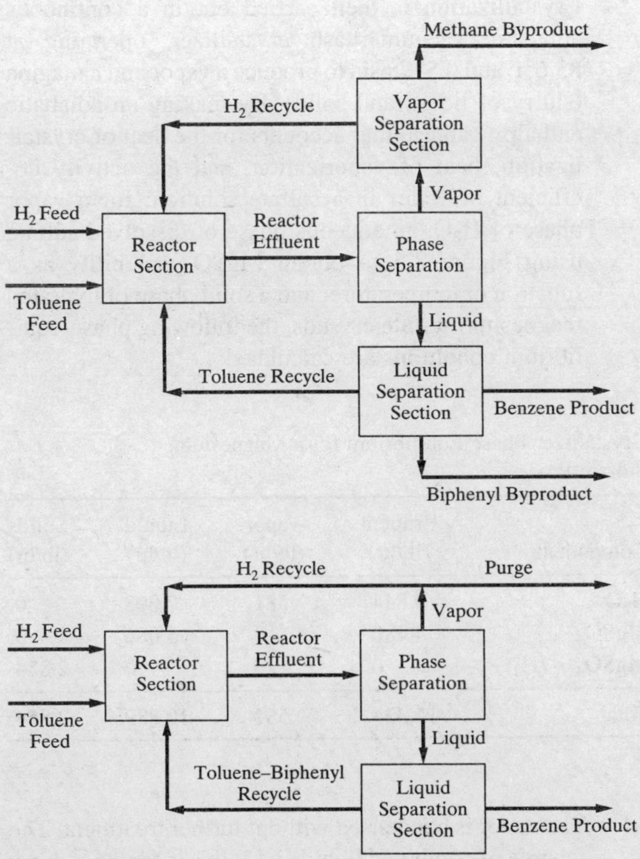

Figure 8.4 Alternative flowsheets for hydrodealkylation of toluene to benzene.

In this case, the vapor is 94 mol% H_2, for which a vapor separation section may not be needed. The organic-rich liquid phase (L1) is sent to a liquid separation section to recover a combined methanol and toluene stream for recycle to the reactor, ethylbenzene as a byproduct, and styrene as the main product. The water-rich liquid phase (L2) is sent to another liquid separation section to recover methanol for recycle to the reactor and water, which is sent to wastewater treatment to remove small quantities of soluble organic components. It is important to note that a two-phase flash calculation would produce erroneous results. If in doubt, perform a three-phase flash calculation, rather than a two-phase flash calculation.

3. ***Vapor-solids case.*** Phthalic anhydride is manufactured mainly by the vapor-phase partial oxidation of orthoxylene with excess air in a shell-and-tube fixed-bed reactor using vanadium pentoxide catalyst packed inside the tubes. Typically the reactor feed is very dilute in the orthoxylene, with only 1.27 moles of orthoxylene per 100 moles of air. The main reaction consumes 80% of the orthoxylene to produce phthalic anhydride and water. The remaining 20% of the orthoxylene is completely and unavoidably oxidized to CO_2 and water vapor. Typical reactor effluent conditions are 660 K and 25 psia. The reactions are exothermic, with heat removal in the reactor by molten salt of the eutectic mixture of sodium and potassium nitrites and nitrates, which recirculates between the

shell side of the reactor and a heat exchanger that produces steam. The reactor effluent is cooled, in a heat exchanger to produce steam from boiler feed water, to 180°C, which is safely above the primary dew point of 140°C, corresponding to condensation of liquid phthalic anhydride. The effluent then passes to one of two parallel desublimation condensers using cooling water, where the effluent is cooled to 70°C at 20 psia. Under these conditions, the phthalic anhydride desublimes on the outside of the extended-surface tubes of the heat exchanger as a solid because the temperature is well below its normal melting point of 131°C. The desublimation temperature of 70°C is safely above the secondary dew point of 36°C for the condensation of water. Two dew points can occur because water and phthalic anhydride are almost insoluble in each other. The water vapor will not begin to condense until its partial pressure in the vapor reaches its vapor pressure. At phase-equilibrium conditions of 70°C and 20 psia (1,034 torr), a two-phase flash calculation on the reactor effluent gives the following results when the Clausius–Clapeyron vapor-pressure equation of Crooks and Feetham (1946) is used for solid phthalic anhydride:

$$\text{Log}_{10} P^s = 12.249 - 4,632/T$$

where vapor pressure, P^s, is in torr and temperature, T, is in K. This equation is valid for temperatures in the range of 30°C to the normal melting point of 131°C and predicts a vapor pressure of 0.000517 torr at 25°C, which is in good agreement with the often-quoted value of 0.000514 torr. Assuming that the solid phase is pure phthalic anhydride, its partial pressure in the equilibrium vapor phase is equal to its vapor pressure.

Reactor Effluent Phase Equilibrium for a Phthalic Anhydride Process-Basis: 100 Moles of Reactor Effluent

Component	Effluent (moles)	Vapor (moles)	Solids (moles)
N_2	77.70	77.70	0.00
O_2	15.05	15.05	0.00
Orthoxylene	0.00	0.00	0.00
CO_2	2.00	2.00	0.00
H_2O	4.25	4.25	0.00
Phthalic anhydride	1.00	0.005	0.995
Total	100.0	99.005	0.995

At these equilibrium conditions of 70°C and 1,034 torr, the partial pressure of phthalic anhydride in the vapor is (0.005/99.005)1,034 = 0.05 torr, which is equal to its vapor pressure. The partial pressure of water in the vapor is (4.25/99.005)1,034 = 44.4 torr, which is well below its vapor pressure of 234 torr at 70°C. Thus, water does not condense at these conditions. The amount of solids in the table above corresponds to a 99.5% desublimation of phthalic anhydride. At 85°C the percent desublimation is only 98%, while at 96.4°C it is only 95%. Thus, the recovery of phthalic anhydride from the reactor effluent is sensitive to the desublimation condenser temperature.

While one desublimation condenser is removing 99.5% of the phthalic anhydride from the effluent, phthalic anhydride in the other condenser is melted with hot water at 160°C flowing inside the tubes, and sent to a liquid separation section for the removal of small amounts of any impurities. Thus, the reactor effluent gas is switched back and forth between the two parallel cooling-water condensers. The vapor leaving the desublimation condenser is sent to a vapor separation section.

4. ***Vapor-liquid-solids case.*** Magnesium sulfate as Epsom salts ($MgSO_4 \cdot 7H_2O$) is produced by the reaction of solid $Mg(OH)_2$ with an aqueous solution of sulfuric acid. A typical reactor effluent is a 10 wt% aqueous solution of $MgSO_4$ at 70°F and 14.7 psia. The effluent is concentrated to 37.75 wt% $MgSO_4$ in a double-effect evaporation system with forward feed, after which filtrate from a subsequent filtering operation is added. Crystallization is then carried out in a continuous adiabatic vacuum flash crystallizer, operating at 85.6°F and 0.577 psia to produce a vapor and a magma (slurry of liquid and solid). By making an adiabatic enthalpy balance that accounts for the heat of crystallization, heat of vaporization, and the activity coefficient of water in a sulfate solution; for a vapor phase of H_2O, an aqueous phase of dissolved sulfate using Figure 8.5 to obtain $MgSO_4$ solubility as a function of temperature; and a solid phase of hydrated magnesium sulfate crystals, the following phase-equilibrium conditions are calculated:

Crystallizer Phase Equilibrium for a Magnesium Sulfate Process

Component	Effluent (lb/hr)	Vapor (lb/hr)	Liquid (lb/hr)	Solids (lb/hr)
H_2O	9,844	581	7,803	0
$MgSO_4$	4,480	0	3,086	0
$MgSO_4 \cdot 7H_2O$	0	0	0	2,854
Total	14,324	581	10,889	2,854

The vapor is condensed without further treatment. The magma of combined liquid and solids is sent to a slurry separation system to obtain a product of dry crystals of $MgSO_4 \cdot 7H_2O$.

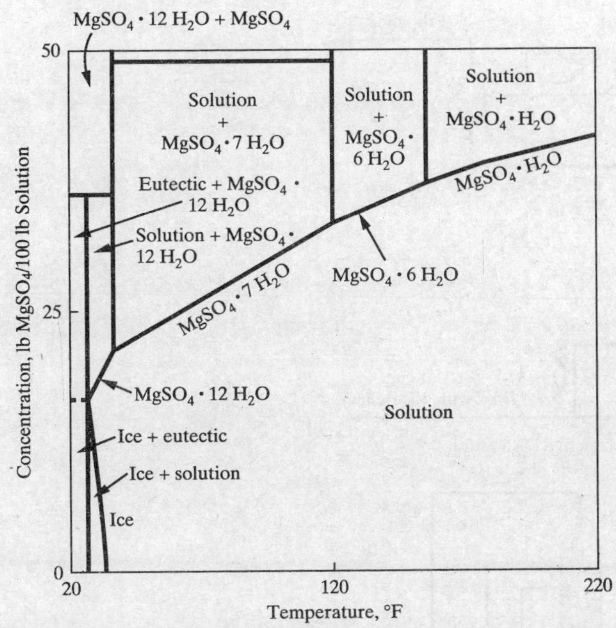

Figure 8.5 Phase diagram for the $MgSO_4 \cdot H_2O$ system.

Industrial Separation Operations

Following phase separation, the individual vapor, liquid, solids, and/or slurry streams are sent to individual separation systems, the most common of which is the liquid separation system. When the feed to a vapor or liquid separation system is a binary mixture, it may be possible to select a separation method that can accomplish the separation task in just one piece of equipment. In that case, the separation system is relatively simple. More commonly, however, the feed mixture involves more than two components. Although some progress is being made in devising multicomponent separation systems involving a single piece of equipment, most systems involve a number of units in which the separations are sequenced, with each unit separating its feed stream into two effluent streams of different composition. The separation in each piece of equipment (unit) is made between two components designated as the key components for that particular separation unit. Each effluent is either a final product or a feed to another separation device. The synthesis of a multicomponent separation system can be very complex because it involves not only the selection of the separation method(s), but also the manner in which the pieces of separation equipment are sequenced. This chapter deals with both aspects of the synthesis problem.

As an example of the complexity of a multicomponent separation system, consider the synthesis of a separation system for the recovery of butenes from a C_4 concentrate from the catalytic dehydrogenation of *n*-butane. The specifications for the separation process are taken from Hendry and Hughes (1972), and are shown in Figure 8.6. The process feed, which contains propane, 1-butene, *n*-butane, *trans*-2-butene, *cis*-2-butene, and *n*-pentane, is to be separated into four fractions: (1) a propane-rich stream containing 99% of the propane in the feed, (2) an *n*-butane-rich stream containing 96% of the nC_4 in the feed, (3) a stream containing a mixture of the three butenes, at 95% recovery, and (4) an *n*-pentane-rich stream containing 98% of the nC_5 in the feed. The C_3 and nC_5 streams are final products, the nC_4 stream is recycled to the catalytic dehydrogenation reactor, and the butenes stream is sent to another dehydrogenation reactor to produce butadienes.

Many different types of separation devices and sequences thereof can accomplish the separations specified in Figure 8.6. In general, the process design engineer seeks the most economical system. One such system, based on mature technology and the availability of inexpensive energy, is shown in Figure 8.7. The system involves two separation methods, distillation and extractive distillation. The process feed from the butane dehydrogenation unit is sent to a series of two distillation columns (1-butene columns, C-1A and C-1B), where the more volatile propane and 1-butene are removed as distillate and then separated in a second distillation column (depropanizer, C-2) into propane and 1-butene.

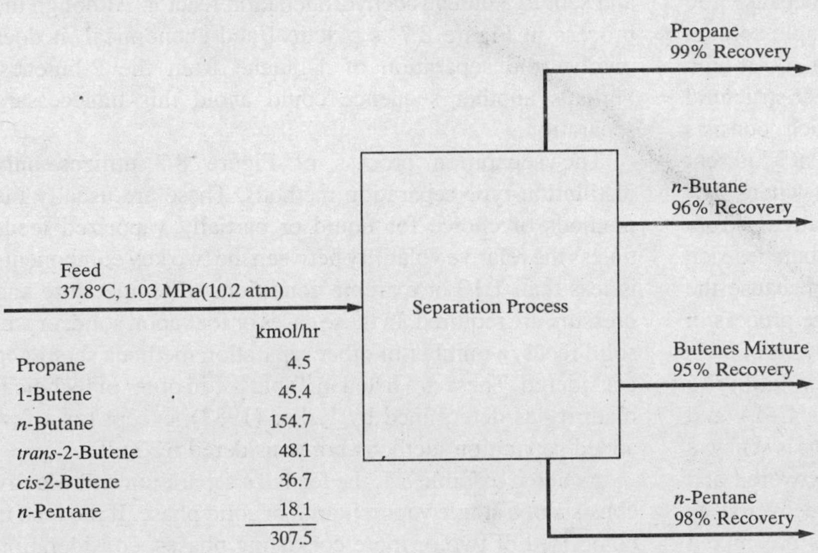

Figure 8.6 Specification for butenes recovery system.

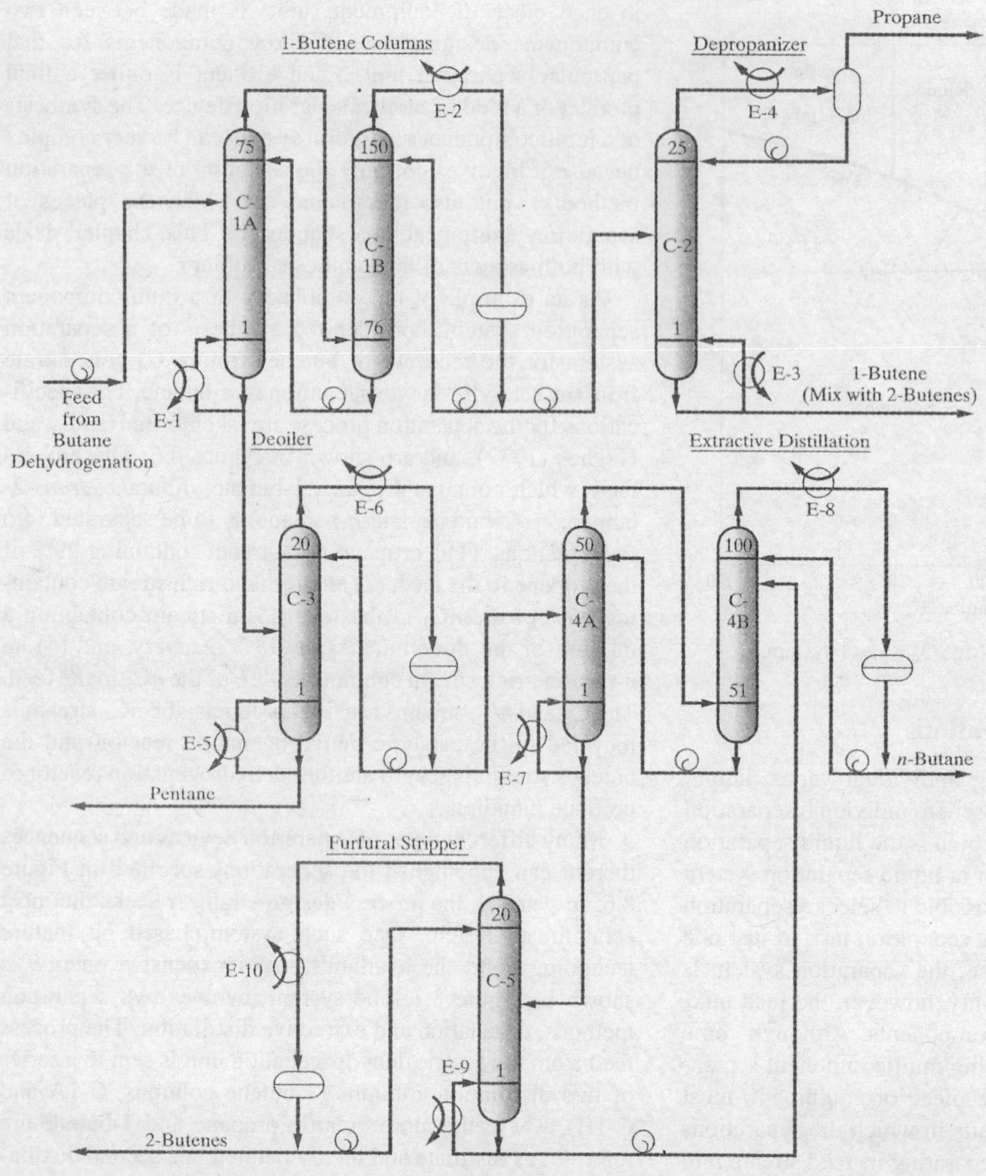

Figure 8.7 Process for butenes recovery: C = distillation column; E = heat exchanger.

Distillation unit C-1 consists of two columns because 150 trays are required, which are too many for a single column (since the tray spacing is typically 2 ft, giving a 300-ft high tower, while most towers do not exceed 200 ft for structural reasons). The bottoms from unit C-1A, which consists mainly of *n*-butane, the 2-butene isomers, and nC_5, is sent to another distillation column (deoiler, C-3), where nC_5 product is removed as bottoms. The distillate stream from unit C-3 cannot be separated into nC_4-rich and 2-butenes-rich streams economically by ordinary distillation because the relative volatility is only about 1.03. Instead, the process in Figure 8.7 uses extractive distillation with a solvent of 96% furfural in water, which enhances the relative volatility to about 1.17. The separation occurs in columns C-4A and C-4B, with nC_4 taken off as distillate. The bottoms is sent to a furfural stripper (C-5), where the solvent is recovered and recycled to unit C-4 and the 2-butenes are recovered as distillate. The 1-butene and 2-butenes streams are mixed and sent to a butenes dehydrogenation reactor. Although the process in Figure 8.7 is practical and economical, it does involve the separation of 1-butene from the 2-butenes. Perhaps another sequence could avoid this unnecessary separation.

The separation process of Figure 8.7 utilizes only distillation-type separation methods. These are usually the methods of choice for liquid or partially vaporized feeds unless the relative volatility between the two key components is less than 1.10 or extreme conditions of temperature and pressure are required. In those cases or for vapor, solid, or wet solid feeds, a number of other separation methods should be considered. These are listed in Table 8.1 in order of technical maturity as determined by Keller (1987), except for a few added separation methods not considered by Keller.

As noted in Table 8.1, the feed to a separation unit usually consists of a single vapor, liquid, or solid phase. If the feed is comprised of two or more coexisting phases, consideration

Table 8.1 Common Industrial Separation Methods

Separation Method	Phase Condition of Feed	Separating Agent(s)	Developed or Added Phase	Separation Property
Flash	L and/or V	Pressure reduction or heat transfer ESA	V or L	Volatility
Distillation (ordinary)	L and/or V	Heat transfer or shaft work ESA	V or L	Volatility
Gas absorption	V	Liquid absorbent MSA	L	Volatility
Stripping	L	Vapor stripping agent MSA	V	Volatility
Extractive distillation	L and/or V	Liquid solvent and heat transfer MSA	L and V	Volatility
Azeotropic distillation	L and/or V	Liquid entrainer and heat transfer MSA	L and V	Volatility
Liquid–liquid extraction	L	Liquid solvent MSA	Second L	Solubility
Crystallization	L	Heat transfer ESA	S	Solubility or melting point
Gas adsorption	V	Solid adsorbent MSA	S	Adsorbability
Liquid adsorption	L	Solid adsorbent MSA	S	Adsorbability
Membrane	L or V	Membrane ESA	Membrane	Permeability and/or solubility
Supercritical extraction	L or V	Supercritical solvent MSA	Supercritical fluid	Solubility
Leaching	S	Liquid solvent MSA	L	Solubility
Drying	S and L	Heat transfer ESA	V	Volatility
Desublimation	V	Heat transfer ESA	S	Volatility

should be given to separating the feed stream into two phases by some mechanical means, of the type shown in Figure 8.2, and then sending the separated phases to different separation units, each appropriate for the phase condition of the stream.

The separation of a feed mixture into streams of differing chemical compositions is achieved by forcing individual species into different spatial locations. This is accomplished by any one or a combination of four common industrial techniques: (1) the creation by heat transfer, shaft work, or pressure reduction of a second phase; (2) the introduction into the system of a second fluid phase; (3) the addition of a solid phase on which selective adsorption can occur; and (4) the placement of a selective membrane barrier. Unlike the mixing of chemical species, which is a spontaneous process, the separation of a mixture of chemicals requires an expenditure of some form of energy. In the first technique, no other chemicals are added to the feed mixture and the separation is achieved by an energy-separating agent (ESA), usually heat transfer, which causes the formation of a second phase. The components are separated by differences in volatility, thus causing each species to favor one phase over another. In the second technique, a second phase is added to the separation unit in the form of a solvent as a mass-separating agent (MSA) that selectively dissolves or alters the volatility of certain species of the mixture. A subsequent separation step is usually required to recover the solvent for recycle. The third technique involves the addition of solid particles that selectively adsorb certain species of the mixture. Subsequently, the particles must be treated by another separation method to recover the adsorbed species and regenerate the adsorbent for further use. Thus, the particles act as an MSA. The fourth technique imposes a barrier that allows the permeation of some species over others. A mechanical energy loss accompanies the permeation. Thus, this technique involves an ESA. For all four techniques, mass transfer controls the rate of migration of species from one phase to another. Except for the fourth technique, the extent of mass transfer is limited by thermodynamic equilibrium between the phases. In the case of membrane separations, the exiting phases do not approach equilibrium; rather the separation occurs strictly because of differences in the rates of permeation through the membrane.

8.2 CRITERIA FOR SELECTION OF SEPARATION METHODS

The development of a separation process requires the selection of (1) separation methods, (2) ESAs and/or MSAs, (3) separation equipment, (4) the optimal arrangement or sequencing of the equipment, and (5) the optimal operating conditions of temperature and pressure for the equipment.

When the process feed is a binary mixture and the task is to separate that mixture into two products, a single separation device may suffice if an ESA is used. If an MSA is necessary, an additional separation device will be required to recover the MSA for recycle. For a multicomponent feed that is to be separated into nearly pure components and/or one or more multicomponent products, more than one separation device is usually required. Not only must these devices be selected, an optimal arrangement of the devices must be sought. In devising such a separation sequence, it is preferable not to separate components that must be blended later to form desired multicomponent products. However, many exceptions exist to this rule. For example, in Figure 8.6, a six-component mixture is separated into four products, one of which contains 1-butene and *cis*- and *trans*-2-butene. However, the process in Figure 8.7 shows the separation of 1-butene from the 2-butenes and subsequent blending to obtain the desired olefin mixture. The unnecessary separation is carried out because the volatility of *n*-butane is intermediate between that of 1-butene and the two 2-butene isomers. The process shown in Figure 8.7 is the most economical one known. In a multicomponent separation process, each separation operation generally separates between two components, in which case the minimum number of operations is one less than the number of products. However, there are a growing number of exceptions to this rule, and cases are described later for which a single separation operation may produce only a partial separation.

Phase Condition of the Feed as a Criterion

When selecting a separation method from Table 8.1, the phase condition of the feed is considered first.

Vapor Feeds

If the feed is a vapor or is readily converted to a vapor, the following operations from Table 8.1 should be considered: (1) partial condensation (the opposite of a flash or partial vaporization), (2) distillation under cryogenic conditions, (3) gas absorption, (4) gas adsorption, (5) gas permeation with a membrane, and (6) desublimation.

Liquid Feeds

If the feed is a liquid or is readily converted to a liquid, a number of the operations in Table 8.1 may be applicable: (1) flash or partial vaporization, (2) (ordinary) distillation, (3) stripping, (4) extractive distillation, (5) azeotropic distillation, (6) liquid–liquid extraction, (7) crystallization, (8) liquid adsorption, (9) dialysis, reverse osmosis, ultrafiltration, and pervaporation with a membrane, and (10) supercritical extraction. A flash and the different types of distillation are also applicable for feeds consisting of combined liquid and vapor phases.

Slurries, Wet Cakes, and Dry Solids

Slurry feeds are generally separated first by filtration or centrifugation to obtain a wet cake, which is then separated into a vapor and a dry solid by drying. Feeds consisting of dry solids can be leached with a selective solvent to separate the components.

Separation Factor as a Criterion

The second consideration for the selection of a separation method is the *separation factor*, SF, that can be achieved by the particular separation method for the separation between two key components of the feed. This factor, for the separation of key component 1 from key component 2 between phases I and II, for a single stage of contacting, is defined as

$$\mathrm{SF} = \frac{C_1^{\mathrm{I}}/C_2^{\mathrm{I}}}{C_1^{\mathrm{II}}/C_2^{\mathrm{II}}} \tag{8.1}$$

where C_j^i is a composition (expressed as a mole fraction, mass fraction, or concentration) of component j in phase i. If phase I is to be rich in component 1 and phase II is to be rich in component 2, then SF must be large. The value of SF is limited by thermodynamic equilibrium, except for membrane separations that are controlled by relative rates of mass transfer through the membrane. For example, in the case of distillation, using mole fractions as the composition variable and letting phase I be the vapor and phase II be the liquid, the limiting value of SF is given in terms of vapor and liquid equilibrium ratios (K-values) by

$$\mathrm{SF} = \frac{y_1/y_2}{x_1/x_2} = \frac{y_1/x_1}{y_2/x_2} = \frac{K_1}{K_2} = \alpha_{1,2} \tag{8.2}$$

where α is the relative volatility. In general, components 1 and 2 are designated in such a manner that SF > 1.0. Consequently, the larger the value of SF, the more feasible is the particular separation operation. However, when seeking a desirable SF value, it is best to avoid extreme conditions of temperature that may require refrigeration or damage heat-sensitive materials; pressures that may require gas compression or vacuum; and MSA concentrations that may require expensive means to recover the MSA. In general, operations employing an ESA are economically feasible at a lower value of SF than are those employing an MSA. In particular, provided that vapor and liquid phases are readily formed, distillation should always be considered first as a possible separation operation if the feed is a liquid or partially vaporized.

When a multicomponent mixture forms nearly ideal liquid and vapor solutions, and the ideal gas law holds, the K-values and relative volatility can be readily estimated from vapor pressure data. Such K-values are referred to as ideal or Raoult's law K-values. Then, the SF for vapor–liquid separation operations employing an ESA (partial evaporation, partial condensation, or distillation) is given by

$$\mathrm{SF} = \alpha_{1,2} = \frac{P_1^s}{P_2^s} \tag{8.3}$$

where P_i^s is the vapor pressure of component i. When the components form moderately nonideal liquid solutions (hydrocarbon mixtures or homologous series of other organic compounds) and/or pressures are elevated, an equation-of-state, such as Soave–Redlich–Kwong (SRK) or Peng–Robinson (PR), may be necessary for the estimation of the separation factor, using

$$\mathrm{SF} = \alpha_{1,2} = \frac{\bar{\phi}_1^L/\bar{\phi}_1^V}{\bar{\phi}_2^L/\bar{\phi}_2^V} \tag{8.4}$$

where $\bar{\phi}_i$ is the mixture fugacity coefficient of component i.

For vapor–liquid separation operations (e.g., azeotropic and extractive distillation) that use an MSA that causes the formation of a nonideal liquid solution but operate at near-ambient pressure, expressions for the *K*-values of the key components are based on a modified Raoult's law that incorporates liquid-phase activity coefficients. Thus, the separation factor is given by

$$\mathrm{SF} = \alpha_{1,2} = \frac{\gamma_1^L P_1^s}{\gamma_2^L P_2^s} \tag{8.5}$$

where γ_i is the activity coefficient of component *i*, which is estimated from the Wilson, NRTL, UNIQUAC, or UNIFAC equations and is a strong function of mixture composition.

If an MSA is used to create two liquid phases, such as in liquid–liquid extraction, the SF is referred to as the relative selectivity, β:

$$\mathrm{SF} = \beta_{1,2} = \frac{\gamma_1^{\mathrm{II}}/\gamma_2^{\mathrm{II}}}{\gamma_1^{\mathrm{I}}/\gamma_2^{\mathrm{I}}} \tag{8.6}$$

where phase II is usually the MSA-rich phase and component 1 is more selective for the MSA-rich phase than is component 2.

In general, MSAs for extractive distillation and liquid–liquid extraction are selected according to their ease of recovery for recycle and to achieve relatively large values of SF. Such MSAs are often polar organic compounds (e.g., furfural), used in the example earlier to separate *n*-butane from 2-butenes. In some cases, the MSA is selected in such a way that it forms one or more homogeneous or heterogeneous azeotropes with the components in the feed. For example, the addition of *n*-butyl acetate to a mixture of acetic acid and water results in a heterogeneous minimum-boiling azeotrope of the acetate with water. The azeotrope is taken overhead, the acetate and water layers are separated, and the acetate is recirculated.

Although the degree of separation that can be achieved for a given value of SF is almost always far below that required to attain necessary product purities, the application of efficient countercurrent-flow cascades of many contacting stages, as in distillation operations, can frequently achieve sharp separations. For example, consider a mixture of 60 mol% propylene and 40 mol% propane. It is desired to separate this mixture into two products at 290 psia, one containing 99 mol% propylene and the other 95 mol% propane. By material balance, the former product would constitute 58.5 mol% of the feed and the latter 41.5 mol%. From equilibrium thermodynamics, the relative volatility for this mixture is approximately 1.12. A single equilibrium vaporization at 290 psia to produce 58.5 mol% vapor results in products that are far short of the desired compositions: a vapor containing just 61.12 mol% propylene and a liquid containing just 51.36 mol% propane at 51.4°C. However, with a countercurrent cascade of such stages in a simple (single-feed, two-product) distillation column with reflux and boilup, the desired products can be achieved with 200 stages and a reflux ratio of 15.9.

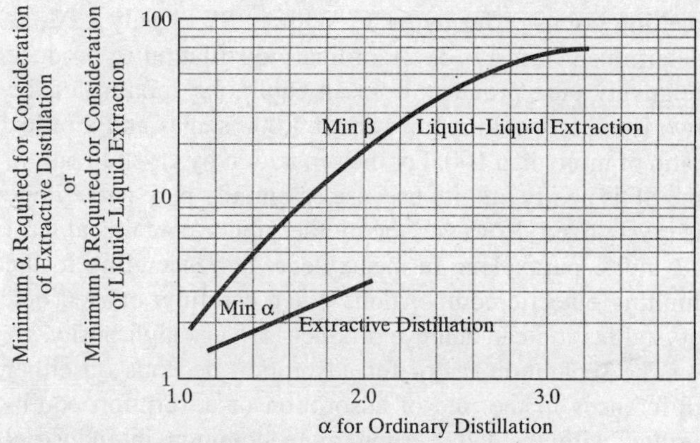

Figure 8.8 Relative selectivities for equal-cost separators (Souders, 1964).

Single-stage operations (e.g., partial vaporization or partial condensation with the use of an ESA) are utilized only if SF between the two key components is very large or if a rough or partial separation is needed. For example, if SF = 10,000, a mixture containing equimolar parts of components 1 and 2 could be partially vaporized to give a vapor containing 99 mol% of component 1 and a liquid containing 99 mol% of component 2. At low values of SF, lower than 1.10 but greater than 1.05, ordinary distillation may still be the most economical choice. However, an MSA may be able to enhance the value of SF for an alternative separation method to the degree that the method becomes more economical than ordinary distillation. As illustrated in Figure 8.8, from Souders (1964), extractive distillation or liquid–liquid extraction may be preferred if the SF can be suitably enhanced. If SF = 2 for ordinary distillation, it must be above 3.3 for extractive distillation to be an acceptable alternative, and above 18 for liquid–liquid extraction.

Unless values of SF are about 10 or above, absorption and stripping operations cannot achieve sharp separation between two components. Nevertheless, these operations are used widely for preliminary or partial separations where the separation of one key component is sharp, but only a partial separation of the other key component is adequate. The degree of sharpness of separation is given by the recovery factor RF,

$$\mathrm{RF} = \frac{n_i^I}{n_i^F} \tag{8.7}$$

where *n* is moles or mass, *I* is the product rich in *i*, and *F* is the feed.

The separation of a solid mixture may be necessary when one or more (but not all) of the components is (are) not readily melted, sublimed, or vaporized. Such operations may even be preferred when boiling points are close but melting points are far apart, as is the case with many isomeric pairs. The classic example is the separation of metaxylene from paraxylene, whose normal boiling points differ only by 0.8°C, but whose melting points differ by 64°C. With an SF of only 1.02, as determined from Eq. (8.2), ordinary distillation to produce relatively pure products from an equimolar mixture of the two isomers would require about 1,000 stages and a reflux ratio of more than 100. For the separation by crystallization, the SF is nearly infinity because essentially pure paraxylene is crystallized. However, the mother liquor contains at least 13 mol% paraxylene in metaxylene, corresponding to the limiting eutectic composition. When carefully carried out, crystallization can achieve products of very high purity.

The separation factor for adsorption depends on either differences in the rate of adsorption or adsorption equilibrium, with the latter being more common in industrial applications. For equilibrium adsorption, Eq. (8.1) applies, where the concentrations are those at equilibrium on the adsorbed layer within the pores of the adsorbent and in the bulk fluid external to the adsorbent particles. High selectivity for adsorbents is achieved either by sieving, as with molecular-sieve zeolites or carbon, or by large differences in adsorbability. For example, in the case of molecular-sieve zeolites, aperture sizes of 3, 4, 5, 8, and 10 Å are available. Thus, nitrogen molecules, with a kinetic diameter of about 3.6 Å, can be separated from ammonia, with a kinetic diameter of about 2.6 Å, using a zeolite with an aperture of 3 Å. Only the ammonia is adsorbed. Adsorbents of silica gel and activated alumina, having wide distributions of pore diameters in the range of 20 to 100 Å, are highly selective for water, while activated carbon with pore diameters in the same range is highly selective for organic compounds. When adsorption is conducted in fixed beds, essentially complete removal from the feed of those components with high selectivity can be achieved until breakthrough occurs. Before breakthrough, regeneration or removal of the adsorbent is required.

If only a small amount of one component is present in a mixture, changing the phase of the components in high concentrations should be avoided. In such a case, absorption, stripping, or selective adsorption best removes the minor component. Adsorption is particularly effective because of the high selectivity of adsorbents and is widely used for purification, where small amounts of a solute are removed from a liquid or vapor feed.

For membrane separation operations, SF may still be defined by Eq. (8.1). However, SF is governed by relative rates of mass transfer, in terms of permeabilities, rather than by equilibrium considerations. For the ideal case where the downstream concentration is negligible compared to the upstream concentration, the separation factor reduces to:

$$\mathrm{SF} = \frac{P_{M_1}}{P_{M_2}} \tag{8.8}$$

where P_{M_i} is the permeability of species i. In most cases, the value of SF must be established experimentally. In general, membrane separation operations should be considered whenever adsorption methods are considered. Membranes are either porous or nonporous. If porous, the permeability is proportional to the diffusivity through the pore. If nonporous, the permeability is the product of the solubility of the molecule in the membrane and its diffusivity for travel through the membrane. An example of the use of membranes is gas permeation with nonporous hollow fibers to separate hydrogen, helium, carbon dioxide, and/or water vapor from gases containing oxygen, nitrogen, carbon monoxide, and/or light hydrocarbons. For a typical membrane, the SF between hydrogen and methane is 6. Because it is difficult to achieve large numbers of stages with membranes, an SF of this magnitude is not sufficient to achieve a sharp separation, but is widely used to make a partial separation. Sharp separations can be achieved by sieving when the kinetic molecular diameters of the components to be separated differ widely, and when membrane pore diameter lies between those kinetic diameters.

Supercritical extraction utilizes the solvent power of a gas at near-critical conditions. It is the preferred method for the removal of undesirable ingredients from foodstuffs with carbon dioxide. The separation factor, which is given by Eq. (8.1), is difficult to estimate from equations of state using Eq. (8.4) and is best determined by experiment. Equation (8.1) also applies for leaching (solid–liquid extraction), often using a highly selective solvent. As with supercritical extraction, the value of SF is best determined by experiment. Because mass transfer in a solid is very slow, it is important to preprocess the solid to drastically decrease the distance for diffusion. Typical methods involve making thin slices of the solid or pulverizing it. Desublimation is best applied when a sublimable component is to be removed from noncondensable components of a gas stream, corresponding to a very large separation factor.

Reason for the Separation as a Criterion

A final consideration in the selection of a separation method is the reason for the separation. Possible reasons are (1) purification of a species or group of species, (2) removal of undesirable constituents, and (3) recovery of constituents for subsequent processing or removal. In the case of purification, the use of an MSA method may avoid exposure with an ESA method to high temperatures that may cause decomposition. In some cases, removal of undesirable species together with a modest amount of desirable species may be economically acceptable. Likewise, in the recovery of constituents for recycle, a high degree of separation from the product(s) may not be necessary.

8.3 SELECTION OF EQUIPMENT

Only a very brief discussion of equipment for separation operations is presented here. Much more extensive presentations, including drawings and comparisons, are given in *Perry's Chemical Engineers' Handbook* (Green and Perry, 2008) and by Kister (1992), Walas (1988), Seader and Henley (2006), and in *Visual Encyclopedia of Chemical Engineering Equipment for MacIntosh and Windows 95/NT* by Montgomery, as described at www.engin.umich.edu/labs/mel/equipflyer/equip.html. In general, equipment selection is based on stage or mass-transfer efficiency, pilot-plant tests, scale-up feasibility, investment and operating cost, and ease of maintenance.

Absorption, Stripping, and Distillation

For absorption, stripping, and all types of distillation (i.e., vapor–liquid separation operations), either trayed or packed columns are used. Trayed columns are usually preferred for initial installations, particularly for columns 3 ft or more in diameter. However, packed columns should be given serious consideration for operation under vacuum or where a low-pressure drop is desired. Other applications favoring packed columns are corrosive systems, foaming systems, and cases where low liquid holdup is desired. Packing is also generally specified for revamps. Applications favoring trayed columns are feeds containing solids, high liquid-to-gas ratios, large-diameter columns, and where operation over a wide range of conditions is necessary. The three most commonly used tray types are sieve, valve, and bubble-cap. However, because of high cost, the latter is specified only when a large liquid holdup is required on the tray, for example, when conducting a chemical reaction simultaneously with distillation. Sieve trays are the least expensive and have the lowest pressure drop per tray, but they have the narrowest operating range (turndown ratio). Therefore, when flexibility is required, valve trays are a better choice. Many different types of packings are available. They are classified as random or structured. The latter are considerably more expensive than the former, but the latter have the lowest pressure drop, the highest efficiency, and the highest capacity compared to both random packings and trays. For that reason, structured packings are often considered for column revamps.

Liquid-Liquid Extraction

For liquid–liquid extraction, an even greater variety of equipment is available, including multiple mixer-settler units or single countercurrent-flow columns with or without mechanical agitation. Very compact, but expensive, centrifugal extractors are also available. When the equivalent of only a few theoretical stages is required, mixer-settler units may be the best choice because efficiencies approaching 100% are achievable in each unit. For a large number of stages, columns with mechanical agitation may be favored. Packed and perforated tray columns can be very inefficient and are not recommended for critical separations.

Membrane Separation

Most commercial membrane separations use natural or synthetic glassy or rubbery polymers. To achieve high permeability and selectivity, nonporous materials are preferred, with thicknesses ranging from 0.1 to 1.0 micron, either as a surface layer or film onto or as part of much thicker asymmetric or composite membrane materials, which are fabricated primarily into spiral-wound and hollow-fiber-type modules to achieve a high ratio of membrane surface area to module volume.

Adsorption

For commercial applications, an adsorbent must be chosen carefully to give the required selectivity, capacity, stability, strength, and regenerability. The most commonly used adsorbents are activated carbon, molecular-sieve carbon, molecular-sieve zeolites, silica gel, and activated alumina. Of particular importance in the selection process is the adsorption isotherm for competing solutes when using a particular adsorbent. Most adsorption operations are conducted in a semicontinuous cyclic mode that includes a regeneration step. Batch slurry systems are favored for small-scale separations, whereas fixed-bed operations are preferred for large-scale separations. Quite elaborate cycles have been developed for the latter.

Leaching

Equipment for leaching operations is designed for either batchwise or continuous processing. For rapid leaching, it is best to reduce the size of the solids by grinding or slicing. The solids are contacted by the solvent using either percolation or immersion. A number of different patented devices are available.

Crystallization

Crystallization operations include the crystallization of an inorganic compound from an aqueous solution (solution crystallization) and the crystallization of an organic compound from a mixture of organic chemicals (melt crystallization). On a large scale, solution crystallization is frequently conducted continuously in a vacuum evaporating draft-tube baffled crystallizer to produce crystalline particles, whereas the falling-film crystallizer is used for melt crystallization to produce a dense layer of crystals.

Drying

A number of factors influence the selection of a dryer from the many different types available. These factors are dominated by the nature of the feed, whether it be granular solids, a paste, a slab, a film, a slurry, or a liquid. Other factors include the need for agitation, the type of heat source (convection, radiation, conduction, or microwave heating), and the degree to which the material must be dried. The most commonly employed continuous dryers include tunnel, belt, band, turbo-tray, rotary, steam-tube rotary, screw-conveyor, fluidized-bed, spouted-bed, pneumatic-conveyor, spray, and drum dryers.

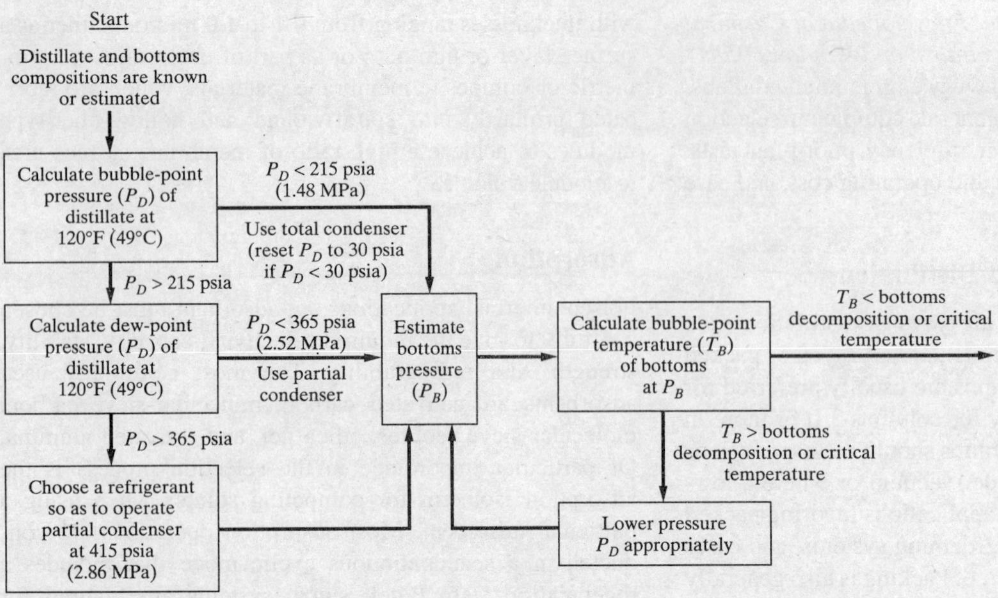

Figure 8.9 Algorithm for establishing distillation column pressure and condenser type.

8.4 SEQUENCING OF ORDINARY DISTILLATION COLUMNS FOR THE SEPARATION OF NEARLY IDEAL FLUID MIXTURES

Multicomponent mixtures are often separated into more than two products. Although one piece of equipment of complex design might be devised to produce all the desired products, a sequence of two-product separators is more common.

For nearly ideal feeds such as hydrocarbon mixtures and mixtures of a homologous series, for example, alcohols, the most economical sequence will often include only ordinary distillation columns, provided that the following conditions hold:

1. The relative volatility between the two selected key components for the separation in each column is >1.05.
2. The reboiler duty is not excessive. An example of an excessive duty occurs in the distillation of a mixture with a low relative volatility between the two key components, where the light key component is water, which has a very high heat of vaporization.
3. The tower pressure does not cause the mixture to approach its critical temperature.
4. The overhead vapor can be at least partially condensed at the column pressure to provide reflux without excessive refrigeration requirements.
5. The bottoms temperature at the tower pressure is not so high that chemical decomposition occurs.
6. Azeotropes do not prevent the desired separation.
7. Column pressure drop is tolerable, particularly if operation is under vacuum.

Column Pressure and Type of Condenser

During the development of distillation sequences, it is necessary to make at least preliminary estimates of column operating pressures and condenser types (total or partial). The estimates are facilitated by the use of the algorithm in Figure 8.9, which is conservative. Assume that cooling water is available at 90°F, sufficient to cool and condense a vapor to 120°F. The bubble-point pressure is calculated at 120°F for an estimated distillate composition. If the computed pressure is less than 215 psia, use a total condenser unless a vapor distillate is required, in which case use a partial condenser. If the pressure is less than 30 psia, set the condenser pressure to 30 psia and avoid near-vacuum operation. If the distillate bubble-point pressure is greater than 215 psia, but less than 365 psia, use a partial condenser. If it is greater than 365 psia, determine the dew-point pressure for the distillate as a vapor. If the pressure is greater than 365 psia, operate the condenser at 415 psia with a suitable refrigerant in place of cooling water. For the selected condenser pressure, add 10 psia to estimate the bottoms pressure, and compute the bubble-point temperature for an estimated bottoms composition. If that temperature exceeds the decomposition or critical temperature of the bottoms, reduce the condenser pressure appropriately.

Number of Sequences of Ordinary Distillation Columns

Initial consideration is usually given to a sequence of ordinary distillation columns, where a single feed is sent to each column and the products from each column number just two, the distillate and the bottoms. For example, consider a mixture of benzene, toluene, and biphenyl. Because the

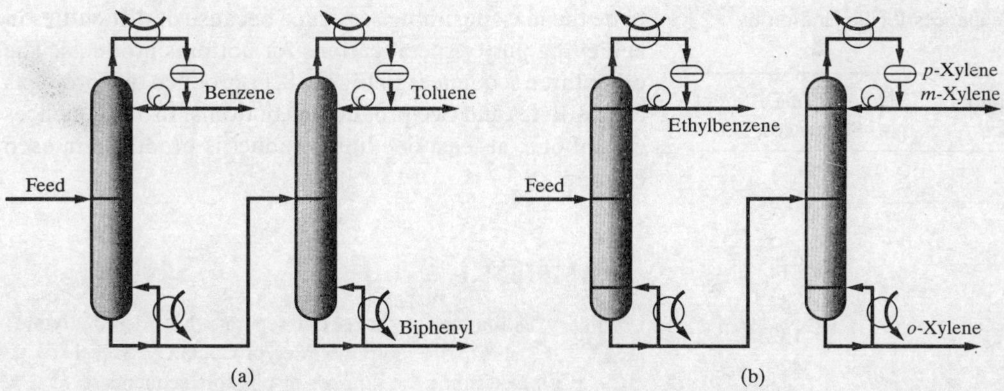

Figure 8.10 Distillation configurations for separation of ternary mixtures: (a) separation of a benzene-toluene-biphenyl mixture; (b) separation of xylene isomers.

normal boiling points of the three components (80.1, 110.8, and 254.9°C, respectively) are widely separated, the mixture can be conveniently separated into three nearly pure components by ordinary distillation. A common process for separating this mixture is the sequence of two ordinary distillation columns shown in Figure 8.10a. In the first column, the most volatile component, benzene, is taken overhead as a distillate final product. The bottoms is a mixture of toluene and biphenyl, which is sent to the second column for separation into the two other final products: a distillate of toluene and a bottoms of biphenyl, the least volatile component.

Even if a sequence of ordinary distillation columns is used, not all columns need give nearly pure products. For example, Figure 8.10b shows a distillation sequence for the separation of a mixture of ethylbenzene, *p*-xylene, *m*-xylene, and *o*-xylene into only three products: nearly pure ethylbenzene, a mixture of *p*- and *m*-xylene, and nearly pure *o*-xylene. The para and meta isomers are not separated because the normal boiling points of these two compounds differ by only 0.8°C, making separation by distillation impractical.

Note in Figure 8.10 that it takes a sequence of two ordinary distillation columns to separate a mixture into three products. Furthermore, other sequences can produce the same final products. For example, the separation of benzene, toluene, and biphenyl, shown in Figure 8.10a, can also be achieved by removing biphenyl as bottoms in the first column, followed by the separation of benzene and toluene in the second column. However, the separation of toluene from benzene and biphenyl by ordinary distillation in the first column is impossible, because toluene is intermediate in volatility. Thus, the number of possible sequences is limited to two for this case of the separation of a ternary mixture into three nearly pure products.

Now consider the more general case of the synthesis of all possible ordinary distillation sequences for a multicomponent feed that is to be separated into *P* final products that are nearly pure components and/or multicomponent mixtures. The components in the feed are ordered by volatility, with the first component being the most volatile. This order is almost always consistent with that for normal boiling point if the mixture forms nearly idéal liquid solutions, such that Eq. (8.3) applies. Assume that the order of volatility of the components does not change as the sequence proceeds. Furthermore, assume that any multicomponent products contain only components that are adjacent in volatility. For example, suppose that the previously cited mixture of benzene, toluene, and biphenyl is to be separated into toluene and a multicomponent product of benzene and biphenyl. With ordinary distillation, it would be necessary first to produce products of benzene, toluene, and biphenyl, and then blend the benzene and biphenyl.

An equation for the number of different sequences of ordinary distillation columns, N_s, to produce a number of products, P, can be developed in the following manner. For the first separator in the sequence, $P - 1$ separation points are possible. For example, if the desired products are A, B, C, D, and E in order of decreasing volatility, then the possible separation points are $5 - 1 = 4$, as follows: A–B, B–C, C–D, and D–E. Now let j be the number of final products that must be developed from the distillate of the first column. For example, if the separation point in the first column is C–D, then $j = 3$ (A, B, C). Then $P - j$ equals the number of final products that must be developed from the bottoms of the first column. If N_i is the number of different sequences for i final products, then, for a given separation point in the first column, the number of sequences is $N_j N_{P-j}$. But in the first separator, $P - 1$ different separation points are possible. Thus, the number of different sequences for P products is the following sum:

$$N_s = \sum_{j=1}^{P-1} N_j N_{P-j} = \frac{[2(P-1)]!}{P!(P-1)!} \tag{8.9}$$

Application of Eq. (8.9) gives results shown in Table 8.2 for sequences producing up to 10 products. As shown, the number of sequences grows rapidly as the number of final products increases.

Equation (8.9) gives five possible sequences of three columns for a four-component feed. These sequences are shown in Figure 8.11. The first, where all final products but one are distillates, is often referred to as the *direct sequence*, and is widely used in industry because distillate final products are more free of impurities such as objectionable high-boiling compounds and solids. If the purity

Table 8.2 Number of Possible Sequences for Separation by Ordinary Distillation

Number of Products, P	Number of Separators in the Sequence	Number of Different Sequences, N_s
2	1	1
3	2	2
4	3	5
5	4	14
6	5	42
7	6	132
8	7	429
9	8	1,430
10	9	4,862

of the final bottoms product (D) is critical, it may be produced as a distillate in an additional column called a *rerun* (or *finishing*) column. If all products except one are bottoms products, the sequence is referred to as the *indirect sequence*. This sequence is generally considered to be the least desirable sequence because of difficulties in achieving purity specifications for bottoms products. The other three sequences in Figure 8.11 produce two products as distillates and two products as bottoms. In all sequences except one, at least one final product is produced in each column.

EXAMPLE 8.1

Ordinary distillation is to be used to separate the ordered mixture C_2, $C_3^{=}$, C_3, $1\text{–}C_4^{=}$, nC_4 into the three products C_2; ($C_3^{=}$, $1\text{–}C_4^{=}$); (C_3, nC_4). Determine the number of possible sequences.

SOLUTION

Neither multicomponent product contains adjacent components in the ordered list. Therefore, the mixture must be completely separated with subsequent blending to produce the ($C_3^{=}$, $1\text{–}C_4^{=}$) and (C_3, nC_4) products. Thus, from Table 8.2 with P taken as 5, $N_s = 14$.

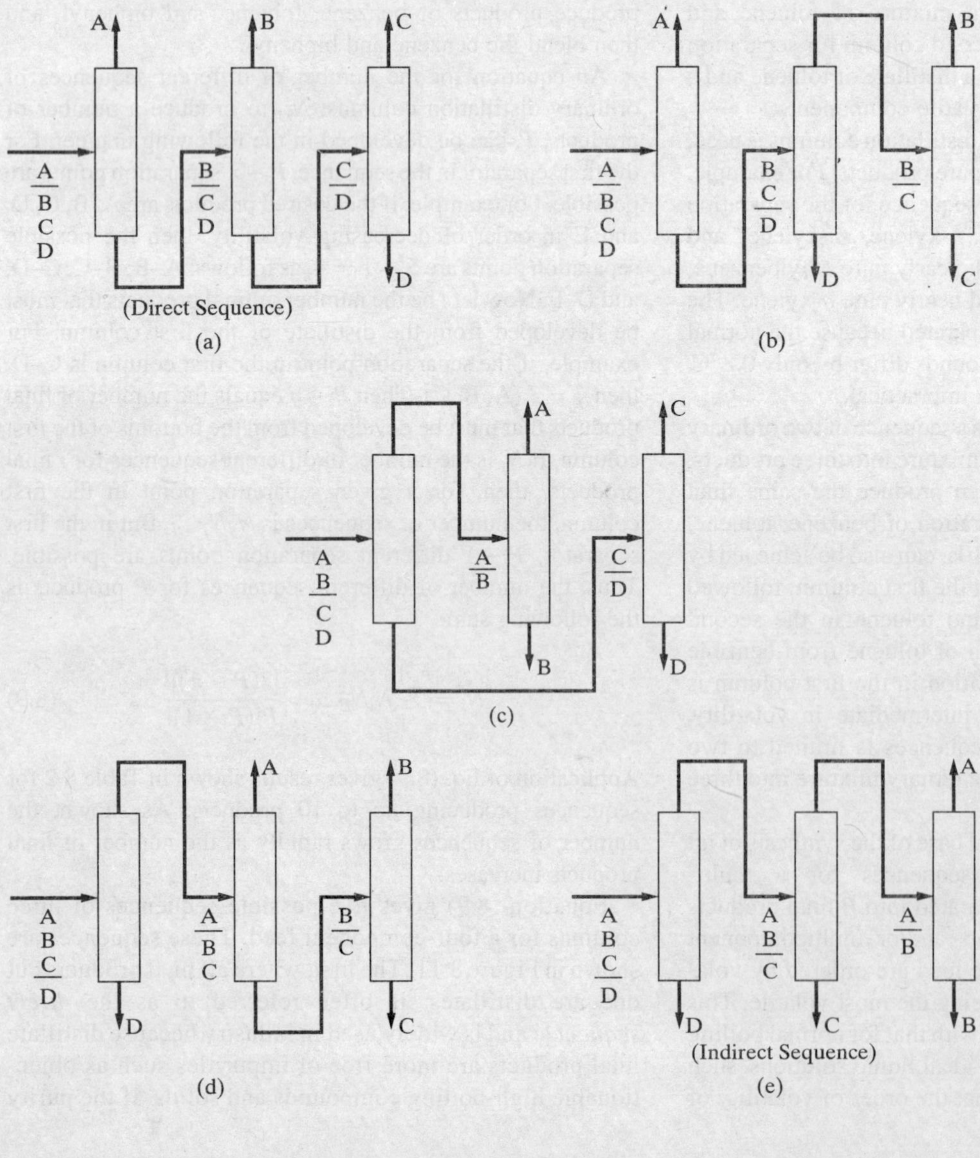

Figure 8.11 The five sequences for a four-component feed.

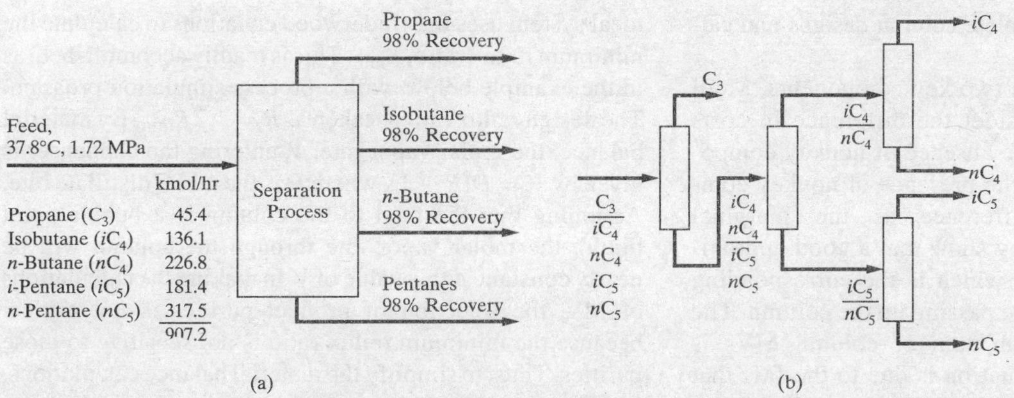

Figure 8.12 Synthesis problem and separation train for Example 8.2: (a) paraffin separation problem; (b) sequence developed from heuristics.

Heuristics for Determining Favorable Sequences

When the number of products is three or four, designing and costing all possible sequences can best determine the most economical sequence. Often, however, unless the feed mixture has a wide distribution of component concentrations or a wide variation of relative volatilities for the possible separation points, the costs will not vary much and the sequence selection may be based on operation factors. In that case, the direct sequence is often the choice. Otherwise, a number of heuristics that have appeared in the literature, starting in 1947, have proved useful for reducing the number of sequences for detailed examination. The most useful of these heuristics are:

1. Remove thermally unstable, corrosive, or chemically reactive components early in the sequence.
2. Remove final products one-by-one as distillates (the direct sequence).
3. Sequence separation points to remove, early in the sequence, those components of greatest molar percentage in the feed.
4. Sequence separation points in the order of decreasing relative volatility so that the most difficult splits are made in the absence of the other components.
5. Sequence separation points to leave last those separations that give the highest-purity products.
6. Sequence separation points that favor near equimolar amounts of distillate and bottoms in each column.

None of these heuristics require column design and costing. Unfortunately, however, these heuristics often conflict with each other. Thus, more than one sequence will be developed, and cost and other factors will need to be considered to develop an optimal final design. When energy costs are relatively high, the sixth heuristic often leads to the most economical sequence. Heuristics 2–6 are consistent with observations about the effect of the nonkey components on the separation of two key components. These nonkey components can increase the reflux and boilup requirements, which, in turn, increase column diameter and reboiler operating cost. These, and the number of trays, are the major factors affecting the investment and operating costs of a distillation operation.

EXAMPLE 8.2

Consider the separation problem shown in Figure 8.12a, except that separate isopentane and n-pentane products are also to be obtained with 98% recoveries. Use heuristics to determine a good sequence of ordinary distillation units.

SOLUTION

Approximate relative volatilities for all adjacent pairs are

Component pair	Approximate α at 1 atm
C_3/iC_4	3.6
iC_4/nC_4	1.5
nC_4/iC_5	2.8
iC_5/nC_5	1.35

For this example, there are wide variations in both relative volatility and molar percentages in the process feed. The choice is Heuristic 4, which dominates over Heuristic 3 and leads to the sequence shown in Figure 8.12b, where the first split is between the pair with the highest relative volatility. This sequence also corresponds to the optimal arrangement.

Marginal Vapor Rate Method

When application of the above heuristics for sequencing ordinary distillation columns is uncertain or conflicting results are obtained, it is preferable to employ sequencing methods that rely on column design and, in some cases, cost estimation. Exhaustive search to calculate the annualized cost of every sequence can determine the optimal sequence, provided that column-operating conditions are optimized, and may be justified for sequences involving just three or possibly four products. However, less rigorous methods are available that can produce good, although not always optimal, sequences. These methods, which attempt to reduce the search space, include those of Hendry and Hughes (1972), Rodrigo and Seader (1975), Gomez and Seader (1976), Seader and Westerberg (1977), and the marginal vapor rate (MV) method of Modi and Westerberg (1992). The latter method outperforms the other methods and can be applied without the necessity of complete column designs and calculations of costs.

For a given split between two key components, Modi and Westerberg (1992) consider the difference in costs between the separation in the absence of nonkey components and the separation in the presence of nonkey components, defining this difference as the marginal annualized cost (MAC). They show that a good approximation of MAC is the MV, which is the corresponding difference in molar vapor rate passing up the column. The sequence with the minimum sum of column MVs is selected. The good approximation is due to the fact that vapor rate is a good measure of cost because it is a major factor in determining column diameter as well as reboiler and condenser areas (thus, column and heat-exchanger capital costs) and reboiler and condenser duties (thus, heat-exchanger annual operating costs).

A convenient method for determining the molar vapor rate in an ordinary distillation column separating a nearly ideal system uses the Underwood equations to calculate the minimum reflux ratio, R_{min}. This is readily accomplished, as in the example below, with a process simulation program. The design reflux ratio is taken as $R = 1.2\,R_{min}$. By material balance, the molar vapor rate, V, entering the condenser is given by $V = D(R+1)$, where D is the molar distillate rate. Assuming that the feed to the column is a bubble-point liquid, the molar vapor rate through the column will be nearly constant at this value of V. In making the calculations of MV, the selection of product purities is not critical because the minimum reflux ratio is not sensitive to those purities. Thus, to simplify the material balance calculations, it is convenient to assume nearly perfect separations, with the light and lighter-than-light key components leaving in the distillate and the heavy and heavier-than-heavy key components leaving in the bottoms. Column top and bottom pressures are estimated with Figure 8.9. The column feed pressure is taken as the average of the top and bottom pressures.

EXAMPLE 8.3

Use the marginal vapor rate (MV) method to determine a sequence for the separation of light hydrocarbons specified in Figure 8.12a, except: (1) remove the propane from the feed, (2) ignore the given temperature and pressure of the feed, and (3) strive for recoveries of 99.9% of the key components in each column. Use a process simulation program, with the Soave–Redlich–Kwong equation of state for K-values and enthalpies, to set top and bottom column pressures and estimate the reflux ratio with the Underwood equations.

Table 8.3 Calculations of Marginal Vapor Rate, MV

Separation	Column Top Pressure (kPa)	Distillate Rate, D (kmol/hr)	Reflux Ratio ($R = 1.2\,R_{min}$)	Vapor Rate, $V = D(R+1)$ (kmol/hr)	Marginal Vapor Rate (kmol/hr)
A/B	680	136.2	10.7	1,594	0
A/BC	680	136.2	11.9	1,757	163
A/BCD	680	136.2	13.2	1,934	340
B/C	490	226.8	2.06	694	0
AB/C	560	362.9	1.55	925	231
B/CD	490	226.8	3.06	921	227
AB/CD	560	362.9	2.11	1,129	435
C/D	210	181.5	13.5	2,632	0
BC/D	350	408.3	6.39	3,017	385
ABC/D	430	544.4	4.96	3,245	613

SOLUTION

To produce four nearly pure products from the four-component feed, five sequences of three ordinary distillation columns each are shown in Figure 8.11. Let A = isobutane, B = n-butane, C = isopentane, and D = n-pentane. A total of 10 unique separations are embedded in Figure 8.11. These are listed in Table 8.3, together with the results of the calculations for the top column pressure, P_{top}, in kPa; the molar distillate rate, D, in kmol/hr; and the reflux ratio, R, using the shortcut (Fenske–Underwood–Gilliland or FUG) distillation model of the CHEMCAD process simulation program. This model applies the Underwood equations to estimate the minimum reflux ratio, as described by Seader and Henley (2006). Column feeds were computed as bubble-point liquids at $P_{top} + 35$ kPa. Also included in Table 8.3 are values of the column molar vapor rate, V, in kmol/hr and marginal vapor rate, MV, in kmol/hr.

From Table 8.3, the sum of the marginal vapor rates is calculated for each of the five sequences in Figure 8.11. The results are given in Table 8.4.

Table 8.4 Marginal Vapor Rates for the Five Possible Sequences

Sequence in Figure 8.11	Marginal Vapor Rate, MV (kmol/hr)
(a) Direct	567
(b)	725
(c)	435
(d)	776
(e) Indirect	844

Table 8.4 shows that the preferred sequence is the one that performs the two most difficult separations, A/B and C/D, in the absence of nonkey components. These two separations are far more difficult than the separation B/C. The direct sequence is the next best.

Complex and Thermally Coupled Distillation Columns

Following the development of an optimal or near-optimal sequence of simple, two-product distillation columns, revised sequences involving complex, rather than simple, distillation columns should be considered. Some guidance is available from a study by Tedder and Rudd (1978a, b) of the separation of ternary mixtures (A, B, and C in order of decreasing volatility) in which eight alternative sequences of one to three columns were considered, seven of which are shown in Figure 8.13. The configurations include the direct and indirect sequences (I and II), two interlinked cases (III and IV), five cases that include the use of sidestreams (III, IV, V, VI, and VII), and one case (V) involving a column with two feeds. All columns in Cases I, II, V, VI, and VII have condensers and reboilers. In Cases III and IV, the first column has a condenser and reboiler. In Case III, the rectifier column has a condenser only, while the stripper in Case IV has a reboiler only. The interlinking streams that return from the second column to the first column thermally couple the columns in Cases III and IV.

As shown in Figure 8.14, optimal regions for the various configurations depend on the process feed composition and on an ease-of-separation index (ESI), which is defined as the relative volatility ratio, $\alpha_{A,B}/\alpha_{B,C}$. It is interesting to note that a ternary mixture is separated into three products with just one column in Cases VI and VII in Figure 8.13, but the reflux requirement is excessive unless the feed contains a large amount of B, the component of intermediate volatility, and little of the component that is removed from the same section of the column as B. Otherwise, if the feed is dominated by B but also contains appreciable amounts of A and C, the prefractionator case (V) is optimal. Perhaps the biggest surprise of the study is the superiority of distillation with a vapor sidestream rectifier, which is favored for a large region of the feed composition when ESI >1.6. The results of Figure 8.14 can be extended to multicomponent separation problems involving more than three components, if difficult ternary separations are performed last.

Case V in Figure 8.13 consists of a prefractionator followed by a product column, from which all three final products are drawn. Each column is provided with its own condenser and reboiler. As shown in Figure 8.15, eliminating the condenser and reboiler in the prefractionator and providing, instead, reflux and boilup to that column from the product column can thermally couple this arrangement, which is referred to as a Petlyuk system after its chief developer and is described by Petlyuk et al. (1965). The prefractionator separates the ternary-mixture feed, ABC, into a top product containing A and B and a bottom product of B and C. Thus,

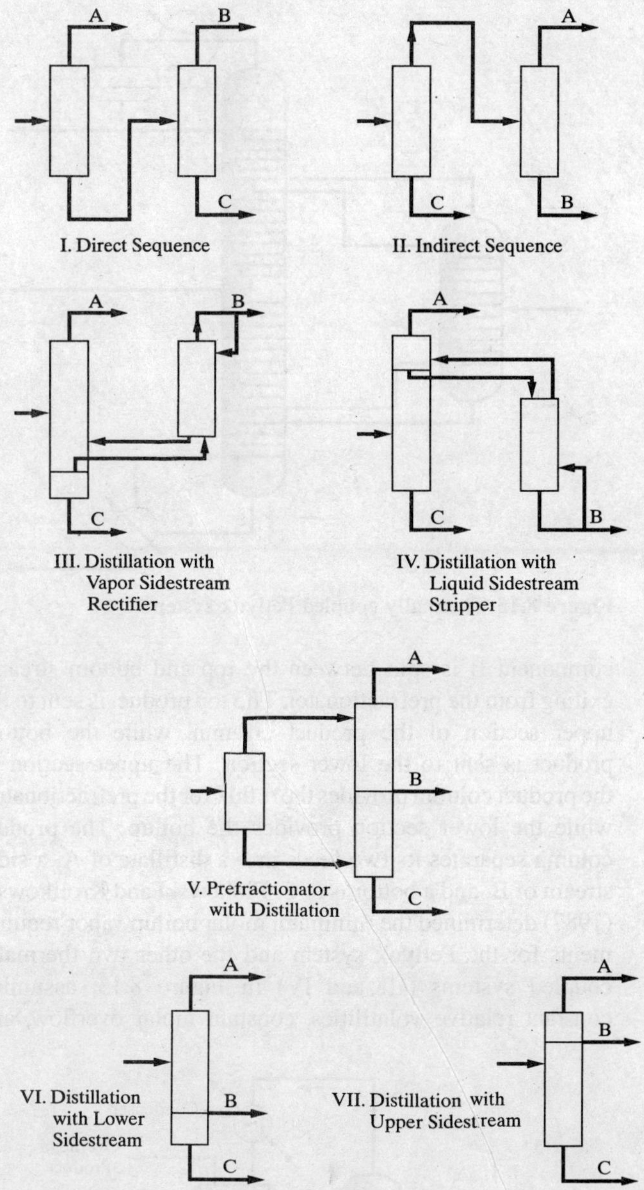

Figure 8.13 Configurations for ternary distillation.

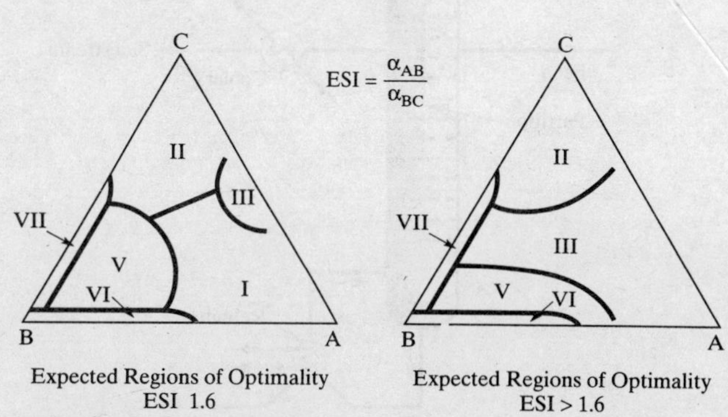

Figure 8.14 Regions of optimality for ternary distillation configurations (Tedder and Rudd, 1978a, b).

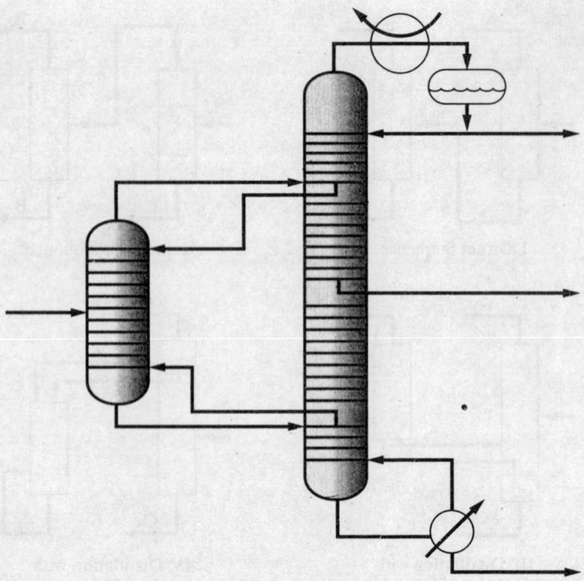

Figure 8.15 Thermally coupled Petlyuk system.

component B is split between the top and bottom streams exiting from the prefractionator. The top product is sent to the upper section of the product column, while the bottom product is sent to the lower section. The upper section of the product column provides the reflux for the prefractionator, while the lower section provides the boilup. The product column separates its two feeds into a distillate of A, a sidestream of B, and a bottoms of C. Fidkowski and Krolikowski (1987) determined the minimum molar boilup vapor requirements for the Petlyuk system and the other two thermally coupled systems (III and IV) in Figure 8.13, assuming constant relative volatilities, constant molar overflow, and

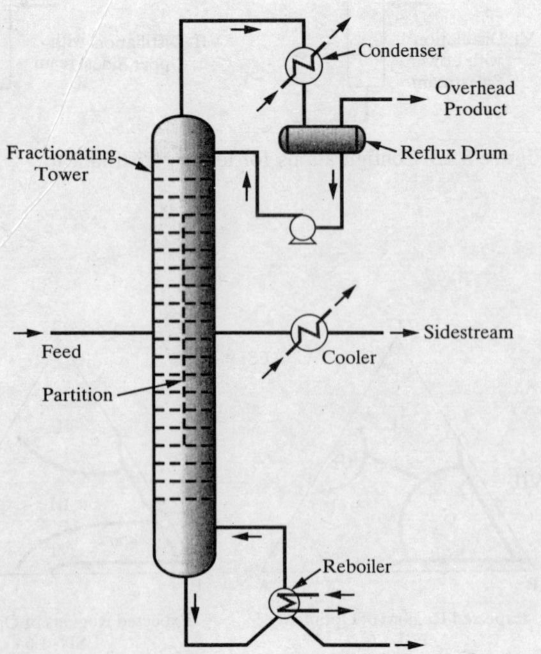

Figure 8.16 Dividing-wall (partition) column of Wright.

bubble-point liquid feed and products. They compared the requirements to those of the conventional direct and indirect sequences shown as Cases I and II in Figure 8.13 and proved that for all combinations of feed flow rates of the components A, B, and C, as well as all values of relative volatilities, that: (1) the Petlyuk system has the lowest minimum molar boilup vapor requirements, and (2) Cases III and IV in Figure 8.13 are equivalent and have lower minimum molar boilup vapor requirements than either the direct or indirect sequence.

Despite its lower vapor boilup requirements, no industrial installations of a two-column Petlyuk system have been reported. Two possible reasons for this, as noted by Agrawal and Fidkowski (1998), are: (1) an unfavorable thermodynamic efficiency when the three feed components are not close-boiling because all of the reboiler heat must be supplied at the highest temperature and all of the condenser heat must be removed at the lowest temperature; and (2) the difficulty in controlling the fractions of vapor and liquid streams in the product column that are returned to the prefractionator as boilup and reflux, respectively. The Petlyuk system can be embodied into a single column, with a significantly reduced capital cost, by using a dividing-wall column (also called divided wall and column-in-column), a concept described in a patent by Wright (1949) and shown by his patent drawing in Figure 8.16. Because the dividing-wall column makes possible savings in both energy and capital, and because control difficulties appear to have been solved, it is attracting much attention. The first dividing-wall column was installed by BASF in 1985. A number of such columns using packing have been installed in the past 15 years and the first dividing-wall column using trays was recently announced. Agrawal and Fidkowski (1998) present other thermally fully coupled (FC) systems of distillation columns that retain the benefit of a minimum vapor requirement and afford easier control. Energy savings can also be achieved by heat-integrating the two columns in a direct sequence. In Figure 8.17, Column 2 is

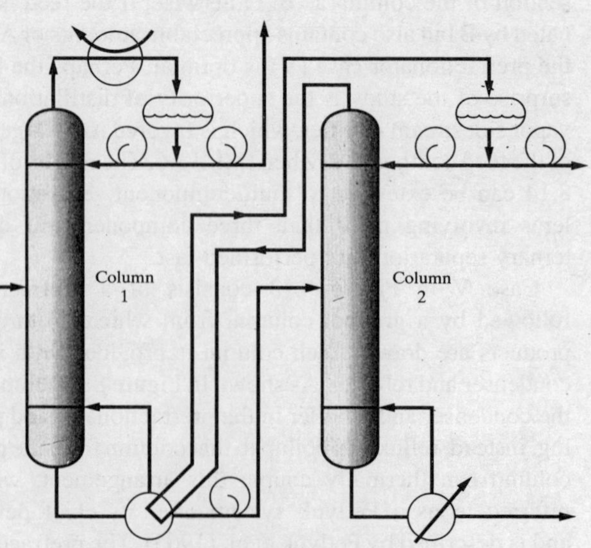

Figure 8.17 Heat-integrated direct sequence of two distillation columns.

operated at a higher pressure than Column 1, such that the condenser duty of Column 2 can provide the reboiler duty of Column 1. Rev et al. (2001) show that heat-integrated systems are often superior in annualized cost to the Petlyuk system. For further discussion of heat-integrated distillation columns, see Sections 9.9, 12.1, and 12S.3.

8.5 SEQUENCING OF OPERATIONS FOR THE SEPARATION OF NONIDEAL FLUID MIXTURES

When a multicomponent fluid mixture is nonideal, its separation by a sequence of ordinary distillation columns will not be technically and/or economically feasible if relative volatilities between key components drop below 1.05 and, particularly, if azeotropes are formed. For such mixtures, separation is most commonly achieved by sequences comprised of ordinary distillation columns, enhanced distillation columns, and/or liquid–liquid extraction equipment. Membrane and adsorption separations can also be incorporated into separation sequences, but their use is much less common. Enhanced distillation operations include extractive distillation, homogeneous azeotropic distillation, heterogeneous azeotropic distillation, pressure-swing distillation, and reactive distillation. These operations are considered in detail in *Perry's Chemical Engineers' Handbook* (Green and Perry, 2008) and by Seader and Henley (2006), Stichlmair and Fair (1998), and Doherty and Malone (2001). A design-oriented introduction to enhanced distillation is presented here.

In many processes involving oxygenated organic compounds such as alcohols, ketones, ethers, and acids, often in the presence of water, distillation separations are complicated by the presence of azeotropes. Close-boiling mixtures of hydrocarbons (e.g., benzene and cyclohexane, whose normal boiling points only differ by 1.1°F) can also form azeotropes. For these and other mixtures, special attention must be given to the *distillation boundaries* in the composition space that confine the compositions for any one column to lie within a bounded region of the composition space. To introduce these boundaries, leading to approaches for the synthesis of separation trains, several concepts concerning azeotropes and *residue curves* and *distillation lines* are reviewed in the subsections that follow.

Azeotropy

Azeotrope is an ancient Greek word that is translated "to boil unchanged," meaning that the vapor emitted has the same composition as the liquid (Swietoslawski, 1963). When classifying the many azeotropic mixtures, it is helpful to examine their deviations from Raoult' s law (Lecat, 1918).

When two or more fluid phases are in physical equilibrium, the chemical potential, fugacity, or activity of each species is the same in each phase. Thus, in terms of species mixture fugacities for a vapor phase in physical equilibrium with a single liquid phase,

$$\bar{f}_j^V = \bar{f}_j^L \qquad j = 1, \ldots, C \tag{8.10}$$

Substituting expressions for the mixture fugacities in terms of mole fractions, activity coefficients, and fugacity coefficients,

$$y_j \bar{\phi}_j^V P = x_j \gamma_j^L f_j^L \qquad j = 1, \ldots, C \tag{8.11}$$

where $\bar{\phi}$ is a mixture fugacity coefficient, γ is a mixture activity coefficient, and f is a pure-species fugacity.

For a binary mixture with an ideal liquid solution ($\gamma_j^L = 1$) and a vapor phase that forms an ideal gas solution and obeys the ideal gas law ($\bar{\phi}_j^V = 1$ and $f_j^L = P_j^s$), Eq. (8.11) reduces to the following two equations for the two components 1 and 2:

$$y_1 P = x_1 P_1^s \tag{8.12a}$$

$$y_2 P = x_2 P_2^s \tag{8.12b}$$

where P_j^s is the vapor pressure of species j.

Adding Eqs. (8.12a and 8.12b), noting that mole fractions must sum to one,

$$\begin{aligned}(y_1 + y_2)P = P &= x_1 P_1^s + x_2 P_2^s \\ &= x_1 P_1^s + (1 - x_1) P_2^s \\ &= P_2^s + (P_1^s - P_2^s) x_1 \end{aligned} \tag{8.13}$$

This linear relationship between the total pressure, P, and the mole fraction, x_1, of the most volatile species is a characteristic of Raoult's law, as shown in Figure 8.18a for the benzene-toluene mixture at 90°C. Note that the bubble-point curve $(P - x)$ is linear between the vapor pressures of the pure species (at $x_1 = 0, 1$), and the dew-point curve $(P - y_1)$ lies below it. When the (x_1, y_1) points are graphed at different pressures, the familiar vapor–liquid equilibrium curve is obtained, as shown in Figure 8.18b. Using McCabe–Thiele analysis, it is shown readily that for any feed composition, there are no limitations to the values of the mole fractions of the distillate and bottoms products from a distillation tower.

However, when the mixture forms a nonideal liquid phase and exhibits a positive deviation from Raoult's law ($\gamma_j^L > 1$, $j = 1, 2$), Eq. (8.13) becomes

$$P = x_1 \gamma_1^L P_1^s + (1 - x_1) \gamma_2^L P_2^s \tag{8.14}$$

Furthermore, if the boiling points of the two components are close enough, the bubble- and dew-point curves may reach a maximum at the same composition, which by definition is the azeotropic point. Such a situation is illustrated in Figure 8.19a

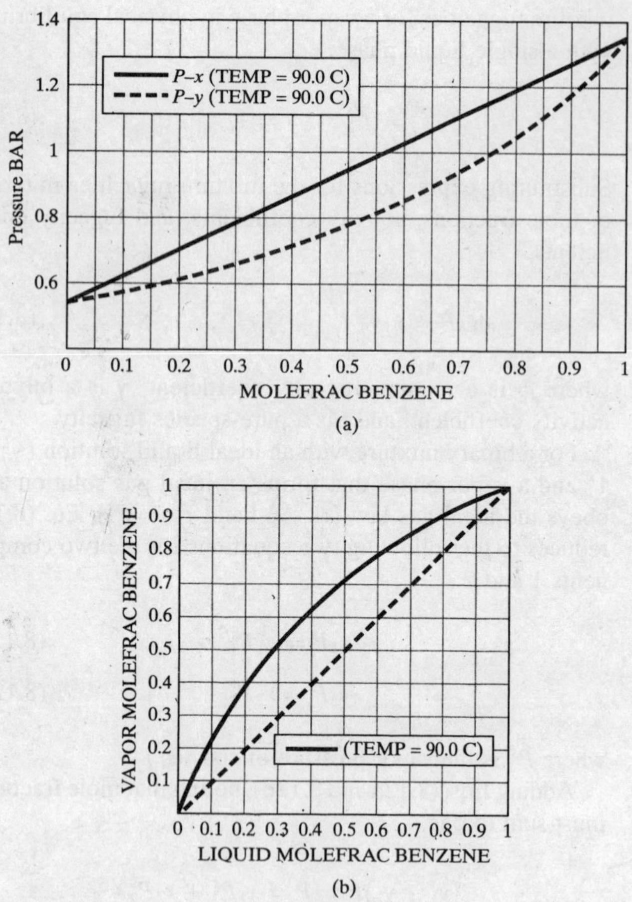

Figure 8.18 Phase diagrams for the benzene–toluene mixture at 90°C, calculated using ASPEN PLUS: (a) P–x–y diagram: (b) x–y diagram.

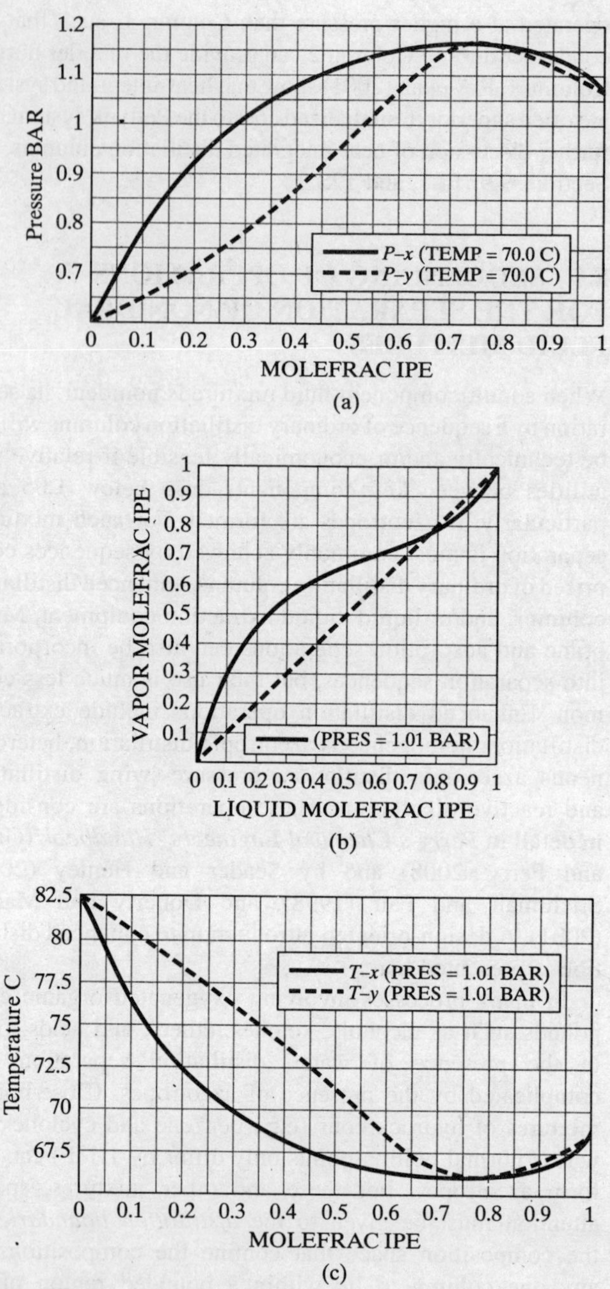

Figure 8.19 Phase diagrams for the isopropyl ether-isopropyl alcohol binary computed using ASPEN PLUS: (a) P–x–y diagram at 70°C; (b) x–y diagram at 101 kPa; (c) T–x–y diagram at 101 kPa.

for the isopropyl ether (1)–isopropyl alcohol (2) binary at 70°C. Figure 8.19b shows the corresponding x–y diagram, and Figure 8.19c shows the bubble- and dew-point curves on a T–x–y diagram at 101 kPa. Note the *minimum-boiling azeotrope* at 66°C, where $x_1 = y_1 = 0.76$. Feed streams having lower isopropyl ether mole fractions cannot be purified beyond 0.76 in a distillation column, and streams having higher isopropyl ether mole fractions have distillate mole fractions that have a lower bound of 0.76. Consequently, the azeotropic composition is commonly referred to as a *distillation boundary.*

Similarly, when the mixture exhibits the less-common negative deviation from Raoult's law ($\gamma_j^L < 1,\ j = 1, 2$), both the bubble- and dew-point curves drop below the straight line that represents the bubble points for an ideal mixture, as anticipated by examination of Eq. (8.14). Furthermore, when the bubble- and dew-point curves have the same minimum, an azeotropic composition is defined, as shown in Figure 8.20a for the acetone–chloroform binary at 64.5°C, where $x_1 = y_1 = 0.35$. For this system, Figures 8.20b and 8.20c show the corresponding x–y diagram and T–x–y diagram at 101 kPa. On the latter diagram, the azeotropic point is at a maximum temperature, and consequently, the system is said to have a *maximum-boiling azeotrope*. In this case, feed streams having lower acetone mole fractions cannot be purified beyond 0.35 in the bottoms product of a distillation column, and streams having higher acetone mole fractions have a lower bound of 0.35 in the acetone mole fraction of the bottoms product.

In summary, at a homogeneous azeotrope, $x_j = y_j$, $j = 1, \ldots, C$, the expression for the equilibrium constant, K_j, for species j becomes unity. Based on the general phase equilibria Eq. (8.11), the criterion for azeotrope formation is:

$$K_j = \frac{y_j}{x_j} = \frac{\gamma_j^L f_j^L}{\overline{\Phi}_j^V P} = 1 \qquad j = 1, \ldots, C \tag{8.15}$$

where the degree of nonideality is expressed by the deviations from unity of the activity coefficients and fugacities for the liquid phase and the fugacity coefficients for the vapor phase. At low pressure, $\overline{\Phi}_j^V = 1$ and $f_j^L = P_j^s$ so that Eq. (8.15) reduces to

$$K_j = \frac{y_j}{x_j} = \gamma_j^L \frac{P_j^s}{P} = 1 \qquad j = 1, \ldots, C \tag{8.16}$$

Because the K-values for all of the species are unity at an azeotrope point, a simple distillation approaches this point, at which no further separation can occur. For this reason, an azeotrope is often called a *stationary* or *fixed* or *pinch point*.

For a minimum-boiling azeotrope, when the deviations from Raoult' s law are sufficiently large ($\gamma_j^L \gg 1.0$, usually > 7), splitting of the liquid phase into two liquid phases (phase splitting) may occur, and a minimum-boiling, heterogeneous azeotrope may form that has a vapor phase in equilibrium with the two liquid phases. A heterogeneous azeotrope occurs when the vapor–liquid envelope overlaps with the liquid–liquid envelope, as illustrated in Figure 8.21b. For a homogeneous azeotrope, when $x_1 = x_{1,\text{azeo}} = y_1$, the mixture boils at this composition, as shown in Figure 8.21a; whereas for a heterogeneous azeotrope, when the overall liquid composition of the two liquid phases, $x_1 = x^0_{1,\text{azeo}} = y_1$, the mixture boils at this overall composition, as illustrated in Figure 8.21b, but the three coexisting phases have distinct compositions.

Residue Curves

To understand better the properties of azeotropic mixtures that contain three chemical species, it helps to examine the properties of *residue curves* on a ternary diagram. A collection of residue curves, which is called a *residue curve map*, can be computed

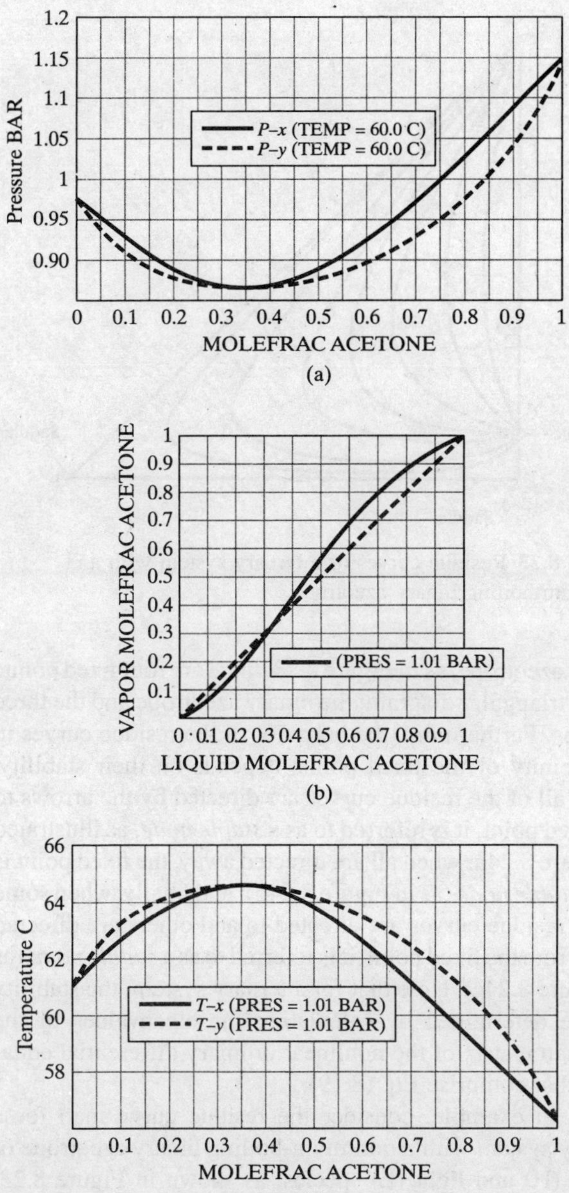

Figure 8.20 Phase diagrams for the acetone–chloroform binary computed using ASPEN PLUS: (a) P–x–y diagram at 60°C; (b) x–y diagram at 101 kPa; (c) T–x–y diagram at 101 kPa.

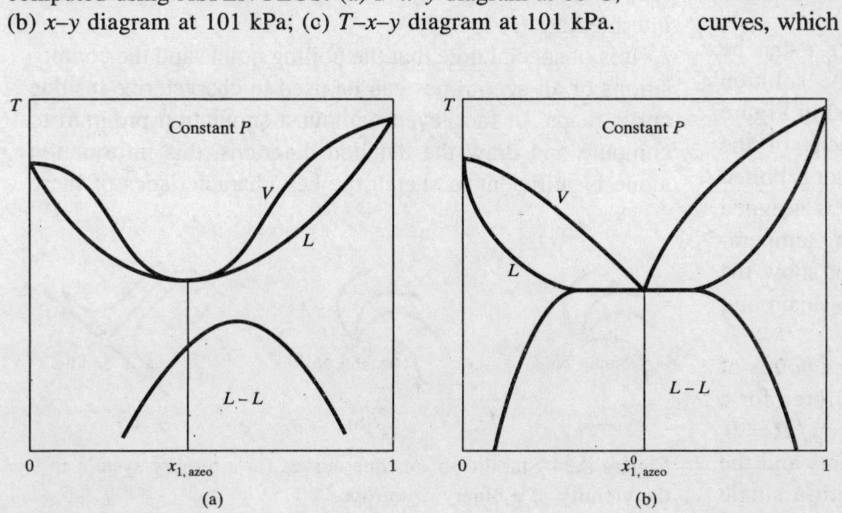

Figure 8.21 Binary phase diagram at a fixed pressure for: (a) homogeneous azeotrope; (b) heterogeneous azeotrope.

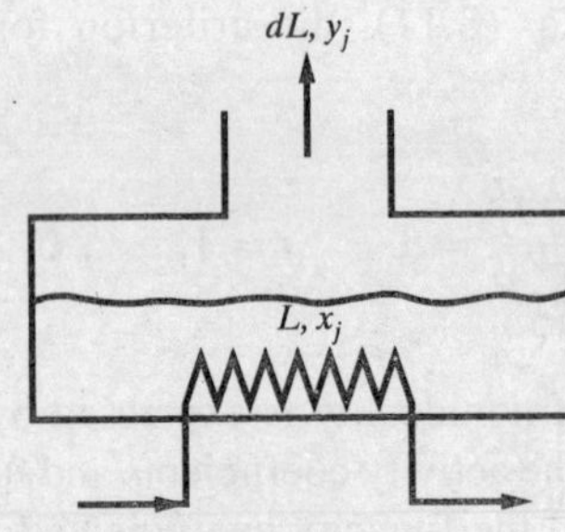

Figure 8.22 Simple distillation still.

and drawn by any of the major simulation programs. Each residue curve is constructed by tracing the composition of the equilibrium liquid residue of a simple (Rayleigh batch) distillation in time, starting from a selected initial composition of the charge to the still, using the following numerical procedure.

Consider L moles of liquid with mole fractions $x_j(j = 1, \ldots, C)$ in a simple distillation still at its bubble point, as illustrated in Figure 8.22. Note that the still contains no trays and that no reflux is provided. As heating begins, a small portion of this liquid, ΔL moles, is vaporized. The instantaneous vapor phase has mole fractions $y_j(j = 1, \ldots, C)$, assumed to be in equilibrium with the remaining liquid. Since the residual liquid, $L - \Delta L$ moles, has mole fractions $x_j + \Delta x_j$, the mass balance for species j is given by

$$Lx_j = (\Delta L)y_j + (L - \Delta L)(x_j + \Delta x_j) \quad j = 1, \ldots, C - 1 \tag{8.17}$$

In the limit, as $\Delta L \rightarrow 0$,

$$\frac{dx_j}{dL/L} = x_j - y_j = x_j(1 - K_j\{T, P, \underline{x}, \underline{y}\}) \quad j = 1, \ldots, C - 1 \tag{8.18}$$

and setting $d\hat{t} = dL/L$,

$$\frac{dx_j}{d\hat{t}} = x_j - y_j \qquad j = 1, \ldots, C - 1 \tag{8.19}$$

where K_j is given by Eq. (8.15). In Eq. (8.19), $\hat{t}$ can be interpreted as the dimensionless time, with the solution defining a family of residue curves, as illustrated in Figure 8.23. Note that each residue curve is the locus of the compositions of the residual liquid in time, as vapor is boiled off from a simple distillation still. Often, an arrow is assigned in the direction of increasing time (and increasing temperature). Note that the residue curve map does not show the equilibrium vapor composition corresponding to each point on a residue curve.

Yet another important property is that the *fixed points* of the residue curves are points where the driving force for a change in the liquid composition is zero; that is, $d\underline{x}/d\hat{t} = 0$. This condition is satisfied at the azeotropic points and the pure-species vertices. For a ternary mixture with a single

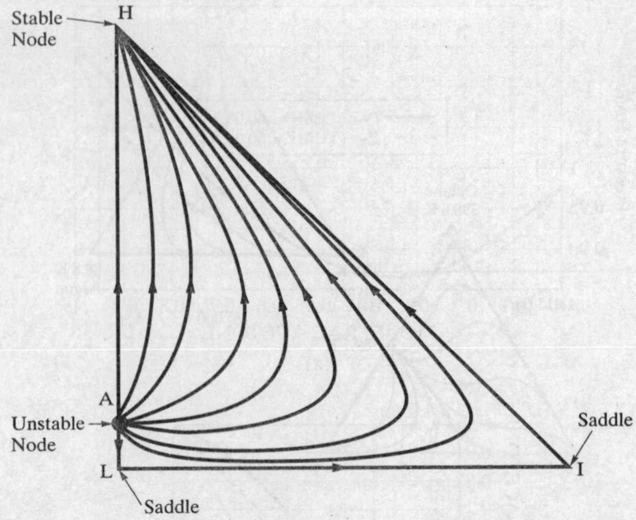

Figure 8.23 Residue curves of a ternary system with a minimum-boiling binary azeotrope.

binary azeotrope, as in Figure 8.23, there are four fixed points on the triangular diagram: the binary azeotrope and the three vertices. Furthermore, the behavior of the residue curves in the vicinity of the fixed points depends on their stability. When all of the residue curves are directed by the arrows to the fixed point, it is referred to as a *stable node*, as illustrated in Figure 8.24a; when all are directed away, the fixed point is an *unstable node* (as in Figure 8.24b); and finally, when some of the residue curves are directed to and others are directed away from the fixed point, it is referred to as a *saddle point* (as in Figure 8.24c). Note that for a ternary system, the stability can be determined by calculating the eigenvalues of the Jacobian matrix of the nonlinear ordinary differential equations that comprise Eq. (8.19).

As an example, consider the residue curve map for a ternary system with a minimum-boiling binary azeotrope of heavy (H) and light (L) species, as shown in Figure 8.23. There are four fixed points: one unstable node at the binary azeotrope (A), one stable node at the vertex for the heavy species (H), and two saddles at the vertices of the light (L) and intermediate (I) species.

It is of special note that the boiling points and the compositions of all azeotropes can be used to characterize residue curve maps. In fact, even without a simulation program to compute and draw the detailed diagrams, this information alone is sufficient to sketch the key characteristics of these

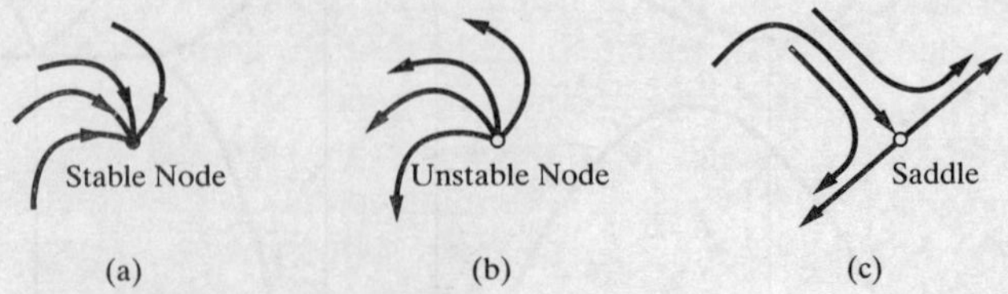

Figure 8.24 Stability of residue curves for a ternary system in the vicinity of a binary azeotrope.

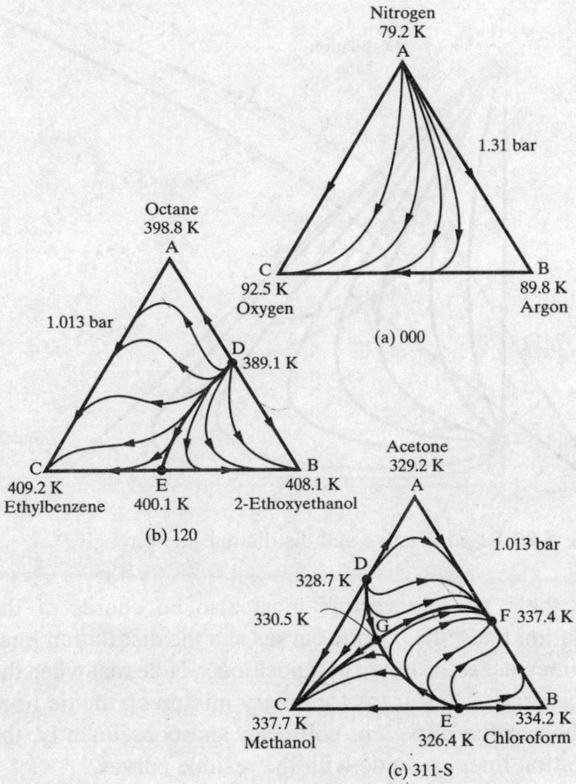

Figure 8.25 Maps of residue curves or distillation lines: (a) system without azeotropes; (b) system with two binary azeotropes; (c) system with binary and ternary azeotropes (Stichlmair et al., 1989).

diagrams using the following procedure. First, the boiling points of the pure species are entered at the vertices. Then the boiling points of the binary azeotropes are positioned at the azeotropic compositions along the edges, with the boiling points of any ternary azeotropes positioned at their compositions within the triangle. Arrows are assigned in the direction of increasing temperature in a simple distillation still. As examples, typical diagrams for mixtures involving binary and ternary azeotropes are illustrated in Figure 8.25. Figure 8.25a is for a simple system, without azeotropes, involving nitrogen, oxygen, and argon. In this mixture, nitrogen is the lowest-boiling species (L), argon is the intermediate boiler (I), and oxygen is the highest-boiling species (H). Thus, along the oxygen–argon edge the arrow is pointing to the oxygen vertex, and on the remaining edges the arrows point away from the nitrogen vertex. Since these arrows point away at the nitrogen vertex, it is an unstable node, and all of the residue curves emanate from it. At the argon vertex, the arrows point to and away from it. Since the residue curves turn in the vicinity of this vertex, it is not a terminal point. Rather, it is referred to as a saddle point. All of the curves end at the oxygen vertex, which is a terminal point or stable node. For this ternary mixture, the map shows that pure argon, the intermediate boiler, cannot be obtained in a simple distillation.

Simple Distillation Boundaries

The graphical approach described here is effective in locating the starting and terminal points and the qualitative locations of the residue curves. As illustrated in Figures 8.25b and 8.25c, it works well for binary and ternary azeotropes that exhibit multiple starting and terminal points. In these cases, one or more *simple distillation boundaries* called *separatrices* (e.g., curved line DE in Figure 8.25b) divide these diagrams into regions with distinct pairs of starting and terminal points. For the separation of homogeneous mixtures by simple distillation, these separatrices cannot be crossed unless they are highly curved. A feed located in region ADECA in Figure 8.25b has a starting point approaching the composition of the binary azeotrope of octane and 2-ethoxyethanol and a terminal point approaching pure ethylbenzene, whereas a feed located in region DBED has a starting point approaching the same binary azeotrope but a terminal point approaching pure 2-ethoxyethanol. In this case, a pure octane product is not possible. Figure 8.25c is even more complex. It shows four distillation boundaries (curved lines GC, DG, GF, and EG), which divide the diagram into four distillation regions.

Distillation Towers

When tray towers are modeled assuming vapor–liquid equilibrium at each tray, the residue curves approximate the liquid composition profiles at *total reflux*. To show this, a species balance is performed for the top n trays, counting down the tower, as shown in Figure 8.26:

$$L_{n-1}\underline{x}_{n-1} + D\underline{x}_D = V_n\underline{y}_n \tag{8.20}$$

where D and $\underline{x}_D$ are the molar flow rate and vector of mole fractions of the distillate. Similarly, L_{n-1} and $\underline{x}_{n-1}$ are for the liquid leaving tray $n-1$, and V_n and $\underline{y}_n$ are for the vapor leaving tray n. Defining h as the dimensionless distance from

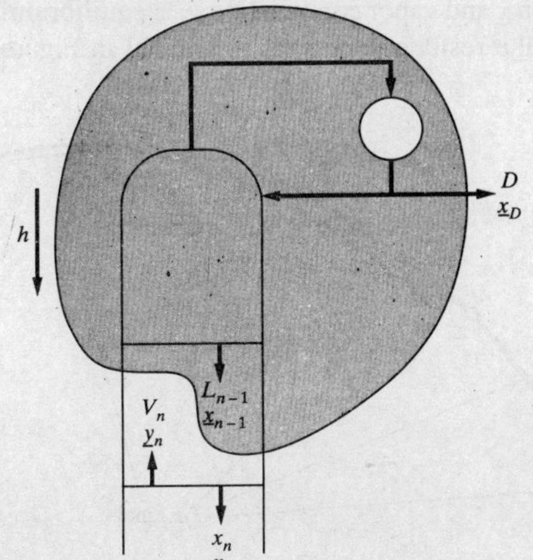

Figure 8.26 Schematic of rectifying section.

the top of the tower, a backward-difference approximation at tray n is

$$\left.\frac{d\underline{x}}{dh}\right|_n \cong \underline{x}_n - \underline{x}_{n-1} \qquad (8.21)$$

Rearranging Eq. (8.20) and substituting in Eq. (8.21),

$$\left.\frac{d\underline{x}}{dh}\right|_n \cong \underline{x}_n - \frac{V_n}{L_{n-1}}\underline{y}_n + \frac{D}{L_{n-1}}\underline{y}_D \qquad (8.22)$$

At total reflux, with $D = 0$ and $V_n = L_{n-1}$, Eq. (8.22) becomes

$$\left.\frac{d\underline{x}}{dh}\right|_n \cong \underline{x}_n - \underline{y}_n \qquad (8.23)$$

Hence, Eq. (8.23) approximates the operating lines at total reflux and, because $\hat{t}$ and h are dimensionless variables and Eq. (8.19) is identical in form, the residue curves approximate the operating lines of a distillation tower operating at total reflux.

Distillation Lines

An exact representation of the operating line for a distillation tower at total reflux, also known as a *distillation line* [as defined by Zharov (1968) and Zharov and Serafimov (1975)], is shown in Figure 8.27. Note that, at total reflux,

$$\underline{x}_n = \underline{y}_{n+1} \quad n = 0, 1, \ldots \qquad (8.24)$$

Furthermore, assuming operation in vapor-liquid equilibrium, the mole fractions on trays n, $\underline{x}_n$, and $\underline{y}_n$, lie at the ends of the equilibrium tie lines.

To appreciate better the differences between distillation lines and residue curves, consider the following observations. First, Eq. (8.19) requires the tie line vectors connecting liquid composition $\underline{x}$ and vapor composition $\underline{y}$, at equilibrium, to be tangent to the residue curves, as illustrated in Figure 8.28.

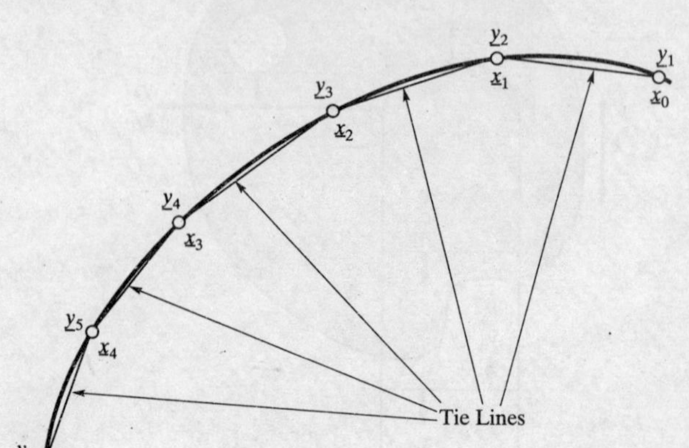

Figure 8.27 Distillation line and its tie lines.

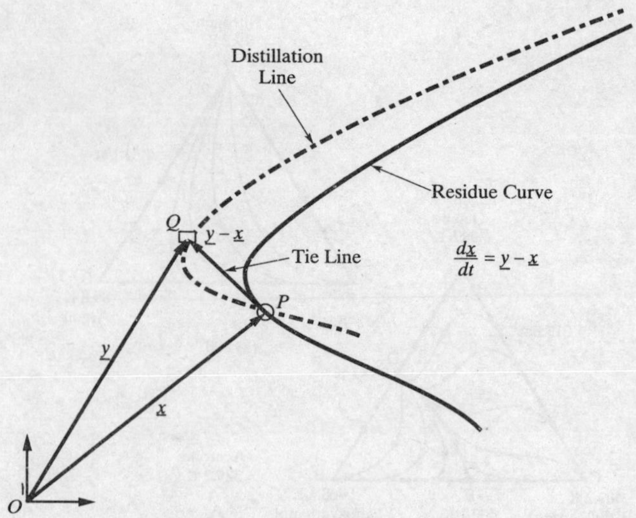

Figure 8.28 Residue curve and distillation line through P.

Since these tie line vectors must also be chords of the distillation lines, the residue curves and the distillation lines must intersect at the liquid composition $\underline{x}$. Note that when the residue curve is linear (as for binary mixtures), the tie lines and the residue curve are collinear, and consequently, the distillation lines coincide with the residue curves.

Figure 8.29a shows two distillation lines (δ_1 and δ_2) that intersect a residue curve at points A and B. As a consequence of Eq. (8.19), their corresponding vapor compositions at equilibrium, a and b, lie at the intersection of the tangents to the residue curves at A and B with the distillation lines δ_1 and δ_2. Clearly, the distillation lines do not coincide with the residue curves, an assumption that is commonly made but that may produce significant errors. In Figure 8.29b, a single distillation line connects the compositions on four adjacent trays (at C, D, E, F) and crosses four residue curves (ρ_C, ρ_D, ρ_E, ρ_F) at these points.

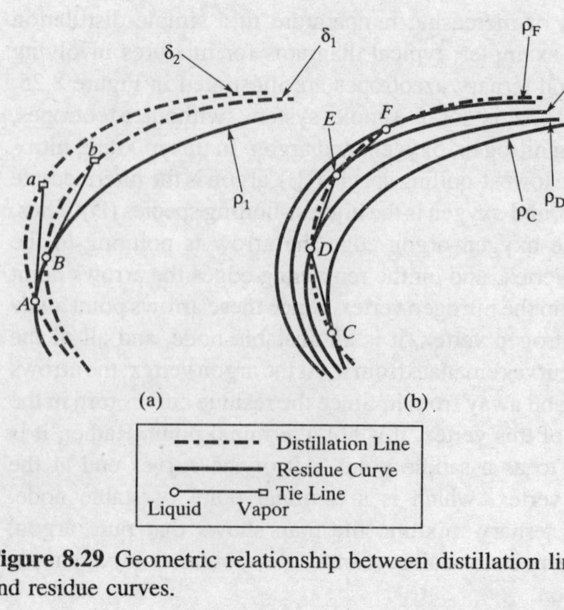

Figure 8.29 Geometric relationship between distillation lines and residue curves.

Note that distillation lines are generated by computer as easily as residue curves and, because they do not involve any approximations to the operating line at total reflux, are preferred for the analyses to be performed in the remainder of this section. However, simulation programs compute and plot only residue curves. It can be shown that distillation lines have the same properties as residue curves at fixed points, and hence, both families of curves are sketched similarly. Their differences are pronounced in regions that exhibit extensive curvature.

Computing Azeotropes for Multicomponent Mixtures

Gmehling (1994) provides data on more than 15,000 binary azeotropes and 900 ternary azeotropes. Undoubtedly, many more ternary azeotropes exist, as well as untold numbers of azeotropes involving more than three components. When a process simulation program is used to compute a residue curve map for a ternary system at a specified pressure, compositions and temperatures of all azeotropes are automatically estimated. The results depend, of course, on the selected vapor pressure and liquid-phase activity coefficient correlations. For quaternary and higher systems, the arclength homotopy-continuation method of Fidkowski, Malone, and Doherty (1993) can be used for homogeneous systems to estimate all azeotropes. They find all roots to the following equations, which define a homogeneous azeotrope:

$$y_j - x_j = 0, \qquad j = 1, 2, \ldots, C - 1 \tag{8.25}$$

$$y_j = \left(\frac{\gamma_j^L P_j^s}{\overline{\phi}_j^V P}\right) x_j, \qquad j = 1, 2, \ldots, C \tag{8.26}$$

$$\sum_{j=1}^{C} x_j = 1 \tag{8.27}$$

$$\sum_{j=1}^{C} y_j = 1 \tag{8.28}$$

$$x_j \geq 0, \qquad j = 1, 2, \ldots, C \tag{8.29}$$

To find the roots, they construct the following homotopy to replace Eqs. (8.25) and (8.26), based on gradually moving from an ideal K-value based on Raoult's law to the more rigorous expression of Eq. (8.26):

$$y_j - x_j = \left[(1-t) + t\frac{\gamma_j^L}{\overline{\phi}_j^V}\right]\frac{P_j^s}{P} x_j = H(t, x_j) = 0, \qquad j = 1, 2, \ldots, C \tag{8.30}$$

Initially, the homotopy parameter, t, is set to 0 and all values of x_j are set to 0 except for one, which is set to 1.0. Then t is gradually and systematically increased until a value of 1.0 is obtained. With each increase, the temperature and mole fractions are computed. If the resulting composition at $t = 1.0$ is not a pure component, it is an azeotrope. By starting from each pure component, all azeotropes are computed. The method of Fidkowski, Malone, and Doherty is included in many of the process simulation programs. Eckert and Kubicek (1997) extended the method of Fidkowski, Malone, and Doherty to the estimation of heterogeneous multicomponent azeotropes.

Distillation-Line Boundaries and Feasible Product Compositions

Of great practical interest is the effect of distillation boundaries on the operation of distillation towers. To summarize a growing body of literature, it is well established that the compositions of a distillation tower operating at total reflux cannot cross the distillation-line boundaries, except under unusual circumstances, where these boundaries exhibit a high degree of curvature. This provides the total-reflux bound on the possible (feasible) compositions for the distillate and bottoms products.

As shown in Figure 8.30a, at total reflux, $\underline{x}_B$ and $\underline{y}_D$ reside on a distillation line. Furthermore, these compositions lie collinear with the feed composition, $\underline{x}_F$, on the overall material balance line. As the number of stages increases, the operating curve becomes more convex and in the limit approaches the two sides of the triangle that meet at the intermediate boiler. As an example, an operating line at total reflux (minimum stages) is the curve AFC in Figure 8.31a. At the other extreme, as the number of stages increases, the operating curve becomes more convex approaching ABC,

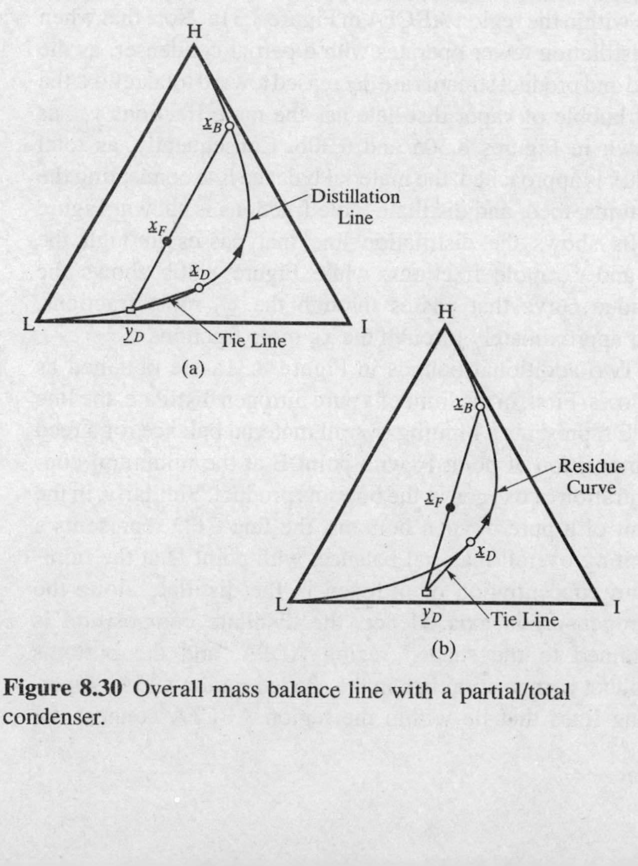

Figure 8.30 Overall mass balance line with a partial/total condenser.

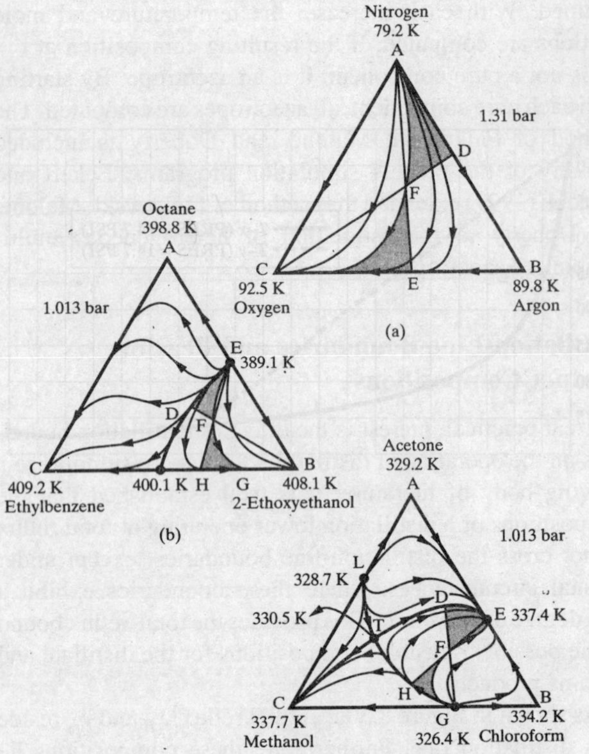

Figure 8.31 Regions of feasible distillate and bottoms product compositions (shaded) for a ternary mixture: (a) system without azeotropes; (b) system with two binary azeotropes; (c) system with binary and ternary azeotropes (Stichlmair et al., 1989).

where the number of stages approaches infinity (corresponding to minimum reflux). Hence, the operating line for a distillation tower that operates within these limiting regimes lies within the region ABCFA in Figure 8.31a. Note that when a distillation tower operates with a partial condenser, as the feed and product streams are decreased toward total reflux, the last bubble of vapor distillate has the mole fractions $\underline{y}_D$, as shown in Figures 8.30a and 8.30b. Consequently, as total reflux is approached, the material balance line connecting the bottoms, feed, and distillate mole fractions is shown. Figure 8.30a shows the distillation line that passes through the $\underline{x}_B$ and $\underline{y}_D$ mole fractions, while Figure 8.30b shows the residue curve that passes through the $\underline{x}_D$ mole fractions, and approximately through the $\underline{x}_B$ mole fractions.

Two additional bounds in Figure 8.31a are obtained as follows. First, in the limit of a pure nitrogen distillate, the line AFE represents a limiting overall material balance for a feed composition at point F, with point E at the minimum concentration of oxygen in the bottoms product. Similarly, in the limit of a pure oxygen bottoms, the line CFD represents a limiting overall material balance, with point D at the minimum concentration of nitrogen in the distillate along the nitrogen–argon axis. Hence, the distillate composition is confined to the shaded region ADFA, and the bottoms product composition lies in the shaded region CEFC. Operating lines that lie within the region ABCFA connect the distillate and bottoms product compositions in these shaded regions. At best, only one pure species can be obtained. In addition, only those species located at the end points of the distillation lines can be recovered in high purity, with one exception to be noted. Hence, the end points of the distillation lines determine the potential distillate and bottoms products for a given feed. This also applies to the complex mixtures in Figures 8.31b and 8.31c. Here, the location of the feed point determines the *distillation region* in which the potential distillate and bottoms product compositions lie. For example, in Figure 8.31b, for feed F, only pure 2-ethoxyethanol can be obtained. When the feed is moved to the left across the distillation-line boundary, pure ethylbenzene can be obtained. In Figure 8.31c, only methanol can be recovered in high purity for feeds in the region LTGCL. For a feed in the region EDTHGBE, no pure product is possible. Before attempting rigorous distillation calculations with a simulation program, it is essential to establish, with the aid of computer-generated residual curve maps, regions of product-composition feasibility such as shown in Figure 8.31. Otherwise it is possible to waste much time and effort in trying to converge distillation calculations when specified product compositions are impossible.

Heterogeneous Distillation

In heterogeneous azeotropic distillation, an entrainer is utilized that concentrates in the overhead vapor and, when condensed, causes the formation of a second liquid phase that can be decanted and recirculated to the tower as reflux. The other liquid phase as well as the bottoms are the products from the distillation. This is possible when the entrainer forms a heterogeneous azeotrope with one or more of the species in the feed. Figure 8.32a shows one possible configuration, with an accompanying triangular diagram in Figure 8.32b for the dehydration of ethanol using toluene as an entrainer. In Column C-1, the feed is preconcentrated in ethanol. Column C-2 is the azeotropic tower. Unfortunately, both products B1 and B2 are bottoms. Ethanol and water form a minimum-boiling azeotrope at 89 mol% ethanol and 1 atm, as shown in Figures 8.32c and 8.32d, which were prepared by ASPEN PLUS. Although toluene is the highest-boiling species, it is an appropriate entrainer because it forms minimum-boiling azeotropes with both water and ethanol. Hence, the arrows on the residue curves point toward both the ethanol and water vertices, allowing ethanol to be recovered in a high-purity bottoms product. Since toluene forms a ternary, minimum-boiling, heterogeneous azeotrope (point D2 in Figure 8.32b), the overhead vapor approaches this composition and condenses into two liquid phases, one rich in toluene (point S2 in Figure 8.32b) and the other rich in water (point S1 in Figure 8.32b), which are separated in the decanter. The former is recycled to the azeotropic tower, while the latter is recycled to the preconcentrator. All column product compositions are shown in Figure 8.32b. A binodal curve for the distillate temperature of the azeotropic tower is

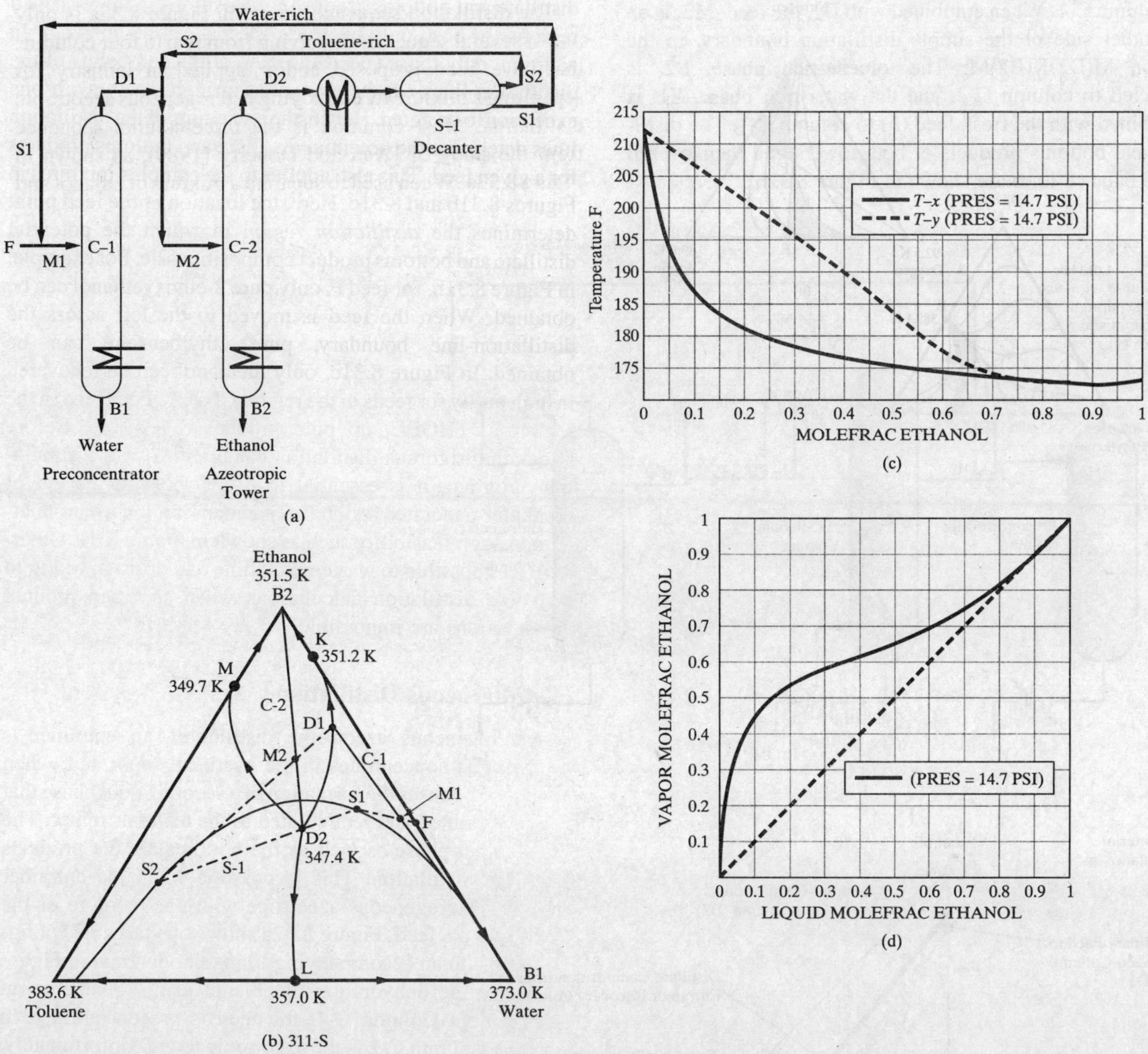

Figure 8.32 Dehydration of ethanol using toluene as an entrainer: (a) process flow diagram; (b) ternary composition diagram; (c) $T-x-y$ diagram at 1 atm; (d) $x-y$ diagram at 1 atm (Stichlmair et al., 1989).

included in Figure 8.32b, together with a tie line through the azeotropic composition of D2 to show the phase split of condensed overhead D2 into liquid phases S1 and S2.

When residue curve and distillation-line maps are constructed for heterogeneous systems using process simulation programs, the composition spaces are also divided into regions with simple distillation boundaries. However, the residue curve and distillation-line maps for systems containing heterogeneous azeotropes are far more restricted. Their azeotropic points can only be minimum-boiling saddles or unstable nodes. More importantly, the compositions of the two liquid phases lie within different distillation regions. This unique property, which is not shared by homogeneous systems, enables the decanter to bridge the distillation regions. This is the key that permits the compositions of a single distillation column to cross from one distillation region into another, as illustrated in Figures 8.32a and 8.32b. In this system, for the dehydration of ethanol using toluene, the preconcentrator, C-1, with mixed feed, M1, removes water, B1, as the bottoms product. Its distillate, at D1, lies just to the right of the simple distillation boundary, K(D2)L, as shown in Figure 8.32b. The addition of entrainer S2 to the distillate, D1, produces a C-2 feed stream, M2, that crosses this boundary into the distillation region just to the left of boundary K(D2), where high-purity ethanol, B2, is obtained as the bottoms product of the azeotropic tower, C-2. Its overhead vapor, D2, is in the vicinity of the heterogeneous ternary azeotrope, and when condensed and subcooled forms two liquid phases that are decanted easily. The organic phase, at S2, lies in a different distillation region than the feed, M1, to column C-1. When combined with D1, the feed, M2, is on the other side of the simple distillation boundary, in the region M(D2)K(B2)M. The toluene-rich phase, S2, is recycled to column C-2, and the water-rich phase, S1, is combined with the fresh feed (F) to column C-1. The distillate and bottoms products of both towers and their overall mass balance lines are shown in Figure 8.32b.

The distillation sequence shown in Figure 8.32a is only one of several sequences involving from two to four columns that have been proposed and/or applied in industry for separating a mixture by employing heterogeneous azeotropic distillation. Most common is the three-column sequence from the study of Ryan and Doherty (1989), as shown in Figure 8.33a. When used to separate a mixture of ethanol and

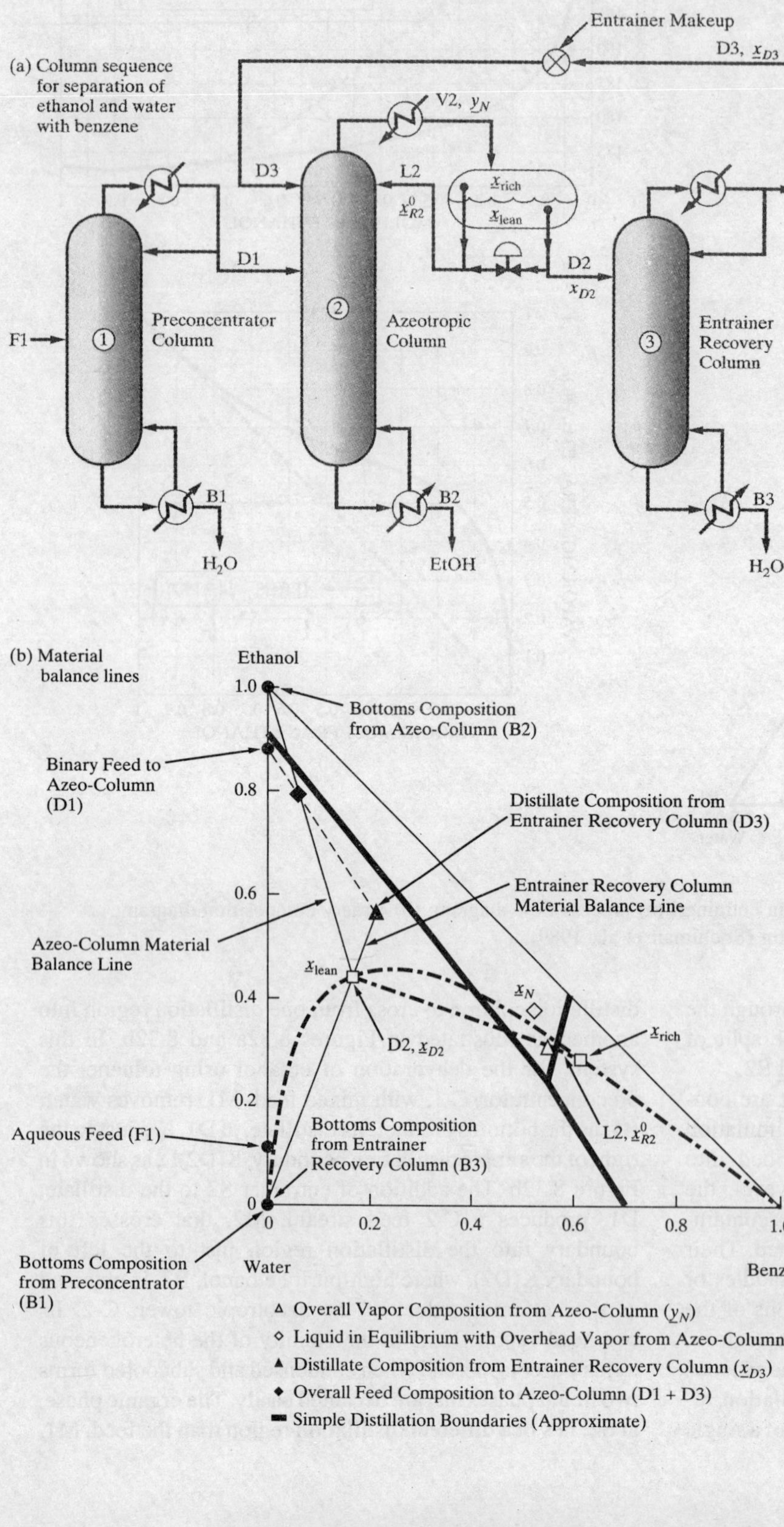

Figure 8.33 Kubierschky three-column system.

water using benzene as the entrainer, the three columns perform the separation in the following manner, where the material-balance lines for Columns 2 and 3 are shown in Figure 8.33b. The aqueous feed, F1, dilute in ethanol, is preconcentrated in Column 1 to obtain a pure water bottoms, B1, and a distillate, D1, whose composition approaches that of the homogeneous minimum-boiling binary azeotrope. The distillate becomes the feed to Column 2, the azeotropic column, where nearly pure ethanol, B2, is removed as bottoms. The overhead vapor from Column 2, V2, is close to the composition of the heterogeneous ternary azeotrope of ethanol, water, and benzene. When condensed, it separates into two liquid phases in the decanter. Most of the organic-rich phase, L2, is returned to Column 2 as reflux. Most of the water-rich phase, D2, is sent to Column 3, the entrainer recovery column. Here, the distillate, D3, consisting mainly of ethanol but with appreciable amounts of benzene and water, is recycled to the top of Column 2. The bottoms, B3, from Column 3 is nearly pure water. All columns operate at close to 1 atm pressure.

Multiple Steady States

The occurrence of multiple steady states in chemical reactors has been well recognized for at least 50 years. The most common example is an adiabatic CSTR, for which in some cases, for the same feed and reactor size, three possible products may be obtained, two of which are stable and one unstable, as shown in Case Study 12S.1. The product obtained in actual operation depends upon the startup procedure for the reactor. Only in the past 25 years has the existence of multiple steady states in distillation towers been shown by calculations and verified by experimental data from tower operation. In particular, azeotropic distillation is especially susceptible to multiple steady states. Disturbances during operation of an azeotropic tower can cause it to switch from one steady state to another, as shown by Prokopakis and Seider (1983). Methods for computing multiple steady states for homogeneous and heterogeneous azeotropic distillation are presented in a number of publications. Kovach and Seider (1987) computed, by an arclength homotopy-continuation method, five steady states for the ethanol–benzene–water distillation. Bekiaris et al. (1993, 1996, 2000) studied multiple steady states for ternary homogeneous- and ternary heterogeneous-azeotropic distillation, respectively. Using the distillate flow rate as the bifurcation parameter, they found conditions of feed compositions and distillation-region boundaries for which multiple steady states can occur in columns operating at total reflux (infinite reflux ratio) with an infinite number of equilibrium stages (referred to as the ∞–∞ case). They showed that their results have relevant implications for columns operating at finite reflux ratios with a finite number of stages. Vadapalli and Seader (2001) used ASPEN PLUS with an arclength continuation and bifurcation method to compute all stable and unstable steady states for azeotropic distillation under conditions of finite reflux ratio and finite number of equilibrium stages. Specifications for their heterogeneous azeotropic distillation example, involving the separation of an ethanol–water mixture using benzene, are shown in Figure 8.34a. The total feed rate to the column is 101.962 kmol/hr. The desired bottoms product is pure ethanol. Using the bottoms flow rate as the bifurcation parameter, computed results for the mole fraction of ethanol in the bottoms are shown in Figure 8.34b as a function of the bifurcation parameter. In the range of bottoms flow rate from approximately 78 to 96 kmol/hr, three steady states exist, two stable and one unstable. For a bottoms rate equal to the flow rate of ethanol in the feed (89 kmol/hr), the best stable solution is an ethanol mole fraction of 0.98; the inferior stable solution is only 0.89. Figure 8.34b shows the computed points. In the continuation method, the results of one point are used as the initial guess for obtaining an adjacent point.

While heterogeneous azeotropic distillation towers are probably used more widely than their homogeneous counterparts, care must be taken in their design and operation. In addition to the possibility of multiple steady states, most azeotropic distillation towers involve sharp fronts as the temperatures and compositions shift abruptly from the vicinity of one fixed point to the vicinity of another. Furthermore, in heterogeneous distillations, sharp fronts often accompany the interface between trays having one and two liquid phases as well. Consequently, designers must select carefully the number of trays and the reflux rates to prevent these fronts from exiting the tower with an associated deterioration in the product quality. While these and other special properties of azeotropic towers (e.g., maximum reflux rates above which the separation deteriorates, and an insensitivity of the product compositions to the number of trays) are complicating factors, fortunately, they are usually less important when synthesizing separation trains, and consequently they are not discussed further here. For a review of the literature on this subject, see the article by Widagdo and Seider (1996).

Pressure-Swing Distillation

In some situations, azeotropic points are sensitive to moderate changes in pressure. When this is the case, pressure-swing distillation can be used in place of azeotropic distillation to permit the recovery of two nearly pure species that are separated by a distillation boundary. This section introduces pressure-swing distillation.

The effect of pressure on the temperature and composition of the ethanol–water and ethanol–benzene azeotropes, two minimum-boiling binary azeotropes, is shown in Figure 8.35. For the first, as the pressure is decreased from 760 to 100 torr, the mole fraction of ethanol increases from 0.894 to 0.980. Although not shown, at a lower pressure, below 70 torr, the azeotrope disappears entirely. While the temperature changes are comparable for the ethanol–benzene azeotrope, the composition is far more sensitive. Many other binary azeotropes are pressure-sensitive, as discussed by Knapp and

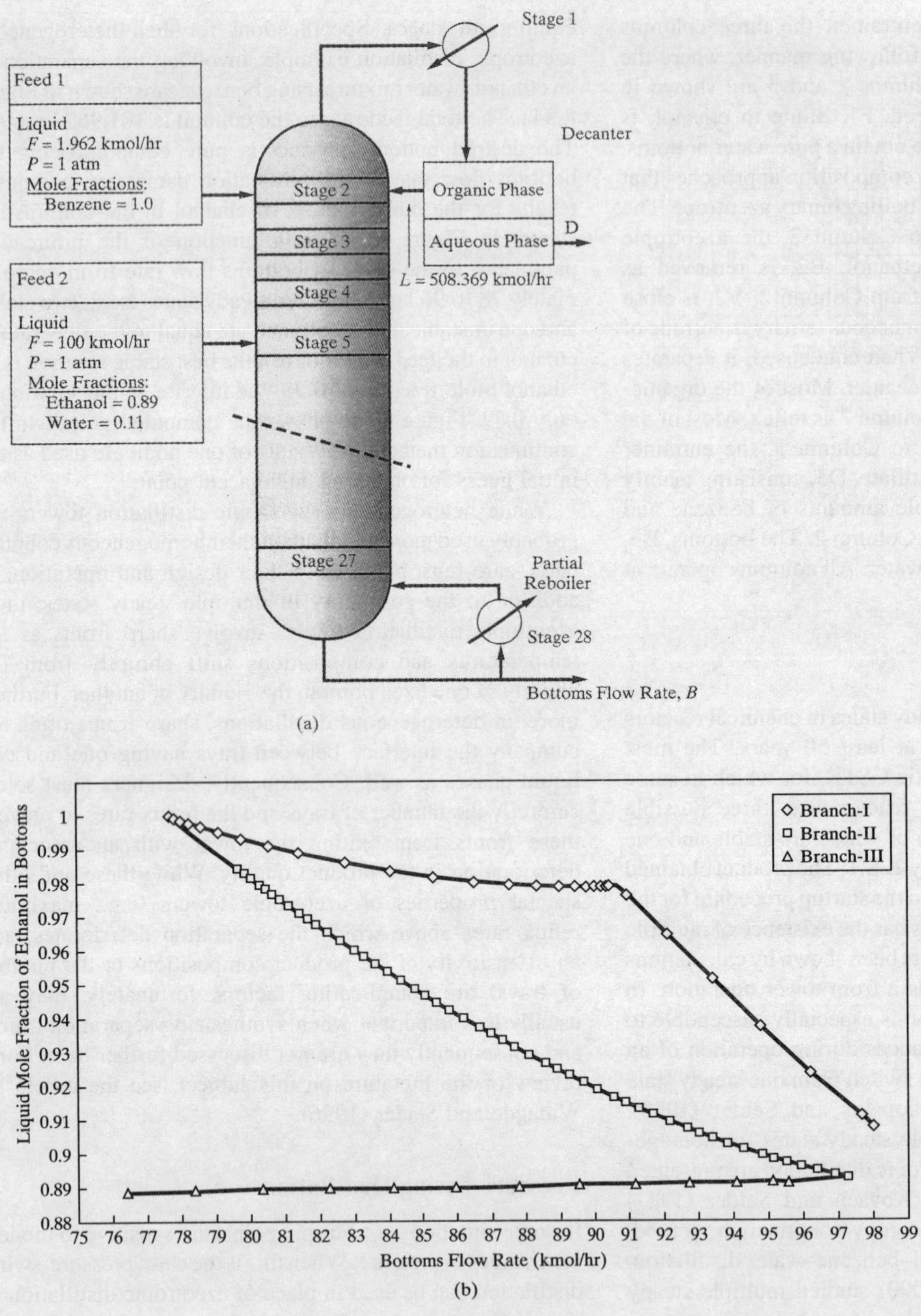

Figure 8.34 Heterogeneous azeotropic distillation: (a) specifications, (b) bifurcation diagram; branches I and III—stable, branch II—unstable.

Doherty (1992), who list 36 systems taken from the compilation of azeotropic data by Horsley (1973).

An example of pressure-swing distillation, described by Van Winkle (1967), is provided for the mixture A–B, having a minimum-boiling azeotrope, with the *T–x–y* curves at two pressures shown in Figure 8.36a. To take advantage of the decrease in the composition of A as the pressure decreases from P_2 to P_1, a sequence of two distillation towers is shown in Figure 8.36b. The total feed to column 1, F_1, operating at the lower pressure, P_1, is the sum of the fresh feed, F, whose composition is richer in A than the azeotrope, and the distillate, D_2, whose composition is close to that of the azeotrope at P_2, and which is recycled from column 2 to column 1. The compositions of D_2, and consequently F_1, are richer in A than the azeotropic composition at P_1. Hence, the bottoms product, B_1, that leaves column 1 is nearly pure A. Since the distillate, D_1, which is slightly richer in A than the azeotropic composition, is less rich in A than the azeotropic composition at P_2, when it is fed to column 2, the bottoms product, B_2, is nearly pure B. Yet another example is provided by Robinson and Gilliland (1950) for the dehydration of ethanol, where the fresh-feed composition is less rich

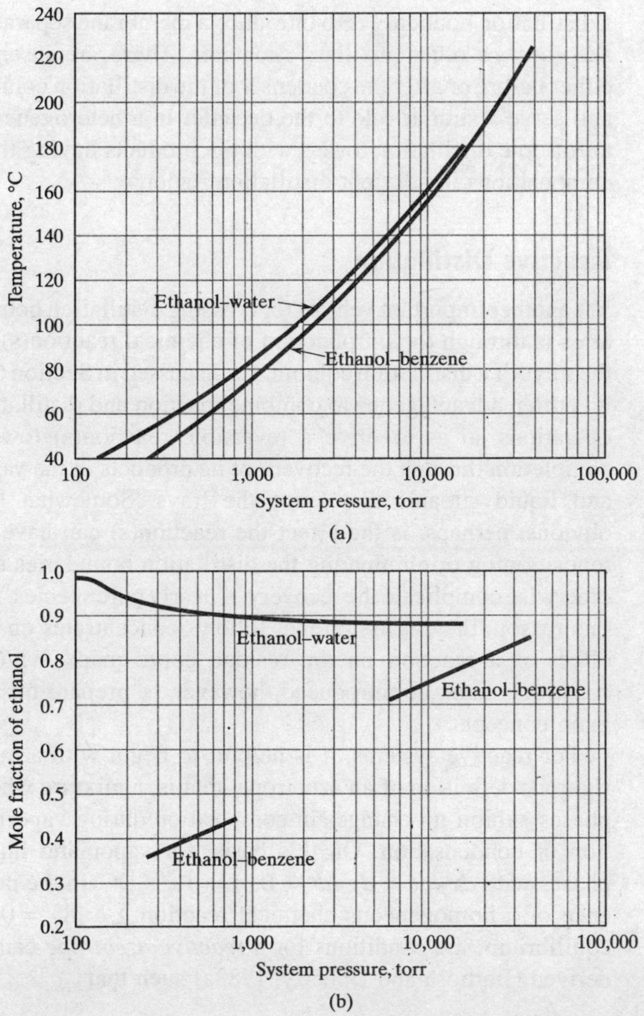

Figure 8.35 Effect of pressure on azeotrope conditions: (a) temperature of azeotrope; (b) composition of azeotrope.

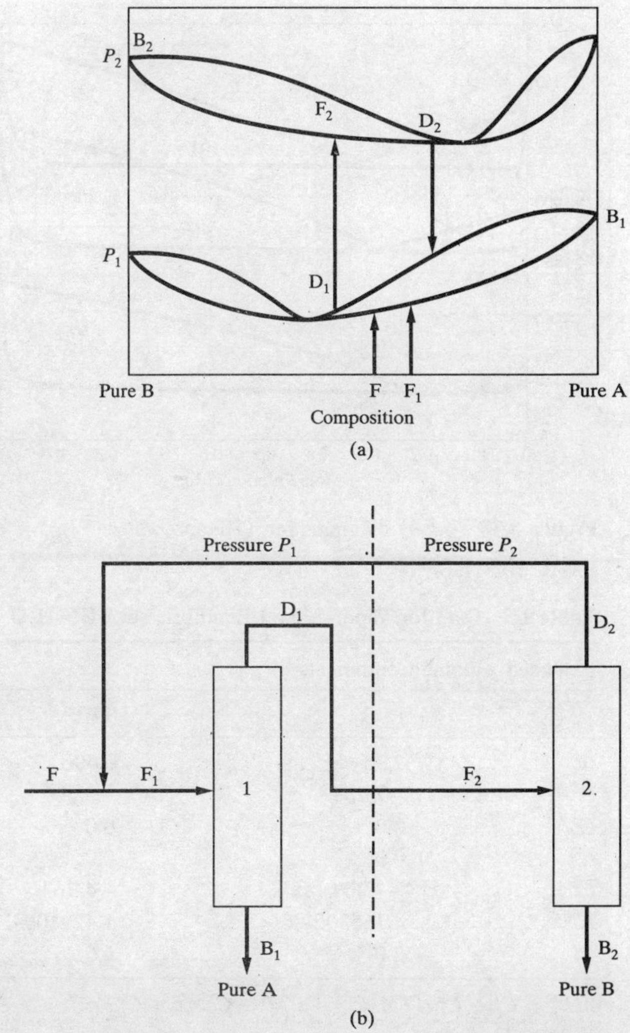

Figure 8.36 Pressure-swing distillation for the separation of a minimum-boiling azeotrope: (a) T–x–y curves at pressures P_1 and P_2 for minimum-boiling azeotrope; (b) distillation sequence for minimum-boiling azeotrope.

in ethanol than the azeotrope. In this case, ethanol and water are removed as bottoms products also, but nearly pure B (water) is recovered from the first column and A (ethanol) is recovered from the second. Similar pressure-swing distillations are designed to separate maximum-boiling binary azeotropes, which are less common.

When designing pressure-swing distillation sequences, the recycle ratio must be adjusted carefully. Note that it is closely related to the differences in the compositions of the azeotrope at P_1 and P_2. Horwitz (1997) illustrates this for the dehydration of ethylenediamine.

EXAMPLE 8.4

Consider the separation of 100 kmol/hr of an equimolar stream of tetrahydrofuran (THF) and water using pressure-swing distillation, as shown in Figure 8.37. The tower T1 operates at 1 bar, with the pressure of tower T2 increased to 10 bar. As shown in the T–x–y diagrams in Figure 8.38, the binary azeotrope shifts from 19 mol% water at 1 bar to 33 mol% water at 10 bar. Assume that the bottoms product from T1 contains pure water and that from D2 contains pure THF. Also, assume that the distillates from T1 and T2 are at their azeotropic compositions. Determine the unknown flow rates of the product and internal streams. Note that data for the calculation of vapor–liquid equilibria are provided in Table 8.5.

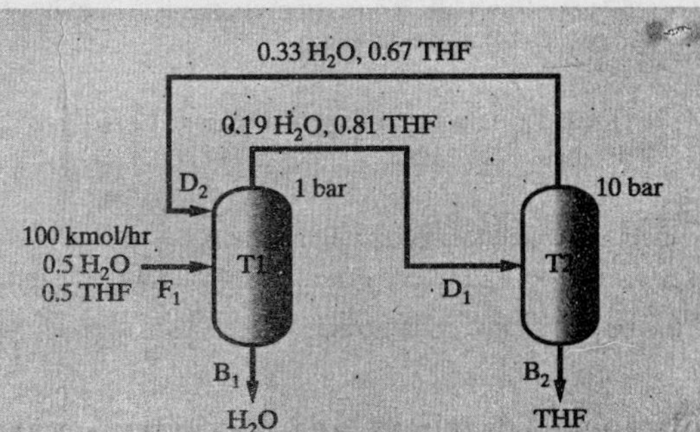

Figure 8.37 Pressure-swing distillation for dehydration of THF with stream compositions in mole fractions.

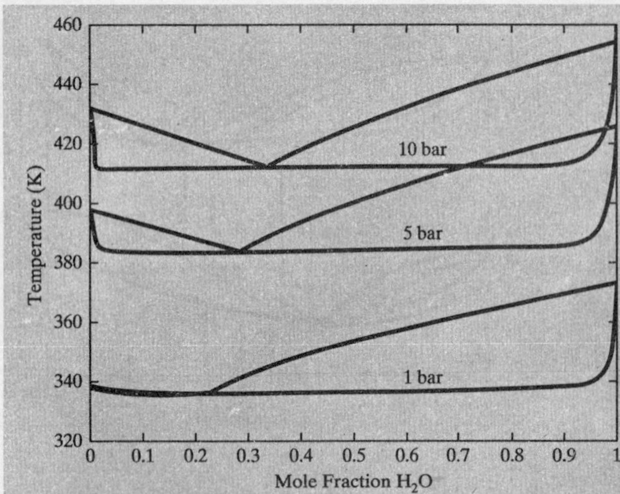

Figure 8.38 T–x–y diagrams for THF and water.

Table 8.5 Data for Vapor–Liquid Equilibria for THF–H_2O

Extended Antoine Coefficients

	H_2O	THF
C^1	7.36	5.490
C^2	−7,258	−5,305
C^3	0.0	0.0
C^4	0.0	0.0
C^5	−7.304	−4.763
C^6	4.16530E-06	1.42910E-17
C^7	2.0	6.0

$\ln P_i^s = C_i^1 + C_i^2/(T + C_i^3) + C_i^4 T + C_i^5 \ln T + C_i^6 T^{C_i^7}$;
P^s, Pascal

Wilson Interaction Coefficients

A_{ij}	H_2O	THF	B_{ij}	H_2O	THF
H_2O	0.0	−23.709	H_2O	0.0	7,500
THF	−2.999	0.0	THF	−45.07	0.0

SOLUTION

Since the bottoms products are pure, $B_1 = 50$ kmol/hr H_2O and $B_2 = 50$ kmol/hr THF. To determine the distillate flow rates, the following species balances apply.

$$\text{H}_2\text{O balance on column T2}: \quad 0.19 D_1 = 0.33 D_2$$
$$\text{THF balance on column T1}: \quad 0.81 D_1 = 0.67 D_2 + 50$$

Solving these two equations simultaneously, $D_1 = 117.9$ kmol/hr and $D_2 = 67.9$ kmol/hr. Exercise 8.19 examines the effect of pressure on the internal flow rates.

Membranes, Adsorbers, and Auxiliary Separators

When operating homogeneous azeotropic distillation towers, a convenient vehicle for permitting the compositions to cross a distillation boundary is to introduce a membrane separator, adsorber, or other auxiliary separator. These are inserted either before or after the condenser of the distillation column and serve a similar role to the decanter in a heterogeneous azeotropic distillation tower, with the products having their compositions in adjacent distillation regions.

Reactive Distillation

Yet another important vehicle for crossing distillation boundaries is through the introduction of chemical reaction(s) on the trays of a distillation column. As discussed in Section 6.3, it is often advantageous to combine reaction and distillation operations so as to drive a reversible reaction(s) toward completion through the recovery of its products in the vapor and liquid streams that leave the trays. Somewhat less obvious, perhaps, is the effect the reaction(s) can have on repositioning or eliminating the distillation boundaries that otherwise complicate the recovery of nearly pure species. For this reason, the discussion that follows concentrates on the effect of a reaction on the residue curve maps. Several constructs must be introduced, however, to prepare for the main concepts.

For reactive systems, it is helpful to begin with a more rigorous definition of an azeotrope, that is, a mixture whose phases exhibit no changes in composition during vaporization or condensation. On this basis, for vapor and liquid phases with $dx_j/dt = dy_j/dt = 0,\ j = 1, \ldots, C$, in the presence of a homogeneous chemical reaction $\Sigma_j \nu_j A_j = 0$, at equilibrium, the conditions for a *reactive azeotrope* can be derived (Barbosa and Doherty, 1988a) such that

$$\frac{y_j - x_j}{\nu_j - x_j \nu_T} = \frac{d\xi}{d\upsilon} = \kappa \qquad j = 1, \ldots, C \tag{8.31}$$

where ν_j is the stoichiometric coefficient of species j, $\nu_T = \Sigma_j \nu_j$, ξ is the extent of the reaction, υ is the moles of vapor, and κ is a constant. Furthermore, it can be shown that the mass balances for simple distillation in the presence of a chemical reaction can be written in terms of transformed variables (Barbosa and Doherty, 1988b):

$$\frac{dX_j}{d\tau} = X_j - Y_j \qquad j = 1, \ldots, C-1;\ j \neq j' \tag{8.32a}$$

where

$$X_j = \frac{x_j/\nu_j - x_{j'}/\nu_{j'}}{\nu_{j'} - \nu_T x_{j'}} \tag{8.32b}$$

$$Y_j = \frac{y_j/\nu_j - y_{j'}/\nu_{j'}}{\nu_{j'} - \nu_T y_{j'}} \tag{8.32c}$$

$$\tau = \frac{H}{\upsilon}\left(\frac{\nu_{j'} - \nu_T y_{j'}}{\nu_{j'} - \nu_T x_{j'}}\right) t \tag{8.32d}$$

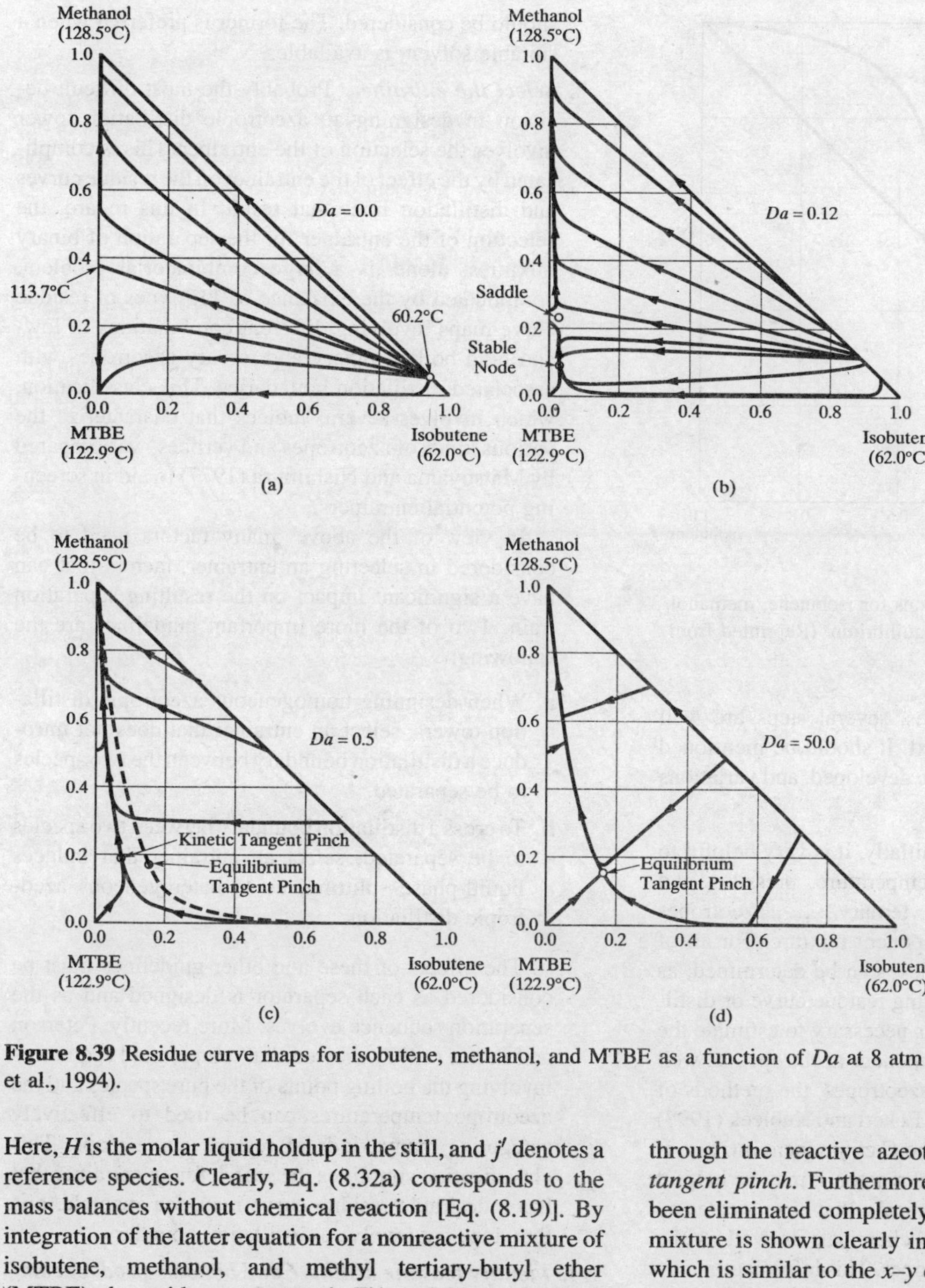

Figure 8.39 Residue curve maps for isobutene, methanol, and MTBE as a function of *Da* at 8 atm (Reprinted from Venimadhavan et al., 1994).

Here, H is the molar liquid holdup in the still, and j' denotes a reference species. Clearly, Eq. (8.32a) corresponds to the mass balances without chemical reaction [Eq. (8.19)]. By integration of the latter equation for a nonreactive mixture of isobutene, methanol, and methyl tertiary-butyl ether (MTBE), the residue curve map in Figure 8.39a is obtained. There are two minimum-boiling binary azeotropes and a distillation boundary that separates two distillation regions.

When the chemical reaction is turned on and permitted to equilibrate, Eq. (8.32a) is integrated and at long times,

$$X_j = Y_j \quad j = 1, \ldots, C \tag{8.33}$$

define the fixed point and are the conditions derived for a reactive azeotrope [Eq. (8.31)]. At shorter times, *reactive residue curves* are obtained, as shown in Figure 8.39d, where the effect of the chemical reaction can be seen. It is clear that the residue curves have been distorted significantly and pass through the reactive azeotrope, or so-called *equilibrium tangent pinch*. Furthermore, the distillation boundary has been eliminated completely. The reactive azeotrope of this mixture is shown clearly in an X–Y diagram (Figure 8.40), which is similar to the x–y diagram when reaction does not occur. Finally, through the use of a kinetic model involving a well-stirred reactor, it is possible to show the residue curves as a function of the residence time (that is, the Damkohler number, Da). Figures 8.39b and 8.39c show how the residue curves change as the residence time increases (Venimadhavan et al., 1994).

Separation Train Synthesis

Beginning with the need to separate a C-component mixture into several products, alternative sequences of two-product distillation towers are considered in this section. Although the synthesis strategies are not as well defined for highly

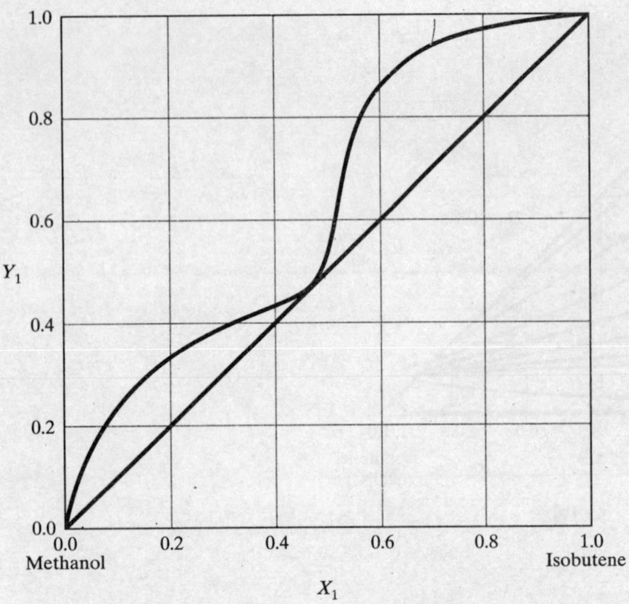

Figure 8.40 Transformed compositions for isobutene, methanol, and MTBE in chemical and phase equilibrium. (Reprinted from Doherty and Buzad, 1992).

nonideal and azeotropic mixtures, several steps are well recognized and are described next. It should be mentioned that these strategies continue to be developed, and variations are not uncommon.

1. ***Identify the azeotropes.*** Initially, it is very helpful to obtain estimates of the temperature, pressure, and composition of the binary, ternary, . . . , azeotropes associated with the C-component mixture. For all of the ternary submixtures, these can be determined, as described above, by preparing residue curve or distillation-line maps. When it is necessary to estimate the quaternary and higher-component azeotropes as well as the binary and ternary azeotropes, the methods of Fidkowski et al. (1993) and Eckert and Kubicek (1997) are recommended. When the C-component mixture is the effluent from a chemical reactor, it may be helpful to include the reacting chemicals, that is, to locate any azeotropes involving these chemicals as well as the existence of reactive azeotropes. This information may show the potential for using reactive distillation operations as a vehicle for crossing distillation boundaries that complicate the recovery of nearly pure species.
2. ***Identify alternative separators.*** Given estimates for the azeotropes, the alternatives for the separators involving all C species are identified. These separate two species that may or may not involve a binary azeotrope. When no binary azeotrope is involved, a normal distillation tower may be adequate, unless the key components are close boiling. For close-boiling binary pairs, or binary pairs with an azeotrope separating the desired products, the design of an extractive distillation tower or an azeotropic distillation tower should be considered. The former is preferred when a suitable solvent is available.
3. ***Select the entrainer.*** Probably the most difficult decision in designing an azeotropic distillation tower involves the selection of the entrainer. This is complicated by the effect of the entrainer on the residue curves and distillation lines that result. In this regard, the selection of the entrainer for the separation of binary mixtures, alone, is a large combinatorial problem, complicated by the existence of 113 types of residue curve maps involving different combinations of low- and high-boiling binary and ternary azeotropes with associated distillation boundaries. This classification, which involves several indices that characterize the various kinds of azeotropes and vertices, was prepared by Matsuyama and Nishimura (1977) to aid in screening potential entrainers.

   In view of the above, many factors need to be considered in selecting an entrainer, factors that can have a significant impact on the resulting separation train. Two of the more important guidelines are the following:

   a. When designing homogeneous azeotropic distillation towers, select an entrainer that does not introduce a distillation boundary between the two species to be separated.

   b. To cross a distillation boundary between two species to be separated, select an entrainer that induces liquid-phase splitting, as in heterogeneous azeotropic distillation.

   The effects of these and other guidelines must be considered as each separator is designed and as the separation sequence evolves. More recently, Peterson and Partin (1997) showed that temperature sequences involving the boiling points of the pure species and the azeotrope temperatures can be used to effectively categorize many kinds of residue curve maps. This classification simplifies the search for an entrainer that has a desirable residue curve map, for example, one that does not involve a distillation boundary.
4. ***Identify feasible distillate and bottoms-product compositions.*** When positioning a two-product separator, it is usually an objective to recover at least one nearly pure species, or at least to produce two products that are easier to separate into the desired products than the feed mixture. To accomplish this, it helps to know the range of feasible distillate and bottoms-product compositions. For a three-component feed stream, the feed composition can be positioned on a distillation-line map and the feasible compositions for the distillate and bottoms product identified using the methods described above in the subsection on distillation-line boundaries and feasible product compositions. For feed mixtures containing four or more species ($C > 3$), a

common approach is to identify the three most important species that are associated with the separator being considered. Note, however, that the methods for identifying the feasible compositions assume that they are bounded by the distillation line, at total reflux, through the feed composition. For azeotropic distillations, however, it has been shown that the best separations may not be achieved at total reflux. Consequently, a procedure has been developed to locate the bounds at finite reflux. This involves complex graphics to construct the so-called *pinch-point trajectories*, which are beyond the scope of this presentation but are described in detail by Widagdo and Seider (1996). Because the composition bounds at finite reflux usually include the feasible region at total reflux, the latter usually leads to conservative designs.

Having determined the bounds on the feasible compositions, the first separator is positioned usually to recover one nearly pure species. At this point in the synthesis procedure, the separator can be completely designed (to determine number of trays, reflux ratio, installed and operating costs, etc.). Alternatively, the design calculations can be delayed until a sequence of separators is selected, with its product compositions positioned. In this case, Steps 2–4 are repeated for the mixture in the other product stream. Initially, the simplest separators are considered, that is, ordinary distillation, extractive distillation, and homogeneous azeotropic distillation. However, when distillation boundaries are encountered and cannot be eliminated through the choice of a suitable entrainer, more complex separators are considered, such as heterogeneous azeotropic distillation; pressure-swing distillation; the addition of membranes, adsorption, and auxiliary separators; and reactive distillation. Normally, a sequence is synthesized involving many two-product separators without chemical reaction. Subsequently, after the separators are designed completely, steps are taken to carry out task integration as described in Section 4.4. This involves the combination of two or more separators and seeking opportunities to combine the reaction and separation steps in reactive distillation towers. As an example, Siirola (1995) describes the development of a process for the manufacture of methyl acetate and the dehydration of acetic acid. Initially, a sequence was synthesized involving a reactor, an extractor, a decanter, and eight distillation columns incorporating two mass-separating agents. The flowsheet was reduced subsequently to four columns, using evolutionary strategies and task integration, before being reduced finally to just two columns, one involving reactive distillation.

As illustrated throughout this section, process simulators have extensive facilities for preparing phase-equilibrium diagrams ($T-x-y$, $P-x-y$, $x-y$, $\ldots$), and residue curve maps and binodal curves for ternary systems. In addition, related but independent packages have been developed for the synthesis and evaluation of distillation trains involving azeotropic mixtures. These include SPLIT™ by Aspen Technology, Inc., and DISTIL™ by Hyprotech (now Aspen Technology, Inc., which contains MAYFLOWER developed by M.F. Doherty and M.F. Malone at the University of Massachusetts).

EXAMPLE 8.5 *Manufacture of Di-Tertiary-Butyl Peroxide*

This example involves the manufacture of 100 million pounds per year of di-tertiary-butyl peroxide (DTBP) by the catalytic reaction of tertiary-butyl hydroperoxide (TBHP) with excess tertiary-butyl alcohol (TBA) at 170°F and 15 psia according to the reaction

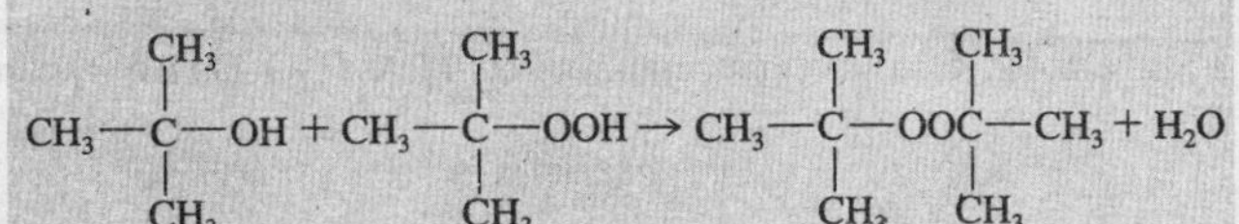

Tertiary-butyl alcohol (TBA) + Tertiary-butyl hydroperoxide (TBHP) → DTBP + H_2O

Assume that the reactor effluent stream contains

	lbmol/hr	Mole Fraction
TBA	72.1	0.272
H_2O	105.6	0.398
DTBP	87.7	0.330
		1.000

and small quantities of isobutene, methanol, and acetone, which can be disregarded. A separation sequence is to be synthesized to produce 99.6 mol% pure DTBP, containing negligible water. It may be difficult to separate TBA and water. Therefore, rather than recovering and recycling the unreacted TBA, the conversion of TBA to isobutene and water in the separation sequence should be considered. In the catalytic reactor, the TBA dehydrates to isobutene, which is the actual molecule that reacts with TBHP to form DTBP. Thus, isobutene, instead of TBA, can be recycled to the catalytic reactor.

SOLUTION

A residue curve map at 15 psia, prepared using ASPEN PLUS (with the NRTL option set and proprietary interaction coefficients), is displayed in Figure 8.41a. There are three minimum-boiling binary azeotropes:

	T, °F	
DTBP–TBA	177	$x_{TBA} = 0.82$
TBA–H_2O	176	$x_{H_2O} = 0.38$
H_2O–DTBP	188	$x_{DTBP} = 0.47$

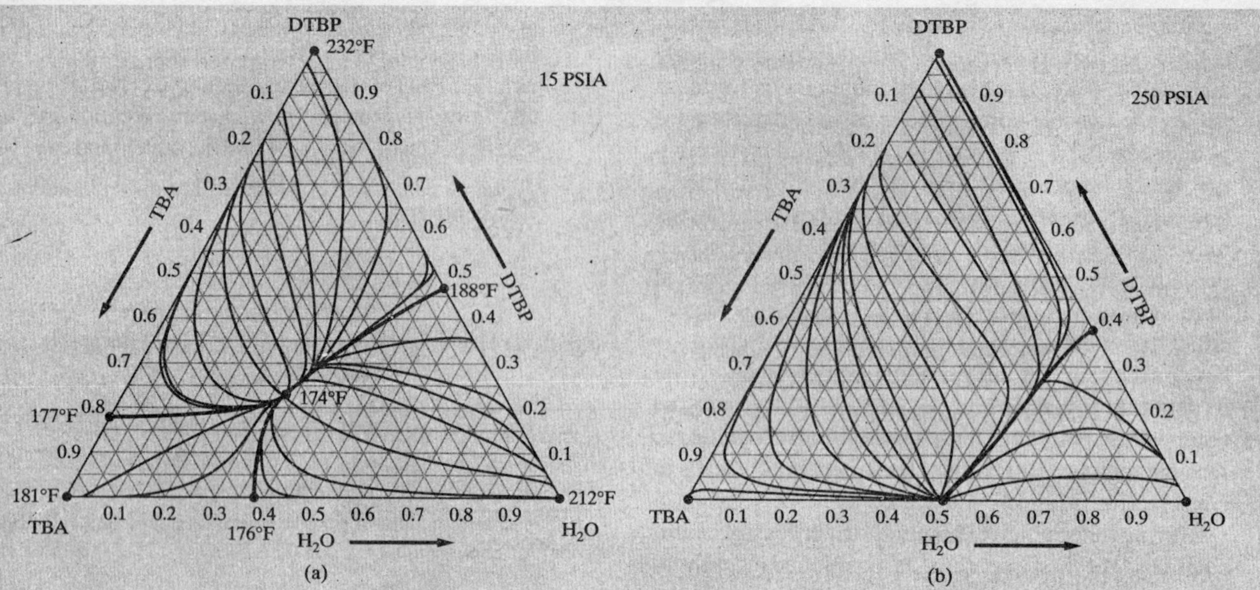

Figure 8.41 Residue curve map for the TBA–H_2O–DTBP system. (a) 15 psia; (b) 250 psia.

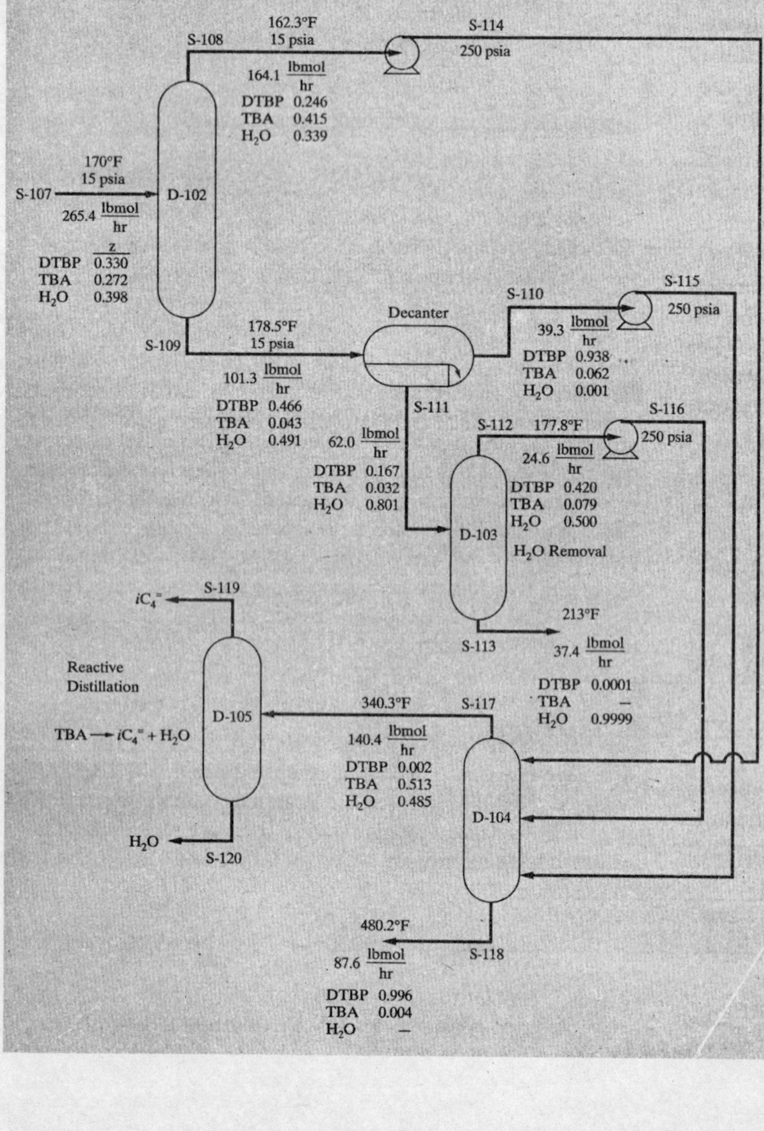

Figure 8.42 Process flowsheet for the DTBP process.

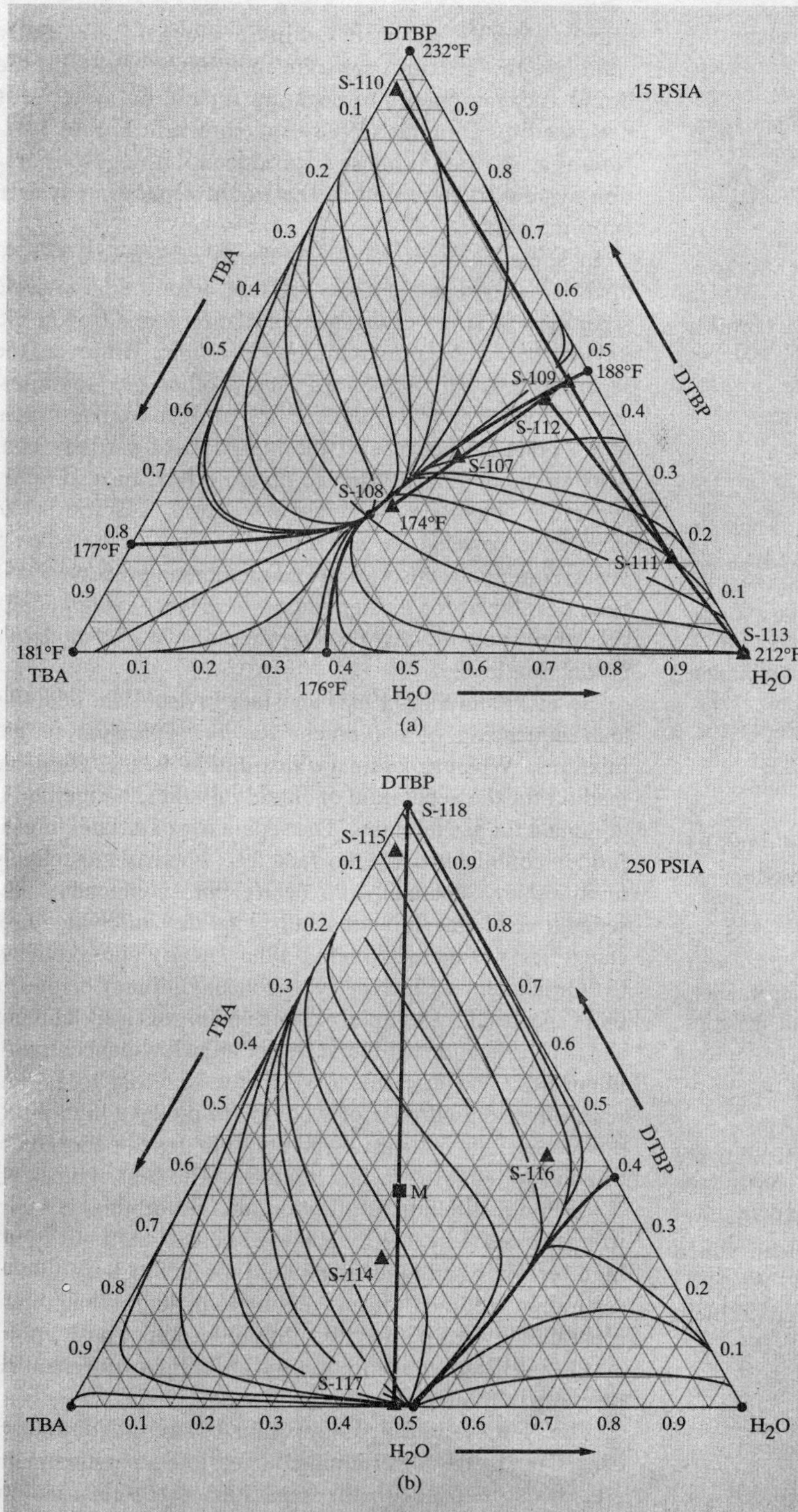

Figure 8.43 Distillation boundaries and material balance lines for the TBA–H_2O–DTBP system: (a) 15 psia; (b) 250 psia.

and the boiling points of the pure species are 181, 212, and 232°F, for TBA, H_2O, and DTBP, respectively. In addition, there is a minimum-boiling ternary azeotrope at $x_{TBA} = 0.44$, $x_{H_2O} = 0.33$, and $x_{DTBP} = 0.23$, and 174°F. Consequently, there are three distinct distillation regions, with the feed composition in a region that does not include the product vertex for DTBP.

To cross the distillation boundaries, it is possible to take advantage of the partial miscibility of the DTBP–H_2O system, as well as the disappearance of the ternary azeotrope at 250 psia as illustrated in Figure 8.41b. One possible design is shown in Figure 8.42, where the reactor effluent is in stream S-107. Column D-102 forms a distillate in stream S-108 whose composition is very close to the ternary azeotrope, and a bottoms product in stream S-109, as shown on the ternary diagram in Figure 8.43a. The latter stream, containing less than 5 mol% TBA, is split into two liquid phases in the decanter. The aqueous phase in stream S-111 enters the distillation tower, D-103, which forms nearly pure water in the bottoms product, stream S-113. The distillate from tower D-102, S-108; the organic phase from the decanter, S-110; and the distillate from tower D-103, S-112, are pumped to 250 psia and sent to the distillation tower, D-104, where they enter on stages that have comparable compositions. The compositions of the streams at elevated pressure, S-114, S-115, and S-116, and the mix point, M, are shown in Figure 8.43b. Note that at 250 psia, M lies in the distillation region that contains the DTBP vertex. Consequently, tower D-104 produces nearly pure DTBP in the bottoms products, S-118, and its distillate, S-117, is sent to the reactive distillation tower, D-105, where the TBA is dehydrated according to the reaction

$$\text{TBA} \rightarrow i\text{-butene} + H_2O$$

with *i*-butene recovered in the distillate, S-119, which is recycled to the catalytic reactor, and water in the bottoms product, S-120. As seen in Figures 8.43a and 8.43b, the material balance lines associated with the distillation towers lie entirely within separate distillation regions. The process works effectively because of the phase split and because the distillation boundaries are repositioned at the elevated pressure. Note, however, that the material balance line for the tower, D-102, would preferably be positioned farther away from the distillation boundary to allow for inaccuracies in the calculation of the distillation boundary.

Since this design was completed, the potential for DTBP to decompose explosively at temperatures above 255°F was brought to our attention. At 250 psia, DTBP is present in the bottoms product of tower D-104 at 480.2°F. Given this crucial safety concern, a design team would seek clear experimental evidence. If positive, lower pressures, with corresponding lower temperatures, would be explored, recognizing that the distillation boundaries are displaced less at lower pressures.

www.wiley.com/go/global/seider

For additional details of this process design, see the design report by Lee et al. (1995). Also, see Problem A-IIS.1.10 in the Supplement_to_Appendix_II.pdf (in the PDF Files folder, which can be downloaded from the Wiley Web site associated with this book) for the design problem statement that led to this design.

8.6 SEPARATION SYSTEMS FOR GAS MIXTURES

Sections 8.4 and 8.5 deal primarily with the synthesis of separation trains for liquid–mixture feeds. The primary separation techniques are ordinary and enhanced distillation. If the feed consists of a vapor mixture in equilibrium with a liquid mixture, the same techniques and synthesis procedures can often be employed. However, if the feed is a gas mixture and a wide gap in volatility exists between two groups of chemicals in the mixture, it is often preferable, as discussed in Section 8.1, to partially condense the mixture, separate the phases, and send the liquid and gas phases to separate separation systems as discussed by Douglas (1988) and shown in Figure 8.44. Note that if a liquid phase is produced in the gas separation system, it is routed to the liquid separation system and vice versa.

In some cases, it has been found economical to use distillation to separate a gas mixture, with the large-scale separation of air by cryogenic distillation into nitrogen and oxygen being the most common example. However, the separation by distillation of many other gas mixtures, such as hydrogen from methane or hydrogen from nitrogen, is not practical because of the high cost of partially condensing the overhead vapor to obtain reflux. Instead, other separation methods, such as absorption, adsorption, or membrane permeation, are employed. In just the past 25 years, continuous adsorption and membrane processes have been developed for the separation of air that economically rival the cryogenic distillation process at low to moderate production levels.

Barnicki and Fair (1992) consider in detail the selection and sequencing of equipment for the separation of gas mixtures. Whereas ordinary distillation is the dominant method for the separation of liquid mixtures, no method is dominant for gas mixtures. The separation of gas mixtures is further complicated by the fact that whereas most liquid mixtures are separated into nearly pure components, the separation of gas mixtures falls into the following three categories: (1) sharp splits to produce nearly pure products, (2) enrichment to increase the concentration(s) of one or more species, for example, oxygen and nitrogen enrichment, and (3) purification to remove one or more low-concentration impurities. The first category is often referred to as *bulk separation*, the purpose of which is to produce high-purity products at high recovery. Separations in this category can be difficult to achieve for gas mixtures. The best choices are cryogenic distillation, absorption, and adsorption. By contrast, the second category achieves neither high purity nor high recovery and is ideally suited for any of the common separation methods for gas mixtures, including membrane separation by gas permeation. To produce high-purity products by purification, adsorption and absorption with chemical reaction are preferred.

The synthesis of a separation train for a gas mixture can be carried out by first determining the feasible separation methods, which depend on the separation categories and the separation factors, and then designing and costing systems involving these methods to determine the optimal train. The design of equipment for absorption, adsorption, distillation, and membrane separations is covered by Seader and Henley (2006). Besides the separation category and separation factor, the production scale of the process is a major factor in determining the optimal train because economies of scale are most pronounced for cryogenic distillation and absorption,

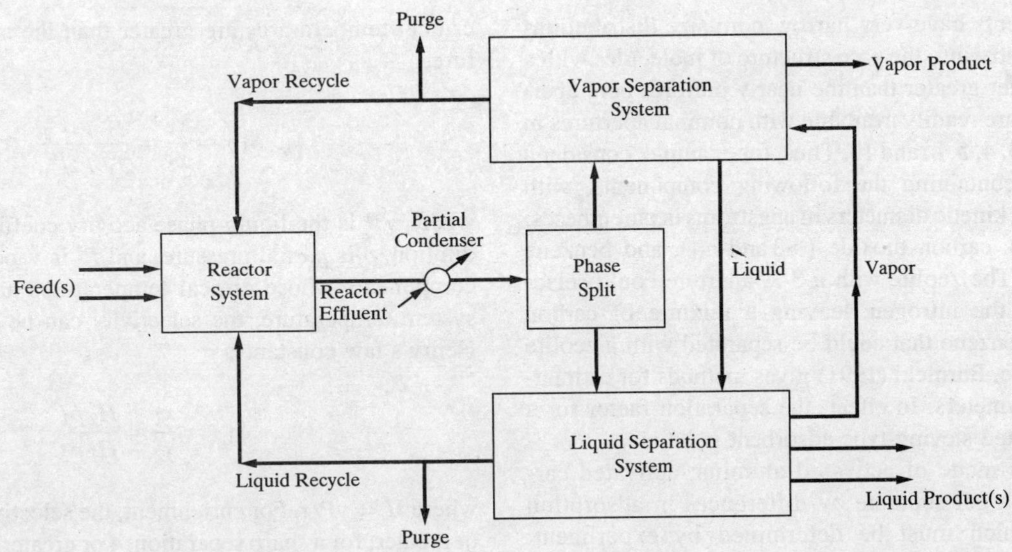

Figure 8.44 Process with vapor and liquid separation systems. (Modified and reprinted with permission from Douglas, 1988).

and least pronounced for adsorption and membrane separations. For example, for the separation of air into nitrogen- and oxygen-enriched products, membrane separations are most economical at low production rates, adsorption at moderate rates, and cryogenic distillation at high rates.

Membrane Separation by Gas Permeation

In gas permeation, the gas mixture is compressed to a high pressure and brought into contact with a very thin membrane to produce two products: (1) a permeate that passes through the membrane and is discharged at a low pressure, and (2) a *retentate* that does not pass through the membrane and is maintained at close to the high pressure of the feed. The separation factor defined by Eqs. (8.2) and (8.8) can be applied to gas permeation, when the retentate-side pressure is much greater than the permeate-side pressure, if y is the mole fraction in the permeate and x is the mole fraction in the retentate. The relative volatility is replaced by the ratio of the membrane permeabilities for the two key components of the feed–gas mixture, sometimes called the permselectivity. Most commercial membranes for gas permeation are nonporous (dense) amorphous or semicrystalline polymers. To pass through such polymers, the gas molecules first dissolve in the polymer and then pass through it by diffusion. Thus, the permeability depends on both solubility and diffusivity in the particular polymer at the conditions of temperature and pressure. The permeability is the product of the solubility and diffusivity. Permeabilities are best determined by laboratory measurements. However, a predictive method given by Barnicki (1991) for a number of glassy and rubbery polymers, which depends on species van der Waals volume and critical temperature, can be applied in the absence of data. In general, gas permeation is commercially feasible when the ratio of permeabilities (permselectivity) for the two components is greater than 15. However, some processes that require only rough enrichments use membranes having permselectivities of only 5. Commercial applications include the recovery of carbon dioxide from hydrocarbons, the adjustment of the hydrogen-to-carbon monoxide ratio in synthesis gas, the recovery of hydrocarbons from hydrogen, and the separation of air into nitrogen- and oxygen-enriched streams.

Adsorption

Adsorption differs from the other techniques in that it is a cyclic operation with adsorption and desorption steps. However, adsorption is a very versatile separation technique. To be economical, the adsorbent must be regenerable. This requirement precludes the processing of gas mixtures that contain (1) high-boiling organic compounds, because they are preferentially adsorbed and are difficult to remove during the regeneration part of the cycle, (2) lower-boiling organic compounds that may polymerize on the adsorbent surface, and (3) highly acidic or basic compounds that may react with the adsorbent surface. In some cases, such compounds can be removed from the gas mixture by guard beds or other methods prior to entry into the adsorption system.

Selectivity in adsorption is controlled by (1) molecular sieving or (2) adsorption equilibrium. When components differ significantly in molecular size and/or shape, as characterized by the kinetic diameter, zeolites and carbon molecular-sieve adsorbents can be used to advantage because of the strong selectivity achieved by molecular sieving.

These adsorbents have very narrow pore-size distributions that prevent entry into the pore structure of molecules with a kinetic diameter greater than the nearly uniform pore aperture. Zeolites are readily available with nominal apertures in angstroms of 3, 4, 5, 8, and 10. Thus, for example, consider a gas mixture containing the following components, with corresponding kinetic diameters in angstroms in parentheses: nitrogen (<3), carbon dioxide (>3 and <4), and benzene (>7 and <8). The zeolite with a 3-Å aperture could selectively adsorb the nitrogen, leaving a mixture of carbon dioxide and benzene that could be separated with a zeolite of 4-Å aperture. Barnicki (1991) gives methods for estimating kinetic diameters. In effect, the separation factor for a properly selected sieving-type adsorbent is infinity.

Adsorbents made of activated alumina, activated carbon, and silica gel separate by differences in adsorption equilibria, which must be determined by experiment. Equilibrium-limited adsorption can be applied to all three categories of separation, but is usually not a favored method when the components to be selectively adsorbed constitute an appreciable fraction of the feed gas. Conversely, equilibrium-limited adsorption is ideal for the removal of small quantities of selectively adsorbed impurities. At a given temperature, the equilibrium loading of a given component, in mass of adsorbate per unit mass of adsorbent, depends on the component partial pressure and to a lesser extent on the partial pressures of the other components. For equilibrium-limited adsorption to be feasible, Barnicki and Fair (1992) suggest that the ratio of equilibrium loadings of the two key components be used as a separation factor. This ratio should be based on the partial pressures in the feed gas. A ratio of at least 2, and preferably much higher, makes equilibrium adsorption quite favorable. However, two other conditions must also be met: (1) the more highly adsorbed component should have a concentration in the feed of less than 10 mol% and (2) for an adsorption time of 2 hr, the required bed height should not exceed 20 ft. Equilibrium-limited adsorption is usually the best alternative for the removal of water and organic chemicals from mixtures with light gases, and should also be considered for enrichment applications..

Absorption

Absorption of components of a gas mixture into a solvent may take place by physical or chemical means. When no chemical reaction between the solute and absorbent occurs (physical absorption), the separation factor is given by Eq. (8.2). Thus, if component 1 is to be selectively absorbed, a small value of SF is desired. Alternatively, Barnicki and Fair (1992) suggest that consideration of physical absorption should be based on a selectivity, $S_{1,2}$, defined as the ratio of liquid-phase mole fractions of the two key components in the gas mixture. This selectivity can be estimated from the partial pressures of the two components in the gas feed and their K-values for the given solvent. For components whose critical temperatures are greater than the system temperature,

$$S_{1,2} = \frac{x_1}{x_2} = \frac{\gamma_2^\infty p_1 P_2^s}{\gamma_1^\infty p_2 P_1^s} \tag{8.34}$$

where γ^∞ is the liquid-phase activity coefficient at infinite dilution, p is partial pressure, and P^s is vapor pressure. For components whose critical temperatures are less than the system temperature, the selectivity can be estimated from Henry's law constants:

$$S_{1,2} = \frac{x_1}{x_2} = \frac{H_2 p_1}{H_1 p_2} \tag{8.35}$$

where $H = yP/x$. For enrichment, the selectivity should be 3 or greater; for a sharp separation, 4 or greater. The number of theoretical stages should be at least 5. For the removal of readily soluble organic compounds from light gases, Douglas (1988) recommends the use of 10 theoretical stages and a solvent molar flow rate, L, based on an absorption factor, A, for solute of 1.4, where

$$A = \frac{L}{KV} \tag{8.36}$$

with V = gas molar flow rate. When the partial pressure, in the gas feed, of the component to be absorbed is very small and a high percentage of it is to be removed, physical absorption may not be favorable. Instead, particularly if the solute is an acid or base, chemical absorption may be attractive.

Partial Condensation and Cryogenic Distillation

The previously discussed separation techniques for gas mixtures all involve a mass separating agent. Alternatively, thermal means are employed with partial condensation and cryogenic distillation. Barnicki and Fair (1992) recommend that partial condensation be considered for enrichment when the relative volatility between the key components is 7. For large-scale (>10–20 tons/day of product gas) enrichment and sharp separations, cryogenic distillation is feasible when the relative volatility between the key components is greater than 2. However, if the feed gas contains components, such as carbon dioxide and water that can freeze at the distillation temperatures, those components must be removed first.

8.7 SEPARATION SEQUENCING FOR SOLID–FLUID SYSTEMS

The final product from many industrial chemical processes is a solid material. This is especially true for inorganic compounds, but is also common for a number of moderate- to high-molecular-weight organic compounds. Such processes involve the separation operations of leaching, evaporation, solution crystallization (solutes with high melting points that are crystallized from a solvent), melt crystallization (crystallization from a mixture of components with low to moderate melting points), precipitation (rapid crystallization from a solvent of nearly insoluble compounds that are usually formed by a chemical reaction), desublimation, and/or drying, as well as the phase-separation operations of filtration, centrifugation, and cyclone separation. In addition, because specifications for solid products may also include a particle-size distribution, size-increase and size-reduction operations may also be necessary. If particle shape is also a product specification, certain types of crystallizers and/or dryers may be dictated. Even when the final product is not a solid, solid–liquid or solid–gas separation operations may be involved. For example, liquid mixtures of *meta*- and *para*-xylene cannot be separated by distillation because their normal boiling points differ by only 0.8°C. Instead, because their melting points differ by 64°C, they are separated industrially by melt crystallization. Nevertheless, the final products are liquids. Another example is phthalic anhydride, which, although a solid at room temperature, is usually shipped in the molten state. It is produced by the air oxidation of napthalene or *ortho*-xylene. The separation of the anhydride from the reactor effluent gas mixture is accomplished by desublimation, followed by distillation to remove impurities and produce a melt.

A common flowsheet for the separation section of a process for the manufacture of inorganic salt crystals from their aqueous solution is shown in Figure 8.45. If the feed is aqueous $MgSO_4$, a typical process proceeds as follows. A 10 wt% sulfate feed is concentrated, without crystallization, to 30 wt% in a double-effect evaporation system. The concentrate is mixed with recycled mother liquors from the hydroclone and centrifuge before being fed to an evaporative vacuum crystallizer, which produces, by solution crystallization, a magma of 35 wt% crystals of $MgSO_4 \cdot 7H_2O$, the stable hydrate at the temperature in the crystallizer. The magma is thickened to 50 wt% crystals in a hydroclone, and then sent to a centrifuge, which discharges a cake containing 35 wt% moisture. The cake is dried to 2 wt% moisture in a direct-heat rotary dryer. Approximately 99 wt% of the dried crystals are retained on a 100-mesh screen and 30 wt% are retained on a 20-mesh screen. The crystals are bagged for shipment. Rossiter (1986) presents a similar flowsheet for the separation of aqueous NaCl, where a fluidized-bed dryer replaces the rotary dryer in Figure 8.45.

When a solid mixture of two components is to be separated, the process is more complicated. Such a process, using solution crystallization and shown in Figure 8.46, is considered by Rajagopal et al. (1988) for the production of crystalline potash (KCl) from sylvinite ore (mixture of 40 wt% KCl and 60 wt% NaCl). The separation scheme is feasible because KCl is less soluble than NaCl in water, and the solubility of KCl in water decreases with decreasing temperature, whereas the reverse is true for NaCl. In the first step of the process, KCl is completely dissolved (leached) by a mixture of makeup water and filtrate from the second filter. The NaCl in the ore is not dissolved because conditions are selected so that the water in the dissolver is saturated with NaCl. Thus, a slurry of solid (undissolved) NaCl and aqueous KCl–NaCl leaves the dissolver. The slurry is filtered in a rotary vacuum filter, which sends the wet cake of NaCl to further processing and the mother liquor to an evaporative crystallizer. There, evaporation lowers the temperature below that in the dissolver, causing crystallization only of the KCl. The magma from the crystallizer is sent to a rotary vacuum filter, from which the mother liquor is recycled to the dissolver, and the filter cake is sent to a direct-heat rotary dryer to produce crystalline potash. Other sequences for multicomponent mixtures are considered by Rajagopal et al. (1991), Cisternas and Rudd (1993), and Dye and Ng (1995b).

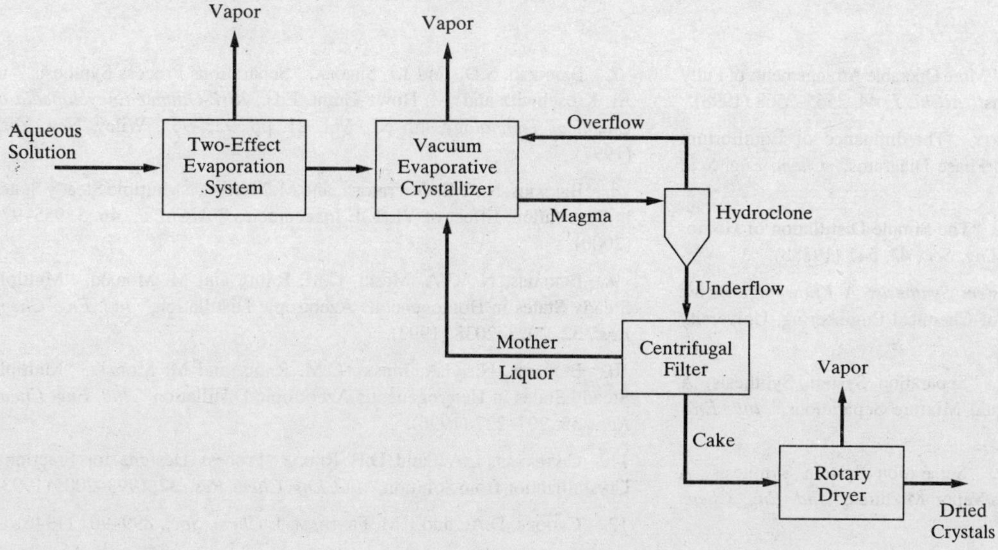

Figure 8.45 Process for producing inorganic salt crystals.

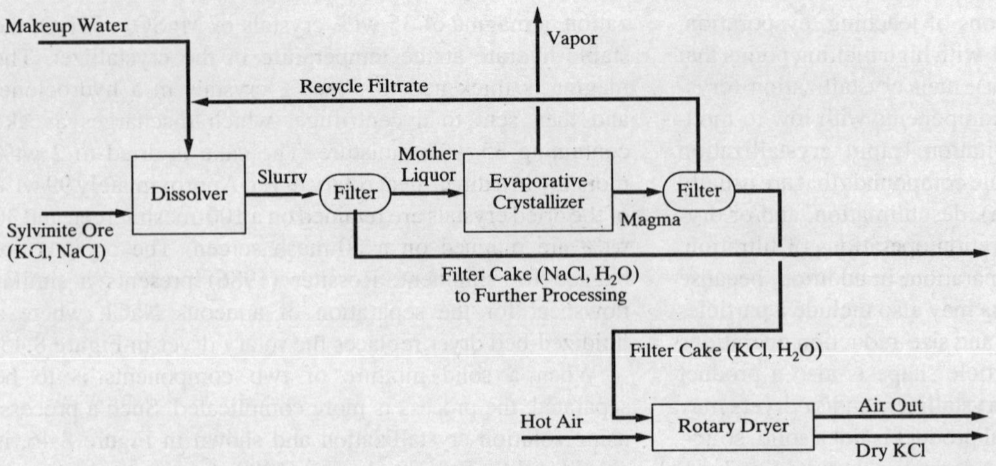

Figure 8.46 Process for separating a solid mixture.

In both of the processes just described, a crystallizer produces a solid and, following a solid–liquid phase separation, a dryer removes the moisture. In some cases, all three of these operations can be carried out in a single piece of equipment, a spray dryer or a drum dryer, but at the expense of increased utility cost because all of the solvent is evaporated. Such dryers are used extensively to produce dried milk and detergents. For these products, spray dryers are particularly desirable, because the drying process produces porous particles that are readily dissolved in water. Spray dryers can also handle slurries and pastes.

As discussed by Barnicki and Fair (1990), melt crystallization is an alternative to other separation techniques for liquid mixtures, including ordinary distillation, enhanced distillation, liquid–liquid extraction, adsorption, and membrane permeation. Melt crystallization should be considered only when ordinary distillation is not feasible, but may be an attractive alternative when the melting-point difference between the two key components exceeds 20°C and a eutectic is not formed. If a eutectic is formed, high recovery may not be possible, as discussed by King (1980). Methods for circumventing the eutectic limitation are discussed by Dye and Ng (1995a).

8.8 SUMMARY

Having studied this chapter, the reader should

1. Know how each of the important industrial separation methods can be applied to the separation of multicomponent mixtures.
2. Know the importance of the separation factor.
3. Know how to determine near-optimal and optimal distillation sequences for nearly ideal systems.
4. Know how to develop separation sequences for nonideal systems that involve the formation of azeotropes.
5. Know how to develop a sequence for separating a gas mixture.
6. Know how to separate solid–fluid and multicomponent solid mixtures.

REFERENCES

1. Agrawal, R., and Z.T. Fidkowski, "More Operable Arrangements of Fully Thermally Coupled Distillation Systems," *AIChE J.*, **44**, 2565–2568 (1998).

2. Barbosa, D., and M.F. Doherty, "The Influence of Equilibrium Chemical Reactions on Vapor-Liquid Phase Diagrams," *Chem. Eng. Sci.*, **43**, 529 (1988a).

3. Barbosa, D., and M.F. Doherty, "The Simple Distillation of Homogeneous Reactive Mixtures," *Chem. Eng. Sci.*, **43**, 541 (1988b).

4. Barnicki, S.D., *Separation System Synthesis: A Knowledge-Based Approach*, Ph.D. dissertation, Dept. of Chemical Engineering, University of Texas at Austin (1991).

5. Barnicki, S.D., and J.R. Fair, "Separation System Synthesis: A Knowledge-Based Approach. 1. Liquid Mixture Separations," *Ind. Eng. Chem. Res.*, **29**, 421–432 (1990).

6. Barnicki, S.D., and J.R. Fair, "Separation System Synthesis: A Knowledge-Based Approach. 2. Gas/Vapor Mixtures," *Ind. Eng. Chem. Res.*, **31**, 1679–1694 (1992).

7. Barnicki, S.D., and J.J. Siirola, "Separations Process Synthesis," in J.I. Kroschwitz and M. Howe-Grant, Eds., *Kirk-Othmer Encyclopedia of Chemical Technology*, 4th ed., Vol. 21, pp. 923–962, Wiley, New York (1997).

8. Bekiaris, N., T.E. Guttinger, and M. Morari, "Multiple Steady States in Distillation: Effect of VL(L)E Inaccuracies," *AIChE J.*, **46**, 5, 955–979 (2000).

9. Bekiaris, N., G.A. Meski, C.M. Radu, and M. Morari, "Multiple Steady States in Homogeneous Azeotropic Distillation," *Ind. Eng. Chem. Res.*, **32**, 2023–2038 (1993).

10. Bekiaris, N., G.A. Meski, C.M. Radu, and M. Morari, "Multiple Steady States in Heterogeneous Azeotropic Distillation," *Ind. Eng. Chem. Res.*, **35**, 207–227 (1996).

11. Cisternas, L.A., and D.F. Rudd, "Process Designs for Fractional Crystallization from Solution," *Ind. Eng. Chem. Res.*, **32**, 1993–2005 (1993).

12. Crooks, D.A., and F.M. Feetham, J. *Chem. Soc.*, 899–901 (1946).

13. Doherty, M.F., and G. Buzad, "Reactive Distillation by Design," *Trans. IChemE*, **70**, 448–458 (1992).

14. Doherty, M.F., and M.F. Malone, *Conceptual Design of Distillation Systems*, McGraw-Hill, Boston (2001).

15. Douglas, J.M., *Conceptual Design of Chemical Processes*, McGraw-Hill, New York (1988).

16. Douglas, J.M., "Synthesis of Separation System Flowsheets," *AIChE J.*, **41**, 2522–2536 (1995).

17. Dye, S., and K.M. Ng, "Bypassing Eutectics with Extractive Crystallization: Design Alternatives and Tradeoffs," *AIChE J.*, **41**, 1456–1470 (1995a).

18. Dye, S., and K.M. Ng, "Fractional Crystallization: Design Alternatives and Tradeoffs," *AIChE J.*, **41**, 2427–2438 (1995b).

19. Eckert, E., and M. Kubicek, "Computing Heterogeneous Azeotropes in Multicomponent Mixtures," *Comput. Chem. Eng.*, **21**, 347–350 (1997).

20. Fidkowski, Z., and L. Krolikowski, "Minimum Energy Requirements of Thermally Coupled Distillation Systems," *AIChE J.*, **33**, 643–653 (1987).

21. Fidkowski, Z.T., M.F. Malone, and M.F. Doherty, "Computing Azeotropes in Multicomponent Mixtures," *Comput. Chem. Eng.*, **17**, 1141 (1993).

22. Gmehling, J., *Azeotropic Data*, VCH Publishers, Deerfield Beach, Florida (1994).

23. Gomez, A., and J.D. Seader, "Separation Sequence Synthesis by a Predictor Based Ordered Search," *AIChE J.*, **22**, 970–979 (1976).

24. Green, D.W., and R.H. Perry Ed., *Perry's Chemical Engineers' Handbook*, 8th ed., McGraw-Hill, New York (2008).

25. Hendry, J.E., and R.R. Hughes, "Generating Separation Process Flowsheets," *Chem. Eng. Prog.*, **68**(6), 69 (1972).

26. Horsley, L.H., *Azeotropic Data–III, Advances in Chemistry Series No. 116*, American Chemical Society, Washington, DC (1973).

27. Horwitz, B.A., "Optimize Pressure-Sensitive Distillation," *Chem. Eng. Prog.*, **93** (4), 47 (1997).

28. Keller, G.M., "Separations: New Directions for an Old Field," *AIChE Monogr. Ser. No. 17*, **83** (1987).

29. King, C.J., *Separation Processes*, 2nd ed., McGraw-Hill, New York (1980).

30. Kister, H.Z., *Distillation Design*, McGraw-Hill, New York (1992).

31. Knapp, J.P., and M.F. Doherty, "A New Pressure-Swing Distillation Process for Separating Homogeneous Azeotropic Mixtures," *Ind. Eng. Chem. Res.*, **31**, 346 (1992).

32. Kovach III, J.W., and W.D. Seider, "Heterogeneous Azeotropic Distillation: Homotopy-Continuation Methods," *Comput. Chem. Eng.*, **11**, 593 (1987).

33. Lecat, M., *L'azcotropism: la tension de vapeur des melanges de liquides. 1. ptie., Donnes experimentales. Bibliographie*, A. Hoste, Gand, Belgium (1918).

34. Lee, K.-S., C. Levy, and N. Steckman, *Ditertiary-Butyl Peroxide Manufacture*, Towne Library, University of Pennsylvania, Philadelphia (1995).

35. Matsuyama, H., and H. Nishimura, "Topological and Thermodynamic Classification of Ternary Vapor–Liquid Equilibria," *J. Chem. Eng. Jpn.*, **10** (3), 181 (1977).

36. Modi, A.K., and A.W. Westerberg, "Distillation Column Sequencing Using Marginal Price," *Ind. Eng. Chem. Res.*, **31**, 839–848 (1992).

37. Peterson, E.J., and L.R. Partin, "Temperature Sequences for Categorizing All Ternary Distillation Boundary Maps," *Ind. Eng. Chem. Res.*, **36**, 1799–1811 (1997).

38. Petlyuk, F.B., V.M. Platonov, and D.M. Slavinskii, "Thermodynamically Optimal Method for Separating Multicomponent Mixtures," *Int. Chem. Eng.*, **5**, 555 (1965).

39. Prokopakis, G.J., and W.D. Seider, "Dynamic Simulation of Azeotropic Distillation Towers," *AIChEJ.*, **29**, 1017–1029 (1983).

40. Rajagopal, S., K.M. Ng, and J.M. Douglas, "Design of Solids Processes: Production of Potash," *Ind. Eng. Chem. Res.*, **27**, 2071–2078 (1988).

41. Rajagopal, S., K.M. Ng, and J.M. Douglas, "Design and Economic Trade-offs of Extractive Crystallization Processes," *AIChE J.*, **37**, 437–447 (1991).

42. Rev, E., M. Emtir, Z. Szitkai, P. Mizsey, and Z. Fonyo, "Energy Savings of Integrated and Coupled Distillation Systems," *Comput. Chem. Eng.*, **25**, 119–140 (2001).

43. Robinson, C.S., and E.R. Gilliland, *Elements of Fractional Distillation*, McGraw-Hill, New York (1950).

44. Rodrigo, B.F.R., and J.D. Seader, "Synthesis of Separation Sequences by Ordered Branch Search," *AIChEJ.*, **21** (5), 885 (1975).

45. Rossiter, A.P., "Design and Optimisation of Solids Processes. Part 3—Optimisation of a Crystalline Salt Plant Using a Novel Procedure," *Chem. Eng. Res. Des.*, **64**, 191–196 (1986).

46. Ryan, P.J., and M.F. Doherty, "Design/Optimization of Ternary Heterogeneous Azeotropic Distillation Sequences," *AIChE J.*, **35**, 1592–1601 (1989).

47. Seader, J.D., and E.J. Henley, *Separation Process Principles*, 2nd ed., Wiley, New York (2006).

48. Seader, J.D., and A.W. Westerberg, "A Combined Heuristic and Evolutionary Strategy for Synthesis of Simple Separation Sequences," *AIChE J.*, **23**, 951 (1977).

49. Siirola, J.J., "An Industrial Perspective of Process Synthesis," in L.T. Biegler and M.F. Doherty, Eds., *Foundations of Computer-Aided Process Design, AIChE Symp. Ser.* No. 304, **91**, 222–233 (1995).

50. Souders, M., "The Countercurrent Separation Process," *Chem. Eng. Prog.*, **60** (2), 75–82 (1964).

51. Stichlmair, J.G., and J.R. Fair, *Distillation–Principles and Practice*, Wiley-VCH, New York (1998).

52. Stichlmair, J.G., J.R. Fair, and J.L. Bravo, "Separation of Azeotropic Mixtures via Enhanced Distillation," *Chem. Eng. Prog.*, **85** (63), 1 (1989).

53. Swietoslawski, W., *Azeotropy and Polyazeotropy*, Pergamon Press, New York (1963).

54. Tedder, D.W., and D.F. Rudd, "Parametric Studies in Industrial Distillation: I. Design Comparisons," *AIChEJ.*, **24**, 303 (1978a).

55. Tedder, D.W., and D.F. Rudd, "Parametric Studies in Industrial Distillation: II. Heuristic Optimization," *AIChEJ.*, **24**, 316 (1978b).

56. Vadapalli, A., and J.D. Seader, "A Generalized Framework for Computing Bifurcation Diagrams Using Process Simulation Programs," *Comput. Chem. Eng.*, **25**, 445–464 (2001).

57. Van Winkle, M., *Distillation*, McGraw-Hill, New York (1967).

58. Venimadhavan, G., G. Buzad, M.F. Doherty, and M.F. Malone, "Effect of Kinetics on Residue Curve Maps for Reactive Distillation," *AIChEJ.*, **40** (11), 1814 (1994).

59. Walas, S.M., *Chemical Process Equipment–Selection and Design*, Butterworth, Boston (1988).

60. Widagdo, S., and W.D. Seider, "Azeotropic Distillation," *AIChE J.*, **42** (1), 96 (1996).

61. Wright, R.O., "Fractionation Apparatus," U.S. Patent 2,471,134 (May 24, 1949).

62. Zharov, V.T., "Phase Representations and Rectification of Multicomponent Solutions," *J. Appl. Chem. USSR*, **41**, 2530 (1968).

63. Zharov, V.T., and L.A. Serafimov, *Physicochemical Fundamentals of Distillations and Rectifications*, (in Russian), Khimiya Leningrad, (1975).

EXERCISES

8.1 Stabilized effluent from a hydrogenation unit, as given below, is to be separated by ordinary distillation into five relatively pure products. Four distillation columns will be required. According to Eq. (8.9) and Table 8.2, these four columns can be arranged into 14 possible sequences. Draw sketches, as in Figure 8.11, for each of these sequences.

Component	Feed Flow Rate (lbmol/hr)	Approximate Relative Volatility Realative to C5
Propane (C3)	10.0	8.1
Butene-1 (B1)	100.0	3.7
n-Butene (NB)	341.0	3.1
Butene-2 isomers (B2)	187.0	2.7
n-Pentane (C5)	40.0	1.0

8.2 The feed to a separation process consists of the following species:

Species Number	Species
1	Ethane
2	Propane
3	Butene-1
4	*n*-Butane

It is desired to separate this mixture into essentially pure species. The use of two types of separators is to be explored:

1. Ordinary distillation
2. Extractive distillation with furfural (species 5)

The separation orderings according to relative volatility are

	Separator Type	
	1	2
Species number	1	1
	2	2
	3	4
	4	3
		5

Notice that the addition of furfural causes *n*-butane (4) to become more volatile than butene-1 (3). Determine the number of possible separation sequences.

8.3 Thermal cracking of naphtha yields a gas that is to be separated by a distillation train into the products indicated in Figure 8.47. If reasonably sharp separations are to be achieved, determine by the heuristics of Section 8.4 two good sequences.

8.4 Investigators at the University of California at Berkeley have studied all 14 possible sequences for separating the following mixture at a flow rate of 200 lbmol/hr into its five components at about 98% purity each (D.L. Heaven, M.S. Thesis in Chemical Engineering, University of California, Berkeley, 1969).

Species	Symbol	Feed Mole Fraction	Approximate Volatility Relative to *n*-Pentane
Propane	A	0.05	8.1
Isobutane	B	0.15	4.3
n-Butane	C	0.25	3.1
Isopentane	D	0.20	1.25
n-Pentane	E	0.35	1.0
		1.00	

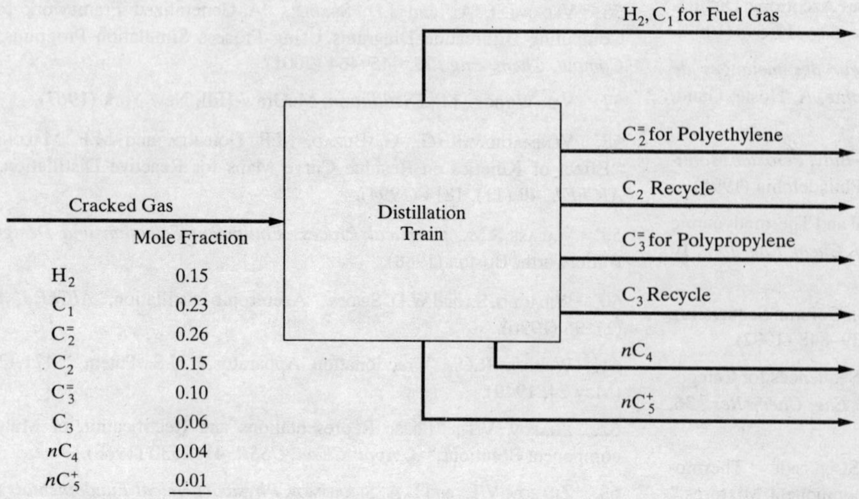

Figure 8.47 Thermal cracking of naphtha.

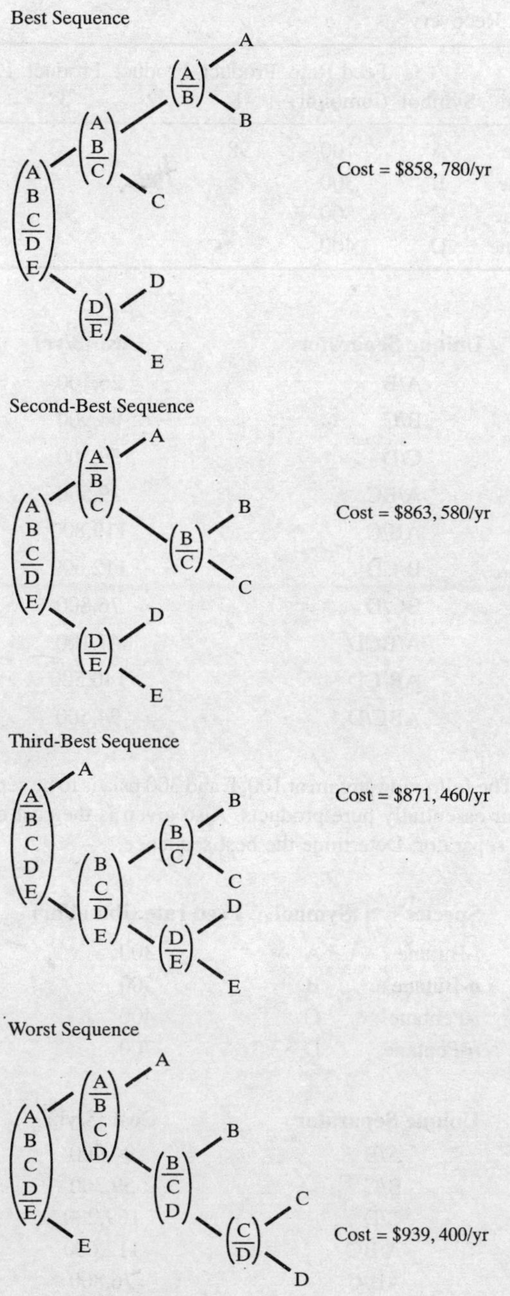

Figure 8.48 Cost data for Exercise 8.4.

For each sequence, they determined the annual operating cost, including depreciation of the capital investment. Cost data for the best three sequences and the worst sequence are in Figure 8.48.

Explain in detail, as best you can, why the best sequences are best and the worst sequence is worst using the heuristics. Which heuristics appear to be the most important?

8.5 The effluent from a reactor contains a mixture of various chlorinated derivatives of the hydrocarbon RH_3, together with the hydrocarbon itself and HCl. Based on the following information and the heuristics of Section 8.4, devise the best two feasible separation sequences. Explain your reasoning. Note that HCl may be corrosive.

Species	lbmol/hr	α Realative to RCl_3	Purity Desired
HCl	52	4.7	80%
RH_3	58	15.0	85%
RCl_3	16	1.0	98%
RH_2Cl	30	1.9	95%
$RHCl_2$	14	1.2	98%

8.6 The following stream at 100°F and 20 psia is to be separated into the four indicated products. Determine the best distillation sequence by the heuristics of Section 8.4. Compare your result with the result obtained by applying the Marginal Vapor Rate method.

		Percent Recovery			
Species	Feed (lbmol/hr)	Product 1	Product 2	Product 3	Product 4
Benzene	100	98			
Toluene	100		98		
Ethylbenzene	200			98	
p-Xylene	200			98	
m-Xylene	200			98	
o-Xylene	200				98

8.7 The following cost data, which include operating cost and depreciation of capital investment, pertain to Exercise 8.1. Determine by finding the total cost for each of the 14 possible sequences:

a. The best sequence

b. The second-best sequence

c. The worst sequence

Are the heuristics of Section 8.4 in agreement with the results based on costs?

Split	Cost ($/yr)
C3/B1	15,000
B1/NB	190,000
NB/B2	420,000
B2/C5	32,000
C3, B1/NB	197,000
C3/B1, NB	59,000
B1, NB/B2	500,000
B1/NB, B2	247,000
NB, B2/C5	64,000
NB/B2, C5	460,000
C3, B1, NB/B2	510,000
C3, B1/NB, B2	254,000
C3/B1, NB, B2	85,000
B1, NB, B2/C5	94,000
B1, NB/B2, C5	530,000
B1/NB, B2, C5	254,000
C3, B1, NB, B2/C5	95,000
C3, B1, NB/B2, C5	540,000
C3, B1/NB, B2, C5	261,000
C3/B1, NB, B2, C5	90,000

8.8 A hypothetical mixture of four species, A, B, C, and D, is to be separated into the four separate components. Two different separator types are being considered, neither of which requires a mass-separating agent. The order of separation for each of the two types are

Separator Type I	**Separator Type II**
A	B
B	A
C	C
D	D

Annual cost data for all the possible splits are given below. Determine by considering each possible sequence:

a. The best sequence

b. The second-best sequence

c. The worst sequence

For each answer, draw a diagram of the separation train, being careful to label each separator as to whether it is type I or II.

Subgroup	Split	Type Separator	Annual Cost × \$10,000
(A, B)	A/B	I	8
		II	15
(B, C)	B/C	I	23
		II	19
(C, D)	C/D	I	10
		II	18
(A, C)	A/C	I	20
		II	6
(A, B, C)	A/B, C	I	10
	B/A, C	II	25
	A, B/C	I	25
		II	20
(B, C, D)	B/C, D	I	27
		II	22
	B, C/D	I	12
		II	20
(A, C, D)	A/C, D	I	23
		II	10
	A, C/D	I	11
		II	20
(A, B, C, D)	A/B, C, D	I	14
	B/A, C, D	II	20
	A, B/C, D	I	27
		II	25
	A, B, C/D	I	13
		II	21

8.9 The following stream at 100°F and 250 psia is to be separated into the four indicated products. Also given is the cost of each of the unique separators. Determine:

a. The best sequence

b. The second-best sequence

Percent Recovery

Species	Symbol	Feed Rate (lbmol/hr)	Product 1	Product 2	Product 3	Product 4
Propane	A	100	98			
i-Butane	B	300		98		
n-Butane	C	500			98	
i-Pentane	D	400				98

Unique Separator	**Cost (\$/yr)**
A/B	26,100
B/C	94,900
C/D	59,300
A/BC	39,500
AB/C	119,800
B/CD	112,600
BC/D	76,800
A/BCD	47,100
AB/CD	140,500
ABC/D	94,500

8.10 The following stream at 100°F and 300 psia is to be separated into four essentially pure products. Also given is the cost of each unique separator. Determine the best sequence.

Species	**Symbol**	**Feed rate (lbmol/hr)**
i-Butane	A	300
n-Butane	B	500
i-Pentane	C	400
n-Pentane	D	700

Unique Separator	**Cost (\$/yr)**
A/B	94,900
B/C	59,300
C/D	169,200
A/BC	112,600
AB/C	76,800
B/CD	78,200
BC/D	185,300
A/BCD	133,400
AB/CD	94,400
ABC/D	241,800

8.11 Consider the problem of separation, by ordinary distillation, of propane, A; isobutane, B; *n*-butane, C; isopentane, D; and *n*-pentane, E. Using the heuristics of Section 8.4, develop flowsheets for:

a. Equimolal feed with product streams A, (B, C) and (D, E) required

b. Feed consisting of A = 10, B = 10, C = 60, D = 10, and E = 20 (relative moles) with products A, B, C, D, and E

Component	Average Relative Volatility
A > B	2.2
B > C	1.44
C > D	2.73
D > E	1.25

8.12 Derive the right-hand side of Eq. (8.9).

8.13 **a.** Consider binary mixtures of acetone and chloroform at 101 kPa, with vapor–liquid equilibria in Figure 8.20.

Using distillation, identify the maximum and minimum mole fractions of acetone in the product streams for feed streams containing:

1. 90 mol% acetone
2. 25 mol% acetone

What are the bubble-point temperatures of the associated distillate and bottoms products?

b. Repeat (a) for isopropyl ether (IPE) and isopropyl alcohol (IPA), using Figure 8.19, with mole fractions of IPE replacing those of acetone.

8.14 A multicomponent mixture is boiled in a flask at 1 atm. The vapors are condensed and recovered as a liquid product. It is desired to examine the mole fractions of the residual liquid in the flask as vaporization proceeds. Although sketches of the residue curve maps are called for in (b)–(d), a process simulator can be used to prepare the drawings accurately.

a. For a mixture of 60 mol% *n*-butane (1), and 40 mol% *n*-pentane (2), determine the residual mole fraction of *n*-butane after 10% of the liquid has vaporized.

b. Consider mixtures of *n*-butane (1), *n*-pentane (2), and *n*-hexane (3). For three typical feed compositions:

	Mole Fractions		
Component	I	II	III
1	0.7	0.15	0.15
2	0.15	0.7	0.15
3	0.15	0.15	0.7

sketch the residue curves (solutions of the ODEs—do not solve them analytically or numerically) on triangular graph paper. Use arrows to show the direction along the trajectories in time.

c. Repeat (b) for mixtures of acetone (1), chloroform (2), and benzene (3). Note that the acetone-chloroform binary exhibits a maximum-boiling azeotrope (64°C) at 35 mol% acetone, with no other azeotropes existing. Sketch any boundaries across which the residue curves cannot traverse.

d. Repeat (c) for mixtures of methyl acetate (1), methanol (2), and *n*-hexane (3). Note the existence of four azeotropes, where compositions are in mol%.

	T°C
Methyl acetate (65%)–methanol (35%)	53.5
Methanol (51%)–*n*-hexane (49%)	50.0
Methyl acetate (60%)–*n*-hexane (40%)	51.8
Methyl acetate (31%)–*n*-hexane (40%)–methanol (29%)	49.0

8.15 Prepare residue curve maps using a process simulation program for the following mixtures at 1 atm. Identify any distillation boundaries.

a. Acetone, *n*-heptane, toluene

b. Methanol, ethanol, water

c. Acetone, chloroform, ethanol

8.16 For a mixture of 70 mol% chloroform, 15 mol% acetone, and 15 mol% ethanol at 1 atm, show on a residue curve map the feasible compositions of the distillate and bottoms product.

8.17 Consider the process for the dehydration of ethanol using toluene in Figure 8.32. Estimate the ratios of the flow rates in the following streams:

a. S1 and S2

b. S2 and D1

c. S1 and F

d. B1 and D1

e. B2 and D2

8.18 For the manufacture of di-tertiary-butyl peroxide in Example 8.5 synthesize an alternative process and show the flow rate and composition of each stream.

8.19 For the pressure-swing dehydration of THF, determine the internal flow rates when the high-pressure column is at 5 bar.

Chapter 9

Heat Exchanger Networks

9.0 OBJECTIVES

This chapter introduces several *algorithmic* approaches that have been developed for process integration to satisfy the cooling, heating, and power demands of a process. After studying this material, the reader should

1. Be able to determine minimum energy requirement (MER) targets; that is, to compute the minimum usage of heating and cooling utilities when exchanging heat between the hot and cold streams in a process. Three methods are introduced: the temperature-interval (TI) method, a graphical approach, and the formulation and solution of a linear program (LP).
2. Be able to design a network to meet the MER targets; that is, to position heat exchangers in a network, assuming overall heat-transfer coefficients. Two methods are introduced: a unit-by-unit method beginning at the closest-approach temperature difference (the *pinch*), and the formulation and solution of a mixed-integer linear program (MILP).
3. Be able to reduce the number of heat exchangers in MER networks, by relaxing the MER target and *breaking the heat loops* (i.e., allowing heat to flow across the pinch), or alternatively, by employing *stream splitting.*
4. Be able to design a network when the minimum approach temperature is below a threshold value, at which either heating or cooling utility is used, but not both.
5. Be able to use the grand composite curve to assist in the selection and positioning of appropriate types of hot and cold utilities in the network.
6. Understand the importance of the specified minimum approach temperature difference on the design of a heat exchanger network (HEN).
7. Understand how to set up a superstructure for the design of a HEN that minimizes the annualized cost and how to formulate and solve its nonlinear program (NLP) using the General Algebraic Modeling System (GAMS).
8. Understand several approaches to designing energy-efficient distillation trains, including the adjustment of tower pressure, multiple-effect distillation, and heat pumping, vapor recompression, and reboiler flashing.
9. Understand the need to position *heat engines* to satisfy power demands of processes, and the need to position *heat pumps* to accomplish refrigeration to reduce power requirements. A methodology is introduced that does not require the usage of formal optimization methods.

9.1 INTRODUCTION

At the start of the task-integration step in process synthesis, the source and target temperatures, T^s and T^t, and power demands for pumping and compression of all streams are known. Heat and power integration seeks to utilize the energy in the high-temperature streams that need to be cooled and/or condensed to heat and/or vaporize the cold streams, and provide power to compressors from turbines and heat engines where possible. In most designs, it is common initially to disregard power demands in favor of designing an effective network of heat exchangers by heat integration, without using the energy of the high-temperature streams to produce power. To accomplish this, N_H hot process streams, with specified source and target temperatures $T^s_{h_i}$ and $T^t_{h_i}$, $i = 1, \ldots, N_H$, are cooled by N_C cold process streams, with specified source and target temperatures, $T^s_{c_j}$ and $T^t_{c_j}, j = 1, \ldots, N_C$, as shown schematically in Figure 9.1a. When either: (a) the sum of the heating requirements does not equal the sum of the cooling requirements; or (b) some source temperatures may not be sufficiently high or low to achieve some target temperatures through heat exchange; or (c) when other restrictions exist, as discussed in Section 9.2, it is always necessary to provide one or more auxiliary heat exchangers for heating or cooling through the use of utilities such as steam and cooling water. It is common to refer to the heat exchangers between the hot and cold process streams as comprising the *interior network,* and those between the hot or cold streams and the utilities as comprising the *auxiliary network,* as shown schematically in Figure 9.1b.

When carrying out the design given the states of the source and target streams (flow rates of the species, temperature, pressure, and phase), it is desired to synthesize the most economical network of heat exchangers. Several measures of economic goodness are possible, as discussed in Section 23.4. Usually, when generating and comparing alternative flowsheets, an approximate profitability measure is sufficient, such as the annualized cost:

$$C_A = i_m(C_{TCI}) + C \tag{9.1}$$

where C_{TCI} is the total capital investment, as defined in Table 22.9, i_m is a reasonable return on investment annually (i.e., when $i_m = 0.33$, a chemical company charges itself annually for one-third of the cost of the capital invested), and C is the annual cost of sales, as defined in the *cost sheet* of Table 23.1. In Tables 22.9 and 23.1, many factors are involved, most of

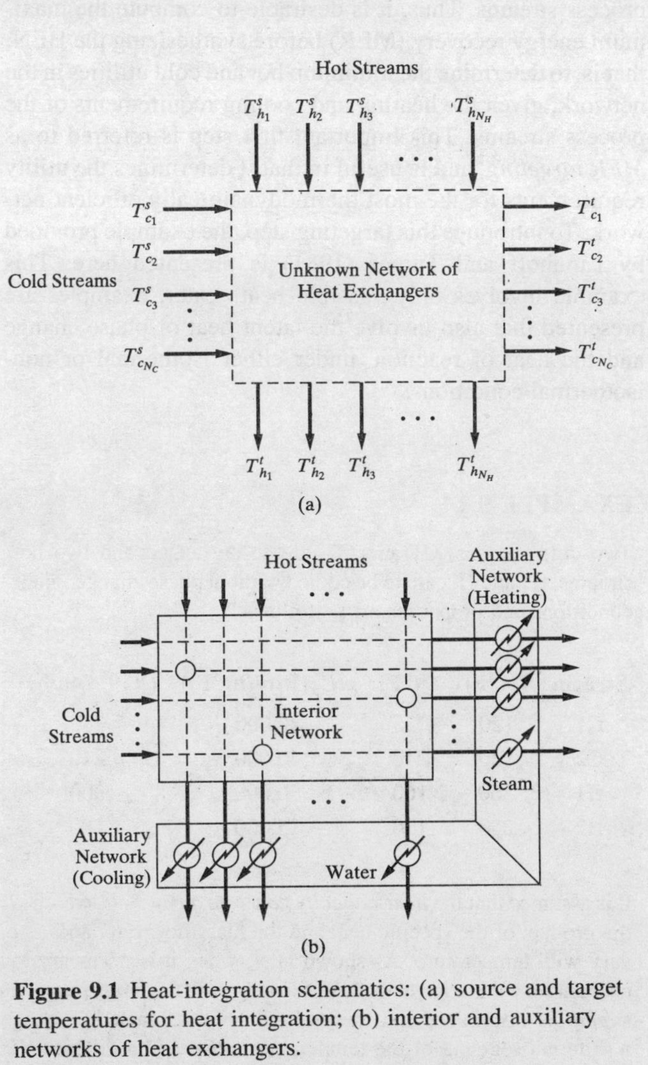

Figure 9.1 Heat-integration schematics: (a) source and target temperatures for heat integration; (b) interior and auxiliary networks of heat exchangers.

which are necessary for a detailed profitability analysis. However, to estimate an approximate profitability measure for the comparison of alternative flowsheets, it is adequate to approximate C_{TCI} as the sum of the purchase costs for each of the heat exchangers (without including installation costs and other capital investment costs). The purchase costs can be estimated based on the area for heat transfer, A, estimated from the heat-transfer rate equation discussed in Section 18.2 [Eq. (18.7)]:

$$A = Q/(UF_T \Delta T_{LM}) \tag{9.2}$$

where Q is the heat duty, U is the overall heat-transfer coefficient, F_T is the correction factor for a multiple-pass exchanger, and ΔT_{LM} is the log-mean temperature-driving force for countercurrent flow based on the approach-temperature differences at the two ends. Equation (9.2) must be used with care because of its restrictions, as discussed in Chapter 18. If both a phase change and a significant temperature change occur for one or both streams, U is not constant and a ΔT_{LM} is not appropriate. Furthermore, multiple-pass exchangers may be required, for which F_T is in the range 0.75–0.9. Nevertheless, to develop a reasonably optimal heat exchanger network, it is common to apply Eq. (9.2) with $F_T = 1.0$. It is adequate to approximate C as the annual cost of the utilities for heating and cooling, typically using steam and cooling water. In summary, with these approximations, Eq. (9.1) is rewritten as

$$C_A = i_m \left[\sum_i C_{P,I_i} + \sum_j C_{P,A_j} \right] + sF_s + (cw)F_{cw} \tag{9.3}$$

where C_{P,I_i} and C_{P,A_j} are the purchase costs of the heat exchangers in the interior and auxiliary networks, respectively, F_s is the annual flow rate of steam (e.g., in kilograms per year), s is the unit cost of steam (e.g., in dollars per kilogram), F_{cw} is the annual flow rate of cooling water, and cw is the unit cost of cooling water. Clearly, when other utilities such as fuel, cool air, boiler feed water, and refrigerants are used, additional terms are needed.

Many approaches have been developed to optimize Eq. (9.3) and similar profitability measures, several of which are presented in this chapter. Probably the most widely used, an approach developed immediately after the OPEC oil embargo in 1973 that triggered a global energy crisis, utilizes a two-step procedure:

1. A network of heat exchangers is designed having the *minimum usage of utilities* (i.e., an MER network), usually requiring a large number of heat exchangers. However, when the cost of fuel is extremely high, as it was in the late 1970s and the mid-2000s, a nearly optimal design is obtained.

2. The number of heat exchangers is reduced toward the minimum, possibly at the expense of increasing the consumption of utilities.

Clearly, as step 2 is implemented, one heat exchanger at a time, capital costs are reduced due to the economy-of-scale in equations of the form:

$$C_P = KA^n \tag{9.4}$$

where K is a constant and n is less than unity, typically 0.6. As each heat exchanger is removed, with the total area for heat transfer approximately constant, the area of each of the remaining heat exchangers is increased, and because $n < 1$, the purchase cost per unit of area is decreased. In addition, as step 2 is implemented, the consumption of utilities is normally increased. At some point, the increased cost of utilities overrides the decreased cost of capital and C_A increases beyond the minimum. When the cost of fuel is high, the minimum C_A is not far from that for a network of heat exchangers using the minimum utilities. Finally, note that Eqs. (22.38)–(22.44) provide more accurate estimates than Eq. (9.4), but are not commonly used when comparing alternative heat exchanger networks during process synthesis.

The selection of the minimum approach temperature, $\Delta T_{\min}$, for the heat exchangers is a key design variable in the synthesis of heat exchanger networks (HENs), because of its impact on lost work associated with heat transfer. Consider the heat transfer between the high- and low-temperature reservoirs in Figure 9.2. Equation (9S.27), which is the result of combining the first and second laws of thermodynamics for the general process in Figure 9S.18, can be simplified to eliminate the term involving the flowing streams, the work term, and the term for unsteady operation, to give

$$LW = \left(1 - \frac{T_0}{T_1}\right)Q + \left(1 - \frac{T_0}{T_2}\right)(-Q) \tag{9.5a}$$

$$= Q\left(\frac{T_0}{T_2} - \frac{T_0}{T_1}\right) \tag{9.5b}$$

$$= QT_0\left(\frac{T_1 - T_2}{T_1 T_2}\right) \tag{9.5c}$$

$$= QT_0\frac{\Delta T}{T_1 T_2} \tag{9.5d}$$

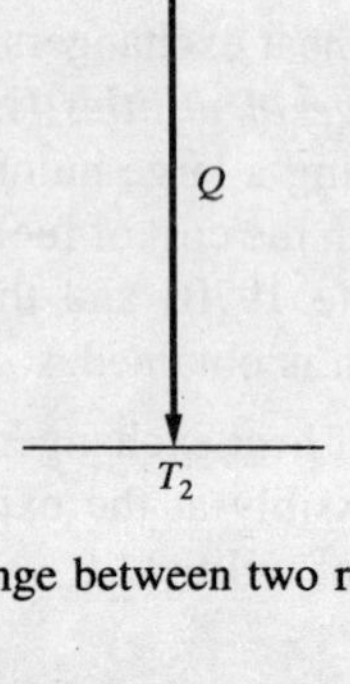

Figure 9.2 Heat exchange between two reservoirs.

where LW is the rate of lost work and T_0 is the absolute temperature of the environment. Note that a simpler notation suffices in this chapter, in which all analysis is in the steady state; hence, $LW \equiv L\dot{W}$, $Q \equiv \dot{Q}$, and $m \equiv \dot{m}$. It can be seen that, for a given rate of heat transfer and a given ΔT approach, the rate of lost work increases almost inversely with the decrease in the square of the absolute temperature level. Thus, as the temperature levels move lower into the cryogenic region, the approach temperature difference, ΔT, must decrease approximately as the square of the temperature level to maintain the same rate of lost work. This explains the need to use very small approach temperature differences, on the order of 1°C, in the cold boxes of cryogenic processes. If the approach temperature differences were not reduced, the large increases in the rate of lost work would sharply increase the energy requirements to operate these processes, especially the operating and installation costs for compressors.

9.2 MINIMUM UTILITY TARGETS

A principal objective in the synthesis of HENs is the efficient utilization of energy in the hot process streams to heat cold process streams. Thus, it is desirable to compute the maximum energy recovery (MER) before synthesizing the HEN; that is, to determine the minimum hot and cold utilities in the network, given the heating and cooling requirements of the process streams. This important first step is referred to as *MER targeting*, and is useful in that it determines the utility requirements for the most thermodynamically efficient network. To introduce this targeting step, the example provided by Linnhoff and Turner (1981) is presented here. This example involves only sensible heat. Later, examples are presented that also involve the latent heat of phase change and the heat of reaction, under either isothermal or nonisothermal conditions.

EXAMPLE 9.1

Two cold streams, C1 and C2, are to be heated and two hot streams, H1 and H2, are to be cooled without phase change. Their conditions and properties are as follows:

Stream	T^s(°F)	T^t(°F)	mC_p[Btu/(hr°F)]	Q(10^4 Btu/hr)
C1	120	235	20,000	230
C2	180	240	40,000	240
H1	260	160	30,000	300
H2	250	130	15,000	180

It is assumed that the *heat-capacity flow rate*, $mC_P = C$, which is the product of the specific heat and the mass flow rate, does not vary with temperature. As shown later, when it is necessary to account for a variation in the heat capacity with the temperature, a stream is *discretized* into several substreams, each involving a different segment of the temperature range and a different C

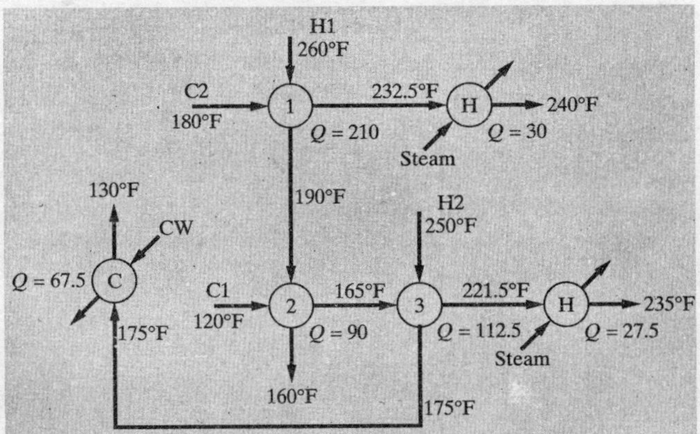

Figure 9.3 Proposed HEN for Example 9.1 with ΔT_{min} = 10°F, showing interior heat exchangers (1–3) and auxiliary heat exchangers (H, C). Multiply heat duties, Q, by 10^4 Btu/hr.

(see Example 9.5). Design a HEN that uses the smallest amounts of heating and cooling utilities possible, such that the closest approach temperature differences never fall below a minimum value. For the temperature range in this example, a reasonable assumption is $\Delta T_{\text{min}} = 10$°F.

SOLUTION

For this system, a total of 480 × 10^4 Btu/hr must be removed from the two hot streams, but only a total of 470 × 10^4 Btu/hr can be consumed by the two cold streams. Hence, from the first law of thermodynamics, a minimum of 10 × 10^4 Btu/hr must be removed by a cold utility such as cooling water. As shown below, this is *not* the minimum utility usage, which, from the second law of thermodynamics, depends on ΔT_{min}. One possible HEN is shown in Figure 9.3, involving six heat exchangers and 57.5 × 10^4 and 67.5 × 10^4 Btu/hr of hot and cold utilities, respectively. Note that the *difference* between the hot and cold utility duties equals that given by the first law of thermodynamics.

Since the design of this HEN has neither considered the MER targets nor utilized procedures for optimal HEN synthesis, its assessment focuses on two questions: (a) How do the heating and cooling utility duties of 57.5 × 10^4 and 67.5 × 10^4 Btu/hr compare with the MER targets? (b) Is it possible to synthesize a HEN with fewer heat exchangers, and if so, what are its utility requirements? In this section, the first of these two questions is addressed, with the second, which involves higher utility costs and decreased capital costs, postponed until later. Three methods are introduced to estimate MER targets: (1) the temperature-interval method, (2) a graphical method using *composite heating and cooling curves*, to be defined, and (3) the formulation and solution of a linear programming (LP) problem.

Temperature-Interval (TI) Method

The *temperature-interval method* was developed by Linnhoff and Flower (1978a, b) following the pioneering work of Hohmann (1971). The method is applied to the hot and cold streams introduced in Example 9.1, and as will be seen, a systematic procedure unfolds for determining the minimum utility requirements over all possible HENs, given just the heating and cooling requirements for the process streams and the minimum approach temperature in the heat exchangers, ΔT_{min}.

EXAMPLE 9.2 ***(Example 9.1 Revisited)***

Returning to Example 9.1, the temperature-interval (TI) method is used for the calculation of MER targets for $\Delta T_{\text{min}} = 10$°F.

SOLUTION

The first step in the TI method is to adjust the source and target temperatures using ΔT_{min}. Somewhat arbitrarily, this is accomplished by reducing the temperatures of the hot streams by ΔT_{min}, while leaving the temperatures of the cold streams untouched as follows:

			Adjusted Temps		
	T^s(°F)	T^t(°F)	T^s(°F)	T^t(°F)	
C1	120		120		T_5
		235		235	T_2
C2	180		180		T_3
		240		240	T_1
H1	260		250		T_0
		160		150	T_4
H2	250		240		T_1
		130		120	T_5

The adjustment of the hot stream temperatures by subtracting ΔT_{min} brings both the hot and cold streams to a common frame of reference so that when performing an energy balance involving hot and cold streams at the same temperature level, the calculation accounts for heat transfer with at least a driving force of ΔT_{min}. Next, the adjusted temperatures are rank-ordered, beginning with T_0, the highest temperature. These are used to create a cascade of *temperature intervals* within which energy balances are carried out. As shown in Table 9.1 and Figure 9.4, each interval, i, displays the enthalpy difference, ΔH_i, between the energy to be removed from the hot streams and the energy to be taken up by the cold streams in that interval. For example, in interval 1, 240°F to 250°F ($\Delta T = 10$°F), only stream H1 is involved. Hence, the enthalpy difference is:

$$\Delta H_1 = (\Sigma C_h - \Sigma C_c)_1 (T_0 - T_1) = (3 \times 10^4) \times (250 - 240)$$
$$= 30 \times 10^4 \text{ Btu/hr}$$

It is noted that initially, no energy is assumed to enter this interval from a hot utility, such as steam at a higher temperature; that is, $Q_{\text{steam}} = 0$. Hence, 30 × 10^4 Btu/hr are available and flow down as a residual, R_1, into the next lowest interval 2; that is, $R_1 = 30 \times 10^4$ Btu/hr. Interval 2 involves streams H1, H2, and

Table 9.1 Enthalpy Differences for Temperature Intervals

Interval, i	$T_{i-1} - T_i$, °F	$\Sigma C_h - \Sigma C_c$, 10^4 Btu/hr-°F	ΔH_i, 10^4 Btu/hr
1	$250 - 240 = 10$	3	30
2	$240 - 235 = 5$	$3 + 1.5 - 4 = 0.5$	2.5
3	$235 - 180 = 55$	$3 + 1.5 - 4 - 2 = -1.5$	−82.5
4	$180 - 150 = 30$	$3 + 1.5 - 2 = 2.5$	75
5	$150 - 120 = 30$	$1.5 - 2 = -0.5$	−15

C2 between 235°F and 240°F($\Delta T = 5$°F), and hence, the enthalpy difference is

$$\Delta H_2 = (\Sigma C_h - \Sigma C_c)_2(T_1 - T_2) = [(3 + 1.5 - 4) \times 10^4] \times (240 - 235) = 2.5 \times 10^4 \text{ Btu/hr}$$

When this is added to the residual from interval 1, R_1, this makes the residual from interval 2, R_2, equal to 32.5× 10^4 Btu/hr. Note that no temperature violations of $\Delta T_{\min}$ occur when the streams are matched in interval 2 because the hot stream temperatures are reduced by $\Delta T_{\min}$. Interval 3, 180°F to 235°F ($\Delta T = 55$°F), involves all four streams, and hence, the enthalpy difference is -82.5×10^4 Btu/hr as detailed in Table 9.1, making the residual from interval 3 equal to $(32.5 - 82.5) \times 10^4 = -50 \times 10^4$ Btu/hr. Similarly, the enthalpy differences in intervals 4 and 5 are 75×10^4 and -15×10^4 Btu/hr, respectively, with the residuals leaving these intervals being 25×10^4 and 10×10^4 Btu/hr. Note that for $Q_{\text{steam}} = 0$, the largest negative residual is from interval 3, $R_3 = -50 \times 10^4$ Btu/hr. Clearly, to satisfy the second law of thermodynamics, all negative residuals must be removed because heat cannot flow from a low- to a high-temperature interval. The only way to avoid negative residuals is to add energy at higher temperatures. This is achieved using low-pressure steam above 250°F. In Figure 9.4, note that when $Q_{\text{steam}} = 50 \times 10^4$ Btu/hr, R_1 becomes 80×10^4 Btu/hr, $R_2 = 82.5 \times 10^4$ Btu/hr, $R_3 = 0$, $R_4 = 75 \times 10^4$ Btu/hr, and $R_5 = Q_{\text{cw}} = 60 \times 10^4$ Btu/hr. Thus, $Q_{\text{steam}} = 50 \times 10^4$ Btu/hr is the smallest amount of energy that must be added above 180°F, and hence, it becomes a lower bound on the hot utility duty, just as $Q_{\text{cw}} = 60 \times 10^4$ Btu/hr becomes the lower bound on the cold utility duty. These are referred to as the *MER targets;* evidently, the HEN in Figure 9.3 exceeds these targets, by 7.5×10^4 Btu/hr each.

Note that $Q_{\text{steam}} - Q_{\text{cw}} = -10 \times 10^4$ Btu/hr, which is consistent with the first law. Furthermore, at minimum utilities, no energy flows between intervals 3 and 4. This is referred to as the *pinch,* with associated temperatures of 180°F for the cold streams and $180 + \Delta T_{\min} = 180 + 10 = 190$°F for the hot streams. To maintain minimum utilities, it is recognized that *no energy is permitted to flow across the pinch.* If, as in the HEN in Figure 9.3, Q_{steam} were increased to 57.5×10^4 Btu/hr, R_3, the transfer of heat across the pinch, would be 7.5×10^4 Btu/hr, and Q_{cw} would increase to 67.5×10^4 Btu/hr.

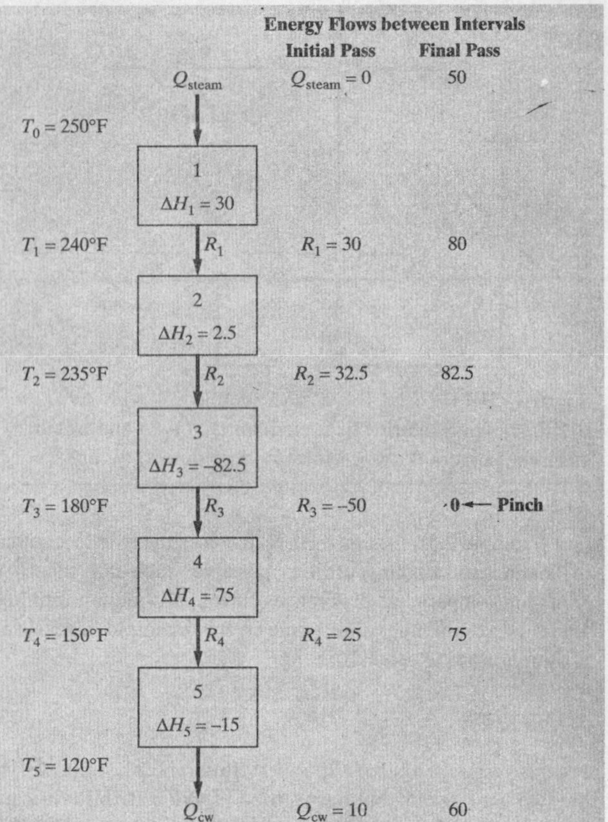

Figure 9.4 Cascade of temperature intervals, energy balances, and residuals; multiply ΔH_i and R_i by 10^4 Btu/hr.

Table 9.2 summarizes the cooling and heating loads in each interval, with the actual temperature ranges shown. Included are the cumulative loads starting from the lowest temperatures.

HEN synthesis is facilitated using the stream representation in Figure 9.5, in which arrows moving from left to right denote the hot streams, while arrows moving from right to left denote the cold streams. The arrows for the hot and cold streams either pass through or begin at the pinch temperatures. To maintain minimum

Table 9.2 Interval Heat Loads (Multiply Q by 10^4 Btu/hr)

	Cooling			Heating		
Interval i	Temperature Range (°F)	Q	Cum. Q	Temperature Range (°F)	Q	Cum. Q
1	250–260	30	480.0	240–250	0	—
2	245–250	22.5	450.0	235–240	20	470.0
3	190–245	247.5	427.5	180–235	330	450.0
4	160–190	135	180.0	150–180	60	120.0
5	130–160	45	45.0	120–150	60	60.0

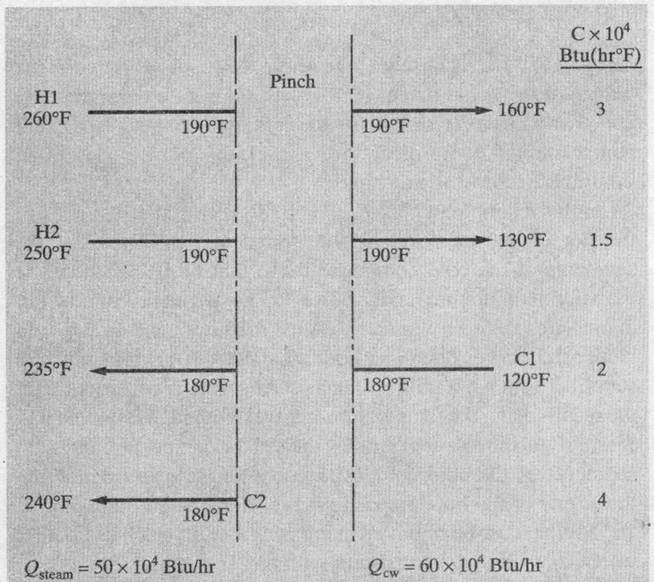

Figure 9.5 Pinch decomposition of the hot and cold streams for Example 9.2.

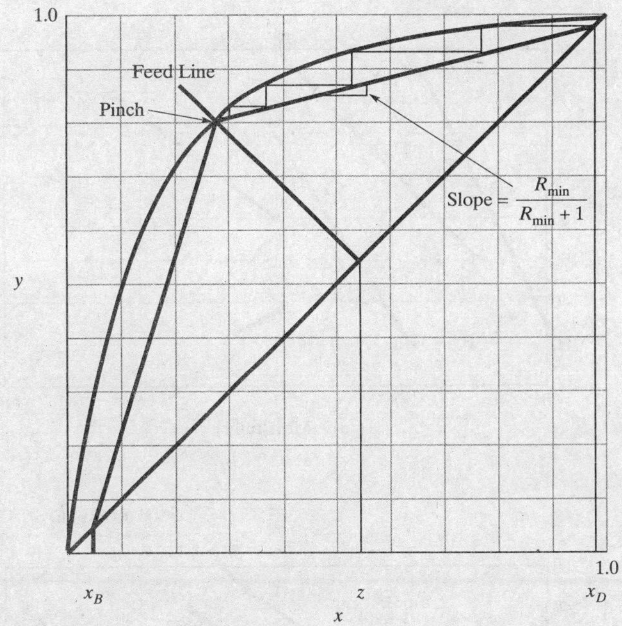

Figure 9.6 McCabe–Thiele diagram showing a pinch at minimum reflux in binary distillation.

utilities, two separate HENs *must* be designed, one on the hot side and one on the cold side of the pinch. Energy is added from hot utilities on the hot side of the pinch (50×10^4 Btu/hr), and energy is removed using cold utilities on the cold side of the pinch (60×10^4 Btu/hr), and no energy is permitted to flow across the pinch. If energy were exchanged between a hot stream on the hot side of the pinch and a cold stream on the cold side of the pinch, this energy would not be available to heat the cold streams on the hot side of the pinch, and additional energy from the hot utilities would be required. Similarly, the cold stream on the cold side of the pinch would not have the ability to remove this energy from the hot streams on the cold side of the pinch, and the same amount of additional energy would have to be removed from the cold streams on the cold side of the pinch using cold utilities.

Composite Curve Method

The terminology *pinch* is understood more clearly in connection with a graphical display, introduced by Umeda et al. (1978), in which composite heating and cooling curves are positioned no closer than ΔT_{min}. As $\Delta T_{min} \rightarrow 0$, the curves *pinch* together and the area for heat exchange approaches infinity. In this respect, there is a close parallel to the graphical approach introduced by McCabe and Thiele (1925) in their classic method for the design of distillation towers to separate binary mixtures. It should be recalled that, on the McCabe–Thiele diagram, a *pinch* occurs when the operating lines intersect the equilibrium curve and the feed line. This occurs at the minimum reflux ratio, R_{min}, where an infinite number of stages accumulate in the vicinity of the pinch point, as shown in Figure 9.6. The parallel is illustrated further in Example 9.3.

EXAMPLE 9.3 ***(Example 9.1 Revisited)***

In this example, the minimum utility requirements for a HEN involving the four streams in Example 9.1 are determined using the graphical approach by Umeda et al. (1978).

SOLUTION

For each of the streams, the temperature, *T*, is graphed on the ordinate as a function of the enthalpy or heat transferred on the abscissa, with the slope being the inverse of the heat-capacity flow rate, *C*. When *C* is constant (i.e., not a function of *T*), the curves are straight lines. For the hot streams, they are *cooling* curves that begin at the highest temperature and finish at the lowest temperature after the energy has been removed. For the cold streams, they are *heating curves* that begin at the lowest temperature and finish at the highest temperature after heat has been added. In Figure 9.7a, the two heating and two cooling curves are displayed, with each of the lines positioned arbitrarily along the abscissa to avoid intersections and crowding.

To display the results of the TI method graphically, Table 9.2 is used to prepare *hot composite* and *cold composite* curves, which combine curves H1 and H2 in Figure 9.7a into one hot composite curve, and curves C1 and C2 into one cold composite curve. First, the hot composite curve is graphed starting with an enthalpy datum of 0 at 130°F, the lowest temperature of a hot stream. From Table 9.2, the hot composite enthalpies are

T(°F)	130	160	190	245	250	260
H (Btu/hr × 10^{-4})	0	45	180	427.5	450	480

These points form the hot composite curve in Figure 9.7b. As seen in the figure, the hot composite curve has a segment from

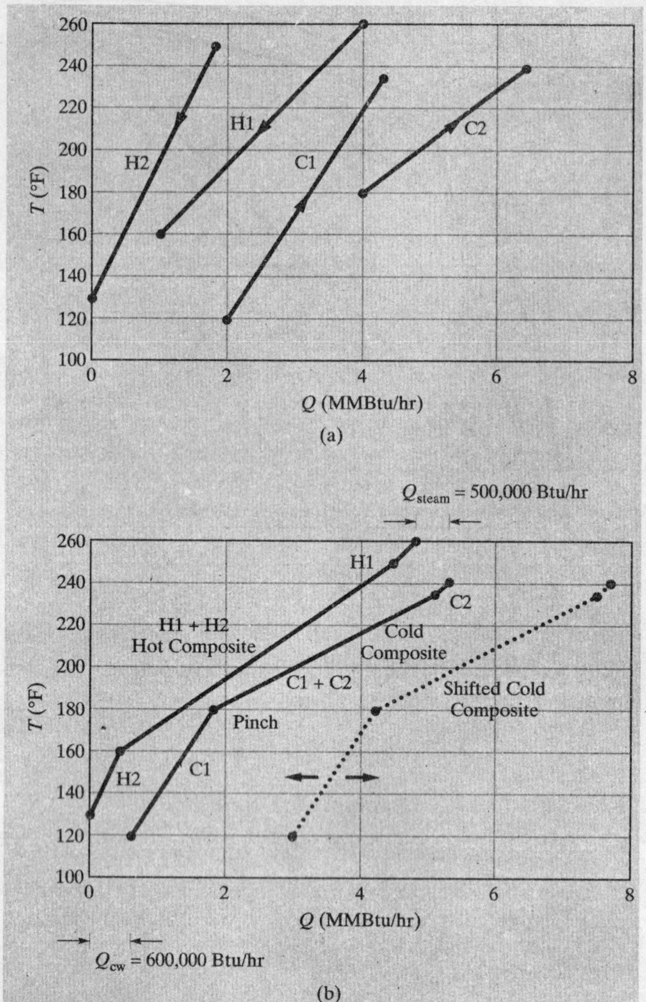

Figure 9.7 Graphical method to determine MER targets: (a) heating and cooling curves for the streams; (b) composite hot and cold curves.

stream H1 between 260 and 250°F. From 250 to 160°F, both streams H1 and H2 coexist, and hence, their cooling requirements are combined. Note that the combined heat-capacity flow rate is increased, and consequently, the slope of the hot composite curve is reduced. Finally, from 160 to 130°F, only stream H2 appears.

Next, the cold composite curve is graphed. For $\Delta T_{\min} = 10$°F, the TI method determined a minimum cooling utility of 60×10^4 Btu/hr. Therefore, the graph begins with an enthalpy datum of that value. From Table 9.2, the cold composite enthalpies are

T(°F)	120	150	180	235	240
H (Btu/hr $\times 10^{-4}$)	60	120	180	510	530

These points form the cold composite curve in Figure 9.7b. From 120 to 180°F, only stream C1 appears in the cold composite curve. From 180°F to 235°F, the streams C1 and C2 coexist and their heating curves are combined. Finally, from 235 to 240°F, only stream C2 exists.

As shown by the solid lines in Figure 9.7b, the composite curves have a closest point of approach of $\Delta T_{\min} = 10$°F at the point where stream C2 begins along the cold composite curve; that is, 180°F. The corresponding temperature on the hot composite curve is 190°F. Consequently, these two points provide the temperatures at the pinch. If $\Delta T_{\min}$ is reduced to zero, the cold composite curve is shifted to the left until it touches the hot composite curve. As mentioned earlier, this corresponds to an infinite area for heat exchange.

In this example, with $\Delta T_{\min} = 10$°F, heat in the segments of the hot composite curve is transferred vertically to heat the segments of the cold composite curve that lie below them. At the high-temperature ends, however, no segments of the hot composite curve lie vertically above the upper end of the cold composite curve. There, an additional 50×10^4 Btu/hr must be supplied from a hot utility such as steam. This is consistent with the results using the temperature-interval method in Example 9.2. Similarly, at the low-temperature ends of the composite curves, no segments of the cold composite curve lie vertically below the lower end of the hot composite curve. Here, an additional 60×10^4 Btu/hr must be removed using a cold utility such as cooling water, a result again consistent with the TI analysis.

Figure 9.7b also includes a dotted, cold composite curve, shifted to the right to give a $\Delta T_{\min}$ of 65°F. Corresponding minimum utilities increase to $Q_{cw} = 300 \times 10^4$ Btu/hr and $Q_{steam} = 290 \times 10^4$ Btu/hr. Although not shown in Figure 9.7b, if the cold composite curve is shifted further to the right so that $\Delta T_{\min}$ is increased to 140°F, all heat would have to be transferred from steam and to cooling water.

Many additional observations are noteworthy in connection with the hot and cold composite curves. One is that the slopes of the composite curves *always* decrease at the inlet temperature of a stream and increase at the outlet temperature of a stream. It follows that points at which the slope decreases are candidate pinch points, and furthermore, when a pinch temperature exists, one of the inlet temperatures is *always* a pinch temperature. Hence, to locate a potential pinch temperature, one needs only to examine the inlet temperatures of the streams.

Yet another observation is that for some $\Delta T_{\min}$ there are no pinch temperatures. In such cases, either hot or cold utilities (not both) are required in an amount equal to the difference between the total energy to be removed from the hot streams and that to be added to the cold streams. The $\Delta T_{\min}$ at which the pinch disappears is referred to as the *threshold* $\Delta T_{\min}$, as discussed in Section 9.5.

Linear Programming Method

A closer examination of the temperature-interval (TI) method shows that the minimum hot and cold utilities can be calculated by creating and solving a linear programming (LP) problem, as discussed in Section 24.4. This approach is illustrated in the example that follows.

EXAMPLE 9.4

It is desired to determine the minimum hot and cold utilities for a HEN involving the four streams in Example 9.1 by creating and solving a linear programming problem, using the energy balance for each interval in the cascade of Figure 9.4.

SOLUTION

The LP is formulated:

Minimize Q_{steam}
With respect to (w.r.t):
Q_{steam}
Subject to (s.t.):

$$Q_{steam} - R_1 + 30 = 0 \qquad \text{(LP.1)}$$

$$R_1 - R_2 + 2.5 = 0 \qquad \text{(LP.2)}$$

$$R_2 - R_3 - 82.5 = 0 \qquad \text{(LP.3)}$$

$$R_3 - R_4 + 75 = 0 \qquad \text{(LP.4)}$$

$$R_4 - Q_{cw} - 15 = 0 \qquad \text{(LP.5)}$$

$$Q_{steam}, Q_{cw}, R_1, R_2, R_3, R_4 \geq 0 \qquad \text{(LP.6)}$$

Note that only Q_{steam} is needed in the objective function because when Q_{steam} is at its minimum, Q_{cw} is also at its minimum. Furthermore, the equality constraints are the energy balances for each of the temperature intervals in Figure 9.4. These must be satisfied at the solution of the LP.

Using the General Algebraic Modeling System (GAMS), the following linear programming problem is defined. Note that the solution is equivalent to that obtained using the TI method, as expected. For an introduction to GAMS, see the file GAMS.pdf in the Program and Simulation Files folder, which can be downloaded from the Wiley Web site associated with this book.

GAMS Program

```
VARIABLES
  Qs, Qcw, R1, R2, R3, R4
  Z          min. util.;
POSITIVE VARIABLE Qs, Qcw, R1, R2, R3, R4;

EQUATIONS
  COST          define objective function
  T1, T2, T3, T4, T5;

  COST . . Z =E= Qs;
  T1 . . Qs - R1 + 30     =E= 0;
  T2 . . R1 - R2 + 2.5    =E= 0;
  T3 . . R2 - R3 - 82.5 =E= 0;
  T4 . . R3 - R4 + 75     =E= 0;
  T5 . . R4 - Qcw - 15    =E= 0;
MODEL HEAT/ALL/;
SOLVE HEAT USING LP MINIMIZING Z;
DISPLAY R1.L, R2.L, R3.L, R4.L, Qs.L, Qcw.L;
```

GAMS Solution

```
**** SOLVER STATUS 1 NORMAL COMPLETION
**** MODEL STATUS 1 OPTIMAL
**** OBJECTIVE VALUE 50.0000
  VARIABLE R1 =   80.000
  VARIABLE R2 =   82.500
  VARIABLE R3 =    0.000
  VARIABLE R4 =   75.000
  VARIABLE QS =   50.000
  VARIABLE QCW = 60.000
```

Note that the residual across the pinch temperatures, $R_3 = R_p$, is zero, as must be the case when the utilities are minimized. These results can be reproduced using the GAMS input files, CASC.1 or CASC.2, in the Program and Simulation Files folder, which can be downloaded from the Wiley Web site associated with this book.

Thus far, only sensible heat changes have been considered. Furthermore, the specific heat or heat capacity has been assumed constant over the range between the source and target temperatures so that the stream heat-capacity flow rates are constant. However, in many processes, latent heat of phase change, heat of reaction, and heat of mixing may also be involved under isothermal or nonisothermal conditions. In addition, the specific heat may not be constant, or sensible heat may be combined with latent heat, heat of reaction, or heat of mixing, such as for multicomponent mixtures passing through condensers, vaporizers, re-boilers, and nonadiabatic reactors and mixers. In such cases, a fictitious heat-capacity flow rate can be used based on the change in enthalpy flow rate due to all applicable effects divided by a temperature range. In general, a plot of stream enthalpy flow rate as a function of temperature is curved, but can be discretized into straight-line segments. Particular attention should be paid to the accuracy of the discretization in the vicinity of the pinch temperatures. Note that a conservative approach is recommended, in which the linear approximations provide a bound for the cold stream temperature–enthalpy curves from above, and for the hot stream curves from below. This conservative approach ensures that the true temperature approach is greater than that computed in terms of the linear approximations. The following example shows how linear piece-wise approximations are used to generate stream data for a design problem involving both the vaporization and condensation of process streams.

EXAMPLE 9.5 *MER Targeting for a Process Exhibiting Phase Changes and Variable Heat Capacities*

Figure 9.8 shows a process for the manufacture of toluene by the dehydrogenation of *n*-heptane. Note that a furnace, E-100, heats the feed stream of pure *n*-heptane, S1, at 65°F, to the reactor feed, S2, at 800°F. Furthermore, the reactor effluent, S3, containing a multicomponent mixture of *n*-heptane, hydrogen, and toluene at 800°F is cooled to 65°F and fed to a separator as stream S4. The pressure is 1 atm throughout the process. It is planned to install a heat exchanger to heat the feed stream, S1, using the hot reactor effluent, S3, and thus reduce the required duty of the preheater, E-100. (a) Generate stream data using piece-wise linear approximations for the heating and cooling curves for the reactor feed and effluent streams. (b) Using the stream data, compute the MER targets for $\Delta T_{min} = 10°F$.

SOLUTION

(a) Generation of stream data

HYSYS is used to simulate the process in Figure 9.8, as shown in the multimedia modules, which can be

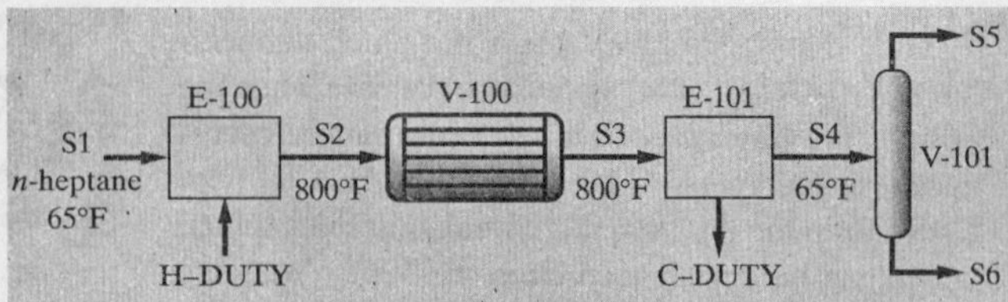

Figure 9.8 Process flow diagram for dehydrogenation of *n*-heptane.

downloaded from the Wiley Web site associated with this book (see HYSYS → Tutorials → Heat Transfer → Toluene Manufacture). The Peng–Robinson equation of state is used to estimate thermodynamic properties. Sensitivity analyses are performed, in which enthalpies for S2 and S4 are computed as a function of temperature, giving the temperature-enthalpy diagrams in Figures 9.9 and 9.10. In Figure 9.9, which is for pure *n*-heptane feed, only liquid sensible heat is involved from 65°F to 209°F, where isothermal vaporization occurs, as represented by the horizontal line. From there until 800°F, only vapor sensible heat is involved. In Figure 9.10, which is for the ternary reactor effluent, only vapor sensible heat is involved from 800°F to 183°F, which is the dew point. Then, condensation occurs, involving both latent heat and sensible heat, to the target temperature of 65°F. Shown in the diagrams are the piece-wise linear approximations, defined by critical coordinates on the original heating and cooling curves. The piece-wise linear approximations are defined in terms of temperature-enthalpy coordinates, (h_k, T_k) and (h_{k+1}, T_{k+1}) through which linear arcs are drawn to approximate the true heating or cooling curves. Each arc represents a new stream, with the source and target temperatures being the abscissa coordinates, T_k and T_{k+1}, and the heat-capacity flow rate being

$$C_k = \frac{h_{k+1} - h_k}{T_{k+1} - T_k} \tag{9.6}$$

where h is the enthalpy flow rate and MM stands for million. This is the inverse of the slope of each linear segment in Figures 9.9 and 9.10. Reasonably accurate linear approximations are obtained using four segments for stream S2 and six for stream S4, whose

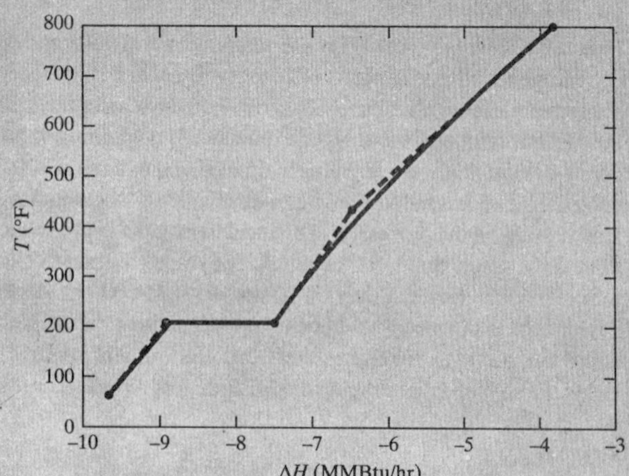

Figure 9.9 Temperature–enthalpy diagram for the cold stream, S2, showing simulation results (solid line) and piece-wise linear approximation (dashed line).

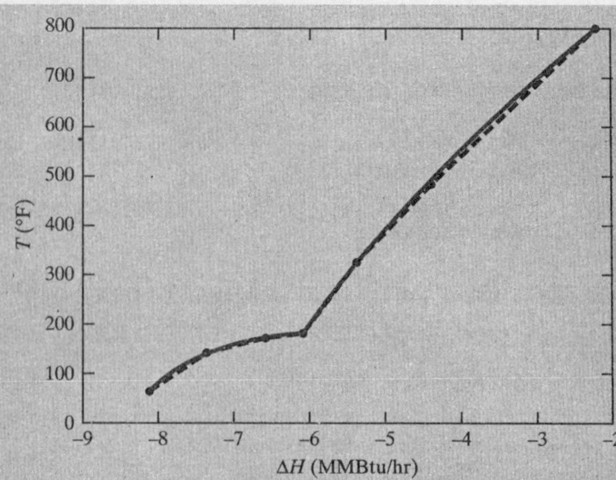

Figure 9.10 Temperature–enthalpy diagram for the hot stream, S4, showing simulation results (solid line) and piece-wise linear approximation (dashed line).

coordinates are positioned to ensure accuracy in the vicinity of the pinch temperatures. The temperature coordinates are determined to the nearest degree. Thus, in Figure 9.9, the horizontal line for the vaporization of *n*-heptane at 209°F is taken to occur over a 1°F interval from 209 to 210°F, giving a fictitious heat-capacity flow rate, C, of 1.4282 MMBtu/hr/1°F = 1.4282 MMBtu/hr–°F.

Having determined the coordinate positions, the stream data for the four cold and six hot streams are computed directly, where the heat-capacity flow rate for the k th stream is given by Eq. (9.6), as shown in Table 9.3. Note the increased values of the heat-capacity flow rate in the region where streams exhibit phase change, and in particular, the large value for the cold stream (pure *n*-heptane), where vaporization is assumed to occur over 1°F.

Table 9.3 Stream Data for Example 9.5

(a) Cold Streams

Stream	T^s(°F)	T^t(°F)	Duty (MMBtu/hr)	C (MMBtu/hr–°F)
S2A	65	209	0.7446	0.5171×10^{-2}
S2B	209	210	1.4282	1.4282
S2C	210	435	1.0492	0.4663×10^{-2}
S2D	435	800	2.6446	0.7245×10^{-2}

(b) Hot Streams

Stream	T^s(°F)	T^t(°F)	Duty (MMBtu/hr)	C (MMBtu/hr–°F)
S4A	800	485	2.1745	0.6903×10^{-2}
S4B	485	326	0.9823	0.6178×10^{-2}
S4C	326	183	0.7304	0.5108×10^{-2}
S4D	183	172	0.4863	4.421×10^{-2}
S4E	172	143	0.7780	2.683×10^{-2}
S4F	143	65	0.7652	0.9810×10^{-2}

Table 9.4 Computation of MER Targets Using the TI Method

	(°F)		(MMBtu/hr)		
Interval	T_i	$T_{i-1}-T_i$	ΔH_i	Q ($Q_{H,min}=0$)	Q ($Q_{H,min}=1.4208$)
T_0	800			0	1.4208
T_1	790	10	−0.0725	−0.0725	1.3483
T_2	475	315	−0.1077	−0.1802	1.2406
T_3	435	40	−0.0427	−0.2229	1.1979
T_4	316	119	0.1803	−0.0426	1.3782
T_5	210	106	0.0472	0.0046	1.4254
T_6	209	1	−1.4231	−1.4185	0.0023
T_7	173	36	−0.0023	−1.4208	0.0000 ← Pinch
T_8	162	11	0.4294	−0.9913	0.4294
T_9	133	29	0.6281	−0.3633	1.0575
T_{10}	65	68	0.3155	−0.0478	1.3729
T_{11}	55	10	0.0981	0.0502	1.4710

(b) Computing MER targets

The TI method is applied using the data in Table 9.3, with the hot temperatures reduced by ΔT_{min}. The results, summarized in Table 9.4, indicate that the cold pinch temperature is 173°F, giving the hot and cold utility targets of $Q_{H,min} = 1.421$ MMBtu/hr and $Q_{H,min} = 1.471$ MMBtu/hr. The location of the cold pinch temperature in the HYSYS simulation of the heat-integrated process designed for $\Delta T_{min} = 10$°F is 172°F, with the heat duties of the preheater and cooler being 1.396 MMBtu/hr and 1.446 MMBtu/hr, respectively. These differences are the result of the linear approximations for the heating and cooling curves.

9.3 NETWORKS FOR MAXIMUM ENERGY RECOVERY

Having determined the minimum utilities for heating and cooling, it is common to design two networks of heat exchangers, one on the hot side and one on the cold side of the pinch, as shown in Figure 9.5. In this section, two methods are presented for this purpose. The first, introduced by Linnhoff and Hindmarsh (1983), places emphasis on positioning the heat exchangers by working out from the pinch. The second is an algorithmic strategy that utilizes a mixed-integer linear program (MILP), which was introduced by Papoulias and Grossmann (1983b) and is solved with GAMS.

Stream Matching at the Pinch

To explain the approach of Linnhoff and Hindmarsh (1983), it helps to refer to a diagram showing the *pinch decomposition* of the hot and cold streams, as shown in Figure 9.5 for the four streams in Example 9.1. Attention is focused at the pinch, where the temperatures of the hot and cold streams

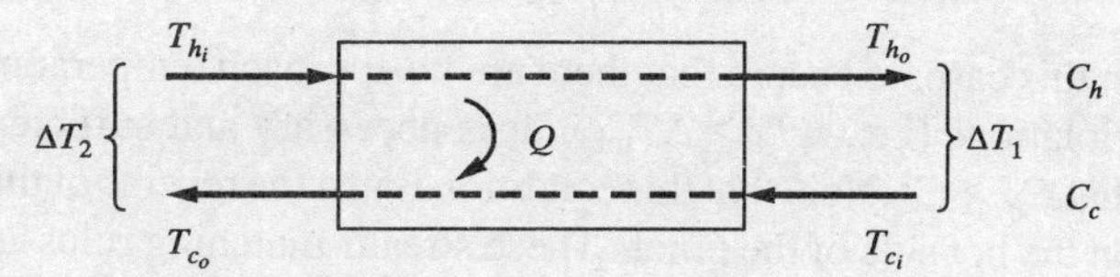

Figure 9.11 Schematic of a countercurrent heat exchanger.

are separated by ΔT_{min}. This, of course, is the location of the closest temperature approach.

Consider the schematic of a countercurrent heat exchanger in Figure 9.11. The hot stream, having a heat-capacity flow rate of C_h, enters at T_{h_i} and exits at T_{h_0}. It transfers heat, Q, to the cold stream that has a heat-capacity flow rate of C_c, entering at T_{c_i} and exiting at T_{c_0}. On the cold end of the heat exchanger, where the temperatures of the hot and cold streams are the lowest, the approach temperature difference is ΔT_1. On the hot end, where the temperatures are the highest, the approach temperature difference is ΔT_2. Carrying out energy balances for the hot and cold streams:

$$Q = C_h(T_{h_i} - T_{h_0}) \quad \text{or} \quad T_{h_i} - T_{h_0} = \frac{Q}{C_h} \tag{9.7}$$

$$Q = C_c(T_{c_0} - T_{c_i}) \quad \text{or} \quad T_{c_0} - T_{c_i} = \frac{Q}{C_c} \tag{9.8}$$

and subtracting Eq. (9.8) from Eq. (9.7):

$$(T_{h_i} - T_{c_0}) - (T_{h_0} - T_{c_i}) = Q\left(\frac{1}{C_h} - \frac{1}{C_c}\right) \tag{9.9}$$

or:

$$\Delta T_2 - \Delta T_1 = \frac{Q(C_c - C_h)}{C_h C_c} \tag{9.10}$$

Following the approach introduced by Linnhoff and Hindmarsh (1983), the potential locations for the heat exchangers at the pinch are considered next. When a heat exchanger is positioned on the hot side of the pinch, which is arbitrarily considered first, $\Delta T_1 = \Delta T_{min}$, and Eq. (9.10) becomes:

$$\Delta T_2 - \Delta T_{min} + \frac{Q(C_c - C_h)}{C_h C_c} \tag{9.11}$$

Then, to ensure that $\Delta T_2 \geq \Delta T_{min}$, since $Q > 0$ and the heat-capacity flow rates are positive, it follows that $C_c \geq C_h$ is a necessary and sufficient condition. That is, for a match to be *feasible* at the pinch, on the hot side, $C_c \geq C_h$ must be satisfied. If two streams are matched at the pinch with $C_c < C_h$, the heat exchanger is *infeasible* because $\Delta T_2 < \Delta T_{min}$.

When a heat exchanger is positioned on the cold side of the pinch, $\Delta T_2 = \Delta T_{min}$, and Eq. (9.10) becomes:

$$\Delta T_1 = \Delta T_{min} - \frac{Q(C_c - C_h)}{C_h C_c} \tag{9.12}$$

In this case, to ensure that there are no approach temperature violations (i.e., $\Delta T_1 \geq \Delta T_{\text{min}}$), it is necessary and sufficient that $C_h \geq C_c$. Note that this condition is just the reverse of that on the hot side of the pinch. These stream-matching rules are now applied to design a HEN for Example 9.1.

EXAMPLE 9.6 *(Example 9.1 Revisited)*

Design a HEN to meet the MER targets for Example 9.1: $Q_{\text{H,min}} = 50 \times 10^4$ Btu/hr and $Q_{\text{C,min}} = 60 \times 10^4$ Btu/hr, where $Q_{\text{H,min}}$ and $Q_{\text{C,min}}$ are the minimum hot and cold utility loads, respectively.

SOLUTION

As stated previously, when designing a HEN to meet MER targets, no heat is transferred across the pinch, and hence, two HENs are designed, one on the hot side and one on the cold side of the pinch, as shown in Figure 9.12. Arbitrarily, the HEN on the hot side of the pinch is designed first. At the pinch, an appropriate match with stream H1 is sought. Since $C_{\text{H1}} = 3 \times 10^4$ Btu/(hr°F), stream C2 must be selected with $C_{\text{C2}} = 4 \times 10^4$ Btu/(hr°F), to ensure that $C_c \geq C_h$. Note that if C1 were selected, $C_{\text{C1}} < C_{\text{H1}}$, and an approach temperature violation occurs, $\Delta T_2 < \Delta T_{\text{min}}$. Consequently, interior heat exchanger 1 is installed, with a heat duty equal to 210×10^4 Btu/hr, the entire cooling requirement of stream H1 on the hot side of the pinch. Similarly, because $C_{\text{C1}} \geq C_{\text{H2}}$, streams H2 and C1 are matched on the hot side of the pinch, using interior heat exchanger 2, with a heat duty equal to 90×10^4 Btu/hr, the entire cooling requirement of H2 on the hot side of the pinch. Since these two heat exchangers bring streams C1 and C2 to 225 and 232.5°F, respectively, utility heaters (labeled "H") are added, to complete the design on the hot side of the pinch, with a total duty of 50×10^4 Btu/hr, which matches the MER heating target. Note that each unit exchanges heat between two process streams in countercurrent flow, with the inlet and outlet temperatures for each stream shown on either side of circles, identified by the heat exchanger number, connected by a vertical line to represent the match, and the heat duty annotated below the circle associated with the cold stream.

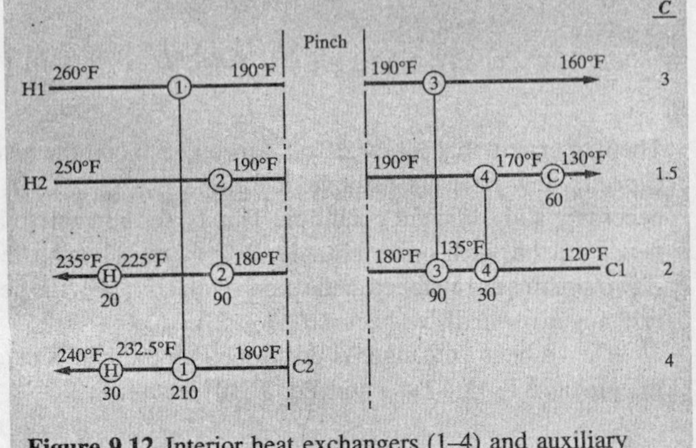

Figure 9.12 Interior heat exchangers (1–4) and auxiliary heat exchangers (H, C). Multiply heat duties by 10^4 Btu/hr and heat-capacity flow rates by 10^4 Btu/(hr°F).

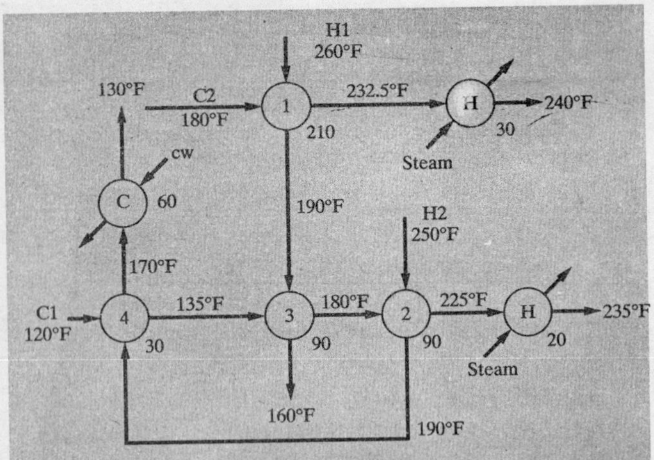

Figure 9.13 Flowsheet for HEN in Figure 9.12.

On the cold side of the pinch, only streams H1 and C1 can be matched, since $C_{\text{H1}} \geq C_{\text{C1}}$. Note that $C_{\text{H2}} < C_{\text{C1}}$, and consequently, if streams H2 and C1 are matched, an approach temperature violation occurs, $\Delta T_1 < \Delta T_{\text{min}}$. Thus, interior heat exchanger 3 is installed, with a heat duty equal to 90×10^4 Btu/hr, the entire cooling requirement of stream H1 on the cold side of the pinch. By energy balance, the temperature of stream C1 entering heat exchanger 3 is 135°F. This allows an additional internal heat exchanger to be positioned to pair streams H2 and C1, noting that the pairing rule $C_h \geq C_c$ applies only *at the pinch.* Heat exchanger 4 is installed with a heat duty equal to 30×10^4 Btu/hr, the remaining heating requirement of stream C1 on the cold side of the pinch. The HEN on the cold side of the pinch is completed by installing a cooler on stream H2 (labeled "C") with a heat duty of 60×10^4 Btu/hr, which matches the MER cooling target. The final design, shown in Figure 9.12, meets the MER targets with a total number of seven heat exchangers. These are displayed in the flowsheet in Figure 9.13, which should be compared with the HEN in Figure 9.3. Note that the former meets the MER energy targets, while both the cold and hot utility targets are exceeded by 7.5×10^4 Btu/hr in the latter. In contrast, the latter involves only six units, compared to seven utilized in the MER design. Cost estimates are needed to select between these and other alternatives. As will be seen in later examples, the tradeoff between capital and operating costs is at the heart of HEN synthesis.

In summary, the HEN design procedure to meet MER targets consists of the following steps:

1. *MER Targeting:* The pinch temperatures are determined, together with minimum hot and cold utility targets, $Q_{\text{H,min}}$ and $Q_{\text{C,min}}$, respectively. Either the temperature-interval or the composite curve method is used, or a linear programming problem is formulated and solved.
2. The synthesis problem is decomposed at the pinch, yielding two independent HENs to be designed, using the representation shown in Figure 9.5. It is helpful to place the heat-capacity flow rates for each stream in a column to the right, for reference.
3. The HEN is designed on the hot side of the pinch, *starting at the pinch,* and working outward. *At the*

pinch, streams are paired such that $C_c \geq C_h$. In general, the heat duty of each interior heat exchanger is selected to be as large as possible, to reduce the total number of exchangers. In some cases (e.g., Example 9.7), duties are selected to retain sufficient temperature driving forces for additional matches. Finally, hot utilities are added to meet cold temperature targets (up to a total of $Q_{H,min}$). *Cold utilities are not used on the hot side of the pinch.*

4. The HEN is designed on the cold side of the pinch, *starting at the pinch,* and working outward. *At the pinch,* streams are paired such that $C_h \geq C_c$. In general, the duty of each interior heat exchanger is selected to be as large as possible, to reduce the total number of exchangers. As mentioned above, it may be necessary to select duties to retain sufficient temperature driving forces for additional matches. Finally, cold utilities are added to meet cold temperature targets (up to a total of $Q_{C,min}$). *Hot utilities are not used on the cold side of the pinch.*

In this synthesis procedure, the designer positions the first heat exchangers *at the pinch,* where the approach temperature difference at one end of each heat exchanger is constrained at ΔT_{min}. Then, working outward, the utility exchangers are positioned last. In cases that do not require stream splitting, to be considered in Section 9.4, this simple procedure is sufficient to guarantee compliance with the MER targets. However, it often leads to designs with a large number of heat exchangers, as illustrated in the following example.

EXAMPLE 9.7

Consider the design of a HEN for the four streams below in a problem presented by Linnhoff and Flower (1978a, b)

Stream	T^s(°C)	T^t(°C)	C (kW/°C)	Q(kW)
C1	60	180	3	360
C2	30	130	2.6	260
H1	180	40	2	280
H2	150	40	4	440

Let $\Delta T_{min} = 10$°C, with the following specifications:

Cooling water(cw):

$T^s = 30$°C, $T^t \leq 80$°C, cost of cw $= 0.00015$ \$/kg

Steam(sat'd., s):

$T = 258$°C, $\Delta H^v = 1,676$ kJ/kg, cost of s $= 0.006$ \$/kg

Overall heat-transfer coefficients:

$U_{heater} = 1$ kW/m²–°C, $U_{cooler} = U_{exch} = 0.75$ kW/m²–°C

Purchase cost of heat exchangers:

$$C_P = 3,000A^{0.5} \quad (\$, \text{m}^2)$$

Equipment operability $= 8,500$ hr/yr, return on investment $= i_m = 0.1$ [for Eq. (9.3)]

SOLUTION

First, the MER targets are computed using the TI method, as summarized in the cascade diagram of Figure 9.14, where the pinch temperatures are 140 and 150°C, and the minimum hot and cold utility duties are 60 kW and 160 kW, respectively.

Next, the MER synthesis procedure is used to design the HEN. Since only streams H1 and C1 appear on the hot side of the pinch, and $C_{H1} < C_{C1}$, a heat exchanger is installed between streams H1 and C1, as shown in Figure 9.15a, with a heat duty equal to 60 kW, the cooling demand of H1 on the hot side of the pinch. Finally, a 60-kW heater is installed to complete heating stream C1.

On the cold side, all four streams are present, but only streams H1, H2, and C1 exist at the pinch, since the target temperature of stream C2 is 130°C, below 140°C, the pinch temperature of the cold streams. Only the H2–C1 match is feasible since $C_{H2} > C_{C1}$, while $C_{H1} < C_{C1}$. Following the guidelines, one would be tempted to install an internal heat exchanger with a duty of

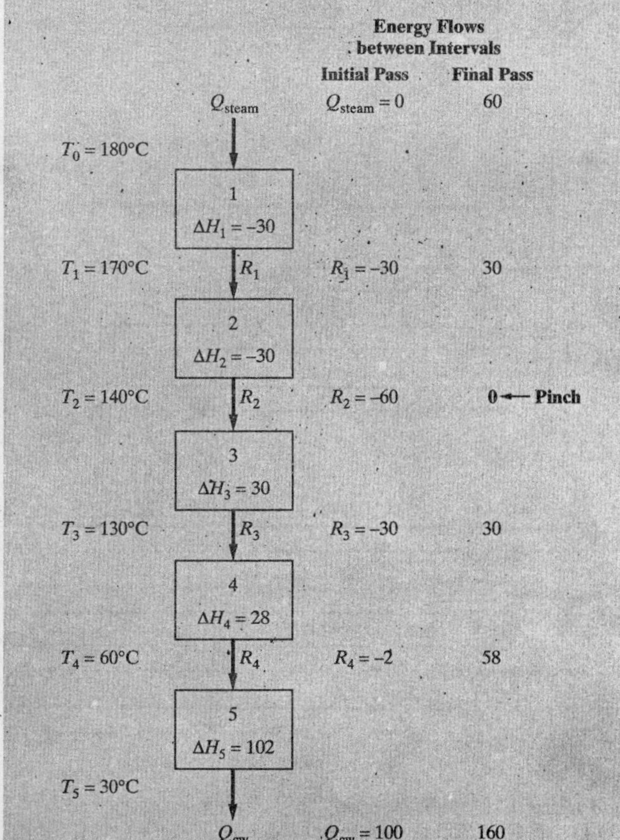

Figure 9.14 Cascade diagram for Example 9.7, showing temperature intervals, heat balances, and residuals ΔH_i and R_i in kilowatts.

240 kW, as shown in Figure 9.15b, which would satisfy the entire energy requirement of stream C1 on the cold side of the pinch. Furthermore, the H1–C2 match is possible, even though $C_{H1} < C_{C2}$, because stream C2 doesn't reach the pinch. Thus, a heat exchanger can be installed between these streams, but since $\Delta T_1 < \Delta T_2$ and $\Delta T_1 \geq \Delta T_{min} = 10°C$, only 86.58 kW can be transferred, as shown in Figure 9.15b. Unfortunately, this requires

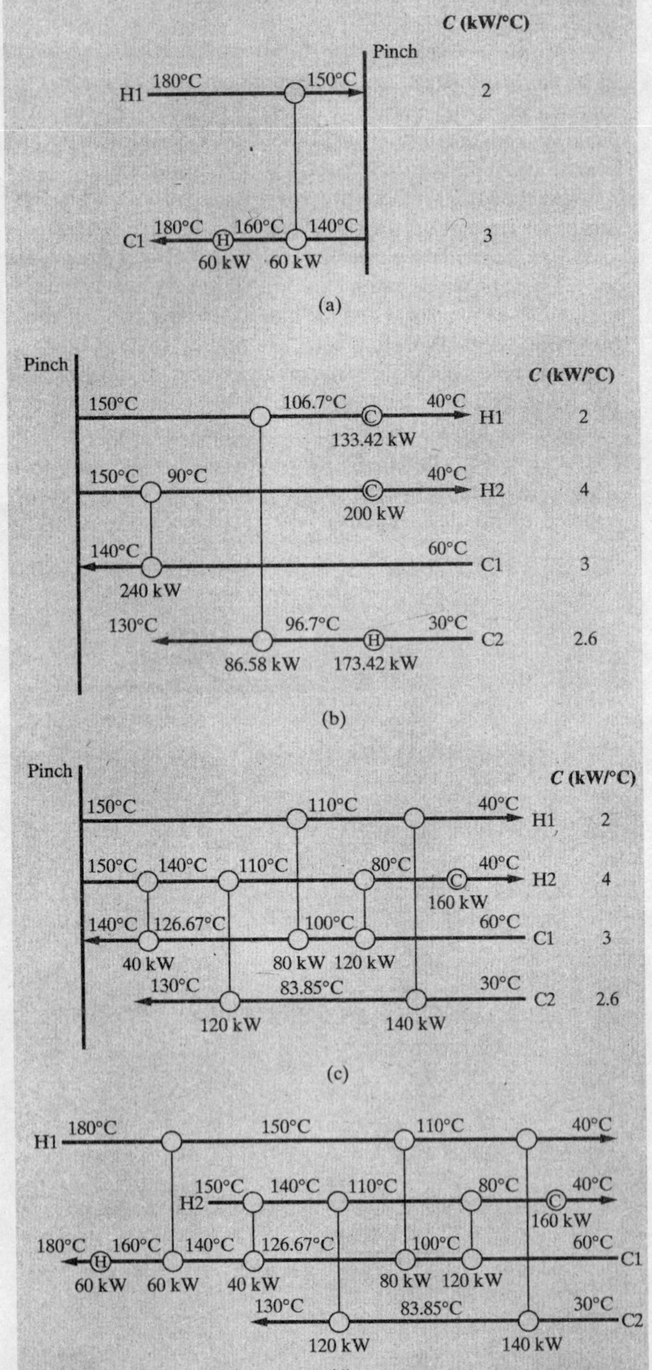

Figure 9.15 MER design for Example 9.7: (a) network on the hot side of the pinch; (b) network on the cold side of the pinch—additional utilities required; (c) network on the cold side of the pinch—minimum utilities; (d) combined network involving eight heat exchangers.

that a portion of stream C1 be heated with hot utility (173.42 kW). The addition of heat from a hot utility on the cold side of the pinch requires that an equivalent amount of heat be removed from the hot streams using cooling water, thereby exceeding the minimum cold utility (160 kW).

Alternatively, a more careful design is performed in which heat exchangers are added with duties assigned such that sufficient temperature differences are retained for additional matches, as shown in Figure 9.15c. The first match, H2–C1, is assigned a heat duty of only 40 kW, so that the H2 effluent temperature is reduced to only 140°C. This allows it to be used to heat stream C2, with a heat duty of 120 kW, which reduces the effluent temperature of H2 to 110°C, allowing its subsequent use to heat stream C1, and so on. As seen in Figure 9.15c, this rather complicated design, involving six heat exchangers, meets the MER cooling target of 160 kW.

When combined with the network in Figure 9.15a, the resulting HEN involves six interior heat exchangers, an auxiliary heater, and an auxiliary cooler, as shown in Figure 9.15d. Note that this network utilizes the minimum hot and cold utilities, but involves eight heat exchangers, three above the minimum, as discussed in Section 9.4. The next step is to consider the possibilities for reducing the number of heat exchangers by removing the *heat loops,* or to introduce stream splitting, as demonstrated later in Section 9.4.

Mixed-Integer Linear Programming

The *transshipment model* for heat transfer, introduced by Papoulias and Grossmann (1983b), provides a more systematic method for stream matching. In this model, the hot streams and hot utilities are viewed as the sources of energy that distribute among the temperature intervals, and that in turn become a source of energy for the sinks (or destinations), that is, the cold streams and cold utilities. The temperature levels of the sources and sinks establish the temperature intervals within which the sources can provide energy, and similarly, within which the sinks can receive energy. Energy can also be carried by the hot streams, as *residuals*, to the adjacent intervals at a lower temperature. Such a transshipment model is illustrated schematically in Figure 9.16a, in which hot streams H1 and H2 potentially exchange energy with cold stream C1 and cooling water W in interval k. As shown, energy from the hot streams in interval k, $Q^{H}_{H1,k}$ and $Q^{H}_{H2,k}$, combines with residual energy in the hot streams from interval $k-1$, $R_{H1,k-1}$ and $R_{H2,k-1}$, to be transferred to cold stream C1 or to cooling water W or rejected to the next interval as residual energy, $R_{H1,k}$ and $R_{H2,k}$. Stated differently, Figure 9.16a shows a superstructure in which all possible exchanges of heat between the hot streams, cold streams, and cold utility are included. Using this superstructure to design a network with a low capital cost having the minimum utilities, Papoulias and Grossmann created a MILP that minimizes the number of matches (i.e., the number of heat exchangers). For this MILP, Figure 9.16b defines the nomenclature, where Q_{ijk} is the heat exchanged between hot stream i and cold stream j in temperature interval k, Q^{H}_{ik} is the heat available

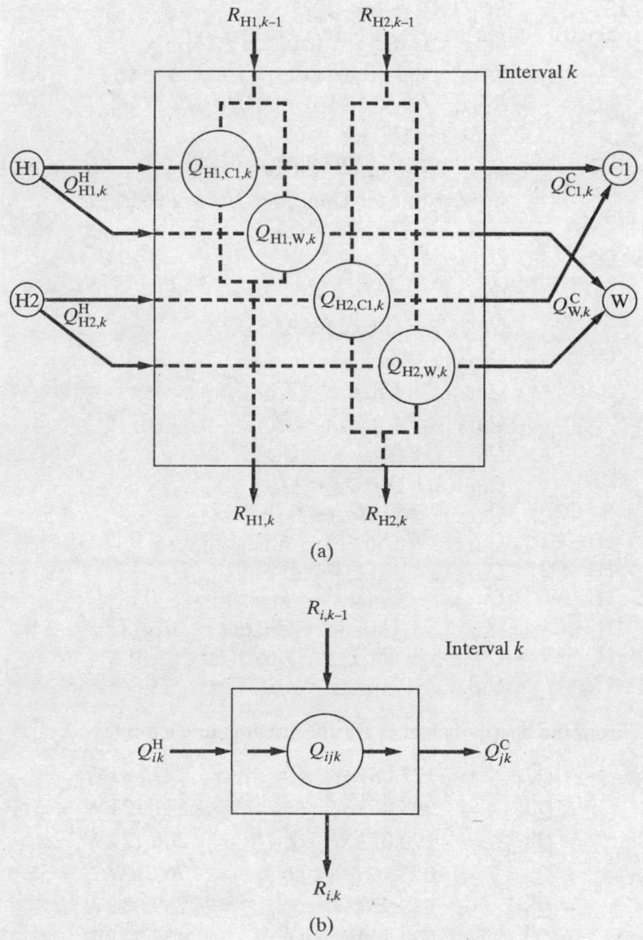

Figure 9.16 Transshipment model for stream matching: (a) superstructure; (b) nomenclature.

from hot stream i in interval k, $R_{i,k-1}$ is the residual heat in hot stream i that is transferred from the adjacent hotter interval $k-1$, $R_{i,k}$ is the residual heat in hot stream i that is transferred to the adjacent colder interval $k+1$, and Q^C_{jk} is the heat that must be transferred to cold stream j in interval k. Then, using this nomenclature, the MILP takes the form:

$$\text{Minimize } z = \sum_i \sum_j w_{ij} y_{ij}$$

w.r.t

$$Q_{ijk},\ y_{ij}$$

s.t.

$$R_{ik} - R_{i,k-1} + \sum_{j \in C_k} Q_{ijk} = Q^H_{ik}, \quad i \in H_k, \quad k = 1, \ldots, K \qquad \text{(MILP.1)}$$

$$\sum_{i \in H_k} Q_{ijk} = Q^C_{ik}, \quad j \in C_k, \quad k = 1, \ldots, K \qquad \text{(MILP.2)}$$

$$\sum_k Q_{ijk} - y_{ij} U_{ij} \leq 0, \quad i \in H, \quad j \in C \qquad \text{(MILP.3)}$$

$$R_{ik} \geq 0, \quad Q_{ijk} \geq 0, \quad y_{ij} \in 0, 1 \qquad \text{(MILP.4)}$$

$$R_{i0} = R_{iK} = 0 \qquad \text{(MILP.5)}$$

Here, y_{ij} is a binary variable that equals unity when a match exists between hot stream i and cold stream j, and is zero otherwise. The objective is to minimize the number of matches, and hence, the objective function sums over all of the possible matches, with the weighting coefficient, w_{ij}, increased as certain matches become less desirable. Constraints (MILP.1) and (MILP.2) are the energy balances for each of the K temperature intervals where i, the hot stream index, belongs to the set of hot streams in interval k, H_k, and j, the cold stream index, belongs to the set of cold streams in interval k, C_k.

Constraints (MILP.3) place bounds on the heat to be transferred when hot stream i and cold stream j are matched. Of course, when $y_{ij} = 0$, there is no match. However, when $y_{ij} = 1$, there is an upper bound on the heat that can be transferred between the two streams in all of the temperature intervals ($\sum Q_{ijk}$). This upper bound, U_{ij}, is the minimum of the heat that can be released by hot stream $i[Q_i = C_i(T^s_i - T^t_i)]$ and that which can be taken up by cold stream $j[Q_j = C_j(T^t_j - T^s_j)]$; that is, $U_{ij} = \min\{Q_i, Q_j\}$. Note that H and C are the sets of the hot and cold streams, respectively.

Constraints (MILP.4) ensure that all of the residuals and rates of heat transfer are greater than or equal to zero and define the binary variables. Finally, constraints (MILP.5) indicate that the residuals at the lower and upper bounds of the temperature intervals are zero, and hence, all streams that exchange energy must have their temperatures within the bounds of the temperature intervals, including the hot and cold utilities. This is one of the principal departures from the temperature intervals in the previous analysis (see Examples 9.2 and 9.6), where the hot and cold utilities are not included in the temperature intervals.

The next example is provided to illustrate the creation and solution of a typical MILP for MER design.

EXAMPLE 9.8

In this example, taken from Linnhoff and Flower (1978a, b), a network of heat exchangers is to be synthesized for two hot and two cold streams, with steam and cooling water as the utilities, as shown below. Note that the minimum utilities have been determined for $\Delta T_{min} = 10°C$, using one of the methods in Section 9.2, and that the temperatures and heat duties of the utilities are given as well:

Stream	T^s(°C)	T^t(°C)	C(kW/°C)	Q(kW)
C1	60	160	7.62	762
C2	116	260	6.08	875.52
H1	160	93	8.79	588.93
H2	249	138	10.55	1171.05
S	270	270	—	127.68
W	38	82	5.685	250.14

The pinch temperatures are 249°C and 239°C. Note also that demineralized water would be required to achieve a target temperature of 82°C without scaling the heat exchanger tubes.

SOLUTION

First, the temperature intervals are identified, somewhat differently than when implementing the TI method to determine the minimum utilities, as shown in Figure 9.17. In this case, the *inlet* (or source) temperatures denote the bounds on the temperature intervals. These are circled in the figure. At each interval boundary, both the hot and cold stream temperatures, on the left and right, respectively, are shown, separated by $\Delta T_{\min} = 10°C$. The steam temperature is selected to be 10°C higher than the highest temperature of the cold streams, 260°C. On the left side of the temperature intervals, the hot streams are shown so that it is clear within which intervals each hot stream appears. Similarly, on the right side, the cold streams are shown. Note that the target temperatures of the hot and cold streams could be used to define additional temperature intervals, but these do not affect the results of the MILP.

Using Figure 9.17, the streams in each interval are

Interval	Stream
1	S, C2
2	H2, C1, C2
3	H1, H2, C1, C2
4	H1, H2*, C1,W
5	H1*, H2*, W

where the asterisks denote that the hot streams can exchange heat in these intervals through their residuals. This is the basis for the construction of the MILP that follows, for the case where all of the weighting coefficients are unity.

$$\text{Minimize } z = y_{S,C2} + y_{H1,C1} + y_{H1,C2} + y_{H1,W} + y_{H2,C1} + y_{H2,C2} + y_{H2,W}$$

w.r.t.

y_{ij}

s. t.

S: $R_{S,1} + Q_{S,C2,1} = Q^H_{S,1}$

H2: $R_{H2,2} + Q_{H2,C1,2} + Q_{H2,C2,2} = Q^H_{H2,2}$

$R_{H2,3} - R_{H2,2} + Q_{H2,C1,3} + Q_{H2,C2,3} = Q^H_{H2,3}$

$R_{H2,4} - R_{H2,3} + Q_{H2,C1,4} + Q_{H2,W,4} = Q^H_{H2,4}$

$-R_{H2,4} + Q_{H2,W,5} = Q^H_{H2,5}$

H1: $R_{H1,3} + Q_{H1,C1,3} + Q_{H1,C2,3} = Q^H_{H1,3}$

$R_{H1,4} - R_{H1,3} + Q_{H1,C1,4} + Q_{H1,W,4} = Q^H_{H1,4}$

$-R_{H1,4} + Q_{H1,W,5} = Q^H_{H1,5}$

C2: $Q_{S,C2,1} = Q^C_{C2,1}$

$Q_{H2,C2,2} = Q^C_{C2,2}$

$Q_{H1,C2,3} + Q_{H2,C2,3} = Q^C_{C2,3}$

C1: $Q_{H2,C1,2} = Q^C_{C1,2}$

$Q_{H1,C1,3} + Q_{H2,C1,3} = Q^C_{C1,3}$

$Q_{H1,C1,4} + Q_{H2,C1,4} = Q^C_{C1,4}$

W: $Q_{H1,W,4} + Q_{H2,W,4} = Q^C_{W,4}$

$Q_{H1,W,5} + Q_{H2,W,5} = Q^C_{W,5}$

S–C2: $Q_{S,C2,1} - y_{S,C2} U_{S,C2} \le 0$

H1–C1: $Q_{H1,C1,3} + Q_{H1,C1,4} - y_{H1,C1} U_{H1,C1} \le 0$

H1–C2: $Q_{H1,C2,3} - y_{H1,C2} U_{H1,C2} \le 0$

H1–W: $Q_{H1,W,4} + Q_{H1,W,5} - y_{H1,W} U_{H1,W} \le 0$

H2–C1: $Q_{H2,C1,2} + Q_{H2,C1,3} + Q_{H2,C1,4} - y_{H2,C1} U_{H2,C1} \le 0$

H2–C2: $Q_{H2,C2,2} + Q_{H2,C2,3} - y_{H2,C2} U_{H2,C2} \le 0$

H2–W: $Q_{H2,W,4} + Q_{H2,W,5} - y_{H2,W} U_{H2,W} \le 0$

From the energy balances for the streams in each interval:

$$
\begin{aligned}
Q^H_{S,1} &= 127.68\,\text{kW} & Q^C_{C2,1} &= 127.68\,\text{kW}\\
Q^H_{H1,3} &= 298.86\,\text{kW} & Q^C_{C2,2} &= 541.12\,\text{kW}\\
Q^H_{H1,4} &= 290.07\,\text{kW} & Q^C_{C2,3} &= 206.72\,\text{kW}\\
Q^H_{H1,5} &= 0 & Q^C_{C1,2} &= 76.2\,\text{kW}\\
Q^H_{H2,2} &= 938.95\,\text{kW} & Q^C_{C1,3} &= 259.08\,\text{kW}\\
Q^H_{H2,3} &= 232.1\,\text{kW} & Q^C_{C1,4} &= 426.72\,\text{kW}\\
Q^H_{H2,4} &= 0 & Q^C_{W,4} &= 125.07\,\text{kW}\\
Q^H_{H2,5} &= 0 & Q^C_{W,5} &= 125.07\,\text{kW}
\end{aligned}
$$

Furthermore, the upper bounds, U_{ij}, for the potential matches are

S–C2: $U_{S,C2} = \min\{127.68, 127.68\} = 127.68$

H1–C1: $U_{H1,C1} = \min\{762, 588.93\} = 588.93$

H1–C2: $U_{H1,C2} = \min\{875.52, 588.93\} = 588.93$

H1–W: $U_{H1,W} = \min\{250.14, 588.93\} = 250.14$

H2–C1: $U_{H2,C1} = \min\{1,171.05, 762\} = 762$

H2–C2: $U_{H2,C2} = \min\{1,171.05, 747.84\} = 747.84$

H2–W: $U_{H2,W} = \min\{1,171.05, 250.14\} = 250.14$

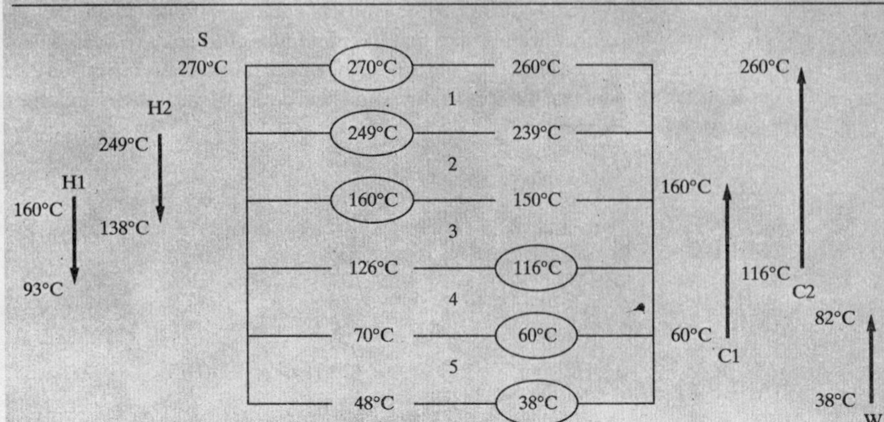

Figure 9.17 Temperature intervals for stream matching.

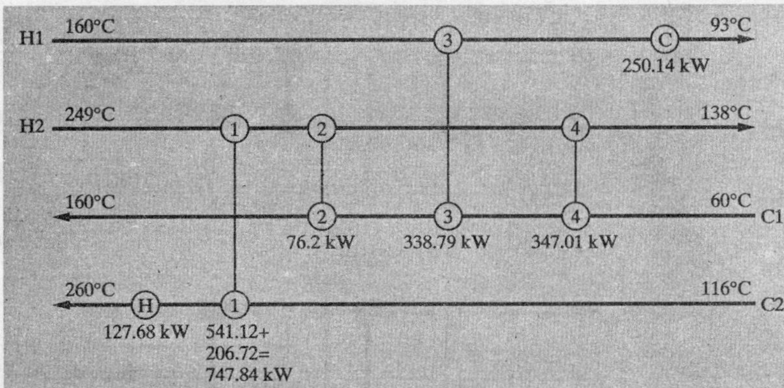

Figure 9.18 HEN for Example 9.8.

When determining U_{ij} to maintain the minimum utilities, heat cannot be exchanged across the pinch. Hence, the heat loads for each of the streams are for the segments above or below the pinch temperatures (249°C and 239°C) depending upon whether the match occurs above or below the pinch.

Substituting the above numbers into the MILP and using the GAMS input file MATCH.1 in the Program and Simulation Files folder, which can be downloaded from the Wiley Web site associated with this book, $z = 5$ at the minimum, with:

$$y_{S,C2} = y_{H1,C1} = y_{H1,W} = y_{H2,C1} = y_{H2,C2} = 1$$

and:

$$\text{S–C2}: \; Q_{S,C2,1} = 127.68$$
$$\text{H1–C1}: \; Q_{H1,C1,3} = 259.08, \quad Q_{H1,C1,4} = 79.71$$
$$\text{H1–W}: \; Q_{H1,W,4} = 125.07, \quad Q_{H1,W,5} = 125.07$$
$$\text{H2–C1}: \; Q_{H2,C1,2} = 76.2, \quad Q_{H2,C1,3} = 0, \quad Q_{H2,C1,4} = 347.01$$
$$\text{H2–C2}: \; Q_{H2,C2,2} = 541.12, \quad Q_{H2,C2,3} = 206.72$$

with:

$$\text{H1}: \; R_{H1,3} = 39.78 \qquad R_{H1,4} = 125.07$$
$$\text{H2}: \; R_{H2,2} = 321.63 \qquad R_{H2,3} = 347.01 \qquad R_{H2,4} = 0$$

As can be seen, these results satisfy all of the equality and inequality constraints. There are five matches ($z = 5$), but these translate into six heat exchangers, as shown in Figure 9.18. Streams H2 and C2 exchange heat in two adjacent intervals, 2 and 3, and hence, heat exchanger 1 is sufficient for this purpose. On the other hand, streams H2 and C1 exchange heat in intervals 2 and 4, with stream C1 exchanging heat with stream H1 in interval 3. Hence, two heat exchangers, 2 and 4, are installed, separated by heat exchanger 3.

In summary, six heat exchangers is the minimum for this network when it is required that the hot and cold utilities be minimized as well. As discussed in Section 9.4, the minimum number of heat exchangers for this system is five, which can be achieved either by *breaking heat loops,* usually at the price of exceeding the MER targets, or by *stream splitting.*

To complete this section, it is important to note that only an introduction to the application of MILP models for stream matching has been presented. A presentation that is far more complete is provided by Floudas (1995) in *Nonlinear and Mixed-integer Optimization: Fundamentals and Applications*, which discusses the theory behind the MILP formulations and shows more advanced techniques for excluding specific matches, matching with multiple hot and cold utilities, and allowing matches between two hot or two cold streams, when this is desirable.

9.4 MINIMUM NUMBER OF HEAT EXCHANGERS

Having designed a HEN that meets the MER targets, it is common to reduce the number of heat exchangers toward the minimum while permitting the consumption of utilities to rise, particularly when small heat exchangers can be eliminated. In this way, lower annualized costs may be obtained, especially when the cost of fuel is low relative to the purchase cost of the heat exchangers. Alternatively, an attempt can be made to design the HEN to minimize the number of heat exchangers *and satisfy the MER target,* by invoking stream splitting.

Reducing the Number of Heat Exchangers—Breaking Heat Loops

Before proceeding, it is important to recognize that, as pointed out by Hohmann (1971), the minimum number of heat exchangers in a HEN is

$$N_{HX,min} = N_s + N_U - 1 \tag{9.13}$$

where N_S is the number of streams and N_U is the total number of distinct hot and cold utility sources. Thus, for hot utilities: fuel, steam at high pressure (hps), intermediate pressure (ips), and low pressure (lps); and for cold utilities: boiler feed water (bfw), cooling water (cw), and refrigeration; each count as distinct utility sources. Hence, for the networks in Examples 9.7 and 9.8, which involve four streams and two utilities (steam and cooling water), $N_{HX,min} = 5$. Equation (9.13) indicates that the minimum number of heat exchangers, and with it the minimum capital cost of the HEN, increases with each distinct utility source utilized in a design. However, as shown in Section 9.8, the operating costs of a HEN are reduced by replacing a portion of the utility duty requirement provided by a costly utility (e.g., refrigeration) by one at a lower cost (e.g., cooling water).

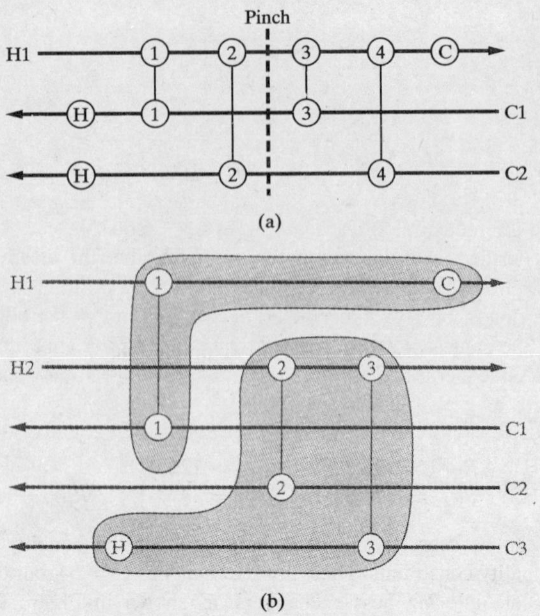

Figure 9.19 HENs involving: (a) one network; (b) two independent networks.

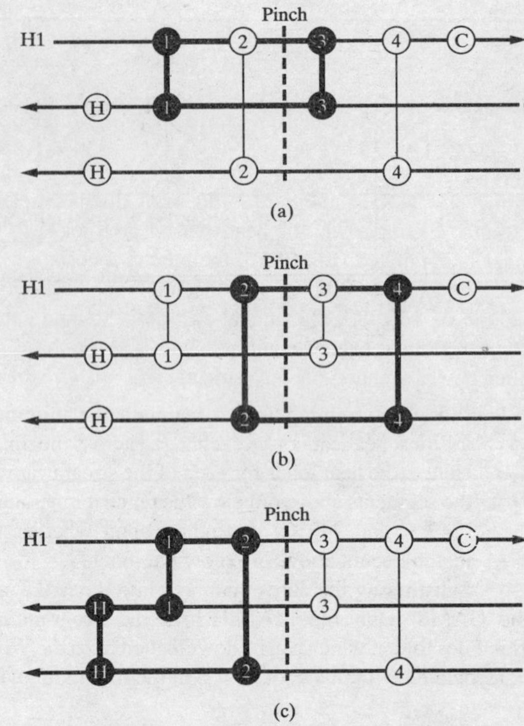

Figure 9.20 The three heat loops in the HEN in Figure 9.19a.

In a more general result, Douglas (1988) shows that the minimum number of heat exchangers is also dependent on the number of *independent networks*, N_{NW}; that is, the number of subnetworks consisting of linked paths between the connected streams:

$$N_{HX;min} = N_s + N_U - N_{NW} \tag{9.14}$$

When all streams in a process are connected directly or indirectly by heat exchangers, as in Figure 9.19a, $N_{NW} = 1$, and Eq. (9.14) reverts to Eq. (9.13), with $N_{HX,min} = 4$, noting that the HEN in Figure 9.19a has seven heat exchangers, three in excess of the minimum possible. In contrast, Figure 9.19b illustrates a network of five streams and two utilities, comprised of two independent subsystems: one connecting streams H1, C1, and the cold utility, and the second connecting streams H2, C2, C3, and the hot utility. In this case, $N_{HX,min} = N_S + N_U - N_{NW} = 5 + 2 - 2 = 5$, which is the number of heat exchangers in the HEN shown.

In one of the first studies of the methods for heat integration, Hohmann (1971) showed that in a HEN with N_{HX} heat exchangers, $N_{HX} - N_{HX,min}$ independent *heat loops* exist, that is, subnetworks that exhibit cyclic heat flows between two or more streams. The simplest case of a heat loop is shown in Figure 9.20a, noting that streams H1 and C1 are matched twice, with heat exchangers 1 and 3. A similar heat loop is shown in Figure 9.20b, where heat flows between streams H1 and C2 in two matches, in heat exchangers 2 and 4. A more complex heat loop is shown in Figure 9.20c, in which heat flows between streams H1 and C1 in heat exchanger 1, between H1 and C2 in heat exchanger 2, and between C1 and C2 and the hot utility, forming a closed heat cycle. Note that if C1 and C2 were to be serviced by two distinct sources of hot utility, there would be no heat loop in Figure 9.20c. Note also that the three heat loops identified in Figure 9.20 explain why the HEN has three heat exchangers more that $N_{HX,min}$.

Since a pinch exists in Figure 9.20, a HEN that meets the MER targets consists of two *independent* networks, one on each side of the pinch, so that: $N^{+}_{HX,min} = N^{-}_{HX,min} = N_S + N_U - 1 = 3 + 1 - 1 = 3$, noting that the "+" and "−" superscripts indicate the hot and cold side of the pinch, respectively. Thus, the HEN in Figure 9.20 can be designed to meet the MER targets with just six heat exchangers, ($N^{MER}_{HX,min} = N^{+}_{HX,min} + N^{-}_{HX,min} = 6$). The simplest change is to eliminate heat exchanger 1, transferring its duty to heat exchanger 2, decreasing the duty of the heater on stream C2 by this amount, and increasing the duty of the heater on stream C1 by this amount. This is referred to as *loop breaking*. Often, the smallest heat exchanger in the heat loop is eliminated, because the cost of the area saved by eliminating a small exchanger is usually more than the cost incurred in increasing the area of a large exchanger by the same amount.

As will be shown in Example 9.9, each heat loop is broken by removing a heat exchanger and adjusting the heat loads accordingly, which often leads to heat flow across the pinch, moving the HEN away from its MER targets.

EXAMPLE 9.9 ***(Example 9.7 Revisited)***

Return to the HEN in Figure 9.15d, which was designed to meet its MER targets, and involves eight heat exchangers, three more than $N_{HX,min}$ given by Eq. (9.13). In this example, the heat loops are identified and removed, one-by-one, taking note of the impact on the capital cost, the cost of the utilities, and the annualized cost, C_A, in Eq. (9.3).

SOLUTION

Beginning with the HEN in Figure 9.15d, using the specifications at the start of Example 9.6, the heat-transfer area for each heat exchanger is computed [Eq. (9.2)], the purchase costs are estimated using $C_P = 3,000A^{0.5}$, and the total purchase cost is computed to be \$66,900. The annual cost of steam and cooling water is \$10,960/yr, which combines with the purchase cost, multiplied by a return on investment ($i_m = 0.1$), to give $C_A = \$17,650/\text{yr}$.

To eliminate the first heat loop, in which heat is exchanged in two heat exchangers between streams H1 and C1, as shown in Figure 9.21a, the 80-kW exchanger is combined with the 60-kW exchanger, as shown in Figure 9.21b. This causes an approach temperature violation (110°C − 113.33°C = −13.33°C), which must be eliminated by transferring less heat, x, in this heat exchanger. This amount of heat is computed so as to give $\Delta T_{\min} = 10°C$ on the cold side; that is, the temperature of stream H1 is reduced to 123.33°C. Then, by heat balance,

$$140 - x = 2(180 - 123.33)$$

and the reduction in the heat load is $x = 26.66$ kW.

Furthermore, to account for this reduction, the steam consumption must be increased by x to heat stream C1 to 180°C, and the loads of five other heat exchangers must be adjusted by the same amount, as shown in Figure 9.21c. This includes the same increase in the consumption of cooling water because x units of heat are transferred across the pinch. After the heat loads are adjusted, the resulting network is shown in Figure 9.21d. A glance at Table 9.5 shows that the total purchase cost of the seven heat exchangers is reduced to \$57,470, but the cost of utilities is increased to \$13,250/yr, and hence, C_A is increased to \$19,000/yr.

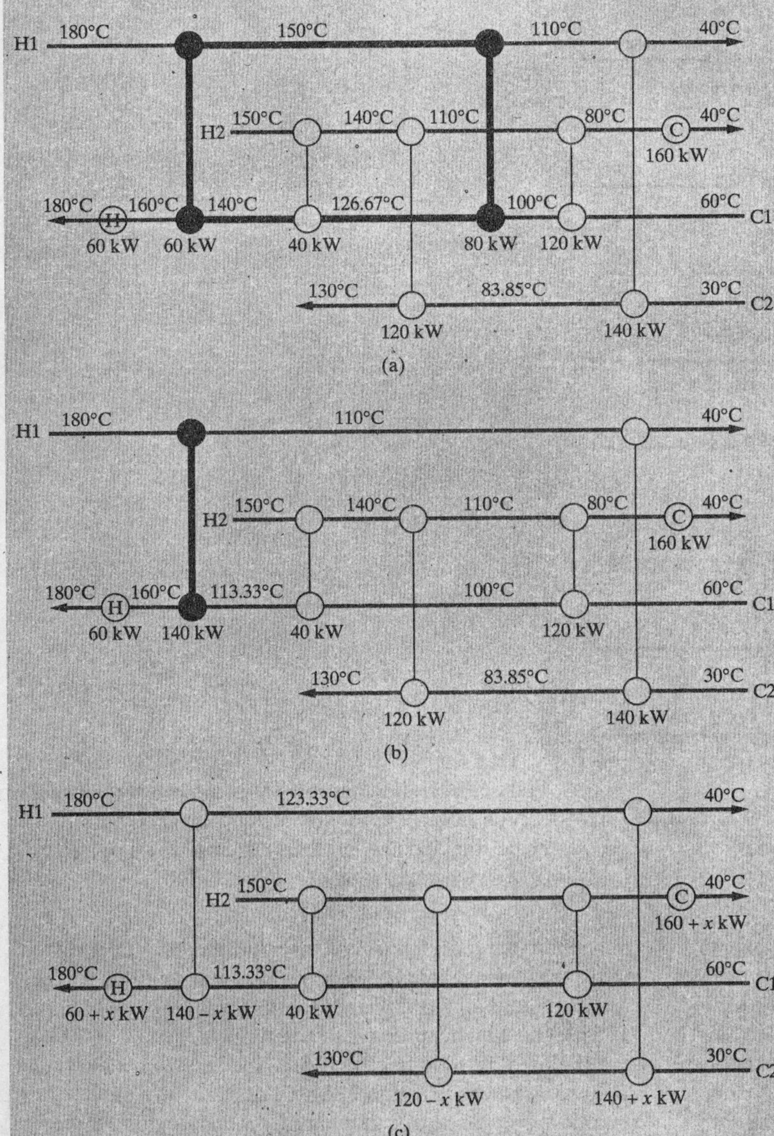

Figure 9.21 Breaking heat loops—Example 9.9: (a) eight heat exchangers, first heat loop; (b) seven heat exchangers, $\Delta T_{\min}$ violation; (c) seven heat exchangers, shifting heat loads (*continued*).

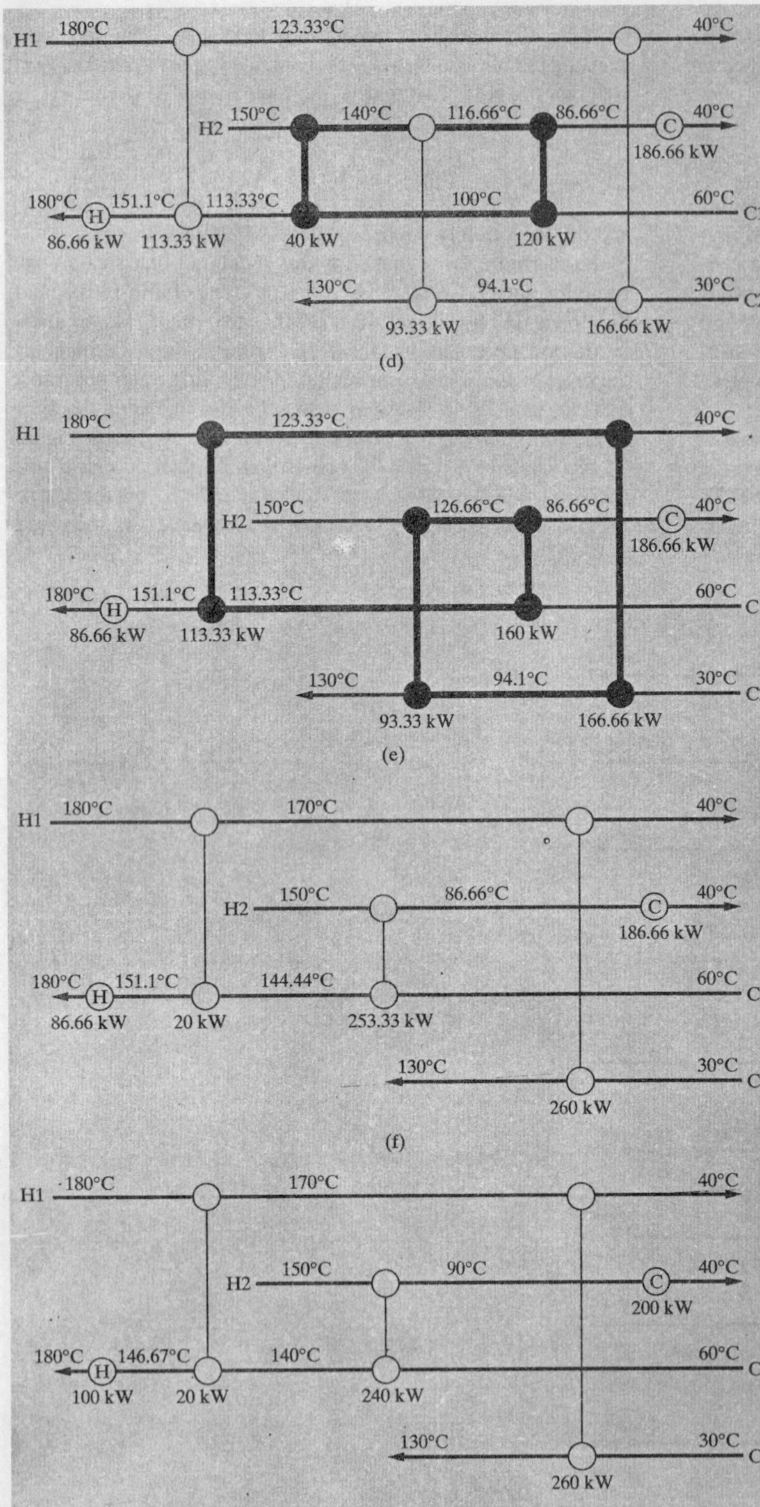

Figure 9.21 (Continued) (d) seven heat exchangers, second heat loop; (e) six heat exchangers, third heat loop; (f) five heat exchangers, ΔT_{min} violation; (g) five heat exchangers.

To eliminate the second heat loop, in which heat is exchanged in two heat exchangers between streams H2 and C1, as shown in Figure 9.21d, the 40-kW exchanger is combined with the 120-kW exchanger, as shown in Figure 9.21e. Normally, the heat exchanger with the smaller load is eliminated and its load transferred to the large one, unless the loads are comparable (as in the case of the first heat loop). In this case, since there are no temperature-approach violations, no adjustments in the loads of the heat exchangers are needed. Hence, as shown in Table 9.5, the cost of utilities is unchanged and the total purchase cost is reduced to $54,430, which reduces C_A to $18,690/yr.

The final heat loop has four heat exchangers involving four streams, as shown in Figure 9.21e. One of these can be eliminated by shifting the load of the smallest heat exchanger around the heat

Table 9.5 Cost Comparison for Example 9.9

Network	Utilities Cost ($/yr)	C_P Purchase Cost ($)	C_A, Annualized Cost ($/yr)
Design for MER: 8 HXs	10,960	66,900	17,650
7 HXs	13,250	57,470	19,000
6 HXs	13,250	54,430	18,690
Design for $N_{HX,min}$: 5 HXs	15,340	45,930	19,930
Design for MER: stream-split design, 6 HXs[a]	10,960	54,890	16,450

[a]See Example 9.11 for details of the stream-split design to meet $N_{HX,min}^{MER}$

loop. The network in Figure 9.21f results, but it has a temperature approach violation ($150 - 144.44 = 5.56 \leq 10°C$). To eliminate this, the 253.33-kW heat load is reduced by y, where:

$$253.33 - y = 3(140 - 60)$$

to enable stream C1 to leave the heat exchanger at 140°C. This amount of heat ($y = 13.33$ kW) must be supplied and removed by additional steam and cooling water, respectively, as shown in Figure 9.21g, where the final HEN is displayed, with only five heat exchangers. In Table 9.5, for this network, the total purchase cost is reduced to $45,930, but the cost of utilities is increased to $15,340/yr, resulting in the largest C_A at $19,930. Hence, for this system, the HEN with eight heat exchangers has the lowest C_A when $\Delta T_{min} = 10°C$. Of course, the tradeoff between the total purchase cost and the cost of utilities shifts as the cost of fuel decreases and the capital cost or the return on investment increases. Under different conditions, another configuration can have the lowest annualized cost.

As seen in Table 9.5, an alternative design involving stream splitting gives the lowest annualized cost. Indeed, as will be shown in the next section, the use of stream splitting often enables the design of HENs that satisfy MER targets with the minimum number of heat exchangers.

Reducing the Number of Heat Exchangers—Stream Splitting

When designing a HEN to meet its MER targets, stream splitting *must* be employed if: (a) the number of hot streams *at the pinch,* on the cold side, is less than the number of cold streams; or (b) the number of cold streams *at the pinch,* on the hot side, is less than the number of hot streams. In this way, parallel pairings that fully exploit the temperature differences between energy sources and sinks are possible. Moreover, stream splitting helps to reduce the number of heat exchangers in a HEN without increasing the utility duties, which often occurs when heat loops are broken. The following example illustrates the use of stream splitting to achieve simultaneously the MER and the minimum units for a HEN involving two hot streams and one cold stream.

EXAMPLE 9.10

It is required to design a HEN, with a minimum number of heat exchangers that satisfy $\Delta T_{min} = 10°C$ and a hot utility MER target of 300 kW, for three streams on the hot side of the pinch:

Stream	T^s(°C)	T^t(°C)	C(kW/°C)	Q(kW)
H1	200	100	5	500
H2	150	100	4	200
C1	90	190	10	1,000

SOLUTION

Since no cold utility is allowed, Eq. (9.13) gives $N_{HX,min} = 3$. Furthermore, because the two hot streams need to be cooled entirely by the cold stream, to avoid an approach-temperature violation (i.e., $\Delta T_1 < \Delta T_{min}$), the cold stream must be split, as shown in Figure 9.22a. Note that the duties of all three heat exchangers have been set to satisfy the MER targets, but the portion of the

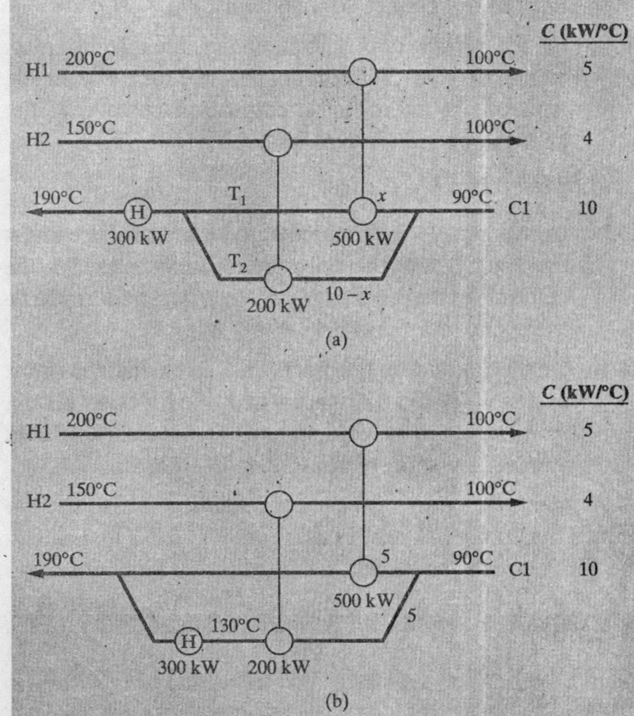

Figure 9.22 MER design for Example 9.10: (a) split determination; (b) improved design.

heat-capacity flow rate assigned to the first stream, x, must be determined by solving the energy balances for the split streams:

$$x(T_1 - 90) = 500 \qquad (9.15)$$
$$(10 - x)(T_2 - 90) = 200 \qquad (9.16)$$

subject to the constraints:

$$200 - T_1 \geq \Delta T_{min} = 10°C \qquad (9.17)$$
$$150 - T_2 \geq \Delta T_{min} = 10°C \qquad (9.18)$$

To minimize lost work, as explained in Section 9S.4, it is desirable to mix the split streams isothermally; that is, to select $T_1 = T_2 = 160°C$, with $x = 7.143$ kW/°C. Unfortunately, isothermal mixing is infeasible since inequality (9.18) limits T_2 to be less than or equal to 140°C. Arbitrarily, the equality is set *active;* that is, $T_2 = 140°C$, giving $x = 6$ kW/°C and $T_1 = 173.33°C$. Note that by making a minor structural change in which the heater is moved upstream of the mixing junction, as in Figure 9.22b, the design is improved to provide isothermal mixing, as desired.

As illustrated in Example 9.10 and introduced by Linnhoff and Hindmarsh (1983), generalized rules for stream splitting on both sides of the pinch to satisfy MER requirements are:

Hot Side of Pinch

H-1. Let N_H and N_C be the number of hot and cold streams at the pinch. Noting that cold utilities cannot be used on the hot side of the pinch, if $N_H > N_C$, *a cold stream must be split,* since for feasibility, $N_H \leq N_C$.

H-2. On the hot side of the pinch, all feasible matches must ensure $C_h \leq C_c$. If this is not possible for every match, split one or more hot streams as necessary. If hot streams are split, return to step H-1 above.

Cold Side of Pinch

C-1. Let N_H and N_C be the number of hot and cold streams at the pinch. Noting that hot utilities cannot be used on the cold side of the pinch, if $N_C > N_H$, *a hot stream must be split,* since for feasibility, $N_C \leq N_H$.

C-2. On the cold side of the pinch, all feasible matches must ensure $C_c \leq C_h$. If this is not possible for every match, split one or more cold streams as necessary. If cold streams are split, return to step C-1 above.

As mentioned previously, stream splitting is also used to reduce the number of heat exchangers for HENs that satisfy the MER targets, as shown in the following example.

EXAMPLE 9.11 ***(Example 9.9 Revisited)***

Return to Example 9.9 and recall that, as heat loops are broken to reduce the number of heat exchangers, heating and cooling utilities are normally increased. Here, stream splitting is used to reduce the number of heat exchangers while satisfying MER targets.

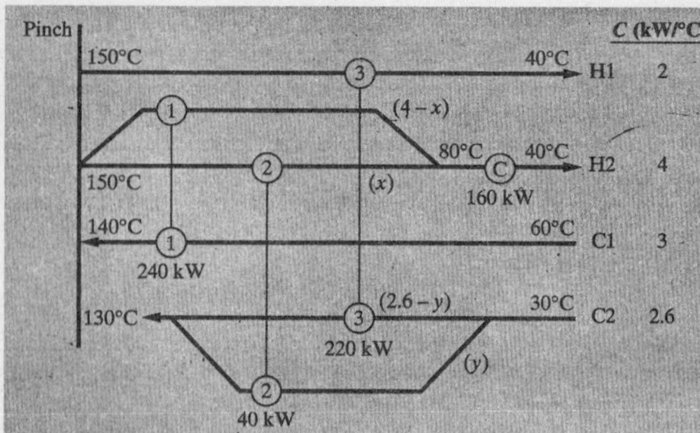

Figure 9.23 Stream splitting on the cold side of the pinch to achieve $N^{-}_{HX,\,min}$ for Example 9.11.

SOLUTION

Since an MER design implies separate HENs on both sides of the pinch, it is helpful to compute the minimum number of heat exchangers in each HEN. On the hot side of the pinch, only streams H1 and C1 exist, and so $N^{+}_{HX,min} = 2$, while on the cold side of the pinch, all streams participate, and $N^{-}_{HX,min} = 4$. Thus, $N^{MER}_{HX,min} = N^{+}_{HX,min} + N^{-}_{HX,min} = 6$. The HEN in Figure 9.15a has the minimum number of heat exchangers, while that in Figure 9.15c exceeds the minimum by two heat exchangers.

Figure 9.23 shows a possible design, in which stream H2 is split *at the pinch,* ensuring that $C_c \leq C_h$, with the largest portion of its heat-capacity flow rate, $4 - x$, paired with stream C1 using heat exchanger 1. The remaining branch is paired with a portion of stream C2, y, using heat exchanger 2. The remaining portion of stream C2, with a heat-capacity flow rate of $2.6 - y$, is paired with stream H1. To determine x and y, the energy balances associated with the stream splits are solved to ensure isothermal mixing, giving $x = 40/70 = 0.57$ kW/°C and $y = 40/100 = 0.4$ kW/°C. Note that $C_c \leq C_h$ must be satisfied only in match 1, because it is the only heat exchanger having both streams at the pinch. The overall HEN, a combination of Figures 9.15a and 9.23, satisfies the MER targets and has the minimum number of heat exchangers (six). Consequently, the cost of utilities is at the minimum, \$10,960, with the total purchase cost, \$54,890, providing an annualized cost of \$16,450, the lowest in Table 9.5.

While stream splitting provides these advantages, its use should be considered with caution because it reduces flexibility and complicates process operability. When possible, it is usually preferable to break heat loops without stream splitting, as shown later in Example 9.17.

9.5 THRESHOLD APPROACH TEMPERATURE

In many cases, the selected ΔT_{min} is such that no pinch exists, and MER design calls for either hot or cold utility to be used, but not both. The critical ΔT_{min} below which no pinch exists is referred to as the *threshold approach temperature difference*, ΔT_{thres}. The following two examples illustrate how this arises and demonstrate how the guidelines presented previously are adapted for HEN design.

EXAMPLE 9.12

Compute MER targets for the following streams as a function of the minimum approach temperature:

Stream	T^s(°C)	T^t(°C)	C(kW/°C)	Q (kW)
H1	300	200	1.5	150
H2	300	250	2	100
C1	30	200	1.2	204

SOLUTION

For $\Delta T_{\min} = 10°\text{C}$, no pinch exists, as shown in Figure 9.24a using the TI method. Note that 46 kW of cold utility are required. When the analysis is repeated as $\Delta T_{\min}$ is varied, for $\Delta T_{\min} > 100°\text{C}$, a pinch exists, as illustrated in Figure 9.24b for $\Delta T_{\min} = 105°\text{C}$. Figure 9.25 shows the threshold, $\Delta T_{\min} = \Delta T_{\text{thres}}$, at which the pinch appears. Since no heating utility is required for $\Delta T_{\min} < \Delta T_{\text{thres}}$, where the cooling requirement is constant at 46 kW, no energy is saved while capital costs are increased as $\Delta T_{\min}$ is decreased. The impact of $\Delta T_{\min}$ on the economics of HENs is considered in Section 9.6.

EXAMPLE 9.13

For the following streams:

Stream	T^s(°C)	T^t(°C)	C(kW/°C)	Q (kW)
H1	590	400	2.376	451.4
H2	471	200	1.577	427.4
H3	533	150	1.320	505.6
C1	200	400	1.600	320.0
C2	100	430	1.600	528.0
C3	300	400	4.128	412.8
C4	150	280	2.624	341.1

design a HEN at the threshold approach temperature difference, $\Delta T_{\min} = \Delta T_{\text{thres}} = 50°\text{C}$, to meet its MER targets: $Q_{\text{H,min}} = 217.5\,\text{kW}$ and $Q_{\text{C,min}} = 0\,\text{kW}$, as well as its $N_{\text{HX,min}}$ target of seven exchangers.

SOLUTION

Although no pinch exists, the MER design procedure is used, starting at the cold end, where matches are placed with $\Delta T_1 = \Delta T_{\min} = 50°\text{C}$, and reserving the allocation of the utility heaters until last. Furthermore, matches at the limiting $\Delta T_{\min}$ must have $C_h \leq C_c$, as occurs on the hot side of the pinch in the MER design procedure.

As seen in Figure 9.26, the first match is made bearing in mind that no cooling utility is allowed, and stream H3 must be cooled to 150°C. Since C2 is the only cold stream that can be used, internal

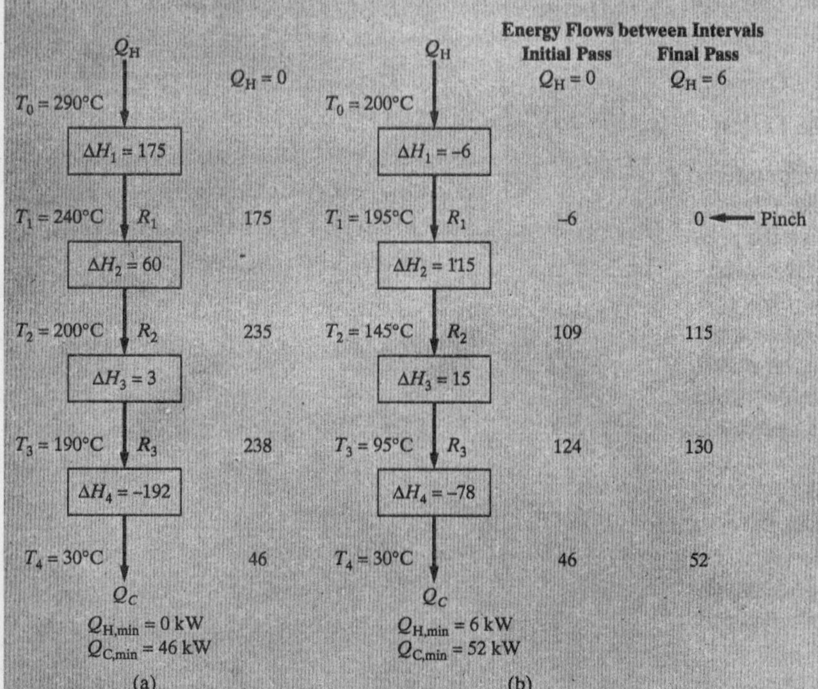

Figure 9.24 Temperature intervals, energy balances, and residuals for Example 9.12 at: (a) $\Delta T_{\min} = 10°\text{C}$; (b) $\Delta T_{\min} = 105°\text{C}$.

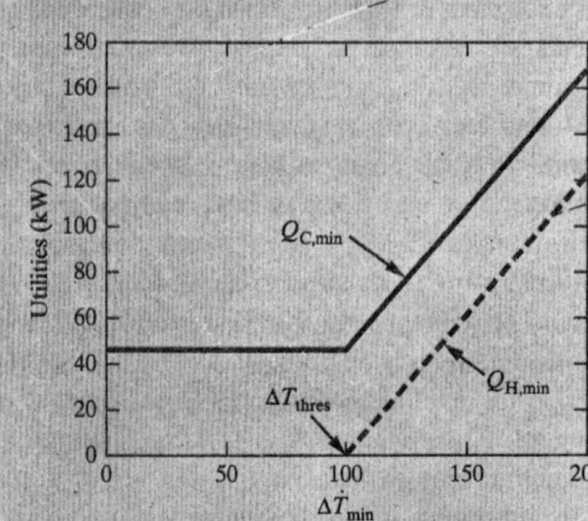

Figure 9.25 Minimum utility requirements as a function of $\Delta T_{\min}$ for Example 9.12.

heat exchanger 1 is installed, with a heat duty of 505.6 kW, which completes the cooling requirement of stream H3. Similarly, the second match is between streams H2 and C4, with a heat duty of 341.1 kW, which completes the heating requirement of stream C4. The remaining matches are far less constrained, and are positioned with relative ease. Note that the last units positioned are the auxiliary heaters, with a total duty equal to the MER target. Also, seven heat exchangers are utilized (five internal heat exchangers and two auxiliary heaters). Note that this network has no heat loops and that stream splitting is not employed.

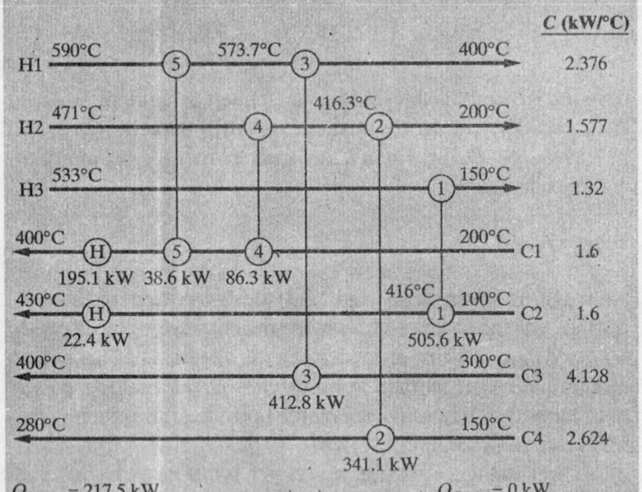

Figure 9.26 HEN with minimum number of exchangers to meet MER target at $\Delta T_{min} = \Delta T_{thres} = 50°C$ in Example 9.13.

9.6 OPTIMUM APPROACH TEMPERATURE

The importance of the minimum approach temperature, ΔT_{min}, has been emphasized in the previous sections. Clearly, as $\Delta T_{min} \rightarrow 0$, the true pinch is approached at which the area for heat transfer approaches infinity, while the utility requirements are reduced to the absolute minimum. At the other extreme, as $\Delta T_{min} \rightarrow \infty$, the heat transfer area approaches zero and the utility requirements are increased to the maximum, with no heat exchange between the hot and cold streams. The variations in heat transfer area and utility requirements with ΔT_{min} translate into the variations in capital and operating costs shown schematically in Figure 9.27. As discussed in the previous section, as ΔT_{min} decreases, the cost of utilities decreases linearly until a threshold temperature, ΔT_{thres}, is reached, below which the cost of utilities is not reduced. Thus, when $\Delta T_{min} \leq \Delta T_{thres}$, the tradeoffs between the capital and utility costs do not apply.

In summary, when designing a HEN, it is important to consider the effect of ΔT_{min}. The next example illustrates this effect, while applying the techniques for stream matching described in Section 9.3.

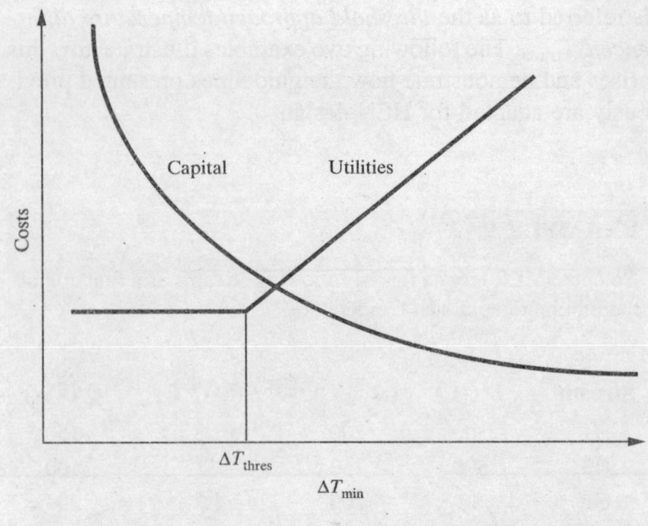

Figure 9.27 Tradeoff between capital and utilities costs as a function of ΔT_{min}.

EXAMPLE 9.14

Consider the design of a HEN for four streams in a problem generated initially by Nishida et al. (1977):

Stream	T^s (K)	T^t (K)	C(kW/K)	Q(kW)
C1	269.3	488.7	36.93	8,102.44
H1	533.2	316.5	10.55	2,286.19
H2	494.3	383.2	26.38	2,930.82
H3	477.5	316.5	15.83	2,548.63

The following specifications apply:

$$\text{Cooling water (cw)}: T^s = 310.9\,\text{K}, T^t \leq 355.4\,\text{K},$$
$$\text{cost of cw} = 0.11023\,\$/1,000\,\text{kg}$$
$$\text{Steam (sat'd., s)}: T = 508.7\,\text{K}, \Delta H^v = 1,785.2\,\text{kJ/kg},$$
$$\text{cost of s} = 2.2046\,\$/1,000\,\text{kg}$$

Overall heat-transfer coefficients:

$U_{heater} = 0.3505\,\text{kW/m}^2\,\text{K}$, $U_{cooler} = U_{exch} = 0.2629\,\text{kW/m}^2\,\text{K}$

Purchase cost of heat exchangers:

$$C_P = 1,456.3A^{0.6}(\$, \text{m}^2)$$

Equipment operability = 8,500 hr/yr

Return on investment = i_m = 0.1 [for Eq.(9.3)]

SOLUTION

For this example, when $\Delta T_{min} \geq \Delta T_{thres} = 25.833$ K, two pinches exist. This is because the heat-capacity flow rates of streams H1 and H2 sum to the heat-capacity flow rate of stream C1. Consequently, when $\Delta T_{min} = 30$ K, at the first pinch the hot and cold stream temperatures are 494.3 K and 464.3 K, and at the second pinch the temperatures are 477.5 K and 447.5 K. The temperature interval between these temperatures involves just streams H1, H2, and C1, and hence, $\Delta T =$ 30 K throughout this interval. Above the pinches, steam provides 490.7 kW and below the pinches, cooling water removes 153.89 kW. These minimum utility requirements, at $\Delta T_{min} = 30$ K, are shown in Figure 9.28a.

When $\Delta T_{min} = \Delta T_{thres} = 25.833$ K, the temperatures of the hot streams at the pinches are unchanged, but the temperatures of the cold streams are 468.5 K and 451.7 K. In this case, on the low-temperature side of the pinches, no cooling water is required, because the composite cold curve begins where the composite hot curve ends, as shown in Figure 9.28b.

Figure 9.29 shows the designs for three HENs having minimum utilities, at three values of ΔT_{min} : 30, 25.833 K, and 16.9 K. For the first two, in Figures 9.29a and 9.29b, using the method of Linnhoff and Hindmarsh (1983), stream splitting is required below the first pinch. Stream C1 is split into two streams between the pinches because just streams H1 and H2 are present. Below the second pinch, stream C1 is split into three streams because all three hot streams are present. As shown in Table 9.6, both the purchase and utility costs are lower at ΔT_{thres}. The former is reduced because the cooling water exchanger is no longer needed. At $\Delta T_{min} = 16.9$ K, the HEN is much simpler because no pinches exist, as shown in Figure 9.29c. Hence, the purchase cost is lower and the cost of utilities is equal to that at ΔT_{thres}. When compared with the other two networks having minimum utilities, the latter has a lower annualized cost. For a more realistic comparison, the heat loops in Figures 9.29a and 9.29b should be broken to determine whether a lower annualized cost is possible.

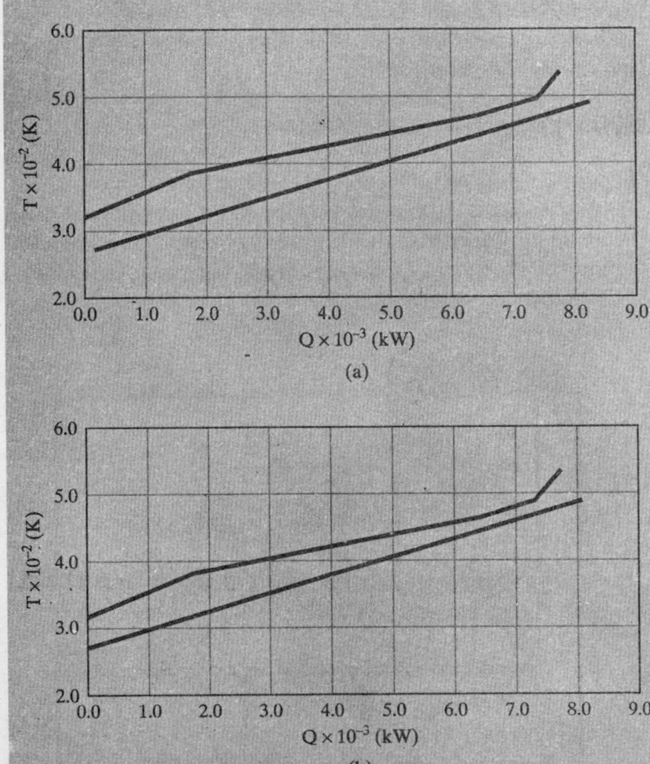

Figure 9.28 Composite heating and cooling curves for Example 9.14: (a) $\Delta T_{min} = 30$ K; (b) $\Delta T_{min} = \Delta T_{thres} =$ 25.833 K.

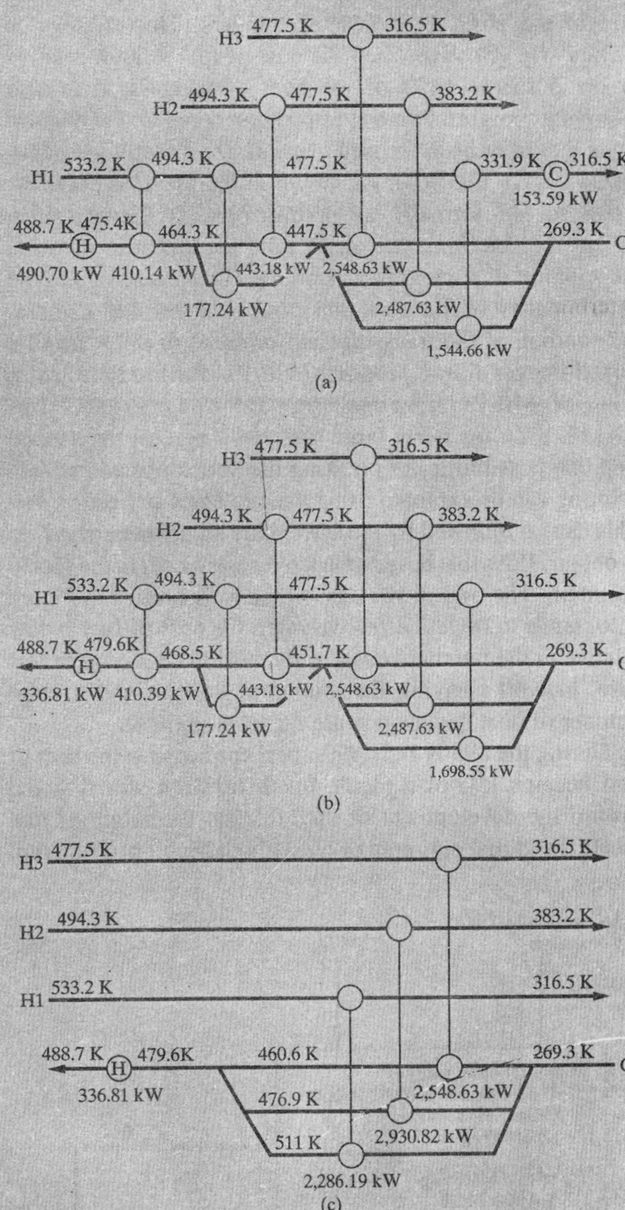

Figure 9.29 HENs for Example 9.14: (a) $\Delta T_{min} = 30$ K; (b) $\Delta T_{min} = \Delta T_{thres} = 25.833$ K; (c) $\Delta T_{min} = 16.9$ K.

Table 9.6 Cost Comparison for Example 9.14

Network	Utilities Cost ($/yr)	C_P, Purchase Cost ($)	C_A, Annualized Cost ($/yr)
$\Delta T_{min} = 30$ K	27,050	82,740	35,320
$\Delta T_{min} = 25.833$ K	12,730	79,410	20,670
$\Delta T_{min} = 16.9$ K	12,730	68,440	19,580

9.7 SUPERSTRUCTURES FOR MINIMIZATION OF ANNUAL COSTS

Thus far, emphasis has been placed on the two-step procedure introduced in Section 9.1, in which HENs are designed initially to have the minimum utilities, followed by a reduction in the number of heat exchangers. This strategy is particularly effective when the cost of fuel is high relative to the purchase costs of the heat exchangers. It is also relatively straightforward to implement compared with strategies that have been devised to minimize directly the annualized cost in Eq. (9.3). As shown in the previous sections, when solved formally as optimization problems, given $\Delta T_{\min}$, the determination of the minimum utilities involves the solution of a linear programming (LP) problem, and the determination of the minimum number of matches involves the solution of the transshipment problem in the form of a mixed-integer linear program (MILP). Furthermore, solutions of MILPs are straightforward using systems like GAMS, even for fairly large systems. Then, in the second step, the possibilities of breaking the heat loops and stream splitting can be explored using the strategies in Section 9.4. This can be followed by the systematic adjustment of $\Delta T_{\min}$ to obtain HENs that have annualized costs closer to the global optimum. The overall strategy, which is illustrated in Figure 9.30, tends to be less effective when the cost of fuel is low relative to the purchase costs of the heat exchangers. In this case, networks having the minimum utilities and a large number of heat exchangers are far from optimal.

During the 1980s and 1990s, perhaps because the cost of fuel became less of a factor, much research was directed toward the development of optimization formulations that locate the global optimum of C_A without decomposition into a two-step strategy. In their most general form, the formulations are mixed-integer nonlinear programs (MINLPs), involving nonlinear terms in the objective function (e.g., $A^{0.7}$) and in the constraints {e.g., $\Delta T_{\text{LM}} = [(T_{h,i} - T_{c,o}) - (T_{h,o} - T_{c,i})]/\ln[(T_{h,i} - T_{c,o})/(T_{h,o} - T_{c,i})]$}. While the MINLPs can be formulated rather easily, unfortunately the solution algorithms available today, in systems such as GAMS, are limited in their ability to obtain converged solutions. As computational speeds increase, along with the size of computer memories, algorithms that are more effective are being developed, and more complex and larger mathematical programs are being solved.

It is beyond the scope of this text to cover in detail the latest approaches to formulating and solving MINLPs. Instead, following the approach of Floudas (1995), a superstructure is presented in Examples 9.15 and 9.16 to introduce several aspects of the approaches. Then, a brief review of the approaches is presented, with the Floudas text referred to for the details. MINLP formulations for HEN synthesis can also be solved reliably using stochastic optimization. Lewin et al. (1998) and Lewin (1998) present an approach using genetic algorithms for the synthesis of large-scale HENs.

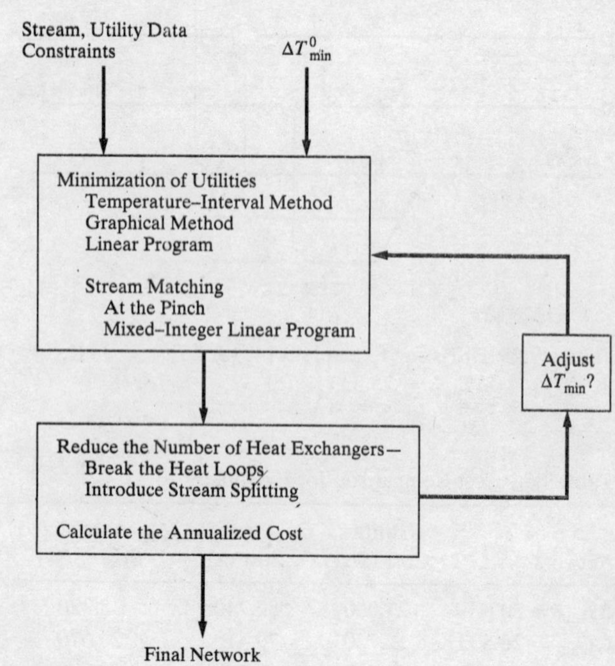

Figure 9.30 Two-step strategy for design of HENs.

EXAMPLE 9.15

In this example, a superstructure is to be created that embeds all of the alternative HENs involving one cold stream, C1, and three hot streams, H1, H2, and H3, as presented by Floudas (1995). The elements of the HENs include (1) heat exchangers, (2) stream mixers, and (3) stream splitters.

SOLUTION

Figure 9.31 shows all of the possible heat exchangers, mixers, and splitters embedded within a single superstructure. A close examination of this superstructure identifies five alternative embedded substructures, including sequences in parallel, series, and combinations thereof, including the possibility of bypasses.

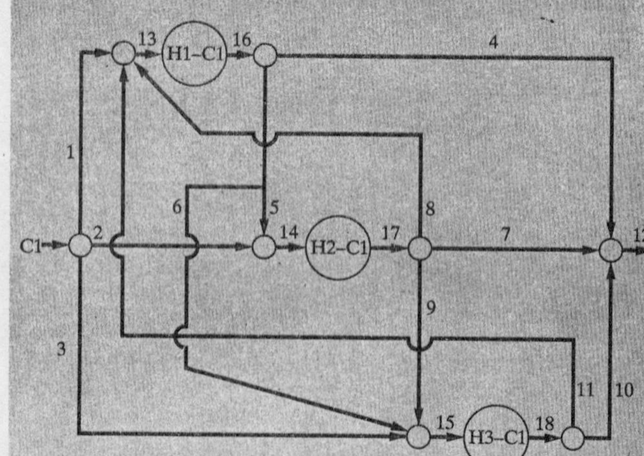

Figure 9.31 Superstructures for Example 9.15 (Reproduced from Floudas, 1995, with permission).

EXAMPLE 9.16

In this example, a superstructure is created for the potential HENs that involve hot stream H1 and cold streams C1 and C2, as specified below:

Stream	T^s(K)	T^t(K)	C(kW/K)	Q(kW)	h[kW/(m²K)]
H1	440	350	22	1,980	2
C1	349	430	20	1,620	2
C2	320	368	7.5	360	0.6667

Note that there is no pinch and that no hot or cold utilities are required. This example, taken from Floudas (1995), illustrates the formulation of the nonlinear program (NLP) to locate the HEN that minimizes the annualized cost. The purchase cost of a heat exchanger is given by $C_P = 13,000A^{0.6}$($, m²), and the return on investment is 0.1.

SOLUTION

For this system, the superstructure that contains all possible HENs is shown in Figure 9.32. Annotated are the unknown variables, $F_1, \ldots, F_8$, T_3, T_4, T_{56}, T_{78}, $A_{\mathrm{H1,C1}}$, and $A_{\mathrm{H1,C2}}$. Note that in this formulation, the heat-capacity flow rates are denoted by F.

The NLP is formulated as follows:

$$\text{Minimize } C_A = 1,300A_{\mathrm{H1,C1}}^{0.6} + 1,300A_{\mathrm{H1,C2}}^{0.6}$$
$$\text{w. r. t.}\quad F_i$$
$$\text{s. t.}$$

$F_1 + F_2 = 22$ Mass balances for splitters

$F_3 - F_5 - F_6 = 0$

$F_4 - F_7 - F_8 = 0$

$F_3 - F_1 - F_8 = 0$ Mass balances for mixers

$F_4 - F_2 - F_6 = 0$

$440\ F_1 - F_8 T_{78} - F_3 T_3 = 0$ Energy balances for mixers

$440\ F_2 - F_6 T_{56} - F_4 T_4 = 0$

$F_3(T_3 - T_{56}) = 1,620$ Energy balances for heat exchangers

$F_4(T_4 - T_{78}) = 360$ Feasibility constraints

$T_3 - 430 \geq 10$

$T_{56} - 349 \geq 10$

$T_4 - 368 \geq 10$

$T_{78} - 320 \geq 10$

$F_1, F_2, \ldots, F_8 \geq 0$ Non-negativity constraints

where the areas in the objective function are defined by

$$A_{\mathrm{H1,C1}} = \frac{1,620}{(1.0)(\Delta T_{\mathrm{LM}})_{\mathrm{H1,C1}}},\ A_{\mathrm{H_1,C2}} = \frac{360}{(0.5)(\Delta T_{\mathrm{LM}})_{\mathrm{H1,C2}}}$$

$$(\Delta T_{\mathrm{LM}})_{\mathrm{H1,C1}} = \frac{(T_3 - 430) - (T_{56} - 349)}{\ln\{(T_3 - 430)/(T_{56} - 349)\}}$$

$$(\Delta T_{\mathrm{LM}})_{\mathrm{H1,C2}} = \frac{(T_4 - 368) - (T_{78} - 320)}{\ln\{(T_4 - 368)/(T_{78} - 320)\}}$$

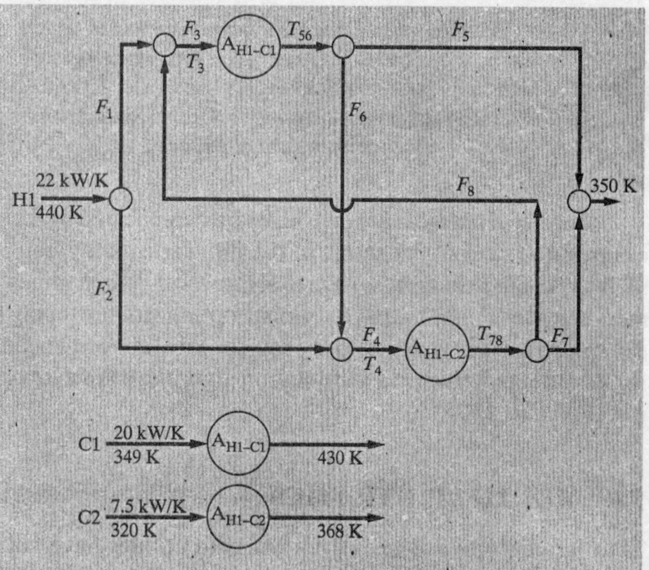

Figure 9.32 Superstructure for Example 9.16 (Reproduced from Floudas, 1995, with permission).

Note that there are eight heat-capacity flow rates, four temperatures, and nine equality constraints, and hence, three decision variables (e.g., F_1, F_6, and F_8). Also, $\Delta T_{\min}$ is set at 10 K. In a more general formulation, it could be adjusted as an optimization variable.

As reported by Floudas (1995), the solution is:

$F_1 = 20$ kW/K $F_2 = 2$ kW/K $F_3 = 20$ kW/K

$F_4 = 8.125$ kW/K $F_5 = 13.875$ kW/K $F_6 = 6.125$ kW/K

$F_7 = 8.125$ kW/K $F_8 = 0$

$T_3 = 440$ K $T_4 = 378.9$ K $T_{56} = 359$ K $T_{78} = 334.64$ K

$A_{\mathrm{H1,C1}} = 162\ \mathrm{m}^2$ $A_{\mathrm{H1,C2}} = 56.717\ \mathrm{m}^2$

with the corresponding HEN shown in Figure 9.33. It includes a splitter associated with the H1–C1 exchanger, a mixer associated with the H1-C2 exchanger, and a stream that connects a portion of the effluent from the former to the latter. Stream 2 bypasses the H1–C1 exchanger and stream 5 bypasses the H1–C2 exchanger.

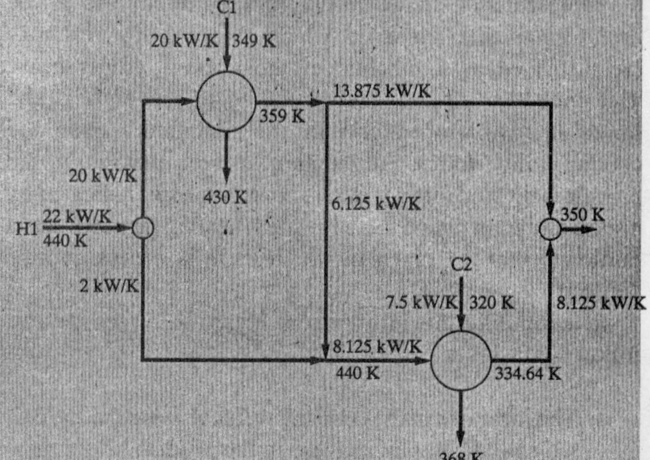

Figure 9.33 Optimal HEN for Example 9.16 (Reproduced from Floudas, 1995, with permission).

See the Program and Simulation Files folder, which can be downloaded from the Wiley Web site associated with this book, for the GAMS input file, COST.1, which obtains a solution using the Chen approximation to the log-mean temperature difference.

Solutions of the most general NLPs are complicated by the existence of nonconvex terms in the bilinear equality constraints. These lead most solvers to locate local optima rather than the global optimum. As global optimizers are being developed that are guaranteed to provide global convergence, improved superstructures and solution algorithms are evolving (Floudas, 1995).

9.8 MULTIPLE UTILITIES

Thus far, multiple sources of hot and cold utilities have not been considered. For example, steam is normally available at several pressures, with its cost a function of the pressure (and temperature) level, as discussed in Section 23.1 (Table 23.1). To reduce the cost of utilities, as well as the lost work, this section shows how to construct and use the *grand composite curve* (GCC) to reduce the temperature driving force in the auxiliary heat exchangers.

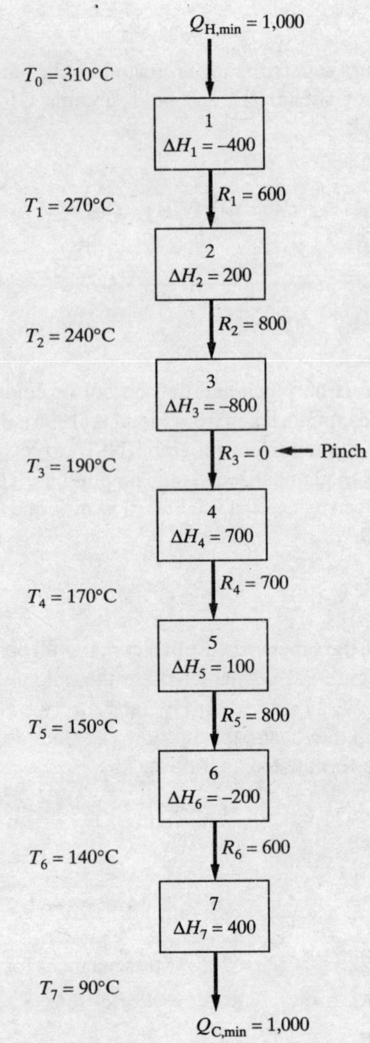

Figure 9.34 Temperature intervals, energy balances, and residuals; ΔH_i, R_i, Q_{Hmin}, and Q_{Cmin} in kilowatts.

Designing HENs Assisted by the Grand Composite Curve

As described in Section 9.2, temperature intervals are identified and residuals computed to estimate the minimum heating and cooling utilities and to locate the pinch temperatures. For example, in the cascade diagram of Figure 9.34, computed for $\Delta T_{\min} = 10°C$, the cold pinch temperature is 190°C, with the minimum hot and cold utility levels being 1,000 kW each. The residual enthalpies leaving each temperature interval indicate the excess heating or cooling capacity of the cascade above and including the temperature interval. This suggests the representation shown in Figure 9.35, referred to as the *grand composite curve,* in which the enthalpy residuals are displayed as a function of the interval temperatures. The enthalpy residuals corresponding to the highest and lowest interval temperatures are the minimum heating and cooling utility duties. Furthermore, as seen previously, the process is divided into a portion above the pinch (indicated by a solid circle) in which there is excess heating demand, and a portion below the pinch in which there is excess cooling demand.

In addition to this information, Figure 9.35 provides the following insights:

a. The local rise in the enthalpy residual along the arc BC, which indicates an increase in the residual heat available, can be used to supply the increased demand denoted by the decrease in the enthalpy residual along the arc CD. This is accomplished using internal heat exchange.

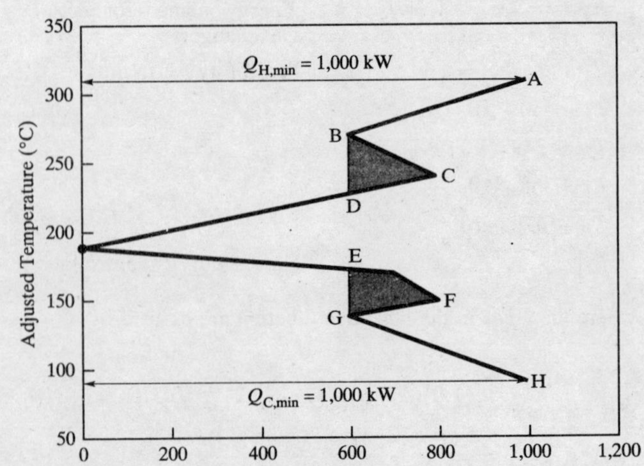

Figure 9.35 Grand composite curve for the temperature intervals in Figure 9.34.

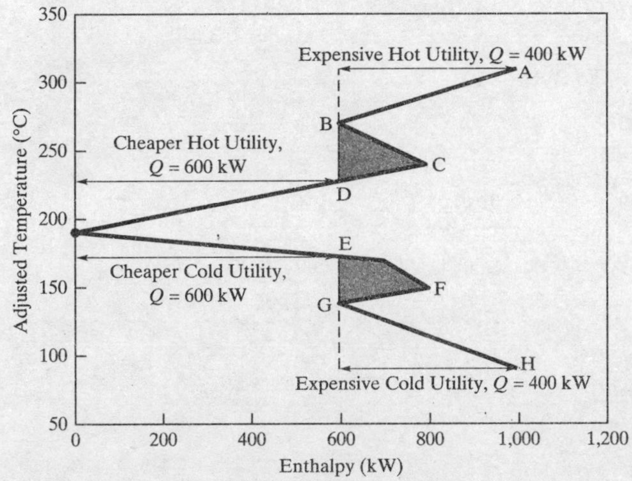

Figure 9.36 Grand composite curve for the temperature intervals in Figure 9.34, showing possible savings by utilizing HPS and BFW.

Similarly, the heat demand along the arc FG can be supplied by internal heat exchange with the arc EF.

b. When 1,000 kW are provided at 330°C, a fired-heater is required that burns fuel. Alternatively, the GCC replotted in Figure 9.36 shows that up to 600 kW can be provided at 230°C using less expensive high-pressure steam at 400 psi, noting that the alternative utility temperature level and duty are limited by the intercept D. This leaves only 400 kW to be supplied in the fired-heater. Similarly, instead of removing 1,000 kW at 90°C using cooling water, up to 600 kW can be removed at 170°C by heating boiler feed water, leading to further savings in operating costs. As before, the temperature level and duty are limited by intercept E. Evidently, significant savings in operating costs are revealed by the GCC.

The grand composite curve helps to better utilize utility resources when designing HENs, as illustrated in the following example.

EXAMPLE 9.17

Consider the design of a HEN for the four streams below, with $\Delta T_{\min} = 10°C$:

Stream	T^s (°C)	T^t (°C)	C(kW/°C)
H1	180	40	20
H2	160	40	40
C1	60	220	30
C2	30	180	22

To reduce operating costs, the design should consider alternative utility sources: high-pressure steam (hps) and intermediate-pressure steam (ips), boiler feed water (bfw), and cooling water.

SOLUTION

The TI method is used to construct the grand composite curve shown in Figure 9.37a, which indicates that MER targets are $Q_{\mathrm{H,min}} = 2,360$ kW and $Q_{\mathrm{H,min}} = 1,860$ kW, with pinch temperatures at 150 and 160°C. Since stream H2 does not appear on the hot side of the pinch, the minimum number of heat exchangers that meets MER targets, $N_{\mathrm{HX,min}}^{\mathrm{MER}} = N_{\mathrm{HX,min}}^{+} + N_{\mathrm{HX,min}}^{-} = 3+4=7$. A HEN designed for $N_{\mathrm{HX,min}}^{\mathrm{MER}} = 7$ is shown in Figure 9.38a, in which cooling water and hps are used. A simpler design, obtained by eliminating heat exchanger 3 and thus breaking one of the heat loops in Figure 9.38a, is shown in Figure 9.38b, noting that this involves 240 kW of additional heating and cooling utilities.

An alternative design is suggested in Figure 9.37b, in which a portion of the hot utility, 1,310 kW, is supplied as ips at 195°C (at an adjusted temperature of 185°C). Furthermore, a portion of the cold utility duty, 320 kW, is used to generate steam from boiler feed water (bfw) at 110°C. These substitutions lead to the more complex design in Figure 9.39, involving three pinches: a process pinch at 150 and 160°C and utility pinches at 110 and 120°C, and

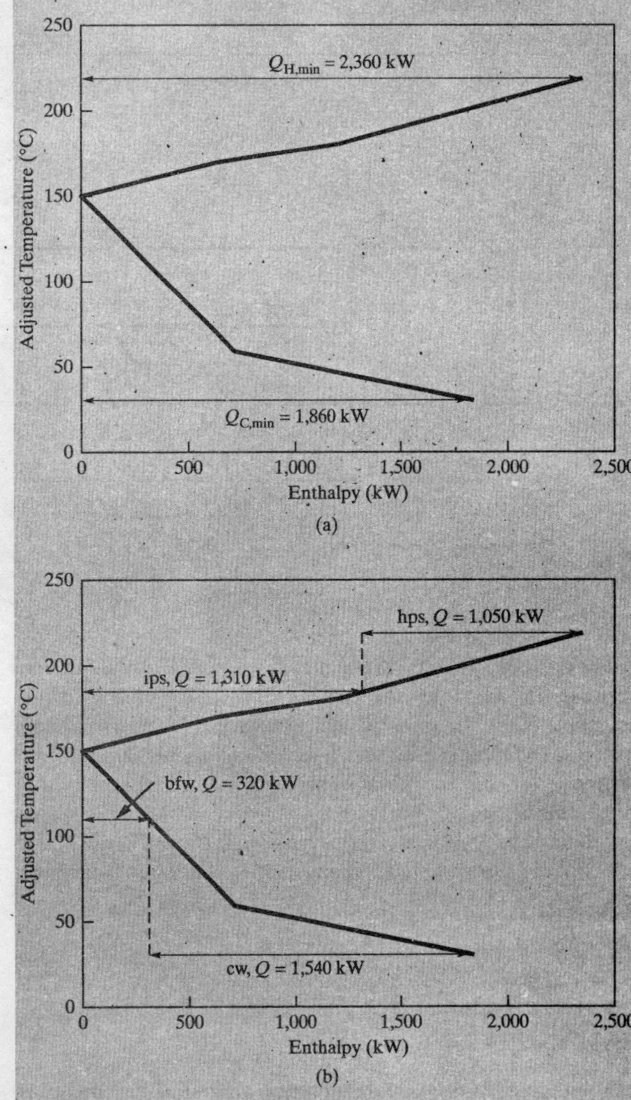

Figure 9.37 Grand composite curve for Example 9.17: (a) MER targets; (b) possible positioning of multiple utilities.

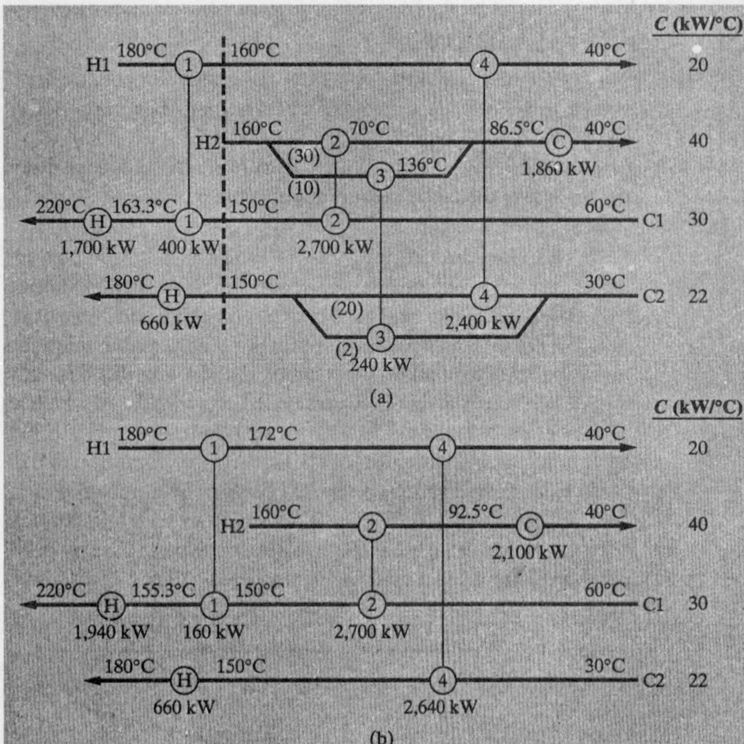

Figure 9.38 HENs for Example 9.17: (a) design to meet the $N_{HX,min}^{MER}$ target; (b) simplified design after breaking a heat loop.

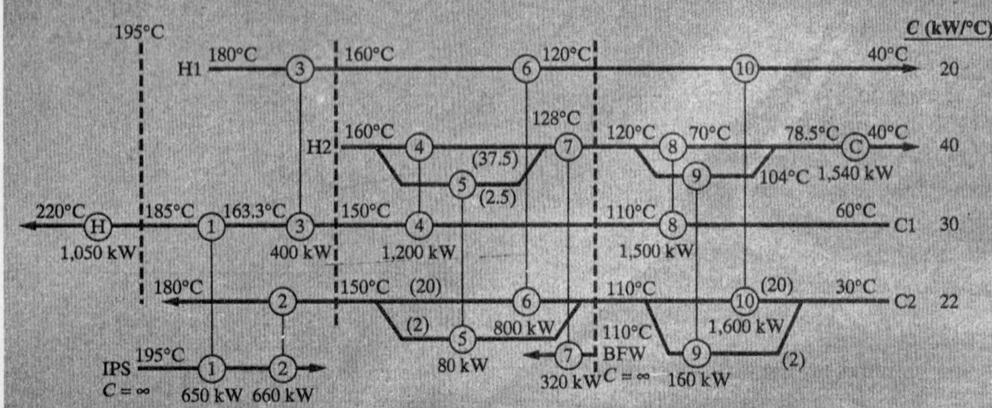

Figure 9.39 HEN for Example 9.17 utilizing cheaper utilities while meeting the $N_{HX,min}^{MER}$ target.

185 and 195°C. The utility pinches arise because of the infinite heat-capacity flow rates associated with the utility streams. Note that the $N_{HX,min}^{MER}$ target of 12 heat exchangers is met by stream splitting and careful matching to permit a feasible design. Furthermore, the number of heat exchangers can be reduced by: (a) combining heat exchangers 6 and 10; (b) eliminating exchanger 1, reducing the capital costs at the expense of shifting 650 kW of hot utility duty from ips to hps; (c) eliminating heat loops, usually at the expense of additional utilities (see Exercise 9.20).

9.9 HEAT-INTEGRATED DISTILLATION TRAINS

Although distillation is highly energy-intensive, having a low thermodynamic efficiency (less than 10% for a difficult separation, as shown in Example 9S.4), it continues to be widely used for the separation of organic chemicals in large-scale chemical processes. As discussed in Chapter 8, the designer normally seeks to utilize more effective separation processes, but in many cases has no choice but to resort to distillation because it is more economical, especially in the manufacture of commodity chemicals.

In the previous sections of this chapter, methods have been discussed for exchanging heat between high-temperature sources and low-temperature sinks. When distillation operations are present in a process flowsheet, it is particularly important to consider the heating requirements in the reboilers and the cooling requirements in the condensers as HENs are designed. Furthermore, over the past three decades, several approaches have been suggested for the energy-efficient incorporation of distillation columns into a process flowsheet. This section is intended to present some of these approaches.

Impact of Operating Pressure

As discussed in Section 8.4, the column pressure of a distillation column is a key design variable because it determines the temperature levels in the reboiler and condenser, and consequently, the possible heating and cooling media. Earlier, when synthesizing the vinyl-chloride process in Figure 4.8, it was noted that: "... heat is needed to drive the reboiler in the first distillation column at 93°C, but the heat of reaction (available from the direct chlorination of ethylene at 90°C) cannot be used for this purpose unless the temperature levels are adjusted." Furthermore, the question was raised: "How can this be accomplished?"

The principal vehicle for enabling these kinds of energy exchanges in most processes is through the adjustment of the pressure of the distillation towers, although in many cases it is possible to operate the reactor at a higher temperature. For the direct chlorination of ethylene in the liquid phase, it should be possible to increase the reactor temperature without increasing the rate of undesirable side reactions, and consequently, this alternative should be considered. The other alternative is to reduce the pressure of the distillation tower, thereby reducing the reboiler temperature and permitting the condensation of the dichloroethane product in the reboiler of the first tower. This reduces the usage of cooling water and steam, or eliminates them entirely, depending upon the cooling and heating demands. Also, as the distillation pressure is reduced, the separation is made easier and the number of stages is decreased. On the down side, however, the temperature of the condenser is reduced. If it is reduced below the temperature at which cooling water can be used, the cost of refrigeration becomes a significant cost factor (largely through an increased compression load). Furthermore, the integrated process is more difficult to control. In many cases, however, the combined savings in the utilities and the capital cost of the heat exchangers and the column exceeds the added refrigeration costs, and a reliable control system can be designed for the integrated process, as discussed in Chapter 12.

When working with the composite heating and cooling curves for a process, it helps to examine the heating and cooling requirements for a distillation column using a T–Q diagram, as shown in Figure 9.40. This diagram shows the heat provided to the reboiler, Q_{reb}, at temperature T_{reb}. Similarly, the diagram shows the heat removed from the condenser, Q_{cond}, at temperature T_{cond}. The principal assumption in Figure 9.40 is that $Q_{reb} \approx Q_{cond}$. At first, this assumption may appear unjustified, but is reasonable for most distillation towers, especially when the feed and product streams are saturated liquids. To justify this, for the distillation tower in Figure 9.41, the energy balance in the steady state is

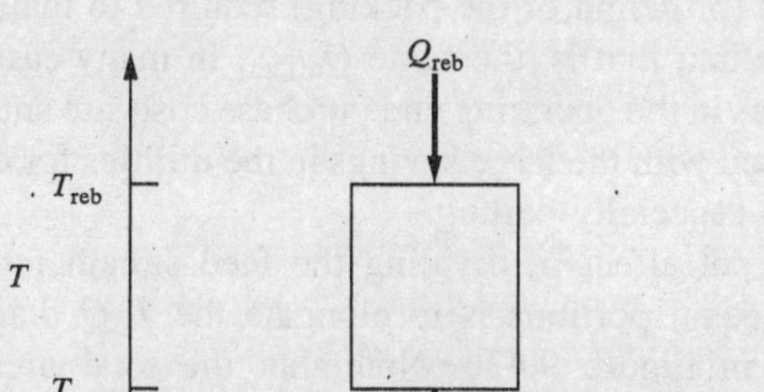

Figure 9.40 T–Q diagram for a distillation column.

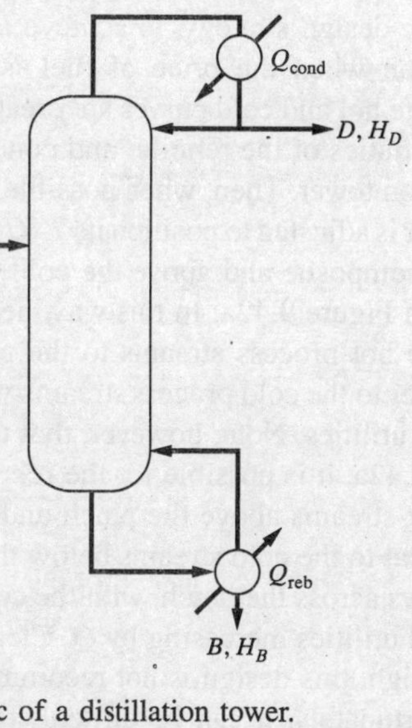

Figure 9.41 Schematic of a distillation tower.

$$FH_F - DH_D - BH_B + Q_{reb} - Q_{cond} = 0 \qquad (9.19)$$

Then, for saturated liquids, where $T_{cond} < T_F < T_{reb}$, a reasonable approximation is:

$$FH_F - DH_D - BH_B \approx 0 \qquad (9.20)$$

and consequently,

$$Q_{reb} \approx Q_{cond} \qquad (9.21)$$

As shown in the multimedia modules, which can be downloaded from the Wiley Web site associated with this textbook, the approximations in Eqs. (9.20) and (9.21) apply for the separation of propane from n-butane in a mixture of five normal paraffins, from ethane to n-hexane. Using both the RADFRAC subroutine in ASPEN PLUS (*SEPARATIONS → Distillation → MESH Equations → RADFRAC*) and the **Column** object in HYSYS (*SEPARATIONS → Distillation → Column Setup*), the approximation in Eq. (9.21) applies. Also, as shown in Dhole and Linnhoff (1993), the grand composite curve provides helpful insights when the column feeds are not saturated liquids.

www.wiley.com/go/global/seider

Returning to Figure 9.40, it is important to note that the heat duty, $Q \approx Q_{reb} \approx Q_{cond}$, is related directly to the reflux ratio. When Q is reduced, because the cost of fuel is high, the number of trays (or the height of the packing) increases. Clearly, the tradeoffs between operating and capital costs significantly influence the optimal design.

In one design strategy, to achieve a high degree of heat integration when the price of fuel is relatively high, the composite hot and cold curves are created without including the heat duties of the reboiler and condenser for a potential distillation tower. Then, when possible, the pressure level of the tower is adjusted to position its T–Q rectangle to lie below the hot composite and above the cold composite curves, as shown in Figure 9.42a. In this way, heat can be transmitted from the hot process streams to the reboiler and from the condenser to the cold process streams without increasing the external utilities. Note, however, that unlike the example in Figure 9.42a, it is possible for the reboiler to accept energy from hot streams above the pinch and for the condenser to reject heat to the cold streams below the pinch. In this case, heat flows across the pinch, with the consumption of the hot and cold utilities increasing by Q. Clearly, when the cost of fuel is high, this design is not recommended.

Unfortunately, it can be difficult to position a tower as shown in Figure 9.42a unless the chemical species being separated are close-boiling (e.g., the split between propylene and propane in Example 9S.4). Alternatively, the pressure of the tower can be adjusted and the tower positioned so that the reboiler receives its energy from a hot utility and the condenser rejects its energy to the cold process streams above the pinch, as shown in Figure 9.42b.

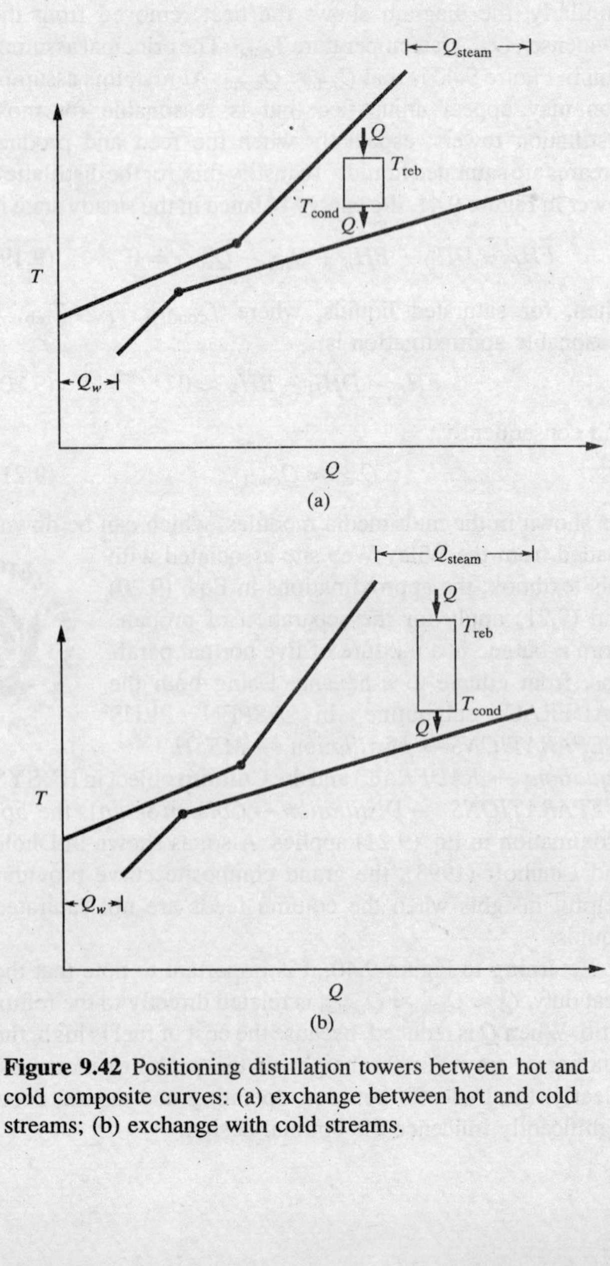

Figure 9.42 Positioning distillation towers between hot and cold composite curves: (a) exchange between hot and cold streams; (b) exchange with cold streams.

This reduces the load of the hot utilities required by the HEN for the remainder of the plant, while satisfying the heating demand of the reboiler. When the cost of fuel is high, this provides an attractive design.

Multiple-Effect Distillation

For separations where the T–Q rectangle for a distillation tower cannot be positioned under the hot composite and above the cold composite curves, as in Figure 9.42b, several possibilities exist for creating a more energy-efficient distillation operation when the price of fuel is high. One widely used configuration for distilling water is *multiple-effect distillation,* in which the feed stream is split into approximately equal portions and sent to the same number of separate distillation towers, each operating at a different pressure, as illustrated in Figure 9.43a for a cascade of two effects. The pressures in the towers decrease from the bottom to the top of the cascade so that the temperatures of the adjacent condensing vapors and boiling liquids differ by ΔT_{min}. In this way, heat from the condensing vapor in the tower below, at a pressure of P_2, is transferred to boil the bottoms liquid from the tower above, at the lower pressure of P_1. Note that the flow rates of the feed streams are adjusted to equate the duties of the adjacent condensing and boiling streams. In this way, the heat duties associated with condensation and boiling, Q_{effect}, are approximately equal to the heat duty for a single effect, Q, divided by the number of effects, N_{effect}; that is, $Q_{effect} \approx Q/N_{effect}$. When the price of fuel is high, this represents a substantial reduction in the operating costs. On the down side, however, the pressure level is increased in $N_{effect} - 1$ of the distillation towers, and multiple towers are needed. While pumping costs to increase the pressure of liquid feed streams are low, the tower walls are thicker, increasing the purchase costs; and the relative volatility is decreased at higher pressures, increasing the number of trays (or height of the packing) required to maintain the same reflux; that is, the same Q_{effect}. In many cases, these increases in the operating and purchase costs are small when compared with the large savings in the utilities for condensing and especially boiling.

The net effect of dividing the feed stream into N_{effect} nearly equal portions is to elongate the T–Q diagram, as shown in Figure 9.43b. Note that the total area is not conserved because the pressure level in each effect determines the bubble-point temperatures of its condensing vapor and boiling liquid streams.

Two variations on multiple-effect distillation do not involve *feed-splitting* (FS), as described above and shown in Figure 9.44a. For these configurations, as shown in Figures

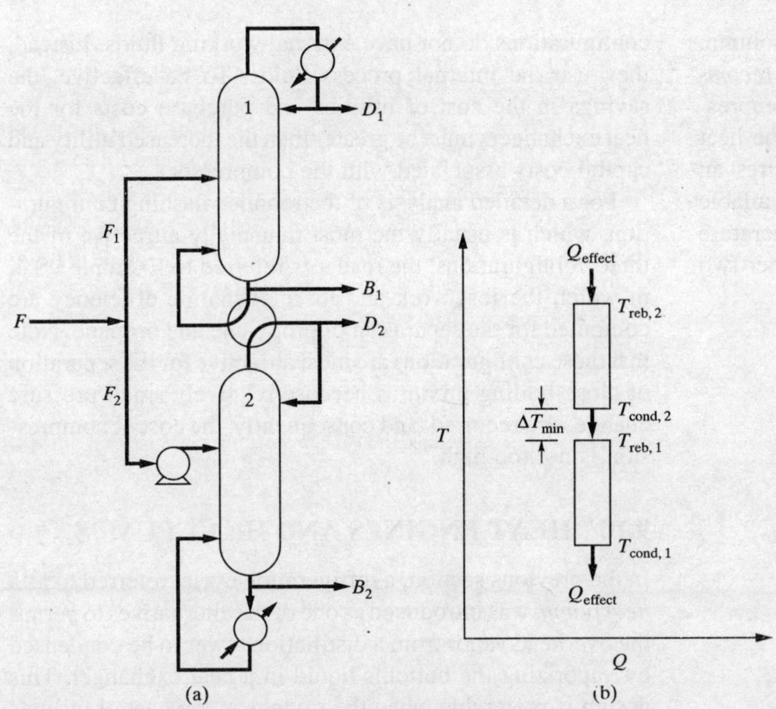

Figure 9.43 Two-effect distillation: (a) tower and heat exchanger configuration; (b) T–Q diagram.

9.44b and 9.44c, the entire feed stream is sent to the first tower. In the *light split/forward heat-integration* (LSF) configuration (Figure 9.44b), the feed is pumped and sent to the high-pressure column. About half of the light key component is removed in the distillate at high purity. The bottoms product, which contains the remainder of the light key component, is fed to the low-pressure column. In this case, the heat integration is in the direction of the mass flow. For the other variation, Figure 9.44c, referred to as the *light split/reverse heat-integration* (LSR) configuration, the feed is sent to the low-pressure column. Here, also, about half of the light key component is removed in the distillate, with the bottoms product pumped and sent to the high-pressure column. In this case, the heat integration is in the reverse direction of the mass flow. Note that these configurations are compared among themselves and with a single column in Example 12S.8, where the dehydration of methanol is examined. First, the configurations are compared when operation is in the steady state. Then, the controllability and resiliency (C&R) of each configuration is assessed in response to typical disturbances, and verified by dynamic simulations of the FS and LSR configurations using HYSYS, confirming the findings of the C&R analysis.

Heat Pumping, Vapor Recompression, and Reboiler Flashing

Other, more sophisticated configurations, designed to increase the thermodynamic efficiency when the price of fuel is high, permit the vapor overhead to be condensed

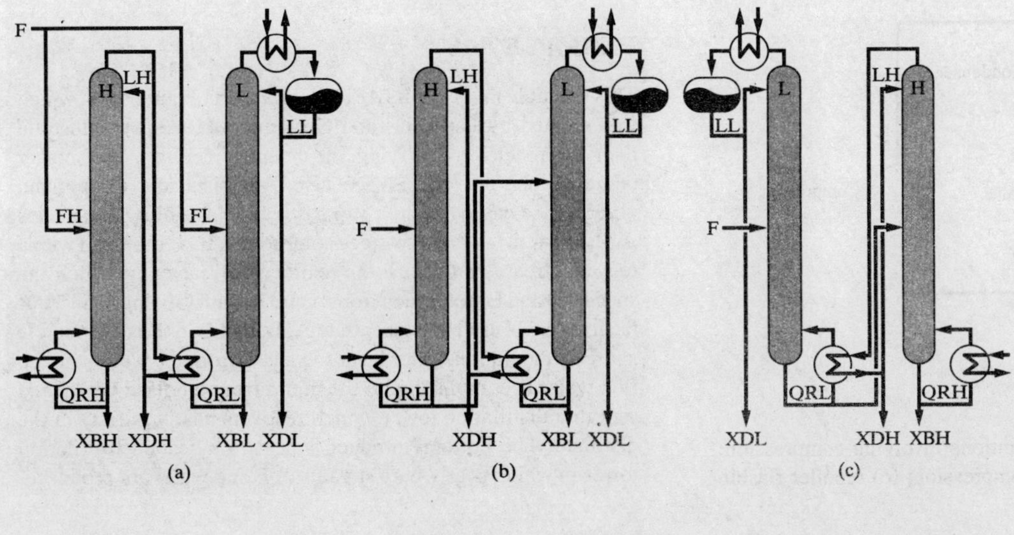

Figure 9.44 Variations on two-effect distillation: (a) FS; (b) LSF; (c) LSR.

with the bottoms liquid from the *same* distillation column. Each of three configurations, *heat pumping, vapor recompression,* and *reboiler flashing,* involves expensive compression of a vapor stream, as shown in Figure 9.45. The heat pump operates like a refrigeration cycle and requires an external fluid as the working medium. It pumps available heat from a low-temperature level up to a higher temperature level where it can be used more effectively. The other two configurations do not have external working fluids. Instead, they use the internal process fluids. To be effective, the savings in the cost of utilities and purchase costs for the heat exchangers must be greater than the increased utility and capital costs associated with the compressor.

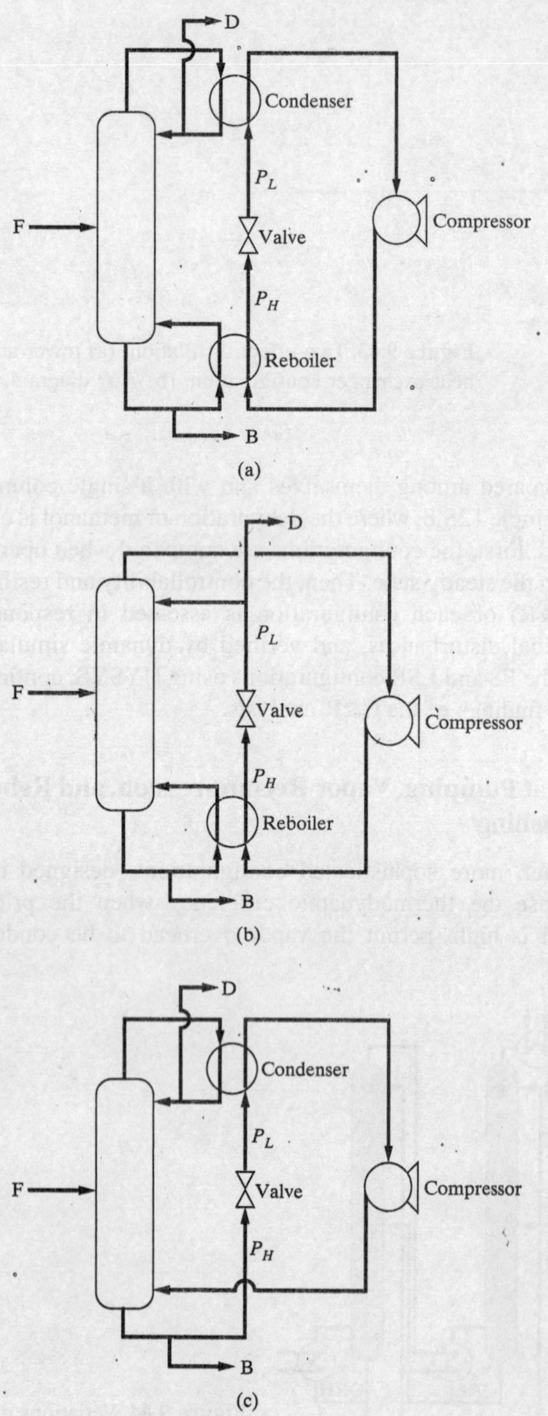

Figure 9.45 Distillation configurations involving compression: (a) heat pumping: (b) vapor recompression; (c) reboiler flashing.

For a detailed analysis of the reboiler-flashing configuration, which is usually the most financially attractive of the three configurations, the reader is referred to Example 9S.4, in which the lost work and thermodynamic efficiency are computed for the separation of propylene and propane. Note that these configurations are most attractive for the separation of close-boiling mixtures because relatively small pressure changes are required, and consequently, the cost of compression is not too high.

9.10 HEAT ENGINES AND HEAT PUMPS

In the previous section, a refrigeration cycle, referred to as a *heat pump,* was introduced as one of the alternatives to permit the overhead vapor from a distillation tower to be condensed by vaporizing the bottoms liquid in a heat exchanger. This design is preferable when the savings in the cost of utilities for condensation and boiling and the purchase cost of the heat exchangers are greater than the costs of operation and installation of a compressor. It illustrates the use of *heat and power integration* to achieve a more profitable design.

The objective of this section is to consider the general role of heat engines, which convert heat to shaft work, and heat pumps in satisfying the heating, cooling, and power demands within a chemical process. In other words, after the source and target temperatures of the streams to be heated and cooled have been established, together with the power requirements for the pumps and compressors, heat exchangers, heat engines, and heat pumps are inserted to satisfy these demands in a profitable manner. The section begins with two examples that show typical processes for which these demands need to be satisfied. Then, some of the important considerations in positioning heat engines and heat pumps are considered.

EXAMPLE 9.18

This example involves the *ABCDE process* in Figure 9.46, which was created by Papoulias and Grossmann (1983c) to illustrate their approach to satisfying the heating, cooling, and power demands using a MILP. See also Papoulias and Grossmann (1983a). As can be seen, a vapor stream containing the species A, B, and C is fed to a two-stage compressor. It is combined with a recycle stream and sent to an exothermic reactor in which the species D and E are formed from A and B, with C being an inert. A flash vessel is used to concentrate A, B, and C in the vapor and D and E in the liquid. Water scrubs D and E from the vapor and the rich water is combined with the liquid from the flash vessel and sent to a distillation tower, which recovers nearly pure D in the distillate. The bottoms product is sent to a second distillation tower in which nearly pure streams of E and water are produced.

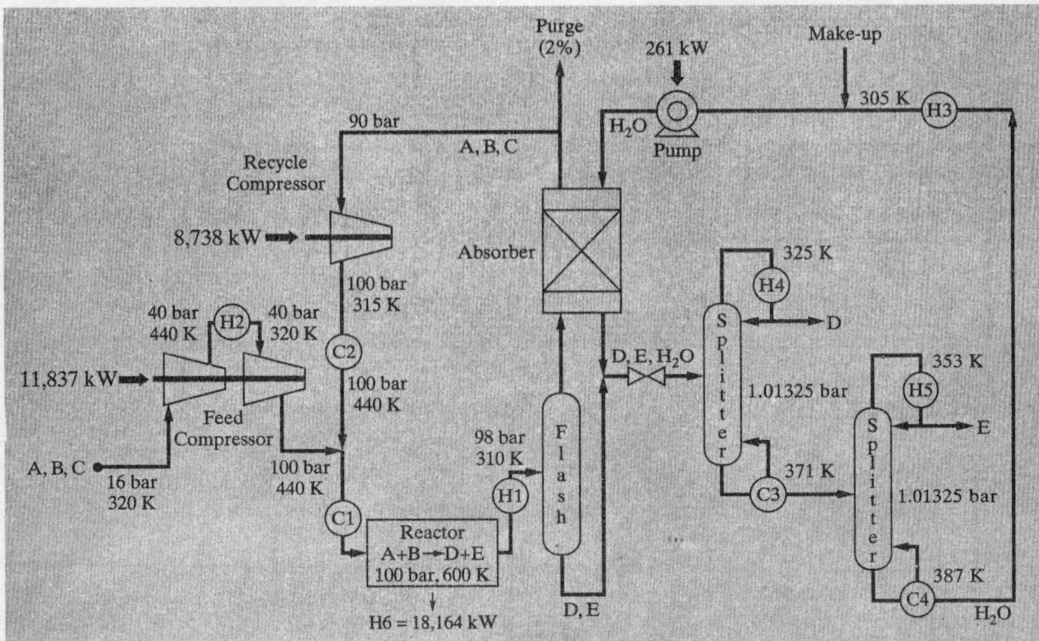

Figure 9.46 ABCDE process.

The water is combined with a make-up stream, pumped, and recycled to the absorber. Inert species C is removed in a small purge stream from the lean vapor from the absorber. The remainder is compressed and recycled. Note that the conditions shown in Figure 9.46 are typical of those produced by a process simulator during the fourth step in process synthesis in which the heating and cooling requirements are established, as well as the power demands for pumping and compression. The latter are annotated on the flowsheet and the former are as follows. Note that values of C (heat-capacity flow rates) for streams undergoing an isothermal phase change or an isothermal reaction are listed as infinite in the table.

Stream	T^s(K)	T^t(K)	C(kW/K)	Q(kW)
H1	600	310	901.0	261,300
H2	440	320	49.36	5,900
H3	387	310	36.44	2,800
H4	325	325	∞	65,300
H5	353	353	∞	171,000
H6	600	600	∞	18,200
C1	440	600	889.6	142,300
C2	315	440	840.5	105,100
C3	371	371	∞	119,200
C4	387	387	∞	90,800

SOLUTION

As can be seen, the heat recovered from the reactor, shown as stream H6, and the heat to be removed from the reactor effluent stream, H1, contain considerable energy at elevated temperatures. In synthesizing a profitable design, an important question concerns whether it is advantageous to utilize a heat engine(s) to generate the power required by the compressors and pump. This question will be addressed after several principles are established for the proper placement of heat engines.

EXAMPLE 9.19

This example involves the ethylene plant in Figure 9.47, which was synthesized by Lincoff (1983) and used by Colmenares and Seider (1989b) to illustrate their method for designing cascade refrigeration systems. The feedstock to the process is a pyrolysis gas containing a mixture of water, hydrogen, methane, ethane, ethylene, propane, propylene, butadiene, butylenes, and steam-cracked naphtha (SCN). This is the quenched, gaseous product of a pyrolysis reactor, in which a mixture of paraffins and steam is cracked at 1,100 K. The pyrolysis gas enters a five-stage compression train at 333 K and 136.5 bar in which it is compressed to 350 bar. After each compression stage, the gas is cooled, condensed water is removed, and a vapor–liquid mixture is separated in a flash vessel. The liquid streams from the flash vessels are fed to a condensate splitter whose overhead vapors are recycled to the fourth stage of compression and the bottoms are sent to a depropanizer. The vapor effluent from the fifth flash vessel is dried in a bed of zeolite molecular sieves and sent to a separation train, where low-temperature refrigeration is required to separate the light products (ethylene, propylene, etc.).

In the separation train, the gas stream is partially liquefied before entering the demethanizer at 320 bar. The overhead vapor, containing methane and hydrogen, is sent to a membrane separator in which these products are separated. The pressure of the bottoms product is reduced to 270 bar and fed to the deethanizer. In this column, the ethylene and ethane are removed in the distillate, whose pressure is reduced to 160 bar before the species are separated in the C-2 splitter. The bottoms product from the deethanizer, containing propylene, propane, and the heavier species, is throttled to 190 bar, mixed with the bottoms product from the condensate splitter, and fed to the depropanizer. The overhead product of the depropanizer is a mixture of propane and propylene and the bottoms product is throttled to 50 bar and sent to the debutanizer. In this column, the butylenes and butadiene are separated from the SCN.

The conditions shown in Figure 9.47 are typical of those produced by a process simulator during the fourth step in process synthesis, in which the heating and cooling requirements are

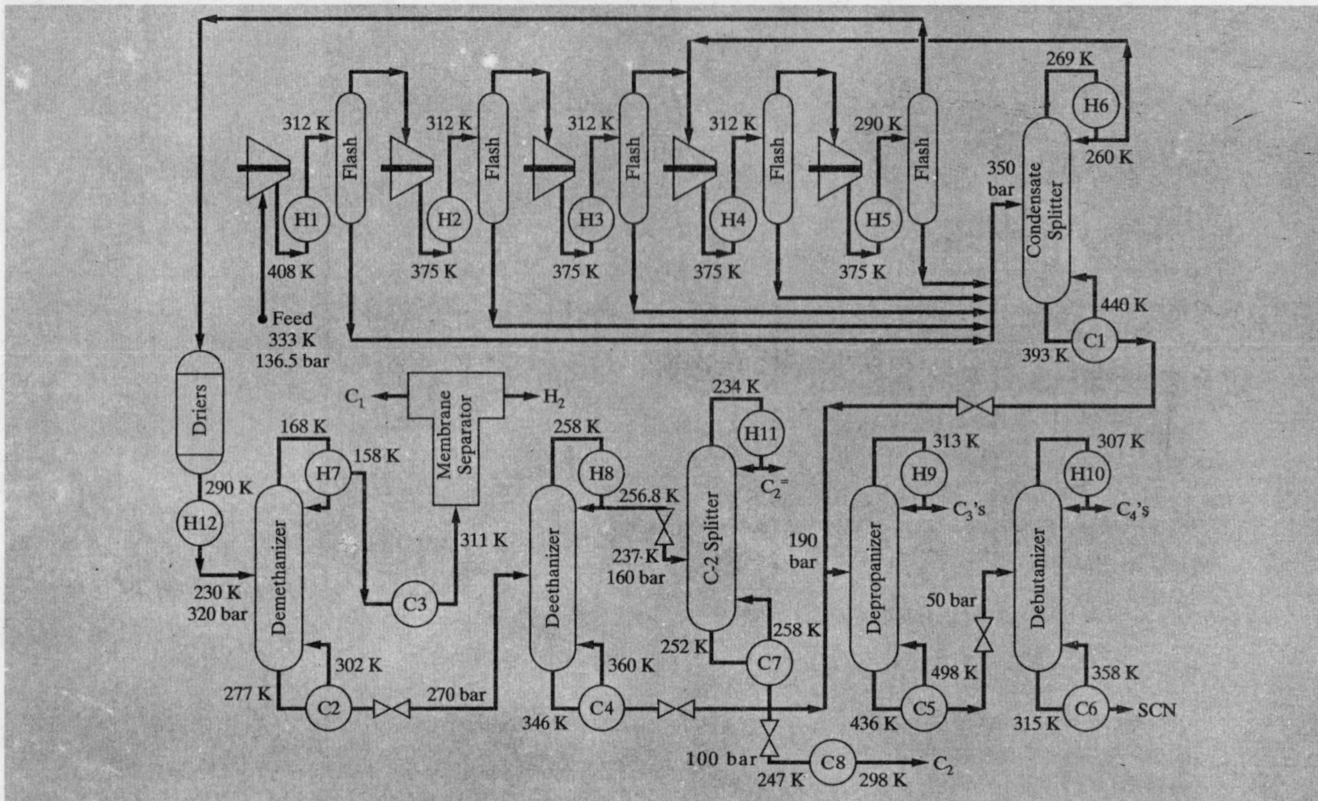

Figure 9.47 Ethylene process.

established, as well as the power demands for the five-stage compression train. The former are tabulated below; the latter are omitted because they do not affect the positioning of the heat pumps. Values of C for streams H9, H10, and H11 are infinite because of isothermal phase change.

Stream	T^s (K)	T^t (K)	C(kW/K)	Q (kW)
H1	408	312	12.35	1,186
H2	375	312	7.397	466
H3	375	312	6.143	387
H4	375	312	6.032	380
H5	375	290	6.729	572
H6	269	260	2.222	20
H7	168	158	15.70	157
H8	258	256.8	317.5	381
H9	313	313	∞	224
H10	307	307	∞	141
H11	234	234	∞	1,081
H12	290	230	7.517	451
C1	393	440	2.362	111
C2	277	302	6.960	174
C3	158	311	1.360	208
C4	346	360	36.86	516
C5	436	498	7.226	448
C6	315	358	2.956	133
C7	252	256	280.0	1.120
C8	247	298	1.882	96

SOLUTION

As can be seen, refrigeration is needed to condense the overhead vapor streams at low temperatures, with the lowest temperatures in the condenser of the demethanizer. In synthesizing a profitable design, an important question concerns how to position the heat pumps (refrigeration cycles) to satisfy the heating and cooling demands. This question will be addressed after several principles are established for the proper placement of heat pumps.

Positioning Heat Engines and Heat Pumps

When processes have significant power demands, usually in compressor loads, it is normally sound practice to operate at or near the minimum utilities for heating and cooling. This is because the annualized cost is dominated often by the operating and capital costs associated with satisfying these power demands. Given the desirability of operating these processes at minimum utilities, Townsend and Linnhoff (1983a, b) make recommendations concerning the positioning of heat engines and heat pumps relative to the pinch, discussed next.

Figure 9.48 shows the temperature intervals for the streams to be heated and cooled in a chemical process, separated into two sections, above (a) and below (c) the pinch temperatures, T_p. Consider the three alternatives for positioning a typical heat engine, as shown in Figure 9.49. The latter is a closed cycle in which condensate, at T_1 and P_c, is pumped to T_2 and P_b, and sent to a boiler, where it leaves as a superheated vapor

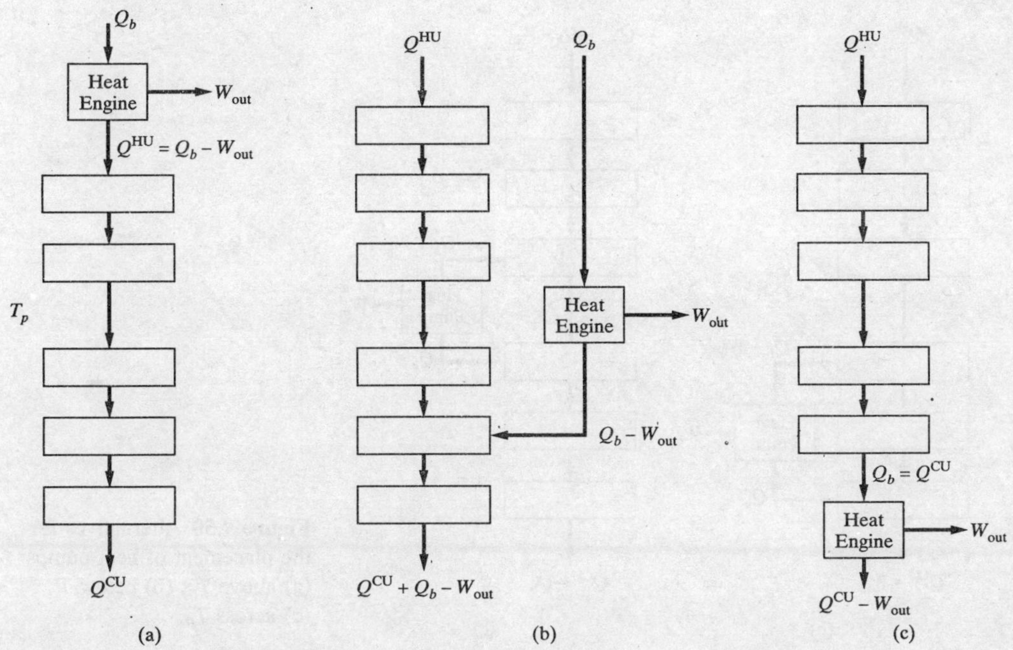

Figure 9.48 Alternatives for the placement of heat engines: (a) above T_p; (b) across T_p; (c) below T_p.

at T_3 and P_b. The boiler effluent is expanded across a turbine to T_4 and P_c, before it is condensed. The net heat energy, $Q_b - Q_c$, is converted to the net power, $W_{out} - W_{in}$, typically at a thermodynamic efficiency (see Section 9S.7) of about 35%. Returning to Figure 9.48a, where the heat engine is positioned above the pinch, to satisfy the demand for hot utilities, Q^{HU}, Q_b is required by the boiler and the net power produced is W_{out}, neglecting the small power requirement of the pump. Hence, W_{out} is produced by adding the equivalent heat to Q^{HU}. In Figure 9.48c, where the heat engine is positioned below the pinch, the heat that would be rejected to cold utilities, Q^{CU}, is sent to the boiler. W_{out} is recovered from the turbine and the remainder is rejected to the cold utilities. The alternative to these two placements, in Figure 9.48b, has the heat engine accepting heat above the pinch and rejecting heat below the pinch. As shown, the total utilities above the pinch are incremented by Q_b, and below the pinch, the cold utilities are incremented by $Q_b - W_{out}$. Clearly, when the heat engine is positioned across the pinch, both the hot and cold utility loads are incremented. When the cost of fuel is

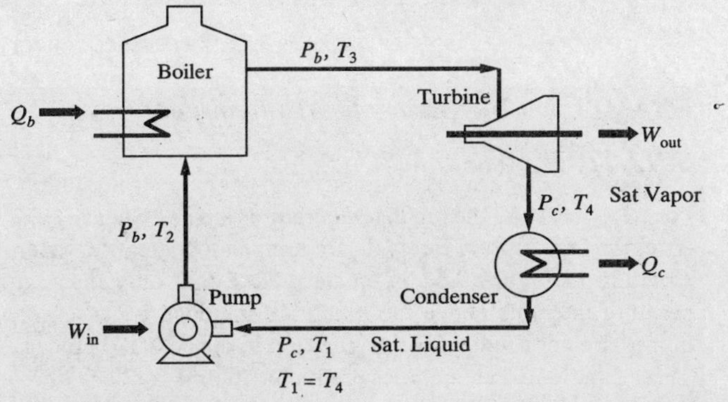

Figure 9.49 Heat engine.

high, this configuration is less profitable than the configurations with the heat engine entirely above or below the pinch.

Similarly, the three alternatives for positioning heat pumps relative to the pinch are shown in Figure 9.50. A typical heat pump is shown in Figure 9.51, where saturated vapor, at T_1 and P_b, is compressed to T_2 and P_c, and condensed, by rejecting its heat (often to the environment, but possibly to the boiler of another heat pump at lower temperature and pressure). The condenser effluent, at T_3 and P_c, is expanded across a valve to reduce its pressure and temperature to T_4 and P_b, while flashing some of the liquid. The remaining liquid is vaporized in the boiler. In Figure 9.50c, where the heat pump is positioned across the pinch, heat is removed from a temperature interval below the pinch and rejected to a temperature interval above the pinch, causing a reduction in both the hot and cold utility loads but at the expense of shaft work. Alternatively, when the heat pump is positioned above the pinch, as in Figure 9.50a, its compression load, W_{in}, reduces the hot utility load by this amount but does not reduce the cold utility load below the pinch. In this case, expensive power is converted directly to less valuable heat to reduce the hot utility load. Finally, when the heat pump is positioned below the pinch, as in Figure 9.50b, its compression load, W_{in}, increases the cold utility load by the same amount without affecting the hot utility load. Clearly, this is less desirable than when the heat pump is positioned across the pinch, where both the hot and cold utilities are decreased.

In summary, two heuristics result:

Townsend and Linnhoff Heuristics

1. ***When positioning heat engines, to reduce the total utilities, place them entirely*** **above** ***or*** **below** ***the pinch.***
2. ***When positioning heat pumps, to reduce the total utilities, place them*** **across** ***the pinch.***

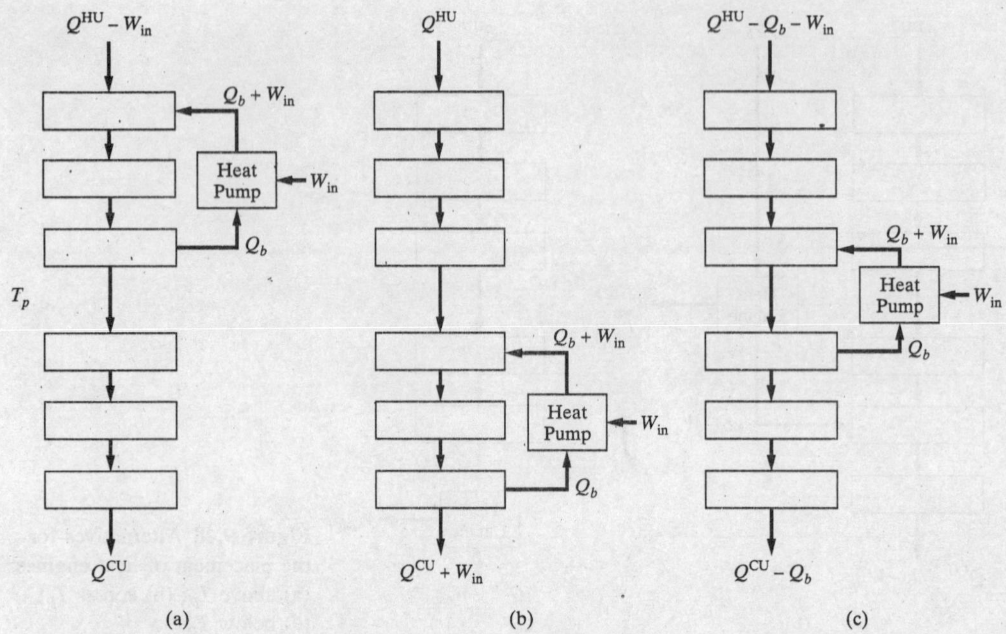

Figure 9.50 Alternatives for the placement of heat pumps: (a) above T_p; (b) below T_p; (c) across T_p.

Optimal Design

Following the heuristics of Townsend and Linnhoff, optimal design strategies have been developed (Colmenares and Seider, 1989a, b). These involve the following steps:

1. Carry out the temperature-interval method to locate the pinch temperatures and the minimum hot and cold utility loads.
2. Lump the temperature intervals together above the pinch to create intervals having heat deficits; that is, intervals in which more heat is required to heat the cold streams than is available from the hot streams to be cooled. Similarly, lump the temperature intervals together below the pinch to create intervals having negative heat deficits; that is, intervals in which more heat must be removed from the hot streams than can be consumed by the cold streams.
3. Create a superstructure that includes the candidate heat engines and heat pumps. These can add heat to the lumped intervals above the pinch and remove heat from the lumped intervals below the pinch.
4. Formulate an NLP to minimize the total annualized cost of the heat engines and heat pumps. The design variables include the pressure levels and the flow rates of the working fluids in the heat engines and the heat pumps. The constraints include the heat balances for the condensers and boilers, the heat balances for the temperature intervals, an energy balance to satisfy the power demand, bounds on the heat removed or added to the temperature intervals, and bounds on the temperatures and pressures.
5. Solve the NLP using a solver such as MINOS within GAMS.

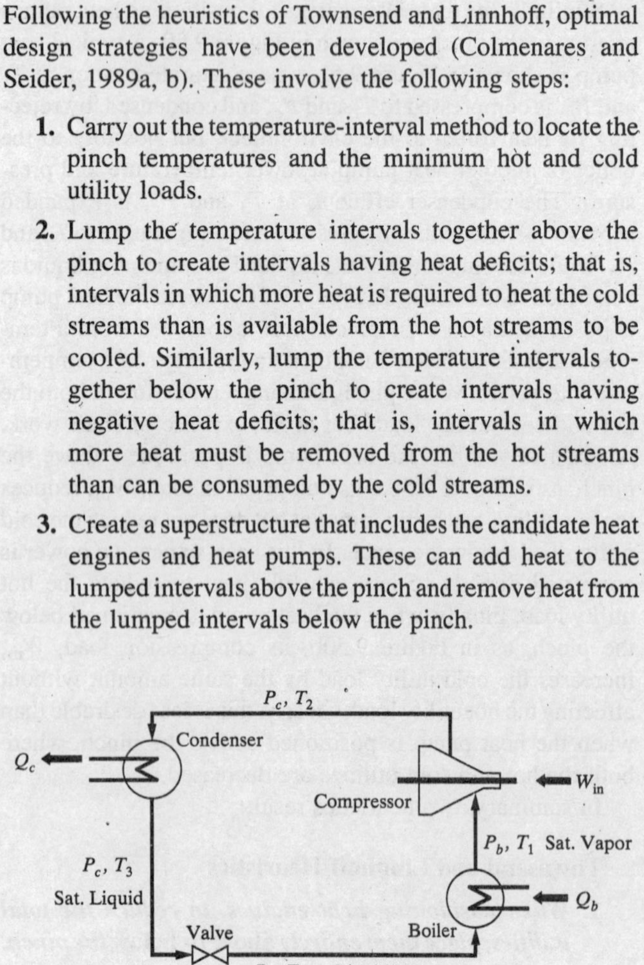

Figure 9.51 Heat pump.

It should be pointed out that appropriate placement of heat pumps and heat engines should also account for the utility pinches, and not just the process pinch as stated above. However, even the simplified approach of Colmenares and Seider cannot be described in detail in the limited space available in this section. Instead, the highlights are summarized in the solutions to Examples 9.18 and 9.19, which are completed below.

EXAMPLE 9.20 ***(Example 9.18 Revisited)***

SOLUTION ***(continued)***

For $\Delta T_{min} = 10$ K, the pinch temperatures are 381 K and 371 K. After the temperature intervals are lumped together, the superstructure in Figure 9.52 is created. Note that only the temperature intervals above the pinch are included because heat pumps are not needed for refrigeration in the ABCDE process. Three candidate heat engines are included, one for each of the temperature intervals having a large heat deficit, and one that provides power without providing heat to satisfy the heating requirements of the process streams. Also, three hot utilities are considered, provided by high-, medium-, and low-pressure steam. The lowest temperature interval is associated with the heating requirement in the reboiler of the first distillation column. Here, vaporization occurs at 371 K, and hence, $T^s = 371^- $ K and $T^t = 371^+$ K.

The annualized cost was minimized given the following data for the utilities:

Steam

	T (K)	P (bar)	\$/ton	ΔH (kJ/kg)	\$/kW
HPS	672.4	68.95	6.5	1,895.2	103.7
MPS	605.4	17.24	5.4	2,149.0	73.32
LPS	411.0	3.45	4.0	2,227.2	56.29

Cooling Water (cw)

$T^s = 300\,K$, $T^t = 322$ K, cost of cw = \$0.07/1,000 gal (= \$6.011/kW)

Fuel (f)

$\Delta H^c = 43.95$ kJ/kg, cost of f = \$143/ton (= \$109.3/kW)

At the minimum, the annualized cost of the heat engines and utilities is \$17,942,000/yr. Only Heat Engine 2 is utilized with $P_b = 53.6$ bar and $P_c = 1.26$ bar. The steam loads are $Q_1^{HU} = 4,092$ kW, $Q_2^{HU} = 4,804$ kW, and $Q_3^{HU} = 147,451$ kW. Note that the details of this solution, together with solutions for more general superstructures involving heat engines that exchange mass and energy, are presented by Colmenares and Seider (1989c).

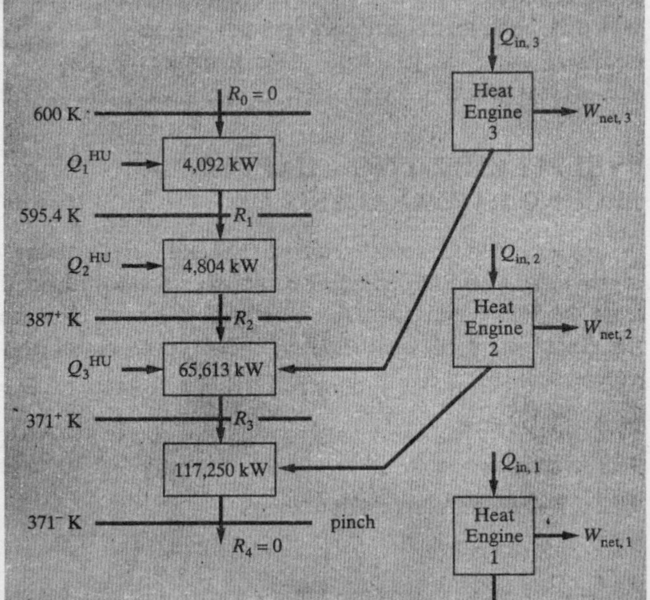

Figure 9.52 Superstructure for integration of ABCDE process with heat engines.

EXAMPLE 9.21 *(Example 9.19 Revisited)*

SOLUTION (continued)

For $\Delta T_{min} = 10$ K, the pinch temperatures are 408 K and 398 K. After the temperature intervals are lumped together, the superstructure in Figure 9.53 is created. Note that only the temperature intervals below the pinch are included, because the power demands for the five-stage compression system do not influence the design of the refrigeration system. As seen, the superstructure involves six potential heat pumps, each having a different working fluid, including commercial refrigerants R-13, R-22, and R-115. Ethane and ethylene are the candidates for removal of heat from the lowest temperature interval, with R-13 and propylene reserved for the intermediate temperature interval, and R-22 and R-115 reserved for the next highest temperature interval. Note that the highest temperature interval below the pinch can reject its heat to cooling water. It should also be noted that in the cascade structure, the heat pumps at the lower temperature levels can reject heat from their condensers to the boilers of heat pumps at the next higher temperature levels. The variables Q_{ij} denote the rate of heat transfer between heat pumps i and j. These variables, together with the pressure levels and the flow rates of the working fluids, are adjusted to minimize the annualized cost of the refrigeration system.

The annualized cost was minimized, including the installed cost of the compressors only, which was estimated using $C_{comp} = 1,925\, W_{in}^{0.963}$, where the power required is in kilowatts. The cost of electricity was estimated to be \$0.04/kWh, and the cost of cooling water was estimated to be \$0.07/1,000 gal. At the minimum, computed by MINOS and shown in Figure 9.54, three heat pumps, involving ethane, propylene, and R-22, are in a cascade. All of the heat from the temperature interval below the pinch is rejected to cooling water, as is the heat from the condenser of the R-22 heat pump. Note that no residual heat flows between the temperature intervals. For the complete details of this solution, showing how the temperature intervals are lumped and the NLP is formulated, see Colmenares and Seider (1989b).

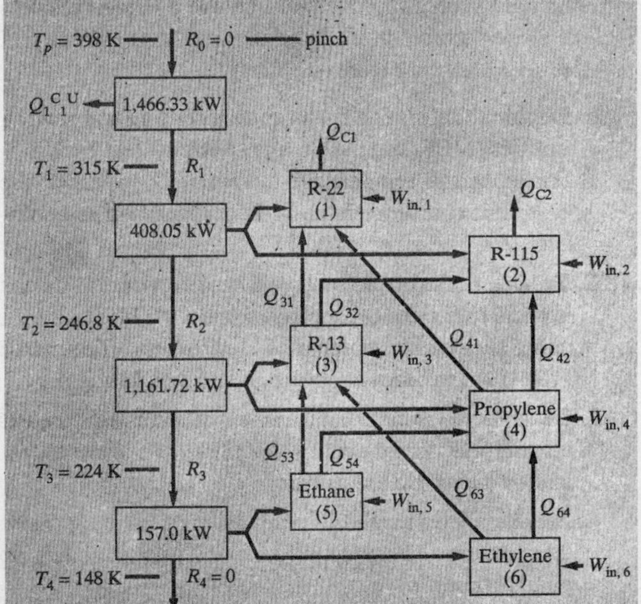

Figure 9.53 Superstructure for integration of the ethylene process with the cascade of heat pumps.

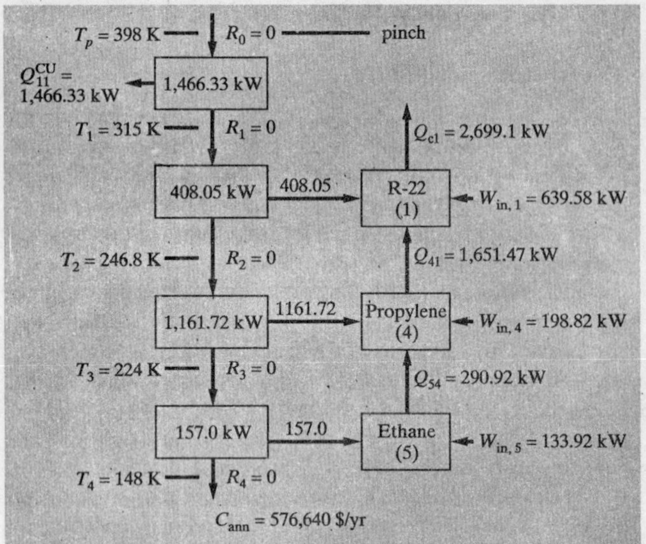

Figure 9.54 Cascade refrigeration system for the ethylene process at the minimum annualized cost.

9.11 SUMMARY

Having studied this chapter, and having solved many of the exercises, the reader should have learned how to achieve effective heat integration using several approaches. In one of the approaches, optimization problems are formulated and solved using GAMS, including linear programs (LPs), mixed-integer linear programs (MILPs), and nonlinear programs (NLPs). Although the chapter cannot provide a comprehensive treatment of the methods for achieving heat-integrated distillation trains and for satisfying power and refrigeration demands with heat engines and heat pumps, the introduction to these topics should enable a design team to include these features in its designs more systematically.

More specifically, the reader should

1. Be able to determine the minimum cooling and heating utilities (MER targets) for a network of heat exchangers using the temperature-interval (TI) method, the composite curve method, or the formulation and solution of a linear program (LP).
2. Be able to design networks of heat exchangers on the hot and cold sides of the pinch to meet the MER targets, using the heuristic method of Linnhoff and Hindmarsh (1983) or the transshipment model in a MILP.
3. Be able to reduce the number of heat exchangers toward the minimum by breaking the heat loops, and/or using stream splitting.
4. Be able to design a HEN when the minimum approach temperature difference is below ΔT_{thres}, at which no pinch occurs.
5. Understand the key role of the minimum approach temperature and appreciate the need to adjust it to achieve more optimal designs.
6. Be able to use the grand composite curve to consider the use of multiple hot and/or cold utilities and to find their optimal locations in the network.
7. Recognize the advantages and disadvantages of formulating superstructures for the design of HENs having the minimum annualized cost.
8. Be able to adjust the pressure in distillation columns to achieve heat integration and to consider the usage of multiple-effect distillation and compression to achieve designs that are more profitable.
9. Be able to insert turbines and heat engines—and compressors, refrigerators, and heat pumps—to satisfy both the heating and power demands for a process.

For additional coverage, based upon the pioneering research of Linhoff and coworkers, see Linhoff et al. (1994).

Heat-Integration Software

Most of the methods introduced in this chapter, especially those for the design of HENs, are implemented in commercial software. Of special note are ASPEN PINCH by Aspen Technology, Inc. (in the Aspen Engineering Suite), HX-NET by Hyprotech, HEXTRAN by Simulation Sciences, Inc., and TARGET by the Linnhoff–March Corp. As discussed in the chapter, many of the methods involve the solution of linear and nonlinear programs (LPs and NLPs), whose fundamentals are introduced in Chapter 24, and their mixed-integer counterparts (MILPs and MINLPs). While solutions using the General Algebraic Modeling System (GAMS) are presented herein and in the Program and Simulation Files folder, which can be downloaded from the Wiley Web site associated with this book, the former packages provide, in addition, excellent graphical user interfaces (GUIs) that simplify their usage.

www.wiley.com/go/global/seider

9S SUPPLEMENT TO CHAPTER 9—SECOND LAW ANALYSIS

A supplement to Chapter 9 entitled "Second Law Analysis" is provided in the PDF Files folder, which can be downloaded from the Wiley Web site associated with this book. See the file Supplement_to_Chapter 9.pdf. The contents for this supplement are:

9S.0 Objectives

9S.1 Introduction

9S.2 The System and Surroundings

9S.3 Energy Transfer

9S.4 Thermodynamic Properties

Typical Entropy Changes

Thermodynamic Availability

Typical Availability Changes

REFERENCES

1. Colmenares, T.R., and W.D. Seider, "Synthesis of Utility Systems Integrated with Chemical Processes," *I&EC Res.*, **28**, 84 (1989a).

2. Colmenares, T.R., and W.D. Seider, "Synthesis of Cascade Refrigeration Systems Integrated with Chemical Processes," *Comput. Chem. Eng.*, **13** (3), 247–258 (1989b).

3. Colmenares, T.R., and W.D. Seider, "Heat and Power Integration of Chemical Processes," *AIChE J.*, **33**, 898 (1989c).

4. Dhole, V.R., and B. Linnhoff, "Distillation Column Targets," *Comput. Chem. Eng.*, **17** (5–6), 549 (1993).

5. Douglas, J.M. *Conceptual Design of Chemical Processes*, McGraw-Hill, New York (1988).

6. Floudas, C.A., *Nonlinear and Mixed-integer Optimization: Fundamentals and Applications*, Oxford University Press, Oxford (1995).

7. Hohmann, E.C., *Optimum Networks for Heat Exchange*, Ph.D. dissertation, University of Southern California, Los Angeles (1971).

8. Lewin, D.R., "A Generalized Method for HEN Synthesis Using Stochastic Optimization: (II) The Synthesis of Cost-Optimal Networks," *Comput. Chem. Eng.*, **22** (10), 1387–1405 (1998).

9. Lewin, D.R., H. Wang, and O. Shalev, "A Generalized Method for HEN Synthesis Using Stochastic Optimization: (I) General Framework and MER Optimal Synthesis," *Comput. Chem. Eng.*, **22** (10), 1503–1513 (1998).

10. Lincoff, A.M., *Separation System for Recovery of Ethylene and Light Products from a Naphtha-pyrolysis Gas Stream*, Process Design Case Study, CACHE Corp., Austin, Texas (1983).

11. Linnhoff, B., and J.R. Flower, "Synthesis of Heat Exchanger Networks: I. Systematic Generation of Energy Optimal Networks," *AIChE* J., **24**, 633 (1978a).

12. Linnhoff, B., and J.R. Flower, "Synthesis of Heat Exchanger Networks: II. Evolutionary Generation of Networks with Various Criteria of Optimality," *AIChE J.*, **24**, 642 (1978b).

13. Linnhoff, B., and E. Hindmarsh, "The Pinch Design Method for Heat Exchanger Networks," *Chem. Eng. Sci.*, **38**, 745 (1983).

14. Linnhoff, B., and J.A. Turner, "Heat Exchanger Network Design: New Insights Yield Big Savings," *Chem. Eng.*, **77**, 56, November (1981).

15. Linnhoff, B., D.W. Townsend, D. Boland, G.E. Hewitt, B.E.A. Thomas, A.R. Guy, and R.H. Marsland. *A User Guide on Process Integration for the Efficient Use of Energy*, Revised 1st ed., The Institution of Chemical Engineers (IChemE), Rugby, England (1994).

16. McCabe, W., and E. Thiele, "Graphical Design of Fractionating Towers," *Ind. Eng. Chem.*, **17**, 605 (1925).

17. Nishida, N., Y.A. Liu, and L. Lapidus, "Studies in Chemical Process Design and Synthesis: III. A Simple and Practical Approach to the Optimal Synthesis of Heat Exchanger Networks," *AIChE J.*, **23**, 77 (1977).

18. Papoulias, S., and I.E. Grossmann, "A Structural Optimization Approach in Process Synthesis—I: Utility Systems." *Comput. Chem. Eng.*, **7**, 695 (1983a).

19. Papoulias, S., and I.E. Grossmann, "A Structural Optimization Approach in Process Synthesis—II: Heat Recovery Networks," *Comput. Chem. Eng.*, **7**, 707 (1983b).

20. Papoulias, S., and I.E. Grossmann, "A Structural Optimization Approach in Process Synthesis—III: Total Processing Systems," *Comput. Chem. Eng.*, **7**, 723 (1983c).

21. Townsend, D.W., and B. Linnhoff, "Heat and Power Networks in Process Design. 1. Criteria for Placement of Heat Engines and Heat Pumps in Process Networks," *AIChE J.*, **29**, 742 (1983a).

22. Townsend, D.W., and B. Linnhoff, "Heat and Power Networks in Process Design. II. Design Procedure for Equipment Selection and Process Matching," *AIChE J.*, **29**, 748 (1983b).

23. Umeda, T., J. Itoh, and K. Shiroko, "Heat Exchange System Synthesis," *Chem. Eng. Prog.*, **74**, 70, July (1978).

24. Yee, T.F., and I.E. Grossmann, "Simultaneous Optimization Models for Heat Integration—II: Heat Exchanger Network Synthesis," *Comput. Chem. Eng.*, **10**, 1165 (1990).

EXERCISES

9.1 Four streams are to be cooled or heated:

Stream	T^s(°C)	T^t(°C)	C(kW/°C)
H1	180	60	3
H2	150	30	1
C1	30	135	2
C2	80	140	5

(a) For $\Delta T_{\min} = 10$°C, find the minimum heating and cooling utilities. What are the pinch temperatures?

(b) Design a heat exchanger network for MER both on the hot and cold sides of the pinch.

9.2 **(a)** For $\Delta T_{\min} = 10$°C, find the minimum utility requirements for a network of heat exchangers involving the following streams:

Stream	T^s(°C)	T^t(°C)	C(kW/°C)
C1	60	180	3
C2	30	105	2.6
H1	180	40	2
H2	150	40	4

(b) Repeat (a) for the following streams:

Stream	T^s(°C)	T^t(°C)	C(kW/°C)
C1	100	430	1.6
C2	180	350	3.27
C3	200	400	2.6
H1	440	150	2.8
H2	520	300	2.38
H3	390	150	3.36

(c) For (a) and (b), design HENs that require the minimum utilities.

9.3 To exchange heat between four streams with $\Delta T_{\min} = 20°C$, the HEN in Figure 9.55 is proposed. Determine if the network has the minimum utility requirements. If not, design a network with the minimum utility requirements. As an alternative, design a network with the minimum number of heat exchangers. Using the specifications in Example 9.7, which alternative is preferred?

9.4 Consider the network of heat exchangers in Figure 9.56:

(a) Determine $N_{HX,\min}$.

(b) Identify the heat loop.

(c) Show one way to break the heat loop using $\Delta T_{\min} = 10°F$. For the resulting network, prepare a revised diagram showing all temperatures and heat duties.

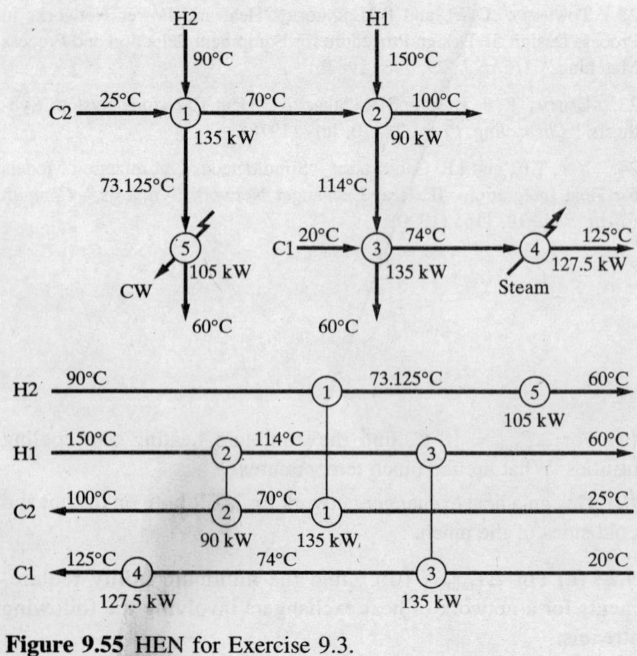

Figure 9.55 HEN for Exercise 9.3.

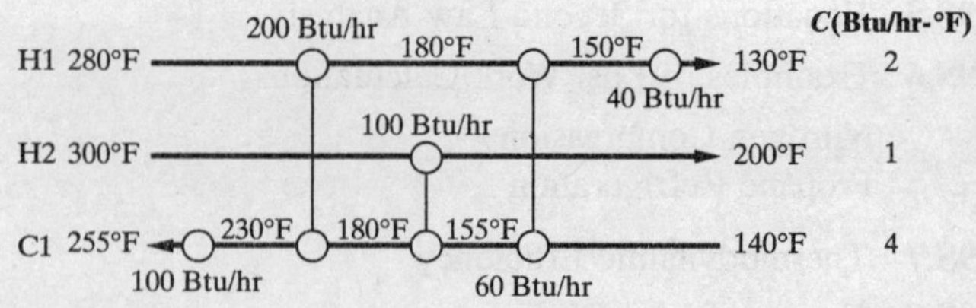

Figure 9.56 HEN for Exercise 9.4.

9.5 For the "pinch match" in Figure 9.57, show that to have a feasible match, that is $T_{h2} - T_{c1} \geq \Delta T_{\min}$, the heat-capacity flow rate of the streams must satisfy $C_h \geq C_c$.

9.6 Consider the design of a network of heat exchangers that requires the minimum utilities for heating and cooling. Is it true that a pinch temperature can occur *only* at the inlet temperature of a hot or cold stream? *Hint:* Sketch typical composite hot and cold curves for two hot and two cold streams.

9.7 Consider Test Case No. 2 by Linnhoff and Flower (1978a):

Stream	T^s(°C)	T^t(°C)	C(kW/°C)
C1	60	180	3
C2	30	130	2.6
H1	180	40	2
H2	150	40	4

(a) Use Figure 9.58 to find the minimum hot and cold utility loads when $\Delta T_{\min} = 10°C$ and 50°C.

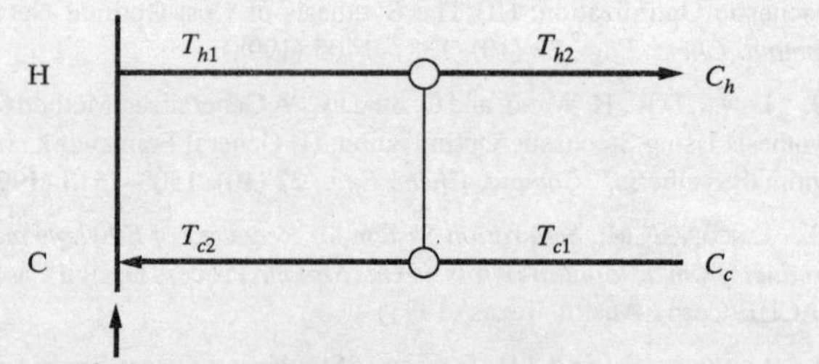

Figure 9.57 A heat exchanger positioned on the cold side of the pinch.

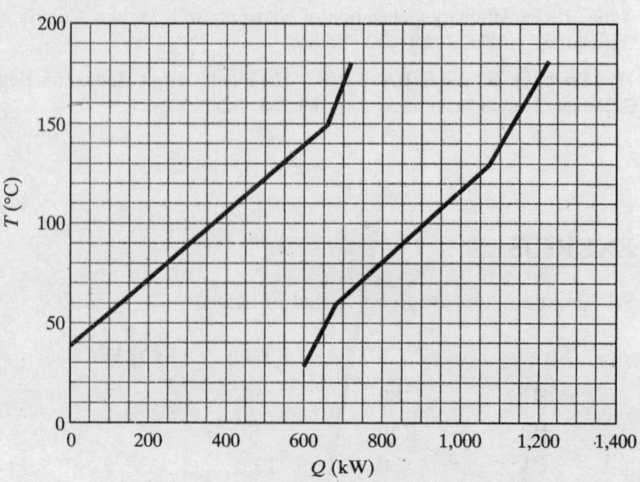

Figure 9.58 Composite hot and cold curves for Exercise 9.7.

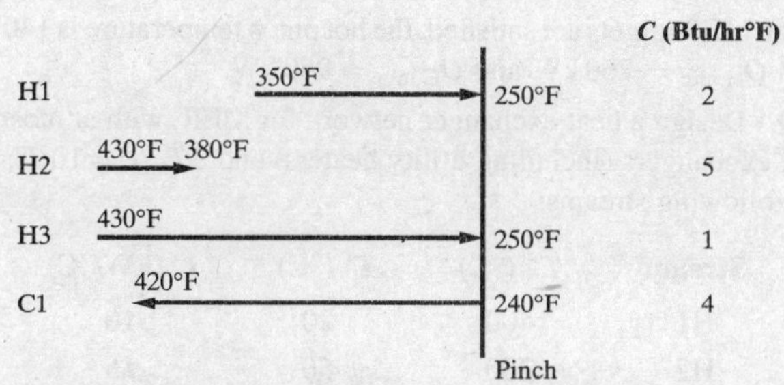

Figure 9.59 Streams for Exercise 9.10.

(b) For $\Delta T_{\min} = 50°C$, design a network of heat exchangers having minimum utilities.

9.8 In Example 9.6, a HEN is designed to meet the MER targets for the problem defined in Example 9.1.

(a) Determine the number of heat loops in the design in Figures 9.12 and 9.13.

(b) Systematically remove the loops, one at a time, and adjust the heat duties of the internal and auxiliary heat exchangers as needed.

(c) How does your design for part (b) compare with that in Figure 9.3?

9.9 Design a HEN for the streams in Exercise 9.4 that meets the MER targets with the minimum number of heat exchangers.

9.10 A process has streams to be heated and cooled above its pinch temperatures, as illustrated in Figure 9.59. Complete a design that satisfies the MER targets with the minimum number of heat exchangers.

9.11 Design a HEN to meet the MER targets for $\Delta T_{\min} = 10°C$ and $N_{HX,\min}$ for a process involving five hot streams and one cold stream, as introduced by Yee and Grossmann (1990):

Stream	T^s(K)	T^t(K)	C(kW/K)
H1	500	320	6
H2	480	380	4
H3	460	360	6
H4	380	360	20
H5	380	320	12
C1	290	670	18

9.12 Consider a process with the following streams:

Stream	T^s(°C)	T^t(°C)	C(kW/°C)
H1	500	50	5
H2	400	100	4
H3	400	100	3
H4	200	50	2
C1	250	450	10
C2	30	430	6

(a) Determine MER targets for $\Delta T_{\min} = 10°C$.

(b) Design a HEN for MER using no more than 10 heat exchangers (including auxiliary heaters and coolers).

(c) Add a stream to your HEN, without increasing the total number of exchangers. The data for the additional stream are

Stream	T^s (°C)	T^t (°C)	C(kW/°C)
C3	40	200	4

9.13 Consider a process with the following streams:

Stream	T^s(°C)	T^t(°C)	C(kW/°C)
H1	350	160	3.2
H2	400	100	3
H3	110	60	8
C1	50	250	4.5
C2	70	320	2
C3	100	300	3

When $\Delta T_{\min} = 10°C$, the minimum utilities for heating and cooling are 237 kW and 145 kW, respectively, with pinch temperatures at 110°C and 100°C. Design a HEN that satisfies the MER targets and has the minimum number of heat exchangers, $N_{HX,\min}^{MER}$. Show the heat duties and temperatures for each heat exchanger.

9.14 Consider the following heating and cooling demands:

Stream	T^s(°C)	T^t(°C)	C(kW/°C)
H1	525	300	2
H2	500	375	4
H3	475	300	3
C1	275	500	6

A HEN is to be designed with $\Delta T_{\min} = 30°C$:

(a) Find the MER targets.

(b) Design a subnetwork of heat exchangers below the pinch that meets the MER targets.

9.15 Design a HEN with $N_{HX,\min}^{MER}$ heat exchangers for Example 9.8. *Hint:* The solution requires stream splitting.

9.16 Consider a process with the following streams:

Stream	T^s(°F)	T^t(°F)	C ($\times 10^{-4}$ Btu/hr°F)
H1	480	250	2.0
H2	430	180	3.0
C1	100	400	2.5
C2	150	360	2.5
C3	200	400	2.5

(a) Compute ΔT_{thres}, as well as the minimum external heating and cooling requirements as a function of $\Delta T_{\min}$.

(b) Design a HEN to meet the MER targets with $N_{HX,\min}^{MER}$ heat exchangers, for $\Delta T_{\min} = 20°F$. Show the heat duties and temperatures for each heat exchanger.

9.17 Consider the process flowsheet in Figure 9.60, where the duties required for each heat exchanger are in MW, and the source and target stream temperatures are

Process Stream	T^s(°C)	T^t(°C)
Feed	25	200
Effluent	260	40
Recycle 1	40	200
Flash liquid	40	100
Recycle 2	50	200
Product	120	40

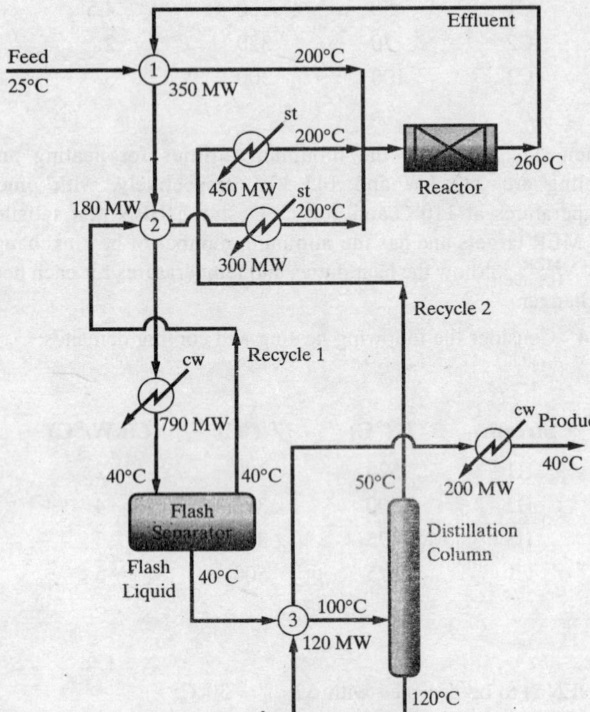

Figure 9.60 Process flowsheet for Exercise 9.17.

(a) The flowsheet calls for 990 MW to be removed by cooling water and 750 MW to be provided by steam. It is claimed that this design does not meet MER targets for $\Delta T_{\min} = 10$°C. Verify or refute this claim.

(b) If verified, design a HEN to meet MER targets for $\Delta T_{\min} = 10$°C.

9.18 Design a heat exchanger network for MER, with at most 15 heat exchangers (including utility heaters) and $\Delta T_{\min} = 10$°C, for the following streams:

Stream	T^s(°C)	T^t(°C)	C(kW/°C)
H1	140	50	10
H2	320	20	9
H3	370	20	8
C1	50	130	10
C2	130	430	8
C3	100	300	6
C4	30	230	5
C5	30	130	4
C6	30	430	1

When MER targets are satisfied, the hot pinch temperature is 140°C, with $Q_{H,\min} = 760$ kW and $Q_{C,\min} = 960$ kW.

9.19 Design a heat exchanger network for MER, with at most 18 heat exchangers (including utility heaters) and $\Delta T_{\min} = 10$°C, for the following streams:

Stream	T^s (°C)	T^t (°C)	C (kW/°C)
H1	400	20	10
H2	200	50	15
H3	350	230	5
H4	400	100	8
C1	80	450	7
C2	20	320	10
C3	50	450	5
C4	50	350	4
C5	100	500	1

When MER targets are satisfied, the hot pinch temperature is 200°C, with $Q_{H,\min} = 1,170$ kW and $Q_{C,\min} = 1,030$ kW.

9.20 In Example 9.17, HENs are designed for a process involving two hot and two cold streams. Note that three designs are proposed: (1) involving only HPS and cooling water that meets the $N_{HX,\min}^{MER}$ target (shown in Figure 9.38a); (2) involving HPS and cooling water with no stream splitting and one less heat exchanger (shown in Figure 9.38b); (3) utilizing HPS and IPS, cooling water, and boiler feed water (shown in Figure 9.39). Which of these designs has the lowest annualized cost, given the following specifications:

Cooling water (cw): $T^s = 30$°C, $T^t \leq 80$°C,
cost of cw = 0.00015 \$/Kg

Boiler feed water (bfw): $T = 110$°C, $\Delta H^v = 2,230$ kJ/kg,
revenue on bfw = 0.001 \$/kg

Saturated ips: $T = 195$°C, $\Delta H^v = 1,958$ kJ/kg,
cost of ips = 0.003 \$/kg

Saturated hps: $T = 258$°C, $\Delta H^v = 1,676$ kJ/kg,
cost of hps = 0.006 \$/kg

Overall heat-transfer coefficients: $U_{\text{heater}} = U_{\text{cooler}} = U_{\text{exch}}$
$= 1$ kW/m²°C

Purchase cost of heat exchangers: $C_P = 3,000A^{0.5}$(\$, m²)

Return on investment, $i_m = 0.1$

Bonus

Adapt the design in Figure 9.39 to produce a cheaper HEN.

9.21 *Flowsheet analysis and HEN synthesis problem.* A material balance has been completed for a process to manufacture styrene and an ethylbenzene byproduct from reactions involving methanol and toluene. See Figure 9.61 for a block flow diagram of the process with the results of material-balance calculations. You are to develop an optimal heat exchanger network for this process. Note that:

1. Stream 1 is fresh methanol feed, which enters at 25°C and 600 kPa.
2. Stream 2 is fresh toluene feed, which enters at 25°C and 600 kPa.

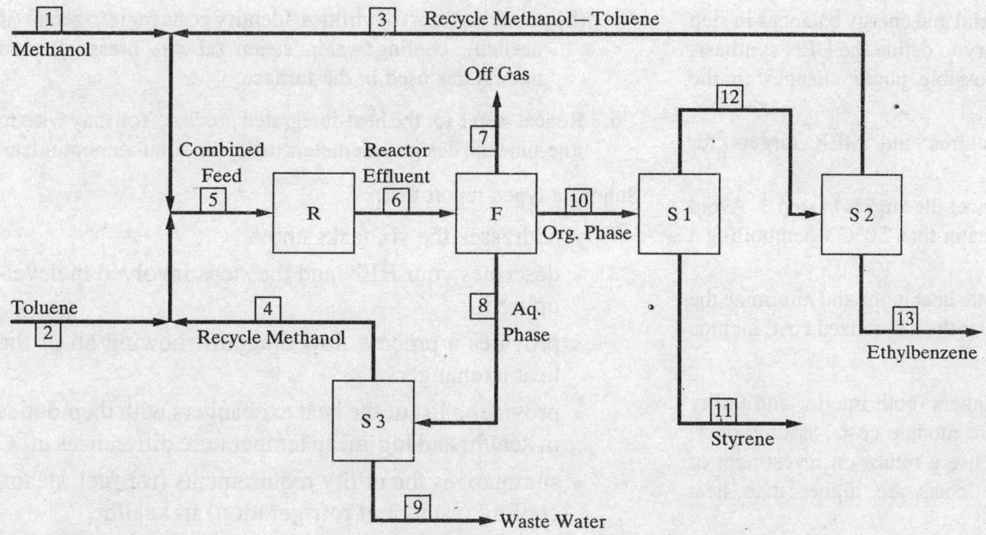

Material Balance for Styrene Process										
	Streams with flow rates in kmol/hr:									
Component	1	2	4	3	6	7	10	8	11	13
Hydrogen					352.2	352.2				
Methanol	493.4		37.0	66.0	107.3	4.3	66.0	37.0		
Water					489.1	7.9		481.2		
Toluene		491.9		104.5	107.3	1.5	105.8			1.3
Ethylbenzene				3.8	140.7	0.7	140.0			136.2
Styrene					352.2	1.6	350.6		346.7	3.9
Total	493.4	491.9	37.0	174.3	1,548.8	368.2	662.4	518.2	346.7	141.4

Figure 9.61 Styrene process.

3. Stream 4 is methanol vapor recycle from the aqueous methanol recovery, S3, system. This stream returns at 113°C at saturation pressure.
4. Stream 3 is an organic recycle from the distillation section. This stream returns at 175°C and saturation pressure as a vapor.
5. Streams 1 and 2 must be brought to the bubble point in separate heat exchangers and then vaporized in separate heat exchangers. These streams can then be mixed as desired with streams 3 and/or 4 to obtain the combined feed to the reactor, which must enter the reactor at 500°C and 400 kPa. A furnace must be used to heat streams from 400 to 500°C.
6. Stream 6, the reactor effluent, leaves the adiabatic reactor as vapor at 425°C and 330 kPa and must be cooled and partially condensed at 278 kPa before entering the three-phase separator. During that condensation, primary and secondary dew points are observed, with the final effluent comprised of a vapor phase and two liquid phases. Be sure to determine a cooling curve for cooling and partially condensing the reactor effluent.
7. Column S1 has a reboiler duty of 2.83×10^7 kcal/hr to reboil the bottoms in the temperature range of 144–145°C.
8. Column S2 has a reboiler duty of 4.64×10^6 kcal/hr to reboil the bottoms in the temperature range of 153–154°C.
9. Column Sl needs a condenser duty of 2.59×10^7 kcal/hr to condense the overhead in the temperature range of 103–80°C.
10. Column S2 needs a condenser duty of 2.62×10^6 kcal/hr to condense the overhead in the temperature range of 108–89°C.
11. Stream 11, the liquid styrene product, needs to be cooled from 145 to 38°C before being sent to storage.
12. Stream 13, the liquid ethylbenzene byproduct, needs to be cooled from 153 to 38°C before being sent to storage.

Your tasks:

1. Solve the material and energy balances for the flowsheet in Figure 9.61 using a process simulator. Adjust the pressure drops in each equipment item to satisfy the pressure specifications above. Two parallel reactions occur in the reactor R, which may be modeled using a **Conversion Reactor** in HYSYS, with the conversion in each reaction specified to closely match the material balances in Figure 9.61. Furthermore, for simplicity, you may use **Component Splitters** for units S1, S2, and S3. Finally, note that it may be necessary to install additional equipment items between units F and S1 and between units S2 and the mixer for streams 1 and 3.

2. Using the solution of the material and energy balances in step 1, extract information necessary to define the HEN synthesis problem. Pay attention to possible phase changes in the streams.
3. Compute the pinch temperatures and MER targets for $\Delta T_{min} = 10°C$.
4. Carry out an MER design to meet the targets in step 3. Avoid temperature-driving forces greater than 50°C when boiling a pure species or a mixture.
5. Refine your solution to eliminate heat loops and minimize the annualized cost of the HEN. For the annualized cost, include estimates for:

   (a) the cost of all heat exchangers (both interior and utility exchangers). Estimate bare module costs, assume operation 330 days/year, and use a return on investment of 20%. Note that furnace costs are higher than heat exchanger costs.

   (b) the annual cost of utilities. Identify costs for refrigerant (if needed), cooling water, steam (at one pressure), and natural gas used in the furnace.
6. Repeat step 1 for the heat-integrated process. You may wish to fine-tune the design parameters using the simulator optimizer.

Submit a typed report that:

- addresses the six tasks above
- describes your HEN and the steps involved in developing it
- provides a process flow diagram showing all of the heat exchangers
- provides a list of the heat exchangers with their duties in kcal/hr and log-mean temperature differences in °C
- summarizes the utility requirements (for fuel, steam, cooling water, and refrigeration) in kcal/hr.

Chapter 10

Mass Exchanger Networks

10.0 OBJECTIVES

This chapter extends the strategies for heat and power integration in Chapter 9 to apply to the mass integration of chemical processes during process synthesis. In Chapter 9, procedures for developing heat exchanger networks (HENs) were presented. In this chapter, analogous procedures for developing mass exchanger networks (MENs) are discussed. Mass exchangers use mass-separating agents (MSAs) to transfer solutes from solute-rich streams to solute-lean streams. Mass integration takes place after the demands for this transfer have been specified.

After studying this chapter, the reader should

1. Be able to compute the minimum usage of external mass-separating agents (MSAs) to determine the minimum operating cost (MOC) target. Two methods are introduced: the composition interval (CI) method (analogous to the TI method for HENs) and a graphical method known as the composite curve method (analogous to the use of temperature-composite curves in synthesizing HENs).
2. Be able to design a mass exchanger network (MEN) that meets the MOC targets. A method is introduced that inserts mass exchangers, one at a time, beginning at the closest approach mass-fraction difference, referred to as the *pinch*.
3. Be able to reduce the number of mass exchangers in MENs by relaxing the MOC target and *breaking mass loops* (i.e., allowing solute to be exchanged across the pinch).

10.1 INTRODUCTION

Almost all commercial operations for separating chemical mixtures utilize either an energy-(heat or shaft work) separating agent (ESA) as in distillation and certain high-pressure membrane separations, or an MSA, as in absorption, stripping, liquid–liquid extraction, solid–liquid extraction, adsorption, ion exchange, and membrane separations using a sweep fluid. With MSAs, solutes in so-called rich process streams are transferred into MSA streams, referred to as lean streams. The solute may then be removed from the MSA to permit its reuse. The network of equipment used to transfer solutes into MSAs is called a mass exchanger network (MEN). In general, it is assumed that the equipment used in the MEN employs countercurrent flow of the rich and lean streams. This is analogous to the assumption of the use of countercurrent flow heat exchangers in HENs. Major goals in the development of a MEN are to find and minimize the need for MSAs.

In Chapter 9, procedures for carrying out heat and power integration were discussed. These procedures are often implemented during the fourth step in process synthesis, where the differences in temperature, pressure, and phase are eliminated—when the source and target temperatures, T^s and T^t, for the streams to be heated and cooled, as well as power demands—have been specified. Emphasis is placed on determining the minimum hot and cold utility requirements [so-called MER (maximum energy recovery) targets for the HEN to be synthesized], and stream matching when positioning heat exchangers in the HEN. These procedures are normally carried out before the detailed design of the individual heat exchangers and turbines, for which techniques are discussed in Chapters 18 and 20.

In this chapter, similar procedures are introduced for carrying out mass integration, which are often implemented during the third step in process synthesis, where the differences in composition are eliminated by introducing separation operations. These differences may be eliminated by the use of ESAs or MSAs. The use of ESAs is usually considered first. When not feasible, MSAs are used and mass integration becomes an important consideration. In this chapter, the separations are achieved by MSAs and the goal is to synthesize an efficient MEN. The development begins with specification of the source and target concentrations, c^s and c^t, of rich and lean streams. Emphasis is placed on determining the minimum amounts of MSAs to be introduced, and stream matching when positioning separators in the MEN. These procedures are normally carried out before the detailed design of the individual separators, for which techniques are discussed in Chapter 19 and *Perry's Chemical Engineers' Handbook* (Green and Perry, 2008). While the first procedures for heat integration appeared in the early 1970s, parallel procedures for MENs were not introduced until two decades later by El-Halwagi and Manousiouthakis (1989). Major applications of MENs have been made in pollution abatement and waste minimization.

To define the MEN synthesis problem, N_R rich streams at mass flow rates F_{R_i}, with specified source and target compositions, say mass fractions y_i^s and y_i^t, $i = 1, \ldots, N_R$, have their solutes removed by N_L lean streams at mass flow rates F_{L_i}, with specified source and target mass fractions, x_i^s and x_i^t, $i = 1, \ldots, N_L$, as shown schematically in Figure 10.1. Notice the similarity between Figure 10.1 for mass integration and Figure 9.1 for heat integration. The streams may be gas, liquid, or solid. Often, the lean streams are already present in the process flowsheet. These are referred to as the N_{LP} *process* mass-separating agents (MSAs). For example, monochlorobenzene (MCB) is the MSA used to separate HCl from benzene in the absorber of Figure 5.23. The remaining N_{LE} mass-separating agents are transferred to the process from *external* sources (like heating and cooling utilities in HEN synthesis). The external source might be steam when being used to strip volatile organic compounds (VOCs) from wastewater. However, note that the auxiliary network of Figure 9.1 for heat integration does not appear in Figure 10.1 for mass integration. To achieve sufficiently low concentrations, the lean stream concentrations must be sufficiently low or external mass-separating agents must be acquired. When it is desired to concentrate the lean streams above the inlet concentrations of the rich streams, more concentrated rich streams must be acquired, although this is not common and not necessary in waste-removal operations.

In aqueous waste-removal operations, the solute is often an undesirable species to be removed from wastewater. After

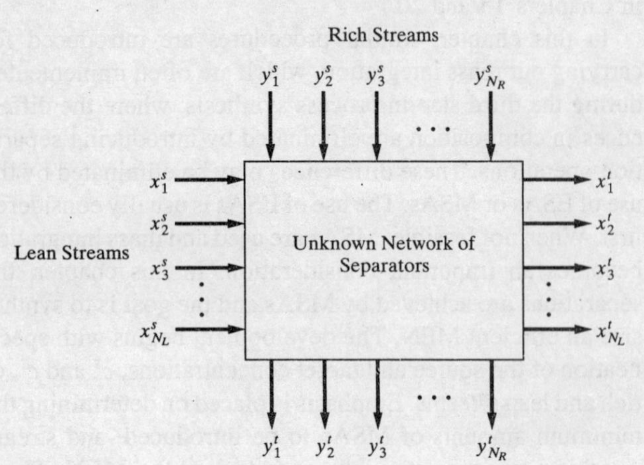

Figure 10.1 Mass-integration schematic with source and target concentrations of rich and lean streams.

solute recovery, the rich streams, with solute in low concentrations, are disposed of, returned to the environment, recycled, or reused. Clearly, when returned to the environment, these streams must meet the latest federal, state, or local regulations. When recycled or reused, solute concentrations must be sufficiently low to meet the requirements of sinks elsewhere in the process.

When carrying out the design given the states of the source and target streams (flow rates and compositions of the solute), it is desired to synthesize the most economical network of mass exchangers. Several measures of economic goodness are possible, as discussed in Section 23.4. Usually, when generating and comparing alternative flowsheets, an approximate profitability measure is sufficient, such as the annualized cost:

$$C_A = i_m(C_{\text{TCI}}) + C \tag{10.1}$$

where C_{TCI} is the total capital investment, as defined in Table 22.9, i_m is a reasonable return on investment annually (i.e., when $i_m = 0.33$, a chemical company charges itself annually for one-third of the cost of the capital invested), and C is the annual cost of sales, as defined in the *cost sheet* of Table 23.1. In Tables 22.9 and 23.1, many factors are involved, most of which are necessary for a detailed profitability analysis. However, to estimate an approximate profitability measure for the comparison of alternative flowsheets, it is adequate to approximate C_{TCI} as the sum of the purchase costs for each of the separators (without including installation costs and other capital investment costs). The purchase costs can be estimated based on the diameter, height, and weight of the process vessels, which are estimated using the procedures in Chapters 19 and 22. It is adequate to approximate C as the annual cost of the external mass-separating agents. In summary, with these approximations, Eq. (10.1) is rewritten as:

$$C_A = i_m\left(\sum_i C_{P_i}\right) + \sum_i e_i F_{E_i} \tag{10.2}$$

where C_{P_i} is the purchase cost of separator i, F_{E_i} is the annual flow rate of external mass-separating agent i (e.g., in kilograms per year), and e_i is the unit cost of external mass-separating agent i (e.g., in dollars per kilogram). Clearly, when utilities such as cooling water, air, steam, fuel, and refrigerants are used, additional terms are needed.

As in the synthesis of heat exchanger networks, two principal steps are typically carried out when synthesizing MENs:

1. A network of mass exchangers is designed having the *minimum amounts of mass-separating agents*, usually requiring a large number of mass exchangers. When the unit costs of the mass-separating agents are high, a nearly optimal design is obtained.
2. The number of mass exchangers is reduced toward the minimum, possibly at the expense of increasing the consumption of MSAs.

As step 2 is implemented, one mass exchanger at a time, capital costs are reduced due to the economy-of-scale in Eqs. (22.49) – (22.58). As each exchanger is removed, the diameters, heights, and weights of the exchangers are increased, and because the slope of the curves in Figure 22.13 are less than unity, the purchase costs per unit volume are decreased. Also, as step 2 is implemented, the consumption of MSAs is normally increased. At some point, the increased cost of MSAs overrides the decreased cost of capital and C_A increases beyond the minimum. When the costs of MSAs are high, the minimum C_A is not far from C_A for a network of mass exchangers using the minimum MSAs.

10.2 MINIMUM MASS-SEPARATING AGENT

When minimizing the utilities in heat integration, the approach temperature difference is the key specification. As ΔT_{min} decreases, the utilities decrease, but the heat-exchange area increases in inverse proportion. Similarly, when minimizing the flow rates of MSAs in mass integration, an approach composition difference must be specified. Here, it is convenient to specify the compositions of the rich and lean streams on the same scale. Commonly used compositions are mass fractions, mole fractions, and parts per million (ppm) on a volume basis for gases and a mass basis for liquids and solids.

Approach to Phase Equilibrium

Beginning arbitrarily with the rich phase, having solute mass fraction y, the composition of the solute in the lean phase that approaches equilibrium with the rich phase is denoted x^*. Here, the approach to phase equilibrium, Δx_{min}, can be specified, where:

$$x^* = x + x_{min} \tag{10.3}$$

Then, on an x–y diagram, in a dilute region where the equilibrium curve is linear, the x mass fraction is displaced to the left by Δx_{min}, as shown in Figure 10.2. Let the equation of the equilibrium curve be a straight line given by:

$$y = mx^* + b \tag{10.4}$$

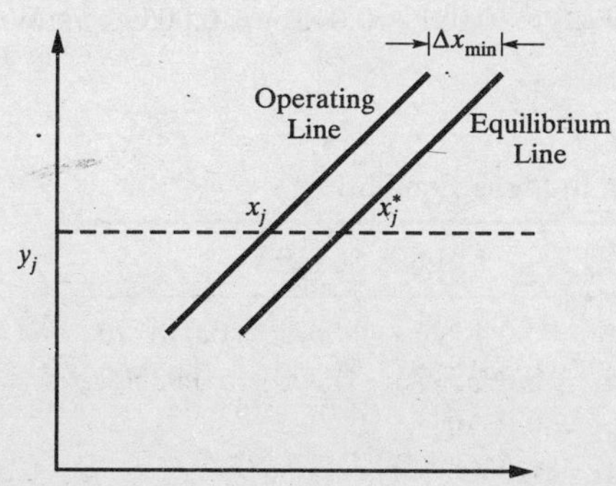

Figure 10.2 Equilibrium curve and approach composition difference.

where m is the slope and b is the ordinate intercept. Substituting Eq. (10.3) in Eq. (10.4) and rearranging,

$$x = \frac{y - b}{m} - \Delta x_{min} \tag{10.5}$$

Now, consider a countercurrent, direct-contact mass exchanger such as a packed column. The packed height of the column is the product of the height of a transfer unit (HTU) and the number of transfer units (NTU). As Δx_{min} decreases, the NTU increases and, in turn, the height of the column and its capital cost increases. However, as will be shown, the amounts of the MSAs decrease.

Concentration-Interval (CI) Method

Consider the following example, which is similar to one introduced by El-Halwagi and Manousiouthakis (1989) in their pioneering paper. To determine the minimum flow rate of an MSA, the concentration-interval method is introduced first, after which, in the next subsection, the composite curve method is introduced.

EXAMPLE 10.1 ***H_2S Removal from Sour Coke Oven Gas***

The process in Figure 10.3 is being designed to remove H_2S from sour coke oven gas (COG), which is a mixture of H_2, CH_4, CO, N_2, NH_3, CO_2, and H_2S. The removal is necessary because H_2S is corrosive and becomes the pollutant SO_2 when the gas is combusted. It is proposed to remove the H_2S and send it to a Claus unit to convert it to sulfur. However, because the conversion of the H_2S is incomplete, the tail gases must be recycled for H_2S removal. Distillation to remove the H_2S is not feasible, but absorption is feasible. Thus, it is proposed to design a MEN based on absorption. One possible MSA is aqueous ammonia, noting that ammonia is already present in the COG and that the flow rate and composition of the recycle stream are specified before the MEN is designed. An alternative MSA is chilled methanol, which is an external MSA. Both ammonia and chilled methanol are to be considered as possible absorbents for the removal of H_2S from the COG and the tail gas. As shown in Figure 10.3, the rich absorbent streams are regenerated by stripping to recover the acid gases, which are sent to the Claus unit.

To begin the development of the MEN, the sour COG and the tail gases are not mixed, and absorption can utilize ammonia, methanol, or both. Mass transfer in all mass exchangers is from the gas phase to the liquid phase.

The specifications for the rich and lean streams are as follows, where compositions y for gases and x for liquids are in mass fractions, F is the stream mass flow rate, and n is the mass flow rate of H_2S transferred to or from the stream:

Stream	y^s or x^s	y^t or x^t	F (kg/s)	n (kg/s)
R1 (COG)	0.0700	0.0005	0.9	0.06255
R2 (Tail gases)	0.0510	0.0003	0.1	0.00507
L1 (Aq. NH_3)	0.0008	0.0310	2.3	0.06946
L2 (Methanol)	0.0001	0.0035	Unlimited	Unlimited

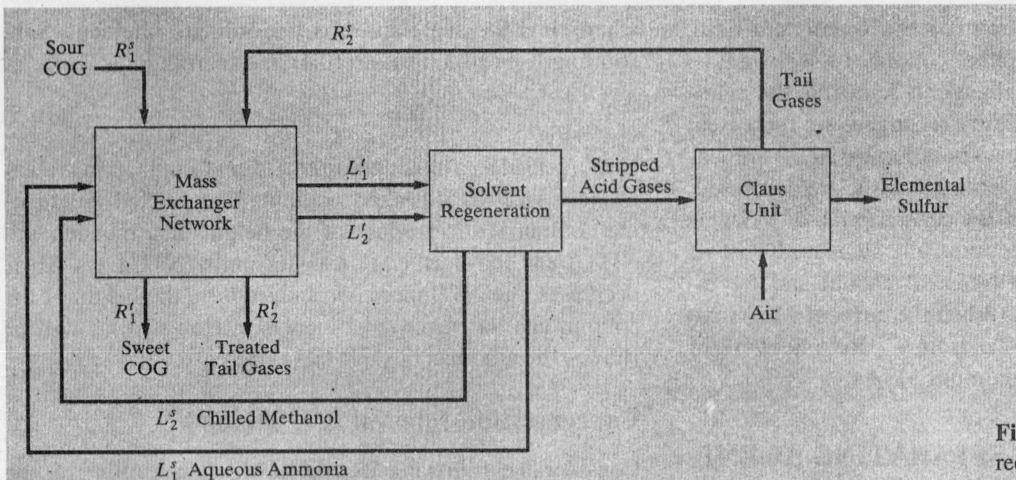

Figure 10.3 Process for recovery of H_2S.

Note that the flow rate of aqueous ammonia is limited, but chilled methanol is considered to be available in unlimited amounts. Note also that the total amount of H_2S to be transferred to the absorbent(s) is $0.06255 + 0.00507 = 0.06762$ kg/s. This is less than the capacity of the aqueous ammonia. However, as in heat exchange, where a driving force is necessary to transfer the heat, mass exchange also requires a driving force and, at this point in the synthesis, it is not known whether sufficient mass-transfer driving forces exist to utilize the capacity of the aqueous ammonia. If not, then the use of chilled methanol must be considered.

All conditions in the above specifications table are considered to be dilute in the solute, H_2S. Therefore, stream flow rates are assumed constant, and at the expected operating conditions of temperature and pressure, the following linear equilibrium equations apply:

$$\text{Aqueous ammonia (1)}, \; y = m_1 x = 1.45x$$

$$\text{Chilled methanol (2)}, \; y = m_2 x = 0.26x \; .$$

For concentrated solutes, it is preferable to use solute-free flow rates and the mass ratios of solute to solute-free solvent.

At this stage in process synthesis, it is desired to determine, by the CI method, the minimum amount of chilled methanol required for a MEN involving these four streams, noting that it may be possible to eliminate the need for chilled methanol. In the solution, the COG and lean gas streams are first matched with the aqueous ammonia stream. Alternatively, it may be preferable to consider first matches with the chilled methanol stream.

SOLUTION

The first step in the CI method is to rank-order the source and target mass fractions of streams R_1, R_2, and L_1, regardless of whether they are associated with the rich or lean phase. This includes computing the mass fractions in the corresponding phase, accounting for an approach to phase equilibrium using Eq. (10.5) with an assumed value of Δx_{min}, taken for this example as 0.0001. Thus, for the rich vapor streams, $x = y/1.45 - 0.0001$, and for the lean liquid streams, $y = 1.45(x + 0.0001)$. The results are given in Table 10.1, where specified values are in boldface and the two columns are rank-ordered, starting with the largest mass fraction at the top.

In the second step, the rank-ordered mass fractions are used to create a cascade of composition intervals, established in Table 10.2, within which mass balances are carried out. This is analogous to the cascade diagram in the TI method for HENs. As shown in Figure 10.4, each interval i displays the difference, Δn_i, between the mass to be removed from the rich streams and the mass to be taken up by the lean streams in the interval. For example, in interval 1 ($0.051 \leq y \leq 0.07$), only R1 is involved. Hence, as shown in Table 10.2, the difference, Δn_1, is $F_{R1} \times (0.07 - 0.051) = 0.0171$ kg/s. Since no excess mass, n_{Excess}, is assumed to enter this interval, 0.0171 kg/s are available and flow down into interval 2; that is, the residual from interval 1 is $R_1 = 0.0171$ kg/s. Interval 2 involves both rich streams, but the lean stream is not present. As shown in Table 10.2, its difference, Δn_2, is $(F_{R1} + F_{R2}) \times (0.051 - 0.0451) = 0.0059$ kg/s. When added

Table 10.1 Rank-Ordered Compositions Including the Approach to Phase Equilibrium

Rich Streams	Lean Streams
$y_0 = \mathbf{0.0700}$	$x_0 = 0.048276 - 0.0001 = 0.048176$
$y_1 = \mathbf{0.0510}$	$x_1 = 0.0035172 - 0.0001 = 0.035072$
$y_2 = 1.45(0.0310 + 0.0001) = 0.0451$	$x_2 = \mathbf{0.0310}$
$y_3 = 1.45(0.0008 + 0.0001) = 0.001305$	$x_3 = \mathbf{0.0008}$
$y_4 = \mathbf{0.0005}$	$x_4 = 0.000345 - 0.0001 = 0.000245$
$y_5 = \mathbf{0.0003}$	$x_5 = 0.000207 - 0.0001 = 0.000107$

Table 10.2 Internal Mass Loads

Interval	From Rich Streams (kg/s)	To Lean Streams (kg/s)	Excess (kg/s)	Residual, R (kg/s)
1	$(0.07 - 0.051) \times 0.9 = 0.0171$	—	0.0171	0.0171
2	$(0.051 - 0.0451) \times 0.9 +$ $(0.051 - 0.0451) \times 0.1 = 0.0059$	—	0.0059	0.0230
3	$(0.04505 - 0.001305) \times 0.9 +$ $(0.04505 - 0.001305) \times 0.1 = 0.04375$	$(0.031 - 0.0008) \times 2.3 = 0.06946$	−0.025665	−0.002665
4	$(0.001305 - 0.0005) \times 0.9 +$ $(0.001305 - 0.0005) \times 0.1 = 0.00080$	—	0.00080	−0.001865
5	$(0.0005 - 0.0003) \times 0.1 = 0.00002$	—	0.00002	−0.001845

to the residual from interval 1, R_1, the residual from interval 2, R_2, $= 0.0230$ kg/s. Interval 3 involves both rich streams ($0.001305 \leq y \leq 0.0451$) and stream L1 ($0.0008 \leq x \leq 0.0310$). Its difference, shown in Table 10.2, is negative, -0.025665 kg/s, which when added to R_2 gives $R_3 = -0.002665$ kg/s. In this interval, the lean stream requires $\Delta n_3 = 0.025665$ kg/s more solute than is available from the rich streams. The residual, R_2, provides 0.0230 kg/s, but this is insufficient, and consequently, the residual from interval 3 is negative; that is, $R_3 = -0.002665$ kg/s. Clearly, the H_2S solute cannot be transferred from the rich streams in interval 4, which are in a lower concentration range. The only way to avoid a negative residual is to add solute at a higher concentration. Note that the least amount of solute to be added is 0.002665 kg/s. Turning next to interval 4, only the rich streams are present ($0.0005 \leq y \leq 0.001305$) and $\Delta n_4 = 0.00080$ kg/s, which when added to R_3, gives $R_4 = -0.001865$ kg/s. Finally, only stream R2 is present in interval 5 ($0.0003 \leq y \leq 0.0005$) and $\Delta n_5 = 0.00002$ kg/s, which when added to R_4, gives $R_5 = -0.001845$ kg/s. Residuals R_3, R_4, and R_5 are negative and infeasible. Note that $n_{LE} = R_5$ is the minimum amount of lean external MSA (chilled methanol) required to remove the H_2S solute from streams R1 and R2. Clearly, it cannot be negative.

Table 10.2 summarizes the solute loads to be removed and added in each interval, as well as the difference, or excess, and the residuals. These results constitute the "Initial Pass" in Figure 10.4.

Clearly, all negative residuals must be removed through the addition of solute at higher concentrations because the solute

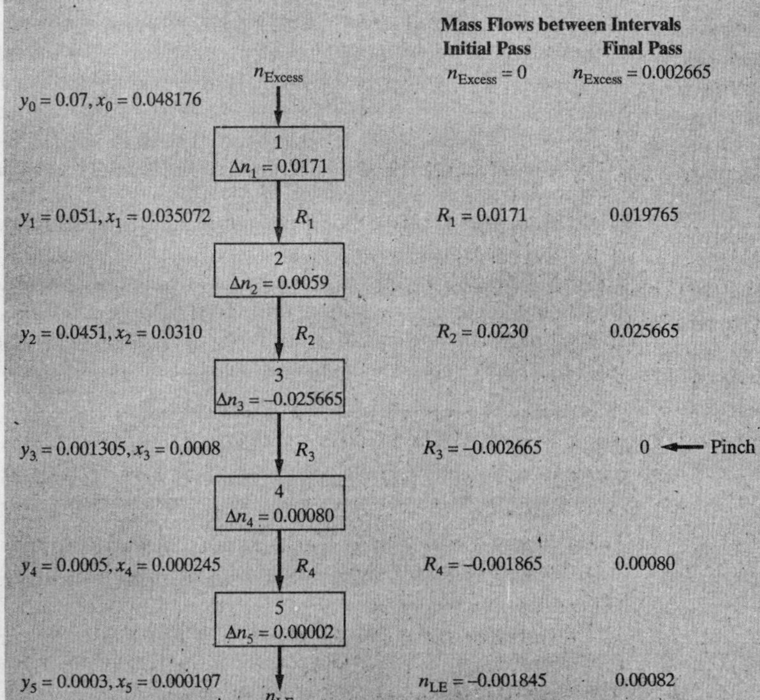

Figure 10.4 Cascade of composition intervals, mass balances, and residuals. R_i, n_{Excess}, and n_{LE} are in kilograms per second; x and y are mass fractions of the lean and rich phases.

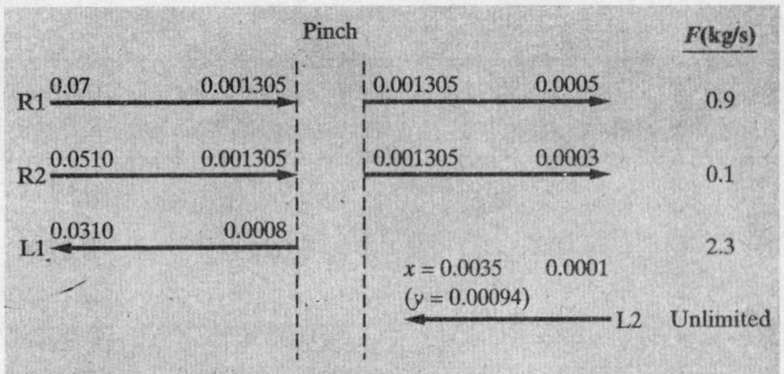

Figure 10.5 Pinch decomposition of the rich and lean streams.

cannot be transferred from a low to a high concentration.[1] In this example, the largest negative residual is −0.002665 kg/s, and consequently, no negative residuals remain when $n_{Excess} =$ 0.002665 kg/s is added at the top of the cascade, as shown in the "Final Pass" column of Figure 10.4. Alternatively, the internal MSA, aqueous ammonia, must be adjusted to reduce its consumption of solute, H_2S, by reducing its flow rate and/or its target mass fraction. By revising values of R_1 to R_4, a final residual of 0.00082 kg/s is found at the bottom of the cascade. This is the minimum amount of the solute that must be removed by the external MSA. To accomplish this, 0.00082/(0.0035 − 0.0001) = 0.2412 kg/s of chilled methanol is required. At the minimum usage of external MSA, no solute flows between intervals 3 and 4. This is referred to as the *pinch*, as shown in Figure 10.4, with associated mass fractions of 0.001305 for the rich streams and 0.0008 for the lean stream. Assuming that the mass fractions satisfy phase equilibrium, this is the location of the closest approach mass fraction, $\Delta x_{min} = 0.0001$. To maintain the minimum external MSA, *no solute is permitted to flow across the pinch.* Should additional solute, say 0.0002 kg/s, be added to the rich streams above the pinch, 0.0002 kg/s would be transferred across the pinch, and the amount of solute to be removed by the external MSA would be increased to 0.00082 + 0.0002 = 0.00102 kg/s.

In Figure 10.5, which helps to define the design requirements when the external MSA is at a minimum, arrows moving from left to right denote the rich streams; lean streams are denoted by arrows moving from right to left. The arrows for the rich and lean streams either pass through or begin at the pinch mass fractions. Note that the smallest mass fraction of the rich streams that can enter a countercurrent mass exchanger with the external MSA is, from Equation (10.5), 0.00094[=0.26 (0.0035 + 0.0001)]. It lies below the mass fraction of the rich streams at the pinch, 0.001305. To maintain the minimum external MSA, two separate MENs *must* be designed, one on the rich side and one on the lean side of the pinch mass fractions. Solute is not permitted to flow across the pinch. On the rich side of the pinch, the flow rate of the internal MSA has been reduced to permit all of the solute in the rich streams to be removed while meeting the target mass fraction of the internal MSA. On the lean side of the pinch, 0.00082 kg/s of H_2S solute are removed by the minimum amount of the external MSA. If solute from the rich streams on the rich side of the pinch were removed by a lean stream on the lean side of the pinch (in this case, the external MSA), solute would flow across the pinch and the amount of the external MSA would be increased above the minimum.

[1] Actually, in a nonideal multicomponent system, as shown by Toor (1957), it is possible for mass transfer of a component to occur against a composition driving force because of cross-coupling effects.

In the next section, methods for inserting the mass exchangers *(stream matching)* are described. Before this, a graphical approach, the *composite curve method*, is discussed in the next subsection.

Composite Curve Method

Similar to the discussion of heat integration in Section 9.2, the terminology *pinch* is understood more clearly in connection with a graphical display, as introduced by El-Halwagi and Manousiouthakis (1989) for mass integration, in which composite rich and lean curves are positioned no closer than the phase-equilibrium departure plus Δx_{min}. As $\Delta x_{min} \rightarrow 0$, the curves pinch together toward the compositions at phase equilibrium and the area for mass transfer approaches infinity. The use of these curves is illustrated next in the composite curve method.

EXAMPLE 10.2 ***H_2S Removal from Sour Coke Oven Gas (Example 10.1 Revisited)***

In this example, the minimum external MSA requirement for a MEN involving the four streams in Example 10.1 is determined using the graphical approach of El-Halwagi and Manousiouthakis (1989).

SOLUTION

For each of the rich streams and the internal MSA lean stream (aqueous ammonia), y or x is graphed on the ordinate as a function of the mass of solute transferred on the abscissa, with the slope being the inverse of the mass flow rate, F (kg/s). For mixtures dilute in the solute, F is nearly constant, and consequently, the curves are approximately straight lines. For the rich streams, the curves begin at the highest mass fractions and finish at the lowest mass fractions after the solute has been removed. For the lean streams, they begin at the lowest mass fractions and finish at the highest mass fractions after the solute has been added. In Figure 10.6a, the three curves are displayed, with each of the lines positioned arbitrarily along the abscissa to avoid intersections and crowding.

To display the results of the CI method graphically, Table 10.2 is used to prepare the *rich composite curve*, which combines curves R1 and R2 in Figure 10.6a into one composite curve. Beginning with zero mass of solute at $y = 0.0003$, the lowest mass fraction of a rich stream, and using Table 10.2, the cumulative mass of solute removed at the interval mass fractions are

y	0.0003	0.0005	0.001305	0.0451	0.051	0.07
n_{cum}	0	0.00002	0.00082	0.04457	0.05047	0.06757

These points form the rich composite curve in Figures 10.6b and 10.6c. Note that the low-concentration end is expanded in the *area of detail* into Figure 10.6c.

Normally, the lean composite curve is prepared in the same way. However, in this example, the curve is that for the one internal MSA in stream L1. It is simply copied from Figure 10.6a, but shifted to the left so as to begin at $n = 0.00082$, the minimum amount of solute that must be removed by the external MSA, chilled methanol.

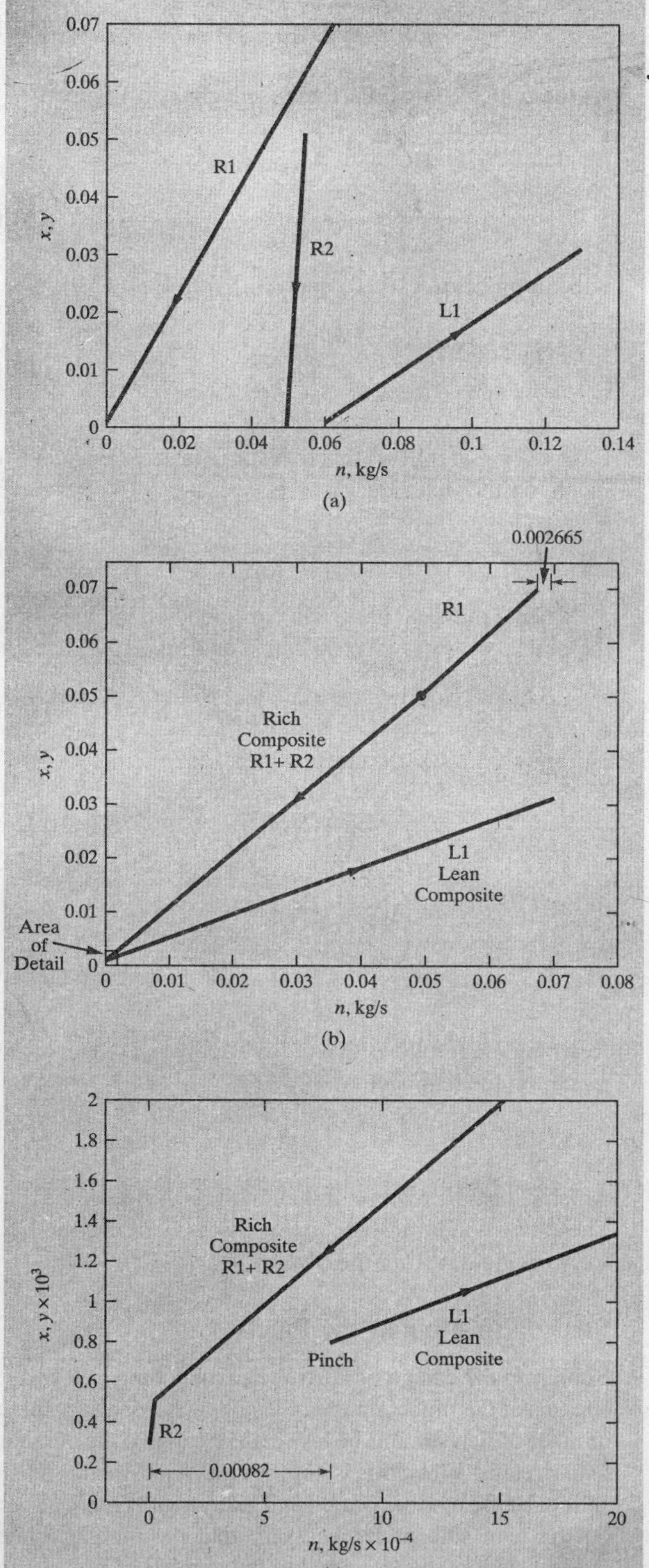

Figure 10.6 Graphical method to locate the minimum external MSA: (a) mass exchange curves for each stream, (b) and (c) composite rich and lean curves.

As shown in Figure 10.6c, the composite curves have a closest point of approach at the point where stream L1 begins along the lean composite curve; that is, $x = 0.0008$. The corresponding mass fraction on the rich composite curve is $y = 0.001305$, with a corresponding equilibrium mass fraction in the aqueous ammonia liquid phase of $0.001305/1.45 = 0.0009$. Thus, the approach is $0.0009 - 0.0008 = 0.0001$, which is $\Delta x_{\min}$. Consequently, these two points provide the mass fractions at the pinch. If $\Delta x_{\min}$ is reduced to zero, the lean composite curve is shifted to the left until $y = 0.00116$ $(= 1.45 \times 0.0008)$, the mass fraction at equilibrium with $x = 0.0008$. As mentioned earlier, this corresponds to an infinitely large mass exchanger.

In this example, with $\Delta x_{\min} = 0.0001$, mass in the segments of the rich composite curve is transferred down vertically to the segments of the lean composite curve that lie below them. At the high-concentration ends, however, no segments on the rich composite curve lie vertically above the upper end of the lean composite curve. There, as computed previously and shown in the "Final Pass" of Figure 10.4, an additional 0.002665 kg/s of solute must be added. Alternatively, the flow rate of L1 can be reduced or its target mass fraction reduced from 0.0310 to eliminate the need for additional solute. This is consistent with the results using the composition-interval method in Example 10.1. Similarly, at the low-concentration ends of the composite curves, no segments of the lean composite curve lie vertically below the lower end of the rich composite curve. Here, 0.00082 kg/s must be removed by the external MSA, a result again consistent with the CI analysis.

As for heat integration in Section 9.2, many additional observations are noteworthy in connection with the rich and lean composite curves. One is that the slopes of the composite curves *always* decrease at the inlet concentration of a stream and increase at the outlet concentration of a stream. It follows that points at which the slope decreases are candidate pinch points, and furthermore, that one of the inlet concentrations is *always* a pinch concentration, when a pinch exists. Hence, to locate a potential pinch concentration, one needs only to examine the inlet concentrations of the streams.

Similarly, for some $\Delta x_{\min}$, there are no pinch concentrations, which is analogous to HENs. In such cases, either excess internal MSAs exist or external MSAs are required, but not both. Their amounts equal the difference between the mass of solute to be removed from the rich streams and that required by the lean streams. In these cases, $\Delta x_{\min} \leq \Delta x_{\text{thres}}$, where Δx_{thres} is the threshold approach concentration difference, above which a pinch exists. The threshold condition for HENs, which is similar to that for MENs, is discussed in Section 9.5.

In other cases, two or more pinch points arise. This occurs when the total flow rates of the rich and lean streams in a concentration interval are equal and the interval contains a pinch point.

10.3 MASS EXCHANGE NETWORKS FOR MINIMUM EXTERNAL MSA

Having determined the minimum flow rate of an external MSA (that is, the MOC target) using one of the two methods above, or using a linear program (similar to that for heat integration in Section 9.2), it is common to design two networks of mass exchangers, one on the rich side and one on the lean side of the pinch, as shown in Figure 10.5. In this section, a method introduced by El-Halwagi and Manousiouthakis (1989) and similar to the method of Linnhoff and Hindmarsh (1983) for heat integration is presented that places emphasis on positioning the mass exchangers *(stream matching)* by working out from the pinch.

Stream Matching at the Pinch

To explain the approach of El-Halwagi and Manousiouthakis (1989), it helps to refer to Figure 10.5, which shows the *pinch decomposition* of the rich and lean streams in Example 10.1. Attention is focused at the pinch where the mass fractions of the rich and lean streams are those determined as in Table 10.1 using Eq. (10.5) and the assumed value of $\Delta x_{\min}$.

Consider the schematic of a countercurrent mass exchanger in Figure 10.7. The rich stream, having a flow rate of F_R, enters at y_i and exits at y_o. It transfers mass, n, to the lean stream that has a flow rate of F_L, enters at x_i, and exits at x_o. Carrying out mass balances for the solute in the rich and lean streams:

$$n = F_R(y_i - y_o) \tag{10.6}$$

$$n = F_L(x_o - x_i) \tag{10.7}$$

$$F_R(y_i - y_o) = F_L(x_o - x_i) \tag{10.8}$$

When a mass exchanger is positioned on the rich side of the pinch, which is considered first arbitrarily, conditions at the lower end of the exchanger in Figure 10.7 become $y_o = y_{\text{pinch}}$, $x_i = x_{\text{pinch}}$, and Eq. (10.8) becomes:

$$F_R(y_i - y_{\text{pinch}}) = F_L(x_o - x_{\text{pinch}}) \tag{10.9}$$

Figure 10.7 Schematic of a countercurrent mass exchanger.

Furthermore, accounting for the approach to phase equilibrium at the pinch, Eq. (10.5) becomes:

$$y_{\text{pinch}} = m(x_{\text{pinch}} + \Delta x_{\min}) + b \tag{10.10}$$

At the upper end of the mass exchanger in Figure 10.7, it follows that the mass fractions at equilibrium must be separated by greater than $\Delta x_{\min}$; that is,

$$y_i \geq m(x_o + \Delta x_{\min}) + b \tag{10.11}$$

Substituting Eqs. (10.10) and (10.11) in Eq. (10.9):

$$\begin{aligned} &F_R[m(x_o + \Delta x_{\min}) + b - m(x_{\text{pinch}} + \Delta x_{\min}) - b] \\ &\quad \leq F_L(x_o - x_{\text{pinch}}) \end{aligned} \tag{10.12}$$

Rearranging, it follows that to install a mass exchanger at the pinch, on the rich side:

$$F_R \leq \frac{F_L}{m} \tag{10.13}$$

When a mass exchanger is positioned on the lean side of the pinch, $y_i = y_{\text{pinch}}$, $x_o = x_{\text{pinch}}$, and Eq. (10.8) becomes:

$$F_R(y_{\text{pinch}} - y_o) = F_L(x_{\text{pinch}} - x_i) \tag{10.14}$$

Substituting Eq. (10.10) and

$$y_o \geq m(x_i + \Delta x_{\min}) + b \tag{10.15}$$

in Eq. (10.14):

$$\begin{aligned} &F_R[m(x_{\text{pinch}} + \Delta x_{\min}) + b - m(x_i + \Delta x_{\min}) - b] \\ &\quad \geq F_L(x_{\text{pinch}} - x_i) \end{aligned} \tag{10.16}$$

Rearranging, it follows that to install a mass exchanger at the pinch, on the lean side:

$$F_R \geq \frac{F_L}{m} \tag{10.17}$$

which is just the reverse of that on the rich side of the pinch. These two conditions are analogous to those used in HENs for working out from the pinch.

Stream Splitting at the Pinch

Similar to the design of HENs when matching rich and lean streams at the pinch, on the rich side, it is necessary that the number of rich streams be less than or equal to the number of lean streams. When this is not the case, lean streams must be split until the number of rich and lean streams is equal. Also, on the lean side, it is necessary that the number of lean streams be less than or equal to the number of rich streams. Here, rich streams must be split until the number of rich and lean streams is equal. The development of MENs above and below the pinch, including the need for stream splitting, is illustrated in the next example.

EXAMPLE 10.3 H_2S Removal from Sour Coke Oven Gas (Examples 10.1 and 10.2 Revisited)

Returning to Figure 10.5, it is desired to synthesize a network of mass exchangers that utilizes the minimum external MSA. Two MENs are needed, one on the rich side and one on the lean side of the pinch.

SOLUTION

As discussed earlier, the capacity of stream L1 to remove solute exceeds the solute in steams R1 and R2 by 0.002665 kg/s. Rather than add solute to the rich streams, it is assumed that the flow rate of stream L1 can be reduced accordingly. Consequently, stream L1 is adjusted to remove 0.06946 − 0.002665 = 0.06680 kg/s of solute. Its adjusted flow rate is

$$F_{\text{L1}} = \frac{0.06680}{0.031 - 0.0008} = 2.2119\,\text{kg/s}$$

which is 2.3 − 2.2119 = 0.0881 kg/s smaller.

Next, inequality (10.13) is checked for each potential match on the rich side of the pinch:

Match	$F_R \leq \frac{F_L}{m}$?
R1–L1	$0.9 \leq \frac{2.2119}{1.45} = 1.525$
R2–L1	$0.1 \leq \frac{2.2119}{1.45} = 1.525$

Both inequalities are satisfied, but stream L1 must be split to permit matches with both streams R1 and R2 at the pinch. This is accomplished as shown in Figure 10.8, based on the amounts of the solute to be removed above the pinch from R1 and R2. Using Table 10.2, 0.9(0.07 − 0.001305) = 0.06183 kg/s of solute is to be removed from R1, and 0.1 (0.051 − 0.001305) = 0.00497 kg/s from R2. Therefore, stream L1 is split into the following portions to remove the entire amounts of solute from streams R1 [0.06183/(0.0310 − 0.0008) = 2.0474 kg/s] and R2 [0.00497/(0.0310 − 0.0008) = 0.1645 kg/s].

On the lean side of the pinch, stream L1 does not appear. Thus, the chilled methanol, stream L2, must be used. Its target mass fraction is $x = 0.0035$, which can be contacted by rich streams

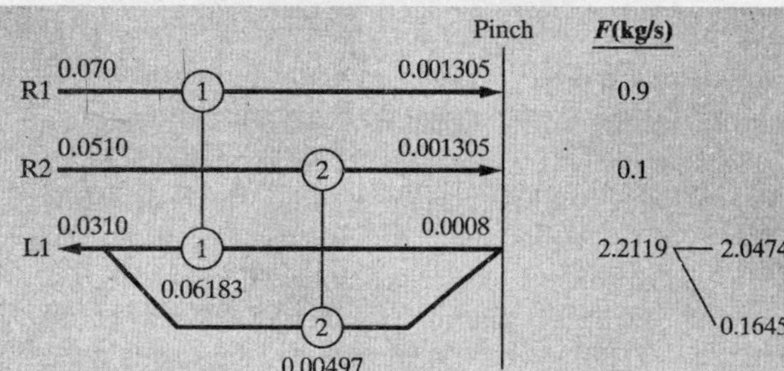

Figure 10.8 MEN on rich side of the pinch.

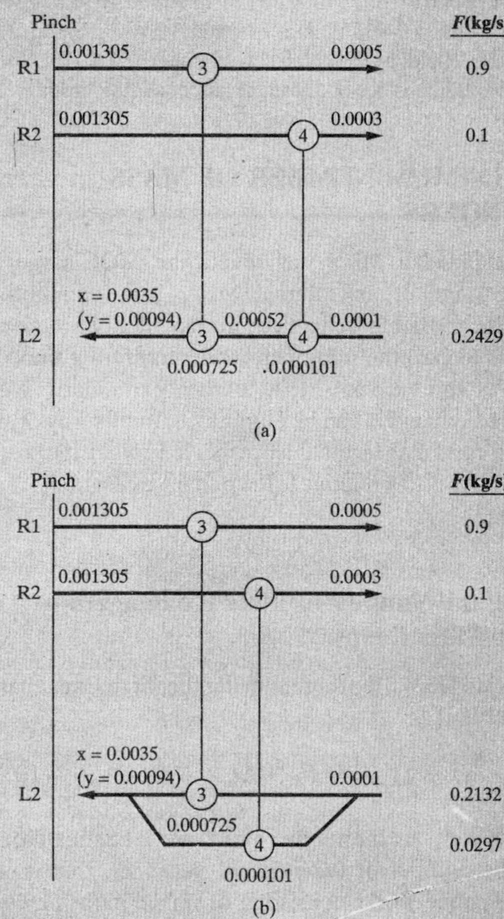

Figure 10.9 MENs on lean side of the pinch.

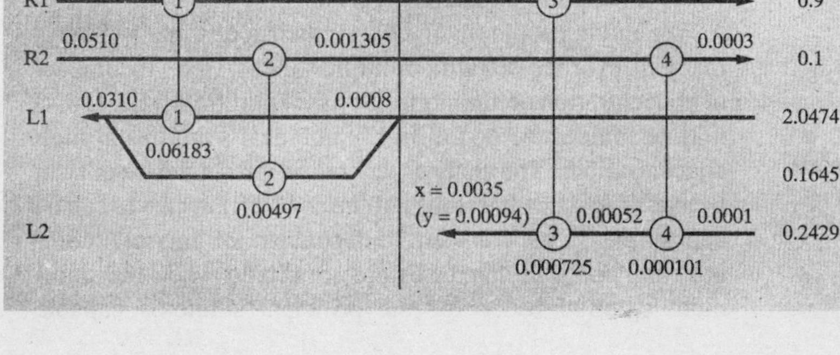

Figure 10.10 MEN having minimum number of mass exchangers, while satisfying the minimum usage of external MSA.

having mass fractions greater than or equal to $y = 0.00094$ [$= 0.26(0.0035 + 0.0001)$], which, as shown in Figure 10.5, is less than the mass fraction at the pinch ($y = 0.001305$). Thus, Eq. (10.17) does not apply. Two potential MENs are shown in Figure 10.9, one of which utilizes stream splitting. In both MENs, 0.000725 kg/s of solute must be removed from R1 and 0.000101 kg/s of solute from R2, as shown in Figure 10.9. In both cases, the equivalent value of y for the entering methanol $= 0.26(0.0001 + 0.0001) = 0.000052$. In both cases, the flow rate of methanol $= (0.000725 + 0.000101)/(0.0035 - 0.0001) = 0.2429$ kg/s, the minimum amount of external MEA.

Finally, the MEN on the rich side of the pinch in Figure 10.8 is combined in Figure 10.10 with the MEN in Figure 10.9a to give a MEN that utilizes the minimum external MSA. Alternatively, Figure 10.9b could be combined with Figure 10.8. The final selection would consider both capital and operating costs.

10.4 MINIMUM NUMBER OF MASS EXCHANGERS

Having designed a MEN that meets the MOC target of minimum usage of an external MSA, it is common to consider the possible reduction in the number of mass exchangers toward the minimum while permitting the consumption of the external MSA to rise, particularly when small mass exchangers can be eliminated. In this way, lower annualized costs may be obtained, especially when the cost of external MSAs is low relative to the purchase cost of the mass exchangers.

Reducing the Number of Mass Exchangers—Breaking Mass Loops

By analogy to HENs, the minimum number of mass exchangers in a MEN is

$$N_{\mathrm{MX,min}} = N_{\mathrm{R}} + N_{\mathrm{L}} - N_{\mathrm{NW}} \qquad (10.18)$$

where N_{R} and N_{L} are the number of rich and lean streams, and N_{NW} is the number of *independent networks;* that is, the number of subnetworks consisting of linked paths between the connected streams. Also, when the number of mass exchangers in a MEN exceeds the minimum, the difference, $N_{\mathrm{MX}} - N_{\mathrm{MX,min}}$, equals the number of independent *mass loops*. As the mass loops are broken by combining mass exchangers, it is common for the amount of external MSA to increase. This is illustrated in the next example.

EXAMPLE 10.4 ***H_2S Removal from Sour Coke Oven Gas (Example 10.3 Revisited)***

The MEN in Figure 10.10 contains the minimum number of mass exchangers with the minimum usage of external MSA, assumed equivalent to the minimum operating costs (MOC), given by:

$$N^{\mathrm{MOC}}_{\mathrm{MX,min}} = N^{+}_{\mathrm{MX,min}} + N^{-}_{\mathrm{MX,min}} \qquad (10.19)$$

where $N^{+}_{\mathrm{MX,min}}$ and $N^{-}_{\mathrm{MX,min}}$ are the minimum number of mass exchangers on the rich and lean side of the pinch, keeping the external MSA at a minimum. For these streams, $N^{+}_{\mathrm{MX,min}} = N^{-}_{\mathrm{MX,min}} = 2 + 1 - 1 = 2$, and consequently, $N^{\mathrm{MOC}}_{\mathrm{MX,min}} = 4$. Since, according to Eq. (10.18), treating the MEN as a whole, $N_{\mathrm{MX,min}} = 2 + 2 - 1 = 3$, one mass loop exists. To reduce to the minimum number of mass exchangers, this mass loop must be broken.

SOLUTION

Suppose the mass loop is broken by eliminating the smallest mass exchanger on the rich side of the pinch and shifting its mass load around the loop. The resulting MEN is shown in Figure 10.11, where the smallest mass exchanger is assumed to be the one with the smallest amount of solute transferred. Observe that the mass of the solute to be removed by the external MSA is increased in exchanger 4 by the amount removed by the internal MSA in mass exchanger 2, or 0.00497 kg/s, giving a new total for that exchanger of 0.00507 kg/s. This corresponds to a substantial increase in the flow rate of the external MSA to 1.7044 kg/s [$= (0.000725 + 0.00507)/(0.0035 - 0.0001)$] from 0.2429 kg/s.

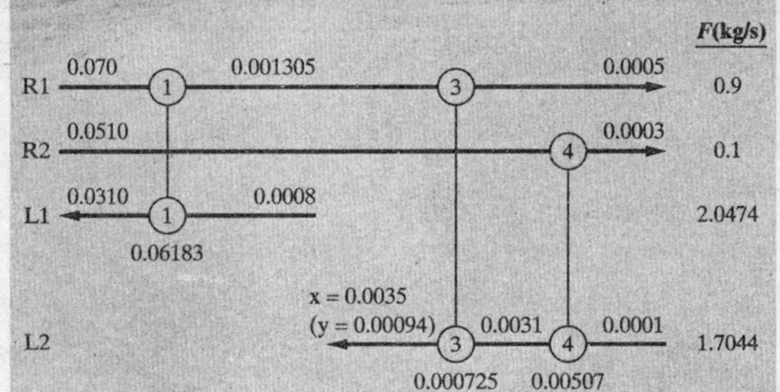

Figure 10.11 MEN having the minimum number of mass exchangers.

10.5 ADVANCED TOPICS

Many of the concepts covered in Chapter 9 on heat and power integration apply in the design of MENs. These include the threshold $\Delta x_{\mathrm{thres}}$ and the optimal Δx_{min}, as well as strategies for mathematical programming, using MILP and MINLP formulations (El-Halwagi and Manousiouthakis, 1990; Papalexandri et al., 1994). For coverage of the former, see the book by El-Halwagi (1997). In addition, Hallale and Fraser (2000a, b) show how to extend the methods with simple approximations to obtain capital and operating cost estimates when calculating annualized costs.

Yet another topic involves the extension of the synthesis methods to processes with multiple solutes. Here, the impact of concentration on the slope of phase equilibrium Eq. (10.4) may become a factor for highly nonideal solutions at high concentrations. The analysis techniques presented herein can be extended when the slopes of the equilibrium curves can be approximated as constant, independent of mixture composition. Also, the analyses are simplified when the minimum external MSA for the principal solute is capable of removing the other solutes as well (El-Halwagi and Manousiouthakis, 1989).

Finally, it is possible to synthesize HENs and MENs simultaneously, for example, in the design of heat-integrated distillation networks (Bagajewicz and Manousiouthakis, 1992; Bagajewicz et al., 1998).

10.6 SUMMARY

Having studied this chapter, the reader should understand the parallels between heat and mass integration and be prepared to carry out analyses to identify the minimum external mass-separating agent. In addition, the reader should be able to position mass exchangers in a MEN beginning at the pinch and working outward. Lastly, the reader should have learned a strategy for stream splitting and for breaking mass loops while allowing solute to be exchanged across the pinch.

REFERENCES

1. Bagajewicz, M.J., and V. Manousiouthakis, "Mass/Heat-Exchange Network Representation of Distillation Networks," *AIChE J.*, **38** (11), 1769 (1992).

2. Bagajewicz, M.J., R. Pham, and V. Manousiouthakis, "On the State Space Approach to Mass/Heat Exchanger Network Design," *Chem. Eng. Sci.*, **53** (14), 2595–2621 (1998).

3. El-Halwagi, M.M., *Pollution Prevention Through Process Integration: Systematic Design Tools*, Academic Press, San Diego (1997).

4. El-Halwagi, M.M., and V. Manousiouthakis, "Synthesis of Mass Exchange Networks," *AIChE J.*, **35** (8), 1233–1244 (1989).

5. El-Halwagi, M.M., and V. Manousiouthakis, "Automatic Synthesis of Mass Exchange Networks with Single-component Targets," *Chem. Eng. Sci.*, **45**, 2813–2831 (1990).

6. Green, D.W., and Perry, R.H., Ed., *Perry's Chemical Engineers' Handbook*, 8th ed., McGraw-Hill, New York (2008).

7. Hallale, N., and D.M. Fraser, "Capital and Total Cost Targets for Mass Exchange Networks, Part 1: Simple Capital Cost Models," *Comput. Chem. Eng.*, **23**, 1661–1679 (2000a).

8. Hallale, N., and D.M. Fraser, "Capital and Total Cost Targets for Mass Exchange Networks, Part 2: Detailed Capital Cost Models," *Comput. Chem. Eng.*, **23**, 1681–1699 (2000b).

9. Linnhoff, B., and E. Hindmarsh, "The Pinch Design Method for Heat Exchanger Networks," *Chem. Eng. Sci.*, **38**, 745 (1983).

10. Papalexandri, K.P., E.N. Pistikopoulos, and C.A. Floudas, "Mass Exchange Networks for Waste Minimization: A Simultaneous Approach," *Trans. Inst. Chem. Eng.*, **72**, 279–293 (1994).

11. Toor, H.L., "Diffusion in Three-component Gas Mixtures," *AIChE J.*, **3**, 198–207 (1957).

EXERCISES

10.1 A copolymerization plant uses benzene solvent. Benzene must be recovered from its gaseous waste stream. Two lean streams in the process, an additive stream and a catalytic solution, are potential process MSAs. Organic oil, which can be regenerated using flash separation, is the external MSA. The stream data are shown below:

Stream	F (kmol/s)	y^s or x^s	y^t or x^t
R1 (off-gas)	0.2	0.0020	0.0001
L1 (additives)	0.08	0.003	0.006
L2 (catalytic soln.)	0.05	0.002	0.004
L3 (organic oil)	Unlimited	0.0008	0.0100

In these concentration ranges, the following equilibrium equations apply.

Additives

$$y = 0.25x$$

Catalytic solution

$$y = 0.5x$$

Organic oil

$$y = 0.1x$$

(a) Show how to utilize the process MSAs and minimize the amount of the external MSA required to remove benzene from the rich stream. Let $\Delta x_{\min} = 0.0001$.

(b) Design a MEN. Assuming that the operating cost of recirculating oil (including pumping, makeup, and regeneration) is \$0.05/kg oil, calculate the annual cost of the oil.

10.2 An oil-recycling plant is shown in Figure 10.12 (El-Halwagi, 1997). Gas oil and lube oil streams are deashed and demineralized. Atmospheric distillation provides light gases, gas oil, and a heavy product, which is distilled under vacuum to produce lube oil. Subsequently, the gas oil is steam-stripped to remove light and sulfur impurities. Similarly, the lube oil is dewaxed and deasphalted before it is steam-stripped to remove light and sulfur impurities. The two wastewater streams contain phenol, a toxic pollutant that depletes oxygen, causes turbidity, and potentially causes objectionable taste and odor in fish and potable water. A MEN is to be designed involving the following internal streams:

Stream		y^s or x^s	y^t or x^t	F (kg/s)
R1	Condensate from Stripper 1	0.050	0.010	2
R2	Condensate from Stripper 2	0.030	0.006	1
L1	Gas oil	0.005	0.015	5
L2	Lube oil	0.010	0.030	3

Potential separations include solvent extraction using gas oil or lube oil. Note that phenol acts as an oxygen inhibitor, improves color stability, and reduces sediment formation in the oils. External MSAs include air, adsorption on activated carbon, and ion exchange on a polymeric resin. The following phase equilibrium data are provided for Eq. (10.4), with $b = 0$:

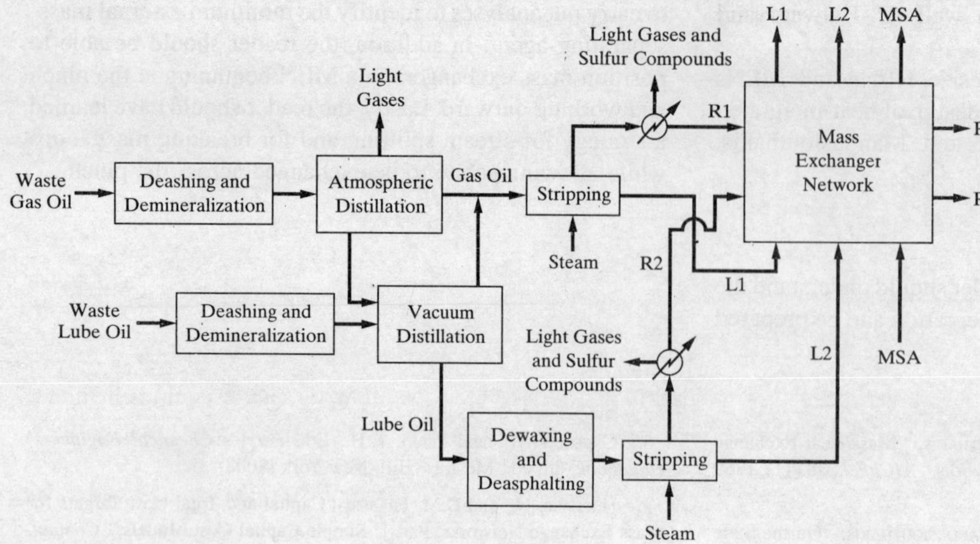

Figure 10.12 Oil-recycling plant.

MSA	m
Gas oil	2.00
Lube oil	1.53
Air	0.04
Activated carbon	0.02
Ion exchange resin	0.09

Let $\Delta x_{\min} = 0.001(\text{kg phenol/kg MSA})$.

(a) Locate the minimum amount of external MSA for a MEN. Use either the composition interval method or the composite curve method.

(b) Using activated carbon as the external MSA, design a MEN that has the minimum number of mass exchangers while utilizing the minimum amount of activated carbon.

(c) Identify any mass loops in the solution to part b. Break these mass loops to design a MEN that has the minimum number of mass exchangers.

10.3 Ammonium nitrate is a fertilizer that is also used in the production of explosives and other chemicals. It is commonly manufactured by neutralizing ammonia with nitric acid. Aqueous waste streams containing even low concentrations of ammonia and ammonium nitrate are toxic to aquatic life and lead to eutrophication of lakes.

Two rich waste streams, R1 and R2, are produced: (1) wastewater containing ammonia, and (2) condensate, from the off-gas condenser downstream from the nitric-acid ammonia reactor, containing ammonium nitrate and ammonia. The treated condensate can be used as boiler feed water. Two potential recovery operations are air stripping and ion exchange. The following table provides data for the rich waste streams (R1 and R2) and the potential lean streams, L1 (containing air) and L2 (containing ion exchange resin).

Stream	F (kg/s)	Species	y^s or x^s (kg/kg)	y^t or x^t (kg/kg)
R1	2.6	NH_3	0.006	0.004
(wastewater)		NH_4NO_3	0.0	0.0
R2	0.8	NH_3	0.02	0.001
(condensate)		NH_4NO_3	0.05	0.002
L1	Unlimited	NH_3	0.0	0.005
(air)		NH_4NO_3	0.0	0.005
L2	Unlimited	NH_3	0.0	0.005
(ion exchange resin)		NH_4NO_3	0.0	0.04

where the flow rates are expressed on a solute-free basis and the compositions are in kg of solute per kg of solute-free solvent. In these concentration ranges, the following equilibrium equations apply.

NH_3

$$y_{\text{air}} = 0.788(x_{\text{air}} + 0.001) - 0.0002$$
$$y_{\text{resin}} = 0.11x_{\text{resin}} - 0.0006$$

NH_4NO_3

$$y_{\text{air}} = 0.98x_{\text{air}}$$
$$y_{\text{resin}} = 0.168x_{\text{resin}} - 0.0001$$

Let $\Delta x_{\min} = 0.0001$.

(a) Use NH_3 as the MSA. Find the minimum amount of NH_3 to achieve the separation.

(b) Use NH_4NO_3 as the MSA. Find the minimum amount of NH_4NO_3 to achieve the separation.

(c) For unit costs of air, 0.001 \$/kg, and ion exchange resin, 0.05 \$/kg, design MENs for parts (a) and (b).

(d) Repeat part (c) for the cheapest network using NH_3 and NH_4NO_3.

Chapter 11

Batch Process Design

11.0 OBJECTIVES

This chapter introduces strategies for designing and scheduling batch processes. It begins with single equipment items, focusing on methods for achieving the optimal batch time and batch size. Then, reactor–separator processes are examined, where tradeoffs exist between the reaction conversion, as it varies with reaction time; the cost of separation, which decreases with conversion; and the batch cycle time. Subsequently, methods of scheduling batch processes with recipes having numerous tasks and equipment items are considered. Initially, schedules are considered for plants involving a single product, produced in production trains that are repeated from batch to batch. The chapter concludes with strategies for designing multiproduct batch plants.

After studying this chapter, the reader should

1. Be knowledgeable about process units operated in batch mode and approaches for optimizing their designs and operations.
2. Know how to determine the optimal reaction time for a batch reactor–separator process.
3. Be able to schedule recipes for the production of a single chemical product.
4. Understand how to schedule batch plants for the production of multiple products.

11.1 INTRODUCTION

Continuous processes are dominant in the chemical process industries for the manufacture of commodity chemicals, plastics, petroleum products, paper, etc. When production rates are low, however—say, in the manufacture of specialty chemicals, pharmaceuticals, and electronic materials—it is difficult to justify the construction of a continuous plant comprised of small vessels and pipes. In these cases, it is common to design *batch processes* or *semicontinuous processes* that are hybrids of batch and continuous processes. The alternatives are illustrated schematically in Figure 11.1, with a continuous process shown in Figure 11.1a. In the batch process of Figure 11.1b, the chemicals are fed *before* (step 1) and the products are removed *after* (step 3) the processing (step 2) occurs. *Fed-batch processes* combine the first two steps with some or all chemicals being fed continuously during the processing. Then, when the processing is finished, the products are removed batch-wise, as shown in Figure 11.1c. In *batch-product removal*, the chemicals are fed to the process before processing begins and steps 2 and 3 are combined; that is, the product is removed continuously as the processing occurs, as shown in Figure 11.1d. In effect, fed-batch and batch-product removal processes are semicontinuous processes.

The challenge in designing a batch, fed-batch, or batch-product removal process is in deciding on the size of the vessel and the processing time. This is complicated for the latter two processes, where the flow rate and concentration of the feed stream or the flow rate of the product stream as a function of time strongly influences the performance of the process. Note that the determination of optimal operating profiles is referred to as the solution to the *optimal control problem*. This subject is introduced in the next section.

Batch and semicontinuous processes are utilized often when production rates are small, residence times are large, and product demand is intermittent, especially when the demand for a chemical is interspersed with the demand for one or more other products and the quantities needed and the timing of the orders are uncertain. Even when the demand is continuous and the production rates are sufficiently large to justify continuous processing, batch and semicontinuous processes are often designed to provide a reliable, though inefficient, route to the production of chemicals. For example, in the emulsion polymerization of resins, large batch reactors are installed, often to avoid

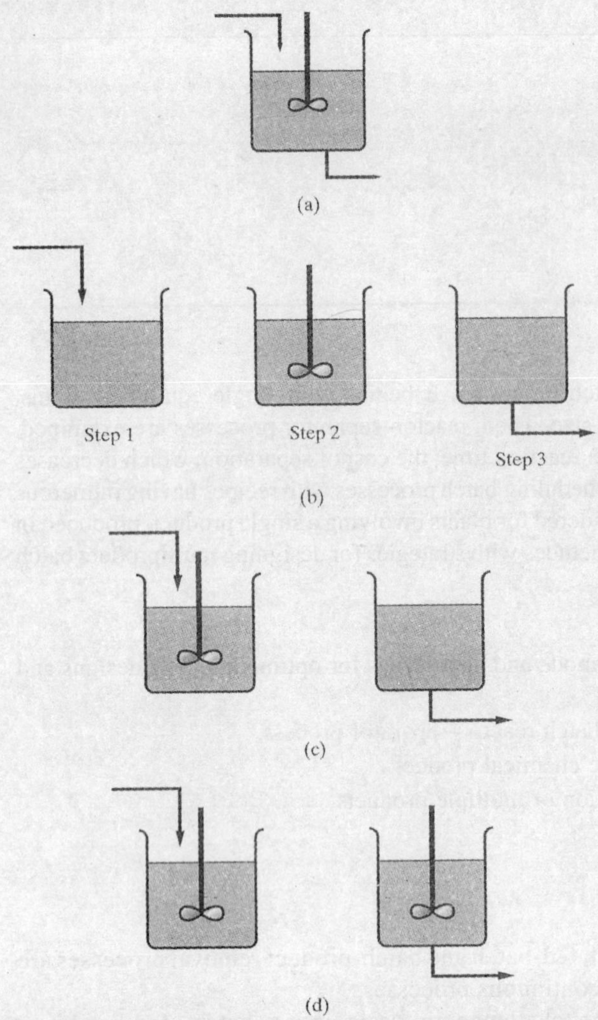

Figure 11.1 Continuous and batch processes: (a) continuous process; (b) batch process; (c) fed-batch process; (d) batch-product removal process.

carrying out these highly exothermic reactions in continuous stirred-tank reactors. Note, however, that while operation at a low-conversion steady state is often less profitable than batch or semicontinuous processing, operation at an open-loop, unstable steady state is often more profitable. Rather than install a control system to stabilize the operation, many companies prefer to operate in batch or semicontinuous mode. Similarly, design teams often opt for batch and semicontinuous processes when the chemicals are hazardous or toxic or when safety aspects are of great concern.

Because the designs for continuous and batch processes are usually very different, the choice of processing mode is made commonly during process synthesis, in the task-integration step, as discussed in Section 4.4. At this stage, the decision to reject continuous processing is based upon rules of thumb rather than a detailed comparison of the alternatives. Through process simulation, as discussed in Chapter 5, and the optimization methods presented in this chapter, more algorithmic methods are available for selecting from among the various batch and continuous processes.

Usually, for the production of small quantities of high-priced chemicals, such as in the manufacture of pharmaceuticals, foods, electronic materials, and specialty chemicals, batch, fed-batch, and batch-product removal processes are preferred. This is often the case in bioprocessing, for example, when drugs are synthesized in a series of chemical reactions, each having small yields, and requiring difficult separations to recover small amounts of product. This is also the case for banquet facilities in hotels, which prepare foods in batches, and for many unit operations in the manufacture of semiconductors. As discussed in Chapters 4 and 5, these processes usually involve a *recipe*, that is, a sequence of *tasks*, to be carried out in various items of equipment. In the latter sections of this chapter, variations on batch process schedules are discussed, as well as methods for optimizing the schedules.

11.2 DESIGN OF BATCH PROCESS UNITS

When designing a process unit to operate in batch mode, it is usually desired to determine the *batch time*, τ, and the *size factor*, *S*, which is usually expressed as the volume per unit mass of product, that maximize an objective like the amount of product. To accomplish this, a dynamic model of the process unit is formulated and the degrees of freedom adjusted, as illustrated in the examples that follow. As will be seen, there are many ways to formulate this *optimal control problem*. To simplify the discussion, models are presented and studied for various input profiles, to see how they affect the objectives. Emphasis is not placed on the formal methods of optimization.

Batch Processing

For conventional batch processing, with no material transfer to or from the batch, performance is often improved by adjusting the operating variables such as temperature and agitation speed. Through these adjustments, reactor conversion is improved, thereby reducing the batch time to achieve the desired conversion. An example is presented next that shows how to achieve this objective by optimizing the temperature during batch processing.

EXAMPLE 11.1 ***Exothermic Batch Reactor***

Consider a batch reactor to carry out the exothermic reversible reaction:

$$n_1 A \rightleftharpoons n_2 B$$

where the rate of consumption of A is:

$$r\{c_A, c_B, t\} = c_A^{n_1} k_1^o e^{\frac{-E_1}{RT}} - c_B^{n_2} k_2^o e^{\frac{-E_2}{RT}} \tag{11.1}$$

and where $E_1 < E_2$ for the exothermic reaction. The reaction is charged initially with A and B at concentrations c_{A_o} and c_{B_o}. To achieve a specified fractional conversion of A, $X = (c_{A_o} - c_A)/c_{A_o}$, determine the profile of operating temperature in time that gives the minimum batch time. This example is based upon the development by Denn (1969).

SOLUTION

The minimum batch time, τ_{min}, is achieved by integrating the mass balances:

$$\frac{dc_A}{dt} = -r\{c_A, c_B, t\} \tag{11.2}$$

$$c_B\{t\} = c_{B_o} + \frac{n_1}{n_2}[c_{A_o} - c_A\{t\}] \tag{11.3}$$

while adjusting T at each point in time to give the maximum reaction rate.

The temperature at the maximum reaction rate is obtained by differentiation of Eq. (11.1) with respect to T:

$$\frac{dr}{dT} = 0 \tag{11.4}$$

Rearranging:

$$T_{opt} = \frac{E_2 - E_1}{R \ln \dfrac{c_B^{n_2} k_2^o E_2}{c_A^{n_1} k_1^o E_1}} \tag{11.5}$$

When an upper bound in temperature, T^U, is assigned, the typical solution profile is as shown in Figure 11.2. Initially, when $T_{opt} > T^U$, the reactor temperature is adjusted to the upper bound, T^U. Then, as conversion increases, the reactor temperature decreases, leveling off to the equilibrium conversion. In practice, this optimal temperature trajectory is approached using feedback control, with the coolant flow rate adjusted to give temperature measurements that track the optimal temperature trajectory.

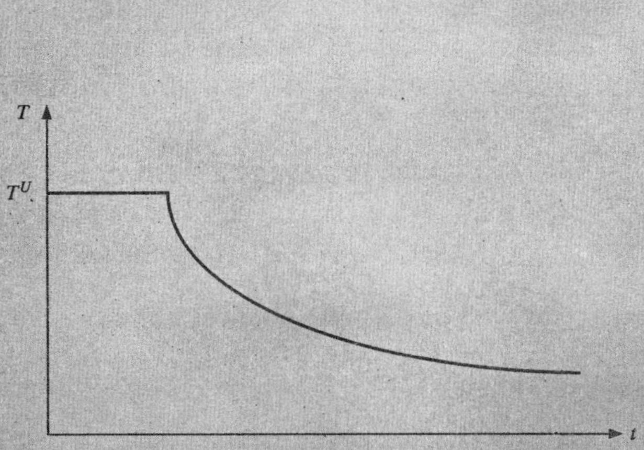

Figure 11.2 Temperature profile to minimize batch reactor time.

Fed-Batch Processing

Fermentation processes for the production of drugs are usually carried out in fed-batch reactors. In these reactors, it is desirable to find the best profile for feeding substrate into the fermenting broth, as illustrated in the next example.

EXAMPLE 11.2 *Biosynthesis of Penicillin*

Consider the fed-batch reactor in Figure 11.3. Initially, the reactor is charged with an aqueous volume, $V\{0\}$, containing *E. coli* cells (referred to as biomass) in concentration $X\{0\}$. Then, an aqueous solution of sucrose (referred to as the substrate; i.e., the substance being acted on) at a concentration S_f(g/L) is fed to the reactor at a variable flow rate, $F\{t\}$(g/hr). The reactor holdup, $V\{t\}$, contains *E. coli* cells in concentration $X\{t\}$ (g/L), penicillin product in concentration $P\{t\}$ (g/L), and sucrose in concentration $S\{t\}$ (g/L). Using Monod kinetics, the specific growth-rate of the cell mass (g cell growth/g cell) is

$$\mu = \mu_{max}\left(\frac{S}{K_x X + S}\right)$$

Lim and co-workers (1986) developed the following expressions for the specific rate of penicillin production (g penicillin/g cell):

$$\rho = \rho_{max}\left(\frac{S}{K_p + S(1 + S/K_{in})}\right)$$

and for the specific consumption rate of substrate (g substrate/g cell):

$$s = m_s\left(\frac{S}{K_m + S}\right)$$

Using mass balances for the cell mass, penicillin, and substrate, as well as the overall mass balance, the following rate equations can be derived (see Exercise 11.2):

$$\dot{X}\{t\} = \mu\{X, S\}X - \frac{X}{S_f V}F$$

$$\dot{P}\{t\} = \rho\{S\}X - K_{deg}P - \frac{P}{S_f V}F$$

$$\dot{S}\{t\} = \mu\{X, S\}\frac{X}{Y_{X/S}} - \rho\{S\}\frac{X}{Y_{P/S}} - s\{S\}X + \left(1 - \frac{S}{S_f}\right)\frac{F}{V}$$

$$\dot{V}\{t\} = \frac{F}{S_f}$$

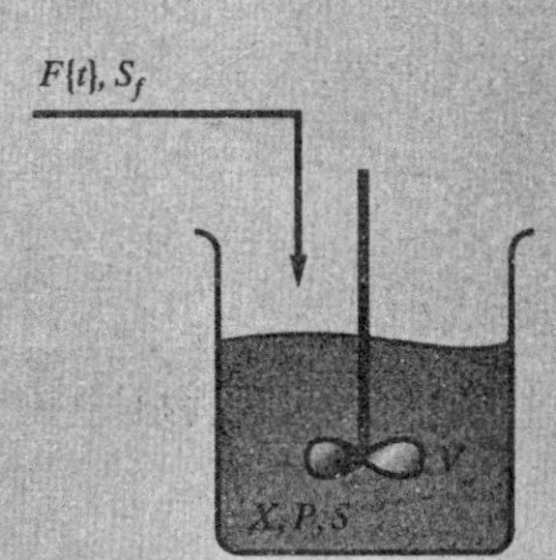

Figure 11.3 Batch penicillin reactor.

where $Y_{X/S}$ and $Y_{P/S}$ are the yield coefficients that relate the rate of substrate consumption to the rates of cell growth and penicillin production, respectively.

Using the feed concentration $S_f = 500g$ S/L, with the kinetic parameters by Lim and co-workers (1986)—$\mu_{max} = 0.11\ \text{hr}^{-1}$, $K_x = 0.006$ g S/g X, $K_P = 0.0001$ g S/L, $\rho_{max} = 0.0055$ g P/(g X · hr), $K_{in} = 0.1$ g S/L, $K_{deg} = 0.01\ \text{hr}^{-1}$, $m_s = 0.029$ g S/(g X · hr), $K_m = 0.0001$ g S/L, $Y_{X/S} = 0.47$ g X/g S} and $Y_{P/S} = 1.2$ g P/g S—for the initial conditions $V\{0\} = 7$ L, $X\{0\} = 1.5$ g/L, and $P\{0\} = S\{0\} = 0$ and for the constraints

$$
\begin{gathered}
0 \leq X\{t\} \leq 40 \\
0 \leq S\{t\} \leq 100 \\
0 \leq V\{t\} \leq 10 \\
0 \leq F\{t\} \leq 50 \\
72 \leq X\{\tau\} \leq 200
\end{gathered}
$$

Cuthrell and Biegler (1989) maximize the production of penicillin, $P\{T\}V\{T\}$, where τ is the batch time, using variational calculus (Pontryagin maximum principle) to obtain the solution in Figure 11.4. As seen, at the optimum, the batch time is 124.9 hr, with the production of 87.05 g of penicillin. It is notable that the optimal feed flow rate is 50 g/hr for the first 11.21 hr, after which the feed stream is turned off until 28.79 hr, when it is held constant at 10 g/hr. This "on–off" control strategy is commonly referred to as *bang–bang control*. To confirm the cell mass, penicillin, and substrate concentration profiles, the differential equations can be integrated using a mathematical software package such as MATLAB. Furthermore, Cuthrell and Biegler (1989) show how to solve numerically for the optimal solution, which is often referred to as the "optimal control profile," using orthogonal collocation on finite elements to discretize the differential equations, and successive quadratic programming (SQP).

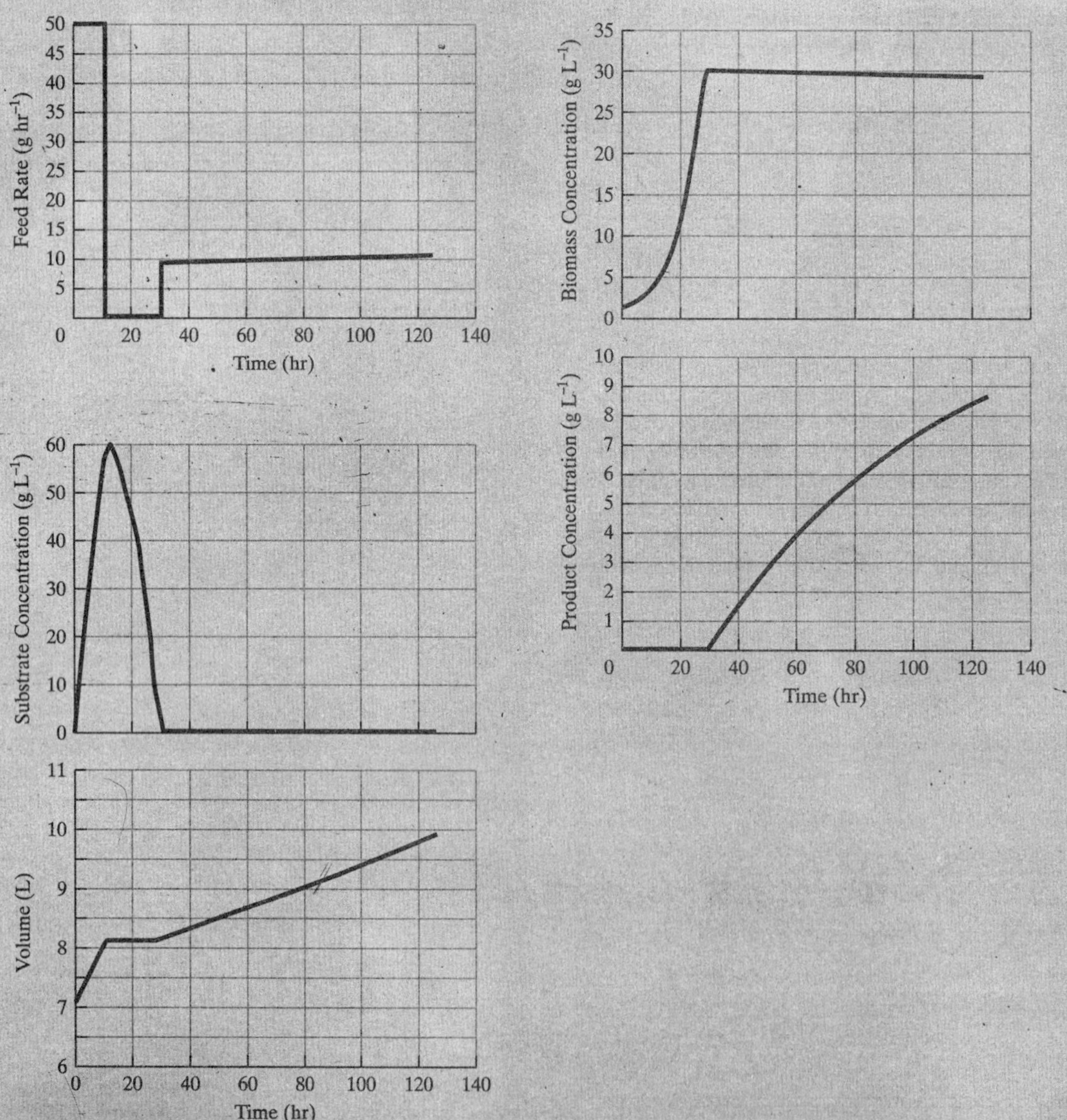

Figure 11.4 Optimal profiles for the penicillin reactor.

Batch-Product Removal

When distillations are carried out in batch mode, the still is charged with the feed mixture and the heat is turned on in the reboiler. The lightest species concentrate in the distillate, which is condensed and recovered in *batch-product removal* mode. As the light species is recovered, it is accompanied by increasing fractions of the heavier species unless a strategy is applied to maintain a high concentration of light species.

For multicomponent separations, to simplify operation it is often satisfactory to adjust the reflux rate once or twice while recovering each species. When the purity of a species that is being collected in the product accumulator drops below its specification, the contents of the product accumulator are dumped into its product receiver, and the reflux rate is adjusted. Stated differently, the reflux rate is increased as the difficulty of the separation between the light and heavy key components increases. This is illustrated in the next example.

EXAMPLE 11.3 *Batch Distillation*

A 100-lb mole mixture of methanol, water, and propylene glycol, with mole fractions 0.33, 0.33, and 0.34, is separated using a 15-tray batch distillation operation, as shown in Figure 11.5. Assume operation at a nominal pressure of 1 atm, realizing that the pressure in the still will have to be somewhat higher than this to avoid a vacuum in the reflux accumulator. The tray and condenser liquid holdups are 0.1 ft^3/tray and 1.0 ft^3, respectively.

In an attempt to devise a satisfactory operating strategy, the following recipe (also called a campaign) was proposed:

Methanol Recovery

1. Bring the column to total reflux operation, with the distillate valve closed.
2. Using a reflux ratio of 3, send 5 lbmol/hr of distillate continuously to the product accumulator. Continue until the mole fraction of water in the instantaneous distillate reaches 0.001.
3. Bring the column to total reflux.
4. Using a reflux ratio of 5, send 2.5 lbmol/hr of distillate continuously to the product accumulator. Continue until the mole fraction of water in the instantaneous distillate reaches 0.001, at which point the distillate valve is closed. Dump the contents of the product accumulator into the methanol product receiver.

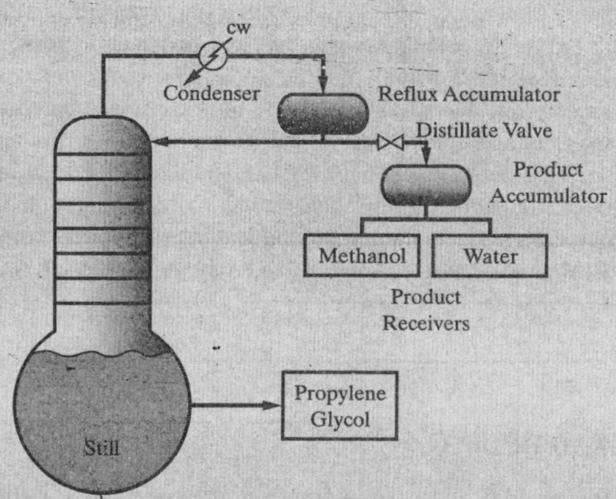

Figure 11.5 Batch distillation operation.

Propylene Glycol Recovery

5. The column is now at total reflux.
6. Using a reflux ratio of 3, send 20 lbmol/hr of distillate continuously to the product accumulator. Continue until the mole fraction of propylene glycol in the instantaneous distillate reaches 0.001. Dump the contents of the product accumulator into the water product receiver.
7. Pump the contents of the still pot into the propylene glycol product receiver.

To examine the performance of this recipe, it is helpful to use a batch distillation program in a process simulator, such as BATCHSEP by Aspen Technology, Inc. Then, a processing objective can be specified (such as minimum batch time, energy consumption, or reject chemicals, or some combination of these), and variations on the recipe can be explored in an effort to achieve more optimal operation, as in Exercise 11.3.

SOLUTION

Using the BATCHSEP simulator, the results are as follows, where Distil. and Accum. refer to the instantaneous distillate and product accumulator, respectively:

After Step		Methanol	Water	Propylene Glycol	Total Amount (lbmol)	Step Time (hr)	Total Time (hr)
1	Charge	0.3300	0.3300	0.3400	100		0
2	Distil.	0.9990	0.0010	—	—	5.63	5.63
	Accum.	0.9999	0.0001	—	28.14		
	Still^x	0.0676	0.4592	0.4732	71.86		
3	Total reflux						
4*	Distil.	0.9990	0.0010	—	—	0.69	6.32
	Accum.	0.9999	0.0001	—	29.88		
	Still^x	0.0446	0.4705	0.4849	70.12		
5	Total reflux						
6*	Distil.	—	0.9990	0.0010	—	1.65	7.97
	Accum.^{x}	0.0947	0.9053	—	37.09		
	Still	—	0.0002	0.9998	33.03		

*Before the reflux accumulator is dumped.

^{x}Includes tray and condenser liquid holdups

As can be seen, the total batch time is nearly 8 hr, and the amounts of 99.9 mol% methanol and 99.98 mol% propylene glycol products are 29.88 lbmol and 33.03 lbmol, respectively. Note that after step 4 the methanol product accumulator contains $0.999 \times 29.88 = 29.85$ lbmol methanol. The remainder, $33.33 - 29.85 = 3.48$ lbmol methanol, is recovered initially in the water product accumulator during step 6. Hence, the water product accumulator contains a "slop cut" of water. Nearly all of the propylene glycol is recovered in the still. These results can be reproduced using the BATCHSEP file, EXAM11-3.bspf, in the Program and Simulation Files folder, which can be downloaded from the Wiley Web site associated with this book.

www.wiley.com/go/global/seider

11.3 DESIGN OF REACTOR–SEPARATOR PROCESSES

In this section, an approach to solving the optimal control problem is introduced for reactor-separator processes. The approach involves the simultaneous determination of the batch times and size factors for both of the process units. Furthermore, the interplay between the two units involves tradeoffs between them that are adjusted in the optimization. It should be noted that simpler models than in normal practice are used here to demonstrate the concept and, in the first example, to provide an analytical solution that is obtained with relative ease.

EXAMPLE 11.4

Consider the batch reactor–separator combination in Figure 11.6, initially presented by Rudd and Watson (1968). In the reactor, the isothermal irreversible reaction A → B is carried out. The separator recovers product B from unreacted A. The rate constant for the reaction is $k = 0.534\ \text{day}^{-1}$. It is assumed that the cost of A is negligible compared to the value of product B, $C_B = \$2.00/\text{lb}$; that the operating costs of the reactor are proportional to the batch time (that is, $C_o = \alpha t$, where $\alpha = \$100/\text{day}$); and that the cost of separation per batch, C_S, is inversely proportional to the conversion of A in the reactor, X (that is, $C_S = KV/X$), where K is the proportionality constant and V is the holdup volume in the reactor. Find the batch time for the reactor when the gross profit in \$/day is maximized. Let $c_{Ao} = 10\ \text{lb/ft}^3$, $V = 100\ \text{ft}^3$, and $K = 1.0\ \$/(\text{ft}^3\cdot\text{batch})$. Note that as the conversion of A to product B increases, the cost of separation decreases, as shown in Figure 11.7b.

SOLUTION

For the first-order reaction, it can be shown (Exercise 11.4) that:

$$X = 1 - e^{-kt} \tag{11.6}$$

as illustrated in Figure 11.7a, and

$$c_B = c_{Ao}X \tag{11.7}$$

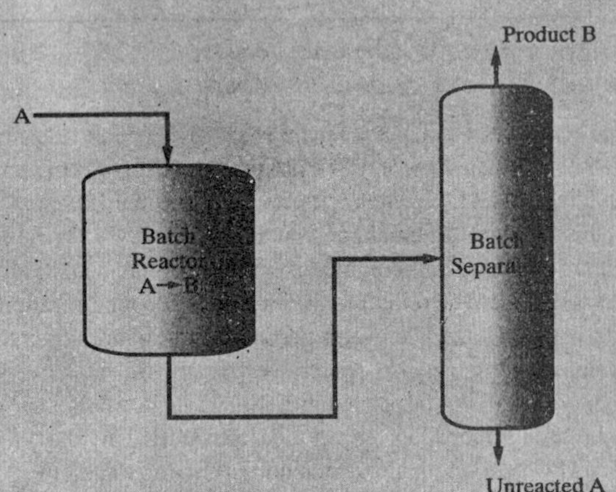

Figure 11.6 Reactor–separator process.

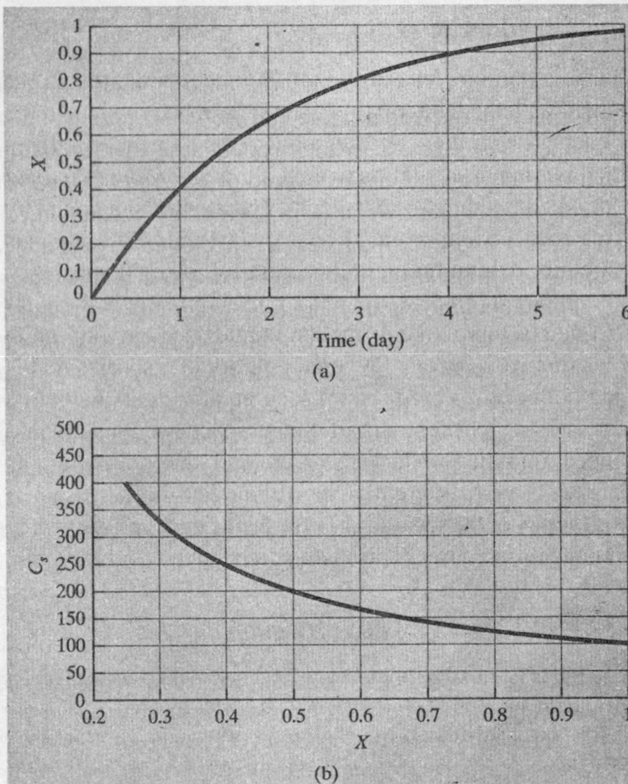

Figure 11.7 Conversion and cost of separation. First-order response of conversion; (b) cost of separation as a function of conversion.

It is desired to maximize the gross profit, GP, in \$/day; that is,

$$\max_t \text{GP} = \frac{1}{t}\left[C_B V c_B\{t\} - C_O\{t\} - C_S\{X\{t\}\}\right] \tag{11.8}$$

To locate the maximum, substitute the equations above and differentiate:

$$\frac{d\text{GP}}{dt} = 0$$

It can be shown (Exercise 11.4) that at the maximum, the reactor batch time is $\tau_{\text{opt}} = 1.35$ day, with a fractional conversion, x_{opt}, of 0.514. At longer times, while the revenues increase due to increased conversion and the separation costs decrease, the gross profit decreases due to the increase in the batch time in the denominator of Eq. (11.8). At shorter times, the revenues decrease due to smaller conversion, and separation costs increase due to a more difficult separation, more than countering the decrease in the denominator and leading to smaller gross profit. At the maximum, $\text{GP}_{\text{opt}} = \$516.8/\text{day}$. Note that GP < 0 when $t < 0.507$ day.

EXAMPLE 11.5

Figure 11.8 shows an isothermal batch reactor in which the irreversible reactions A → B → C take place, with B the desired

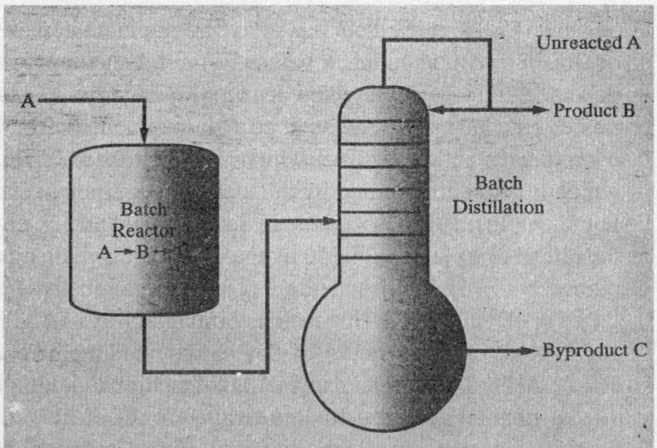

Figure 11.8 Reactor–distillation process.

product and C an undesired byproduct. The reactions are irreversible and first-order in the reactants. The reaction rate constants are $k_1 = 0.628\,\text{hr}^{-1}$ and $k_2 = 0.314\,\text{hr}^{-1}$. The reactor products are fed from an intermediate storage tank (not shown) to a batch distillation column, from which the most volatile species, A, is removed first in the distillate, followed by product B, which is the intermediate boiler. For specified reactor volume V_r, and column volume V_c, it is desired to determine the batch times for the reactor and column that minimize the cost of producing the desired amount of product B, B_{tot} moles, in a single campaign. Following Barrera and Evans (1989), the analysis is simplified by assuming that the distillation column produces pure A until it is depleted from the still, followed by pure B. Furthermore, specifications include the total molar concentration in the reactor, C, the distillate flow rate, F_d, the time horizon within which the campaign must be completed, τ_{hor}, and the cleaning times between batches for the reactor and distillation column, t_{cr} and t_{cc}. In addition, several cost coefficients are specified, including the cost of fresh feed, P_A (\$/mol); recycling credit for A, P_{rA}; cost or credit for the byproduct C, P_C; the rental rates of the reactor, r_r (\$/hr), intermediate storage, r_s, and distillation column, r_c; the costs of cleaning the reactor, C_{clr} (\$/batch), and distillation column, $C_{c\ell c}$; and the utility cost per mole of distillate, P_u. These specifications are given in Table 11.1.

Table 11.1 Specifications for Example 11.5

$V_r = 3{,}503\,\text{L},\ V_c = 784\,\text{L}$
$B_{tot} = 800\,\text{mol}$
$C = 0.5\,\text{mol/L}$
$F_d = 206.2\,\text{mol/hr}$
$\tau_{hor} \leq 24\,\text{hr}$
$t_{cr} = 0.8\,\text{hr},\ t_{cc} = 0.5\,\text{hr}$
$P_A = 1\,\$/\text{mol},\ P_{rA} = 0.2\,\$/\text{mol},\ P_C = 0.4\,\$/\text{mol}$
$r_r = 100\,\$/\text{hr},\ r_s = 20\,\$/\text{hr},\ r_c = 100\,\$/\text{hr}$
$C_{clr} = 100\,\$/\text{batch},\ C_{clc} = 100\,\$/\text{batch}$
$P_u = 0.45\,\$/\text{mol}$

SOLUTION

By integration of the kinetic rate equations, it can be shown (Exercise 11.5) that the mole fraction profiles in time are given by:

$$x_A = e^{-k_1 t_r} \tag{11.9}$$

$$x_B = \frac{k_1(e^{-k_1 t_r} - e^{-k_2 t_r})}{k_2 - k_1} \tag{11.10}$$

$$x_C = 1 - x_A - x_B \tag{11.11}$$

These profiles are graphed in Figure 11.9.

The minimization of the campaign costs is expressed as:

$$\min_{t_r, T_c} \phi = \phi_{rm} + \phi_{eq} + \phi_{cl} + \phi_u$$

where t_r is the reaction time (reactor operating time), τ_c is the batch time for the distillation column, and the campaign costs for the raw materials, equipment rental, cleaning, and utilities, are

$$\phi_{rm} = CV_r[P_A + P_{rA}x_A + P_C x_C]\tau_{tot}/\tau_r$$
$$\phi_{eq} = [r_r + r_s + r_c]\tau_{tot}$$
$$\phi_{cl} = [C_{clr}/\tau_r + C_{clc}/\tau_c]\tau_{tot}$$
$$\phi_u = P_u F_d t_c \tau_{tot}/\tau_c$$

Note that the quantity $P_u F_d$ is \$/hr charge for utilities, t_c/τ_c is the fraction of column batch time during which the column is operational (and using utilities), and τ_{tot} is the total batch time. These quantities are multiplied to give ϕ_u, the cost of utilities per batch. This assumes that the utilities are applied during a t_c/τ_c fraction of the reactor operation. The minimization is carried out subject to a mass balance for the reactor:

$$B_{tot} = CV_r x_B \tau_{tot}/\tau_r,$$

an equation that ensures periodicity in the storage tank (i.e., equal volumes processed per unit time):

$$\frac{V_r}{\tau_r} = \frac{V_c}{\tau_c},$$

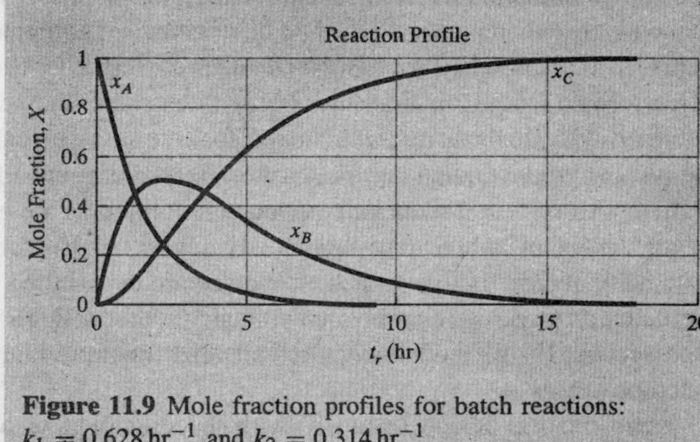

Figure 11.9 Mole fraction profiles for batch reactions: $k_1 = 0.628\,\text{hr}^{-1}$ and $k_2 = 0.314\,\text{hr}^{-1}$.

a mass balance for the distillation column:

$$t_c = V_r C(1 - x_C)/F_d$$

and the total campaign time:

$$\tau_{tot} = \tau_r + \tau_c$$

where τ_r is the batch time for the reactor and t_c is the column operating time. In addition, as the optimization proceeds, it is necessary to satisfy the following inequality constraints:

$$\tau_{tot} \leq \tau_{hor}$$
$$\tau_r - (t_r + t_{cr}) \geq 0$$
$$\tau_c - (t_c + t_{cc}) \geq 0$$

For the specifications in Table 11.1, the following solution was obtained using successive quadratic programming (SQP) in GAMS (see Section 24.6). The reactor operates for $t_r = 4.44$ hr to produce a product containing 0.062, 0.373, and 0.565 mole fractions of A, B, and C. For periodic cycling, its batch time is $\tau_r = 18.74$ hr, which exceeds the total of the reactions and cleaning times. The distillation column operates for $t_c = 3.69$ hr, which together with its cleaning time, $t_{cc} = 0.5$ hr, gives a batch time for the column, $\tau_c = 4.19$ hr. The total time, or cycle time, is $\tau_{tot} = 22.93$ hr, which falls within the horizon time specified, $\tau_{hor} = 24$ hr. These times correspond to the minimum cost, $\phi = \$10,240$. By varying t_r, it can be shown that this is the minimum cost.

11.4 DESIGN OF SINGLE-PRODUCT PROCESSING SEQUENCES

Having examined small optimal control problems for batch process units in Section 11.2 and for reactor–separator sequences in Section 11.3, it should be clear that the determination of optimal batch times, given batch sizes expressed as batch volumes per unit mass of product, can be demanding computationally. Since most processes, in practice, have recipes with numerous tasks and a comparable number of processing units (e.g., the tPA process in Chapters 4 and 5), it is normally not practical to optimize the batch times for the individual processing units when preparing a schedule of tasks and equipment items for the manufacture of a product. Consequently, when preparing a schedule of tasks and equipment items, it is common to specify batch times for tasks to be performed in specific units, usually with batch sizes, and to optimize cycle times for a specific recipe. In some cases, using the rates of production and yields, the vessels are designed as well; that is, vessel sizes are determined to minimize the cost of the plant while determining the cycle times for a specific recipe. In this section, schedules are determined for the batch processes that involve only single products. In the next section, the methodology is extended for multiproduct batch processes.

Batch process design begins with the specification of a *recipe* of *tasks* to produce a product. In continuous processing, each task is carried out in a specific equipment item, with one-to-one correspondence between them, and shown on a flowsheet that remains fixed in time. Similarly, in batch processes, the tasks are assigned to equipment items, but over specific intervals of time, which vary with batch size, which is often determined by the available equipment sizes. For example, in the tPA process in Sections 4.4 and 5.5, given the rate of tPA production [50 pg tpa/(cell-day), where pg are picograms $= 10^{-12}$g] and the cell concentration (between 0.225×10^6 and 3×10^6cell/mL), the availability of a 5,000-L cultivator determines the 14-day batch time and the batch size (2.24 kg of tPA, produced in 4,000 L of medium, yielding 1.6 kg of final product) for the cultivator. As discussed in Section 4.4, process synthesis involves the creation of a sequence or flowsheet of operations, which can be referred to as a *recipe* of operations or tasks. During the *task-integration* step, tasks are often combined to be carried out in a single equipment item, for example, heating and reaction in a pyrolysis furnace. Also during this step, the decision to use continuous or batch processing is made. At this point, the available equipment sizes often determine the batch sizes and times.

Batch Cycle Times

When scheduling and designing batch processes, several formalisms are widely used, as reviewed by Reklaitis (1995). In this section, and those that follow, portions of the presentation are derived from his article.

In batch processes, it is common for a task to consist of a sequence of *steps* to be carried out in the same equipment unit. For example, Figure 11.10 shows a typical recipe with its tasks and steps. Note that each step involves a *batch time,* which is determined by the processing rates and the *batch size,* that is, the amount of the *final* product manufactured in one batch. Furthermore, a *production line* is a set of equipment items assigned to the tasks in a recipe to produce a

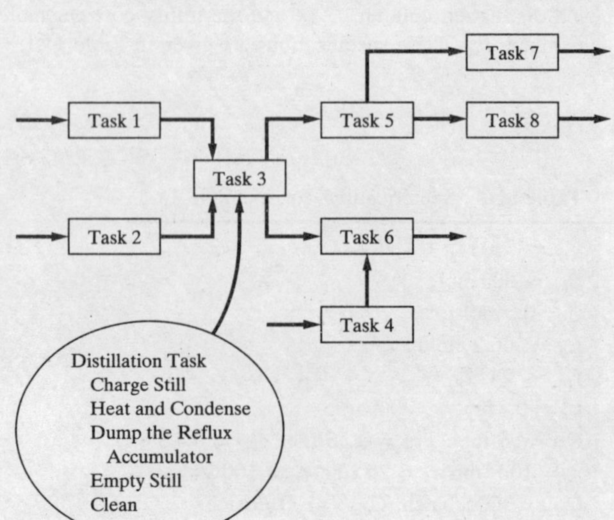

Figure 11.10 Recipes, tasks, and subtasks.

product. When a production line is used to produce a sequence of identical batches, the *cycle time* is the time between the completions of batches. To better visualize the schedule of production, an equipment occupation diagram known as a *Gantt chart* is prepared, showing the periods of time during which each equipment item is utilized, as shown in Figure 11.11a. Note that the unit having the longest batch time (6 hr), U2, is the *bottleneck* unit, as it is always in operation. Note also that the second batch is begun in time to produce the feed to unit U2 when the latter becomes available after processing the first batch. In this diagram, the batches are transferred from unit-to-unit immediately (so-called *zero-wait* strategy, with no intermediate storage utilized). Clearly, the cycle time, 6 hr, is the batch time of U2.

In the schedule in Figure 11.11a, the serial process has a distinct task assigned to each equipment item. Often, to utilize the equipment more efficiently, it is possible to use an equipment item to carry out two or more tasks. Note that this may not be possible when manufacturing specialty chemicals that are very sensitive to contamination, as in

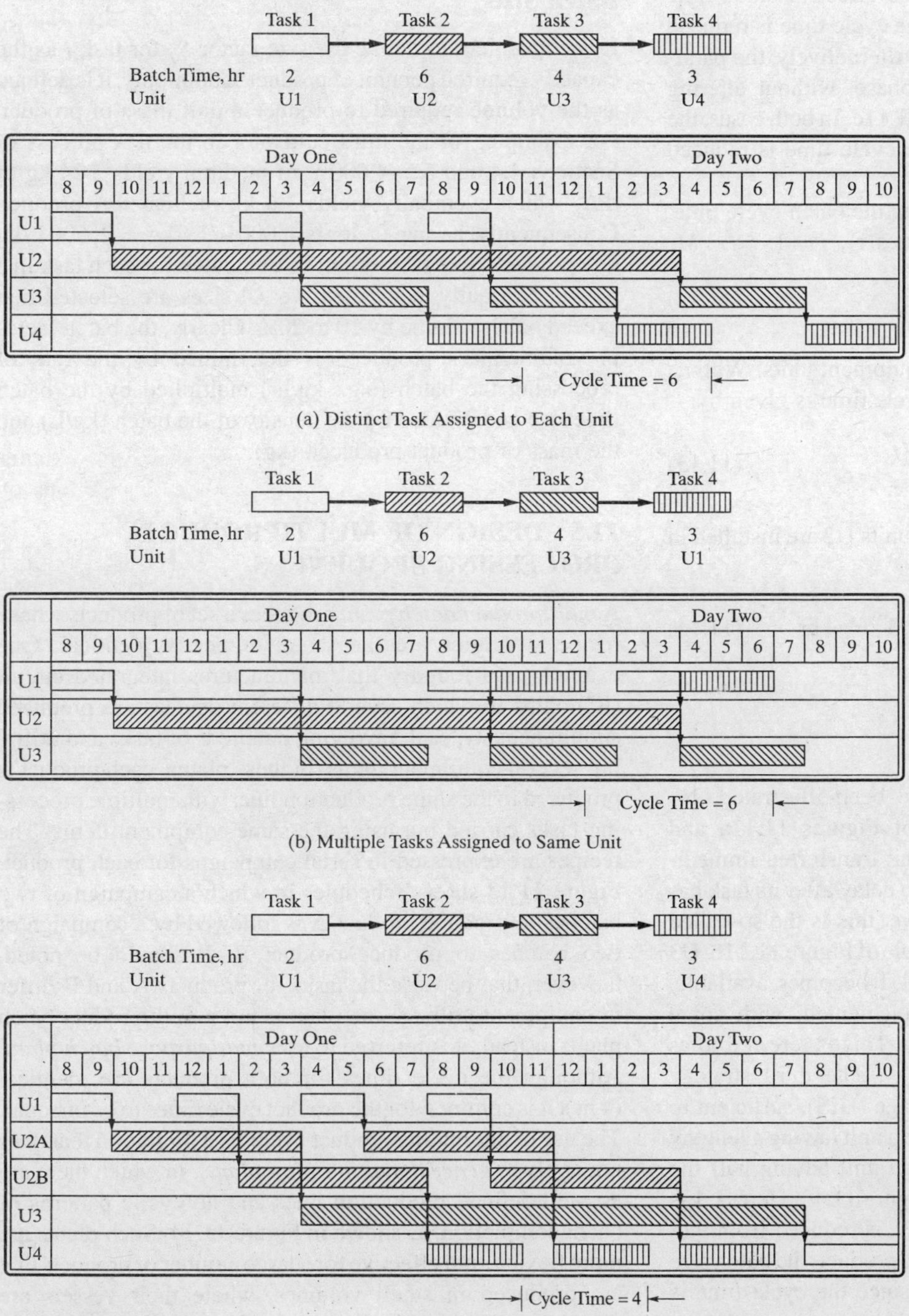

(c) Parallel Units

Figure 11.11 Serial recipe and Gantt charts.

the manufacture of pharmaceuticals. Returning to the schedule in Figure 11.11a, when the fourth task can be carried out in U1, this unit is better utilized and U4 can be released for production elsewhere in the batch plant, as shown in Figure 11.11b. Note that to achieve this schedule without adding intermediate storage, it is necessary to retain the batch within U3 until U1 becomes available. Furthermore, to increase the efficiency of the schedule, that is, to reduce the cycle time, it is common to add one or more units in parallel. When in phase, it is clear that the batch time for the unit is reduced to τ_j/n_j, where n_j is the number of units in parallel for task j. For example, when two U2 units, each half-size, are installed in parallel, the effective batch time for unit U2 is reduced to 3 hr, and the cycle time is reduced to 4 hr, with U3 the bottleneck unit. Alternatively, the parallel units can be sequenced out-of-phase, without altering their batch time, as shown in Figure 11.11c. In both cases, the U2 bottleneck is eliminated and the cycle time is reduced to 4 hr.

Clearly, without parallel operation, the batch cycle time, CT, is the maximum of the batch times τ_j, $j = 1, \ldots, M$:

$$CT = \max_{j=1,\ldots,M} \tau_j \tag{11.12}$$

where M is the number of unique equipment units. With n_j units in parallel and in phase, the cycle time is given by:

$$CT = \max_{j=1,\ldots,M} \frac{\tau_j}{n_j} \tag{11.13}$$

Returning to the example, when two units U2 are installed in parallel to perform task 2:

$$CT = \max_{j=1,\ldots,M} \frac{\tau_j}{n_j} = \max\{2, \tfrac{6}{2}, 4, 3\} = 4\,\text{hr} \tag{11.14}$$

Intermediate Storage

Thus far, two storage options have been illustrated. No storage is used in the schedules of Figures 11.11a and 11.11c, with the contents of each unit transferred immediately to the next unit, experiencing no delay after its task has been completed. As mentioned above, this is the so-called *zero-wait* (ZW) strategy. In the schedule of Figure 11.11b, U3 provides intermediate storage until U1 becomes available. Hence, a zero-wait strategy is implemented, with some intermediate storage when necessary. This is referred to as an *intermediate storage* (IS) strategy. The third strategy involves unlimited intermediate storage (UIS), sufficient to hold the contents of the products from a unit having a lengthy batch time to be used repeatedly in a unit having half the batch time or less, as illustrated in Figure 11.12. Here, U1 is utilized at all times and the cycle time is reduced from 9 to 3 hr. To produce a specified amount of product, the batch size is reduced by a factor of one-third since the cycle time is divided by three.

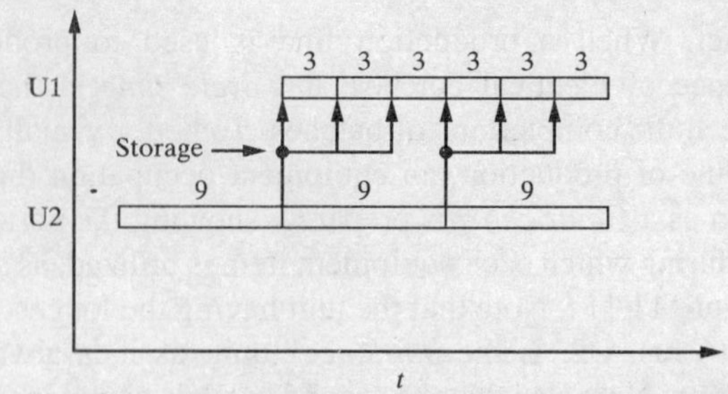

Figure 11.12 Gantt chart with unlimited intermediate storage (UIS).

Batch Size

It is convenient to define the size factor S_j for task j as the capacity required per unit of product. Commonly, it is defined as the volume required to produce a unit mass of product. For example, for the third cultivator in the tPA process of Sections 4.4 and 5.5, 4,000 L of medium yields 2.24 kg of tPA, which eventually yields 1.6 kg of final tPA product. Consequently, its size factor is $4,000\,\text{L}/1.6\,\text{kg} = 2,500\,\text{L/kg}$ tPA product. Size factors can be computed for each task in a recipe. Normally, equipment vessel sizes are selected that exceed batch volume by 10 to 20%. Clearly, the batch factor in volume/mass produced is determined by the rate of processing the batch (e.g., kg/hr) multiplied by the batch time (hr) and divided by the density of the batch (kg/L) and the mass of product produced (kg).

11.5 DESIGN OF MULTIPRODUCT PROCESSING SEQUENCES

A *multiproduct batch plant* produces a set of products whose recipe structures are the same, or nearly identical. One example is a foundry that manufactures integrated circuit (IC) chips in which several different devices are produced simultaneously, each involving hundreds of tasks and utilizing several equipment items. In these plants, each product is produced in the same production line, with multiple processing tasks carried out using the same equipment items. The recipes are expressed in serial campaigns for each product. Figure 11.13 shows schedules in which a campaign of two batches to produce product A is followed by a campaign of two batches to produce product B. It should be noted, however, that because the tasks for products A and B differ in equipment utilized, the plant is not a multiproduct batch plant; instead, it is referred to as a *multipurpose batch plant.* Although the cycle times for both products are identical (4 hr), it is common for the product cycle times to be unequal. The use of alternating product cycles is a limitation that does not apply to *general multipurpose plants,* in which there are no well-defined production lines and no cyclic patterns of batch completion, as shown in Figure 11.14. Such plants are more flexible and effective for a large number of products that are produced in small volumes, where their vessels are cleaned easily and the presence of trace contaminants in

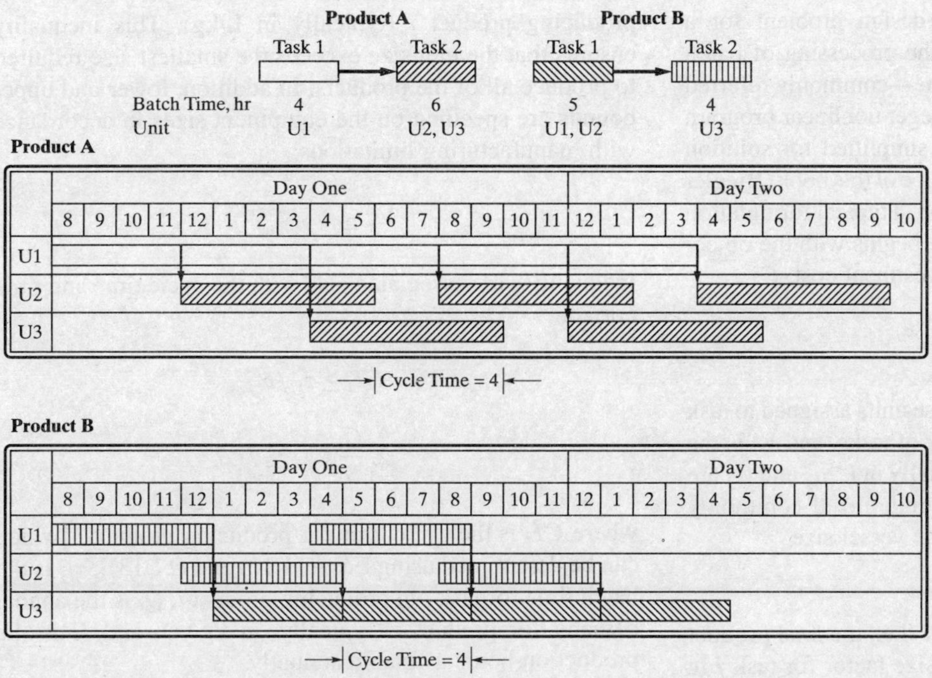

Figure 11.13 Gantt charts for a multipurpose plant.

the products is not a concern. Their equipment items are utilized more completely, without the idle-time gaps in plants with cyclic campaigns for each product. Consequently, multiproduct batch plants are used for larger-volume products having similar recipes, as is often the case for plants that produce a family of grades of a specific product.

Scheduling and Designing Multiproduct Plants

For an existing plant, the scheduling problem involves a specification of the: (1) product orders and recipes, (2) number and capacity of the equipment items, (3) listing of the equipment items available for each task, (4) limitations on the shared resources (e.g., involving the usage of utilities and manpower), and (5) restrictions on the use of equipment due to operating or safety considerations. In solving the problem, that is, determining an optimal schedule, the order in which tasks use the equipment and resources is determined, with specific timings of the tasks provided that optimize the plant performance (which can be specified in many ways, e.g., to maximize the gross profit).

When the plant does not exist, that is, when a new plant is to be designed, the product orders are usually not well defined. Otherwise, the specifications are identical. In fact, the design problem encompasses the scheduling problem in that its solution involves determining the number and capacity of the equipment items in addition to the optimal schedule. For the design problem, these are determined to optimize an objective that includes the investment costs of the equipment, such as the annualized cost. Because the product orders are not as well known during the design stage, it is common to solve the scheduling problem less rigorously.

As mentioned earlier, it is common to specify size factors and input/output ratios as known constants when defining recipes. Also, batch times for each task are often specified as constant, or as known functions of the batch size. These can be determined by optimizing the operation of each equipment item, as discussed in Section 11.2.

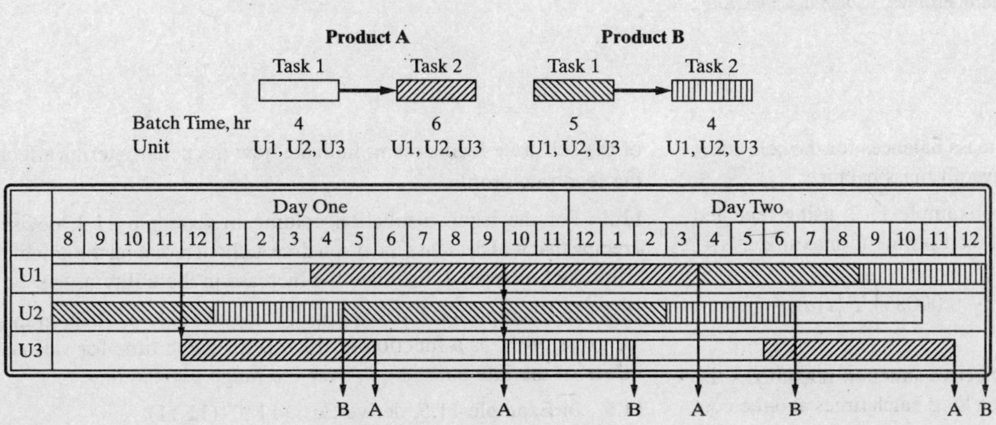

Figure 11.14 Gantt chart for a general multipurpose plant.

It is common to formulate the design problem for a multiproduct batch plant involving the processing of batch campaigns in series (i.e., one-at-a-time—commonly referred to as a *flowshop plant)* as a mixed-integer nonlinear program (MINLP). Then, the formulation is simplified for solution using strategies that are beyond the scope of this book (Biegler et al., 1998). Herein, as an introduction, a typical formulation is presented without simplification. It begins with the objective, that is, to minimize the total investment cost, C:

$$\min C = \sum_{j=1}^{M} m_j a_j V_j^{\alpha_j}$$

where m_j is the number of out-of-phase units assigned to task j (integer variables), M is the number of tasks, and V_j is the size of the unit assigned to task j (usually in L; a_j and α_j are cost coefficients). This objective is minimized commonly subject to inequalities that involve the vessel size:

$$V_j \geq B_i S_{ij}$$

where B_i is the batch size of product i (i.e., the final product size, typically in kg), and S_{ij} is the size factor for task j in producing product i (typically in L/kg). This inequality ensures that the unit size exceeds the smallest size required to produce all of the products. In addition, lower and upper bounds are specified on the equipment sizes in accordance with manufacturing limitations:

$$V_j^L \leq V_j \leq V_j^U$$

Inequalities are associated also with the cycle time and time horizon:

$$CT_i \geq \tau_{ij}/m_j$$

$$\sum_{i=1}^{N} \frac{Q_i}{B_i} CT_i \leq H$$

where CT_i is the cycle time for producing product i [which can be determined using Eqs. (11.12) and (11.13)], τ_{ij} is the batch time for task j in producing product i, Q_i is the annual demand for product i (typically, in kg/yr), and H is the production hours available annually.

11.1 SUMMARY

Initially, this chapter focuses on the optimal control of batch processing units, with emphasis on reducing the batch time and batch size. Then, the batch times for reactor–separator processes are optimized with emphasis on the interactions between the process units and the tradeoffs in adjusting their batch times.Finally, the problem of determining operating schedules for single- and multiproduct batch plants, involving the possibility of intermediate storage and complex recipes with numerous tasks in numerous process units, is examined.

REFERENCES

1. BARRERA, M.D., and L.B. EVANS, "Optimal Design and Operation of Batch Processes," *Chem. Eng. Comm.*, **82**, 45–66 (1989).

2. BIEGLER, L.T., I.E. GROSSMANN, and A.W. WESTERBERG, *Systematic Methods of Chemical Process Design*, Prentice-Hall, Englewood Cliffs, New Jersey (1998).

3. BOSTON, J.F., H.I. BRITT, S. JIRAPONGPHAN, and V.B. SHAH, "An Advanced System for Simulation of Batch Distillation Operations," in R.S.H. MAH and W.D. SEIDER, eds., *Foundations of Computer-aided Process Design*, Vol. 2, AIChE, New York (1981).

4. CUTHRELL, J.E., and L.T. BIEGLER, "Simultaneous Optimization and Solution Methods for Batch Reactor Control Profiles," *Comput. Chem. Eng.*, **13**, 49–62 (1989).

5. DENN, M.M., *Optimization by Variational Methods*, McGraw-Hill (1969).

6. LIM, H.C., Y.J. TAYEB, J.M. MODAK, and P. BONTE, "Computational Algorithms for Optimal Feed Rates for a Class of Fed-batch Fermentations: Numerical Results for Penicillin and Cell Mass Production," *Biotechnol. Bioeng.*, **28**, 1408–1420 (1986).

7. REKLAITIS, G.V., "Computer-aided Design and Operation of Batch Processes," *Chem. Eng. Ed.*, 76–85, Spring (1995).

8. RUDD, D.F., and C.C. WATSON, *Strategy of Process Engineering*, Wiley, New York (1968).

EXERCISES

11.1 In Example 11.2, derive the mass balances for the cell mass, penicillin, and substrate, and the overall mass balance.

11.2 For the penicillin reactor in Example 11.2, using repeated simulations, search for the optimal feed profile to maximize:

$$0.025P\{\tau\}V\{\tau\} - 1.68\tau - 0.00085\int_0^{\tau} F\{t\}dt$$

where τ is the batch time. This objective function maximizes the penicillin produced while penalizing long batch times and the cost of the substrate feed stream. Indicate how the penalty terms affect the feed rate profile.

11.3 For the batch distillation column in Example 11.3, devise a recipe that will decrease the batch time without reducing the amount of product recovered. Estimate the increase in the utility usage.

11.4 In Example 11.4, derive Eqs. (11.6) and (11.7). Then, graph the gross profit as a function of the reactor batch time for various values of the rate constant, k, over the range 0.4–0.6 day^{-1}.

11.5 In Example 11.5, derive Eqs. (11.9)–(11.11).

11.6 For the reactor–distillation process in Example 11.5, recompute the solution when the reactor and column volumes are decreased by 20%.

11.7 Construct a Gantt chart for the general multipurpose plant in Figure 11.14, but with the unit assignments specified in Figure 11.13.

11.8 A batch process requires the following operations to be completed in sequence: 3 hr of mixing, 5 hr of heating, 4 hr of reaction, 7 hr of purification, and 2 hr of transfer.

(a) When the five operations are carried out in vessels U1, U2, U3, U4, and U5, respectively, determine the cycle times, and construct Gantt charts corresponding to the zero-wait, intermediate storage, and unlimited intermediate storage inventory strategies.

(b) When a new purification vessel U4A is purchased, so that two 7-hr purifications can take place in parallel, determine the system bottleneck using the intermediate storage inventory strategy.

Chapter 12

Plantwide Control

12.0 OBJECTIVES

In this chapter, the importance of considering controllability and operability issues early in the design process is demonstrated by showing how controllability considerations can help to differentiate between processes that are easy, and processes that are difficult, to control. It provides a recommended methodology to initiate the design of attractive plantwide control systems.

After reading this chapter, the student should be able to:

1. Identify potential control problems in a process flowsheet.
2. Classify and select controlled and manipulated variables for a plantwide control system.
3. Perform a conceptual synthesis of plantwide control structures (pairings) based on degrees-of-freedom analysis and qualitative guidelines.

12.1 INTRODUCTION

The design of a continuous chemical process is usually carried out at steady state for a given operating range, it being assumed that a control system can be designed to maintain the process at the desired operating level and within the design constraints. However, unfavorable process static and dynamic characteristics could limit the effectiveness of the control system, leading to a process that is unable to meet its design specifications. A related issue is that, usually, alternative designs are judged based on economics alone, without taking controllability and resiliency into account. This may lead to the elimination of easily controlled, but slightly less economical, alternatives in favor of slightly more economical designs that may be extremely difficult to control. It is becoming increasingly evident that design based on steady-state economics alone is risky, because the resulting plants are often difficult to control (i.e., inflexible, with poor disturbance-rejection properties), resulting in off-specification product, excessive use of fuel, and associated profitability losses.

Consequently, there is a growing recognition of the need to consider the *controllability and resiliency* (C&R) of a chemical process during its design. *Controllability* can be defined as the ease with which a continuous plant can be held at a specific steady state. An associated concept is *switchability*, which measures the ease with which the process can be moved from one desired stationary point to another. *Resiliency* measures the degree to which a processing system can meet its design objectives despite external disturbances and uncertainties in its design parameters. Clearly, it would be greatly advantageous to be able to predict as early as possible in the design process how well a given flowsheet meets these dynamic performance requirements.

Table 12.1 summarizes the four main stages in the design of a chemical process. In the conceptual and preliminary stages, a large number of alternative process flowsheets, in the steady state (SS), are generated. Subsequent stages involve more detailed analysis in the steady state, followed by the testing of the dynamic (Dyn) performance of the controlled flowsheets. Here, considerably more engineering effort is expended than in the preliminary stages. Therefore, far fewer designs are considered, with many of the initial flowsheets having been eliminated from further consideration by screening in the preliminary stages.

The need to account for the controllability of competing flowsheets in the early design stages is an indication that simple screening measures, using the limited information available, should be employed to select from among the flowsheets. Here, if high-fidelity, closed-loop, dynamic modeling were used, the engineering effort and time for development and analysis would *slow* the design process significantly. The right-hand-side columns in Table 12.1 show that the shortcut C&R tools provide a bridge between steady-state simulation for process design and the rigorous dynamic simulation required to verify switchability and other attributes of the closed-loop dynamics of the final design.

Table 12.1 Process Design Stages, Issues, and Tools

Design Stage (see Figure PI.1)	Issues	What Gets Fixed	Tools SS	Tools C&R	Tools Dyn
1. Process creation	Selecting between alternative material pathways and flowsheets	Material pathways	■		
2. Development of base-case design	Feasibility studies based on fixed material pathways Unit operations selection Heat integration superstructure	Flowsheet structure	■	■	
3. Detailed design	Optimization of key process variables Analysis of process sensitivity to disturbances and uncertainties	Optimal flowsheet parameters	■	■	
4. Plantwide controllability assessment	Flowsheet controllability Dynamic response of the process to disturbances Selection of the control system structure and its parameters	Control structure and its parameters			■

ASPEN PLUS, PRO/II, ASPEN HYSYS, UNISIM, and CHEMCAD are commonly used simulation packages, all of which enable both steady-state and dynamic simulation.

In the following examples, the impact of design decisions on controllability and resiliency is introduced for four processes. The supplement to this chapter (Section 12S in the file, Supplement_-to_Chapter_12.pdf, in the PDF Files folder, which can be downloaded from the Wiley Web site associated with this book) expands upon this introduction and shows how to eliminate the less desirable alternatives and validate the performance of the most promising designs.

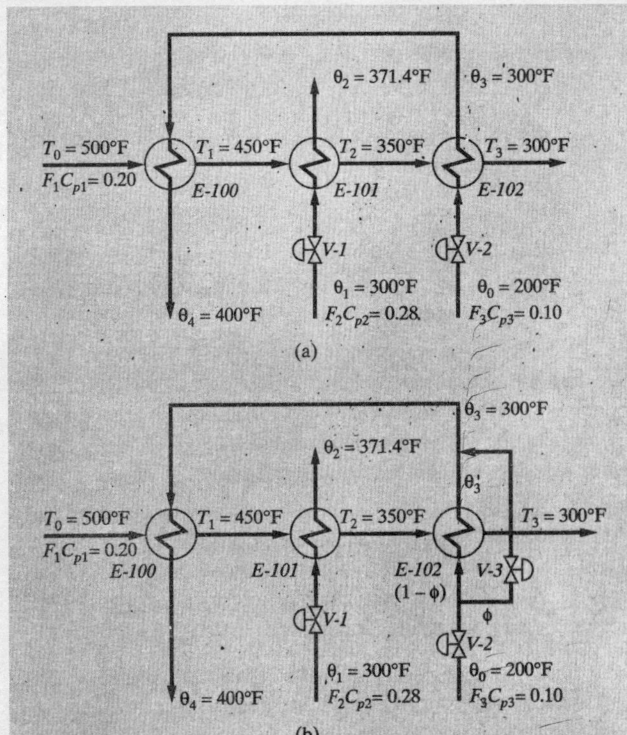

Figure 12.1 Heat exchanger network: (a) original configuration; (b) modification with bypass.

EXAMPLE 12.1 *Heat Exchanger Networks*

The network shown in Figure 12.1a, which was introduced by McAvoy (1983), cools hot stream 1 from 500 to 300°F using cold streams 2 and 3, having feed temperatures of 300 and 200°F and corresponding target temperatures of 371.4 and 400°F, respectively, with the heat-capacity flow rates in MMBtu/hr-°F. Furthermore, the feed rate and temperature of the hot stream are considered as disturbances.

As shown in Figure 12.1a, two of the target temperatures can be controlled by manipulating the flow rates of the two cold streams. This means that one of the target temperatures is left uncontrolled in the face of disturbances in the hot stream. An alternative design, involving a bypass around exchanger E-102, is illustrated in Figure 12.1b. As shown, this simple modification allows all three target temperatures to be regulated. Because the selection of the appropriate bypass flow fraction, ϕ, and of the most effective control configuration is not trivial, controllability analysis should be carried out on the alternative networks and their candidate control structures. This will assist in selecting one of the two designs, as shown in the supplement to this chapter. Furthermore, Example 25.1 shows how a six-sigma approach can be used to significantly improve the resiliency of the network.

EXAMPLE 12.2 *Heat-Integrated Distillation Columns*

The production of methanol is carried out in a moderate-pressure synthesis loop by the direct hydrogenation of carbon dioxide,

$$CO_2 + 3H_2 \rightleftharpoons CH_3OH + H_2O \quad (12.1)$$

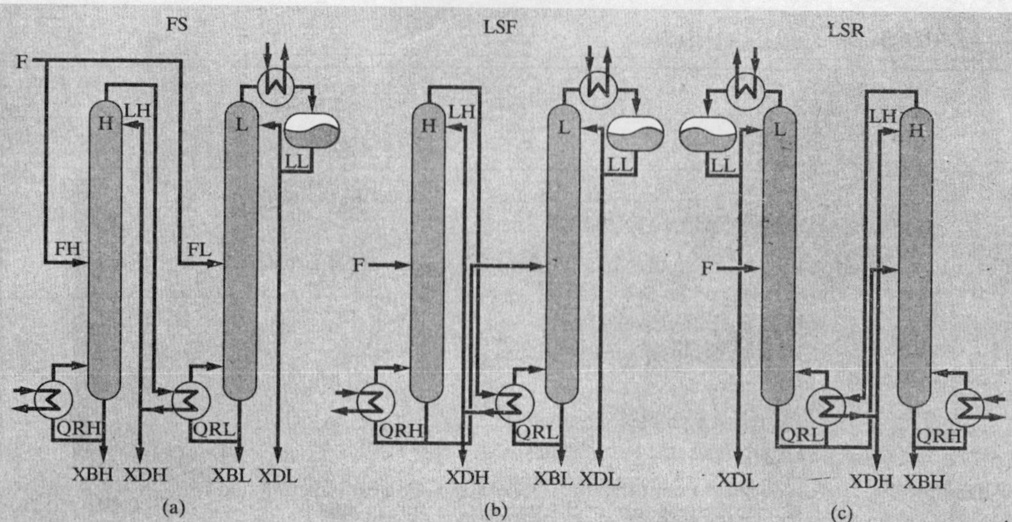

Figure 12.2 Three heat-integrated alternatives to a single distillation column.

which generates a liquid product that contains a binary mixture of methanol and water in approximately equal proportions. To provide commercial methanol that is nearly free of water, dehydration is achieved commonly by distillation. To reduce the sizable energy costs, three double-effect, heat-integrated configurations, shown in Figure 12.2, are commonly considered as alternatives to a single distillation column (SC):

FS (Feed Split). The feed is split nearly equally ($F_H \approx F_L$) between two columns to achieve optimal operation. The overhead vapor product of the high-pressure column supplies the heat required in the low-pressure column.

LSF (Light-Split/Forward Heat Integration). The entire feed is fed to the high-pressure column. About half of the methanol product is removed in the distillate from the high-pressure column, and the bottoms product is fed into the low-pressure column. In this configuration, heat integration is in the same direction as the mass flow.

LSR (Light-Split/Reverse Heat Integration). The entire feed is fed to the low-pressure column, with the bottoms product from the low-pressure column fed into the high-pressure column. Here, heat integration is in the direction opposite to that of the mass flow.

As discussed under "Multiple-Effect Distillation" in Section 9.9, these configurations reduce the energy costs by using the heat of condensation of the overhead stream from the high-pressure column (H) to supply the heat of vaporization of the boilup in the low-pressure column (L). Although they are more economical, assuming steady-state operation, they are more difficult to control because the configurations (1) are more interactive, and (2) have one less manipulated variable for process control, since the reboiler duty in the low-pressure column can no longer be manipulated independently.

To show the energy savings, the four flowsheets were simulated on the basis of an equimolar feed of 2,700 kmol/hr, producing 96 mol% methanol in the distillate and 4 mol% methanol in the bottoms product, assuming 75% tray efficiency and no heat loss to the surroundings, and using UNIFAC to estimate the liquid-phase activity coefficients. The total energy requirements for the four alternatives were computed as follows:

SC	2.12×10^7 kcal/hr	LSR	1.23×10^7 kcal/hr
LSF	1.33×10^7 kcal/hr	FS	1.23×10^7 kcal/hr

Clearly, the LSR and FS configurations save the most energy, although the energy consumption in the LSF configuration is only 8% higher. Based on steady-state economics alone, one of these three configurations would be selected. However, disturbance resiliency analysis (Chiang and Luyben, 1988; Weitz and Lewin, 1996) shows that either the LSR or LSF configurations are preferred for disturbance rejection, as they provide performance only slightly worse than that of a single column, SC. The FS configuration, on the other hand, does considerably worse. The supplement to this chapter shows how to obtain this information when selecting from among these alternatives.

EXAMPLE 12.3 *Heat Recovery from an Exothermic Reactor*

Often, the heat from an exothermic reactor is used to preheat the reactor feed, thus saving energy, as discussed in Section 6.5. Figure 12.3b shows a configuration using a feed/product heat exchanger that is commonly preferred to the configuration with independent preheat in Figure 12.3a. However, the heat-integrated configuration shares the same disadvantages as the heat-integrated distillation systems discussed in the previous example (i.e., one less manipulated variable and possibly unfavorable dynamic interactions). Furthermore, the feed-effluent heat exchanger introduces positive feedback, and the possibility of thermal runaway.

EXAMPLE 12.4 *Reactor-Flash-Recycle System*

Reactor design for complete conversion may be impossible thermodynamically or undesirable because of reduced yields

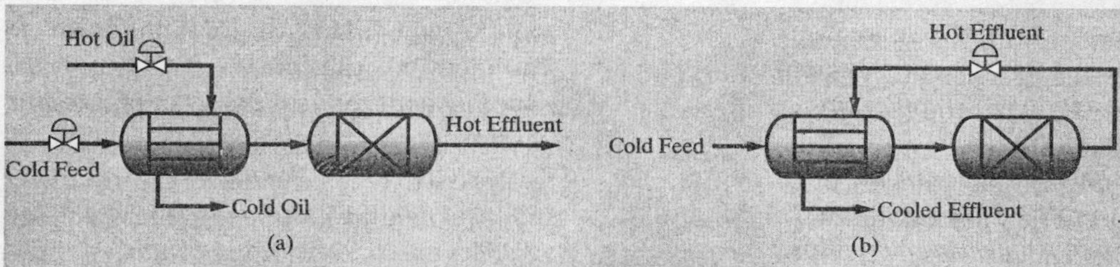

Figure 12.3 Two configurations for an exothermic reactor requiring feed preheating: (a) reactor with independent preheat; (b) heat-integrated reactor.

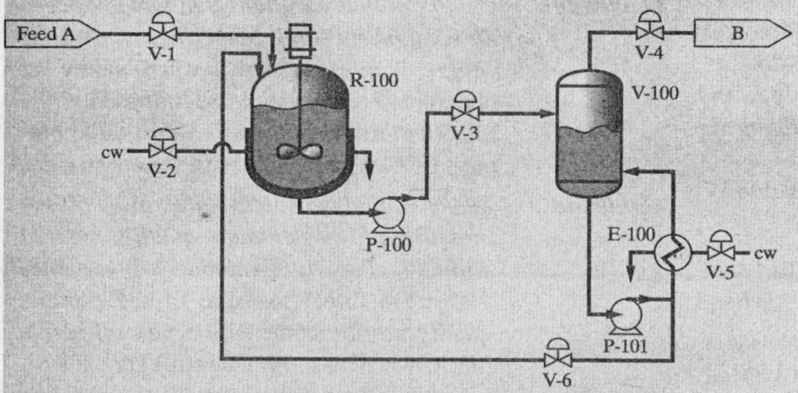

Figure 12.4 Reactor-flash-recycle system for the production of B.

when byproducts are formed. In such cases, an economic alternative is to design a combined reactor-separator-recycle system, as illustrated in the simple example in Figure 12.4 and discussed in Section 7S.1. Here, the reaction A → B is carried out in a CSTR, whose liquid feed is a stream containing pure A. In the event that B is sufficiently more volatile than A, the separation can be performed using a flash vessel, with unreacted A recycled to the reactor. As will be seen in the plantwide control examples at the end of this chapter and in the quantitative analysis in the supplement to this chapter, the presence of the recycle complicates control of the process, and requires special attention.

To enable the evaluation of the controllability and resiliency of alternative process configurations, it is important to consider two aspects of the design of plantwide control systems:

1. The classification and selection of controlled and manipulated variables
2. The qualitative synthesis of plantwide control structures based on degrees-of-freedom analysis and qualitative guidelines. These are examined in the next two sections.

12.2 CONTROL SYSTEM CONFIGURATION

The design of a control system for a chemical plant is guided by the objective to maximize profits by transforming raw materials into useful products while satisfying product specifications, safety and operational constraints, and environmental regulations. All four constraints require special consideration.

1. ***Product specifications.*** To satisfy customer expectations, it is important that product quality and production rate meet specifications. This has been the driving force for the implementation of online, optimal process control in the chemical industry. More recently, statistics-based approaches such as Six-Sigma methodologies have been harnessed for this purpose, as discussed in Chapter 25.
2. ***Safety.*** The plant must be operated safely to protect the well-being of plant personnel and nearby communities. As an example, a typical safety-driven constraint requires that the temperature and pressure of a steel vessel not exceed upper limits dictated by metallurgy. For other examples, see Section 1.5.
3. ***Operational constraints.*** These constraints, which are often not associated directly with safety, place upper bounds on the vapor velocity to avoid flooding in distillation columns, and place upper bounds on the reactor temperatures to prevent degradation of the catalyst or the onset of undesirable side reactions.
4. ***Environmental regulations.*** These require that processing plants comply with constraints on air and water quality as well as waste disposal. Many examples are discussed in Section 1.4.

Classification of Process Variables

When designing a plantwide control system, it is common to view the process in terms of its input and output variables. These variables include flow rates of streams entering and leaving process equipment, and temperatures, pressures, and compositions in entering and leaving streams and/or within equipment.

Process *output variables* are those that give information about the state of the process. They are usually associated with streams leaving the process or with measurements inside a process vessel. When designing a control system, output variables are usually referred to as *controlled variables*, which are measured (online or off-line).

Process *input variables* are independent variables that affect the output variables of a process. They can be subdivided into two subgroups: (1) *manipulated variables* (also called *control variables*), which can be adjusted freely by an operator or a control mechanism, and (2) *disturbance variables* (also called *externally defined variables*), which are subject to the external environment and thus cannot be controlled. These variables are associated typically with the inlet and outlet streams. In a control system, manipulated variables cause changes to controlled variables.

There are three main reasons why it may be impossible to control all of the output variables of a process.

1. It may not be possible to measure online all of the output variables, especially compositions. Even when it is possible, it may be too expensive to do so.
2. By a degrees-of-freedom analysis, described below, there may not be enough manipulated variables available to control all of the output variables.
3. Potential control loops may be impractical because of slow dynamics, low sensitivity to the manipulated variables, or interactions with other control loops.

Qualitative criteria have been suggested by Newell and Lee (1988) to guide the selection of controlled and manipulated variables that are suitable for an initial analysis in the design of a plantwide control system. These guidelines, which are presented next, are driven by the plant and control objectives and should not be applied without due consideration. When two guidelines conflict, the most important of the two should be adopted. In critical cases, the more reliable quantitative screening approaches, discussed in the supplement to this chapter, should be considered. Following presentation of the guidelines, examples of the selection of variables are given.

Selection of Controlled (Output) Variables

Guideline 1: *Select output variables that are either non-self-regulating or unstable.* A *self-regulating process* is one that is described by a state equation of the form $\dot{x} = f\{x, u\}$, where x is an output variable and u is an input variable. A change in u will result in the process moving to a new steady state. A *non-self-regulating* process is described by $\dot{x} = f\{u\}$. As a result, changes in the input variable, u, affect the process output as a pure integrator. An example of a non-self-regulating output variable is the liquid level of a surge tank, whose effluent feeds a pump followed by a control valve. Clearly, if the control valve is left uncontrolled, a positive feed disturbance to the surge drum may cause the vessel to overflow. When the process is unstable in the open loop (that is, in the absence of feedback control), a change in the input variable causes the system to become unstable. Clearly, *non-self-regulating* and *unstable* process output variables must be selected as controlled variables.

Guideline 2: *Choose output variables that would exceed the equipment and operating constraints without control.* Clearly, when safety or operational constraints are imposed, it is important to measure and control these output variables to comply with the constraints.

Guideline 3: *Select output variables that are a direct measure of the product quality or that strongly affect it.* Examples of variables that are a direct measure of the product quality are the composition and refractive index, whereas those that strongly affect it are temperature and pressure. This guideline helps the control system to ensure that the product specifications are regulated and met.

Guideline 4: *Choose output variables that exhibit significant interactions with other output variables.* Plantwide control must handle the potential interactions in the process. Improved closed-loop performance is achieved by stabilizing output variables that interact significantly with each other.

Guideline 5: *Choose output variables that have favorable static and dynamic responses to the available manipulated variables.* All other things being equal, this guideline should be applied.

Selection of Manipulated Variables

Guideline 6: *Select manipulated variables that significantly affect the controlled variables.* For each control loop, select an input variable with as large a steady-state gain as possible and sufficient range to adjust the controlled variable. For example, when a distillation column operates with a large reflux ratio, that is, values greater than four (Luyben et al., 1999), it is much easier to control the level in the reflux drum using the reflux flow rate rather than the distillate flow rate.

Guideline 7: *Select manipulated variables that rapidly affect the controlled variables.* This precludes the selection of inputs that affect the outputs with large delays or time constants.

Guideline 8: *Select manipulated variables that affect the controlled variables directly rather than* *indirectly.* For example, when appropriate for the design of an exothermic reactor, it is preferable to inject a coolant directly rather than use a cooling jacket.

Guideline 9: *Avoid recycling disturbances.* It is usually better to eliminate the effect of disturbances by allowing them to leave the process in an exiting stream rather than having them propagate through the process by the manipulation of a feed or recycle stream.

Selection of Measured Variables

Both input and output variables may be measured variables, with online measurement preferred to off-line measurement. Seborg et al. (1989) discuss the importance of measurements in control, and provide three guidelines for the selection of variables to be measured and the location of the measurements.

Guideline 10: *Reliable, accurate measurements are essential for good control.* An example of a poorly designed measurement would be an orifice positioned to measure flow rate with an insufficient entry length of piping.

Guideline 11: *Select measurement points that are sufficiently sensitive.* Consider, for example, the indirect control of the product compositions from a distillation column by the regulation of a temperature near the end of the column. In high-purity distillation columns, where the terminal temperature profiles are almost flat, it is preferable to move the temperature measurement point closer to the feed tray.

Guideline 12: *Select measurement points that minimize time delays and time constants.* Large time delays and dynamic lags in the process limit the achievable closed-loop performance. These should be reduced, whenever possible, in the process design and the selection of measurements.

Degrees-of-Freedom Analysis

Before selecting the controlled and manipulated variables for a control system, one must determine the number of manipulated variables permissible. As discussed under "Degrees of Freedom" in Section 5.2, the number of manipulated variables cannot exceed the number of degrees of freedom, which are determined using a process model according to

$$N_D = N_{\text{Variables}} - N_{\text{Equations}} \tag{12.2}$$

where N_D is the number of degrees of freedom, $N_{\text{Variables}}$ is the number of process variables, and $N_{\text{Equations}}$ is the number of independent equations that describe the process. However, the number of manipulated variables is generally less than the number of degrees of freedom, since one or more variables may be externally defined (disturbed); that is, $N_D = N_{\text{Manipulated}} + N_{\text{Externally Defined}}$. Consequently, the number of manipulated variables can be expressed in terms of the number of externally defined variables:

$$N_{\text{Manipulated}} = N_{\text{Variables}} - N_{\text{Externally Defined}} - N_{\text{Equations}} \tag{12.3}$$

The number of manipulated variables equals the number of controlled variables that can be regulated. When a manipulated variable is paired with a regulated-output variable, its degree of freedom is transferred to the output's setpoint, which becomes the new independent variable.

Next, degrees-of-freedom analyses are carried out and their implications for control system design are considered for heat exchanger networks, jacketed stirred-tank reactors, a utility system, a flash vessel, and a distillation column.

EXAMPLE 12.5 ***Control Configurations for Heat Exchanger Networks (Example 12.1 Revisited)***

Referring to Figure 12.1a, the process can be described in terms of 15 variables: $F_1, F_2, F_3, T_0, T_1, T_2, T_3, \theta_0, \theta_1, \theta_2, \theta_3, \theta_4, Q_1, Q_2,$ and Q_3. Of these, assume that four variables can be considered to be externally defined: $F_1, T_0, \theta_0,$ and θ_1. A steady-state model for the process consists of three equations for each heat exchanger. For example, for the first heat exchanger, the following equations apply:

$$Q_1 = F_1 C_{P1}(T_0 - T_1) \tag{12.4}$$

$$Q_1 = F_3 C_{P3}(\theta_4 - \theta_3) \tag{12.5}$$

$$Q_1 = U_1 A_1 \frac{(T_0 - \theta_4) - (T_1 - \theta_3)}{\ln\left\{\dfrac{T_0 - \theta_4}{T_1 - \theta_3}\right\}} \tag{12.6}$$

In these equations, Q_i, U_i, and A_i are the heat duty, heat-transfer coefficient, and heat-transfer area, respectively, for heat exchanger i. Values for the latter two are assumed known, so they are not process variables. Similar equations are written for the other two heat exchangers, making a total of nine equations. Consequently, the number of manipulated variables is computed: $N_{\text{Manipulated}} = N_{\text{Variables}} - N_{\text{Externally Defined}} - N_{\text{Equations}} = 15 - 4 - 9 = 2$.

Thus, two variables can be manipulated. Two candidates are the flow rates of the two cold streams: F_2 and F_3. Ideally, for the selection of the controlled variables, it would be desirable to regulate all three target temperatures: T_3, θ_2, and θ_4. However, with only two manipulated variables, only two controlled variables can be selected. The guidelines presented above are

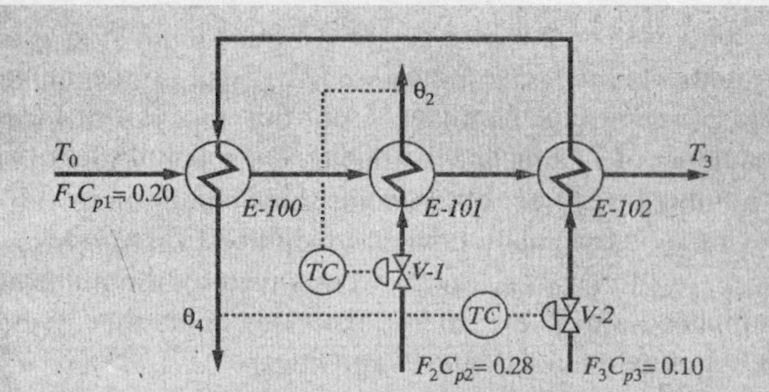

Figure 12.5 Control system for original heat exchanger network.

insufficient to select which two of the three should be picked, since all three provide a direct measure of the product quality (Guideline 3), and there are clearly significant interactions among all three of the variables (Guideline 4). Without quantitative analysis, one cannot gauge which of the three have the most favorable static and dynamic responses to the manipulated variables. If only T_3, θ_2, and θ_4 are considered as potential controlled variables, three possible control systems should be investigated. As an illustration, Figure 12.5 shows one possible configuration of two control loops. One loop adjusts the flow rate, F_2, to control the temperature, θ_2, while the other loop adjusts F_3 to control θ_4. An alternative configuration with reversed pairings (i.e., $\theta_4 - F_2$, $\theta_2 - F_3$) is unstable, as shown in Case Study 12S.2.

The design in Figure 12.1b, involving a bypass on exchanger E-102, permits the regulation of all three target temperatures. In this case, the number of variables is increased by two (the bypass flow fraction, ϕ, and the temperature, θ_3'), giving a total of 17 variables, with the same four disturbance variables. For constant heat capacities and no phase change, the process is modeled by one additional energy balance for the mixer,

$$\theta_3 = (1 - \phi)\theta_0 + \phi\theta_3' \tag{12.7}$$

where ϕ is the E-102 bypass flow fraction and θ_3' is the temperature leaving heat exchanger E-102. Since $N_{\text{Manipulated}} = N_{\text{Variables}} - N_{\text{Externally Defined}} - N_{\text{Equations}} = 17 - 4 - 10 = 3$, three variables can be manipulated, namely, F_2, F_3, and ϕ. The flow rate of the second cold stream, F_3, affects two of the three heat exchangers, whereas F_2 affects only the second one directly, and ϕ affects T_3 directly (Guidelines 6, 7, and 8). The control structure shown in Figure 12.6 is the most resilient and controllable regulatory structure, as is demonstrated in Case Study 12S.2 and Example 25.1 using quantitative analysis.

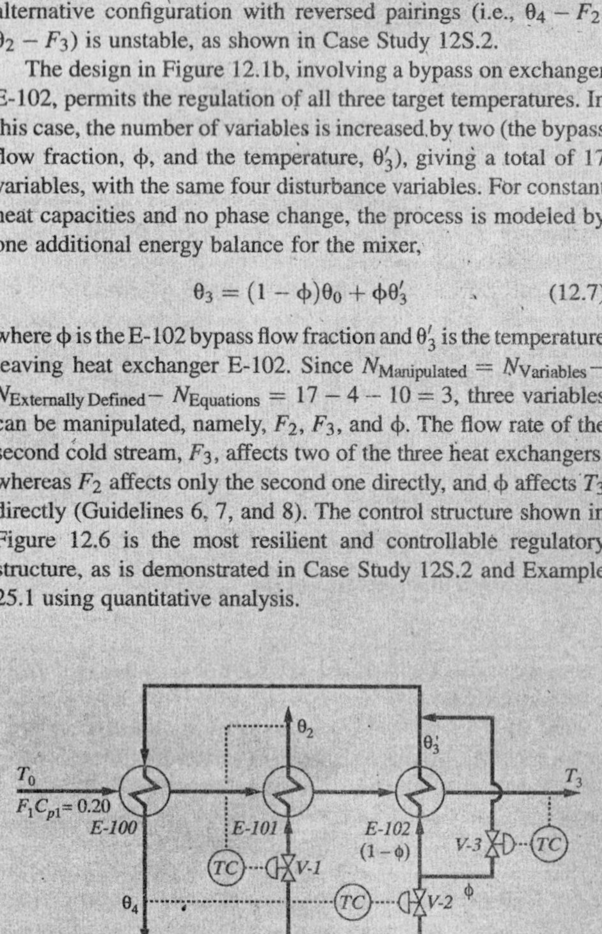

Figure 12.6 Control system for the modified heat exchanger network.

EXAMPLE 12.6 *Control Configuration for a Jacketed CSTR*

Consider the control of a jacketed continuous-stirred-tank reactor (CSTR) in which the exothermic reaction A → B is carried out. This system can be described by 10 variables, as shown in Figure 12.7: h, T, C_A, C_{Ai}, T_i, F_i, F_o, F_c, T_c, and T_{co}, three of which are considered to be externally defined: C_{Ai}, T_i, and T_{co}. Its model involves four equations, assuming constant fluid density.

1. Overall mass balance:

$$A\frac{dh}{dt} = F_i - F_o \tag{12.8}$$

2. Mass balance on component A:

$$A\frac{d}{dt}(hC_A) = F_iC_{Ai} - F_oC_A - Ah \cdot r\{C_A, T\} \tag{12.9}$$

3. Energy balance on the reacting mixture:

$$A\rho c_p \frac{d}{dt}(h \cdot T) = F_i\rho c_p T_i - F_o\rho c_p T + Ah \cdot r\{C_A, T\}(-\Delta H) - UA_s(T - T_c) \tag{12.10}$$

4. Energy balance on the jacket coolant:

$$V_c\rho_c c_{pc}\frac{dT_c}{dt} = F_c c_{pc} T_{co} - F_c c_{pc} T_c + UA_s(T - T_c) \tag{12.11}$$

where A is the cross-sectional area of the vessel, h is the liquid level in the reactor, A_s is the area for heat transfer, U is the overall heat-transfer coefficient, C_{Ai} and C_A are the inlet and reactor concentrations of A, T_i and T are the inlet and reactor temperatures, F_i and F_o are the inlet and outlet volumetric flow rates, ρ is the fluid density, F_c is the coolant mass flow rate, ρ_c is the coolant density, T_{co} and T_c are the inlet coolant and jacket temperatures, V_c is the volume of fluid in the cooling jacket, r is the intrinsic rate of reaction, ΔH is the heat of reaction, and c_p and c_{pc} are the specific heats of the reacting mixture and coolant, respectively.

Here, the number of variables that can be manipulated independently is $N_{\text{Manipulated}} = N_{\text{Variables}} - N_{\text{Externally Defined}} - N_{\text{Equations}} = 10 - 3 - 4 = 3$.

Selection of controlled variables. C_A should be selected because it affects the product quality directly (Guideline 3). T should be selected because it must be regulated properly to avoid safety problems (Guideline 2) and because it interacts with C_A (Guideline 4). Finally, h must be selected as a controlled output because it is non-self-regulating (Guideline 1).

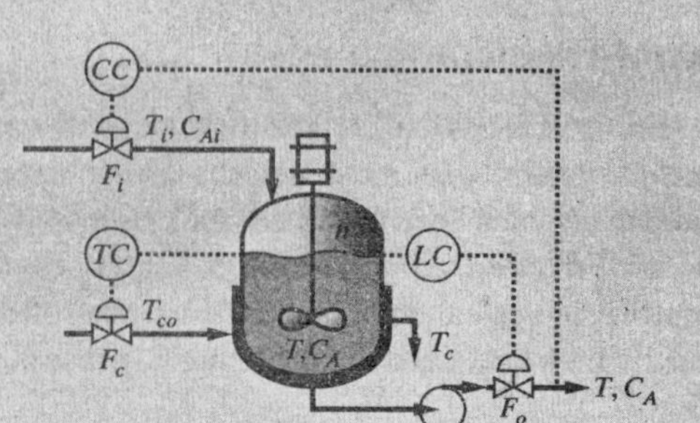

Figure 12.7 Control system for a jacketed CSTR.

Selection of manipulated variables. The volumetric feed flow rate, F_i, should be selected because it directly and rapidly affects the conversion (Guidelines 6, 7, and 8). Using the same reasoning, F_c is selected to control the reactor temperature, T; and the flow rate of the reactor effluent, F_o, is selected to control h. This configuration, which is included in Figure 12.7, should be compared with other pairings using the quantitative analysis presented in the supplement to this chapter, it being noted that there are several opportunities for improvement.

EXAMPLE 12.7 *Control Configuration for a Utilities Subsystem*

Often, the contents of a chemical batch reactor are heated initially to achieve ignition, and then cooled to remove the heat generated in reaction. In such cases, it is common to install a jacket supplied with both cooling and heating utility streams, as shown in Figure 12.8. The utilities subsystem involves eight variables: P_{cf}, T_{cf}, F_{c1}, T_{c1}, F_{c2}, T_{c2}, F_c, and T_{co}. Of these, two are externally defined, and constitute disturbance variables: P_{cf} and T_{cf}. Four material and energy balances relate the subsystem variables: (1) an energy balance for the cooling branch, (2) an energy balance for the heating branch, (3) an energy balance for the mixing junction, and (4) a mass balance for the mixing junction. Hence, the number of variables to be manipulated independently is $N_{\text{Manipulated}} = N_{\text{Variables}} - N_{\text{Externally Defined}} - N_{\text{Equations}} = 8 - 2 - 4 = 2$. This is also the number of subsystem variables that can be controlled independently.

Selection of controlled variables: The guidelines presented earlier are not helpful because no output variable has a direct effect on the product quality, all are self-regulating, and none are directly associated with equipment or operating constraints. Nonetheless, F_c and T_{co} are obvious choices for the controlled variables because the objective of this subsystem is to control the temperature and flow rate of the utility stream fed to the reactor jacket.

Selection of manipulated variables: The two obvious candidates are F_{c1} and F_{c2}, since both affect the two outputs directly and rapidly (Guidelines 7 and 8). However, linear and nonlinear combinations of these flow rates are also possible. As shown in Example 12S.4, a quantitative analysis is needed to make the best selection.

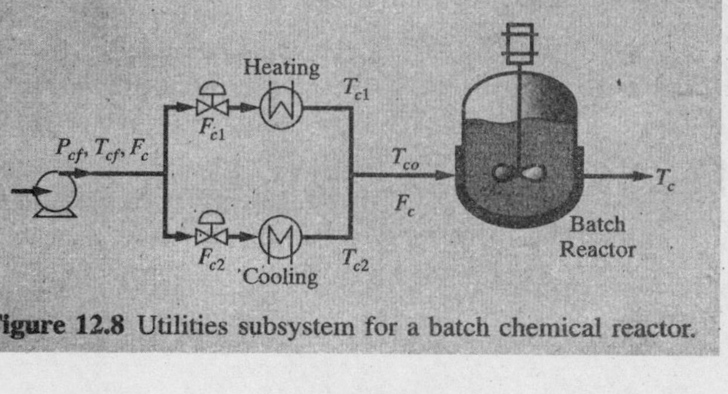

Figure 12.8 Utilities subsystem for a batch chemical reactor.

EXAMPLE 12.8 *Control Configuration for a Flash Drum*

The flash drum in Figure 12.9 illustrates a situation where a stream containing a binary mixture of two components, A and B, is flashed through a valve and separated in a flash drum into an overhead vapor stream and a residual liquid product stream. External heat exchange with liquid recycle is provided with a fixed recycle ratio. This process is modeled with 11 variables: F_i, T, C_A, F_W, P_f, h, T_f, F_V, y_A, F_L, and x_A. Two variables are considered to be externally defined, T and C_A. The model involves five equations: a total mass balance, a mass balance for component A, an overall energy balance, and a vapor–liquid equilibrium equation for each component. Thus, the number of variables to be manipulated independently is $N_{\text{Manipulated}} = N_{\text{Variables}} - N_{\text{Externally Defined}} - N_{\text{Equations}} = 11 - 2 - 5 = 4$.

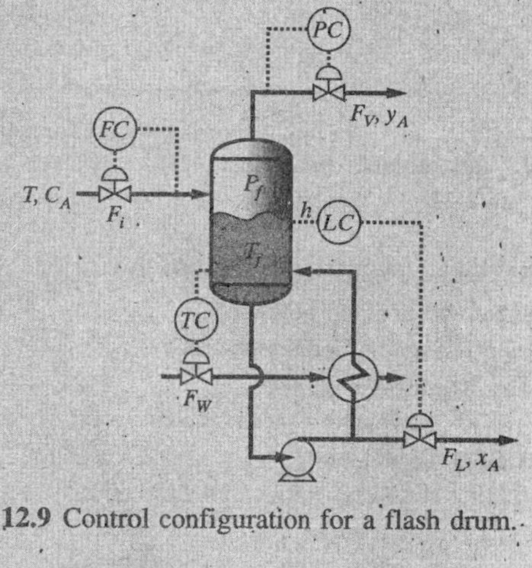

Figure 12.9 Control configuration for a flash drum.

Selection of controlled variables: P_f is selected because of the potential safety problems (Guideline 2) and because it affects the product concentrations (Guideline 3). T_f should be selected because it directly affects the product quality (Guideline 3). The liquid height in the drum, h, must be selected because it is non-self-regulating (Guideline 1), and F_i is selected because it controls the product flow rate directly, one of the overall control objectives (Guideline 3). Note that all of these outputs exhibit significant interaction.

Selection of manipulated variable: F_i is adjusted to achieve its setpoint (Guideline 8), defined as the *production handle*. F_V has a rapid, direct effect on the vessel pressure, P_f, and almost no effect on any other output (Guidelines 7 and 8). For similar reasons, F_L is selected to control the liquid level, h. F_W is selected because it directly controls the flash temperature, T_f (Guideline 8).

EXAMPLE 12.9 *Control Configuration for a Binary Distillation Column*

This analysis, for the distillation operation in Figure 12.10, is based on the following assumptions and specifications: (a) constant relative volatility, (b) saturated liquid distillate, (c) negligible vapor holdup in the column, (d) constant tray pressure drops

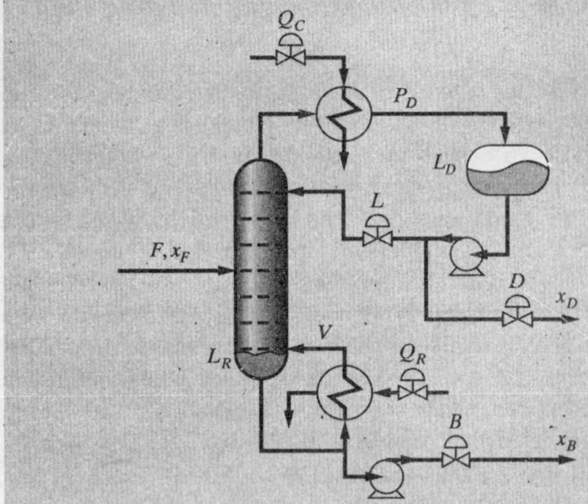

Figure 12.10 A distillation column with two liquid products.

(Luyben, 1990), and (e) negligible heat losses except for the condenser and reboiler. When using N_T trays, the column is modeled with $4N_T + 13$ variables:

Vapor and liquid compositions on each tray:	$2N_T$
Tray liquid flow rates and holdups:	$2N_T$
Reflux drum holdup and composition:	2
Reflux and distillate flow rates:	2
Column sump vapor and liquid compositions:	2
Column sump liquid holdup:	1
Bottoms product and reboiler vapor flow rates:	2
Feed flow rate and composition:	2
Condenser pressure:	1
Condenser duty:	1

The system is described by $4N_T + 6$ equations:

Species mass balances (trays, sump, reflux drum):	$N_T + 2$
Total mass balances (trays, sump, reflux drum):	$N_T + 2$
Vapor–liquid equilibrium (trays, sump):	$N_T + 1$
Tray hydraulics for tray holdup:	N_T
Total vapor dynamics:	1

Assuming that the feed flow rate and composition are externally defined, the number of variables to be manipulated independently is $N_{\text{Manipulated}} = N_{\text{Variables}} - N_{\text{Externally Defined}} - N_{\text{Equations}} = 4N_T + 13 - 2 - (4N_T + 6) = 5$.

Selection of controlled variables: The condenser pressure, P_D, should be regulated, since it strongly affects the product compositions (Guidelines 3 and 4). The reflux drum and sump liquid inventory levels, L_D and L_R, need to be regulated, since they are not self-regulating (Guideline 1). This leaves two additional variables that can be regulated. When distillate and bottoms streams are product streams, their compositions, x_D and x_B, are often selected as controlled variables (Guideline 3). However, as pointed out by Luyben et al. (1999), in plantwide control, it is often sufficient to control one composition, for example, when the composition of a recycle stream is not regulated. Since significant delay times are often associated with composition measurements, tray temperatures (which are measured without delay times) are often used to infer compositions (Guideline 12). In this regard, temperatures must be measured on trays that are sensitive to column upsets (Guideline 11).

Selection of manipulated variables: As shown in Figure 12.10 by labels on the valves, the five manipulated variables are the flow rates of the reflux, distillate, and bottoms streams L, D, and B, and the cooling and heating duties, Q_C and Q_R. It is most common to control P_D in columns with a liquid overhead product by manipulating Q_C. Since it is impossible to control the product compositions while setting the product flow rates, the latter are commonly used to regulate the liquid inventory levels (that is, the liquid levels in the reflux drum and the sump). This leaves two variables to control the product compositions. Figure 12.11a shows the so-called LV-configuration, where L controls x_D, Q_R

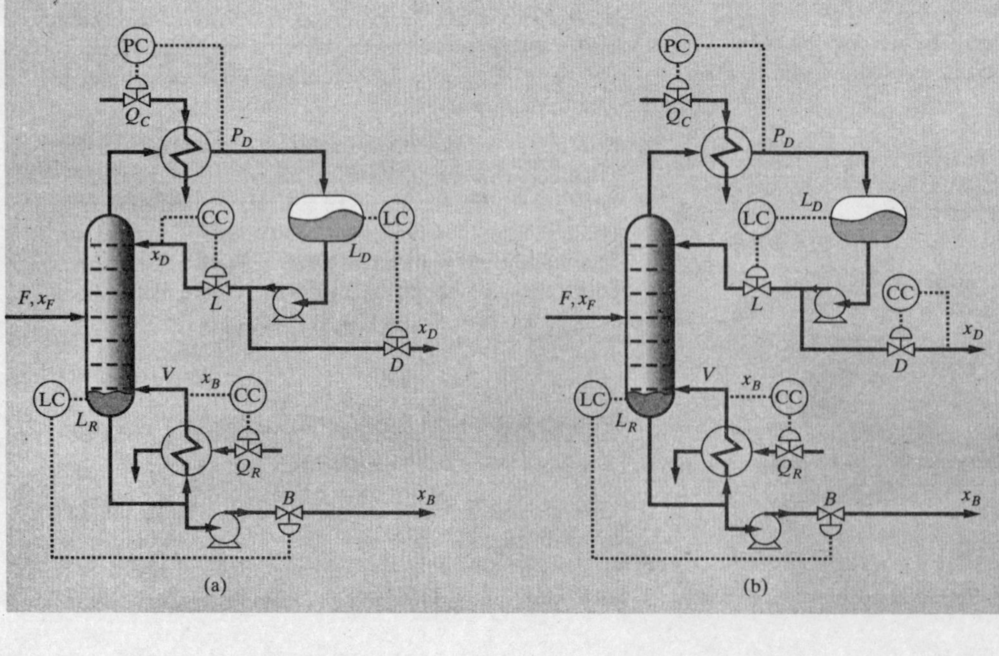

Figure 12.11 Two control configurations for a binary distillation column: (a) LV; (b) DV.

(which is closely related to V) controls x_B, D controls L_D, B controls L_R, and Q_C controls P_D. In columns operating with large reflux ratios, it is advisable to use L to control L_D (Guideline 6). Then, D regulates x_D, in the so-called DV-configuration, shown in Figure 12.11b. Alternative configurations involve ratios of manipulated variables, intended to decouple the control loops by reducing the interaction between them (Shinskey, 1984; Luyben et al., 1999). The dynamic performance of a column control system should be verified using the quantitative methods described in the supplement to this chapter.

12.3 QUALITATIVE PLANTWIDE CONTROL SYSTEM SYNTHESIS

As pointed out by Luyben et al. (1999), the design of a plantwide control system should be driven by the objectives of the overall process rather than by considerations of the individual processing units as in the preceding section. Their strategy for control system design utilizes the available degrees of freedom to achieve these objectives in order of importance, by adopting a "top-down" approach, in common with successful programming practice. Alternatively, in a simpler "bottom-up" approach (Stephanopoulos, 1984), the process is divided into subsystems, with each subsystem often comprised of several process units that share a common processing goal. Then, a control system is formulated for each subsystem, relying on the qualitative guidelines in Section 12.2 or the quantitative analysis to be described in the supplement to this chapter. Finally, an integrated system is synthesized by eliminating possible conflicts among the subsystems. The main disadvantage of this "bottom-up" approach is that good solutions at the subsystem level may not satisfy the process objectives. This can occur when manipulated variables are assigned to meet the control objectives of a subsystem, leaving less attractive inputs to satisfy those of the overall process. Furthermore, interactions among subsystems, such as those resulting from heat integration and material recycle, are not addressed in this decomposition approach, which often leads to unworkable solutions, as will be shown.

The qualitative design procedure for plantwide control by Luyben et al. (1999) consists of the following steps:

Step 1: ***Establish the control objectives.*** As mentioned above, these are related closely to the process objectives. For example, one may wish to impose a given production rate, while ensuring that the products satisfy the quality specified by the market, and guaranteeing that the process meets environmental and safety constraints.

Step 2: ***Determine the control degrees of freedom.*** In practice, the degrees-of-freedom analysis in Section 12.2 may be too cumbersome for the synthesis of plantwide control systems. In a more direct approach, the number of control valves in the flowsheet equals the degrees of freedom (Luyben et al., 1999). As the valves are positioned on the flowsheet, care must be taken to avoid the control of a flow rate by more than one valve. In cases where the degrees of freedom are insufficient to meet all of the control objectives, it may be necessary to add control valves, for example, by adding bypass lines around heat exchangers, as shown in Example 12.5, or by adding trim-utility heat exchangers.

Step 3: ***Establish the energy-management system.*** In this step, control loops are positioned to regulate exothermic and endothermic reactors at desired temperatures. In addition, temperature controllers are positioned to ensure that disturbances are removed from the process through utility streams rather than recycled by heat-integrated process units.

Step 4: ***Set the production rate.*** This is accomplished by placing a flow control loop on the principal feed stream (referred to as *feed flow control*), or on the principal product stream (referred to as *on-demand product flow*), noting that these two options lead to very different plantwide control configurations. Alternatively, the production rate is controlled by regulating the reactor operating conditions, for example, temperature and feed composition.

Step 5: ***Control the product quality and handle safety, environmental, and operational constraints.*** Having regulated the production rate and the effect of temperature disturbances, secondary objectives to regulate product quality and satisfy safety, environmental, operational, and process constraints are addressed in this step.

Step 6: ***Fix a flow rate in every recycle loop and control vapor and liquid inventories (vessel pressures and levels).*** Process unit inventories, such as liquid holdups and vessel pressures (measures of vapor holdups), are relatively easy to control. While vessel holdups are usually non-self-regulating (Guideline 1), the *dynamic performance* of their controllers is usually less important. In fact, level controllers are usually detuned to allow the vessel accumulations to dampen disturbances in the same way that shock absorbers cushion an automobile, as demonstrated in the supplement to this chapter. Less obvious is the need to handle plantwide holdups in recycle loops. As will be shown qualitatively in several examples that follow, and quantitatively in the supplement to this chapter, failure to impose flow control on each recycle stream can result in the loss of control of the process.

Step 7: ***Check component balances.*** In this step, control loops are installed to prevent the accumulation of individual chemical species in the process. Without control, chemical species often build up, especially in material recycle loops.

Step 8: ***Control the individual process units.*** At this point, the remaining degrees of freedom are assigned to ensure that adequate local control is provided in each process unit. Note that this comes *after* the main plantwide issues have been handled.

Step 9: ***Optimize economics and improve dynamic controllability.*** When control valves remain to be assigned, they are utilized to improve the dynamic and economic performance of the process.

Next, the above procedure, which focuses the control system design on meeting the process objectives, is demonstrated on three processes of increasing complexity: (a) an acyclic process; (b) the reactor-flash-recycle process in Example 12.4; and (c) the vinyl-chloride process discussed in Sections 4.4 and 4.5. The last two examples feature recycle loops.

EXAMPLE 12.10 ***Plantwide Control System Configuration for an Acyclic Process***

The chemical process shown in Figure 12.12 is based on an example by Stephanopoulos (1984). It consists of a CSTR in which species A reacts to form B in an exothermic reaction. The reactor effluent is fed to a flash vessel, where the heavier product B is concentrated in the liquid stream, and unreacted A is discarded in the vapor stream. A preheater recovers heat from the hot reactor effluent, with a so-called trim heater installed to ensure that the liquid reactor feed is at the desired temperature. To ensure that the reactor temperature remains on target, the CSTR is equipped with a jacket fed with cooling water to attenuate the heat released. Seven flow-control valves are shown. An eighth valve could be placed on the liquid recycle to V-100, but is not.

Applying the nine-step design procedure for plantwide control by Luyben and co-workers (1999), and using, where possible, the guidelines of Section 12.2:

Step 1: *Set objectives.* The control objectives for this process are as follows:

1. Maintain the production rate of component B at a specified level.
2. Keep the conversion of the plant at its highest permissible value.
3. Achieve constant composition in the liquid effluent from the flash drum.

The structure of the plantwide control system depends on the primary control objective: that is, to maintain a desired production rate. The two possible interpretations of this goal are: (a) to ensure a desired flow rate of the product stream by flow control using valve V-7, which leads to the "on-demand product flow" configuration shown in Figure 12.13, or (b) to ensure a desired production level by "feed flow rate control" using valve V-1, which leads to the control configuration shown in Figure 12.14. The *on-demand product flow* configuration is considered first.

Step 2: *Define control degrees of freedom.* As shown in Figure 12.12, the process has seven degrees of freedom for manipulated variables, with three valves controlling the flow rates of the utility streams (V-2, V-3, and V-6), one controlling the feed flow rate (V-1), two controlling the product flows (V-5 and V-7), and one controlling the reactor effluent flow rate (V-4). Having decided to design an *on-demand product flow* configuration, the valve controlling the B product flow rate (V-7) is reserved for independent flow control (i.e., it directly controls the flow rate). This leads to the control system shown in Figure 12.13.

Step 3: *Establish the energy-management system.* The critical energy management for the CSTR is handled next, since loss of control of the reactor would have serious plantwide consequences. Using the guidelines for controlled and manipulated variable selection, the reactor feed and effluent temperatures are identified as critical for safety (Guideline 2) and quality assurance (Guideline 3). The obvious choices for valves to control these two temperatures are V-2, the jacket coolant valve, and V-3, the steam valve for the trim heater, both of which have a direct effect (Guidelines 6 and 7). These are assigned to temperature control loops.

Step 4: *Set the production rate.* As mentioned above, the B product valve, V-7, is assigned to a flow controller, whose setpoint directly regulates the production rate.

Step 5: *Control product quality, and meet safety, environmental, and operational constraints.* The product

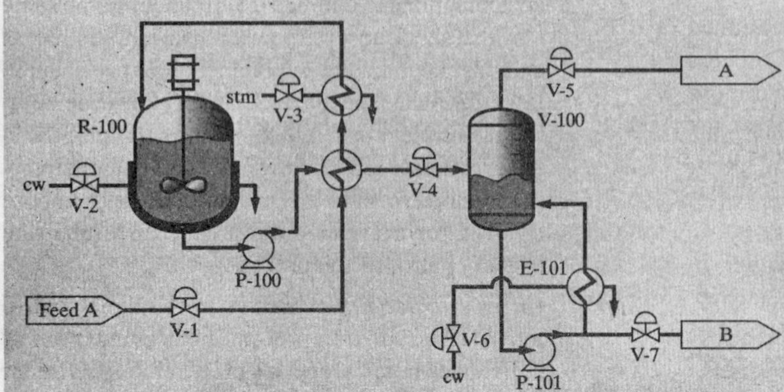

Figure 12.12 Process flowsheet for the acyclic process.

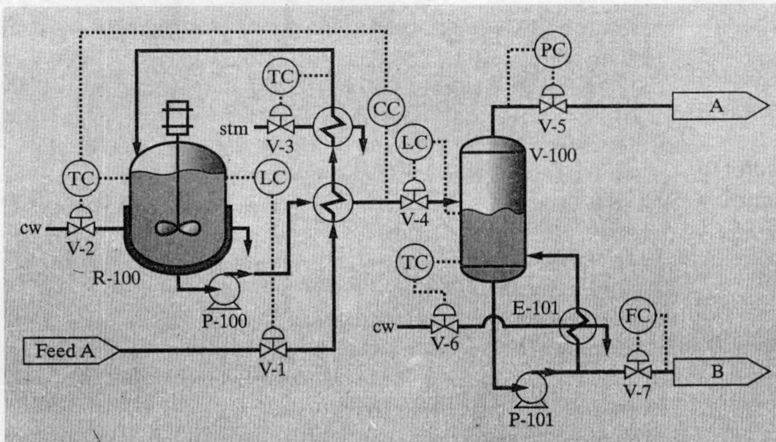

Figure 12.13 Control structure for on-demand product flow.

quality is controlled by maintaining the operating temperature and pressure in the flash vessel at setpoints (Guideline 3). The former is regulated by adjusting the coolant water flow rate through V-6, while the latter is controlled by adjusting the flow rate through overhead valve V-5. These valves are selected because of their rapid and direct effect on the outputs (Guidelines 6, 7, and 8). In addition, these two control loops satisfy the third control objective; that is, to provide tight product-quality control.

Step 6: *Fix recycle flow rates and vapor and liquid inventories.* The liquid inventories in the flash vessel and reactor are non-self-regulating, and therefore, need to be controlled (Guideline 1). Since the liquid product valve from the flash vessel has been assigned to control the product flow rate, the inventory control must be in the reverse direction to the process flow. Thus, the reactor effluent valve, V-4, controls the flash vessel liquid level, and the feed valve, V-1, controls the reactor liquid level. Both of these valves have rapid, direct effects on the liquid holdups (Guidelines 6, 7, and 8). The vapor product valve, V-5, which has been assigned to control the pressure in V-100, thereby controls the vapor inventory.

Step 7: *Check component balances.* With the controllers assigned above, A and B cannot build up in the process, and consequently, this step is not needed.

Step 8: *Control the individual process units.* Since all of the control valves have been assigned, no additional control loops can be designed for the process units.

Step 9: *Optimize economics and improve dynamic controllability.* While a temperature control system for the CSTR is in place, its setpoint needs to be established. To meet the second control objective, which seeks to maximize conversion, a cascade controller is installed, in which the setpoint of the reactor temperature controller (TC on V-2) is adjusted to control the concentration of B (CC) in the reactor effluent. If the reaction is irreversible, conversion is maximized by operating the reactor at the highest possible temperature, making this controller unnecessary.

This completes the control system design for the *on-demand product flow* configuration in Figure 12.13. The performance of the control system needs to be verified by using controllability and resiliency assessment and by applying dynamic simulation as described in the supplement to this chapter. As an alternative to the *on-demand product flow* configuration, the production level can be maintained by fixing the feed flow rate, which leads to the control system shown in Figure 12.14. This control configuration is derived using the same procedure as with Figure 12.13, with the only difference being in Step 6, where the liquid levels are controlled in the direction of the process flow. For the fixed-feed configuration, reaction kinetics may dictate that the reactor

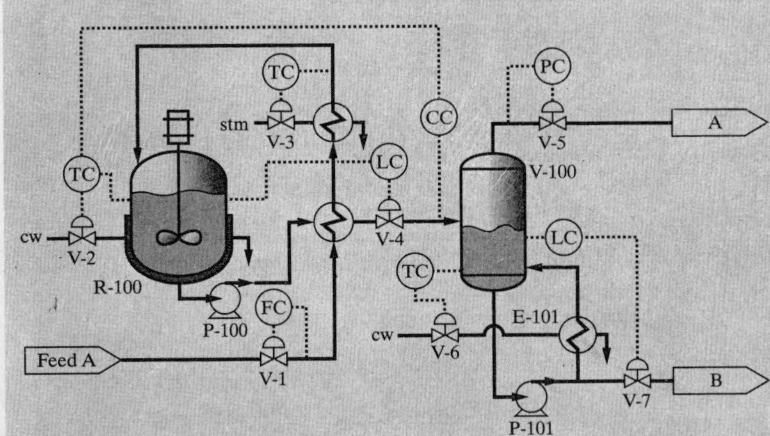

Figure 12.14 Control structure for fixed fresh feed.

holdup be manipulated in concert with throughput changes. In this case, it may be necessary to coordinate the reactor level setpoint with the feed flow rate.

Processes that involve significant heat integration and/or material recycle present more challenging plantwide control problems. The next two examples involve material recycle loops.

EXAMPLE 12.11 ***Plantwide Control System Configuration for Reactor-Flash-Recycle Process (Example 12.4 Revisited)***

For the reactor-flash-recycle process introduced in Example 12.4, Figure 12.15 shows a control system with the control objectives to:

1. Maintain the production rate of component B at a specified level.
2. Keep the conversion of the plant at its highest permissible value.

This controller configuration results from using a unit-by-unit design approach, with each vessel inventory controlled by manipulation of its liquid effluent flow. Although the control pairings are acceptable for each process unit in isolation, the overall control system does not establish flow control of the recycle stream. Consequently, a change in the desired feed rate, keeping the reactor inventory constant with level control, causes an excessive increase in the reactor effluent flow, which is transferred rapidly to the recycle flow by the flash level controller. This undesirable positive feedback is referred to as the "snowball effect" by Luyben and co-workers (1999), and is the consequence of not ensuring flow control of the recycle stream.

Since Luyben identified the snowball effect (Luyben, 1994), the sensitivity of reactor-separator-recycle systems to external disturbances has been the subject of several studies (e.g., Wu and Yu, 1996; Skogestad, 2002). Recent work by Bildea and co-workers (Bildea et al., 2000; Kiss et al., 2002) has shown that a critical reaction rate can be defined for each reactor-separator-recycle system using the Damköhler number, *Da* (dimensionless rate of reaction, proportional to the reaction rate constant and the reactor holdup). When the Damköhler number is below a critical value, Bildea et al. show that the conventional unit-by-unit approach in Figure 12.15 leads to the loss of control. Furthermore, they show that controllability problems associated with exothermic CSTRs and PFRs are often resolved by controlling the total flow rate of the reactor feed stream.

The extent of the snowball effect is shown next by analysis of the controlled process in Figure 12.15. The combined feed of pure A and recycle is partially converted to B in reactor R-100 by the isothermal, liquid-phase, irreversible reaction A → B, which has first-order kinetics. The reactor effluent is flashed across valve V-3, to give a vapor-product stream, assumed to be pure B, and a liquid-product stream, assumed to be pure A. The liquid stream is recycled to the reactor, where it is mixed with fresh feed A to give the combined feed. The control configuration consists of six control loops: (1) production rate F_0, controlled using valve V-1 on the fresh feed stream; (2) temperature control using valve V-2 to ensure isothermal operation of R-100; (3) level control in R-100 holdup using valve V-3; (4) level control in V-100 using valve V-6; (5) pressure control in V-100 using the vapor-product valve, V-4; and (6) temperature control in V-100 (controlling product quality) using the coolant valve, V-5.

What happens when the fresh feed flow rate changes or is disturbed? Note that even though the flow controller fixes the fresh feed rate in Figure 12.15, it can still be disturbed. The equations that apply are:

Combined molar feed to the CSTR :	$F_0 + B$
Molar material balance around the flash vessel :	$F_0 + B = D + B$
Overall molar material balance :	$F_0 = D$

where F_0, D, and B are the molar flow rates of the feed, flash vapor, and flash liquid streams. Finally, the rate of consumption of A in the reactor is:

$$r_A = kc_A \tag{12.12}$$

where r_A is the intrinsic rate of reaction, k is the first-order rate constant, and c_A is the molar concentration of A in the reactor

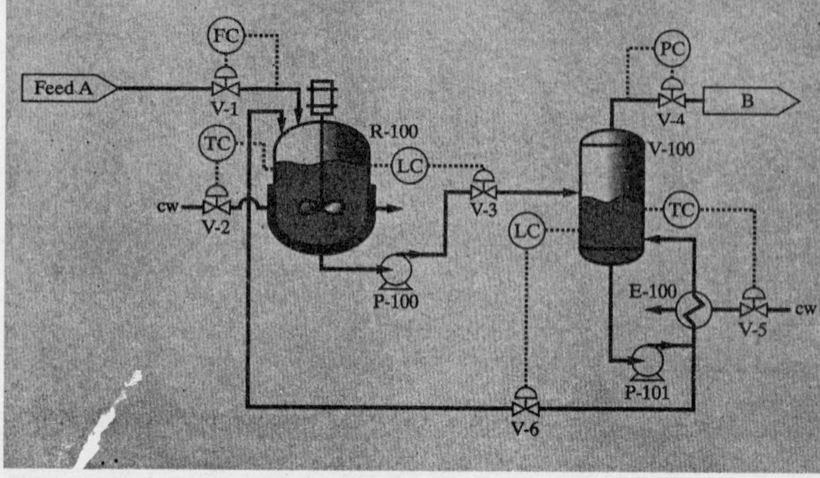

Figure 12.15 Control structure based on unit-by-unit design approach.

effluent. Using c_{total} and x_A, the total molar concentration and the mole fraction of A in the reactor effluent, Eq. (12.12) becomes:

$$r_A = kx_A c_{total} \tag{12.13}$$

The molar flow rate of B in the reactor effluent is:

$$(1 - x_A)(F_0 + B) = kx_A c_{total} V_R \tag{12.14}$$

where V_R is the volume of the reactor holdup. Then, substituting $c_{total} V_R = n_T$.

$$(1 - x_A)(F_0 + B) = kx_A n_T \tag{12.15}$$

where n_T is the total molar holdup in the reactor. Rearranging Eq. (12.15) for the flash liquid stream flow rate (recycled to the reactor), B:

$$B = \frac{x_A(F_0 + kn_T) - F_0}{1 - x_A} \tag{12.16}$$

With the reactor temperature and holdup fixed, a change to the fresh feed flow rate by a disturbance causes the mole fraction of A in the reactor effluent to change. Therefore, to obtain the effect of a change in F_0 on B, x_A must be eliminated from Eq. (12.16). For a perfect separation, an overall balance on the disappearance of A gives:

$$F_0 = kx_A n_T \tag{12.17}$$

Rearranging Eq. (12.17) for x_A and substituting in Eq. (12.16) gives:

$$B = \frac{F_0^2}{kn_T - F_0} \tag{12.18}$$

Equation (12.18) shows clearly that the numerator increases as the square of F_0 while the denominator decreases with increasing F_0. Thus, B increases by more than the square of F_0! As an example, consider F_0 in the range of 50 to 150, with $kn_T = 200$. Then, Eq. (12.18) gives the recycle rate, B:

F_0	B
50	16.7
75	45
100	100
125	208
150	450

Thus, when the feed rate is tripled from 50 to 150, the recycle rate increases by a factor of 450/16.7 = 27. This result assumes a fixed Kn_T. A more general result relies on reformulating Eq. (12.18) in terms of the Damköhler number, $Da = kn_T/F_0$, giving:

$$B = \frac{F_0}{Da - 1} \tag{12.19}$$

Equation (12.19) shows that for values of Da much larger than unity, no snowball effect is expected. The snowball effect occurs as Da approaches a critical value of 1, and is eliminated by controlling the recycle flow rate, as shown next.

To generate a workable plantwide control system, as shown in Figure 12.16, the design procedure for plantwide control by Luyben and co-workers (1999) is applied:

Step 1: *Set objectives.* To achieve the primary control objective, the production level is maintained by flow control of the feed stream using valve V-1.

Step 2: *Define control degrees of freedom.* As shown in Figure 12.4, the process has six degrees of freedom with two valves controlling the flow rates of the utility streams (V-2 and V-5), one controlling the feed flow rate (V-1), one controlling the product flow (V-4), one controlling the reactor effluent flow rate (V-3), and one controlling the recycle flow rate (V-6). Having chosen *constant feed flow* in Step 1, the feed valve (V-1) is reserved for independent flow control.

Step 3: *Establish the energy-management system.* The reactor temperature, which affects the process yield and stability (Guidelines 2 and 3), is controlled by adjusting the coolant flow rate, using valve V-2.

Step 4: *Set the production rate.* As stated previously, the feed valve, V-1, is assigned to a flow controller, whose setpoint regulates the production rate.

Step 5: *Control product quality, and meet safety, environmental, and operational constraints.* A conventional pressure and temperature control system is set up for the flash vessel, as in the previous example.

Step 6: *Fix recycle flow rates and vapor and liquid inventories.* To eliminate the snowball effect, the recycle flow rate must be controlled by installing a flow controller, either on the reactor effluent or on the flash liquid effluent. As shown in Figure 12.16, the second option forces the reactor effluent valve, V-3, to control the flash vessel liquid inventory, in the absence of other alternatives. Then, to regulate the reactor inventory, a cascade control system is designed in which the reactor level controller (LC) adjusts the setpoint of the feed flow controller (FC on V-1). This does not conflict with the objective to set the production rate by fixing the feed flow rate because, in stable operation, the reactor level and feed flow rate vary proportionally, through higher conversion in the CSTR. The vapor product valve, V-4, which has been assigned to control the pressure, thereby controls the vapor inventory.

Steps 7 and 8: *Check component balances and control individual process units.* As in Example 12.10, the controllers assigned thus far prevent the buildup of A and B in the process. Furthermore, no valves are available to improve control of either process unit.

Step 9: *Optimize economics and improve dynamic controllability.* To maximize conversion, a cascade controller is installed as in the previous example, in which the

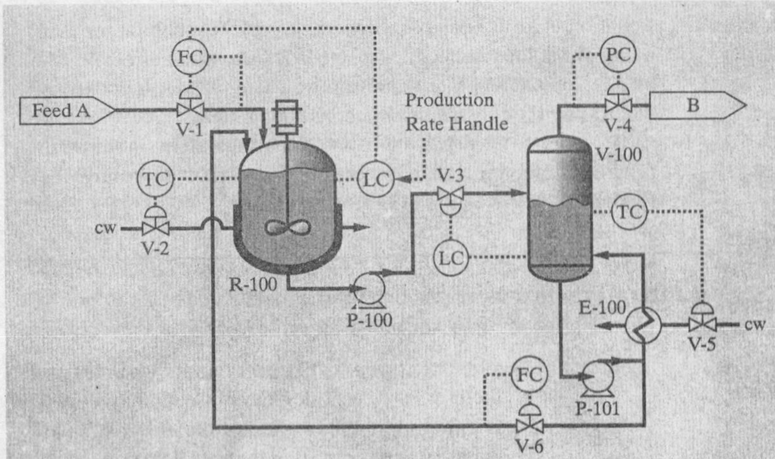

Figure 12.16 Workable plantwide control structure.

setpoint of the reacto. temperature controller (TC on V-2) is adjusted to control the concentration of B in the reactor effluent. Again, for an irreversible reaction, it is enough to operate the reactor at the highest possible temperature.

EXAMPLE 12.12 *Plantwide Control System Configuration for the Vinyl-Chloride Process*

For the vinyl-chloride process synthesized in Section 4.4, a preliminary design of its plantwide control system helps to assess the ease of maintaining the desired production level. As shown in Figures 12.17 and 12.18, this is achieved following the design procedure of Luyben and co-workers (1999):

Step 1: *Set objectives.* Note that nearly 100% conversion is achieved in the dichloroethane reactor (R-100). Assuming that the conversion in the pyrolysis furnace (F-100) cannot be altered, the production level must be maintained by flow control of the ethylene feed flow rate using valve V-1.

Step 2: *Determine the control degrees of freedom.* Twenty control valves have been positioned in the PFD, as shown in Figure 12.17.

Step 3: *Establish the energy-management system.* The coolant valve, V-3, in the overhead condenser of the exothermic dichloroethane reactor, R-100, is used for temperature control. The yield in the pyrolysis furnace, F-100, is controlled by maintaining the outlet temperature at 500°C using the fuel gas valve, V-6. To attenuate the effect of temperature disturbances, the flow rates of the utility streams are adjusted to regulate effluent temperatures in the evaporator, E-101 (using V-5); the quench tank, V-100 (using the cooler E-102 and manipulating V-7); the partial condenser, E-103 (using V-8); and the recycle cooler, E-108 (using V-20). All of these valves act rapidly and directly on the controlled outputs (Guidelines 6, 7, and 8). Note that the temperature control loops using utility exchangers ensure that temperature disturbances are not recycled (Guideline 9).

Step 4: *Set the production rate.* As stated previously, the feed valve, V-1, is assigned to a flow controller, whose setpoint regulates the production rate.

Step 5: *Control product quality, and meet safety, environmental, and operational constraints.* The overhead product compositions in both distillation columns are regulated by adjusting the reflux flow rates using valves V-11 and V-16, both of which provide fast, direct control action (Guidelines 6, 7, and 8). The bottoms product compositions are controlled using the reboiler steam valves, V-13 and V-18. Thus, the control systems for both columns are in the LV-configuration. The process design calls for pressure reduction at the feed of each distillation column, using valves V-9 and V-14. However, these valves are not appropriate for the control of the column pressures, which are regulated in Step 6 by manipulation of downstream, rather than upstream, valves. Thus, V-9 is maintained constant at its design position, while V-14 is utilized for inventory control in the sump of column T-100 using Step 6.

Step 6: *Fix recycle flow rates and vapor and liquid inventories.* The recycle flow rate must be held constant by flow control. However, since two-point composition control has been installed in T-101, it is not possible to also fix its bottoms flow rate by manipulating V-19. Thus, a flow controller is installed to fix the combined recycle and feed flow rates using V-4. In addition, liquid inventory control must be installed for the reactor, R-100, as well as for the sumps and reflux drums of the two columns, with vapor inventory control needed in the two columns. Having assigned V-4 for recycle flow rate control, a level controller for R-100 is cascaded with the ethylene flow controller, making the level setpoint the production handle, as in Example 12.11 (Figure 12.16). Inventory control of T-101 is assigned first. With the total recycle flow rate under control, the bottoms flow rate in T-101, adjusted by valve V-19, is used for sump level control. The liquid level in reflux drum V-102 is controlled by

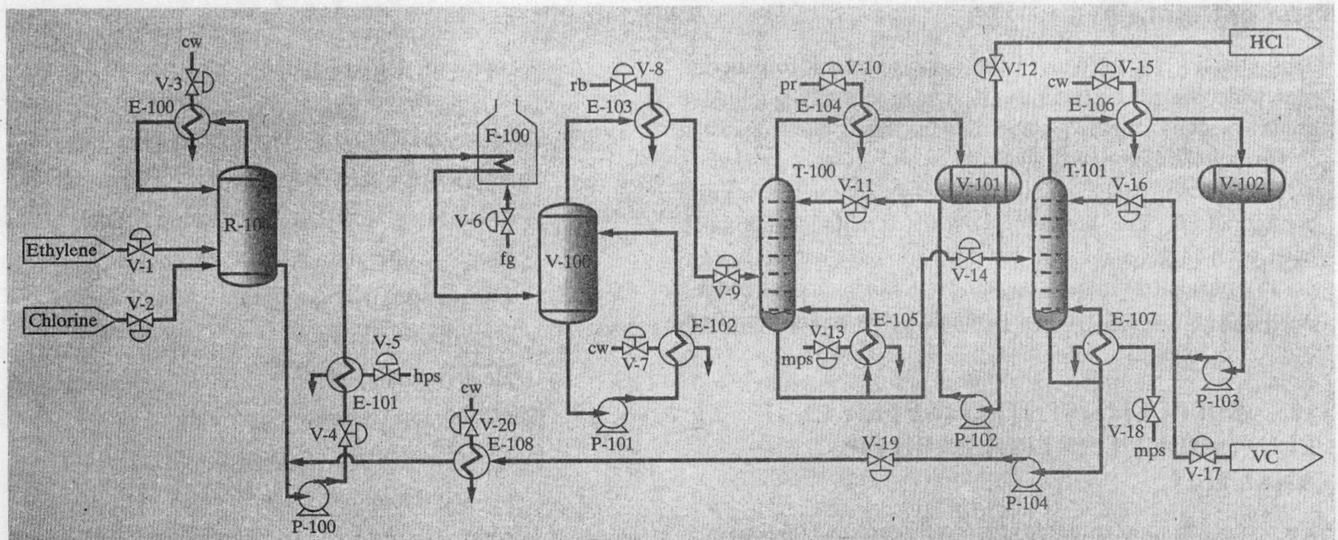

Figure 12.17 Control valve placement for the vinyl-chloride process.

manipulating the distillate valve, V-17. Inventory control for T-101 is completed by controlling the overhead pressure using the coolant valve, V-15. Turning to the HCl column, T-100, the bottoms product valve, V-14, is assigned to control the sump level. Since the overhead product is vapor, the condenser pressure is regulated using the distillate valve, V-12. This frees the condenser coolant valve, V-10, to regulate the reflux drum liquid level.

Steps 7 and 8: *Check component balances and control individual unit operations.* At this point, all but one of the valves (V-2) has been assigned. To ensure a stoichiometric ratio of reagents entering reactor R-100, the chlorine feed is adjusted to ensure complete conversion of the ethylene, using a composition controller on the reactor effluent.

Step 9: *Optimize economics and improve dynamic controllability.* As in the previous example, to improve the range of production levels that can be tolerated, the setpoint of the recycle flow controller is set in proportion to the feed flow rate, suitably lagged for synchronization with the propagation rate through the process. For clarity, this is not shown in Figure 12.18.

The complete control system is shown in Figure 12.18. Many of the qualitative decisions need to be checked by quantitative analysis or by simulation. For example, the interaction between the control systems of the two columns may require careful controller tuning. These refinements are discussed in the supplement to this chapter.

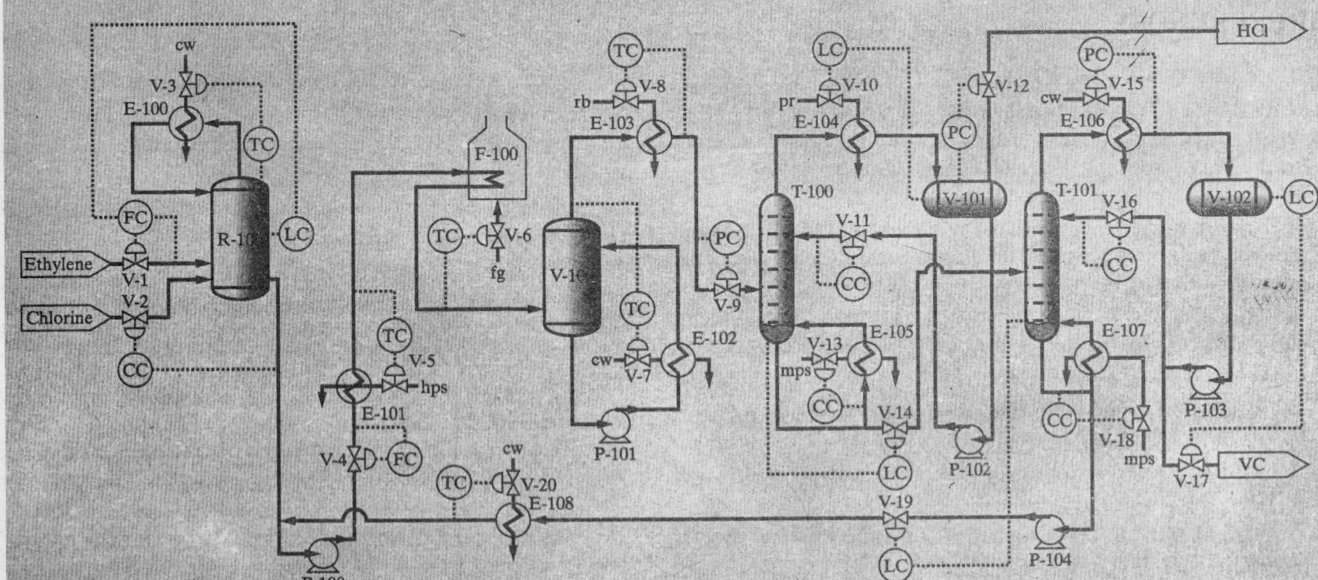

Figure 12.18 Control system for the vinyl-chloride process.

12.4 SUMMARY

This chapter has introduced the importance of considering plantwide control early in the design process. A qualitative control synthesis method, combining the approaches suggested by Newell and Lee (1988) and Luyben and co-workers (1999), was presented to show how to generate alternative control configurations. The limitations of this qualitative approach have been highlighted, and the need for the quantitative approach presented in the supplement to this chapter, which involves analysis and dynamic simulation, has been established.

12S SUPPLEMENT TO CHAPTER 12—FLOWSHEET CONTROLLABILITY ANALYSIS

A supplement to Chapter 12, entitled "Flowsheet Controllability Analysis," is provided in the PDF Files folder, which can be downloaded from the Wiley Web site associated with this book. See the file Supplement_to_Chapter 9.pdf. The contents of this supplement are:

12S.0 Objectives

12S.1 Generation of Linear Models in Standard Forms

12S.2 Quantitative Measures for Controllability and Resiliency

- Relative-Gain Array (RGA)
- Properties of Steady-State RGA
- Dynamic RGA (McAvoy, 1983)
- The RGA as a Measure of Process Sensitivity to Uncertainty
- Using the Disturbance Cost to Assess Resiliency to Disturbances

12S.3 Toward Automated Flowsheet C&R Diagnosis

- Short-Cut C&R Diagnosis
- Generating Low-Order Dynamic Models
- Steady-State Gain Matrix, $\underline{\underline{K}}^c$
- Dynamics Matrix, $\underline{\underline{\psi}}^c\{s\}$
- Distillation Columns
- Heat Exchangers

12S.4 Control Loop Definition and Tuning

- Definition of PID Control Loop
- Controller Tuning
- Model-Based PI-Controller Tuning

12S.5 Case Studies

- Case Study 12S.1 Exothermic Reactor Design for the Production of Propylene Glycol
- Case Study 12S.2 Two Alternative Heat Exchanger Networks
- Case Study 12S.3 Interaction of Design and Control in the MCB Separation Process

12S.6 MATLAB for C&R Analysis

12S.7 Summary

12S.8 References

12S.9 Exercises

REFERENCES

1. AL-ARFAJ, M.A., and W.L. LUYBEN, "Control of Ethylene Glycol Reactive Distillation Column," *AIChE J.*, **48**, 905–908 (2002).

2. BILDEA, C.S., A.C. DIMIAN, and P.D. IEDEMA, "Nonlinear Behavior of Reactor-Separator-Recycle Systems," *Comput. Chem. Eng.*, **24**, 2–7, 209–215 (2000).

3. CHIANG, T., and W.L. LUYBEN, "Comparison of the Dynamic Performances of Three Heat-Integrated Distillation Configurations," *Ind. Eng. Chem. Res.*, **27**, 99–104 (1988).

4. KISS, A.A., C.S. BILDEA, A.C. DIMIAN, and P.D. IEDEMA, "State Multiplicity in CSTR-Separator-Recycle Systems," *Chem. Eng. Sci.*, **57**, 4, 535–546 (2002).

5. LUYBEN, W.L., *Process Modeling, Simulation, and Control for Chemical Engineers*, 2nd ed., McGraw-Hill, New York (1990).

6. LUYBEN, W.L., "Snowball Effects in Reactor/Separator Processes with Recycle," *Ind. Eng. Chem. Res.*, **33**, 299–305 (1994).

7. LUYBEN, W.L., B.D. TYREUS, and M.L. LUYBEN, *Plantwide Process Control*, McGraw-Hill, New York (1999).

8. MCAVOY, T.J., *Interaction Analysis, Instrument Society of America*, Research Triangle Park, North Carolina (1983).

9. NEWELL, R.B., and P.L. LEE, *Applied Process Control*, Prentice-Hall of Australia, Brookvale, NSW (1988).

10. SEBORG, D.E., T.F. EDGAR, and D.A. MELLICHAMP, *Process Dynamics and Control*, Chapter 28, Wiley, New York (1989).

11. SHINSKEY, F.G., *Distillation Control*, 2nd ed., McGraw-Hill, New York (1984).

12. SKOGESTAD, S., "*Plantwide Control: Towards a Systematic Procedure*," in J. GRIEVINK and J. van der Schijndel, Eds., *Computer Aided Chemical Engineering-10, European Symposium on Computer Aided Process Engineering-12*, Elsevier (2002).

13. STEPHANOPOULOS, G., *Chemical Process Control*, Chapter 23, Prentice-Hall, Englewood Cliffs (1984).

14. WEITZ, O., and D.R. LEWIN, "Dynamic Controllability and Resiliency Diagnosis Using Steady State Process Flowsheet Data," *Comput. Chem. Eng.*, **20** (4), 325–336 (1996).

15. WU, K.L., and C.C. YU, "Reactor/Separator Process with Recycle-1. Candidate Control Structures for Operability," *Comput. Chem. Eng.*, **20**, 11, 1,291–1,316 (1996).

EXERCISES

12.1 Perform a degrees-of-freedom analysis for the non-interacting exothermic reactor shown in Figure 12.3a. Suggest an appropriate control structure. Carry out the same exercise for the heat-integrated reactor shown in Figure 12.3b. Compare the results.

12.2 Consider the mixing vessel shown in Figure 12.19. The feed stream flow rate, F_1, and composition, C_1, are considered to be disturbance variables. The feed is mixed with a control stream of flow rate F_2, and constant known composition, C_2. To ensure a product of constant composition, it is also possible to manipulate the flow rate, F_3, of the product stream. Perform a degrees-of-freedom analysis and suggest alternative control system configurations. Note that unsteady-state balances are required.

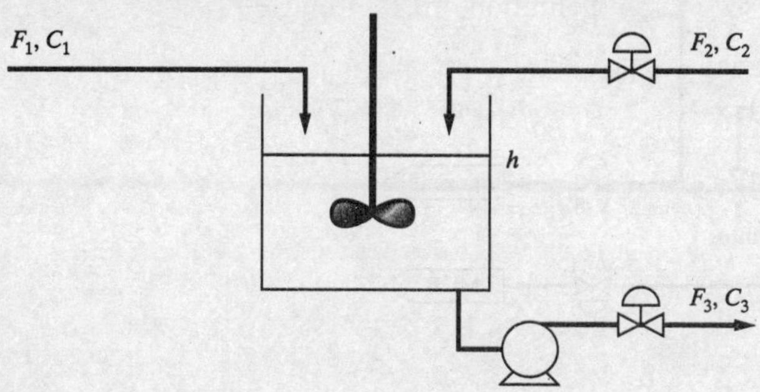

Figure 12.19 Mixing vessel.

12.3 Consider the FS two-column configuration for the separation of methanol and water in Figure 12.2 and (a) determine the number of degrees of freedom for the overall system, (b) determine the number of controlled and manipulated variables, and (c) select a workable control configuration using qualitative arguments.

12.4 Repeat Exercise 12.3 for the LSF configuration in Figure 12.2.

12.5 A control system is suggested for the exothermic reactor in Figure 12.7. Suggest alternative configurations and compare them with the original configuration.

12.6 Figure 12.20 shows a process for the isothermal production of C from A and B ($A + B \rightarrow C$). The two reagents are fed to a CSTR, R-100, where complete conversion of B is assumed. The reactor effluent stream, consisting of C and unreacted A, is separated in a distillation column, T-100, where the more volatile A is withdrawn in the distillate and recycled, and product C is withdrawn in the bottoms stream. Your task is to devise a conceptual plantwide control system for the process. *Hint:* It may be helpful to reposition the feed stream of A.

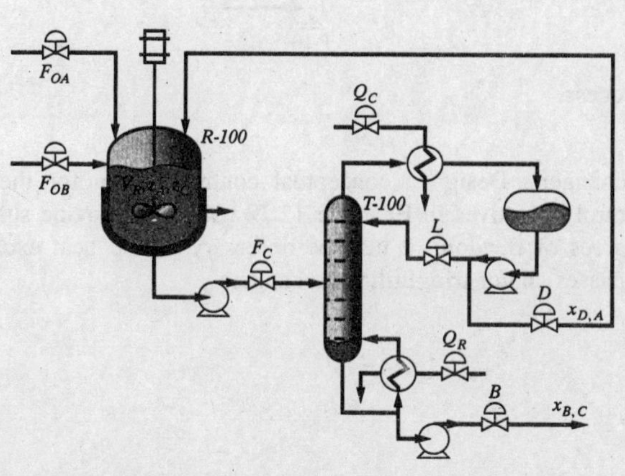

Figure 12.20 Process flowsheet for Exercise 12.6.

12.7 Figure 12.21 shows the flowsheet for a reactive distillation column for the production of ethylene glycol (EG) from ethylene oxide (EO) and water (Al-Arfaj and Luyben, 2002):

$$EO + H_2O \rightarrow EG$$

Note that the reaction proceeds to 100% conversion in the column, with part of the EG undergoing a secondary (undesired) reaction to diethylene glycol (DEG):

$$EO + EG \rightarrow DEG$$

For this reason, the EO is fed to the column in slight excess. The EG product is withdrawn as the bottoms stream, and almost pure water concentrates at the top of the column. Your task is to *use the procedure of Luyben and co-workers, showing all steps*, to devise a conceptual plantwide control system for the process with the following objectives:

(a) Control production rate

(b) Ensure the EG product is at the required concentration

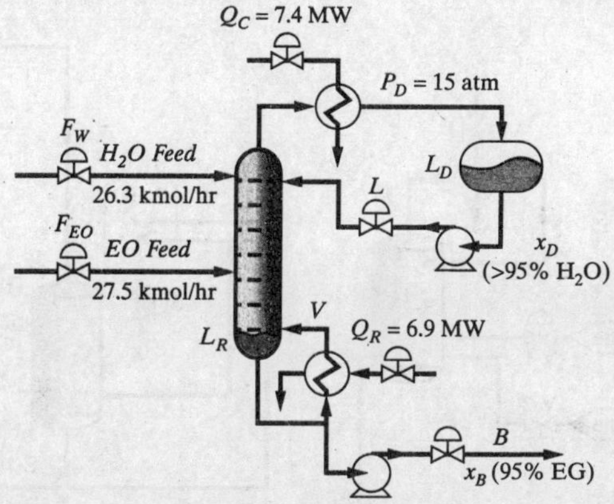

Figure 12.21 Ethylene glycol reactive distillation column.

12.8 Figure 12.22 shows the monochlorobenzene separation process introduced in Section 5.4. The process involves a flash vessel, V-100, an absorption column, T-100, a distillation column, T-101, a reflux drum, V-101, and three utility heat exchangers. As shown in Figure 5.23, most of the HCl is removed at high purity in the vapor effluent of T-100. However, in contrast with the design shown in Chapter 5, that in Figure 12.22 does not include a "treater" to remove the residual HCl; instead, this is purged in a small vapor overhead product stream in T-101. The benzene and monochlorobenzene are obtained at high purity as distillate and bottoms liquid products in T-101. Note that the 12 available control valves are identified. Your task is to design a conceptual control system to ensure that the process provides stable production at a desired level, while meeting quality specifications.

12.9 How would the control configuration for the vinyl-chloride process in Figure 12.18 change if the primary control objective is to provide on-demand vinyl-chloride product?

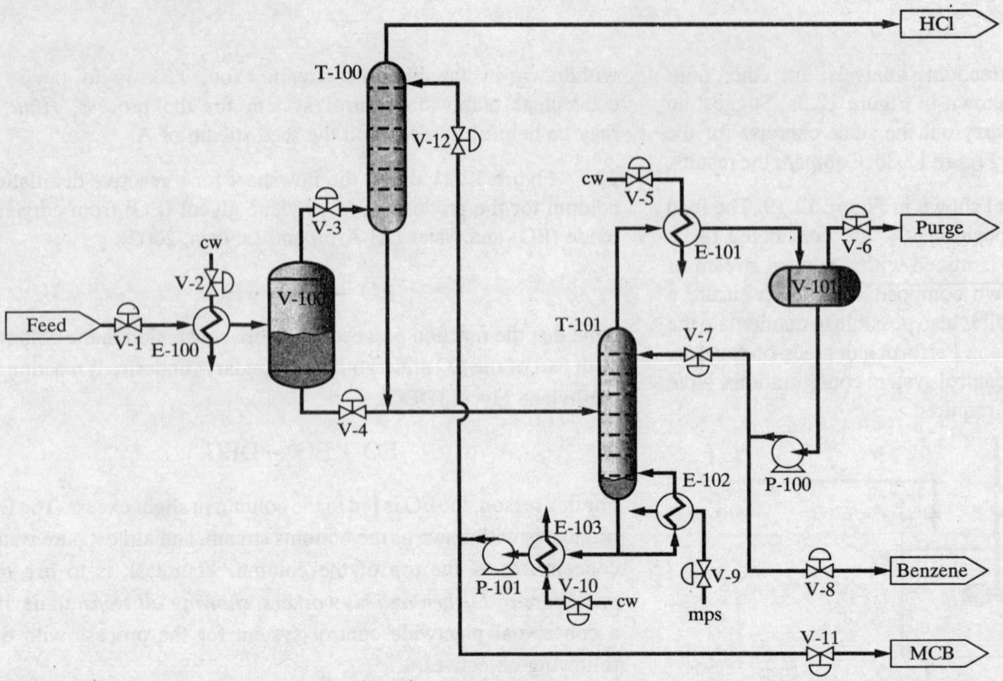

Figure 12.22 Process flowsheet for the MCB separation process.

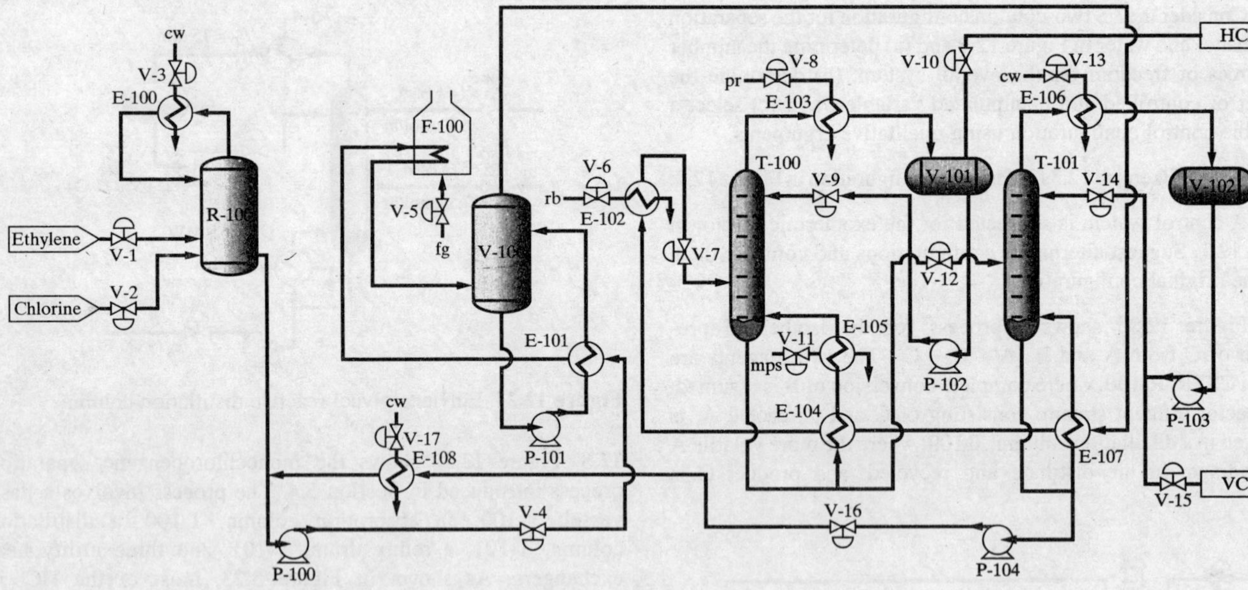

Figure 12.23 Process flowsheet for the heat-integrated vinyl-chloride process.

12.10 Figure 12.23 shows a heat-integrated process for the manufacture of vinyl chloride. As discussed in Section 4.4, this heat integration is possible when the rate of carbon deposition is sufficiently low. This design sharply reduces the utilization of utilities in Figure 12.17, without requiring additional heat exchangers. Design a conceptual control system for the same control objectives in Example 12.12. *Hint:* To provide sufficient degrees of freedom, it may be necessary to add heat exchanger bypasses and/or trim-utility exchangers.

Chapter 13

Product Design Case Studies for Basic Chemicals

13.0 OBJECTIVES

This chapter provides case studies to illustrate the steps in the design of *basic* chemical products using the Stage-Gate^{TM} Product-Development Process. Emphasis is placed on the *concept* and *feasibility* stages in Figure PI.1. Only the key issues are summarized in the *development* stage.

After studying this chapter, the reader should:

1. Be able to use the elements of the Stage-Gate^{TM} Product-Development Process for the design of *basic* chemical products.
2. Have an appreciation of how appropriate design methodology is invoked as needed using the Stage-Gate^{TM} Product-Development Process.

13.1 INTRODUCTION

In this chapter, three case studies are presented involving *basic* chemical products. The first, which involves a plant to produce ammonia, focuses on process design to manufacture this well-understood basic chemical. A number of design innovations are considered in an attempt to reduce costs so as to compete with competitors. For the second, which involves the production of an environmentally friendly refrigerant to replace a refrigerant that is no longer acceptable environmentally, the focus is on the techniques for molecular structure design presented in Chapter 3. Finally, the third involves the design of a water-dispersible β-carotene product for the beverage industry, beginning with the basic chemical β-carotene.

These case studies follow the Stage-Gate^{TM} Product-Development Process, as discussed in Chapter 2 and summarized in Figure PI.1, in solving the three product/process design problems.

13.2 AMMONIA CASE STUDY

Before proceeding with the ammonia case study, the reader should be conversant with the key steps in process design as introduced in Part One, specifically Chapters 4 to 9. Furthermore, since this case study is driven by economics, it is helpful to be familiar also with capital cost-estimation methods, introduced in Chapter 22, and with profitability analysis, introduced in Chapter 23.

Project Charter and New Technologies

As seen in Figure PI.1, it is recommended that the design team begin to develop a new product and process by creating a project charter. Before introducing the initial project charter prepared by the design team, a brief history of the manufacture and purchase of ammonia by Haifa Chemicals is reviewed. Haifa Chemicals is an international corporation, established in 1967, that produces and markets specialty fertilizers, food additives, and technical chemicals. Their production plants are located in Israel at Haifa Bay and in the northern part of the Negev region, and in Lunel, France.

In the 1970s, a 250-ton/day ammonia plant was located at Haifa Bay in close proximity to the Haifa Refinery, which provided naphtha that was converted to hydrogen in a reformer. Then, as demand increased in the late 1980s, the plant was shut down, with ammonia supplied by ship from external producers. To ensure continuous operation, a 12,000-ton storage vessel, containing a month's supply, was installed at Haifa Bay. The principal usage of ammonia in Israel is in the manufacture of $(NH_4)_3PO_3$, that is, fertilizer pellets for use in farming. Haifa Chemicals operates manufacturing facilities for ammonium phosphate in Ramat Hovav, which is in the Negev Desert in southern Israel, and in Haifa Bay, with ammonia being shipped by truck to Ramat Hovav, and phosphate rock being shipped by train to Haifa. As a result of the heightened security threat posed by the Lebanese Hezbollah in the summer of 2006, the State of Israel has been considering alternatives to relocate the storage tank to a remote location, such as the Negev Desert.

Table 13.1 Initial Project Charter

Project Name	Ammonia Production in Israel
Project Champions	Business Director of Haifa Chemicals, Inc.
Project Leaders	Speedy and Ploni Gonzales
Specific Goals	Produce and store NH_3 from synthesis gas beginning with natural gas. Consider relocating the facility to a remote location to reduce the risk to population centers in the event of the release of NH_3 that might result from terrorist rocket fire. Consider the initial design proposed by Emek Projects Ltd (EPL) as a possible starting point. Consider a joint facility with the State of Jordan and the purchase of natural gas from Egypt and/or the Palestinian Authority.
Project Scope	In-scope: • H_2 from natural gas • N_2 from air Out-of-scope: • Production and storage in port cities with large populations (e.g., Haifa, Ashdod)
Deliverables	• Business opportunity assessment • Technical feasibility assessment • Product life-cycle assessment
Time Line	Feasible processing package within 6 months

Consider the example project charter in Table 13.1 assembled by a typical design team in 2007. The goals of this project charter were centered around circumventing the safety hazard created by the storage of toxic ammonia in the Haifa area, which has a metropolitan population of over 1 million persons. At that time, the Israeli demand for ammonia was about 350 metric ton/day. This demand could have been augmented by the joint construction of a manufacturing plant with Jordan, which was storing approximately 30,000 metric tons of ammonia in Akaba that arrived by ship from external suppliers. In 2007, it was also significant that Egypt was constructing a 2, 000-metric-ton/day ammonia plant in the Suez Industrial Zone, scheduled to begin operation in 2008, at an investment cost of $540,000,000. While implicit in satisfying the need for ammonia product, an important objective was to provide guidance and recommendations to the Israeli government regarding policies for the storage of toxic chemicals. To protect the population, given the prices established by external suppliers of natural gas and ammonia, it may have been necessary for the government to consider the provision of tax breaks and low-cost loans. Initially, the preliminary design proposed by Emek Projects Ltd (EPL), an imaginary company, was available.

This initial project charter identified the production of synthesis gas using natural gas as the source of hydrogen and air as the source of nitrogen as *in-scope*. On the other hand, solutions involving the storage of ammonia in port cities, such as Haifa and Ashdod, were considered to be *out-of-scope*.

The first deliverable was a business opportunity assessment involving a profitability analysis. In addition, a technical feasibility assessment was to be provided, including the results of process simulations. And, finally, a product life-cycle assessment was to be conducted that addressed: (1) the danger of an ammonia release from a storage facility, (2) the release of carbon dioxide byproduct into the atmosphere, and (3) the possible conversion of ammonia to urea by reaction with carbon dioxide as a vehicle for curbing the release of the carbon dioxide byproduct. Note, however, that due to space limitations, this life-cycle assessment is the subject of Exercises 13.1–13.3.

Finally, the product time line required the delivery of a feasible process package within six months.

Innovation Map

Having created a project charter, the design team next turned to an examination of the customer needs (that is, the *customer-value proposition*) and the new technologies likely to play an important role in providing the ammonia product, as introduced in Section 1.3. These are shown linked together in the *innovation map* of Figure 13.1.

To construct this *innovation map*, the design team first identified the elements in its four levels, moving from the bottom to the top of the map:

1. *Process/Manufacturing Technology:* Improved heat integration, membrane separation to recover H_2 from the purge stream, and heat and mass exchange (HME) technology to enhance conversion in the NH_3

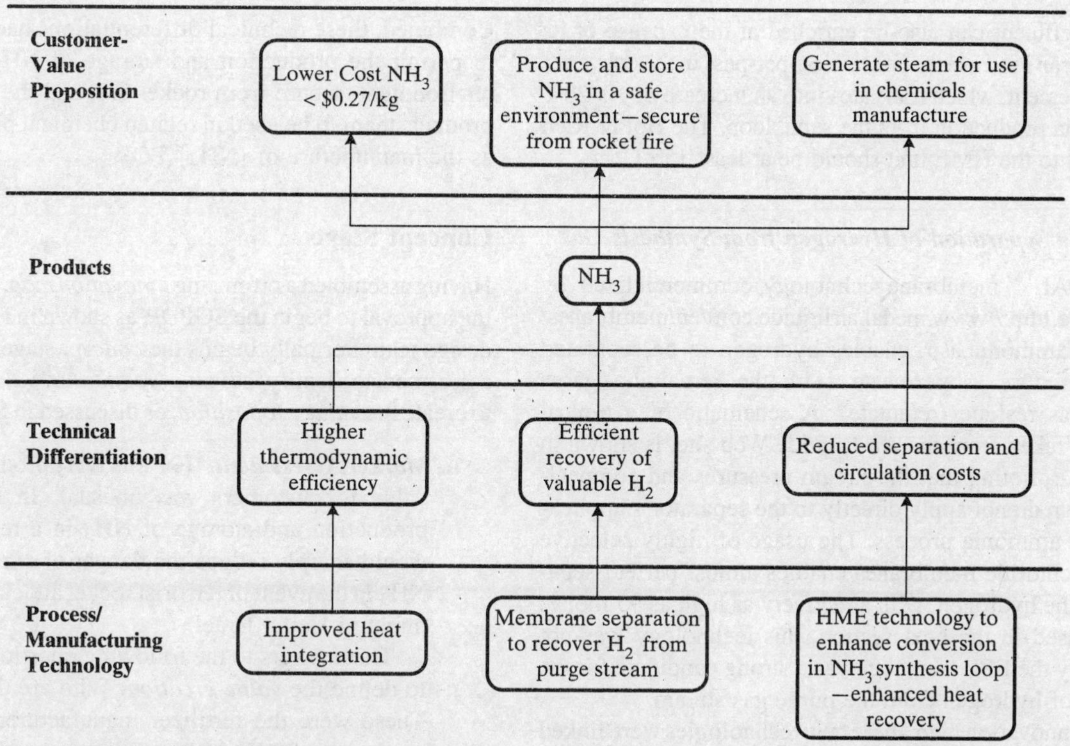

Figure 13.1 Innovation map for ammonia product.

synthesis loop and provide enhanced heat recovery. These were new technologies to be considered to enhance the profitability of the process under development.

2. *Technical Differentiation (Technical-Value Proposition):* Higher thermodynamic efficiency, resulting from improved heat-integration methods; efficient recovery of valuable H_2 by utilizing membrane-separation technology; and reduced separation and recirculation costs in the synthesis loop by incorporating HME technology.
3. *Products:* NH_3.
4. *Customer-Value Proposition:* Produce NH_3 to sell for \$0.27/kg or less, while reducing the potential security impact of a rocket attack.

After identifying the elements at all four levels of the innovation map, their connectivity in the map was added to show the interplay between the technological elements, the *technical-value proposition*, and ultimately the *customer-value proposition*.

In this case, for a well-known commodity chemical product such as NH_3, the new technologies that had the potential to satisfy the customer needs were process/manufacturing technologies. The first was the potential for improved heat integration using the algorithmic methods discussed in Chapter 9; that is, methods to increase the energy recovery. Yet another advance was possible through the use of membranes to recover valuable H_2 from the vapor purge stream that exits from the NH_3 synthesis loop. And, finally, the new heat and mass exchange (HME) technology had the potential to significantly increase the conversion to NH_3 while providing enhanced heat recovery. Both of these separation technologies are discussed next.

Heat and Mass Exchange Technology

HME technology (Lavie, 1987) involves a heat-insulated pair of adsorbing vessels that perform their normal role of adsorption while also transferring heat from the hot regenerant stream to the cold process stream from which one or more species are removed by adsorption, as depicted schematically in Figure 13.2. In NH_3 production, the synthesis gas directed to the converter must be preheated, while the effluent stream from the converter must be cooled to condense the ammonia. This is normally accomplished by heat exchange between the two streams. Also, the maximum concentration of ammonia in the converter effluent is limited by equilibrium and is therefore essentially independent of the ammonia concentration in the feed to the converter. However, by installing an HME unit to completely or partially replace the heat exchanger, the

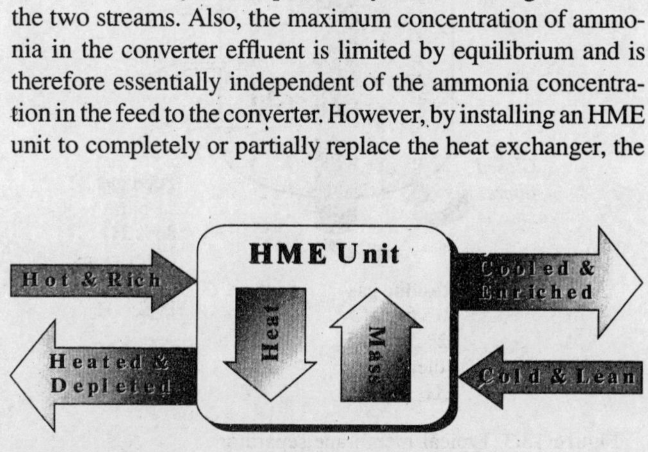

Figure 13.2 HME schematic.

converter effluent can also be enriched at the expense of its feed. This can increase the conversion-per-pass in the converter by a few percent, which translates into an increase of 10–20% in ammonia production from the same loop. The Hot & Rich stream fed to the HME unit should be at least 150°C.

Membrane Separation of Hydrogen from Synthesis Gas

The MEDAL™ membrane technology, commercialized by AirLiquide, http://www.medal.airliquide.com/en/membranes/hydrogen/ammonia.asp, enables hydrogen to be separated from a mixture as permeate, with the remaining gases removed as residue (retentate). A schematic of a typical separator, taken from the Air Liquide Web site, is shown in Figure 13.3, noting that the stream pressures and compositions shown do not apply directly to the separator's application in an ammonia process. The usage of highly selective polyvinylchloride membranes enables almost perfect separation of the hydrogen, with a recovery as high as 95 mol%. As discussed in the next section, this technology was not adopted by the EPL team, but was a strong candidate for the recovery of hydrogen from the purge gas stream.

In the innovation map, these new technologies were linked to their related technical differentiations, which permit the generation of the NH_3 product to satisfy the customer needs identified under the *customer-value proposition*. The first, which is imperative to achieve lower costs in energy-intensive processes that produce commodity chemicals, was to yield a high thermodynamic efficiency through the extensive use of heat integration. The second, which has been enabled by hollow-fiber membranes in recent years, was to recover valuable H_2 in purge streams rather than burn it in a flare device. And, finally, the application of HME technology led directly to reduced separation and circulation costs.

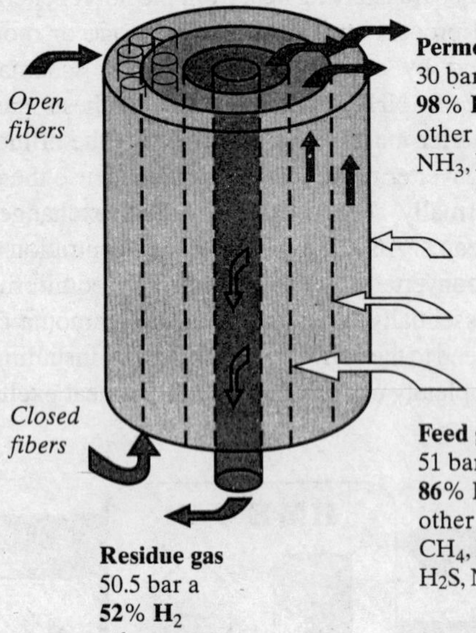

Figure 13.3 Typical membrane separator.

Combined, these technical differentiations had the potential to permit the production and storage of NH_3 in a remote environment, secure from rocket fire, and the generation of product steam to be used in related chemical processes, such as the manufacture of $(NH_4)_3PO_3$.

Concept Stage

Having assembled a promising *innovation map*, and after gaining approval to begin the SGPDP, as shown in Figure PI.1, the design team normally begins the *concept* stage with a market assessment to identify the *value creation* and *value capture*, and to create the *value proposition*, as discussed in Section 2.4.

a. *Market Assessment.* For this NH_3 product, the added value to customers was unusual. In this case, the production and storage of NH_3 in a remote location would sharply reduce the danger of exposure to toxic NH_3 in the event of terrorist rocket attacks, a significant threat in Haifa, Israel.

The answers to the following questions also helped to define the *value creation*: Who are the customers? These were the fertilizer manufacturers, that is, the producers of $(NH_4)_3PO_3$, and the manufacturers of industrial refrigeration units, in both Israel and, possibly, Jordan. Of these customers, who are most likely to buy? Clearly, the Israelis were the most likely customers.

Turning to *value capture*, in this case, there was no clear competition. While the NH_3 could be purchased from external suppliers, especially the new NH_3 plant under construction in Egypt, unfortunately, it was not possible to import and store NH_3 in the Israeli ports Haifa and Ashdod, because they are large population centers.

Finally, the design team identified a concise statement that summarized the need for the new remote facility to produce the product NH_3. Their *value-proposition* statement, was: "To remove its dependence on imported NH_3, the Government of the State of Israel supports the construction of a new plant to produce NH_3 in the Negev Desert. This will permit the dismantling of the hazardous NH_3 storage facility in Haifa Bay."

b. *Customer and Technical Requirements, and Superior Product Concepts.* For this commodity chemical, NH_3, the customer requirements were well established when the project charter was created. These were not further developed in the *concept* stage. Their translation into technical requirements (see Section 2.4) was rather straightforward: that is, to produce 99 mol% NH_3 containing 1 mol% water. The latter is needed to reduce stress-corrosion cracks (SCC) in the stainless steel containers exposed to pure ammonia. Normally, the next steps toward creating superior product concepts are to create a preliminary database and to carry out preliminary process synthesis, as shown in Figure PI.1 and discussed in Section 4.4. In this case, the most promising flowsheets of process operations are well established for the conversion of natural gas to synthesis gas and for the NH_3 synthesis loop. For this reason, as discussed next, the design team opted to adopt, initially, the preliminary process flowsheet created by EPL as their base-case design. This flowsheet, in Figures 13.4 and 13.5, is described next. It was further developed in the *feasibility* stage of the SGPDP.

Table 13.2 Prices of Raw Materials and Products

Commodity	Assumed Price ($/kg).
Ammonia	0.27
Methane	0.20
Process steam	0.02

Feasibility Stage

As discussed above, for this case study a preliminary process flowsheet had been created by EPL. It provided the base-case design to be improved upon by the design team. Note that, in Figure PI.1, it is recommended that a base-case design be developed at the outset of the *feasibility* stage.

Proposed Design by EPL

Figures 13.4 and 13.5, to be described below, show the design proposed by EPL, which assumes a basis of 12,000 kg/hr of methane as feed, with raw material and product prices assumed as given in Table 13.2. Note that the EPL simulations were carried out using UNISIM, the Honeywell version of HYSYS. This design has a return on investment (ROI) of −15.4% and a venture profit (VP, computed as the profit minus a 20% return on investment; see Eq. (23.9)) of −$25,500,000; that is, an annual **loss** of over $25 million. EPL stated in their report that: *"This poor economic evaluation is due to the relatively low market price for ammonia. It is impossible to design a more profitable process without a significant increase in the price of ammonia."* During this *feasibility* stage, the design team disputed this assertion, believing that it was possible to generate an acceptable profit and that poor engineering practice led to EPL's negative assessment.

The popular route to ammonia, adopted by EPL, is from natural gas (largely methane). The process involves two main sections: one for synthesis gas generation (Figure 13.4) and the other for the ammonia synthesis loop (Figure 13.5). Full details of the original EPL design, as well as a complete project tender

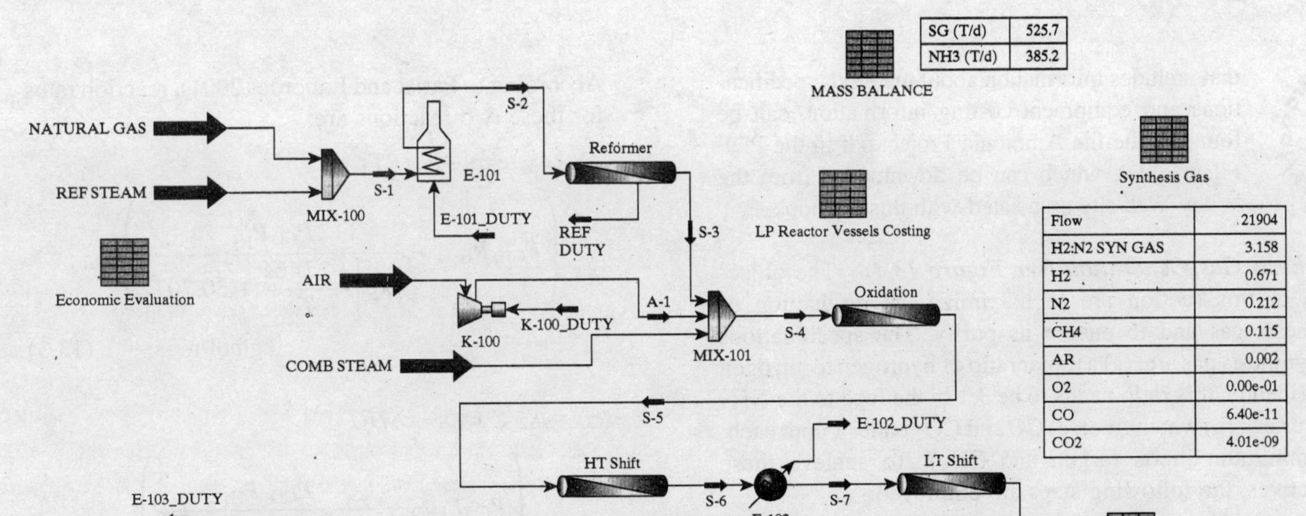

Figure 13.4 UNISIM PFD for the EPL design: Synthesis gas section. In the PFD, mass flows are in kg/hr, temperatures are in °C, pressures are in bar, and compositions are in mole fractions.

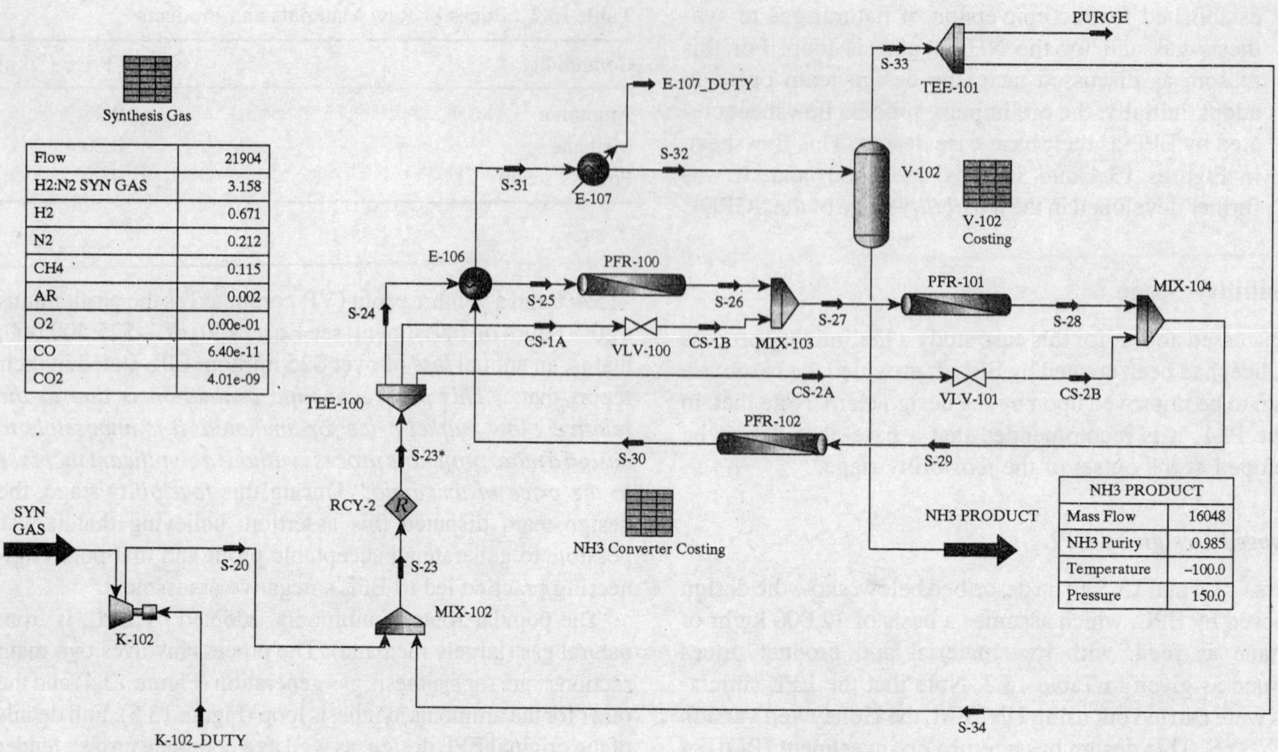

Figure 13.5 UNISIM PFD for the EPL design: Synthesis loop section. Units as in Figure 13.4.

that includes information about materials specifications and equipment costing information, can be found in the file Ammonia Project.pdf in the PDF Files folder, which can be downloaded from the Wiley Web site associated with this textbook.

Synthesis Gas Generation (see Figure 13.4). The objectives of this section are to maximize the production of synthesis gas and to ensure its purity. The specifications for synthesis gas are: (a) a molar ratio of hydrogen to nitrogen of 3 (ideally, this ratio needs to be 3:1 in the feed to the NH_3 converter); (b) water-free; (c) CO and CO_2 under 1 ppm each; (d) minimum inerts (argon and CH_4). To achieve these objectives, the following steps are employed:

a. Methane is combined with reformer steam, preheated in furnace E-101, and fed to the reformer, in which most of the methane is converted to hydrogen. The reformer is a furnace in which the reaction mixture flows through tubes arranged along the furnace wall. It is modeled in UNISIM as an isothermal PFR (with the effluent temperature set at the feed temperature). In the EPL design, the operating temperature is selected as 700°C. The following reactions take place in the reformer:

$$CH_4 + H_2O \leftrightarrow 3H_2 + CO \tag{13.1}$$

$$CO + H_2O \leftrightarrow CO_2 + H_2 \tag{13.2}$$

According to Parisi and Laborde (2001), reaction rates for these two reactions are:

$$-r_{CH_4} = k_{1,\infty} \cdot \exp[-E_1/RT] \left(P_{CH_4} P_{H_2O} - \frac{P_{CO} P_{H_2}^3}{\exp\left[\frac{-27{,}464}{T} + 30.707\right]} \right) \quad [\text{kgmol/m}^3\text{-s}] \tag{13.3}$$

$$-r_{CO} = k_{2,\infty} \cdot \exp[-E_2/RT] \left(P_{CO} P_{H_2O} - \frac{P_{CO_2} P_{H_2}}{\exp\left[\frac{4{,}048}{T} - 3.765\right]} \right) \quad [\text{kgmol/m}^3\text{-s}] \tag{13.4}$$

Note that in the above equations, the species partial pressures are in atm, T is the temperature in K, and Eqs. (13.3) and (13.4) hold for $T > 860$ K. These authors provide the following kinetic parameters: $E_1 = E_2 = 16{,}000$ kJ/kgmol, $k_{1,\infty} = 200$ kgmol/m^3-s, and $k_{2,\infty} = 100$ kgmol/m^3-s.

b. The reformer effluent is combined with air and steam to ensure a 3:1 mixture of hydrogen and nitrogen in the resulting synthesis gas. First, the effluent is reacted in the oxidation reactor, often referred to as a "secondary reformer," modeled in UNISIM as an adiabatic PFR, where the oxygen in the air generates CO, which leads to additional hydrogen. In addition to reaction (13.1) above, the following reaction takes place in the oxidation reactor:

$$CH_4 + 2O_2 \rightarrow CO_2 + 2H_2O \qquad (13.5)$$

According to Wolf et al. (1997), its reaction rate takes the form:

$$-r_{CH_4} = \frac{k_{3,\infty} \cdot \exp[-E_3/RT] \cdot P_{CH_4} P_{O_2}}{(1+K_{CH_4}P_{CH_4}+K_{O_2}P_{O_2}+K_{CO_2}P_{CO_2}+K_{H_2O}P_{H_2O})^2}, \quad [\text{kgmol/m}^3\text{-s}] \qquad (13.6)$$

Note that in the above equations, the partial pressures are in kPa and T is the temperature in K. Wolf et al. (1997) provide the following kinetic parameters:

$$K_{CH_4} = 1.1\times10^{-6}\cdot(-E_{CH_4}/RT), E_{CH_4} = 32{,}200\,\text{kJ/kgmol}$$

$$K_{O_2} = 1.1\times10^{-2}\cdot(-E_{O_2}/RT), E_{O_2} = 28{,}400\,\text{kJ/kgmol}$$

$$K_{CO_2} = 1.5\times10^{-4}\cdot(-E_{CO_2}/RT), E_{CO_2} = 32{,}900\,\text{kJ/kgmol}$$

$$K_{H_2O} = 5.3\cdot(-E_{H_2O}/RT), E_{H_2O}=27{,}300\,\text{kJ/kgmol}$$

The two remaining parameters are selected to be: $E_3 =$ 32,000 kJ/kgmol and $k_{3,\infty} = 1{,}000$ kgmol/m^3-s.

c. Since the first two reaction steps also generate CO, which poisons the ammonia synthesis catalyst, shift reaction steps are employed to convert CO to CO_2. These two reactors are modeled in UNISIM as adiabatic PFRs. In the EPL design, the oxidation reactor effluent is fed to the first shift reactor, HT shift. Then, heat exchanger E-102 reduces its effluent temperature to 500°C before it enters the second shift reactor, LT shift. Note that E-102 generates HT steam, providing a revenue source. The so-called water-gas shift reaction takes place in the shift reactors:

$$CO + H_2O \leftrightarrow CO_2 + H_2 \qquad (13.2)$$

The same kinetic form is used as in Eq. (13.4), but two of the kinetic parameters are slightly different (Parisi and Laborde, 2001) because the reaction occurs at lower temperatures:

$$-r_{CO} = k_{2,\infty}\cdot\exp[-E_2/RT] \cdot \left(P_{CO}P_{H_2O} - \frac{P_{CO_2}P_{H_2}}{\exp\left[\frac{4,577}{T} - 4.33\right]}\right) [\text{kgmol/m}^3\text{-s}] \qquad (13.7)$$

Note that the partial pressures are in atm, T is the temperature in K, and Eq. (13.7) holds for $T < 860$ K. Parisi and Laborde (2001) provide the following kinetic parameters: $E_2 = 16{,}000$ kJ/kgmol, and $k_{2,\infty} = 100$ kgmol/m^3-s.

d. According to reaction (13.1), any remaining CO reacts reversibly to methane in the methanator, which is modeled in UNISIM as an adiabatic PFR. Note that the reaction rate in Eq. (13.3) applies, with the operating temperature sufficiently low to ensure that the reverse reaction dominates. In the EPL design, the methanator feed temperature is selected as 250°C.

e. The water produced in the previous reaction steps is removed next. In the design suggested by EPL, heat exchanger E-104 cools the methanator effluent to 40°C, condensing the water, which is removed largely in the flash vessel, V-100. Residual water is removed using adsorption beds, modeled using a separator, X-100, in UNISIM.

f. The CO_2 produced in the previous steps is removed next. In the EPL design, the heat exchanger, E-105, cools the effluent from X-100 to −120°C, condensing CO_2, most of which is recovered in flash vessel V-101 as byproduct. The residual CO_2 is removed using adsorption beds X-101.

As shown in Figure 13.4, the EPL design produces very clean synthesis gas, easily satisfying the impurity constraints. However, only 526 T/day of synthesis gas is produced, given 12,000 kg/hr of methane feed, and the H_2/N_2 ratio is imprecisely maintained in excess of 3:1.

Ammonia Synthesis Loop (see Figure 13.5). The objectives of this section are to maximize the ammonia production and to ensure its purity ($\geq$98 mol%). To achieve these objectives, the following steps are employed:

a. The synthesis gas is compressed to the operating pressure of the synthesis loop in compressor K-102; that is, 150 bar in the EPL design.

b. The synthesis gas is combined with the recycle stream from the vapor effluent of the flash vessel, V-102.

c. The combined feed from the mixer, MIX-102, is split three ways, with the largest portion entering the ammonia converter through heat exchanger E-106, where it is preheated to the ignition temperature using the hot converter effluent from reactor PFR-102. In this design, the reacting synthesis gas progresses through three adiabatic PFRs, with intercooling provided by two cold shots in streams CS-1A/B and CS-2A/B. The reaction in the adiabatic beds is:

$$0.5\,N_2 + 1.5\,H_2 \leftrightarrow NH_3 \qquad (13.8)$$

The rate of reaction is given by the following kinetic expression (Eq. 7.32):

$$-r_{N_2} = 10^4\exp[-9.1\times10^4/RT]P_{N_2}^{0.5}P_{H_2}^{1.5} - 1.3\times10^{10}\exp[-1.4\times10^5/RT]P_{NH_3} \qquad (13.9)$$

where $-r_{N_2}$ is the rate of nitrogen disappearance in kmol/m^3-s, T is the temperature in K, P_i are the partial pressures of the reacting species in atm, and the activation energies for the forward and reverse reactions are in kJ/kmol.

d. The hot reactor effluent is cooled by exchange with the cold synthesis gas in heat exchanger E-106. It is further cooled in heat exchanger E-107, with the effluent temperature (of stream S-32) low enough to ensure sufficiently pure ammonia product. In the EPL design, cooler E-107 is cooled with expensive methane refrigerant. The cooled converter effluent is flashed in V-102 to liquid ammonia product, with the vapor stream recycled.

e. A small purge stream is split away from the vapor recycle in TEE-101.

Sensitivity Analysis

For a complex process, it helps to identify the decision variables having the greatest effect on the profitability, with a comparable effect on the plant feasibility, like the critical-to-quality (CTQ) variables in product design. These decision variables are:

- ***Production rate.*** The design basis of 12,000 kg/hr methane adopted by EPL produces 385 T/day of ammonia, which meets Israel's needs. However, this scale of operation may not be profitable, as will be shown for the EPL base-case design. Clearly, a well-conceived design that generates a reasonable profit at minimum capacity is sought.
- ***Steam/methane ratio.*** This strongly affects the conversion of methane to hydrogen.
- ***Air/methane ratio.*** This affects the H_2/N_2 ratio in the synthesis gas fed to the synthesis loop. Note that the EPL design generates synthesis gas having a 3.16 H_2/N_2 ratio, which exceeds the stoichiometric ratio.
- ***Feed temperatures of the shift reactors.*** These control the conversion of CO, which poisons the ammonia synthesis catalyst, to CO_2. Note that the EPL design doesn't control the feed temperature to the HT shift but sets the feed temperature to the LT shift to 500°C, leading to excessive CO in the effluent of the shift reactor train. This CO is subsequently converted to methane in the methanator, leading to highly inert (CH_4) concentrations in the synthesis loop, significantly reducing its efficiency and profitability.
- ***Synthesis loop pressure.*** The synthesis loop of the EPL design operates at 150 bar. While the conversion per pass increases with increasing pressure, as shown in the solution to Exercise 6.3, so do the equipment and operating costs. The impact of the synthesis loop pressure on the profitability of the process is considered in Exercise 13.4.
- ***Feed temperature to the ammonia converter.*** This should be adjusted to the lowest possible value that guarantees a reasonable stability margin; that is, one that avoids operation in the vicinity of multiple steady states (see Section 7.2).
- ***Control of "cold-shot" bypasses in the ammonia converter.*** The fractions of feed in the bypasses permit the conversion per pass to be maximized, as discussed in Example 7.3.

In addition to the above, following Chapter 9, MER targeting should be carried out to compute hot and cold utility targets. Also, as will be shown, the grand composite curve helps to distribute the optimal utility sources. A heat exchanger network can then be designed that provides an adequate approach to the MER targets, thereby sharply reducing the usage of external utilities.

Refining the Solution

First, the design basis of 12,000 kg/hr methane feed in the EPL design is retained, with the aim being to estimate the best profitability achievable for this scale of operation. Then, economy-of-scale arguments are employed to estimate the effect of the production rate on the profitability of the process.

Refining the Solution for 12,000 kg/hr Methane Feed

There are several weaknesses in the EPL design: (a) too little hydrogen is produced in the synthesis gas section due to the poor performance of the reformers; (b) the H_2:N_2 ratio of the synthesis gas is 3.158, and 4.255 in the feed to the converter, due to the poor control of the quantity of air fed to the synthesis gas section; (c) the percentage of inerts in the synthesis gas, especially CH_4, is excessive (the composition of methane is 11.5 mol%) due to: (i) the relatively poor conversion of CH_4 in the reformers, and (ii) the relatively large concentration of CO remaining in the synthesis gas effluent from the shift converters, which is subsequently converted to methane in the methanator; (d) the ammonia composition in the NH_3 converter effluent is about 8.5 mol% due to: (i) the excessive amount of inerts in the converter feed (over 44 mol%!), (ii) the high H_2:N_2 ratio in the feed to the converter, which is far above stoichiometric, and (iii) poor performance of the converter, with the bypass fractions suboptimal and the feed temperature too high. All of these weaknesses are dealt with sequentially. Due to space limitations, only the key improvements are highlighted next:

Improving Hydrogen Yield. The main reason for the relatively low hydrogen yield is the low temperature in the reformer, with the yield increasing exponentially with temperature. Setting the operating temperature to 850°C, while

Table 13.3 Sensitivity of Primary and Secondary Reformer Performance to Steam Flow Rates. Outputs — H_2 (in Roman font) and O_2 Mass Flows (in *italics*) — in kg/hr

		Reformer Steam [kgmol/hr]			
		1,500	2,000	2,500	3,000
Combustion Steam [kgmol/hr]	600	4,334	4,449	4,514	4,571
		4.35 × 10^{-6}	*1.15 × 10^{-5}*	*1.20 × 10^{-2}*	*4.27*
	900	4,302	4,421	4,494	4,556
		1.25 × 10^{-3}	*2.02 × 10^{-6}*	*7.52 × 10^{-2}*	*2.47*
	1,200	4,268	4,397	4,476	4,542
		1.05 × 10^{-1}	*3.01 × 10^{-6}*	*2.42 × 10^{-2}*	*1.17*
	1,500	4,234	4,375	4,460	4,530
		2.11 × 10^{-3}	*2.56 × 10^{-6}*	*9.49 × 10^{-4}*	*4.48 × 10^{-1}*

adjusting the air flow rate to ensure a H_2:N_2 ratio of 3, reduces the methane composition in the effluent from the oxidation reactor from 3.6 to a very low 0.04 mol%. Furthermore, to maximize the hydrogen produced, a sensitivity analysis shows the effect of the steam molar flow rate in the reformer and oxidation reactor on the hydrogen mass flow rate in the oxidation reactor effluent. The results are presented in Table 13.3. As seen, the optimal operation is at a reformer steam flow rate of 2,500 kgmol/hr, rather than 2,000 kgmol/hr, and a combustion steam flow rate of 600 kgmol/hr, rather than 1,200 kgmol/hr. These settings ensure the highest possible hydrogen generation rate while keeping the oxygen level sufficiently low. With these settings, the methane composition in the oxidation reactor effluent drops further, from 0.04 to 0.009 mol%.

Minimizing CO Composition in the Shift Reactor Train Effluent. Next, the optimal operation of the two shift reactors is considered. To improve their performance, a cooler is inserted immediately before the HT shift reactor, allowing the feed temperatures to the two reactors to be selected independently. A sensitivity analysis shows their effect on the CO concentration in the effluent from the LT shift reactor as its feed temperature varies from 300 to 500°C. The results in Figure 13.6 show that the minimum CO concentration, at 1.33 mol%, is achieved when the feed temperatures to the HT and LT shift reactors are at 400 and 375°C, respectively. With these settings, the methane composition in the synthesis gas drops to 2.5 mol%. The new cooler also allows a large quantity of high-pressure steam (hps) to be produced, providing significant revenues.

As seen in Table 13.4, with these modifications to the synthesis gas section, the ROI and VP increase to −2.4 and −$20,000,000, respectively. Although the profitability is improved, it remains unacceptable. Additional improvements are needed in the NH_3 synthesis loop.

www.wiley.com/go/global/seider

Increasing Conversion in the Ammonia Converter. First, notice that the conversion in the ammonia reactors is very low. As shown in the multimedia modules, which can be downloaded from the Wiley Web site associated with this textbook (*HYSYS → Tutorials → Process Design Principles → Ammonia Converter Design*), the conversion can be significantly enhanced by adjusting the cold-shot fractions. Consequently, both are increased from 0.1 to 0.2. Furthermore, the reactor feed temperature is reduced from 310°C to just above the extinction temperature, 250°C, further increasing the conversion, which is favored by low temperatures. Because multiple steady states are possible, to avoid operating problems, care is taken to select feed temperatures associated with unique steady states.

Reducing the Cost of Ammonia Separation. Because the ammonia purity is much higher than required, the operating temperature of V-102 can be increased, reducing the refrigeration costs. For example, using ammonia refrigerant at −30°C, the V-102 temperature can be as high as −20°C, assuming a minimum temperature approach of 10°C. With this change, the operating expenses of the process are reduced sharply, at the cost of increasing the NH_3 concentration in the recycle stream.

Hydrogen Recovery from the Purge Stream. The purge stream is at 4.3 T/hr and 61 mol% H_2, which has been generated at considerable expense. As an alternative, hydrogen can be recovered using a membrane-separation unit, which is designed on the assumption of 95% hydrogen recovery. Because the hydrogen recovered is at 50% of the pressure of the synthesis loop, the single-stage compressor must be replaced by a two-stage compressor, including an intercooler, with the recovered hydrogen fed to the suction line of the second compressor. The additional equipment costs must be justified by the increased NH_3 production rates.

Figures 13.7 and 13.8 show the PFDs for the improved synthesis gas generation and synthesis loop sections. To achieve profitability, it is necessary to reduce the H_2:N_2 ratio of the synthesis gas, such that after the recovered hydrogen is recycled, the H_2:N_2 ratio entering the converter approaches three. As seen in Table 13.4, with these modifications to the

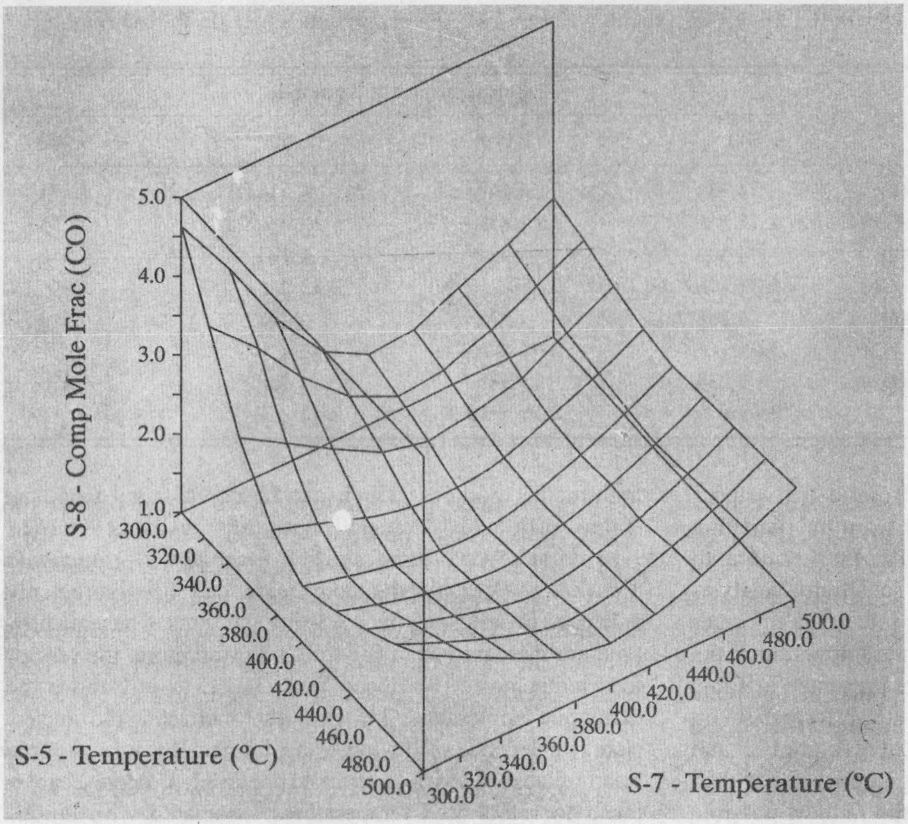

Figure 13.6 Effect of the feed temperatures to the two shift reactors on effluent CO.

Table 13.4 Profitability Analysis by Design Step (Basis: 12,000 kg/hr methane)

Step	NH_3 (T/d)	GP ($MM)	C_{TCI} ($MM)	ROI (%)	PBP (yrs)	VP ($MM)
EPL Design	385.2	−14.8	72.0	−15.4	n/a	−25.5
After modifications to synthesis gas section	530.5	−2.8	89.6	−2.4	n/a	−20.0
After modifications to synthesis loop section	608.9	16.1	96.0	12.5	7.8	−7.2
After implementing heat integration	608.8	17.5	89.3	14.7	6.8	−4.7
After implementing HME	610.2	19.0	88.5	16.1	6.2	−3.4

synthesis loop section, the ROI and VP are increased significantly to 12.5%, and −\$7,200,000. Next, heat integration is considered as a means of further improving the profitability.

MER Targeting, MER Design, and Optimal Energy Management

The objective of this step is to improve the profitability of the process by heat integration, using techniques discussed in Chapter 9. At the outset, note that the primary (NH_3) and secondary (purge) products are currently produced at −20°C, with valuable cooling resources not being exploited. To use these resources, two heaters are inserted into the flowsheet: E-109, which raises the CO_2 product temperature to its bubble point (the highest liquid temperature), and E-110, which raises the ammonia product temperature to its bubble point. It is noted that ammonia storage is designed so that ammonia is stored either at ambient temperature (at about 10 bar) or at atmospheric pressure (thus requiring refrigeration to maintain the ammonia at its bubble-point temperature of −30°C). For this simplified design, it is assumed that the ammonia can be produced at the bubble-point temperature corresponding to the loop pressure.

Next, a preliminary analysis of the opportunities for heat integration is presented; that is, the demands of streams to be heated and cooled are listed below:

COLD_1: 24.02×10^6 kcal/hr is required between 185 and 850°C to preheat the reformer feed. The current design requires \$2,760,000/yr in fuel costs in a furnace (E-101) whose FOB purchase price is \$1,970,000.

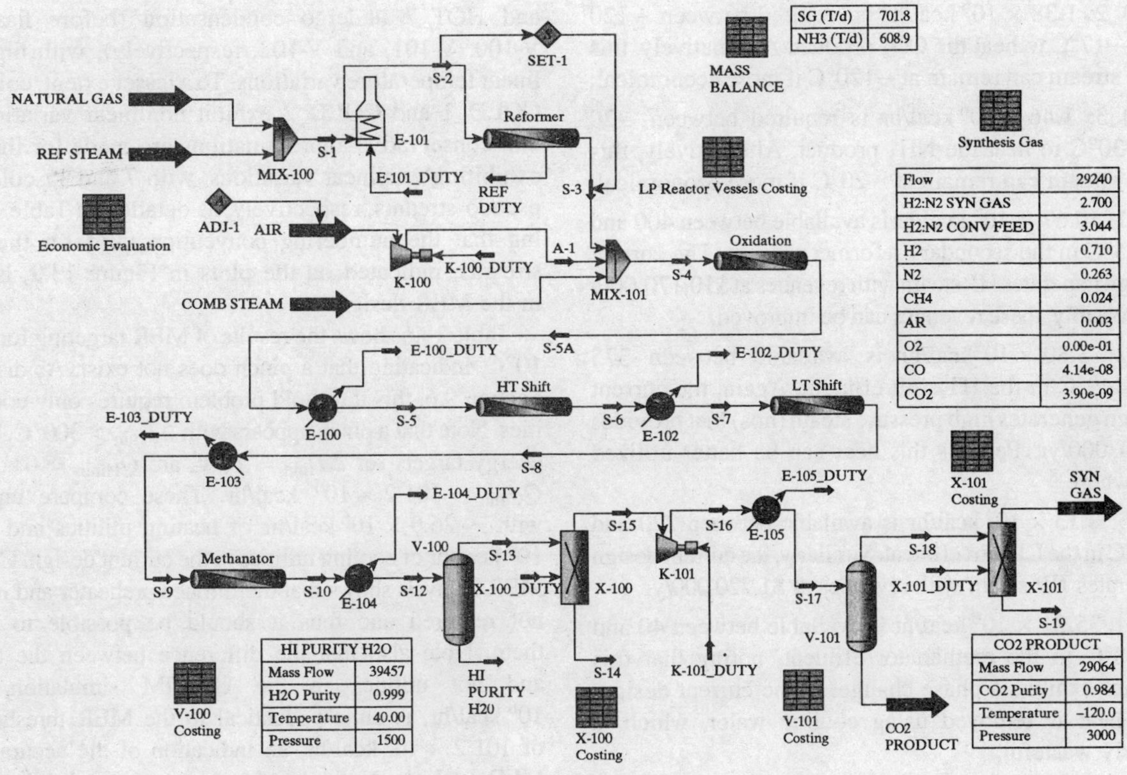

Figure 13.7 UNISIM PFD for the synthesis gas section after modifications. Units as in Figure 13.4.

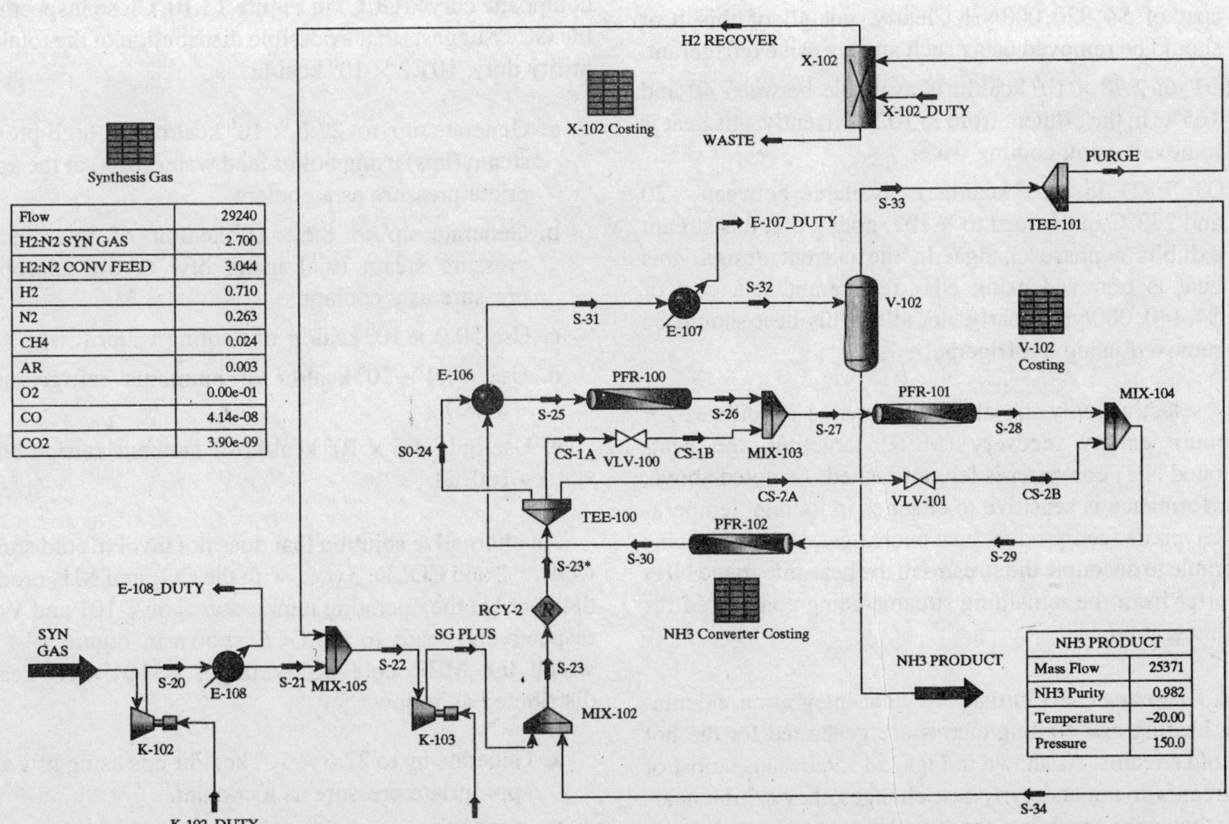

Figure 13.8 UNISIM PFD for the synthesis loop section after modifications. Units as in Figure 13.4.

COLD_2: 1.38×10^6 kcal/hr is required between −120 and −17°C to heat the CO_2 product. Alternatively, this cold stream can remain at −120°C if more economical.

COLD_3: 1.46×10^6 kcal/hr is required between −20 and 30°C to heat the NH_3 product. Alternatively, this cold stream can remain at −20°C if more economical.

HOT_1: 40.67×10^6 kcal/hr is available between 400 and 1,145°C in the secondary reformer effluent. The current design generates HP steam with revenues at $10,170,000/yr. Possibly, these revenues can be improved.

HOT_2: 3.80×10^6kcal/hr is available between 375 and 449°C in the HT shift effluent. Again, the current design generates high pressure steam (hps) that provides $950, 000/yr. Perhaps this heat can be better utilized elsewhere.

HOT_3: 8.13×10^6 kcal/hr is available between 220 and 382°C in the LT shift effluent. Similarly, the current design generates HP steam with revenues at $1,220,000/yr.

HOT_4: 35.21×10^6 kcal/hr is available between 40 and 302.8°C in the methanator effluent, noting that this stream exhibits a phase change. In the current design, this heat is removed using cooling water, which is clearly wasteful.

HOT_5: 9.83×10^6 kcal/hr is available between −120 and 125.7°C in the feed stream to V-101, noting that this stream exhibits a phase change. In the current design, this duty is removed using methane refrigerant (mr), at a cost of $4, 920, 000/yr! Clearly, not all of this heat should be removed using such an expensive refrigerant.

HOT_6: 2.78×10^6 kcal/hr is available between 40 and 165°C in the effluent from K-102. Currently, this heat is removed using cooling water.

HOT_7: 27.58×10^6 kcal/hr is available between −20 and 232°C in the feed to V-102, noting that this stream exhibits a phase change. In the current design, this heat is removed using NH_3 refrigerant at a cost of $4, 140, 000/yr! Clearly, not all of this heat should be removed using a refrigerant.

By selecting only the above streams as candidates for maximum energy recovery (MER) targeting, the heat-integrated NH_3 converter is left untouched. As noted above, its performance is sensitive to changes in its inlet temperature. Given its feed/product heat exchanger, E-106, it seems preferable to decouple the streams in the heat-integrated NH_3 converter from the remaining streams being considered for heat integration.

MER Targeting. To initiate the heat-integration calculations, heating and cooling curves are generated for the hot and cold streams. As shown in Figure 13.9, because most of the streams do not undergo phase changes, they exhibit near-linear behavior, enabling them to be approximated using constant heat capacities, C_P. Only streams HOT_4, HOT_5, and HOT_7 undergo condensation (before flash drums V-100, V-101, and V-102, respectively), with highly non-linear temperature variations. To a lesser extent, cold streams COLD_1 and COLD_2 exhibit nonlinear variations. Note that conservative approximations are made for the streams exhibiting nonlinear variations, with 7 and 17 cold and hot pseudo-streams, respectively, as detailed in Table 13.5, noting that the numbering convention used for the pseudo-streams, indicated on the plots in Figure 13.9, is adopted in the MER design.

Table 13.6 shows the results of MER targeting for $\Delta T_{min} =$ 10°C, indicating that a pinch does not exist. As discussed in Section 9.6, this threshold problem requires only cooling utilities. Note that a pinch appears with $\Delta T_{min} > 300$°C. The MER energy targets for $\Delta T_{min} = 10$°C are $Q_{Hmin} = 0$ kcal/hr and $Q_{Cmin} = 101.2 \times 10^6$ kcal/hr. These compare unfavorably with $\sim 26.9 \times 10^6$ kcal/hr of heating utilities and $\sim 128 \times 10^6$ kcal/hr of cooling utilities in the current design. Clearly, the MER analysis shows that the furnace preheater and its fuel are not required and thus it should be possible to eliminate them. Note also that the difference between the total cold and hot utilities in the UNISIM simulation, 101.1×10^6 kcal/hr, is almost identical to the MER threshold target of 101.2×10^6 kcal/hr, an indication of the accuracy of the MER analysis. Note also that because zero heating utility is required, the residual heat flows between the temperature intervals are identical in columns 3 and 4 of Table 13.6.

The data in Table 13.5 are used to generate the grand composite curve (GCC) in Figure 13.10. Close inspection of the GCC suggests that a possible distribution of the total cold utility duty, 101.2×10^6 kcal/hr, is:

a. Generate up to 27.6×10^6 kcal/hr of high-pressure steam (hps) using boiler feed water (bfw) at the appropriate pressure as a coolant.

b. Generate up to 8.6×10^6 kcal/hr of intermediate-pressure steam (ips) using bfw at the appropriate pressure as a coolant.

c. Use 50.2×10^6 kcal/hr of cooling water (cw).

d. Use 10.2×10^6 kcal/hr of ammonia refrigerant (at −30°C).

e. Use only 4.6×10^6 kcal/hr of methane refrigerant (at −160°C).

An alternative solution that does not involve cold streams COLD_2 and COLD_3 (i.e., with the CO_2 and NH_3 products delivered at the operating temperatures of V-101 and V-102, respectively) leads to the GCC shown in Figure 13.11, in which the MER cold utility target is 104×10^6 kcal/hr, distributed as follows:

a. Generate up to 27.6×10^6 kcal/hr hps using bfw at the appropriate pressure as a coolant.

b. Generate up to 8.5×10^6 kcal/hr ips using bfw at the appropriate pressure as a coolant.

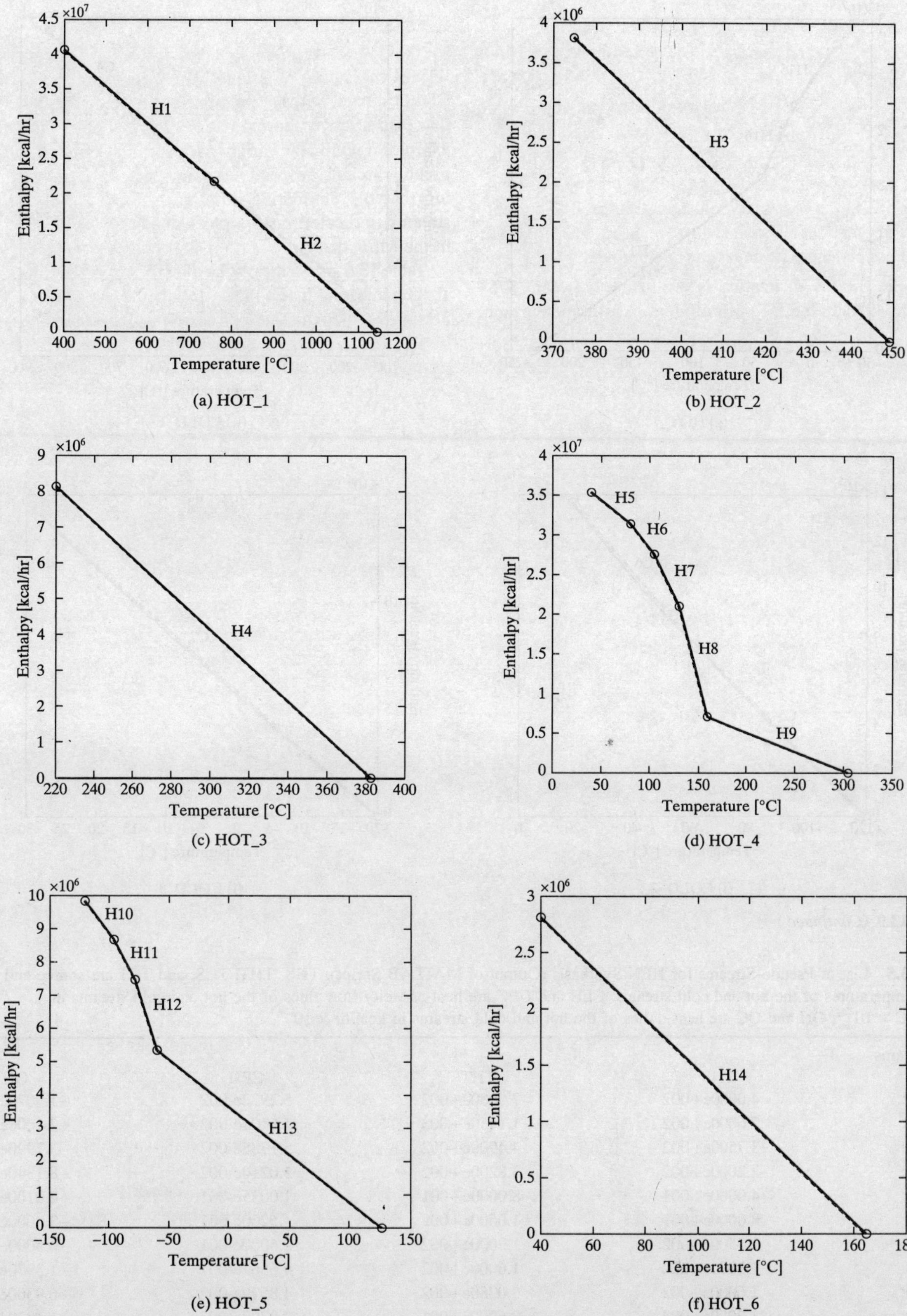

Figure 13.9 Heating and cooling curves for the streams in the HEN (*Continued*).

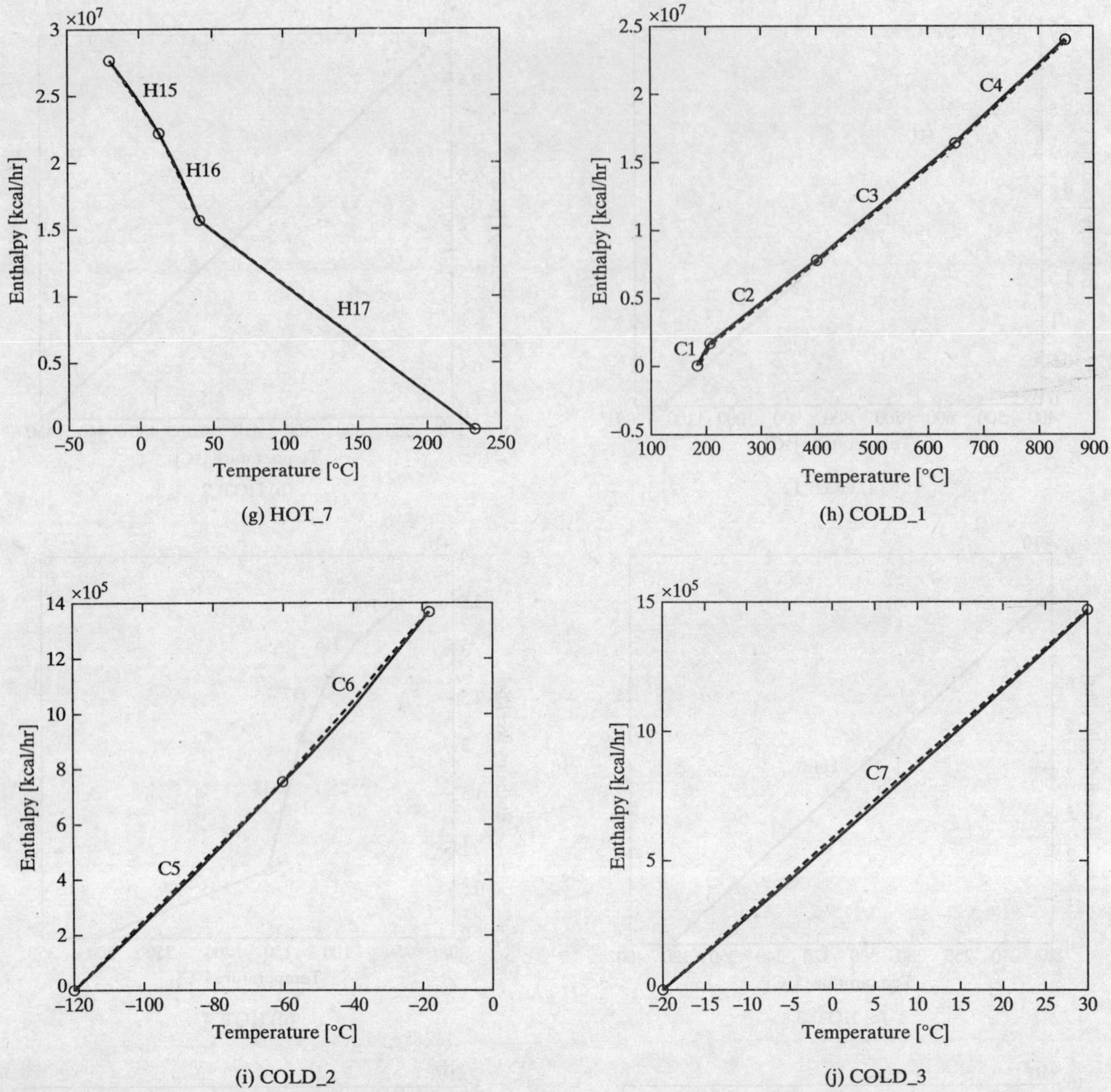

Figure 13.9 *(Continued)*

Table 13.5 List of Pseudo-Streams for HEN Synthesis (Output of MATLAB Script). THS, THT, TCS, and TCT are source and target temperatures of the hot and cold streams. CFH and CPC are heat-capacity flow rates of the hot and cold streams in kcal/hr°C × 10^{-6}. QH and QC are heat duties of the hot and cold streams in kcal/hr × 10^{-6}

Hot Streams	THS	THT	CPH	QH
H1:	4.0000e+002	7.6000e+002	5.2972e-002	−1.9070e+001
H2:	7.6000e+002	1.1450e+003	5.6104e-002	−2.1600e+001
H3:	3.7500e+002	4.4900e+002	5.1338e-002	−3.7990e+000
H4:	2.2000e+002	3.8200e+002	5.0210e-002	−8.1340e+000
H5:	4.0000e+001	8.0000e+001	1.0025e-001	−4.0100e+000
H6:	8.0000e+001	1.0500e+002	1.5200e-001	−3.8000e+000
H7:	1.0500e+002	1.3000e+002	2.6000e-001	−6.5000e+000
H8:	1.3000e+002	1.6000e+002	4.6567e-001	−1.3970e+001
H9:	1.6000e+002	3.0280e+002	4.8529e-002	−6.9300e+000
H10:	−1.2000e+002	−9.6000e+001	4.9708e-002	−1.1930e+000

(Continued)

Table 13.5 (*Continued*)

H11:	−9.6000e+001	−7.8000e+001	6.6667e-002	−1.2000e+000
H12:	−7.8000e+001	−5.9000e+001	1.1526e-001	−2.1900e+000
H13:	−5.9000e+001	1.2570e+002	2.8424e-002	−5.2500e+000
H14:	4.0000e+001	1.6560e+002	2.2237e-002	−2.7930e+000
H15:	−2.0000e+001	1.3900e+001	1.6165e-001	−5.4800e+000
H16:	1.3900e+001	4.1600e+001	2.3466e-001	−6.5000e+000
H17:	4.1600e+001	2.3260e+002	8.1675e-002	−1.5600e+001

Cold Streams

	TCS	TCT	CPC	QC
C1:	1.8500e+002	2.0800e+002	7.4348e-002	1.7100e+000
C2:	2.0800e+002	4.0000e+002	3.1510e-002	6.0500e+000
C3:	4.0000e+002	6.5000e+002	3.4560e-002	8.6400e+000
C4:	6.5000e+002	8.5000e+002	3.8100e-002	7.6200e+000
C5:	−1.2000e+002	−6.0000e+001	1.2583e-002	7.5500e−001
C6:	−6.0000e+001	−1.8000e+001	1.4643e-002	6.1500e−001
C7:	−2.0000e+001	3.0000e+001	2.9200e-002	1.4600e+000

Table 13.6 MER Targeting Results for $\Delta T_{min} = 10°C$ (Output of MATLAB Script)

Interval Temp (°C) 1.0e+003×	ΔH	Energy Flows $Q_H = 0$	Energy Flows $Q_H = 0$
1.1350	0	0	0
0.8500	0.0160	0.0160	0.0160
0.7500	0.0018	0.0178	0.0178
0.6500	0.0015	0.0193	0.0193
0.4390	0.0039	0.0232	0.0232
0.4000	0.0027	0.0259	0.0259
0.3900	0.0007	0.0266	0.0266
0.3720	0.0004	0.0270	0.0270
0.3650	0.0005	0.0275	0.0275
0.2928	0.0014	0.0288	0.0288
0.2226	0.0047	0.0335	0.0335
0.2100	0.0019	0.0354	0.0354
0.2080	0.0002	0.0356	0.0356
0.1850	0.0013	0.0369	0.0369
0.1556	0.0038	0.0407	0.0407
0.1500	0.0009	0.0416	0.0416
0.1200	0.0171	0.0587	0.0587
0.1157	0.0016	0.0602	0.0602
0.0950	0.0081	0.0683	0.0683
0.0700	0.0071	0.0754	0.0754
0.0316	0.0089	0.0844	0.0844
0.0300	0.0006	0.0850	0.0850
0.0039	0.0061	0.0911	0.0911
−0.0180	0.0035	0.0946	0.0946
−0.0200	0.0003	0.0949	0.0949
−0.0300	0.0018	0.0967	0.0967
−0.0600	0.0004	0.0971	0.0971
−0.0690	0.0001	0.0972	0.0972
−0.0880	0.0020	0.0992	0.0992
−0.1060	0.0010	0.1002	0.1002
−0.1200	0.0005	0.1007	0.1007
−0.1300	0.0005	0.1012	0.1012

c. Use approximately 50.2×10^6 kcal/hr of cw.

d. Use approximately 11.5×10^6 kcal/hr of ammonia refrigerant (at −30°C).

e. Use approximately 5.8×10^6 kcal/hr of methane refrigerant (at −160°C).

Since this second alternative is much easier to implement and involves only 3% more cold utilities than the best possible option shown in Figure 13.10, it is selected as the basis for the HEN design.

Design of MER Network. Next, the HEN in Figure 13.12 is designed to meet the MER targets defined above for $\Delta T_{min} = 10°C$. A total of 12 heat exchangers are required, after grouping all contiguous pseudo-streams. Note that:

a. A single process-process heat exchanger is required, in which heat is exchanged from the hot exothermic reactor effluent to the process feed (i.e., heat exchange from pseudo-streams H1–H2 to C1–C4), with a total duty of 24.02×10^6 kcal/hr. This eliminates the furnace heater, although a small heater for process startup may be necessary.

b. The remaining system involves the usage of utility coolers. For process temperatures above 370°C, hps has been raised by using bfw as a coolant. Above 220°C, bfw is used to raise ips.

c. As much as possible, cw has been used. For process temperatures above −20°C, ammonia refrigerant (ar) utility has been used, with methane refrigerant (mr) utility used for colder process temperatures.

d. The above design uses the following distribution of external utilities:

a. $16.65 + 3.8 = 20.45 \times 10^6$ kcal/hr hps is raised using bfw.

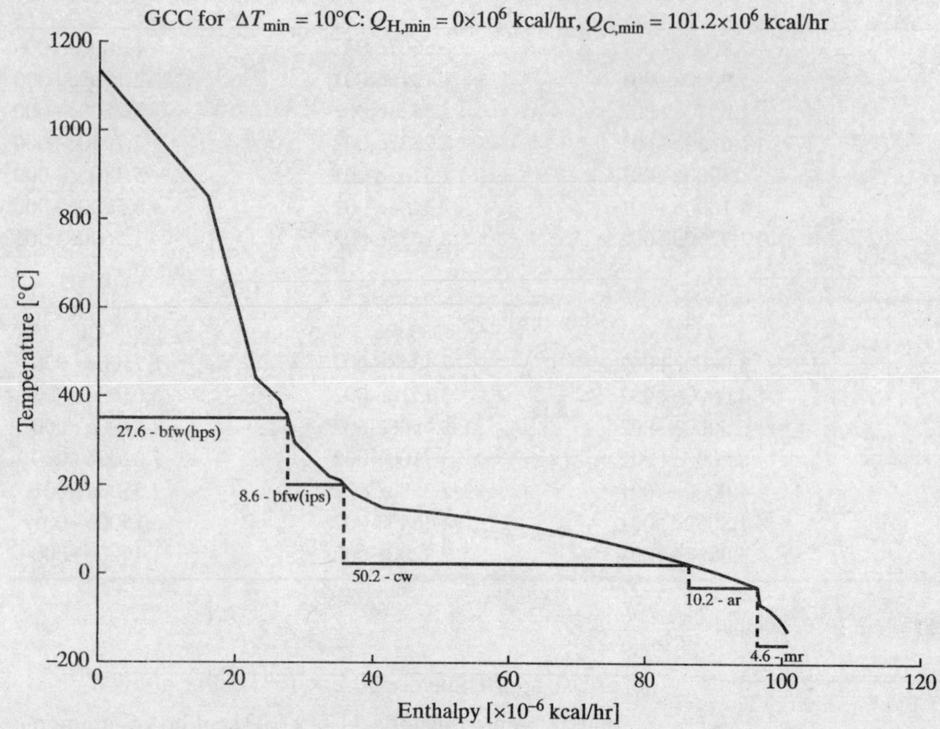

Figure 13.10 Grand composite curve for $\Delta T_{\min} = 10°C$, showing a possible distribution of utilities; $Q_{H,\min} = 0$, $Q_{C,\min} = 101.2 \times 10^6$ kcal/hr.

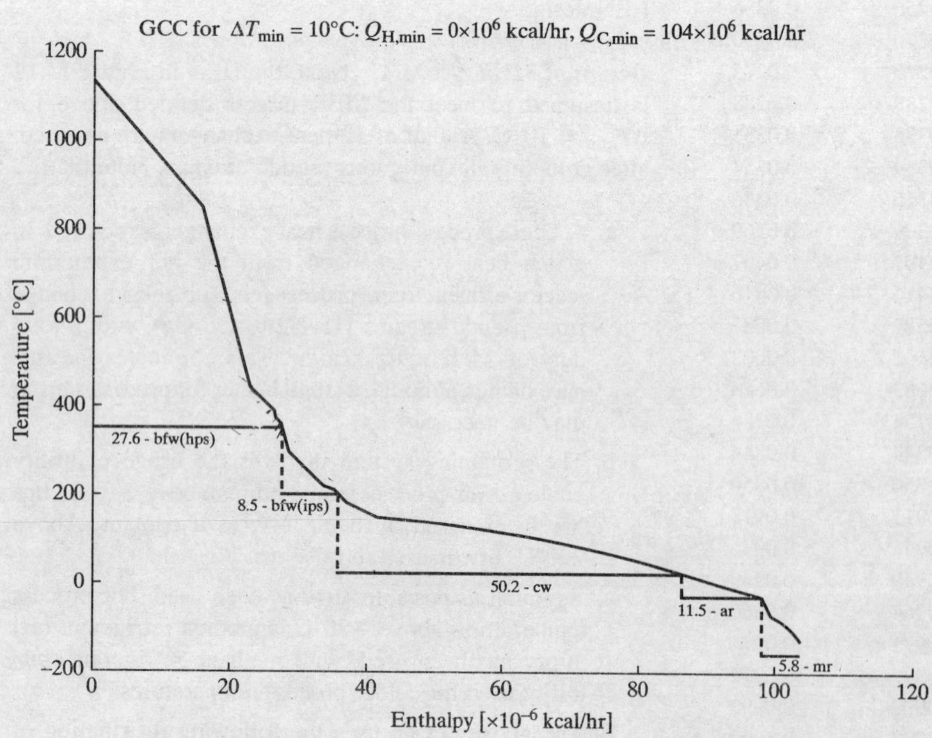

Figure 13.11 Grand composite curve for $\Delta T_{\min} = 10°C$, with C1 as the only cold stream, showing a possible distribution of utilities; $Q_{H,\min} = 0$, $Q_{C,\min} = 104 \times 10^6$ kcal/hr.

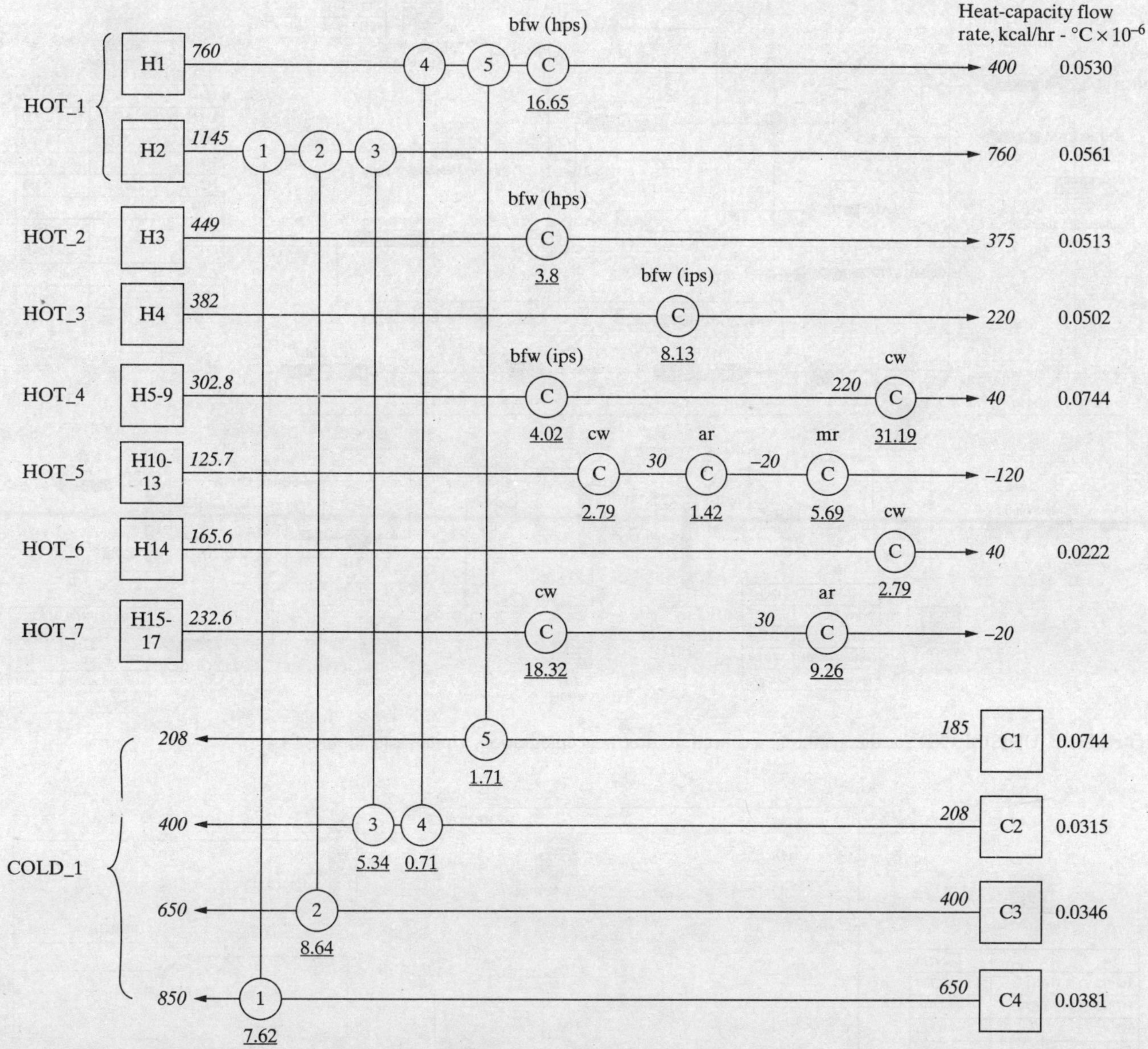

Figure 13.12 MER design for $\Delta T_{\min} = 10°C$. The average heat-capacity flow rates for H10–H13 (HOT_5) and H15–H17 (HOT_7) are not shown. Temperatures in °C (italics). Heat duties in kcal/hr (Roman underlined).

b. $8.13 + 4.02 = 12.15 \times 10^6$ kcal/hr ips is raised using bfw.

c. A total of 55.02×10^6 kcal/hr of cw.

d. 10.68×10^6 kcal/hr of ammonia refrigerant is used (at −30°C).

e. 5.69×10^6 kcal/hr of methane refrigerant is used (at −160°C).

Figures 13.13 and 13.14 show the UNISIM PFD for the heat-integrated process, which implements the HEN in Figure 13.12. As shown in Table 13.4, this heat integration further increases the ROI and VP to 14.7% and −$4,700,000.

Installing a HME into the Synthesis Loop

As shown in Figure 13.15, the heat and mass exchanger (HME) is installed such that the HOT RICH stream is the effluent of a new heat exchanger, E-112, installed to generate ips (S-34B), and the COLD LEAN stream is the combination of the vapor recycled from the flash vessel, V-102, and the makeup synthesis gas stream (Total_SG). The COOLED ENRICHED stream is fed to cooler E-111 (cooled with cw), and the HEATED DEPLETED stream is fed directly to the ammonia reactors. The degrees of freedom for the HME are defined as: (a) 50% of the ammonia in the HOT RICH stream is transferred to the

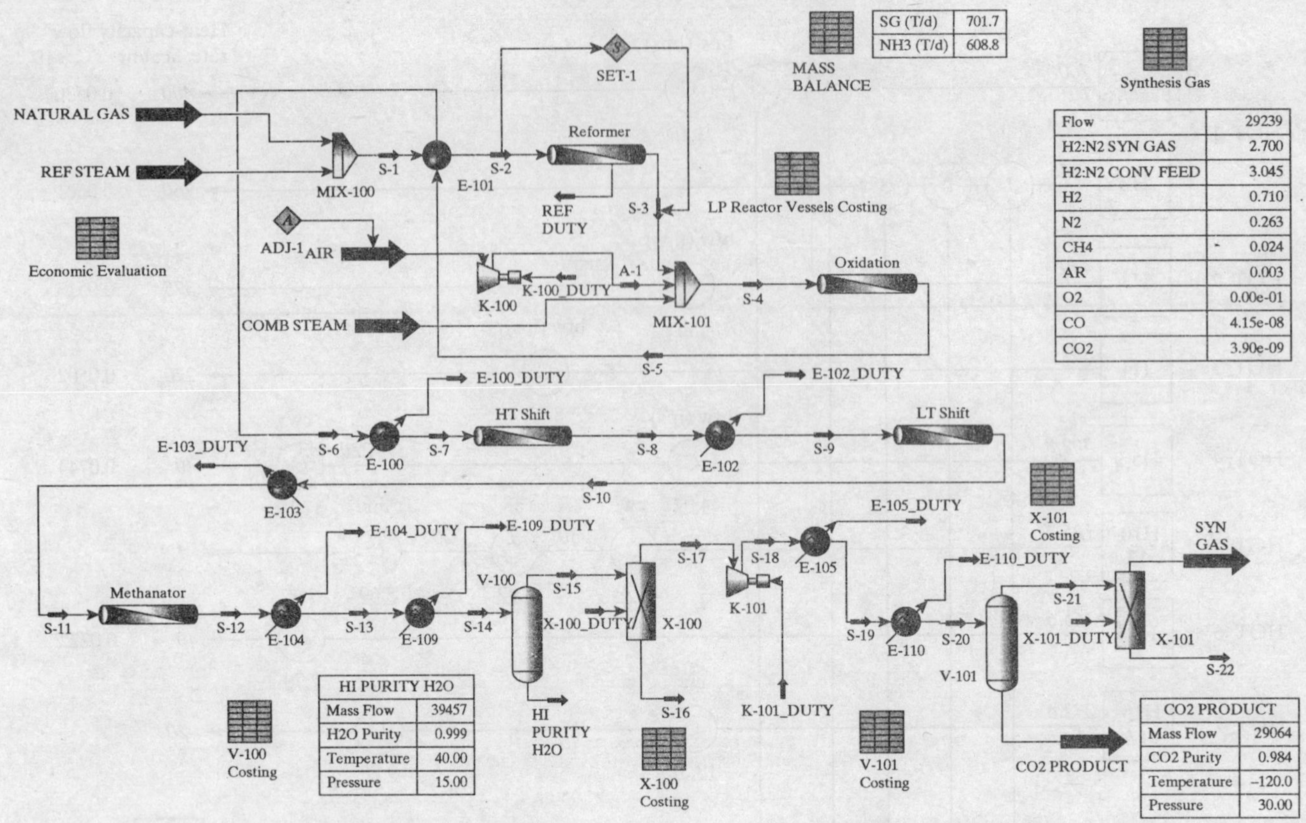

Figure 13.13 UNISIM PFD for the synthesis gas section after heat integration. Units as in Figure 13.4.

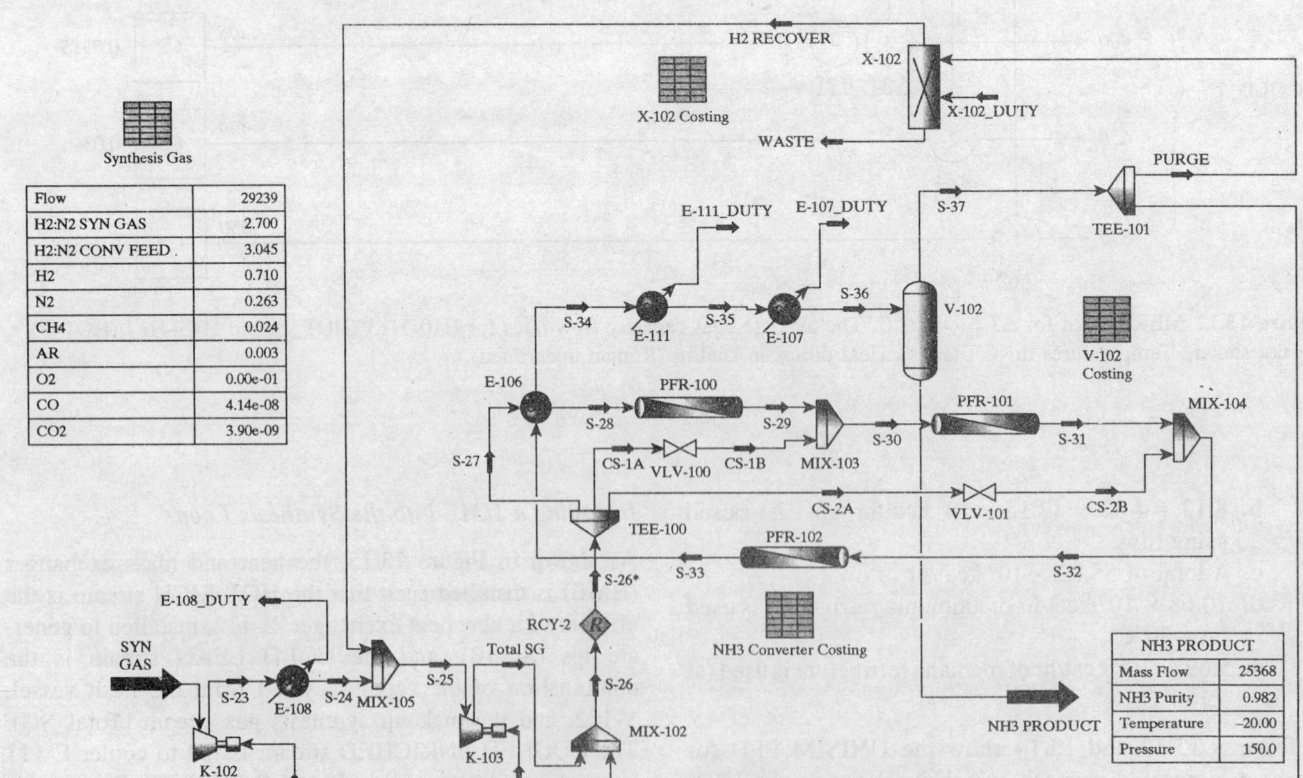

Figure 13.14 UNISIM PFD for the synthesis loop section after heat integration. Units as in Figure 13.4.

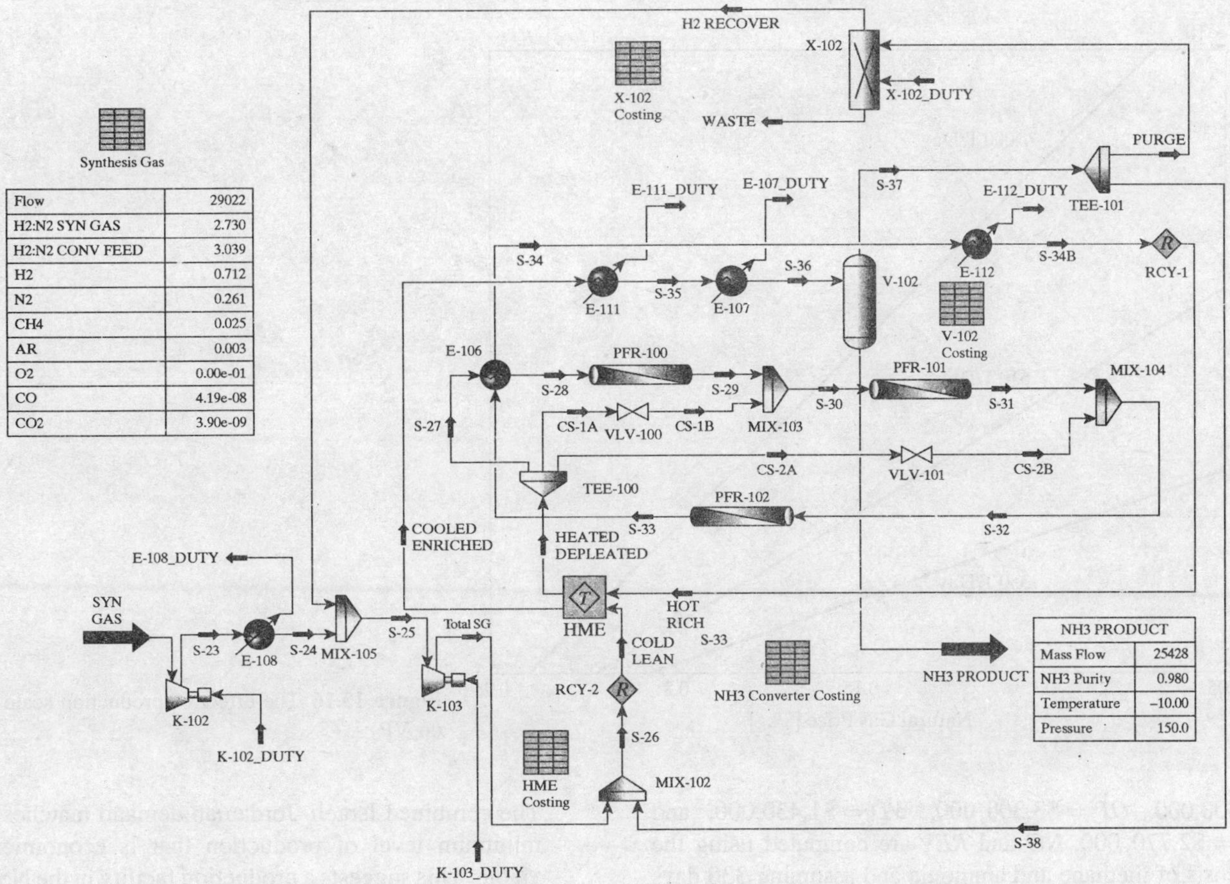

Figure 13.15 UNISIM PFD for the synthesis loop section after installing an HME. Units as in Figure 13.4.

COOLED ENRICHED stream; (b) the temperature of the HEATED DEPLETED stream is 50°C less than that of the HOT RICH stream. Note that the internals of the HME block in the PFD are as given in the project tender in Figure 6 (see the file Ammonia Project.pdf in the PDF Files folder, which can be downloaded from the Wiley Web site associated with this textbook).

The installation of the HME alters the material balance in the synthesis loop, requiring a small adjustment in the H_2:N_2 ratio in the makeup stream to maintain the H_2:N_2 ratio in the reactor feed close to the optimal value of 3. Furthermore, because the reactor feed temperature is increased, the optimal cold-shot fractions move from 0.6, 0.2, and 0.2 to 0.34, 0.33, and 0.33. The reactor feed temperature is also reduced to 240°C. These modifications increase the NH_3 composition in the stream fed to the flash vessel, V-102, to 18.7 mol% (from 16.5 mol%). Also, the feed temperature to the flash vessel is increased to −10°C (thus saving significant refrigeration costs). This, of course, increases the NH_3 composition in the recycle stream, but this is corrected for by the action of the HME. Finally, note that about $1,000,000/yr of revenue is obtained from ips generation in E-112.

As shown in Table 13.4, the introduction of the HME, while not optimally designed, further increases the ROI and VP to above 16% and –$3,400,000, with a payback period that just exceeds 6 years. Clearly, the process at the EPL production rate is only marginally acceptable. To secure an adequate profit, a larger production rate is required, as considered next.

Economy-of-Scale

In this section, the venture profit is estimated at various levels of operation using *economy-of-scale* methods. As defined in Eq. (23.9), the venture profit, VP, is:

$$VP = (1 - t)GP - i_{min}C_{TCI}, \tag{13.10}$$

where t is the income tax rate, GP is the annual pretax earnings (S − C = gross profit), S *is the annual sales,* C *is the annual cost of production,* C_{TCI} *is total capital investment,* and i_{min} is the minimum acceptable return on investment.

For the ammonia process, the annual cost of production is:

$$C = NG + OP + WD + LAB, \tag{13.11}$$

where NG is the annual cost of natural gas, OP is the annual cost of operations, WD is the annual cost of waste disposal, and LAB is the annual cost of labor. Returning to Table 13.4, the most profitable configuration (after implementing the HME) for a feed of 12,000 kg/hr methane has C_{TCI} =

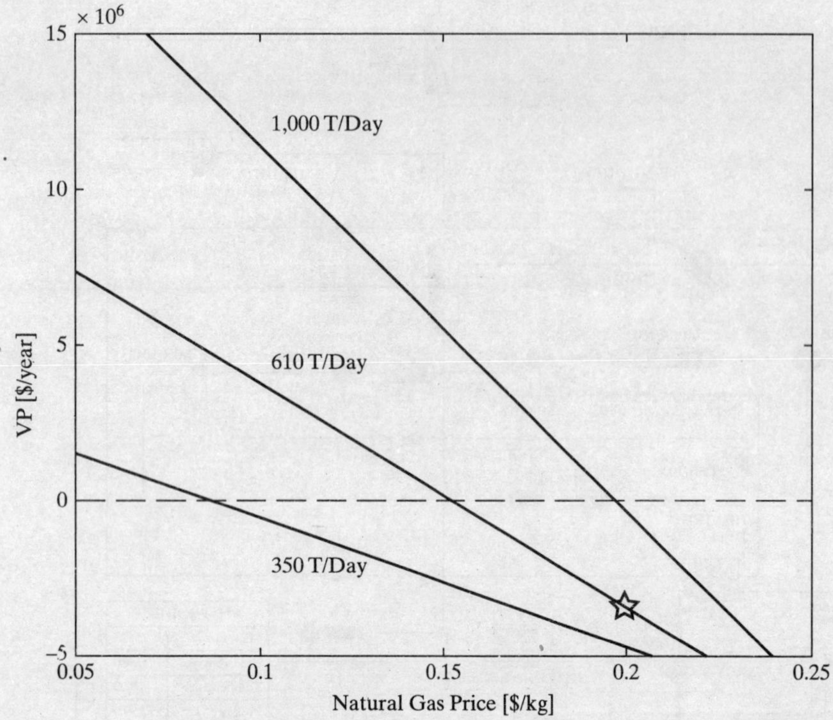

Figure 13.16 The effect of production scale on VP.

\$88,500,000, $OP = \$3,300,000$, $WD = \$1,430,000$, and $LAB = \$2,770,000$. NG and REV are computed using the unit costs of methane and ammonia and assuming 330 days of production per year. A power law is commonly used to scale the C_{TCI} with the production level; that is, for the ammonia process, $C_{TCI} = 0.993P^{0.7}$ million dollars. Taking $i_{min} = 0.2$ and $t = 0.25$, the VP is estimated as a function of the production level, with the results shown in Figure 13.16 as a function of the price of methane. Several obvious conclusions are:

a. Ammonia cannot be produced at a level that satisfies the Israeli national demand unless the cost of natural gas is about half its current level. Given that methane prices are expected to rise, this eliminates the feasibility of production to satisfy the Israeli market demand only. The possible impact of governmental incentives on the feasibility of production at this level is considered in Exercise 13.5.

b. A plant that produces approximately twice the Israeli national demand would be feasible if the cost of methane is subsidized by 25%. However, the surplus ammonia produced would have to be exported, probably by sea, requiring bulk storage facilities in population centers—an infeasible option.

c. A plant to produce three times the Israeli national demand, that is, 1,000 T/day, is feasible economically. If this solution were adopted in Haifa, the storage facilities would be doubled, with two-thirds of the production exported using the city's port facilities, which is also an infeasible option. Alternatively, a joint venture with Jordan, whose ammonia demand is approximately twice that of Israel's, could be encouraged.

The combined Israeli–Jordanian demand matches the minimum level of production that is economically viable. This suggests a production facility in the Negev Desert, close to the Israel–Jordan border.

Development Stage

As indicated in Figure PI.1, the main task in the *development* stage of the SGPDP, as applied to the design of *basic chemical* products, is to carry out a detailed plant design and draw conclusions about the feasibility of the project. This stage remains to be implemented for the ammonia product; see Exercise 13.6.

Postscript

This case study documents the final-year design project undertaken by undergraduate students of the Department of Chemical Engineering at the Technion, Class of 2007. The project tender, provided in the file Ammonia Project.pdf in the PDF Files folder, which can be downloaded from the Wiley Web site associated with this textbook, was addressed to the students. The project attracted considerable media attention in the wake of the 2006 Lebanese War and the missile bombardment of Haifa. The following is an article published in the *Jerusalem Post*:

www.wiley.com/go/global/seider

http://www.jpost.com/servlet/Satellite?cid=1185379003413&pagename=JPost%2FJPArticle%2FShowFull

A joint Israeli-Jordanian factory for the production of ammonia—for making fertilizer and other products—to solve the problems of Haifa Bay's ammonia storage

facility has been proposed by researchers at the Technion-Israel Institute of Technology in Haifa.

Prof. Danny Lewin and students from the chemical engineering faculty were invited by the Haifa City Association to find ways to transfer the ammonia facility from Haifa to the South because of environmental and security risks in its present location. The students' three-semester project was to suggest a way to turn the money-losing facility into a profitable one while ensuring safety and protecting the environment. An ammonia leak could endanger the lungs of neighbors and even pose the danger of death.

Ammonia is not manufactured in Haifa, but imported and stored in the Haifa tanks. During the Second Lebanon War, residents were very fearful that a Hezbollah rocket would hit the facility and set off an environmental disaster.

Lewin said he wanted students to finish their studies less as technocrats and more as involved with people's problems and taking a global approach. They worked for three semesters on a problem that experienced engineers had failed to solve and came up with a way to manufacture ammonia naturally, safely and efficiently.

They reached the conclusion that it was not economically viable to manufacture ammonia for Israel alone, as any facility would produce three times what Israel needs. Instead, they said a partnership should be formed with Jordan to manufacturer the chemical at Rotem in the South, close to the Jordanian border. The area already has a facility to manufacture fertilizers from Haifa Chemicals' ammonia.

13.3 ENVIRONMENTALLY FRIENDLY REFRIGERANT CASE STUDY

In Section 3.2, the issues involved in designing environmentally friendly refrigerants are discussed. These employ molecular-structure design to locate molecules involving C, N, O, S, and H atoms, and the halogens F, Br, and I (and not Cl), that have: (1) a large latent heat of vaporization, to reduce their throughput when removing a specified heat duty; (2) a low viscosity, to reduce the recirculation power; and (3) a low freezing point, to reduce the possibility of freezing. Note that Cl was banned in the Montreal protocol of 1987 because the composition of Cl-containing refrigerants had increased to parts-per-billion in the stratosphere, with the chlorine atoms reacting with ozone, decreasing the earth's ozone layer. Hence, in 1987, the product design problem was to design environmentally friendly molecules, that is, refrigerants that exclude Cl.

For this case study, it is presumed that a design team was assembled in the late 1980s to create a new environmentally friendly refrigerant product. In hindsight, of course, many refrigerants were designed. The purpose of this case study is to trace the steps in product design, following Figure PI.1, as they may have been carried out by design teams at that time. Note that these scenarios are postulated by the authors, who were not involved in the design teams.

Project Charter

As discussed in Section 2.2, and shown in Figure PI.1, most design teams begin to develop a new product by creating a project charter. For an environmentally friendly refrigerant product, a typical project charter—containing specific goals, a project scope, deliverables, and a time line—is given in Table 13.7. Clearly, the project charter summarizes the objectives above and sets the time line with deliverables.

Molecular-Structure Design

The next step in Figure PI.1 requires that the design team answer the question: "Is materials technology invention required?" In this case, of course, it is necessary to "find chemicals or chemical mixtures that have the desired properties." This involves molecular-structure design, as discussed under "Refrigerant Design" in Section 3.5.

In the strategy described, the design team begins by selecting the extreme temperatures in refrigeration cycles for which the refrigerant is needed; that is, the temperatures at which heat is to be absorbed in the evaporator and rejected in the condenser of a refrigerator. For a typical refrigeration cycle, see Figure 9S.20. Then, k atoms and molecular groups are selected to be included in the candidate molecules, each of which may appear n times.

First, Example 3.3 considers 13 atoms and molecular groups—CH_3, CH_2, CH, C, OH, O, NH_2, NH, N, SH, S, F, and Cl—with Cl included in the search to show that molecules containing chlorine are the most desirable when the ozone layer is disregarded. Then, in Example 3.4, when an ozone depletion potential (ODP) is included, refrigerants without Cl are obtained.

For these examples, the refrigerant is designed to absorb heat at temperatures as low as 30°F (−1.1°C) and reject heat at temperatures as high as 110°F (43.3°C). In an optimization formulation, the refrigerants are selected to have:

1. A vapor pressure, $P^s\{-1.1°C\} > 1.4$ bar, to ensure that leaks are from the refrigerant system (rather than from vacuum operation, into which air and water vapor can leak),
2. $P^s\{43.3°C\} < 14$ bar, to keep the compression ratio from exceeding 10,
3. An enthalpy of vaporization, $\Delta H^v\{-1.1°C\} > 18.4$ kJ/mol, to reduce the amount of refrigerant needed (where 18.4 kJ/mol is the latent heat of vaporization of Freon 12, the chlorine-containing refrigerant banned in 1987), and
4. A liquid heat capacity, $c_{pl}\{21.1°C\} < 32.2$ cal/(mol K), to reduce the amount of refrigerant that flashes across the valve (where 32.2 cal/(mol K) is the heat capacity of liquid Freon 12). Note that 21.1°C is the average of the extreme temperatures.

Table 13.7 Possible Project Charter for the Environmentally Friendly Refrigerant

Project Name	Environmentally Friendly Refrigerant
Project Champions	Business Director of the Refrigerant Business
Project Leader	John Doe
Specific Goals	Refrigerant that doesn't react with ozone in the earth's stratosphere
Project Scope	In-scope: • Determination of acceptable molecules involving C, N, O, S, H atoms, and the halogens F, Br, and [illegible]. • Minimal changes in the current manufacturing capabilities Out-of-scope: • Molecules containing Cl atoms • Major manufacturing changes
Deliverables	• Business opportunity assessment • Technical feasibility assessment • Manufacturing capability assessment
Time Line	A process design and product prototypes for market testing within 12 months

Using the group-contribution methods to estimate P^s, ΔH^v, and c_{pl}, as shown in Example 3.3, the results of optimization calculations (to solve a mixed-integer nonlinear program—Duvedi and Achenie (1996)) are shown. Then when the ozone depletion potential (ODP) is defined in Example 3.4, two promising molecules are obtained: CH_3CHF_2 and SF_2. The first is produced from chlorinated hydrocarbons, such as 1,1-dichloroethane (see Section 4.4), chloroethene, and hydrofluoric acid (HF), using various reaction paths and catalysts. While the second is produced by halogen exchange with SCl_2, it decomposes rapidly and is not sufficiently stable to be a refrigerant product.

Innovation Map

The next step in Figure PI.1 is to ask the question: "Is process/manufacturing technology invention required?" Looking ahead to the design of a process for the manufacture of CH_3CHF_2, it seems clear that standard processing operations will be sufficient, and consequently, the design team responds negatively.

At this point, the design team can create an innovation map that links the new technologies with the customer needs, as shown in Figure 3.1, which is repeated here for convenience. Note that, on the right-hand side of the map, the new material technologies in the late 1980s are linked to the customer needs.

Concept Stage

Having completed an innovation map, as shown in Figure PI.1, the design team next seeks approval to initiate the Stage-Gate™ Product-Development Process (SGPDP). When granted approval, the *concept* stage is initiated, beginning with opportunity assessments and the determination of customer and technical requirements, as discussed in Section 2.4. For this case study, these steps are not discussed here. Then, prior to preliminary process synthesis, a preliminary database is assembled, as discussed in Section 4.2. In the *concept* stage for the design of *basic* chemical products such as CH_3CHF_2, preliminary process synthesis is the principal step. As this step has been covered extensively in Section 4.4 for two *basic* chemicals, vinyl chloride and tissue plasminogen activator (tPA), the preliminary process synthesis is not included here for these refrigerants. Instead, the synthesis of a process to manufacture CH_3CHF_2 is the subject of Exercise 13.7 at the end of this chapter. The last step, bench-scale laboratory work, would normally be carried out to verify the principal assumptions in the preliminary process synthesis.

Feasibility Stage

Again turning to Figure PI.1, the steps shown in the *feasibility* stage would be carried out. These begin with developing the base-case design, using one of the most promising flowsheets obtained in preliminary process synthesis. As the base-case design is developed, normally process simulators are used and pilot-plant testing is carried out, as discussed in Section 4.5. Simultaneously, the algorithmic methods for improving the process synthesis (see Chapters 7–11) are normally carried out, usually using approximate economic measures such as the annualized cost. Also, a plantwide controllability assessment is normally carried out, as discussed in Chapter 12.

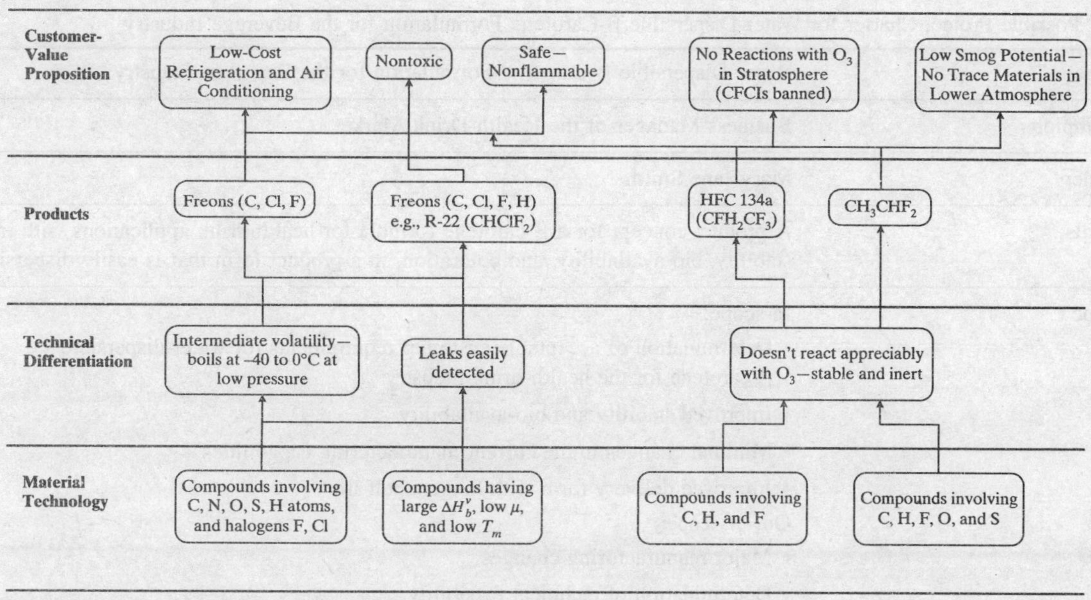

Figure 3.1 Environmentally friendly refrigerant innovation map.

Development Stage

The *development* stage in Figure PI.1 would be carried out next, involving detailed design, equipment sizing, profitability analysis, and optimization, as discussed in Part Four of this textbook. Also included would be the development of startup strategies and safety analysis, often using the HAZOP techniques introduced in Section 1.5.

13.4 WATER-DISPERSIBLE β-CAROTENE CASE STUDY

This case study is presented to illustrate the steps in the Stage-Gate™ Product-Development Process (SGPDP) for designing a *basic* chemical product. Its time frame is assumed to be the mid-1990s, when a design team at a hypothetical food company was developing a carotenoid formulation for the beverage industry.

The carotenoids are common sources of coloring agents for a variety of foods and beverages. Of importance commercially, the carotenoids include β-carotene, lycopene, canthaxantrin, and apocarotenal, which provide color pigments that range from yellow to red. β-carotene is especially important in that it is not only useful as a colorant, but also as a source of Vitamin A. Unfortunately, pure β-carotene crystals do not dissolve readily in water. Rather, they are easily oxidized and, consequently, cannot be used in the beverage industry without further processing and formulation. For this reason, the goal of this case study is to develop a β-carotene product that circumvents these problems and is suitable for the beverage industry.

Many of the considerations in the product-development scenario in this case study are based on the conceptual design article by Leuenberger (2007) and related patents published in the 1990s. Note, however, that the authors were not involved in this development process, and consequently, the scenario presented is of events likely to have taken place. Also, the design team that developed the water-dispersible β-carotene formulation may not have used the Stage-Gate™ Product-Development Process (SGPDP).

Project Charter

As described in Section 2.2, a well-crafted project charter helps a design team develop the desired product on time and at expected cost. The typical elements, that is, specific goals, project scope, deliverables, and time line, are illustrated in Table 13.8 for a water-dispersible β-carotene formulation being developed for the beverage industry. These are discussed next.

Note that the compositions and methods of manufacture of β-carotene crystals were well known in the mid-1990s, with β-carotene, $C_{40}H_{56}$, a well-defined *basic* chemical product having the molecular structure in Figure 13.17.

Because β-carotene is insoluble in water and has very limited solubility in fats, its classical usage was limited to the coloring of fats, such as butter and margarine. Then, in the 1970s, a water-dispersible β-carotene formulation was patented (U.S. Patent 3,655,406) that made it suitable for the coloring of products such as pharmaceuticals, food products, and cosmetics such as lipstick. However, this formulation didn't satisfy the requirements of the beverage industry in the mid-1990s, which sought a stable formulation more readily dispersible in water, having excellent bioavailability, and providing intense coloration at low concentrations. In short, the solid lipophilic β-carotene needed to be transformed into a form that could be used for water-based systems like beverages. Stable formulations were needed to

Table 13.8 Possible Project Charter for Water-Dispersible β-Carotene Formulation for the Beverage Industry

Project Name	Water-Dispersible β-Carotene Formulation for the Beverage Industry
Project Champions	Business Manager of the Health-Drink Market
Project Leader	Mary Jane Smith
Specific Goals	A product concept for a β-carotene formula for health-drink applications with improved stability, bio-availability, and coloration, in a product form that is easily dispersible in water
Project Scope	In-scope: • Determination of acceptable customer requirements for water-dispersible β-carotene for the health-drink industry • Improved stability and bio-availability • Minimal changes in the current manufacturing capabilities • Improved delivery form with longer shelf life Out-of-scope: • Major manufacturing changes • Determination of technical feasibility • Market opportunity assessments
Deliverables	• Voice of the customer • Product concepts
Time Line	Product concepts developed within 3 months

achieve longer shelf lives. Also, bio-availability implies sufficiently small particle sizes to enter cells, at sufficiently high concentrations in the liver, for example, acting as an antioxidant to protect cells' photosynthesis pathways.

As stated, these goals were probably in response to the market trend in the health-drink industry in the mid-1990s. Our hypothetical company would likely have sought to deliver a ready-to-use β-carotene formula having desired coloration capabilities as well as positive health effects. Presumably, its market opportunity assessment had identified a fast-growing market, and a design team was assembled to gather customer requirements and create new product concepts, within 3 months, for lab prototyping and testing.

Regarding the project scope, emphasis was likely placed on determining more explicit customer requirements for improved stability and bio-availability, as well as the preferred form for delivery of the product. And, minimal changes in the manufacturing facilities used for conventional formulations were probably required for producing the new β-carotene formulations.

Innovation Map

Next, following the product design steps in Figure PI.1, having completed its project charter, the design team likely asked whether materials technology and process/manufacturing technology inventions were required. For β-carotene formulations to be used in beverage-industry applications in the mid-1990s, many of the required technologies were in place, as discussed next.

It was well known that beverages containing β-carotene, such as carrot juice and blends of juices containing carrot juice, had very poor shelf life due to their degradation of color and flavor (taste) over time. This is because β-carotene is prone to discolor and to experience oxidation by irradiation; that is, it is both photosensitive and reactive with oxygen. In 1992, however, Ohtaka and Sudo (U.S. Patent 5,153,012) reported that the addition of antioxidative vitamins, such as Vitamins C, B_2, and E, reduce the loss of β-carotene, even at ambient temperatures. Also, in their patent, they show that the loss of β-carotene at room temperature over a 6-month period is less than 10%, while the loss is negligible at 10°C over the same period of time. This is shown in Figure 13.18.

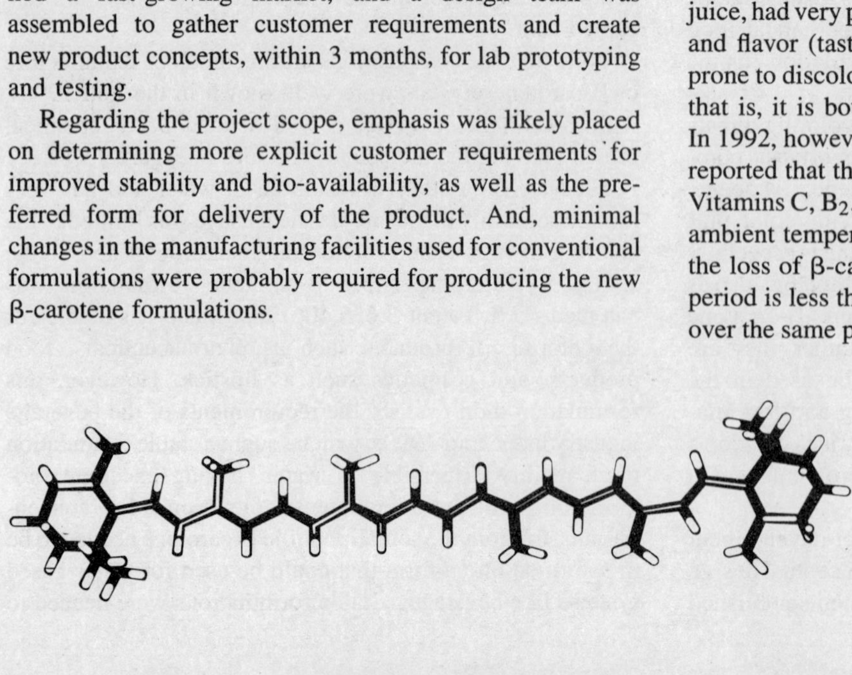

Figure 13.17 β-carotene.

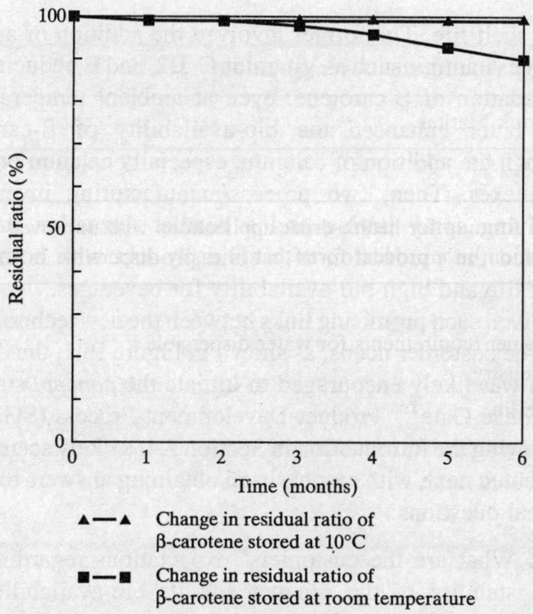

Figure 13.18 Percent loss of β-carotene during storage (U.S. Patent 5,153,012).

Then, in 1996, building on the discovery of Ohtaka and Sudo, Heckert et al. (U.S. Patent 5,516,535) found that the addition of calcium, preferably calcium organic complexes (such as calcium citrate malate, $Ca_6(C_6H_5O_7)_2(C_4H_5O_5)_3 \cdot 6H_2O$ – mol. wt. = 1, 123.1) enhanced the bio-availability of β-carotene. They reported that a mixture of β-carotene and calcium, encapsulated in dextrin, increased the level of Vitamin A, to as high as 35 IU (International Units, where 1 IU of Vitamin A = 0.6 μg of β-carotene), in the livers of laboratory animals. Other encapsulation agents, such as starch and gelatin, were also used, although the latter was found to have adverse effects on the β-carotene-calcium interaction.

In the mid-1990s, it seems clear that dry formulations or colloid emulsions were being considered for product delivery. The latter would have involved β-carotene dissolved in an organic phase, which forms a colloid with water. For the former, Leuenberger (2007) discusses two methods of producing a nano-size β-carotene dispersion, so-called *emulsion powders*: spray drying and the formation of *beadlets*. Note, first, that a typical emulsion powder is comprised of β-carotene dispersed in water, which forms a colloid with gelatin that distributes on the particles of a pulverized carrier material. The latter is insoluble in water with a lipophilic (gelatin-loving) surface.

The spray-drying process involves two steps: (1) the preparation of a β-carotene-water dispersion or an emulsion involving the former and a water-immiscible solvent, and (2) the removal of water by spray drying. This gives stable, nano-particles of β-carotene, which have high bio-availability. The dispersion or emulsion can be produced by different processes, for example:

- β-carotene is dissolved in a water-immiscible solvent, followed by emulsification in water and removal of the solvent—leaving a dispersion of β-carotene nano-particles in water.
- A water-miscible solvent is used, with β-carotene particles dispersed (in a water solution).
- A supersaturated β-carotene solution in oil is prepared and emulsified in water using supercritical gases.

In the beadlet formation process, the same β-carotene dispersion is used, but the spray-drying step is replaced by the beadlet formation process. In the latter, the dispersion or emulsion is sprayed into a fluidized starch bed, with the nano-droplets (100–500 nm) covered by layers of starch. The resulting beadlets are shown schematically in Figure 13.19. Then, the beadlets are dried to the desired moisture content.

The nano-particles of β-carotene in Figure 13.19 are shown encapsulated in a matrix of carbohydrate, ascorbylpalmitate, and gelatin, thus protecting them from oxidation and photodegradation. In the outer layer, the starch skin provides excellent water dispersibility. The overall size of the beadlets, about 0.4 mm, is significantly larger than particles produced by spray drying. In a related 1999 patent, Cox et al. (U.S. Patent 6,007,856) found that beadlets formed

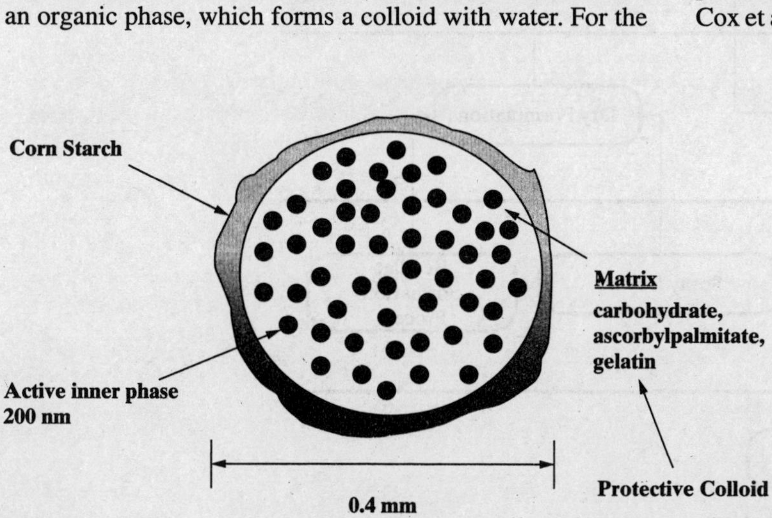

Figure 13.19 Schematic of β-carotene beadlets (Leuenberger, 2007).

using an oil-in-water dispersion of β-carotene are protected (stabilized) from oxidation. These beadlets provide excellent water dispersibility, high coloring intensity even at low concentration, and high bio-availability.

Given these materials and process/manufacturing inventions, as described in Section 1.3, an *innovation map* helps to guide the technology- and product-development processes. Note that the *innovation map* in Figure 13.20 was constructed in hindsight, as it might have been constructed by a design team in the mid-1990s. As seen, it contains elements at five levels, moving from the bottom to the top of the map:

1. *Materials Technology:* antioxidative additives (vitamins), calcium organic complexes
2. *Process/Manufacturing Technology:* spray drying, beadlet formation process
3. *Technical Differentiation (Technical-Value Proposition):* stable formulation, dry formulation
4. *Products:* colloid for pharmaceuticals, cosmetics, etc; solid particles for beverages
5. *Customer-Value Proposition:* long shelf life, high bio-availability

Beginning at the left, the inventions by Ohtaka and Sudo (U.S. Patent 5,153,012) in 1992 and Heckert et al. (U.S. Patent 5,516,535) in 1996 enabled the development of stable β-carotene in colloid products, satisfying customer needs for long shelf life. The former involved the addition of antioxidative vitamins, such as Vitamins C, B2, and E, reducing the degradation of β-carotene, even at ambient temperatures. The latter enhanced the bio-availability of β-carotene through the addition of calcium, especially calcium organic complexes. Then, two process/manufacturing inventions involving spray drying and a beadlet formation process enabled the production of a solid product with both long shelf life and high bio-availability for beverages.

Given such promising links between the new technologies and the customer needs, as shown in Figure PI.1, the design team was likely encouraged to initiate the *concept* stage of the Stage-Gate^{TM} Product-Development Process (SGPDP). Following the introduction in Section 2.4, a likely scenario is presented next, with emphasis on obtaining answers to three typical questions:

1. What are the customers' expectations regarding the stability of the product and its bio-availability for health-drink applications?
2. What is the preferred delivery form?
3. Which product concepts can be identified to satisfy the customer requirements?

Concept Stage

The *concept* stage normally begins with several assessments and activities, which are discussed next as they were likely to

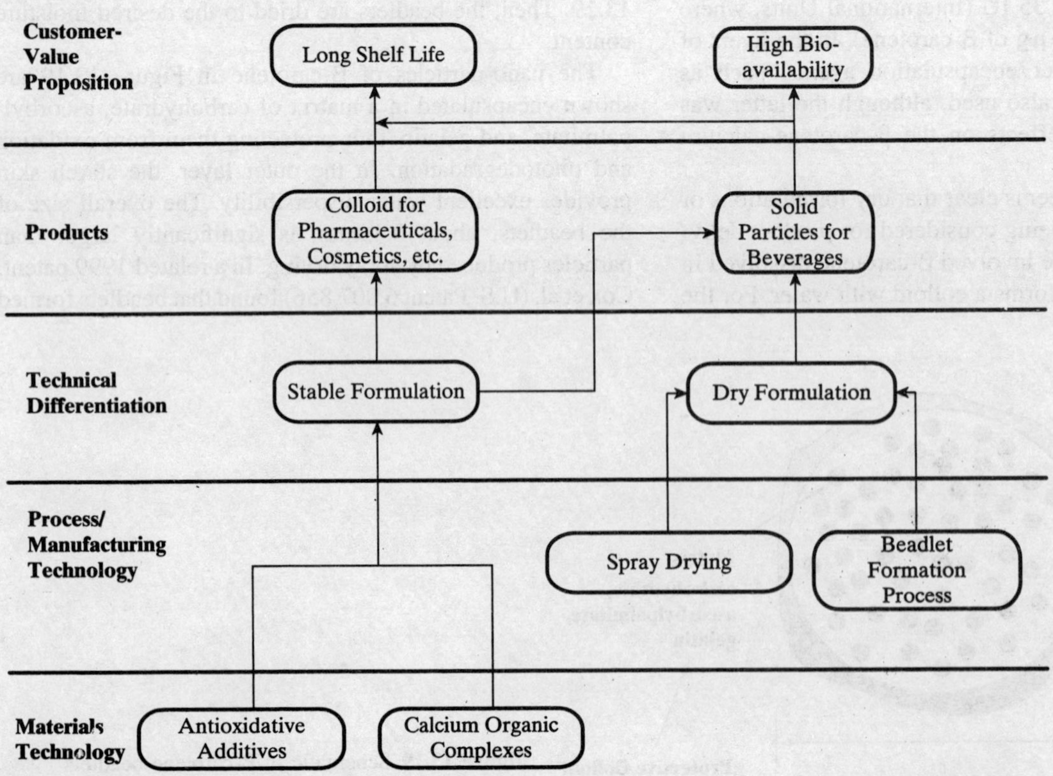

Figure 13.20 Innovation map for water-dispersible β-carotene.

Table 13.9 New-Unique-and-Difficult (NUD) Requirements for Water-Dispersible β-Carotene

Customer Requirement	Product Requirement	Type	Weighting Factor (%)
High bio-availability at point of use	Protected active ingredients (Vitamin A)	NUD	60
Stable during storage	Low β-carotene loss at storage temperatures	NUD	40

have occurred during the development of the β-carotene formulations for the health-drink market in the mid-1990s.

a. ***Opportunity Assessments.*** As mentioned above, in this case the market opportunity assessment was likely considered to be out-of-scope. It had probably been carried out by a business team, which had likely convinced itself that a good opportunity existed for a new product. Given this scenario, the needs to understand better the customer requirements and to identify winning product concepts would have remained.

b. ***Customer Requirements.*** Because β-carotene formulations for health drinks probably presented an emerging opportunity for a new product(s), its customer requirements were probably still being interpreted. Consequently, it would have been important to gather the *voices of the customers* (VOCs), as introduced in Section 2.4.

Following the methodology introduced therein, the design team would likely have created the following objectives to obtain the:

- Requirements for coloration, stability, and bio-availability
- Preferred product form

and would have formulated several questions associated with each of these objectives:

Coloration, Stability, and Bio-Availability

- Under what conditions (temperature, humidity, nitrogen blanket) should the raw materials be stored?
- How long do the raw materials need to be stored?
- What concentration range of β-carotene is desirable in health-drink products?
- What particle size of the dispersions is needed for bio-availability and coloration?

Preferred Delivery Form

- What is the preferred product form—liquid, emulsion (colloidal structure), dispersion (particles in liquid), or solid (powder, beadlets, etc.)? Note that an emulsion involves uniformly distributed small droplets of a water (or organic) phase in a continuous organic (or water) phase. A dispersion involves uniformly distributed, insoluble, small particles in a liquid phase—typically, water.
- What is the preferred particle size or particle-size distribution?
- What temperatures—warm, room temperature, or cold—are desired for dispersions in water?

The value chain of the product likely included the β-carotene formulator, which possibly produced β-carotene crystals, the health-drink manufacturers, retailers, and the end users. Because β-carotene formulators sell their product to beverage manufacturers, clearly the latter were the preferred interviewees, including marketing personnel, product developers, and manufacturing staff. In addition to granting interviews, it is likely that several beverage manufacturers hosted visits to their manufacturing sites, with observations probably recorded as so-called *customer images*. During and after the interviews, the design team probably sought to quantify the relative importance of the various customer voices.

After the interviews, the results were likely analyzed and compiled, with the customer voices and images translated into customer requirements; for example:

- Stability: Less than 10% loss of active β-carotene stored at room temperature over a 6-month period.
- Bio-availability: High bio-availability when particles are dispersed in water.
- A dry delivery form, rather than liquid, is preferred. An excellent dispersion in water is expected.

Table 13.9 shows typical requirements that were likely identified. Note that the product requirements follow directly from the customer requirements.

In addition, *fitness-to-standard* (FTS) requirements that customers expected were probably listed, for example, as shown in Table 13.10, this being an unusual case, with just one FTS requirement.

Table 13.10 Fitness-to-Standard (FTS) Requirements for Water-Dispersible β-Carotene

Customer Requirement	Product Requirement	Type	Weighting Factor (%)
Dry product form	Easy to disperse in water	FTS	100

Table 13.11 Technical NUD Requirements for Water-Dispersible β-Carotene Formulations

Product Requirement	Technical Requirement	Target
Protected active ingredients	Protected active ingredients	Vitamin A level in liver higher than 30 International Units (IU)[§]
Low β-carotene loss at storage temperatures	Low β-carotene loss at storage temperatures	Less than 10% loss of active β-carotene stored at room temperature over 6-month period

[§]IU of Vitamin A = 0.6 μg of β-carotene

As discussed in Section 2.4, in these tables the weighting factors play an important role, especially when the requirements compete and compromises are necessary. Here, among the NUD requirements, the high bio-availability at point of use is normally considered to be more important than stability during storage. In this case, the relative importance between the NUD and FTS requirements would not have been difficult to set. When compromises are difficult to achieve, the design team may try *conjoint* analysis (Bakken and Frazier, 2006), as discussed in the case study in Section 15.3.

c. ***Technical Requirements.*** Because conventional β-carotene products were being extended to water-dispersible β-carotene formulations, most of the technical requirements were likely known, including stability during storage, water dispersibility, and bio-availability properties. The translation of the new NUD customer requirements to their technical requirements for water-dispersible β-carotene formulations is discussed next.

d. ***Determination of Critical-to-Quality (CTQ) Variables.*** Normally, the NUD requirements translate into the *critical-to-quality* variables. And consequently, it is likely that the product requirements *protected active ingredients* and *low β-carotene loss at storage temperatures* were selected during the development of the water-dispersible β-carotene formulations. A typical translation of these requirements into CTQ variables is shown in Table 13.11.

Following the recommendations in Section 2.4, a House of Quality (HOQ) would normally be prepared, relating the various requirements to one another. For this case study, the reader is asked to prepare a HOQ in Exercise 13.8.

e. ***Development of Product Concepts.*** As discussed in Section 2.4, the identification and selection of solution concepts is at the heart of new-product development, especially when several alternatives are available. Furthermore, as the design team generated new formulations, beginning with the new technologies discussed above, the Pugh matrix (Pugh, 1996), discussed in Section 2.4, may have been helpful for screening purposes. In this matrix, as shown in Table 13.12, each potential solution concept is compared with a reference concept. Usually, the reference concept is the best known in the market (or the best potential solution), which in the mid-1990s, was the Heinrich formulation (U.S. Patent 3,655,406). Note that this patent introduces the emulsion powder discussed above, which distributes on the particles of a pulverized carrier. In this patent, Heinrich claims that the composition of these beadlet particles is suitable for coloring products such as pharmaceuticals, food products, and cosmetics such as lipstick, where intense coloring must be achieved in thin coating applications.

For each potential concept, each technical requirement is compared against that for the reference concept and assigned a qualitative valuation of inferior (−), superior (+), or equal (0). Multiple (−) or (+) entries signify decreasing or increasing levels of inferiority or superiority. As seen, Table 13.12 shows the performance comparisons between the Heinrich reference formulation and the Ohtaka and Sudo formulation

Table 13.12 Pugh Matrix Concept Selection

Technical Requirement	Target	Reference Formula	Formula A Ohtaka, Sudo	Formula B Heckert et al.	Formula C Cox
Protected active ingredients	Vitamin A level in liver higher than 30 International Units (IU)[§]	U.S. Patent 3,655,406 Heinrich	0	+	+
Low β-carotene loss at storage temperatures	Less than 10% loss of active β-carotene stored at room temperature over 6-month period		+	+	+
Easy to disperse	Uniform		0	0	+

[§]IU of Vitamin A = 0.6 μg of β-carotene

(Formula A), the Heckert et al. formulation (Formula B), and the Cox formulation (Formula C).

f. ***Selection of Superior Product Concepts.*** In the *concept* stage of the SGPDP, the selection of superior product concepts is based primarily on the satisfaction of the technical requirements, in particular, the *new-unique-and-difficult* (NUD) requirements. Given Table 13.12, clearly Formula C of Cox promises to provide the best performance, with all of the NUD and FTS requirements met.

g. ***Unit-Cost Estimation.*** Normally, having generated superior product concepts, preliminary cost estimates are prepared for the raw materials and products—often using Henderson's law, based upon the price performance of earlier related products, as discussed in Section 2.8. Also, as shown in Figure PI.1, it is common to carry out preliminary process synthesis calculations, often yielding approximate estimates for installation and operating costs. Note that where similar manufacturing processes exist, preliminary estimates for installation and operating costs are often obtained. For this case study, however, insufficient pricing information was available to prepare these estimates, and time wasn't available to synthesize a potential manufacturing process. Normally, such a process would be improved upon during the following steps of the SGPDP, with more detailed estimates prepared for pilot plants (*development* stage) and manufacturing facilities (*manufacturing* stage).

h. ***Gate Review.*** To complete the *concept* stage, a gate review would likely have been carried out in which the design team answered the three questions associated with the deliverables. Such a review is not discussed here. However, based on the discussion above, the design team would likely have had convincing answers for all of its questions. It had clarified its customers' requirements, had considered several promising product formulations, especially those involving β-carotene beadlets, and had selected a superior product concept that satisfied all of its customers' NUD and FTS requirements. Given such positive results, its business decision makers were likely to have responded affirmatively, granting funding to proceed to the *feasibility* stage. Note that more complete coverage of the gate reviews for the *concept* stage is provided in Section 2.4, and a detailed gate review is presented in a case study for the design of a halogen light bulb in Section 17.2. For this case study, the remaining stages of the SGPDP in Figure PI.1 are not presented.

13.5 SUMMARY

Case studies for the design of three *basic* chemical products have been presented.

Emphasis has been placed on the project charter, the role of the innovation map, and the *concept* stage of the Stage-Gate^{TM} Product-Development Process (SGPDP).

REFERENCES

1. Bakken, D., and C.L. Frazier, *Conjoint Analysis–Understanding Consumer Decision Making*, Chapter 15 of *The Handbook of Market Research: Uses, Misuses, and Future Advances*, R. Grover and M. Vriens (eds.), Sage Publications (2006).

2. Duvedi, A.P., and L.E.K. Achenie, "Designing Environmentally Safe Refrigerants Using Mathematical Programming," *Chem. Eng. Sci.*, **51** (15), 3727–3729 (1996).

3. Lavie, R., "Ammonia Synthesis Enhancement Through Heat-Mass-Exchange," *Plant/Operations Progress*, **6** (2), 122–126 Apr. (1987).

4. Leuenberger, B.H., "Conceptual Design of Carotenoid Product Forms," Chapter 11 in B. Ulrich, W. Meier, and G. Wagner, Eds., *Product Design and Engineering*, Vol. 2, Wiley, 2007.

5. Parisi, D. R., and M. A. Laborde, "Modeling Steady-state Heterogeneous Gas-solid Reactors using Feedforward Neural Networks," *Comput. Chem. Eng.*, **25**, 1241–1250 (2001).

6. Pugh, S., *Creating Innovative Products Using Total Design*, Addison-Wesley-Longman, 1996.

7. Wolf, D., M. Höhenberger, and M. Baerns, "External Mass and Heat Transfer Limitations for the Partial Oxidation of Methane over a Pt/MgO Catalyst—Consequences for Adiabatic Reactor Operation," *Ind. Eng. Chem. Res.*, **36**, 3345–3353 (1997).

Patents—Water Dispersible β-carotene

8. U.S. Patent 3,655,406, Heinrich, K.R., *Carotenoid Compositions* (1972).

9. U.S. Patent 5,153,012, Ohtaka, H., and R. Sudo, *Process for Preparing Beverages Containing Beta-Carotene* (1992).

10. U.S. Patent 5,516,535, Heckert, D.C., H. Mehansho, G.R. Hudepohl, and S. Crosby, *Beverage Compositions Having Enhanced Beta-Carotene Bioavailability* (1996).

11. U.S. Patent 6,007,856, Cox, D.J., D.R. Kearney, S.T. Kirksey, and M.J. Taylor, *Oil-in-Water Dispersions of Beta-Carotene and Other Carotenoids Stable Against Oxidation Prepared From Water-Dispersible Beadlets Having High Concentrations of Carotenoid* (1999).

EXERCISES

13.1 As part of an NH_3-product life-cycle assessment, evaluate the danger of an ammonia release, due to a rocket attack, from a 12,000-ton storage tank. Note that the existing facility is a refrigerated tank at atmospheric pressure.

13.2 As part of an NH_3-product life-cycle assessment, when producing synthesis gas using the heat-integrated process in Figure 13.13, consider the impact of releasing the CO_2 byproduct into the atmosphere.

13.3 As part of an NH_3-product life-cycle assessment, consider the alternative of converting ammonia to urea by reaction with CO_2, as a vehicle for curbing the release of the CO_2 byproduct.

13.4 Consider the effect on profitability of changing the operating pressure of the ammonia synthesis loop. What is the optimal synthesis loop operating pressure?

13.5 Consider the effect on profitability of reducing the investor's rate of return to 6%, in line with possible governmental incentives for the ammonia project. Compute the tradeoff line, at which VP is zero, relating the minimum price of ammonia as a function of the cost of methane. In addition, you are requested to estimate the cost of the ammonia produced in a plant that meets the requirement of 350 tonnes/day.

13.6 For the ammonia product, carry out the *development* stage of the SGPDP, as suggested in Section 13.2.

13.7 For the refrigerant CH_3CHF_2, carry out a preliminary process synthesis to create at least one promising process flowsheet. Begin with a literature and patent search to select raw materials and the principal reaction paths.

13.8 Prepare the first House of Quality for a water-dispersible β-carotene product for the health-drink industry.

Part Two

Product Design for Industrial Chemicals

Part Two presents, in two chapters, the underlying technologies and strategies for the design of *industrial chemical* products and the processes to produce them. It follows the Stage-Gate^{TM} Product-Development Process (SGPDP), which was introduced in Chapters 1 and 2 and is presented for the design of *industrial chemical* products in Figure PII.1 As discussed in Section 1.3, while *industrial chemicals* are characterized by thermophysical and transport properties (like *basic chemicals*), other properties are often dominant in satisfying customer needs, including microstructure; particle-size distribution; and functional (e.g., cleansing, adhesion, shape), sensorial (e.g., feel, smell), rheological (non-Newtonian flow), and physical (e.g., stability) properties. The latter are often the focus of engineered materials with specially designed surface coatings such as specialty fibers; engineered plastics; monolayer and multilayer films; creams and pastes; organic polymers tailored for semiconductors (i.e., polymer semiconductors) that enable thin-film transistor technology; specialty glass substrates, designed to match the silicon coefficient of expansion, that enable chip-on-glass technology; and many others that are typical *industrial chemical* products covered in Part Two. As indicated in Figure 1.3, most *industrial chemicals* are not sold to the consumer. Rather, they are the ingredients and building blocks for other *industrial chemicals* and *configured consumer products*, which are covered in Part Three.

Next, the steps for the design of *industrial chemical* products in Figure PII.1 are introduced. Subsequently, they are illustrated in the innovation maps of Chapter 14 and the case studies of Chapter 15.

MATERIALS AND PROCESS/ MANUFACTURING TECHNOLOGIES DEVELOPMENT

After the design team creates its project charter, as discussed in Section 2.2, it seeks to identify appropriate materials technologies to achieve its objectives when they are needed. This is the step at the top left of Figure PII.1, which, for *industrial chemicals*, usually involves a search for the appropriate molecules, engineered polymers and composites, or advanced engineered materials to satisfy the other property specifications, in addition to thermophysical and transport properties that align closely with customer needs. Examples include: (1) pastes and creams, that is, colloids having the functional, sensorial, rheological, and physical properties mentioned above; (2) specialty fibers having, for example, length and diameter distributions, surface coatings, and tensile strength; (3) polymer semiconductors having P- and N-channel characteristics with high charge-carrier mobility, threshold voltage, and on-off current ratio properties; and (4) thin specialty glass substrates for liquid crystal displays (LCDs) having a *green* composition and durability, thermal deformation, and roughness properties.

For many *industrial chemicals* to achieve these desired properties, it becomes necessary to invent or utilize new so-called *process/manufacturing* technologies. For example, as discussed in Section 14.2, Corning® developed the Isopipe^{TM} process for the fusion of thin glass substrates.

The next step is for the design team to formulate an *innovation map*, as discussed in Section 1.3, and to decide whether sufficient new technologies are in place to satisfy the anticipated customer needs, that is, the *voice of the customer*. When these technologies are in place, the design team initiates the SGPDP; otherwise, the project charter is rejected.

CONCEPT STAGE

As discussed in Section 2.4, the *concept* stage focuses on: (1) making opportunity assessments, (2) identifying customer requirements, (3) identifying technical requirements, (4) determining the critical-to-quality

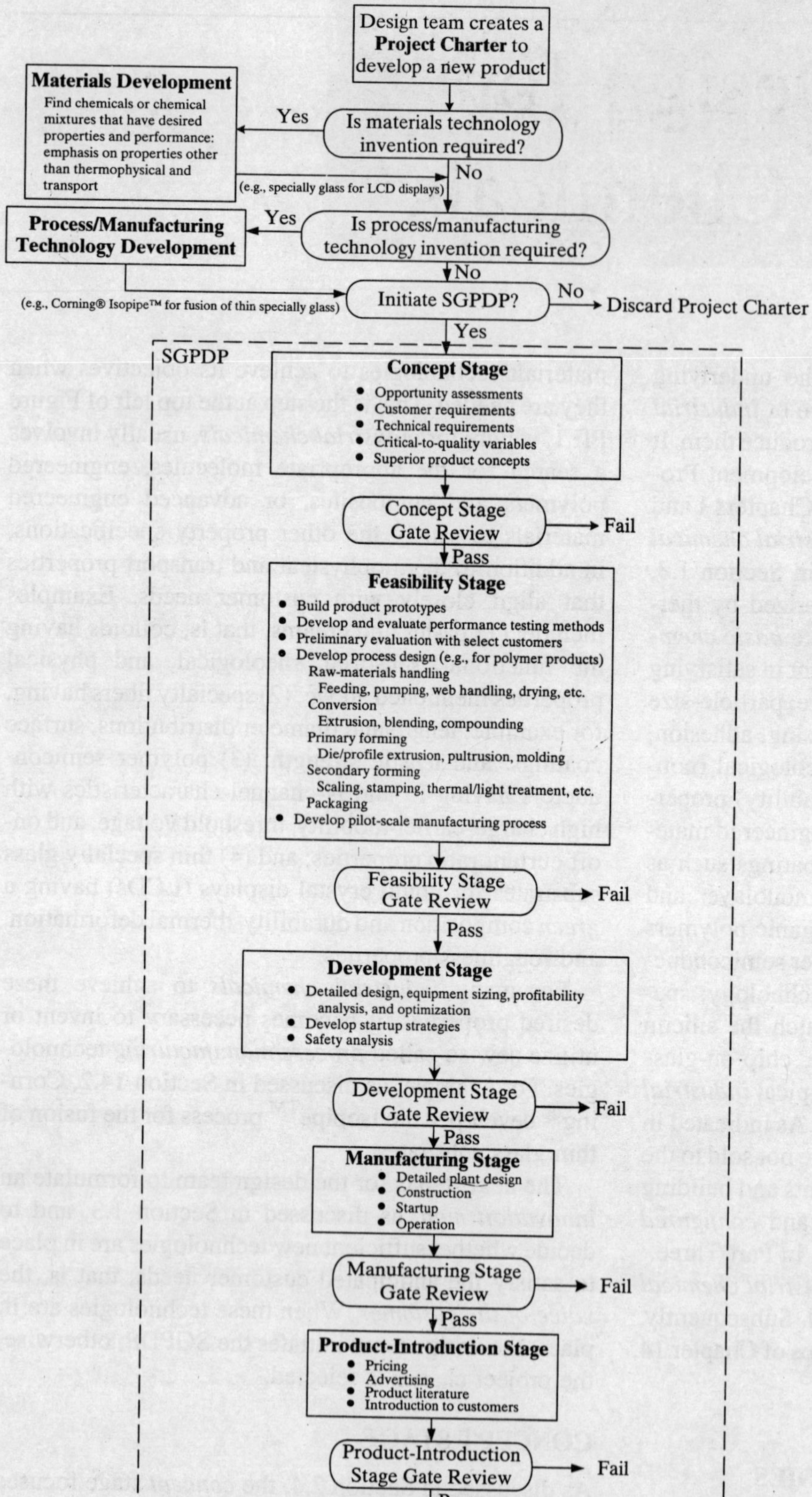

Figure PII.1 Steps in industrial chemical product design.

(CTQ) variables, and (5) determining the superior product concepts. Note that these items are somewhat more difficult to achieve for *industrial chemicals* as compared with *basic chemicals* because the former's customer needs are more difficult to translate into technical requirements, and the generation of superior product concepts is normally more complex. However, the two are similar in that, for *basic chemicals*, preliminary process synthesis leads to a tree of promising flowsheets, while superior product concepts for *industrial chemicals* are selected from those generated earlier in the *concept* stage.

FEASIBILITY STAGE

Normally, product samples (prototypes) are prepared to demonstrate the feasibility of the superior product concepts and performance testing methods are developed and evaluated. Product prototypes are built, preliminary evaluation is carried out to assess their performance, and testing is undertaken with selected customers to obtain their feedback on product performance and compatibility with their manufacturing processes. When promising, a process is designed to manufacture the product. Note that Figure PII.1 illustrates the processing operations utilized in producing many polymer products. For *industrial chemicals*, unlike for *basic chemicals*, these operations depend upon the technology platforms involved. Pastes and creams have colloidal structures that require operations for the generation of microdroplets in continuous phases, micromixing devices, and homogenizers. Epitaxial silicon films for wafer substrates are generated using chemical vapor-deposition reactors, often involving electrodes to create plasmas. Furthermore, processes to produce these products are often not synthesized until the *feasibility* stage—after prototypes have been created in the laboratory. This differs appreciably from the design of most *basic chemical* products, in which preliminary process synthesis is normally carried out in the *concept* stage, primarily because the thermophysical properties of basic chemical products are often achieved using routine processing operations (e.g., continuous-stirred-tank reactors, flash vessels, and distillation columns).

DEVELOPMENT STAGE

Detailed Design, Equipment Sizing, Profitability Analysis, and Optimization

For a new process to produce *industrial chemicals*, the design team usually receives additional assistance in carrying out the detailed process design, equipment sizing and capital-cost estimation, profitability analysis, and optimization of the process. These topics are covered in separate chapters in Part Four, which begins the *development* stage of the SGPDP. Note that the detailed process design for some industrial chemicals involves more specialized processing units such as extruders, micromixers, emulsion guns, glass-fusion processes, clean-room operations, photolithography processes, and nanofabrication processes, for which more specialized techniques to size equipment and estimate capital costs are often required. To illustrate the approaches, Chapter 21, "Polymer Extrusion," discusses the kinds of extruders available, selection considerations, and sizing techniques. Methods for estimating capital and operating costs, and computing profitability measures, are provided in Chapters 22 and 23. Optimization methods are presented in Chapter 24.

When the detailed process design is completed, the economic feasibility of the process is checked to confirm that the company's profitability requirements have been met. If this proves unsatisfactory, the design team determines whether the process is still promising. If so, the team returns to an earlier step to make changes that it hopes will improve the profitability. Otherwise, this process design is rejected.

Develop Startup Strategies

Again, partially due to the more complex properties associated with *industrial chemicals* as compared with *basic chemicals*, simulation methods are less commonly used to develop startup strategies. Often, more experimental approaches are used.

Safety Analysis

Another crucial activity involves a *formal* analysis of the reliability and safety of the proposed process, as discussed in Section 1.5. Note that, as discussed in Section 1.5 and throughout the book, these considerations must be foremost throughout the design process. If not accomplished earlier during process creation, detailed design, and controllability analysis, formal safety analysis usually involves laboratory and pilot-plant testing to confirm that typical faults (valve and pump failures, leaks, etc.) cannot propagate through the plant to create accidents such as explosions, toxic clouds of vapor, or fires. Often, HAZOP (*Haz*ard and *Op*erability) analyses are carried out to check systematically all of the anticipated eventualities. Methods for and examples of HAZOP analysis, together with risk assessment, are presented in Section 1.5. Also, the reader is referred to the text by Crowl and Louvar (1990) and the following books developed by the Center for Chemical Process Safety of the American Institute of Chemical Engineers:

1. *Safety, Health, and Loss Prevention in Chemical Processes: Problems for Undergraduate Engineering Curricula—Student Problems* (1990).
2. *Guidelines for Hazard Evaluation Procedures, Second Edition with Worked Examples* (1992).
3. *Self-Study Course: Risk Assessment* (2002).

The latter reference is particularly noteworthy for instructors because it provides a PowerPoint file that can be integrated into a safety lecture.

MANUFACTURING STAGE

Plant Design, Construction, Startup, and Operation

Detailed plant design, construction, startup, and operation are carried out in the *manufacturing* stage of the SGPDP, as shown in Figure II.1 In creating the plant design for an *industrial chemical* process, much detailed work is done, often by contractors, using many mechanical, civil, and electrical engineers. For processes that produce *industrial chemicals*, engineers complete equipment drawings, piping diagrams, instrumentation diagrams, the equipment layout, the construction of a scale model, and the preparation of bids. Then, the construction phase is entered, in which engineers and project managers play a leading role. The design team often returns to assist in plant startup and operation. Note that the final design and construction activities are usually not the responsibilities of chemical engineers.

PRODUCT-INTRODUCTION STAGE

As the plant comes online, product-launch strategies are normally implemented. These include setting the product price, marketing and advertising to prospective customers, preparing and distributing product literature, and introducing the product to selected customers. These are normally the responsibilities of sales and marketing personnel, many of whom have been trained as chemical engineers.

SUMMARY

This brief introduction to Figure PII.1 should give the reader a good appreciation of the subjects to be learned in the design of industrial chemical products and processes, and how this text is organized to describe the design methodologies.

REFERENCES

1. American Institute of Chemical Engineers, *Safety, Health, and Loss Prevention in Chemical Processes: Problems for Undergraduate Engineering Curricula—Student Problems*, AIChE, New York (1990).

2. American Institute of Chemical Engineers, *Guidelines for Hazard Evaluation Procedures, Second Edition with Worked Examples*, AIChE, New York (1992).

3. American Institute of Chemical Engineers, *Self-Study Course: Risk Assessment*, AIChE, New York (2002).

4. Crowl, D.A., and J.F. Louvar, *Chemical Process Safety: Fundamentals with Applications*, Prentice-Hall, Englewood Cliffs, New Jersey (1990).

Chapter **14**

New Technologies for Industrial Chemical Products

14.0 OBJECTIVES

New product-development programs are often plagued by the need for technological inventions that prolong the product-development process. In these cases, both product and technology developments are often carried out concurrently to reduce the product-development time. Yet, it is recommended that these two activities be decoupled as much as possible. In this chapter, the *innovation map* (introduced in Section 1.3) is used to carry out customer-driven technology development; that is, to show how perceived customer needs are coupled with the development of new technologies.

Herein, *innovation maps* are developed for the industrial chemical products discussed in the two case studies of Chapter 15. The utility of these *innovation maps* is shown for converting materials and process/manufacturing technologies into inventions necessary and sufficient to meet perceived customer requirements. The case studies involve thin-glass substrates for liquid-crystal displays (LCDs) and washable mixtures for crayons.

After studying this chapter, the reader should:

1. Be able to construct an *innovation map* for an industrial chemical product.
2. Be able to identify critical inventions and innovations for materials and process/manufacturing technologies for industrial chemical products.
3. Appreciate the need for technology-protection strategies.

14.1 INTRODUCTION

When developing *industrial chemical* products, chemical engineers in industry often focus on properties beyond the normal thermophysical and transport properties of their pure species and mixtures. Recall that *industrial chemical* products are normally sold to other industrial firms or used downstream in the production of other *industrial chemicals* or *configured consumer* products. The latter are normally sold to consumers.

Consider, next, some common examples of *industrial chemicals* and properties attractive to their users:

1. *Pastes and creams.* These are colloids that have microstructures and particle-size distributions that are characterized by functional (cleansing, adhesion, . . .), sensorial (feel, smell, . . .), rheological (non-Newtonian flow), and physical (stability, . . .) properties. Often these are sold to industrial firms to be blended with other ingredients and packaged for consumer usage.
2. *Woven and nonwoven fibers.* Natural fibers are often produced from vegetables, wood sources, animals (e.g., hair, sinew, . . .), and minerals (asbestos), while man-made fibers are manufactured from natural raw materials or synthetic chemicals (such as polymer fibers—e.g., polyamide nylon, PET polyester, polyvinyl chloride, . . .). These are characterized by diameter distribution, tensile strength, water absorption, etc. Woven fibers are either woven or knit, while nonwoven fibers, for example, felt, are neither woven nor knit. Fibers are industrial chemicals that are produced to be incorporated into numerous consumer products; for example, nonwovens are often used for diapers, surgical gowns, microfilters, tea bags, insulation, . . .
3. *Thin-glass substrates for LCDs.* As introduced in Section 1.3 and expanded upon in Section 14.2, the front and back panels in active-matrix, liquid-crystal displays (AM-LCDs) are thin glass sheets. These are sold to industrial firms that manufacture AM-LCDs

for many applications including laptops, LCD TVs, cell phones, etc. Especially for large-diagonal displays, the array of desirable properties include durability (when exposed to etching chemicals), low deformation (especially when heated), flawless and smooth surfaces, etc.

4. *Marine anti-fouling agents for use in paints.* Biocides are industrial chemicals sold to industrial firms that produce paints for protection of construction steel for ships and prevention of undue hull roughness. These products must be selected to protect against animals and algae for 3 to 5 years, while being nontoxic to humans and remaining active in normal paint formulations. The paints involve selected pigments, solvents, and additives (e.g., anti-foaming agents, corrosion inhibitors, thixotropic agents, . . .)
5. *Monolayer films for food wraps.* Plastic films for food wraps, commercially known as Saran™ or Glad™ wrap, as well as cling film are typically used to seal foods stored in glass or ceramic containers to keep them fresh; that is, to prevent mass transfer with surrounding air. Historically, plastic food wraps have been extruded polyvinylchloride (PVC), but recently there has been a shift to low-density polyethylene (LDPE) due to concerns about the transfer of PVC plasticizers into foods. Cling films are relatively thin (8–15 μm) and cling to various glass, ceramic, stainless steel, etc., containers, while having limited attraction to themselves. Other important properties include a resistance to oxygen and flavor transport, as well as high tensile strength and puncture resistance. For further discussion, see "Plastic Wrap" (Wikipedia, 2007).
6. *Thin layers of epitaxial silicon for integrated circuit chips.* Layers of epitaxial silicon are industrial chemical products used by the manufacturers of integrated circuits to improve the performance of silicon wafers. These layers must be uniform in thickness and roughness, with well-distributed defects and dislocations. Because these layers are deposited by the manufacturers of integrated circuits, the industrial chemical product in this case is the reactor for chemical vapor deposition, to be licensed to manufacturing firms.
7. *Polycarbonate materials for optical applications.* Optical-grade polycarbonate provides balanced properties of transparency, impact strength, heat resistance, and dimensional stability, along with electrical properties. It is widely used for highly demanding applications such as the manufacture of compact discs (CDs), magneto-optical (MO) discs, and DVD optical discs.
8. *Microspheres for controlled release of pesticides.* Microspheres, having diameters on the order of 50–100 μm and nano-liter volumes, are industrial chemicals sold to industrial firms for impregnation with pesticides and other chemicals (e.g., pharmaceuticals) for the controlled release of these chemicals. The latter are products sold to the consumer; that is, configured consumer products. For microspheres, properties like the diameter distribution, molecular-weight distribution, and solute permeability are particularly important. Biodegradable microspheres are most attractive.
9. *Industrial catalysts.* These industrial chemical products, widely used in the chemical industry, include families of zeolite crystals, with channels designed to block the passage of large reactants or products, and ion-exchange resins, such as the Amberlyst™ family. In these cases, the pore-size distribution and diffusivities of the reacting species are important properties.

These and many other *industrial chemical* products are discussed in several articles and books that focus on product design (e.g., Westerberg and Subrahmanian, 2000; Cussler and Moggridge, 2001; Shaeiwitz and Turton, 2001; Cussler et al., 2002; Favre et al., 2002, 2005; Cussler and Wei, 2003; Hill, 2004; Saraiva and Costa, 2004; Seider et al., 2004; Costa et al., 2006; Bröckel et al., 2007a, b; Ng et al., 2007; Wei, 2007). *Industrial chemical* products often involve complex phenomena such as multiphase interactions, amorphous and crystalline structures with dislocations and defects, surface roughness, and stress-strain relationships. Often new materials technologies underlie new *industrial chemical* products; for example, the discovery of chemical mixtures that phase-split to form stable micro-emulsions. And, in many cases, new process/manufacturing technologies underlie the development of new products; for example, an improved extruder that produces grooved fibers.

Note that many other chemical products have properties similar to the nine examples above, but are sold directly to the consumer. These include *configured consumer chemical* products like soap bars, ice cream, cheese substitutes using vegetable oils, and peanut butter. These are discussed in Part Three of this book.

This chapter focuses on the initial steps in the design of *industrial chemical* products, after a design team has created its project charter. These steps, shown in Figure PII.1, involve the selection of materials and process/manufacturing technologies. These are normally new technologies intended to create products that satisfy customer needs while offering a competitive advantage. As discussed in Sections 1.3 and 2.2, to achieve these objectives, it is helpful to create *innovation maps* that show the connections between the two technological components and customer satisfaction, that is, the *customer-value proposition.* The success of new products often relies on careful attention to this interplay. As shown in Figure PII.1, when the *innovation maps* are promising, the design team begins product design following the Stage-Gate™ Product-Development Process (SGPDP).

Chapter 15 presents case studies of *industrial chemical* product design following the SGPDP, beginning with the *innovation maps* discussed in this chapter.

In the sections that follow, *innovation maps* are created for potential product designs involving thin-glass substrates for LCDs and washable mixtures for crayons. In each section, a brief history of the technological developments is presented before the innovation map is created. Note that these innovation maps are prepared in hindsight, while in practice, design teams revise innovation maps multiple times to reflect the progress of their designs.

This chapter also introduces product/process synthesis and simulation for the manufacture of industrial chemical products. Because of the vast number of technology platforms involving a host of industrial chemical products, comprehensive coverage is not attempted. Instead, materials and process/manufacturing technologies are selected that are featured in the case studies of Chapter 15.

14.2 INNOVATION MAP FOR THIN-GLASS SUBSTRATES IN LCDS

Active-matrix, liquid-crystal displays (AM-LCDs) comprise the dominant display technology for desktop, laptop, and notebook computers; and handheld devices such as cellular phones, Personal Digital Assistants (PDAs), calculators, and digital watches. They are increasingly important for large flat-panel televisions, currently challenging the plasma display panel (PDP) technology developed in the mid-2000s. At the heart of an AM-LCD is a sheet of liquid crystals arranged in an active matrix, bounded by a scanning electrode on one side and a color filter on the other. This sandwich sits within thin-glass sheets, the back and front panels, which are surrounded by polarizer layers. On the back panel, thin-film transistor (TFT) circuits are built. On the front panel, a color filter is superimposed, as illustrated in Figure 14.1a.

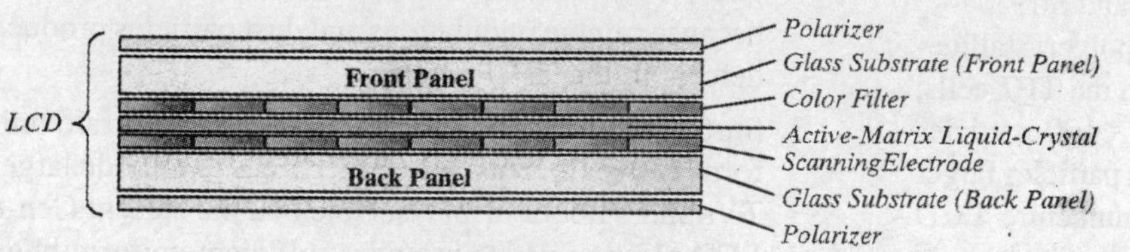

a. Side view of LCD panel.

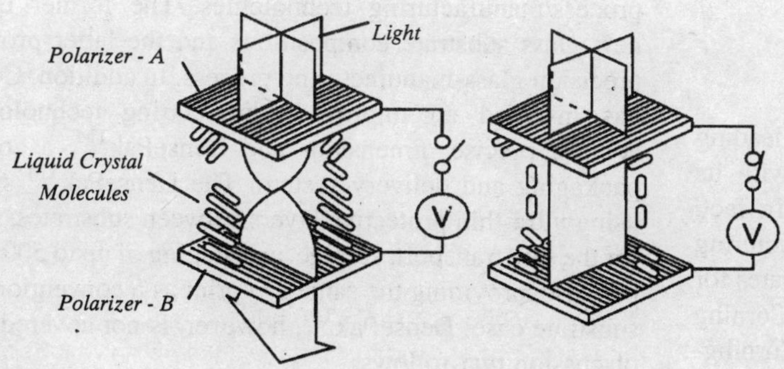

b. Polarization of backlights in a LCD.

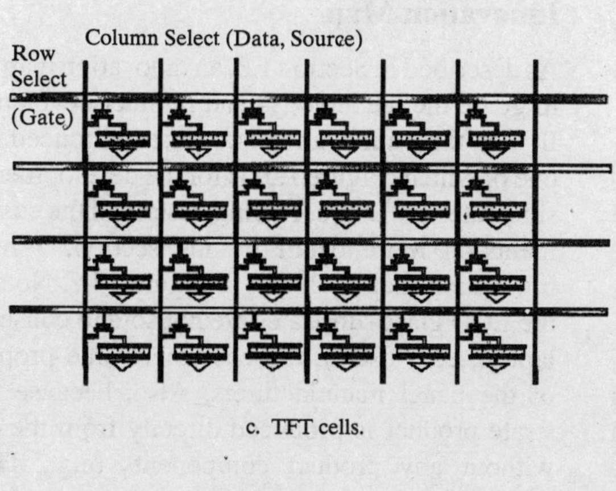

c. TFT cells.

Figure 14.1 Components of liquid-crystal displays (LCDs).

The liquid crystals (LCs), for example, 4-methoxy benzylidene-4′-butylaniline (MBBA), can be reoriented by applying an electric field. As shown in Figure 14.1b, because these materials are optically active, their naturally twisted structure can be used to turn the polarization of light by, for example, 90°. The two polarizers, A and B, transmit light in orthogonal planes. Light that exits polarizer A is naturally twisted 90°, and consequently, it can pass through polarizer B, producing a bright pixel associated with a specific cell. However, when an electric field is applied, the helical structure of light moving between A and B is unwound. As a result, light doesn't transfer through polarizer B, resulting in a dark pixel.

The AM-LCD permits each LC cell to be addressed, with each cell corresponding to one monochrome pixel. In its simplest form, the AM-LCD contains one thin-film transistor for each cell, as shown in Figure 14.1c. A row of pixels is selected by applying a voltage to the selected line connecting the thin-film transistor (TFT) gates for that row of pixels. When a row of pixels is selected, the voltage is adjusted according to the data line. The TFT active matrix can be considered an array of ideal switches that turn a row of pixels on and off.

Commonly, either amorphous-Si (a-Si) or polycrystalline-Si (p-Si) is used for the TFTs. To manufacture the TFT cells, a clean room is required [within Class 100 ($\leq$100 particles larger than 0.5 μm/ft^3 air) to 10,000 ($\leq$10,000 particles larger than 0.5 μm/ft^3 air)]. The processes to manufacture LCD panels often include: (1) plasma-enhanced chemical vapor deposition, (2) sputtering, (3) photolithography, (4) wet processing and cleaning, (5) dry etching, and (6) TFT cell fabrication and assembly.

Thin-Glass Substrates

In the mid-2000s, Corning Incorporated was the leading manufacturer of thin-glass substrates for LCDs, with its development efforts having begun in the 1980s. In fact, nearly all of the new materials and process/manufacturing technologies in the development of thin-glass substrates for LCDs since then are attributed to Corning. In 1987, Corning introduced the first LCD glass substrate, known as Corning-7059, based upon barium boro-silicate chemistry. Since then, a series of improved products has evolved, including Corning-1737, Eagle2000™, and EagleXG™.

Customer specifications for LCD glass substrates are very stringent due to the precise alignments required for the back and front panels in the LCD sandwich. Minute misalignments, on the order of microns, would significantly reduce the image quality and viewing angle. For small displays, especially notebook computers, weight is a key market driver and, consequently, thinner, less dense panels are preferable.

In summary, for these panel manufacturers, the following features and functionalities are desired:

1. *Dimensional stability.* Minute alignment errors in TFT patterning or mismatch between the TFT pattern and the color filter destroys the fidelity of the image.
2. *Surface quality.* Small surface imperfections cause pattern defects during the lithography process that imprints the TFT pattern on the glass substrate.
3. *Surface flatness.* Variable gaps between the back and front panels affect the performance of the liquid crystals, which are sandwiched between the two panels.

To achieve these features and functionalities, the following requirements are imposed upon the thin-glass substrate product:

1. Dimensional stability requires that the glass substrates have minimal thermal shrinkage, low internal stress, and thermal expansion comparable to that of the amorphous silicon used for the TFT pattern. This is because temperatures as high as 400°C are achieved when depositing amorphous silicon on the glass substrate.
2. High surface quality requires high chemical durability to resist the acids used for etching during the photolithography process when building the TFT patterns. Also, very smooth and clean surfaces are required because minute undulations and dust particles produce defects in the TFT patterns.
3. Surface flatness requires extreme glass substrate uniformity over increasingly larger areas to provide larger TVs and outdoor displays. Note that the current Gen-8 LCD glass substrate provides sufficient uniformity up to approximately 7 × 8 ft.

Corning inventions have provided new materials and process/manufacturing technologies. The former involve new glass substrate compositions and the latter provide a precision glass-manufacturing process. In addition, Corning has invented an improved glass-cutting technology to achieve precise dimensions and DensePak™, a compact packaging and delivery system. The DensePak™ system, using ultra-thin protective layers between substrates, allows for the safe transport, storage, and staging of up to 500 sheets per case, providing the same footprint as a conventional 20-substrate case. DensePak™, however, is not covered in the discussion that follows.

Innovation Map

As described in Section 1.3, an innovation map is a useful tool to guide the technology- and product-development process. In fact, three innovation maps are introduced in Section 1.3, one of which (Figure 1.5) is for the development of thin-glass substrates for LCDs. Having described the customer requirements in more detail in this section, a more complete innovation map is shown in Figure 14.2. Note that, because the LCD glass substrates are not sold to consumers (i.e., TV, laptop, etc., users), the customer-value proposition is that of the panel manufacturers. Also, because the glass substrate product is produced directly from the raw materials, without any product components (e.g., TFTs), no new

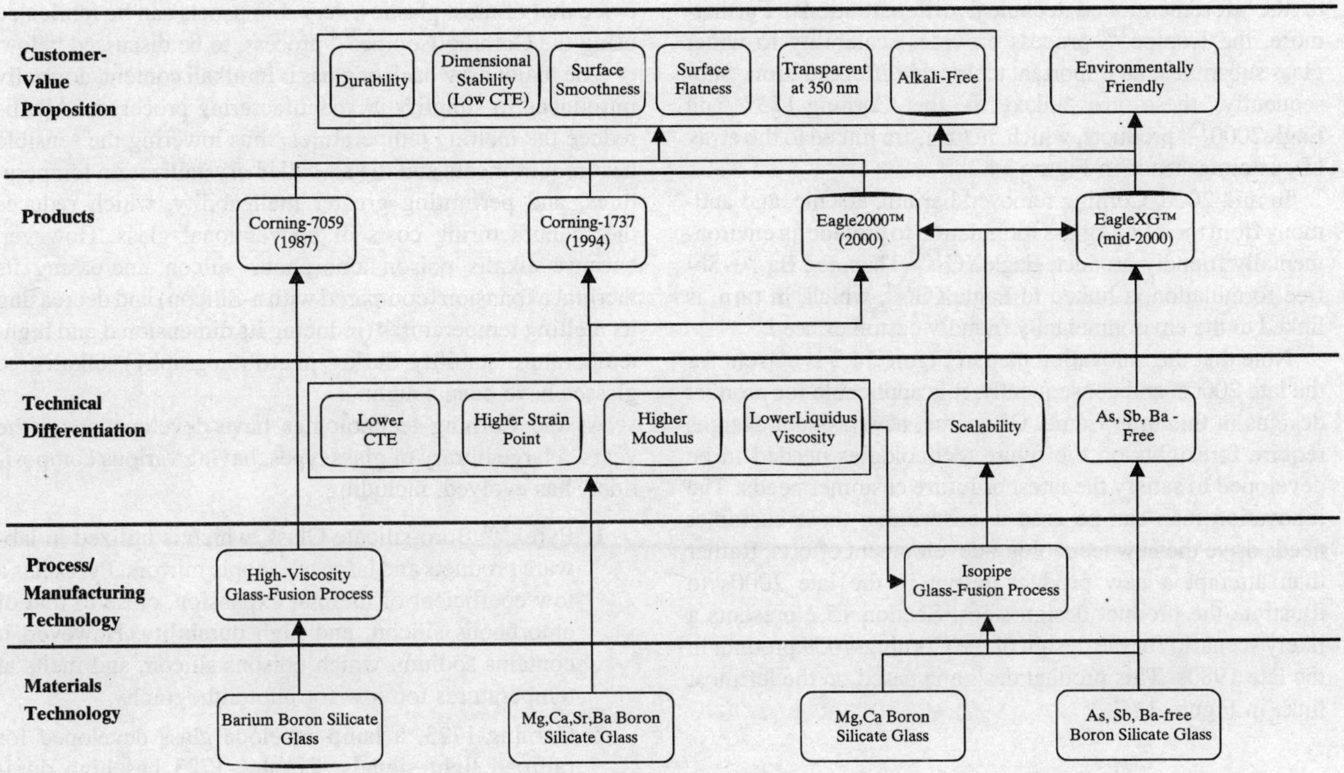

Figure 14.2 Innovation map for Corning LCD glass substrates.

product technologies are required and, consequently, the product technology layer is not included.

To construct the innovation map in hindsight, one must first identify the elements in its five levels, moving from the bottom to the top of the map:

1. *Materials Technology:* Barium boron silicate glass; Mg, Ca, Sr, Ba Boron silicate glass; Mg, Ca Boron silicate glass; As, Sb, Ba-free Boron silicate glass.
2. *Process/Manufacturing Technology:* High-viscosity, glass-fusion process; Isopipe™ glass-fusion process.
3. *Technical Differentiation (Technical-Value Proposition):* Lower CTE manufacturing; higher strain point; higher modulus; scalability; As, Sb, Ba-free.
4. *Products:* Corning-7059; Corning-1737; Eagle2000™; EagleXG™.
5. *Customer-Value Proposition™:* durability; dimensional stability (low CTE); surface smoothness; surface flatness; transparent at 350 nm; alkali-free; environmentally friendly.

Note that these elements are discussed below as well as by Bocko and Mitchell (2003), Bocko et al. (2006), Brody (1997), and Lapp (2004a, 2004b).

After identifying the elements at all five levels of the innovation map, their connectivity is added to show the interplay between the technological elements, the *technical-value proposition*, and ultimately the *customer-value proposition*. These linkages identify the primary inventions needed to satisfy customer needs, possibly providing a competitive advantage such as a price premium over competitive products and/or increased market share. Where customer needs remain unmet, their resolution is often an objective for the next generation of products.

There are two primary inventions that have provided Corning a competitive edge over their competitors: (1) novel glass compositions and the Isopipe™ glass-fusion process. These two new technologies are discussed in the sections that follow on materials technology development and process/manufacturing technology. Next, the linkages in the innovation map are described.

Initially, in the 1980s, the novel barium-boron-silicate glass formulation, together with the high-viscosity fusion process, enabled the product development of the Corning-7059 LCD glass substrate that satisfied the customer requirements at that time. To illustrate this interplay, in Figure 14.2 the barium-boron-silicate glass (new materials technology) and the high-viscosity fusion process (new process/manufacturing technology) are linked to the Corning-7059 product, which, in turn, is linked to the customer-value proposition, including durability, dimensional stability, and surface quality (smoothness and flatness).

The subsequent evolutionary changes in glass formulations—that is, the Mg-Ca-Sr-Ba boron silicate and Mg-Ca boron silicate glasses, together with the Isopipe™ glass-fusion process—improved the coefficient of thermal expansion, the strain point, and the modulus. Consequently, these glass formulations and the Isopipe™ process are linked to the aforementioned technical differentiations. Furthermore, the Isopipe^{TM} process provides scalability to wider glass substrates, an important technical differentiation. Subsequently, these are linked to the Corning-1737 and Eagle2000^{TM} products, which, in turn, are linked to the array of customer needs in Figure 14.2.

In mid-2000, Corning removed barium, arsenic, and antimony from their LCD glass formulation to provide an environmentally friendly product, EagleXG^{TM}. Thus, the Ba-Ar-Sb-free formulation is linked to EagleXG^{TM}, which, in turn, is linked to the environmentally friendly customer need.

Note that the innovation map in Figure 14.2 is current for the late 2000s, and consequently, it is applicable for product designs in this time frame. Of course, new product designs require foresights on the future technologies needed to be developed to satisfy the latest or future customer needs. The innovation map can be used to ensure that these customer needs drive the new technology-development efforts. Rather than attempt a new product design in the late 2000s to illustrate the product design steps, Section 15.2 presents a likely scenario for the design of the Corning-7059 product in the late 1980s. This product design is based on the leftmost links in Figure 14.2.

Materials Technology Development

As mentioned earlier, the active search for suitable substrates for LCDs began in the 1980s. Both glass and plastic substrates were leading candidates, with glass selected for the reasons discussed below. Both the back and front panels in LCDs must meet high dimensional stability and surface-quality (smoothness and flatness) requirements, but the back panel must permit the installation and proper functioning of the TFTs; that is, it must:

1. withstand processing temperatures up to 400°C for up to one hour without any deformation,
2. be transparent at a wavelength of 350 nm for photolithography to build arrays of TFTs,
3. survive the acidic environment during etching in photolithography without deterioration,
4. expand and contract with temperature like the amorphous silicon used to construct the TFTs,
5. not contain elements that contaminate silicon and reduce its electrical performance, especially the alkali cations in conventional glass, and
6. be light.

Compared with plastics, glass deforms at much higher temperatures, is practically inert, and is transparent in the near-UV region. Glass has much lower thermal expansion, but its thermal expansion differs more substantially from that of amorphous silicon. Glass is also impermeable to gas, which permits it to contain the liquid crystals within the LCD sandwich. Although the density of glass is approximately twice that of most plastics, very thin panels can be fabricated using the Corning Isopipe^{TM} process, to be discussed below.

The major drawback of glass is its alkali content, normally introduced to simplify its manufacturing processes. Alkalis reduce the melting temperatures, thus lowering the sensible heat requirements and the so-called crystallization temperatures, and permitting greater malleability, which reduces the manufacturing costs of conventional glass. However, because alkalis poison amorphous silicon, increasing its thermal expansion (compared with a-Silicon) and decreasing its melting temperatures (reducing its dimensional and high-temperature stability during photolithography), alkali-free glasses have been sought.

As the Corning technologies have developed over the years, a large library of glass types, having various compositions, has evolved, including:

1. **Pyrex^{TM} Boro-silicate Glass**, which is utilized in labware products and large telescopic mirrors. Pyrex has a low coefficient of thermal expansion, close to that of amorphous silicon, and high durability. However, it contains sodium, which poisons silicon, and melts at temperatures too low for photolithography.
2. **Corning-1723**, a lamp-envelope glass developed for railroad light signals. Corning-1723 has high durability and stability at high temperatures, and has a low alkali concentration (<2,000 ppm). Consequently, it has a high melting point, which decreases its malleability and increases its thermal expansion (compared to silicon), increasing its manufacturing costs.
3. **Corning-7980**, a fused silica glass that is widely used in UV optical materials and often used as the envelope in halogen light bulbs. Corning-7980 is very stable at extremely high temperatures (>1,000°C) and, consequently, is highly durable. It has an extremely low alkali content (<20 ppm). Compared with Corning-1723, it is very expensive to manufacture, very difficult to manipulate (to achieve high surface qualities), and has thermal expansion coefficients *lower* than desirable.

Because none of these glasses satisfied all of the properties required for LCD substrates, Corning sought to develop a glass having:

1. A coefficient of expansion comparable to that of Pyrex^{TM} glass.
2. A deformation temperature (strain point) comparable to that of lamp-envelope glass (Corning-1723).
3. Low alkali content (<2,000 ppm), comparable to lamp-envelope glass.
4. Low iron content to achieve UV transparency, which is not possible with their existing glass types.

Beginning with Pyrex^{TM} glass, an alkaline-earth aluminosilicate glass ($RO\text{-}Al_2O_3\text{-}SiO_2$, with R, an alkaline-earth element such as Mg, Ca, Sr, or Ba), researchers added boron oxide, B_2O_3, and barium oxide, BaO. They adjusted the composition to meet most of the LCD substrate requirements, but the coefficient of thermal expansion was slightly higher than desired and somewhat improved thermal stability was needed. Given that Pyrex™ glass was invented in 1948, with most patent coverage having expired, this new glass composition had a marginal IP position. Nevertheless, in 1987, Corning introduced this product (Corning-7059) into the market as their first LCD substrate.

Since then Corning has continually adjusted its compositions to achieve improved performance, with the following resulting products:

1. ***Corning-1737.*** In this glass, the BaO in Corning-7059 is replaced with a mixture of four alkaline-earth compounds, MgO, CaO, SrO, and BaO, and the B_2O_3 content is reduced by one-half. This gives an overall composition closer to the eutectic point of the CaO-Al_2O_3-SiO_2 system, a strain point increased from 600 to 666°C, a CTE reduced from 4.6 to 3.8 ppm/°C, and a density reduced from 2.74 to 2.55 g/cc. Its modulus is higher, with less sag (permitting the production of larger sheets), and it has higher durability. For processing, this composition provides a very high liquidus viscosity, permitting more precise sheet forming.
2. ***Eagle2000™.*** While most customers were satisfied with Corning-1737 glass, some sought a lower density and CTE. This was achieved by removing the BaO and replacing most of the SrO with CaO. In addition, an increased strain point was obtained. Furthermore, by increasing the B_2O_3 content by approximately 20%, the liquidus temperature was minimized, giving a slightly reduced melting point.
3. ***EagleXG™.*** To achieve a more environmentally friendly glass, traces of As and Sb were removed, while maintaining the key characteristics of Eagle-2000™ glass.

In the future, as the TV and outdoor display markets move toward even larger displays, glass compositions providing a higher modulus, to reduce sagging, are anticipated. Also, with the shift toward polycrystalline silicon, involving photolithography at higher temperatures, compositions that provide higher strain points are likely to be developed.

Process/Manufacturing Technology: Corning Glass-Fusion Process

To meet the requirements of high surface quality and thickness uniformity, scientists and engineers at Corning developed the Isopipe™ glass-fusion process, shown schematically in Figure 14.3. In this simple, elegant design,

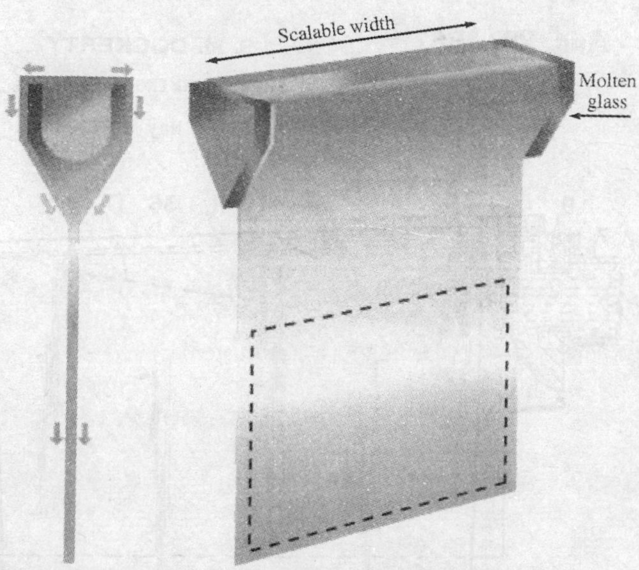

Figure 14.3 Schematic of the Corning Isopipe™ glass-fusion process.

molten glass is side-fed into a trough with a pointed base (see the cross section on the left) and overflows into two separate curtains that join at the vertex of the pointed base, forming a single glass curtain. As it falls by gravity, the glass solidifies and is cut to size.

This process has the following requirements, features, functionalities, and benefits:

1. The glass composition is selected to permit flow without bubble formation and the creation of other defects.
2. As the thin-glass substrates are formed, they are in contact with clean air only. Consequently, high surface smoothness is achieved without downstream polishing, which otherwise imparts surface textures.
3. The Isopipe™ glass-fusion process provides excellent thickness control and is capable of delivering LCD glass substrates with thickness uniformity over the large surface areas required by LCD manufacturers.
4. The Isopipe™ glass-fusion process is scalable for thickness and width. Corning has produced substrates as thin as 0.6 mm and over a width of 2 m for use in Gen-8 LCD manufacturing.

Isopipe™ Process Design

The original design of the Corning Isopipe™ glass-fusion process was based on an apparatus invented by Dockerty in the 1960s and patented by Corning (U.S. Patent 3,338,696). It was intended to produce sheet glass having a uniform thickness across its width and having virgin surfaces; that is, smooth surfaces without scratches and defects. Dockerty based his design on a draw-down glass sheet-making process

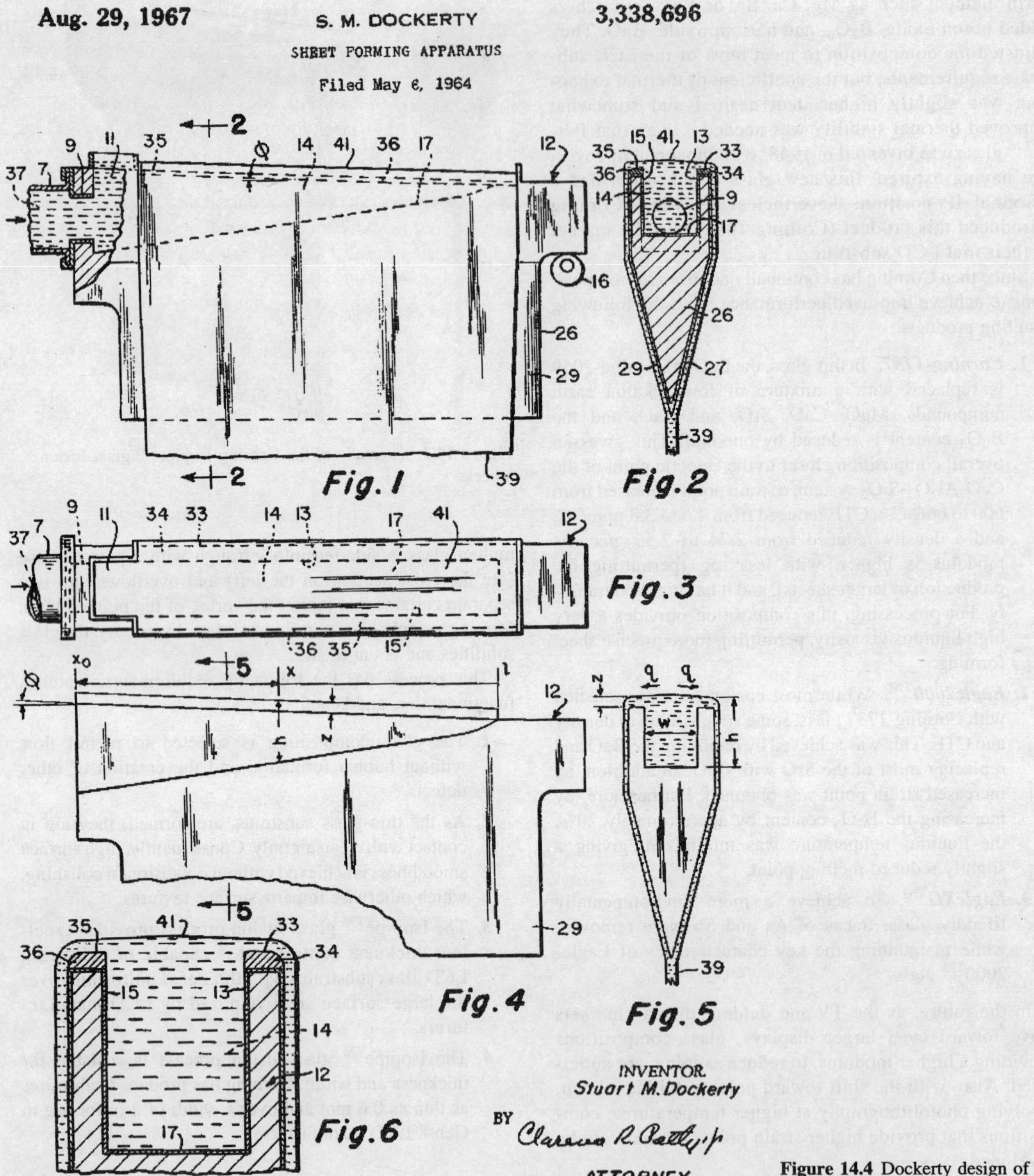

Figure 14.4 Dockerty design of the glass-fusion process.

that was prone to flow fluctuations induced by its molten-glass feeding method. To overcome these fluctuations, Dockerty slightly tilted the feeding trough and computed the shape of a contoured bottom that would provide a uniform flow over its outer walls, which act like weirs.

Figure 14.4 is the cover page of the Dockerty patent, which illustrates many of the features, somewhat exaggerated. In Fig. 1, a side view of the sloped trough is shown, with viscous liquid glass entering through the channel at the left. Note that the bottom of the trough, represented by the dashed line, is sloped upward to reduce the axial flow rate and produce a nearly constant flow rate over the weirs as x varies axially. Item 16 is a roller used to adjust the angle, ϕ, of the trough with the horizontal axis. It is used for flow control of the viscous glass, as discussed below. Fig. 2 is a view taken on the 2-2 line of Fig. 1. Note that it shows a flat trough bottom, rather than the complex contoured bottom designed by Dockerty. The latter is crucial in creating a steady flow devoid of intermittent pulses and surges. Fig. 3 is the top view, and Figs. 4 and 5 annotate the dimensions in the Dockerty model, where l is the trough length, z is the overflow elevation over the weirs, w is the width of the trough, and h is the height of the trough.

Although not derived in the patent, the equations describing the contoured bottom of the trough are provided. For such a design, Dockerty indicates that a uniform flow rate is achieved, given by:

$$\frac{q\,\mu}{\tan\phi} = K \qquad (14.1)$$

where q is the mass flow rate (independent of axial distance, x), μ is the viscosity of glass (~200,000 poise), and K is a constant that depends on h, w, and the axial pressure gradient. Clearly, small adjustments in ϕ can control the overflow rate. Furthermore, this tilt angle can be adjusted to produce glass sheets with a continuous variation of thicknesses, for example, wedge-shaped cross sections.

For more specifics, the reader should consult the Dockerty patent, as well as several patents that describe improvements since then.

14.3 INNOVATION MAP FOR CRAYON MIXTURES

In this section, the new technologies associated with the development of crayon mixtures are discussed in preparation for the case study in Section 15.3 on a new industrial chemical product. Note that the popular Crayola® product, crayons, is a configured consumer product manufactured downstream, beginning with a *crayon mixture*, an industrial chemical product. In the forthcoming case study, the objective will be to create a new crayon mixture that, when processed into crayon-like products, produces water-erasable markings; that is, markings removable without the use of household cleaners. Before considering the new technologies, it is helpful to briefly review the history of crayons in America by Crayola LLC, formerly known as Binney & Smith, Inc. This brief history is taken from the Crayola® Web site and the Wikipedia encyclopedia (Blachford, 2002; Woods, 1999).

History of Crayons

Europe is considered to be the birthplace of the modern crayon, initially a cylindrical stick manufactured from a mixture of charcoal and oil. Over time, colors were introduced as charcoal was replaced with powdered pigments. And, subsequently, it was discovered that wax could be substituted for oil to create sturdier, easier to handle drawing sticks.

In 1864, Joseph W. Binney founded the Peekskill Chemical Company in Peekskill, New York, which produced two pigments: carbon black for increasing the lifetime of car tires and red iron oxide for barn paint. His son, Edwin Binney, and his nephew, C. Harold Smith, formed the Binney & Smith partnership in 1885. They acquired a stone mill in Easton, Pennsylvania, associated with nearby slate quarries, where they created the first slate pencils, introducing them to the educational market.

Building on their core competency in color pigments, Binney and Smith, following the tradition of Joseph Binney, instructed their sales force to listen to their customers' needs. One need involved chalk, which "crumbled easily and was too dusty." In response, Binney and Smith chemists began work on a dustless chalk product formulated in an extrusion process, which was introduced in 1902. As this product met great success in the nation's classrooms, the sales force learned, principally from teachers, that the market was ready for inexpensive, nontoxic, brightly colored crayons. Note that colored crayons were available at the time, but from European manufacturers, with high importation costs.

Subsequently, in 1903, the company chemists created their legendary product, named *Crayola*® by Mrs. Binney, a term derived from the French *craie* (meaning chalk) and the English adjective *oleaginous* (meaning oily).

Over a century later, Crayola® is a household brand name familiar to children around the world. Demand for crayons has continued to grow, creating a half-billion-dollar-per-year business that employs 1,200 men and women who exploit petroleum chemicals to extend the colorful world of children's imaginations. Paraffin wax from distant refineries is delivered in heated railroad tank cars to manufacturing plants in Easton, Pennsylvania. Crayola® has produced over 100 billion crayons since 1903, with current production at about 3 billion crayons annually.

Next, consider the creation of an innovation map that traces the development of crayon mixtures for various Crayola® products created during the past century to satisfy the needs of children and teachers in the educational market.

Innovation Map

As described in Section 1.3, an innovation map is useful in guiding the technology- and product-development process. It shows how new technologies are linked to new products that satisfy customer needs. As such, it is helpful in tracing the development of new technologies as they are incorporated into new products that satisfy customer needs. More importantly, as new products are being developed, it shows how the latest new technologies can potentially be linked to satisfy the latest customer needs.

In this section, the development of an innovation map for crayon mixtures is traced, but not for the final product, crayons. Although crayon mixtures and crayons are produced by the same company, Part Two (Chapters 14 and 15) focuses on industrial chemical products; that is, crayon mixtures. Consequently, discussion is limited to crayon mixtures, the intermediate in producing crayons, and excludes the forming, labeling, and packaging steps. Herein, the new crayon mixtures are viewed as new products that are sold to customers within Crayola,® that is, the crayon product manufacturers.

For the crayon mixtures, the company material scientists, chemists, and chemical engineers developed various material technologies—that is, *manufacturable* formulations—that produce bright colors, a larger selection of colors, easy-to-clean crayons, scented crayons, and color-changing crayons, for example, crayons that exhibit phosphorescence, fluorescence, thermochromic effects, etc.

The basic formulation of crayon mixtures consists of a binder, typically molten wax, and a suitable pigment (U.S. Patents 320,009 and 5,460,647). In addition, various functional additives achieve the aforementioned features, as well as easy and cost-effective manufacture.

Among the most important requirements, the dyes and additives must be nontoxic, especially to avoid the adverse effects of involuntary ingestion by children. In addition, crayon mixtures must meet other specifications including heat stability, absence of odors, moderate viscosities, mold release properties, mechanical strength, color uniformity, color blending, color overlay, UV light stability, and marking (ease of writing, that is, proper frictional-resistance) properties.

Crayon mixtures should also yield crayons with good appearance and mechanical properties. More particularly, they should yield crayons with sufficient mechanical strength to withstand rubbing on surfaces without crumbling or fracturing. Also, the crayon, when applied to a surface, should provide a relatively smooth and uniform layer having uniform color marking. The resulting crayons should not be hygroscopic, that is, they should not absorb water or moisture and thus begin to feel wet and lose mechanical strength.

For such a seemingly simple product, however, satisfying all of the above requirements is not an easy task. Rather, new crayon formulations, especially those with special properties such as being washable, glittering, and changing colors, are often plagued by formulation and manufacturing problems. These are mainly due to new ingredients never before combined with the basic crayon formulation.

To construct the innovation map in hindsight, after a century of product development, one must first identify the elements in its five levels, moving from the bottom to the top of the map. Note that the elements are listed first and discussed below:

1. *Materials Technology:* Wax binder, nontoxic pigments, additives (UV absorbers, mold release, stearic hardener), special effects subtances (reflective materials, perfumes and scents, thermochromic dyes), washable hardener (alkoxylated compounds, polyethylene-glycol-free alkoxylated compounds).
2. *Process/Manufacturing Technology:* Batch mixing, gravity molding, continuous mixing (extrusion compounding), and injection molding.
3. *Technical Differentiation (Technical-Value Proposition):* Easy to mix, easy to mold, high mechanical strength, color stability, more uniform mixing, shorter cooling time, less bubble formation, and aqueous soluble mixture.
4. *Products:* Standard crayon mixture, crayon mixture with special effects (changing color and scented), water-washable crayon mixture.
5. *Customer-Value Proposition:* nontoxic, uniform color, usable on multiple surfaces, strong, long shelf life, wide color selection, special effects (glittering, scented, changing color), and washable.

As shown in Figure 14.5, at each level the elements are placed from left to right as time passes. Initially, at the beginning of the 20th century, crayon mixtures were comprised of a paraffin wax binder and nontoxic pigments, the new materials technologies at the time. These were blended with functional additives, such as talc and UV absorbers, in a batch mixer to achieve the desired mechanical strength and color stability. Note that the batch mixers were the process/manufacturing technology to produce the industrial product, crayon mixtures. The gravity molding units are shown in a dashed envelope because they were used downstream to produce the configured consumer product, crayons. Hence, when designing the industrial product, crayon mixtures, the technical differentiations were the ease of mixing and molding (downstream), the color stability, and the high mechanical strength of crayons formed from these mixtures. This industrial product satisfied the customer-value proposition, that is, the needs for nontoxic, uniform color mixtures that could be formed into crayons that were usable on multiple surfaces, were strong (resistant to fracture), and had a long shelf life.

Gradually throughout the 20th century, the basic crayon mixture formulation was modified to include many new colors, expanding from 8 to over 100 colors, while using the same manufacturing process. This, of course, satisfied customer needs for a wide selection of colors.

As the standard product matured, the crayon formulators added several new features, including reflective materials, perfumes and scents, and thermochromic dyes. These new formulations were processed using the existing manufacturing facility to produce crayon mixtures with special effects, including a glittering appearance, pleasing scents, and color changing (as temperature varies).

The most recent materials technology, in the 1990s, involved a significant change in the basic formulation of the crayon mixtures. Until that time, crayon markings on surfaces and fabrics could not be removed with water or soap

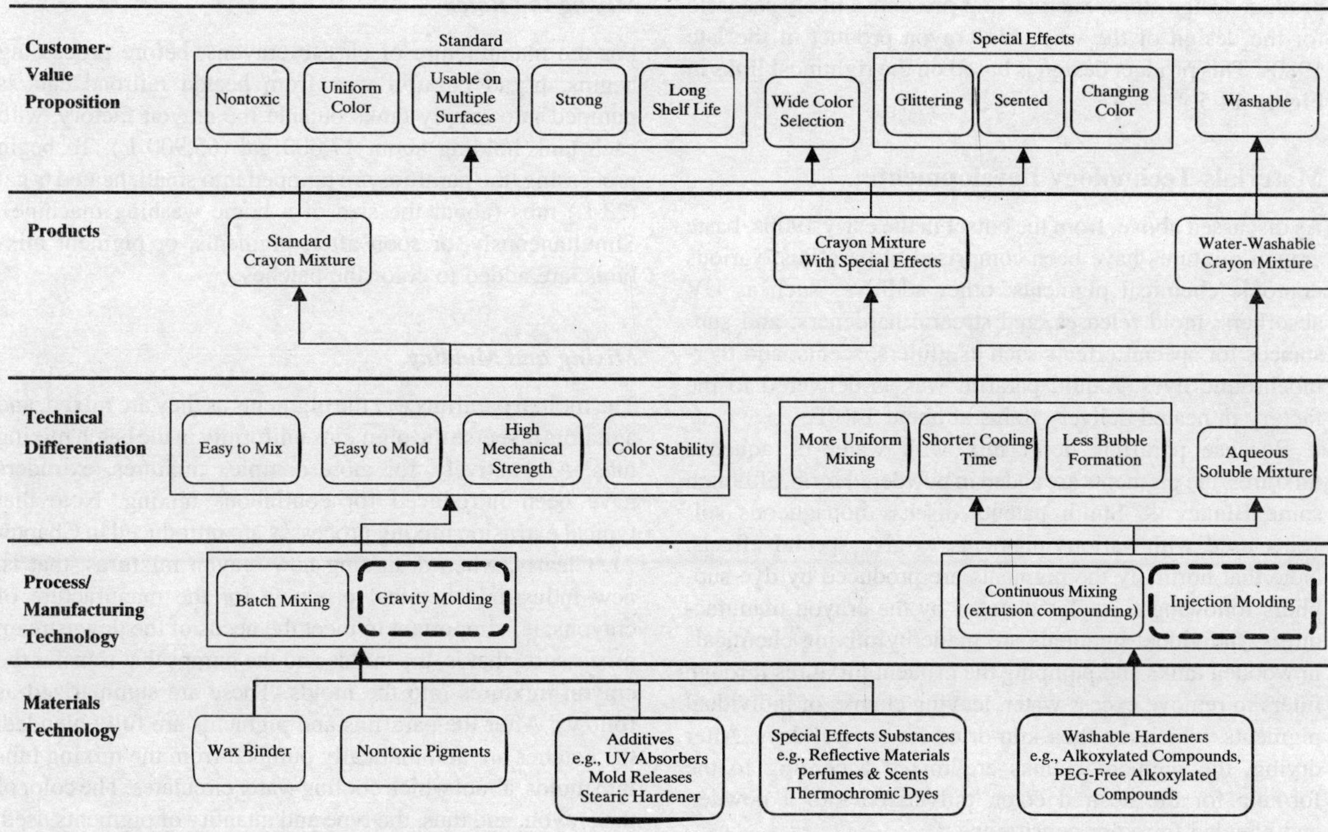

Figure 14.5 Innovation map for crayon mixtures.

and detergent solutions. Although the markings could be removed with various organic solvents, most notably WD-40, a penetrating oil spray consisting mainly of Stoddard solvent (a mixture of C7 to C12 saturated hydrocarbons), LPG, and mineral oil, the dyes left behind required a subsequent removal step with water or soap and detergent solutions. Crayons inadvertently left in fabrics placed into washing machines could deposit on the inside surface of the machine. As cleaning crayon markings using toxic and flammable household cleansers such as WD-40 needed to be avoided, it was deemed desirable to formulate new crayon mixtures that could be easily washed with water for removal from various surfaces (walls, refrigerators, tiles, etc.) and, by washing with detergents, from children's clothing. To accomplish this, the paraffin wax and nontoxic pigments were replaced by water-soluble alkoxylated compounds, in polyethylene glycol (PEG) resins with pigments. According to U.S. Patent 4,978,390, the alkoxylated compounds include viscosity modifiers [stearyl alcohol ($CH_3(CH_2)_{17}OH$) and cetyl alcohol ($CH_3(CH_2)_{15}OH$)], compatible plasticizers [oleyl alcohol ($CH_3(CH_2)_7-CH=CH-(CH_2)_8OH$) and propylene glycol ($CH_3CH(OH)CH_2OH$)], and inexpensive, functional fillers (clay, calcium carbonate, talc, or mica). Note that talc and mica add strength to the eventual crayon product. The alkoxylated compounds provide moisture curing (that is, they react with water, adding strength to crayons). Also, when combined with PEG, they become water-erasable. Because these new formulations are more viscous, as implied by U.S. Patent 6,039,797, they require continuous mixing, using extrusion compounding, as discussed in Chapter 21, and injection molding. Note also that these mixtures are often optimized to obtain uniform mixing and shorter cooling times, and to reduce bubble formation.

Because PEG-based crayon mixtures yield crayons with lower mechanical strength, PEG-free alkoxylated compounds were introduced in the 2000s in U.S. Patent 6,039,797. In this new formulation, PEG is replaced with alkoxylated fatty acids [e.g., ethoxylated stearic acid ($CH_3(CH_2)_{17}OOH$)], which add strength to the resulting crayons while improving the marking properties (reducing crumbling). Also included in the fatty alcohols is glyceryl monostearate ($CH_3(CH_2)_{17}COOCH_2CH(OH)CH_2OH$). These mixtures are melt-blended in extruders, as discussed in Chapter 21, and crayons are formed by injection molding.

This innovation map is current for the late 2000s, and consequently, it is applicable for product designs in this time frame. Of course, new product designs require foresights on the future technologies needed to be developed to satisfy the latest or future customer needs. The innovation map can be used to ensure that these customer needs drive the new technology-development efforts. Rather than attempt a new product design in the late 2000s to illustrate the product design steps, Section 15.3 presents a likely scenario for the design of the washable crayon product in the late 1980s. This product design is based on the rightmost links in Figure 14.5.

Materials Technology Development

As discussed above, from the outset in the early 1900s, basic crayon mixtures have been comprised of paraffins; various nontoxic chemical pigments; other additives such as UV absorbers, mold releases, and strearic hardeners; and substances for special effects such as glitters, scents, and thermochromic dyes. Liquid paraffin wax is delivered to the factory in heated delivery trains at about 140°F.

Because paraffins don't mix with water or aqueous mixtures, the pigments are added in powdered form, although some Binney & Smith patents discuss nonaqueous solvents used with various pigments to give special effects. Note that normally the pigments are produced by dye suppliers following formulas dictated by the crayon manufacturers. Individual pigments are made by mixing chemicals in wooden tanks and pumping the effluent mixtures through filters to remove excess water, leaving chunks of individual pigments, which are then kiln-dried for several days. After drying, the pigment chunks are mixed according to the formula for the desired color, pulverized into a powder, and blended for color consistency.

Over several decades, additional ingredients have been added to crayon mixtures. One of the most popular is glitter, that is, small pieces of reflective material that cause crayon products to both absorb and reflect light. Perfumes and other scents have been added as well. In fact, crayon mixtures were prepared with vegetable and fruit scents, providing crayons that, unfortunately, had to be recalled after children ingested them—as discussed further in the subsection "Environmental Concerns."

Finally, in the latest materials technologies, the paraffins are being replaced by water-soluble compounds designed to perform like paraffin-based crayons, but providing water-washable properties.

Process/Manufacturing Technology

While crayon manufacturing is simple in principle, it can be labor intensive, especially in the transfer of crayon mixtures into molds when producing crayons, the configured consumer product. Note, however, that newer Crayola® plants incorporate some automation in transferring crayon mixtures into molding machines. For manufacture of the intermediates, the most critical processing step involves blending all of the ingredients into crayon mixtures. Eventually, downstream, this mixture must be transferred into the molding machines that form the crayon sticks.

Next, the principal mixing and molding operations are discussed, with emphasis on the new extrusion technologies needed for the more recent crayon products.

Mixing the Batch

For the manufacture of classic crayons, before processing begins, liquid paraffin wax from heated railroad cars is pumped into supply tanks outside the crayon factory, with each tank holding about 17,000 gal (65,900 L). To begin processing, the paraffins are pumped into small, heated 6-gal (23-L) tubs (about the size of a home washing machine). Simultaneously, or soon after, pigments, or pigment mixtures, are added to color the batches.

Mixing and Molding

The molten paraffins *wet* the pigments as they are mixed, and agitators disperse the pigments uniformly in the batch mixing tubs. Alternatively, for more complex mixtures, extruders have been introduced for continuous mixing. Note that typical extrusion mixing processes are introduced in Chapter 21. Clearly, when designing new crayon mixtures, that is, new industrial chemical products for the manufacture of crayons, it is important to meet the needs of the downstream processors; that is, the molds and the pumps that transfer the crayon mixtures into the molds. These are summarized as follows. After the paraffins and pigments are fully blended, the batches are automatically pumped from the mixing tubs into molds, about which cooling water circulates. The color of the crayon, and thus, the type and quantity of pigments used, affect the cooling time, which varies from 4 to 7 minutes. Note that some molds hold as many as 2,400 crayon sticks.

Until recently, the paraffin-pigment mixtures were poured by hand from the tubs into buckets and then into molds. Newer machinery automatically pumps the mixtures into their molds, although in large, older factories, both processes are often found.

With more complex, higher-viscosity formulations, batch mixing and gravity molding are becoming less effective. For washable crayon formulations, U.S. Patent 6,039,797 indicates that extrusion compounding is necessary to mix the more viscous PEG-free alkoxylated compounds. Furthermore, to reduce the cooling time and prevent bubble formation during cooling, U.S. Patent 5,066,216 suggests the use of injection molding for the more viscous crayon mixtures.

Technology Protection

Because research and development of new technologies consumes as much as 10–12% of company revenues, patent protection is important to protect the new technologies prior to introducing a new product. For Binney & Smith Co., that is, Crayola LLC since 2007, formulations and manufacturing processes for crayons have been protected over the years by hundreds of patents. For example, for the washable crayons introduced in 1991, Binney & Smith applied for at least four patents beginning in 1988:

1. U.S. Patent 4,978,390. Snedeker, C.M., *Washable Solid Marking Composition* (1990)—filed in 1988.

2. U.S. Patent 5,460,647. Snedeker, C.M., and D.C. Fistner, *Color-Changing Marking Composition System* (1995)—filed in 1995.
3. U.S. Patent 5,486,228. Miller, R.E., C.R Couch, and C.D. Krieger, *Washable Color-Changing Compositions* (1996)—filed in 1994.
4. U.S. Patent 6,039,797. Fistner, D.C., *Washable Marking Composition* (2000)—filed in 1999.

These were accompanied by three notably competitive patents:

1. U.S. Patent 5,084,098. Olson, J.D., *Water Soluble Crayon Compositions* (1992)—filed in 1989 by an independent investigator.
2. U.S. Patent 5,380,357. Lytton, R.N., *Water Soluble Crayon Composition and Method* (1995)—filed in 1993 by La-Co Industries, Inc.
3. U.S. Patent 5,417,746. Cheng, C. -P., *Washable Crayon Composition* (1995)—filed in 1994 by an independent investigator.

U.S. Patent 5,084,098, the first to discuss the use of alkoxylated compounds to give washability, was referenced by Fistner in U.S. Patent 6,039,797. Note the absence of PEG in the Fistner formulation. In this case, with both the Olson and Fistner patents in the same field, legal professionals may be needed for advice regarding the rights to practice the inventions. In such cases, it may be necessary to obtain a license to practice the technology protected by a patent(s) granted to one's own firm. For this reason, cross licensing between companies is often practiced to exchange the right to use competitive patents in producing a new product.

In the processing area, Binney & Smith protected their molding technologies with at least two patents: U.S. 3,957,408, filed in 1974, and U.S. 5,066,216, filed in 1989. The former discusses their design for the continuous, gravity molding machine for standard crayon formulations, and the latter protects their injection molding apparatus for more viscous crayon formulations such as the washable mixtures. See the former patent for a figure showing the top view of a molding machine that consists of a wax-supply system (WS), which supplies liquid crayon mixtures to a series of mold cavities in a horizontal mold table (CM) that rotates, together with a crayon removal system (CR). Fig. 1 of the latter patent illustrates an injection molding apparatus that includes a vertically adjustable manifold valve (22), positioned above the mold table (26), that controls the flow of viscous crayon mixtures into the mold cavities (30). Fig. 3 shows a close-up top view of the mold cavity with the manifold cover removed.

In many similar cases, patent filings precede the introduction of a product to the market. In such situations, they are a leading indicator of future products in a company's product pipeline.

Environmental Concerns

In the early 1990s, Binney & Smith successfully marketed nontoxic crayons with food scents to match colors and crayon names. Subsequently, in response to concerns that children were eating the crayons, the company restricted the scents to those of nonedible objects like flowers. Note that, although the ingestion of paraffins often results in stomachaches, long-term effects were not anticipated.

In this regard, U.S. law requires that all art materials sold in the United States be nontoxic, with most toxicology evaluations performed under the auspices of the Arts and Crafts Materials Institute in Boston, Massachusetts. The formulas for new art products are submitted to the Institute and evaluated by toxicologists. In addition to the toxicity of an ingredient, a broad range of possible health effects are investigated. Testing is sometimes required, for example, to evaluate the interactions of individual ingredients within a single product or to determine whether a product causes skin irritation. All substances are evaluated at least every five years, with changes in their formulations triggering a new evaluation and approval.

14.4 SUMMARY

In this chapter, *innovation maps* have been presented for two industrial chemical products, LCD glass substrates and crayon mixtures, showing their new materials and product/manufacturing technologies, with links to the new products that satisfy specific customer needs. Having gained familiarity with these innovation maps, the reader is prepared to study their product design case studies in Chapter 15. When undertaking product designs, design teams seek to take advantage of the linkages between the new technologies and the customer-value proposition shown in their innovation maps.

REFERENCES

1. Bröckel, U., W. Meier, and G. Wagner, Eds., *Product Design and Engineering, Volume 1: Basics and Technologies*, Wiley-VCH (2007a).

2. Bröckel, U., W. Meier, and G. Wagner, Eds., *Product Design and Engineering, Volume 2: Raw Materials, Additives, and Applications*, Wiley-VCH (2007b).

3. Costa, R., G.D. Moggridge, and P.M. Saraiva, "Chemical Product Engineering: An Emerging Paradigm within Chemical Engineering," *AIChE J.*, **52**, 6, 1976–1986 (2006).

4. Cussler, E.L., and G.D. Moggridge, *Chemical Product Design*, Cambridge Univ. Press (2001).

5. Cussler, E.L., D.W. Savage, A.P.J. Middelberg, and M. Kind, "Refocusing Chemical Engineering," *CEP*, 26S–31S (Jan., 2002).

6. Cussler, E.L., and J. Wei, "Chemical Product Engineering," *AIChE J.*, **49**, 5, 1072–1075 (2003).

7. Favre, E., L. Marchal-Heusler, A. Durand, N. Midoux, and C. Roizard, "A Graduate-level-equivalent Curriculum in Chemical Product Engineering," *Chem. Eng. Educ.*, 264–271 (Fall 2005).

8. Favre, E., L. Marchal-Heusler, and M. Kind, "Chemical Product Engineering: Research and Educational Challenges," *Trans. I. Chem. E.*, **80**, Part A, 65–74 (Jan., 2002).

9. Hill, M., "Product and Process Design for Structured Products," *AIChE J.*, **50**, 8, 1656–1661 (2004).

10. Ng, K.M., R. Gani, and K. Dam-Johansen, *Chemical Product Design: Toward a Perspective Through Case Studies*, Elsevier (2007).

11. Saraiva, P.M., and R. Costa, "A Chemical Product Design Course with a Quality Focus," *Trans. I. Chem. E.*, **82** (A11), 1476–1484 (Nov., 2004).

12. Seider, W.D., J.D. Seader, and D.R. Lewin, "Chemical Product and Process Design Education," in *Proceedings of FOCAPD-2004 Conference*, Princeton University, Princeton, NJ (2004).

13. Shaeiwitz, J.A., and R. Turton, "Chemical Product Design," *Chem. Eng. Educ.*, 280–285, (Fall 2001).

14. Wei, J., *Product Engineering: Molecular Structure and Properties*, Oxford Univ. Press (2007).

15. Westerberg, A.W., and E. Subrahmanian, "Product Design," *Comput. Chem. Eng.*, **24**, 959–966 (2000).

Plastic Wrap

16. Plastic Wrap, http://en.wikipedia.org/wiki/Plastic-wrap, Wikipedia, May 2007.

Thin-Glass Substrate for LCDs

17. Bocko, P.L., and M.H. Mitchell, "AMLCD Glass Substrates Foundation for HighTech Displays" *Glass Researcher*, **12**, 1/2. 26–28 (2003).

18. Bocko, et al., "Developing Winner," *American Ceramic Soc. Bulletin*, **85**, 35–36 (2006).

19. Brody, T.P., "Birth of the Active Matrix," *Information Display*, **13**, 10, 28–32 (2007).

20. Lapp, J.C., "AMLCD Substrates Trends in Technology," Corning Technical Information Paper, 2004.

21. Lapp, J.C., "Glass Substrates for AMLCD Application; Properties and Implications," Corning Technical Information Paper, 2004.

Patent—Thin-Glass Substrates

22. U.S. Patent 3,338,696, Dockerty, S. M., Sheet Forming Apparatus (1967).

23. Blachford, S.L., Ed., *Crayon: How Products are a Made,* Gale Group, Inc., 2002. eNotes.com, http://www.enotes.com/how-products-encyclopedia/crayon, Dec, 2007.

24. Crayola, http://en.wikipedia.org/wiki/Crayola, Wikipedia (Dec., 2007).

25. History of Crayons, www.crayola.com/mediacenter/HistoryOfCrayons.doc (Dec., 2007).

26. Woods, S.G., *Crayon: from Start to Finish,* Blackbirch Press, CT, 1999.

Patents—Crayon Mixtures

27. U.S. Patent 320,009, Sleeper, J.J., and H.A. Johns, *Composition for Crayons* (1885).

28. U.S. Patent 3,957,408, Clymer, W.L., and C.J. Klara, *Automated Molding Machine* (1976).

29. U.S. Patent 4,978,390, Snedeker, C.M., *Washable Solid Marking Composition* (1990).

30. U.S. Patent 5,066,216, Kowtko, R.A., and R.A. Mellwig, *Apparatus for Injection of Viscous Material* (1991).

31. U.S. Patent 5,084,098, Olson, J.D., *Water Soluble Crayon Compositions* (1992).

32. U.S Patent 5,380,357, Lytton, R.N., *Water Soluble Crayon Composition and Method* (1995).

33. U.S. Patent 5,417,746, Cheng, C.-P., *Washable Crayon Composition* (1995).

34. U.S. Patent 5,460,647, Snedeker, C.M., and D.C. Fistner, *Color-Changing Marking Composition System* (1995).

35. U.S. Patent 5,486,228, Miller, R.E., C.R. Couch, and C.D. Krieger, *Washable Color-Changing Compositions* (1996).

36. U.S. Patent 6,039,797, Fistner, D.C., *Washable Marking Composition* (2000).

EXERCISES

14.1 Develop the value chain for the LCD glass substrate business. You may need to conduct a literature search on the Internet. Note that LCD glass is sold business-to-business; it is not sold to the end users through a retailer.

(a) Identify the elements of the value chain.

(b) Identify the role of each element in the value chain.

(c) Identify the key players for each entity in the value chain.

Hint: You may need to search for the value chain for the LCD polarizer or LCD notebooks, monitors, and TVs.

14.2 Many advances have been made toward developing organic semiconductors, bonded on plastic substrates as replacements for silicon TFTs on glass. Conduct a patent search to assess the viability of plastic substrates.

(a) Identify the leading technological options.

(b) Identify the key players for each technological alternative.

(c) Comment on their strengths and weaknesses.

14.3 Identify two major competitors of Corning LCD glass substrates. Identify their differentiating technologies as compared with the Corning glass technology.

14.4 Identify alternate manufacturing technologies used by the Corning competitors identified in Exercise 14.3.

14.5 As indicated in Chapter 2, the patent portfolio of a company can serve as a leading indicator of their new product-development strategy. Analyze the patent portfolio of Corning for the past five years and identify potential new directions being pursued by Corning.

14.6 Develop an innovation map for color changing and washable color changing crayon mixtures using U.S. Patents 5,460,647 and 5,486,228.

Chapter 15

Product Design Case Studies for Industrial Chemicals

15.0 OBJECTIVES

This chapter provides two case studies to illustrate the steps in designing new *industrial chemical* products using the Stage-Gate^{TM} Product-Development Process. Emphasis is placed on the *concept* and *feasibility* stages. Only the key issues are discussed in the remaining *development*, *manufacturing*, and *product-introduction* stages.

After studying this chapter, the reader should:

1. Be able to use the elements of the Stage-Gate^{TM} Product-Development Process for the design of *industrial chemical* products.
2. Be able to identify and use the necessary technological innovations, discussed in Chapter 14, in the development and design of a new product.

15.1 INTRODUCTION

This chapter is comprised of two case studies that illustrate the principal aspects of *industrial chemical* product design. Before reading these case studies, the reader should be conversant with the key steps in product design, as introduced in Sections 1.2 and 1.3, Chapter 2, and the introduction to Part Two, especially Figure PII.1. For each case study, the history of the technologies involved and an innovation map are presented in Chapter 14. Sections 14.2 and 15.2 are devoted to the design of an LCD glass substrate. Being the first case study, the design scenario for the LCD glass substrate is presented more thoroughly. The second case study, involving washable mixtures for crayons, is introduced in Section 14.3 and discussed in Section 15.3.

15.2 LCD GLASS SUBSTRATE CASE STUDY

This case study is presented in hindsight to illustrate the steps in the Stage-Gate^{TM} Product-Development Process (SGPDP). The time frame for the product development is assumed to be the mid-1980s, when a design team at Corning was concentrating on the development of the LCD glass substrate known as Corning-7059. The development of the underlying new technologies, from the mid-1980s to mid-2000s, was presented in Section 14.2, with each new technology positioned on the *innovation map* for LCD glass substrates in Figure 14.2. The latter shows the role of each new substrate in the Corning LCD glass-product platform, which is comprised of the Corning-7059, Corning-1737, Eagle2000^{TM} and EagleXG^{TM} products. Similar case studies can be created for the other products; that is, Corning-1737 in the early 1990s (see Exercise 15.1) and EagleXG^{TM} in the early 2000s (see Exercise 15.2). Corning-7059 is preferable for this case study because it involved creating a first-generation, *new-to-the-world* product, while the others involved *product extensions* that were less risky.

As Corning introduced its barium boro-silicate glass in the early 1980s, the product development of its first LCD glass substrate coincided with the introduction of the first commercial LCD monitor by Sharp in the late 1980s. In this case study, many of the considerations in the product-development process for the Corning-7059 glass substrate are discussed. Note, however, that the authors were not involved in the product and technology development of glass substrates for LCDs, and consequently, the scenario presented is of the events that were likely to have taken place. Also, the Corning LCD glass-development team might not have used the Stage-Gate^{TM} Product-Development Process (SGPDP).

Project Charter

As described in Section 2.2, a good project charter is the focal point of a product-development effort. Typical elements, that is, specific goals, project scope, deliverables, and time line,

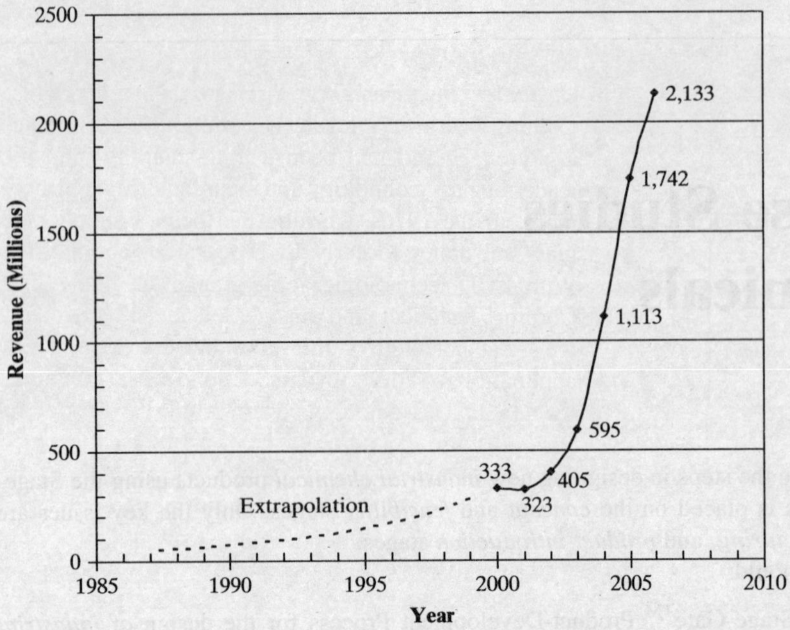

Figure 15.1 Revenues from the Corning LCD glass substrate business. (*Source:* Corning Annual Reports 2000–2006.)

are illustrated for the LCD glass substrates in Table 15.1 and discussed next.

In the late 1970s and early 1980s, Corning was embarking on a new business venture to develop a glass substrate suitable for the emerging LCD technology. At that time, the business opportunity was not clear and, indeed, was far less promising than came to pass 20 years later. Most likely, a small group of technical and business experts envisioned a thriving LCD glass business, with a product-development team formed and charged to develop a glass substrate product suitable for LCD panels.

Table 15.1 Possible Project Charter for the Corning-7059 Glass Substrate*

Project Name	LCD Glass
Project Champions	Business Director of the Specialty Glass Substrate Business
Project Leader	John Doe
Specific Goals	Glass substrate suitable for LCD panel manufacturing, meeting the durability, thermal stability, and surface quality requirements
Project Scope	In-scope: • Determination of acceptable technical requirements for LCD glass substrates • Minimal changes in the current manufacturing capabilities Out-of-scope: • Major manufacturing changes
Deliverables	• Business opportunity assessment • Technical feasibility assessment • Manufacturing capability assessment
Time Line	Product prototypes for market testing within 12 months

*Details of the Corning product-development effort were not available to the authors.

Because active-matrix LCD technology was being developed at that time, the team was likely chartered to determine suitable requirements for LCD glass substrates. Also, it is likely that the team was requested to use current manufacturing capabilities because extensive capital funds may not have accumulated in the preceding years—in which glass profits were stable, but not ignited by breakthrough products.

Finally, the team was probably requested to create a prototype for market testing within a year and to present its technical and manufacturing feasibility.

Next, following the product design steps in Figure PII.1, having completed its project charter, the design team likely asked whether materials technology and process/manufacturing technology invention were required. For the glass substrate products, in the late 1980s when the Corning-7059 product was being developed, the required technologies were in place, as discussed in Section 14.2 (when presenting the innovation map in Figure 14.2). Furthermore, with such promising new technologies, permission to proceed into the SGPDP was probably granted enthusiastically.

Concept Stage

As discussed in Sections 1.2 and 2.4, and shown in Figure PII.1, the SGPDP begins with the *concept* stage. It normally involves five principal assessments and activities, which are discussed next as they were likely to have occurred for the Corning-7059 product.

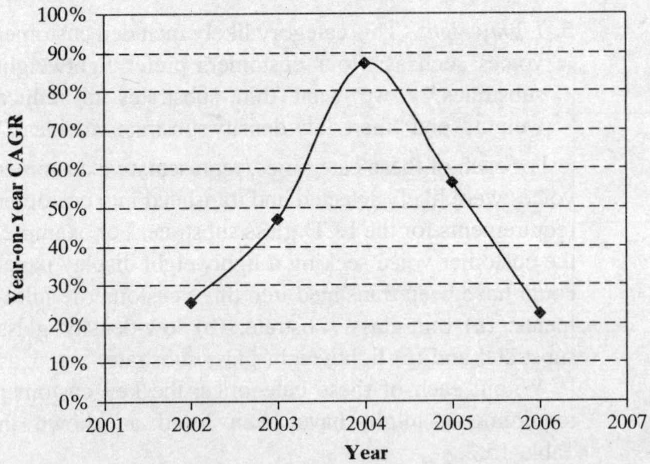

Figure 15.2 Percentage changes in Corning LCD glass substrate revenues. CAGR = Compounded Annual Growth Rate. *(Source:* Corning Annual Reports 2000–2006.)

a. ***Opportunity assessments*** for *new-to-the-world*, first-generation products, such as LCD glass substrates in the 1980s, are very problematic. When introducing these products, most successful companies rely on "mega-market" trends while recognizing principal performance bottlenecks, and their abilities to provide differentiated and sustainable solutions. Furthermore, as exemplified by Corning, in addition to visionary insights, established firms often have a commitment and the means (financial strength) to sustain a business even when the incubation time to achieve satisfactory rewards is longer than anticipated.

For Corning, after entering the market with its first alkali-free glass substrate in 1987, it experienced just modest sales. The market for LCD glass substrates was small in the mid-1980s, but a high-growth potential was likely to have been anticipated. Then, in the early 2000s, as LCD notebooks, monitors, and TV businesses took off exponentially, the revenues of the Corning Display Technology segment, which produced the LCD glass substrates, took off in step. This is shown in Figure 15.1, in which revenue data from Corning Annual Reports (2000 to 2006) are extrapolated back to the mid-1980s (to give approximately \$50 million/yr). In hindsight, this exponential growth was not realized until the early 2000s, peaking in 2004, as shown in Figure 15.2. The decline can be attributed to the shift to commoditization of LCDs accompanied by an inevitable decrease in growth rate.

In summary, for *new-to-the-world* products, companies often rely on "disruptive" innovations and long-term visions rather than on short-term cash-flow projections (commonly used for new product extensions having known or existing markets). In addition, Corning could rely on its core expertise in glass compositions and its novel glass-fusion technology to dominate the market in the sales of its ultra-clean, precision thin-glass sheets.

b. ***Customer Requirements.*** Prior to the early 1980s, Corning had been a leader in commodity glass, such as Pyrex^{TM}, and had been a long-standing industry leader in glass technology and manufacturing. Furthermore, in the 1970s, Corning had been a supplier of glass substrates to early LCD research laboratories. With LCD technologies emerging, it is likely that Corning technical and business leaders recognized the great potential of thin-glass substrates for LCD applications. As described in Section 14.2, at that time, the new LCD technologies were primarily directed to small displays. Their potential customers, the notebook computer manufacturers, were concerned with reducing the displays' weight, and consequently, they recognized the need for thinner and lighter-weight display panels. Also, the development team recognized that the alternate technology, a plastic-based substrate, could not match the thermal stability demanded for TFT processing, improving Corning's likelihood of success.

It is likely that the challenge for the Corning product-development team was that the voices of their customers (the notebook manufacturers), not being expert in photolithography technologies, couldn't prescribe quantitative requirements for the panels other than weight and thickness. To uncover better voices, the team needed to learn the requirements from their direct customers, the panel makers.

Using the methodology described in Section 2.4 for learning the voice of the customer, it is likely that the development team selected several panel-making laboratories and collected the following customer needs (voices), among others:

1. Be able to withstand processing temperatures up to 400°C for up to one hour without deformation or deterioration.
2. Be transparent at a wavelength of 350 nm for photolithography to create transistors using amorphous silicon.
3. Expand and contract with temperature like amorphous silicon.
4. Contain no elements that contaminate silicon, especially alkali cations.
5. Be lightweight.

Because development teams commonly gather hundreds of voices and images from potential customers, often seemingly disjointed voices must be processed into groups of needs having an "affinity" to one another. The standard method of doing so is the so-called KJ, or affinity, analysis. For the thin-glass substrate, using the KJ analysis described in Section 2.4

(see also Creveling et al., 2003 and Belliveau et al., 2004), these voices were likely grouped initially on the basis of their similarities, and then on the basis of their functions, to arrive at the following five categories, some of which were discussed in Section 14.2:

1. *Dimensional stability* to avoid minute alignment errors in the TFT patterns or mismatch between the TFT pattern and the color filter, which destroys the fidelity of the image. This category likely included customer voices such as: "panel substrates ideally should not expand and contract much," "LCD substrates should not change much with temperature," and "panel substrates should behave like amorphous silicon."
2. *Surface quality* to prevent small surface imperfections from causing pattern defects during the photolithography process that imprints the TFT pattern on the glass substrate. In this category, typical customer voices were likely: "we need panel substrates that can withstand chemical treatment," "LCD substrates should be free of defects and dirt," and "we need smooth surfaces."
3. *Surface flatness* to prevent variable gaps between the back and front panels, which adversely affect the performance of the liquid crystals (sandwiched between the two panels). Here, typical customer voices were likely: "we need flat-panel substrates," "the thickness should not vary," and "flatness is important for maintaining uniform distances between the back and front panels."
4. *Optical properties* to enable downstream processing such as photolithography of the TFT layer for the back panel and the imposition of a color filter layer for the front panel. In this category, typical customer voices were likely: "bubbles should be avoided," "transparency at 350 nm is important for TFT processing," and "clarity and inclusion-free substrates are needed."
5. *Lightweight*. This category likely included customer voices such as: "our customers prefer lightweight substrates," "we want thin substrates to reduce weight," and "inert, low-density substrates are ideal."

For each of these categories, representative customer voices were likely selected and translated into customer requirements for the LCD glass substrate. For example, the customer voice seeking a lightweight display panel could have been translated into three customer requirements: (a) thin-glass substrate, (b) low-density glass substrate, and (c) lightweight glass substrate.

Within each of these categories, the key customer requirements might have been listed as shown in Table 15.2.

As discussed in Section 2.4, some requirements were likely identified as *fitness-to-standard* (FTS) and the remainder as *new-unique-and-difficult* (NUD), with each requirement assigned a weighting factor as a measure of its relative importance. These weighting factors play an important role, in particular when there are competing objectives and compromises are necessary. Here, the dimensional stability, surface quality, and flatness requirements were likely the most important NUDs. The lightweight requirement, although classified as a NUD, likely had the lowest weighting factor.

c. *Technical Requirements.* Before product concepts were developed, the glass substrate requirements in Table 15.2 were likely translated into quantifiable technical requirements more amenable to technical development work. For the Corning-7059 glass substrate, the translation into quantifiable technical requirements, such as the coefficient of thermal expansion (CTE) and surface quality, among others, might have been similar to that shown in Table 15.3. In general, this translation occasionally results in multiple technical requirements—for example, the transparency requirement involves two technical requirements.

Table 15.2 Customer Requirements

Customer Requirement Category	Glass Substrate Requirement	Type	Weighting Factor (%)
Dimensional stability	Thermal stress Internal stress Thermal expansion	NUD	25
Surface quality	Chemical resistance Surface smoothness Surface cleanliness	NUD	25
Surface flatness	Surface thickness variation	NUD	25
Optical properties	Transparent No inclusions	FTS	15
Lightweight	Density Thickness	NUD	10

Table 15.3 Technical Requirements

Glass Substrate Requirement	Technical Requirement	Target
Thermal expansion	Coefficient of thermal expansion (CTE)	<5 ppm/°C
Chemical resistance	Alkali content	<2,000 ppm
Surface smoothness	Roughness	$<2.5 \times 10^8$ m
Surface cleanliness	Surface particle count	<1 particle/m^2
Flatness	Thickness variation	<0.5%
Transparent	Optically clear	Transparent at 350 nm
	No inclusions	No bubbles or particles
Density	Specific gravity	<2.8

d. *Determination of Critical-to-Quality Variables (CTQs).* Normally, the NUD requirements are taken as the *critical-to-quality* variables. And consequently, it is likely that the CTE, alkali content, surface roughness, surface particle count, thickness variation, and density were selected during the development of the Corning-7059 glass substrate.

As described in Section 2.4, in the *concept* stage of the SGPDP, the first *House of Quality* (HOQ), also known as the *Quality Function Deployment* (QFD), is normally assembled. It displays the results of the process to obtain, translate, and deploy the *voice of the customer* into technical requirements. This process is repeated at various stages in the product-development

Table 15.4 First House of Quality

								1	Spec. Gravity
							1	+	Inclusion
						1	−		Clarity
					1			−	Thickness Var
				1		−			Surface Particle Count
			1	+	+	−			Roughness
		1	−		−				Alkali Content
	1	+							CTE
Customer Requirement	CTE	Alkali Content	Roughness	Surface Particle Count	Thickness Var	Clarity	Inclusion	Spec. Gravity	Weight
Dimensions Stability									0.25
Thermal Expansion	×								
Surface Quality									0.25
Chemical Resistance		×							
Surface Smoothness			×						
Surface Cleanliness				×					
Surface Flatness					×				0.25
Optical Properties									0.15
Transparent						×			
Inclusion							×		
Light Weight									0.1
Density								×	

process, with the results updated from time to time. Note that while these helpful constructs had not been defined in the 1980s, the relationships shown must have been well understood by the Corning design team.

In Table 15.4, the lower rectangular matrix pairs all of the customer requirements in the first column with at least one quantitative technical requirement or parameter in the adjacent columns. For example, dimensional stability requires a low CTE, closely matching the CTE of a-Si. In other cases, the customer requirement translates directly into the technical requirement; for example, light weight translates into thin glass and low density.

At the top of the house, the *interaction matrix* shows the synergistic technical requirements or parameters, for example, the density and substrate thickness. Note that + indicates that the variables increase or decrease together; and − indicates that when one variable increases, the other decreases, and vice versa; for example, the level of inclusion varies inversely with the optical clarity. A blank entry indicates that no significant relationship exists between the variables.

Because the precision manufacturing process affects the properties of the glass sheets and the glass formulations affect the processability of the molten glass, it is helpful to construct the manufacturing House of Quality, as shown in Table 15.5. This house shows the relationships between the glass technical requirements (or properties) and the materials parameters that impact the rheological and thermal properties of the molten glass. These parameters include the liquidus viscosity, deformation temperature (strain point), and molten glass density.

e. ***Development of Superior Product Concepts.*** The selection of solution concepts is at the heart of new product development, especially when several alternatives are available. Note that these concepts often apply to specific elements of a product assembly. When developing glass substrates for LCDs in the 1980s, as discussed in Section 14.2, the new concepts included barium boro-silicate glass and an improved process for manufacturing thin-glass sheets. Note also that in the *concept* stage, this selection can be risky because, in many cases, a more thorough judgment must be deferred until a prototype is created and tested in the *feasibility* stage. Often, gatekeepers exhibit flexibility when deciding not to eliminate a concept and to carry it through to the *feasibility* stage.

The so-called Pugh matrix (Pugh, 1996), in which each solution concept (partial and complete) is judged against a reference solution, is useful for screening

Table 15.5 Manufacturing House of Quality

			1	Glass Molten Density
		1	+	Liquidus Viscosity
	1	+		Strain Point
Technical Requirements	Strain Point	Liquidus Viscosity	Glass Molten Density	Weight
CTE				0.15
Alkali Content	×	×		0.10
Roughness		×		0.15
Surface Particle Count				0.10
Thickness Var		×		0.25
Clarity				0.10
Inclusion		×		0.05
Spec. Gravity			×	0.10

Table 15.6 Pugh Concept Selection

Technical Requirement	Target	Reference Concept	Concept A: Corning-1723	Concept B: Corning-7980
CTE	<5 ppm/°C		−	−
Alkali content	<2,000 ppm		+	
Roughness	<2.5 × 10^{-8} m			−
Surface particle count	<1 particle/m^2			
Thickness var.	<0.5%	Pyrex™		−
Clarity	Transparent at 350 nm			+
Inclusion	None			
Specific gravity	<2.8			
Strain point	>550°C		−	+
Liquidus viscosity	>3 × 10^5 poise		−	

purposes, as illustrated in Table 15.6 for the Corning-7059 glass substrate. The reference solution is usually the best known in the market (or the best potential solution); in the 1980s, it was Pyrex™ boro-silicate glass. Each potential concept is evaluated against the reference solution and assigned a qualitative valuation of inferior (−), superior (+), or equal (0).

For the Corning-7059 glass substrate, it is likely that the Corning design team initially selected three candidates, Pyrex™ boro-silicate glass, Corning-1723 (lamp-envelope glass), and Corning-7980 (fused-silica glass), from the Corning library of glass formulations. As shown in Table 15.6, none of these formulations met the product and processing requirements.

Next, it is likely that the design team judged other new materials technologies that had been developed at Corning, as described in Section 14.2. The most attractive candidates probably involved various concentrations of boron oxide and barium oxide, with several iterations probably required before the glass formulation known as Corning-7059 was selected, as shown in Table 15.7.

Table 15.7 Subsequent Pugh Matrix

Technical Requirement	Target	Reference Concept	Concept C Corning-7059
CTE	<5 ppm/°C		0
Alkali content	<2,000 ppm		+
Roughness	<2.5 × 10^{-8} m		+
Surface particle count	<1 particle/m^2		+
Thickness var.	<0.5%	Pyrex™	+
Clarity	Transparent at 350 nm		+
Inclusion	None		+
Specific gravity	<2.8		+
Strain point	600°C		0
Liquidus viscosity	>3 × 10^5 poise		+
Glass molten density	2.75 g/cc		0

f. *Selection of Superior Concepts.* In the *concept* stage of the SGPDP, the selection of superior concepts is based primarily on the satisfaction of the technical requirements, in particular, the *new-unique-and-difficult* (NUD) requirements. For the Corning glass substrate, that would have been Concept C (Corning-7059), shown in Table 15.7. Although the Corning-7059 formulation has a slightly higher CTE and slightly lower durability than desired, it was likely selected as the superior concept in the *concept* stage, recognizing that it would be tested extensively in the *feasibility* stage.

g. *Gate Review.* To complete the *concept* stage, a gate review might have been carried out in which the product design team would have answered questions associated with several deliverables. Such a review, which may or may not have been carried out by Corning in the 1980s, is not discussed here. For this case study, the reader may presume that the business decision makers judged the product design team's presentation to be satisfactory, and funded the project to proceed to the *feasibility* stage. Note that the gate review for the *concept* stage is discussed in Section 2.4, and a detailed gate review is presented in the case study for the halogen light bulb (Section 17.2).

For more information, including data used in the *concept* stage, see the Elisson Web site and Lapp (2004a, 2004b).

Feasibility Stage

Following the SGPDP, having received authorization and funding to proceed, the design team would next concentrate on the deliverables to be completed during the *feasibility* stage (see Figure PII.1). Note that in the 1980s, prior to the SGPDP, a similar phase of the project was likely to have been initiated. In this phase, emphasis would have been placed on the technical feasibility of the superior concept(s). In

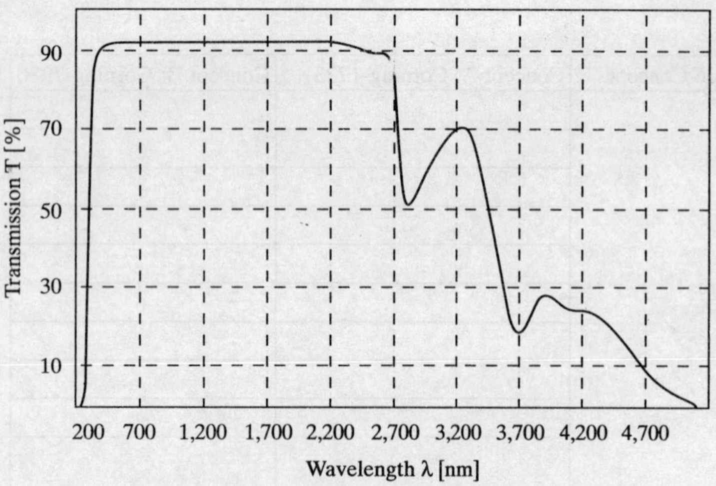

Figure 15.3 Transmission curve for Corning-7059 thin-glass substrate. [*Source:* Präzisions glass & optik (http://www.pgo-online.com/intl/katalog/7059.html).]

addition, several issues would normally have been addressed, arising from a market assessment, a competitive analysis (including IP strategy), and examinations of health-safety-environment concerns and the product life cycle. Again, the reader should view the following discussion as a likely scenario of the events that took place in the 1980s.

a. *Technical Feasibility.* The goal of a technical feasibility assessment would have been to ensure that the superior concept(s), Concept C above, met the customer requirements. For the LCD glass substrate, this was likely to have been accomplished by determining that the superior concept(s) satisfied the technical requirements. In so doing, each critical-to-quality variable was probably checked as follows.

The transparency would have been evaluated at different wavelengths, with typical results shown in Figure 15.3 for 0.5 mm glass. Given that the light transmission exceeds 90% at 350 nm, the design team would likely have been reassured.

Next, it is likely that the coefficient of thermal expansion (CTE) was measured from 0 to 300°C for several prototypes, with typical data for seven prototypes shown in Table 15.8. Clearly, for the data shown, some of the prototypes did not satisfy the dimensional stability test (CTE < 5 ppm/°C), and the variation was large. Of greater concern, the CTE data often exceeded $CTE_{a\text{-}Si}$ = 4.6 ppm/°C.

Table 15.8 Coefficient of Thermal Expansion of Corning-7059 Glass Prototypes

Prototype	CTE (ppm/°C) from 0 to 300°C
A	4.5
B	4.5
C	4.7
D	4.2
E	4.1
F	5.2
G	5.0

For several additional prototypes (five in Table 15.9), measurements were probably collected to evaluate the surface quality; that is, the chemical resistance, surface smoothness, and surface cleanliness. In Table 15.9, typical values for the chemical resistance are displayed as the weight loss in 5 wt% hydrochloric acid solution at 95°C over 24 hours. As shown, weight losses for all of the prototypes fall below 10 mg/day/cm^2, a typical upper bound (U.S. Patent 4,824,808). Table 15.9 also shows typical results for the surface particle counts and the surface smoothness measurements.

Given comparable data, it is likely that the Corning design team prepared a few glass samples for customer testing. Herein, it is presumed that the team measured the CTE, surface chemical resistance, and surface cleanliness and smoothness for three samples. See Table 15.10.

Table 15.9 Surface Resistance, Cleanliness, and Smoothness

Prototype	Weight Loss (mg/day/m^2)	Particle Count (particle/m^2)	Surface Smoothness (10^{-8} m)
D1	7.7	0	1.02
D2	8.0	0	1.07
D3	8.3	0	0.99
D4	8.5	0	1.02
D5	7.0	1	0.96

Table 15.10 Test Results for Three Prototypes Prepared for Customer Feedback

Prototype	CTE (ppm/°C)	Weight Loss (mg/day/m^2)	Particle Count (particle/m^2)	Surface Smoothness (10^{-8} m)
P-1	4.5	8.0	0	1.02
P-2	4.3	8.1	0	1.04
P-3	4.7	7.9	0	1.02

Table 15.11 Customer Feedback on Dimensional Stability Measured by CTE (ppm/°C)

Customer	Prototype P-1	Prototype P-2	Prototype P-3
C-1	4.5	4.3	4.7
C-2	4.6	4.4	4.8
C-3	4.6	4.4	4.8

Table 15.12 Customer Feedback on Surface Durability (mg/day/m^2)

Customer	Prototype P-1	Prototype P-2	Prototype P-3
C-1	8.0	8.1	7.9
C-2	9.2	9.1	9.2
C-3	9.2	9.1	9.2

b. *Customer Verification.* Prior to conducting a field test with potential customers, confidentiality agreements are executed, after which product prototypes are distributed. Typical results of customer tests that were likely obtained by the Corning design team are presented in Tables 15.11–15.14. Note that customer C-1 was a panel-maker laboratory, customer C-2 was a major panel maker, and customer C-3 was the panel integrator (that is, the assembler of the integrated back-front panel LCD unit) that worked with customer C-2. Tables 15.11–15.14 show typical measurements returned by these customers for dimensional stability, surface durability, and surface cleanliness and smoothness. Note that because the LCD industry was new, the Corning team had to collaborate with the development labs of panel makers and integrators.

In all cases, customers C-2 and C-3 provided identical results except for the particle counts, suggesting that as partners, they may have reported the same data. Customer C-2 expressed concerns that the CTE of the

Table 15.13 Customer Feedback on Surface Cleanliness (particle/m^2)

Customer	Prototype P-1	Prototype P-2	Prototype P-3
C-1	1	1	1
C-2	2	3	4
C-3	5	5	7

Table 15.14 Customer Feedback on Surface Smoothness (10^{-8} m)

Customer	Prototype P-1	Prototype P-2	Prototype P-3
C-1	1.02	1.04	1.07
C-2	1.02	1.07	1.04
C-3	1.02	1.07	1.04

Table 15.15 Customer Willingness to Purchase the LCD Glass Substrate

Customer	Customer Comments
C-1	Will not purchase.
C-2	Depending on the price, we will consider purchasing. Can the CTE be further reduced to match $CTE_{a\text{-}Si}$?
C-3	Great performance, contact us when the product is launched. Better dimensional stability is desirable.

prototypes was higher than desired, and asked if it could be lowered. Also, because all particle counts exceeded the team measurements (see Table 15.10), particles may have been acquired during shipping or customer testing.

Finally, it is likely that the design team asked questions concerning the willingness of these customers to purchase the glass substrate, given their performance measurements. Table 15.15 displays typical responses with varying degrees of acceptance. Customer C-1 was unwilling to purchase the product, customer C-2 indicated the price was a deciding factor, and customer C-3 was willing to consider the product, an encouraging indicator. Furthermore, two of the three potential customers requested a lower coefficient of thermal expansion, and consequently, for this response, it is likely that the design team planned to seek the reduction desired and to explore its impact on the product price and manufacturability.

c. *Market Assessment.* As mentioned in the discussion of the *concept* stage, estimates of the market sizes for LCDs in the 1980s were highly uncertain. Figure 15.4 shows the existence of some strong proponents, with these projections probably having a significant influence on the business decision makers at Corning. Although sales in the mid-1980s were nearly nonexistent, Corning firmly pursued this new opportunity.

d. *Competitive Analysis.* The customer feedback in Table 15.15, mostly expressing satisfaction with the technical performance, would have been encouraging. While the design team developed Corning-7059 substrate glass, no significant competitive offerings existed. Furthermore, a patent (U.S. 3,338,696) for the manufacture of precision thin glasses was about to expire, having been granted in 1967. In the mid-1980s, Corning inventors and manufacturing engineers would have sought to improve the process and file additional patents to protect their new substrate glass. For the glass compositions, Corning scientists filed several patents, including U.S. 4,634,683, *Barium and/or Strontium Aluminosilicate Crystal-containing Glasses for Flat Panel Display Devices*, filed in 1985; and U.S. 4,824,808, *Substrate Glass for Liquid Crystal Displays*, filed in 1987.

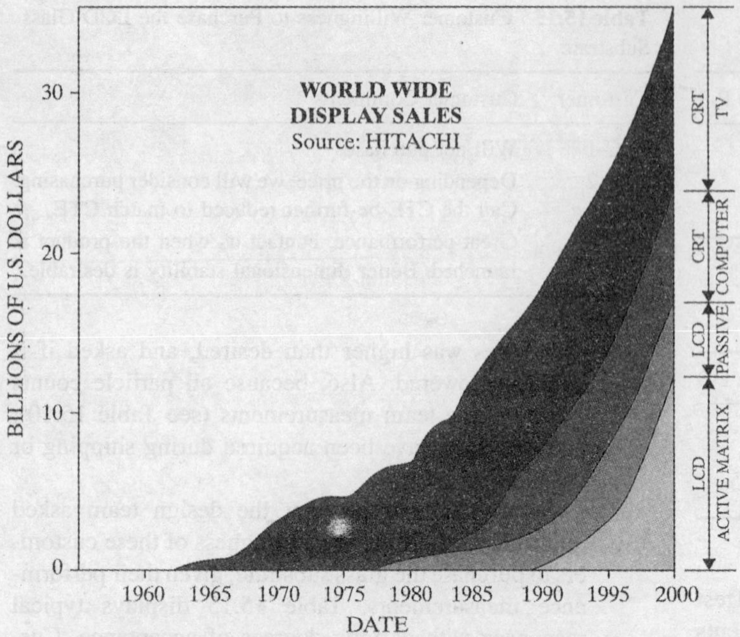

Figure 15.4 Sales projections. (*Source:* Display Technology in Japan (1992)—a report by WTEC, http://www.wtec.org/loyola/dsply_jp/c1_s10.htm.)

Development Stage

The main task in the *development* stage of the SGPDP is to design a manufacturing process at the *pilot-plant* level. Note that in many large manufacturing companies, pilot-plant equipment is often used to produce product at the beginning of the product-introduction period. As the demand grows, production is expanded into a large manufacturing site. As shown in Figure PII.1, this normally involves detailed design, equipment sizing, profitability analysis, and optimization. The *House of Quality* is used to identify the critical process variables for each manufacturing process. In addition, a failure-mode analysis is often performed to identify a plan(s) to return the process to its nominal operating point after abnormal events (severe disturbances or faults) occur. Often, short-term runs are sufficient to check the manufacturing process stability. When necessary, the market assessment and product recycle management are revisited.

For the Corning-7059 glass substrate, it is likely that the Corning glass-fusion process, described in Section 14.2, was designed and pilot-plant–tested for the most promising glass compositions. Normally, several runs are scheduled to check the manufacturability of the superior product concepts.

Manufacturing Stage

In this stage, the manufacturing process for the thin-glass substrate would have been designed, with emphasis on the scale-up of the pilot-plant process. Also, manufacturing costs for the thin-glass substrate would have been estimated. Yet another *House of Quality* would likely have been prepared for the manufacturing process, relating the process variables and materials parameters to the process outputs and product performance. The emphasis would have been on identifying the critical processing and materials parameters that affect the product quality. Furthermore, in recent years, a statistical study using six-sigma methodology, as discussed in Chapter 25, would be conducted to project the capability of the manufacturing process to meet product/customer specifications. Normally, statistical measures are used to ensure stable manufacturing operation and the satisfaction of customer specifications. The manufacturing process is deemed to be stable when it produces consistent products within customer tolerances. When necessary, the market assessment and product recycle management are revisited.

More specifically, for the Corning-7059 glass substrate, it is likely that a manufacturing site was selected to verify the process stability and controllability, and its longer-term behavior. Preferably, several production runs were likely conducted with key processing variables charted to assess the stability and controllability of the process. In addition, the glass quality variation from batch to batch of raw glass materials was probably studied and verified to be sufficiently low.

Typically, demand volumes are estimated based on sales projections, and a 2- to 3-month inventory is created prior to product launch. Note that normally, in an increasingly global economy, manufacturing sites and their *qualification schedules* are planned accordingly. The latter are scheduled manufacturing runs to show long-term capabilities of producing product that meets customer specifications; typically, three runs demonstrate 24-hr continuous, stable manufacturing operation.

Product-Introduction Stage

In the *product-introduction* stage, normally a launch strategy is developed—often including pricing, the launch channel, advertisements, product literature, and early product

Table 15.16 Corning-7059 Product Specifications

Glass Code	Corning-7059
Type	Alkali-Free Boro-silicate
Color	Clear
Principal Use	Optical and Electrical Substrates
Refractive Index at 589.3 nm	1.5333
Density at 20°C	2.76 g/cm^3
Coefficient of Thermal Expansion (20–300°C)	46.0×10^{-7}/°C
Strain Point	539°C
Annealing Point	639°C
Softening Point	844°C
Dielectric Constant at 20°C and 1 MHz	5.84
Dielectric Loss Factor at 20°C and 1 MHz	0.1%
Thickness	0.7 mm +/− 0.127

introduction to a limited market or to selected customers—prior to mass production. For the Corning-7059 glass substrate, as a first-of-a-kind product for flat-panel display manufacturers, pricing was probably difficult to set. Care must have been taken, given that prices rarely increase significantly. It seems clear, however, that Corning used a value-based pricing strategy, as described in Section 2.8, in which the price was set at the perceived value to the customer relative to alternative products. Corning must have chosen this strategy, which proved fruitful, recognizing that there was little competition in the marketplace.

In addition, product literature detailing the properties of Corning-7059 would have been prepared, with a summary shown in Table 15.16. Also, training of the technical-service personnel and the sales force would have been planned. This training might have included the manufacturing of flat-panel displays, in particular the deposition of a-Si on the back-panel displays.

15.3 WASHABLE CRAYON CASE STUDY

This case study is presented to illustrate the steps in the Stage-Gate^{TM} Product-Development Process (SGPDP). Its time frame is assumed to be the mid-1980s, when a design team at Binney & Smith Co. (now Crayola LLC) was developing a washable crayon mixture. As background, the development of the underlying new technologies, from the mid-1980s to mid-2007, was presented in Section 14.3, with each new technology positioned on the *innovation map* for crayon mixtures in Figure 14.5.

Binney & Smith introduced its washable crayon in the early 1990s, in response to requests by parents and teachers to solve the long-standing problem of children using crayons on unintended surfaces, such classroom and home walls, refrigerators, desks and tables, computers, clothing, and other surfaces. The Binney & Smith washable crayon mixture was designed to be wiped off most walls and nonporous household surfaces with just warm water and a sponge, as well as machine-washed from most children's clothing and fabrics. In this case study, many of the considerations in the product-development process are discussed. Note, however, that the authors were not involved in this development process and, consequently, the scenario presented is of the events that were likely to have taken place. Also, the team that developed the washable crayon mixture may not have used the Stage-Gate^{TM} Product-Development Process (SGPDP).

Project Charter

As described in Section 2.2, a well-crafted project charter helps a design team develop the desired product in time and at expected cost. The typical elements, that is, specific goals, project scope, deliverables, and time line, are illustrated in Table 15.17 for the washable crayon mixture, and are discussed next.

Keep in mind that, in 1903, Binney & Smith began with 8 standard crayon colors, which were expanded to 48 in the late 1950s and 64 in the late 1960s. Then, in 1972, 8 fluorescent colors were added, and in 1990, another 8 fluorescent colors were added, to give a total of 80 colors. In 1996, another 16 colors included food colors (e.g., asparagus, Granny Smith apple, and macaroni and cheese), and in 1998, another 24 colors were added, to give the current total of 120 colors. Note that, historically, several colors were dropped and replaced with new colors.

In 1984, Binney & Smith was acquired by Hallmark Card, Inc., which, not surprisingly, expected its new subsidiary to expand its crayon product lines beyond the addition of colors. By the late 1980s, following the introduction of washable Magic Markers in 1987, the Binney & Smith subsidiary embarked on a new product extension to develop a water-washable crayon formula. As discussed in Section 14.3, toxic and flammable household cleansers (such as WD-40) were not acceptable for cleaning crayon markings. Thus, it was deemed desirable to formulate new crayon mixtures that could be easily washed with water for removal from various surfaces (walls, refrigerators, tiles, etc.) and from children's clothing by washing with detergents.

As stated, its specific goals were probably to relieve its customers' grief in dealing with 10-year-olds and younger using crayons beyond their intended surfaces, including walls, desks, and clothing. Because a washable formula was elusive, the design team was probably asked to develop a formula for internal testing and for toxicology certification within a year, as the certification process was anticipated to require considerable time.

Regarding the project scope, emphasis was likely placed on determining more explicit customer requirements for washable crayons. At least for the initial products, it was probably anticipated that just 16 colors would be sufficient. And, the new crayon mixtures were probably to be designed to require minimal changes in the manufacturing facilities for conventional crayons. Finally, the design team was likely

Table 15.17 Possible Project Charter for Washable Crayon Mixtures*

Project Name	Washable Crayon Formula
Project Champions	Business Manager of Crayon Products
Project Leader	Mary Jane Smith
Specific Goals	A washable crayon mixture that meets the current standard product requirements and is water washable (possibly with soap or detergent) from most walls, nonporous household surfaces, and clothing
Project Scope	In-scope: • Determination of acceptable customer requirements for washable crayons • Washable color crayon mixtures for a 16-color set • Minimal changes in the current manufacturing capabilities • Cost target of less than 25 cents per stick or $2 per 8-color box Out-of-scope: • Major manufacturing changes • Nonaqueous solvent for clean wipe off
Deliverables	• Technical feasibility assessment • Manufacturing feasibility assessment • Cost estimation
Time Line	Formula prototypes for internal and toxicology testing within 12 months

*Details of the Binney & Smith product-development effort were not available to the authors.

requested to design a crayon mixture that would keep the price low, say, below $2 per 8-color box (the cost of conventional crayons in the late 1980s).

Furthermore, while the overall customer needs were clear, an acceptable price level was uncertain. For this reason, the development team would have been asked to develop a cost estimate, not a business opportunity assessment. While washable crayons might *cannibalize* the conventional crayon product, this was likely viewed as far more acceptable than a market shift to competitors like Dixon Ticonderoga and Faber-Castell. Note that 15 years later, in 2007, Crayola® sold washable crayons at a 50% price premium [$3 per washable crayon 8-pack compared with $2 for a conventional 8-pack (at Amazon.com)], while Dixon Ticonderoga marketed their washable crayon 8-pack for $1.65. While low-cost competitors offer products with similar or equal quality, clearly, the Crayola® brand name commands a higher price.

Finally, the team was probably requested to demonstrate the technical and manufacturing feasibility of the washable Crayola® mixture.

Turning next to the *concept* stage of the Stage-Gate™ Product-Development Process (SGPDP), which was introduced in Section 2.4, a possible scenario is presented, with emphasis on answering the following four questions:

1. What are the customers' washability expectations?
2. Is it possible to produce a washable crayon mixture that satisfies all of the requirements of conventional crayons?
3. Is it possible to manufacture the washable crayon mixture using the current manufacturing facility? If not, what are the new manufacturing requirements?
4. What is the cost of producing washable crayons?

Concept Stage

The *concept* stage normally begins with several assessments and activities, which are discussed next as they were likely to have occurred during the development of the washable crayon formula in the late 1980s.

a. ***Opportunity Assessments.*** As mentioned above, in this case the market (with its well-defined customer needs and potential response to competitors' pressure) *pushed* the technology and product development, with the market ready for the product when it became available. In such cases, opportunity assessments are less important. Of greater potential consequence in this case, because the washable crayon was likely to command higher prices, was its *cannibalization* of the conventional crayon product, which would yield a lower profit margin. However, note that cannibalization by an internal product is far preferable to the loss of sales to competitors.

In short, for such *market-ready* products, speed to the market is usually more important than optimization of their profitabilities.

b. ***Customer Requirements.*** Although the market was ready for washable crayons, a complete understanding of their *washability* requirements was probably not in hand. Consequently, it was important to capture the *voice of the customers* (VOCs) to understand better their requirements for washable crayons.

Following the methodology in Section 2.4, the design team was likely to have created the following objectives to obtain:

- A definition of washability
- Preferences for cleaning methods
- The nature and types of surfaces to be cleaned

Then, for these objectives, the team probably generated the following kinds of questions:

Washability

- What comes to mind when you clean crayon residues?
- How do you define *washability* when removing crayon residues?
- Which is the biggest challenge in cleaning crayon residues?
- Which crayon color is hardest to remove?
- How do you assess the partial removal of crayon residues? Is *ghosting*, that is, the appearance of a slight residue, acceptable?

Cleaning Methods

- How do you clean crayon residues?
- What cleaning liquids do you use to remove crayon residues?

Nature and Types of Surfaces to Be Cleaned

- On what surfaces do crayon residues appear?
- How would you prefer to wash crayon residues from your children's clothing?
- What objects do your children mark with crayons?
- What is the hardest surface from which to remove crayon residues?

Because most crayon users were children, the customers most likely faced with cleaning crayon residues were parents and teachers. Clearly, these were the preferred interviewees. In addition to interviews, the design team may have arranged visits to schools to observe teachers and custodians as they removed crayon residues from walls, desks, whiteboards, cabinets, etc. It may also have arranged visits to homes to observe parents removing crayon markings. These kinds of observations were probably recorded as so-called *customer images*. During the interviews, the design team probably sought to quantify the relative importance of the customer voices.

After the interviews, the results were likely analyzed and compiled—and the customer voices and images translated into customer requirements, for example:

- Water washable or wipe off—with water being the preferred method of cleaning. Easy wipe off with a wet sponge was frequently cited. Washable with common home laundry detergent was preferred for children's clothing.
- Blue was the most popular color and the most difficult to remove.
- Easy removal from painted walls, wallpaper, metal surfaces such as refrigerators, home and classroom furniture, and children's clothing was most desirable.

Also, the key customer requirements were probably listed, for example, as shown in Table 15.18.

In addition, *fitness-to-standard* (FTS) requirements that customers expected in crayon products were probably listed, for example, as shown in Table 15.19.

Table 15.18 New-Unique-and-Difficult (NUD) Requirements for Washable Crayons

Customer Requirement	Crayon Mixture Requirement	Type	Weighting Factor (%)
Washability	Nontoxic water-washable formula (including the dye and binder)	NUD	70
Washable cleanly from many surfaces	Washable from: • Walls • Wallpaper • Metal surfaces • Wood surfaces • Clothing	NUD	30

Table 15.19 Fitness-to-Standard (FTS) Requirements for Washable Crayons

Customer Requirement	Crayon Mixture Requirement	Type	Weighting Factor (%)
Nontoxic	Nontoxic	FTS	40
Usable on multiple surfaces	Easy marking; good color overlay	FTS	10
Long shelf life	Color stability	FTS	10
Strong	Mechanically strong	FTS	20
Uniform Color	Uniform color	FTS	20

Table 15.20 Technical NUD Requirements for Water-Soluble Crayon Formulation

Crayon Mixture Requirement	Technical Requirement	Target
Water-washable formula	Water-soluble formulation	Low water-insoluble residue (less than 1%)
Washable from: • Walls • Wallpaper • Metal surfaces • Wood surfaces	Clean wipe off from hard surfaces	2 wipes with wet sponge
Washable from clothing	Clean removal from textiles	Clean removal of crayon using regular detergent in a single wash with a gentle cycle without leaving any crayon residue on the surface of the washing machine

In these tables, the weighting factors play an important role, especially when the requirements compete and compromises are necessary. Here, among the NUD requirements, water washability (for both the dyes and crayon binder) is normally considered to be more important than washable cleanly from many surfaces. Among the FTS requirements, nontoxicity is most important, followed by mechanical strength and color uniformity. Note that the relative importance between the NUD and FTS requirements can be difficult to set, as many customers have difficulties assigning priorities. When the design team cannot achieve an ideal product that satisfies all NUD requirements, with compromises needed to be selected from among various product concepts, it may try *conjoint* analysis (Bakken and Frazier, 2006), as discussed next.

Conjoint analysis seeks answers to questions regarding how customers decide on product alternatives with multiple attributes. In this analysis, customers are presented with choices that satisfy several requirements to various degrees at different costs. Using a well-designed survey, customers help the development team decide on the product alternatives to be launched. At one extreme, the best product has a high price, while at the other extreme it barely meets several NUD requirements, but is inexpensive. Given the distribution of votes for the various product alternatives, and depending on the market segment preferred and the potential product positioning, specific products normally emerge as winners.

c. ***Technical Requirements.*** Because washable crayons extend the conventional crayon product, most of the technical requirements are known, including nontoxicity, heat stability, odorless behavior, low viscosity, mold release properties, mechanical strength, color uniformity, color blending, color overlay ability, UV light stability, and marking properties. The translation of the new NUD requirements for washable crayons is discussed next.

d. ***Determination of Critical-to-Quality (CTQ) Variables.*** Normally, the NUD requirements translate into the *critical-to-quality* variables. And consequently, it is likely that *water washability* and *washable cleanly from many surfaces* were selected during the development of the washable crayon mixtures. A typical translation of these requirements into CTQ variables is shown in Table 15.20.

Following the recommendations in Section 2.4, a House of Quality (HOQ) is normally prepared, relating the various requirements to one another. For this case study, the reader is asked to prepare an HOQ in Exercise 15.3.

e. ***Development of Product Concepts.*** As discussed in Section 2.4, the identification and selection of solution concepts is at the heart of new product development, especially when several alternatives are available. In the late 1980s, when the chemists and chemical engineers at Binney & Smith were developing water-soluble crayon mixtures, their new formulations built upon the work of Otis Bill Woolly, who filed for a patent in 1975 on water-soluble coating materials for possible use in crayons (U.S. Patent 3,993,492). Woolly's mixtures included three groupings of polyethylene glycol (PEG) blends having different molecular weight ranges: low (400–800), medium (2,000–4,500), and high (6,000–7,500). Unfortunately, while their washability performance was acceptable, their physical properties, mechanical strength, and marking characteristics were inferior to those of conventional crayons.

Subsequently, in 1988, Colin Snedeker of Binney & Smith filed for a patent (U.S. Patent 4,978,390) that modified the Woolly formulation by excluding PEG resins having an average molecular weight below 7,000. This led to their high-molecular-weight PEG washable crayon mixture, which circumvented lower molecular weight resins with reduced mechanical strength, dimensional stability, and marking performance.

As the design team considered this solution concept, the Pugh matrix (Pugh, 1996), discussed in Section 2.4, may have been helpful for screening purposes. In this matrix, shown in Table 15.21 for the washable crayon mixture, each potential solution concept is judged against

Table 15.21 Pugh Matrix Concept Selection

Technical Requirement	Target	Reference Formula	Formula A Standard Crayon	Formula B-1, Snedeker	Formula B-2, Snedeker
Water-soluble formulation	Low water-insolubles content		−−−	0	0
Clean wipe off from hard surfaces	2 wipes with wet sponge		−−−	+	0
Clean removal from textiles	Clean removal of crayon using regular detergent in a single wash with a gentle cycle		−−−	0	+
Nontoxicity	No toxicity		0	0	0
Heat stability	Below 50°C				
Odor	None	U.S. Patent 3,993,492 —Woolly	0	0	0
Mechanical strength	Below 2,000 g		+++	+	+
Color uniformity	5 = excellent		0	0	0
Color overlay	5 = excellent		0	0	0
UV light stability	5 = excellent		0	0	0
Marking properties	5 = excellent		+++	+	+
Low viscosity	< 10 cp at 190°F and 50 rpm		+++	−	−
Color blending	5 = excellent		+++	0/−	0/−
Mold release	5 = excellent		+++	+	+

a reference solution. Usually, the reference concept is the best known in the market (or the best potential solution), which in the 1980s was the Woolly washable crayon formulation (U.S. Patent 3,993,492). For each potential concept, each technical requirement is compared against that for the reference concept and assigned a qualitative valuation of inferior (−), superior (+), or equal (0). Multiple (−) or (+) entries signify decreasing or increasing levels of inferiority or superiority.

The performance comparisons between the Woolly reference formulation and the conventional crayon formulation (Formula A), as well as the improved Snedeker formulation (Formula B), are shown in Table 15.21. Alternatively, the conventional formulation may have been used as the reference. Note that the standard crayon formulation is included in the Pugh analysis as the control for the physical properties, mechanical strength, marking properties, and ease of manufacture. Also, two variants of the Snedeker formulation are labeled as Formulas B-1 and B-2.

Note that Formulas B-1 and B-2 perform similarly, except that Formula B-1 is more easily removed from hard surfaces and Formula B-2 is more easily removed from textiles.

f. *Selection of Superior Product Concepts.* In the *concept* stage of the Stage-Gate™ Product-Development Process (SGPDP), the selection of superior product concepts is based primarily on the satisfaction of the technical requirements, in particular, the *new-unique-and-difficult* (NUD) requirements. Given Table 15.21, it seems likely that the design team would have sought to create a compromise Formula B*, that provides adequate washability from hard surfaces and textiles. Such a potential product is compared in the improved Pugh matrix of Table 15.22. Note that while Formula B* achieves these objectives, it performs less favorably in mechanical strength and marking properties (giving more flaky and less smooth markings) and is less compatible (due to higher viscosity and higher melting temperatures) with the existing manufacturing processes (batch liquid blending and gravity molding) than conventional crayon formulations. However, the superior concept, Formula B*, being the best possible in the late 1980s, was acceptable as a first-generation product. Often, companies introduce a less-than-optimal product to capture the market before their competitors. Their competitors often face resistance to change on the part of customers that have adopted the first product to market.

g. *Unit-Cost Estimation.* Having selected a superior product concept, it remained to estimate the cost of manufacturing water-washable crayons. Because Binney & Smith had been manufacturing conventional crayons since 1903, most of the cost components were known. Only the effect of replacing paraffin wax with high-molecular-weight PEG resin had to be estimated. This was accomplished by estimating a 25% increase in materials costs and an overall processing yield reduction of 60% due to higher rejection rates (because of increased tip breakages). Note that in the late 1980s, the

Table 15.22 Subsequent Pugh Matrix—Superior Concept Selection

Technical Requirement	Target	Reference Formula	Formula A Standard Crayon	Formula B* Snedeker
Water-soluble formulation	Low water-insolubles content		−−−	0
Clean wipe off from hard surfaces	2 wipes with wet sponge		−−−	+
Clean removal from textiles	Clean removal of crayon using regular detergent in a single wash with a gentle cycle		−−−	+
Nontoxicity	No toxicity		0	0
Heat stability	Below 50°C			
Odor	None	U.S. Patent 3,993,492 —Woolly	0	0
Mechanical strength	Below 2,000 g		+++	+
Color uniformity	5 = excellent		0	0
Color overlay	5 = excellent		0	0
UV light stability	5 = excellent		0	0
Marking property	5 = excellent		+++	+
Low viscosity	< 10 cp at 190°F and 50 rpm		+++	−
Color blending	5 = excellent		+++	0/−
Mold release	5 = excellent		+++	+

unit cost for conventional crayons was 15 cents per crayon, with a 5-cent cost of materials and a 10-cent processing cost. On this basis, the unit cost of washable crayons was estimated, as shown in Table 15.23.

Consequently, an 8-color pack of washable crayons would have been estimated to cost $1.83, just below the target price of $2 per pack, but at a cost exceeding that of conventional crayons by 53%. This would have required a higher selling price to maintain a similar net profit. Note that in the late 1980s, the suggested selling price for an 8-color pack of conventional crayons was $2.

h. *Gate Review.* To complete the *concept* stage, a gate review might have been carried out in which the washable crayon team answered the four questions associated with the deliverables. Such a review, which may or may not have been conducted by Binney & Smith in the 1980s, is not discussed here. However, based on the discussion above, the design team had convincing answers for all of its questions. It had determined its customers' requirements; had developed a viable washable crayon formulation, although its mechanical properties and marking characteristics were inferior compared to conventional crayons; and had met the price target in spite of increased materials cost and loss of productivity due to lower mechanical strength. Given such positive results, the Binney & Smith business decision makers were likely to have responded affirmatively, granting funding to proceed to the *feasibility* stage. Note that more complete coverage of the gate reviews for the *concept* stage is provided in Section 2.4, and a detailed gate review is presented in a case study for the design of a halogen light bulb in Section 17.2.

Table 15.23 Unit Cost for Washable Crayons

Cost Component	Conventional Crayons (cent/stick)	Washable Crayons (cent/stick)
Materials	5	$1.25 \times 5 = 6.25$
Processing	10	$1/0.6 \times 10 = 16.67$
Total Cost	15	22.92

Fifteen years later, it is noteworthy that Binney & Smith launched their first washable crayons in 1991 with a suggested selling price of $2.99 for an 8-color pack. Then, in 1999, they filed for a patent (Fistner, U.S. Patent 6,039,797) that reduced many of the deficiencies in the Snedeker formulation. This formulation provides mechanical strengths equivalent to those of conventional crayons with a much lower viscosity. This improved viscosity, coupled with high mechanical strength, provides a simpler molding process having fewer defects; that is, lower processing costs. However, as introduced in Section 14.3, this new formulation contains alkoxylated fatty acids (e.g., stearic and lauric acids) that are produced in high-pressure reactors, increasing the cost of materials significantly. To estimate the effect on its selling price, the reader is referred to Exercise 15.4.

15.4 SUMMARY

Case studies for the design of two industrial chemical products have been presented. Emphasis has been placed on the project charter, the role of the innovation map, and the *concept* stage of the Stage-Gate™ Product-Development Process (SGPDP).

REFERENCES

1. Bakken, D., and C.L., Frazier, *Conjoint Analysis—Understanding Consumer Decision Making*, Chapter 15 of R. Grover, and M. Vriens, Eds., *The Handbook of Marketing Research: Uses, Misuses, and Future Advances*, Sage Publications (2006).

2. Belliveau, P., A. Griffin, and S. Somermeyer, Eds., *The PDMA ToolBook 2 for New Product Development*, John Wiley & Sons (2004).

3. Corning Financial and Annual Reports 2000–2006, Corning Inc.

4. Creveling, C.M., J.L. Slutsky, and D. Antis, Jr., *Design for Six Sigma in Technology and Product Development*, Pearson Education (2003).

5. Display Technology in Japan, World Technology Evaluation Center, http://www.wtec.org/reports.htm (1992).

6. Ellison, A., "Glass Substrates for Thin Film Transistor Displays," http://www.mse.cornell.edu/courses/mse407/Corning-CU-talk%2009-05.ppt.

7. Lapp, J. C., "AMLCD Substrates Trends in Technology," Corning Technical Information Paper (2004a).

8. Lapp, J. C., "Glass Substrates for AMLCD Applications; Properties and Implications," Corning Technical Information Paper (2004b).

9. Pugh, S., *Creating Innovative Products Using Total Design*, Addison Wesley Longman (1996).

10. 7059 Barium-Borosilicate Glass, Präzisions glass & optik, http://www.pgo-online.com/intl/katalog/7059.html.

Patents—Thin-Glass Substrates

11. U.S. Patent 3,338,696, Dockerty, S.M., *Sheet Forming Apparatus* (1967).

12. U.S. Patent 4,634,683 Dumbaugh, W.H., *Barium and/or Strontium Aluminosilicate Crystal-containing Glasses for Flat Panel Display Devices* (1987).

13. U.S. Patent 4,824,808, Dumbaugh, W.H., *Substrate Glass for Liquid Crystal Displays* (1989).

Patents—Crayon Mixtures

14. U.S. Patent 3,993,492, Woolly, O.B., *Water Soluble Transfer Coating Materials and Articles Incorporating the Same* (1976).

15. U.S. Patent 4,978,390, Snedeker, C.M., *Washable Solid Marking Composition* (1990).

16. U.S. Patent 6,039,797, Fistner, D.C., *Washable Marking Composition* (2000).

EXERCISES

15.1 Design an improved version of the Corning-7059 glass substrate with a coefficient of thermal expansion closer to that of a-Si. Use the technology described in Section 14.2 for the Corning-1737 glass substrate, developed in the early 1990s. You can also make use of patents filed by Corning.

(a) Develop the technical requirements.

(b) Prepare the first House of Quality with typical weighting factors.

(c) Suggest superior product concepts and prepare Pugh matrices.

(d) Complete recommendations for the concept stage review.

15.2 Repeat Exercise 15.1 for development of Eagle™-XG substrate glass, taking into account the changes in the competitive landscape, especially the need for a more environmentally friendly product.

15.3 Develop the first House of Quality for the NUD requirements of the washable crayon formulation given in Table 15.20.

15.4 Repeat the washable Crayola case study in Section 15.3 using the new crayon mixture formulation by Fistner in U.S. Patent 6,039,797. For cost estimation, assume that the overall processing yield is 95% of the conventional crayon yield and that the material cost is doubled. Can a cost target of $2 per 8-pack of washable color crayons be met?

Part Three

Product Design for Configured Consumer Products

Part Three presents, in two chapters, strategies for the design of *configured consumer chemical* products and the processes to produce them. It follows the Stage-Gate™ Product-Development Process (SGPDP), which was introduced in Chapters 1 and 2, and is presented for the design of *configured consumer chemical* products in Figure PIII.1. As discussed in Section 1.3, like *basic* and *industrial chemicals*, *configured consumer chemical* products are characterized by thermophysical and transport properties; and like *industrial chemicals*, their other properties are often dominant in satisfying customer needs, including microstructure; particle-size distribution; and functional (e.g., cleansing, adhesion, shape), sensorial (e.g., feel, smell), rheological (non-Newtonian flow), and physical (e.g., stability) properties. Unlike most *industrial chemicals*, *configured consumer chemical* products are sold directly to the consumer. In some cases, their three-dimensional configurations are crucial in satisfying customer needs; for example, the configurations of a home hemodialysis device, discussed in Sections 16.3 and 17.3, and a lab-on-a-chip for high-throughput screening, discussed in Sections 16.4 and 17.4. Other configured consumer products include halogen light bulbs, discussed in Sections 16.2 and 17.2; espresso machines, discussed in Section 25.3; hand lotions in plastic containers; and hand warmers in polymer sacks.

Next, the steps in Figure PIII.1 for the design of configured consumer chemical products are introduced. Subsequently, they are illustrated in the innovation maps of Chapter 16 and the case studies of Chapter 17.

MATERIALS, PROCESS/MANUFACTURING, AND PRODUCT TECHNOLOGIES DEVELOPMENT

After the design team creates its project charter, as discussed in Section 2.2, it seeks to identify appropriate materials technologies to achieve its objectives when they are needed. This is the step at the top left of Figure PIII.1, which, for *configured consumer chemical* products, usually involves a search for the appropriate molecules or mixtures of molecules to satisfy the other property specifications, in addition to thermophysical and transport properties that align closely with customer needs. Examples include: (1) pastes and creams (packaged in a container for consumer use), that is, colloids having the appropriate functional, sensorial, rheological, and physical properties; and (2) defect-free tungsten for filaments in longer-life incandescent light bulbs.

For many *configured consumer chemical* products to achieve these desired properties, it becomes necessary to invent or utilize new so-called *process/manufacturing* technologies. For example, as discussed in Section 16.2, Coolidge developed a process for producing ductile tungsten rods, which permitted the manufacture of low-cost incandescent light bulbs.

Also, for many *configured consumer chemical* products to satisfy other customer needs, it is necessary to invent or utilize new so-called *product* technologies, which include (1) microstructures that often characterize *industrial chemicals*, as discussed in Part Two; and (2) secondary or supporting devices and the like that are an integral part of the final product construction. For example, *product* technologies include: (1) the quartz

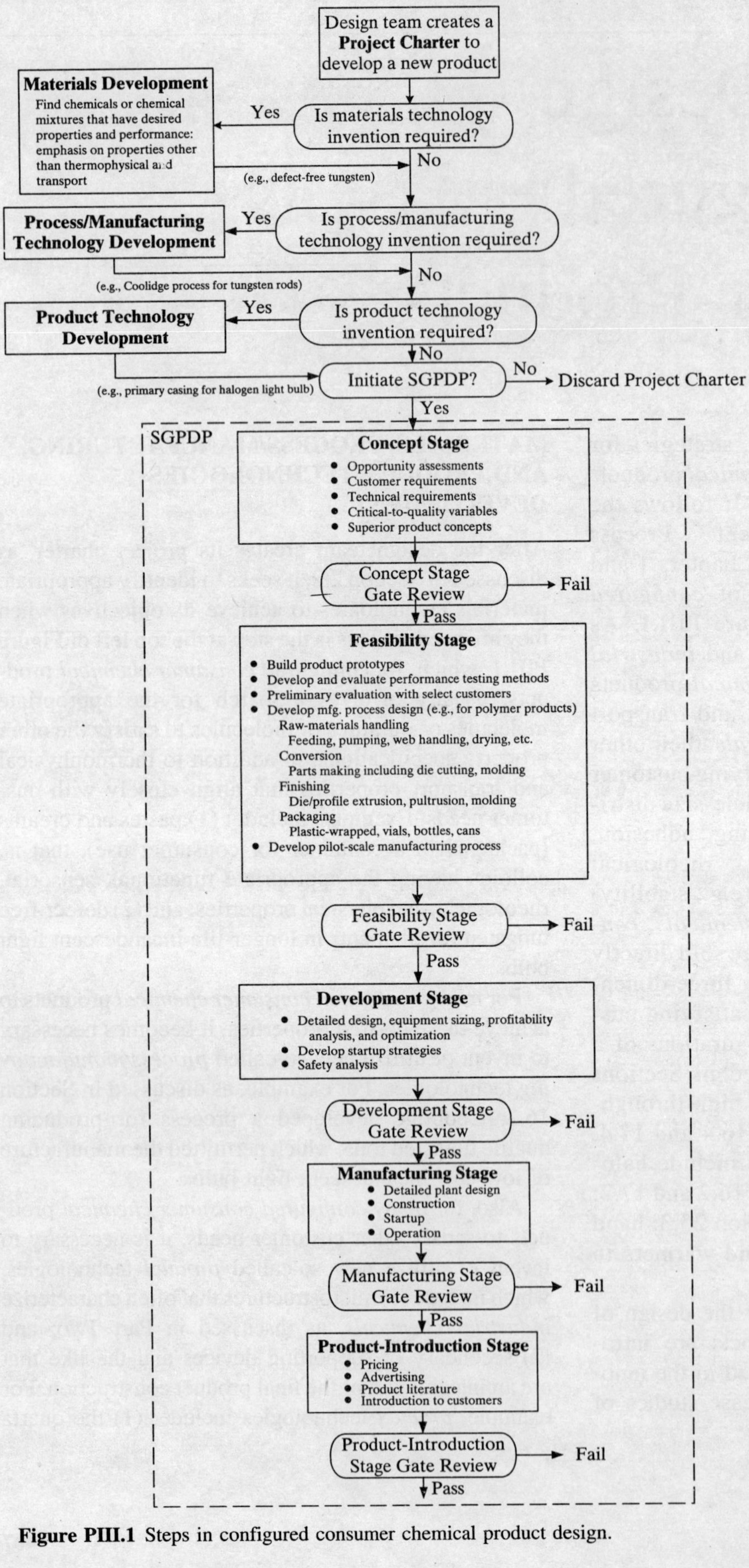

Figure PIII.1 Steps in configured consumer chemical product design.

encasement for a halogen light bulb; (2) Internet communications and alarms to monitor a home hemodialysis product; (3) automated image analysis and database storage for a lab-on-a-chip product; (4) thin-film transistor networks using polymer semiconductors on flexible displays, (5) chip-on-glass structures for LCDs permitted by specialty-glass substrates designed to match the thermal coefficient of expansion of silicon, and (6) multilayer optical films that are the primary building blocks for the Dual Brightness Enhancement Film (DBEF) on LCDs. Note that many of the *product* technologies are not chemically oriented. Yet, they are necessary to permit chemically oriented products to satisfy customer needs. Many *configured consumer chemical* products involve new technologies provided by other companies, often the work of engineers and scientists in other disciplines, often located in geographically distributed locations around the world.

The next step is for the design team to formulate an *innovation map*, as discussed in Section 1.3, and to decide whether sufficient new technologies are in place to satisfy the anticipated customer needs, that is, the *voice of the customer.* When these technologies are in place, the design team initiates the SGPDP; otherwise, the product charter is rejected.

CONCEPT STAGE

As discussed in Section 2.4, the *concept* stage focuses on: (1) making opportunity assessments, (2) identifying customer requirements, (3) identifying technical requirements, (4) determining the critical-to-quality (CTQ) variables, and (5) determining the superior product concepts. Note that these items are somewhat more difficult to achieve for *configured consumer chemical* products (and *industrial chemicals*) as compared with *basic chemicals* because the former's customer needs are more difficult to translate into technical requirements, and the generation of superior product concepts is normally more complex. However, there are similarities in that, for *basic chemicals*, preliminary process synthesis leads to a tree of promising flowsheets, while superior product concepts for *industrial chemicals* and *configured consumer chemical* products are selected from those generated earlier in the *concept* stage.

FEASIBILITY STAGE

Normally, product prototypes are prepared to demonstrate the feasibility of the superior product concepts and performance testing methods are developed and evaluated. Preliminary evaluation is carried out, and customers are selected for testing. When promising, a process is designed to manufacture the product. Given the broader array of new technologies incorporated into *configured consumer chemical* products as compared with industrial chemicals, these steps often require the expertise of mechanical engineers, materials specialists, and others.

When designing the manufacturing process, because the products are configured (at least involving the packaging of three-dimensional containers—e.g., vials, tubes, bottles), an assembly line with robots and advanced control systems is normally needed. Again, the designs of such manufacturing facilities are the responsibilities of mechanical engineers and related technologists.

Note that Figure PIII.1 illustrates the processing operations utilized in producing many polymer products. For *configured consumer chemical* products (and *industrial chemicals*), unlike for *basic chemicals*, these operations depend upon the technology platforms involved. Incandescent light bulbs require the preparation of a coiled tungsten filament and the creation of a blown glass enclosure. Clearly, the operations for raw-materials handling, conversion, finishing, and packaging differ significantly from those for polymer products.

DEVELOPMENT STAGE

Detailed Design, Equipment Sizing, Profitability Analysis, and Optimization

For a new process to produce *configured consumer chemical* products, the design team usually receives additional assistance in carrying out the detailed process design, equipment sizing and capital-cost estimation, profitability analysis, and optimization of the process. These topics are covered in separate chapters in Part Four, which begins the *development* stage of the SGPDP. Methods for estimating capital and operating costs, and computing profitability measures, are provided in Chapters 22 and 23. Optimization methods are presented in Chapter 24.

When the detailed manufacturing process design is completed, the economic feasibility of the process is checked to confirm that the company's profitability requirements have been met. If this proves unsatisfactory, the design team determines whether the process is still promising. If so, the team returns to an earlier step to make changes that it hopes will improve the profitability. Otherwise, this process design is rejected.

Develop Startup Strategies

For *configured consumer chemical* products as compared with *basic chemicals* and *industrial chemicals*, startup strategies are considerably more complex, and consequently, qualitative approaches are often used.

Safety Analysis

Another crucial activity involves a *formal* analysis of the reliability and safety of the proposed manufacturing process, as discussed in Section 1.5. Note that, as discussed in Section 1.5 and throughout the book, these considerations must be foremost throughout the design process. Formal safety analysis usually involves laboratory and pilot-plant testing to confirm that typical faults cannot propagate to create accidents such as explosions or fires. Often, HAZOP (*Haz*ard and *Op*erability) analyses are carried out to check systematically all of the anticipated eventualities. Methods for and examples of HAZOP analysis, together with risk assessment, are presented in Section 1.5. Also, the reader is referred to the text by Crowl and Louvar (1990) and the following books developed by the Center for Chemical Process Safety of the American Institute of Chemical Engineers:

1. *Safety, Health, and Loss Prevention in Chemical Processes: Problems for Undergraduate Engineering Curricula—Student Problems* (1990).
2. *Guidelines for Hazard Evaluation Procedures, Second Edition with Worked Examples* (1992).
3. *Self-Study Course: Risk Assessment* (2002).

The latter reference is particularly noteworthy for instructors because it provides a PowerPoint file that can be integrated into a safety lecture.

MANUFACTURING STAGE

Plant Design, Construction, Startup, and Operation

Detailed plant design, construction, startup, and operation are carried out in the *manufacturing* stage of the SGPDP, as shown in Figure PIII.1. In creating the plant design for a *configured consumer chemical* product, much detailed work is done, often by contractors, using many mechanical, civil, and electrical engineers. For processes that produce *configured consumer chemical* products, engineers complete equipment drawings, piping diagrams, instrumentation diagrams, the equipment layout, the construction of a scale model, and the preparation of bids. Then, the construction phase is entered, in which engineers and project managers play a leading role. The design team often returns to assist in plant startup and operation. Note that the final design and construction activities are usually not the responsibilities of chemical engineers.

PRODUCT-INTRODUCTION STAGE

As the plant comes online, product-launch strategies are normally implemented. These include setting the product price, marketing and advertising to prospective customers, preparing and distributing product literature, and introducing the product to selected customers. These are normally the responsibilities of sales and marketing personnel, many of whom have been trained as chemical engineers.

SUMMARY

This brief introduction to Figure PIII.1 should give the reader a good appreciation of the subjects to be learned in the design of *configured consumer chemical* products and processes, and how this text is organized to describe the design methodologies.

REFERENCES

1. American Institute of Chemical Engineers, *Safety, Health, and Loss Prevention in Chemical Processes: Problems for Undergraduate Engineering Curricula—Student Problems*, AIChE, New York (1990).

2. American Institute of Chemical Engineers, *Guidelines for Hazard Evaluation Procedures, Second Edition with Worked Examples*, AIChE, New York (1992).

3. American Institute of Chemical Engineers, *Self-Study Course: Risk Assessment*, AIChE, New York (2002).

4. Crowl, D.A., and J.F. Louvar, *Chemical Process Safety: Fundamentals with Applications*, Prentice-Hall, Englewood Cliffs, New Jersey (1990).

Chapter 16

New Technologies for Configured Consumer Products

16.0 OBJECTIVES

New product-development programs are often plagued by the need for technological inventions that prolong the product-development process. In these cases, both product and technology developments are often carried out concurrently to reduce the product-development time. Yet, it is recommended that these two activities be decoupled as much as possible. In this chapter, the *innovation map* is used to show how perceived customer needs are coupled to the development of new technologies.

Herein, *innovation maps* are developed for the configured consumer products discussed in the case studies of Chapter 17. The utility of these *innovation maps* is shown for converting material, process/manufacturing, and product technologies into inventions necessary and sufficient to meet perceived customer requirements. The case studies include three new products: halogen light bulbs, home hemodialysis devices, and high-throughput screening devices for kinase inhibitors.

After studying this chapter, the reader should:

1. Be able to construct an *innovation map* for a configured consumer product.
2. Be able to identify critical inventions for materials, process/manufacturing, and product technologies for configured consumer products.
3. Appreciate the need for technology-protection strategies.

16.1 INTRODUCTION

Like *industrial chemical* products, *configured consumer chemical* products usually focus on properties beyond the normal thermophysical and transport properties of their pure species and mixtures. However, unlike *industrial chemical* products, *configured consumer chemical* products are normally sold directly to consumers. Also, in many cases, aspects of their three-dimensional configurations are important in satisfying customer needs. This is also the case for some *industrial chemical* products, for example, thin films with imprinted nano-structures.

A few examples of *configured consumer chemical* products, and their properties attractive to consumers, are:

1. *Halogen light bulbs.* These normally involve bromine, which at high temperature (~3,100 K) reverses the evaporation of tungsten filaments, extending the lifetime of incandescent light bulbs. As discussed in Sections 16.2 and 17.2, tungsten filaments surrounded by a bromine atmosphere are encapsulated in primary and secondary casings to protect consumers from the high-temperature reactions.
2. *Hemodialysis devices.* These are mass exchangers, involving hollow-fiber bundles designed to transfer urea from the blood to a dialysate solution. As discussed in Sections 16.3 and 17.3, the potential for home hemodialysis devices, which operate while the patient sleeps, suggests adjusted configurations for the mass exchanger and the need for automated alarm and communication systems.
3. *Soap bars.* The esterification reactions in the manufacture of soap—involving fatty acids (e.g., stearic and coconut acids) and surfactants (e.g., sodium isethionate) to produce detergents (e.g., sodium acyl isethionate)—are well understood. However, the manufacture of soap bars involves several additional species (e.g., sodium soap, containing 80% tallow and 20% coconut; alkyl benzene sulfonate; water; perfume; salts) to prevent cracking, permit extrusion and stamping, and provide firmness, a rich, creamy lather, absence of grit, no unpleasant odors or colors, a slippery feel when wet, low mush rate, and other so-called secondary properties. Of special note is an article by Hill and Post (2007), in which they indicate that for

over 50 years, the composition of the Dove® soap bar has not changed significantly. Initially, when it was promoted as not leaving a bathtub ring, sales were modest. However, when marketing was changed, approximately 25 years later, to emphasize extreme mildness to the skin, the Dove® Beauty Bar became the best seller in the United States.

4. *Ice cream.* The best-tasting ice creams involve a well-defined blending recipe and careful handling procedures. The ingredients are simple: principally water (from milk and cream), sweeteners (corn syrup or sugar), flavorings, emulsifiers, milk solids, milk fats, and air (between 20–50% by volume). A key to success is in achieving the proper colloidal structure; that is, a dispersion of tiny air bubbles and ice crystals among liquid water and a network of destabilized fat globules. When biting into ice cream, it is thought that the flavor released into one's mouth is a function of the structure. And, consequently, it is important to maintain the structure with proper refrigeration.
5. *Cheese substitutes.* These substitutes, often consisting of caseinates and vegetable oils, have comparable nutritional and textural characteristics to natural cheese, with the caseinates serving as the protein source and polyunsaturated vegetable fats and oils providing a cholesterol-free product. While flavor differences from conventional cheese are usually detectable, when incorporated as an ingredient, for example, in pizza, the distinctions may not be significant.
6. *Labs-on-a-chip.* These are microfluidic devices that create nano-liter droplets of thousands of potential therapeutic drugs to assess their effectiveness in inhibiting the adverse reactions of enzymes and proteins. As discussed in Sections 16.4 and 17.4, these devices currently have the potential to test on the order of 10,000 kinase inhibitors at varying concentrations, as they interact with 100 kinase enzymes in a single day. For these analysis products, their three-dimensional configurations are important to achieve successful designs.

Like *industrial chemical* products, many *configured consumer chemical* products are discussed in several articles and books that focus on product design (e.g., Westerberg and Subrahmanian, 2000; Cussler and Moggridge, 2001; Shaeiwitz and Turton, 2001; Cussler et al., 2002; Favre et al., 2002, 2005; Cussler and Wei, 2003; Hill, 2004; Saraiva and Costa, 2004; Seider et al., 2004a, b; Costa et al., 2006; Bröckel et al., 2007a, b; Ng et al., 2007; Wei, 2007). And, like *industrial chemical* products, *configured consumer chemical* products often involve complex phenomena such as multiphase interactions, amorphous and crystalline structures with dislocations and defects, surface roughness, and stress-strain relationships. Often new materials technologies underlie new *configured consumer chemical* products; for example, the discovery of bromine reactions that reverse the evaporation of tungsten in light bulbs. In many cases, new process/manufacturing technologies underlie the development of new products; for example, an emulsion gun that creates nano-liter aqueous droplets in flowing oil at high frequency (20,000 drop/sec). And, for configured products, whether *industrial* or *configured consumer chemical* products, new product technologies often underlie new products; for example, arrays of thin-film transistors (TFTs) on LCDs.

This chapter focuses on the initial steps in the design of *configured consumer chemical* products, after a design team has created its project charter. These steps, shown in Figure PIII.1, involve the selection of materials, process/manufacturing, and product technologies. These are normally new technologies intended to create products that satisfy customer needs while offering a competitive advantage. As discussed in Sections 1.3 and 2.2, to achieve these objectives, it is helpful to create *innovation maps* that show the connections between the three technological components and customer satisfaction, that is, the *customer-value proposition*. The success of new products often relies on careful attention to this interplay. As shown in Figure PIII.1, when the *innovation maps* are promising, the design team often begins product design following the Stage-Gate™ Product-Development Process (SGPDP). Chapter 17 presents case studies of *configured consumer chemical* product design following the SGPDP, beginning with the *innovation maps* discussed in this chapter.

In the sections that follow, *innovation maps* are created for potential product designs involving halogen light bulbs, home hemodialysis devices, and labs-on-a-chip for high-throughput screening of pharmaceuticals. In each section, a brief history of the technological developments is presented before the innovation map is created. Note that these innovation maps are prepared in hindsight, while in practice, design teams revise their innovation maps as they make progress in their designs.

16.2 INNOVATION MAP FOR THE INCANDESCENT LIGHT BULB

The basic principle in creating incandescent light involves the passing of electrical current through a filament, with the heat generated radiating light in the visible range. Beginning in the early 1800s, Humphrey Davy, an electrochemist, demonstrated this concept by passing electrical current through a platinum strip. Then, in about 1820, a French inventor, Warren de La Rue, made the first incandescent bulb by passing electrical current through a platinum coil in a vacuum glass tube.

Twenty years later, Sir William Robert Grove, the inventor of the fuel cell, used a concept similar to that of de La Rue to light a public theater. In 1878, Joseph Swan replaced the very expensive platinum coil with a much less expensive carbon filament. This was the *turning point* in the mass production of electricity as we know it today.

It was not until 1880 that Edison successfully commercialized the production of incandescent bulbs. Edison was the first to develop an electrical power supply, an electricity delivery system, and a manufacturing process to produce cheaper and longer-lasting bulbs. Because Swan first patented the method of manufacturing carbon filaments, Edison made him a business partner to gain the right to manufacture carbon filaments and provide an electrical power supply system to end users. Edison's contribution to electric lighting wasn't the *invention* of the light bulb, but rather the development of its mass production and an electrical delivery system.

The major disadvantage of the carbon-based filament is the production of combustion products that turn the bulb dark black. This was overcome in 1903, when William Coolidge invented an improved method of making tungsten filaments. The tungsten filament outlasted all other types of filaments and, consequently, Coolidge was able to manufacture affordable bulbs. He succeeded in preparing a ductile tungsten wire by doping tungsten oxide before its reduction to tungsten. The resulting metal powder was pressed, sintered, and forged into thin rods. Very thin wires were then drawn from these rods.

In 1906, the General Electric Company patented a method of making tungsten filaments for use in incandescent light bulbs. Tungsten filaments offer a high melting temperature and low vapor pressures, which translate to a lower evaporation rate of tungsten vapor and reduced *blackening*.

Another GE researcher, Irving Langmuir, took advantage of the ductile tungsten rods produced by the Coolidge process. Prior to the work of Langmuir, GE bulbs were evacuated so that oxygen would not oxidize the filaments. In a vacuum, low-wattage bulbs performed sufficiently well, but the tungsten wires in high-wattage bulbs performed poorly. Tungsten vapor was gradually deposited on the inside walls of the bulbs, turning them black. Langmuir realized that tungsten evaporation could be suppressed by filling the light bulb with an inert gas that wouldn't react with the filament. However, as the inert gas circulated in the bulb, it carried away too much heat, which significantly reduced the brightness of the bulb. In earlier research, Langmuir learned that a filament wound in a tight coil releases significantly less heat. Applying this observation, he invented tightly coiled tungsten filaments using the Coolidge process to produce highly ductile tungsten rods. The tightly coiled tungsten filament in an inert gas was described in his 1916 patent, which continues as the basis for modern incandescent light bulbs. For a more complete description of the invention of the first long-life, high-wattage incandescent light bulb, see this patent (U.S. Patent 1,180,159) and the URL (http://americanhistory.si.edu/lighting/history/patents/mosby1.htm).

Innovation and Product Design of the Incandescent Light Bulb

The preceding subsection provides an historical account of the invention of the incandescent light bulb. Through hindsight, the relationships among the many inventions that led to the successful production of long-life, high-wattage light bulbs have been constructed.

As introduced in Section 1.3, the *innovation map* relates the technological components of product developments to the technical advantages, showing the technical differentiation, ultimately to the satisfaction of the *customer-value proposition.*

For the product development of the incandescent light bulb, consider the *innovation map* in Figure 16.1. Note that while this map is readily generated in hindsight, the generation of such a map can be very helpful during product design. In the early 1900s, the challenge was to produce long-life light bulbs—with vacuum light bulbs having carbon filaments as the state-of-the-art, a technology that served consumer needs well at low-wattage requirements. At the time of Langmuir's invention, tungsten was considered to be the front-runner to replace carbon and platinum filaments.

To construct the *innovation map* of the modern incandescent light bulb in hindsight, one must first identify the elements in its six levels, moving from the bottom to the top of the map:

1. *Materials Technology*: platinum, carbon, tungsten, inert gases, glass (clear, frosted)
2. *Process/Manufacturing Technology*: invention of the Coolidge manufacturing process to produce ductile tungsten rod, the inert-gas-filled bulb, the vacuum bulb
3. *Technical Differentiation (Technical-Value Proposition)*: low-cost manufacturing, higher melting point, lower evaporation rate.
4. *Product Technology*: vacuum bulb, gas-filled bulb, tightly coiled filaments, low- and high-wattage bulbs
5. *Products*: light bulb, long-life incandescent light bulb, coiled light bulb, conical light bulb, frosted light bulb, . . .
6. *Customer-Value Proposition*: long-life light bulb, low cost, versatility of shape, light quality (warm, cool, daylight)

Note that in product design, the time frame is current, with the key inventions normally protected by patents, as discussed under "Technology Protection" later in this section. For the incandescent light bulb, Figure 16.1 shows an innovation map that might have been prepared by the primary manufacturers (e.g., General Electric Co.) in the 1920s.

After identifying the elements at all six levels of the innovation map, their connectivity is added to show the interplay between the technological elements, the *technical-value proposition*, and ultimately the *customer-value proposition.* Where there is an unmet customer need, such as the versatility of shape, its resolution can be an objective for the next generation of products.

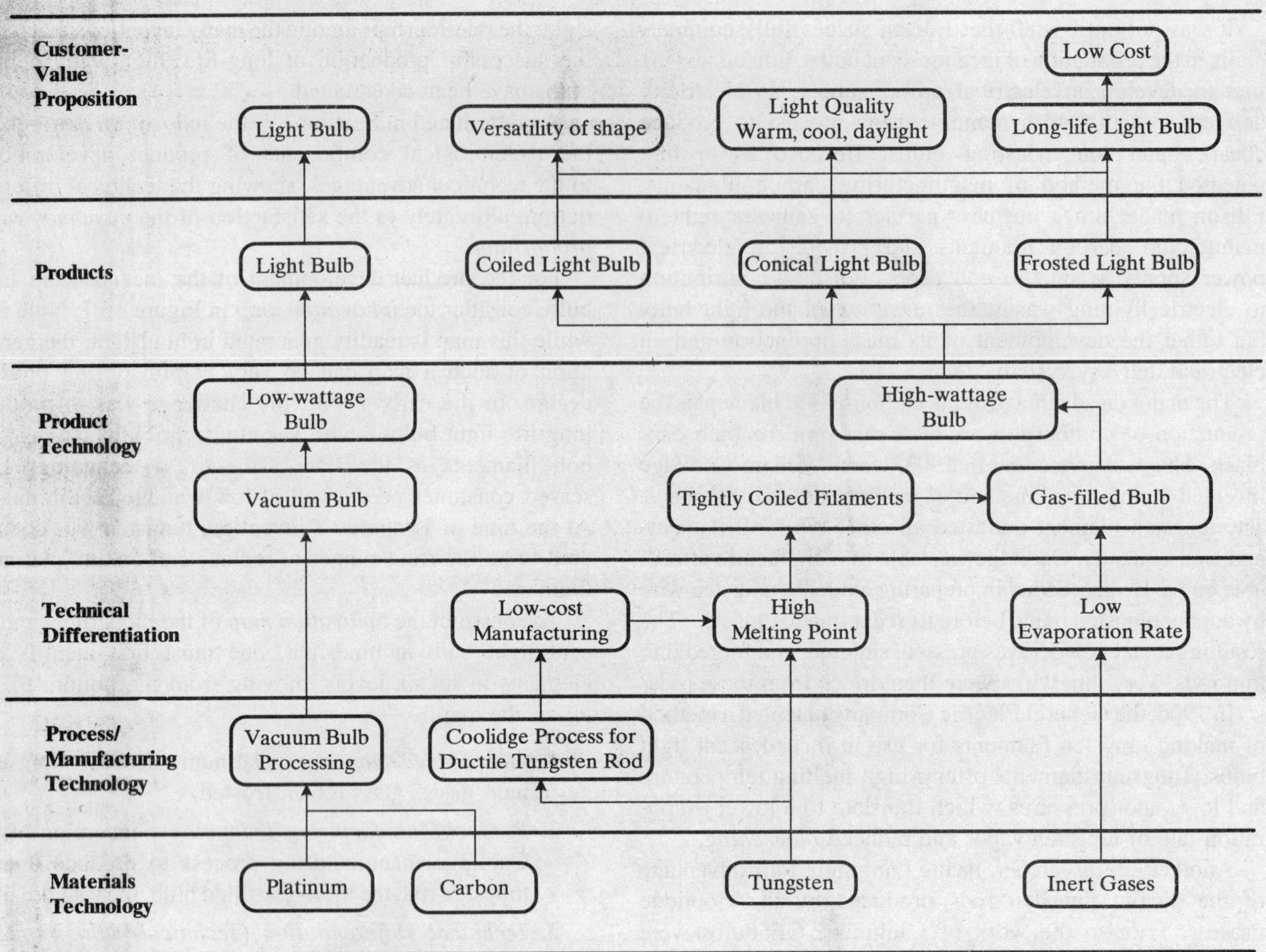

Figure 16.1 Conventional light bulb innovation map.

In the early 1800s, the initial inventions involving *platinum* (materials technology) and *vacuum bulb processing* (process/manufacturing technology) led to *vacuum* and *low-wattage bulbs* (new product technologies). The new product was a *light bulb* that satisfied the *customer-value proposition* of a compact bulb that provided light. Fifty years later, Swan added *carbon* filaments, an important materials technology. These linkages are clearly established in the left-hand column of the innovation map. Indeed, for these conventional incandescent light bulbs, the replacement of *platinum* with *carbon* was a key invention.

Having been introduced to low-wattage, short-life light bulbs, the *customer-value proposition* led to the need for *versatility of shape*, *different light qualities*, *long life*, and *low cost* in new light bulb products. The *Coolidge process* in the early 1900s for producing ductile tungsten rods, shown linked to *tungsten*, provided a *low-cost manufacturing* technique. These were the key materials and process/manufacturing technologies that provided the technical differentiations, that is, a *high-melting-point* material that could be manufactured at low cost.

When Langmuir invented *tightly coiled tungsten filaments*, a key product technology, it became possible to fill bulbs with inert gases (another product technology), lowering the tungsten evaporation rate, another important technical differentiation. These two product technologies led to the *high-wattage bulb*, the key product technology. In turn, these linked to the several new products including *coiled light bulbs*, *conical light bulbs*, and *frosted light bulbs*, which linked to the *customer-value propositions*—that is, the *long-life light bulb*; the *light quality: warm, cool, daylight*; and *low cost*. These linkages are shown to the left on the innovation map.

Halogen Light Bulb Technology

By the mid-1960s, it had been long recognized that a major drawback of incandescent light bulbs is their filament lifetimes, which trade off with their luminous efficiencies. At higher filament temperatures, luminous efficiencies increase, but filaments evaporate faster, reducing their lifetimes. Typically, household incandescent light bulbs had been designed to operate for 750 to 1,000 hours. However, based upon a customer survey, longer lifetimes were desired, at least twice those of the products on the market at that time.

The lifetime of an incandescent bulb depends on the durability of tungsten filaments at high temperature. Thickness variation of the tungsten filaments leads to runaway heating, ultimately causing the filament to fail (Garbe, 1980; Garbe and Hanloh, 1983; Rieck, 1957). Thinner segments, having higher resistance, experience higher temperatures, which lead to higher tungsten evaporation rates and cause the segments to become thinner and hotter, eventually fracturing.

F. Mosby had found (U.S. Patent 3,243,634) that halogen gases react with tungsten in equilibrium at high temperatures, enabling evaporated tungsten to redeposit on the higher-temperature, thinner filaments, interrupting the unstable breakdown and thereby increasing their lifetime. Typically, halogen light bulbs, with a tungsten filament sealed in a small halogen-gas-filled bulb, have twice the lifetime of conventional incandescent light bulbs. In addition, halogen light bulbs have higher efficiencies and whiter color temperatures.

To maintain high reaction temperatures, the halogen light bulb requires a tight (smaller) casing/bulb, typically using high-temperature glass or quartz. A secondary closure is often added to isolate the hot halogen bulb from accidental contact.

Following Mosby, the halogen light bulb technology was developed in the 1970s. Next, in Example 16.1, an innovation map is created based upon the technology developed by the Philips Corporation during the 1970s and 1980s. This precedes a case study for the product design of the halogen light bulb in Section 17.2. By presenting this case study in hindsight, the development and competitive landscape is shown after the fact. Emphasis is placed on the key technology inventions and their primary competing technology, the fluorescent light bulb, during that period. Obviously, if the case study were carried out at the current time (late 2000s), a different competitive landscape would be considered, involving the advancement of the compact fluorescent light bulb (CFL) and the LED (light-emitting diode) technology.

EXAMPLE 16.1 ***Innovation Map for the Halogen Light Bulb***

For the halogen light bulb, the key inventions were in materials (the use of halogen gases), process/manufacturing (quartz primary casing), and product technology (secondary casing). In this example, it is assumed that you are a member of a product-development team in the mid-1980s, and that the team is considering a halogen light bulb product. The objective is to incorporate these inventions into an innovation map. Wherever possible, your design team should provide experimental and theoretical justifications for the promise of these inventions in the *concept* stage. Because much work on halogen light bulbs took place from 1970 to 2000, at this point, you have access to patents and a significant body of technical literature.

Note that the purpose of this example is to show how innovation maps are created with existing technologies. New technologies are more difficult to present, as they are often developed just prior to, or even worse, during the product-development process. Often the product family is derived from new technologies not known *a priori*. In some cases, which are best avoided if possible, the technologies are further developed during the product-development process. To illustrate aspects of a more recent product development, see Exercises 16.3 and 16.4, which involve the creation of an innovation map for the compact fluorescent light bulb (CFL). Also, see Exercises 17.3 and 17.4, involving the product design of CFLs.

SOLUTION

The *innovation map* for the halogen light bulb is shown in Figure 16.2. Unlike Figure 16.1, it does not include the conventional light bulb that stems from the platinum and carbon materials. This is because the low-wattage, short-life light bulb was no longer competitive in the 1980s. Like Figure 16.1, it retains the link to tungsten, as tungsten filaments are incorporated into halogen light bulbs. Its new elements are:

1. *Materials Technology*: halogen gases
2. *Process/Manufacturing Technology*: quartz primary casing
3. *Technical Differentiation (Technical-Value Proposition)*: high-temperature reaction, equilibrium deposition
4. *Product Technology*: secondary casing
5. *Products*: automobile headlight bulb, display lamp bulb, factory lamp bulb, . . .
6. *Customer-Value Proposition*: long-life light bulb, brighter, whiter bulb

Clearly, the *halogen gases* (materials technology) were the drivers of the new innovation. As will be shown, they introduce the reversible deposition of tungsten on the filament, the key technical differentiation. The *quartz primary casing* was the new process/manufacturing technology containing the associated *high-temperature tungsten-halogen reaction*, a related technical differentiation. These two technical differentiations are linked to the existing *gas-filled bulb* and *high-wattage bulb* product technologies. However, a new product technology was added for safety reasons: the *secondary casing* to protect consumers from the high-temperature primary casing. These product technologies are linked to the new products: *automobile headlight bulb, display lamp bulb, factory lamp bulb, . . .*, which are in turn linked to the customer-value propositions, including *brighter, whiter bulb*. For further discussion of this innovation map see Widagdo (2006).

In the remainder of this section, these promising technological inventions, which were the basis for the halogen light bulb design, are discussed in four subsections: (a) Materials Technology, (b) Process/Manufacturing Technology, (c) Product Technology, and (d) Technology Protection.

a. Materials Technology

In this subsection, the key inventions in materials technology are considered, with emphasis on the: (1) halogen lamp failure modes, based upon the key transport mechanisms,

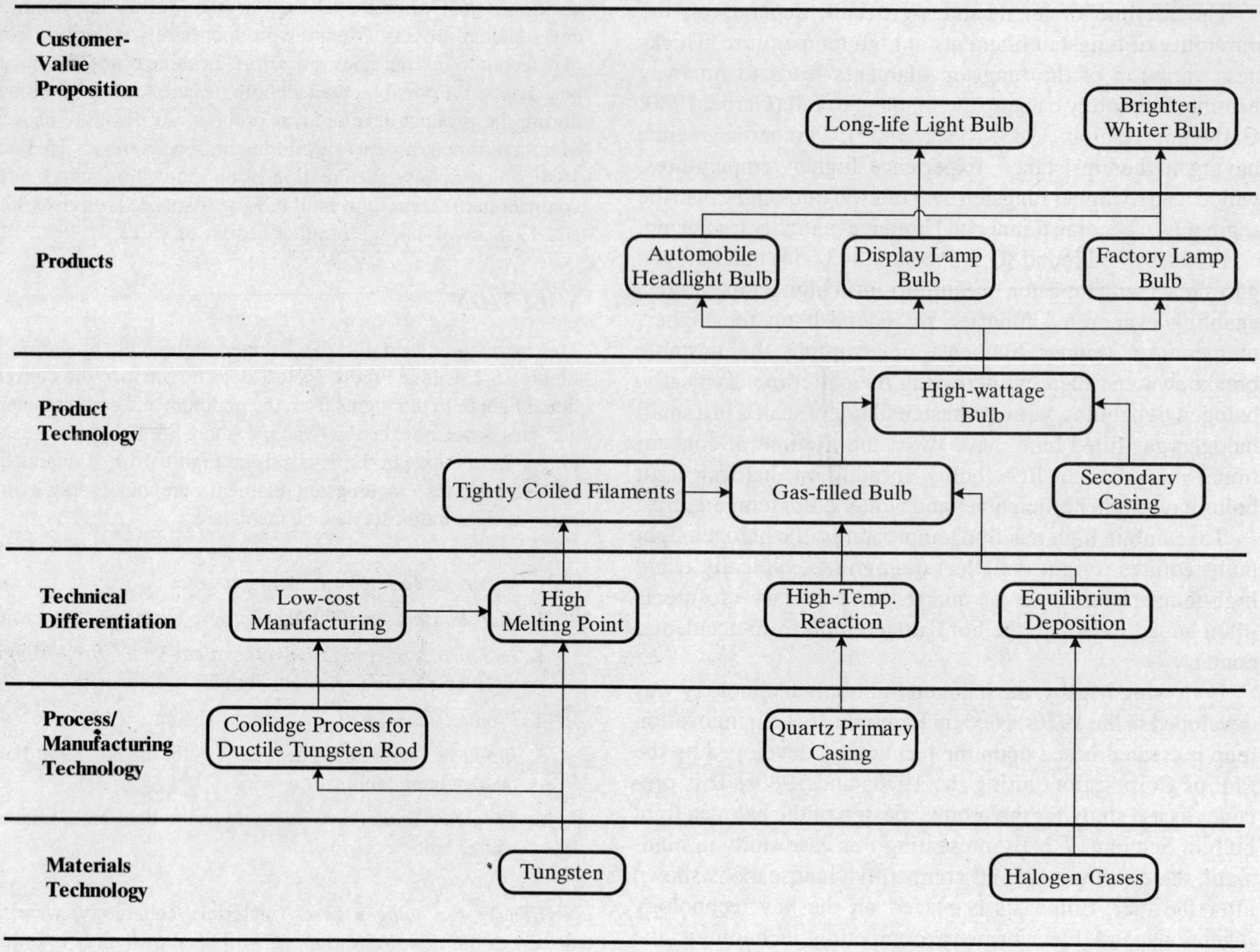

Figure 16.2 Halogen light bulb innovation map.

(2) halogen-tungsten reactions, and (3) surface morphology of the tungsten filaments.

Transport of Tungsten in Halogen Incandescent Light Bulbs

The lifetime of halogen incandescent light bulbs depends on: (1) the halide and inert gas filling, (2) the chemical reactions between the halides, tungsten, bulb wall, and impurities such as oxygen and water, (3) the geometry of the bulb, (4) the gradient between the filament and bulb temperatures, and (5) transport phenomena such as heat conduction, diffusion, and convection within the bulb.

Initially, in a study of the radial transport phenomena around tungsten filaments, Schnedler (1983a, b) showed the importance of thermal diffusion as well as the reactions between the halides and the condensed phases on the bulb inner wall. Because this model could not predict the key parameters such as the lifetime, Schnedler (1985) developed a 3-D model that accounts for the temperature gradient between the filament and the inner wall surface, the mass and heat transfer by diffusion, thermal diffusion, heat conduction in the gas phase, and chemical reactions in the gas phase, both on the filaments and on the bulb surface—with convective transport neglected for small, linear-wall (cylindrical) halogen incandescent light bulbs, as shown schematically in Figure 16.3.

The details of these models involve ordinary and thermal diffusion in energy and mass balances, with the latter accompanied by chemical equilibria involving the species W, Br_2, WBr_2, O_2, and WO_2Br_2. Their solutions provide estimates of the temperatures and partial pressures throughout the cylindrical bulb, as discussed next. The equations for ordinary and thermal diffusion and the energy and mass balances, in the form of partial differential equations, are presented in the file, Supplement_to_Chapter_16.pdf in the PDF Files folder, which can be downloaded from the Wiley Web site associated with this book.

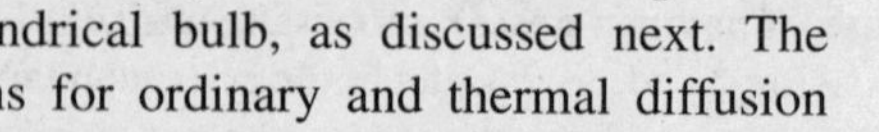

Figure 16.3 Schematic of a linear-wall halogen incandescent light bulb.

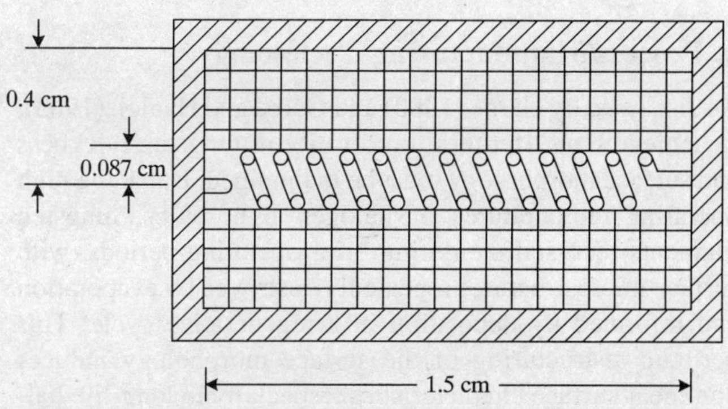

Figure 16.4 Finite difference grid. Reprinted with permission (Schnedler, 1985).

These mass and energy balances were solved by Schnedler (1985) using finite differences and the grid in Figure 16.4. Two simulations were run. First, the tube was filled at 25°C with Br_2 at 0.005 bar and Kr at 5 bar. This mixture has $N = 3$ species (W, Br_2, WBr_2) and one independent reaction at equilibrium, $W + Br_2 \leftrightarrow WBr_2$. For the second simulation, the tube was filled at 25°C with Br_2 at 0.005 bar, O_2 at 0.00005 bar, and Kr at 5 bar. This mixture has $N = 4$ species (W, Br_2, O_2, WO_2Br_2) and one independent reaction at equilibrium, $W + Br_2 + O_2 \leftrightarrow WO_2Br_2$.

Simulation Results

Tungsten-Bromine System To determine whether bromine-filled light bulbs blacken, calculations were carried out for a cylindrical geometry (0.8 cm diameter, 1.5 cm long), as shown schematically in Figure 16.5. At 30V and 250W, the maximum filament temperature is 3,170 K, with the temperature assumed to be 1,000 K on the inner surface of the bulb, based upon experimental measurements. Note that, as shown in the solution to Exercise 16.1, fortunately, the temperature distribution varies only slightly with the bulb temperature.

Figure 16.6 shows the calculated temperature profile, with a maximum temperature of 3,170 K at the centerline of the light bulb (middle of the coiled tungsten filament). Clearly, the temperature gradients are largest at the ends of the filament.

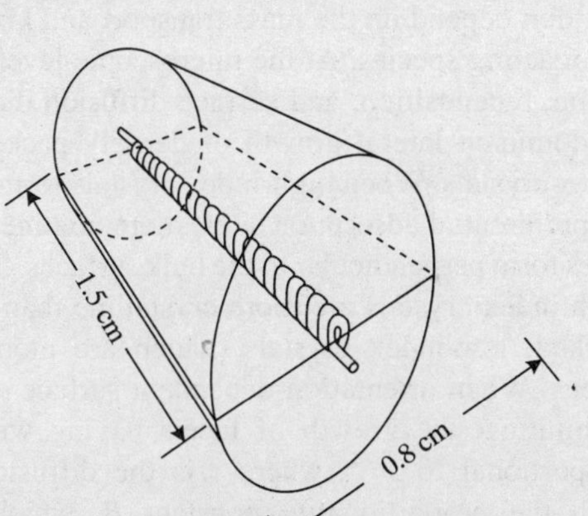

Figure 16.5 Schematic of the bromine-filled light bulb. Reprinted with permission (Schnedler, 1985).

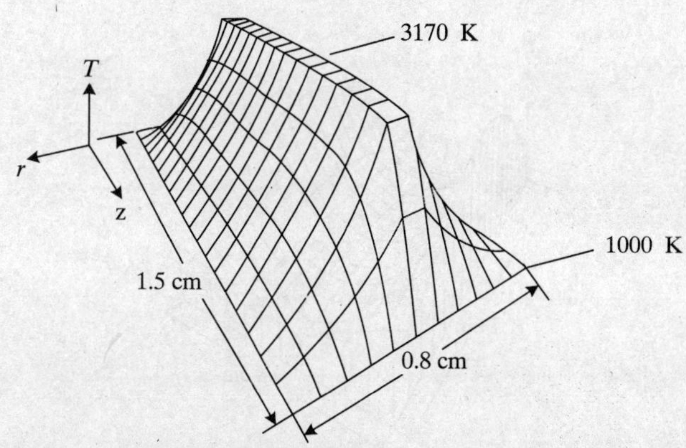

Figure 16.6 Temperature profile within the bromine-filled light bulb. Reprinted with permission (Schnedler, 1985).

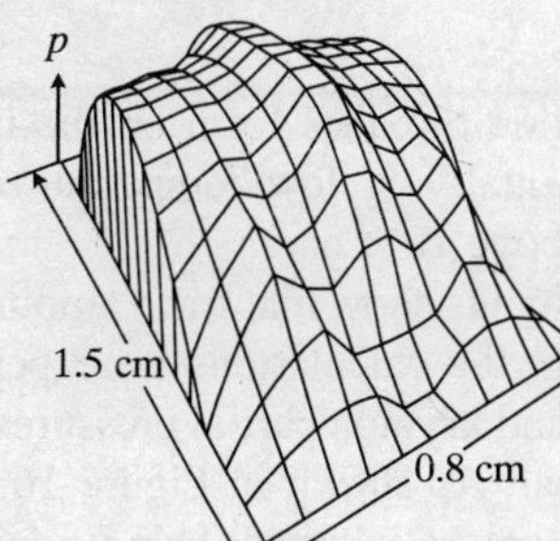

Figure 16.7 Tungsten partial-pressure distribution in the bromine-filled bulb. Reprinted with permission (Schnedler, 1985).

Using the computed temperature distribution, with partial pressures of 0.005 and 5 bar for bromine and krypton at room temperature, solution of the mass balances gives an operating pressure of 24.3 bar. Unfortunately, the solution shows condensation of the tungsten bromides at the inner bulb wall with the bromine partial pressure reduced to 10^{-8} bar, which is insufficient to prevent the bulb from blackening due to tungsten condensation on the inner wall. Figure 16.7 shows that the supersaturated tungsten partial pressure peaks at the center of the filament, where the deposition rates of tungsten are highest. This is consistent with the experimental observation that tungsten is regenerated on the hot filaments.

The experiment and simulation were repeated with bromine and krypton partial pressures of 0.05 and 5 bar; that is, a 10-fold increase in the bromine partial pressure. As expected, a larger amount of bromine deposits prior to the condensation of tungsten on the inner surface of the bulb.

Tungsten-Bromine-Oxygen System In the presence of oxygen or residual water vapor in trace amounts from the tungsten filaments, bromine forms bromide oxides with tungsten (Brongersma et al., 1981; De Maagt and Rouweler, 1980; Dittmer and Niemann, 1981; Eckerlin and Garbe, 1980; Zubler, 1972; Zubler, 1975). Furthermore, small amounts of oxygen and water vapor reside in the bulb itself,

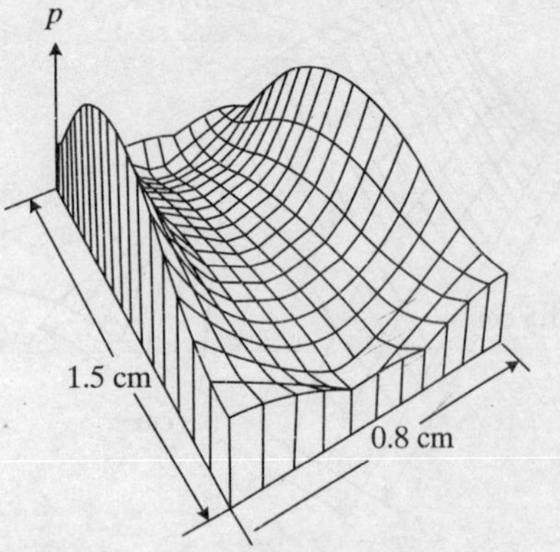

Figure 16.8 Tungsten partial-pressure distribution in the bromine-oxygen-filled bulb. Reprinted with permission (Schnedler, 1985).

and tungsten dihalide dioxides form on the inner wall, as observed experimentally at low temperatures and halide partial pressures above 10^{-2} bar.

Because experiments show that trace amounts of oxygen prevent blackening, the calculations were performed with bromine, oxygen, and krypton partial pressures at 5×10^{-3}, 5×10^{-5}, and 5 bar. As shown in Figure 16.8, the partial pressures of tungsten are inverted, being minimized at the center of the tungsten filament. Similar inversions are computed for bromine and oxygen, but not displayed here, indicating that WBr_2 and WO_3 condense on the inner surface of the bulb, as observed experimentally. However, the remaining bromine and oxygen prevent the condensation of tungsten on the inner surface.

The lines of constant tungsten flux in Figure 16.9 show the transport of tungsten from the hotter portion of the filaments to the cooler portions, with no condensation of tungsten detected at the wall. Notice that the lines of constant tungsten flux run nearly parallel to the wall, indicating that the flux of tungsten to the wall is negligible. Consequently, the bulbs operate without blackening. Instead, the lines of constant tungsten flux impinge nearly orthogonally on the filament, indicating that tungsten redeposits on the filament. Using the order of magnitude of the maximum tungsten flux, the filament lifetime is estimated to be several thousand hours, in agreement with experiments.

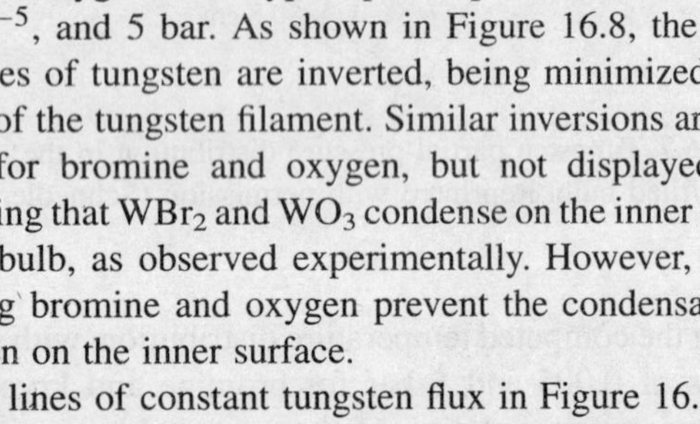

Figure 16.9 Tungsten flux distribution in the bromine-oxygen-filled bulb. Reprinted with permission (Schnedler, 1985).

b. Process/Manufacturing Technology

As discussed by Garbe (1980) and Garbe and Hanloh (1983), to achieve long lifetimes, uniformity in tungsten thickness and surface morphology must be maintained. But at the high operating temperatures of halogen light bulbs, tungsten filaments recrystallize during brief operating periods, with surface microstructures repeatedly destroyed by evaporation and reformed by deposition in a regenerative cycle. This periodic restructuring of the surface morphology induces nonlinear surface characteristics, especially in long-life halogen light bulbs.

This *faceted growth*, which minimizes the surface Gibbs-free energy, forms periodic hill- and valley-like structures over an initially smooth surface in one of two mechanisms (Garbe, 1980):

1. Crystalline tungsten, at equilibrium with its vapor, reduces its surface free energy by rearranging its surface morphology due to the nonuniformity (nonlinearity) of the surface energy of single-crystal surfaces. Figure 16.10 shows a schematic of the rearrangement of a smooth surface into a faceted structure with a minimum surface Gibbs-free energy that favors an increase in the surface area, characterized by hills and valleys.

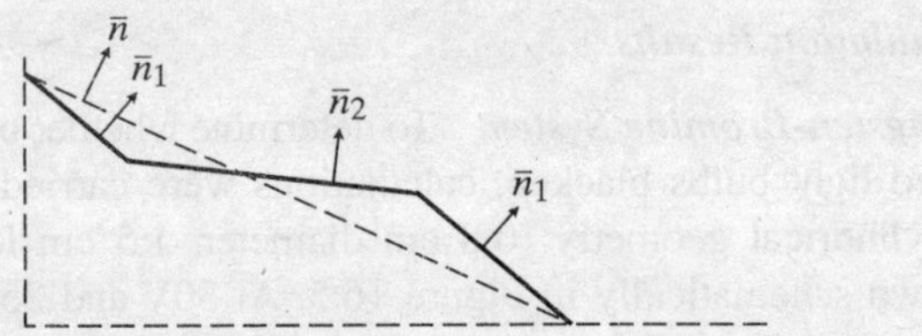

Figure 16.10 Facet growth from a smooth surface. $\bar{n}$ is orthogonal to the surface. Reprinted with permission (Garbe, 1980).

2. Reversible vaporization and redeposition in the tungsten-halide reactions on the surface forms periodic structures that minimize the surface Gibbs-free energy. From a macroscopic view, this vaporization and redeposition depend on the mass transport and kinetics of the reacting species. At the microscopic level, evaporation, redeposition, and surface diffusion dictate the predominant lateral growth of densely packed, low-index crystals. When high-index crystals form, having no preferential adsorption sites, more nonlinear structures form perpendicular to the bulk surfaces. Note that high-index crystals are more crystalline than densely packed, low-index crystals, which are more amorphous. When orientation-dependent surface diffusion is limiting, the growth of facets having width s is proportional to $t^{1/4}$, where t is the diffusion time, with the proportionality constant B, which is the surface-diffusion coefficient (Mullins, 1961):

$$s = Bt^{1/4} \tag{16.1}$$

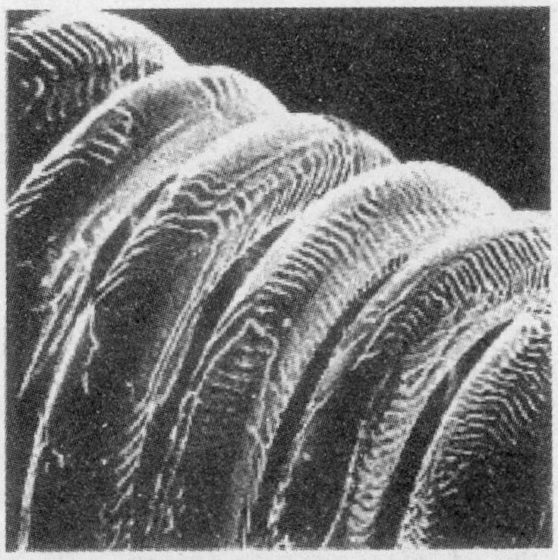

Figure 16.11 Periodic faceting structure after 123 hr at 3,350 K. Reprinted with permission (Garbe, 1980).

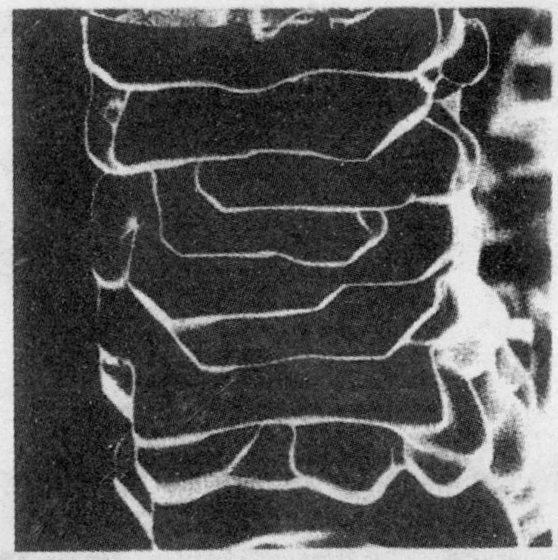

Figure 16.13 Faceted structure after 5,500 hours at 3,000 K. Reprinted with permission (Garbe, 1980).

Initially, the recrystallized tungsten filaments form a single coil having about one grain boundary per turn, resulting in a highly periodic structure early in the service life of a halogen light bulb, as illustrated in Figure 16.11.

These grain boundaries, having nonlinear surface morphology, induce preferential evaporation depending on the surface orientations of the grains. Smooth areas are formed from low-index crystal grains having different radiation characteristics than highly faceted areas, as shown in Figure 16.12. Note that the variations of the surface roughness cause the radiation emissivities to vary.

Very large faceted structures may grow in long-life halogen bulbs, as shown in Figure 16.13, after 5,500 hours of service at 3,000 K. Also, due to bromine bubbles, holes in the tungsten filament often form. For example, Figure 16.14 shows a hole, vertically oriented, on the left of the V-shaped grooves, and Figure 16.15 shows a cross section of two turns. In the right-hand cross section, holes formed by two neighboring bubbles are seen.

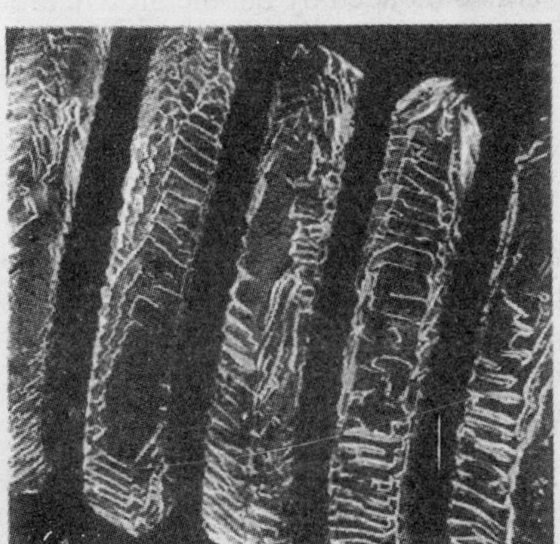

Figure 16.12 Faceted structure after 3,096 hours at 3,000 K. Reprinted with permission (Garbe, 1980).

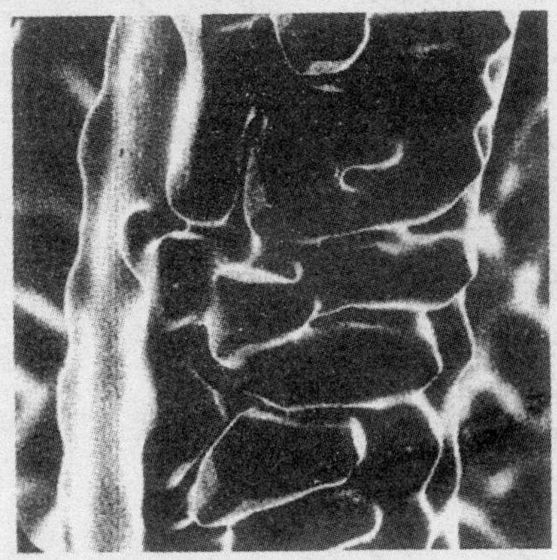

Figure 16.14 V-shaped grooves and holes after 3,500 hours at 3,000 K. Reprinted with permission (Garbe, 1980).

Faceted structures have larger surface areas for both emission and reflection, and consequently, increased effective emissivities. As shown in Figure 16.16, surfaces with V-shaped grooves and holes have increased effective emittance (emissivity), ε, depending on the V-groove apex angle, or the ratio of the hole effective diameter, d_o, to the effective diameter of the coil, D.

Experimental measurements of the effective emissivity, for 650 μm light waves, as a function of service time, are shown in Figure 16.17. Observe the sigmoidal variation, beginning with an effective emissivity of 0.46 and approaching 0.64 asymptotically after 5,000 hours.

Garbe (1980) further observed that this increase in effective emissivity results in: (1) a 150 K reduction of the tungsten filament temperature from 3,000 K, and (2) local microscopic surface variations, induced by facet structures, that may create local temperature differences of 20 to 40 K, which may lead to premature lamp failures. In addition, an increase in the rate of axial transport was observed due to differences in the local faceting patterns. Examination of cross sections of the tungsten filaments at various axial positions,

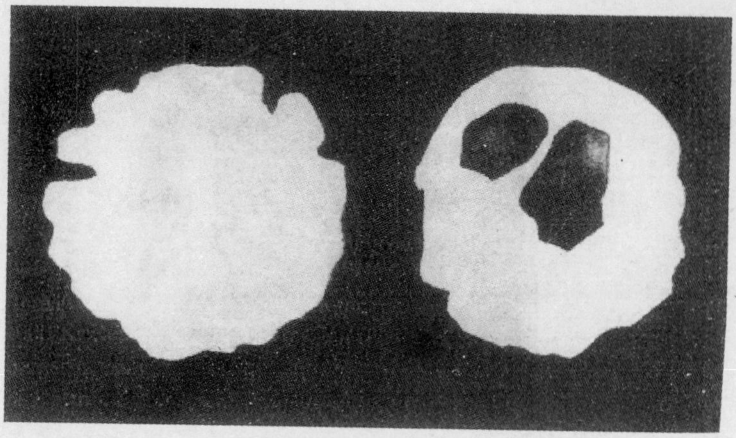

Figure 16.15 Cross section of two turns of a tungsten filament after 3,050 hours at 3,000 K. Reprinted with permission (Garbe and Hanloh, 1983).

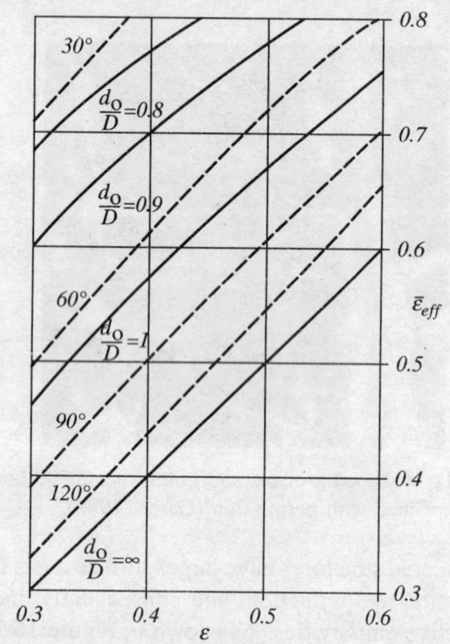

Figure 16.16 Effective emittance for V-shaped grooves or spherical holes. Reprinted with permission (Garbe, 1980).

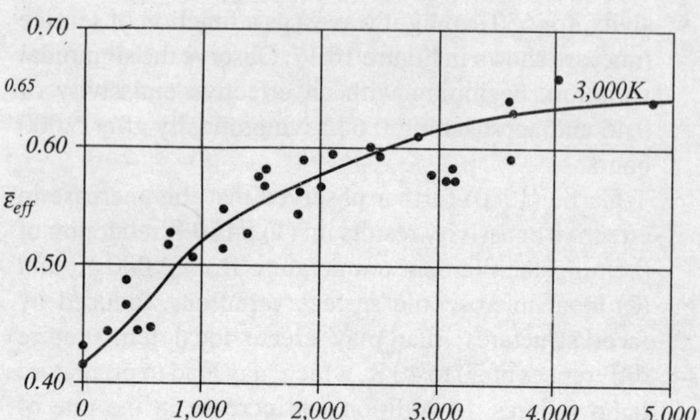

Figure 16.17 Effective emittance for 650 μm light waves at 3,000 K. Reprinted with permission (Garbe, 1980).

with varying faceting patterns, supports the view that the surface-structure-induced, local-emissivity variations lead to temperature gradients and the development of hot spots.

Subsequent to these studies, Garbe and Hanloh (1983) studied the effects of *dopants* used in the manufacture of tungsten filaments on hot-spot development and intergranular fractures. Note that since Coolidge invented the tungsten manufacturing process in 1910, potassium, silicon, aluminum, and various compounds have been added as dopants to give powder metallurgical materials with *non-sag* qualities. Furthermore, some dopants form bubbles that induce the formation of interlocked grain structures. At high operating temperatures, with the associated regenerative process (i.e., repeated evaporation and recrystallization), it is likely that bubbles and their coalescence induce intergranular fractures that reduce the lifetime of halogen light bulbs.

To better understand this mechanism, Garbe and Hanloh (1983) studied the growth of potassium-filled bubbles in tungsten filaments as a function of the operating temperature, temperature gradients, and time. They observed that: (1) local temperature rises are predicted in the vicinity of large voids; (2) the mechanism of bubble-induced hot-spot growth decreases the filament life for wires with larger diameter, larger temperature gradients, and long lifetimes; and (3) creep and fractures induced by bubble growth are predicted after long lifetimes of 3,000–5,000 hours with initial temperatures of 3,000–3,100 K.

c. Product Technology

Because halogen bulbs operate at very high temperatures, the primary enclosure must be made of hard glass or fused quartz instead of ordinary soft glass, which would soften and flow too much at these temperatures.

The primary enclosure materials can be selected and modified (by means of optical coatings) to achieve the required lamp characteristics. Halogen bulbs are increasingly used in automobile headlamps, for example, and because headlamps often contain plastic parts, envelopes for halogen headlamp bulbs are made out of hard glass, or quartz *doped* with additives to block most of the UV output. Note that hard glass blocks UV without the need for dopants.

Conversely, some applications *require* ultraviolet radiation, and in such cases, the lamp envelope is made out of undoped quartz. Thus, the lamp becomes a source of UV-B radiation. Undoped quartz halogen lamps are used in some scientific, medical, and dental instruments as a UV-B source.

Note that, because no significant new technologies are needed, secondary enclosures are not discussed here. See Section 17.2 for a discussion of their characteristics.

Fire and UV Burn Hazards. Because halogen light bulbs operate at very high temperatures, they pose fire and burn hazards. In addition, the possibility of sunburn from excessive exposure to the UV emitted by an undoped quartz halogen lamp is a growing concern. To mitigate the negative effects of unintentional UV exposure, and to contain hot bulb fragments in the event of explosive bulb failure, manufacturers of lamps intended for general-purpose usage usually install UV-absorbing glass filters over or around the bulb. Alternatively, a coating of UV inhibitors that effectively filters UV radiation is installed on the bulb envelope. When implemented correctly, a halogen lamp with UV inhibitors produces less UV than its standard incandescent counterpart.

Handling of the Quartz Primary Casing. Any surface contamination, such as oil from fingerprints, can damage the quartz envelope when it is heated by causing the quartz to change from its vitreous form into a weaker, crystalline form that leaks gas. Consequently, quartz lamps should be handled without touching the clear quartz, either by using a clean paper towel or by carefully holding the porcelain base. When the quartz is contaminated in any way, it must be thoroughly cleaned with rubbing alcohol and dried before use.

d. Technology Protection

Corporations commonly protect their inventions and investments in developing new technologies by filing for patents. When patents cannot be obtained or are deemed to be less than beneficial, companies often rely on trade secrets and public disclosures. Even when filing for a patent, important know-how is frequently omitted. During the development of the halogen light bulb, the Philips Corporation filed for approximately 11 and 36 patents during the 1970s and 1980s, respectively. Beyond that, in the 1990s, Philips filed for another 77 patents. In comparison, the number of patents their key competitors, General Electric, Osram, and Sylvania, filed for is shown in Table 16.1.

During the 1970s, Philips patented the basic technologies, including the manufacturing aspects of the tungsten elements, the bulb geometries, and the composition of the filling gases. By the 1980s, their attention shifted to *article/application patents* that are related to the products themselves, such as the construction of the primary and secondary casings.

Table 16.1 Patent Filing Portfolio (Number of Patents) for Halogen Light Bulb

Company Name	1970s	1980s	1990s
Philips Corporation	11	36	77
General Electric	21	45	48
Osram	—	—	12
Sylvania	10	1	13

Portions of typical *composition of matter* and *product-by-process* patents are shown in Figures 16.18 and 16.19, respectively.

Often a competing technology survey is conducted, which is protected as well, as a method to *picket-fence* the main inventions.

In general, the patent protections for materials and process/manufacturing inventions are filed long before the product technology inventions appear. Consequently, it is important to postulate the potential product technologies before the design is undertaken. This was not the case for the halogen light bulb technologies patented by the Philips Corporation. A patent for the method of manufacture (U.S. Patent 3,932,164) was filed in August 1974, and one for the composition of the halogen gases (U.S. Patent 4,074,168) was filed in October 1974.

During the product design process, additional patents may be filed that would cover the specific product design inventions. For the halogen light bulb, these included patents for an infrared filter (U.S. Patent 4,017,758), a light reflector unit (U.S. Patent 4,081,708), and an automobile headlight (U.S. Patent 4,119,877).

In summary, the elements of the innovation map for the development of long-life incandescent light bulbs as they existed in the mid-1980s, including the key inventions required in the materials, process/manufacturing, and product technologies, have been discussed in this section. The technology protection subsection discussed the competitive landscape filled by the major players such as Philips, GE, Osram, and Sylvania in these technology areas. These technological inventions and their protection became the basis for the development of the halogen light bulb product, as discussed in Section 17.2.

16.3 INNOVATION MAP FOR HOME HEMODIALYSIS DEVICE

Hemodialysis is one of two types of artificial dialysis treatments (the other being peritoneal dialysis) that replace the function of the kidneys, which is to regulate the composition of the bloodstream—containing red cells, white cells, platelets, and plasma—by removing waste products and excess fluids while maintaining the proper chemical balance.

United States Patent [19]

T' Jampens

[11] **4,074,168**

[45] **Feb. 14, 1978**

[54] HALOGEN INCANDESCENT LAMP WHOSE FILLER GAS COMPRISES BROMINE, CHLORINE AND HYDROGEN

[75] Inventor: Germain Reml T'Jampens, Eindboven, Netherlands

[73] Assignee: U.S. Philips Corporation, New York, N.Y.

[21] Appl. No.: 511,056

[22] Filed: Oct. 1, 1974

[30] Foreign Application Priority Data

Oct. 10, 1973 Netherlands 7313906

[51] Int. Cl. H01K 1/50

[52] U.S. Cl. 313/222

[58] Field of Search 313/222, 223

[56] **References Cited**

U.S. PATENT DOCUMENTS

3,418,512	12/1966	T'Jampen et al.	313/222 X
3,484,146	12/1969	Meijer et al.	316/20
3,584,896	6/1971	Meijer et al.	313/222 X
3,719,853	3/1973	Sugano et al.	313/222

Primary Examiner — Palmer C. Demeo

Attorney, Agent, or Firm—Frank R. Trifari; Robert S. Smith

[57] **ABSTRACT**

Halogen incandescent lamp having a gas filling comprising dibromonethane (CH_2Br_2) and carbontetrachloride (CCl_4)

3 Claims, No Drawings

Figure 16.18 Typical *composition of matter* patent.

United States Patent [19]

Gebhardt

[11] **3,932,164**

[45] **Jan. 13, 1976**

[54] METHOD OF MANUFACTURING MINIATURE INCANDESCENT LAMPS

[75] Inventor: Karl Gebhardt, Stolberg-Busbach an Kurths-Muhle, Germany

[73] Assignee: U.S. Philips Corporation, New York, N.Y.

[22] Filed: Aug. 8, 1974

[21] Appl. No.: 495,916

[30] **Foreign Application Priority Data**

Aug. 14, 1973 Germany 2341028

[52] U.S. Cl. 65/32; 65/34; 65/36; 65/152

[51] Int. Cl. C03B 5/16; C03B 23/20

[58] Field of Search 65/34, 43, 32, 36, 152

[56] **References Cited**

UNITED STATES PATENTS

1,398,033	11/1921	Maurer	65/34 X
1,448,351	3/1923	Kirlin	65/34
2,006,568	7/1935	Gustin et al.	65/34 X
2,945,327	7/1960	Malm et al.	65/43 X
3,409,342	11/1968	Anderson et al.	65/32 X
3,505,050	4/1970	Huston, Jr.	65/34
3,688,812	9/1972	Federicks	65/34
3,798,491	3/1974	Malm	65/34 X

FOREIGN PATENTS OR APPLICATIONS

507,999	12/1954	Canada	65/34
261,398	11/1925	United Kingdom	65/34

Primary Examiner—Robert L. Lindsay, Jr.

Assistant Examiner—Frank W. Miga

Attorney, Agent, or Firm—Frank R. Trifari

[57] **ABSTRACT**

In the manufacture of miniature incandescent lamps having a pinched glass cap, an inert cooling gas is blown through the exhaust tube during the pinching operation so as to keep the exhaust tube open.

5 Claims, 3 Drawing Figures

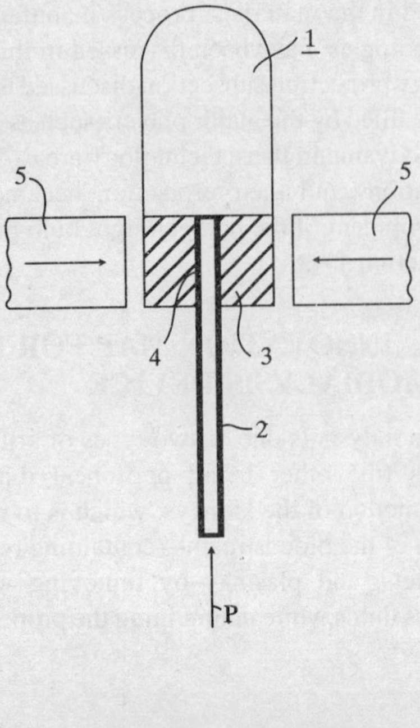

Figure 16.19 Typical *product-by-process* patent.

Plasma, which accounts for about 55 vol % of the blood, consists of about 92% water, with the balance being inorganic salts, organic chemicals, and dissolved gases. A typical male adult has a blood volume of 5 L with a resting cardiac output of 5 L/min and a systemic blood pressure of 120/80 mm Hg. The most common application of hemodialysis is for patients with temporary or permanent kidney failure. It is the only treatment available for patients with end-stage renal disease (ESRD), because their kidneys are no longer capable of their function. This treatment, which is required three times per week for an average of 3–4 hours per dialysis, was performed on more than 200,000 patients in the United States in 1996. It is estimated that in 2005, 300,000 patients used hemodialysis devices.

Hemodialysis Device Inventions

Before creating an *innovation map* for a home hemodialysis device, one of the first steps in product design, it is important to be knowledgeable about the available technologies, including several that are of historical interest.

Enzyme Reactor

One of the first technologies, a column packed with microcapsules containing the enzyme urease and/or aspartase in a membrane gel lattice of semipermeable polymer, was introduced by Chibata et al. (U.S. Patent 3,865,726) in 1975. In the microcapsules, urea is decomposed by the urease into NH_3 and CO_2. The NH_3 reacts with fumarate in the dialysis solution to form aspartic acid by the action of the aspartase. The aspartic acid is nontoxic to human beings and can be readily eliminated using an ionic-exchange resin.

Hollow-Fiber Module

A very common commercial device for hemodialysis is the C-DAK 4000 artificial kidney of Cordis Dow Corporation. This device is covered largely by U.S. Patent 4,276,173, issued on June 30, 1981. This disposable, sterilized membrane module, shown schematically in Figure 16.20, resembles a shell-and-tube heat exchanger. The tubes, which number 10,000, are hollow fibers, 200 microns i.d. by 10 microns wall thickness by 22 cm long, made of hydrophilic microporous cellulose acetate of 15 to 100 Å pore diameter. Alternatively, fibers of polycarbonate, polysulfone, and other polymers are used. The shell, made of acrylonitrile-butadiene-styrene (ABS) plastic with inlet and outlet side ports of polycarbonate, is 24 cm long by 4 cm in outside diameter with centered ports in the heads (caps) at either end of the shell for delivering blood flow into and out of the hemodialyzer. The fibers are potted at each end into polyurethane, sliced at each end to open the fibers, and sealed into the shell heads (caps) to prevent leakage between the tube side and the shell side. The total membrane area, based on the inside area of the hollow fibers, is 1.38 m^2. The packing density is 4,670 m^2 of membrane area per m^3 of membrane module volume. In 1992, 60 million of these units, which weigh less than 100 g each, were sold at 5 to 6 U.S. dollars each. Based on total membrane area used and dollar value, artificial kidneys are the single largest application of membranes.

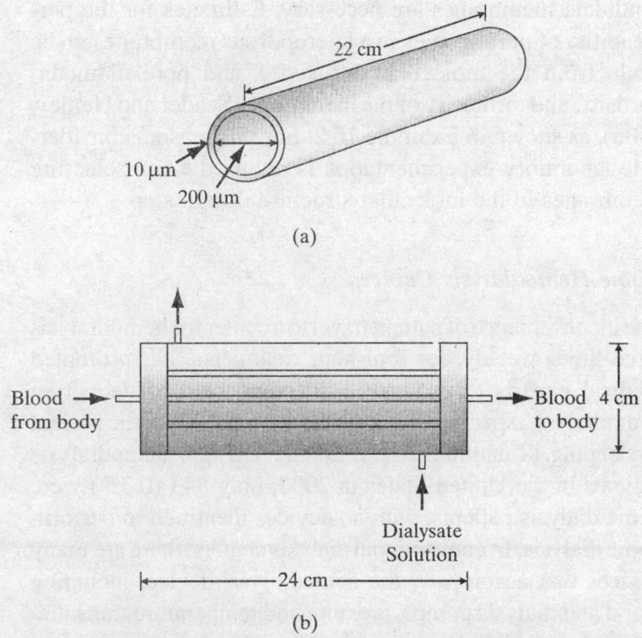

Figure 16.20 Hemodialysis device. (a) Single tube. (b) Complete module.

A hemodialysis treatment with a C-DAK artificial kidney involves the insertion of two needles into the patient's vein, with attachment to plastic tubes to carry the patient's blood to and from the artificial kidney. The blood flows through the hollow fibers at a flow rate monitored by and controlled at 200 ml/min by a dialysis machine. A sweep dialysate solution of water, glucose, and salt passes countercurrently to the blood through the shell side of the artificial kidney at a rate of 500 ml/min. The pressure difference across the membrane from the tube side to the shell side is approximately equal to the diastolic pressure of the patient (e.g., 120 to 150 mmHg), since the dialysate is pulled through the module with suction. The pressure drop on the tube side is 30 mmHg, while that on the shell side is 12 mmHg. The pressure difference across the membrane causes excess fluids in the bloodstream to pass to the dialysate. Concentration differences between the blood and the dialysate cause urea (CH_4ON_2), uric acid ($C_5H_4O_3N_4$), creatinine ($C_4H_7ON_3$), phosphates, and other low-molecular-weight metabolites to transfer by diffusion from the blood to the dialysate, and cause glucose and salts to transfer by diffusion from the dialysate to the blood. When the kidneys of an adult function normally, the urea content of the blood is maintained in the range of 10–20 mg of urea nitrogen per 100 ml (1 deciliter, dl).

For the design of the C-DAK 4000 artificial kidney, and the many similar hemodialysis devices (Daugirdas and Ing, 1988), rates of permeation of the species through the candidate membranes are necessary. Estimates for the permeability of pure species in a microporous membrane can be made from the molecular diffusivity, and pore diameter, porosity, and tortuosity of the membrane (Seader and Henley, 2006), as shown in Example 16.2. For this reason, considerable laboratory experimentation is required when selecting membranes in the molecular-structure design step.

Home Hemodialysis Devices

The inconvenience of patient travel to a center for hemodialysis three times weekly, for four-hour treatments, has prompted medical experts to evaluate a different form of delivering hemodialysis, specifically, at night, at home, while the patient is sleeping (Talamini, 2005). Of the 281,600 hemodialysis patients in the United States in 2002, only 843 (0.3%) were home dialysis patients, with no devices identified to perform home dialysis. In conventional dialysis centers, there are many devices that accompany the hemodialysis device, including blood and dialysis pumps, pressure and temperature transducers, and an air-detection system. Also included are numerous alarms associated with instruments that measure pressure and temperature and the occurrence of blood leaks, air embolism, and vascular access disconnects. For home dialysis, additional devices are envisioned, including sorbent dialysis devices (to recover urea from the dialysate solution), safeguards to prevent blood-access disconnections, alarms to detect fluid (blood or dialysate) leaks, software allowing connection to the Internet for remote monitoring, central monitoring of treatment and patient parameters (e.g., blood pressure, pulse, venous and arterial pressures), and the like. With the large number of home dialysis clinics being established throughout the United States, it is expected that the use of home dialysis will increase rapidly in the coming years.

Innovation Map

For the product development of a hemodialysis device, consider the *innovation map* in Figure 16.21, which is a typical one that might be prepared in the late 2000s. Note that, during product design, an initial innovation map can be prepared when the design team answers the questions concerning the new technologies in the upper portion of Figure PIII.1. Then, the innovation map can be refined as the SGPDP is carried out, following the steps in the product design case study in Section 17.3.

To construct the *innovation map*, the historical information above, coupled with an observation of customer needs, provide the elements to be positioned in its six levels, moving from the bottom to the top of the map:

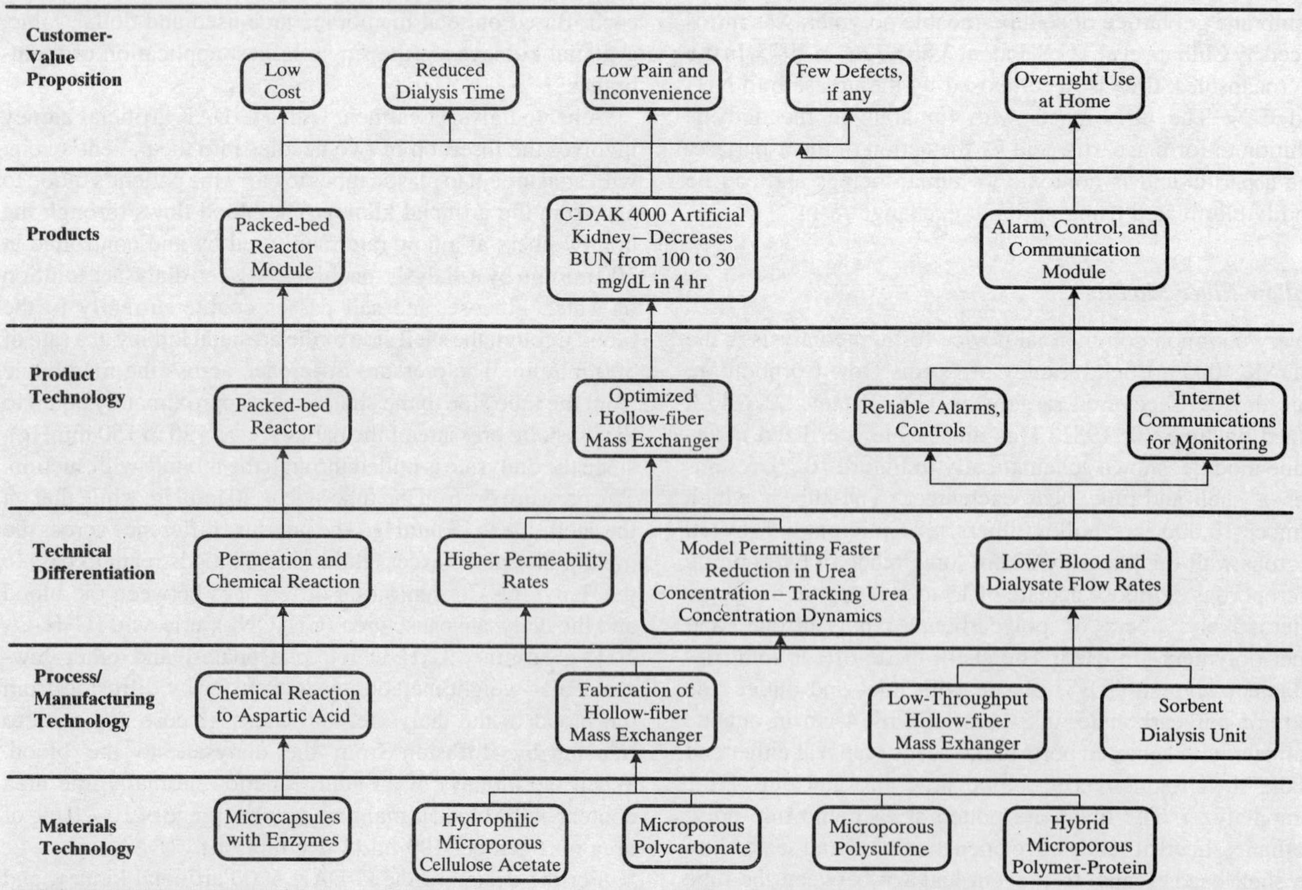

Figure 16.21 Innovation map for home hemodialysis device.

1. *Materials Technology*: microcapsules with enzymes; hydrophobic microporous cellulose acetate, polycarbonate, and polysulfone; hybrid microporous polymer-protein.
2. *Process/Manufacturing Technology*: chemical reaction to aspartic acid, fabrication of hollow-fiber mass exchanger, low-throughput hollow-fiber mass exchanger, and sorbent dialysate unit.
3. *Technical Differentiation (Technical-Value Proposition)*: permeation and chemical reaction, higher permeability rates, model permitting faster reduction in urea concentration—tracking urea concentration dynamics, and lower blood and dialysate flow rates.
4. *Product Technology*: packed-bed reactor, optimized hollow-fiber mass exchanger, reliable alarms and controls, and Internet communications for monitoring.
5. *Products*: packed-bed reactor module; C-DAK 4000 Artificial Kidney; alarm, control, and communication module(s).
6. *Customer-Value Proposition*: low cost, reduced dialysis time, low pain and inconvenience, few defects, overnight use at home.

Given these elements, the connectivity is added to show the interplay between the technological elements, the *technical-value proposition*, and the *customer-value proposition*. As unmet customer needs are encountered, they can be added for the next generation of products.

The initial inventions, in the leftmost column of Figure 16.21, involved the *microcapsules with enzymes* (materials technology) and *chemical reaction to aspartic acid* (process/manufacturing technology) to provide the technical differentiation *permeation and chemical reaction*. This led to the *packed-bed reactor* (product technology) and the *packed-bed reactor module* product, which was created to satisfy the customer needs of *low cost*, *reduced dialysis time*, *low pain and inconvenience*, and *few defects* (customer-value proposition).

Then, in the 1970s, it became possible to extrude hollow fibers reliably. Several materials technologies, including *microporous cellulose acetate*, *polycarbonate*, and *polysulfone*, led to the *fabrication of hollow-fiber mass exchangers* (process/manufacturing technology). Note that a potential new material technology, involving the development of a *hybrid microporous polymer-protein*, is also shown on the innovation map. This process/manufacturing technology led to the technical differentiations *higher permeability rates* and a *model permitting faster reduction in urea concentration—tracking urea concentration dynamics*. In turn, these led to an *optimized hollow-fiber mass exchanger* (product technology) and the *C-DAK 4000 Artificial Kidney* product, which better satisfied the customer-value proposition.

Finally, in the 2000s, moving to the rightmost portion of Figure 16.21, the customer need for *overnight home use* is likely to take advantage of new product/manufacturing technologies, *low-throughput hollow-fiber mass exchanger* and *sorbent dialysis unit*. In turn, their technical differentiation, *lower blood and dialysate flow rates*, might lead to new product technologies: *reliable alarms and controls* and *Internet communications for monitoring*, which may provide a new product(s), that is, an *alarm, control, and communication module(s)*.

In the remainder of this section, aspects of these promising technological inventions are discussed in subsections on: (a) Process/Manufacturing Technology, and (b) Technology Protection.

a. Process/Manufacturing Technology

Hemodialysis Device

Having selected the candidate polymer membranes, as discussed above, that must be available in hollow-fiber membranes, a key challenge in the development of the C-DAK 4000 Artificial Kidney was to fabricate a new hollow-fiber mass exchanger (new process/manufacturing technology). This, together with a model for tracking its urea concentration dynamics, provided the technical differentiation that permitted an optimized configuration (product technology). In practice (for normal hemodialysis, rather than potential home hemodialysis), blood flow rates range from 100 to 400 ml/min, while dialysate flow rates range from 200 to 800 ml/min. Decisions regarding the fiber inside diameter, wall thickness, and length, and the number of fibers influence the pressure of the fluids on each side, the surface area for mass transfer, and the rates of mass transfer. To illustrate the development of this model and demonstrate its use for creating optimized designs, Example 16.2 is provided next.

EXAMPLE 16.2

Develop a design procedure for a hollow-fiber hemodialysis device of the type shown in Figure 16.20. Base the design on a blood flow rate of 200 ml/min and a dialysate flow rate of 500 ml/min. Assume the design will be controlled by mass transfer of one of the blood plasma components to be removed, for example, urea. The blood will flow through the hollow fibers, while the dialysate will flow past the outside surface of the fibers in a direction countercurrent to the flow of the blood plasma. A typical patient will require hemodialysis when the blood reaches a urea nitrogen level (BUN) of 100 mg/dl. A target for the hemodialysis device is to reduce the BUN to 30 mg/dl within 4 hr, which corresponds to normal operation at a hemodialysis center.

SOLUTION

Several key steps in the design procedure are presented next, beginning with the estimation of the mass-transfer coefficient for transport of urea across the membrane. Next, pressure drops are estimated both in and outside of the hollow fibers. Then, a mass-transfer model is solved for the concentration of urea in the bloodstream as a function of time, which assists in sizing the dialysis device.

1. Overall mass-transfer coefficient for urea:

The rate of mass transfer of urea from the blood plasma, through the membrane, and to the dialysate is given by

$$n = K_i A_i \Delta C_{LM} \tag{16.2}$$

where:

- n = rate of mass transfer of urea nitrogen, mg/min
- K_i = overall mass-transfer coefficient based on the inside area, cm/min
- A_i = mass-transfer area based on the inside area of the hollow fibers, cm^2
- ΔC_{LM} = log-mean urea-nitrogen concentration difference for mass transfer

The overall mass-transfer coefficient, which must consider the resistances of the blood plasma, the membrane, and the dialysate, is given by

$$\frac{1}{K_i} = \frac{1}{k_b} + \frac{l_M}{P_M}\frac{A_i}{A_m} + \frac{1}{k_d}\frac{A_i}{A_o} = \frac{1}{k_b} + \frac{l_M}{P_M}\frac{D_i}{D_m} + \frac{1}{k_d}\frac{D_i}{D_o} \tag{16.3}$$

where:

- k_b = mass-transfer coefficient on the inside where the blood flows, cm/min
- k_d = mass-transfer coefficient on the outside where the dialysate flows, cm/min
- A_o = mass-transfer area based on the outside area of the hollow fibers, cm^2
- A_m = arithmetic mean of A_i and A_o, cm^2
- l_M = membrane wall thickness, cm
- P_M = membrane permeability, cm^2/min

The mass-transfer coefficients, k_b and k_d, can be estimated by analogy from available dimensionless empirical correlations for heat transfer, which are taken from Knudsen and Katz (1958). These analogous correlations depend on the flow regime and relate the Sherwood number to the Reynolds and Schmidt numbers.

For the flow of blood plasma inside the hollow fibers, the flow regime will be laminar because of the very small fiber diameter and the need to avoid high flow velocities that would stress the blood cells to destruction. For example, for the C-DAK artificial kidney described above, 200 ml/min (200 cm^3/min), Q_b, of blood flows through 10,000 fibers, each of 200 microns (0.02 cm) inside diameter, D_i, with 20 microns wall thickness, t_w, rather than the 10 micron thickness shown in Figure 16.20, and 22 cm in length, L. The cross-sectional area for flow in each fiber is $3.14(0.02)^2/4 = 0.000314\,\text{cm}^2$. The total flow area, S_i, for 10,000 fibers is $10{,}000(0.000314) = 3.14\,\text{cm}^2$. The average blood velocity, V_b, is $(200)/3.14 = 63.7$ cm/min $= 1.062$ cm/s. Blood has a density, ρ_b, of 1.06 g/cm^3 and a viscosity, μ_b, of approximately 0.014 g/cm-s. This gives a Reynolds number for blood flow through the fibers, $N_{Re_b} = D_i V_b \rho_b / \mu_b = 0.02(1.062)(1.06)/0.014 = 1.608$, which is in the laminar-flow regime. Because the fully developed parabolic velocity profile for laminar flow is obtained by a fiber length $L = 0.05\, D_i N_{Re_b}$, the velocity profile will be developed in less than one fiber diameter. Thus, for mass transfer of urea through the blood plasma, the important criterion is the Peclet number for mass transfer, N_{Pe_M}, which is the product of the Reynolds number and the Schmidt number, $N_{Sc_b} = \mu_b/\rho_b D_{urea}$ or

$$N_{Pe_M} = \frac{D_i V_b}{D_{urea}} \tag{16.4}$$

where D_{urea} = molecular diffusivity of urea in blood plasma, cm^2/s. The diffusivity for urea in blood plasma at a body temperature of 37°C is 0.8×10^{-5} cm^2/s. Thus, the Peclet number $= 0.02(1.062)/(0.8 \times 10^{-5}) = 2{,}650$. The Sherwood number depends on the ratio of N_{Pe_M} to (L/D_i), which equals $2{,}650/(22/0.02) = 2.41$. For this condition, the Sherwood number, $N_{Sh} = D_i k_b / D_{urea}$, is approximately constant at a value of 4.364. Thus, $k_b = 4.364\ (0.8 \times 10^{-5})/0.02 = 0.00175$ cm/s.

Estimation of the mass-transfer coefficient in the dialysate outside the fibers is considerably more difficult because of the complex geometry. In a shell-and-tube heat exchanger, baffles are used to help direct the shell-side fluid to flow back and forth in directions largely normal to the tube length. It would be extremely difficult to include such baffles in a hollow-fiber hemodialysis unit. Instead, as shown in Figure 16.20, the dialysate leaves the inlet port at one end to enter the shell normal to the fibers, turns 90° and flows parallel to the fibers, and then turns 90° to enter the exit port at the other end. Presumably, the dialysate flow is largely parallel to the fibers along their length. For the example being considered here, assume the inside diameter of the shell is 3.8 cm, giving a cross-sectional area of $3.14(3.8)^2/4 = 11.34\,\text{cm}^2$. The 10,000 fibers, with an outside diameter of $200 + 40 = 240$ microns or 0.024 cm, occupy a cross-sectional area of $10{,}000(3.14)\ (0.024)^2/4 = 4.52\,\text{cm}^2$. Thus, the area for flow of the dialysate is $11.34 - 4.52 = 6.82\,\text{cm}^2$. As an approximation, estimate the shell-side mass-transfer coefficient, k_d, from a circular tube inside-flow correlation as for k_b, by replacing tube inside diameter by an equivalent diameter equal to $4r_H$, where r_H is the hydraulic radius, which is equal to the cross-sectional area for flow divided by the wetted perimeter. For 10,000 fibers of 0.024 cm diameter on a square pitch, the wetted perimeter is $3.14(0.024)(10{,}000/4) = 188.4$ cm. This gives a hydraulic radius of $6.82/188.4 = 0.0362$ cm. The equivalent diameter $= 4(0.0362) = 0.145$ cm. Assume dialysate properties at 37°C of $\rho_d = 1.05$ g/cm^3 and $\mu_d = 0.007$ g/cm-s. Take a dialysate flow rate of 500 cm^3/min. The average velocity of the dialysate, V_d, is $(500/60)/6.82 = 1.22$ cm/s. The dialysate Reynolds number, $N_{Re_d} = 4r_H V_d \rho_d / \mu_d$, is $0.145(1.22)(1.05)/\ 0.007 = 26.5$, which is much higher than for the tube side, but is still in the laminar-flow region. The estimated entry length, using the shell-side version of the tube-side equation given above, is $0.05(0.145)(26.5) =$

0.192 cm, making it possible to again assume fully developed flow. The molecular diffusivity of urea through the dialysate solution is 1.8×10^{-5} cm²/s. The Peclet number, using a shell-side version of the equation given above, is $N_{Pe_M} = (0.145)(1.22)/(1.8 \times 10^{-5}) = 9,830$. The ratio of N_{Pe_M} to $(L/4r_H)$ equals 9,830/(22/0.145) = 64.8. For this condition, the Sherwood number, $N_{Sh} = 4r_H k_d / D_{urea}$, is approximately 7. Thus, $k_d = 7(1.8 \times 10^{-5})/0.145 = 0.00087$ cm/s.

For a microporous membrane, the membrane permeability is the effective diffusivity for urea. For the microporous membrane of 20 microns (0.002 cm) wall thickness, assume a porosity, ε, of 0.25 and a tortuosity, τ, of 1.5. The effective diffusivity or permeability in the membrane is given by $(D_{urea})_{eff} = P_M = \varepsilon(D_{urea})_d/\tau = 0.25(1.8 \times 10^{-5})/1.5 = 3 \times 10^{-6}$ cm²/s. The permeance, $P_M/l_M = 3 \times 10^{-6}/0.002 = 0.0015$ cm/s.

The overall mass-transfer coefficient based on the inside area of the hollow fibers is obtained from Eq. (16.3):

$$\frac{1}{K_i} = \frac{1}{0.00175} + \frac{1}{0.0015}\left(\frac{0.020}{0.022}\right) + \frac{1}{0.00087}\left(\frac{0.020}{0.024}\right)$$

$$= 571 + 606 + 958 = 2,135 \text{ s/cm}$$

This result shows that 27% of the mass-transfer resistance is on the blood side, 28% is in the membrane, and 45% is on the dialysate side. These percentages change as the fiber geometry is changed. Taking the reciprocal, the overall mass-transfer coefficient is $K_i = 0.000468$ cm/s.

2. Pressure drop for blood flow through the hollow fibers:

Because the blood flow through the hollow fibers of the hemodialyzer is laminar, the pressure drop is computed from the Hagen–Poiseuille equation:

$$-\Delta P_b = \frac{32\mu_b L V_b}{D_i^2} = \frac{32(0.014)(22)(1.062)}{0.02^2}$$

$$= 26,170 \text{ g/cm-s}^2 \text{ or } 2.617 \text{ kPa or } 19.6 \text{ mmHg}$$

This pressure drop compares well with the pressure drop of 30 mmHg cited above for the commercial C-DAK 4000 artificial kidney.

3. Pressure drop for dialysate flow past the hollow fibers:

The flow on the shell side is also laminar. Using the hydraulic-radius concept, an estimate can be made of the pressure drop by replacing D_i in the above Hagen–Poiseuille equation with $4r_H$. Thus, the pressure drop is

$$-\Delta P_d = \frac{32\mu_d L V_d}{(4r_H)^2} = \frac{32(0.007)(22)(1.22)}{(0.145)^2}$$

$$= 286 \text{ g/cm-s}^2 \text{ or } 0.0286 \text{ kPa or } 0.2 \text{ mmHg}$$

This is far less than the 12 mmHg quoted above for the C-DAK 4000. This difference could be due to the entering and exiting flow of dialysate normal to the fibers, which would increase the pressure drop considerably. A better estimate could be made with a computational fluid dynamics (CFD) program. This pressure-drop calculation also sheds some doubt on the above calculation of the mass-transfer coefficient on the shell side, which could also be greater than that calculated by assuming flow parallel to the length of the fibers.

4. Urea mass transfer and time required to reduce its concentration in the blood:

A compartment model, shown in Figure 16.22, has been successful in following changes in solute concentrations with time for a patient undergoing hemodialysis. The model consists of three perfectly mixed compartments and one membrane separator. Streams that are assumed to have negligible volume connect the compartments and separator. The upper compartment, of volume V_P, represents a patient's body fluid, other than blood. A solute, such as urea, is transferred to the body fluid at the constant mass rate, G. The second compartment, of volume V_B, represents the patient's blood, which circulates between these two compartments at a volumetric rate, Q_P. Below the second compartment is the hemodialysis unit, which transfers solutes such as urea across hollow-fiber membranes to a dialysate. Blood circulates at a volumetric rate, Q_B, between the hemodialyzer and the second compartment. Dialysate circulates at a volumetric rate, Q_D, between the bottom compartment (dialysate holding tank) and the hemodialyzer, through which it flows countercurrently to the blood flow. From the dialysate holding tank, a constant volumetric flow rate of waste dialysate, Q_W, is withdrawn. An equal volumetric flow rate of fresh dialysate of zero waste solute concentrations, for example, urea, is added to the circulating dialysate before it enters the hemodialyzer. Also indicated in Figure 16.22 are symbols, C_j, for solute (for example, urea) concentrations in the various streams in units of mass/unit volume. Because of the assumption of perfectly mixed compartments, concentrations of solutes in the three compartments are equal to the solute concentrations in the streams leaving the corresponding compartments.

Model equations for a system similar to that of Figure 16.22 were developed and solved by Spaeth (1970), whose equations are applied here. Because solute concentrations change with time, the following solute mass balance equations apply to the three compartments:

$$V_P \frac{dC_P}{dt} = G - Q_P(C_P - C_B) \quad (16.5)$$

$$V_B \frac{dC_B}{dt} = Q_P(C_P - C_B) - Q_B(C_B - C_{B,out}) \quad (16.6)$$

$$V_D \frac{dC_W}{dt} = Q_D(C_{D,out} - C_W) \quad (16.7)$$

Equations for the rate of mass transfer of a solute across the walls of the hollow fibers in the membrane unit and a solute mass balance are as follows, where countercurrent flow is assumed in the hemodialyzer, with a corresponding log-mean concentration driving force:

$$Q_B(C_B - C_{B,out}) = K_i A_i \Delta C_{LM}$$

$$= K_i A_i \left[\frac{(C_B - C_{D,out}) - (C_{B,out} - C_D)}{\ln \dfrac{(C_B - C_{D,out})}{(C_{B,out} - C_D)}} \right] \quad (16.8)$$

$$Q_B(C_B - C_{B,out}) = Q_D(C_{D,out} - C_D) \quad (16.9)$$

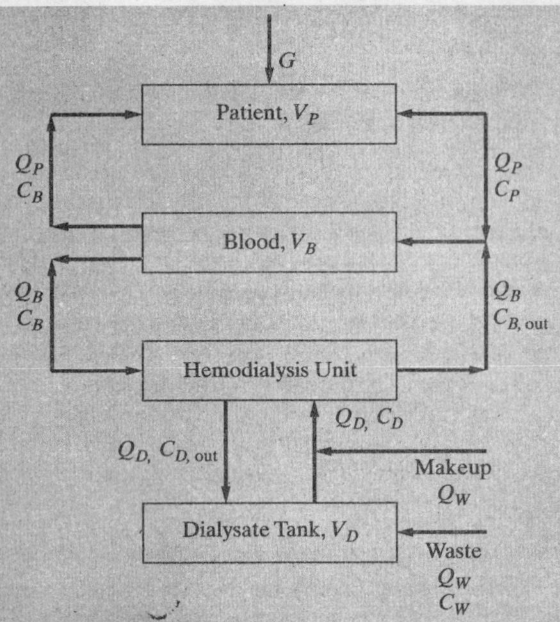

Figure 16.22 Hemodialysis model.

Finally, a solute mass balance around the mixing point, assuming that the makeup dialysate contains no solute, gives

$$Q_D C_D = (Q_D - Q_W) C_W \tag{16.10}$$

Equations (16.5) to (16.10) constitute six equations in the six variables C_P, C_B, C_D, C_W, $C_{B,\text{out}}$, and $C_{D,\text{out}}$, all of which vary with time. The six equations can be reduced to the following three ordinary differential equations in the three variables C_P, C_B, and C_D:

$$\frac{dC_P}{dt} = \frac{G}{V_P} - \left(\frac{Q_P}{V_P}\right)(C_P - C_B) \tag{16.11}$$

$$\frac{dC_B}{dt} = \left(\frac{Q_P}{V_B}\right)(C_P - C_B) - \left(\frac{Q_B}{V_B}\right)E(C_B - C_D) \tag{16.12}$$

$$\frac{dC_D}{dt} = \left[E\left(\frac{Q_B}{V_D}\right)\left(1 - \frac{Q_W}{Q_D}\right)\right]C_B - \left[\left(\frac{Q_W}{V_D}\right) + E\left(\frac{Q_B}{V_D}\right)\left(1 - \frac{Q_W}{Q_D}\right)\right]C_D \tag{16.13}$$

where the parameter, E, is defined by

$$E \equiv \frac{C_B - C_{B,\text{out}}}{C_B - C_D} \tag{16.14}$$

and is computed from

$$E = \frac{1 - \exp\left[\dfrac{K_i A_i}{Q_B}\left(1 - \dfrac{Q_B}{Q_D}\right)\right]}{\dfrac{Q_B}{Q_D} - \exp\left[\dfrac{K_i A_i}{Q_B}\left(1 - \dfrac{Q_B}{Q_D}\right)\right]} \tag{16.15}$$

which is derived from Eqs. (16.8) and (16.9).

Equations similar to Eqs. (16.11) to (16.13), but for a two-compartment model, are solved analytically by Bird et al. (2002). However, a numerical solution suitable for a spreadsheet is used here, beginning with conditions at time $t = 0$ for the three solute concentrations. Once C_P, C_B, and C_D are obtained as functions of time, the other three concentrations, C_W, $C_{B,\text{out}}$, and $C_{D,\text{out}}$ are computed as functions of time from Eqs. (16.10), (16.14), and (16.9), respectively.

As an example of the application of the above equations to urea, the following values are used:

$$V_P = 40{,}000\,\text{cm}^3,\ V_B = 5{,}000\,\text{cm}^3,\ \text{and}$$
$$V_D = 1{,}000\,\text{cm}^3$$
$$Q_P = 5{,}000\,\text{cm}^3/\text{min},\ Q_B = 200\,\text{cm}^3/\text{min},$$
$$Q_D = 500\,\text{cm}^3/\text{min},\ \text{and}\ Q_W = 250\,\text{cm}^3/\text{min}$$
$$G = 5\,\text{mg/min},\ K_i = 0.000468\,\text{cm/s(from above)}$$
$$= 0.0281\,\text{cm/min},$$
$$\text{and}\ A_i = 13{,}800\,\text{cm}^2$$

Initial conditions for the three concentrations are

$$C_P = 1\,\text{mg/cm}^3,\ C_B = 1\,\text{mg/cm}^3,\ \text{and}\ C_D = 0\,\text{mg/cm}^3$$

The numerical solution gives the following results over a 6-hr period, where all three concentrations are given in mg urea/cm^3.

Time (hr)	C_P	C_B	C_D
0	1.00	1.00	0.00
1	0.86	0.84	0.20
2	0.74	0.73	0.18
3	0.64	0.63	0.15
4	0.55	0.54	0.13
5	0.48	0.47	0.11
6	0.42	0.41	0.10

Thus, in 4 hr, which is a typical treatment time, the urea concentrate has been reduced by 45%. It is left to Exercise 16.5 at the end of the chapter for a study of the effect on the rate of urea removal of changing the hemodialyzer geometry, the blood and dialysate flow rates to the dialyzer, the rate of waste withdrawal, the volume of the dialysate tank, and the sensitivity of the rate of urea mass transfer to the mass-transfer coefficient. In particular, the above estimate of the coefficient on the shell side may be low because the entry to and exit from the hemodialyzer of the dialysate is normal to, rather than parallel to, the fibers. This should enhance the shell-side coefficient.

Finally, the model in Example 16.2 can be used to redesign the hemodialysis device for overnight home usage. In this case, dialysis could be accomplished seven nights per week for, say, six hours per night. This would permit much smaller blood and dialysate flow rates, and maintain lower urea concentrations in patients. See Exercise 16.6 at the end of the chapter.

Sorbent Dialysis

At most hemodialysis centers, the rich dialysate from the hemodialysis devices is not regenerated and recycled, but disposed of through drains and replaced by freshly diluted solution. This process, known as single-pass dialysis, requires large volumes of ultra-pure water, typically prepared using reverse osmosis, activated carbon, deionization, and the like. The necessary equipment and holding tanks are expensive and require much space and, consequently, are not well suited for home hemodialysis products.

In the Allient® Sorbent Hemodialysis System (see the *Sorbent Dialysis Primer* and *History of Sorbent Dialysis* Web sites), the spent dialysate is chemically regenerated by passing through a disposable cartridge containing layers of activated carbon, the urease enzyme, and ion-exchange membranes. A small amount of makeup solution, called the infusate, replenishes any essential ions lost in the filter. Thanks to this regeneration process, only 6 L are required for each treatment session, compared with greater than 120 L for single-pass dialysis. In addition to removing urea from the dialysate, the cartridge removes bacteria, heavy metals, and other contaminants from ordinary tap water, eliminating the need for water-treatment equipment. New devices are completely portable, requiring only a source of electricity (see the *Sorbent Dialysis Primer* Web site).

The first sorbent dialysis system was developed in the early 1970s and was sold under the REDY® brand name (for REcirculating DYalysis). Its use continued throughout the 1980s and, in 1993, a portable device was created. FDA approval for this device was not completed, however, and production was suspended until 1999, when SORB™ Technology, Inc., obtained the REDY® brand and renewed interest in sorbent dialysis. After merging with Renal Solutions, Inc., in 2001, they began to develop a system for home use. Renal Solutions, Inc., recently obtained FDA approval to market their product, the Allient® Sorbent Dialysis System, for both in-center and home treatment (see the *History of Sorbent Dialysis* Web site).

As shown in Figure 16.23, the used dialysate from the hemodialysis unit enters a purification layer, consisting of activated carbon, to remove particulate matter, heavy metals, etc. Next, the dialysate passes through a layer of the urease enzyme, which converts urea to ammonium and carbonate ions, that is, ammonium carbonate. It next passes through a zirconium phosphate layer (cation exchanger), where the dangerous ammonium ions are exchanged for hydrogen and sodium. A layer of zirconium oxide (anion exchanger) binds fluorides and phosphates that have been removed from the blood. Because the urease enzyme can break down unlimited amounts of urea, the main limitation of the cartridge is its capacity to bind ammonia. Current cartridges (known as SORB™ and HISORB™) can process 20–30 g of urea nitrogen before losing their ability to bind ammonia. A breakthrough of ammonia, when its concentration exceeds 2 wt%, indicates that the capacity has been exceeded and the cartridge must be changed (see the *Sorbent Dialysis Primer* Web site, page 6).

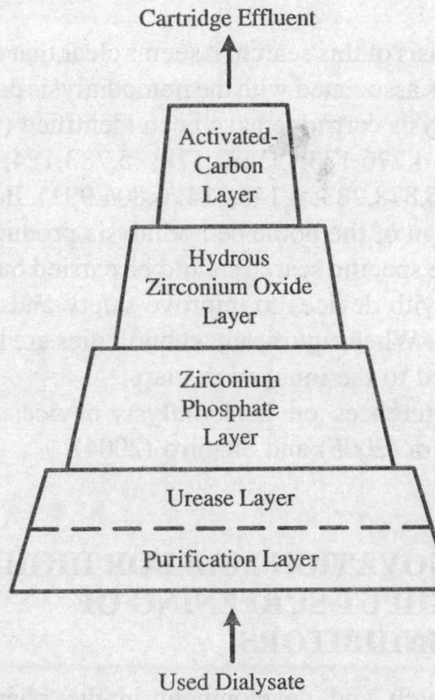

Figure 16.23 Schematic of the Allient® sorbent cartridge.

b. Technology Protection

To protect their technology, prior to designing a new home hemodialysis product, product-development teams need to be aware of existing patents. In July 2007, an advanced Google patent search for the term "home hemodialysis" returned 37 patents. The breakdown by decade is as follows:

Decade	Patents
1960s	1
1970s	1
1980s	0
1990s	33
2000s	2

As seen, far more patents were issued in the 1990s than in the other decades. Many of the inventors cited the low percentage of patients undergoing dialysis at home and offered solutions they believed would increase the popularity of this treatment. The following categories were well represented in the search results:

Category	Patents
Dialysis Machines and Components	12
Dialysate Solution	6

On the basis of this search, it seems clear that the principal technologies associated with the hemodialysis device and the sorbent dialysis cartridge have been identified (U.S. Patents 3,865,726; 4,276,173; D282,578; 5,783,124; 3,669,878; 4,473,449; 6,878,283; 4,148,314; 6,804,991). Before initiating the design of the home hemodialysis product in Section 17.3, a more specific search should be carried out for patents associated with devices to improve safety and for machine disinfection. When significant technologies are located, they can be added to the innovation map.

Other references on hemodialysis devices, in general, include Misra (2005) and Shapiro (2004).

16.4 INNOVATION MAP FOR HIGH-THROUGHPUT SCREENING OF KINASE INHIBITORS

Much research and development in the pharmaceuticals industry involves the invention of new therapeutics. These are often small molecules (< 500 MW) that are synthesized in small mg quantities, often tested in-vitro with small quantities of proteins, peptides, and the like that are produced recombinantly or extracted from natural sources. Because screening tests need to be small, often involving nano-liter samples, and operate quickly, handling many samples in parallel over periods on the order of minutes, there is growing interest in the design of *lab-on-a-chip* products.

This section addresses the potential creation of lab-on-a-chip products for the screening of kinase inhibitors (KIs), which, among other things, are promising cancer therapeutics. These are small molecules that reduce the phosphorylation activity of one or more of the ~500 kinase enzymes in the human body (examples include gleevec, temserolimus, gefitinib, staurosporine, . . .). To be useful, a KI must be highly specific—targeting one or a small number of kinase enzymes without significantly affecting the activity of the others. Typically, these inhibitors compete against ATP in the kinase active site, although alloseriz inhibitors are possible. In 2004, it was estimated that pharmaceutical companies were spending ~ $100 MM/year screening potential therapeutic KI compounds created by their drug-discovery efforts.

One potential lab-on-a-chip product would perform, at high speed, standardized kinase inhibition assays on compounds developed by pharmaceutical companies. All reagents would be purchased commercially or easily synthesized in-house using benchtop-scale equipment. Because of the high cost of the purified kinase enzymes (typically $5,000/mg), the use of at most 10–100 nl total reaction volume per assay is advantageous, with the reagents and samples handled and mixed using microfluidic methods. Ideally, kinase enzyme (KE) function and reactant concentration would be followed in real time using non-destructive, mix-and-read, separation-free methods such as fluorescence polarization (FP), fluorescent resonance energy transfer (FRET), or luminescence, which can be readily parallelized so that a single charge-coupled device (CCD) detector/scanner can monitor several hundred or thousand assays in parallel.

In this section, an *innovation map* is created to examine the key technological inventions that have accompanied chemical products closely related to the lab-on-a-chip being designed. These inventions, either in materials, processing/manufacturing, or product technologies, are critical to attracting customers and satisfying their perceived needs. Note that they are the answers to the three questions raised by design teams in the upper portion of Figure PIII.1. To illustrate the creation of an *innovation map* for a lab-on-a-chip to screen kinase inhibitors, the technologies related to its design are reviewed next. Then, in Section 17.4, the designs for potential lab-on-a-chip products are presented. This review begins with a brief discussion of the normal functioning of kinase enzymes and the role of kinase inhibitors.

Kinase Reactions and Lab-on-a-Chip Inventions

Kinase Enzyme Reactions

In general, each kinase enzyme is a protein that phosphorylates a target protein or set of proteins, affecting their function much like an on–off switch. Along with phosphatases that remove phosphate groups, these enzymes are largely responsible for the biochemical regulation within living cells. In general, the reaction resembles:

$$\begin{aligned} \text{KE} + \text{target protein} + \text{ATP} &\rightarrow \text{KE} \\ &+ (\text{target protein})\text{PO}_4 + \text{ADP} \end{aligned} \qquad \text{(R1)}$$

That is, the kinase enzyme binds adenosine triphosphate (ATP), which is illustrated in Figure 16.24, and an active site on a target protein moves one phosphate group onto the target protein and unbinds, releasing the target protein and ADP. As an enzyme, the kinase is not modified but repeats the reaction as long as target protein and ATP are available. About 250 KEs have been cloned and purified to date and are commercially available.

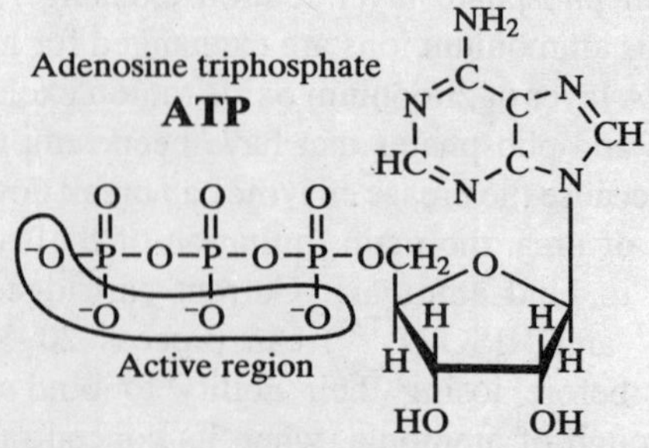

Figure 16.24 Adenosine triphosphate.

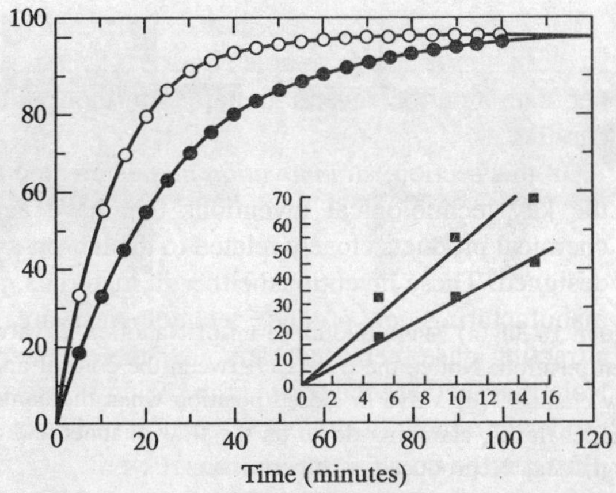

Figure 16.25 Product concentration of the phosphorylation reaction with (o) and without (•) kinase inhibitor (Copeland, 2003).

The function of the KI is to compete with the target protein by binding to the KE's active site. When binding is strong, it effectively blocks access of ATP and significantly slows the above reaction. Figure 16.25 shows the product concentration of the phosphorylation reaction as a function of time, with and without the kinase inhibitor present.

The lab-on-a-chip product would create inhibitory concentration 50% ($IC_{50} = [KI]_{50}$) data; that is, it would determine the KI concentration at which a given KE's activity (rate of reaction, r_P) drops by 50%. This is traditionally determined by performing multiple assays varying the KI concentration over many orders of magnitude, and finding the crossover (if any) in the resulting "S"-shaped inhibition curve (rate of reaction, r_P, as a function of inhibitor concentration), shown schematically in Figure 16.26.

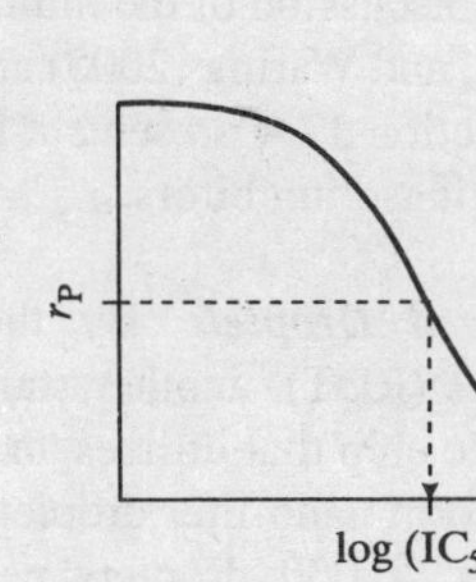

Figure 16.26 IC_{50} curve.

One approach is to use luminescence to report the concentration of ATP at some point in time, say 1 hour or less. Firefly luciferase and similar compounds produce light using an ATP-dependent chemical reaction; the brightness of the light is essentially proportional to the concentration of ATP, [ATP]. Thus, a high-resolution camera, such as a low-light CCD camera, can integrate the light signal for a short period with the brightness of a few pixels proportional to [ATP]. The time rate of change of [ATP] is then proportional to the consumption of ATP by the KE (assuming that the amount consumed by the luciferase is negligible by comparison).

Lab-on-a-Chip Inventions

During the first decade of the 21st century, two promising lab-on-a-chip technologies have been invented, one by the Fluidigm Corporation and the other by the RainDance Technologies and Nano-reactor Company, two successful startup companies.

Fluidigm Two-Layer Soft Lithography In the early 2000s, Fluidigm Corporation designed two-layer chips using poly (dimethylsiloxane), PDMS (whose chemical structure is shown in Figure 16.27), a soft polymer known for its low cost, flexibility, and optical transparency. As shown, PDMS is a repeating polymer consisting of multiple $-OSi(CH_3)_2-$ units in series. Its flexibility allows for deflections that permit the implementation of valves and peristaltic pumps, as described below. To prevent protein denaturation resulting from the hydrophobicity of the $-CH_3$ groups, the PDMS is treated with air plasma, which converts the PDMS surface from hydrophobic to hydrophilic.

Fluidigm used soft lithography to create a PDMS mold. First, the chip designer, using a computer-aided design (CAD) program, devised a master mold that contained all of the microstructures to be placed on the chip. Then, the master mold was created using UV photolithography, as shown in Figure 16.28. To prevent irreversible bonding between the PDMS and the master mold, the latter was coated with fluorinated silanes. And, to complete the process, a liquid PDMS prepolymer was poured onto the master to create the mold. These PDMS molds are highly reproducible and precisely made in less than a day.

To add devices for control (that is, valves and peristaltic pumps) to the chip, an addition-cure process called multilayer soft lithography was developed by Quake et al. (2000).

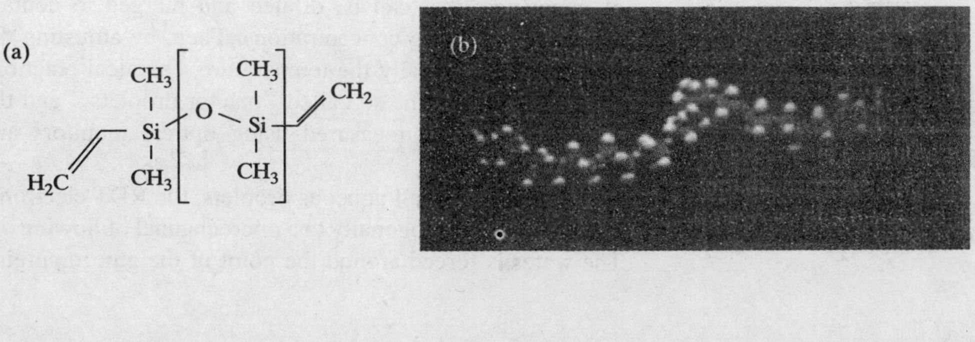

Figure 16.27 (a) Chemical structure of PDMS, comprised of repeating $-OSi(CH_3)_2-$ units. Usual polymer lengths are approximately $n = 60$. (b) A visual model of the PDMS elastomer.

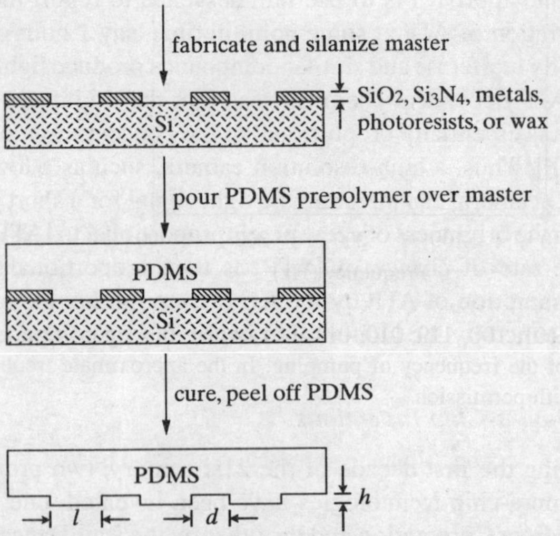

Figure 16.28 Schematic repr sentation of the creation of PDMS molds. Each master molds fabricates 50 PDMS molds. Height, depth, and length range from 0.2–20, 0.5–200, and 0.5–200 μm, respectively (Xia and Whiteside, 1998). Reproduced with permission.

This process bonds two PDMS molds into, effectively, a monolithic mold. The bottom layer is a PDMS mold containing vinyl groups and a platinum catalyst, and the top layer is a PDMS mold containing a cross-linker, with silicon hydride (Si-H) groups. When the two layers are joined, the latter covalently bonds to form a hermetic seal. Since the entire mold is monolithic, interlayer adhesion failures and thermal stress problems are avoided.

The pressure-driven top channels, which run orthogonal to the lower channels, control the fluid flow in the lower channels, as shown in Figure 16.29. The latter contain the reagents for all of the assays to be carried out.

Due to the flexibility of PDMS, as the pressure increases, the channels deflect. Thus, as pressure is exerted on a top

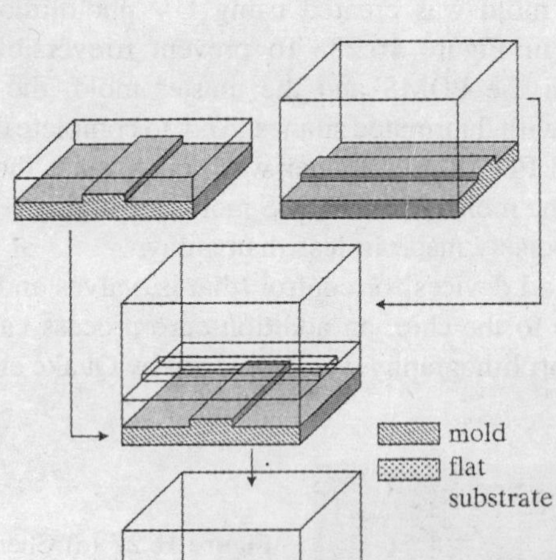

Figure 16.29 Schematic of multilayer soft lithography. The bonding between the bottom and top layers requires 1.5 hours at 80°C. Reproduced with permission.

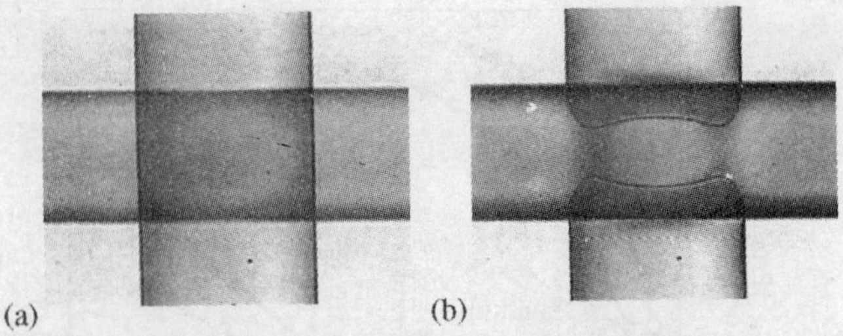

Figure 16.30 (a) Magnified close-up of Nanoflex™ valve in open position. Notice the overlap between the control and the flow channel. (b) Valve in closed position when the control channel flexes, clamping down on the flow channel and creating a tight seal. Reproduced with permission.

channel, it bends downward, by as much as 30 μm, until it touches and even deflects the bottom layer. With an applied pressure of approximately 100 kPa, a typical response time is on the order of 1 ms. Given sufficient pressure, a top channel seals off the bottom channels under it, as shown in Figure 16.30.

To control the fluid movement in the bottom channels, peristalsis is used. Three microfabricated valves are placed in series, generating wavelike contractions and essentially forming a peristaltic pump, as shown in Figure 16.31a. The pump uses the pattern 101, 100, 110, 010, 011, 001, where 0 and 1 represent open and closed valves, respectively.

As shown is Figure 16.31b, the pumping rate for the bottom channels is a function of the pattern-repeat frequency of the top channels. To control the flow rate, the *sensible* region of the graph is used, that is, the leftmost portion, where the flow rate varies linearly for frequencies between 0 and ~75 Hz. In the design herein, a typical frequency is 100 Hz, which corresponds to a flow rate of 2.4 nL/s, or 24 pL per pump cycle. For further discussion of the multi-layer PDMS molds, see Chen, Heend, and Waring (2005) and U.S. Patent 6,951,632. Note that Section 17.4 presents a Fluidigm chip design for the HTS of kinase inhibitors.

RainDance Micron-Sized Droplets By the mid-2000s, RainDance Technologies (RDT), another startup company, introduced a microfluidic chip that utilizes an electronically gated emulsion gun to form nano-liter droplets at very high frequencies, as high as 20,000 droplets per second. As syringe pumps force low-viscosity oil through a micro fabricated chip, referred to as a Personal Laboratory System™ (PLS™), emulsion guns form aqueous droplets containing various reactants. These droplets have volumes on the order of nano-liters that can be diluted and merged to contain reactants at varying concentrations. Then, by adjusting the environment, typically the temperature, chemical reactions are carried out in the so-called "reactor droplets," and the concentrations are measured using optical monitors and cameras.

To form the small aqueous droplets, the RDT electronic gun is pointed orthogonally to a microchannel of flowing oil. The water is forced around the point of the gun impinging

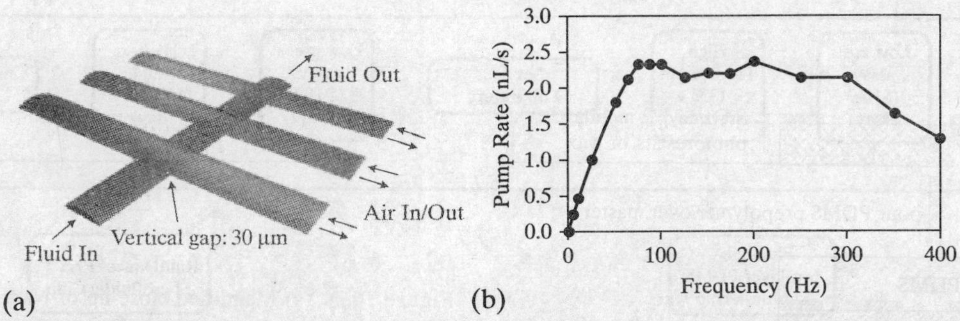

Figure 16.31 (a) When three control channels are actuated by the pattern 101, 100, 110, 010, 011, 001, where 0 = open valve and 1 = closed valve, peristaltic pumping occurs. (b) Pump rate as a function of the frequency of pumping. In the approximate frequency range of 75 Hz–100 Hz, pump rate is constant, ~2.4 nL/s. Reproduced with permission.

into the oil, forming a meniscus and creating charged droplets after each voltage alternation at high frequency (20,000 Hz), yielding as many as 20,000 droplets per second. Often, two guns create two streams of droplets, each containing different reactants and having opposite charges. Then, on the PLS™, as shown in Figure 16.32, the droplets, flowing side by side, are attracted to each other. Eventually, the droplets coalesce to contain two or more reagents; for example, a kinase enzyme (KE) and a potential kinase inhibitor (KI).

To permit the combination of various reagents at differing concentrations, the PLS™ is configured by RainDance personnel for specific applications, involving two or more emulsion guns, mixers, and splitters. The latter allow droplets to be split, with portions (say, one-third volume) recycled to be met by water droplets for dilution purposes.

Because the droplets are formed at high frequencies, and contain different reagents at differing concentrations, it is normally important to label the contents of each droplet using so-called *barcodes*. This led Molecular Probes®, another startup company, to develop TransFluoSphere™ fluorescent microspheres. Using three dyes (say, red, green, and blue), at three significantly different concentrations, each droplet can be assigned a unique code, which is detected by recording the emission wavelengths after laser excitation.

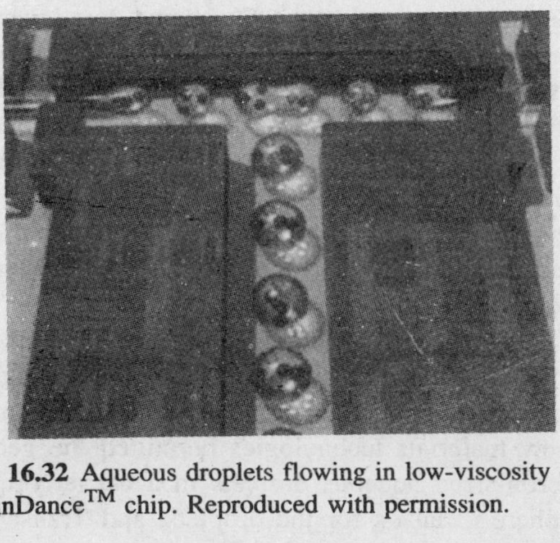

Figure 16.32 Aqueous droplets flowing in low-viscosity oil on the RainDance™ chip. Reproduced with permission.

Yet another advance was necessary to measure the concentration of the reaction products in each reactor droplet, to be related to the rate of reaction. Consider the phosphorylation reaction, R1, with KIs bound to KEs to slow the rate of conversion of ATP to ADP. To measure the ADP concentration, the BellBrook Laboratories, yet another startup company, introduced a Transcreener™ Assay, a fluorescence polarization technique. This assay uses a fluorescent tracer bound to a large antibody that is selected carefully to be specific to ADP. When this complex is exposed to polarized light, the ADP concentration, [ADP], is proportional to the intensity of the depolarized light emitted. To measure the depolarized light emitted, Agilent Technologies developed their Agilent SureScan™ micro-array reader, which uses AlexaFluor 633™, a very bright, stable dye that is excited by the 633 nm line of the He-Ne laser. The latter is included in the Transcreener™ kit, which provides a mixture of ADP and AlexaFluor 66™ to be blended with the kinase enzyme. Note that Section 17.4 presents a product design using a combination of these new technologies.

Innovation Map

For the product development of a lab-on-a-chip for the high-throughput screening of kinase inhibitors, consider the *innovation map* in Figure 16.33, which is typical of one prepared in the mid-2000s.

To construct the *innovation map*, the historical information above, coupled with an observation of customer needs, provide the elements to be positioned in its six levels, moving from the bottom to the top of the map:

1. *Materials Technology*: PDMS soft polymer, 10–100 μm aqueous droplet in carrier oil, Molecular Probes® TransFluoSphere™ fluorescent microspheres, and Transcreener® Assay—fluorescence polarization technique to measure [ADP].
2. *Process/Manufacturing Technology*: Fluidigm® two-layer soft lithography to create PDMS mold, photon generation using luciferase, RainDance® electronically gated emulsion gun—oil wets aqueous droplets, micro-array reader.

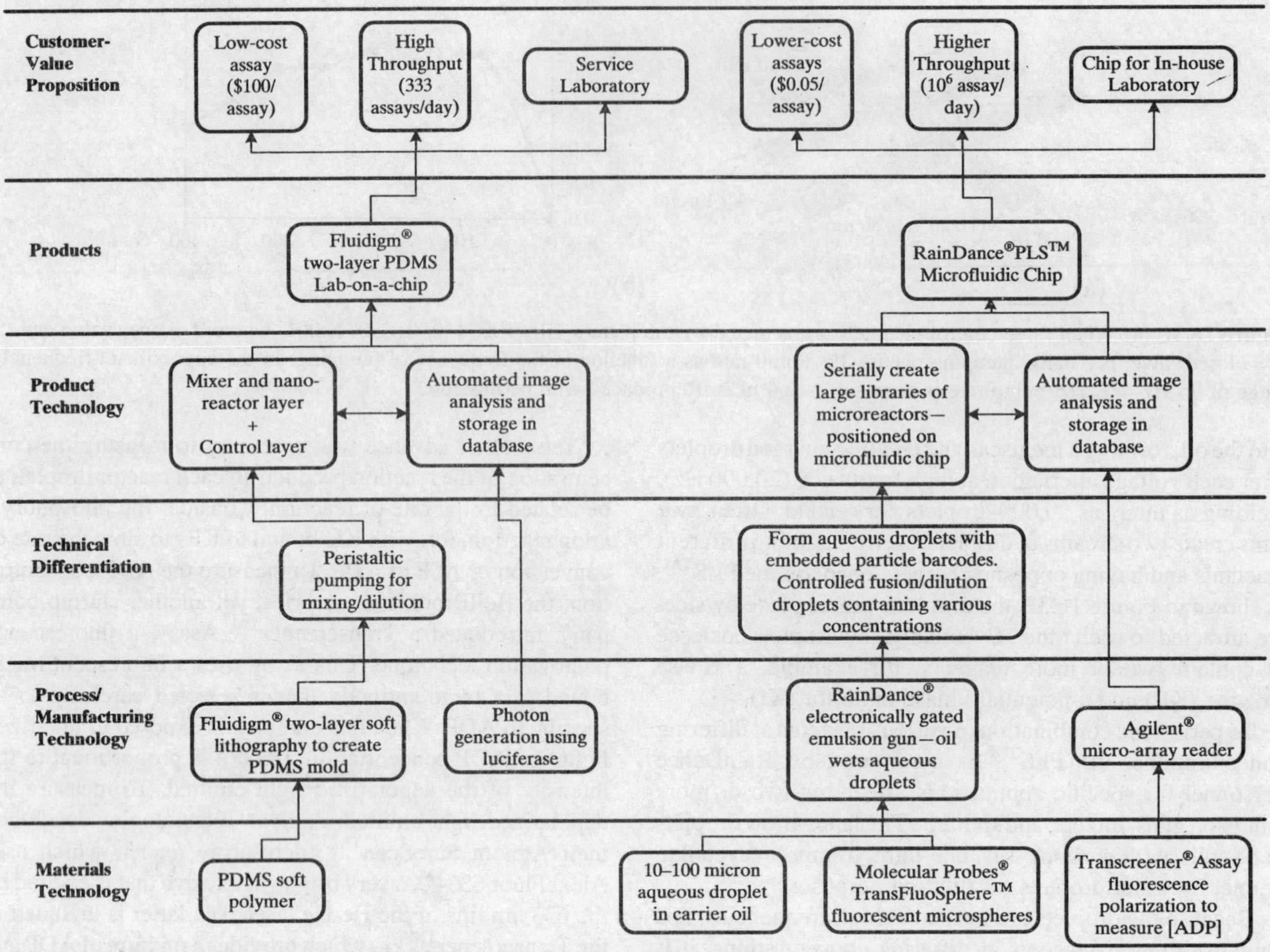

Figure 16.33 Innovation map for the lab-on-a-chip.

3. *Technical Differentiation (Technical-Value Proposition)*: peristaltic pumping for mixing/dilution, form aqueous droplets with embedded particle barcodes—controlled fusion/dilution droplets containing various concentrations.
4. *Product Technology*: mixing and nano-reactor layer + control layer, automated image analysis and storage in database, serially create large libraries of microreactors—positioned on the microfluidic chip.
5. *Products*: Fluidigm® two-layer lab-on-a-chip, RainDance® PLS™ microfluidic chip.
6. *Customer-Value Proposition*: low-cost assays ($100/assay), high throughput (333 assays/day), service laboratory, lower-cost assays ($0.05/assay), higher throughput (10^6 assays/day), chip for in-house laboratory.

Given these elements, the connectivity is added to show the interplay between the technological elements, the *technical-value proposition*, and the *customer-value proposition*. As unmet customer needs are encountered, they can be added for the next generation of products.

In 2000, the initial inventions, shown in the leftmost portion of Figure 16.33, involved the use of *PDMS soft polymer* (materials technology) with *Fluidigm® two-layer soft lithography to create a PDMS mold* (process/manufacturing technology) to provide the technical differentiation, *peristaltic pumping for mixing and dilution*. This led to the *mixer and nano-reactor layer plus a control layer* (product technology) and the *Fluidigm® two-layer PDMS lab-on-a-chip* product, which was created to satisfy the customer needs *low-cost assay ($100/assay), high throughput (333 assays/day)*, and *service laboratory* (customer-value proposition). In parallel, photon generation using luciferase (process/manufacturing technology) permitted the *automated image analysis and storage in database* (product technology), which led to the lab-on-a-chip product.

By 2004, the new materials and process/manufacturing technologies in the innovation map, shown in the right-hand portion of Figure 16.33, led to a key technical differentiation. The new materials technologies permitted the generation of micron-sized aqueous droplets in a carrier oil, TransFluoSphere™ labels for the droplets, and Transcreener®

fluorescence polarization to measure [ADP]. RainDance Technologies® introduced the new processing technology that permitted nano-liter aqueous droplets to be formed at high frequency, with varying compositions of reagents. This, in turn, led to a new product technology involving the creation of large libraries of microreactors on microfluidic chips. Together with new image analysis techniques, the new product, the RainDance PMS™ chip was introduced, which satisfied the customer needs for lower-cost assays, higher throughputs, and a chip configured for in-house laboratories.

In the remainder of this section, aspects of these promising technological inventions are discussed in subsections on: (a) Process/Manufacturing Technology, and (b) Technology Protection.

a. Process/Manufacturing Technology

This section begins with an introduction to enzyme kinetics, with emphasis on the dual-substrate enzyme kinetics and inhibition associated with the kinase enzymes and their inhibitors. This is needed to understand the usage of the new Fluidigm process/manufacturing technology for the lab-on-a-chip product, which will be described in Section 17.4. More specifically, the kinetics model is needed to determine the concentration of ATP, [ATP], to be catalyzed by luciferase to produce sufficient photons for the CCD camera, while maintaining sufficiently high levels of [ATP]; that is, [ATP] $\sim 1\ \mu$M. Also, the method for estimating the droplet reactor volume to produce sufficient photons is discussed, taking into account the incident photon fraction, quantum fraction, quantum efficiency, and similar factors.

Dual-Substrate Enzyme Kinetics

To design the screening device presented in Section 17.4, rate data are needed for the phosphorylation reaction, R1, and for the reaction involving the kinase inhibitor, to be discussed shortly. Before considering this data, which are provided by suppliers of the kinases, such as Caliper Life Sciences, Inc., and applied for the dual-substrate reaction, a brief review of Michaelis–Menton enzyme kinetics is included for reactions involving just a single substrate:

$$E + S \rightleftharpoons E\text{-}S \rightarrow E + P \qquad \text{(R2)}$$

where E is the enzyme, S is the substrate, E-S is the intermediate enzyme-substrate complex, and P is the product. To derive the rate of formation of the product, rate constants are defined for the reversible and irreversible reactions:

$$E + S \xrightarrow{k_1} E\text{-}S \qquad \text{(R3)}$$

$$E\text{-}S \xrightarrow{k_{-1}} E + S \qquad \text{(R4)}$$

$$E\text{-}S \xrightarrow{k_2} E + P \qquad \text{(R5)}$$

Because the enzyme is a catalyst, its total concentration, $[E_T]$, remains constant such that:

$$[E_T] = [E] + [E\text{-}S] \qquad (16.16)$$

where [E] is the enzyme concentration and [E-S] is the concentration of the enzyme bonded to the substrate.

The mass balance for the enzyme-substrate complex gives its rate of formation, $r_{E\text{-}S}$, as a function of its rates of generation and consumption in reactions R3–R5:

$$\begin{aligned} r_{E\text{-}S} &= k_1[E][S] - k_{-1}[E\text{-}S] - k_2[E\text{-}S] \\ &= k_1[E][S] - (k_{-1} + k_2)[E\text{-}S] \end{aligned} \qquad (16.17)$$

Since [E-S] is an intermediate in the reactions, $r_{E\text{-}S}$ is commonly approximated using the pseudo-steady-state assumption, $r_{E\text{-}S} = 0$, which gives:

$$[E][S] = (k_{-1} + k_2)/k_1[E\text{-}S] = K_m[E\text{-}S] \qquad (16.18)$$

where K_m is referred to as the Michaelis–Menten constant. Using Eq. (16.16) to substitute for [E]:

$$([E_T] - [E\text{-}S])[S] = K_m[E\text{-}S] \qquad (16.19)$$

Rearranging for [E-S] and multiplying by k_2 gives the rate of production of the product, P:

$$r_P = k_2[E\text{-}S] = \frac{k_2[S][E_T]}{K_m + [S]} \qquad (16.20)$$

Furthermore, because the maximum rate of production is $r_{P,\max} = k_2[E_T]$, Eq. (16.20) can be written as:

$$r_P = \frac{r_{P,\max}[S]}{K_m + [S]} \qquad (16.21)$$

which is the familiar form of the Michaelis–Menten equation. Note that at low substrate concentrations, where $K_m \gg [S]$, the rate of production is first-order in [S]. At high concentrations, where $K_m \ll [S]$, substrate inhibition occurs and the rate becomes zero-order in [S].

Returning to the phosphorylation reaction, R1, which involves the two substrates ATP and target protein (which is replaced by a polypeptide in the lab), the kinetics mechanism can be represented as:

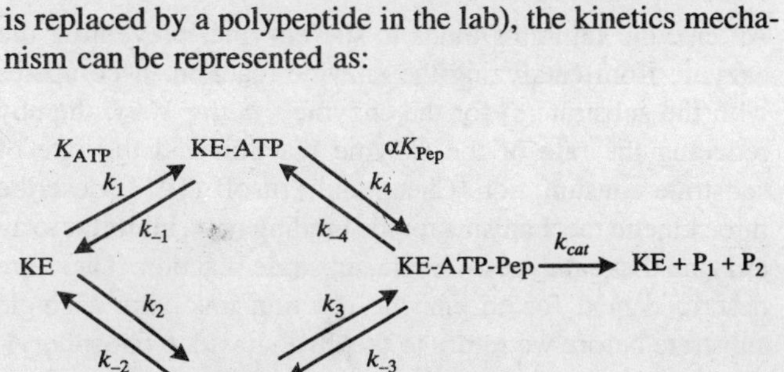

(R6)

where KE is the kinase enzyme, KE-ATP and KE-Pep are single-substrate complexes, KE-ATP-Pep is the dual-substrate complex, and P_1 and P_2 are two reaction products. Assuming *rapid equilibrium binding* (that is, equal forward and backward reaction rates), and defining the dissociation equilibrium constants:

$$K_{\text{ATP}} = \frac{k_{-1}}{k_1}, \quad \alpha K_{\text{ATP}} = \frac{k_{-3}}{k_3}, \quad K_{\text{Pep}} = \frac{k_{-2}}{k_2},$$

$$\alpha K_{\text{Pep}} = \frac{k_{-4}}{k_4} \tag{16.22}$$

and using the pseudo-steady-state assumption for the rates of production of the intermediate complexes, the rate of production of the products is:

$$r_{\text{P}} = \frac{k_{cat}[\text{KE}_\text{T}][\text{ATP}][\text{Pep}]}{\alpha K_{\text{ATP}} K_{\text{Pep}} + \alpha K_{\text{Pep}}[\text{ATP}] + \alpha K_{\text{ATP}}[\text{Pep}] + [\text{ATP}][\text{Pep}]} \tag{16.23}$$

where $[\text{KE}_\text{T}]$ is the total concentration of the kinase enzyme, [ATP] is the concentration of adinosine triphosphate, and [Pep] is the concentration of the polypeptide. Note that α multiplies the equilibrium constant for single-substrate binding to give the equilibrium constant for binding after the other substrate has been bound, it being assumed that α is independent of the binding sequence. Note also that the pseudo-steady-state assumption is redundant when all of the reversible reactions are assumed to be in local equilibrium. Finally, because the maximum rate of production of products, $r_{\text{P,max}}$, is $k_{cat}[\text{KE}_\text{T}]$, Eq. (16.23) can be written:

$$r_{\text{P}} = \frac{r_{\text{P, max}}[\text{ATP}][\text{Pep}]}{\alpha K_{\text{ATP}} K_{\text{Pep}} + \alpha K_{\text{Pep}}[\text{ATP}] + \alpha K_{\text{ATP}}[\text{Pep}] + [\text{ATP}][\text{Pep}]} \tag{16.24}$$

Note that, typically, $K_{\text{ATP}} \cong 1 - 10\ \mu\text{M}$ and $K_{\text{Pep}} \cong 1 - 10\ \mu\text{M}$.

Kinase Inhibition Reactions

An enzyme inhibitor binds to the enzyme, preventing the enzyme from catalyzing the enzyme reaction. It competes with the substrate(s) for the enzyme's *active sites*, thereby reducing the rate of the enzyme reaction and the rate of substrate consumption. Cheng and Prusoff (1973) describe three kinetic mechanisms for the binding of an inhibitor to an enzyme that catalyzes a single-substrate reaction. These are described next for an enzyme reaction involving a single substrate before we return to the dual-substrate phosphorylation reaction.

Single-Substrate Competitive Inhibitor. The first occurs when a *competitive* inhibitor is present; that is, one that competes with the substrate for the binding site on the enzyme:

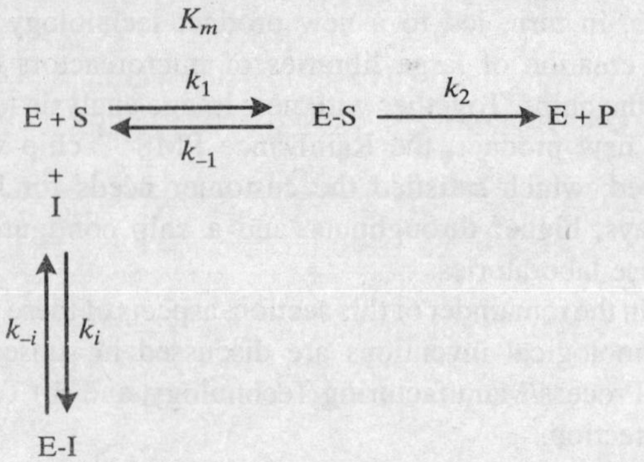

(R7)

Assuming *rapid equilibrium binding* and defining the dissociation equilibrium constant:

$$K_\text{I} = \frac{k_{-i}}{k_i} \tag{16.25}$$

and using the pseudo-steady-state assumption for the rate of production of E-S, the rate of production of the products is:

$$r_{\text{P,I}} = \frac{r_{\text{P, max}}[\text{S}]}{K_m\left(1 + \dfrac{[\text{I}]}{K_\text{I}}\right) + [\text{S}]} \tag{16.26}$$

where $r_{\text{P,I}}$ is the rate of production with the inhibitor present, and $r_{\text{P, max}} = k_2[\text{E}_\text{T}]$. Then, to determine $[\text{I}_{50}]$, set $r_\text{P} = 2r_{\text{P,I}}$, and solve:

$$[\text{I}_{50}] = K_\text{I}\left(1 + \frac{[\text{S}]}{K_m}\right) \tag{16.27}$$

Single-Substrate Non-Competitive Inhibitor. When a *non-competitive* inhibitor is present, it can bind to both the substrate, E, and the enzyme-substrate complex, E-S:

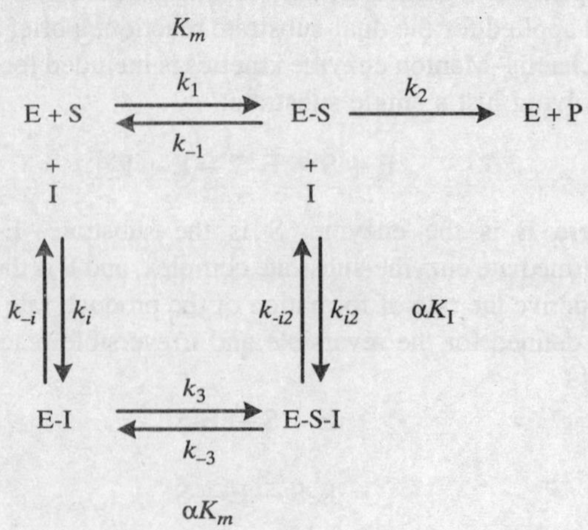

(R8)

Again, assuming *rapid equilibrium binding* and defining the dissociation equilibrium constants:

$$K_I = \frac{k_{-i}}{k_i}, \quad \alpha K_I = \frac{k_{-i2}}{k_{i2}}, \quad \alpha K_m = \frac{k_{-3}}{k_3} \tag{16.28}$$

and using the pseudo-steady-state assumption for the rate of production of E-S, the rate of production of the products is:

$$r_{P,I} = \frac{r_{P,\max}[S]}{K_m\left(1+\frac{[I]}{K_I}\right)+[S]\left(1+\frac{[I]}{\alpha K_I}\right)} \tag{16.29}$$

Then, to determine $[I_{50}]$, set $r_P = 2r_{P,I}$, and solve:

$$[I_{50}] = \frac{K_m+[S]}{\frac{K_m}{K_I}+\frac{[S]}{\alpha K_I}} \tag{16.30}$$

Single-Substrate Uncompetitive Inhibitor. For an *uncompetitive* inhibitor, binding is with the enzyme-substrate complex, E-S, only:

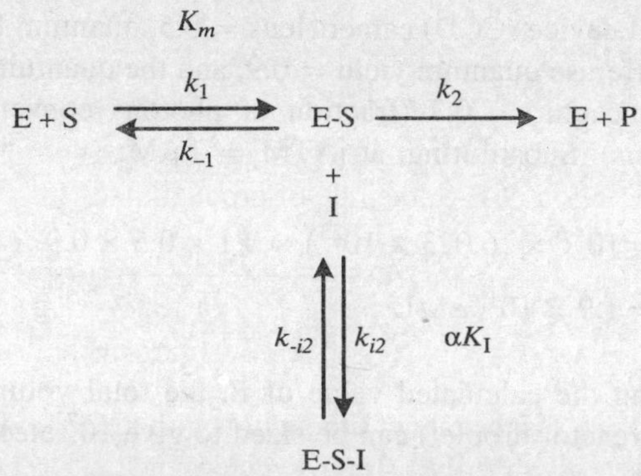

(R9)

Also, assuming *rapid equilibrium binding* and using the pseudo-steady-state assumption for the rate of production of E-S, the rate of production of the products is:

$$r_{P,I} = \frac{r_{P,\max}[S]}{K_m + [S]\left(1+\frac{[I]}{\alpha K_I}\right)} \tag{16.31}$$

and,

$$[I_{50}] = \alpha K_I\left(1+\frac{K_m}{[S]}\right) \tag{16.32}$$

Dual-Substrate Competitive Inhibitor. The competitive kinase inhibitor, KI, binds only with the kinase enzyme, KE:

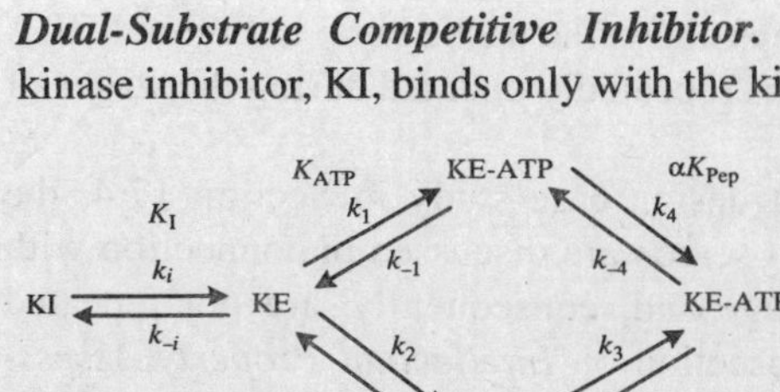

(R10)

Assuming *rapid equilibrium binding* and using the pseudo-steady-state assumption for the rates of production of the intermediate complexes, the rate of production of the products is:

$$r_{P,I} = \frac{r_{P,\max}[ATP][Pep]}{\alpha K_{ATP}K_{Pep}\left(1+\frac{[KI]}{K_I}\right)+\alpha K_{Pep}[ATP]+\alpha K_{ATP}[Pep]+[ATP][Pep]} \tag{16.33}$$

Note that, when designing KIs, it is common to target the ATP binding site on the KE, primarily because the steric structure of the ATP binding sites are well understood. In contrast, the steric structures of the protein binding sites are not known.

Inhibition Detection Methods

For the phosphorylation reaction (R1), the rate loss due to the binding of kinase inhibitors is normally determined at high [ATP], $[ATP_h]$, where the reaction is zero-order in [ATP]. In this case, the rate of production of products, r_{P,I,ATP_h}, in Eq. (16.33) is simplified to:

$$r_{P,I,ATP_h} = \frac{R_{ATP_h}[Pep]}{K_{m.ATP_h}+[Pep]} \tag{16.34a}$$

where

$$R_{ATP_h} = \frac{r_{P,\max}[ATP_h]}{\alpha K_{ATP}+[ATP_h]} \tag{16.34b}$$

$$K_{m,ATP_h} = \frac{\alpha K_{Pep}\left(K_{ATP}\left(1+\frac{[KI]}{K_I}\right)+[ATP_h]\right)}{\alpha K_{ATP}+[ATP_h]} \tag{16.34c}$$

Then, the rate of reaction can be measured by monitoring [Pep]. Alternatively, the rates can be measured at high [Pep], $[Pep_h]$, where the reaction is zero-order in [Pep], by monitoring [ATP]. In this case, the rate of production of products, r_{P,I,Pep_h}, in Eq. (16.33) is simplified to:

$$r_{P,I,Pep_h} = \frac{R_{Pep_h}[ATP]}{K_{m.Pep_h}+[ATP]} \tag{16.35a}$$

where

$$R_{Pep_h} = \frac{r_{P,\max}[Pep_h]}{\alpha K_{Pep}+[Pep_h]}, \tag{16.35b}$$

$$K_{m,Pep_h} = \frac{\alpha K_{ATP}\left(K_{Pep}\left(1+\frac{[KI]}{K_I}\right)+[Pep_h]\right)}{\alpha K_{Pep}+[Pep_h]} \tag{16.35c}$$

At $[ATP_h]$, in the homogeneous, time-resolved, fluorescence (HTRF) assay, antibodies specific to the phosphorylated peptide product are used to generate a fluorescence, resonance, energy-transfer (FRET) signal proportional to [Pep]. This technique has the disadvantage of requiring an antibody specific to each kinase enzyme. Also, the fluorescent group can alter the kinetics of the phosphorylation reaction.

Alternatively, at [Pep_h], a simpler approach involves the use of firefly luciferase, which catalyzes the reaction of luciferin with ATP to yield light whose intensity is proportional to [ATP]. The reaction occurs in two steps:

$$\begin{aligned}&\text{Luciferin} + \text{Luciferase} + \text{ATP} \equiv \text{Luciferin-Luciferase-AMP}\\&\quad + \text{PP}_\text{i}\text{-Luciferin-AMP}\\&\text{Luciferin-Luciferase-AMP} + \text{PP}_\text{i}\text{-Luciferin-AMP} + \text{O}_2 \rightarrow\\&\quad \text{Luciferase} + \text{Oxyluciferin} + \text{AMP} + \text{CO}_2 + \text{Light}\end{aligned} \tag{R11}$$

Yellow-green light, having a maximum wavelength of 575 nm, with a quantum yield of 0.9, permits the luciferase to detect low ATP concentrations. At $[\text{ATP}] > 8\ \mu\text{M}$, there is a flash of light followed by a fast decline in light production. Hence, to avoid rapid signal loss and decreased sensitivity, thermostable versions of firefly luciferase are being marketed (e.g., by Promega), which claim a stable signal for up to four hours. Fairly uniform streams of light are produced with [ATP] on the order of 1 μM.

To obtain an accurate measure of [ATP], the rate of consumption of ATP when generating light through reactions R11 must be small compared with the rate of consumption in the kinase-inhibited phosphorylation reaction, R1. This is accomplished by adjusting the luciferase concentration such that the rate of R11 is one-tenth the rate of R1.

The smallest usable reaction volume is ultimately set by the number of photons emitted—surely [ATP] cannot be accurately followed if only a few photons are emitted by the luciferase reaction. Electronically, this issue is manifested by what is called "shot noise," an irreducible source of noise related to the statistical number of fluctuations in the electrons in a signal. Briefly, the photons that strike an electronic detector are converted to electrons by the photoelectric effect, and the electrons are accumulated during an exposure time interval, amplified by a low-noise amplifier, and digitized. In the same way that 200 flips of a coin result in $100 \pm \sqrt{100}$, rather than exactly 100 heads, any electronic measurement that should yield N_{e^-} electrons will actually yield $N_{e^-} \pm \sqrt{N_{e^-}}$, corresponding to a fractional standard error:

$$\text{Shot noise} = \frac{\sqrt{N_{e^-}}}{N_{e^-}} = \frac{1}{\sqrt{N_{e^-}}}. \tag{16.36}$$

For example, if a 1% error is desired, 10,000 electrons are required, or 10^6 electrons to give a 0.1% expected error. In practice, well-made CCD cameras achieve a noise performance comparable to this irreducible limit. The fact that the light is spread over several pixels is irrelevant, provided that the signals from the pixels are pooled together to give the total light emitted. Thus, if 10 measurements with 0.1% error are desired to follow [ATP] accurately, the reaction must yield $10 \times 10^6 = 10^7$ photoelectrons.

To calculate the photoelectrons collected per unit reaction volume, E:

$$\begin{aligned}E = {}&[\text{ATP}] \times N_A \times (\text{ATP conv.})\\&\times (\text{incid. photon frac.}) \times (\text{quantum frac.})\\&\times (\text{quantum effic.})\end{aligned} \tag{16.37}$$

where N_A = Avogadro's number = 6.023×10^{23} photons/mole, ATP conv. is the fraction of ATP consumed by luciferase = 0.1, incid. photon frac. is the fraction of photons incident upon the Research STL-1001E charge-coupled device (CCD) camera lens = 0.5, quantum frac. is the luciferase quantum yield = 0.9, and the quantum effic. of light capture = 0.7 (fraction of photons converted to electrons). Substituting, at $[\text{ATM}] = 1\ \mu\text{M}$:

$$\begin{aligned}E &= 10^{-6} \times (6.023 \times 10^{23}) \times 0.1 \times 0.5 \times 0.9 \times 0.7\\&= 1.9 \times 10^{16}\, e^-/\text{L}\end{aligned}$$

Using the calculated value of E, the total volume per bolus (reactor droplet) can be sized to give 10^7 electrons:

$$V_{bolus} = \frac{10^7}{1.9 \times 10^{16}} = 0.53 \times 10^{-9}\,\text{L} = 0.53\,\text{nL}$$

Conservatively, the lab-on-a-chip will be designed to accommodate boluses having a 1 nL volume, as shown in Section 17.4.

b. Technology Protection

Before designing a new product for the high-throughput screening of kinase inhibitors, to protect its technology it is important to be aware of existing patents. Normally, an initial patent search is carried out before beginning the SGPDP, in the steps above the dashed box in Figure PIII.1. Then, this search is repeatedly updated during the stages of the SGPDP.

In the product design case study in Section 17.4, the results of a patent search are discussed in connection with the *feasibility* stage and, consequently, are not repeated here. See the subsection on *Intellectual Property Assessment*.

16.5 SUMMARY

In this chapter, *innovation maps* have been presented for three configured consumer chemical products, showing their new materials, product/manufacturing, and product technologies, with links to the new products that satisfy specific consumer needs. Having gained familiarity with these innovation maps, the reader is prepared to study their product design case studies in Chapter 17. When undertaking product designs, design teams seek to take advantage of the linkages between the new technologies and the customer-value proposition shown in innovation maps.

REFERENCES

1. Bröckel, U., W. Meier, and G. Wagner, Eds., *Product Design and Engineering, Volume 1: Basics and Technologies*, Wiley-VCH (2007a).

2. Bröckel, U., W. Meier, and G. Wagner, Eds., *Product Design and Engineering, Volume 2: Raw Materials, Additives, and Applications*, Wiley-VCH (2007b).

3. Costa, R., G.D. Moggridge, and P.M. Saraiva, "Chemical Product Engineering: An Emerging Paradigm within Chemical Engineering," *AIChE J.*, **52**, 6, 1976–1986 (2006).

4. Cussler, E.L., and G.D. Moggridge, *Chemical Product Design*, Cambridge Univ. Press (2001).

5. Cussler, E.L., D.W. Savage, A.P.J. Middelberg, and M. Kind, "Refocusing Chemical Engineering," *CEP*, 26S–31S (Jan., 2002).

6. Cussler, E.L., and J. Wei, "Chemical Product Engineering," *AIChE J.*, **49**, 5, 1072–1075 (2003).

7. Favre, E., L. Marchal-Heusler, A. Durand, N. Midoux, and C. Roizard, "A Graduate-level-equivalent Curriculum in Chemical Product Engineering," *Chem. Eng. Educ.*, 264–271 (Fall 2005).

8. Favre, E., L. Marchal-Heusler, and M. Kind, "Chemical Product Engineering: Research and Educational Challenges," *Trans. I. Chem. E.*, **80**, Part A, 65–74 (Jan., 2002).

9. Hill, M., "Product and Process Design for Structured Products," *AIChE J.*; **50**, 8, 1656–1661 (2004).

10. Hill, M.I., and A.J. Post, "Design of the Dove® Beauty Bar," in K.M. Ng, R. Gani, and K. Dam-Johansen, Eds., *Chemical Product Design: Toward a Perspective Through Case Studies*, Elsevier (2007).

11. Ng, K.M., R. Gani, and K. Dam-Johansen, *Chemical Product Design: Toward a Perspective Through Case Studies*, Elsevier (2007).

12. Saraiva, P.M., and R. Costa, "A Chemical Product Design Course with a Quality Focus," *Trans. I. Chem. E.*, **82** (A11), 1476–1484 (Nov., 2004).

13. Shaeiwitz, J.A., and R. Turton, "Chemical Product Design," *Chem. Eng. Educ.*, 280–285 (Fall 2001).

14. Seider, W.D., J.D. Seader, D.R. Lewin, and S. Widagdo, "Chemical Product and Process Design," in *Proceedings of AspenWorld 2004*, Orlando, FL (2004a).

15. Seider, W.D., J.D. Seader, and D.R. Lewin, "Chemical Product and Process Design Education," in *Proceedings of FOCAPD-2004 Conference*, Princeton University, Princeton, NJ (2004b).

16. Wei, J., *Product Engineering: Molecular Structure and Properties*, Oxford Univ. Press (2007).

17. Westerberg, A.W., and E. Subrahmanian, "Product Design," *Comput. Chem. Eng.*, **24**, 959–966 (2000).

Incandescent Light Bulb—General

18. Widagdo, S., "Incandescent Light Bulb: Product Design and Innovation," *Ind. Eng. Chem. Res.*, **45**, 8231–8233 (2006).

19. http://americanhistory.si.edu/lighting/history/patents/mosby1.htm

Tungsten Filaments

20. Garbe, S. "Morphological Effects of Tungsten Filaments in Halogen Lamps: Their Influence on the Temperature Distribution and Evidence for Hot Spot Growth due to Different Faceting," *Philips J. Res.*, **35**, 326–336 (1980).

21. Garbe, S., and S. Hanloh, "Growth of Potassium-Filled Bubbles in Doped Tungsten and Its Relation to Hot Spot Development and Intergranular Fracture," *Philips J. Res.*, **38**, 248–262 (1983).

22. Rieck, G.D., "The Texture of Drawn Tungsten Wires," *Philips Res. Rep.*, **12**, 423–431 (1957).

Wall Deposition

23. Eckerlin, P., and S. Garbe, "Analysis of Tungsten Compounds in the Wall Region of Halogen Lamps," *Philips J. Res.*, **35**, 320–325 (1980).

24. Mullins, W.W., "Theory of Linear Facet Growth during Thermal Etching," *Phil. Mag.*, **6**, 1313 (1961).

25. Schnedler, E., "Description of Tungsten Transport Processes in Inert Gas Incandescent Lamps," *Philips J. Res.*, **38**, 224–235 (1983a).

26. Schnedler, E., "Description of Tungsten Transport Processes in Halogen Incandescent Lamps," *Philips J. Res.*, **38**, 236–247 (1983b).

Transport Phenomena in Halogen Light Bulb

27. Schnedler, E., "Three Dimensional Model for the Theoretical Description of Tungsten Transports in Halogen Incandescent Lamps," *High Temp. Sci.*, **19**, 237–252 (1985).

Halogen-Tungsten Reaction

28. Dittmer, G., and U. Niemann, "Heterogeneous Reactions and Chemical Transport of Tungsten with Halogens and Oxygen under Steady-State Conditions of Incandescent Lamps," *Philips J. Res.*, **36**, 89–113 (1981).

29. De Maagt, B.J., and G.C.J. Rouweler, "The Reaction of Bromine and Oxygen with a Tungsten Surface," *Philips J. Res.*, **35**, 95–103 (1980).

30. Brongersma, H.H., G.C.J. van der Ligt, and G. Rouweler, "The Reaction fo Bromine and Oxygen with a Tungsten Surface Studied by Means of Low-Energy Ion Scattering," *Philips J. Res.*, **36**, 1–14 (1981).

31. Zubler, E.G., "Kinetics of the Tungsten-Oxygen-Hydrogen Bromide Reaction," *J. Phys. Chem.*, **79**(16), 1703–1706 (1975).

32. Zubler, E.G., "Adsorption Effects in Tungsten-Oxygen-Hydrogen Bromide Reaction," *J. Phys. Chem.*, **76**(3), 320–322 (1972).

Patents—Light Bulbs

33. U.S. Patent 1,180,159, Langmuir, I, *Gaseous Electric Light* (1916).

34. U.S. Patent 3,243,634, Mosby, F., *Electric Lamp and Support Web* (1966).

35. U.S. Patent 3,932,164, Gebhardt, K., *Method of Manufacturing Miniature Incandescent Lamps* (1976).

36. U.S. Patent 4,017,758, Raymund, F.H., *Incandescent Lamp with Infrared Filter* (1977).

37. U.S. Patent 4,074,168, T'Jampens, G.R., *Halogen Incandescent Lamp whose Filler Gas Comprises Bromine, Chlorine, and Hydrogen* (1978).

38. U.S. Patent 4,081,708, Notelteirs, V.R., *Incandescent Lamp-Reflector Unit* (1978).

39. U.S. Patent 4,119,877, Grewe, F.J., H. Rosler, O. Ruhl, K. Stenke, E. van Wersch, and L. Wings, *Electric Lamp Having a Cap Plate* (1978).

Hemodialysis Devices—General

40. Bird, R.B., W.E. Stewart, and E.N. Lightfoot, *Transport Phenomena*, 2nd ed., Wiley, New York (2002).

41. Daugirdas, J.T., and T.S. Ing, eds., *Handbook of Dialysis*, Little, Brown & Co., Boston (1988).

42. *History of Sorbent Dialysis*, Renal Solutions, Inc., Web site: *http://www.renalsolutionsinc.com/*.

43. Knudsen, J.G., and D.L. Katz, *Fluid Dynamics and Heat Transfer* McGraw-Hill, New York (1958).

44. Misra, M., "The Basics of Hemodialysis Equipment," *Hemodialysis Int'l.*, **9**, 30–36 (2005).

45. Seader, J.D., and E.J. Henley, *Separation Process Principles*, Second Edition, Wiley, New York (2006).

46. Shapiro, W.B. "Sorbent Dialysis," in *Clinical Dialysis, Fourth Edition*, (Chapter 38: Sorbent Dialysis), A. R. Nissenson and R.N. Fine, Eds., McGraw-Hill, New York (2004).

47. *Sorbent Dialysis Primer*, SORB Technologies Web site: http://www.sorb.net/.

48. Spaeth, E.E., *Washington University Case Study 9, Analysis and Optimization of an Artificial Kidney System*, Department of Chemical Engineering, Washington University, St. Louis, Missouri (1970).

49. Talamini, M., "Guidance for Nocturnal Home Hemodialysis Devices," prepared for the Nocturnal Home Hemodialysis Devices Panel Meeting, 2005. [Report was found in response to a Google search. It contains a discussion of the key issues and a lengthy list of references.]

Patents—Hemodialysis Devices—General

50. U.S. Patent 3,865,726, Chibata, I, T. Tosa, T. Sato, and T. Mori, *Blood Urea Removal Device* (1975).

51. U.S. Patent D 282,578. Humphreys, L.R., D. Barone, F.C. Baker, III, T.S. Bowers, and J.E. Jamieson. *Processor for a Portable Recirculatory Hemodialysis Unit* (1986).

Patents—Hemodialysis Devices—Hollow-Fiber Membranes

52. U.S. Patent 4,276,173, Kell, M.J., and R.D. Mahoney, *Cellulose Acetate Hollow Fiber and Method of Making the Same* (1981).

53. U.S. Patent 5,783,124. Uenishi, T., I. Yamamoto, K. Okamoto, H. Sido. Y. Shiota, H. Shikurai, S. Watanuki, and M. Suzuki. *Cellulose Acetate Hemodialysis Membrane* (1998).

Patents—Hemodialysis Devices—Dialysate Regeneration

54. U.S. Patent 3,669,878. Marantzm L.B. *Treatment of Dialysate Solution for Removal of Urea* (1972).

55. U.S. Patent 4,473,449. Michaels, A.S., A.J. Appleby, and J.C. Wright. *Flowthrough Electrohemical Hemodialysis Regeneration* (1984).

56. U.S. Patent 6,878,283. Thompson, R.P. *Filter Cartridge Assemblies and Methods for Filtering Fluids* (2005).

Patents—Hemodialysis Devices—Alarms/User Interface

57. U.S. Patent 4,148,314. Yin, C. *Blood Pressure Alarms for Dialysis Machines* (1979).

58. U.S. Patent 6,804,991. Balschat, K., H. Ender, A. Gagel, and R. Spickermann. *Method and Device for Detecting a Leakage in a Fluid System of a Blood Treatment Apparatus* (2004).

High-Throughput Screening—General

59. Chen, Q.-M., L. B. Heend, and J. Waring, *NANOLUX Screening Technologies*, Towne Library, Univ. of Pennsylvania, 2005.

60. Cheng, Y.-C., and W.H. Prusoff, "Relationship Between the Inhibition Constant (K_I) and the Concentration of Inhibitor Which Causes 50 Percent Inhibition (I_{50}) of an Enzyme Reaction," *Biochem. Pharma.*, **22**, 3099–3108 (1973).

61. Copeland, R.A.,"Review: Mechanistic Considerations in High-Throughput Screening," *Anal. Biochem.*, **320**, 1–12 (2003).

62. Quake, S.R., H.P. Chou, M.A. Unger, T. Thorsen, and A. Scherer, "Monolithic Microfabricated Valves and Pumps by Multilayer Soft Lithography," *Science*, **288**, 113–116 (2000).

63. Xia, Y., and G.M. Whiteside, "Soft Lithography," *Angew. Chem. Int. Ed.*, **37**, 550–575 (1998).

Patents—High-throughput Screening

64. U.S. Patent 6,951,632. Unger, M.A., H.-P. Chou, I.D. Manger, D. Fernandes, and Y. Yi, *Microfluidic Devices for Introducing and Dispensing Fluids from Microfluidic Systems* (2005).

EXERCISES

16.1 For the bromine-filled light bulbs, compute the sensitivity of the temperature and tungsten partial-pressure distributions to the bulb temperature. Note that Figures 16.6 and 16.7 are at a bulb temperature of 1,000 K.

16.2 Repeat the mass-transfer calculations assuming that thermal diffusion is negligible. Compare the tungsten compositions at the bulb wall with and without thermal diffusion.

16.3 Carry out a patent search for patents associated with compact fluorescent light bulbs (CFLs).

(a) Identify the principal companies involved and summarize their technology platforms.

(b) Locate patents involving the conversion of fluorescent light to various colors, including soft yellow lighting.

16.4 Develop an *innovation map* for compact fluorescent light bulbs (CFLs), assuming product design in the mid-1990s. These bulbs use electricity to ionize noble gases that produce UV light, which is converted into visible light by fluorescent coatings on the inner surface of the lamp enclosure. These lamps offer longer lifetimes, cooler operating temperatures, and much higher luminous efficiencies compared to standard incandescent light bulbs.

(a) Identify the key components of the materials, process/manufacturing, and product technologies. You can research the open literature for key inventions in each of these areas.

(b) Identify the *customer-value propositions* for the fluorescent light bulbs.

(c) Identify the links between the materials, process/manufacturing, and product technology elements and the *customer-value proposition*. Justify your reasoning in creating these connections.

(d) Draw the *innovation map* for the fluorescent lamps.

16.5 Consider the hemodialysis device in Example 16.2. Examine the effect on the rate of urea removal of changing the hemodialyzer geometry, the blood and dialysate flow rates to the dialyzer, the rate of waste withdrawal, the volume of the dialysate tank, and the sensitivity of the rate of urea mass transfer to the mass-transfer coefficient. In particular, the above estimate of the coefficient on the shell side may be low because the entry to and exit from the hemodialyzer of the dialysate is normal to, rather than parallel to, the fibers. This should enhance the shell-side coefficient.

16.6 Repeat Exercise 16.5 for a hemodialysis device designed for overnight home use. Assume that the device will be used 7 nights/wk for 6 hr/night.

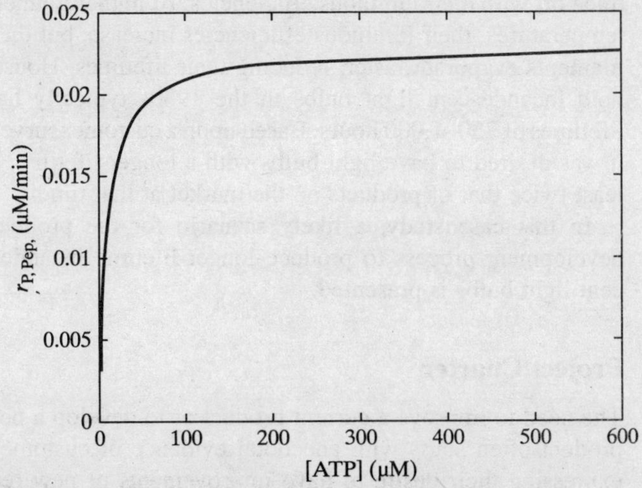

Figure 16.34 Caliper Life Sciences data. [KE] = 2.5 μM, [Pep] = 1.5 μM, R_{Pep_h} = 0.0229 μM/min, and $K_{m.Pep_h}$ = 9.9 μM.

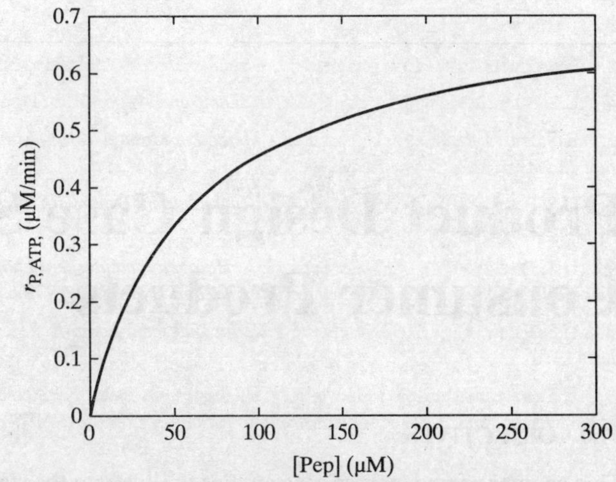

Figure 16.35 Caliper Life Sciences data. [KE] = 2.5 μM, [ATP] = 250 μM, R_{ATP_h} = 0.73 μM/min, and $K_{m.ATP_h}$ = 61.6 μM.

16.7 Data for the phosphorylation reaction, R1, are provided by Caliper Life Sciences, Inc., in Figure 16.34 at high [Pep] = 1.5 μM and in Figure 16.35 at high [ATP] = 250 μM, both with [KE] = 2.5 μM. Using these two curves, find the reaction rate at [Pep] = 60 μM and [ATP] = 1 μM, which are well suited to test typical kinase inhibitors and provide a uniform stream of photons for the CCD camera.

Chapter 17

Product Design Case Studies for Configured Consumer Products

17.0 OBJECTIVES

This chapter provides three case studies to illustrate the steps in designing new *configured consumer chemical* products using the Stage-Gate^{TM} Product-Development Process (SGPDP—Cooper, 2001, 2002, 2005; Cooper et al., 2005). Emphasis is placed on the *concept* and *feasibility* stages. Only the key issues are discussed in the remaining *development*, *manufacturing*, and *product-introduction* stages.

After studying this chapter, the reader should be able to:

1. Use the Stage-Gate^{TM} Product-Development Process (SGPDP) for the design of configured consumer products.
2. Identify and use the necessary technological inventions, discussed in Chapter 16, in the development and design of a new product.

17.1 INTRODUCTION

This chapter is comprised of three case studies that illustrate the principal aspects of *configured consumer chemical* product design. Before reading these case studies, the reader should be conversant with the key steps in product design, as introduced in Sections 1.2 and 1.3, Chapter 2, and the introduction to Part Three, especially Figure PIII.1. For each case study, the history of the technologies involved and an innovation map are presented in Chapter 16. Sections 16.2 and 17.2 are devoted to the design of halogen light bulbs, Sections 16.3 and 17.3 to the design of home hemodialysis devices, and Sections 16.4 and 17.4 to the design of labs-on-a-chip for the high-throughput screening of kinase inhibitors. Being the first case study, the design scenario for halogen light bulbs is presented most thoroughly. For the other products, only the principal steps in their designs are presented in detail.

17.2 HALOGEN LIGHT BULB CASE STUDY

This case study is presented in hindsight to illustrate the steps in the Stage-Gate^{TM} Product-Development Process (SGPDP—Cooper, 2001, 2002, 2005; Cooper et al., 2005). The time frame for the product development is taken to be the mid-1980s. It is assumed that a design team was convened, which elected to concentrate on the development of a *halogen* light bulb product. The justification for this focus is presented in Example 16.1, where the technology platform for halogen light bulbs is positioned on an *innovation map*. An alternate product design, perhaps typical of a project in the 1990s, would probably have involved compact fluorescent light bulbs (CFLs), the subject of Exercises 17.3 and 17.4.

As introduced in Section 16.2, one of the major drawbacks of incandescent light bulbs is their filament lifetimes, which trade off with their luminous efficiencies. At higher filament temperatures, their luminous efficiencies increase, but their filaments evaporate faster, reducing their lifetimes. Household incandescent light bulbs in the 1980s typically had lifetimes of 750–1,000 hours. Based upon a customer survey, it was desired to have light bulbs with a longer lifetime, at least twice that of products on the market at that time.

In this case study, a likely scenario for the product-development process to produce longer-lifetime incandescent light bulbs is presented.

Project Charter

The need to improve a current product or to develop a new product often starts with anecdotal evidence of customers expressing their desire to have improvements or new features, as introduced in Section 2.2. Related requests or ideas can overwhelm the development workforce in a company, and consequently, a prioritization of efforts is a prerequisite to any decision to initiate a new product development, including the improvement of an existing product. Because this prioritization process, and associated product-development portfolio management, are beyond the scope of this case study, it is assumed that the business strategy team had decided in the mid-1980s that the development of longer-lifetime incandescent light bulbs was a priority for the company to meet its sales targets.

Having made this decision, the business strategy team probably formed its multifunctional design team, typically consisting of multidisciplinary personnel (Creveling et al., 2003), including:

- Scientists and engineers, such as chemists, material scientists, and chemical engineers who had developed technologies of potential use in this product. For this case study, members of the team may have been involved in developing the halogen light bulb technologies discussed in Section 16.2.
- Development engineers in related fields such as packaging and thermal engineering.
- Senior manufacturing engineers, such as those who had developed previous generations of the product, that is, the incandescent light bulb.
- Technical- and customer-service personnel who had worked on previous generations of the product, or who had extensive experience in handling technical problems and customer concerns.
- Marketing and sales personnel.
- Supply chain specialists.
- Health, safety, environmental, and regulatory specialists.

Note that such an internally focused team is often complemented with technology-development partners from industry, academia, and/or the government. It may also include selected customers.

As recommended at the top of Figure PIII.1, the team probably began by creating its project charter, recognizing that a good project charter can help to guide a successful product-development effort. As discussed in Section 2.2, its key elements are specific goals, a project scope, deliverables, and a time line, as illustrated in Table 17.1 for the development of halogen light bulbs. Beginning with its specific goals, a likely statement might have been: "To develop new incandescent light bulbs with a lifetime of 2,000 hr, or at least twice that of the current product line, at the same cost."

Returning to Example 2.1, the reader should follow the development of the project scope, expanding upon its initial thoughts in Table 2.1. He or she should also reread the discussion of the four deliverables and the time line.

Note that the project charter, when completed, is the equivalent of a formal *contract* between the project *champions/sponsors* and the product-development team. The former ideally consists of the business, technical, and marketing proponents, to provide a full range of perspectives.

Having created its project charter, it is likely the design team proceeded to consider answers to the three questions near the top of Figure PIII.1. These refer to the potential new technologies at that time, which have been discussed in Section 16.2 and by Widagdo (2006). With these new technologies looking promising, the design team likely received enthusiastic approval to proceed to the *concept* stage of the SGPDP, which is discussed next.

Table 17.1 Project Charter for the Development of Halogen Light Bulbs

Project Name	Longer-Lifetime Incandescent Light Bulb
Project Champions	Business Director of the Home Lighting Business
Project Leader	John Doe
Specific Goals	Incandescent light bulbs with a lifetime of 2,000 hr, or at least twice that of the current product line, at the same cost
Project Scope	In-scope: • Light bulbs for household lighting • Minimal changes to the current manufacturing capability Out-of-scope: • Light bulbs for non-household applications
Deliverables	• Business opportunity assessment • Technical feasibility assessment • Manufacturing capability assessment • Product life-cycle assessment
Time Line	Product prototypes for market testing within 12 months

Concept Stage

As discussed in Sections 1.2 and 2.4, the *concept* stage begins the SGPDP (Cooper, 2001, 2002, 2005; Cooper et al., 2005). It normally involves seven assessments or activities to be completed, which are discussed next for the halogen light bulb product.

a. ***Opportunity assessments*** for halogen light bulbs probably included an examination of the technological challenges and market size, a competitive analysis, and a *value-chain analysis*. The following analyses focus on the *voice of the market*.

Market Analysis: For the purposes of this case study, the available data (Home Channel News, 2001) were projected back to the mid-1980s to illustrate how decisions were made. The market size of the light bulb industry for household usage was approximately \$1.2 billion, with a sales volume of 785 million units. Incandescent light bulbs outsold fluorescent tubes by 233%, down from 255%.

Competitive Analysis: In the mid-1980s, the choices for household lighting included the standard incandescent light bulbs and fluorescent tubes, but not compact fluorescent light bulbs (CFLs). The former have relatively low initial costs, but low energy efficiencies. Only 5–8% of their energy input produces light, with the remainder dissipated as heat. For comparable light outputs, fluorescent tubes use 60–80% less energy and have 10 to 20 times longer life, but they were incompatible with standard fittings (for example, in lamps and lighting fixtures), a limitation that was largely eliminated by CFLs in the 1990s. According to the 2005 Canadian *Survey of Household Energy Use*, in the mid-1980s, almost 80% of the population used standard incandescent light bulbs, with the remainder using fluorescent tubes. Note that the usage of incandescent light bulbs must have been even higher, with just a slow movement toward fluorescent tubes at that time. Consequently, despite their higher energy efficiency and longer lifetime, this survey exposed the disappointingly low level of market penetration for fluorescent lighting. Furthermore, because of their compatibility with standard fittings, and the unavailability of CFLs in the mid-1980s, it was anticipated that longer-lifetime incandescent life bulbs would increase their market share.

Technological Challenge: As mentioned earlier, the lifetime of tungsten filaments in incandescent light bulbs trades off with their luminous efficiencies. At higher filament temperatures, the luminous efficiency increases, but the filaments evaporate faster and their lifetimes are reduced.

Incandescent Light Bulb Output: At a reference supply voltage of 130V, the light output of incandescent light bulbs is shown in Figure 17.1 as a function of the power consumption.

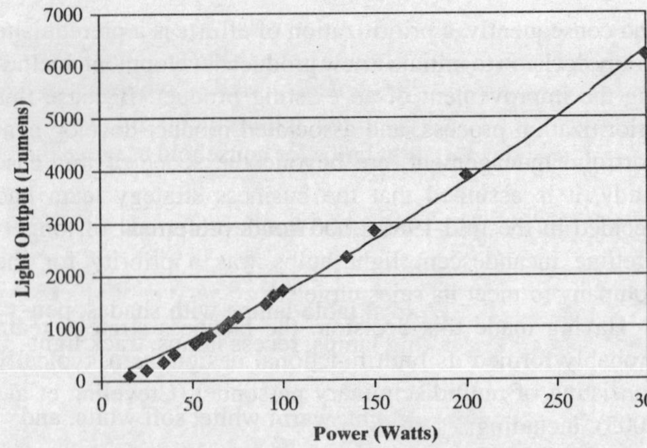

Figure 17.1 Light output of incandescent light bulbs as a function of power consumption at a reference supply voltage of 130V.

Furthermore, incandescent light bulbs are very sensitive to changes in the power input. As shown in the "Incandescent light bulb" Wikipedia reference, for a supply voltage V:

- Light output is proportional to $V^{3.4}$
- Power consumption is proportional to $V^{1.6}$
- Lifetime is proportional to V^{-16}
- Color temperature is proportional to $V^{0.42}$

A nominal-rating 100W conventional incandescent bulb has a light output of 1,700 lumens and is rated to have a lifetime of 750 hours. Consequently, in North America, where the household voltage is normally 115V, the power consumption and light output are reduced by 18% and 34%, respectively, but the lifetime is increased by sevenfold. This is a typical strategy for longer-life incandescent light bulbs. Due to a reduction in 34% of light output, this "longer-life light bulb" becomes equivalent to a 60-Watt bulb at its nominal rating.

EXAMPLE 17.1

Compute the lifetime enhancement for a nominal-rating 750-hour conventional incandescent light bulb when operated at the normal household voltage in North America.

SOLUTION

The nominal voltage, V_1, is 130 volts and the normal household voltage in North America, V_2, is 115 volts. The nominal lifetime, L_1, is 750 hours. As the lifetime is proportional to V^{-16}, the lifetime at 115 volts, L_2, is:

$$L_2 = L_1\left(\frac{V_2}{V_1}\right)^{-16} = 750\left(\frac{115}{130}\right)^{-16} = 750 \times 0.8846^{-16}$$

$$= 5,335\,\text{hr}$$

b. ***Customer requirements*** for halogen light bulbs can be determined by developing *fitness-to-standard* (FTS) and *new-unique-and-difficult* (NUD) requirements. Note that typical customer requirements for longer-life incandescent light bulbs for household usage in the mid-1980s were:

- Lifetime of at least 2,000 hours preferred
- No cost premium
- Fits various fixtures: table lamps with shades, pendant lamps, ceiling lamps, recess lamps, track lighting
- Various colors of light: warm white, soft white, and cool white
- Energy efficient

Of these requirements, some were likely regarded as *fitness-to-standard* and the remainder as *new-unique-and-difficult* requirements, as illustrated in Table 17.2. Note that the FTS and NUD requirements are introduced in Section 2.4. Normally, each requirement is assigned a weighting factor as a measure of its relative importance.

Often, the weighting factors are determined through primary research by surveying prospective customers, or through secondary research involving literature surveys. In a typical customer survey, customers would be given a list of customer requirements and requested to distribute 100 points among them according to their importance. The responses of numerous customers would be averaged to give weighting factors similar to those in Table 17.2. Note that the NUD requirements normally have the highest weighting factors.

c. ***Technical Requirements.*** Before product concepts are developed, customer requirements are translated by the design team into quantifiable technical requirements that are more amenable to technical development work. This translation may result in multiple technical requirements for some of the customer requirements. For halogen light bulbs, a likely translation in the mid-1980s into quantifiable technical requirements such as lifetime and color spectrum, among others, is shown in Table 17.3.

Table 17.2 Customer Requirements for Longer-Lifetime Incandescent Light Bulbs

Customer Requirement	Type	Weighting Factor (%)
Lifetime	NUD	30
No cost premium	FTS	20
Fits various fixtures	FTS	10
Available in various colors of light	FTS	10
Energy efficient	NUD	30

Table 17.3 Technical Requirements for Longer-Lifetime Halogen Light Bulbs

Customer Requirement	Technical Requirement	Target
Lifetime	Lifetime	> 2,000 hr
No cost premium	Cost per watt	1.5 cents
Fits various fixtures	Max. Operating Temperature	
	Lamps with shades	60°C
	Recess lamps	80°C
	Tracking lights	200°C
Available in various colors of light	Color Temperature	
	Warm light	2,700–3,000 K
	Cool light	3,500–6,500 K
Energy efficient	Energy efficiency	> 5%
	Luminous efficacy	> 12 lumen/W

d. ***Determination of Critical-to-Quality Variables (CTQs).*** Usually the NUDs, that is, the differentiators, are taken as the *critical-to-quality* variables. Thus, for the halogen light bulb product design, the lifetime, and the energy efficiency and luminous efficacy, were likely selected in the mid-1980s.

In the *concept* stage of the SGPDP, the first *House of Quality* (HOQ), also known as the *Quality Function Deployment* (QFD), is often assembled by the design team. It displays the results of a process for obtaining, translating, and deploying the *voice of the customer* into technical requirements. This process is repeated at various phases of the product-development process, with the results being updated.

In Table 17.4, a typical HOQ is shown, as likely prepared by a design team in the mid-1980s. Here, the lower rectangular matrix pairs all of the customer requirements in the first column with at least one quantitative technical requirement or parameter in the adjacent columns. For example, the customer requirement that the bulbs fit in lamps with shades, recess lamps, or tracking light fixtures imposes a maximum operating temperature; that is, a technical variable with an upper bound to prevent fires. Similarly, the quality of light, *warm* or *cool*, imposes an operating temperature known as the *color* temperature. In other cases, the customer requirement translates directly into the technical requirement, for example, the lifetime.

At the top of the house, the *interaction matrix* shows the synergistic technical requirements or parameters, for example, the lifetime and color temperature. Note that + (plus) indicates that both variables increase or decrease; – (minus) indicates that when one variable increases, the other decreases, and vice versa; and a blank entry indicates no significant relationship between the variables. For the halogen light bulb, the higher the color temperature, the higher the bulb temperature, and consequently, the shorter the lifetime.

Table 17.4 First House of Quality for Longer-Lifetime Halogen Light Bulbs

						1	Lum. Eff
					1	+	Energy Eff.
				1	+	–	Color Temp
			1				Max Op. Temp
		1	–	–	–	–	Cost $/Watt
	1	+		–			Lifetime (hours)
Customer Requirement	Lifetime (hours)	Cost $/Watt	Max Op. Temp	Color Temp	Energy Eff.	Lum. Eff	Weight
Lifetime	x						0.3
No Cost Premium		x					0.2
Fit Various Fixtures							0.1
Lamps with Shade			x				
Recess Lamps			x				
Tracking Lights			x				
Colors of Light							0.1
Warm Light				x			
Cool Light				x			
Energy Efficient					x	x	0.3

Thus, the – signals the need for a compromise between the color temperature and the lifetime.

e. ***Development of Superior Product Concepts.*** A key activity in the development of a new product is the identification of possible solution concepts. Often these concepts apply to specific elements of a product assembly. For development of a long-life halogen light bulb in the mid-1980s, as discussed in Section 16.2, the new concepts involved a bromine vapor atmosphere and a quartz primary casing to contain the hot reacting gases. Note that these were the riskiest aspects of the *concept* stage of the SGPDP for the halogen light bulb product. Also, in many similar products, it is necessary to defer judgment until a prototype can be created in the *feasibility* stage, that is, the next stage in the SGPDP. Often, flexibility is crucial when deciding when to eliminate a concept or to carry it through to the *feasibility* stage.

The so-called Pugh matrix (Pugh, 1996), in which each solution concept (partial and complete) is judged against a reference solution, is useful for screening purposes. For long-life halogen light bulbs, Table 17.5 is a typical Pugh matrix as likely prepared by a design team in the mid-1980s. Here the reference solution was the best known in the market, in this case, estimated in Example 17.1. Each concept was evaluated against the reference solution and assigned a qualitative valuation of inferior (–), superior (+), or equal (0).

Table 17.5 shows that while Concept A was anticipated to be equivalent to the reference concept, it was expected to be more costly in the mid-1980s. Concept B, with thicker tungsten filaments, which should be more durable, was anticipated to extend the lifetimes. However, thicker filaments have lower resistances, generating less heat, with lower operating temperatures and lower color temperatures. These tradeoffs, involving materials technology, are discussed in Section 16.2.

Because these two concepts didn't meet all of the critical-to-quality requirements—that is, both increased lifetime, and energy efficiency and luminous efficacy—the product-development team likely conducted brainstorming sessions with different functional groups within the company (often represented on the team as discussed above) to seek better solutions. Let's assume that these sessions uncovered a promising

Table 17.5 Pugh Matrix for Longer-Lifetime Incandescent Light Bulbs

Technical Requirement	Target	Reference Concept	Concept A Improved Tungsten Filament	Concept B Thicker Tungsten Filament
Lifetime	> 2,000 hr	5,300 hr	0	0
Cost per watt	1.5 cents	1.5 cents	–	–
Max. operating temperature				
Lamps with shade	60°C		0	+
Recess lamps	80°C	80°C	0	+
Tracking lights	200°C	200°C	0	+
Color temperature				
Warm light	2,700–3,000 K	2,700–3,000 K	0	–
Cool light	3,500–6,500 K	3,500–6,500 K	0	–
Energy efficiency	> 5%	4 %	0	–
Luminous efficacy	> 12 lumen/W	10 lumen/W	0	–

new technology, for example, halogen-filled incandescent light bulbs (see this materials technology in Section 16.2), which had been under development by the corporate R&D group. Having determined that this technology had been well developed, the team would likely have evaluated it in an augmented Pugh matrix, as shown in Table 17.6.

Clearly, Concept C was more promising, as it met all of the critical-to-quality requirements, except that its temperature was expected to be too high (> 60°C) for lamps with shades. Note that in this case, a temperature warning would likely have been issued by the R&D group; that is, a promising technological solution would have been available, but important product design questions remained to be resolved, including:

- Q1: Can a casing be designed for the halogen-filled incandescent bulb to maintain its optimum operating temperature—that is, a high and uniform temperature?
- Q2: Assuming that halogen-filled incandescent bulbs cannot be designed to operate below 60°C, would the size of the remaining market, excluding lamps with shades, be sufficiently large to justify the product?
- Q3: Can the bulb enclosure design address high-temperature concerns (fire, burn hazard, etc.) to pass a UL certification? Note that the Underwriters Laboratories, Inc., (UL) is an independent, not-for-profit product-safety testing and certification organization. The UL tests products for public safety, with a UL certification often required to sell a product in the United States.

Had no technological solutions become available within the company, the team might have recommended termination or postponement of the product-development

Table 17.6 Improved Pugh Matrix for Longer-Lifetime Halogen Light Bulbs

Technical Requirement	Target	Reference Concept	Concept A Improved Tungsten Filament	Concept B Thicker Tungsten Filament	Concept C Halogen-Filled Light Bulb
Lifetime	> 2,000 hr	5,300 hr	0	0	−(4,000 hr)
Cost per watt	1.5 cents	1.5 cents	–	–	0
Max. operating temperature					
Lamps with shade	60°C		0	+	–
Recess lamps	80°C	80°C	0	+	0
Tracking lights	200°C	200°C	0	+	0
Color temperature					
Warm light	2,700–3,000 K	2,700–3,000 K	0	–	0/+
Cool light	3,500–6,500 K	3,500–6,500 K	0	–	0/+
Energy efficiency	> 5%	4 %	0	–	+
Luminous efficacy	> 12 lumen/W	10 lumen/W	0	–	+

effort until a technological solution became available. Note that in some cases, new technology-development efforts are initiated and/or searches for external technological solutions are pursued. Once a technological solution(s) is found, the product-development effort often resumes, assuming that the prior business justification is still valid; that is, the customer needs and business opportunity continue to be promising.

f. *Selection of Superior Concepts.* In the *concept* stage of the SGPDP, the selection of superior concepts is based primarily on the satisfaction of the technical requirements, in particular, the *new-unique-and-difficult* (NUD) requirements; that is, increased lifetime and energy efficiency and luminous efficacy for the incandescent light bulb. It is recognized that superior concepts will be tested extensively in the *feasibility* stage, where more quantitative validation and refinement will be carried out with product prototypes.

When selecting superior concepts for the light bulb product, clearly, it would have been noted that Concept C generated two new product requirements, which were potentially satisfied by the answers to the first and third questions above:

- Q1: A casing design was needed to contain and maintain a uniform high temperature of the halogen gas.
- Q3: UL requirements for the bulb enclosure that include fire hazard and personal injury (thermal burn) prevention had to be satisfied.

Given these two requirements, the steps in the *concept* stage of the SGPDP under b, c, and d would likely have been repeated. With solution concepts generated that satisfied requirements Q1 and Q3, a superior concept would likely have been selected that best satisfied all of the requirements, as discussed next.

Primary Casing Design

As described in Section 16.2, the temperature of the tungsten filaments, and the temperature profile in close proximity to them, are critical. They determine the rates of the regeneration reactions as the tungsten filament decays due to local temperature gradients on the filament surface. And consequently, they would likely have been the basis for new design concepts involving the shapes and materials of construction of the primary casing, as suggested in Table 17.7.

Here, for the reference concept, the shape and materials of construction in a standard incandescent light bulb would likely have been used. Concepts D1 and D2 involve different materials of construction for the primary casings, tempered glass and quartz glass, while concepts E1 and E2 involve two different shapes of the casing, spherical and cylindrical. For each concept, qualitative estimates of their bulk average temperatures and temperature variations [percent deviation of the temperature at distance r from that at the light source ($r = 0$), where $r = 1$ at the wall] would likely have been prepared, as shown in Figures 17.2 and 17.3—with detailed calculations carried out in the *feasibility* stage.

Clearly, on the basis of these estimates, the cylindrical quartz casing would have been considered superior to the tempered glass casing even though quartz is more expensive. In the *concept* stage, however, where emphasis is placed upon identifying the concepts that meet the technical requirements, the cost, being a *fitness-to-standard*, is often ignored.

Secondary Enclosure

In summary, the primary casing would have been designed to provide favorable temperatures for the tungsten-halogen reactions, permitting the tungsten surface to be renewed and thereby increasing the bulb lifetime. The secondary casing, on the other hand, would have been

Table 17.7 Pugh Matrix for Primary Casing

Technical Requirement	Target	Reference Concept	Concept D1 Tempered Glass	Concept D2 Quartz	Concept E1 Spherical	Concept E2 Cylindrical
Lifetime	> 2,000 hr	5,300 hr	0	0	– (4,000 hr)	0
Cost per watt	1.5 cents	1.5 cents	– –	– –	0	0
Max. operating temperature						
Lamps with shade	60°C		–	–	–	–
Recess lamps	80°C	80°C	0	0	0	0
Tracking lights	200°C	200°C	0	0	0	0
Color temperature						
Warm light	2,700–3,000 K	2,700–3,000 K	0	0/+	0/+	0/+
Cool light	3,500–6,500 K	3,500–6,500 K	0	0/+	0/+	0/+
Energy efficiency	> 5%	4 %	0	0	+	+
Luminous efficacy	> 12 lumen/W	10 lumen/W	0	+	+	+
Avg. bulk temperature	2,000–7,000 K	1,500–5,000 K	0	+	+	+
Temperature variations	< 2%	3%	0/+	+	+	+

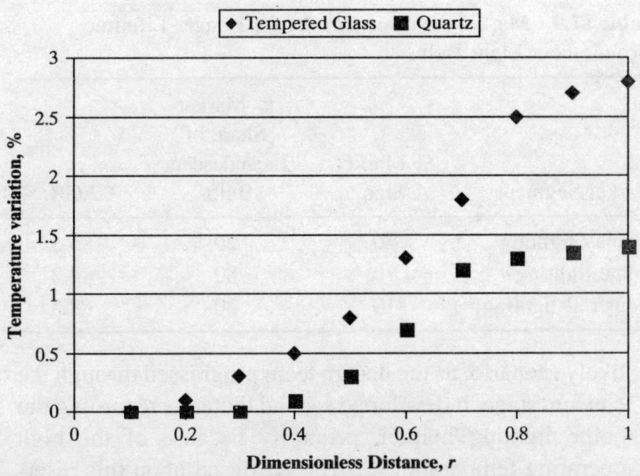

Figure 17.2 Temperature gradient for different materials of construction.

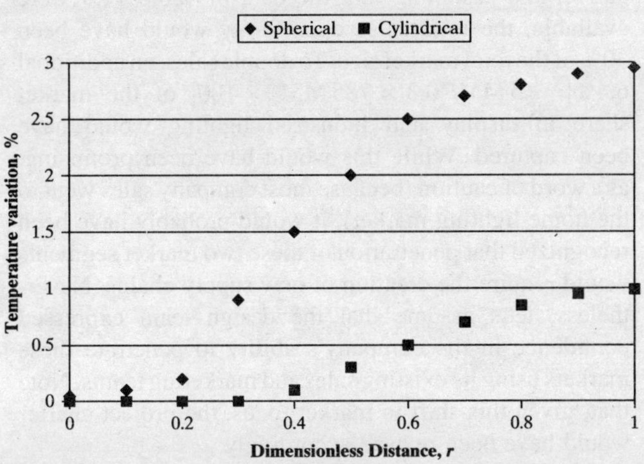

Figure 17.3 Temperature gradient for quartz primary casings.

needed to eliminate fire hazards and safeguard personal safety, it being recognized that the external surface temperature of the quartz casing could potentially exceed 1,500 K.

At this point, it is assumed that brainstorming sessions involving the rapid prototyping and product design teams took place, yielding the following viable concepts:

- Concept F1: A metal open grid over the bulb primary housing
- Concept F2: A clear glass secondary casing over the bulb primary housing
- Concept F3: A clear tempered glass secondary casing over the bulb primary housing.
- Concept F4: A ceramic secondary casing with open design

Given these concepts, it is likely that two new product requirements were identified: (1) the surface temperature must be less than 500 K, and (2) the UL regulations must be satisfied. These would likely have been added to the Pugh matrix, as shown in Table 17.8, and the four concepts would likely have been checked to determine whether they satisfied the augmented requirements. Note that, for the reference concept, no secondary casing would likely have been selected.

While concepts F3 and F4 satisfied all of the product requirements, yet another concept might have been introduced by the design team, that is, a hybrid of the two, F3+F4:

- Concept F3+F4: A ceramic secondary casing with a clear tempered glass window.

To test this concept, the Pugh matrix would have been extended similarly.

Table 17.8 Pugh Matrix for Secondary Enclosure

Technical Requirement	Target	Reference Concept	Concept F1	Concept F2	Concept F3	Concept F4
Lifetime	> 2,000 hr	5,300 hr	0	0	0	0
Cost per watt	1.5 cents	1.5 cents	0	0	0	+
Max. operating temperature						
Lamps with shade	60°C					
Recess lamps	80°C	80°C	0	0	0	0
Tracking lights	200°C	200°C	0	0	0	0
Color temperature						
Warm light	2,700–3,000 K	2,700–3,000 K	0	0	0	0
Cool light	3,500–6,500 K	3,500–6,500 K	0	0	0	0
Energy efficiency	> 5%	4%	0	0	0	+
Luminous efficacy	> 12 lumen/W	10 lumen/W	0	0	0	+
Avg. bulk temperature	2,000–7,000 K	1,500–5,000 K	0	0	0	0
Temperature variations	< 2%	3%	0	0	0	+
Surface temperature	< 500 K	1,500 K	0	+ (800 K)	+ (500 K)	+ (350 K)
UL testing	Pass	No	No	No	Yes	Yes

At this point, three superior concepts would have been identified:

- Concept C: Halogen-filled incandescent bulb
- Concept D2+E2: Quartz bulb enclosure with a cylindrical shape
- Concepts F3, F4, or F3+F4: Secondary casing design

In summary, Concepts C and D2 would have satisfied the primary customer requirements, while concepts F3, F4, and F3+F4 would have satisfied the requirements for UL certification.

g. ***Gate Review.*** At the completion of the *concept* stage of the SGPDP, answers to the critical questions raised in the project charter must be presented to the *business decision makers* at the gate review. In this subsection, for each of the principal deliverables, a list of typical associated questions is presented before their answers are addressed.

- ***Business opportunity assessment:*** What is the size and scope of the opportunity?
- ***Technical feasibility assessment:*** How innovative are the solution concepts? How much confidence is there in a *win* in the marketplace?
- ***Manufacturing capability assessment:*** Do the company's manufacturing capabilities support this product? Is a significant investment required?
- ***Product life-cycle assessment:*** Are there health, safety, environmental, and regulatory issues?

Under each deliverable, these questions are answered by addressing several key issues. Typical issues and answers for halogen incandescent light bulbs, as they were likely prepared in the mid-1980s, are presented next.

Business Opportunity Assessment

Customer Requirements. These are the results of the market study projected to the mid-1980s, as presented in Table 17.2, with the lifetime and energy efficiency as the major requirements. A lifetime that exceeds 2,000 hours was desired.

Preliminary Opportunity Analysis. Referring to the market analysis described earlier, the market size was quite large, approximately \$1.2 billion, with a compounded annual growth rate (CAGR) of 6.9%. Given current company sales of conventional incandescent light bulbs at 15 million units per year, let's assume that sales were projected to double with the new halogen light bulb concept, capturing an additional 1.9%[= 15 MM / 785 MM × 100] of the total market opportunity (~785 million units per year).

Although based upon this assumption the market potential for sales of halogen light bulbs was high, in one likely scenario, as the design team progressed through the *concept* stage, it developed second thoughts regarding the home lighting market, primarily because of the high operating temperatures, 3,100 K. Based upon this reservation, a shift to the display and industrial lighting market might have been more promising. Considering this alternative, and assuming that the data in Table 17.9 were available, the total sales opportunity would have been 30% of the total market size. To double sales, an additional 6.4%[= 15 MM/(0.3 × 785 MM) × 100] of the market share in display and industrial lighting would have been captured. While this would have been promising, as a word of caution (because most company sales went to the home lighting market), it would probably have been recognized that penetration of these two market segments would require the creation of new supply chains. Nevertheless, let's assume that the design team expressed confidence in the company's ability to penetrate these markets using its existing sales and marketing teams. Note that, given this shift in market focus, the project charter would have been revised accordingly.

Table 17.9 Market Analysis Results for Longer-Lifetime Incandescent Light Bulbs

Market Segment	% Market Size	% Market Share of Incandescent Bulbs	CAGR, %
Display lighting	20	20	5.5
Home lighting	70	80	4.2
Industrial lighting	10	20	8.5

Technical Feasibility Assessment

Competitive Analysis. In the mid-1980s, it was probably well recognized that cool fluorescent light bulbs provided the only significant competition on the market, their main *value proposition* being in their high luminous efficiencies, which significantly reduced energy costs to their end users. At that time, however, because energy efficiency was less important (fuel was relatively cheap) and the business opportunities in the display and industrial lighting markets were so promising, fluorescent technologies were probably not anticipated to be competitive. Furthermore, it was likely recognized that, to attain a large fraction of the home lighting market, compact fluorescent light bulbs (CFLs) that could be inserted into standard home lighting fixtures would have to be created. Given that a mechanism for converting from white (daylight) to yellowish color, comparable to that emitted from most incandescent light bulbs, had not been developed, it was probably recognized that customer interests in CFLs would be slow to develop. In summary, these factors likely led to the decision to refocus on the display and industrial lighting markets.

Superior Concept. Having shifted its focus to the display and industrial lighting markets and to satisfy the lifetime and luminous efficiency requirements, the preliminary technical analysis by the design team probably led it to concentrate on developing halogen-tungsten light bulbs—rather than CFLs, which were perceived to have color limitations at that time. However, halogen light bulbs had one principal drawback—operation at higher temperatures, which created safety and hazard concerns (to be described under Product Life-Cycle Assessment).

Probability of Technical Success. The design team likely expressed confidence in the high probability of technical success. Critical parameters to quality (lifetime and energy efficiency) had been identified, along with appropriate methods for satisfying them. Concerns remained involving the production of tungsten filaments having uniform thickness, but these were probably anticipated to be resolved during the manufacturing stage.

IP Assessment. The patent landscape for halogen light bulbs was not crowded in the mid-1980s. Based upon a preliminary patent analysis, problems in protecting its technologies were not likely to have been anticipated. Patents would have been sought to protect the new tungsten-halogen technologies and the halogen light bulb designs, as well as their application areas.

Manufacturing Capability Assessment

As mentioned above, it would have remained to address the manufacturing of tungsten filaments having uniform thickness. During the *manufacturing* stage, it would likely have been anticipated that the production of tungsten rods, prior to stretching into coils, could be perfected.

Product Life-Cycle Assessment

During the health, safety, environmental, and regulatory assessment, two safety and hazard issues—fire and personal burn injuries—would likely have been addressed. Although less significant in the display and industrial market segments, a secondary enclosure was probably designed as a protective measure. Also during safety and hazard testing, another issue involved UV light emission from the high-temperature halogen bulb. For certain applications, such as lighting in the art display market, it was likely recognized that exposure to UV light had to be eliminated or minimized. Note that it would also have been necessary to seek UL approval for home usage of products intended for the display and industrial markets.

Recommendations

On the basis of these four assessments, it is likely that the design team recommended that authorization and funding be granted to proceed to the *feasibility* stage of the SGPDP, where the issues identified in the next subsection would have been addressed. In so doing, it likely presented a plan for the *feasibility* stage, which likely included the following deliverables:

1. A refined business opportunity that addressed the supply chain for display and industrial applications.
2. Refined technical solutions that addressed UV emissions. These likely involved halogen light bulb prototypes and a market study conducted with selected customers.
3. A refined manufacturing analysis that addressed tungsten-thickness uniformity and surface texture.
4. An assessment that addressed UL requirements.

Feasibility Stage

Having received authorization and funding to proceed, the design team would probably have concentrated next on the deliverables to be completed during the *feasibility* stage. Emphasis would likely have been placed on the technical feasibility of the superior concept(s). In addition, several issues would likely have been addressed, arising from a market assessment, a competitive analysis (including IP strategy), and examinations of health-safety-environment concerns and the product life cycle. For this case study, to illustrate the steps in the *feasibility* stage, hypothetical data, representative of that likely to have been reported by customers, are presented. The actions of the design team are in response to this hypothetical, but representative, data.

a. *Technical Feasibility.* The goal of the technical feasibility assessment likely was to ensure that the superior concept(s) C, D2+E2, and F3+F4 met the *customer* requirements. For the halogen light bulb, this was likely accomplished by checking that the superior concepts satisfied the *technical* requirements.

First, the critical-to-quality variables, the lifetime and the energy efficiency, had been checked in the *concept* stage, with simulations carried out showing that thermal diffusion, the halogen reactions with condensed phases at the bulb wall, and the oxygen concentration were critical to the lifetime. For details, see Section 16.2 and the file, Supplement_to_Chapter_16.pdf, in the PDF Files folder, which can be downloaded from the Wiley Web site associated with this book.

In the *feasibility* stage, the results of these simulations, together with the predicted solution to Exercise 16.1, were likely confirmed with prototype light bulbs. Table 17.10 displays typical, but hypothetical, experimental data, showing that all of the prototypes satisfied the lifetime and energy efficiency requirements, although the variation in lifetime was large. At this stage, however, it would likely have been sufficient that the prototypes met the customer requirements. More

Table 17.10 Hypothetical Lifetime and Energy Efficiency Data for Tungsten-Bromine Prototypes

Prototype	Lifetime (hr)	Energy Efficiency (%)
A	5,000	5.5
B	6,500	5.0
C	4,500	5.3
D	5,500	5.8
E	6,000	5.2
F	7,000	5.0
G	5,000	6.0

uniform performance would likely have been sought in the *manufacturing* stage.

It is likely that primary and secondary casing prototypes were also tested experimentally. Because several preliminary tests failed to satisfy the maximum temperature specification of 350 K, the casing subteam likely selected new ceramic materials for the primary casing. Table 17.11 shows hypothetical surface temperature data for several prototypes using light bulb prototype D. Still, four of the prototypes didn't meet specifications, and consequently, additional prototypes would likely have been built for customer testing—with hypothetical data shown in Table 17.12.

b. *Customer Verification.* The results of hypothetical customer tests are presented in Table 17.13, with the customer profiles summarized as follows. Customer C-1 was the best customer that sold to industrial end users, Customer C-2 was the best customer with the largest global distribution in home applications, and Customer C-3 was the smallest customer selling to various market segments globally. Also shown in Tables 17.13–17.15 are hypothetical lifetime, energy efficiency, and surface temperature data typical of that reported by these customers.

Note that while small customers are often excluded, they frequently provide the most useful feedback for the evaluation of new products. Even better evaluations are provided by potential customers who are currently buying the products of competitors. In addition, it is very helpful when a working relationship with customers has been cultivated over several product cycles.

Finally, the design team likely questioned the willingness of these customers to purchase the light bulbs based upon their technical performance, with typical results of these inquiries displayed in Table 17.16. Clearly, the degrees of acceptance of the halogen light

Table 17.11 Hypothetical Surface Temperature Data for Casings with Light Bulb Prototype D

Prototype	Surface Temperature (K)
D1	350
D2	360
D3	350
D4	345
D5	355

Table 17.12 Hypothetical Test Results from Prototypes Prepared for Customer Feedback

Prototype	Lifetime (hr)	Energy Efficiency (%)	Surface Temperature (K)
P-1	5,500	5.8	350
P-2	6,500	5.5	325
P-3	6,000	5.7	340

Table 17.13 Customer Feedback—Hypothetical Lifetime (hr) Data

Customer	Prototype P-1	Prototype P-2	Prototype P-3
C-1	5,500	6,500	6,000
C-2	5,500	6,000	6,000
C-3	5,000	6,500	6,000

Table 17.14 Customer Feedback—Hypothetical Energy Efficiency Data

Customer	Prototype P-1	Prototype P-2	Prototype P-3
C-1	5.8	5.5	5.7
C-2	5.7	5.5	5.6
C-3	5.5	5.6	5.5

Table 17.15 Customer Feedback—Hypothetical Surface Temperature (K) Data

Customer	Prototype P-1	Prototype P-2	Prototype P-3
C-1	350	325	340
C-2	375	350	360
C-3	350	300	320

Table 17.16 Customer Willingness to Buy

Customer	Typical Customer Comments
C-1	Definitely will purchase. When will it be available?
C-2	Depending on the price, we will consider purchasing. Can the surface temperature be further reduced?
C-3	Great performance, contact us when the product is launched. Further reduction in surface temperature will be more attractive to us.

Table 17.17 Revised Market Analysis

Market Segment	% Market Size	% Market Share of Incandescent Bulbs	CAGR, %
Display Lighting	20	20	5.5
Industrial Lighting	10	20	8.5

bulb concept differ. Customer C-1 was eager to buy, Customer C-2 indicated the price was a deciding factor, and Customer C-3 was willing to consider, an encouraging indicator. Furthermore, two of the three potential customers requested a lower surface temperature, and consequently, the design team planned to explore the reduction desired and its impact on product price.

c. ***Market Assessment.*** With the potential introduction of the compact fluorescent light bulb (CFL), the marketing team likely anticipated that the halogen light bulb would face severe competition. Table 17.17 shows how, in the *feasibility* stage, their market assessment might have been adjusted.

Not shown is the portion associated with the home lighting business, which was about 70% of the entire market (approximately \$1.2 billion/yr) in the mid-1980s; that is, $0.7 \times 1,200\text{MM} = \$840\text{MM}$. On the basis of the estimates in Table 17.17, the remaining market segments, display and industrial lighting, would have been projected to be approximately $0.2 \times 1,200\text{MM} = \$240\text{MM/yr}$ and \$120MM/yr, respectively. Consequently, in these segments, the corresponding sales of incandescent light bulbs would have been projected to be $0.2 \times 240\text{MM} = \$48\text{MM/yr}$ and $0.2 \times 120\text{MM} = \$24$ million/yr, respectively, totaling \$72MM/yr. Because the remaining \$288MM/yr$(= 360 - 72)$ would have been more attractive, providing new sales for the company, the marketing team would likely have decided to focus on the larger opportunity; that is, to replace fluorescent light bulbs with halogen light bulbs in the display and industrial market segments.

d. ***Competitive Analysis.*** The customer feedback, expressing satisfaction with the technical performance, would likely have been considered to be very encouraging. The most likely concerns would have involved pricing and increasing competition from fluorescent light bulbs. At the time, fluorescent light bulb manufacturers had exhibited prototypes of compact fluorescent light bulbs (CFLs), but their potential home usage was not considered to be significant in the immediate future.

e. ***Product Life-Cycle Management.*** With prototype testing data showing that the primary and secondary enclosures performed well, the design team would likely have concluded that health, safety, and environmental concerns had been resolved. While UL certification might not have been completed in the *feasibility* stage, it likely would have been in progress.

f. ***Gate Review.*** The *feasibility* Stage-Gate review, often known as the *Money* gate, is the last gate prior to the commitment of significant resources, both personnel and financial. The focus of this review is normally on demonstrating that the superior concepts satisfy the customer requirements, both through the demonstration of the technical performance of a product prototype(s) in the laboratory and customer feedback and verification. Equally important is the business case for the proposed new product(s) that should include sales potential and market segmentation.

For the halogen light bulb product, the hypothetical assessments provided in this section would likely have been sufficiently promising to complete the *feasibility* stage. Regarding the business case, projections of the market share for the large existing market would likely have been considered adequate to proceed to the *development* stage.

Development Stage

As introduced in Section 2.6, the main task in the *development* stage of the SGPDP is to design a manufacturing process at the *pilot-plant* level. The *House of Quality* is normally used to identify the critical process variables for each manufacturing process. In addition, a failure-mode analysis is often performed to identify a reaction plan(s) to abnormal events (severe disturbances or faults) that return the process to its nominal operating point. Often, short-term runs are sufficient to check the manufacturing process stability. When necessary, the market assessment and product recycle management are revisited.

For the halogen light bulb, the manufacturing process would have required careful design to produce tungsten rods having uniformly smooth surface structures. In addition, it would have been necessary to design the manufacturing process of the primary enclosure to provide a favorable thermal environment ensuring the longevity of the tungsten coils, with a well-constructed reaction chamber that prevented early burnout.

Manufacturing Stage

In this stage, the manufacturing process for the halogen light bulb would have been designed, with emphasis on the scale-up of the pilot-plant process. Manufacturing costs for the halogen light bulbs would also have been estimated. Yet another *House of Quality* would likely have been prepared for the manufacturing process, relating the process variables and materials parameters to the process outputs and product performance. The emphasis would likely have been on identifying the critical processing and materials parameters that affect the product quality. Note that today, a statistical study using six-sigma methodology, as discussed in Chapter 25, would be conducted to project the capability of the manufacturing process to meet product/customer specifications. When necessary, the market assessment and product recycle management would be revisited.

For the halogen light bulb, the manufacture of tungsten wire was likely to have been the critical processing step. As discussed in Section 16.2, the design for uniform surface structure would have been key to ensuring high product quality.

Product-Introduction Stage

As discussed in Section 2.8, in the *product-introduction* stage, normally a launch strategy is developed—often including pricing, the launch channel, advertisements, product literature, and early product introduction to a limited market or to selected customers—prior to mass production. For the halogen light bulb product, because the company had been selling incandescent light bulbs to the selected market segment, the launch channel would likely already have been considered as well as established. Instead, the product literature, pricing, and the inventory were likely to have received the most attention.

The product literature, in addition to providing product specifications, would likely have included a section on product usage and high operating temperatures, with warnings of potential IR and UV exposure. The training of technical-service personnel and the sales force would likely have been planned as well.

In addition, a pricing analysis, taking into consideration the prices of existing products, would likely have been undertaken at this stage. With the likely objective to migrate users toward the new halogen light bulbs, a small price reduction on the order of 5% would probably have been considered sufficient. Alternatively, had the objective been to differentiate between the halogen and standard incandescent light bulbs—and prevent product cannibalization—a slight price premium might have been set.

Depending on the supply-chain infrastructure, it may have been necessary to create a one- or two-month inventory prior to launching the new halogen light bulb product.

17.3 HOME HEMODIALYSIS DEVICE CASE STUDY

As discussed in Section 16.3, this case study is based upon a document prepared for a panel of medical experts assembled to evaluate the potential for delivering hemodialysis at home, specifically, at night, while the patient is sleeping (Talamini, 2005). The underlying new technologies developed during the 1970s and 1980s and presented in Section 16.3 were positioned in the *innovation map* of Figure 16.21, which shows the linkages between these new technologies and the products that satisfy the needs of hemodialysis patients; that is, the *customer-value proposition.* In this case study, the Stage-Gate^{TM} Product-Development Process (SGPDP) is applied to design a home hemodialysis product, beginning with the formation of a product design team. Emphasis is placed on developing the project charter and the *concept* stage of the SGPDP.

Project Charter

Following the steps at the top of Figure PIII.1, initially the design team prepares a project charter, as shown in Table 17.18. Clearly, the challenge is to develop a product that will permit the safe use of a hemodialysis device at home, conveniently at night while patients sleep. Under the Project Scope, the design team recognizes the need to set home hemodialysis schedules and to reconfigure the hemodialysis device to adhere to these schedules. While a product involving the hemodialysis device and associated operating equipment (pumps, holding tanks, sorbent dialysis, etc.), sensors, and alarms is envisioned, the installation of the operating system is not included in the product package. Installation instructions would be provided, but not an installation service. Furthermore, the retrofitting of a bedroom to provide high standards of cleanliness is also not covered by the cost of the new product. The design team envisions the demonstration of a working prototype system within 12 months.

Having created a promising project charter and an acceptable innovation map, it is next assumed that the design team is authorized to proceed to the *concept* stage of the SGPDP.

Concept Stage

The concept stage begins with an opportunity assessment, which identifies the potential market. In 2002, 281,600 patients visited hemodialysis centers in the United States, three times weekly, to receive four-hour hemodialysis treatments, with only 843 patients (0.3%), equipped to receive hemodialysis treatments at home. Considering the inconvenience of traveling to hemodialysis centers to receive time-consuming treatments during daylight hours, competing with other activities, customer interviews indicate that low-cost home treatment, while patients sleep, would be far preferable. It is expected that with an easy-to-use and safe home hemodialysis system, the usage rate will be significantly higher than the current 0.3%.

According to Talamini (2005), these potential customers need hemodialysis systems that operate reliably at home in the absence of medical personnel, with comfortable, less-disruptive treatments desired. Potentially, these should involve lower blood and dialysate flow rates, with no noticeable effects (that is, smooth, continuous operation without surging) during sleep. Because the treatments would be administered over longer periods, as many as seven days

Table 17.18 Project Charter of the Home Hemodialysis Device

Project Name	Home Hemodialysis Device
Project Champions	Business Director of the Hemodialysis Business
Project Leader	John Doe
Specific Goals	Hemodialysis devices that operate at home while patients sleep. The product involves an array of devices to be installed at home, including alarms with sensors to detect and provide notifications of malfunctions—with alerts to medical personnel at hemodialysis centers.
Project Scope	In-scope: • Determination of home hemodialysis schedules • Reconfiguration of the hemodialysis device to adhere to the adjusted schedules Out-of-scope: • Installation of the equipment, instrumentation, and alarms • Retrofitting of bedrooms to achieve high standards of cleanliness
Deliverables	• Business opportunity assessment • Technical feasibility assessment
Time Line	Product prototypes for market testing within 12 months

weekly, concentrations of urea in the blood should be lower on average, thus requiring smaller amounts of dialysate solution per treatment. If possible at low cost, it would be desirable to recover urea from the dialysate before recirculation to the hemodialysis device, rather than discarding the spent solution without recycling.

The most obvious customer needs are summarized in Table 17.19. Note that these needs are difficult to satisfy, and consequently are classified as *new-unique-and-difficult* (NUD) rather than *fitness-to-standard* (FTS).

Next, these customer requirements are translated into a set of technical requirements in Table 17.20. Note that the latter are normally adjusted by the design team as the product-development process proceeds.

Reliable home operation requires careful monitoring of many variables, such as the pressure, temperature, and flow rates, throughout the treatment session. The critical-to-quality (CTQ) variables selected must be monitored and controlled within the ranges shown in Table 17.20. The dialysate temperature is an important measure that should be kept near body temperature. Temperatures above 42°C may lead to protein denaturation while temperatures above 45°C can cause hemolysis (rupturing of red blood cells).

Table 17.19 Customer Requirements for the Home Hemodialysis Product

Customer Requirement	Type	Weighting Factor (%)
Reliable operation at home	NUD	50
More comfortable, less disruptive treatments	NUD	50
No noticeable disturbances while sleeping		

Pressures are typically measured at two points: upstream of the blood pump (arterial pressure) and downstream of the dialyser (venous pressure). Safety alarms should be triggered when the pressure moves ±10% from the setpoint value, P^{SP} (which depends on the flow rate). Pressures are monitored similarly in the dialysate circuit. Finally, monitors must detect air embolisms and blood leakage.

For the extended dialysis schedule, 7 day/wk, 6 hr/day, the urea concentration in a patient's blood, $C_{P,i}$, builds to just 0.6 mg/cm^3, as seen by solving the following equation:

$$\frac{1.0\,\text{mg/cm}^3 - 0.30\,\text{mg/cm}^3}{56\,\text{hr}} = \frac{C_{P,i} - 0.30\,\text{mg/cm}^3}{24\,\text{hr}}$$

Most sources of home hemodialysis products suggest lower flow rates than those at dialysis centers; that is, blood and dialysate flow rates at 200 and 300 cm^3/min, respectively. Note, however, that the model in Example 16.2, with the same specifications except for the initial urea concentration in the blood reduced to 0.6 mg/cm^3, indicates that the urea concentration in the blood can be reduced to 0.3 mg/cm^3 in six hours, using flow rates of 125 and 150 cm^3/min, respectively. In this regard, blood and dialysate pumps are available for the lower flow rates.

To keep patients comfortable during dialysis, surges in the blood flow rate must be controlled. In the *concept* stage, bounds at ±10% of the setpoint flow rate seem reasonable. This range must be carefully assessed in the *feasibility* stage, when prototype products are constructed and tested.

Given the more frequent dialysis schedule, the average urea concentration is reduced to 0.45 mg/cm^3, midway between 0.6 and 0.3 mg/cm^3.

Table 17.20 Technical Requirements for the Home Hemodialysis Product

Customer Requirement	Technical Requirement	Typical Values
Reliable operation at home	Lower average urea concentration	≤ 0.45 mg/cm^3
	Remove urea and recirculate dialysate	Residual urea concentration ≤ 1 ppm
	Monitoring Systems—*P*, *T*, air detection, blood-leak detection Alarms—*P*, *T*, blood leak, air embolism, vascular access	$36 \leq T \leq 40$°C $0.9P^{SP} \leq P \leq 1.1P^{SP}$ $0.25 \leq$ BLS* ≤ 0.35 mL/L blood Alarms ≥ 70 decibel
More comfortable, less disruptive treatments	Lower blood and dialysate flow rates	$Q_B = 200$ cm^3/min $Q_D = 300$ cm^3/min
No noticeable disturbances while sleeping	Surge-free pumps	$\Delta Q_B \leq \pm 0.1 Q_B^{SP}$ $\Delta Q_D \leq \pm 0.1 Q_D^{SP}$
	Dialysis schedule	7 night/wk, 6 hr/night

*Blood-leak sensitivity—mL/L blood; Q_B is the blood flow rate; Q_D is the dialysate flow rate; T is the temperature; P is the pressure

Finally, when the dialysate is recycled, in the *concept* stage an upper bound on the urea concentration of 1 ppm seems reasonable. This upper bound must also be assessed in the *feasibility* stage. Also, when sorbent dialysis is implemented, urea decomposes to ammonia, which must not exceed 2 wt % in the recycled dialysate. Note that the cartridges hold up to 20–30 g of urea, more than accumulated in a single dialysis treatment. When full, the ammonia concentration in the dialysate effluent builds as ammonia begins to break through. Yet another consideration is a bound on the dialysate concentration of bacteria, which must not exceed 200 colony forming units of bacteria per milliliter (200 cfu/cm^3).

Another important aspect of reliable home operation is the user interface, which will typically be a computer or touch-sensitive screen. During normal operation, this displays details such as pressure, temperature, and flow rates. When an alarm is triggered, the user interface must clearly communicate the problem to the patient and provide instructions for correcting it.

Having defined the CTQ variables, it remains to generate the product concepts and evaluate these concepts against the best existing solution (Hemodialysis Center). This leads to the selection of the superior concept. In this case, the most promising product would include sorbent dialysis to eliminate the need for water-treatment equipment and dialysate storage.

Feasibility Stage

The three principal activities in the *feasibility* stage involve:

1. Adjusting the design of the dialysis unit to improve its performance. See Exercise 16.6.
2. Constructing and testing prototype products.
3. Creation of a business plan. See Exercise 17.5.

Having obtained promising results, the design team would normally pass the gate review and proceed to the *development* stage of the SGPDP. Note that the authors have completed solutions to Exercises 16.6 and 17.5. The product design and business plan, including profitability analyses, are very promising.

Development Stage

This stage would normally include an optimized design. In addition, designs of the monitoring systems would be completed with failure-mode analyses, to ensure the home hemodialysis product is safe and can handle various failure modes.

17.4 HIGH-THROUGHPUT SCREENING OF KINASE INHIBITORS CASE STUDY

A design team has been formed with the goal to create a lab-on-a-chip for the high-throughput screening of kinase inhibitors. The chip would be available in a commercial laboratory, which carries out assays of kinase inhibitors sent by pharmaceutical companies, or the chip would be licensed to pharmaceutical companies to be installed in their in-house laboratories. This case study begins with the *innovation map* presented earlier in Figure 16.33, which shows consumer needs and two principal new technologies for the lab-on-a-chip product. Note that other technologies are available, but these two are the technologies to be considered in this limited case study.

Due to space limitations, only the principal steps in this design are presented. In brief, the project charter requires that a design be completed in six months. The time frame for the design project is the mid-2000s. The design team has received endorsement for its *innovation map* and is proceeding to the SGPDP.

Concept Stage

In the mid-2000s, it is estimated that pharmaceutical companies are spending approximately \$200MM/yr for screening potential therapeutic kinase inhibitors. The market and competitive analyses suggest that the design team should strive to achieve the highest-speed throughput practical, performing on the order of 1 MM IC_{50} assays/day, which on a 8-hr day, 5-day/wk schedule, corresponds to 250 MM assays/yr. On the order of 10,000 kinase inhibitors would be tested daily against 100 kinase enzymes, giving 1 MM IC_{50} assays/day.

The customers (pharmaceutical companies) seek low-cost, high-throughput screening products available in a commercial service laboratory or, preferably, in their in-house laboratories. The innovation map in Figure 16.33 shows rough price expectations associated with the Fluidigm® and RainDance® technologies. Reliable assays are required, involving considerable redundancy to achieve six-sigma performance; that is, 3.4 defective assays per million. These customer requirements are translated into product requirements as follows. First, for both Fluidigm® and RainDance® chips, configurations are needed to create nano-liter boluses (i.e., reactor droplets) involving 100 kinase enzymes and 10,000 kinase inhibitors at different concentrations. Then, to determine the rates of the phosphorylation reaction (R1 in Chapter 16), an optical means of measuring the concentrations [ATP] or [ADP] is needed. As shown on the innovation map, two different technologies are selected for use with the Fluidigm® and RainDance® chips.

Several CTQ variables are identified next. For both chips, these include:

1. The number of reactor droplets, at different [KI], to achieve accurate IC_{50} concentrations, $[KI_{50}]$. Six is the minimum commonly selected.
2. The number of redundant droplets, having the same [KI], to achieve accurate IC_{50} concentrations, $[KI_{50}]$. Five is selected.
3. The ranges of [ATP] and [Pep] concentrations in the reactor droplets, which influence the kinetics of the phosphorylation reaction and the rate of photon generation. In Section 16.4, these concentrations are set at $[ATP] = 1\ \mu M$ and $[Pep] = 60\ \mu M$.
4. The volume of the reactor droplets. In Section 16.4, 1 nL is selected.
5. The reaction, or incubation, time at 37°C. In Section 16.4, 1 hr is selected.

Using these CTQ variables, two product concepts are considered involving two chip designs, which are described next.

Fluidigm® Chip

Figure 17.4 shows a typical configuration of the bolus (reactor droplet) preparation system on the Fluidigm® chip. For a discussion of the components of Fluidigm® chips, see Section 16.4. Two reservoirs (A and B) on the lower PDMS mold, that is, the *process layer*, contain the kinase enzyme, ATP (1 μM), luciferase, luciferin, substrate peptide, fluorophore (Alexa Fluor® 350), and buffer (to maintain pH = 7.5). In addition, reservoir A contains the kinase inhibitor. Each reservoir is 10 μm high, with a diameter of 100 μm. Reservoir B is referred to as *without inhibitor* and reservoir A as *with inhibitor.* The two reservoirs are connecting to channels 10 μm high and 100 μm wide, each of which is crossed by three orthogonal channels in the upper PDMS mold, that is, the *control layer.* Each set of three channels, 50 μm wide per channel, serves as a peristaltic pump. Note that the two channels are separated by 100 μm. In addition, a third line, 50 μm wide, lies 50 μm below the kinase inhibitor line. For this line, a peristaltic pump is sufficient to insert perfluorodecalin (PFD), a spacer, to separate the mixed boluses. Note also that these three lines are repeated 48 times in the process layer, permitting assays to be obtained in parallel for 48 kinase inhibitors.

Returning to the effluent lines from reservoirs A and B, these pass through valves into a 3-nL Fluidigm® mixer. Note the dimensions in Figure 17.4: all lines are 10 μm high and 100 μm wide, giving a volume of $(2 \times 1,200 \times 100 + 2 \times 300 \times 100) \times 10\ \mu m^3 = 3 \times 10^6\ \mu m^3 = 3,000\ pL = 3\ nL$. For the initial bolus, both inlet valves are open and the effluent valve is closed as the solutions are pumped into the mixer. Then, the inlet valves are closed and the mixer pump is operated for approximately 3 seconds of mixing time, after which the three valves are opened and additional fluid from reservoirs A and/or B is pumped to displace the

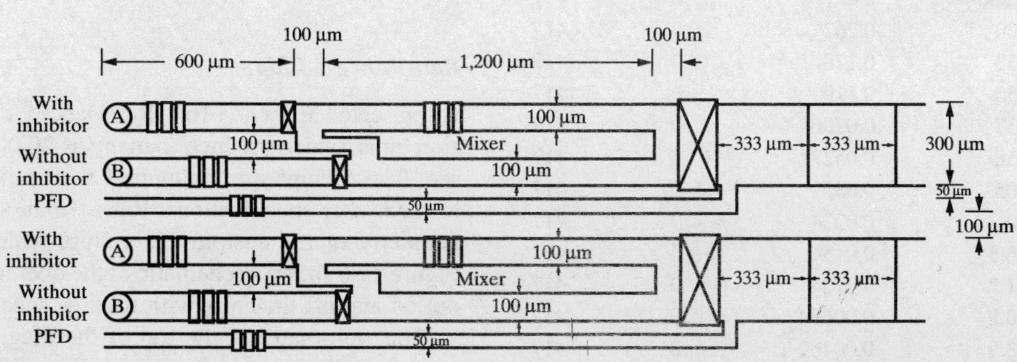

Figure 17.4 Schematic of bolus preparation system.

fluid into the effluent line. Just prior to the next mixing operation, with the effluent valve closed, 1 nL of PFD spacer is pumped to displace the first bolus in the effluent line. This process is repeated to form 15 reacting boluses in each effluent line, each having a different inhibitor concentration. Note that because the effluent line is 300 μm wide, the length of each bolus and spacer is 333 μm, to give 1 nL volume.

The dilution process is accomplished by varying the number of cycles for either or both of the pumps associated with reservoirs A and B, thereby changing the concentrations in each bolus. In each pumping cycle, 1 nL of fresh reactants displaces 1 nL of fluid from the mixer. See Table 17.21 for a typical dilution schedule; that is, for each pumping cycle, the volumes of inhibitor (from reservoir A) in the mixer after pumping, the pumping times for the two pumps, and the inhibitor/total volume ratio in the mixer. As shown, in the initial cycle (1), 2,700 pL of fluid from reservoir A are pumped in 1,125 ms, and 300 pL from reservoir B are pumped in 125 ms. During cycle 2, 1,000 pL of fluid are displaced from the mixer (containing 900 pL of fluid A and 100 pL of fluid B) and replaced by 1,000 pL of fluid B. The resulting contents of the mixer are 1,800 pL of fluid A and 1,200 pL of fluid B; that is, 60% of fluid A. To pump 1,000 pL of fluid B, 416.7 ms are required. During each of the remaining 13 pumping cycles, an additional 1,000 pL of fluid B are added.

After 15 pumping cycles, which take approximately 1 minute, a heating plate is activated that heats the 15 boluses to 37°C, the temperature at which the phosphorylation reaction takes place. For the next hour, photons generated by the luciferase reaction (R11 in Section 16.4) are collected

Table 17.21 Typical Dilution Schedule

Cycle	Inhibitor Volume (pL)	Inhibitor/Mixer Volume Ratio	Pump A (ms)	Pump B (ms)
1	2,700	0.9	1,125	125
2	1,800	0.6	0	417
3	1,200	0.4	0	417
4	800	0.267	0	417
5	533	0.178	0	417
6	355	0.119	0	417
7	237	0.0790	0	417
8	158	0.0527	0	417
9	105	0.0351	0	417
10	70.2	0.0234	0	417
11	46.8	0.0156	0	417
12	31.2	0.0104	0	417
13	20.8	0.00694	0	417
14	13.9	0.00462	0	417
15	9.25	0.00308	0	417

by a charge-coupled device (CCD) detector/scanner to measure [ATP] and the related rate of the phosphorylation reaction (R1).

Figure 17.5 shows a top view of the chip design discussed thus far. There are two sets of 48 three-channel devices, side by side, requiring 2.54 cm = 1 in. These two sets are separated by a trough. For a given kinase enzyme, each three-channel device involves 48 different kinase inhibitors. For each inhibitor, on each side of the trough, 15 boluses at different inhibitor concentrations are generated, giving a total of 30 boluses for each inhibitor. The length of the lines on each side of the trough is determined by adding the length of the entrance region (600 μm), the length of the mixer (1,400 μm), and the length of the bolus region ($2 \times 333 \times 15 = 10,000$ μm), to give a total length of 12,000 μm = 1.2 cm. Hence, 2.54 cm = 1 in. is sufficient to contain the lines on both sides of the trough and the trough, where spent reagents are accumulated and removed.

In summary, using two channels for each inhibitor, six decades of dilution are generated. On a single chip, there are 96 three-channel devices, providing a total of 48 IC_{50}'s that are generated every ninety minutes (30 min for preparation and one hour for reaction). With one CCD camera over an 8½ hour day, seven chips are processed, giving some time to spare. Consequently, 10,080 boluses per day, providing 336 IC_{50} curves per day (that is, 30 boluses per IC_{50} curve), are carried out. Each chip is discarded after one utilization, providing 48 IC_{50} curves. Hence, 336 IC_{50} curves are produced using seven chips. This product design is presented by Chen, Heend, and Waring (2005).

Clearly, this product concept can be improved significantly—and this would be the case if it were selected as the superior product concept. If the number of boluses per IC_{50} was reduced to six, each 1-inch × 1-inch chip would produce $5 \times 48 = 240\ IC_{50}$/chip and $7 \times 240 = 1,680\ IC_{50}$/day

Note, however, that the RainDance® emulsion guns, introduced three years after the Fluidigm® chips, produce reactor droplets on the order of 1,000 times faster than the Fluidigm® mixer. This has led to the customer request for 1 MM IC_{50}/day. For this reason, the Fluidigm® chip is rejected in favor of the RainDance® chip.

RainDance® Chip

As presented in Section 16.4, RainDance® chips utilize emulsion guns, which produce as many as 20,000 microdroplets/sec. These chips are configured with mixers and splitters to meet user specifications. Rather than show the detailed geometry of the custom-made chip, which is proprietary, Figure 17.6 shows a schematic of the operations to be carried out on and off the chip, with timings specified.

The well-plates at the top of the diagram are received, often in mailers, from external suppliers. When operating as a commercial service laboratory, one well-plate containing

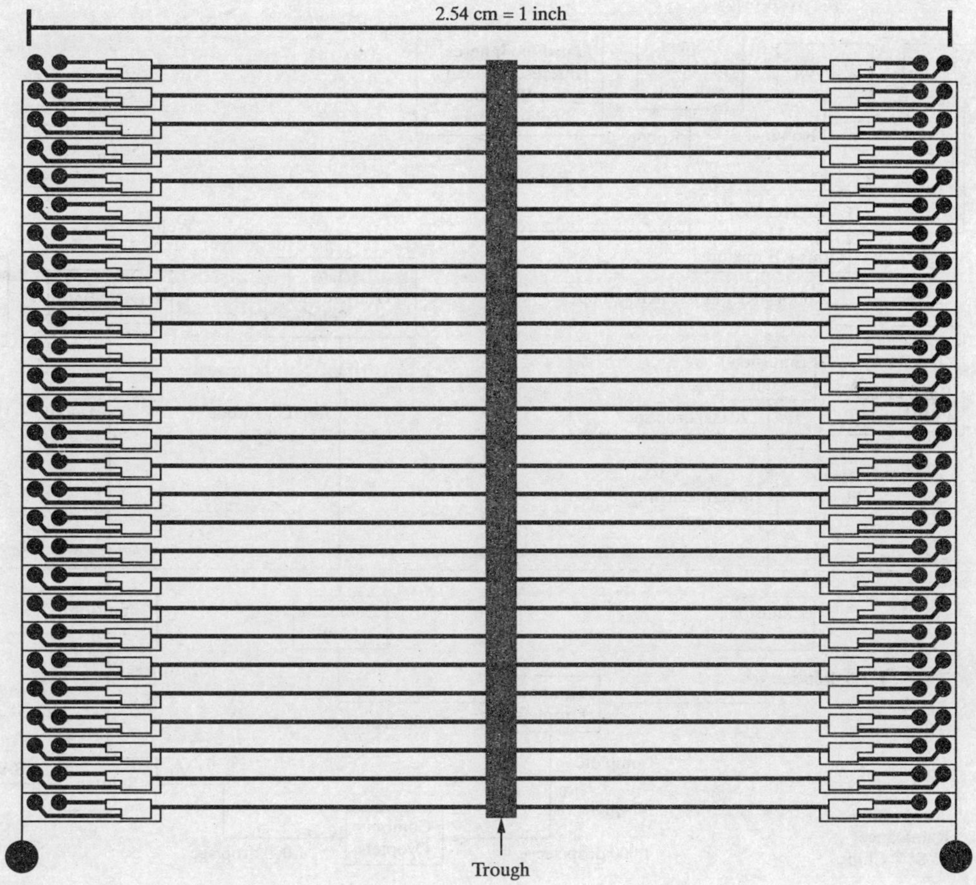

Figure 17.5 Top view of chip layout. Two arrays of 48 three-channel devices, each containing 15 boluses on each side of the trough. The PFD spacer is stored in one large reservoir on each side of the trough.

10,000 KIs is received daily from a pharmaceutical company. In the design shown, the first kinase enzyme, KE_1, is transferred by a JANUS robot arm from its well-plate to a Labcyte® ECHO 555™ liquid handler, where it is mixed with peptide and buffer solution from their well-plates. Note that typical transfer times for the JANUS robot are 1.5 sec, but the robot is washed and evacuated after each kinase enzyme is transferred, requiring an additional 1.65 sec. In parallel, the JANUS robot transfers a mixture of TransFluoSpheres™ (fluorescent beads), which serves as a *barcode* for KE_1. The resulting mixture, KE_1/Pep, is transferred by the JANUS robot to an emulsion gun on the first RainDance® PLS™ chip, which produces nano-liter droplets at 20,000 drop/sec, as discussed in Section 16.4, that are conveyed to a microcentrifuge tube.

Note that the enzymes are mixed one at a time, each enzyme being labeled with a unique combination of three of four colors at three concentration levels; that is, 12C3 = 220. Because 112 combinations involve different concentrations of the same color that cannot be distinguished, these are discarded, leaving 108 unique labels (for 100 kinase enzymes). The resulting mixtures, KE_1/Pep, . . . , KE_{100}/Pep, are transferred one-by-one by the JANUS robot to the emulsion gun on the first RainDance® PLS™ chip.

As discussed below, sufficient droplets are generated to mix with droplets of the 10,000 kinase inhibitors. For each KI, five copies (for redundancy to counter measurement errors) of six droplets, at different KI concentration levels, are needed. Hence, for each kinase enzyme, 5 drops/conc. $\times$ 6 conc./KI $\times$ $10,000\,\text{KI} = 300,000$ droplets are generated at 20,000 drops/sec, requiring 15 sec. In addition, 3.15 sec/KE are required for robot transfer, washing, and evacuation. Consequently, the total time to create 30,000,000 droplets of KE_1/Pep, . . . , KE_{100}/Pep is $(15 + 3.15)$ sec/KE $\times$ 100 KE $= 1,815\,\text{sec} \approx 30\,\text{min}$, which can be prepared one day in advance. The barcoded droplets in the microcentrifuge tube can be randomized by agitation overnight. Note that 1.65 sec are also required to wash and evacuate the emulsion gun before the next KE is loaded, but this can be accomplished during the 3.15 sec/KE required for robot transfer, washing, and evacuation.

Subsequently, these randomized droplets are transferred by the JANUS robot to an emulsion gun on the second RainDance® PLS™ chip, where they are released, carefully synchronized at 1,000 drops/sec, to be mixed with droplets of the ADP/Transcreener® assay solution. The latter is transferred by robot to an emulsion gun, which also operates at 1,000 drops/sec. The two trains of droplets are combined, as

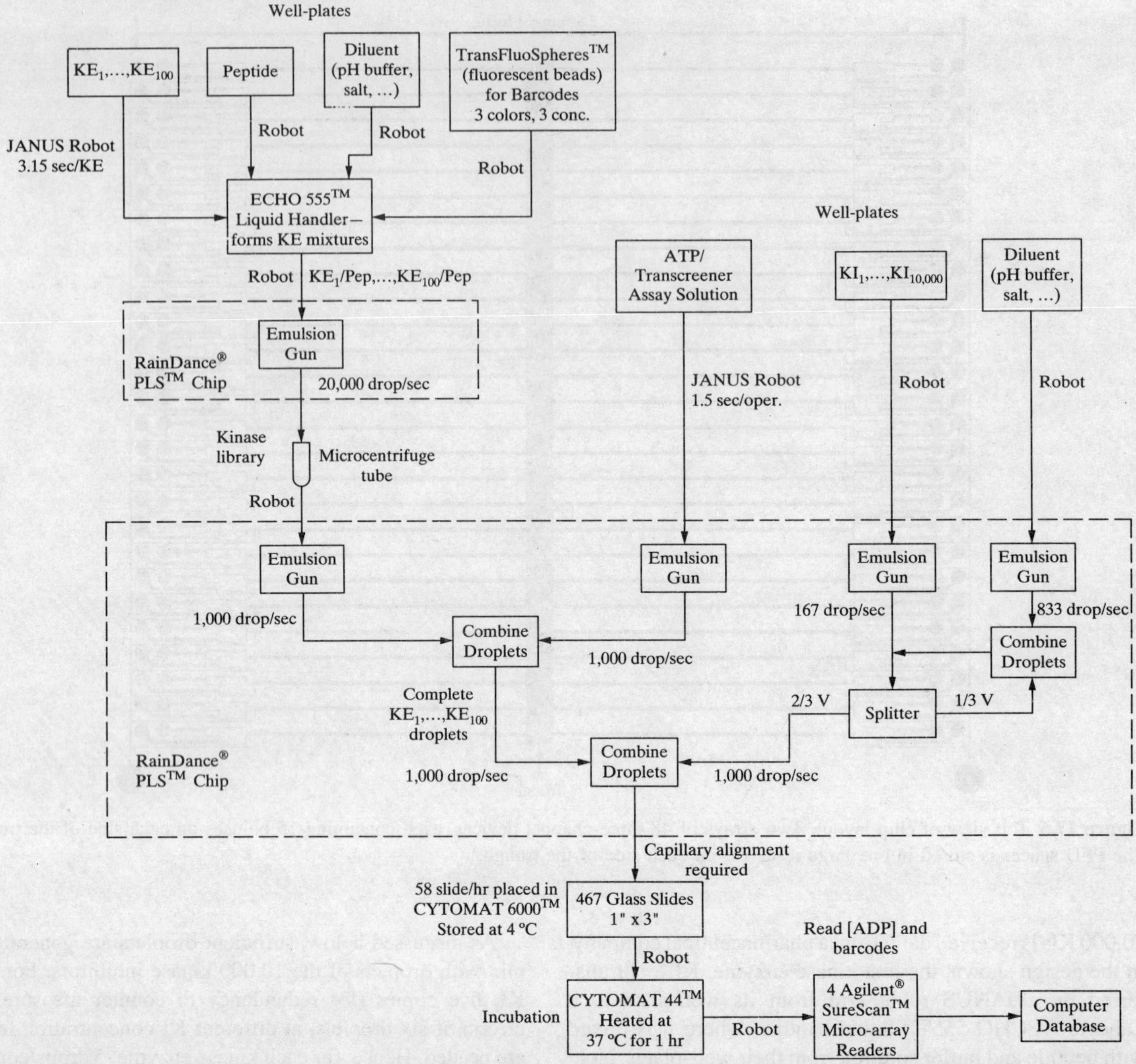

Figure 17.6 Schematic of operations on and off RainDance® PLS™ chip.

shown in Figure 16.32, to give complete (randomized and barcoded) kinase enzyme droplets.

The product concept in Figure 17.6 shows an approach for generating kinase inhibitor droplets serially without labels. In this approach, the 10,000 kinase inhibitors are transferred from their well-plates to the emulsion gun, one-by-one, by the JANUS robot. For each KI, six droplets are generated at 167 drops/sec. In a splitter, one-third of each droplet is recovered and combined with droplets from a diluent emulsion gun generated at 833 drop/sec. These are also sent to the splitter. Hence, for each KI droplet, five additional droplets at decreasing [KI] are generated in series. Note, however, that the KI emulsion gun releases its droplets slightly delayed, to avoid intersecting with the last diluted droplet, and the splitter has a gate that turns off after the lowest concentration droplet is created. This is repeated for each of the droplets from the KI emulsion gun. To provide redundancy to overcome experimental errors, five sets of six droplets are created, that is, 30 drops/KI. Because 30 drops/KI are needed for each KE, a string of 30 drops/KI $\times$ 100 KE/KI $=$ 3,000 drops/KI at 1,000 drops/sec, requiring 3 sec/KI. The emulsion gun is loaded (1.5 sec/KI), washed (1.5 sec/KI), and evacuated (0.15 sec/KI), totaling 3.15 sec/KI. Consequently, to generate the string of droplets for 10,000 KIs, 31,500 sec/day $=$ 8.75 hr/day are required. These KI droplets are combined with the randomized, complete KE droplets at 1,000 drops/sec.

The resulting droplets are transmitted to 1-inch $\times$ 3-inch glass slides, which have been etched with 100 μm channels (10 μm high), permitting 127 channels in parallel. Each channel receives 485 droplets, giving 61,600 drops/slide. Hence, with 467 slides, 30 MM drops/day are stored. Note that a precise robot is required to align rapidly the outlet capillary from the PLS™ chip with the capillaries on the glass slides, moving from capillary to capillary in sequence.

At a rate of 58 slides/hr, the JANUS robot places the loaded slides into a Heraeus CYTOMAT 6000 for storage at 4°C. These, in turn, are transferred by the robot to a CYTOMAT 44, where they are heated at 37°C for 1 hr, the incubation time. Then, the slides are transferred, one-by-one, to Agilent® SureScan™ micro-array readers, which require 1 min to scan each slide; that is, 467 slide/day $\times$ 4 $\times$ 4 scan/slide $\times$ 1 min/scan $= 1,868$ min/day $= 31.1$ hr/day. Note that one scan records [ADP] for each droplet and a second scan detects its barcode. Two additional scans are performed per slide to counter errors by the micro-array readers. Consequently, four micro-array readers are required to complete the scanning during an 8-hr day. The resulting data are transmitted to a computer database for determination of the IC_{50} for each of the 10,000 KIs. For more specifics of this product concept, the reader is referred to the design report by Levin et al. (2007).

With all operations having batch times on the order of 8 hr or less, it is possible to design a serial train of operations with a cycle time of 9–10 hr, which can be accomplished in one working day. In one design, two KI emulsion guns are installed in parallel, permitting one gun to undergo loading, washing, and evacuation (3.15 sec/KI) while the other generates droplets (3 sec/KI). The two emulsion guns handle alternate KIs.

One possible schedule of operations for this product concept is shown in Figure 17.7. Operations begin with the preparation of the kinase enzyme library, as shown in Figure 17.7a. The first kinase enzyme, KE_1, is loaded by a JANUS robot arm onto the first PLS™ chip in 1.5 sec, after which the emulsion gun forms 300,000 droplets in 15 sec. Then, the robot arm loads the washout fluid onto the chip in 1.5 sec and the chip is evacuated for 0.15 sec. This sequence is repeated for the remaining 99 enzymes. For the entire KE library, $1,815 \text{ sec} = \sim 30 \text{ min} = 0.5 \text{ hr}$ are required, which can be prepared during the preceding afternoon.

The schedule for the preparation of KI droplets at six different concentrations is shown in Figure 17.7b. Here, the first robot arm loads KI_1 onto the second chip in 1.5 sec, after which the first emulsion gun forms 500 KI_1 droplets at 167 drops/sec in 3 sec. Each of these droplets is split and merged with droplets from the diluent emulsion gun at 833 drops/sec, giving a total of 3,000 droplets of KI_1. Then, the robot arm loads the washout fluid onto the chip in 1.5 sec, after which the KI emulsion gun is evacuated. The second kinase inhibitor, KI_2, is loaded by the second robot arm, after which the second emulsion gun forms KI_2 droplets. While the latter are being formed, the first robot arm loads KI_3 onto the chip, after which the first emulsion gun forms KI_3 droplets. Note that the two robots and emulsion guns alternate between the kinase inhibitors. While one inhibitor is being loaded and emulsified, the other robot arm and emulsion gun are being loaded with washout fluid and evacuated. Meanwhile, the diluent emulsion gun forms droplets of diluent nearly continuously, except for brief periods in which neither KI emulsion gun is in operation. Note that 6.15 sec are required to form droplets for two KIs, and consequently, the time for the preparation of droplets for 10,000 KIs is:

$$\frac{6.15 \text{ sec}}{2 \text{ KI}} \times 10,000 \frac{\text{KI}}{\text{day}} = 30,750 \frac{\text{sec}}{\text{day}} = 8.5 \frac{\text{hr}}{\text{day}}$$

Finally, the Gantt chart in Figure 17.7c shows that the slide-loading robot arm operates with negligible delay following the formation of reaction droplets and loading onto the glass slides. Note that the CYTOMAT 6000 is delayed by the 1-hr incubation time and that the Agilent Micro-array readers begin shortly after the first batch of glass slides leave the incubator. Consequently, 1 MM IC_{50} assays are completed in $\sim$9.5 hr.

Superior Product Concept

Clearly, the RainDance® chip concept, having been developed three years after the Fluidigm® chip concept, is preferable. In short, the generation of droplets by emulsion guns is approximately three orders of magnitude faster than by mixers driven by peristaltic pumps. If the performances were less differentiated, the design team would generate approximate cost estimates and profitability analyses to guide its decision.

Feasibility Stage

In practice, the *feasibility* stage would begin with a technical feasibility analysis to show that the RainDance® PLS™ chip can deliver the desired performance requirements. This would involve the creation of a prototype and testing. In addition, alternate configurations for the PLS™ would be suggested and tested. For example, as proposed in Exercise 17.6, strategies would be considered involving: (1) no barcodes, that is, sequential generation of both kinase enzymes and inhibitors, and (2) barcodes for both the kinase enzymes and inhibitors.

Note that the product concept in Figure 17.7 presumes that the combined reaction droplets can be loaded onto the glass slides without excess pressure drops. This can be checked using estimates of the Darcy friction factor, as requested in Exercise 17.7.

Business Case

For the product concept shown in Figure 17.6, it is proposed to operate a commercial service laboratory approximately 10 hr daily, five days per week, in the United States. At full production, customers will send 1 well-plate/day containing 10,000 kinase inhibitors to be tested against 100 kinase enzymes. Each IC_{50} assay will be priced at \$0.05. A scenario

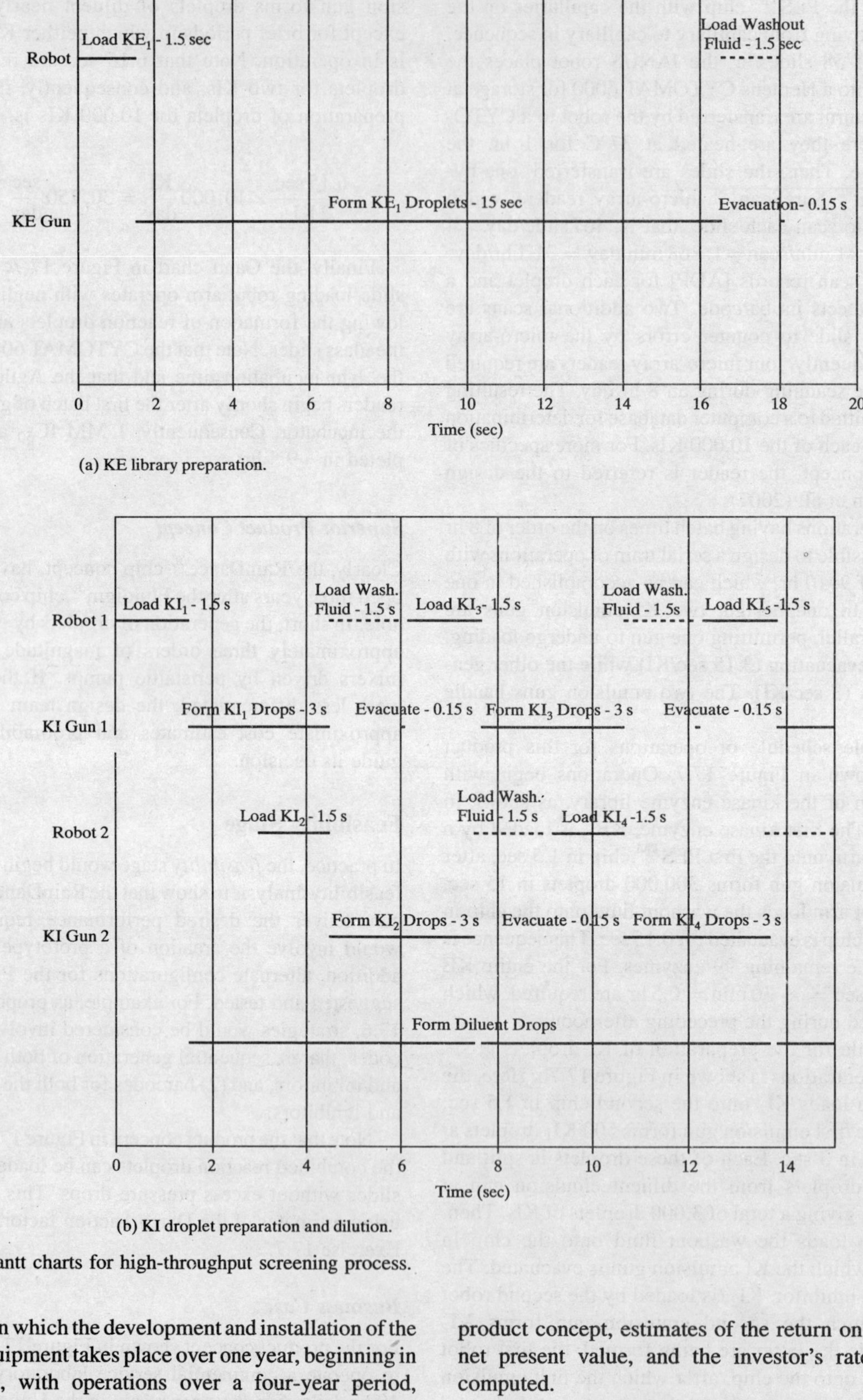

(a) KE library preparation.

(b) KI droplet preparation and dilution.

Figure 17.7 Gantt charts for high-throughput screening process.

is envisioned in which the development and installation of the processing equipment takes place over one year, beginning in January 2009, with operations over a four-year period, beginning in January 2010. During the first year of operation, only 20% of the design production capacity is assumed, increasing to 60% in the second year, and 100% in the third and fourth years. To judge the potential profitability of the product concept, estimates of the return on investment, the net present value, and the investor's rate of return are computed.

Table 17.22 provides a listing of estimates of the installed equipment costs, most of which have been obtained from equipment and software vendors. All items will be purchased and installed in 2009 except for two Agilent Microarray

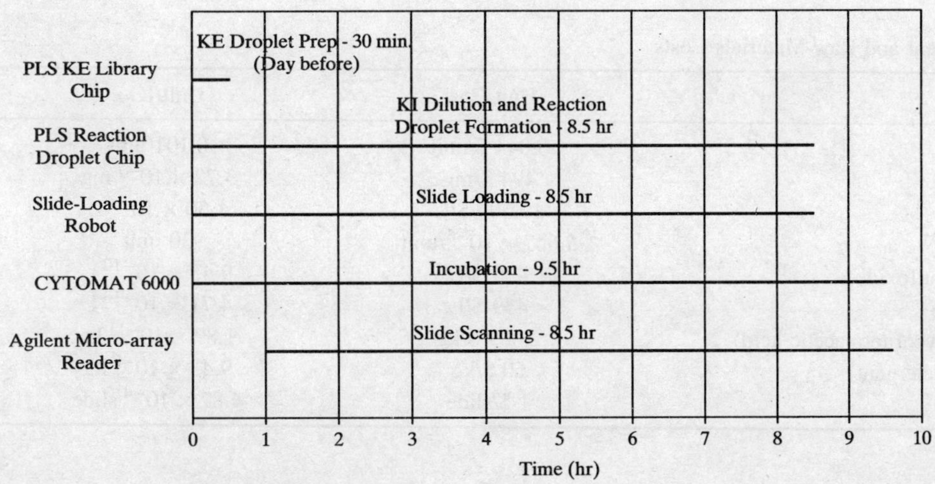

(c) Overall process.

Figure 17.7 (*Continued*)

Readers, which will be added at the beginning of the second production year, 2011.

Variable operating costs include the costs of reagents and raw materials, which are presented in Table 17.23. Here, a unit of kinase enzyme is the amount of enzyme needed to transfer *x* mol of phosphate to enzyme per unit time, where *x* is specific to each enzyme. Given 100 kinase enzymes, a representative unit price was selected. For full production (1 MM IC_{50} assays/day), the total cost of reagents and raw materials is 0.0056 \$/IC_{50} or \$1.358 MM/yr. Other variable costs include the cost of sales and research and development, estimated at 6% and 5% of the total sales; that is, \$750,000/yr and \$625,000/yr at full production. These sum to \$2,732,600/yr of variable costs.

The fixed operating costs include operations, maintenance, lab and office space rental, licensing and security, and property insurance and taxes. Three operators are assumed at \$85,000/oper/yr (40 hr/wk×50 wk/yr×\$42.50/hr) and five professionals at \$125,000/prof/yr. Due to the maintenance of expensive automated equipment, approximately \$175,000/yr is allocated for maintenance. Lab space is

Table 17.22 Equipment and Software Costs

	Installed Cost, \$
RainDance® PLS™ chips	200,000
Computer system	5,000
4 Agilent Micro-array readers	504,000
Labcyte® ECHO™ liquid handler	330,750
Perkin-Elmer JANUS Robotic Arm	150,000
Heraeus CYTOMAT 6000 Incubator	20,000
Refrigerator	18,000
Heraeus CYTOMAT 44	35,000
Lab Management Software	100,000
Security System	2,000
Total	1,364,750

estimated at \$35/ft^2, and consequently, for 2,500 ft^2, the cost is \$87,500/yr during the four operating years. For office space, the figures are \$25/ft^2, 2,000 ft^2, and \$50,000/yr. The costs of licensing and security are estimated at \$100,000/yr and \$50,000/yr. At 2% of the total depreciable capital, insurance and tax costs are approximately \$34,000/yr. These fixed costs sum to approximately \$1.4 MM/yr.

Added to the total cost of equipment and software, \$1,365,000, is the cost of lab preparation; that is, 5% of the total cost of equipment and software, or \$68,200. This gives a total permanent investment of \$1,433,000. Adding 18% for the cost of contingencies and fees gives a total depreciable capital of \$1,690,900. Another \$250,000 is added for startup costs, to give \$1,940,900. Finally, the cost of working capital is estimated as the accounts receivable for 21 days of sales, that is, \$1,050,000. This gives a total capital investment of \$2,990,900.

Three profitability measures are computed, all showing very favorable results. An approximate measure, the return on investment estimated in the third production year, is 181.4%. Two more rigorous measures, based upon the time-value of money, are the investor's rate of return (IRR) at 90.01%, and the net present value (NPV) at \$5.795 MM, based upon a 19% annual interest rate. On this basis, it seems clear that the business case for further developing this lab-on-a-chip product is very strong. In preparing for the gate review, it is helpful to examine the price sensitivity of these profitability measures. This is the purpose of Exercise 17.8.

Intellectual-Property Assessment

Having prepared a strong business case for the lab-on-a-chip product, the patent landscape was examined in July 2007. At that time, an advanced Google patent search using the key words "microfluidic" and "lab-on-a-chip," returned 194 patents, only 12 of which were filed prior to 2000. These

Table 17.23 Reagent and Raw-Materials Costs

	Unit Cost	Unit/IC_{50}	Cost/IC_{50}, \$
Kinase enzymes	0.044 \$/unit	0.101 unit	4.44×10^{-3}
Peptide	494 \$/mg	3.72×10^{-7} mg	1.84×10^{-4}
Transcreener® kits	19,500 \$/L	1.50×10^{-8} L	2.92×10^{-4}
TransFluoSpheres™	5.667×10^{-6}/unit	30 unit	1.70×10^{-4}
DMSO (dimethyl sulfoxide)	439 \$/L	9.43×10^{-10} L	4.14×10^{-7}
HEPES (pH buffer)	439 \$/L	4.71×10^{-10} L	2.07×10^{-7}
EGTA (ethylene glycol tetraacetic acid)	2,089 \$/kg	1.89×10^{-11} kg	3.95×10^{-8}
Brij-35 (nonionic detergent)	60 \$/kg	9.43×10^{-8} kg	5.66×10^{-6}
Glass slides	1 \$/slide	4.87×10^{-4} slide	4.87×10^{-4}

Table 17.24 Results of Google Patent Search for "Microfluidic" and "Lab-on-a-Chip"

Category	Patents
Fluid Movement and Mixing	44
Microfluidic Devices and Applications	47
Fabrication	21
Scanning, Detection, and Measurement	16
Sorting and Separation	10

patents are classified roughly in the categories shown in Table 17.24, with some patents having entries in more than one category.

Patents under fluid movement and mixing introduce pumps, valves, and other methods of moving fluids on chips. Those under microfluidic devices and applications concentrate on the chip architecture (substrate, channels, pumps, etc.) and potential applications. Under fabrication, the emphasis is on the methods of manufacture, such as micro-contact printing and photochemistry. Those under scanning, detection, and measurement emphasize optical probes and electrical instruments to measure activities on chips. Finally, those under sorting and separation introduce devices for separation of particles based upon size and other properties.

While several companies had filed for several patents [Syrrx, Inc. (9), Caliper Life Science (8), Caliper Technologies Corp. (7), Nanostream, Inc. (5), and Fluidigm (4)], no patents appeared to threaten the foundations of the business venture proposed. Clearly, there has been much patent activity since 2000, and consequently, it is advisable to project only four operating years, assuming that new technologies will lead to more competitive products.

It is noteworthy that no patents filed by RainDance Technologies were identified in the Google search. In view of this, before signing a contract with RainDance to produce their PLS™ chips to the specifications of Figure 17.6, it is important to check that RainDance has not violated patents filed by others—given the danger that the proposed commercial laboratory could be shut down due to IP violation claims.

In addition, it is important to confirm that all of the technology inventions in the *innovation map* (Figure 16.33) are protected by patents as well. See Exercise 17.9. Patent protection strengthens the business position of the commercial laboratory by preventing competition from firms that provide comparable products at lower fees.

Development Stage

In the development stage, the key design variables that can be adjusted to optimize the design are adjusted. These include the emulsion gun frequencies, as they influence the throughput. By optimizing the operating variables, it should be possible to significantly increase the throughput and profitability.

Also, in the development stage, more advanced prototypes are developed, with more extensive testing. More quantitative performance measures are examined to determine the extent of redundancy necessary to reduce the errors in high-throughput screening.

17.5 SUMMARY

Case studies for the design of three configured consumer chemical products have been presented. Emphasis has been placed on the project charter, the role of the innovation map, and the *concept* and *feasibility* stages of the Stage-Gate™ Product-Development Process (SGPDP).

REFERENCES

1. Chen, Q.–M., L.B. Heend, and J. Waring, *NANOLUX Screening Technologies*, SEAS Library, Univ. of Pennsylvania, 2005.

2. Cooper, R.G., *Winning at New Products: Accelerating the Process from Idea to Finish, Third Ed.*, Perseus Publ., Cambridge, MA. (2001).

3. Cooper, R.G., *Product Leadership: Creating and Launching Superior New Products*, Perseus Publ., Cambridge, MA. (2002).

4. Cooper, R.G., *Product Leadership: Creating and Launching Superior New Products*, 2nd Edition, Basic Books, Cambridge, MA. (2005).

5. Cooper, R.G., S. Edgett, and E. Kleinschmidt, *Portfolio Management for New Products*, Second Ed., Perseus Publishing (2005).

6. Creveling, C.M., J.L. Slutsky, and D. Antis, Jr., *Design for Six Sigma in Technology and Product Development*, Pearson Education (2003).

7. Home Channel News, Light Bulb Market Information, http://www.findarticles.com/p/articles/mi_m0VCW/is_16_27/ai_78399529 (Sept. 3, 2001).

8. Incandescent light bulb—Wikipedia, http://en.wikipedia.org/wiki/Light_bulb.

9. Levin, J., J. Chung, and J. Irving, *Novel High Throughput Screening of Kinase-Inhibitors Using Raindance Technologies' Personal Laboratory System*, Univ. of Pennsylvania (2007).

10. Pugh, S., *Creating Innovative Products Using Total Design*, Addison Wesley Longman (1996).

11. *Survey of Household Energy Use* (SHEU), Office of Energy Efficiency—Natural Resources Canada (Dec., 2005).

12. Talamini, M., "Guidance for Nocturnal Home Hemodialysis Devices," prepared for the Nocturnal Home Hemodialysis Devices Panel Meeting, 2005. [Report was found in response to a Google search. It contains a discussion of the key issues and a lengthy list of references.]

13. Widagdo, S., "Incandescent Light Bulb: Product Design and Innovation," *Ind. Eng. Chem. Res.*, **45**, 8231–8233 (2006).

EXERCISES

17.1 Perform a market analysis for the light bulb industry. Examine the segments for household, display, and industrial lighting. For household lighting, what fraction is associated with compact fluorescent light bulbs (CFLs)?

17.2 In Europe, electrical power is provided to household users at 220 volts. Estimate the lifetime of a nominal 750-hour conventional incandescent light bulb. How do its light output and power consumption compare with those for the conventional bulb?

17.3 Design a compact fluorescent light bulb (CFL) product for the home lighting market. Assume that the product is being designed in 1995. Note that the patent search and innovation map are prepared as the solutions to Exercises 16.3 and 16.4.

(a) Select the technical requirements.

(b) Prepare the first House of Quality with typical weighting factors.

(c) Suggest superior product concepts and prepare Pugh matrices.

(d) Complete recommendations for concept development stage review.

(e) Prepare a market assessment in the feasibility stage.

(f) Prepare a business case analysis for the feasibility stage review.

(g) Carry out a pricing analysis.

17.4 Repeat Exercise 17.3 for a product design in the mid-2000s, taking into account the changes in the competitive landscape.

17.5 Consider the home hemodialysis product.

(a) Complete Exercise 16.6; that is, design a hemodialysis device that lowers the urea concentration in the blood to 0.3 mg/cm^3, operating 7 nights/wk for 6 hr/night.

(b) Complete a product design including all units—that is, the hemodialysis device, sorbent recovery unit, pumps, holding tanks, sensors, alarms, computer, and display. Estimate their purchase costs.

(c) Set a product price and complete a business plan, including estimates of the return on investment, net present value (at 15% annual interest rate), and investor's rate of return.

17.6 Consider the product concept in Figure 17.6 for the high-throughput screening of kinase inhibitors. Two competitive concepts are proposed:

(a) Use no barcodes; that is, use sequential generation of both the kinase enzymes and inhibitors.

(b) Use barcodes for both the kinase enzymes and inhibitors. Determine the number of colors and concentrations necessary to provide barcodes for 10,000 kinase inhibitors.

Estimate the operation times and compare the purchase costs and net present values of the three concepts over three years.

17.7 Given the 1-inch × 3-inch glass slides upon which 1,000 aqueous drops/sec are loaded into 100 × 10 μm parallel channels, estimate the pressure drop. Is it sufficiently low to permit usage of the glass slides? *Hint:* Use the Darcy friction factor.

17.8 Reestimate the profitability measures in Section 17.4 as the price of IC_{50} assays varies. Use a spreadsheet to increase and decrease the price from \$0.05/IC_{50} assay.

17.9 The lab-on-a-chip product in Figure 17.6 incorporates the inventions in the innovation map of Figure 16.33. Carry out a patent search to determine whether all of the technological inventions in Figure 16.33 have been protected by patents.

Part Four

Detailed Design, Equipment Sizing, Optimization, and Product-Quality Analysis

This part covers the steps of detailed design in (1) determining equipment sizes, (2) carrying out economic calculations, (3) optimizing designs and flowsheets, and (4) achieving product-quality control. In all of the discussions, approximate and rigorous methods are presented. The approximate methods are used throughout the *concept* and *feasibility* stages of the Stage-Gate™ Product-Development Process as discussed in Parts One, Two, and Three. Consequently, this part of the book should be covered long before Parts One, Two, and Three are completed. In most cases, the rigorous methods are used in the *development* stage of the SGPDP.

Chapter 18, on the design of heat exchangers, contains some basic introductory material that may be helpful before or during the study of Chapter 9 on heat and power integration. As mentioned therein, the emphasis in that chapter is on the design of energy-efficient heat exchanger networks (HENs) and not on the details of individual heat exchangers. This is the subject of Chapter 18. Students who have had a strong course in heat transfer are likely to find much of this material a review. They should, however, find the many recommendations for design helpful, including the selection of heating and cooling media, the avoidance of temperature crossovers, and the frequent need for multiple shell-and-tube passes in heat exchangers.

Chapters 19 and 20 provide similar coverage on the details of the design of multistage and packed towers, and of pumps, compressors, and expanders, respectively. While some of this material may be a review for some students, the recommendations for equipment design should be helpful.

Chapter 21 provides detailed design techniques for polymer-compounding devices, that is, extruders. These are particularly important in the manufacture of many *industrial* and *configured consumer* products. Because many product designers have had little, if any, exposure to extruder design, Chapter 21 provides guidelines for characterizing the materials and the required processing steps. Then, heuristics are presented for selecting a feeding strategy, for screw design, and for setting the operating parameters.

Chapters 22 and 23, on cost accounting and capital cost estimation, and profitability analysis, respectively, provide a comprehensive treatment of these subjects. Both approximate and rigorous methods are presented. Equations are provided for estimating the purchase cost of a broad array of process equipment. Also, instruction is provided in the use of the Aspen Icarus Process Evaluator (IPE), together with the process simulators, for the estimation of purchase costs and the total permanent investment. In addition, an EXCEL spreadsheet is presented for calculation of profitability analyses.

Chapter 24 presents a general discussion of optimization methods as applied to the optimization of product and process designs. Subsequently, it shows how to use the process simulators to optimize a process flowsheet by adjusting the continuous variables such as the purge/recycle ratio and the reflux ratio.

Finally, Chapter 25 provides an introduction to six-sigma methodology for product-quality analysis. This chapter shows how to analyze product quality for existing products and processes and how to use six-sigma analysis in the design of new products and processes.

Chapter 18

Design of Heat Exchangers

18.0 OBJECTIVES

Storage tanks, reactors, and separation units in a chemical process are operated at specified temperatures, pressures, and phase conditions. In continuous processes, pressure conditions are established by valves and pumps for liquids, and valves, compressors, and turbines or expanders for gases. Valves are also used to partially or completely convert liquids to gases. Temperature and phase conditions are established mainly by heat exchangers, which are the subject of this chapter.

After studying this chapter, and the multimedia materials on heat exchangers, which can be downloaded from the Wiley Web site associated with this book, the reader should

1. Understand how the temperature and phase conditions of a stream can be changed by using a heat exchanger.
2. Be able to specify a heat exchanger when modeling just one side.
3. Be able to select heat-transfer media for the other side of the exchanger.
4. Understand the importance of heating and cooling curves and how to generate them and use them to avoid temperature crossover violations of the second law of thermodynamics.
5. Be familiar with the major types of heat exchange equipment and how they differ in flow directions of the two streams exchanging heat, and the corresponding effect on the temperature-driving force for heat transfer.
6. Be able to specify a heat exchanger when modeling both sides.
7. Know how to estimate overall heat-transfer coefficients, including the effect of fouling.
8. Understand the limitations of boiling heat transfer.
9. Be able to design a shell-and-tube heat exchanger with the help of a simulator.

18.1 INTRODUCTION

This chapter begins with consideration of the effects of changing temperature, pressure, and phase condition, for a single stream, on stream enthalpy and heat duty. Then heating and cooling media are discussed, and the temperature-driving force for effecting a desired change in stream conditions is considered. Selection of heat exchange equipment is followed by a discussion of methods of determining exchanger sizes from estimates of overall heat-transfer coefficients. The chapter concludes with a comprehensive design problem for a shell-and-tube heat exchanger. In addition, the multimedia modules, which can be downloaded from the Wiley Web site associated with this book show how to model heat exchangers using ASPEN PLUS and HYSYS; see *ASPEN → Heat Exchangers* and *HYSYS→Heat Exchangers*.

Heat Duty

In the early stages of process design, heating and cooling of solids, liquids, and vapors, partial and complete vaporization of liquids, partial and complete condensation of vapors, and sensible and latent heat changes for streams containing solids are treated without regard to (1) the source or sink of thermal energy transferred to or from the stream, (2) the rate at which the energy is transferred, or (3) the type and size of heat exchanger needed. Only the overall enthalpy change (*heat duty*) of the stream for the specified heat exchanger inlet and outlet conditions, and the variation of enthalpy with intermediate conditions in the exchanger, are of interest. The variation is represented most conveniently by *heating and cooling curves*. The heat duty and these curves are most easily obtained, especially for streams that are multicomponent mixtures undergoing phase change, with a steady-state process simulator. The calculations are not simple because effects of temperature, pressure, and composition on enthalpy are taken into account, and the phase condition is established by a phase-equilibrium calculation.

Consider the heat exchanger in Figure 18.1. The continuous, steady-state heat duty is given by

$$Q = m(H_{out} - H_{in}) \tag{18.1}$$

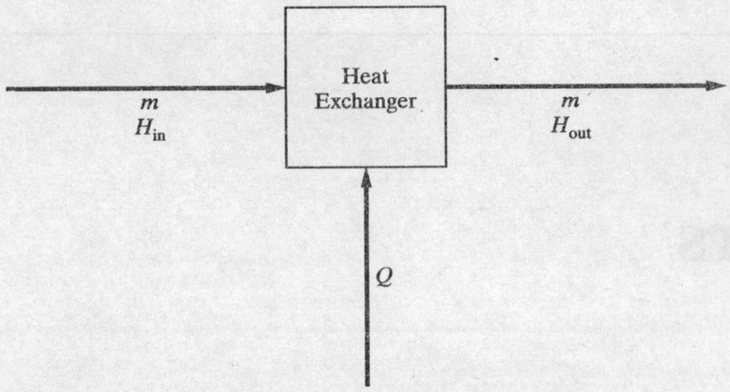

Figure 18.1 One-sided heat exchanger.

where Q is the heat duty (rate of heat transfer), m is the flow rate of the stream (mass or molar), H_{in} is the enthalpy of the stream entering (per unit mass or mole), and H_{out} is the enthalpy of the stream leaving (per unit mass or mole). Simulation programs refer to this type of model as a *one-sided heat exchanger* because only one of the two streams exchanging heat is considered. The calculations are illustrated in the following example.

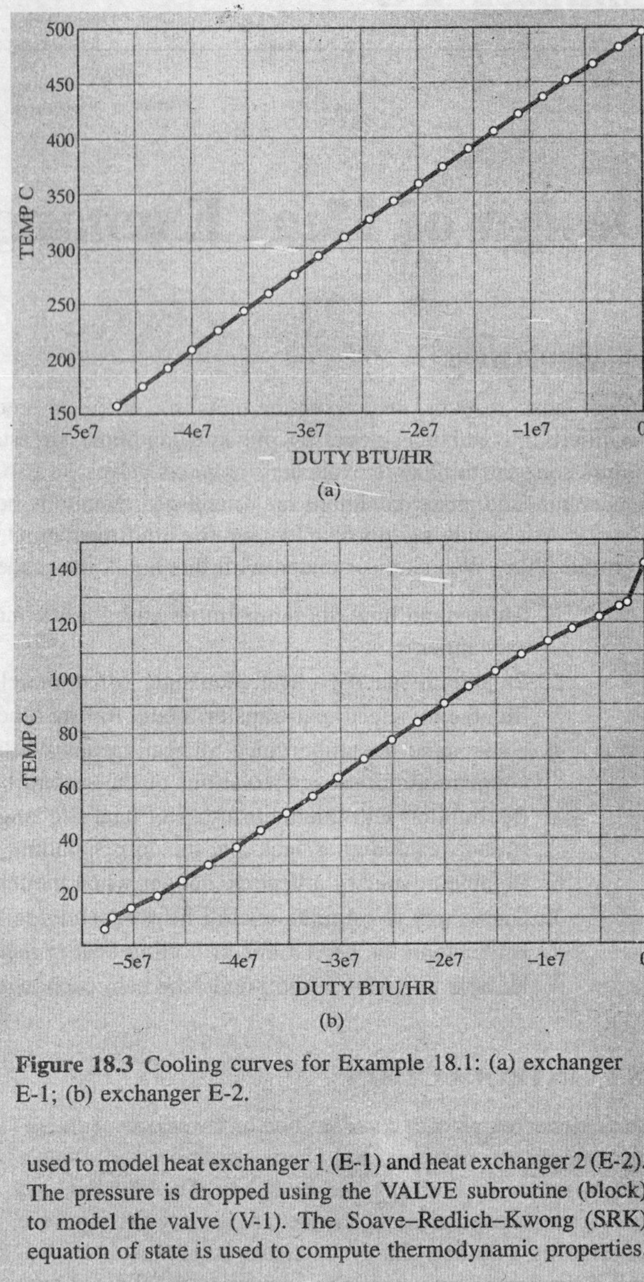

Figure 18.3 Cooling curves for Example 18.1: (a) exchanger E-1; (b) exchanger E-2.

EXAMPLE 18.1

In Figure 4.7, the reactor effluent from the pyrolysis reactor consists of 58,300 lb/hr of HCl, 100,000 lb/hr of vinyl chloride, and 105,500 lb/hr of 1,2-dichloroethane at 500°C and 26 atm. Before entering a distillation section, this stream is cooled and condensed to 6°C at 12 atm. Assume that this is to be done in three steps: (1) cooling in heat exchanger 1 at 26 atm to the dew-point temperature, (2) adiabatic expansion across a valve to 12 atm, and (3) cooling and condensation in heat exchanger 2 at 12 atm to 6°C. Determine the heat duties and cooling curves for each heat exchanger. Note that the pressure drop in each of the two exchangers is neglected.

SOLUTION

This example was solved using ASPEN PLUS. The flowsheet is shown in Figure 18.2, where the HEATER subroutine (block) is used to model heat exchanger 1 (E-1) and heat exchanger 2 (E-2). The pressure is dropped using the VALVE subroutine (block) to model the valve (V-1). The Soave–Redlich–Kwong (SRK) equation of state is used to compute thermodynamic properties.

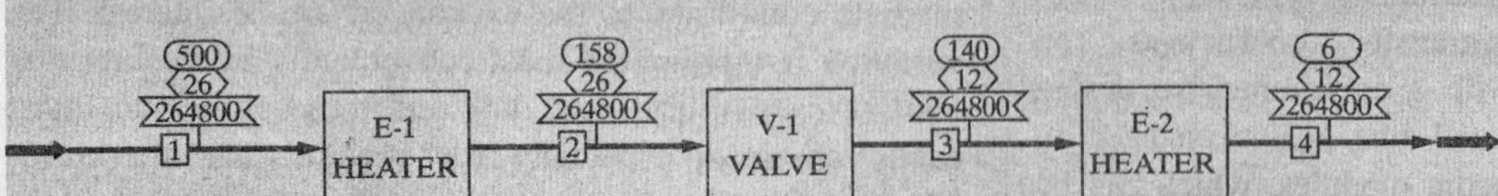

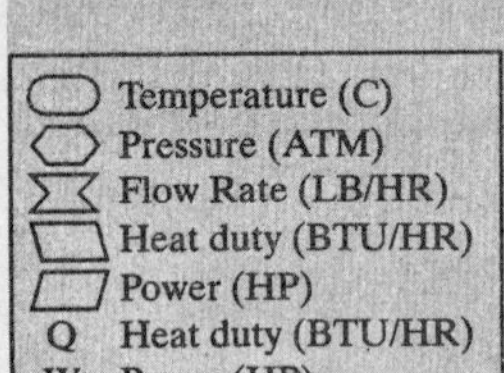

Heat and Material Balance Table

Stream ID		1	2	3	4
Temperature	C	500.0	157.6	140.2	6.00
Pressure	ATM	26.00	26.00	12.00	12.00
Vapor Frac		1.000	1.000	1.000	0.000
Mole Flow	LBMOL/HR	4275.224	4275.224	4275.224	4275.224
Mass Flow	LB/HR	264800.000	264800.000	264800.000	264800.000
Volume Flow	CUFT/HR	165314.984	77820.844	177819.281	3933.389
Enthalpy	MMBTU/HR	-47.128	-93.912	-93.912	-146.911
Mass Flow	LB/HR				
HCL		58300.000	58300.000	58300.000	58300.000
VC		100000.000	100000.000	100000.000	100000.000
DCE		106500.000	106500.000	106500.000	106500.000

Figure 18.2 ASPEN PLUS flowsheet for Example 18.1.

The results of the simulation are included in Figure 18.2, where the heat duties, computed from Eq. (18.1), are shown to be 46,780,000 Btu/hr for E-1 and 53,000,000 Btu/hr for E-2. Stream conditions leaving E-1 are at the dew-point temperature of 157.6°C at 26 atm. The stream leaves valve V-1 as a vapor at 140.2°C and 12 atm. Thus, the adiabatic expansion lowers the temperature by 17.4°F. Stream conditions leaving E-2 are liquid at 6°C and 12 atm. The cooling curve for E-1 is given in Figure 18.3a. Vapor conditions persist throughout E-1; thus, the enthalpy change is all sensible heat. Because the vapor heat capacity varies only slightly with temperature, the graph of the temperature as a function of the enthalpy change is almost linear. The cooling curve for E-2 is given in Figure 18.3b. Entering E-2, the stream is slightly superheated at 140.2°F, with the dew point occurring at 126°C, as seen by the significant change in the slope of the curve in Figure 18.3b. Another significant change in slope occurs at 10°C, which is the bubble point. Between the dew point and the bubble point, both sensible and latent heat changes occur, with the curve deviating somewhat from a straight line.

Heat-Transfer Media

Heat is transferred to or from process streams using other process streams or *heat-transfer media.* In a final process design, every effort is made to exchange heat between process streams and thereby minimize the use of heat-transfer media (usually referred to as *utilities*). Inevitably, however, some use of media, mostly cooling water, steam, and the products of combustion, is necessary. When media must be used, the heat exchangers are called *utility exchangers*.

Heat-transfer media are classified as *coolants* (heat sinks) when heat is transferred to them from process streams, and as *heat sources* when heat is transferred from them to process streams. Process design includes the selection of appropriate heat-transfer media, data for which are listed in Table 18.1, where the media are ordered by temperature range of application.

The most common coolant, by far, is cooling water, which is circulated through a cooling tower. As indicated in Heuristic 27 of Chapter 6, the water typically enters the utility exchanger at 90°F and exits at no higher than 120°F. The cooling tower restores the cooling water temperature to 90°F by contacting the water with air, causing evaporation of a small amount of the water. The enthalpy of evaporation is supplied mainly from the water, causing it to cool. The evaporated water is replaced by treated water. With cooling water, process streams can be cooled and/or condensed to temperatures as low as about 100°F (depending on seasonal temperatures). When the plant is located near an ocean or river, that water is sometimes used for cooling without using a cooling tower. When water is scarce at the plant location, air is used for cooling, but air can only cool process streams economically to about 120°F.

When exchanger inlet temperatures of process streams to be cooled are higher than 250°F, consideration is given to transferring at least some of the heat to treated boiler feedwater to produce steam. The steam is produced at as high a pressure, and corresponding saturation temperature, as possible, subject to a reasonable temperature-driving force for heat transfer in the utility exchanger. For process design purposes, the boiler feedwater enters the utility exchanger as a saturated liquid at the selected pressure, and exits without temperature change as a saturated vapor. The steam is available for use elsewhere in the process. For process stream temperatures above the critical temperature of water, supercritical water is sometimes used as the coolant.

Table 18.1 Heat-Transfer Media

Medium	Typical Temperature Range (°F)	Mode
Coolants:		
Ethylene	−150 to −100	Vaporizing
Propylene	−50 to 10	Vaporizing
Propane	−40 to 20	Vaporizing
Ammonia	−30 to 30	Vaporizing
Tetrafluoroethane	−15 to 60	Vaporizing
Chilled brine	0 to 60	Sensible
Chilled water	45 to 90	Sensible
Cooling water	90 to 120	Sensible
Boiler feedwater	220 to 450	Vaporizing
Heat sources:		
Hot water	100 to 200	Sensible
Steam	220 to 450	Condensing
Heating oils	30 to 600	Sensible
Dowtherm A	450 to 750	Condensing
Molten salts	300 to 1,100	Sensible
Molten metals	100 to 1,400	Sensible
Combustion gases	30 to 2,000	Sensible

When process streams must be cooled below 100°F in utility exchangers, refrigerants are used, which are designated with an R number by the American Society of Heating, Refrigerating and Air-Conditioning Engineers (ASHRAE). When the process involves light hydrocarbons, the refrigerant can be one of the hydrocarbons, for example, propane (R-290). Otherwise, a commercial refrigerant, for example, R-717 (ammonia) or R-134a (tetrafluoroethane), is selected. A widely used refrigerant, R-12 (dichlorodifluoromethane), is being phased out because of the accepted hypothesis that chlorine and bromine, but not fluorine, atoms from halocarbons, when released to the air, deplete ozone in the atmosphere. Feasible refrigerants are included in Table 18.1 for a range of coolant temperatures. When the refrigerant is a pure compound, as it often is, the process design calculation assumes that the refrigerant enters the utility exchanger, at a specified pressure, as a saturated liquid and exits, without temperature change, as a saturated vapor. The refrigerant is circulated through a refrigeration cycle, often consisting of a compressor (to increase the pressure), a condenser (to condense the compressed vapor), a throttle valve (to reduce the pressure), and the utility exchanger (also called the refrigerant boiler), as discussed in Example 9S.2. The refrigerant boiling temperature is chosen to avoid freezing the process stream at the wall of the exchanger, unless it is a crystallizer. When process streams are to be cooled to temperatures between 45 and 90°F, chilled water is often used as the coolant rather than a boiling refrigerant. Chilled aqueous brines can be used to temperatures as low as 0°F. Extensive information on refrigerants is given in the *ASHRAE Handbook*.

The most common heat source for heating and/or vaporizing process streams in a utility exchanger is steam, which is available in most chemical plants from a boiler, at two, three, or more pressure levels. For example, the available levels might be 50, 150, and 450 psig, corresponding to saturation temperatures of 298, 366, and 459°F for a barometric pressure of 14.7 psia. For process design purposes, the steam enters the utility exchanger as a saturated vapor, and exits without pressure change as a saturated liquid (condensate), which is returned to the boiler. Uncondensed steam is prevented from leaving the utility exchanger by a steam trap.

Although condensing steam can be used as a heat source to temperatures as high as about 700°F (critical temperature = 705.4°F), steam pressures become very high at high temperatures (3,206 psia at the critical temperature). It is more common to use other media for temperatures above about 450°F. As listed in Table 18.1, these include the diphenyl (26.5 wt%)–diphenyloxide (73.5 wt%) eutectic (Dowtherm A) for temperatures from 450 to 750°F, and various heating oils, molten salts, and molten metals for higher temperatures. Alternatively, as indicated in Heuristic 25 of Chapter 6, a furnace (fired heater), burning gas, fuel oil, or coal is often used in place of a utility heat exchanger when the chemicals being heated are not subject to decomposition and heating is required above 750°F.

Temperature-Driving Force for Heat Transfer

When streams on both sides of a heat exchanger are considered in process design with a simulation program, a two-sided heat exchanger model is used. The model applies Eq. (18.1) to each side under conditions of equal heat-transfer rates, assuming that the exchanger is well insulated such that heat losses are negligible. Thus, all of the heat released by one side is taken up by the other side. In addition, a transport equation is applied:

$$Q = UA\,\Delta T_m \tag{18.2}$$

where U is the overall heat-transfer coefficient, A is the area for heat transfer, and ΔT_m is the mean temperature-driving force for heat transfer.

The driving force is a critical component of Eq. (18.2). For a given heat exchange task, the rate of heat transfer, Q, is computed from Eq. (18.1). Depending on the geometry and extent of fouling of the heat exchanger, and the conditions of the streams passing through the exchanger, the overall coefficient, U, can be computed from correlations described later in this chapter. The mean driving force, ΔT_m, then determines the heat exchanger area, A. The driving force depends on the entering and exiting stream temperatures, the variation of enthalpy with temperature and pressure of each of the two streams as they pass through the exchanger (as given by the heating and cooling curves), and the stream flow patterns in the exchanger. The latter requires careful consideration.

Examples of a few standard flow patterns are shown in Figure 18.4. The standard and most efficient pattern is countercurrent flow of the two streams. For this case, reference temperature-driving forces are those at the two ends of the exchanger. At one end, ΔT is the difference between the temperatures of the entering hot stream and exiting cold stream. At the other end, ΔT is the difference between the temperatures of the exiting hot stream and the entering cold stream. The smallest of the two differences is called the *closest* or *minimum temperature approach*. It is common to specify the design of a two-sided heat exchanger in terms of inlet conditions for each stream, the pressure drop across the exchanger for each stream, and a minimum approach temperature that reflects economics, as shown in Section 9.6. The simulation program determines to which end of the exchanger the minimum applies, and then calculates the exiting stream temperatures and the heat duty.

The optimal minimum approach temperature is a function mainly of the temperature levels of the two streams, as indicated in Heuristic 25 of Chapter 6, and the lost work analysis in Section 9.2. At temperatures below ambient, it is less than 10°F and may be only 1–2°F at highly cryogenic conditions. At ambient temperature it is about 10°F. At temperatures above ambient, up to 300°F, it is about 20°F. At higher temperatures it may be 50°F. In a furnace, the flue gas temperature may be 250 to 350°F above the inlet process stream temperature. When one stream is boiled, a special

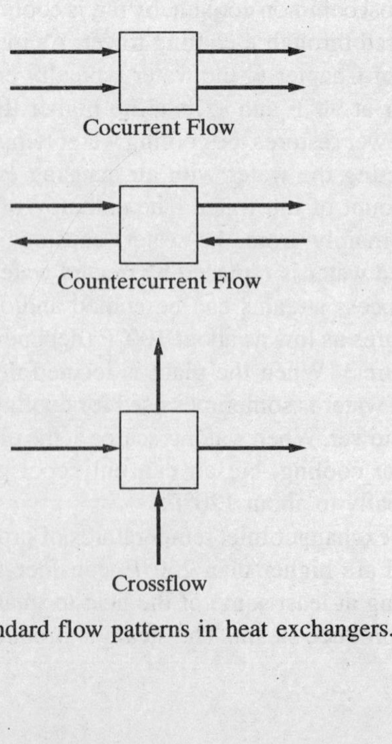

Figure 18.4 Standard flow patterns in heat exchangers.

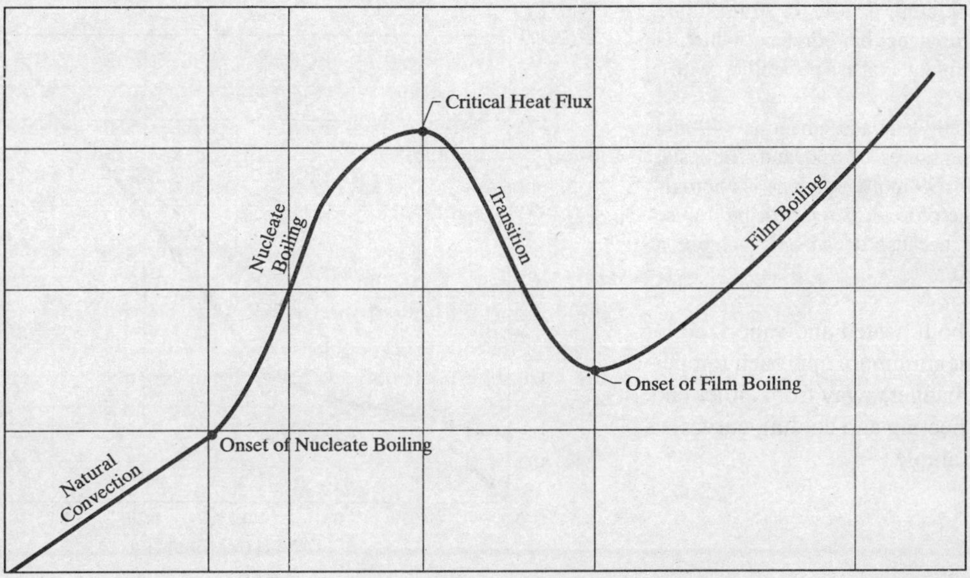

Figure 18.5 Modes in boiling heat transfer.

consideration is necessary. Evaporation can take place in any of four different modes, as shown in Figure 18.5. At temperature-driving forces on the boiling side of less than about 10°F, natural convection is dominant and heat-transfer rates are low. At driving forces between about 20 and 45°F, *nucleate boiling* occurs, with rapid heat-transfer rates because of the turbulence generated by the bubbles. For driving forces above about 100°F, *film boiling* takes place and heat-transfer rates are again low because the mechanism is conduction through the gas film. The region between about 50 and 100°F is in transition. Heat exchangers for vaporization and reboiling avoid film boiling and are designed for the nucleate boiling region to maximize heat-transfer rates. A conservative rule of thumb is to employ Heuristic 28 of Chapter 6, which suggests using a mean overall temperature-driving force of 45°F. This driving force can be achieved by adjusting the pressure at which boiling takes place or the temperature of the heating medium.

EXAMPLE 18.2

Toluene is converted to benzene by hydrodealkylation. Typically, a 75% conversion is used in the reactor, which necessitates the recovery and recycle of unreacted toluene. In addition, a side reaction occurs that produces a small amount of a biphenyl byproduct, which is separated from the toluene. A hydrodealkylation process is being designed that includes a distillation column for separating toluene from biphenyl. The feed to the column is 3.4 lbmol/hr of benzene, 84.6 lbmol/hr of toluene, and 5.1 lbmol/hr of biphenyl at 264°F and 37.1 psia. The distillate is to contain 99.5% of the toluene and 2% of the biphenyl. If the column operates at a bottoms pressure of 38.2 psia, determine the bottoms temperature and select a suitable heat source for the reboiler. Steam is available at pressures of 60, 160, and 445 psig. The barometer reads 14 psia.

SOLUTION

Assume that no benzene is present in the bottoms because it has a much higher vapor pressure than toluene, and a sharp separation between toluene and biphenyl is specified. By material balance, the bottoms contains 0.423 lbmol/hr of toluene and 4.998 lbmol/hr of biphenyl. A bubble-point calculation for this composition at 38.2 psia, using ASPEN PLUS with the SRK equation of state for *K*-values, gives a temperature of 510.5°F. The highest-pressure steam available is at 459 psia, with a saturation temperature of 458°F. Thus, steam cannot be used as the heat source for the reboiler. Instead, Dowtherm A is specified. It enters the exchanger as a saturated vapor and exits as a saturated liquid. To ensure nucleate boiling, the overall temperature-driving force for reboiling the biphenyl bottoms is taken as 45°F. Thus, the condensing temperature for the Dowtherm A is 555.5°F. From data supplied by Dow Chemical Co., the saturation pressure at this temperature is only 28.5 psia, and the heat of vaporization is 116 Btu/lb. If saturated steam at 555.5°F were available, the pressure would be 1,089 psia with a heat of vaporization of 633 Btu/lb. Thus, the use of Dowtherm A at high temperatures results in much lower pressures, but its low heat of vaporization requires a higher circulation rate.

EXAMPLE 18.3

A mixture of 62.5 mol% ethylene and 37.5 mol% ethane is separated by distillation to obtain a vapor distillate of 99 mol% ethylene with 98% recovery of ethylene. When the pressure in the reflux drum is 200 psia, determine the distillate temperature and select a coolant for the condenser. What pressure is required to permit the use of cooling water in the condenser?

SOLUTION

Using the CHEMCAD simulator, the dew-point temperature for 99 mol% ethylene and 1 mol% ethane at 200 psia is −42°F. Assuming a minimum approach temperature of 5°F and a boiling refrigerant, the refrigerant temperature is −47°F. From Table 18.1, a suitable refrigerant is propylene, but ethylene, which is available at 99 mol% purity in the plant, is also a possibility, with a boiling pressure of 185 psia.

The critical temperatures of ethylene and ethane are 49 and 90°F, respectively, at critical pressures of 730 and 708 psia, respectively. The critical point for 99 mol% ethylene is approximately at 50°F and 729 psia. Therefore, it is not possible to use cooling water in the condenser because it can only achieve a condensing temperature of 100°F.

When a process stream is both heated and vaporized, or both cooled and condensed, the minimum approach temperature can occur within the exchanger, away from either end. This can be determined from heating and cooling curves, as illustrated in the following example.

EXAMPLE 18.4

A mixture of 100 lbmol/hr of ethyl chloride and 10 lbmol/hr of ethanol at 200°F and 35 psia is cooled with 90 lbmol/hr of ethanol at 90°F and 100 psia in a countercurrent heat exchanger. Determine the stream outlet conditions and the heat duty for a minimum approach temperature of 10°F. Assume a pressure drop of 5 psi on the hot side and 10 psi on the cold side.

SOLUTION

The calculations are made with the CHEMCAD program using the HTXR model with the UNIFAC method for computing *K*-values. The hot stream is found to enter the exchanger as superheated vapor and exit partially condensed. The cold stream is found to be a liquid throughout the exchanger. The *plot* menu is used to generate heating and cooling curves, which are shown in Figure 18.6a. It is seen that the minimum approach temperature of 10°F is placed by the HTXR model at the 200°F hot stream feed end to give a cold stream outlet temperature of 190°F. At the other end of the exchanger, the hot stream exits at 105.5°F, so the driving force at that end is 105.5 − 90 = 15.5°F. The heat duty is 277,000 Btu/hr. However, Figure 18.6a shows a temperature crossover within the exchanger, which violates the second law of thermodynamics. This crossover is caused by the condensation of the hot stream, which begins at a dew-point temperature of approximately 120°F. This results in a sharp change in slope of the temperature-enthalpy curve for the hot stream. From 120°F to the exit temperature, the hot stream undergoes partial condensation to an exit condition of 93 mol% vapor. The HTXR model has an option that can be used to detect a crossover during execution. This option, which is suggested in Heuristic 29 of Chapter 6, is a zone analysis called "No. of Zones." If, for example, 20 zones are specified, stream temperatures are computed at 19 intermediate points in the exchanger. From these temperatures, the intermediate temperature-driving forces for the heat exchanger are checked to determine if any are negative. If so, the HTXR model terminates, with a warning to the user.

When a crossover occurs, a trial-and-error procedure can be applied to place the minimum approach temperature within the exchanger. This involves increasing the specified minimum approach temperature, which, as mentioned above, is placed at one end or the other. For this example, the result is shown in Figure 18.6b, where it is seen that the minimum approach temperature occurs at the dew-point temperature of the hot stream. This is achieved by specifying a minimum approach temperature of 23°F, which is placed by the HTXR model at the hot stream exit end of the exchanger. Now the hot stream is cooled only to 113°F and the cold stream is heated only to 161°F. The heat duty is reduced to 190,000 Btu/hr. The hot stream exits with 97.8 mol% vapor.

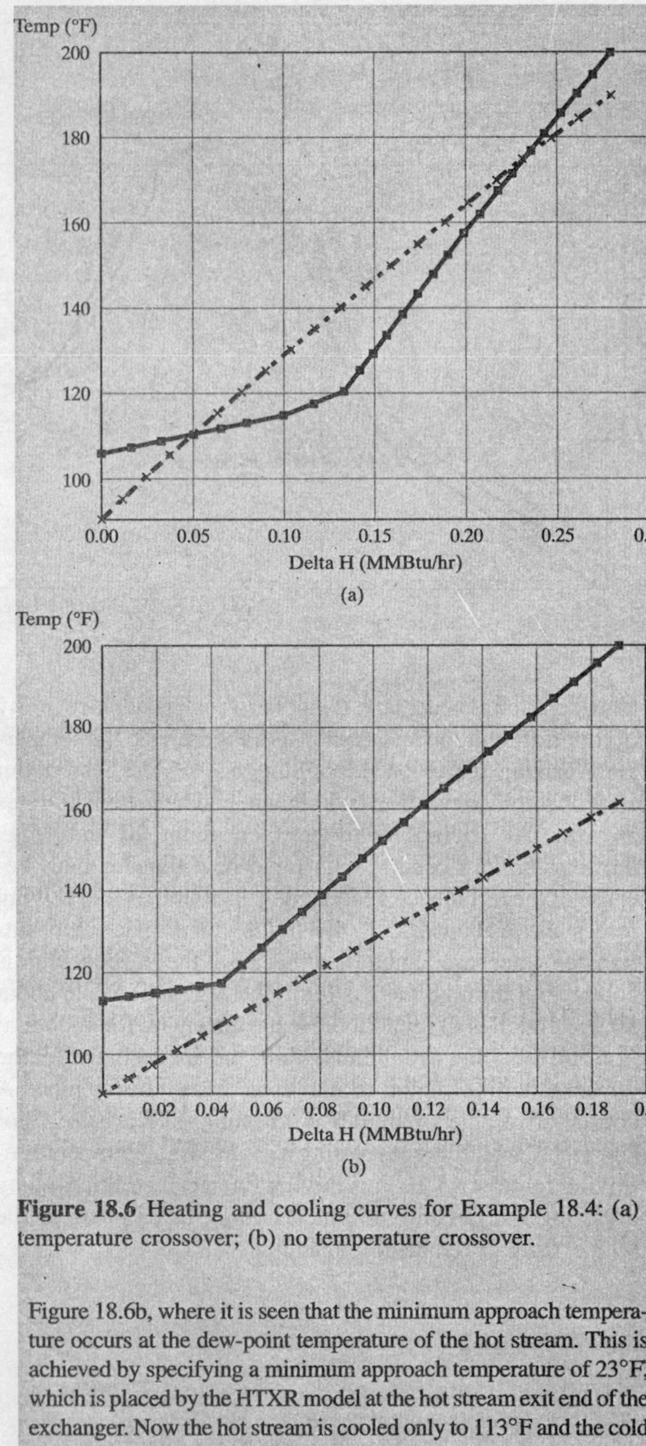

Figure 18.6 Heating and cooling curves for Example 18.4: (a) temperature crossover; (b) no temperature crossover.

To see how ASPEN PLUS and HYSYS are used to model a heat exchanger in which both streams undergo phase changes, see *ASPEN PLUS→ Tutorials→ Heat Transfer→Toluene Manufacture* and *HYSYS→ Tutorials→Heat Transfer→Toluene Manufacture* in the multimedia modules, which can be downloaded from the Wiley Web site associated with this book.

www.wiley.com/go/global/seider

Pressure Drop

The final design of a heat exchanger includes pressure-drop calculations on each side. For process design when using a simulation program, preliminary conservative estimates of pressure drops due to friction are as follows as suggested in Heuristic 31 of Chapter 6. An additional pressure change occurs if the exchanger is placed vertically, due to energy conversions between pressure head and potential energy.

	Pressure Drop	
Liquid streams with no phase change	5 to 9 psi	35 to 62 kPa
Vapor streams with no phase change	3 psi	21 kPa
Condensing streams	1.5 psi	10 kPa
Boiling streams	1.5 psi	10 kPa
Process stream passing through a furnace	20 psi	140 kPa

Methods for determining pressure drop when heat exchanger geometry is known are discussed in Section 18.3.

18.2 EQUIPMENT FOR HEAT EXCHANGE

As listed in Table 18.2, a wide variety of equipment is available for conducting heat exchange. Commercial units range in size from very small, *double-pipe heat exchangers*, with less than 1 ft^2 of heat-transfer surface, to large, air-cooled units called *fin-fan heat exchangers* because they consist of tubes with external peripheral fins and fans to force air past the tubes. Finned area in a single unit is as large as 20,000 ft^2. By far the most common units are *shell-and-tube heat exchangers*, which come in a variety of configurations in sizes from 50 to 12,000 ft^2. For specialized applications, *compact heat exchangers* are challenging shell-and-tube units.

Table 18.2 Heat Exchange Equipment

Double-pipe
Shell-and-tube
Countercurrent flow
Parallel (cocurrent) flow
Crossflow
1-2, 1-4, 1-6, 1-8
2-4, 2-8
3-6
4-8
6-12
Air-cooled (fin-fan)
Compact
Plate-and-frame
Spiral-plate
Spiral-tube
Plate-fin

Double-Pipe Heat Exchangers

A typical double-pipe unit is shown in Figure 18.7a. In its simplest form, it consists of an inner straight pipe of circular cross section, concentric to and supported within an outer straight pipe by means of packing glands. One stream flows through the inner pipe, while the other stream flows countercurrently through the annular passage between the outer wall of the inner pipe and the inner wall of the outer pipe. When the inner pipe is 12-ft-long, $1\frac{1}{4}$-in., schedule 40 pipe, the heat-transfer area from Table 18.3 is 5.22 ft^2 based on the outside wall of the inner pipe. When the inner pipe is 20-ft-long, 3-in., schedule 40 pipe, the heat-transfer area is 18.34 ft^2. When more heat-transfer area is needed, return bends and heads are used with additional pipes to build a hairpin unit, as shown in Figure 18.7b. Hairpin units are available in sizes up to about 200 ft^2 of heat-transfer area, and are competitive with shell-and-tube exchangers in the range of 100 to 200 ft^2. To prevent sagging of the inner pipe with a resulting distortion of the annular cross section, pipe length is limited to 20 ft. Therefore, a 200-ft^2 unit of 3-in.-diameter inner pipes requires 10 hairpin connections. When one stream is at high temperature and/or high pressure, and/or is corrosive, it is passed through the inner pipe. If the other stream is a gas, longitudinal fins can be added to the outside surface of the inner pipe to help balance the inner and outer heat-transfer resistances. If crystallization occurs from a liquid stream flowing through the inner pipe, scrapers can be added inside that pipe to prevent buildup of crystals on the inner wall. Double-pipe exchangers are not recommended for use in boiling or vaporization services.

Shell-and-Tube Heat Exchangers

Heat-transfer area per unit volume is greatly increased by placing a large number of small-diameter tubes inside a shell, that is, a pressure vessel. Shell-and-tube heat exchangers, whose design is standardized by the Tubular Exchanger Manufacturers Association (TEMA) and has changed little in almost 70 years, is shown in one configuration in Figure 18.8a. Data for heat exchanger tubes are given in Table 18.4.

The following heuristic is useful in making preliminary calculations:

Heuristic 54: ***For shell-and-tube heat exchangers, tubes are typically $\frac{3}{4}$-in. O.D., 16 ft long, and on 1-in. triangular spacing. A single-tube-pass shell of 1-ft inside diameter accommodates a tube outside area of approximately 300 ft^2; 2-ft inside diameter, 1,330 ft^2; and 3-ft inside diameter, 3,200 ft^2.***

As a further example of this type of heat exchanger, a standard 37-in. I.D. shell can accommodate 1,074 $\frac{3}{4}$-in. O.D., 16 BWG (Birmingham wire gauge, which determines the tube wall thickness) tubes on a 1-in. triangular pitch (tube center-to-center distance). When the tubes are 20 ft long, the

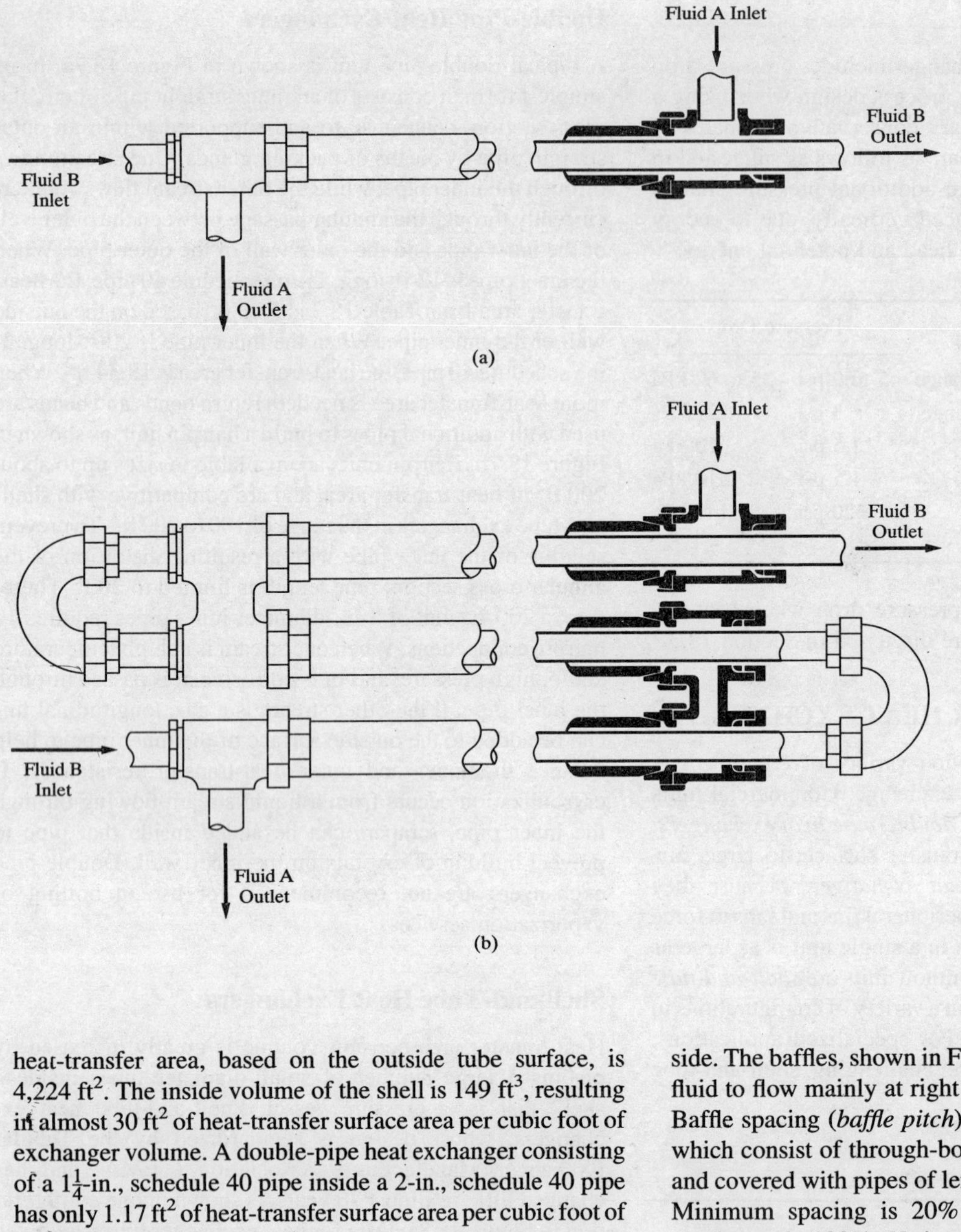

Figure 18.7 Double-pipe heat exchangers: (a) single unit; (b) hairpin unit.

heat-transfer area, based on the outside tube surface, is 4,224 ft^2. The inside volume of the shell is 149 ft^3, resulting in almost 30 ft^2 of heat-transfer surface area per cubic foot of exchanger volume. A double-pipe heat exchanger consisting of a $1\frac{1}{4}$-in., schedule 40 pipe inside a 2-in., schedule 40 pipe has only 1.17 ft^2 of heat-transfer surface area per cubic foot of exchanger volume.

Many configurations of shell-and-tube heat exchangers are available, with Figure 18.8a being the simplest. It is a one-tube-pass, one-shell-pass, fixed (stationary)-head heat exchanger. One stream (tube-side fluid) flows through the tubes; the other (shell-side fluid) flows through the shell, across the outside of the tubes. The exchanger consists of a shell (1), to which are attached an inlet nozzle (2) and an outlet nozzle (3) for the shell-side fluid. At either end of the shell are tube sheets (4), into which tubes are expanded to prevent leakage of streams between the tube side and the shell side. Attached to the tube sheets are channels (5) with inlet and outlet nozzles (6, 7) for the tube-side fluid. Attached to the channels are covers (8, 9). To induce turbulence and increase the velocity of the shell-side fluid, transverse baffles (10), through which the tubes pass, are employed on the shell side. The baffles, shown in Figure 18.8b, cause the shell-side fluid to flow mainly at right angles to the axes of the tubes. Baffle spacing (*baffle pitch*) is fixed by baffle spacers (11), which consist of through-bolts screwed into the tube sheets and covered with pipes of length equal to the baffle spacing. Minimum spacing is 20% of the shell inside diameter; maximum is 100%. Various types of baffles are available, but the segmental is the most common, with a segment height equal to 75% of the shell inside diameter. This is often referred to as a *baffle cut* of 25%. Maximum baffle cut is 45%. It is not practical to fit the baffles snugly to the inside surface of the shell. Instead, there is a shell-to-baffle clearance, which depends on shell inside diameter. The diametric shell-to-baffle clearance (twice the clearance) varies from approximately $\frac{1}{8}$ to $\frac{3}{8}$ in. for shell inside diameters of 12 to 84 in.

Several different tube layout patterns are used, four of which are shown in Figure 18.9. Tube spacing is characterized by the *tube pitch*, which is the closest center-to-center distance between the adjacent tubes; or *tube clearance*, which is the shortest distance between two adjacent tube holes. The most common tube layouts are

Table 18.3 Steel Pipe Data

Nominal Pipe Size (in.)	O.D. (in.)	Schedule No.	I.D. (in.)	Flow Area per Pipe (in.^{2})	Surface per Linear Foot (ft^2)		Weight per Linear Foot (lb steel)
					Outside	Inside	
$\frac{1}{8}$	0.405	40†	0.269	0.058	0.106	0.070	0.25
		80‡	0.215	0.036	0.106	0.056	0.32
$\frac{1}{4}$	0.540	40	0.364	0.104	0.141	0.095	0.43
		80	0.302	0.072	0.141	0.079	0.54
$\frac{3}{8}$	0.675	40	0.493	0.192	0.177	0.129	0.57
		80	0.423	0.141	0.177	0.111	0.74
$\frac{1}{2}$	0.840	40	0.622	0.304	0.220	0.163	0.85
		80	0.546	0.235	0.220	0.143	1.09
$\frac{3}{4}$	1.05	40	0.824	0.534	0.275	0.216	1.13
		80	0.742	0.432	0.275	0.194	1.48
1	1.32	40	1.049	0.864	0.344	0.274	1.68
		80	0.957	0.718	0.344	0.250	2.17
$1\frac{1}{4}$	1.66	40	1.380	1.50	0.435	0.362	2.28
		80	1.278	1.28	0.435	0.335	3.00
$1\frac{1}{2}$	1.90	40	1.610	2.04	0.498	0.422	2.72
		80	1.500	1.76	0.498	0.393	3.64
2	2.38	40	2.067	3.35	0.622	0.542	3.66
		80	1.939	2.95	0.622	0.508	5.03
$2\frac{1}{2}$	2.88	40	2.469	4.79	0.753	0.647	5.80
		80	2.323	4.23	0.753	0.609	7.67
3	3.50	40	3.068	7.38	0.917	0.804	7.58
		80	2.900	6.61	0.917	0.760	10.3
4	4.50	40	4.026	12.7	1.178	1.055	10.8
		80	3.826	11.5	1.178	1.002	15.0
6	6.625	40	6.065	28.9	1.734	1.590	19.0
		80	5.761	26.1	1.734	1.510	28.6
8	8.625	40	7.981	50.0	2.258	2.090	28.6
		80	7.625	45.7	2.258	2.000	43.4
10	10.75	40	10.02	78.8	2.814	2.62	40.5
		60	9.75	74.6	2.814	2.55	54.8
12	12.75	30	12.09	115	3.338	3.17	43.8
16	16.0	30	15.25	183	4.189	4.00	62.6
20	20.0	20	19.25	291	5.236	5.05	78.6
24	24.0	20	23.25	425	6.283	6.09	94.7

†Schedule 40 designates former "standard" pipe.
‡Schedule 80 designates former "extra-strong" pipe.

Layout	Tube O.D. (in.)	Tube Pitch (in.)
Square	$\frac{3}{4}$	1
Square	1	$1\frac{1}{4}$
Triangular	$\frac{3}{4}$	$1\frac{5}{16}$
Triangular	$\frac{3}{4}$	1
Triangular	1	$1\frac{1}{4}$

It is not practical to fit tubes tightly to the baffles. Accordingly, some shell-side fluid leaks through the clearance between the tubes and the baffle holes. This leakage is in addition to the leakage through the clearance between the shell and the baffles. Although tubes can completely fill the shell, there must be a clearance between the outermost tubes and the shell. Typical clearance between the outer-tube limit (OTL) and the shell inside diameter is $\frac{1}{2}$ in. Common tube lengths are 8,12, 16, and 20 ft.

The 1-1 fixed-head shell-and-tube heat exchanger of Figure 18.8a has several limitations:

1. The inside surfaces of the tubes can be cleaned, when necessary, by removing the end covers of the shell and reaming out the tubes, but the outside surfaces of the tubes cannot be cleaned because the tube bundle is fixed inside the shell.

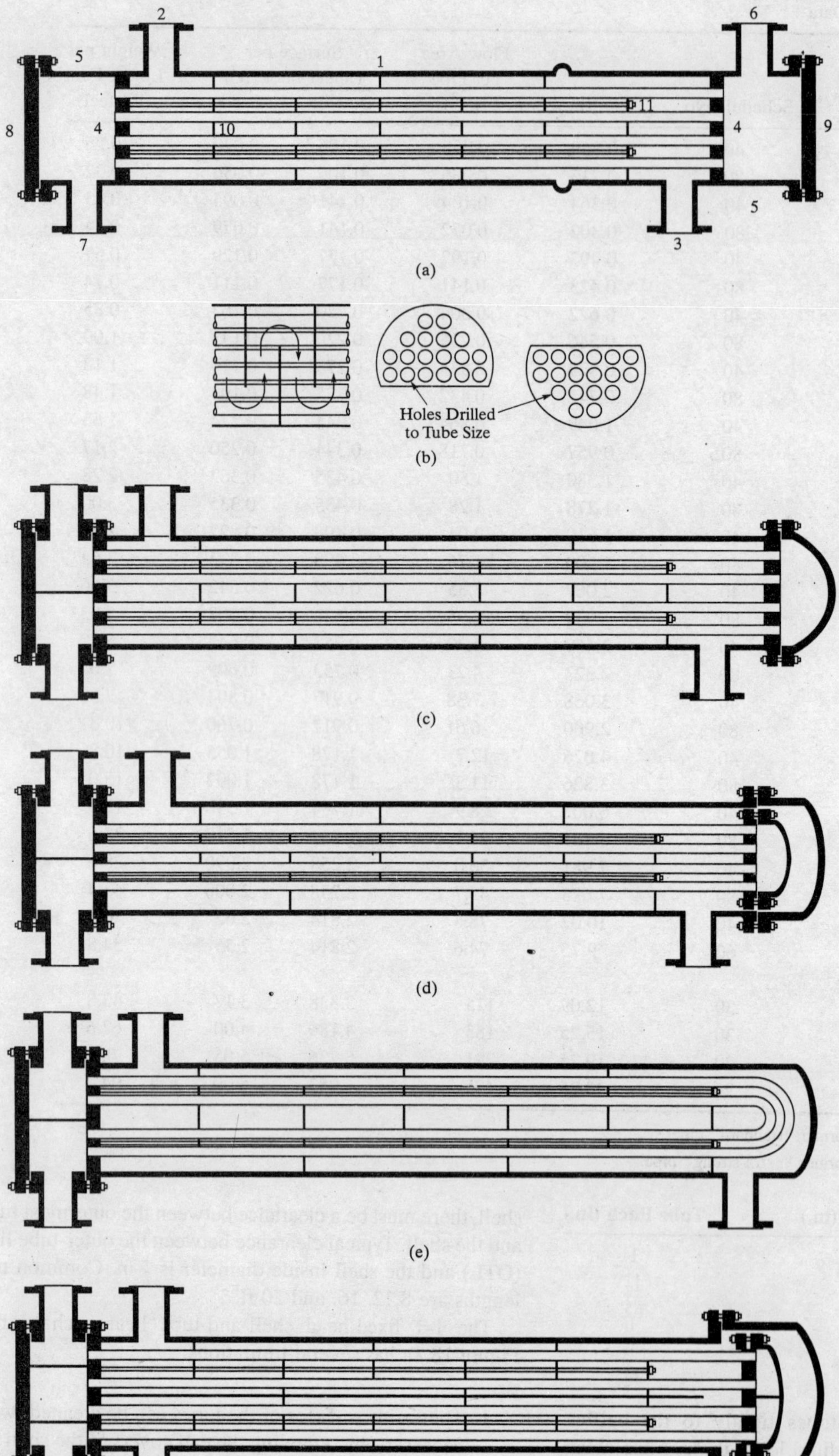

Figure 18.8 Shell-and-tube heat exchangers: (a) 1-1 fixed head; (b) segmental baffles; (c) 1-2 fixed head; (d) 1-2 floating head. (e) 1-2 U-tube; (f) 2-4 floating head.

Table 18.4 Heat Exchanger Tube Data

Tube O.D. (in.)	BWG	Wall Thickness (in.)	I.D. (in.)	Flow Area per Tube (in.^{2})	Surface per Linear Foot (ft^2) Outside	Surface per Linear Foot (ft^2) Inside	Weight per Linear Foot (lb steel)
$\frac{1}{2}$	12	0.109	0.282	0.0625	0.1309	0.0748	0.493
	14	0.083	0.334	0.0876	0.1309	0.0874	0.403
	16	0.065	0.370	0.1076	0.1309	0.0969	0.329
	18	0.049	0.402	0.127	0.1309	0.1052	0.258
	20	0.035	0.430	0.145	0.1309	0.1125	0.190
$\frac{3}{4}$	10	0.134	0.482	0.182	0.1963	0.1263	0.965
	11	0.120	0.510	0.204	0.1963	0.1335	0.884
	12	0.109	0.532	0.223	0.1963	0.1393	0.817
	13	0.095	0.560	0.247	0.1963	0.1466	0.727
	14	0.083	0.584	0.268	0.1963	0.1529	0.647
	15	0.072	0.606	0.289	0.1963	0.1587	0.571
	16	0.065	0.620	0.302	0.1963	0.1623	0.520
	17	0.058	0.634	0.314	0.1963	0.1660	0.469
	18	0.049	0.652	0.334	0.1963	0.1707	0.401
1	8	0.165	0.670	0.335	0.2618	0.1754	1.61
	9	0.148	0.704	0.389	0.2618	0.1843	1.47
	10	0.134	0.732	0.421	0.2618	0.1916	1.36
	11	0.120	0.760	0.455	0.2618	0.1990	1.23
	12	0.109	0.782	0.479	0.2618	0.2048	1.14
	13	0.095	0.810	0.515	0.2618	0.2121	1.00
	14	0.083	0.834	0.546	0.2618	0.2183	0.890
	15	0.072	0.856	0.576	0.2618	0.2241	0.781
	16	0.065	0.870	0.594	0.2618	0.2277	0.710
	17	0.058	0.884	0.613	0.2618	0.2314	0.639
	18	0.049	0.902	0.639	0.2618	0.2361	0.545
$1\frac{1}{4}$	8	0.165	0.920	0.665	0.3271	0.2409	2.09
	9	0.148	0.954	0.714	0.3271	0.2498	1.91
	10	0.134	0.982	0.757	0.3271	0.2572	1.75
	11	0.120	1.01	0.800	0.3271	0.2644	1.58
	12	0.109	1.03	0.836	0.3271	0.2701	1.45
	13	0.095	1.06	0.884	0.3271	0.2775	1.28
	14	0.083	1.08	0.923	0.3271	0.2839	1.13
	15	0.072	1.11	0.960	0.3271	0.2896	0.991
	16	0.065	1.12	0.985	0.3271	0.2932	0.900
	17	0.058	1.13	1.01	0.3271	0.2969	0.808
	18	0.049	1.15	1.04	0.3271	0.3015	0.688
$1\frac{1}{2}$	8	0.165	1.17	1.075	0.3925	0.3063	2.57
	9	0.148	1.20	1.14	0.3925	0.3152	2.34
	10	0.134	1.23	1.19	0.3925	0.3225	2.14
	11	0.120	1.26	1.25	0.3925	0.3299	1.98
	12	0.109	1.28	1.29	0.3925	0.3356	1.77
	13	0.095	1.31	1.35	0.3925	0.3430	1.56
	14	0.083	1.33	1.40	0.3925	0.3492	1.37
	15	0.072	1.36	1.44	0.3925	0.3555	1.20
	16	0.065	1.37	1.47	0.3925	0.3587	1.09
	17	0.058	1.38	1.50	0.3925	0.3623	0.978
	18	0.049	1.40	1.54	0.3925	0.3670	0.831

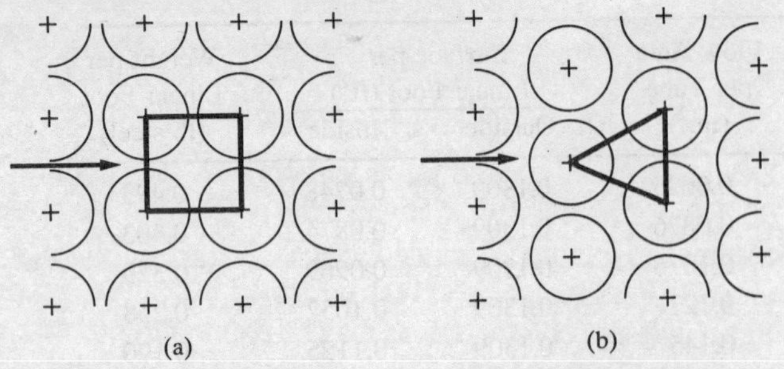

Figure 18.9 Tube layout patterns: (a) square pitch; (b) triangular pitch; (c) square pitch rotated; (d) triangular pitch with cleaning lanes.

2. If large temperature differences exist between the shell-side and tube-side fluids, differential expansion between the shell and tubes may exceed limits for bellows or expansion joints.
3. The velocity of the tube-side fluid may be too low to obtain a reasonable heat-transfer coefficient.

These limitations are avoided by other configurations in Figure 18.8. The floating-head unit of Figure 18.8d eliminates the differential expansion problem. Also, the pull-through design permits removal of the tube bundle from the shell so that the outside surfaces of the tubes can be cleaned. The square-pitch tube layout is preferred for cleaning.

To increase the tube-side fluid velocity, a one-shell-pass, two-tube-pass (1-2) exchanger—shown in Figures 18.8c, 18.8d, and 18.8e, respectively, fixed-head, floating-head, and U-tube units—is used. A disadvantage of the U-tube unit is the inability to clean the insides of the tubes completely.

With the one-tube-pass exchangers of Figures 18.8a and 18.8b, efficient countercurrent flow between the tube-side and shell-side fluids is closely approximated. This is not the case with the 1-2 exchangers of Figures 18.8c, 18.8d, and 18.8e because of the reversal of the tube-side fluid flow direction. The flow is countercurrent in one tube pass and cocurrent (parallel) in the other. As shown later in this section, this limits heat recovery because of the reduction in the mean temperature-driving force for heat transfer. Note that a video of an industrial 1-2 exchanger is provided in the multimedia modules, which can be downloaded from the Wiley Web site associated with this text. See *ASPEN→Heat Exchangers → Introduction with Video* or *HYSYS→Heat Exchangers→Theory.*

The shell-side fluid velocity is increased and the exchanger heat recovery is improved with the two-shell-pass, four-tube-pass (2-4) configuration shown in Figure 18.8f, where a longitudinal baffle creates the two shell passes in a single shell. Alternatively, two exchangers in series, each with a single shell pass and two tube passes, can be employed. Further improvements are achieved with 3-6 and 4-8 exchangers, but at the cost of more complexity in the exchanger design. Customarily, not more than two shell passes are provided in a single shell. Thus, a 3-6 pass exchanger would consist of three shells (exchangers) in series, each with two tube passes. When even higher tube-side velocities are desired, 1-4, 1-6, or 2-8 exchangers can be specified. Heat recovery for these various combinations of shell-and-tube passes is considered in detail later in this section.

The exchangers in Figure 18.8 are suitable for heating, cooling, condensation, and vaporization. However, a special design, the *kettle reboiler*, shown in Figure 18.10, is also in

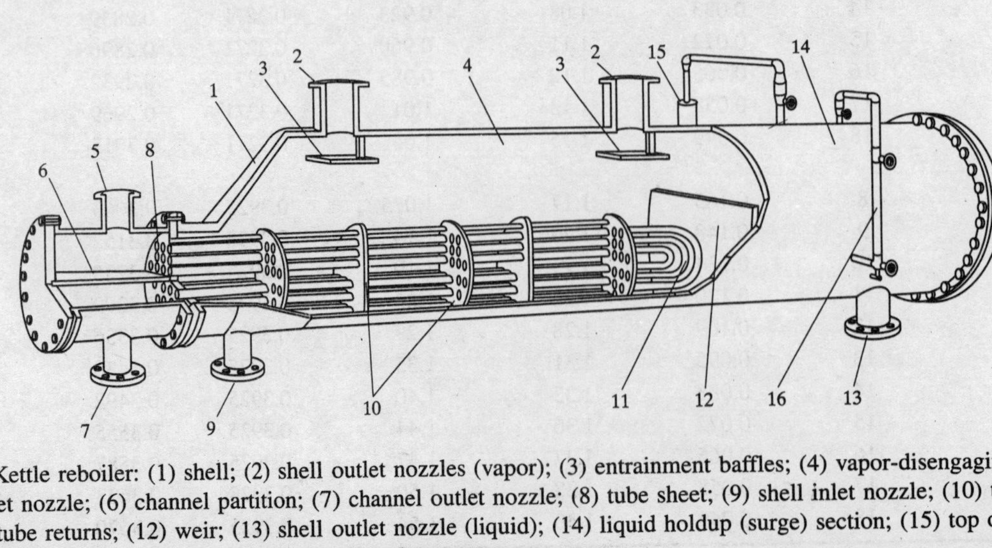

Figure 18.10 Kettle reboiler: (1) shell; (2) shell outlet nozzles (vapor); (3) entrainment baffles; (4) vapor-disengaging space; (5) channel inlet nozzle; (6) channel partition; (7) channel outlet nozzle; (8) tube sheet; (9) shell inlet nozzle; (10) tube support plates; (11) U-tube returns; (12) weir; (13) shell outlet nozzle (liquid); (14) liquid holdup (surge) section; (15) top of level—instrument housing (external displacer); (16) liquid level gauge.

common use for vaporization or boiling. Compared to a 1-2 exchanger, the kettle reboiler has a weir to control the liquid level in the shell and a disengagement region in the space above the liquid level. In a typical service, steam is condensed inside the tubes and liquid is vaporized from the pool of liquid outside the tubes.

When employing a shell-and-tube heat exchanger, a decision must be made as to which fluid passes through the tubes (tube side) and which passes through the shell outside the tubes (shell side). The following heuristic is useful in making this decision:

Heuristic 55: ***The tube side is for corrosive, fouling, scaling, hazardous, high-temperature, high-pressure, and more expensive fluids. The shell side is for more viscous, cleaner, lower-flow rate, evaporating, and condensing fluids.***

Air-Cooled Heat Exchangers

When cooling water is scarce, air is used for cooling and condensing liquid streams in fin-fan heat exchangers. A common configuration is shown in Figure 18.11. See also a video of an industrial fin-fan cooler in the multimedia modules, which can be downloaded from the Wiley Web site associated with this text. The liquid to be cooled and/or condensed passes through the inside of the tubes. Peripheral fins on the outside of the tubes, across which the air flows, increase the outside heat-transfer area and thereby lower the outside thermal resistance so that it approaches the tube inside resistance. The tubes are arranged in banks, with the air forced across the tubes in crossflow by one or more fans. No shell is needed, fouling on the outside of the tubes does not occur, and inside tube cleaning is readily accomplished. For initial design, the following heuristic is useful:

www.wiley.com/go/global/seider

Heuristic 56: ***For an air-cooled exchanger, the tubes are typically 0.75–1.00 in. in outside diameter. The ratio of fin surface area to tube outside bare area is large at 15–20. Fan power requirement is in the range of 2–5 Hp per million Btu/hr transferred, or about 20 Hp per 1,000 ft^2 of tube outside bare surface (fin-free) area. Minimum approach temperature is about 50°F, which is much higher than with water-cooled exchangers. Without the fins, overall heat-transfer coefficients would be about 10 Btu/hr ft^2 °F. With the fins, $U = 80$–100 Btu/hr ft^2 °F, based on the tube outside bare surface area.***

Design is usually based on an entering air temperature of 90°F (hot summer day), for which the process stream can be assumed to exit the air-cooled heat exchanger at 140°F. For air at 70°F, a process stream can be cooled typically to 120°F. Special design considerations may be required for the use of air coolers in the Middle East, where air temperatures may vary from 130°F during the day to 35°F at night. Overhead condensers sometimes combine an air cooler with a cooling-water condenser to reduce the cooling-water load.

Compact Heat Exchangers

Compact heat exchangers have been available for more than a century, but have been slow to replace shell-and-tube exchangers. This has been due to the standards established by TEMA for shell-and-tube exchangers and their applicability to high pressures and temperatures, and to streams containing particulate matter. Nevertheless, for nondemanding services, compact exchangers can offer significant economies and deserve consideration.

When the two fluids exchanging heat must be kept clean, *plate-and-frame heat exchangers* made of stainless steel are commonly used. A typical configuration, shown in Figure 18.12a, consists of a series of pressed corrugated plates on close spacing. Hot and cold fluids flow on opposite sides of a plate. Heat-transfer coefficients are high because of the enhancement of turbulence by the corrugations. Fouling of the surfaces is low, and the heat-transfer surfaces are readily cleaned. Because gasket seals are necessary in the grooves around the periphery of the plates to contain and direct the fluids, operating pressures and temperatures are limited to 300 psig and 400°F. Plate-and-frame units with as much as 16,000 ft^2 of heat-transfer surface area are available.

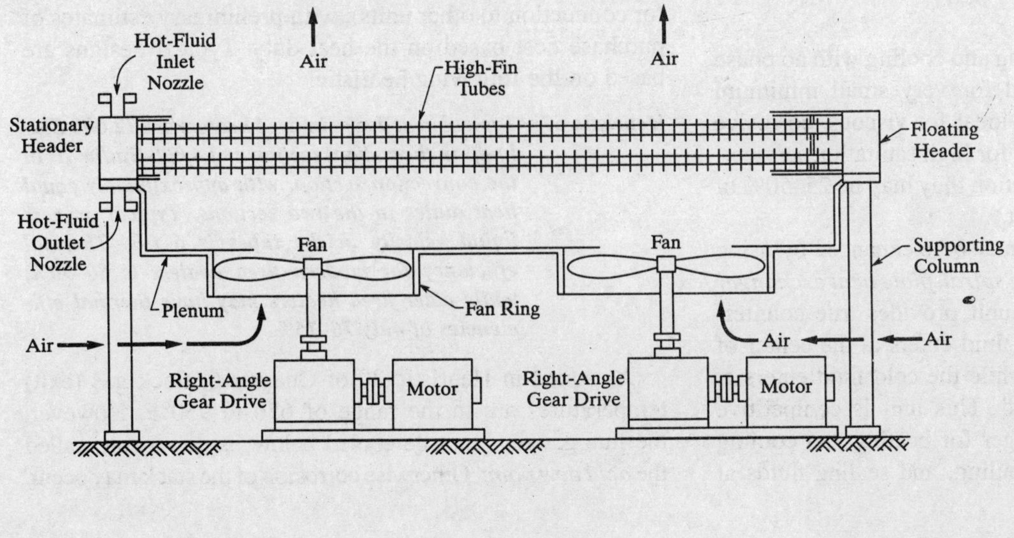

Figure 18.11 Fin-fan heat exchanger.

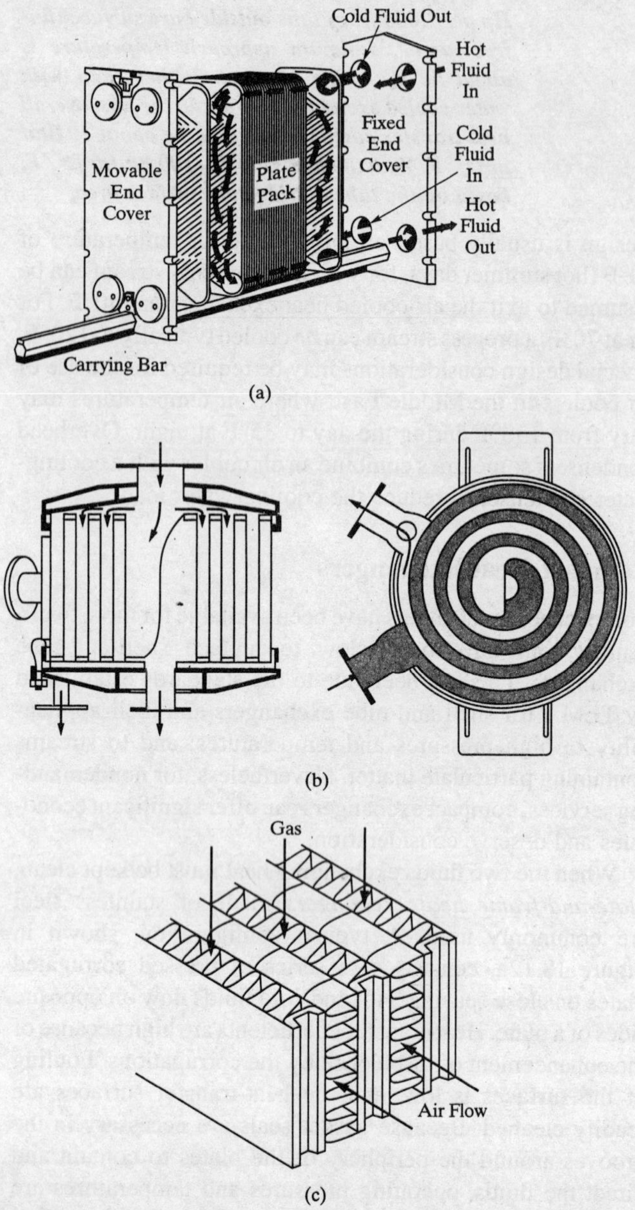

Figure 18.12 Compact heat exchangers: (a) plate-and-frame; (b) spiral-plate; (c) plate-fin.

They are suitable only for heating and cooling with no phase change. They can be designed for very small minimum approach temperatures and are ideal for viscous, corrosive fluids. They are also well suited for high sanitation services, where in stainless steel construction they may be 25–50% of the cost of a shell-and-tube unit.

Heat-transfer coefficients can also be enhanced by using spiral flow passageways as in the *spiral-plate heat exchanger* shown in Figure 18.12b. This unit provides true countercurrent flow. Typically, the hot fluid enters at the center of the spiral and flows outward, while the cold fluid enters at the periphery and flows inward. This unit is competitive with the shell-and-tube exchanger for heating and cooling of highly viscous, corrosive, fouling, and scaling fluids at ambient to moderate pressures. Units with up to 2,000 ft^2 of heat-transfer surface area are available.

For operation at high pressures, a spiral of adjacent tubes can be used. One fluid flows through the tube coil, while the other fluid flows countercurrently in the spiral gap between turns of the coil. The shell side is readily cleaned, but the tube side is not. Sizes of the *spiral-tube heat exchanger* are limited to 500 ft^2 of heat-transfer surface area.

When sensible heat is to be exchanged between two gases, extended heat-transfer surface in the form of fins is desirable on both sides. This is accomplished by *plate-fin heat exchangers*, an example of which is shown in Figure 18.12c. These compact units achieve heat-transfer surface areas of 350 ft^2/ft^3 of unit, which is much higher (up to 4 times) than that for shell-and-tube heat exchangers. The fins consist of corrugated surfaces of 0.2- to 0.6-mm thickness and 3.8- to 11.8-mm height. Fin density is 230–700 fins/m. Plate-fin units can be designed for high pressures, and for countercurrent or crossflow. Two, three, or more streams can exchange heat in a single unit.

Furnaces

Furnaces (also called fired heaters) are often used to heat, vaporize, and/or react process streams at high temperatures and high flow rates. Heat duties of commercial units are in the range of 3 to 100 MW (10,000,000 to 340,000,000 Btu/hr). A number of different designs exist, using either rectangular or cylindrical steel chambers, lined with firebrick. The process fluid flows through tubes that are arranged in a so-called radiant section around the inside wall of the furnace enclosure. In this section, heat transfer to the outer surface of the tubes is predominantly by radiation from combustion gases resulting from burning of the furnace fuel with air. To recover as much energy as possible from the combustion gases, a so-called convection section, where the gases flow over a bank of extended-surface tubes, surmounts the radiant section. In this section, heat transfer from the gases to the tubes is predominantly by forced convection. In some cases, plain tubes are placed in the bottom part of the convection section to shield the extended-surface tubes from excessive radiation. Furnaces are purchased as package units (complete units ready for connection to other units), with preliminary estimates of purchase cost based on the heat duty. Typical designs are based on the following heuristic:

Heuristic 57: ***Typical heat fluxes in fired heaters are 12,000 Btu/hr-ft^2 in the radiant section and 4,000 Btu/hr-ft^2 in the convection section, with approximately equal heat duties in the two sections. Typical process liquid velocity in the tubes is 6 ft/s. Thermal efficiency for modern fired heaters is 80–90%, while older fired heaters may have thermal efficiencies of only 70–75%.***

As stated in Heuristic 30 of Chapter 6, stack gas (exit) temperatures are in the range of 650 to 950°F. However, the flue gas must not be cooled below its dew point, called the *acid dew point*. Otherwise corrosion of the stack may occur.

Temperature-Driving Forces in Shell-and-Tube Heat Exchangers

The rate of heat transfer between two streams flowing through a heat exchanger is governed by Eq. (18.2). Except for a few simple, idealized cases, the mean temperature-driving force, ΔT_m, is a complicated function of the exchanger flow configuration and the thermodynamic and transport properties of the fluids. When a phase change occurs, an additional complication enters into its determination.

The simplest expression for ΔT_m is determined when the following assumptions hold:

1. Stream flows are at steady state.
2. Stream flows are countercurrent or cocurrent to each other.
3. The overall heat-transfer coefficient is constant throughout the exchanger.
4. Each stream undergoes only sensible enthalpy changes (heating or cooling), with constant specific heat.
5. Heat losses are negligible.

For these assumptions, changes in the stream temperatures with distance through the exchanger, or with stream enthalpy, are linear, as shown in the heating and cooling curves of Figure 18.13a for countercurrent flow and Figure 18.13b for cocurrent flow. The ΔT_m is then a function only of the driving forces at the two ends of the exchanger, ΔT_1 and ΔT_2, in the form of a log mean:

$$\Delta T_{\text{LM}} = \frac{\Delta T_1 - \Delta T_2}{\ln(\Delta T_1/\Delta T_2)} \tag{18.3}$$

If one or both of the streams undergo isothermal condensation or boiling, the specific heats are constant, and the above assumptions 1, 3, and 5 apply, then the log-mean temperature difference applies to all heat exchanger configurations, including multiple tube- or shell-pass arrangements.

When shell-and-tube exchangers with multiple-tube passes, or multiple shell-and-tube passes are used, the flow directions of the two fluids are combinations of countercurrent and cocurrent flow. The resulting ΔT_m for given values of ΔT_1 and ΔT_2, based on countercurrent flow, is less than the ΔT_{LM} given by Eq. (18.3). For assumptions 1, 3, 4, and 5 above, the true mean temperature-driving force for a 1-2 exchanger was derived by Nagle (1933) and Underwood (1934). The resulting equation is commonly expressed in terms of the ratio F_T = correction factor = $\Delta T_m/\Delta T_{\text{LM}}$:

$$F_T = \frac{\sqrt{R^2+1}\,\ln[(1-S)/(1-RS)]}{(R-1)\ln\left[\dfrac{[2-S(R+1-\sqrt{R^2+1})]}{[2-S(R+1+\sqrt{R^2+1})]}\right]} \tag{18.4}$$

where

$$R = \frac{T_{\text{hot in}} - T_{\text{hot out}}}{T_{\text{cold out}} - T_{\text{cold in}}} \tag{18.5}$$

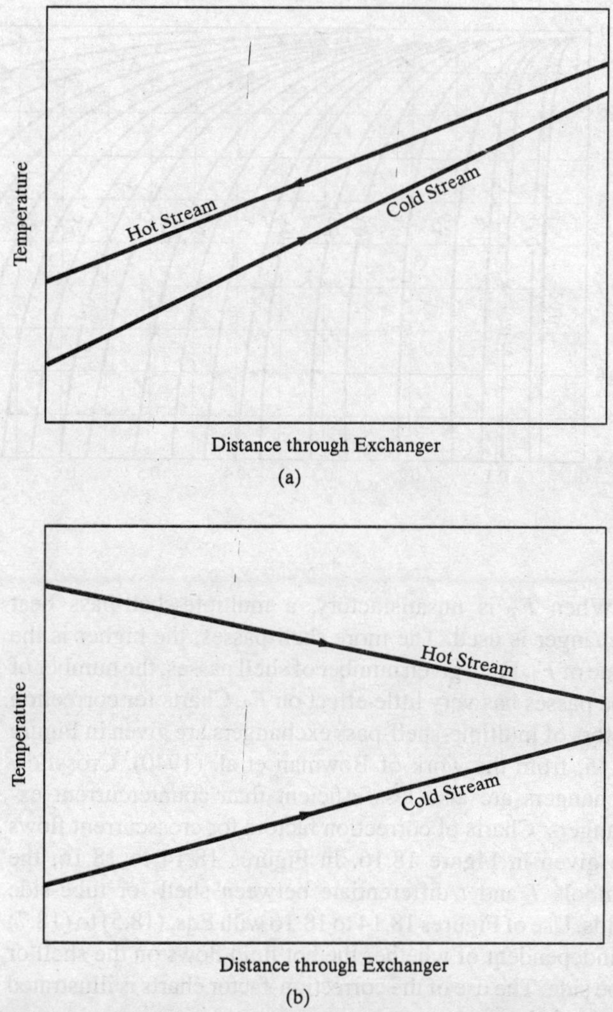

Figure 18.13 Ideal heating and cooling curves: (a) countercurrent flow; (b) cocurrent flow.

$$S = \frac{T_{\text{cold out}} - T_{\text{cold in}}}{T_{\text{hot in}} - T_{\text{cold in}}} \tag{18.6}$$

The rate of heat transfer in multipass exchangers then becomes

$$Q = UAF_T\Delta T_{\text{LM for countercurrent flow}} \tag{18.7}$$

A graph of Eq. (18.4) appears in Figure 18.14, with F_T as a function of S and R as a parameter. Values of F_T are always less than 1. In heat exchanger applications, it is desirable to have a value of F_T of 0.85 or higher. Values of less than 0.75 are generally unacceptable because below this value, the curves in Figure 18.14 turn sharply downward. Thus, small errors in R and S, or small deviations from the above assumptions, can result in values of F_T much lower than anticipated. Values of F_T are not significantly decreased further by using exchangers with additional tube passes, such as 1-4, 1-6, or 1-8. At most, F_T for a 1-8 exchanger differs by less than 2% from that for a 1-2 exchanger.

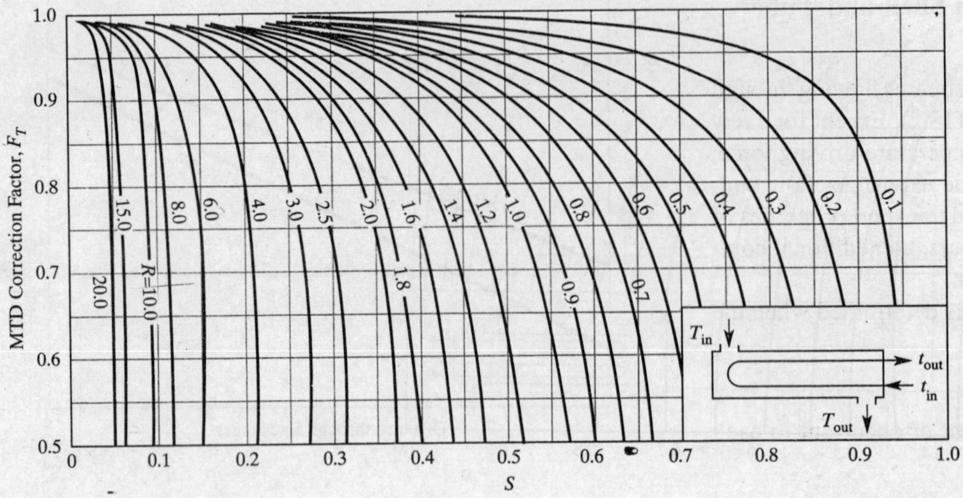

Figure 18.14 Temperature-driving-force correction factor for 1-2 shell-and-tube exchanger. [Adapted from Bowman et al., *Trans. ASME*, **62**, 283 (1940).]

When F_T is unsatisfactory, a multiple-shell-pass heat exchanger is used. The more shell passes, the higher is the value of F_T. For a given number of shell passes, the number of tube passes has very little effect on F_T. Charts for correction factors of multiple-shell-pass exchangers are given in Figure 18.15, from the work of Bowman et al. (1940). Crossflow exchangers are also less efficient than countercurrent exchangers. Charts of correction factors for crosscurrent flows are given in Figure 18.16. In Figures 18.14 to 18.16, the symbols T and t differentiate between shell- or tube-side fluids. Use of Figures 18.14 to 18.16 with Eqs. (18.5) to (18.7) is independent of whether the hot fluid flows on the shell or tube side. The use of the correction-factor charts is illustrated by the following example.

EXAMPLE 18.5

A hot stream is being cooled from 200°F to 140°F by a cold stream that enters the exchanger at 100°F and exits at 190°F. Determine the true mean temperature-driving force for multiple-tube-pass shell-and-tube exchangers.

SOLUTION

For countercurrent flow, the temperature-driving forces at the two ends of the exchanger are $200 - 190 = 10$°F and $140 - 100 = 40$°F. The log-mean driving force, using Eq. (18.3), is

$$\Delta T_{LM} = \frac{40 - 10}{\ln(40/10)} = \frac{30}{1.386} = 21.6°\text{F}$$

For multiple-pass exchangers, using Eqs. (18.5) and (18.6),

$$R = \frac{200 - 140}{190 - 100} = \frac{60}{90} = 0.667 \text{ and } S = \frac{190 - 100}{200 - 100} = \frac{90}{100} = 0.9$$

For a 1-2 exchanger, using Figure 18.14, the value of F_T cannot be read because it is less than 0.5. When it is computed from Eq. (18.4), the argument of the ln term in the denominator of Eq. (18.4) is negative. Thus, a real value of F_T cannot be computed. This indicates that a temperature crossover occurs in a 1-2 exchanger.

For a 2-4 exchanger, using Figure 18.15a, F_T is again less than 0.5. For a 3-6 exchanger, using Figure 18.15b, $F_T = 0.7$, which is risky. For a 4-8 exchanger, using Figure 18.15c, $F_T = 0.85$, which is satisfactory. The mean temperature-driving force is $F_T\,\Delta T_{LM} = 0.85(21.6) = 18.4$°F.

18.3 HEAT-TRANSFER COEFFICIENTS AND PRESSURE DROP

To determine the heat-transfer area of a heat exchanger from Eq. (18.7), an overall heat-transfer coefficient is required. It can be estimated from experience or from the sum of the individual thermal resistances. For double-pipe and shell-and-tube heat exchangers, the area for heat transfer increases across the pipe or tube wall from the inside to the outside surface. Accordingly, the overall heat-transfer coefficient is based on the inner wall, i, the outer wall, o, or, much less frequently, a mean, m. The three coefficients are related by

$$\frac{1}{UA} = \frac{1}{U_oA_o} = \frac{1}{U_iA_i} = \frac{1}{U_mA_m} \tag{18.8}$$

When the outer wall is used, the area is A_o and

$$U_o = \frac{1}{R_{f,o} + \left(\dfrac{1}{h_o}\right) + \left(\dfrac{t_wA_o}{k_wA_m}\right) + \left(\dfrac{A_o}{h_iA_i}\right) + R_{f,i}\left(\dfrac{A_o}{A_i}\right)} \tag{18.9}$$

where $R_{f,o}$ is the outside fouling factor, $R_{f,i}$ is the inside fouling factor, h is the individual heat-transfer coefficient, k_w is the thermal conductivity of the cylindrical wall, t_w is the thickness of the cylindrical wall,

$$A_o = \pi D_oL \quad A_i = \pi D_iL \quad A_m = \frac{\pi L(D_o - D_i)}{\ln(D_o/D_i)}$$

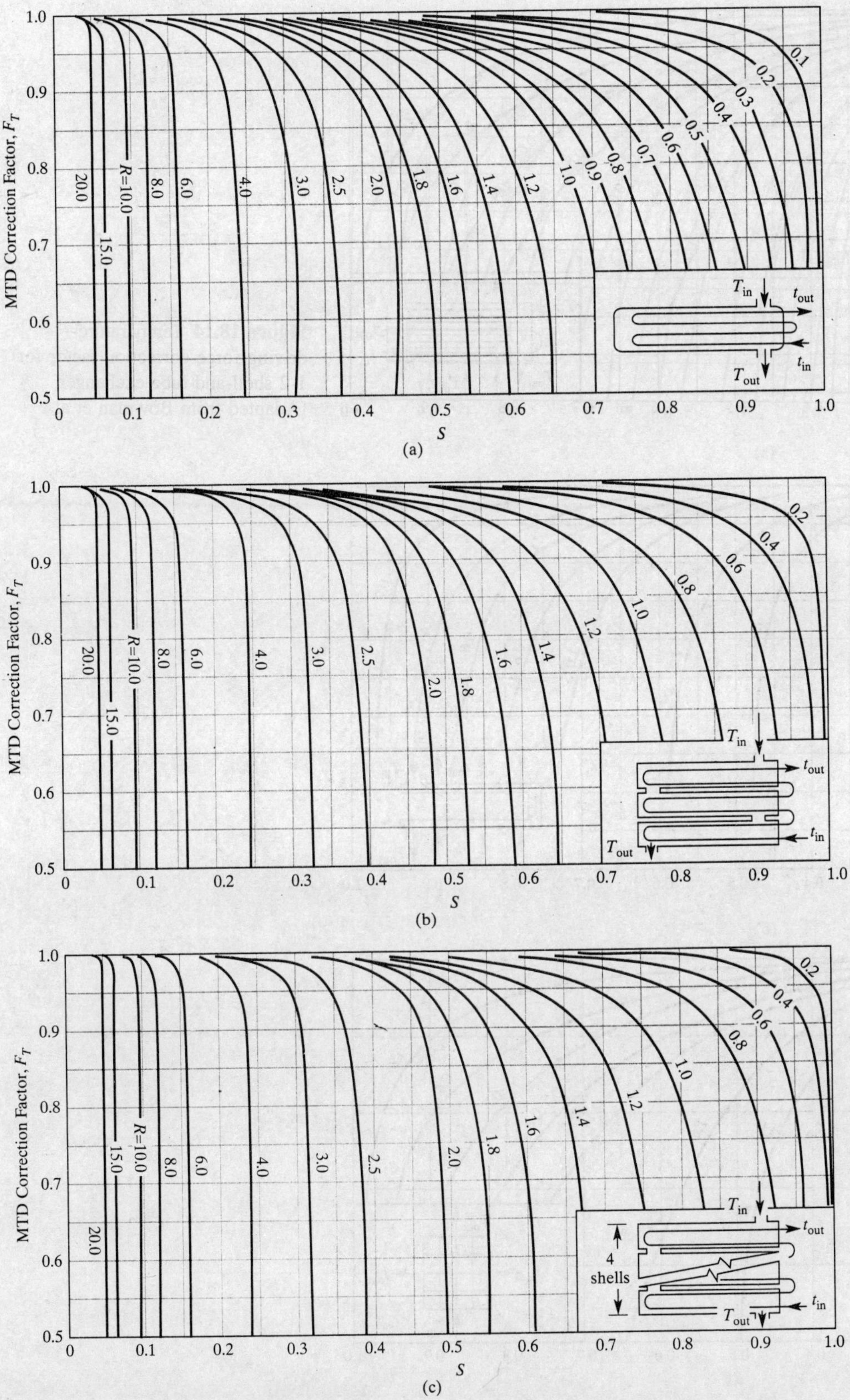

Figure 18.15 Temperature-driving-force correction factor for multiple-shell-pass heat exchangers: (a) 2-4 exchanger; (b) 3-6 exchanger; (c) 4-8 exchanger. [Adapted from Bowman et al., *Trans. ASME*, **62**, 283 (1940).]

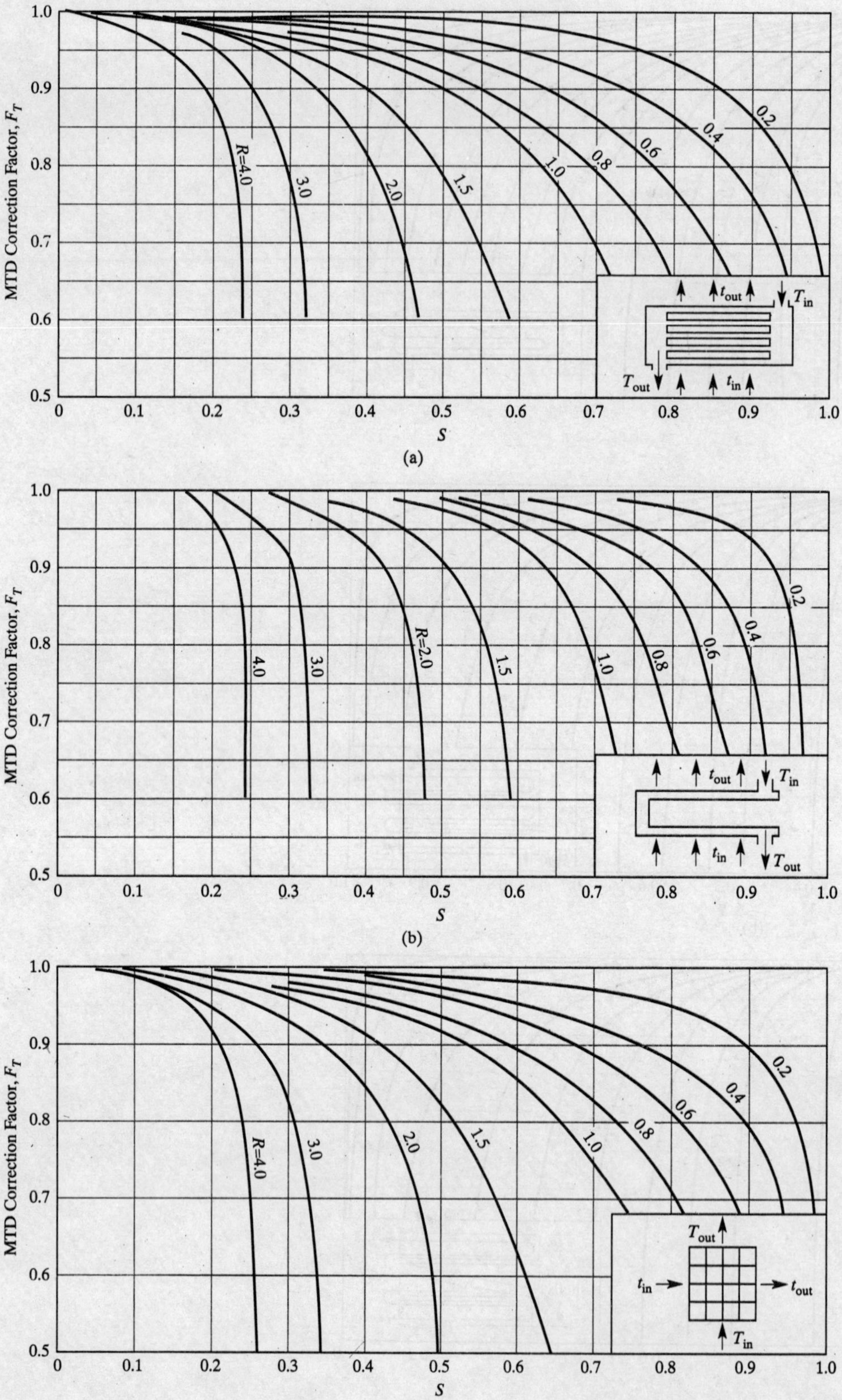

Figure 18.16 Temperature-driving-force correction factor for crossflow heat exchangers: (a) one shell pass, one or more parallel rows of tubes; (b) two shell passes, two rows of tubes (for more than two passes, use $F_T = 1$); (c) one shell pass, one tube pass, both fluids unmixed. [Adapted from Bowman et al., *Trans. ASME*, **62**, 283 (1940).]

D is the tube or pipe diameter, and L is the tube or pipe length. When the inner wall is used, the area is A_i and

$$U_i = \cfrac{1}{R_{f,o}\left(\frac{A_i}{A_o}\right) + \left(\frac{1}{h_o}\right)\left(\frac{A_i}{A_o}\right) + \left(\frac{t_w A_i}{k_w A_m}\right) + \left(\frac{1}{h_i}\right) + R_{f,i}} \tag{18.10}$$

Estimation of Overall Heat-Transfer Coefficients

For preliminary design, the heat-transfer area is computed from Eq. (18.7) using a rough estimate of the overall heat-transfer coefficient, U, based on the service. Because the values are rough, the basis for the area is of no concern. Typical values of U for shell-and-tube heat exchangers are given in Table 18.5. The values include a fouling-factor contribution referred to as *total dirt*, equal to $R_{f,o} + R_{f,i}$. For example, for gasoline on the shell side and water in the tubes, U is given as 60–100 Btu/°F-ft^2-hr with total dirt of 0.003(hr-ft^2-°F)/Btu. The U in Table 18.5 can be referred to as U_{dirty}. Thus, $1/U_{\text{dirty}} = 0.010\text{–}0.017$ (hr- ft^2-°F)/Btu. For a clean exchanger, $1/U_{\text{clean}} = 1/U_{\text{dirty}} - (R_{f,o} + R_{f,i}) = 0.007\text{–}0.014$ (hr-ft^2-°F)/Btu or $U_{\text{clean}} = 70\text{–}140$ Btu/°F-ft^2-hr.

EXAMPLE 18.6

A mixture of 60 mol% propylene and 40 mol% propane at a flow rate of 600 lbmol/hr is distilled at 300 psia to produce a distillate of 99 mol% propylene and a bottoms of 95 mol% propane. The bottoms temperature is 138°F and the heat duty of the reboiler, Q, is 33,700,000 Btu/hr. When waste heat, consisting of saturated steam at 220°F, is used as the heating medium in the reboiler, estimate the area of a shell-and-tube reboiler.

SOLUTION

Assume that the bottoms is on the shell side and steam is inside the tubes. Because the bottoms is almost pure, assume that it vaporizes at 138°F, whereas the steam condenses at 220°F. Therefore, $\Delta T_{\text{LM}} = \Delta T_m = 220 - 138 = 82°\text{F}$. From Table 18.5, under vaporizers, with propane on the shell side and steam condensing on the tube side, $U = 200\text{–}300$ Btu/°F-ft^2-hr. Note that this includes a fouling resistance of 0.0015(hr-ft^2-°F)/Btu. The correction factor, F_T, is 1, regardless of the number of passes or flow directions, because at least one fluid is at a constant temperature in the exchanger. From Eq. (18.7), using 200 Btu/°F-ft^2-hr for U,

$$A = \frac{Q}{UF_T\Delta T_{\text{LM}}} = \frac{33,700,000}{(200)(1.0)(82)} = 2,050\ \text{ft}^2$$

The heat flux in the reboiler is

$$\frac{Q}{A} = \frac{33,700,000}{2,050} = 16,400\ \text{Btu/ft}^2\text{-hr}$$

Note that ΔT_m greatly exceeds the maximum value of 45°F suggested earlier for reboilers. However, that value pertains to just the portion of the ΔT on the boiling side of the exchanger. In this example, when the total driving force of 82°F is divided among the five resistances, it is possible that the maximum value might not be exceeded. Alternative limits on reboilers for the vaporization of organic chemicals are maximum heat fluxes of 12,000 Btu/ft^2-hr for natural circulation and 20,000 Btu/ft^2-hr for forced circulation. Therefore, with a heat flux of 16,400 Btu/ft^2-hr, a kettle reboiler should not be specified. Instead, a pump-through reboiler should be used to pump the bottoms through the shell side of the reboiler. Alternatively, the heating steam temperature could be reduced. However, this would result in vacuum steam, which is very undesirable because air that leaks into the steam can interfere with condensation.

Estimation of Individual Heat-Transfer Coefficients and Frictional Pressure Drop

An enormous amount of experimental work on convective heat transfer and skin-friction pressure drop was reported during the 20th century. This was accompanied by theoretical developments. For laminar flow, heat transfer-coefficients and friction factors for well-defined, simple geometries can be accurately predicted from theory. For turbulent flow, both theoretical equations and empirical correlations of data are available. No attempt is made in the brief space permitted here to present recommended methods for predicting convective heat-transfer coefficients and friction factors for the wide variety of commercial heat exchanger geometries. Instead, the reader is referred to the *Handbook of Heat Exchanger Design*, edited by G.F. Hewitt (1992), which provides a comprehensive coverage by experts in the field. A brief discussion is given here of turbulent-flow convective heat transfer and skin friction without phase change. In general, turbulent flow is preferred in heat exchangers because of the higher heat-transfer coefficients that can be achieved.

Turbulent Flow in Straight, Smooth Ducts, Pipes, and Tubes of Circular Cross Section

In double-pipe and shell-and-tube heat exchangers, fluids flow through straight, smooth pipes and tubes of circular cross section. Many correlations have been published for the prediction of the inside-wall, convective heat-transfer coefficient, h_i when no phase change occurs. For turbulent flow, with Reynolds numbers $N_{\text{Re}} = D_i G/\mu$ greater than 10,000, three empirical correlations have been widely quoted and applied. The first is the Dittus–Boelter equation (Dittus and Boelter, 1930) for liquids and gases in fully developed flow ($D_i/L < 60$), and with Prandtl numbers $N_{\text{Pr}} = C_P\mu/k$ between 0.7 and 100:

$$N_{\text{Nu}} = \frac{h_i D_i}{k_b} = 0.023\left(\frac{D_i G}{\mu_b}\right)^{0.8}\left(\frac{C_{P_b}\mu_b}{k_b}\right)^n \tag{18.11}$$

where D_i is the inside duct, pipe, or tube diameter, G is the fluid mass velocity (flow rate/cross-sectional area for flow),

Table 18.5 Typical Overall Heat-Transfer Coefficients for Shell-and-Tube Heat Exchangers [U = Btu/(°F-ft^2-hr)]

Shell Side	Tube Side	Design U	Includes Total Dirt
Liquid–liquid media			
Aroclor 1248	Jet fuels	100–150	0.0015
Cutback asphalt	Water	10–20	0.01
Demineralized water	Water	300–500	0.001
Ethanol amine (MEA or DEA) 10–25% solutions	Water or DEA, or MEA solutions	140–200	0.003
Fuel oil	Water	15–25	0.007
Fuel oil	Oil	10–15	0.008
Gasoline	Water	60–100	0.003
Heavy oils	Heavy oils	10–40	0.004
Heavy oils	Water	15–50	0.005
Hydrogen-rich reformer stream	Hydrogen-rich reformer stream	90–120	0.002
Kerosene or gas oil	Water	25–50	0.005
Kerosene or gas oil	Oil	20–35	0.005
Kerosene or jet fuels	Trichlorethylene	40–50	0.0015
Jacket water	Water	230–300	0.002
Lube oil (low viscosity)	Water	25–50	0.002
Lube oil (high viscosity)	Water	40–80	0.003
Lube oil	Oil	11–20	0.006
Naphtha	Water	50–70	0.005
Naphtha	Oil	25–35	0.005
Organic solvents	Water	50–150	0.003
Organic solvents	Brine	35–90	0.003
Organic solvents	Organic solvents	20–60	0.002
Tall oil derivatives, vegetable oil, etc.	Water	20–50	0.004
Water	Caustic soda solutions (10–30%)	100–250	0.003
Water	Water	200–250	0.003
Wax distillate	Water	15–25	0.005
Wax distillate	Oil	13–23	0.005
Condensing vapor–liquid media			
Alcohol vapor	Water	100–200	0.002
Asphalt (450°F)	Dowtherm vapor	40–60	0.006
Dowtherm vapor	Tall oil and derivatives	60–80	0.004
Dowtherm vapor	Dowtherm liquid	80–120	0.0015
Gas-plant tar	Steam	40–50	0.0055
High-boiling hydrocarbons V	Water	20–50	0.003
Low-boiling hydrocarbons A	Water	80–200	0.003
Hydrocarbon vapors (partial condenser)	Oil	25–40	0.004
Organic solvents A	Water	100–200	0.003
Organic solvents high NC, A	Water or brine	20–60	0.003
Organic solvents low NC, V	Water or brine	50–120	0.003
Kerosene	Water	30–65	0.004
Kerosene	Oil	20–30	0.005
Naphtha	Water	50–75	0.005
Naphtha	Oil	20–30	0.005
Stabilizer reflux vapors	Water	80–120	0.003
Steam	Feed water	400–1,000	0.0005
Steam	No. 6 fuel oil	15–25	0.0055
Steam	No. 2 fuel oil	60–90	0.0025
Sulfur dioxide	Water	150–200	0.003
Tall oil derivatives, vegetable oils (vapor)	Water	20–50	0.004
Water	Aromatic vapor-stream azeotrope	40–80	0.005

(Continued)

Table 18.5 (Continued)

Shell Side	Tube Side	Design U	Includes Total Dirt
Gas-liquid media			
Air, N_2, etc. (compressed)	Water or brine	40–80	0.005
Air, N_2, etc., A	Water or brine	10–50	0.005
Water or brine	Air, N_2 (compressed)	20–40	0.005
Water or brine	Air, N_2, etc., A	5–20	0.005
Water	Hydrogen containing natural-gas mixtures	80–125	0.003
Vaporizers			
Anhydrous ammonia	Steam condensing	150–300	0.0015
Chlorine	Steam condensing	150–300	0.0015
Chlorine	Light heat-transfer oil	40–60	0.0015
Propane, butane, etc.	Steam condensing	200–300	0.0015
Water	Steam condensing	250–400	0.0015

NC = noncondensable gas present.

V = vacuum.

A = atmospheric pressure.

Dirt (or fouling factor) units are (hr-ft^2-°F/Btu).

To convert British thermal units per hour-square foot-degrees Fahrenheit to joules per square meter-second-Kelvin, multiply by 5.6783; to convert hr-ft^2-°F/Btu to sec-m^2-K/joule, multiply by 0.1761.

Source: From Green, D.W. and R.H. Perry. Perry's Chemical Engineers' Handbook, 8th ed., McGraw-Hill, New York (2008).

k is the fluid thermal conductivity, C_p is the fluid specific heat, μ is the fluid viscosity, subscript b refers to average bulk fluid conditions, and exponent $n = 0.4$ for heating the fluid and 0.3 for cooling.

The Colburn equation (Colburn, 1931) also applies to liquids and gases and is almost identical to the Dittus–Boelter equation, but is usually displayed in a j-factor form in terms of a Stanton number, $N_{St} = h_i/GC_p$. It is considered valid to a Prandtl number of 160:

$$\frac{h_i}{GC_{p_b}}\left(\frac{C_{p_f}\mu_f}{k_f}\right)^{2/3} = 0.023\left(\frac{D_i G}{\mu_f}\right)^{-0.2} \tag{18.12}$$

where the subscript f refers to a film temperature midway between the wall and bulk condition.

The Sieder–Tate equation (Sieder and Tate, 1936) is specifically for liquids, especially viscous liquids where the viscosities at the wall and in the bulk may be considerably different. It is claimed to be valid for very high Prandtl numbers. In Nusselt number form, it is

$$N_{Nu} = \frac{h_i D_i}{k_b} = 0.027\left(\frac{D_i G}{\mu_b}\right)^{0.8}\left(\frac{C_{p_b}\mu_b}{k_b}\right)^{1/3}\left(\frac{\mu_b}{\mu_w}\right)^{0.14} \tag{18.13}$$

where the subscript w refers to the temperature at the wall.

In Section 2.5.1 of Hewitt (1992), prepared by Gnielinski, a more accurate and more widely applicable correlation is given that accounts for tube diameter-to-tube length ratio for $0 < D_i/L < 1$, and is applicable to wide ranges of Reynolds and Prandtl numbers of 2,300 to 1,000,000 and 0.6 to 2,000, respectively. The correlation has a semitheoretical basis in the Prandtl analogy to skin friction in terms of the Darcy friction factor, f_D:

$$N_{Nu} = \frac{h_i D_i}{k_b} = \frac{(f_D/8)(N_{Re} - 1{,}000)N_{Pr}}{1 + 12.7\sqrt{f_D/8}\left(N_{Pr}^{2/3} - 1\right)} \times \left[1 + \left(\frac{D_i}{L}\right)^{2/3}\right] \tag{18.14}$$

where

$$f_D = (1.82\log_{10}N_{Re} - 1.64)^{-2} \tag{18.15}$$

The Darcy friction factor is related to the Fanning friction factor by $f_D = 4f$. The application of Eq. (18.14) is made easy because all properties are evaluated at the bulk fluid conditions. However, for viscous liquids, the right-hand side is multiplied by a correction factor K, where

$$K = \left(\frac{N_{Pr_b}}{N_{Pr_w}}\right)^{0.11} \tag{18.16}$$

For gases being heated, a different correction factor is employed:

$$K = \left(\frac{T_b}{T_w}\right)^{0.45} \tag{18.17}$$

where T is absolute temperature. The Gnielinski equations are preferred for computer calculations in heat exchanger design programs.

The pressure drop for the flow of a liquid or gas under isothermal conditions without phase change through a straight circular tube or pipe of constant cross-sectional area is given by either the Darcy or Fanning equation:

$$-\Delta P = \frac{f_D G^2 L}{2 g_c \rho D_i} = \frac{2 f G^2 L}{g_c \rho D_i} \tag{18.18}$$

where:

$-\Delta P = P_{\text{in}} - P_{\text{out}}$ = pressure drop

L = length of tube or pipe

g_c = conversion factor = 32.17 ft-lbm/lbf-s^2
= 1 in SI units

For turbulent flow at $N_{\text{Re}} > 10,000$ with a smooth wall, f_D is given by Eq. (18.15), or a Fanning friction factor chart can be used to obtain f.

Equation (18.18) accounts for only skin friction at the inside wall of the tube or pipe. Pressure drop also occurs as the fluid enters (by contraction) or leaves (by expansion) the tube or pipe from or to, respectively, the header, and as the fluid reverses flow direction in exchangers with multiple-tube passes. In addition, pressure drop occurs as the fluid enters the exchanger from a nozzle and passes out through a nozzle. For nonisothermal flow in a multitube-pass exchanger, Eq. (18.18) is modified to:

$$-\Delta P_i = K_P \frac{N_P f_D G^2 L}{2 g_c \rho D_i \phi} = K_P \frac{2 N_P f G^2 L}{g_c \rho D_i \phi} \tag{18.19}$$

where:

K_P = correction factor for contraction, expansion, and reversal losses

N_P = number of tube passes

ϕ = correction factor for nonisothermal turbulent flow = $1.02(\mu_b/\mu_w)^{0.14}$, where subscript w refers to the average inside wall temperature

A reasonable value for K_P is 1.2. If the exchanger is vertical and flow is upward, the outlet pressure is further reduced by the height of the heat exchanger times the fluid density. If the flow is downward, the outlet pressure is increased by the same amount.

Turbulent Flow in the Annular Region Between Straight, Smooth Concentric Pipes of Circular Cross Section

In double-pipe heat exchangers, one fluid flows through the annular region between the inner and outer pipes. To predict the heat-transfer coefficient at the outside of the inner pipe, Eqs. (18.14) and (18.15), with the K corrections, can be used by replacing D_i by $D_2 - D_1$, where D_2 is the inside diameter of the outer pipe and D_1 is the outer diameter of the inner pipe. Then the following correction is made:

$$\frac{N_{\text{Nu, annulus}}}{N_{\text{Nu, tube}}} = 0.86\left(\frac{D_1}{D_2}\right)^{-0.16} \tag{18.20}$$

When the flow is through the annulus of a double-pipe heat exchanger, Eqs. (18.15) and (18.19) can be used to estimate the frictional pressure drop, provided that the inside diameter, D_i, of the tube or pipe is replaced by the hydraulic diameter, D_H, which is defined as 4 times the channel cross-sectional area divided by the wetted perimeter. For an annulus, $D_H = D_2 - D_1$.

Turbulent Flow on the Shell Side of Shell-and-Tube Heat Exchangers

Accurate predictions of the shell-side heat-transfer coefficient and pressure drop are difficult because of the complex geometry and resulting flow patterns. A number of correlations are available, none of which is as accurate as those above for the tube side. All are based on crossflow past an ideal tube bank, either staggered (triangular pitch pattern) or inline (square pitch pattern). Corrections are made for flow distortion due to baffles, leakage, and bypassing. From 1950 to 1963, values of h_o, the shell-side convective heat-transfer coefficient, were most usually predicted by the correlations of Donohue (1949) and Kern (1950), which are suitable for hand calculations. Both of these correlations are of the general Nusselt number form

$$N_{\text{Nu}} = \frac{h_o D}{k_b} = C\left(\frac{DG}{\mu_b}\right)^n \left(\frac{C_{P_b}\mu_b}{k_b}\right)^{1/3} \left(\frac{\mu_b}{\mu_w}\right)^{0.14} \tag{18.21}$$

The two correlations differ in how D and G are defined, and in how C and n are determined. For D, Donohue uses the tube outside diameter, whereas Kern uses the hydraulic diameter. For the mass velocity, G, Donohue uses a geometric mean of (1) the mass velocity in the free area of the baffle window, parallel with the tubes, and (2) the mass velocity normal to the tubes for the row closest to the centerline of the exchanger; Kern uses just the latter mass velocity. Donohue uses $n = 0.6$ and $C = 0.2$; Kern uses 0.55 and 0.36, respectively. Kern's correlation is valid for N_{Re} from 2,000 to 1,000,000. Donohue's correlation is considered to be conservative.

For flow of a gas or liquid across the tubes on the shell side of a shell-and-tube heat exchanger, a preliminary estimate of the shell-side pressure drop can be made by the method of Grimison (1937). The pressure drop is given by a modified Fanning equation:

$$-\Delta P_t = K_S \frac{2 N_R f' G_S^2}{g_c \rho \phi} \tag{18.22}$$

where K_S is a correction factor for friction due to inlet and outlet nozzles and to the presence of shell-side baffles that cause reversal of the flow direction, recrossing of tubes, and variation in cross-sectional area for flow. K_S may be taken as approximately 1.10 times (1 + number of baffles). N_R is the number of tube rows across which the shell fluid flows, which equals the total number of tubes at the center plane minus the number of tube rows that pass through the cut portions of the baffles. For 25% cut baffles, N_R may be taken as 50% of the number of tubes at the center plane. For example, if the inside shell diameter is 25 in., the tube outside diameter is 0.75 in., and the tube clearance is 0.25 in. (pitch = 1 in.), the number of tubes in the row at the center plane is 25. With 25% cut baffles, $N_R = 0.5 \times 25 \cong 13$. G_S is the fluid mass velocity based on the flow area at the center plane, which equals the distance between baffles times the tube clearance times the number of tubes at the center plane. f' is the modified friction factor such that:

$$f' = b\left(\frac{D_o G_S}{\mu_b}\right)^{-0.15} \tag{18.23}$$

where b for triangular pitch (staggered tubes) is

$$b = 0.23 + \frac{0.11}{(x_T - 1)^{1.08}} \tag{18.24}$$

and for tubes in line, for example, square pitch, b is

$$b = 0.044 + \frac{0.08x_L}{(x_T - 1)^{0.43+1.13/x_L}} \tag{18.25}$$

Here, x_T is the ratio of the pitch transverse to flow-to-tube outside diameter, and x_L is the ratio of pitch parallel to flow-to-tube outside diameter. For square pitch, $x_T = x_L$.

In 1963, Bell and co-workers at the University of Delaware published a comprehensive method for predicting the shell-side pressure drop and convective heat-transfer coefficient. This method is often referred to as the Bell–Delaware method, and is described in detail in Section 11 of *Perry's Chemical Engineers' Handbook* (1997). Experts in Hewitt (1992) consider it to be the best method available. To use the method, geometric and construction details of the exchanger must be known. The calculations are best carried out with a computer. The method considers the effects of tube layout, bypassing, tube-to-baffle leakage, shell-to-baffle leakage, baffle cut, baffle spacing, and adverse temperature gradients. These effects are applied as corrections to an equation of the form of Eq. (18.21). However, the exponent n on the Reynolds number depends on the Reynolds number.

When making estimates of the heat-transfer coefficients and pressure drop for shell-and-tube heat exchangers, using the methods discussed previously or the more accurate methods in *Perry's Chemical Engineers' Handbook* (Green and Perry, 2008), tubesheet layouts must be known as a function of shell-and-tube diameters. Typical layouts are given in Table 18.6 for shell diameters ranging from 8 to 37 in., and for $\frac{3}{4}$- and 1-in. O.D. tubes.

Heat-Transfer Coefficients for Laminar-Flow, Condensation, Boiling, and Compact Heat Exchangers

Correlations are available for predicting pressure drops and convective heat-transfer coefficients for laminar flow inside and outside of ducts, tubes, and pipes; for pipes with longitudinal and peripheral fins; for condensation and boiling; and for several different geometries used in compact heat exchangers. No attempt is made to discuss or summarize these correlations here. They are presented by Hewitt (1992).

Table 18.6 Tube Sheet Layouts

	One-Pass		Two-Pass		Four-Pass	
Shell I.D., in.	Square Pitch	Triangular Pitch	Square Pitch	Triangular Pitch	Square Pitch	Triangular Pitch
			$\frac{3}{4}$-in. O.D. Tubes on 1-in. Pitch			
8	32	37	26	30	20	24
12	81	92	76	82	68	76
$15\frac{1}{4}$	137	151	124	138	116	122
$21\frac{1}{4}$	277	316	270	302	246	278
25	413	470	394	452	370	422
31	657	745	640	728	600	678
37	934	1,074	914	1,044	886	1,012
			1-in. O.D. Tubes on $1\frac{1}{4}$-in. Pitch			
8	21	21	16	16	14	16
12	48	55	45	52	40	48
$15\frac{1}{4}$	81	91	76	86	68	80
$21\frac{1}{4}$	177	199	166	188	158	170
25	260	294	252	282	238	256
31	406	472	398	454	380	430
37	596	674	574	664	562	632

18.4 DESIGN OF SHELL-AND-TUBE HEAT EXCHANGERS

The design of a shell-and-tube heat exchanger is an iterative process because heat-transfer coefficients and pressure drop depend on many geometric factors, including shell and tube diameters, tube length, tube layout, baffle type and spacing, and the numbers of tube and shell passes, all of which are initially unknown and are determined as part of the design process.

A procedure for an iterative design calculation is as follows, where it is assumed that the inlet conditions (temperature, pressure, composition, flow rate, and phase condition) are known for the two streams entering the heat exchanger and that an exit temperature or some equivalent specification is given for one of the two streams. If a heating or cooling utility is to be used for one of the two streams, it is selected from Table 18.1, together with its entering and leaving temperatures. A decision is made as to which stream will flow on the tube side and which will flow on the shell side. Shell-and-tube side pressure drops are estimated using the values suggested at the end of Section 18.1. With this information, an overall energy balance is used, as discussed in Section 18.1, to calculate the heat duty and the remaining exiting conditions for the two streams. If a heating or cooling utility is to be used, its flow rate is calculated from an energy balance.

A one-tube-pass, one-shell-pass, countercurrent-flow exchanger is assumed. A check is made to make sure that the second law of thermodynamics is not violated and that a reasonable temperature-driving force exists at the two ends of the exchanger, as discussed in Section 18.1. If a phase change occurs on either side of the exchanger, a heating and/or cooling curve is calculated as discussed in Section 18.1, and a check is made to make sure that a temperature crossover is not computed within the exchanger.

A preliminary estimate of the heat exchanger area is made by using Table 18.5 to estimate first the overall heat-transfer coefficient and then using the heating and/or cooling curves or Eq. (18.3) to compute the mean driving force for heat transfer, followed by Eq. (18.7) to estimate the heat exchanger area, with $F_T = 1$. If the area is greater than 8,000 ft^2, multiple exchangers of the same area are used in parallel. For example, if an area of 15,000 ft^2 is estimated, then two exchangers of 7,500 ft^2 each are used.

From the estimated heat-transfer area, preliminary estimates are made of the exchanger geometry. A tube-side velocity in the range of 1 to 10 ft/s is selected, with a typical value being 4 ft/s. The total inside-tube cross-sectional area is then computed from the continuity equation. A tube size is selected, for example, $\frac{3}{4}$-in. O.D., 14 BWG, which, from Table 18.4, has an inside diameter of 0.584 in. and an inside flow area, based on the inside cross-sectional area, of 0.268 in.^{2}. From this, the number of tubes per pass per exchanger is calculated. A tube length is selected, for example, 16 ft, and the number of tube passes per exchanger is calculated. The tube-side velocity and tube length are adjusted, if necessary, to obtain an integer number for the number of tube passes.

If more than one tube pass is necessary, the log-mean temperature-driving force is corrected, using Figures 18.14 through 18.16. This may require using more than one shell pass, as discussed in Section 18.2 and illustrated in Example 18.5. A tube-sheet layout is then selected from Table 18.6, and a baffle design and spacing is selected for the shell side. This completes a preliminary design of the heat exchanger.

A revised design is made next by using the geometry of the preliminary design to estimate an overall heat-transfer coefficient from calculated individual heat-transfer coefficients and estimated fouling factors, as well as pressure drops, using the methods discussed in Section 18.3. Then the entire procedure for sizing the heat exchanger is iterated until changes to the design between iterations are within some tolerance.

The previous procedure is tedious if done by hand calculations. Therefore, it is more convenient to conduct the design with available computer programs. For example, the HEATX subroutine of the ASPEN PLUS simulator computes heat-transfer coefficients, pressure drops, and outlet conditions for a shell-and-tube heat exchanger of known geometry, as illustrated in Example 18.7. It can be used by trial and error with the iterative procedure to design an exchanger.

EXAMPLE 18.7

An existing 2-8 shell-and-tube heat exchanger in a single shell (equivalent to two shells in series with 4 tube passes in each shell) is to be used to transfer heat to a toluene feed stream from a styrene product stream. The toluene enters the exchanger on the tube side at a flow rate of 125,000 lb/hr at 100°F and 90 psia. The styrene enters on the shell side at a flow rate of 150,000 lb/hr at 300°F and 50 psia. The exchanger shell and tubes are carbon steel. The shell has an inside diameter of 39 in. and contains 1,024 $\frac{3}{4}$-in., 14 BWG, 16-ft-long tubes on a 1-in. square pitch. Thirty-eight segmental baffles are used with a baffle cut of 25%. Shell inlet and outlet nozzles are 2.5-in., schedule 40 pipe, and tube-side inlet and outlet nozzles are 4-in., schedule 40 pipe. Fouling factors are estimated to be 0.002(hr-ft^2-°F)/Btu on each side. Determine the exit temperatures of the two streams, the heat duty, and the pressure drops.

SOLUTION

The HEATX subroutine (block) of the ASPEN PLUS simulator is used to make the calculations. It has built-in correlations of the type described above for estimating shell-side and tube-side heat-transfer coefficients and pressure drops. The following results are obtained (both streams are liquid):

Toluene exit temperature = 257.4°F
Styrene exit temperature = 175.9°F
Tube-side tube pressure drop = 3.59 psi
Tube-side nozzle pressure drop = 0.56 psi
Toluene exit pressure = 85.85 psia
Shell-side baffled pressure drop = 4.57 psia
Shell-side nozzle pressure drop = 4.92 psia
Styrene exit pressure = 40.52 psia
Heat-transfer area (tube outside) = 3,217 ft^2

Heat duty = 8,775,000 Btu/hr

Estimated overall heat-transfer coefficient, U_o, clean = 101.6 Btu/(hr-ft^2-°R)

Estimated overall heat-transfer coefficient, U_o, dirty = 69.4 Btu/(hr-ft^2-°R)

Log-mean temperature difference based on countercurrent flow = 57.6°F

Correction factor for 2-8 exchanger, $F_T = 0.682$

Maximum velocity in the tubes = 2.90 ft/s

Maximum Reynolds number in the tubes = 34,000

Maximum crossflow velocity in the shell = 2.36 ft/s

Maximum crossflow Reynolds number in the shell = 32,400

Flow regime on tube and shell sides = turbulent

Note that the file EXAM18-7.bkp in the Program and Simulation Files folder, which can be downloaded from the Wiley Web site associated with this book, can be used to reproduce these results.

Of even greater utility are the B-JAC programs of Aspen Technology, Inc., which are a suite of three programs: (1) HETRAN for the detailed thermal design, rating, and simulation of shell-and-tube heat exchangers, including sensible heating and cooling, condensation, and vaporization; (2) AEROTRAN for the detailed design, rating, and simulation of air-cooled heat exchangers; and (3) TEAMS for the mechanical design of shell-and-tube heat exchangers, using the pressure vessel code. Of particular importance is HETRAN, which can determine the optimal geometry for a shell-and-tube heat exchanger. This program evaluates all possible baffle and shell- and tube-pass arrangements, and seeks the exchanger with the smallest shell diameter, shortest tube length, minimum reasonable baffle spacing, and maximum reasonable number of tube passes, subject to allowable shell- and tube-side pressure drops. The result is a complete TEMA (Tubular Exchanger Manufacturers Association) specification sheet.

EXAMPLE 18.8

Design a new shell-and-tube heat exchanger for the conditions of Example 18.7, but with maximum shell-side and tube-side pressure drops of 10 psi each.

SOLUTION

In this case, it is convenient to use a heat exchanger design program such as HETRAN in B-JAC. For this example, the inlet conditions for the toluene and styrene streams are taken from Example 18.7. That example used physical properties of ASPEN PLUS. In this example, the physical property correlations provided with the B-JAC programs were used. The results showed less than a 5% difference. The current versions of ASPEN PLUS and B-JAC now show no difference. To provide the best comparison with Example 18.7, the same two exit temperatures computed in Example 18.7 (165.2°F for styrene and 268.7°F for toluene) were specified. To maintain an energy balance, the toluene flow rate was increased by 4.5%. The computed heat duty was 9,970,000 Btu/hr compared to 9,472,000 Btu/hr in Example 18.7. HETRAN considered 17 designs, with up to three shell passes in series and total tube passes ranging from two to eight. Maximum tube length was limited to 20 ft. Most designs resulted in pressure drops that exceeded the 10-psi maximum. The recommended design was a 3-12 exchanger with three exchangers in series, each with one shell pass and 4 tube passes. Tubes of 0.75-inch O.D., 0.065-inch thickness, 16-ft long, and on 0.9375-inch triangular spacing were selected. Other results are as follows in the same order as Example 18.7:

Toluene exit temperature = 268.7°F

Styrene exit temperature = 165.2°F

Tube-side tube pressure drop = 5.37 psi

Tube-side nozzle pressure drop = 1.02 psi

Toluene exit pressure = 83.61 psia

Shell-side baffled pressure drop = 8.02 psi

Shell-side nozzle pressure drop = 1.16 psi

Styrene exit pressure = 40.82 psia

Heat-transfer area (tube outside) = 3,663.2 ft^2 or 1,221.1 ft^2 in each of the three shells

Heat duty = 9,970,000 Btu/hr

Estimated heat-transfer coefficient on the tube side = 304 Btu/hr-ft^2-°F

Estimated heat-transfer coefficient on the shell side = 344 Btu/hr-ft^2-°F

Estimated overall heat-transfer coefficient, clean = 140 Btu/hr-ft^2-°F

Estimated overall heat-transfer coefficient, fouled = 86.6 Btu/hr-ft^2-°F

Log-mean temperature difference based on countercurrent flow = 46.4°F

Correction factor for 3-12 exchanger, $F_T = 0.75$

Velocity in the tubes = 3.49 ft/s

Nominal Reynolds number in the tubes = 44,000

Velocity in the shell = 1.67 ft/s

Nominal Reynolds number in the shell = 29,000

Flow regime on tube and shell sides = turbulent

Additional results were 20 baffles in each shell on an 8.5-inch spacing and with a 25% baffle cut, 392 tubes in each shell for a total of 1,176 tubes, and a shell inside diameter of 21.25 inches. The setting plan and tube layout for each of the three shells in series is shown in Figure 18.17, while the heat exchanger specification sheet is shown in Figure 18.18. In Figure 18.17, the dimensions are in inches and the line marked "O" is a reference line for dimensions to the left and right of it. Note that the file EXAM18-8.bjt in the Program and Simulation Files folder, which can be downloaded from the Wiley Web site associated with this book, can be used to reproduce these results.

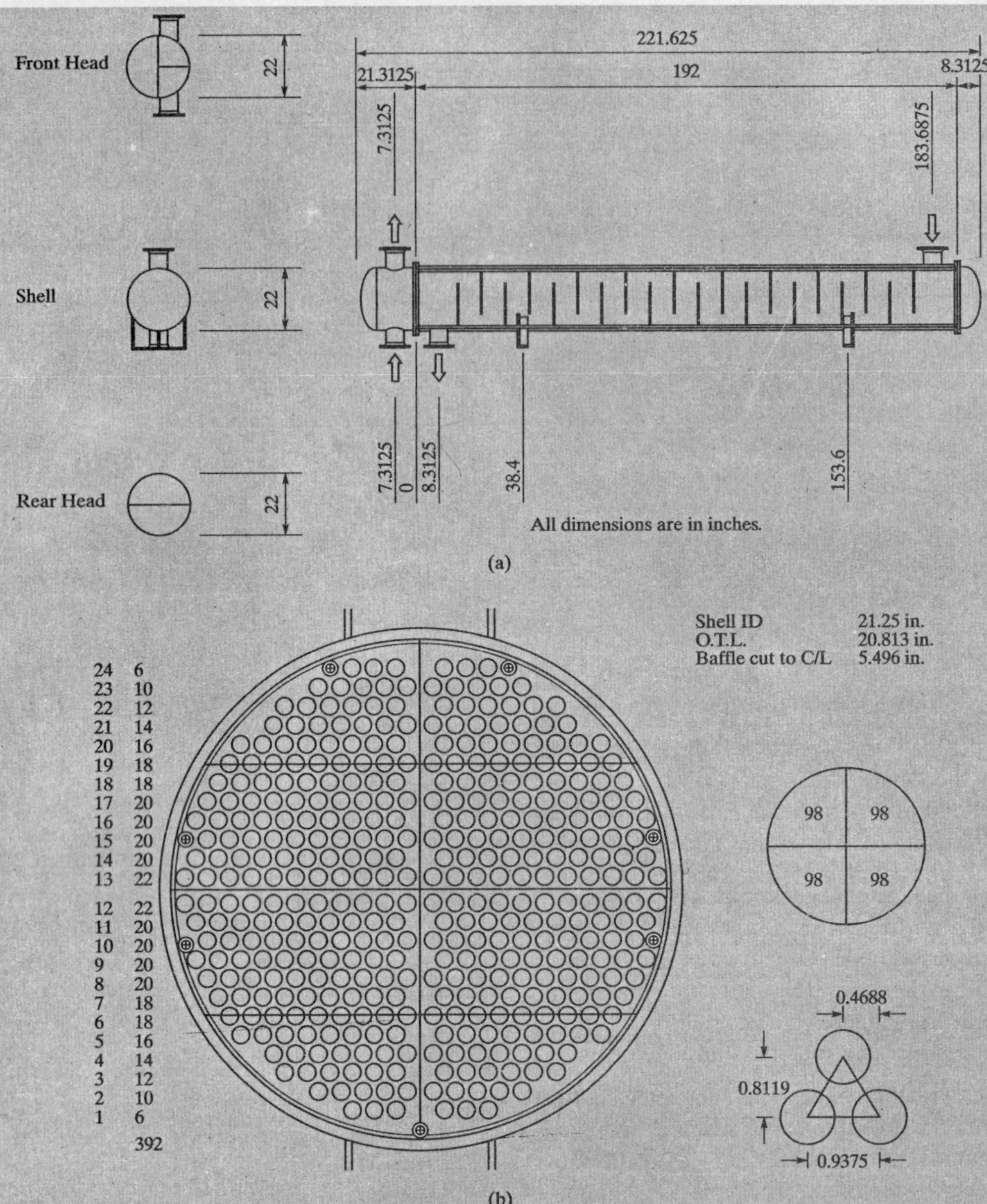

Figure 18.17 Heat exchanger for Example 18.8: (a) setting plan, (b) tube layout.

1	Company:							
2	Location:							
3	Service of Unit:	Our reference:						
4	Item No.:	Your reference:						
5	Date:	Rev No.:	Job No.:					
6	Size	21–192	in	Type BEM	hor	Connected in	1 parallel	3 series
7	Surf./unit(eff.)	3663.2	ft2	Shells/unit 3	Surf/shell (eff.)	1221.1	ft2	
8	**PERFORMANCE OF ONE UNIT**							
9	Fluid allocation		Shell Side		Tube Side			
10	Fluid name							
11	Fluid quantity, Total	lb/h	150000		130714			
12	Vapor (In/Out)	lb/h						
13	Liquid	lb/h	150000	150000	130714	130714		
14	Noncondensable	lb/h						
15								
16	Temperature (In/Out)	F	300	165.2	100	268.7		
17	Dew / Bubble point	F						
18	Density	lb/ft3	48.617	53.007	53.284	47.245		
19	Viscosity	cp	0.214	0.381	0.478	0.217		
20	Molecular wt, Vap							
21	Molecular wt, NC							
22	Specific heat	BTU/(lb*F)	0.5491	0.447	0.4234	0.4855		
23	Thermal conductivity	BTU/(ft*h*F)	0.066	0.074	0.077	0.061		
24	Latent heat	BTU/lb						
25	Inlet pressure (abs)	psi	50		90			
26	Velocity	ft/s	1.67		3.49			
27	Pressure drop, allow./calc.	psi	10	9.177	10	6.392		
28	Fouling resist. (min)	ft2*h*F/BTU	0.002		0.002			
29	Heat exchanged	9969642	BTU/h	MTD corrected	34.83	F		
30	Transfer rate, Service	78.14	Dirty	86.58	Clean	140.24	BTU/(h*ft2*F)	
31	**CONSTRUCTION OF ONE SHELL**					Sketch		
32			Shell Side		Tube Side			
33	Design /Test pressure	psi	75/	/Code	90/	/Code		
34	Design temperature	F	360		330			
35	Number passes per shell		1		4			
36	Corrosion allowance	in	0.0625		0.0625			
37	Connections	In	6 / 150 ANSI		6 /150 ANSI			
38	Size/rating	Out	6 / 150 ANSI		6 / 150 ANSI			
39	in/	Intermediate	/ 150 ANSI		/ 150 ANSI			
40	Tube No.	OD 0.75	Tks-avg 0.065	in Length 16	ft	Pitch 0.9375	in	
41	Tube type	Material CS	Tube pattern 30					
42	Shell CS	ID	OD22	in	Shell cover			
43	Channel or bonnet CS				Channel cover			
44	Tubesheet-stationary CS				Tubesheet-floating			
45	Floating head cover				Impingement protection	None		
46	Baffle-crossing CS	Type single seg	Cut(%d) 24	hor	Spacing: c/c 8.5	in		
47	Baffle-long	Seal type	Inlet 14.4375	in				
48	Supports-tube	U-bend	Type					
49	Bypass seal	Tube-tubesheet joint	groove/expand					
50	Expansion joint	Type						
51	RhoV2-Inlet nozzle	927	Bundle entrance 489	Bundle exit 448	lb/(ft*s2)			
52	Gaskets - Shell side	Tube Side						
53	Floating head							
54	Code requirements	ASME Code Sec VIII Div 1	TEMA class B					
55	Weight/Shell	5400.9	Filled with water 7927	Bundle 3600.5	lb			
56	Remarks							
57								
58								

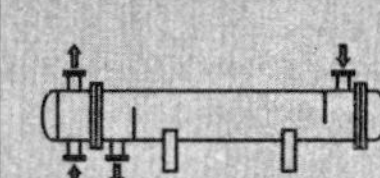

Figure 18.18 Heat exchanger specification sheet for Example 18.8.

18.5 SUMMARY

Having studied this chapter, the reader should

1. Know how the temperature and phase condition of a stream can be changed by using a heat exchanger.
2. Be able to specify and use a simulation program to calculate a heat exchanger when modeling just one side.
3. Be able to select heat-transfer media for the other side of the exchanger.
4. Know the importance of heating and cooling curves, how to generate them with a simulation program, and how to use them to avoid crossover violations of the second law of thermodynamics.
5. Know the major types of heat exchange equipment and how they differ in flow directions of the two fluids exchanging heat, and how to determine the corrected temperature-driving force for heat transfer.
6. Know how to specify a heat exchanger when modeling both sides with a simulation program.
7. Know how to estimate overall heat-transfer coefficients, including the effect of fouling.
8. Know the limitations of boiling heat transfer.
9. Be able to design a shell-and-tube heat exchanger with the help of a simulator.

REFERENCES

1. BOWMAN, R.A., A.C. MUELLER, and W.M. NAGLE, "Mean Temperature Difference in Design," *Trans. ASME*, **62**, 283–293 (1940).

2. COLBURN, A.P., *Trans. AIChE*, **29**, 166 (1931).

3. DITTUS, F.W., and L.M.K. BOELTER, *Univ. Calif. (Berkeley) Pub. Eng.*, **2**, 443 (1930).

4. DONOHUE, D.A., *Ind. Eng. Chem.*, **41**, 2,499 (1949).

5. GREEN, D.W., and R.H. PERRY, *Perry's Chemical Engineers' Handbook*, 8th ed., McGraw-Hill, New York (2008).

6. GRIMISON, E.D., *Trans. ASME*, **59**, 583 (1937).

7. HEWITT, G.F., Ed., *Handbook of Heat Exchanger Design*, Begell House, New York (1992).

8. KERN, D.Q., *Process Heat Transfer*, McGraw-Hill, New York (1950).

9. NAGLE, W.M., "Mean Temperature Differences in Multipass Heat Exchangers," *Ind. Eng. Chem.*, **25**, 604–609 (1933).

10. SIEDER, E.N., and G.E. TATE, "Heat Transfer and Pressure Drops of Liquids in Tubes," *Ind. Eng. Chem.*, **28**, 1,429–1,436 (1936).

11. UNDERWOOD, A.J.V., "The Calculation of the Mean Temperature Difference in Multipass Heat Exchangers," *J. Inst. Petroleum Technol.*, **20**, 145–158 (1934).

EXERCISES

18.1 In Example 18.7, an existing exchanger is used to transfer sensible heat between toluene and styrene streams. A minimum approach temperature of 31.3°F is achieved. Design a new shell-and-tube heat exchanger for a 10°F minimum approach temperature.

18.2 A heat exchange system is needed to cool 60,000 lb/hr of acetone at 250°F and 150 psia to 100°F. The cooling can be achieved by exchanging heat with 185,000 lb/hr of acetic acid, which is available at 90°F and 75 psia and needs to be heated. Four 1-2 shell-and-tube heat exchangers are available. Each has an inside shell diameter of 21.25 in. and contains 270$\frac{3}{4}$-in. O.D., 14 BWG, 16-ft-long carbon steel tubes in a square layout on a 1-in. pitch. Segmental baffles with a 25% baffle cut are spaced 5 in. apart. Determine whether one or more of these exchangers can accomplish the task. Note that if two, three, or four of the exchangers are connected in series, they will be equivalent to one 2-4, 3-6, or 4-8 exchanger, respectively. If the exchangers are not adequate, design a new exchanger or exchanger system that is adequate. Assume a combined fouling factor of 0.004(hr-ft^2-°F)/Btu.

18.3 A trim heater is to be designed to heat 116,000 lb/hr of 57 wt% ethane, 25 wt% propane, and 18 wt% *n*-butane from 80 to 96°F. The stream will enter the exchanger at 520 psia and must not reach the bubble point in the exchanger. The stream will be heated with gasoline, which will enter at 240°F and 95 psia, with a flow rate of 34,000 lb/hr. Standard practice of the company is to use 1-2 shell-and-tube heat exchangers with $\frac{3}{4}$- in., 16 BWG carbon steel tubes, 20 ft long, 1-in. square pitch. Tube count depends on shell diameter, with the following diameters available:

Shell I.D. (in.)	Tube Count
10	52
12	78
13.25	96
15.25	136
17.25	176
19.25	224

The gasoline will flow on the shell side. Assume a combined fouling factor of 0.002(hr-ft^2-°F)/Btu. Design a suitable heat exchange system, assuming a 25% overdesign factor.

18.4 Design a shell-and-tube heat exchanger to cool 60,000 lb/hr of 42° API (American Petroleum Institute) kerosene from 400 to 225°F by heating a 35° API distillate from 100 to 200°F under the following specifications. Allow a 10-psi pressure drop for each stream and a combined fouling factor of 0.004(hr-ft^2-°F)/Btu. Neglect the tube-wall resistance. Use $\frac{3}{4}$-in., 16 BWG tubing, O.D. = 0.75 in., I.D. = 0.620 in., flow area/tube = 0.302 in.^{2}, surface/linear foot = 0.1963 ft^2 outside and 0.1623 ft^2 inside. Use 1-in. square pitch. Place kerosene on the shell side. If necessary, change the configuration to keep the tube lengths below 20 ft and the pressure drops below 10 psi.

DATA

	42° API		35° API	
	400°F	225°F	200°F	100°F
C_p, Btu/lb°F	0.67	0.56	0.53	0.47
μ, cP	0.20	0.60	1.3	3.4
k, Btu/hr-ft-°F	0.074	0.078	0.076	0.078
Sp. gr.	0.685	0.75	0.798	0.836

18.5 Hot water at 100,000 lb/hr and 160°F is cooled with 200,000 lb/hr of cold water at 90°F, which is heated to 120°F in a countercurrent shell-and-tube heat exchanger. The exchanger has 20-ft steel tubes with 0.75-in. O.D. and 0.62-in. I.D. The tubes are on a 1-in. square pitch. The thermal conductivity of steel is 25.9 Btu/(ft-hr-°F). The mean heat-transfer coefficients are estimated as h_i = 200 Btu/(ft^2-hr-°F) and h_o = 200 Btu/(ft^2-hr-°F). Estimate:

(a) The area for heat transfer

(b) The diameter of the shell

18.6 A horizontal 1-4 heat exchanger is used to heat gas oil with saturated steam. Assume that h_o = 1,000 Btu/(ft^2-hr-°F) for condensing steam and the fouling factor = 0.004 ft^2-hr-°F/Btu[1 bbl(barrel) = 42 gal].

(a) For a tube-side velocity of 6 ft/s, determine the number and length of tubes and the shell diameter.

(b) Determine the tube-side pressure drop.

	Shell Side		Tube Side	
	Inlet	Outlet	Inlet	Outlet
Fluid	Steam Condensate		Gas oil	
Flow rate (bbl/hr)			1,200	
Temperature (°F)			60	150
Pressure (psig)	50	50	60	
Viscosity (cP)			5.0	1.8
Sp. gr.			0.840	0.810
Thermal conductivity [Btu/(ft-hr-°F)]			0.078	0.083
Heat capacity (Btu/lb-°F)			0.480	0.461

The tubes are 1-in. O.D. by 16 BWG on a 1.25-in. square pitch.

18.7 An alternative heating medium for Exercise 18.6 is a distillate:

	Shell Side	
	Inlet	Outlet
Fluid	— 35° API distillate	—
Flow rate		
Temperature (°F)	250	150
Pressure (psig)	80	
Viscosity (cP)	1.3	3.4
Sp. gr.	0.798	0.836
Thermal conductivity (Btu/hr-ft-°F)	0.076	0.078
Heat capacity (Btu/lb-°F)	0.53	0.47

Determine the tube-side velocity, number and length of tubes, and shell diameter for a 1-6 shell-and-tube heat exchanger using the 1-in. O.D. by 16 BWG tubes on a 1.25-in. square pitch. Design to avoid pressure drops greater than 10 psia. If necessary, change the configuration to keep the tube length below 20 ft.

18.8 Ethylene glycol at 100,000 lb/hr enters the shell of a 1-6 shell-and-tube heat exchanger at 250°F and is cooled to 130°F with cooling water heated from 90 to 120°F. Assume that the mean overall heat-transfer coefficient (based on the inside area of the tubes) is 100 Btu/(ft^2-hr-°F) and the tube-side velocity is 5 ft/s. Use $\frac{3}{4}$-in. 16 BWG tubing (O.D. = 0.75 in., I.D. = 0.62 in.) arranged on a 1-in. square pitch.

(a) Calculate the number of tubes, length of the tubes, and tube-side heat-transfer coefficient.

(b) Calculate the shell-side heat-transfer coefficient to give an overall heat-transfer coefficient of 100 Btu/(ft^2-hr-°F).

DATA

	Ethylene Glycol	Water
	190°F	105°F
C_p, Btu/lb-°F	0.65	1.0
μ, cP	3.6	0.67
k, Btu/hr-ft-°F	0.154	0.363
Sp. gr.	1.110	1.0

Chapter 19

Design of Separation Towers

19.0 OBJECTIVES

The most commonly used separation method in industrial chemical processes is distillation, including enhanced distillation (extractive, azeotropic, and reactive), which is carried out in towers of cylindrical shape containing either plates or packing for contacting the vapor flowing up the tower with the liquid flowing down. The process design of such towers consists of a number of calculations, which are described and illustrated in this chapter. Most of these calculations are readily made with a simulator. The same calculations apply to any multistage separation involving mass transfer between vapor and liquid phases, including absorption and stripping.

After studying this chapter and the materials on distillation on the multimedia modules, which can be downloaded from the Wiley Web site associated with this book, the reader should

1. Be able to determine the tower operating conditions of pressure and temperature and the type of condenser to use.
2. Be able to determine the number of equilibrium stages and reflux required.
3. Be able to select an appropriate contacting method (plates or packing).
4. Be able to determine the number of actual plates or packing height required, together with feed and product locations.
5. Be able to determine the tower diameter.
6. Be able to determine other factors that may influence tower operation.

19.1 OPERATING CONDITIONS

Multistage towers for separations involving mass transfer between vapor and liquid phases can operate anywhere within the two-phase region, but proximity to the critical point should be avoided. Typical operating pressures for distillation range from 1 to 415 psia. For temperature-sensitive materials, vacuum distillation is very common, with pressures as low as 5 mm Hg. Except for low-boiling components and cases where a vapor distillate is desired, a total condenser is used. Before determining a feasible and, hopefully, a near-optimal operating pressure, a preliminary material balance must be made to estimate the distillate and bottoms product compositions. As a starting point for establishing a reasonable operating pressure and a type of condenser, the graphical algorithm in Figure 8.9 can be applied in the following manner, noting that it is based on the use of cooling water that enters the condenser at 90°F and exits at 120°F. The pressure at the exit of the condenser (or in the reflux drum), P_D, is determined so as to permit condensation with cooling water, if possible. This pressure is computed as the bubble-point pressure at 120°F. If this pressure is less than 215 psia, a total condenser is used. However, if the pressure is less than 30 psia, the condenser outlet pressure is set at 20 to 30 psia to avoid vacuum operation. If the pressure at 120°F is greater than 215 psia, the dew-point pressure of the distillate is calculated at 120°F. If that pressure is less than 365 psia, a partial condenser is used; if it is greater than 365 psia, a refrigerant is selected that gives a minimum approach temperature of 5 to 10°F, in place of cooling water for the partial condenser, such that the distillate dew-point pressure does not exceed 415 psia. Up to this point, the tower operating pressure has been determined by the composition of the distillate. Conditions based on the composition of the bottoms product must now be checked. Using the determined condenser outlet pressure, assume a condenser pressure drop in the range of 0–2 psia. Assume a tower pressure drop of from 5 to 10 psia. This will give a pressure at the bottom of the column, P_B, in the range of 5–12 psia higher than the condenser outlet pressure. Almost all reboilers that produce a bottoms product at or close to the bubble point are partial reboilers. Therefore, the bottoms temperature, T_B, is determined by a bubble-point calculation based on the estimated bottoms composition and the bottoms pressure. If this exceeds the decomposition, polymerization, or critical temperature of the bottoms, then a bottoms pressure is computed based on a bottoms temperature safely below the limiting temperature. Then using the assumed pressure drops, a new condenser outlet pressure and corresponding temperature are calculated. This may require a change in the coolant used in the condenser and the type of condenser. Also, the new condenser outlet pressure may be less than about 15 psia, in which case a vacuum system for the tower will be necessary, as discussed in Section 22.5.

In some distillations, the overhead vapor may contain components covering a wide range of volatility. For example, the overhead vapor from a vacuum tower will contain air from leakage into the tower mixed with other components that could be condensed with cooling water or a modest refrigerant. In other cases, the overhead vapor may contain hydrogen and other light gases mixed with easily condensable components. In those cases, neither a total nor a partial condenser is used. Instead, the condenser is designed to produce both a vapor distillate and a liquid distillate. The latter has the same composition as the reflux. For vacuum operation, the vapor distillate is sent to a vacuum pump. To determine the pressure, P_D, compositions of the vapor distillate and the liquid distillate are calculated for a series of pressures at a temperature of 120°F for cooling water, or at a lower temperature if a refrigerant is necessary to recover a higher percentage of the less-volatile components in the liquid distillate. When using a refrigerated condenser, one should always consider placing a water-cooled partial condenser ahead of it. From the results of the calculations, a reasonable pressure is selected.

For extractive and azeotropic distillation, the condenser outlet pressure is usually near ambient pressure, in the range of 20–30 psia, and a total condenser is used. An exception is azeotropic distillation when a low-molecular-weight entrainer is used that necessitates a higher pressure. For reactive distillation, the pressure must be sufficiently high to give corresponding temperatures in the range of reasonable reaction rates.

Absorbers and strippers usually involve components that cover a very wide range of volatility. For example, an absorber might have a feed gas that contains methane, while the absorbent might be an oil of 150 molecular weight. For these two separation operations, which frequently do not utilize either a condenser or reboiler, the tower operating pressure cannot be determined from bubble- and/or dew-point calculations because they can be extremely sensitive to assumed product vapor and/or liquid compositions. Instead, the following rules may apply:

Absorption favors high pressures and low temperatures. Therefore, cool the feed gas and absorbent with cooling water or a refrigerant. If the internal temperature rise in an absorption column is large, interstage coolers can be added. However, because of the high cost of gas compression, it may not be economical to increase the pressure of the feed gas. But do not decrease the pressure of the feed gas.

Stripping favors low pressures and high temperatures. Therefore, heat the liquid feed and stripping agent and lower the pressure to near ambient, but not to a vacuum.

19.2 FENSKE–UNDERWOOD–GILLILAND (FUG) SHORTCUT METHOD FOR ORDINARY DISTILLATION

For ordinary distillation of a single feed to give only distillate and bottoms products, the FUG method, which is included in the library of equipment models of all simulators, is useful for making an initial estimate of the reflux ratio, the number of equilibrium stages, and the location of the feed stage. The method is quite accurate for ideal mixtures of a narrow-boiling range. However, for nonideal mixtures, particularly those that form azeotropes, and for wide-boiling feeds, the FUG method can be quite inaccurate. Therefore, before applying the method, the vapor-liquid equilibrium of the feed should be carefully examined for the magnitude of liquid-phase activity coefficients and the possibility of azeotropes over the range of possible compositions. Note that for nonideal mixtures especially, design engineers often skip this approximate method, in preference to running a few iterations using a rigorous model, as discussed in Section 19.4. Often, reasonable guesses can be provided for the number of stages and reflux ratio to achieve a satisfactory simulation that can be fine-tuned to satisfy product specifications.

The FUG method, which applies to binary and multicomponent feeds, is described in detail by Seader and Henley (2006) and in *Perry's Chemical Engineers' Handbook* (Green and Perry, 2008). Only the procedure is discussed here. The method involves five steps based on the desired separation of two key components in the feed. It includes an estimation of the separation of the nonkey components.

Step 1: *Estimation by the Fenske equation of the minimum number of equilibrium stages, $N_{\min}$(corresponding to total reflux or infinite reflux ratio), needed to separate the two key components.* The Fenske equation is simple and readily applied, even by hand. It involves only one assumption, that of an average relative volatility, $\alpha_{\mathrm{LK,HK}}$, between the two key components, throughout the tower. This may be the geometric average of the distillate and bottoms, or the geometric average of the feed, distillate, and bottoms. The Fenske equation may be written as follows:

$$N_{\min} = \frac{\log\left[\left(\frac{d_{\mathrm{LK}}}{b_{\mathrm{LK}}}\right)\left(\frac{b_{\mathrm{HK}}}{d_{\mathrm{HK}}}\right)\right]}{\log(\alpha_{\mathrm{LK,HK}})} \tag{19.1}$$

where d is a component flow rate in the distillate and b is a component flow rate in the bottoms product.

Step 2: *Estimation by the Fenske equation [Eq. (19.1)] of the distribution, d/b, of the nonkey components between distillate and bottoms at total reflux using the value of $N_{\min}$ computed in Step 1, the b/d ratio for the heavy key, and the relative volatility between the nonkey and the heavy key, $\alpha_{\mathrm{NK,HK}}$.*

Although this estimate is for total reflux conditions, it is a surprisingly good estimate for the distribution of the nonkey components at finite reflux conditions for nearly ideal mixtures.

Step 3: *Estimation by the Underwood equations of the minimum reflux ratio, R_{min}(corresponding to an infinite number of equilibrium stages), needed to separate the two key components*. This calculation is complicated because it involves the solution of nonlinear equations and requires a calculation of the distribution of the nonkey components at minimum reflux even though that distribution is not used for any other purpose. The application of the Underwood equations involves two serious assumptions: (1) The molar liquid flow rate is constant throughout the rectifying section, and (2) the relative volatility is constant in the pinch region. When these assumptions are not valid, the estimated minimum reflux ratio can be less than the true value, making the method nonconservative. More details of the use of the Underwood equations are given by Seader and Henley (2006).

Step 4: *Estimation by the Gilliland correlation of the actual number of equilibrium stages, N, for a specified ratio of actual reflux ratio, R, to minimum reflux ratio, R_{min}*. The Gilliland correlation, which is shown in Figure 19.1, has no theoretical foundation, but is an empirical fit of many rigorous binary and multicomponent calculations when plotted as $(N - N_{min})/(N + 1)$ as a function of $(R - R_{min})/(R + 1)$. The accuracy of the Gilliland method is limited because it ignores the effect of the feed condition (from subcooled to superheated), and can be badly in error when stripping is much more important in the separation than rectification. For optimal design, the recommended value of R/R_{min} to use with the Gilliland method is typically in the range of 1.1 to 1.5, with the lower value for difficult separations requiring more than 100 equilibrium stages and the higher value for easy separations of less than 10 equilibrium stages. At $R/R_{min} = 1.3$, N/N_{min} is often equal to approximately 2.

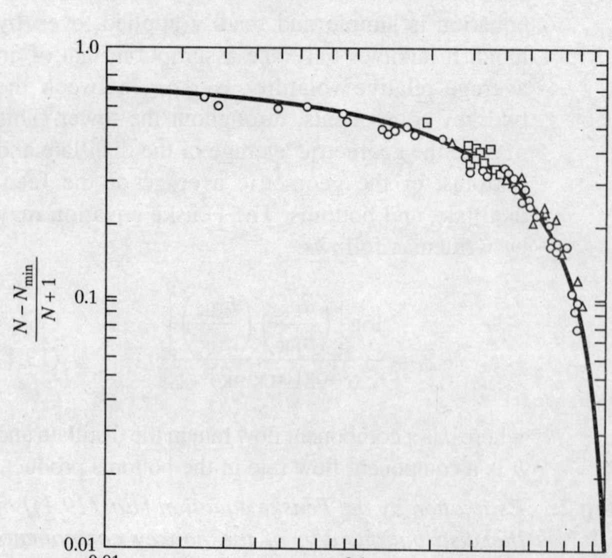

Figure 19.1 Gilliland correlation for ordinary distillation.

Step 5: *Estimation of the feed-stage location by the Fenske equation*. The calculation is made with Eq. (19.1) by applying it to the section of stages between the feed composition and the distillate composition to obtain the minimum number of rectification stages, $N_{R,min}$, and then to the section of stages between the feed and bottoms product to obtain the minimum number of stripping stages, $N_{S,min}$. The ratio of $N_{R,min}$ to $N_{S,min}$ is assumed to be the same as the ratio of N_R to N_S at finite reflux conditions. Alternatively, the empirical, but often more accurate, Kirkbride equation can be applied.

19.3 KREMSER SHORTCUT METHOD FOR ABSORPTION AND STRIPPING

For adiabatic absorbers and strippers with one feed, one absorbent or stripping agent, and two products, a simple and useful shortcut method for estimating the minimum absorbent or stripping agent flow rate is the Kremser method. It applies in the limit of an infinite number of equilibrium stages for the specified absorption or stripping of one component, the key component, from the feed. It also applies for a finite number of equilibrium stages, *N*. Although the method is not included in the library of equipment models of most simulators, it is quite straightforward to apply the Kremser method using hand calculations or a spreadsheet. The derivation of the equations is presented in detail by Seader and Henley (2006) and in *Perry's Chemical Engineers' Handbook* (Green and Perry, 2008).

The separation factor in the Kremser method is an effective absorption factor, A_e, for absorption and a stripping factor, S_e, for stripping, rather than a relative volatility as in the FUG method for distillation. These two factors, which are different for each component, are defined by:

$$A_e = L/KV \tag{19.2}$$

$$S_e = KV/L \tag{19.3}$$

The total molar liquid rate down the tower, *L*, the total molar vapor rate up the tower, *V*, and the *K*-value all vary from the top stage to the bottom stage of the tower. However, sufficiently good estimates by the Kremser method can be achieved by using average values based on the flow rates and temperatures of the two streams entering the tower.

For an absorber, the design basis is the tower pressure; the flow rate, composition, temperature, and pressure of the entering vapor feed; the composition, temperature, and pressure of the absorbent; and the fraction to be absorbed of one key component. The minimum molar absorbent flow rate is estimated from:

$$L_{\min} = K_K V_{\text{in}}(1 - \phi_{A_K}) \tag{19.4}$$

where K_K is the K-value of the key component computed at the average temperature and pressure of the two entering streams and $(1 - \phi_{A_K})$ is the fraction of the key component in the feed gas that is to be absorbed. Typically, the operating absorbent rate is 1.5 times the minimum value. Then, the following equation, due to Kremser and shown in Figure 19.2, is used to compute the number of equilibrium stages required. This equation assumes that the absorbent does not contain the key component.

$$\phi_{A_K} = \frac{A_{e_K} - 1}{A_{e_k}^{N+1} - 1} \tag{19.5}$$

With the value of N computed for the key component, Eq. (19.5) is then used to compute the values of ϕ_A for the other components in the feed gas using their absorption factors. From this, a material balance around the tower can be completed.

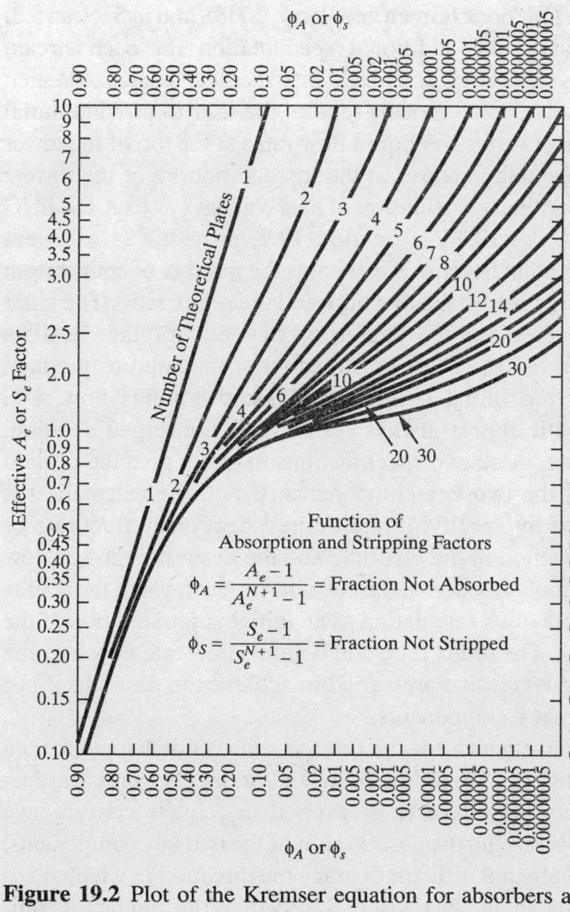

Figure 19.2 Plot of the Kremser equation for absorbers and strippers.

For a stripper, the design basis is the tower pressure; the flow rate, composition, temperature, and pressure of the entering liquid feed; the composition, temperature, and pressure of the stripping agent; and the fraction of the key component to be stripped. The minimum molar stripping agent flow rate is estimated from:

$$V_{\min} = \frac{L_{\text{in}}}{K_K}(1 - \phi_{S_K}) \tag{19.6}$$

where K_K is the K-value of the key component computed at the average temperature and pressure of the two entering streams and $(1 - \phi_{S_K})$ is the fraction of the key component in the feed gas that is to be stripped. Typically, the stripping agent rate is 1.5 times the minimum value. Then, the following equation is used to compute the number of equilibrium stages required. This equation assumes that the stripping agent does not contain the key component.

$$\phi_{S_K} = \frac{S_{e_K} - 1}{S_{e_K}^{N+1} - 1} \tag{19.7}$$

With the value of N computed for the key component, Eq. (19.7) is then used to compute the values of ϕ_S for the other components in the feed liquid using their stripping factors. From this, a material balance around the tower can be completed.

EXAMPLE 19.1

The feed gas to an absorber at 105°F and 400 psia contains 150 kmol/hr of methane, 350 kmol/hr of ethane, 250 kmol/hr of propane, and 50 kmol/hr of n-butane. The absorber is to absorb 90% of the n-butane with an oil at 90°F and 50 psia. Estimate, with the Kremser equation, the number of stages required and the amounts absorbed of the other three components in the feed gas.

SOLUTION

Set the absorber pressure at the feed gas pressure of 400 psia and neglect the pressure drop in the absorber. Use a pump to increase the pressure of the absorbent to 400 psia. The entering vapor rate is $V = 150 + 350 + 250 + 50 = 800$ kmol/hr. The average temperature of the two entering streams is $(105 + 90)/2 = 97.5$°F. The K-value for the key component, n-butane, at 400 psia and 97.5°F, is 0.22 by the SRK equation of state. Using Eq. (19.4) with $(1 - \phi_{A_K}) = 0.90$, the minimum absorbent rate is

$$L_{\min} = 0.22(800)(0.90) = 158 \text{ kmol/hr}$$

Select an operating absorbent flow rate of $L = 1.5 L_{\min} = 1.5(158) = 237$ kmol/hr. The absorption factor for n-butane, from Eq. (19.2), $= A_{e_K} = 237/[0.22(800)] = 1.35$. This is close to 1.40, which is often quoted as the optimal value of the absorption factor. Equation (19.5), which is nonlinear in N, is now applied with $(1 - \phi_{A_K}) = 0.90$, which gives $\phi_{A_K} = 0.10$.

$$0.10 = \frac{1.35 - 1}{1.35^{N+1} - 1}$$

Solving, $N = 4$ equilibrium stages. The result for this exercise is very useful as a first approximation for a rigorous equilibrium-stage method using a simulator, as described in the next section.

19.4 RIGOROUS MULTICOMPONENT, MULTI-EQUILIBRIUM-STAGE METHODS WITH A SIMULATOR

Almost all multistage, multicomponent vapor-liquid separation towers, whether plate or packed, are routinely designed with simulators. The calculations are usually based on the assumption of equilibrium stages, but more realistic mass-transfer models are also available (e.g., see Chapter 12 of Seader and Henley, 2006). The equilibrium-stage calculations apply component mole balances, enthalpy balances, and vapor-liquid phase equilibrium at each stage, and utilize any of a number of reasonably rigorous thermodynamic correlations based on equations of state or liquid-phase activity coefficients to estimate K-values and enthalpies. The resulting large set of equations is nonlinear and is solved iteratively for stagewise profiles of vapor flows and compositions, liquid flows and compositions, and temperatures, from a set of starting guesses by either an *inside-out method* or a *Newton method*, both of which are described in some detail by Seader and Henley (2006) and in *Perry's Chemical Engineers' Handbook* (Green and Perry, 2008). The inside-out method is fast and is the most widely used, but Newton's method is sometimes preferred for highly nonideal systems. However, convergence of the solution of the nonlinear equations is not guaranteed for either method. When a method fails to converge within the default number of iterations (usually 20): (1) more iterations can be specified, (2) a damping factor can be applied to limit the changes made by the method to the guesses of the unknowns between iterations to prevent wild swings, and/or (3) the initial guesses of the unknowns can be changed. In this way, most problems, unless infeasibly specified, can be converged. Infeasible specifications include those where an inadvertent attempt is made to violate the order of volatility of the components.

Equilibrium-stage methods are usually adequate for nearly ideal distillation systems when coupled with calculations of plate efficiency to estimate actual trays or, in the case of packed towers, when HETS (height equivalent of a theoretical stage) or HETP (height equivalent to a theoretical plate) values are known from experience or from experiment to enable the estimation of packed height. For absorbers, strippers, and nonideal distillation systems, mass-transfer models are preferred, but their use requires a value for the tower diameter and a tray layout or type and size of packing. Even when mass-transfer models are preferred, initial calculations are usually made with equilibrium-stage models. Also, note that data for reliable mass-transfer coefficients are often difficult to obtain.

Both of the equilibrium-stage methods can handle almost any tower configuration, including multiple feeds, vapor and liquid sidestreams, and interheaters and intercoolers. Some of these methods can also handle pumparounds (liquid side draws returned to the column at a higher tray after heat exchange with other streams), bypasses, two liquid phases, chemical reaction, interlinked towers, and specified plate efficiencies. Thus, these models can be applied to ordinary and complex distillation, extractive distillation, homogeneous azeotropic distillation, heterogeneous azeotropic distillation, reactive distillation, absorption, stripping, reboiled stripping, and reboiled absorption.

When using an equilibrium-stage model, the following must be specified: (1) all stage pressures; (2) type of condenser (total, partial, or mixed) and type of reboiler; (3) all tower feed streams and feed stage locations, including total feed flow rate, composition, temperature, and pressure; (4) and number of equilibrium stages. In addition, stage locations for sidestreams, intercoolers, and interheaters are necessary. From a degrees-of-freedom analysis, as discussed by Seader and Henley (2006), in *Perry's Chemical Engineers' Handbook* (Green and Perry, 2008), and in Section 5.2, this leaves one additional specification for each stream leaving the tower and each intermediate heat exchanger. In addition, some models require the user to provide initial guesses of vapor and liquid flow rates at the top of the tower and stage temperatures at the top and bottom of the tower.

For ordinary distillation of nearly ideal systems, the FUG method, described in Section 19.2, provides an excellent starting point because it estimates the number of equilibrium stages, the feed stage location, and the reflux ratio. The latter can be used for the degree of freedom for the distillate product. For the degree of freedom of the bottoms product, a preferred initial specification is the bottoms flow rate, because it almost always results in a converged solution. However, these two specifications may not give the desired split of the two key components. If not, the calculation is repeated by specifying the desired heavy-key flow rate or mole fraction in the distillate and the desired light-key flow rate or mole fraction in the bottoms product, using the results of the previous calculation as an initial approximation of the solution. The reflux ratio and bottoms flow rate now become initial guesses that are varied to achieve the desired split of the two key components.

If convergence for the desired split is not achieved, then estimates of the reflux ratio and/or the bottoms product flow rate may have to be revised to achieve convergence when specifying the desired split of the two key components. It is usually not difficult to judge the direction in which these estimates should be revised. Rarely does the number of equilibrium stages have to be increased or decreased. However, the degree of separation as high purities are approached is more sensitive to the number of stages than to the reflux ratio. Finally, it is useful to vary the feed stage location to determine its optimal value, which corresponds to the lowest necessary reflux ratio.

For converged calculations, simulators can provide tables and graphs of temperature, vapor and liquid flow rates, and vapor and liquid compositions as a function of stage number. These profiles should be examined closely to detect the existence of any pinch points where little or no change occurs over a section of stages. If a pinch point is found, say over a region of 4 stages, then the number of stages in that section of the column can probably be reduced by 4 without changing the degree of separation. This should be confirmed by calculations.

For simple absorbers and strippers, the Kremser method described in Section 19.3 can be used to obtain an initial approximation to the number of equilibrium stages and the flow rate of the absorbent or stripping agent. Then, with the rigorous method, the latter can be varied to achieve the desired separation of the key component for a fixed number of stages.

When the FUG method is not valid for obtaining initial estimates for use with the rigorous methods, the following procedure may be useful. It focuses on an attempt to at least estimate the number of equilibrium stages required for each section of stages bounded by feeds and/or products. These estimates are provided by the Fenske equation, applied to key-component concentrations at either end of the section, where the computed $N_{\min}$ is multiplied by 2 to approximate the necessary N. This is illustrated in the following example.

EXAMPLE 19.2

A distillation column for the separation between propane and n-butane is to have the following two feeds:

	Upper Feed	Lower Feed
Temperature, °F	170	230
Pressure, psia	245	245
Component feed rates, lbmol/hr:		
Ethane	2.5	0.5
Propane	14.0	6.0
n-Butane	10.0	18.0
n-Pentane	5.0	30.0
n-Hexane	0.5	4.5

Use the Fenske equation to estimate the number of stages that should be placed between the two feeds.

SOLUTION

First compute the relative volatility between propane and n-butane at 245 psia and the average temperature of the two feeds of (170 + 230)/2 = 200°F. The respective average K-values by the SRK equation of state are 1.76 and 0.84, giving $\alpha_{\mathrm{LK,HK}} = 1.76/0.84 = 2.10$. Applying the Fenske equation [Eq. (19.1)] between the two feeds, using the key component feed flow rates, gives:

$$N_{\min} = \frac{\log\left[\left(\frac{14}{6}\right)\left(\frac{18}{10}\right)\right]}{\log(2.10)} = \frac{0.623}{0.322} = 1.93$$

Therefore, $N = 2(1.93) = 3.86$. If this is rounded up to a value of 4, then four equilibrium stages should be placed between the two feed stages.

Rigorous calculations for extractive distillation are usually readily converged once the user determines which components the solvent forces to the bottom of the tower. The Fenske equation can be applied, in a manner similar to that in Example 19.2, to determine at which stage down from the top to bring in the solvent so as to minimize its loss to the distillate. Rigorous calculations for azeotropic distillation are another matter. Before even attempting a rigorous calculation, a triangular residue curve map, which can be drawn by the simulators, should be used to determine feasible entrainer flow rates and product compositions, as described in Section 8.5. In addition, for heterogeneous azeotropic distillation, a triangular liquid-liquid phase equilibrium diagram should be used to determine preliminary values for the flows and compositions of the phase split that occurs in the overhead decanter. Failure to make these preliminary studies can result in much time and effort spent in trying to converge an infeasible tower specification. Most difficult of all are reactive distillation calculations. Again, preliminary calculations are necessary, including (1) independent reactor calculations, with a CSTR model, to determine an operating temperature range that gives reasonable reaction rates, and (2) flash calculations to determine component volatilities of reaction mixtures.

19.5 PLATE EFFICIENCY AND HETP

If a mass-transfer (rate-based) model is used, of the type described by Seader and Henley (2006), the stages will be actual trays or packed height in the case of packings. If an equilibrium-stage model is used, plate efficiencies for tray towers or HETP values for packed towers must be estimated to convert equilibrium stages to actual trays or to packed height. One of the major factors that influences mass transfer is the viscosity of the liquid phase. In distillation, liquid viscosities are generally low, often in the range of 0.1 to 0.2 cP, and overall plate efficiencies, E_o, are relatively high, in the range of 50 to 100%. Because of a liquid crossflow effect in large-diameter distillation towers, efficiencies even higher than 100% have been measured. Liquid viscosity in absorbers and some strippers is often in the range of 0.2 to 2.0 cP, and overall plate efficiencies are in the range of 10 to 50%. Very approximate estimates that are sometimes used are 70% for

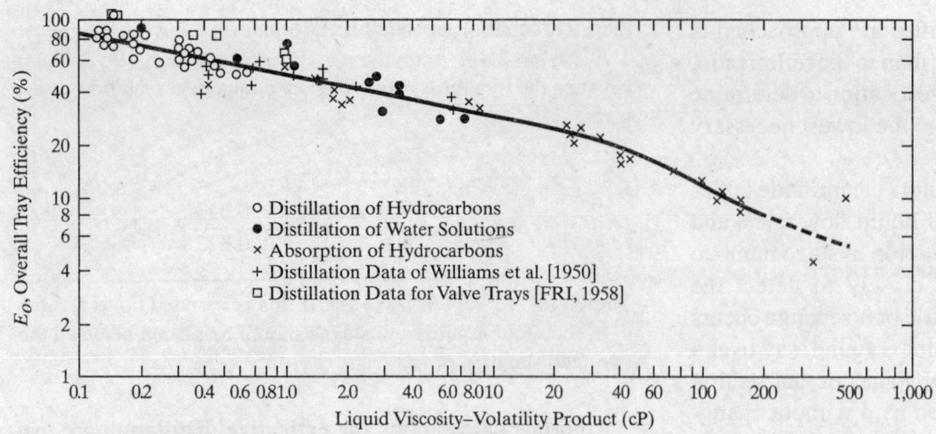

Figure 19.3 Lockhart and Leggett version of O'Connell correlation for plate efficiency.

distillation, 50% for strippers, and 30% for absorbers. The number of actual plates required is

$$N_{\text{actual}} = N_{\text{equilibrium}}/E_o \qquad (19.8)$$

A better estimate of overall plate efficiency can be made with the Lockett and Leggett version of the empirical O'Connell correlation, as shown in Figure 19.3. In this plot, the overall plate efficiency depends on the product of the average liquid-phase viscosity in cP and a dimensionless volatility factor. For distillation, the volatility factor is the average relative volatility between the light and heavy key components, $\alpha_{\text{LK,HK}}$. For absorbers and strippers, the volatility factor is 10 times the average *K*-value of the key component. If an even better estimate of the plate efficiency is desired, and in particular one that depends on plate location and component, a semitheoretical method developed by Chan and Fair (1984a,b), based on the definition of the Murphree vapor-phase efficiency, can be applied, as discussed by Seader and Henley (2006).

For packed columns, HETP values are usually used to convert equilibrium stages to packed height even though the alternative concept of HTU (height of a transfer unit) together with NTU (number of transfer units) is on a more firm theoretical foundation. Values of HETP are generally derived from experimental data for a particular type and size of packing, and are often available from packing vendors. Typically cited, in the absence of data, is an HETP of 2 ft for modern random packings, and 1 ft for structured packings. However, Kister (1992) suggests the following, where D_P is the nominal diameter of random packings and a is the specific surface area of structured packings:

1. For modern random packings with low-viscosity liquids:

$$\text{HETP, ft} = 1.5(D_p, \text{ in.})$$

2. For structured packings at low-to-moderate pressures and low-viscosity liquids:

$$\text{HETP, ft} = 100/a, \text{ ft}^2/\text{ft}^3 + 0.333$$

3. For absorption with a viscous liquid:

$$\text{HETP} = 5 \text{ to } 6 \text{ ft}$$

4. For Vacuum service:

$$\text{HETP, ft} = 1.5(D_p, \text{ in.}) + 0.50$$

5. For high-pressure service with structured packings:

$$\text{HETP, ft} > 100/a, \text{ ft}^2/\text{ft}^3 + 0.333$$

6. For small-diameter towers less than 2 ft in diameter: HETP, ft = tower diameter in feet, but not less than 1 ft

The packed height is given by:

$$\text{Packed height} = N_{\text{equilibrium}}(\text{HETP}) \qquad (19.9)$$

If a more accurate estimate of packed height is desired, correlations of experimental mass-transfer coefficients or heights of transfer units should be used for the particular packing selected. Some of these correlations are provided in simulators, and the method of calculation is given in detail by Seader and Henley (2006).

19.6 TOWER DIAMETER

The tower diameter depends on the vapor and liquid flow rates and their properties up and down the tower. The tower diameter is computed to avoid flooding, where the liquid begins to fill the tower and leave with the vapor because it cannot flow downward at the required rate.

Tray Towers

For a given vapor flow rate in a tray tower, *downcomer flooding* occurs when the liquid rate is increased to the point where the liquid froth in the downcomer backs up to the tray above. This type of flooding is not common, because most tray towers have downcomers with an adequate cross-sectional area for liquid flow. A common rule is to compute the height of clear liquid in the downcomer. At low to moderate pressures, if the height is less than 50% of the tray spacing, it is unlikely that downcomer flooding will occur. However, at high pressures, this value may drop to 20–30%. Another rule is to provide a downcomer cross-sectional area of at least 10–20% of the total tower cross-sectional area, with the larger percentage pertaining to high pressure.

More commonly, the diameter of a tray tower is determined to avoid *entrainment flooding.* For a given liquid rate, as the vapor rate is increased, more and more liquid droplets are carried by the vapor to the tray above. Flooding occurs when the liquid entrainment by the vapor is so excessive that column operation becomes unstable.

The tower inside cross-sectional area, A_T, is computed at a fraction f (typically 0.75 to 0.85) of the vapor flooding velocity, U_f, from the continuity equation for one-dimensional steady flow, applied to the vapor flowing up to the next tray through area $(A_T - A_d)$:

$$\dot{m}_V = G = (fU_f)(A_T - A_d)\rho_G \qquad (19.10)$$

where G = mass flow rate of vapor, A_d = downcomer area, and ρ_G = vapor density. Substituting $A_T = \pi(D_T)^2/4$ for a circular cross section into Eq. (19.10) and solving for the tower inside diameter, D_T, gives

$$D_T = \left[\frac{4G}{(fU_f)\pi\left(1 - \dfrac{A_d}{A_T}\right)\rho_G}\right]^{1/2} \qquad (19.11)$$

The flooding velocity is computed from an empirical capacity parameter, C, based on a force balance on a suspended liquid droplet:

$$U_f = C\left(\frac{\rho_L - \rho_G}{\rho_G}\right)^{1/2} \qquad (19.12)$$

The capacity parameter is given by:

$$C = C_{SB}F_{ST}F_F F_{HA} \qquad (19.13)$$

The parameter C_{SB}, for towers with perforated (sieve) plates, is given by the correlation of Fair (1961), based on data from commercial-size towers, covering tray spacings from 6 to 36 in. A revision by Fair of the original correlation, shown in Figure 19.4, applies to all common crossflow plates (sieve, valve, and bubble-cap), with tray spacing, T_S, in mm, from 150 to 900 and C_{SB} in m/s. The abscissa in Figure 19.4 is a flow ratio parameter, $F_{LG} = (L/G)(\rho_G/\rho_L)^{1/2}$, where both the liquid rate, L, and vapor rate, G, are mass flow rates. The surface tension factor, F_{ST}, is equal to $(\sigma/20)^{0.20}$, where the surface tension, σ, is in dyne/cm. The foaming factor, F_F, is 1 for non-foaming systems, typical of distillation, and 0.5 to 0.75 for foaming systems, typical of absorption with heavy oils. The hole-area factor, F_{HA}, is 1 for valve and bubble-cap trays. For sieve trays, it is 1 for $(A_h/A_a) \geq 0.10$, and $[5(A_h/A_a) + 0.5]$ for $0.06 \leq (A_h/A_a) \leq 1.0$, where A_h is the total hole area on a tray and A_a is the active area of the tray = $(A_T - 2A_d)$ where bubbling occurs.

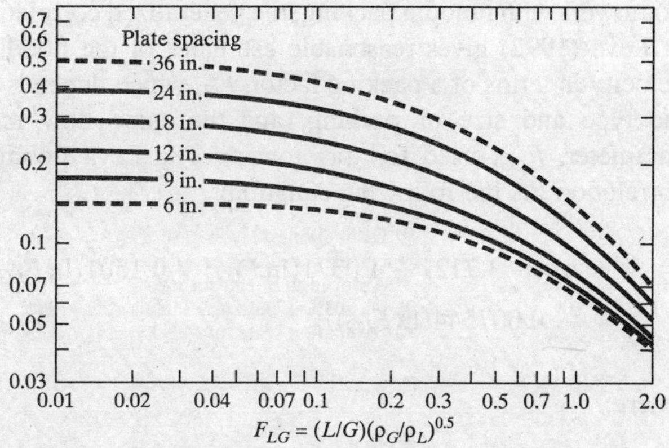

Figure 19.4 Flooding correlation for sieve, valve, and bubble-cap trays.

In Eq. (19.11), the ratio (A_d/A_T) may be estimated by

$$\frac{A_d}{A_T} = \begin{Bmatrix} 0.1, & F_{LG} \leq 0.1 \\ 0.1 + \dfrac{(F_{LG} - 0.1)}{9}, & 0.1 \leq F_{LG} \leq 1.0 \\ 0.2, & F_{LG} \geq 1.0 \end{Bmatrix}$$

Example 19.3 below illustrates the calculation of the tower diameter for a sieve tray.

Packed Towers

If a packed tower is irrigated from a good distributor by a downflow of liquid, the liquid flows over the packing surface and a volumetric holdup of liquid in the tower is observed. As vapor is passed up the tower at low flow rates, countercurrent to the liquid, little or no drag is exerted by the vapor on the liquid and the liquid holdup is unchanged. The liquid has no difficulty leaving the tower as fast as it enters. However, if the gas flow rate is increased, a point is eventually reached where, because of drag, the liquid holdup begins to increase significantly with increasing vapor rate. This is called the loading point. Further increases in the vapor rate eventually reach the point where liquid begins to fill the tower, causing a rapid increase in pressure drop. The flooding point can be defined as the point where the pressure drop rapidly increases with a simultaneous decrease in mass-transfer efficiency Typically, the flooding point is accompanied by a pressure head of approximately 2 in. of water/ft of packing. For a given liquid flow rate, the loading gas flow rate, which is typically 70% of the flooding gas flow rate, is often used to compute the tower inside diameter.

The diameter of a packed tower is calculated from an estimated flooding velocity with a continuity equation similar to Eq. (19.11) for tray towers:

$$D_T = \left[\frac{4G}{(fU_f)\pi\rho_G}\right]^{1/2} \qquad (19.14)$$

For towers with random packing, the generalized correlation of Leva (1992) gives reasonable estimates of the flooding velocity in terms of a packing factor, F_P, which depends on the type and size of packing, and the same flow ratio parameter, F_{LG}, used for tray towers. The Leva flooding correlation fits the following equation:

$$Y = \exp[-3.7121 - 1.0371(\ln F_{LG}) - 0.1501(\ln F_{LG})^2 - 0.007544(\ln F_{LG})^3] \quad (19.15)$$

where:

$$Y = \frac{U_f^2 F_P}{g}\left(\frac{\rho_G}{\rho_{H_2O(L)}}\right) f\{\rho_L\} f\{\mu_L\} \quad (19.16)$$

The flooding velocity factor Y is dimensionless, with U_f in ft/s, F_P in ft²/ft³, and $g = 32.2\,\text{ft/s}^2$. Values of F_P for several representative packings are listed in Table 19.1. Equation (19.15) is valid for $Y = 0.01$ to 10.

The density function is given by:

$$f\{\rho_L\} = -0.8787 + 2.6776\left(\frac{\rho_{H_2O(L)}}{\rho_L}\right) - 0.6313\left(\frac{\rho_{H_2O(L)}}{\rho_L}\right)^2 \quad (19.17)$$

for density ratios from 0.65 to 1.4.

Table 19.1 Packing Factors for Calculating Flooding Velocity

Type Packing	Material	Nominal Diameter, D_P(in.)	Packing Factor, F_P(ft²/ft³)
Raschig rings	Ceramic	1.0	157
		2.0	58
		3.0	33
Raschig rings	Metal	1.0	165
		2.0	71
		3.0	40
Intalox saddles	Ceramic	1.0	92
		2.0	30
		3.0	15
Intalox saddles	Plastic	1.0	36
		2.0	25
Pall rings	Metal	1.0	56
		1.5	29
		2.0	27
		3.5	16
Pall rings	Plastic	1.0	53
		2.0	25
		3.5	15

For random packings of 1 in. or greater nominal diameter, the viscosity function is

$$f\{\mu_L\} = 0.96\,\mu_L^{0.19} \quad (19.18)$$

for liquid viscosities from 0.3 cP to 20 cP.

For a value of F_{LG}, Y is computed from Eq. (19.15), and U_f is then computed from Eq. (19.16) for a given packing type and size, with F_P from Table 19.1 and using Eqs. (19.17) and (19.18). Then for $f = 0.7$, the tower diameter is computed from Eq. (19.14). The tower inside diameter should be at least 10 times the nominal packing diameter and preferably closer to 30 times.

The determination of the flooding velocity in structured packings is best carried out by using interpolation of the flooding and pressure-drop charts for individual structured packings in Chapter 10 of Kister (1992).

19.7 PRESSURE DROP AND WEEPING

In general, pressure drop per unit height is least for towers with structured packings and greatest with tray towers, with randomly packed towers in between. For sieve trays, the components of the pressure drop are (1) pressure drop through the holes in the tray, which depends on the hole diameter, hole area, and vapor volumetric flow rate; (2) pressure drop due to surface tension; and (3) the head of equivalent clear liquid on the tray, which depends on the weir height, weir length, and froth density. Detailed methods of calculation of tray pressure drop are presented by Kister (1992), Seader and Henley (2006), and in *Perry's Chemical Engineers' Handbook* (Green and Perry, 2008). Most simulators perform this calculation. However, the user should minimize the hydraulic gradient of the liquid flowing across the tray before requesting the calculation, by considering the number of liquid passes to use. Columns of diameter larger than 4 ft and operating with liquid rates greater than 500 gal/min frequently employ multipass trays to increase weir length and shorten the liquid flow path across the tray. Figure 19.5 shows three multipass arrangements and a correlation for selecting the number of passes to use. For preliminary design, a pressure drop of 0.10 psi/tray can be assumed for columns operating at ambient pressure or higher. For vacuum operation, trays should be designed so as not to exceed 0.05 psi/tray, or packing should be considered as a substitute for trays to give an even lower pressure drop. Methods for estimating pressure drop in packed towers are found in Kister (1992), Seader and Henley (2006), and in *Perry's Chemical Engineers' Handbook* (Green and Perry, 2008) and are performed by simulators.

For sieve trays, the possibility of weeping of liquid through the holes in the trays should be checked, particularly when the vapor flow rate is considerably below the flooding point. Methods for checking this are used by the simulators. Note that, in general, weeping rates as high as 10% do not affect the tray efficiency, primarily because the weeping liquid is in contact with the vapor as it falls to the tray below.

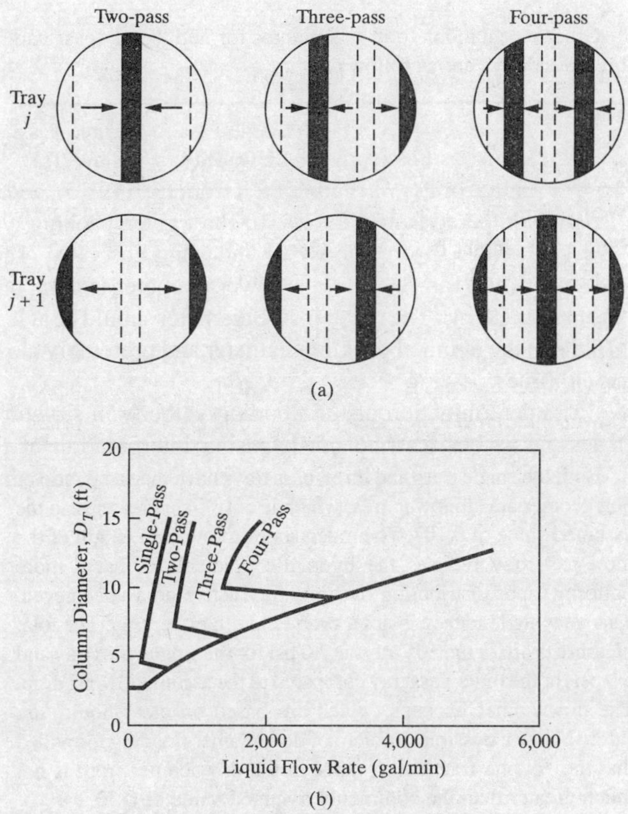

Figure 19.5 Selection of multipass trays. (a) Multipass trays: (1) two-pass; (2) three-pass; (3) four-pass, (b) Flow pass correlation. (Derived from *Koch Flexitray Design Manual, Bulletin 960*, Koch Engineering Co., Inc., Wichita, Kansas, 1960.)

EXAMPLE 19.3

Several alternative distillation sequences are being examined for the separation of a mixture of light hydrocarbons. The sequences are to be compared on the basis of annualized cost, discussed in Chapter 23 and given by Eq. (23.10). This requires estimates of the total capital cost and the annual operating cost of the columns, trays, condensers, reboilers, and reflux accumulators. To estimate these costs, equipment sizes must be determined. In this example, calculations of the height and diameter are illustrated for one column in one of the sequences.

The column to be sized is a deisobutanizer with a saturated liquid feed of 500 lbmol/hr of isobutane and 500 lbmol/hr of *n*-butane. The distillate is to be 99 mol% isobutane and the bottoms 99 mol% *n*-butane. The column shell is carbon steel, with carbon-steel sieve trays on 24-in. spacing. The trays have 0.25-in. diameter holes with a hole-area-to-active-area ratio of 0.1. The weir height is 2 in.

SOLUTION

Following the procedure outlined above, the following results are obtained:

1. Using a simulator, with the Soave–Redlich–Kwong (SRK) equation of state for thermodynamic properties, a bubble-point pressure of 98 psia is computed at 120°F for the distillate composition. Therefore, from Figure 8.9, a total condenser should be used with cooling water. Assuming a pressure drop of 2 psia across the condenser, the pressure at the top of the column is 100 psia. Assuming a 10-psi drop across the tower, the tower bottoms pressure is 110 psia. This gives a bubble-point bottoms temperature of 152°F, which is far below the decomposition temperature of *n*-butane. The assumed tower pressure drop is checked by a simulator after the column diameter is determined.
2. Using the Fenske–Underwood–Gilliland shortcut model with a process simulator, for a reflux-to-minimum reflux ratio of 1.10 (because this is a difficult separation with a relative volatility predicted by the SRK equation of state of approximately 1.30), gives 36.4 minimum stages, a minimum reflux ratio of 6.6, 85.6 equilibrium stages at a reflux ratio of 7.25, and a feed-stage location of 43 stages from the top (approximately at the middle stage). Using these results as a first approximation, a rigorous equilibrium-stage calculation for 84 equilibrium stages in the column, an equilibrium-stage reboiler, and a total condenser (86 stages in all) with a feed stage at the middle, gives a reflux ratio of 7.38 (only 2% greater than the FUG value) to achieve the specified distillate and bottoms purities. Thus, for this nearly ideal system, the FUG method is in close agreement with a rigorous method. The computed condenser duty is 31,600,000 Btu/hr and the reboiler duty is 31,700,000 Btu/hr.
3. Use Figure 19.3 to estimate the plate efficiency for average conditions in the tower. Using a simulator, the estimated average liquid viscosity = 0.12 cP, while the average relative volatility = 1.30. Using the product of these two factors, 0.12(1.3) = 0.156, Figure 19.3 predicts E_O = 0.80. Therefore, the number of actual trays = 84/0.80 = 105, with the partial reboiler counted as an additional stage.
4. For a 24-in. tray spacing, allowing a 10-ft-high liquid bottoms storage (sump) below the bottom tray, and a 4-ft disengagement height above the top tray, the tower height is 222 ft (tangent-to-tangent, i.e., not including the top and bottom tower heads).
5. Assume that the tower diameter will be determined from the entrainment flooding velocity rather than by downcomer flooding. The clear liquid height in the downcomer is one of a number of items computed by a simulator when a tray design is specified. That height should be checked to determine if it is less than 50% of the tray spacing. If not, to prevent downcomer flooding, the downcomer cross-sectional area should be increased. For conditions at the top stage of the column, a process simulation program gives the following results.

Liquid phase:

Surface tension = 7.1 dyne/cm

Flow rate = 215,000 lb/hr

Density = 32.4 lb/ft^3 or 4.33 lb/gal

Molecular weight = 58.12

Vapor phase:

Flow rate = 244,000 lb/hr

Density = 1.095 lb/ft^3

Molecular weight = 58.12

The flow ratio parameter $= F_{LG} = (215,000/244,000) \times (1.095/32.4)^{0.5} = 0.162$

From Figure 19.4 for 24-in. (approximately 600-mm) tray spacing, $C_{SB} = 0.09$ m/s

The surface-tension factor $= F_{ST} = (7.1/20)^{0.2} = 0.81$. Assume $F_F = 1$. Also, $F_{HA} = 1$. Therefore, from Eq. (19.13), $C = 0.09(0.81)(1)(1) = 0.073$ m/s. From Eq. (19.12), $U_f = 0.073[(32.4 - 1.095)/1.095]^{0.5} = 0.390$ m/s = 4,610 ft/hr. Assume operation on the top tray of 80% of flooding ($f = 0.80$). To determine the ratio A_d/A_T, $0.1 + (F_{LG} - 0.1)/9 = 0.1 + 0.062/9 = 0.107 = A_d/A_T$. From Eq. (19.11),

$$D_T = \left[\frac{4(244,000)}{0.80(4,610)(3.14)(1 - 0.107)(1.095)}\right]^{1/2} = 9.3 \text{ ft}$$

For this large a tower diameter, the need for a multipass tray needs to be considered, using Figure 19.5. The volumetric liquid flow rate = (215,000/60)/4.33 = 828 gpm. For this diameter and liquid flow rate, a three-pass tray is indicated. For a one-pass tray, a simulator gives a tower diameter of 9.5 ft, when the diameter is restricted to increments of 0.5 ft. For a three-pass tray, the tower diameter remains at 9.5 ft.

Other calculations from a simulator for both single-pass and three-pass trays are as follows:

	Single-Pass Sieve Tray	Three-Pass Sieve Tray
Weir length, ft	7.3	23.3
Flow path length, ft	6.1	2.2
Active area, ft^2	70.9	70.9
Weeping tendency	Barely	No
Pressure drop, psi	0.067	0.056
Downcomer backup, ft	0.70	0.54
Downcomer area/Tower area	0.122	0.122

Both the single-pass and three-pass trays have the same ratio of downcomer area to tower area, which is only 14% greater than the assumed value of 0.107. The much shorter flow path length of the three-pass tray reduces the hydraulic gradient so that a more uniform vapor distribution over the tray active area is achieved. The weeping tendency is not a problem with either tray. The total pressure drop for the 105 trays is 7.0 psi for the single-pass tray and 5.9 psi for the three-pass tray compared to the assumed 10-psi drop. The downcomer backups, which are based on clear liquid, are safely below a possible problem of downcomer flooding, provided that the volume fraction of vapor in the downcomer froth is not much greater than the commonly assumed value of 0.50.

19.8 SUMMARY

After studying this chapter and completing a few exercises, the reader should have learned to

1. Select an appropriate operating pressure for a multistage tower and a condenser type for distillation.
2. Determine the number of equilibrium stages required for a separation and a reasonable reflux ratio for distillation.
3. Determine whether trays, packing, or both should be considered.
4. Determine the number of actual trays or packing height required.
5. Estimate the tower diameter.
6. Consider other factors for successful tower operation.

REFERENCES

1. Chan, H., and J.R. Fair, "Prediction of Point Efficiencies on Sieve Trays. 1. Binary Systems," *Ind. Eng. Chem. Process Des. Dev.*, **23**, 814–819 (1984a).

2. Chan, H., and J.R. Fair, "Prediction of Point Efficiencies on Sieve Trays. 2. Multicomponent Systems," *Ind. Eng. Chem. Process Des. Dev.*, **23**, 820–827 (1984b).

3. Fair, J.R., *Petro./Chem. Eng.*, 33, 211–218 (Sept., 1961).

4. FRI (Fractionation Research Institute) report of Sept. 3, 1958, *Glitsch Ballast Tray*, published as Bulletin No. 159 of Fritz W. Glitsch and Sons, Inc., Dallas, Texas (1958).

5. Green, D.W., and R.H. Perry, Eds. *Perry's Chemical Engineers' Handbook*, 8th ed., McGraw-Hill, New York (2008).

6. Kister, H.Z., Distillation Design, McGraw-Hill, New York, (1992).

7. *Koch Flexitray Design Manual, Bulletin 960*, Koch Engineering Co., Inc., Wichita, Kansas, 1960.

8. Leva, M., "Reconsider Packed-Tower Pressure-Drop Correlations," *Chem. Eng. Prog.*, **88** (1), 65–72 (1992).

9. Seader, J.D., and E.J. Henley, Separation Process Principles, 2nd ed., John Wiley & Sons, New York (2006).

10. Williams, G.C., E.K. Stigger, and J.H. Nichols, "A Correlation of Plate Efficiencies in Fractionating Columns," *Chem. Eng. Prog.*, **46** 1, 7–16 (1950).

EXERCISES

19.1 In Example 19.1, an absorber with an absorbent rate of 237 kmol/hr and 4 equilibrium stages absorbs 90% of the entering *n*-butane. Repeat the calculations for:

(a) 474 kmol/hr of absorbent (twice the flow) and 4 equilibrium stages.

(b) 8 equilibrium stages (twice the stages) and 237 kmol/hr of absorbent.

Which case results in the most absorption of *n*-butane? Is this result confirmed by the trends of the curves in the Kremser plot of Figure 19.2?

19.2 The feed to a distillation tower consists of 14.3 kmol/hr of methanol, 105.3 kmol/hr of toluene, 136.2 kmol/hr of ethylbenzene, and 350.6 kmol/hr of styrene. The bottoms product is to contain 0.1 kmol/hr of ethylbenzene and 346.2 kmol/hr of styrene. Determine a suitable operating pressure at the top of the tower, noting that the bottoms temperature is limited to 145°C to prevent the polymerization of styrene.

19.3 A mixture of benzene and monochlorobenzene is to be separated into almost pure products by distillation. Determine an appropriate operating pressure at the top of the tower.

19.4 In a reboiled absorber, operating as a deethanizer at 400 psia to separate a light hydrocarbon feed, conditions at the bottom tray are

Liquid phase:

Molar flow = 1,366 lbmol/hr

MW = 91.7

Density = 36.2 lb/ft^3

Surface tension = 10.6 dyne/cm

Vapor phase:

Molar flow = 735.2 lbmol/hr

MW = 41.2

Density = 2.83 lb/ft^3

If sieve trays are used with hole area of 10% and a 24-in. tray spacing, determine the tower diameter. Assume 80% of flooding and a foaming factor of 0.75.

19.5 A distillation tower with sieve trays is to separate benzene from monochlorobenzene. Conditions at a plate near the bottom of the column are

Vapor phase:

Mass flow rate = 24,850 lb/hr

Density = 0.356 lb/ft^3

Liquid phase:

Mass flow rate = 41,850 lb/hr

Density = 59.9 lb/ft^3

Surface tension = 24 dyne/cm

Determine a reasonable tower diameter.

19.6 Water is to be used to absorb acetone from a dilute mixture with air in a tower packed with 3.5-in. metal Pall rings. Average conditions in the tower are

Temperature = 25°C

Pressure = 110 kPa

Liquid phase:

Water = 1,930 kmol/hr

Acetone = 5 kmol/hr

Density = 62.4 lb/ft^3

Surface tension = 75 dyne/cm

Vapor phase:

Air = 680 kmol/hr

Water = 13 kmol/hr

Acetone = 5 kmol/hr

Determine the column diameter for operation at 70% of flooding.

Chapter 20

Design of Pumps, Compressors, and Expanders

20.0 OBJECTIVES

This chapter presents brief descriptions and some theoretical background of the most widely used pumps for liquids, and compressors and expanders for gases, all of which are modeled in simulators. Heuristics for the application of these devices during the synthesis of a chemical process are presented in Chapter 6. Further information on their selection and capital cost estimation is covered in Chapter 22. More comprehensive coverage of the many types of pumps, compressors, and expanders available is presented in Sandler and Luckiewicz (1987) and in *Perry's Chemical Engineers' Handbook* (Green and Perry, 2008). After studying this chapter and the materials on pumps, compressors, and turbines on the multimedia modules, which can be downloaded from the Wiley Web site associated with this book, the reader should be able to explain how the more common types of pumps, compressors, and expanders work and how a simulator computes their power input or output.

20.1 PUMPS

The main purpose of a pump is to provide the energy needed to move a liquid from one location to another. The net result of the pumping action may be to increase the elevation, velocity, and/or pressure of the liquid. However, in most process applications, pumps are designed to increase the pressure of the liquid. In that case, the power required is

$$\dot{W} = F\text{v}(\Delta P), \qquad (20.1)$$

where F is the molar flow rate, v is the molar volume, and P is pressure. Because the liquid molar volume is usually much smaller than that of a gas, pumps require relatively little power compared to gas compressors for the same molar flow rate and increase in pressure. Therefore, when a vapor stream is produced from a liquid stream with increased pressure and temperature, it is generally more economical to increase the pressure while the stream is a liquid. Except for very large changes in pressure, the temperature of the liquid being pumped increases only slightly.

The main methods used to move a liquid are centrifugal force, displacement, gravity, electromagnetic force, and transfer of momentum from another fluid, with the first two methods being the most common for chemical processes. Pumps that use centrifugal force are sometimes referred to as kinetic pumps, but more commonly as *centrifugal pumps*. Displacement of one part of a fluid with another part takes place in so-called *positive-displacement pumps*, whose action is either reciprocating or rotary. The use of electromagnetic force is limited to fluids that can conduct electricity. Jet pumps, either eductors or injectors, are simple devices that transfer momentum from one fluid to another. Their application is also limited because the motive and pumped fluids contact each other and may mix together, and the efficiency of transfer is very low.

The two most important characteristics of a pumping operation are the *capacity* and the *head*. The capacity refers to the flow rate of the fluid being pumped. It may be stated as a mass flow rate, a molar flow rate, or a volumetric flow rate. Most common is the volumetric flow rate, Q, in units of either m^3/hr or gal/min (gpm). The head, or pump head, H, refers to the increase in total head across the pump from the suction, s, to the discharge, d, where the head is the sum of the velocity head, static head, and pressure head. Thus,

$$H, \text{pump head} = \left(\frac{V_d^2}{2g} + z_d + \frac{P_d}{\rho_d g}\right) - \left(\frac{V_s^2}{2g} + z_s + \frac{P_s}{\rho_s g}\right) \qquad (20.2)$$

where V is the average velocity of the liquid, z is the elevation, P is the pressure of the liquid, g is the gravitational acceleration (32.2 ft/s^2, 9.81 m/s^2), and ρ is the liquid density. The head is expressed in units of ft or m of liquid. The required pump head or pressure increase is determined by an energy balance, as discussed with a heuristic and example in Section 6.7.

Centrifugal Pumps

As shown in Figure 20.1, a centrifugal pump consists of an impeller, mounted on a shaft and containing a number of blades rotating within a stationary casing that is provided with an inlet and outlet for the liquid being pumped. Power,

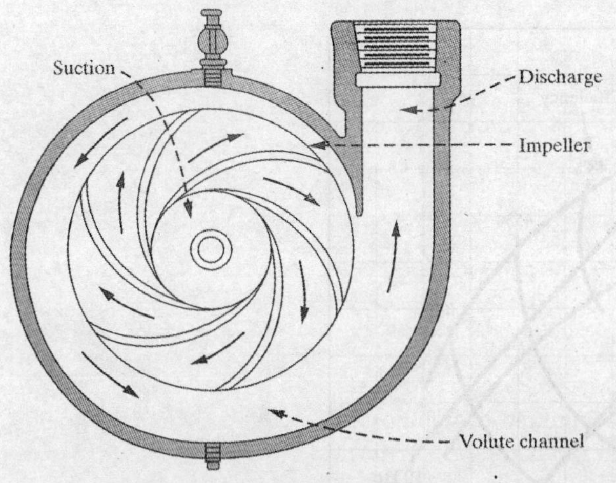

Figure 20.1 Schematic of centrifugal pump.

usually from an electric motor, rotates the shaft, which rotates the impeller. The rotating blades reduce the pressure at the inlet or eye of the impeller, causing liquid to enter the impeller from the suction of the pump. This liquid is forced outward along the blades to the blade tips at an increasing tangential velocity. At this point the liquid has acquired an increased velocity head from the power input to the pump. The velocity head is then reduced and converted to a pressure head as the liquid passes into the annular (volute) chamber within the casing and beyond the blades, and thence to the pump outlet or discharge.

When a centrifugal pump is installed in a pumping system and operated at a particular rotational rate, N (usually 1,750 to 3,450 rpm), the flow rate can be varied by changing the opening on a valve located in the pump discharge line. The variation of H with Q defines a unique *characteristic curve* for the particular pump operating at N with a fluid of a particular viscosity. Each make and model of a centrifugal pump is supplied by the manufacturer with a characteristic curve determined by the manufacturer when pumping water. Corrections are necessary when other fluids are pumped. Corresponding to the variation of H with Q, curves representing the effect of Q on the brake horsepower, P, and the pump efficiency, η, are shown in Figure 20.2. Typically, the pump head decreases with increasing flow rate, while the brake horsepower increases with increasing flow rate. The pump efficiency passes through a maximum. The pump will only operate at points on the characteristic curve. Therefore, for a particular pumping task, the required head-volumetric flow rate point must lie somewhat below the characteristic curve. The difference between the two heads, (pump head – required head), can be throttled across a control valve in the discharge line. Ideally, a centrifugal pump should be selected so that the operating point is located on the characteristic curve at the point of maximum efficiency.

For a given centrifugal pump, the characteristic curve moves upward with increasing rate of rotation, N, as shown in Figure 20.3. Similarly, for a pump of a particular design, the characteristic curve moves upward with increasing impeller

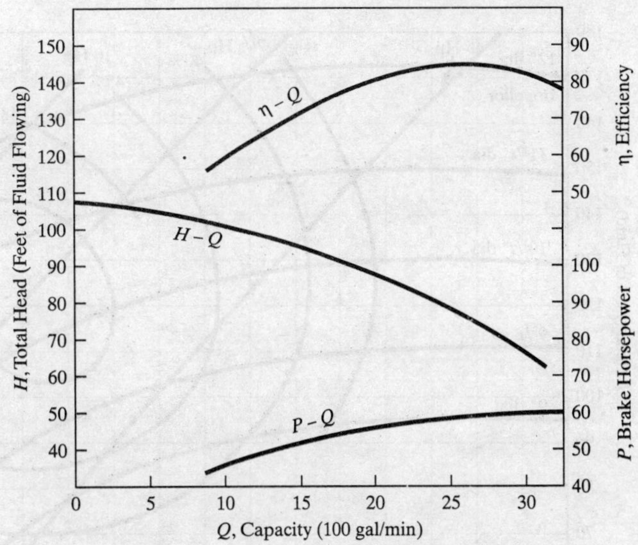

Figure 20.2 Characteristic curves for a centrifugal pump.

diameter, D, as shown in Figure 20.4. When a characteristic curve for just one rotation rate and/or impeller diameter is available and an approximate characteristic curve is desired for another rotation rate and/or impeller diameter, the affinity laws for a centrifugal pump can be applied:

$$Q_2 = Q_1\left(\frac{N_2}{N_1}\right) \tag{20.3}$$

$$H_2 = H_1\left(\frac{N_2}{N_1}\right)^2 \tag{20.4}$$

$$Q_2 = Q_1\left(\frac{D_2}{D_1}\right) \tag{20.5}$$

$$H_2 = H_1\left(\frac{D_2}{D_1}\right)^2 \tag{20.6}$$

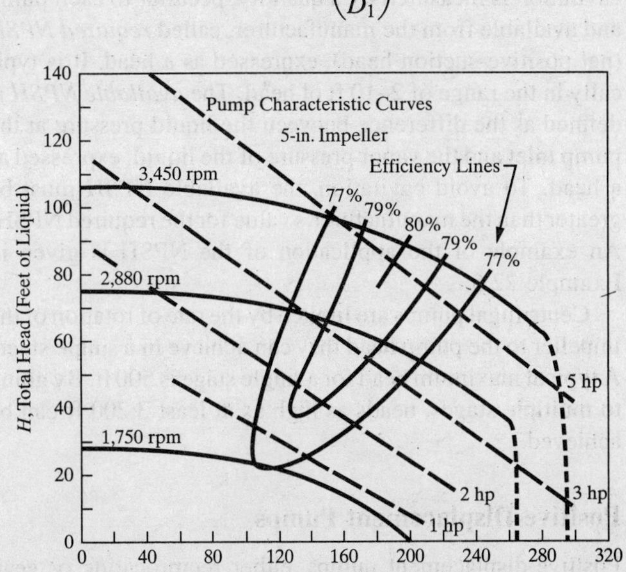

Figure 20.3 Effect of rate of rotation on characteristic curves.

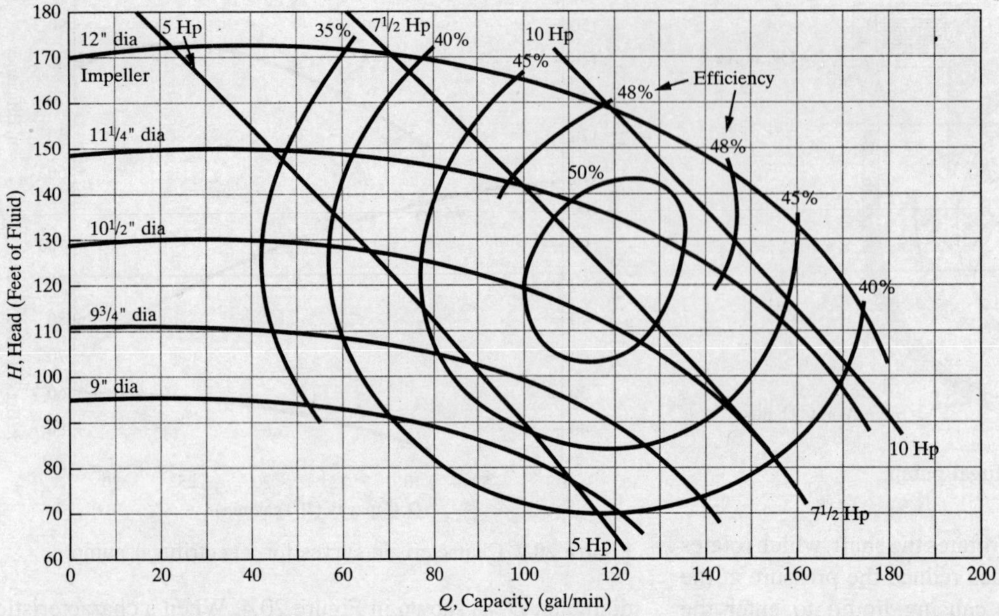

Figure 20.4 Effect of impeller diameter on characteristic curves.

More difficult is the correction for viscosity. In general, increasing viscosity for a fixed capacity, Q, decreases the pump head and the pump efficiency, and increases the brake horsepower. Typical effects of viscosity are shown in Figure 20.5. As seen, the effect of viscosity can be substantial.

Because centrifugal pumps operate at high rates of rotation, the imparted high liquid velocities can lower the local pressure. If that pressure falls below the vapor pressure of the liquid, vaporization will produce bubbles that may collapse violently against surfaces where a higher pressure exists. This phenomenon is called *cavitation* and must be avoided. Otherwise, besides a lowering of efficiency and flow rate, the pump may be damaged. The tendency for cavitation is measured by a quantity, peculiar to each pump and available from the manufacturer, called *required NPSH* (net positive-suction head), expressed as a head. It is typically in the range of 2–10 ft of head. The *available NPSH* is defined as the difference between the liquid pressure at the pump inlet and the vapor pressure of the liquid, expressed as a head. To avoid cavitation, the available NPSH must be greater than the manufacturer's value for the required NPSH. An example of the application of the NPSH is given in Example 22.5.

Centrifugal pumps are limited by the rate of rotation of the impeller to the pump head they can achieve in a single stage. A typical maximum head for a single stage is 500 ft. By going to multiple stages, heads as high as at least 3,200 ft can be achieved.

Positive-Displacement Pumps

Positive-displacement pumps, either reciprocating or gear, are essentially metering pumps designed to deliver a volumetric flow rate, Q, that is independent of the required pump head, H. Thus, the characteristic curve of a positive-displacement pump, if it can be called that, is a vertical line on a plot of Q as a function of H. The pump head is limited only by the Hp of the driver, the strength of the pump, and/or possible leakage through clearances between moving pistons, plungers, gears, or screws, and stationary cylinders or casings. Unlike centrifugal pumps, where the flow rate can be changed (while staying on the characteristic curve) by adjusting a valve on the discharge line, the flow rate of a positive-displacement pump must be changed by a bypass or with a speed changer on the motor. The efficiency of positive-displacement pumps is greater than for centrifugal pumps because less friction occurs in the former, and cavitation is not usually a concern with positive-displacement pumps.

The three main classes of reciprocating pumps are piston, plunger, and diaphragm, which are shown schematically in Figure 20.6. They all contain valves on the inlet and outlet. During suction, a chamber is filled with liquid, with the inlet valve open and the outlet valve closed. During discharge of the liquid from the chamber, the inlet valve is closed and the outlet valve opened. This type of action causes pressure pulsations, which cause a fluctuating flow rate and discharge pressure. These fluctuations can be reduced by employing a gas-charged surge chamber in the discharge line and/or by using multiple cylinders in parallel. In addition, if pistons are used, the pump can be double-acting, with chambers on either side of the piston. With a plunger, only a single-action is used. Reciprocating pumps with a flexible diaphragm of metal, rubber, or plastic eliminate packing and seals, making them useful for hazardous or toxic liquids.

Rotary pumps include gear pumps and screw pumps, which are shown schematically in Figure 20.7. These must be designed to tight tolerances to avoid binding and excessive

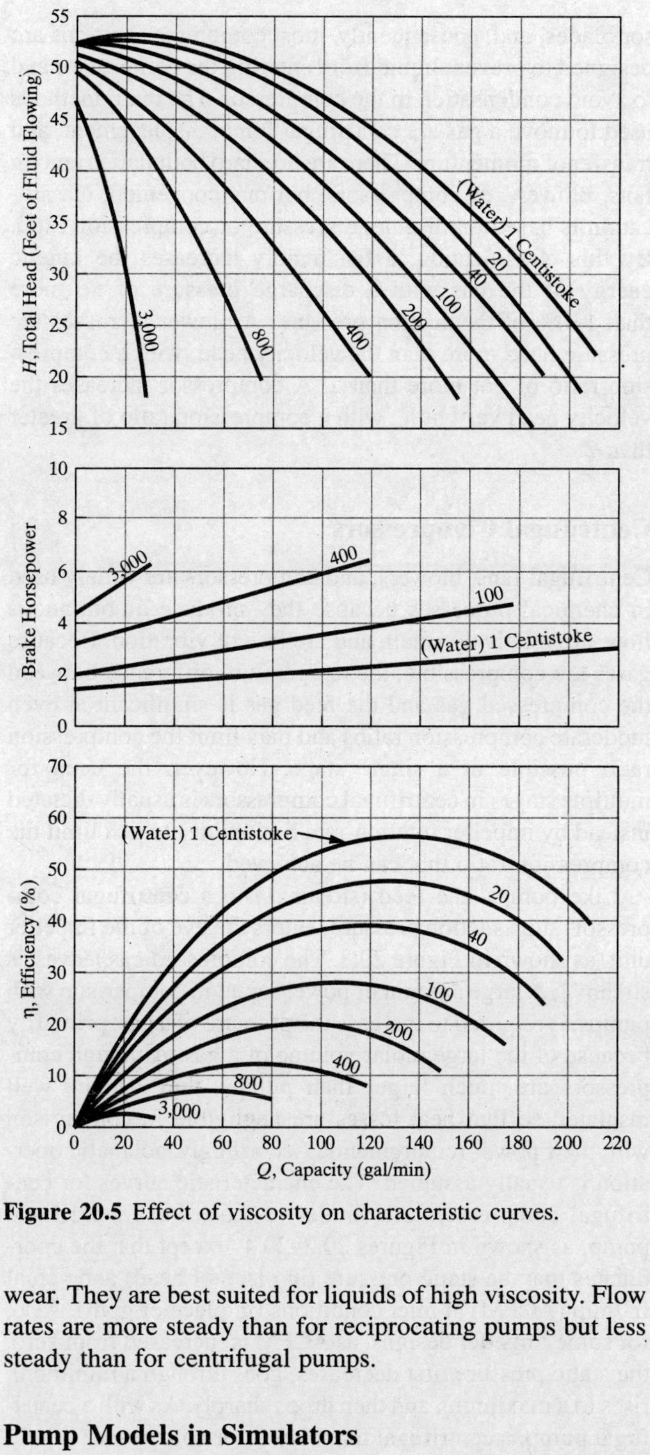

Figure 20.5 Effect of viscosity on characteristic curves.

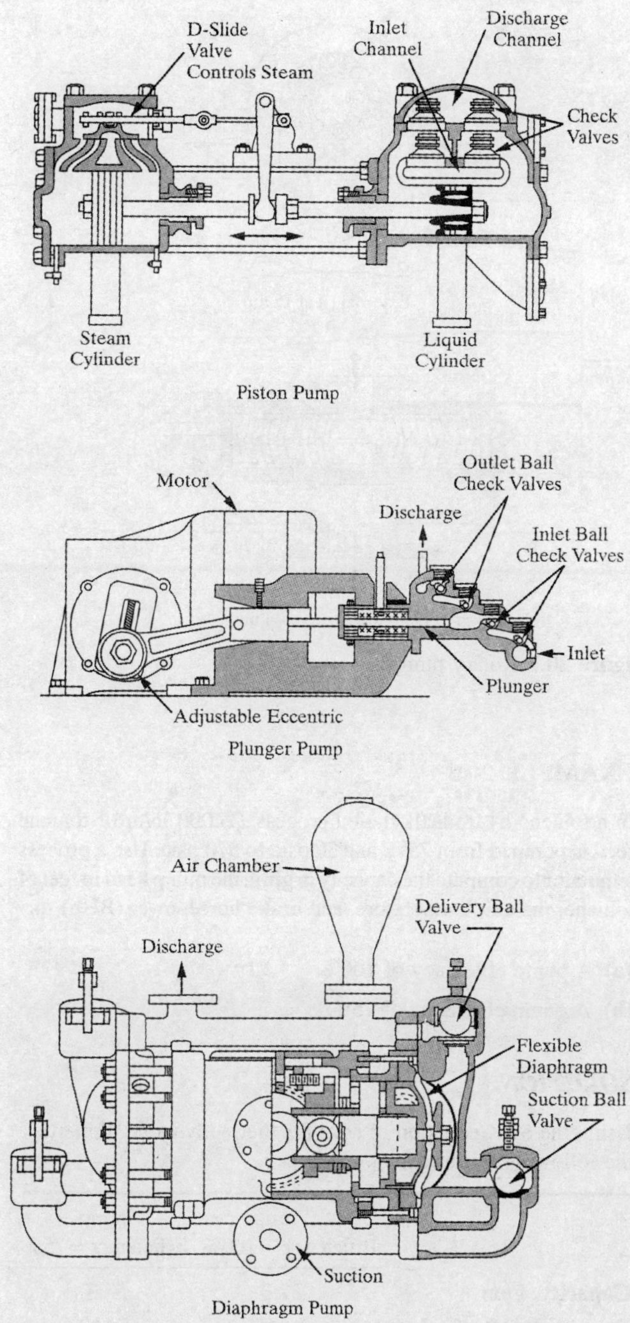

Figure 20.6 Reciprocating pumps.

wear. They are best suited for liquids of high viscosity. Flow rates are more steady than for reciprocating pumps but less steady than for centrifugal pumps.

Pump Models in Simulators

The pump models in process simulators do not differentiate between centrifugal pumps and positive-displacement pumps when calculating theoretical power requirement from the product of the capacity and required head [Eq. (20.1)]. In some cases the models do utilize built-in efficiency equations, which differentiate between the two types of pumps when computing the brake horsepower. Most models calculate a discharge temperature, which is based on the small variation of density with temperature and an assumption that all of the pump inefficiency produces friction that causes an increase in the liquid temperature. In most cases, the temperature and enthalpy changes of the liquid across the pump are small. Simulators do not provide built-in characteristic curves to help select a suitable centrifugal pump nor do they consider multiple stages or cylinders. Pump subroutines are discussed further in the Program and Simulation Files folder, which can be downloaded from the Wiley Web site associated with this book, accompanied by a video of an industrial-scale centrifugal pump (*ASPEN → Pumps, Compressors & Expanders → Pumps* and *HYSYS → Pumps, Compressors & Expanders → Pumps*).

www.wiley.com/go/global/seider

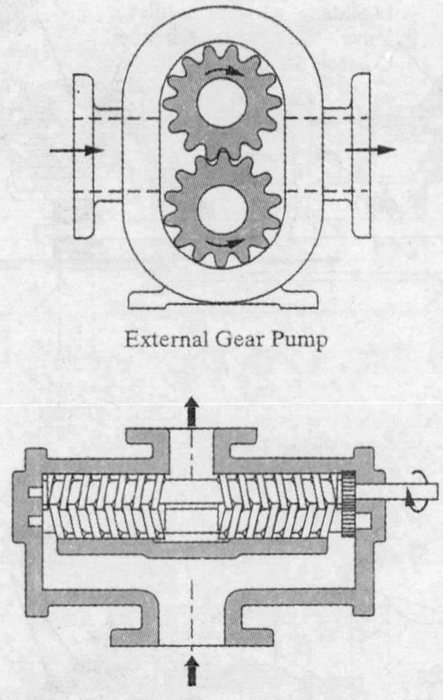

Figure 20.7 Rotary pumps.

EXAMPLE 20.1

In a toluene hydrodealkylation process, 25,000 lb/hr of toluene feed is pumped from 75°F and 30 psia to 570 psia. Use a process simulator to compute the capacity in gpm, the pump head in feet of toluene, the exit temperature, and brake horsepower (BHp) for:

(a) A pump efficiency of 100%.

(b) A pump efficiency of 75%.

SOLUTION

Using the SRK equation of state for thermodynamic properties, the following results are obtained

	Pump Efficiency = 100%	Pump Efficiency = 75%
Capacity, gpm	57.3	57.3
Pump head, ft of toluene	1,440	1,440
Outlet temperature, °F	75.78	77.37
Brake horsepower, BHp	18.2	24.3

The temperature rise is very small even for the 75% efficiency case. The pump head is well above the limit of 500 ft for a single-stage centrifugal pump. Therefore, a multistage centrifugal pump would be required.

20.2 COMPRESSORS AND EXPANDERS

Gas compressors (including fans and blowers), unlike pumps, are designed to increase the velocity and/or pressure of gases rather than liquids. In fact, small amounts of liquid can cause significant amounts of degradation to the compressor blades, and, consequently, most compressor systems are designed to prevent liquid from entering the compressor and to avoid condensation in the compressor. The main methods used to move a gas are centrifugal force, displacement, and transfer of momentum. There are no sharp boundaries among fans, blowers, or compressors, but one convenient classification is based on discharge pressure or compression ratio. By this classification, a fan mainly increases the kinetic energy of the gas with a discharge pressure of no more than 110% of the suction pressure. A blower increases the pressure head more than the velocity head, with a compression ratio of not more than 2. A compressor increases the velocity head very little, with a compression ratio of greater than 2.

Centrifugal Compressors

Centrifugal fans, blowers, and compressors are widely used in chemical processes because they produce a continuous flow, are relatively small, and are free of vibration. Because gases are compressible, the temperature difference between the compressed gas and the feed gas is significant at even moderate compression ratios and may limit the compression ratio possible in a single stage. However, the need for multiple stages in centrifugal compressors is usually dictated instead by impeller rotation-rate limitations, which limit the compression ratio that can be achieved.

Like pumps, the feed (stream 1) to a centrifugal compressor, at its suction pressure, enters the eye of the impeller unit, as shown in Figure 20.8. The compressed gas leaves as stream 2. A large amount of power input, in comparison with pumps, is required to increase the pressure of a gas, primarily because of the large molar volume of a gas. Although compressors are much larger than pumps, they can be well insulated so that heat losses are negligible in comparison with their power requirements. Accordingly, adiabatic operation is usually assumed. The characteristic curves for centrifugal compressors are similar to those for a centrifugal pump, as shown in Figures 20.2–20.4, except that the coordinates may be static pressure (in place of head) and actual ft^3/min (ACFM) at inlet conditions (in place of gpm). Also, for some impeller designs, as ACFM is increased from zero, the static pressure first decreases, goes through a minimum, rises to a maximum, and then drops sharply. As with a centrifugal pump, a centrifugal fan, blower, or compressor should be selected for operation at the point of maximum efficiency on the characteristic curve.

Positive-Displacement Compressors

Positive-displacement fans, blowers, and compressors are similar in action to positive-displacement pumps, and include reciprocating compressors, two- or three-lobe blowers, and screw compressors. However, with gases, the almost vertical characteristic curves bend to the left more than for liquids because of the greater tendency for slip.

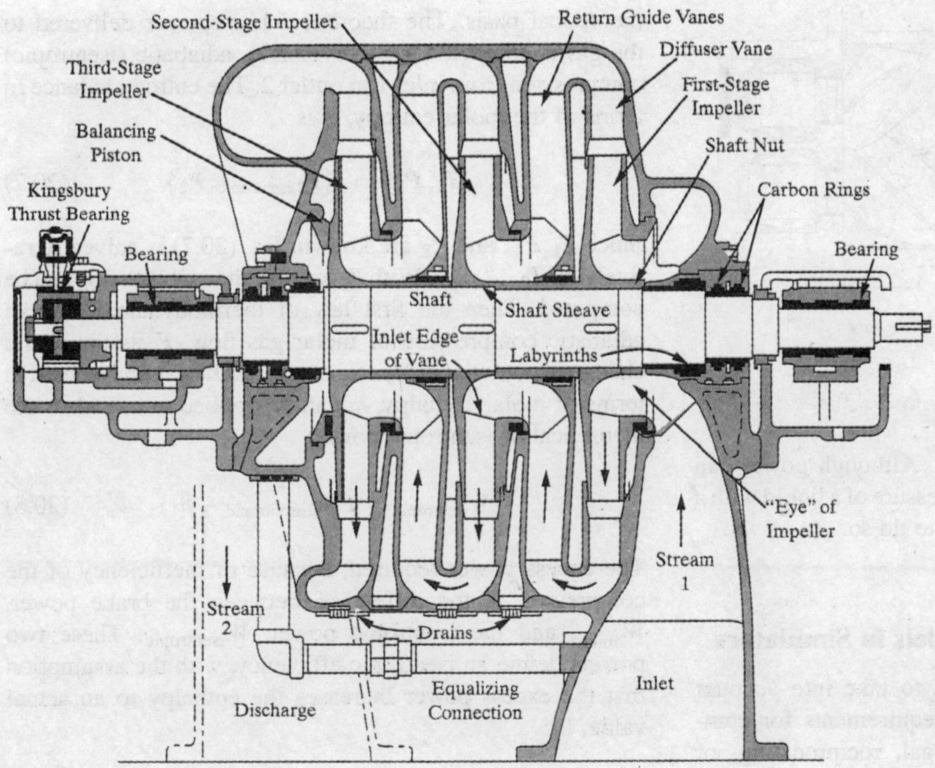

Figure 20.8 Cross section of a three-stage centrifugal compressor.

Reciprocating compressors use pistons with either single- or double-action. As discussed in Section 22.5, compression ratios in a single stage are limited to a discharge temperature of 400°F. This corresponds to compression ratios about 2.5–6 as the specific heat ratio of the gas decreases from 1.67 (monotomic gas) to 1.30 (methane). Compression ratios of even 8 are possible with high-molecular-weight gases. If higher compression ratios are needed, a multistage reciprocating compressor is used with intercooling, usually by water. See, for example, the video of a two-stage reciprocating compressor with an intercooler in the multimedia modules *(ASPEN → Pumps, Compressors & Expanders → Compressors & Expanders* and *HYSYS → Pumps, Compressors & Expanders → Compressors & Expanders).* Reciprocating compressors must be protected by knock-out drums to prevent the entry of liquid.

A lobed blower, shown in Figure 20.9, is similar to a gear pump. Both two- and three-lobe units are common. They are limited to low capacity and low heads because shaft deflection must be kept small to maintain clearance between the rotating lobes and the casing. If higher compression ratios are required, multiple stages can be used. A screw compressor, as shown in Figure 20.10, with two screws, male and female, that rotate at speeds typical of centrifugal pumps, can operate at higher capacities to give higher compression ratios that may be limited by temperature. If so, higher compression ratios can be achieved with multiple stages separated by intercoolers. Screw compressors can run dry or can be flooded with oil.

Expanders

Expanders (also called turboexpanders and expansion turbines) are often used in place of valves to recover power from a gas when its pressure must be decreased. At the same time, the temperature of the gas is reduced, and often the chilling of the gas is more important than the power recovery. Most common is the radial-flow turbine, as shown in Figure 20.11, which resembles a centrifugal pump and can handle inlet pressures up to 3,000 psi and temperatures up to 1,000°F. With an impeller tip speed of 1,000 ft/s, a single stage of expansion can reduce the enthalpy of the gas by as much as 50 Btu/lb (116 kJ/kg). When calculations show that condensation may occur during the expansion, the expander must be designed to avoid erosion of the impeller. Expanders are

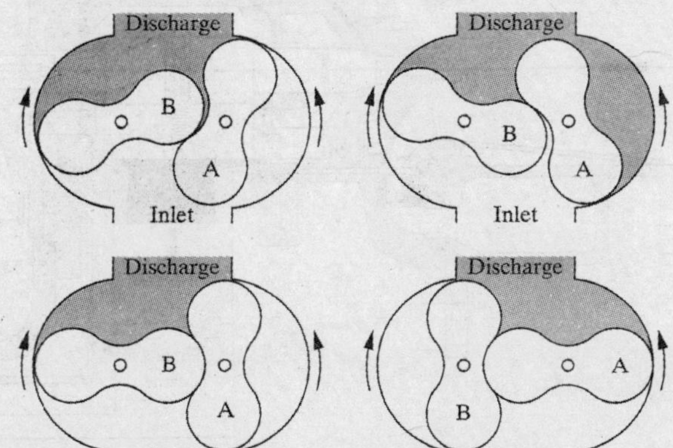

Figure 20.9 Lobed blower.

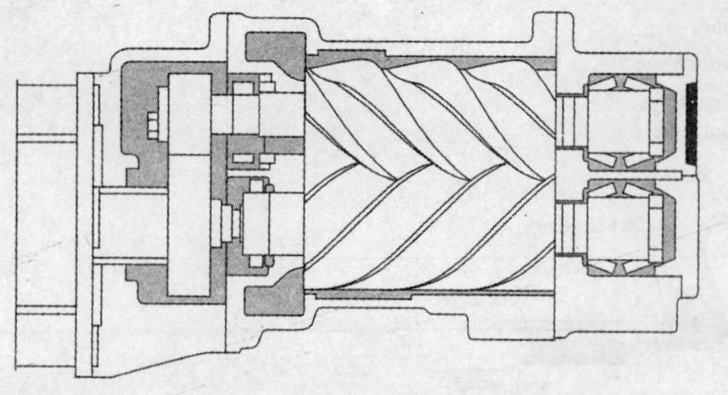

Figure 20.10 Screw compressor.

widely used at cryogenic conditions. Although power can also be recovered by decreasing the pressure of a liquid with a turbine, it is usually not economical to do so.

Compressor and Expander Models in Simulators

Either of two methods can be used to take into account efficiency when calculating power requirements for compressors, whether they are centrifugal, reciprocating, or screw. One method is the *polytropic method*, based on the expression PV^n = constant during compression, where V is the gas volume and n is the polytropic coefficient, which lies between 1 and the specific heat ratio.

Since the advent of simulation programs that routinely calculate entropy, the second method, called the *isentropic method*, has become preferred, because it has a sound theoretical basis. The theoretical horsepower delivered to the gas is computed for a reversible, adiabatic (isentropic) compression from inlet 1 to outlet 2. The entropy balance in terms of the molar entropy, s, is

$$s\{T_1, P_1\} = s\{T_{2,\text{isentropic}}, P_2\} \tag{20.7}$$

Since T_1, P_1, and P_2 are known, Eq. (20.7) is solved iteratively for $T_{2,\text{isentropic}}$. With P_2 known, the exit enthalpy can be computed. Then the first law of thermodynamics for an adiabatic compression of molar gas flow, F, assuming no change in potential or kinetic energy of the gas and written in terms of molar enthalpy, h, can be applied to calculate the theoretical or isentropic power:

$$\dot{W}_{\text{isentropic}} = F(h_{2,\text{isentropic}} - h_1) \tag{20.8}$$

The excess power required, because of inefficiency of the compressor, is the difference between the brake power, $\dot{W}_{\text{brake}}$, and the isentropic power, $\dot{W}_{\text{isentropic}}$. These two powers define an isentropic efficiency, with the assumption that the excess power increases the enthalpy to an actual value, h_2:

$$\eta_s = \frac{\dot{W}_{\text{isentropic}}}{\dot{W}_{\text{brake}}} = \frac{h_{2,\text{isentropic}} - h_1}{h_2 - h_1} \tag{20.9}$$

The actual temperature of the discharged compressed gas, T_2, is then computed iteratively from the actual enthalpy, h_2. The actual temperature T_2 can be significantly higher than the isentropic temperature $T_{2,\text{isentropic}}$.

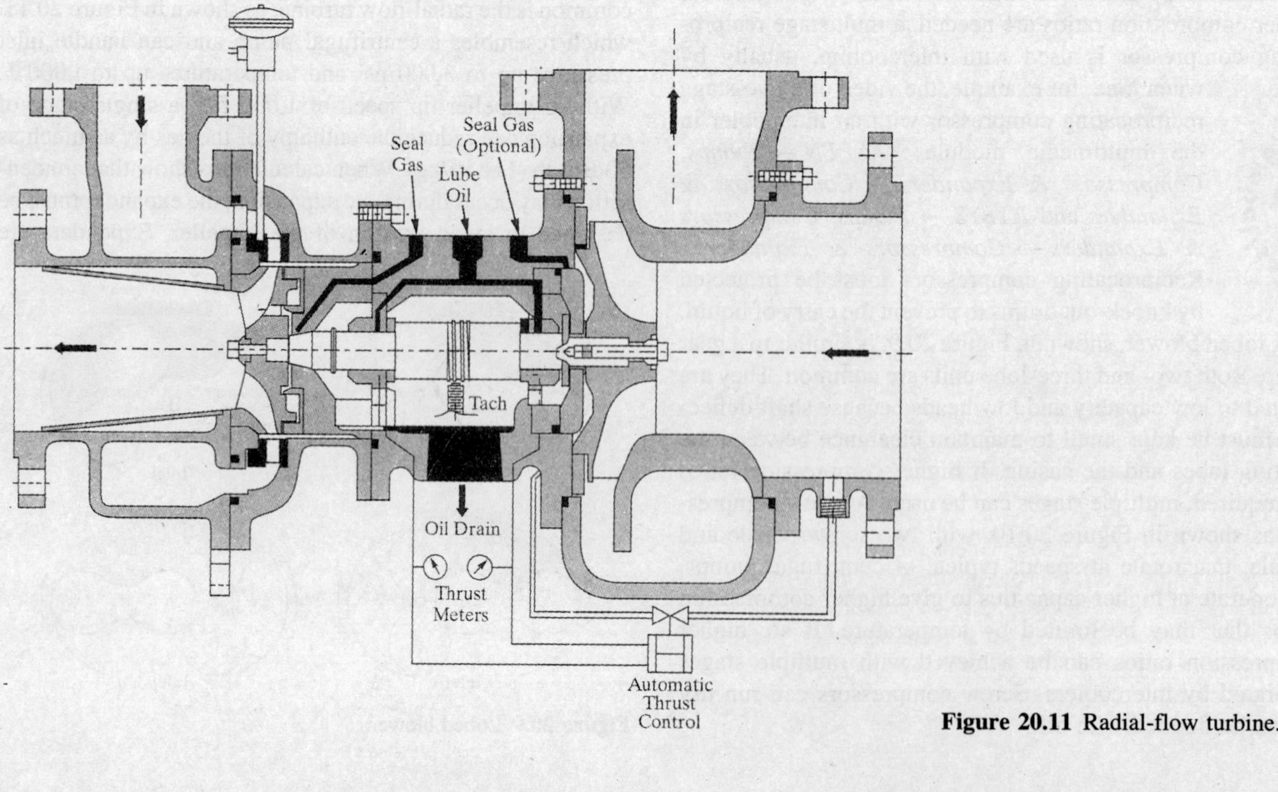

Figure 20.11 Radial-flow turbine.

The isentropic method is also applied to an expander. Eq. (20.7) is used for calculating the isentropic exit temperature, but takes into account possible condensation of the gas. Like the exit pressure, the exit temperature will be less than the inlet value. Then, the exit isentropic enthalpy is computed, from which Eq. (20.8) is used to calculate the power recovered, which will be a negative value. The effect of the expander efficiency is just the opposite of the compressor efficiency, as indicated by a revision of Eq. (20.9) for applicability to expanders:

$$\eta_s = \frac{\dot{W}_{\text{brake}}}{\dot{W}_{\text{isentropic}}} = \frac{h_1 - h_2}{h_1 - h_{2,\text{isentropic}}} \qquad (20.10)$$

Because of inefficiency, the brake horsepower recovered is less than the isentropic horsepower, and the exit temperature is higher than the isentropic exit temperature. Thus, inefficiency will reduce the tendency for condensation to occur.

EXAMPLE 20.2

A natural gas stream of 5,000 kmol/hr at 25°C and 1,500 kPa contains 90% methane, 7% ethane, and 3% propane. Currently this gas is expanded adiabatically across a valve to 300 kPa. Use a process simulator to determine the exit temperature and recovered power if the valve is replaced with:

(a) an isentropic expansion turbine, and

(b) an expansion turbine with an isentropic efficiency of 75%.

SOLUTION

Using the SRK equation of state for thermodynamic properties, the following results are obtained.

	Valve	Isentropic Expander	Expander
Isentropic efficiency, η_s	—	1.00	0.75
Exit temperature, °C	18.5	−69.7	−47.1
Power recovered, kW	0	4,480	3,360
Power recovered, BHp	0	6,010	4,510

The results show that not only does the expander recover a significant amount of power, but it is also very effective in reducing the temperature, compared to the valve. However, the actual exit temperature is almost 20°C higher than the isentropic value. In all cases, no condensation is found to occur, since the dew point of the exit gas at 300 kPa is computed to be −83.2°C.

20.3 SUMMARY

Having studied this chapter, the reader should

1. Be able to explain how the more common types of pumps, compressors, and expanders work.
2. Understand the types of calculations made by a simulator for pumps, compressors, and expanders.

REFERENCES

1. GREEN, D.W., and R.H. PERRY, Eds., *Perry's Chemical Engineers' Handbook*, 8th ed., McGraw-Hill, New York, 2008.
2. SANDLER, H.J., and E.T. LUCKIEWICZ, *Practical Process Engineering*, McGraw-Hill, New York, 1987.

EXERCISES

20.1 Liquid oxygen is stored in a tank at −298°F and 35 psia. It is to be pumped at 100 lb/s to a pressure of 300 psia. The liquid oxygen level in the tank is 10 ft above the pump, and friction and acceleration losses from the tank to the pump suction are negligible. If the pump efficiency is 80%, calculate the BHp, the oxygen discharge temperature, and the available NPSH using a simulator to make the calculations.

20.2 Use a simulator to design a compression system with intercoolers to compress 600 lb/hr of a mixture of 95 mol% hydrogen and 5 mol% methane at 75°F and 20 psia to a pressure of 600 psia, if the maximum exit temperature from a compressor stage is 400°F and compressor efficiency is 80%. Assume gas outlet temperatures from the intercoolers at 120°F. For each compressor stage, compute the BHp. For each intercooler, compute the heat duty in Btu/hr.

20.3 Superheated steam, available at 800 psia and 600°F, is to be expanded to a pressure of 150 psia at the rate of 100,000 lb/hr. Calculate, with a simulator, the exit temperature, phase condition, and Hp recovered for:

(a) an adiabatic valve,

(b) an isentropic expansion turbine, and

(c) an expansion turbine with an isentropic efficiency of 75%.

20.4 Propane gas at 300 psia and 600°F is sent to an expansion turbine with an efficiency of 80%. What is the lowest outlet pressure that can be achieved without condensing any of the propane?

Chapter 21

Polymer Extrusion

21.0 OBJECTIVES

Polymer compounding, involving extrusion devices, is often required for the manufacture of *industrial* and *configured consumer* products. This chapter provides detailed design techniques for twin-screw polymer compounders. Guidelines are provided for characterizing the materials and the required processing steps. Then, heuristics are presented for selecting a feeding strategy, for screw design, and for setting the operating parameters.

After completing this chapter, the reader should be able to

1. Select and configure extruders for polymer-compounding applications.
2. Estimate the key performance parameters for specific designs.

21.1 INTRODUCTION

Compounding often refers to the mixing and blending of various ingredients into homogeneous and uniform mixtures. In this chapter, as illustrated in Figure 21.1, compounding is limited to the continuous mixing of non-reactive, melt-processable polymers and functional additives and/or fillers into homogenous and uniform mixtures, to be subsequently formed into useful articles such as films, sheets, pellets, foams, nonwoven fibers, molded parts, pharmaceutical tablets, pasta, breakfast cereals, snacks, and candies. Compounding has been used widely in the food and plastics industries, and recently it has been applied to the mixing and shaping of pharmaceutical products. For reactive-extrusion processes, the reader should refer to an excellent discussion by Xanthos (1992).

A compounding process involves not only mixing, but also feeding, solids conveying, melting, melt-pumping, and devolatilization/degassing steps. Its success depends on understanding the response of materials to the conditions imposed by the compounding process and the operation of its extruder. A goal of this chapter is to convey a basic understanding of these interactions to achieve successful compounding processes. The reader will find excellent descriptions, with models, of the compounding processes in books by Tadmor and Klein (1978), Agassant and co-workers (1991), Todd (1998), and Tadmor and Gogos (2006). Also, excellent discussions of devolatilization are given by Welling (1980), Biesenberger (1983), and Albalak (1996). Finally, the development of polymer structures during compounding is discussed by Wilkinson and Ryan (1998).

This chapter is presented in six sections that focus on:

a. Compounding technologies

b. Compounding machinery

c. Understanding polymeric materials

d. Feeding protocols

e. Screw design

f. Setting the processing conditions

21.2 COMPOUNDING TECHNOLOGIES

Mixing can be divided into two types: *distributive* and *dispersive*. Distributive mixing refers to the blending of components into spatially uniform mixtures without size changes of the ingredients. On the other hand, dispersive mixing is distributive mixing accompanied by the breakup of the ingredients into sufficiently small particles. For immiscible polymer blends, dispersive mixing involves the breakup of immiscible polymeric domains into the desired domain size or morphology. These types are illustrated schematically in Figure 21.2.

The quantification of the goodness of mixing remains an active research area, with microscopic analysis often used to determine the morphology of a mixture and to judge the degree of mixing. While the major drawback of microscopic analysis is its local examination area, usually too small to represent the degree of mixing throughout, microscopic analysis continues to be the most frequently used measure of mixing. Recently, however, chaos theory has been applied

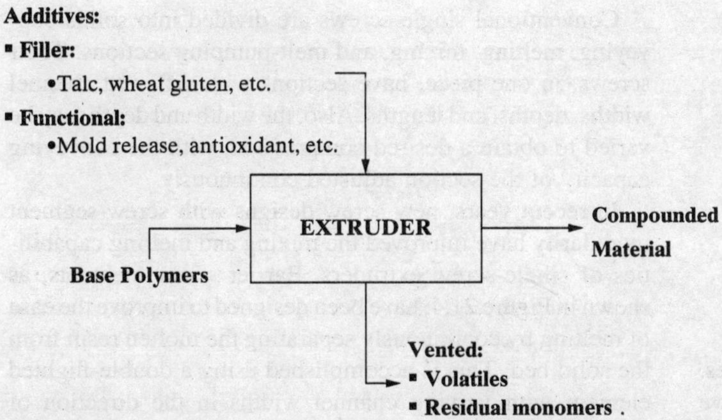

Figure 21.1 Schematic of compounding processes.

to characterize simple flows in simple geometries. But, for complex flows and geometries, these characterizations are quite limited. Rauwendaal (1991, 1998) discusses mixing in polymer processing.

The choice of the best compounding equipment for a given mixing task depends on many factors such as the:

- Number of and physical form of the raw materials
- Physical, thermal, and rheological properties of the ingredients
- Loading levels and productivity constraints
- Kind of and desired degree of mixing

Extruders are commonly used to perform the compounding of foods (biopolymers) and melt-processable polymers. Compounding refers to the mixing and blending of many ingredients to produce a uniform product with desired properties. It involves the following steps: pretreatment, feeding, melting, mixing (dispersive and distributive), venting (degassing/devolatilation), and pressure generation for downstream shaping processes.

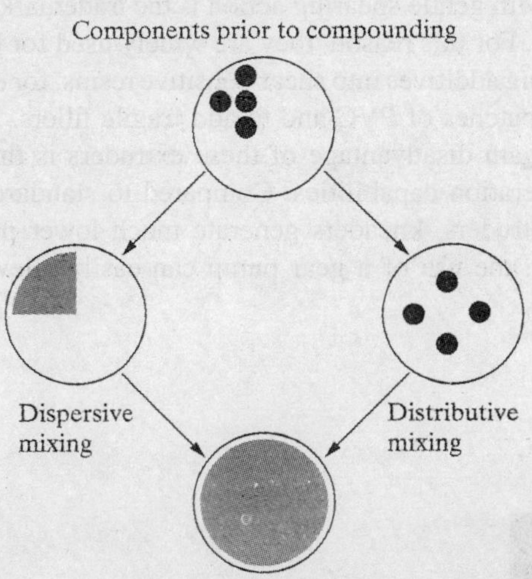

Figure 21.2 Distributive and dispersive mixing (Martin, 2008). Reprinted with permission.

Pretreatment, such as the drying or pre-blending of raw materials, may be necessary. For the former, fluctuations in temperature or humidity during storage may introduce unwanted moisture into hydrophilic polymers or additives. This can degrade polymers like polyamides and polyesters, which are particularly sensitive to moisture in the molten state. For the latter, hoppers equipped with a nitrogen blanket may be necessary to prevent oxidative degradation.

Feeding is often accomplished using multiple feeders, or the ingredients may be dry-blended and fed using a single feeder. Gear pumps are used to feed liquid ingredients into their injection ports.

After melting and mixing in extruders, venting may be necessary to remove the air or moisture often associated with powdered fillers, often in the form of residual monomers and other volatiles. Venting requires a melt seal upstream of the vent port to prevent air from entering the feed port; and to achieve venting, the screw is designed to run partially full of polymer in the venting zone. Also, the volatiles are removed under vacuum to prevent condensate from returning to the extruder. Venting is needed to prevent bubbles from forming in the polymer, which can result in pellets, for example, having non-uniform density.

At the end of the extruder, pressure builds to overcome the pressure drop in downstream equipment (e.g., pelletizing dies). Because most polymers are incompressible, temperatures rise in the pressure-development zone. Also, because polymers are poor heat conductors, heat is often removed inefficiently, possibly causing undesirable degradation. Furthermore, extruders with diameters larger than six inches normally operate near adiabatic conditions, with the rate of heat transfer less than the rate of heat generation.

The classification of extrusion devices commonly used for polymer compounding is shown in Figure 21.3.

Non-reciprocating single-screw extruders are the least favored for difficult-to-mix materials, although the recent development of advanced single-screw devices has broadened their usage. Normally, these are cost-effective and have the ability to generate appreciable pressure increases.

Reciprocating single-screw extruders are used more widely than non-reciprocating single-screw extruders, as

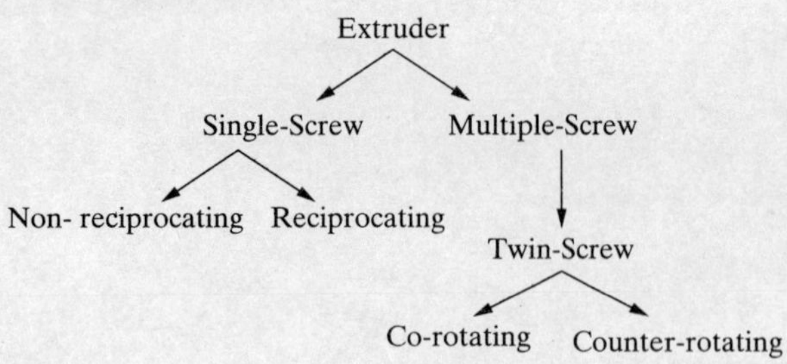

Figure 21.3 Classification of compounding extruders.

they normally provide good mixing at low shear energies. Consequently, these are often used for shear-sensitive materials, such as PVC, and conductive carbon for static-dissipation applications. Their main limitations often stem from their ineffectiveness in pressure generation and devolatilization.

Twin-screw extruders, especially co-rotating devices, are the most commonly used for compounding due to their versatility and their mixing effectiveness, even when the most difficult mixing requirements must be satisfied.

Recently, multi-screw extruders have become available, with the promise to provide better mixing, increased productivity, and better devolatilization capabilities.

21.3 COMPOUNDING MACHINERY

In this section, three extruder types, single-screw, reciprocating single-screw, and twin-screw, are discussed. Both reciprocating and non-reciprocating single-screw extruders usually provide sufficient distributive mixing. Non-reciprocating single-screw extruders excel in providing distributive mixing action at low shearing forces. Twin-screw extruders are preferred to accomplish dispersive mixing.

Single-Screw Extruder

Single-screw extruders are commonly used to add colorants, fillers, and additives into resins with loading levels up to 30 wt%. They are also commonly used for compounding reinforced composites, foams, and for plastics recycling. The advantages of single-screw extruders over other continuous systems include lower costs, less maintenance, and simpler operation. The main disadvantage is their limited dispersive mixing capability. Pre-compounded master batches are often a viable solution to this shortcoming.

Conventional single screws are divided into solids conveying, melting, mixing, and melt-pumping sections. Their screws, in one piece, have sections with different channel widths, depths, and lengths. Also, the width and depth may be varied to obtain a desired compression, with the conveying capacity of the section adjusted continuously.

In recent years, new screw designs with screw-segment modularity have improved the mixing and melting capabilities of single-screw extruders. Barrier screw elements, as shown in Figure 21.4, have been designed to improve the ease of melting by continuously separating the molten resin from the solid bed. This is accomplished using a double-flighted element with varying channel widths in the direction of material flow; that is, with the width of the solid channel decreasing and the width of the melt channel increasing along the length of the screw.

Mixing screws in single-screw extruders are classified according to their distributive and dispersive capabilities. An example of a distributive mixing element is the pin mixer shown in Figure 21.5. Dispersive mixing elements include the Maddox mixer and the cavity-transfer mixer, also shown in Figure 21.5.

In the late 1990s, their leading manufacturers included Davis Standard, Killion, and Black Clausen. Todd (1998) provides an extensive discussion and description of the devices from each manufacturer.

Reciprocating Single-Screw Extruder

Reciprocating single-screw extruders are also known as co-kneaders, as shown in Figure 21.6. Typical is the Buss Kneader, which involves a single screw that reciprocates. Unlike standard single-screw extruders, the barrel of the kneaders has a series of teeth that match the wedge-shaped channels on the screw. During each complete rotation, the teeth travel back and forth through the wedge-shaped channels providing an intense, yet gentle, mixing action. Effective mixing with gentle shearing action is the trademark of these kneaders. For this reason, they are widely used for blending and mixing additives into shear-sensitive resins, for example, to color batches of PVC and to add fragile fillers.

The main disadvantage of these extruders is their pressure-generation capabilities. Compared to standard single-screw extruders, kneaders generate much lower pressures. However, the use of a gear pump can easily alleviate this drawback.

Figure 21.4 Barrier screw. From www.rdray.com/BarrierScrews.htm. Reprinted with permission.

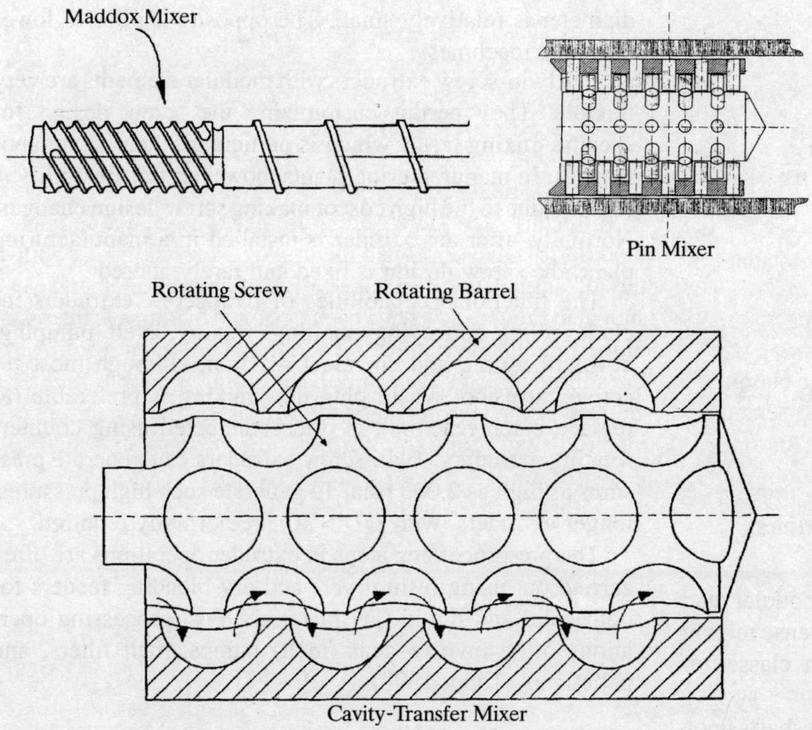

Figure 21.5 Distributive and dispersive mixing elements (Rauwendaal, 2004; Dray, 2006).

Twin-Screw Extruder

Twin-screw extruders are either conical or non-conical and intermeshing or non-intermeshing. The conical machines yield higher pressures and are often used for higher-density materials in the melt rather than in the solid state. The intermeshing machines have tight shaft-to-shaft clearances and, consequently, provide excellent wiping action. However, when the wiping action generates shear-induced degradation, it is reduced by increasing the shaft-to-shaft clearance, which reduces the wiping action—leading toward non-intermeshing performance. Three configurations, two intermeshing and one non-intermeshing, are shown in Figure 21.7. Herein, only fully intermeshing extruders are discussed.

Twin-screw extruders are often considered to be the most effective and flexible continuous mixers. They are regarded as general-purpose and are the most popular in industry.

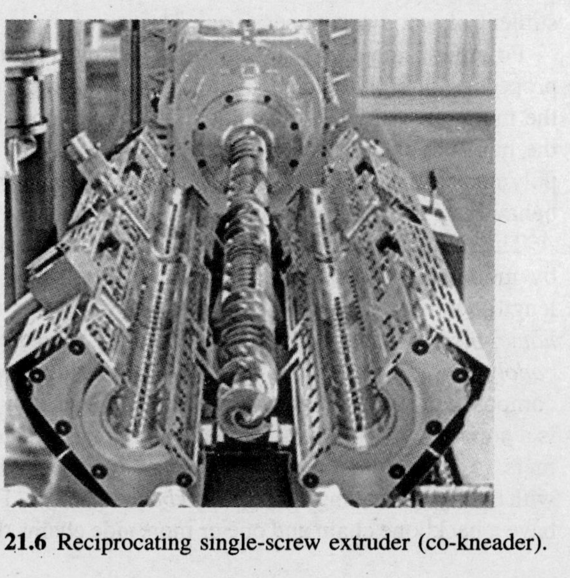

Figure 21.6 Reciprocating single-screw extruder (co-kneader).

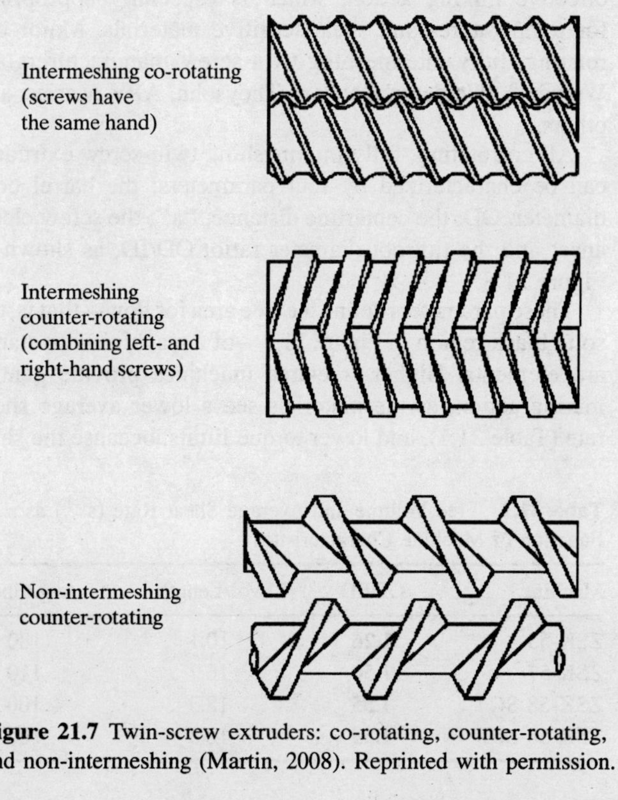

Figure 21.7 Twin-screw extruders: co-rotating, counter-rotating, and non-intermeshing (Martin, 2008). Reprinted with permission.

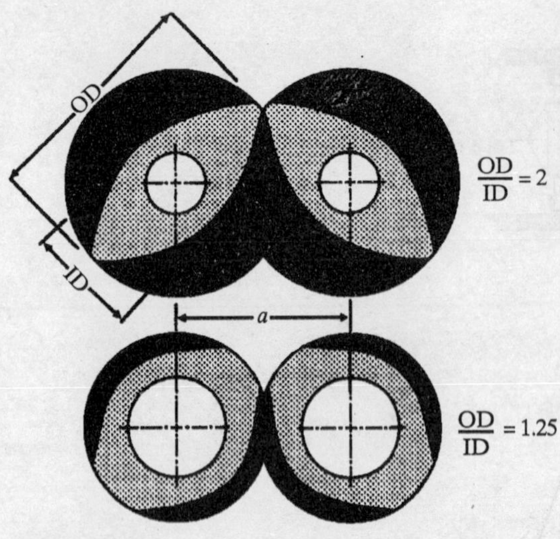

Figure 21.8 Twin-screw extruder machine characteristics.

Their flexibility is closely related to their modular and exchangeable screw elements, which provide intense mixing action, so intense that it can degrade certain classes of materials.

Twin-screw extruders are equipped with twin shafts upon which modular screw elements can be arranged. The elements provide either conveying, kneading, or specialty mixing. Most elements are double flighted (bi-lobal), but single- and tri-lobal elements are available from various manufacturers. A double-flighted (bi-lobal) element, involving two tips per screw, is shown in Figure 21.8, and a tri-lobal element, involving three tips per screw, is shown later in Figure 21.21. The latter is said to provide gentler, yet effective mixing action, which is especially appropriate for temperature- and shear-sensitive materials. Major co-rotating, fully intermeshing twin-screw manufacturers are Werner & Pfleiderer, Berstorf, Theysohn, APV, Maries, and others.

All co-rotating, fully intermeshing twin-screw extruders can be characterized by four parameters: the barrel bore diameter, OD; the centerline distance, "a"; the screw clearance; and the tip/root diameter ratio, OD/ID, as shown in Figure 21.8.

These parameters define the free area for flow—that is, the solid black region in Figure 21.8—of a machine with standard elements. Higher free-area machines provide gentler mixing action, as the materials see a lower average shear rate (Table 21.1), and lower torque limits, because the shaft diameter is relatively small. The opposite applies to lower free-area machines.

Table 21.1 Free Volume and Average Shear Rate (s^{-1}) as a Function of Machine Characteristics

Machine	OD/ID	Vol/Length	Avg. Shear
ZSK-53	1.26	10.1	180
ZSK-57	1.50	16.7	110
ZSK-58 SC	1.55	18.3	100
ZSK-58 MC	1.55	18.3	100

The twin-screw extruders with modular elements are very flexible. They permit customizing the screw design for specific mixing tasks, which is particularly helpful in laboratories. In manufacturing plants, however, this flexibility is reduced due to the high cost of making screw design changes. Normally, after an extruder is installed in a manufacturing plant, the screw design is fixed and rarely altered.

The functional capabilities of twin-screw extruders include solids conveying, melting, mixing, melt pumping, devolatilization, and chemical reaction, although most reactive extrusion, where plug-flow mixing is preferable (to replace batch reactors), is often conducted using counter-rotating extruders. Twin-screw extruders can generate pressures as high as 2,000 psia. To generate such high pressures, longer extruders, with $L/D > 30$, are normally required.

The pre-processing steps in extruder operations are often carried out using resin dryers and dry-blending feeders for both solid and liquid raw materials. Post-processing operations often involve gear (melt) pumps, melt filters, and pelletizers (Todd, 1998).

21.4 UNDERSTANDING POLYMERIC MATERIALS

As discussed in Sections 3.4 and 3.5, polymers are molecules having high molecular weights and composed of repeating units (monomers) connected by covalent chemical bonds. Polymeric materials can be considered to be mixtures of polymers with various chain lengths or molecular weights, as illustrated in Figure 21.9. While the term "polymer" in popular usage suggests "plastic," polymers comprise a large class of natural and synthetic materials with a variety of properties and purposes. Natural polymer materials such as shellac and amber have been in use for centuries. Biopolymers such as proteins (for example, hair, skin, and bone components) play crucial roles in biological processes. Other natural polymers include cellulose, which is the main constituent of wood and paper.

Polymeric materials can be described by: (1) structural properties directly related to the physical arrangement of the monomers along the backbone of the polymer chain at the nano- or microscopic scale, (2) the morphology of the polymer matrix at the mesoscopic scale, and (3) the bulk behavior at the macroscopic scale.

The structural properties of a polymer can be characterized by monomer type(s), and the chain linearity and size (or length). Polymers that contain a single monomer type are *homopolymers*, while those with two or more types are *copolymers*. For example, polystyrene is a homopolymer composed of styrene monomers, while ethylene-vinyl acetate is a copolymer containing ethylene and vinyl acetate monomers. Also, polymers are classified as *linear* or *branched*, with the former having a *linear backbone* structure. The latter have a backbone chain and one or more side chains that form

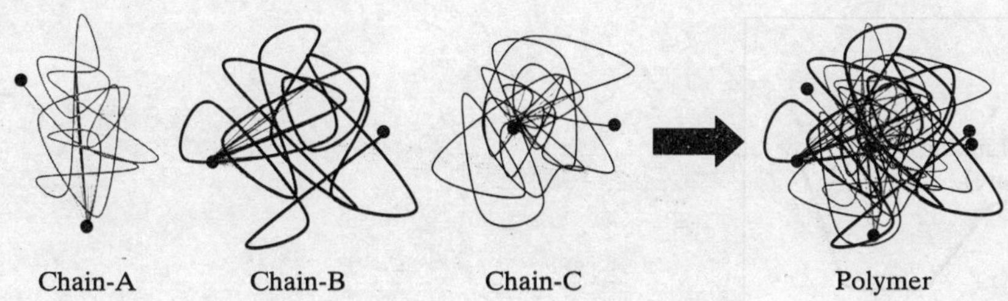

Figure 21.9 Polymeric material.

star, brush, or comb structures. These are often highly cross-linked polymers involving four or more distinct polymeric chains, often referred to as polymer networks. At an extreme, as networks become more complex, polymer *gels* are formed, where all chains have links into a single molecule. One fascinating example is an *aerogel,* whose refractive index approaches that of air, due to its largely hollow structure.

The degree of polymerization determines the chain lengths, that is, the number of monomer units that form the chain. Because polymer chains are not homogeneous in length, their molecular weights are expressed statistically using weight- or number-average, molecular-weight distributions. The ratio of these two molecular-weight measures is the so-called polydispersity index, which describes the breadth of the distribution.

A synthetic polymer may contain both crystalline and amorphous phases, which can coexist in regions of the polymer, as shown in Figure 21.10. The crystalline phase is composed of a three-dimensional ordering at the atomic length scales, as a result of intramolecular folding or stacking of the adjacent chains.

The degree is crystallinity is expressed as a weight or volume fraction of the crystalline phase. Amorphous polymers contain no crystalline regions. Most synthetic polymers are either amorphous or semi-crystalline.

For non-reactive compounding processes, the bulk properties of interest are the glass-transition temperature (T_g); the melting point (T_m); the degradation profile (that is, the weight loss during heating in thermo-gravimetric analysis (TGA), discussed later in this section; and the rheological properties.

The glass-transition temperature is the temperature at which a brittle, glassy amorphous solid undergoes a second-order phase transition into a rubbery, viscous amorphous phase. This is also known as the softening temperature. The glass-transition temperature of polymeric materials can be altered by adding a plasticizer or by changing the degree of branching or cross-linking. Plasticizers tend to reduce the glass-transition temperature, while branching and cross-linking tend to increase it.

The melting point of a polymer refers to the transition temperature between the crystalline phase and the polymer melt. Consequently, only crystalline or semi-crystalline thermoplastic polymers have melting temperatures, while amorphous polymers just soften continuously with temperature increases. So-called thermosets decompose at high temperatures rather than melt.

Differential scanning calorimetry (DSC) is widely used to determine the glass-transition and melting-point temperatures. DSC measures the energy or temperature change as a function of temperature, or time, for a specified temperature profile. A schematic of the sequence of phase/state changes that often occur during heating or cooling is shown in Figure 21.11. As the temperature increases during heating, the first phase transition takes place at the transition temperature at which the polymer begins to soften. For crystalline or semi-crystalline materials, heat is generated during the crystallization period. As the specimen is further heated, melting occurs. For certain polymers, cross-linking may occur, with heat generated during curing. Finally, as the specimen is further heated, oxidation and/or decomposition often take place.

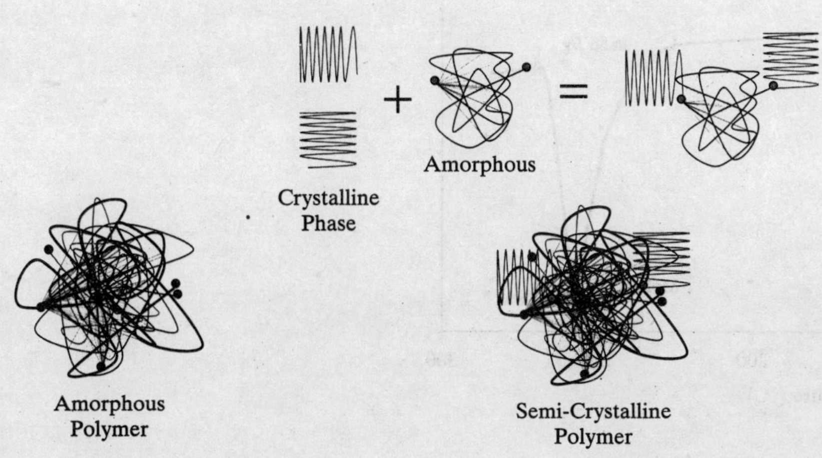

Figure 21.10 Polymer crystallinity.

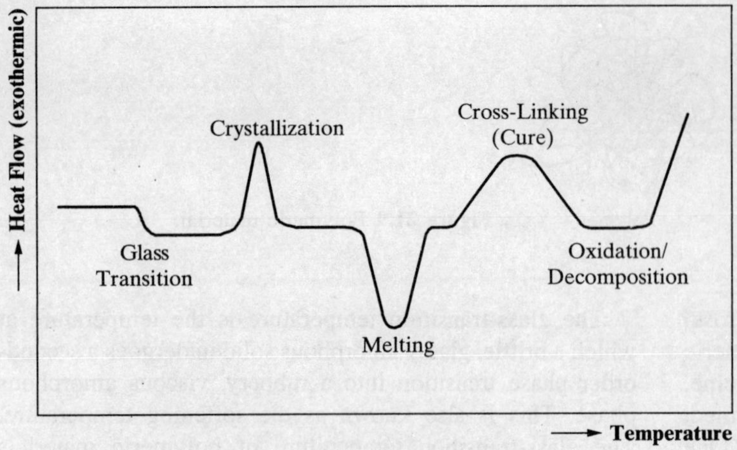

Figure 21.11 Phase changes during heating or cooling in a differential scanning calorimeter (DSC).

As an example, polyethylene terephthalate (PET) is a thermoplastic polymer of the polyester family used widely in beverage bottles, food containers, and synthetic fibers. In its natural state, it is semi-crystalline and opaque. When the melt is cooled rapidly it becomes clear and amorphous. The DSC data in Figure 21.12 show a glass-transition temperature at 75.35°C, a crystallization temperature at 139.93°C, and a melting-point temperature at 256.78°C. Note that glass-transition and melting-point temperatures are estimated for many polymers using the group-contribution methods discussed in Sections 3.4 and 3.5 (Bicerano, 1993).

Furthermore, to determine the onset of softening and/or the melting-point temperature, it is important to determine the temperature at which the polymer begins to degrade—as these two temperatures bound the compounding temperature. To determine decomposition temperatures, as well as the degassing profile, thermo-gravimetric analysis (TGA) is widely used. TGA measures the weight loss of a sample (in nitrogen or air) as a function of temperature and time for a given heating profile. As an example, Figure 21.13 shows TGA data for a typical rubber product. The left ordinate represents the remaining weight percentage in the sample and the right ordinate shows the temperature during the heating profile.

As heating occurs, at ~18 min, with the temperature at 188°C, just 1 percent of the weight of the sample has been lost; that is, the degradation of the rubber has just begun. About 25 min later, at about 450°C, the weight of the sample has decreased by about 43.25%. Then, at ~75 min, having reached 800°C, the carbon black and inorganic fillers have decomposed. Note that the onset of degradation, 188°C, provides a likely upper bound on the compounding temperature. Yet another example involves TGA data for cellulose acetate, shown in Figure 21.14, which shows the onset of degradation at ~208°C.

The viscosity of polymeric materials is a measure of the resistance to deformation by either shear or extensional stress. Whereas the viscosity of Newtonian fluids is independent

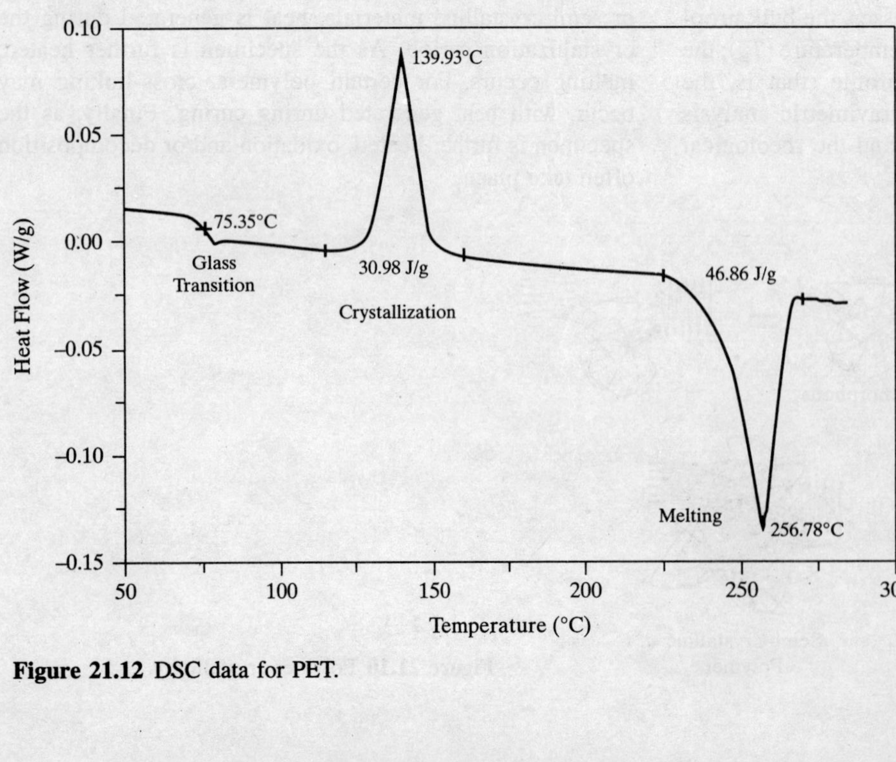

Figure 21.12 DSC data for PET.

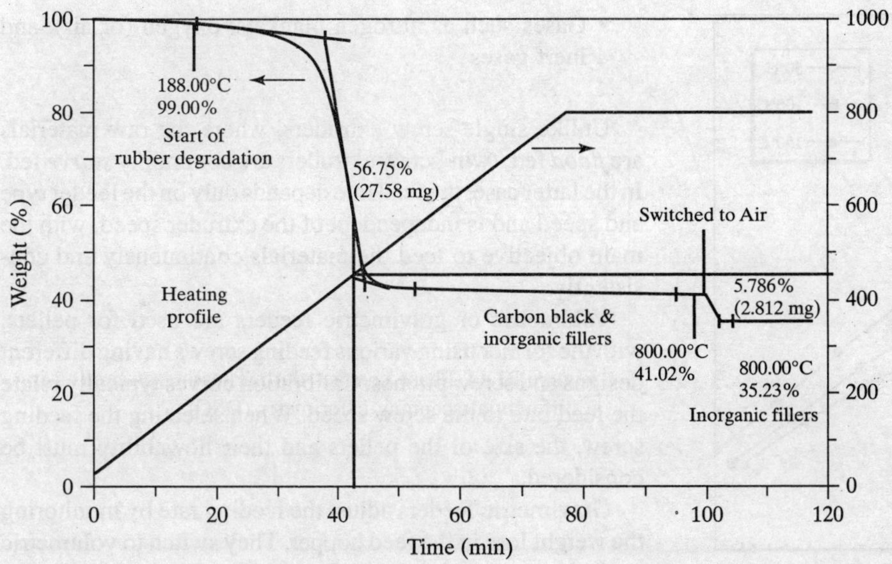

Figure 21.13 Thermo-gravimetric analysis of a rubber product.

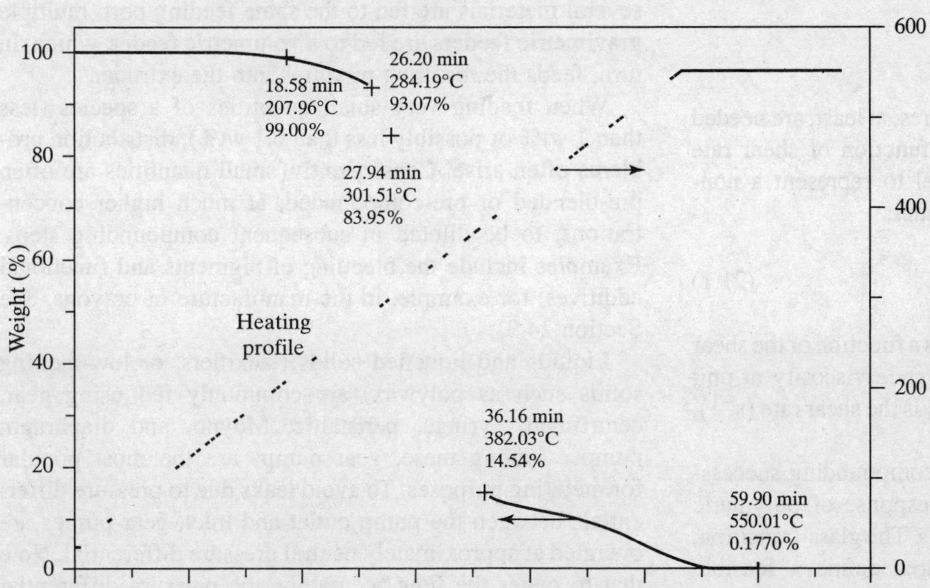

Figure 21.14 Thermo-gravimetric analysis (TGA) for cellulose acetate.

of the shear rate, in contrast, the viscosity of non-Newtonian fluids depends on the shear rate—with most polymers showing non-Newtonian behavior. In Figure 21.15, at lower shear rates, in the plateau region of the viscosity curve, the polymer behaves as a Newtonian fluid. At higher shear rates, the viscosity decreases within the shear-thinning regime.

Viscosity is commonly perceived as "thickness" or resistance to flow of a polymer in the molten or softened state. In compounding operations, the polymeric materials are subjected to shear rates on the order of a few hundred to a few thousand reciprocal seconds, the range within which viscosity measurements are needed. Most polymeric viscosity measurements are taken with capillary rheometers. As an example, for a typical polymer melt, its viscosity is displayed as a function of the shear rate and temperature in Figure 21.15. As the temperature increases from 50°C to 150°C, at a shear rate of 100 s^{-1}, the viscosity-shear rate relationship changes from power-law-like to Newtonian, where the viscosity is independent of shear rate. At low shear rates, less than 10 s^{-1}, the behavior is Newtonian, with shear thinning observed at higher shear rates. The threshold shear rate at which the polymer begins to shear thin changes with temperature. Note that the viscosity curves approach a common asymptote at high shear rates.

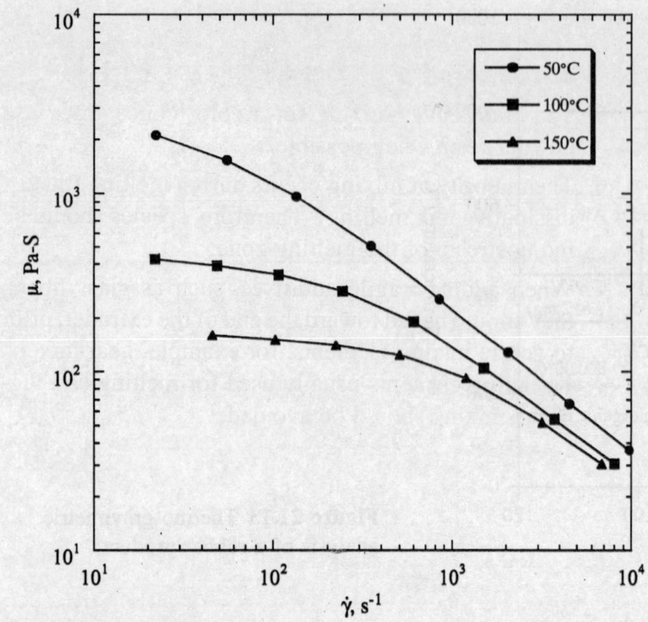

Figure 21.15 Rheology curves.

Viscosity curves at three temperatures, at least, are needed to correlate this viscosity data as a function of shear rate and temperature. The simplest model to represent a non-Newtonian fluid is the power-law model:

$$\mu(\dot{\gamma}, T) = \mu_0 \dot{\gamma}^{n-1} \tag{21.1}$$

where $\mu(\dot{\gamma}, T)$ is the viscosity (Pa-s) as a function of the shear rate and temperature, μ_0 is the reference viscosity at unit shear rate and reference temperature, $\dot{\gamma}$ is the shear rate (s^{-1}), and n is the power-law index.

In summary, to perform polymer compounding successfully, one needs to understand the response of polymeric materials to their processing conditions. The glass-transition, melting, and degradation temperatures define a feasible range for operations, while the viscosity curves characterize the flow behavior of the polymers at various shear rates stresses and temperatures.

21.5 FEEDING PROTOCOLS

In this section, feeding strategies, that is, feeding methods and locations, for compounding processes are discussed. The various materials fed into extruders include:

- Base polymers in the form of pellets, powder, flakes, crumbs, or ropes
- Solid fillers and functional additives in the form of powders (course, fine, and nano-particles), fibers (chopped or continuous), or spheres (solid, hollow, and expandable)
- Liquid additives with viscosities ranging from 1 to 10,000 cP
- Gases such as nitrogen blankets, oxygen (or air), and inert gases

Unlike single-screw extruders, where the raw materials are *flood* fed, twin-screw extruders are commonly *starve* fed. In the latter case, the feed rate depends only on the feeder type and speed and is independent of the extruder speed, with the main objective to feed the materials continuously and consistently.

Volumetric or gravimetric feeders are used for pellets, with the former using various feeding screws having different designs and screw pitches. Calibration curves typically relate the feed rate to the screw speed. When selecting the feeding screw, the size of the pellets and their flowability must be considered.

Gravimetric feeders adjust the feeding rate by monitoring the weight loss in the feed hopper. They switch to volumetric operation when the feed hopper is refilled. Gravimetric feeders are widely used in large manufacturing plants. When several materials are fed to the same feeding port, multiple gravimetric feeders are fed to a volumetric feeder, which, in turn, feeds the resulting mixtures into the extruder.

When feeding very small quantities of a species (less than 1 wt% or possibly less than 0.1 wt%), distribution problems often arise. Consequently, small quantities are often pre-blended or pre-compounded, at much higher concentrations, to be diluted in subsequent compounding steps. Examples include the blending of pigments and functional additives, for example, in the manufacture of crayons. See Section 14.3.

Liquids and liquefied solids (tackifiers, or low-melting solids such as polywax) are commonly fed using gear, centrifugal, syringe, peristaltic, Moyno, and diaphragm pumps. Among these, gear pumps are the most popular for metering purposes. To avoid leaks due to pressure differentials between the pump outlet and inlet, gear pumps are operated at approximately neutral pressure differential. Note that to meter the flow accurately, the pressure differential must be approximately zero; otherwise the pressure differential will create additional flow.

Several heuristics help to establish feeding protocols. These include:

1. When there are low-melting solids in the mixture, such as tackifiers or polyethylene wax, they should be liquefied before feeding.
2. When liquid components comprise more than 40 vol% of the mixture, they should be split and fed at two or three locations.
3. Liquid must be incorporated into the melt immediately after feeding to prevent pooling. Gear-like mixing elements, as shown in Figure 21.16, are often used because they achieve better distributive mixing.
4. For low-bulk-density pellets (with high void volumes), use the following feeding protocol to remove air prior

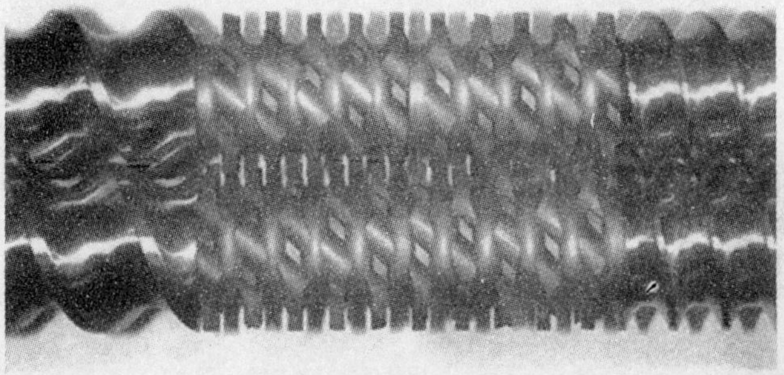

Figure 21.16 Gear mixer, Martin (2008). Reprinted with permission.

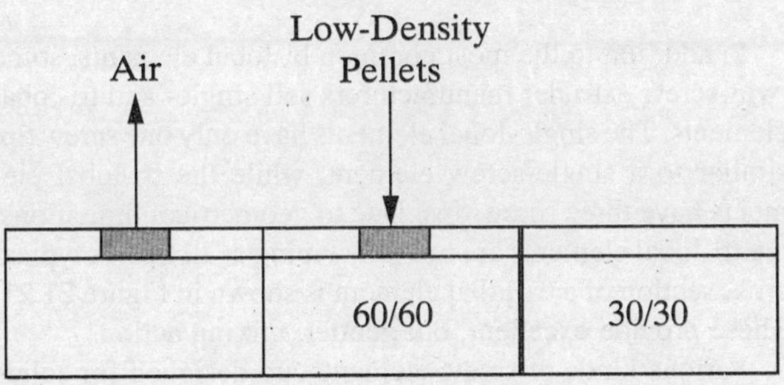

Figure 21.17 Feeding protocol for pellets with low bulk density. Flow direction is from left to right.

to melting, thereby avoiding bubbles. This involves reducing the pitch of the conveying element, which removes the air as the pellets are compressed. Figure 21.17 shows a schematic of the protocol using an element that reduces the pitch from 60 mm to 30 mm, where 60/60 refers to a conveying element 60 mm long with a 60-mm pitch.

5. To obtain better mixing for a color concentrate, use powder resin when possible.
6. The majority of mixing occurs during melting (that is, dissipative mix melting). Therefore, species should be fed upstream of the melting zone.
7. When adding fragile additives, such as glass fibers, they should be fed toward the end of the extruder, prior to gentle mixing elements, for example, gear mixers. Kneading elements usually used for melting and vigorous mixing should be avoided.

EXAMPLE 21.1

Synthetic rubber adhesives, such as those commonly used to send membership cards in the mail, are composed of synthetic rubber resins (such as polystyrene rubber), antioxidant (to prevent degradation during processing), and liquid tackifiers. For a mixture containing 50 wt% rubber, 5 wt% antioxidant, and 45 wt% tackifier, propose a feeding strategy for efficient compounding.

SOLUTION

A schematic of a proposed unit operation for compounding is shown in Figure 21.18a. The synthetic rubber and the powder antioxidant are fed at the beginning of the extruder prior to the melting section. The liquid tackifier is divided evenly and fed at two locations, followed immediately by mixing zones. Finally, the compounded mixture is conveyed and pumped using a gear pump to the next downstream processing station. This implemented feeding protocol is shown in Figure 21.18b.

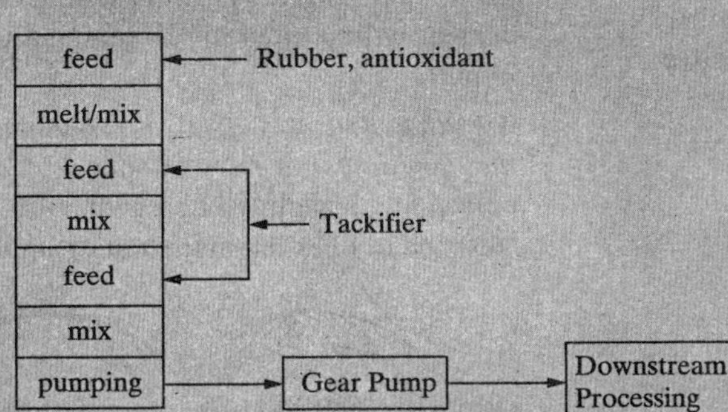

(a) Processing unit operations

(b) Extruder screw. Drawn using TXS Software.

Figure 21.18 Synthetic rubber compounding.

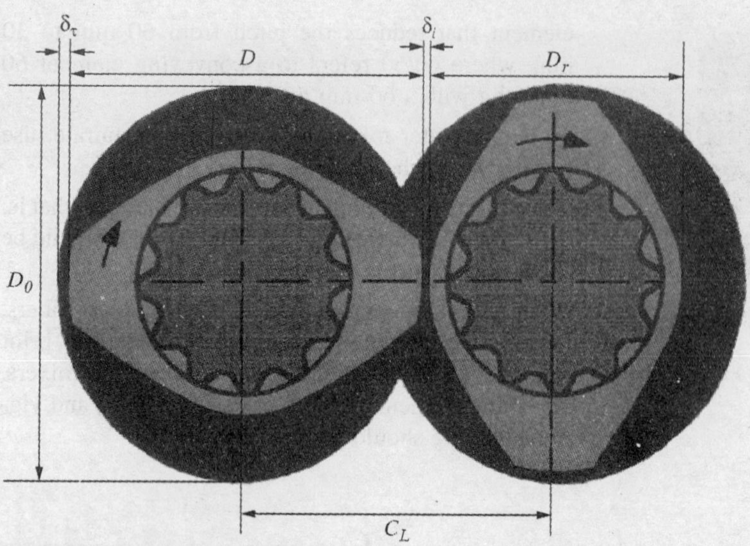

Figure 21.19 Flow cross section of bi-lobal twin-screw elements. Adapted from Fig. 3.14 of Todd (1998). Reprinted with permission.

21.6 SCREW DESIGN

In this section, the types of twin-screw elements, their functions, and their flow characteristics are discussed. The main objective is to cover techniques for configuring the screws to accomplish compounding tasks.

The cross section of twin-screw elements (conveying and kneading blocks) are identical except for special elements, such as gear mixers. One typical flow cross section, for bi-lobal twin-screw elements, is shown in Figure 21.19. Here, D_O is the screw bore diameter, D is the screw tip diameter, D_r is the screw root diameter, δ_τ is the screw tip clearance, δ_i, is the intermeshing clearance, and C_L is the centerline distance, with the latter expressed as:

$$C_L = (D + D_r)/2 + \delta_i \tag{21.2a}$$

The tip angle is defined as:

$$\cos\beta = \frac{C_L}{D} \tag{21.2b}$$

The longitudinal cross section of these elements is shown in Figure 21.20. Here, L is the pitch of the element, which consists of two flow channels, with the pitch being the distance between two tips across two complete rotations.

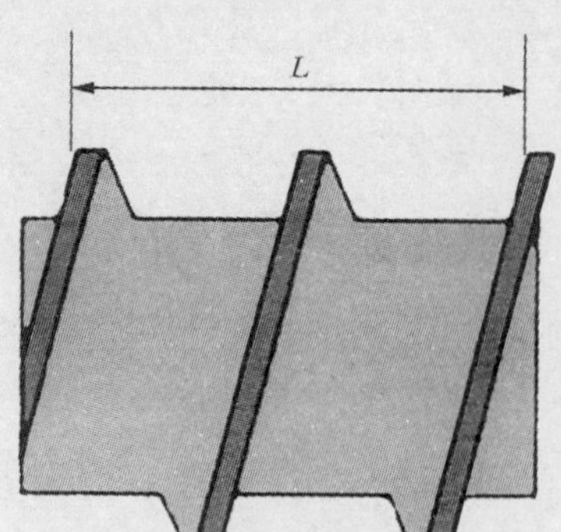

Figure 21.20 Longitudinal cross section of a bi-lobal twin-screw element.

In addition to the most common bi-lobal elements, some twin-screw extruder manufacturers sell single- and tri-lobal elements. The single-lobal elements have only one screw tip, similar to a single-screw element, while the tri-lobal elements have three screw tips. Due to geometrical limitations, the tri-lobal elements are not self-wiping at all tips. A typical cross section of a tri-lobal element is shown in Figure 21.21. These provide excellent, but gentler, mixing action.

Various kinds of screw elements are designed for solids conveying, melting, mixing, melt pumping, devolatilization, and chemical reaction. These include kneading blocks, conveying elements, and special elements such as gear mixers, as shown in Figure 21.22. These screw elements are further differentiated into *forwarding* and *reversing* types. The forwarding element has a forwarding pitch, while the reversing elements act as a flow barrier by providing a reversing flow.

Furthermore, these screw elements are named using the following conventions:

1. *Conveying elements: Total pitch/length/F or R*. F is for forwarding and R is for reversing elements. For example, 60/60/F represents a forwarding conveying screw element with a pitch of 60 mm and a total length of 60 mm.
2. *Kneading blocks: Offset angle/number of kneaders/total length/F or R*. For example, 45/5/42/R represents a reversing kneading block with a 45° offset between subsequent lobes having a total length of 42 mm.

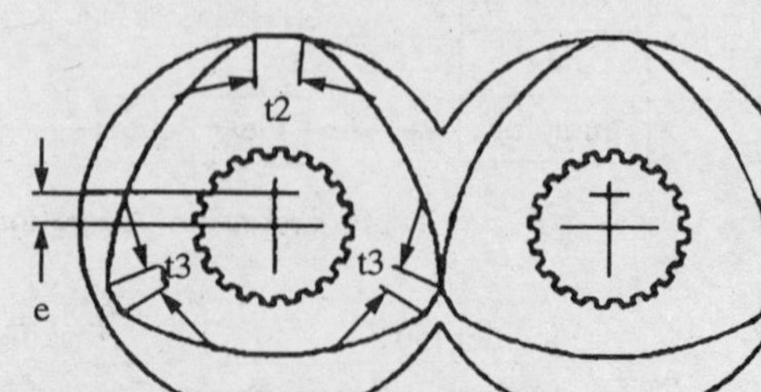

Figure 21.21 Flow cross section of a tri-lobal element (Padmanabhan, 2004).

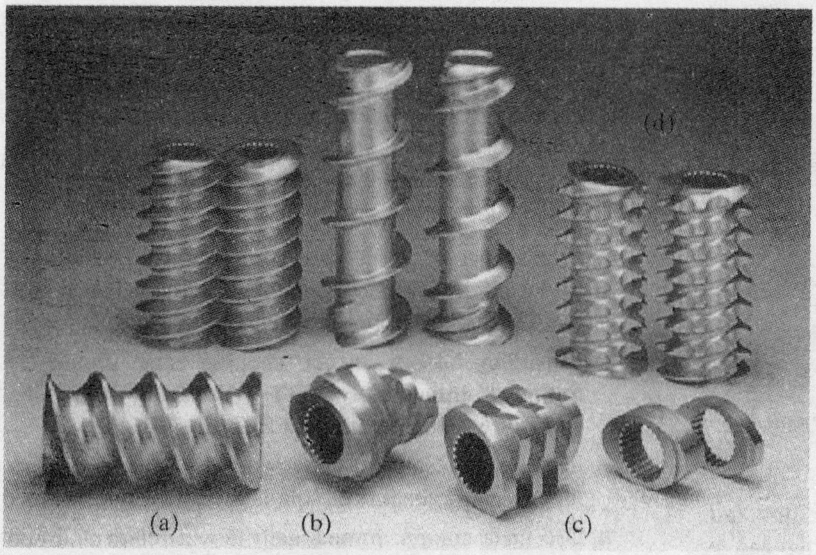

Figure 21.22 Twin-screw elements: (a) conveying element, (b) kneading block, (c) neutral kneading block, (d) gear mixer (Martin, 2008). Reprinted with permission.

3. *Gear mixers: Number of gear teeth/total length/F or R.* For example, 12/10/F is for a forwarding gear with 12 teeth and a total length of 10 mm.

Note that naming conventions vary among twin-screw extruder manufacturers. Each manufacturer may add a prefix, such as ZSK by Werner and Pfleiderer.

Unlike for single-screw extruders, where the screws are full over their entire length, twin-screw extruders are starve fed, leaving distinct regions completely full and others partially filled. The degree of filling, f, in a twin-screw extruder varies as a function of the screw design, that is, the sequence of screw elements. It is defined as the ratio of the net flow rate, Q, to the drag flow rate, Q_d:

$$f = \frac{Q}{Q_d} \tag{21.3}$$

The drag flow rate is the maximum volume that can be conveyed in one rotation:

$$Q_d = \frac{G_D}{2}\left(\pi(D/2)^2\right)LN \tag{21.4}$$

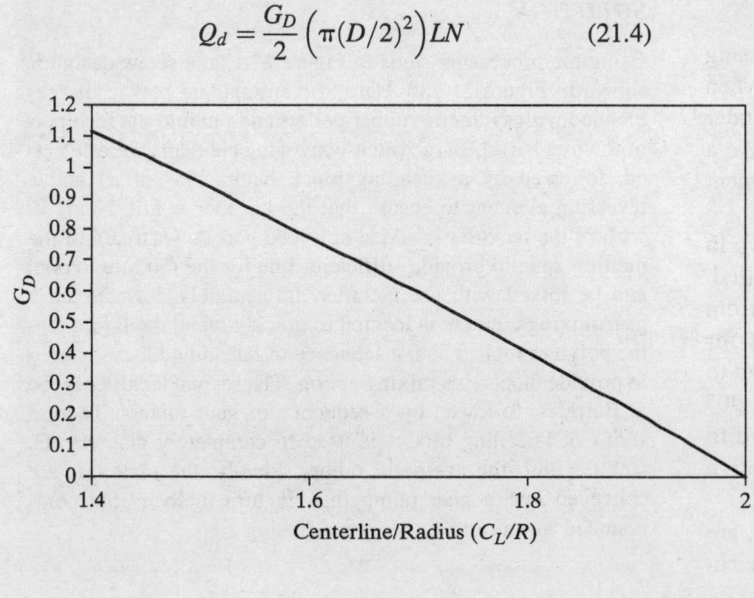

Figure 21.23 Free cross-sectional area correction factor.

The geometric factor, G_D, is the ratio of the free area for flow and the cross-sectional area of a tube with a diameter equal to the screw tip diameter. L is the length of the element, and N is the screw speed (e.g., revolutions per second). For bi-lobal twin-screw elements, it can be shown (Booy, 1978) that:

$$G_D = \frac{\text{Free area}}{\pi(D/2)^2} = \frac{1}{\pi}\left[6\beta - 4\pi\left(\frac{C_L}{D}\right)^2 + 4(\pi - 4\beta)\left(\frac{C_L}{D}\right) + 10\sin(\beta)\left(\frac{C_L}{D}\right)\right] \tag{21.5}$$

where G_D is displayed as a function of $\left(\frac{C_L}{R}\right)$ in Figure 21.23. Note that Eq. (21.5) was derived for bi-lobal elements with zero clearance.

The net volumetric flow rate in a twin-screw extruder is commonly expressed as:

$$Q = Q_d - Q_p - Q_l \tag{21.6}$$

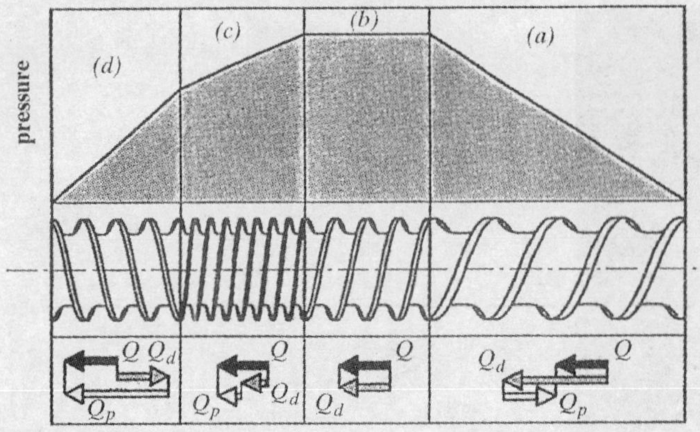

Figure 21.24 Pumping characteristics of conveying elements.

where Q_d is the drag flow rate, Q_p is the pressure flow rate, and Q_l is the leakage flow rate. While the latter can be significant, it is difficult to model, and consequently, neglected in many cases, giving:

$$Q = Q_d - Q_p \tag{21.7}$$

Depending on the relative values of the drag and the net flow rate, conveying elements may increase or decrease the pressure, as shown in Figure 21.24. Note that the direction of the flow is from right to left.

- Section *a*: The drag capacity is higher than the net throughput; pressure is generated.
- Section *b*: The drag capacity is equal to the net throughput; no pressure is generated.
- Section *c*: The drag capacity is smaller than the net throughput; pressure is decreasing.
- Section *d*: The drag capacity of the reversing conveying element is in the opposite direction of the net flow rate; a large pressure decrease is generated.

Note that the pressure flow direction in a reversing element is the same as that for a forwarding element when the drag capacity is smaller than the net throughput. Under this condition, a forwarding conveying element acts like a reversing element. The degree of fill is unity and the element is fully filled.

The kneading block elements with 45° offset, shown in Figure 21.22b, are efficient melters when fully filled. They also provide excellent dispersive mixing action. Thicker kneading block elements are more efficient for mixing. Other offset angles such as 30° or 60° are used to provide less vigorous dispersive mixing action. The 90° kneading blocks, shown in Figure 21.22c and referred to as neutral kneading blocks, can provide good distributive mixing.

The gear-mixing elements, shown in Figure 21.22d, are primarily used to provide distributive mixing. As discussed in Section 21.5, gear mixers are used widely to prevent pooling when blending liquids into molten polymers.

In designing the screw configuration, one must decide on the sequence of unit operations needed before selecting the appropriate screw elements to accomplish the operations. The following heuristics should be considered when designing the screw configuration:

1. A sequence of conveying elements with decreasing screw pitch is recommended to increase the degree of fill before the polymer pellets enter the melting zone.
2. Kneading blocks in the melting zone should be followed immediately with a reversing conveying element to ensure complete filling.
3. Use the largest pitch-conveying elements to accommodate the transport of solids during feeding.
4. Use gear mixers immediately downstream of liquid feeding.
5. Use kneading blocks for dispersive mixing.
6. Use gear-mixing elements or neutral kneading blocks to accomplish distributive mixing.
7. Use a reversing element immediately downstream of a degassing port to ensure a good melt seal.
8. Use high free-volume elements underneath the degassing port to provide better conveyance (to prevent flooding) and higher surface renewals for degassing operation.
9. Use small pitch elements to generate pressure efficiently.

EXAMPLE 21.2

Propose a screw design for the synthetic rubber compounding in Example 21.1.

SOLUTION

Using the processing steps in Figure 21.18a, a screw design is shown in Figure 21.18b. Here, the antioxidant powder is preblended with synthetic rubber pellets and fed into the feed port of the first barrel. Large pitch-conveying elements were selected, followed by a kneading block having 45° offset and a reversing element to ensure that the kneader is full. Next, 50 wt% of the tackifier is added at a feed port downstream of the melting zone to provide sufficient time for the mixture to cool and be mixed with the tackifier. Immediately downstream a gear-mixing element is located to quickly blend the liquid into the polymer melt. Then, a sequence of kneading blocks is used to provide dispersive mixing action. The second tackifier is fed at Barrel-7, followed by a sequence of gear mixers. Then, a series of kneading blocks is used to completely disperse the tackifier into the synthetic rubber. Finally, the materials are conveyed into a gear pump that, in turn, delivers the compounded mixture to the next processing step.

21.7 SETTING THE PROCESSING CONDITIONS

In this section, techniques for setting the processing conditions during melting, mixing, and conveying or pumping are discussed. The objective is to accomplish complete melting and sufficient mixing without degradation, either thermal or mechanical.

Degradation in compounding operations is primarily caused by viscous energy dissipation when shearing highly viscous materials, coupled with poor heat transfer in the melt phase as well as through the barrel walls. The viscous energy dissipation (*VED*) is given by:

$$VED = \mu(\dot{\gamma}, T)\dot{\gamma}_{avg}^2 \tag{21.8}$$

where *VED* is the rate of viscous energy dissipation per unit volume, $\mu(\dot{\gamma}, T)$ is the melt viscosity at the shear rate and melt temperature, and $\dot{\gamma}_{avg}$ is the average shear rate.

The average shear rate in the channel of a standard element (conveying or kneading block) in a twin-screw extruder is given by:

$$\dot{\gamma}_{avg} = \frac{\pi D N}{\bar{h}} \tag{21.9}$$

with the screw channel divided into three regions: the tip, Booy's region, and the waist region, as shown in Figure 21.25 (Booy, 1978). Here, h is channel depth at angle θ from the abscissa, and $\bar{h}$ is the average channel depth, where $0 \le \theta \le \pi/2$. Also, D is the screw bore diameter and N is the screw speed.

The following derivation assumes zero screw clearances. In the tip and waist regions, the channel depth is constant. In the waist region:

$$h = H;\ 0 \le \theta \le \frac{\alpha}{2} \tag{21.10}$$

In the tip region:

$$h = 0;\ \frac{\pi}{2} - \frac{\alpha}{2} \le \theta \le \frac{\pi}{2} \tag{21.11}$$

In Booy's region, the channel depth is given by:

$$h\{\theta_b\} = R(1 + \cos\theta_b) - \sqrt{C_L^2 - R^2\sin^2\theta_b};\quad \theta_b = \theta - \frac{\alpha}{2} \text{ and } \frac{\alpha}{2} \le \theta \le \frac{\pi}{2} - \frac{\alpha}{2} \tag{21.12}$$

where:

$$\frac{\alpha}{2} = \frac{\pi}{4} - \cos^{-1}\left(\frac{C_L}{D}\right) \tag{21.13}$$

The average channel depth is:

$$\bar{h} = \frac{\int_0^{\pi/2} h\{\theta\}\, d\theta}{\int_0^{\pi/2} d\theta} = \frac{\int_0^{\alpha/2} h_{waist}\{\theta\}\, d\theta + \int_{\alpha/2}^{\pi/2-\alpha/2} h_{Booy}\{\theta_b\}\, d\theta + \int_{\pi/2-\alpha/2}^{\pi/2} h_{tip}\{\theta\}\, d\theta}{\int_0^{\pi/2} d\theta} \tag{21.14}$$

Substituting Eqs. (21.10) to (21.12) into Eq. (21.14) gives:

$$\bar{h} = G_H H = \left(\frac{2}{\pi}\right)\left[\frac{\alpha}{2} + \frac{1}{1 - R_r/R}\int_0^{\pi/2-\alpha}(1 + \cos\theta_b - \sqrt{(1 + R_r/R)^2 - \sin^2\theta_b})\, d\theta_b\right] H \tag{21.15a}$$

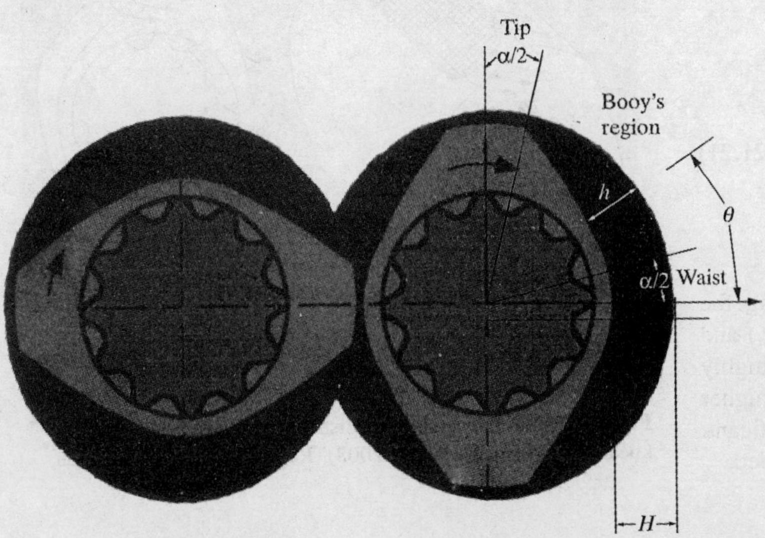

Figure 21.25 Cross-sectional area of twin-screw elements showing the tip, Booy's, and waist regions. Adapted from Fig 3.14 of Todd (1998). Reprinted with permission.

where the geometric factor, G_H, for the channel depth is:

$$G_H = \left(\frac{2}{\pi}\right)\left[\frac{\alpha}{2} + \frac{1}{1 - R_r/R}\int_0^{\pi/2-\alpha}(1 + \cos\theta_b - \sqrt{(1 + R_r/R)^2 - \sin^2\theta_b})\,d\theta_b\right] \quad (21.15b)$$

EXAMPLE 21.3

Compute the geometric factor, G_H, for the channel depth of a W&P Extruder with a tip-to-root radius ratio, R_r/R, of 1.55.

SOLUTION

Using Eq. (21.15):

$$G_H = 0.6477 \quad (21.16)$$

The adiabatic temperature rise due to *VED*, ΔT, can be estimated using:

$$VED = \frac{Q_m C_p \Delta T}{AL} \quad (21.17)$$

where Q_m is the mass flow rate, C_p is the heat capacity of the polymer, A is the free area for flow, and L is the element length. Substituting $A = G_D \pi D^2/4$ from Eq. (21.5):

$$VED = \frac{Q_m C_p \Delta T}{G_D\left(\frac{\pi D^2}{4}\right)L} \quad (21.18)$$

Substituting Eq. (21.9) into Eq. (21.8):

$$VED = \mu\frac{\pi^2 D^2 N^2}{\overline{h}^2} \quad (21.19)$$

Substituting $\overline{h} = G_H H$ from Eq. (21.15a):

$$VED = \mu\frac{\pi^2 D^2 N^2}{G_H^2 H^2} \quad (21.20)$$

Equating Eq. (21.18) and Eq. (21.20) and rearranging:

$$\Delta T = \underbrace{\left(\frac{\pi}{4}\frac{G_D}{G_H^2}\frac{D^4 L}{H^2}\right)}_{\text{Geometric factor}}\ \underbrace{\left(\frac{\mu}{C_p}\right)}_{\text{Material properties}}\ \underbrace{\left(\frac{N^2}{Q_m}\right)}_{\text{Operating conditions}} \quad (21.21)$$

The effects of *VED* can be reduced by lowering the residence time (i.e., providing higher throughput, Q_m) and reducing the screw speed, N. When compounding highly filled materials, such as those with very high loadings (higher than 60 wt%), the effect of increased viscosity is significant. In that case, reducing the degree of fill is recommended.

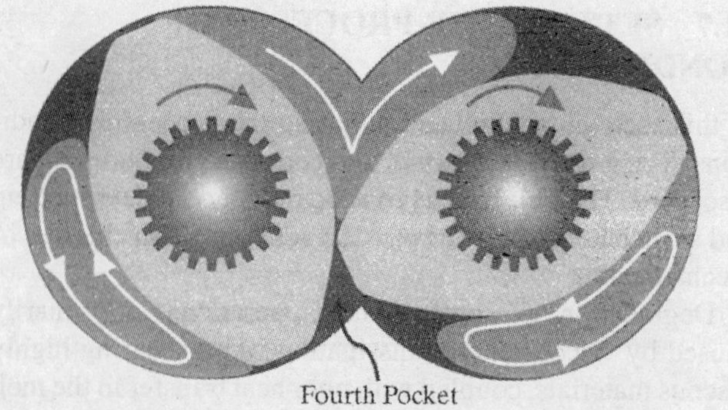

Figure 21.26 The fourth pocket phenomena in kneading elements. Source: Coperion Corp. Reprinted with permission.

As a rule of thumb, the barrel temperature in the mixing zone should be set approximately 50°C higher than the glass-transition temperature or the melting temperature. To achieve complete melting, it is often necessary to minimize the barrel temperature in the melting zone. Melting phenomena in twin-screw extruders are now understood to result from solid compression in the fourth pocket region, shown in solid grey in the lower apex of Figure 21.26. This region is formed when the leading tip of the right screw element is at its lower apex. A similar region can be formed at the upper apex. The volume of the fourth pocket depends on the thickness of the kneading element. The fourth pocket is formed and disappears as the screw rotates, expanding or contracting. It is during this expansion and compression that melting of the solid takes place as well as mixing in the axial direction.

This is the main reason the kneading block provides good melting and dispersive mixing action. It supplements the well-known flow over the screw tip clearance and around the thickness of the kneading element, as illustrated in Figure 21.27.

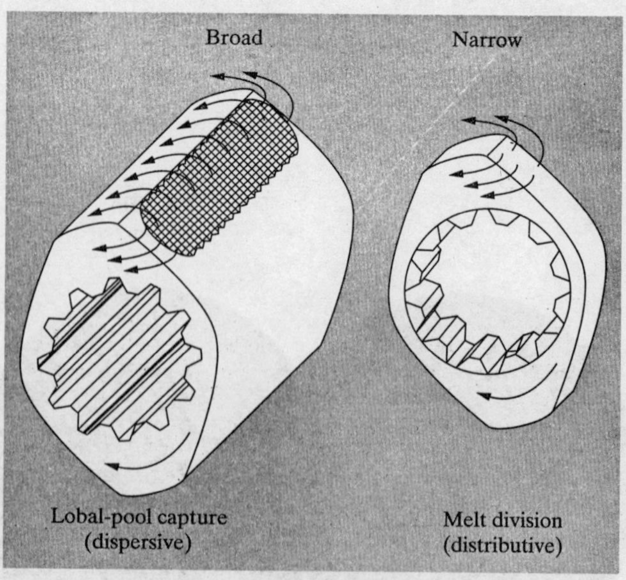

Figure 21.27 Dispersive and distributive mixing action in a kneading element (Martin, 2008). Reprinted with permission.

21.8 SUMMARY

In this chapter, polymer blending and mixing is discussed, with the focus on the compounding of polymeric materials in twin-screw extruders. A framework is provided to first study the characteristics of the materials, followed by the unit operations, to determine the processing steps. Using heuristics, the feeding strategy and extruder screws are designed, and the operating conditions are set. Simple models are presented for estimating the degree of fill and the degradation represented by viscous energy dissipation.

REFERENCES

1. AGASSANT, J.F., P. AVENAS, and J.P. SERGENT, *Polymer Processing—Principles and Modeling*, Hanser, Munich (1991).

2. ALBALAK, R., Ed., *Polymer Devolatilization*, Marcel Dekker, New York (1996).

3. BICERANO, J., *Prediction of Polymer Properties*, Marcel Dekker, New York (1993).

4. BIESENBERGER, J.A., Ed., *Devolatilization of Polymers*, Hanser, New York (1983).

5. BOOY, M.L., *Polym. Eng. Sci.*, **18**, 973 (1978).

6. DRAY, R.F., Plastics Screw with Barrier Members, U.S. Patent 6,988,821 (2006).

7. MARTIN, C., "In the Mix: Continuous Compounding Using Twin-Screw Extruders," http://www.devicelink.com/mddi/archive/00/04/010.html (Jan., 2008).

8. PADMANABHAN, B., Fractional and Higher Lobed Co-Rotating Twin-Screw Elements, U.S. Patent 6,783,270 (2004).

9. RAUWENDAAL, C., Ed., *Mixing in Polymer Processing*, Marcel Dekker, New York (1991).

10. RAUWENDAAL, C., *Polymer Mixing—A Self Study Guide*, Hanser, Munich (1998).

11. RAUWENDAAL, C.J., Intermeshing Element Mixer, U.S. Patent 6,709,147 (2004).

12. TADMOR, Z., and C.G. GOGOS, *Principles of Polymer Processing*, 2nd ed., John Wiley & Sons, New York (2006).

13. TADMOR, Z., and I. KLEIN, *Engineering Principles of Plasticating Extrusion*, Krieger, New York (1978).

14. TODD, D.B., Ed., *Plastics Compounding—Equipment and Processing*, Hanser, Munich (1998).

15. WELLING, S., *Devolatilisation of Plastics*, English Translation, VDI Verlach, Dusseldorf (1980).

16. WILKINSON, A.N., and A.J. RYAN, *Polymer Processing and Structure Development*, Kluwer, Dordrecht (1998).

17. XANTHOS, M., Ed., *Reactive Extrusion—Principles and Practice*, Hanser, New York (1992).

EXERCISES

21.1 Propose a compounding design, including the feeding location and screw design, for a polystyrene-polyethylene blend having a 40/60 weight ratio. The glass-transition temperature of polystyrene is 100°C and the melting point of polyethylene is 135°C. The viscosity of both polymers at a shear rate of 300 s^{-1} is approximately 2,000 Pa-s.

21.2 Show that the free area in a conveying and kneading element is constant along the length of the element.

21.3 Compare the relative effectiveness of the pressure generation of two conveying elements with different screw pitches; that is, a 60/60 conveying element compared with a 30/60 element.

21.4 Compute the average shear rate experienced by the polymer in a conveying element of 60/60 mm in a W&P extruder with a tip-to-root diameter ratio of 1.55, run at 100 rpm.

Chapter 22

Cost Accounting and Capital Costs

22.0 OBJECTIVES

Throughout the design process, as discussed in the previous chapters, estimates of the cost of equipment and other costs related to the capital investment play a crucial role in selecting from among the design alternatives. This chapter presents the various methods in common use for making preliminary estimates of capital costs of ventures for new products and processing plants or revamps of existing plants and should be studied in connection with the other chapters as needed. Some readers may prefer to study Sections 22.1 to 22.4 even before working with the previous chapters, especially when studying the techniques for process synthesis that require estimates of capital costs. In Chapter 23, capital cost estimates prepared according to the information presented in this chapter are combined with process operating costs and other expenses to determine the profitability of a proposed venture. However, even though a venture is predicted to be profitable, the financial condition of the company exploring the venture may not be sufficient to justify a decision to proceed with the venture, or competing ventures may be more attractive. In the former case, it is important to understand measures used to determine the financial condition of a company. These measures are intimately tied to the field of cost accounting, which is the subject of the first section of this chapter.

After studying this chapter, the reader should

1. Be able to assess the financial condition of a company from its annual report.
2. Be able to use the equations provided to estimate the purchase cost of representative types of process equipment and, when applicable, components of basic, industrial, and configured-consumer chemical products.
3. Be able to estimate the cost of installation of the equipment units, including the cost of chemical materials, labor, and indirect costs.
4. Be able to estimate the total capital investment for the process, including the costs of spare equipment, storage tanks, surge vessels, site preparation and service facilities, allocated costs for utilities and related facilities, contingencies, land, royalties, startup, and working capital. Understand the need to reestimate the working capital after preparing the cost sheet, as discussed in Chapter 23.

22.1 ACCOUNTING

Accounting is the systematic recording, reporting, and analysis of the financial transactions of a business. Accounting is necessary and valuable to a company because it:

1. Provides a record of property owned, debts owed, and money invested.
2. Provides a basis for the preparation of a report that gives the financial status of the company.
3. Gives assistance and direction to those managing the affairs of the company.
4. Provides a basis for stockholders and others to evaluate management of the company.

Debits and Credits

By tradition, since the 15th century, the recording of financial transactions by accountants is carried out by the *double-entry method* of *debits* and *credits*. Surprisingly, debits are increases (not decreases) in assets, where an *asset* is anything of economic value that is owned by the company. Credits are just the opposite; that is, they are decreases in assets. One possible explanation for the definitions of debits and credits is that the giver receives a credit while the receiver acquires a debit. By custom, all transactions are initially recorded chronologically, in terms of debits and credits, in *a journal*, where for every debit, there is a credit of equal amount. The debits and credits apply to different accounts (cash, land,

Table 22.1 Typical Journal Page

JOURNAL					Page 43
2008				Debit ($)	Credit ($)
June	3	Heat Exchanger	15	80,450	
		Cash	11		80,450
		Purchase of a heat exchanger for ammonia plant			
	4	Cash	11	125,000	
		Ammonia product	12		125,000
		Sales of product from ammonia plant to ABC			
	6	Land	20	265,000	
		Cash	11		265,000
		Purchase of land in Iowa from XYZ			

equipment, bank account, etc.), which are maintained in separate *ledgers* for each account. Entries in the journal are posted to the ledgers, usually with the debit entry going to one ledger and the corresponding credit entry going to a different ledger. Although the journal might seem superfluous, it serves two useful purposes besides being a chronological record: (1) It reduces the possibility of error because for each transaction, the debit and corresponding credit are recorded together, and (2) if desired, a detailed explanation for a transaction is easily entered into the journal. In both the journal and the separate account ledgers, debits are entered to the left of the credits. At any point in time, the sum of debit entries for all ledgers must equal the sum of credit entries for all ledgers. Although it is not necessary for engineers to be accountants, it is important for engineers to understand what accountants do and why they do it.

Tables 22.1 and 22.2 show an example of double-entry bookkeeping, with a journal and ledgers, for the following transactions. The company purchases a heat exchanger for $80,450, paid for by a check. The next day the company sells $125,000 of ammonia product, with payment by check. Two more days later, the company purchases land for $265,000, paid for by check. In all cases, the checks are handled with the same bank account. Note that four separate accounts are involved. Suppose they have been assigned the following account numbers: Bank Account, 11; Plant Equipment, 15; Sales of Products, 12; and Land, 20. The journal page (say page 43) is shown in Table 22.1, where the ledger account numbers are added in the column to the left of the debits column when the journal entries are posted to the ledgers. Postings to the ledger page (page 5 of this ledger) for the Bank Account are shown in Table 22.2. Included to the left of the debits and credits columns are the corresponding pages in the journal, in this case just page 43. This ledger page is for the month of June, for which the initial balance in the bank account, page 42, was an amount of $500,000. At the end of June 6, the bank account balance was $500{,}000 + 125{,}000 - 80{,}450 - 265{,}000 = \$279{,}550$. Balances may be entered into the ledger at the end of each month.

The Annual Report (Form 10-K)

Every publicly held company in the United States is required by the Securities and Exchange Commission (SEC) to submit an annual report, referred to as Form 10-K.

Table 22.2 Typical Ledger Page

BANK ACCOUNT, LEDGER 11									Page 5
2008				Debit ($)	2008				Credit ($)
June	2	Balance forward	42	500,000	June				
	4	Sales	43	125,000		3	Purchase	43	80,450
						6	Purchase	43	265,000

In addition to the 10-K, most companies also issue a more user-friendly, attractive annual report written primarily for stockholders that typically provides the following useful information:

1. Nature of the company's business.
2. Summary of important events and new developments of the year.
3. New acquisitions and formation of partnerships.
4. Plans for the near future.
5. Summary of concerns that might influence the company's business.
6. Collection of financial statements from Form 10-K, including:
   (a) Balance Sheet
   (b) Income Statement
   (c) Cash Flow Statement

The current financial condition of a company can be assessed by an analysis of the financial statements in its annual report. Annual reports of more than 11,000 U.S. companies can be viewed at the Web site www.irin.com.

The Balance Sheet

The *balance sheet*, also called the *consolidated balance sheet* or *statement of consolidated financial condition*, is a quantitative summary of a company's financial condition at a specific point in time (at the end of the calendar or fiscal year), including *assets, liabilities*, and *net worth* (share owners' equity, stockholders' equity, or proprietorship). *Equity* means ownership, generally in the form of common stock or as a holding company. The balance sheet is prepared from balances in the ledger accounts. The overall entries in the balance sheet must conform to the fundamental accounting equation:

$$\text{Assets} = \text{Liabilities} + \text{Net Worth} \qquad (22.1)$$

For publicly held companies, the net worth is the stockholders' equity. Bankers and other grantors of credit to companies are concerned with the margin of security for their loans. The balance sheet provides them with two important measures: (1) The assets owned by the company, and (2) the liabilities owed by the company. A representative consolidated balance sheet for a large fictitious corporation, U.S. Chemicals, is given in Table 22.3 for the calendar year ending December 31, 2006. In the United States, a corporation is the most common form of business organization, one that is chartered by a state and given many legal rights as an entity separate from its owners. It is characterized by the limited liability of its owners, the issuance of shares of easily transferable stock, and existence as a going concern. The balance sheet is divided, according to Eq. (22.1), into three sections: Assets, Liabilities, and, in this case, Shareholders' Equity (in place of Net Worth). Each section is divided into accounts, where the entries are the balances in the ledger accounts as of the date of the balance statement. All numbers in the table represent millions of U.S. dollars.

As shown in Table 22.3, assets for this corporation are divided into Current Assets, Investments, Property, and Other Assets. *Current assets* are items of economic value that could be converted to cash in less than one year, including cash and cash equivalents, marketable securities, accounts receivable, inventories, prepaid expenses, and deferred income taxes. The current assets total \$4,630,000,000. *Investments* pertain to investments in companies in which ownership interest by U.S. Chemicals is 50% or less, but where U.S. Chemicals exercises significant influence over operating and financial policies. *Property* constitutes fixed assets, including land, buildings, machinery, equipment, and software, and is listed at its *book value*, which is the original cost (the so-called basis), corrected for accumulated depreciation. *Depreciation* is the allocation of the cost of an asset over a period of time for accounting and tax purposes. It accounts for a decline in the value of a property due to general wear and tear or obsolescence. Property still in use remains on the books and the balance sheet even after it is completely depreciated. *Goodwill* is an intangible asset that provides to a company a competitive advantage, such as a well-known and accepted brand name, reputation, or high employee morale. In addition to goodwill, other intangible assets may be listed, such as patents and trademarks. The total assets are shown as \$14,211,000,000.

The second part of the balance sheet in Table 22.3 lists the liabilities and stockholders' equity. *Current liabilities* include all payments that must be made by the company within one year. The total for U.S. Chemicals is \$4,153,000,000. *Long-term debts*, often in the form of bonds, are due after more than one year from the date of the balance sheet. They total \$3,943,000,000. *Other noncurrent liabilities* total \$1,754,000,000 and include pension and other post-retirement benefits as well as reserves for any company operations that are discontinued. Total liabilities are \$9,850,000,000. We note that liabilities are less than assets by \$4,361,000,000. Thus, by Eq. (22.1), this difference must be the stockholders' equity. This equity includes the par value of issued common stock, which totals \$1,000,000,000. The par value of a share of stock is an arbitrary amount that has no relationship to the market value of the stock, but is used to determine the amount credited to the stock account. If the stock is issued for more than its par value, the excess is credited to the account shown as *capital in excess of par value*. In Table 22.3, the par value is \$1.00 per share but the stock was issued at \$4.23 per share. Companies frequently repurchase shares of their common stock, resulting in a reduction of stockholders' equity. Because the shares are placed in a treasury, the transaction appears as *treasury stock at cost*. In Table 22.3, that amount is \$3,428,000,000. The other account under stockholders' equity is *retained earnings*, which is the accumulated retained earnings that are increased each year by net income. The amount of this entry must be such that Eq. (22.1)

Table 22.3 Consolidated Balance Sheet for U.S. Chemicals in Millions of Dollars as of 31 December 2006

ASSETS		
Current Assets		
Cash and cash equivalents	107	
Marketable securities	45	
Accounts receivable	2,692	
Inventories:		
Finished products and work in progress	1,420	
Materials and supplies	312	
Deferred income tax assets	54	
Total current assets		4,630
Investments		
In nonconsolidated affiliates	544	
Other	1,476	
Total investments		2,020
Property		
Land	200	
Buildings	2,190	
Plant machinery and equipment	7,684	
Office equipment	645	
Computer software	242	
Less accumulated depreciation	(6,006)	
Net property		4,955
Other Assets		
Goodwill	952	
Other intangible assets	1,654	
Total other assets		2,606
TOTAL ASSETS		14,211
LIABILITIES		
Current Liabilities		
Short-term debt (payable within one year)	150	
Accounts payable	2,773	
Income taxes payable	130	
Deferred income tax liabilities	21	

(Continued)

Table 22.3 (*Continued*)

Dividends payable	104		
Accrued current liabilities	975		
Total current liabilities		4,153	
Long-Term Debt		3,943	
Other Noncurrent Liabilities			
Pension and other post-retirement benefits	892		
Reserve for discontinued operations	78		
Other noncurrent obligations	784		
Total other noncurrent liabilities		1,754	
TOTAL LIABILITIES			9,850
STOCKHOLDERS' EQUITY			
Common stock (authorized 2,000,000,000 shares at $1.00 par value; 1,000,000,000 issued)	1,000		
Capital in excess of par value of common stock	4,230		
Retained earnings	2,559		
Less treasury stock at cost, 300,000,000 shares	(3,428)		
NET STOCKHOLDERS EQUITY			4,361
TOTAL LIABILITIES + STOCKHOLDERS' EQUITY			14,211

is satisfied. This is seen to be the case in Table 22.3, where the net stockholders' equity is $4,361,000,000, giving total liabilities plus stockholders' equity as $14,211,000,000, which is equal to total assets by Eq. (22.1).

The Income Statement

An annual report also contains *the income statement*, also called the *statement of consolidated income* (loss) or *statement of consolidated operations*, which is an accounting of sales, expenses, and net profit (same as net earnings and net income) for a given period. In the annual report, the period is for one calendar or fiscal year. However, many companies also issue quarterly statements. Bankers, other grantors of credit, and investors and speculators pay close attention to the income statement because it provides the net profit of the company, which is an indication of the ability of the company to pay its debts and grow. *Net profit* is defined as *revenues* (sales) minus cost of sales, operating expenses, and taxes over a given period of time, with *gross profit* (gross earnings or gross income) being revenues minus just cost of sales and operating expenses (i.e., profit before taxes).

A representative consolidated income statement for the large fictitious corporation U.S. Chemicals is given in Table 22.4 for the calendar year 2006. *Net sales* is gross sales minus returns, discounts, and allowances. The *cost of goods sold* (cost of sales) is the cost of purchasing the necessary *raw materials* to produce the goods plus the *cost of manufacturing* the finished products. Operating expenses are expenses other than those of manufacture and include research and development expenses; selling, general, and administrative expenses; insurance and finance company operations; and amortization and adjustments of goodwill. *Amortization* is the gradual elimination of a liability, such as a mortgage, in regular payments over a specified period of time, where the payments are sufficient to cover both principal and interest. Income from operations equals gross profit minus operating expenses. From this, interest expenses are subtracted to give gross income (sometimes called net profit before income taxes). Interest expenses pertain to interest payments to bond holders and interest on loans. Subtraction of income taxes gives net income. Table 22.4 shows that from an annual net sales of $11,504,000,000, the net profit is $803,000,000 or 6.98%.

Table 22.4 Consolidated Income Statement for U.S. Chemicals in Millions of Dollars for the Calendar Year 2006

Net sales	11,504	
Cost of goods sold	9,131	
GROSS PROFIT		2,373
OPERATING EXPENSES		
Research and development expenses	446	
Selling, general, and administrative expenses	439	
Insurance and finance company operations	34	
Amortization and adjustments of goodwill	64	
TOTAL OPERATING EXPENSES		983
INCOME FROM OPERATIONS		1,390
Interest expense	185	
GROSS INCOME		1,205
Provision for income taxes	402	
NET INCOME		803

Table 22.5 Consolidated Cash Flow Statement for Chicago Chemicals in Millions of Dollars for the Calendar Year 2006

OPERATING ACTIVITIES		
Net income available for common stockholders	3,151	
Adjustments to reconcile net income to net cash:		
Depreciation	675	
Depletion	383	
Amortization	486	
Provision for deferred income tax	125	
Net gain (loss) on sales of property	(103)	
Changes in assets and liabilities involving cash:		
Accounts and notes receivable	(441)	
Inventories	(389)	
Accounts payable	315	
CASH PROVIDED BY OPERATING ACTIVITIES		4,202
INVESTING ACTIVITIES		
Capital expenditures	(1,227)	
Sales of property	231	
Sales (purchases) of investments	2,221	
CASH PROVIDED IN INVESTING ACTIVITIES		1,225
FINANCING ACTIVITIES		
Payments on long-term debt	(524)	
Purchases of treasury stock	(15)	
Proceeds from sales of preferred stock	620	
Dividends paid to stockholders	(485)	
CASH PROVIDED (USED) IN FINANCING ACTIVITIES		(404)
INCREASE (DECREASE) IN CASH AND CASH EQUIVALENT		5,023

The Cash Flow Statement

The *cash flow statement*, also called the *consolidated statement of cash flow* or *statement of consolidated cash flow*, is a summary of the cash flow of a company over a given period of time. The *cash flow* equals cash receipts minus cash payments over a given period of time or, equivalently, net profit plus amounts charged off for depreciation, depletion, and amortization. These latter three items are added back because they do not represent any cash transactions. *Depletion*, which is similar to depreciation, accounts for the exhaustion of natural resources owned by the company, such as oil, timber, and minerals. The cash flow statement is a measure of a company's financial health and, in recent years, has become a very important feature of the annual report.

A representative consolidated cash flow statement for a fictitious company, Chicago Chemicals, is given in Table 22.5 for the calendar year 2006. The statement is divided into three parts: operating activities, investing activities, and financing activities. Cash flows are either into or out of the company. In this statement, cash flows out of the company are stated in parentheses. Under operating activities, to the net income available for holders of common stock is added depreciation, depletion, amortization, and provision for deferred income tax; subtracted is a net loss on sales of property. Since the cash is not yet in hand, receivables and inventory are subtracted, but accounts payable (not yet paid) is added. The resulting cash flow into the company for operating activities for the year 2006 is $4,202,000,000. Under investing activities, capital expenditures by the company are subtracted from the sum of sales of property and sales of investments to give a cash flow of $1,225,000,000 into the company. Under financing activities, cash flows out of the company due to payments of long-term debt, purchases of treasury stock, and dividends paid to stockholders are partially offset by proceeds to the company from sales of preferred stock to give a cash flow out of the company of $404,000,000. For the combined three activities, the cash flow into the company is $5,023,000,000.

Financial Ratio Analysis

The analysis of the performance and financial condition of a company is carried out by computing several ratios from information given in its annual report. Such analysis must be done carefully because seemingly good performance might be due more to such factors as inflation and reduction of inventory than to improvements in company operations.

Current Ratio

The *current ratio* is defined as current assets divided by current liabilities. It is an indication of the ability of a company to meet short-term debt obligations. The higher the current ratio, the more liquid the company is. However, too high a ratio may indicate that the company is not putting its cash or equivalent cash to good use. A reasonable ratio is two, but it is better to compare current ratios of companies in a similar business. From Table 22.3, the current assets ratio of U.S. Chemicals is $4,630/4,153 = 1.11$, which is a low value. On August 31, 2006, Monsanto Company had a much better current ratio of 2.40.

Acid-Test Ratio

The *acid-test ratio*, also called the *quick ratio*, is a modification of the current ratio with the aim of obtaining a better measure of the liquidity of a company. In place of current assets, only assets readily convertible to cash, called *quick assets*, are used. Thus, it is defined as the ratio of current assets minus inventory to current liabilities. Marketable securities, accounts receivable, and deferred income tax assets are considered to be part of quick assets. From Table 22.3, the quick assets for U.S. Chemicals, in millions of dollars, are $4,630 - 1,420 - 312 = 2,898$. This gives an acid-test ratio of $2,898/4,153 = 0.70$, which is not a desirable ratio, since it is less than 1. As of August 31, 2006, Monsanto Company had a much better acid-test ratio of 1.66.

Equity Ratio

The *equity ratio* is defined as the ratio of stockholders' equity to total assets. It measures the long-term financial strength of a company. From Table 22.3, the equity ratio for U.S. Chemicals is $4,361/14,211 = 0.31$, which, again, is too low a value. This ratio should be about 0.50. If the equity ratio is too high, the company may be curtailing its growth. As of August 31, 2006, Monsanto Company had a satisfactory equity ratio of 0.56.

The above three financial ratios use only data from the balance sheet. We next consider two ratios that use both the balance sheet and the income statement, followed by two ratios that use data from the income statement only. These ratios are particularly susceptible to economic conditions, which can, sometimes, change quickly from year to year. In the United States, a brief recession took place between March and November of 2001, but economic growth was positive during the years 2002 through 2006.

Return on Total Assets (ROA)

One measure of how a company uses its assets is the *return on total assets*, which is defined as the ratio of income before interest and taxes to total assets. Using data from Tables 22.3 and 22.4, this ratio for U.S. Chemicals is $1,390/14,211 = 0.098$ or 9.8%, which is higher than the 9.00% achieved by Monsanto Company in the fiscal year of 2006.

Return on Equity (ROE)

A more widely quoted return measure is the *return on stockholders' equity*, which measures the ability of a company to reinvest its earnings to generate additional earnings. It is identical to ROA except that the divisor in the ratio is common stockholders' equity instead of total assets. For U.S. Chemicals, the ROA from data in Tables 22.3 and 22.4 is $1,390/4,361 = 0.319$ or 31.9%, which, again, is higher than the 16.17% achieved by Monsanto Company in the year 2006. Historically, the norm for ROE in the United States is approximately 11%. In 2005, the 30 U.S. companies that make up the Dow Jones Industrial Average (DJIA) had an average ROE of 17.4%.

Operating Margin

The operating margin is defined as the ratio of income from operations (called revenues) to net sales. For some industrial groups, it is highly susceptible to general economic conditions. For U.S. Chemicals, Table 22.4 gives an operating margin of $1,390/11,504 = 12.1\%$, which is somewhat lower than the 14.4% achieved by Monsanto Company in the year 2006.

Profit Margin

The *profit margin*, also called the *net profit ratio*, is defined as the ratio of net income after taxes to net sales. It is more widely quoted than the operating margin and is also more susceptible to general economic conditions. When used over a period of years, it is a very useful measure of the growth of a company. Using the data of Table 22.4, the profit margin for U.S. Chemicals was $803/11,504 = 0.070$ or 7.0%. In the third quarter of the recession year 2001, the average operating margin for 900 companies in the United States was only 3.0%, with the 25 largest publicly held chemical companies averaging a dismal 0.9%. A year earlier, when general economic conditions were much more favorable, these two values were higher, at 6.8% and 6.6%, respectively. For some companies, operating margins can be quite high. From 2002 to 2006, Microsoft had an average profit margin of 25.3%, compared to an industry average of 18.3% and an S&P 500 company average of 11.7%. In 2006, Pfizer achieved 25.2%, while 3M, Questar Gas, DuPont, EXXON/Mobil, IBM, and Starbucks had lower profit margins of 17.0%, 14.2%, 10.8%, 10.5%, 10.3%, and 7.5%, respectively.

Cost Accounting

Cost accounting is a branch of accounting that deals specifically with the identification, recording, tracking, and control of costs. Accountants allocate these costs among: (1) *direct costs*, both labor and materials, (2) *indirect costs or overhead*, and (3) other miscellaneous expenses. Direct costs are those costs directly attributable to a project such as the construction of a new plant or the operation of an existing plant. Indirect costs or overhead are costs that are generally shared among several projects and are allocated to the individual projects by a formula or some other means. Miscellaneous expenses include administration, distribution and selling, research, engineering, and development. Direct costs can be more accurately identified, measured, and controlled, and are generally the largest fraction of the total cost.

Cost accounting is of great interest in plant construction and plant operation. It is also of importance when making an economic evaluation of a process design to determine whether a plant should be built. By studying company cost accounting records for existing manufacturing plants, chemical engineers preparing economic evaluations of proposed new processing plants or revamps of existing processes are less likely to omit or neglect costs that may have a significant influence on estimated process profitability measures. Large companies that engage in a number of plant construction and plant operation projects use cost accounting records to make comparisons of costs. These records are invaluable to company process design engineers when preparing, for new projects, estimates of investment and operating costs.

Cost accounting for direct costs is accomplished in terms of unit cost and quantity. The product of these two is the cost. For example, an existing process may have used 11.2 million kg/yr of a raw material with an average unit cost of \$0.52/kg. The cost is then \$5,824,000/yr. At the beginning of the year, a standard cost and a standard quantity are established for the year, for example, 10,500,000 kg and \$0.51/kg. The budgeted cost for the year is $10,500,000 \times 0.51 = \$5,355,000$. The differences between the actual and standard unit costs and quantities are called variances. Accountants can set variance flags to warn process managers of possible cost overruns. In this example, the quantity variance is 11.2 − 10.5 or 0.7 million kg/yr. This is a % variance of $0.7/10.5 \times 100\% = 6.7\%$. The cost variance is \$0.01/kg or a % variance of $0.01/0.51 \times 100\% = 2.0\%$.

A *variance in quantity* may reveal the extent of waste. For example, suppose a plant is scheduled to produce 22,700,000 kg for the coming year. Design calculations indicate that 1.2 kg of raw material is required to produce each kilogram of product. The actual production rate for the year is 21,800,000 kg. The design rate for the raw material at the actual production rate is $1.2 \times 21,800,000 = 26,160,000$ kg/yr. However, the accounting records show for the raw material a beginning inventory of 1,070,000 kg, an ending inventory of 1,120,000 kg, and a purchase of 26,980,000 kg. Thus, the amount of raw material used is $26,980,000 + 1,070,000 - 1,120,000 = 26,930,000$ kg. If the design calculations are accepted as the basis, the waste is $26.93 - 26.16 = 0.77$ million kg. At \$0.52/kg, this represents a loss for the year of $\$0.52 \times 770,000 = \$400,400$, a significant amount of money and an incentive to find the reasons for the waste and eliminate it. Similar calculations can be made for utility usage by expressing both the design and the actual values for the quantities of utilities on a per-kilogram production of product basis.

The analysis of *variance in cost* is complicated when the price of the material, whether it be the raw material or the finished product, varies during the year. This is because of the need for the company to maintain inventories of raw materials and finished products. Two methods of costing in common use are: (1) *first-in, first-out*, and (2) *last-in, first-out*. In the first method, abbreviated as *fifo*, the cost of the oldest material in the inventory is used first. In the second method, abbreviated as *lifo*, the cost of the most recent material in the inventory is used first; that is, the older material is kept in the inventory. Companies may also use any of several average costing methods. To illustrate the possible significance to the company of choosing between the fifo and lifo methods, consider the following example. At the beginning of the year 2005, ABC Oil Producing Company has an inventory of 100,000 barrels (bbl) of crude oil with a unit cost of \$50/barrel. During the first quarter of 2005, purchases of crude oil are made at three different prices as follows:

Month	Barrels Purchased	Cost (\$/bbl)
January	80,000	50
February	100,000	60
March	150,000	30

At the end of March, the inventory is 75,000 barrels. Determine the cost of the barrels sold during the quarter. The total barrels sold during the quarter is $100,000 - 75,000 + 80,000 + 100,000 + 150,000 = 355,000$ barrels.

Fifo method:

Cost of barrels sold by first-in, first-out is

$$100,000 \times 50 = \$5,000,000$$
$$80,000 \times 50 = 4,000,000$$
$$100,000 \times 60 = 6,000,000$$
$$75,000 \times 30 = 2,250,000$$
$$\text{Total barrels sold} = 355,000$$
$$\text{Total cost of barrels sold} = \$17,250,000$$

Lifo method:

Cost of barrels sold by first-in, first-out is

$$150,000 \times 30 = \$4,500,000$$
$$100,000 \times 60 = 6,000,000$$
$$80,000 \times 50 = 4,000,000$$
$$25,000 \times 50 = 1,250,000$$
$$\text{Total barrels sold} = 355,000$$
$$\text{Total cost of barrels sold} = \$15,750,000$$

The total cost of barrels sold by the lifo method is $1,500,000 less than that of the fifo method. The unit costs are $48.59/bbl for the fifo method and $44.37/bbl for the lifo method.

22.2 COST INDEXES AND CAPITAL INVESTMENT

In all stages of the design process, estimates of both the total capital investment (TCI) and the annual cost of manufacture (COM) are crucial for the evaluation of product and processing alternatives. As discussed in Chapters 6, 7, and 8, many heuristics have been developed to create process flowsheets that reduce costs and increase the profitability of the processes being designed. In Chapters 8 and 9, approximate measures are used, such as the annualized cost (involving both the capital investment and the annual manufacturing cost), for the comparison of alternative process flowsheets. In some cases, when the manufacturing costs, especially the cost of fuel, are high, it is possible to compare the alternatives on the basis of the lost work or thermodynamic efficiency. This is the subject of Chapter 9, in which several considerations are presented for adjusting the minimum approach temperature in heat exchangers, replacing valves with turbines, and reducing pressure drops in pipelines.

In this chapter and the next, commonly used methods are developed for assessing the profitability of product and process designs. This chapter focuses on the so-called *direct permanent* (capital) *investment*, C_{DPI}, that is, the estimation of the purchase cost of required equipment and the cost of its installation in a potential chemical process. To this is added a contingency, the cost of land, any applicable royalties, and the cost of starting up the plant, to give the *total permanent investment*, C_{TPI}. The *contingency* accounts for uncertainty in the estimate and the possibility of not accounting for all of the costs involved. *Royalties* are payments made for the use of property, especially a patent, copyrighted work, franchise, or natural resource through its use. In Chapter 23, the *annual manufacturing costs*—together with the general annual expenses such as administration and marketing, which are listed in a *Cost Sheet*—are considered. These costs are the basis for an estimate of the *working capital* needed to compute the *total capital investment* for a chemical process. Then, together with depreciation and tax schedules, cash flows are computed that lead to profitability measures such as the *investor's rate of return* (IRR), also known as the *discounted cash flow rate of return* (DCFRR).

Cost Indexes

As discussed in Section 22.5, the purchase cost of processing equipment is generally obtained from charts, equations, or quotes from vendors. However, costs are not static. Because of inflation, they generally increase with time. Thus, charts and equations apply to a particular date, usually month and year, or to an average for a particular year. Quotes from vendors are often applicable only for a month or two. An estimate of the purchase cost at a later date is obtained by multiplying the cost from an earlier date by the ratio of a *cost index*, I, at that later date to a base cost index, I_{base}, that corresponds to the date that applies to the purchase cost:

$$Cost = Base\ Cost\left(\frac{I}{I_{base}}\right) \tag{22.2}$$

The indexes most commonly considered by chemical engineers are

1. ***The Chemical Engineering (CE) Plant Cost Index*** It is published in each monthly issue of the magazine *Chemical Engineering*, with $I = 100$ for 1957–1959. A complete description of the index appears in *Chemical Engineering*, **109** (1), 62–70 (2002) in an extensive revision by W.M. Vatavuk.
2. ***The Marshall & Swift (MS) Equipment Cost Index*** It is published in each monthly issue of the magazine *Chemical Engineering*, with $I = 100$ for 1926. A complete description of the index appears in *Chemical Engineering*, **54** (11), 124 (1947); **85** (11), 189 (1978); and **92** (9), 75 (1985).
3. ***The Nelson–Farrar (NF) Refinery Construction Cost Index*** It is published in the first issue each month of the magazine *Oil & Gas Journal*, with $I = 100$ for 1946. A complete description of the index appears in the magazine *Oil & Gas Journal*, **63** (14), 185 (1965); **74** (48), 68 (1976); and **83** (52), 145 (1985).
4. ***The Engineering News-Record (ENR) Construction Cost Index*** It is published each week in the magazine *Engineering News-Record* and in each monthly issue of

Chemical Engineering, with $I = 100$ for 1967. A complete description of the index appears in *Engineering News-Record*, **178** (11), 87 (1967).

The CE and NF indexes pertain to the entire processing plant, taking into account labor and materials to fabricate the equipment, deliver it, and install it. However, the NF index is restricted to the petroleum industry, while the CE index pertains to an average of all chemical processing industries. The ENR index, which is a more general index that pertains to the average of all industrial construction, is a composite of the costs of structural steel, lumber, concrete, and labor. The MS index pertains to an all-industry average equipment purchase cost. However, it is accompanied by a more useful process industries average equipment cost index, averaged mainly for the chemicals, petroleum products, paper, and rubber industries. The CE and NF indexes also provide cost indexes for only the purchase cost of several categories of processing equipment, including heat exchangers, pumps and compressors, and machinery.

Figure 22.1 compares, on a semilogarithmic plot, the values of the CE Plant Cost Index, MS Process Industries Average Cost Index, NF Refinery Cost Index, and ENR Construction Cost index for the period of 1980 to 2006. The same values are tabulated in Table 22.6. It can be seen that costs increased at a more rapid annual rate from 2003 to 2006 than from 1990 to 2003. In the 26-year period of Table 22.6, the cost indexes increased by factors of 1.916 for CE, 2.022 for MS, 2.440 for NF, and 2.383 for ENR. These factors correspond to the following respective increases per year: 2.53%, 2.75%, 3.49%, and 3.40%. Included in Figure 22.1 and Table 22.6 are values for the U.S. Consumer Price Index (CPI), published by the federal government and used to measure the rate of inflation, with a basis of 10.0 for the year 1914. In the 26-year period from 1980 to 2006, the CPI increased by a factor of 2.447, giving an average consumer inflation rate of 3.50% per year, very close to the ENR index rate.

Commodity Chemicals

Manufactured chemicals, also referred to as basic chemical products, can be classified as either (1) primary chemicals, usually referred to as *commodity chemicals* or *bulk chemicals;* or (2) secondary chemicals, which include *fine chemicals* and *specialty chemicals*. Commodity chemicals have a known chemical structure, are most often produced by continuous processing, and are characterized by high production rates (typically more than 10 million lb/yr), high sales volumes, low selling prices, and global competition. Secondary chemicals are most often produced by batch processing, and are characterized by low production rates (typically less than 1 million lb/yr), low sales volume, and high selling prices.

Many of the high-volume commodity chemicals are listed in Table 22.7, including U.S. production rates in the year 2005 from the July 10, 2006, issue of *Chemical & Engineering News*, typical sales price in the year 2006 from the *Chemical Market Reporter*, and required raw materials. Note that 14 commodity chemicals are produced at total rates of more than 10 billion lb/yr. Typical large plants produce 1 billion lb/yr, equivalent to about 125,000 lb/hr. A large petroleum refinery, producing many products, feeds 100,000 barrels of crude oil per day, equivalent to about 1,300,000 lb/hr. Prices for the commodity chemicals in Table 22.7 range from as low as \$0.04/lb for sulfuric acid to \$0.80/lb for ethylene oxide. These prices may be compared to regular-grade gasoline at the pump before state and federal taxes are added, at \$2.50/gal, equivalent to about \$0.42/lb. Most fine and specialty chemicals cost much more than commodity chemicals. For example, cetyl (palmityl) alcohol in crystalline flake form, which is used as a textile conditioner,

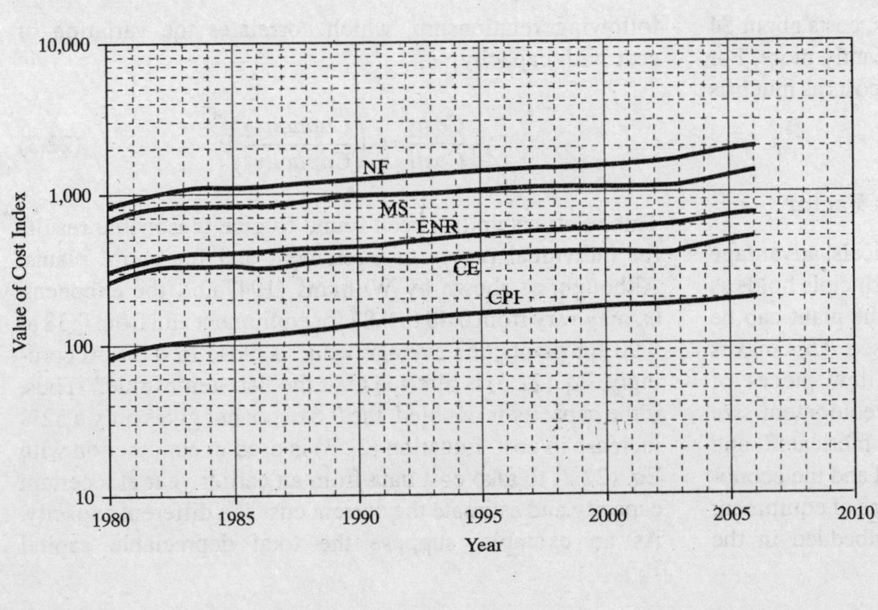

Figure 22.1 Comparison of annual average cost indexes.

Table 22.6 Comparison of Annual Average Cost Indexes

Year	CE Chemical Engineering Plant Cost Index Years 1957–1959 = 100	MS Marshall-Swift Process Industry Index Year 1926 = 100	NF Nelson-Farrar Refinery Construction Index Year 1946 = 100	ENR Engineering News-Record Construction Index Year 1967 = 100	CPI U.S. Federal Government Consumer Price Index Year 1914 = 10
1980	261	675	823	303	82.4
1981	297	745	904	330	90.9
1982	314	774	977	357	96.5
1983	317	786	1,026	380	99.6
1984	323	806	1,061	387	103.9
1985	325	813	1,074	392	107.6
1986	318	817	1,090	401	109.6
1987	324	830	1,122	412	113.6
1988	343	870	1,165	422	118.3
1989	355	914	1,196	429	124.0
1990	358	935	1,226	441	130.7
1991	361	952	1,253	450	136.2
1992	358	960	1,277	464	140.3
1993	359	975	1,311	485	144.5
1994	368	1,000	1,350	503	148.2
1995	381	1,037	1,392	509	152.4
1996	382	1,051	1,419	524	156.9
1997	387	1,068	1,449	542	160.5
1998	390	1,075	1,478	551	163.0
1999	391	1,083	1,497	564	166.6
2000	394	1,110	1,543	579	172.2
2001	394	1,109	1,580	590	177.1
2002	396	1,121	1,642	609	179.9
2003	402	1,143	1,710	623	184.0
2004	444	1,202	1,834	662	188.9
2005	468	1,295	1,919	693	195.3
2006	500	1,365	2,008	722	201.6

an emulsifier, and a component of cosmetics, costs about $4 for 1 lb. Many pharmaceuticals cost significantly more. For example, tPA, a drug to degrade blood clots, costs as much as $3,000 for a 100-mg dose.

Economy-of-Scale and the Six-Tenths Factor

When demand is high for commodity chemicals, advantage can be taken of the *economy-of-scale*. This principle holds as long as each major piece of equipment in the plant can be made larger as the production rate is increased. This makes possible a single-train plant, with no or few pieces of equipment duplicated. However, when the equipment size exceeds the maximum size that can be fabricated and shipped, then equipment must be duplicated and the economy-of-scale is lost because two or more trains of equipment are required. The economy-of-scale is embedded in the following relationship, which correlates the variation of cost with capacity:

$$\frac{Cost_2}{Cost_1} = \left(\frac{Capacity_2}{Capacity_1}\right)^m \tag{22.3}$$

This relationship has been found to give reasonable results for individual pieces of equipment and for entire plants. Although, as shown by Williams (1947a,b), the exponent, m, may vary from 0.48 to 0.87 for equipment and from 0.38 to 0.90 for plants, the average value is close to 0.60. Accordingly, Eq. (22.3) is referred to as the "six-tenths rule." Thus, if the capacity is doubled, the 0.6 exponent gives only a 52% increase in cost. Equation (22.3) is used in conjunction with Eq. (22.2) to take cost data from an earlier year at a certain capacity and estimate the current cost at a different capacity. As an example, suppose the total depreciable capital

Table 22.7 Major U.S. Commodity Chemicals

Chemical	U.S. Production in 2005 (Millions of Pounds)	Typical Price in 2006 ($/lb)	Typical Raw Materials Required
Sulfuric acid	80,512	0.04	Sulfur dioxide, oxygen, water
Ethylene	52,853	0.55	Petroleum
Propylene	33,803	0.45	Petroleum
Phosphoric acid	25,571	0.23	Phosphorun, oxygen, water
Ethylene dichloride	24,930	0.19	Ethylene, chlorine
Chlorine	22,432	0.17	Sodium chloride, water
Ammonia	21,550	0.16	Nitrogen, hydrogen
Sodium hydroxide	18,483	0.05	Sodium chloride, water
Benzene	14,635	0.32	Coal tar, petroleum
Ammonium nitrate	14,006	0.09	Ammonia, nitric acid
Nitric acid	13,951	0.11	Ammonia, oxygen
Urea	12,789	0.12	Carbon dioxide, ammonia
Ethylbenzene	11,576	0.49	Benzene, ethylene
Styrene	11,116	0.70	Ethylbenzene
Hydrochloric acid	9,713	0.06	Byproduct of chemical processes
Cumene	7,736	0.37	Benzene, propylene
Ethylene oxide	6,980	0.80	Ethylene, oxygen
Ammonium sulfate	5,683	0.07	Ammonia, sulfuric acid
Vinyl acetate	2,926	0.56	Ethylene, acetic acid, oxygen
Acrylonitrile	2,917	0.48	Propylene, ammonia, oxygen
Sodium chlorate	2,577	0.24	Chlorine, sodium hydroxide

investment for a plant to produce 1,250 tonnes/day (1 tonne = 1,000 kg) of ammonia was $140 million in 1990. In the year 2006, the estimated investment for a 2,500 tonnes/day plant is as follows, where the CE index in Table 22.6 was used:

Estimated investment, millions of U.S.$

$$= 140\left(\frac{2,500}{1,250}\right)^{0.6}\left(\frac{500}{358}\right) = 140(1.52)(1.40) = 297$$

As discussed below in Section 22.7, the Aspen Engineering Suite provides methods more accurate than the six-tenths factor method of Equation (22.3) for determining the effect of scale on capital cost. The Aspen methods apply engineering-based scale-up rules to each piece of process equipment and to buildings, site development, and other items of capital cost. See Section 22.7; that is, Section 22S.1 in the file, Supplement_to_Chapter_22.pdf (in the PDF File folder, which can be downloaded from the Wiley Web site associated with this book.).

www.wiley.com/go/global/seider

Typical Plant Capacities and Capital Investments for Commodity Chemicals

Because of the economy-of-scale, one might ask: How large are the capacities of the plants used to produce commodity chemicals and what are the corresponding capital investments? Haselbarth (1967) presented investment and plant capacity data for 60 types of chemical plants. This was followed by a more extensive compilation by Guthrie (1970, 1974) for 54 chemical processes. Unfortunately, because competition for the manufacture of commodity chemicals has become so keen in recent years, companies are now reluctant to divulge investment figures for new plants. However, Table 22.8 presents recent data for many of the commodity chemicals in Table 22.7. Plant production rates are large, 0.360 to 4.0 billion lb/yr. Corresponding total depreciable capital investments are also large, ranging from 20 million to 400 million U.S. dollars in 1995.

Note that in several cases, the plants produce combined products. Both ethylene and propylene are produced from a naphtha cut obtained from the fractionation of crude oil. A combined ammonia-urea fertilizer plant is common. The electrolysis of a brine solution produces both chlorine and sodium hydroxide. Recent literature data are usually given for plant capacities in tonnes per year (1 tonne = 1,000 kg) or tons per day (1 ton = 2,000 lb) of product, but the capacity data in Table 22.8 are given in pounds of product per year. Also included in the table is the value of C_b for use in the following modification of Eq. (22.3):

$$C_{\text{TDC}} = C_b\left(\frac{\text{pounds/year}}{\text{production rate in Table 22.8}}\right)^{0.6}\left(\frac{I}{I_b}\right) \quad (22.4)$$

where C_{TDC} is the total depreciable capital (TDC) investment, for the desired production rate and year. The data in Table 22.8 are indexed to the year 1995, when, according to

Table 22.8 Representative Plant Capacity and Capital Investment for Some Commodity Chemicals

Commodity Chemical(s)	Production Rate(s) (Millions of Pounds/Year)	Capital Investment Factor [C_b in Eq. (22.4) for 1995]
Ethylene and propylene	1,200 and 600	$300,000,000
Sulfuric acid	4,000	$30,000,000
Ethylene dichloride	1,000	$80,000,000
Ammonia and urea	400 and 1,500	$400,000,000
Chlorine and sodium hydroxide	360 and 400	$80,000,000
Ethylbenzene	2,800	$80,000,000
Phosphoric acid	3,200	$50,000,000
Styrene	2,500	$200,000,000
Nitric acid	1,400	$50,000,000
Ethylene oxide	600	$80,000,000
Cumene	600	$30,000,000
Ammonium nitrate	800	$20,000,000

Table 22.6, the Chemical Engineering Plant Cost Index was 381. Thus, for the year 2006, the right side of Eq. (22.4) would include a cost index ratio of 500/381 = 1.312.

EXAMPLE 22.1

Estimate the total depreciable capital investment in the year 2006 for a plant to produce 90 ton/day of chlorine and 100 ton/day of sodium hydroxide. Assume that the plant will operate continuously for 330 days of the 365-day year.

SOLUTION

The production rate can be based upon either the chlorine or the sodium hydroxide since both are produced in the same plant. The annual production rate of, say, chlorine in lb/year = $90(2,000)(330) = 59,400,000$ lb/year. In Table 22.8, $C_b =$ \$80,000,000 for a chlorine production rate of 360,000,000 lb/yr. Using Eq. (22.4),

$$C_{TDC} = \$80 \text{ million} \left(\frac{59.4}{360}\right)^{0.6} \times 1.312 = \$35.6 \text{ million}$$

22.3 CAPITAL INVESTMENT COSTS

The total capital investment (TCI) of a chemical plant or a chemical product manufacturing facility is a one-time expense for the design, construction, and startup of a new plant or a revamp of an existing plant. It is analogous to the purchase price of a new house, where the price includes purchase of the land, building-permit fees, excavation of the land, improvements to the land to provide utilities and access, preparation of architectural and construction drawings, construction of the house, landscaping, and contractor's fee. For convenience in cost accounting and cost estimation, Busche (1995) divides the TCI into the components listed in Table 22.9.

Before Table 22.9 is discussed, it is important to make a few distinctions. A new chemical processing plant may be an addition to an existing *integrated complex*, such as the addition of a polyethylene plant to a petroleum refinery that produces ethylene as one of its products; or it may be a *grass-roots plant*, with no other chemical plants nearby. In both cases, a new plant requires *auxiliary facilities*, including utilities such as steam, cooling water, and electricity; and other services, such as waste treatment and railroad facilities. A grass-roots plant may also require other new facilities, such as a cafeteria and a maintenance shop. In the integrated complex, the auxiliary facilities may be shared among the various plants in the complex. For either an integrated complex or a grass-roots plant, it is customary to separate the processing equipment directly associated with the manufacturing process from the auxiliary facilities by an imaginary fence that defines so-called *battery limits*, with the chemical processing plant inside the limits in an *on-site* area. The utilities and other services are outside the battery limits and are referred to as *offsite facilities*. Figure 22.2 shows typical offsite auxiliary facilities that might be associated with a grass-roots plant. Depending on the extent of the offsite facilities, they can be a significant fraction of the total capital investment.

Table 22.9 begins with the sums of so-called bare-module costs for fabricated process equipment and process machinery. These refer to the on-site part of the plant, which can be divided into modules, each of which contains a processing unit that may be a piece of *fabricated process equipment*, such as a heat exchanger, vessel, or distillation column; or an item of *process machinery*, such as a pump, compressor, centrifuge, conveyor, or robot arm. Fabricated equipment is custom-designed, usually according to the pressure vessel code and other design standards, for any size and shape that can be shipped. Process machinery is selected from a vendor-supplied list of standard sizes and often includes a driver, such as an electric motor. A module contains not only the

Table 22.9 Components of Total Capital Investment (TCI)

Total bare-module costs for fabricated equipment	C_{FE}			
Total bare-module costs for process machinery	C_{PM}			
Total bare-module costs for spares	C_{spare}			
Total bare-module costs for storage and surge tanks	$C_{storage}$			
Total cost for initial catalyst charges	$C_{catalyst}$			
Total bare-module costs for computers and software, including distributed control systems, instruments, and alarms	C_{comp}			
Total bare-module investment, TBM		C_{TBM}		
Cost of site preparation		C_{site}		
Cost of service facilities		C_{serv}		
Allocated costs for utility plants and related facilities		C_{alloc}		
Total of direct permanent investment, DPI			C_{DPI}	
Cost of contingencies and contractor's fee			C_{cont}	
Total depreciable capital, TDC				C_{TDC}
Cost of land				C_{land}
Cost of royalties				C_{royal}
Cost of plant startup				$C_{startup}$
Total permanent investment, TPI				C_{TPI}
Working capital				C_{WC}
Total capital investment, TC				C_{TCI}

piece of equipment or machinery, but also all other materials for installing it (setting it up and connecting it to equipment in other modules), including the piping to and from other modules; the concrete (or other) foundation; ladders and other steel supporting structures; the instruments, controllers, lighting, and electrical wiring; insulation; and painting. Also, depending on plant location and size, some equipment may be housed in process buildings or shelters.

Given the purchase cost of a process unit, the installed cost is obtained by adding the cost of installation. It is common to estimate the cost of installation using *factored-cost methods* based on the f.o.b. purchase cost of the process equipment.

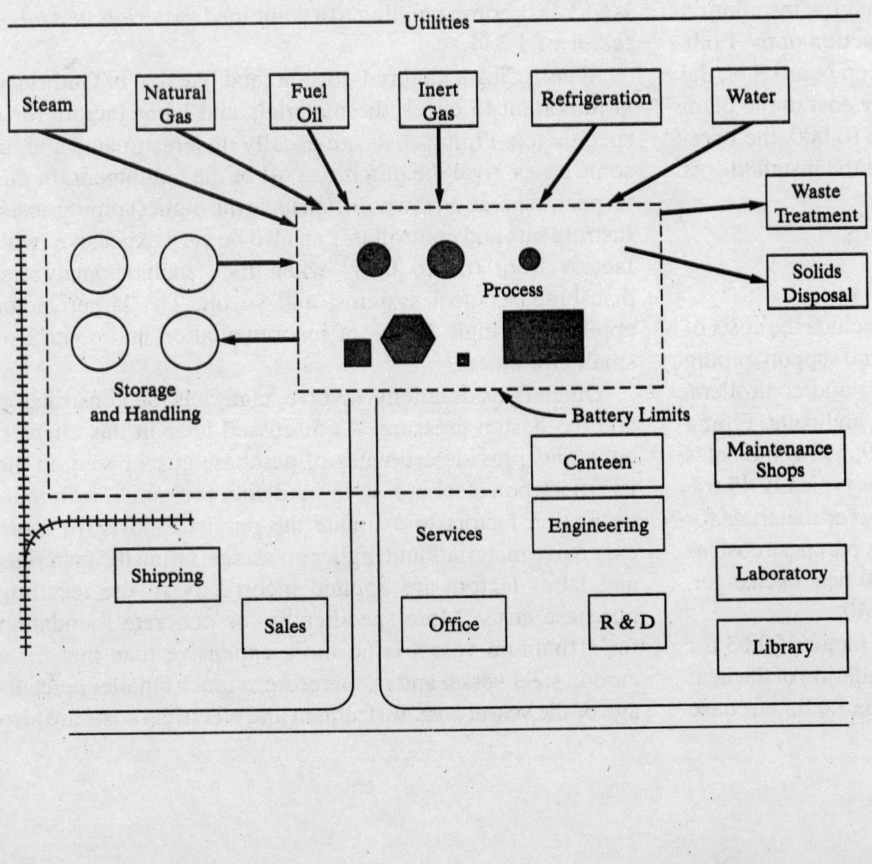

Figure 22.2 Plant services outside of process battery limits (courtesy of C.A. Miller).

Table 22.10 Example of Installation Costs for a Heat Exchanger to Give the Bare-Module and Total-Module Costs

	Cost ($)	Total Costs ($)	Fraction of f.o.b. Purchase Cost (C_P)
Direct module expenses			
Equipment purchase price, f.o.b., C_P		10,000	1.00 C_P
Field materials used for installation			
Piping	4,560		
Concrete	510		
Steel	310		
Instruments and controllers	1,020		
Electrical	200		
Insulation	490		
Paint	50		
Total of direct field materials, C_M		7,140	$C_M = 0.714\ C_P$
Direct field labor for installation			
Material erection	5,540		
Equipment setting	760		
Total of direct field labor, C_L		6,300	$C_L = 0.63\ C_P$
Indirect module expenses			
Freight, insurance, taxes, C_{FIT}	800		$C_{FIT} = 0.08\ C_P$
Construction overhead, C_O	5,710		$C_O = 0.571\ C_P$
Contractor engineering expenses, C_E	2,960		$C_E = 0.296\ C_P$
Total indirect expenses, C_{IE}		9,470	$C_{IE} = 0.947\ C_P$
Bare-module cost, C_{BM}		32,910	$C_{BM} = 3.291\ C_P$
			$F_{BM} = 3.291$

For each piece of equipment, Guthrie (1969, 1974) provides factors to estimate the direct costs of materials and labor, as well as indirect costs involved in the installation procedure. When these costs are added to the purchase cost, Guthrie calls the result the *bare-module* cost instead of the installed cost. As an illustration, see Table 22.10, in which the installation costs for a heat exchanger are given as a fraction of the f.o.b. purchase cost, C_P, where f.o.b. means free on board (i.e., the purchase cost does not include the delivery cost to the plant site). Although the f.o.b. purchase cost is $10,000, the bare-module cost is $32,910. The components of the installed cost are as follows.

Direct Materials and Labor (M&L)

The costs of materials, C_M, for installation include the costs of concrete for the foundations, steel for structural support, piping from and to the other modules, instruments and controllers, lighting and electrical materials, insulation, and paint. Piping costs can be very substantial. Guthrie (1969, 1974) indicates that the cost of piping for a heat exchanger is typically 45.6% of the f.o.b. purchase cost, while the total cost of materials for installation is estimated at 71.4% of the f.o.b. purchase cost, as shown in Table 22.10. Hence, for a $10,000 heat exchanger, the cost of materials for installation is $7,140.

Similarly, Guthrie provides a field labor factor of 0.63 for installation; that is, the cost of labor for installation of the heat exchanger in Table 22.10, C_L, is 63% of the f.o.b. purchase cost or $6,300. The field labor cost accounts for the setting in place of the exchanger, installation of the associated piping, and all other labor costs utilizing the field materials. Combined with the cost of materials, the total cost of direct materials and labor for installation of the heat exchanger is $13,440, corresponding to a combined *materials and labor factor* of 1.344.

When using a factored-cost method like that of Guthrie, it is important to check the materials and labor factors for a specific job. Piping costs are usually underestimated and, in some cases, rival the purchase cost of the equipment. In this respect, separation towers often have the highest piping costs. Instruments and controllers can also be very expensive (with factors from 0.1 to 0.75) when they include analyzers, distributed control systems, and so on. The larger factor applies to a high degree of instrumentation and control of small equipment.

Other considerations involve materials of construction and the design pressure. As discussed later in this chapter, equations provide estimates of purchase costs based on the use of carbon steel at low-to-moderate pressures, with multiplication factors to estimate the purchase costs for more expensive materials and higher pressures. Often the materials and labor factors are applied incorrectly to the resulting purchase costs. More specifically, the concrete foundation for a titanium vessel is no more expensive than that for a carbon steel vessel and is, therefore, a much smaller percentage of the vessel cost. Instrument and electrical costs are also a smaller percentage of the vessel cost. An example of the correct application of materials and labor factors is given later in Example 22.4.

The above discussion supposes that the vendor assembles all fabricated process equipment before shipping it to the plant for installation. In some cases, equipment cannot be shipped to the plant site in one piece and pre-installation field assembly will be required. Examples are furnaces and very large distillation columns and other vessels that cannot be trucked, barged, or sent by rail in one piece to the plant site. Large columns may be fabricated in sections in the shops of the vendor and transported to the plant site, where the sections are welded in a horizontal orientation before the column is erected to a vertical position. In this chapter, the purchase cost of field-assembled equipment includes the cost of pre-installation assembly at the plant site. Field-assembly costs are usually included in the purchase-cost quote from a vendor.

Indirect Costs

Other costs, such as the costs of freight to deliver the equipment to the plant site, with associated insurance and taxes, are considered to be *indirect expenses*, C_{FIT}. As shown in Table 22.10, these are estimated to be approximately 8% of C_P, that is, \$800 for the heat exchanger. These are accompanied by construction overhead, C_O, which includes fringe benefits for the workers (health insurance, vacation pay, sick leave, etc.), so-called *burden* (Social Security taxes, unemployment insurance, etc.), and salaries, fringe benefits, and burden for the supervisory personnel. The construction overhead also includes the costs of temporary buildings, roads, parking areas, cranes and machinery (purchased or rented), job site cleanup, security costs, and so on. These costs are estimated at approximately 57.1% of the f.o.b. purchase cost of the equipment, or \$5,710 for a \$10,000 heat exchanger. Contractor engineering expenses, C_E, are also included in the indirect expense category. This covers the costs of engineering, including salaries for project and process engineers, designers, procurement expenses, home office expenses, and so on. They are estimated to be 29.6% of C_P, that is, \$2,960 for a \$10,000 heat exchanger.

Summing the indirect module expenses (\$9,470), and combining the result with the cost of materials and labor, the *bare-module cost* for the \$10,000 heat exchanger, C_{BM}, is \$32,910, and, hence, the *bare-module factor*, F_{BM}, is 3.291.

Guthrie also computes and lists factors for starting with the f.o.b. purchase cost and arriving at the bare-module cost. For the heat exchanger example of Table 22.10, the factors are derived as follows:

Start with the equipment purchase price, f.o.b. of \$10,000.

The direct field materials, including the equipment, total $(10,000 + 7,140) = \$17,140$.

Guthrie defines the direct materials factor as $(17,140)/10,000 = 1.714$.

The direct field labor totals \$6,300.

Guthrie defines the direct labor factor as $(6,300)/10,000 = 0.630$.

The combined direct field materials and labor totals \$23,440.

Guthrie defines the combined direct field materials and labor factor as $(23,440/10,000) = 2.344$.

The indirect module expenses total \$9,470.

Guthrie defines the indirect factor as $(23,440 + 9,470)/23,440 = 1.404$.

The total bare-module factor is 2.344 times $1.404 = 3.291$, the same as above.

Bare-module factors vary among the various types of fabricated equipment and process machinery, decreasing somewhat with increasing size. The extent of this variation for ordinary materials of construction and low-to-moderate pressures can be seen in Table 22.11, which is taken from Guthrie (1974), based on single units of smaller size, where the factors are as much as 10% lower for multiple units of the same type. For the solids-handling equipment, the indirect factor is taken from Guthrie (1969) as 1.29. The bare-module factors vary from a value of 1.39 for crushers to separate solid particles by size to 4.16 for vertical pressure vessels,

Table 22.11 Bare-Module Factors of Guthrie (1974) for Ordinary Materials of Construction and Low-to-Moderate Pressures

	Bare-Module Factor (F_{BM})
Furnaces and direct-fired heaters, Shop-fabricated	2.19
Furnaces and direct fired heaters, Field-fabricated	1.86
Shell-and-tube heat exchangers	3.17
Double-pipe heat exchangers	1.80
Fin-tube air coolers	2.17
Vertical pressure vessels	4.16
Horizontal pressure vessels	3.05
Pumps and drivers	3.30
Gas compressors and drivers	2.15
Centrifuges	2.03
Horizontal conveyors	1.61
Bucket conveyors	1.74
Crushers	1.39
Mills	2.30
Crystallizers	2.06
Dryers	2.06
Evaporators	2.45
Filters	2.32
Flakers	2.05
Screens	1.73

which are widely used for distillation, absorption, stripping, and flash drums. All of the equipment for handling solids and fluid-solids mixtures have factors less than 2.45. In Table 22.9, the sum of the bare-module costs for all items of fabricated equipment is C_{FE}, while the sum of the bare-module costs for all items of process machinery is C_{PM}.

Other Investment Costs

In addition to the bare-module costs associated with the process equipment in a flowsheet, capital costs are incurred for spare items of equipment, C_{spare}; for storage and surge tanks, $C_{storage}$; for initial charges of catalyst, $C_{catalyst}$; and for computers and software, including instruments, distributed control systems, and alarms, C_{comp}. As shown in Table 22.9, these costs are added to the bare-module costs for the on-site equipment to give the *total bare-module investment*, C_{TBM}. Other investment costs include site preparation or development, C_{site}; service facilities (e.g., utility lines and industrial buildings), C_{serv}; and allocated costs to purchase or upgrade the utility plants and other offsite facilities (e.g., for steam and electricity generation and waste disposal), C_{alloc}, shown in Figure 22.2. These are added to C_{TBM} to give the *direct permanent investment*, C_{DPI}. After adding funds (typically 18% of C_{DPI}) to cover contingencies and a contractor fee, the *total depreciable capital*, C_{TDC}, is obtained. Depreciation is very important to companies that use equipment to manufacture goods because it permits companies to reduce their taxes. As will be discussed in detail in the next chapter, depreciation is the allocation of the cost of an asset, such as a processing plant, over a period of time for accounting and tax purposes. Depreciation accounts for the decline in the value of equipment due to general wear and tear. To the C_{TDC} is added the investment in nondepreciable items—including the cost of land, the cost of royalties for the use of processes patented by others, and costs for plant startup—to give the *total permanent investment*, C_{TPI}, also referred to as *total fixed capital*. After the working capital is added, the *total capital investment*, C_{TCI}, is obtained.

The working capital is described in detail in the next chapter. It includes the initial investment in temporary and consumable materials, as well as cash for initial payments of salaries and other operating expenses prior to the receipt of payments for plant products. These other costs, which represent a significant fraction of the total capital investment, are considered next.

Spares, Storage Tanks, Surge Vessels, Catalyst Costs, and Computer Costs

In addition to the bare-module costs for each process unit in the flowsheet as discussed above, it is often recommended to provide funds for *spares*, C_{spare}, especially for liquid pumps, to permit uninterrupted operation when a process unit becomes inoperable. Pumps are relatively inexpensive but require frequent maintenance to prevent leaks. Funds are also provided for storage units and surge tanks, $C_{storage}$, to provide improved control and intermediate storage between sections of the plant, so that one section can continue operation when an adjoining section is down or operating under capacity. The amount of storage depends on the anticipated periods of downtime and the importance of maintaining steady, uninterrupted operation. In addition, it is common to include the cost of the initial charge of catalyst, $C_{catalyst}$, which is a sizable investment cost in some plants. Finally, the cost of computers and software, with associated instruments, distributed control systems, and alarms, C_{comp}, is gaining importance in many plants, especially those that manufacture industrial and configured consumer products. These costs are included in C_{TBM}, as indicated in Table 22.9.

Site Preparation

Site preparation typically involves making land surveys, dewatering and drainage, surface clearing, rock blasting, excavation, grading, piling; and addition of fencing, roads, sidewalks, railroad sidings, sewer lines, fire protection facilities, and landscaping. Costs for site preparation and development, C_{site}, can be quite substantial for grass-roots plants, in the range of 10–20% of the total bare-module cost of the equipment. For an addition to an existing integrated complex, the cost may only be in the range of 4–6% of the total bare-module cost of the equipment.

Service Facilities

Costs for service facilities, C_{serv}, include utility lines, control rooms, laboratories for feed and product testing, maintenance shops, and other buildings. Service facility costs can be substantial for a grass-roots plant when administrative offices, medical facilities, cafeterias, garages, and warehouses are needed.

Allocated Costs for Utilities and Related Facilities

Allocated costs, C_{alloc}, are included to provide or upgrade offsite utility plants (steam, electricity, cooling water, process water, boiler feed water, refrigeration, inert gas, fuels, etc.) and related facilities for liquid waste disposal, solids waste disposal, off-gas treatment, and wastewater treatment. Some typical capital investment costs for utility plants are shown in Table 22.12. Cogeneration plants can provide both steam and electricity by burning a fuel. When utilities such as electricity are purchased from vendors at so many cents per kilowatt-hour, that cost includes the vendor investment cost. Thus, a capital cost for the plant is then not included in the capital cost estimate.

Contingencies and Contractor's Fee

Contingencies are unanticipated costs incurred during the construction of a plant. To account for the cost of contingencies, it is common to set aside 15% of the direct permanent investment, C_{DPI}, which is comprised of the components in

Table 22.12 Allocated Capital Investment Costs for Utility Plants (CE cost index = 500)

Utility	Size Factor, *S*	Range of *S*	Allocated Cost, $
Steam	Flow rate, lb/hr	20,000–1,000,000 lb/hr	$C_{alloc} = 820\,S^{0.81}$
Electricity	Power, MW	0.5–1,000 MW	$C_{alloc} = 2,600,000\,S^{0.83}$
Cooling water	Flow rate, gpm	1,000–200,000 gpm	$C_{alloc} = 1,000\,S^{0.68}$
Process water	Flow rate, gpm	5–10,000 gpm	$C_{alloc} = 1,500\,S^{0.96}$
Refrigeration (evaporator temperature = −20°F)	Tons	3–1,000 tons	$C_{alloc} = 11,000\,S^{0.77}$

Table 22.9. In addition, Guthrie (1969) adds a contractor fee of 3% of the direct permanent investment. When this total of 18%, designated C_{cont}, is added, the total depreciable capital, C_{TDC}, is obtained.

The cost of contingencies varies considerably, with 15% being a useful estimate when the design team is unable to make a better estimate. With processes for which the company has considerable experience, the cost of contingency is much lower than when a plant is being designed to produce a new chemical product just discovered by a research group. When deciding on a contingency cost, the design team should address the following three groups of questions:

1. How well is the product or process known? Has the process been demonstrated commercially, in a pilot plant, or in a laboratory? How long has the process run? Are the corrosion rates associated with the equipment well established? Has a demonstration of the process included all of the recycle streams?
2. How complete is the design? Has a simulation model been prepared? Is the detailed design complete? How much is known about the plant site?
3. How accurate are the estimates? Is the equipment conventional, or are there new and complex equipment items with which the company has little experience and cost history? In most cases, a 15% contingency is low, except when the experience factor is very great. Typical designs that have not entered the *development* (final design) stage in Figures PI.1, PII.1, and PIII.1, which are representative of most designs by student groups at universities, are likely to require 35% for contingency costs. When the chemistry is new and not well understood, 100% might be more realistic.

Land

The cost of land, C_{land}, is nondepreciable, since land rarely decreases in value, and in the absence of data can be taken as 2% of the total depreciable capital, C_{TDC}.

Royalties

When a company desires to use a product or process that is covered by patents owned by another company, a license may sometimes be negotiated. The license fee may be a one-time fee, in which case that fee is included in the capital investment as a one-time royalty or paid-up license, C_{royal}. A more common arrangement is to pay an initial license fee, included in the capital investment, and an annual royalty based on the amount or dollar value of product sold, as discussed in Section 23.2 under *Licensing Fees*. The amount of the annual royalty depends on the uniqueness of the process and the chemical being produced, with a range of 1–5% of product sales. In the absence of data, an initial royalty fee of 2% of C_{TDC} may be assumed together with an annual royalty of 3% of product sales.

Startup

The cost of plant startup, $C_{startup}$, is typically estimated as 10% of C_{TDC}. However, according to Feldman (1969), if the process and equipment are well known to skilled operators and the new process is not dependent on the operation of another plant, the startup cost may be as low as 2% of C_{TDC}. At the other extreme, if the process and the equipment are radically new, and the new process is dependent on another plant, the startup cost may be as high as 30% of C_{TDC}. In this latter case, it may be necessary to modify the process and add more equipment. With respect to startup, it is important that the process design include additional equipment, such as heat exchangers, to achieve the startup. This is particularly important for processes involving significant recycle streams and/or a high degree of energy integration. Startup time depends on the same factors as startup costs and is generally taken as a percentage of construction time, varying from 10 to 40%.

Some company accountants may prefer to divide plant startup costs into two categories: (1) those costs incurred by the contractor in checking equipment performance, calibrating controllers and other plant equipment, and commissioning the plant, and (2) those costs incurred by plant operating personnel when starting up and shutting down the plant. The former costs are included in the capital cost while the latter are considered operating costs.

Investment Site Factors

In many companies, *investment site factors*, F_{ISF}, are used to multiply the total permanent investment, C_{TPI}, to account

Table 22.13 Typical Investment Site Factors, F_{ISF}

U.S. Gulf Coast	1.00
U.S. Southwest	0.95
U.S. Northeast	1.10
U.S. Midwest	1.15
U.S. West Coast	1.25
Western Europe	1.20
Mexico	0.95
Japan	1.15
Pacific Rim	1.00
India	0.85

for different costs in different localities based on the availability of labor, the efficiency of the workforce, local rules and customs, union status, and other items. Typical factors in recent use by one of the major chemical companies are provided in Table 22.13, where a plant in the U.S. Gulf Coast area is given a base factor of 1.0. The factors range from 0.85 in India to 1.25 in the U.S. West Coast area. The corrected total permanent investment is computed as:

$$C_{TPI_{corrected}} = F_{ISF} C_{TPI}$$

Working Capital

Working capital funds, C_{WC}, are needed to cover operating costs required for the early operation of the plant, including the cost of the inventory and funds to cover accounts receivable. Because they involve the costs of the raw materials and the values of the intermediates, products, and byproducts, the working capital is normally estimated in connection with the calculation of the operating "Cost Sheet," which is presented in Table 23.1 and discussed in Section 23.3. Note that funds are usually allocated for a spare charge of catalyst, often kept in a warehouse, as a backup in case an operating problem causes the catalyst to become ineffective.

Table 22.14 Capital Cost Estimate of Bare-Module Equipment Cost for an Ammonia Plant—Costs in Millions of U.S. Dollars (Year 2006)

	C_P	F_{BM}	C_{BM}
Fabricated equipment			
Heat exchangers	6.67	3.3	22.01
Flash drum	0.01	4.3	0.04
Distillation column	0.09	4.3	0.38
Adsorbers	0.23	4.3	0.98
Absorber	0.25	4.3	1.09
Membrane separators	4.52	3.2	14.46
Reactor	0.43	4.3	1.86
Process machinery			
Gas compressors	27.72	3.5	97.00
Pumps	0.09	3.4	0.30
Total bare-module cost for on-site equipment			138.12

Example of an Estimate of Capital Investment

An example of an estimate of the total capital investment for a processing plant is given in Tables 22.14 and 22.15 for an ammonia plant producing 1 billion lb/yr. The costs are for the year 2000 at a U.S. Midwest location. The plant is part of an integrated complex. The process involves a variety of

Table 22.15 Total Capital Investment for an Ammonia Plant—Costs in Millions of U.S. Dollars (Year 2006)

Total bare-module cost for on-site equipment	138.12				
Cost for spares	0.66				
Cost for storage and surge tanks	0.57				
Cost for initial catalyst charge	0.63				
Cost of computers, software, and associated items					
Total bare-module investment		139.98			
Cost of site preparation		4.20			
Cost of service facilities		2.09			
Allocated costs for utility plants and related facilities		19.61			
Direct permanent investment			165.88		
Cost of contingencies and contractor's fee			29.86		
Total depreciable capital				195.74	
Cost of land				3.91	
Cost of plant startup				15.63	
Total permanent investment					215.28
Working capital					12.80
Total capital investment					228.08

Note: In Table 22.15, the cost of computers, software, and associated items is included in the total bare-module cost for on-site equipment.

equipment, including gas compressors, pumps, heat exchangers, a catalytic reactor, a distillation column, an absorber, a flash drum, a gas adsorber, and gas permeation membrane separators. The material of construction is almost exclusively carbon steel.

For the ammonia process, which operates at high pressure (200 atm) mostly in the gas phase, the total f.o.b. purchase cost of the on-site process equipment is $40,010,000. Installation costs boost this amount by a factor of 3.452 to a total bare-module cost of $138,120,000. As seen in Table 22.14, this cost is dominated by the gas compressors, with significant contributions from the heat exchangers and the membrane separators. Surprisingly, the reactor cost is a small fraction of the total cost. This is often the case for chemical plants. The reactor may not cost much, but it is the heart of the process and it had better produce the desired results.

Table 22.15 continues the cost estimate to obtain the total capital investment. After all other investment costs are added to the total bare-module cost of the on-site equipment, the total capital investment becomes $228,080,000. The total permanent investment is $215,280,000, which is a factor of 5.38 times the total f.o.b. purchase cost of the on-site process equipment. The startup cost here is taken as 8% of C_{TDC}.

22.4 ESTIMATION OF THE TOTAL CAPITAL INVESTMENT

As the project for manufacturing a new or existing chemical by a new process progresses from laboratory research through pilot-plant development to a decision for plant construction, a number of process design studies of increasing complexity may be made, accompanied at each step by capital cost estimates of increasing levels of accuracy as follows:

1. *Order-of-magnitude estimate* based on bench-scale laboratory data sufficient to determine the type of equipment and its arrangement to convert the feedstock(s) to product(s).
2. *Study estimate* based on a preliminary process design.
3. *Preliminary estimate* based on detailed process design studies leading to an optimized process design.
4. *Definitive estimate* based on a detailed plant design, including detailed drawings and cost estimates, sufficient to apply cost accounting.

If the process is well known and has been verified by one or more commercial operating plants, only estimate levels 3 and 4 are necessary. Methods for making capital investments at the first three levels are discussed next. This chapter is concluded with an example of a definitive estimate using the Aspen Icarus Process Evaluator (IPE), which is part of the Aspen Engineering Suite that includes ASPEN PLUS.

Currently, Aspen Icarus software systems are the only commercially available systems that are recognized as the standard of the industry for estimating process costs. These systems, which are discussed in Section 22.7 (i.e., Section 22S.1 in the file, Supplement_to_Chapter_ 22.pdf in the PDF File folder, which can be downloaded from the Wiley Web site associated with this book), have more than 35 years of field-testing on commercial plants and are in use worldwide by owner companies and engineering design and construction firms. In our experience, application of the Aspen Icarus Process Simulator (IPE) is rather quickly understood and applied by chemical engineering students and practitioners after having studied the simpler, but less accurate, costing methods presented here in Sections 22.1 to 22.6.

www.wiley.com/go/global/seider

Method 1. Order-of-Magnitude Estimate (Based on the Method of Hill, 1956)

This estimation method can be applied rapidly and is useful in determining whether a new process is worth pursuing, especially when there are competing routes. The method is particularly useful for low-pressure petrochemical plants, where it has an accuracy of approximately ±50%. For moderate-to-high pressure processes, the actual cost may be as much as twice the estimate. To produce the estimate, only two things are needed, a production rate in pounds per year and a flowsheet showing the gas compressors, reactors, and separation equipment required. Heat exchangers and liquid pumps are not considered in making the estimate. Also, it is not necessary to compute a mass and energy balance or to design or size the equipment, but it is important that the process has been sufficiently studied that the flowsheet is complete with all the major pieces of gas movement, reactors, and separation equipment and their required materials of construction. Another important factor in making the estimate is the design pressure of each major piece of equipment if it is greater than 100 psi. The method proceeds as follows based on a year 2006 Marshall and Swift Process Industries Average Cost Index of 1,365, a base production rate of 10,000,000 lb/yr for the main product, carbon steel construction, and a design pressure of less than 100 psi.

Step 1: Establish the production rate of the main product in pounds per year. Compute a production rate factor, F_{PR}, using the six-tenths rule:

$$F_{PR} = \left(\frac{\text{Main product flow rate, lb/yr}}{10,000,000}\right)^{0.6} \tag{22.5}$$

Step 2: Using a process flowsheet, calculate from the following equation a module cost, C_M, for purchasing, delivering, and setting in place each major piece of equipment, including gas compressors and blowers (but not low-compression ratio recycle compressors and blowers); reactors; separators such as distillation columns, absorbers, strippers, adsorbers, membrane units, extractors, electrostatic precipitators, crystallizers, and evaporators;

but not heat exchangers, flash and reflux drums, or liquid pumps:

$$C_M = F_{PR} F_M \left(\frac{\text{design pressure, psia, if} > 100\,\text{psi}}{100} \right)^{0.25} \times (\$160,000) \tag{22.6}$$

where F_M is a material factor, as follows:

Material	F_M
Carbon steel	1.0
Copper	1.2
Stainless steel	2.0
Nickel alloy	2.5
Titanium clad	3.0

Step 3: Sum the values of C_M; multiply the sum by the factor F_{PI} to account for piping, instrumentation and automatic controls, and indirect costs; and update with the current MS cost index, giving the total bare-module investment, C_{TBM}:

$$C_{TBM} = F_{PI} \left(\frac{\text{MS index}}{1,365} \right) \Sigma C_M \tag{22.7}$$

where the factor F_{PI} depends on whether the plant processes solids, fluids, or a mixture of the two, as follows:

Type of Process	F_{PI}
Solids handling	1.85
Solids-fluids handling	2.00
Fluids handling	2.15

Step 4: To obtain the direct permanent investment, C_{DPI}, multiply C_{TBM} by the following factors to account for site preparation, service facilities, utility plants, and related facilities:

$$C_{DPI} = (1 + F_1 + F_2) C_{TBM} \tag{22.8}$$

where the factors F_1 and F_2 are

	F_1
Outdoor construction	0.15
Mixed indoor and outdoor construction	0.40
Indoor construction	0.80

	F_2
Minor additions to existing facilities	0.10
Major additions to existing facilities	0.30
Grass-roots plant	0.80

Outdoor construction is common except where winters are very severe and/or solids handling is critical.

Step 5: Obtain the total permanent investment and the total capital investment by the following equations, where a large contingency of 40% is used because of the approximate nature of the capital cost estimate, and the costs of land, royalties, and plant startup are assumed to add an additional 10%. Working capital is taken as 15% of the total permanent investment.

$$C_{TPI} = 1.50 C_{DPI}$$

$$C_{TCI} = 1.15 C_{TPI}$$

EXAMPLE 22.2

Make an order-of-magnitude estimate of the total capital investment, as of early 2007 with MS = 1,400, to produce benzene according to the toluene hydrodealkylation process shown in Figure 5.20. Assume an overall conversion of toluene to benzene of 95% and 330 days of operation per year. Also, assume the makeup gas enters at the desired pressure and a clay adsorption treater must be added to the flow sheet after the stabilizer. The treater removes contaminants that would prevent the benzene product from meeting specifications. In addition, in order for the reactor to handle the high temperature, it must have a brick lining on the inside, so take a material factor of $F_M = 1.5$. Otherwise, all major equipment is constructed of carbon steel. The plant will be constructed outdoors with major additions to existing facilities.

SOLUTION

Step 1: The plant will operate $330(24) = 7,920$ hr/yr. For a toluene feed rate of 274.2 lbmol/hr, the annual benzene production rate is $0.95(274.2)(78.11)(7,920) = 161,000,000$ lb/yr. Thus,

$$F_{PR} = \left(\frac{161,000,000}{10,000,000} \right)^{0.6} = 5.3$$

Step 2: The flowsheet includes one reactor (with $F_M = 1.5$) operating at 570 psia, three distillation columns operating at pressures less than 100 psia, one compressor operating at 570 psia, and one adsorption tower, assumed to operate at less than 100 psia. Therefore, the sum of the C_M values is

$$\Sigma C_M = 5.3 \left[1.5 \left(\frac{570}{100} \right)^{0.25} + 3 + 1 \left(\frac{570}{100} \right)^{0.25} + 1 \right] (\$160,000) = \$6,670,000$$

Step 3: From Eq. (22.7), the total bare-module investment for a fluids handling process is

$$C_{TBM} = 2.15 \left(\frac{1,400}{1,365} \right) (\$6,650,000) = \$14,700,000$$

Steps 4 and 5:

$$C_{DPI} = (1 + 0.15 + 0.30)(\$14,700,000) = \$21,300,000$$

$$C_{TPI} = 1.50(\$21,400,000) = \$32,100,000$$

$$C_{TCI} = 1.15(\$32,100,000) = \$36,900,000$$

Table 22.16 Original and Recommended Lang Factors

	Original Lang Factors, *Not* Including Working Capital	$f_{L_{TPI}}$ Recommended Lang Factors of Peters, Timmerhaus, and West, *Not* Including Working Capital	$f_{L_{TCI}}$ Recommended Lang Factors of Peters, Timmerhaus, and West, Including Working Capital
Solids processing plant	3.10	3.97	4.67
Solids-fluids processing plant	3.63	4.28	5.03
Fluids processing plant	4.74	5.04	5.93

Method 2. Study Estimate (Based on the Overall Factor Method of Lang, 1947a, b, and 1948)

In a series of three articles from 1947 to 1948, Lang developed a method for estimating the capital cost of a chemical plant using overall factors that multiply estimates of the delivered cost of the major items of process equipment. This method requires a process design, complete with a mass and energy balance, and equipment sizing. In addition, materials of construction for the major items of equipment, including the heat exchangers and pumps, must be known. Considerably more time is required for making a study estimate than for the preceding order-of-magnitude estimate. But, the accuracy is improved to ±35%. To apply the method, the f.o.b. purchase cost of each piece of major equipment must be estimated. F.o.b purchase costs of a wide range of chemical processing equipment are given in the next section of this chapter. The Lang method proceeds by steps as follows:

Step 1: From the process design, prepare an equipment list, giving the equipment title, label, size, material of construction, design temperature, and design pressure.

Step 2: Using the data in Step 1 with f.o.b. equipment cost data, add to the equipment list the cost and the corresponding cost index of the cost data. Update the cost data to the current cost index, sum the updated purchase costs to obtain the total f.o.b. purchase cost, C_P, and multiply by 1.05 to account for delivery of the equipment to the plant site. Then, multiply the result by an appropriate Lang factor, f_L, to obtain the total permanent investment (fixed capital investment), C_{TPI} (i.e., without the working capital), or the total capital investment, C_{TCI} (i.e., including an estimate of the working capital at 15% of the total capital investment or 17.6% of the total permanent investment):

$$C_{TPI} = 1.05\, f_{L_{TPI}} \sum_i \left(\frac{I_i}{I_{b_i}}\right) C_{P_i} \tag{22.9}$$

$$C_{TCI} = 1.05\, f_{L_{TCI}} \sum_i \left(\frac{I_i}{I_{b_i}}\right) C_{P_i} \tag{22.10}$$

The original Lang factor, based on capital costs for 14 different chemical plants, was found to depend on the extent to which the plant processes solids or fluids. Lang's factors, which at that time did not account for working capital, are given in the second column of Table 22.16.

A more detailed development of the Lang factors, based on an analysis of 156 capital cost estimates, was published by the editors of *Chemical Engineering* magazine in the September 30, 1963, issue on pages 120 and 122, as "Cost File 81." A further refinement, carried out by Peters, Timmerhaus, and West (2003), gives the most widely accepted values of the Lang factors, which are included in Table 22.16 and are the factors recommended here for use in Eqs. (22.9) and (22.10). The detailed breakdown of costs by Peters et al. is given in Table 22.17, which assumes that major plant additions are made to an existing site. The numbers in the table are based on a value of 100 for the total delivered cost of the process equipment. Here, the delivered cost is estimated as 1.05 times the f.o.b. purchase cost. The Lang factors apply to total permanent investments of up to approximately $100 million U.S. dollars. Note that the combined contractor's fee and legal expenses, as well as the contingency, are quite generous. If Eq. (22.9) is used, a detailed estimate of the working capital should be made according to the method presented in Section 23.3. No provision is made in the Lang-factor estimates for spares, storage and surge tanks, initial catalyst charge, royalties, or plant startup. However, these additional items can be added when desired. The fixed capital investment in Table 22.17 is the same as the total permanent investment in Table 22.9.

EXAMPLE 22.3

Use the Lang-factor method to estimate the total capital investment, as of the year 2006 (MS = 1,365), to produce cyclohexane according to the benzene hydrogenation process shown in Figure 9S.24. However, the makeup H_2 feed is not available at 335 psia, but at 75 psia. Therefore, a feed-gas compressor, K2, has been added. Also two heat exchangers have been added. Reactor effluent stream S7 now enters new exchanger H2, which cools the effluent to 260°F by producing 10 psig steam from boiler feed water. The effluent leaves H2 as stream S7A and enters new exchanger H3, where it is heat-exchanged with the feed benzene, heating the benzene to 235°F while being cooled to 201°F and

Table 22.17 Breakdown of Lang Factors by Peters, Timmerhaus, and West (2003)

	Percent of Delivered-Equipment Cost for		
	Solids Processing Plant	Solids-Fluids Processing Plant	Fluids Processing Plant
Delivered cost of process equipment	100	100	100
Installation	45	39	47
Instrumentation and control	18	26	36
Piping	16	31	68
Electrical	10	10	11
Buildings (including services)	25	29	18
Yard improvements	15	12	10
Service facilities	40	55	70
Total direct plant cost	269	302	360
Engineering and supervision	33	32	33
Construction expenses	39	34	41
Total and indirect plant costs	341	368	434
Contractor's fee and legal expenses	21	23	26
Contingency	35	37	44
Fixed capital investment	397	428	504
Lang factor, f_{LTPI}, for use in Eq. (22.9)	3.97	4.28	5.04
Working capital	70	75	89
Total capital investment	467	503	593
Lang factor, f_{LTCI}, for use in Eq. (22.10)	4.67	5.03	5.93

leaving as stream S7B, which now enters existing exchanger H1. The process design has been completed, with the equipment sizes in the "Equipment List" of Table 22.18. Also included in Table 22.18 are estimates of the f.o.b. purchase costs in the year 1977 (MS index of 514). All equipment is fabricated from carbon steel.

SOLUTION

Referring to Table 22.18, the total f.o.b. purchase cost corresponding to an MS index of 514 is \$176,900. However, if we provide spares for the two pumps, the total becomes \$178,600.

Table 22.18 Equipment List, Including Purchase Costs, for Cyclohexane Process

Equipment Name	Equipment Label	Size	Design Temperature (°F)	Design Pressure (psia)	C_P, f.o.b. Purchase Cost (MS Index = 514)
Recycle compressor	K1	3 Hp	120	350	2,000
Feed-gas compressor	K2	296 Hp	450	350	80,000
Benzene feed pump	P1	4 Hp	120	350	1,200
Recycle pump	P2	1 Hp	120	350	500
Cooler	H1	210 ft^2	210	300	4,000
Effluent-BFW HX	H2	120 ft^2	400	320	2,500
Effluent-benzene HX	H3	160 ft^2	270	310	3,200
High-pressure flash	F1	2 ft diam. × 8 ft height	120	300	5,000
Low-pressure flash	F2	2 ft diam. × 8 ft height	120	20	3,500
Reactor	R1	8 ft diam. × 30 ft height	400	330	75,000

From Eq. (22.10), using a Lang factor for fluids processing of 5.93 from Table 22.16 and an updated MS index of 1,365, the estimated total capital investment is

$$C_{\mathrm{TCI}} = 1.05(5.93)\left(\frac{1{,}365}{514}\right)\$178{,}600 = \$2{,}950{,}000$$

For this example, the order-of-magnitude estimate of Method 1 gives $4,700,000.

Method 3. Preliminary Estimate (Based on the Individual Factors Method of Guthrie, 1969, 1974)

This method is best carried out after an optimal process design has been developed, complete with a mass and energy balance, equipment sizing, selection of materials of construction, and development of a process control configuration as incorporated into a P&ID. More time is required for making a preliminary estimate than for the preceding study estimate, but the accuracy is improved to perhaps $\pm 20\%$. To apply the method, the f.o.b. purchase cost of each piece of major equipment must be estimated, as was the case with the Lang method. However, instead of using an overall Lang factor to account for installation of the equipment and other capital costs, individual factors for each type of equipment, given in Table 22.11 above, are used, as developed first by Hand (1958) and later in much more detail by Guthrie (1969, 1974), who introduced the bare-module concept. Furthermore, the Guthrie method takes into account the fact that the installation cost of equipment made of stainless steel or other expensive materials and/or for operation at a high pressure is not as large a factor of the purchase cost as it is with carbon steel and/or at near ambient pressure because with stainless steel and/or at high pressure, the foundation, supporting structures and ladders, electrical, insulation, and paint will be the same cost as when carbon steel is used and/or the design pressure is near ambient. The only difference is in the cost of the attached piping, because it is stainless steel, for example, instead of carbon steel, and may have to be thicker for high pressure. This is illustrated by an example below. F.o.b. purchase costs of a wide range of chemical processing equipment are given in the next section. The Guthrie method involves the summation of estimates of module costs for the four different modules listed below. To this summation is added a contingency and contractor fee, in terms of a factor, to obtain the total permanent investment. An appropriate estimate of the working capital is added to obtain the total capital investment. Thus, the components of the total permanent investment are accounted for in a manner somewhat different from Table 22.9, but the overall result is the same. The equation for the total capital investment by the Guthrie method is

$$\begin{aligned} C_{\mathrm{TCI}} &= C_{\mathrm{TPI}} + C_{\mathrm{WC}} \\ &= 1.18(C_{\mathrm{TBM}} + C_{\mathrm{site}} + C_{\mathrm{buildings}} \\ &\quad + C_{\mathrm{offsite\ facilities}}) + C_{\mathrm{WC}} \end{aligned} \tag{22.11}$$

Equation (22.11) does not account for royalties or plant startup. These additional costs should be added if they are known or can be estimated.

The total bare-module cost, C_{TBM}, refers to the summation of bare-module costs for all items of process equipment, including fabricated equipment, process machinery, spares, storage tanks, surge tanks, and computers and software. The initial charge of catalyst is included with the corresponding catalytic reactor cost. As shown in the heat exchanger example of Table 22.10, the bare-module cost is based on the f.o.b. equipment purchase cost, to which is factored in direct field materials and labor, and indirect expenses such as freight, insurance, taxes, overhead, and engineering.

Site development costs, C_{site}, are discussed above in the section on capital investment costs. In lieu of a detailed estimate, which is not normally prepared at this stage of cost estimation, a value of 10–20% of C_{TBM} may be assigned for a grass-roots plant and 4–6% for an addition to an integrated complex.

Building costs, $C_{\mathrm{buildings}}$, are also discussed above in the section on capital investment costs. In the Guthrie method, buildings include process buildings and non-process buildings. Again, a detailed estimate is not generally made at this stage of cost estimation. Instead, an approximate estimate is sufficient, but must consider whether some or all the process equipment must be housed in buildings because of weather or other conditions, and whether a grass-roots location or an addition to an integrated complex is being considered. If the equipment is housed, the cost of process buildings may be estimated at 10% of C_{TBM}. If a grass-roots plant is being considered, the non-process buildings may be estimated at 20% of C_{TBM}. If the process is to be an addition to an integrated complex, the non-process buildings may be estimated at 5% of C_{TBM}.

Offsite facilities include utility plants when the company provides its own utilities, pollution control, ponds, waste treatment, offsite tankage, and receiving and shipping facilities. The utility plants may be estimated with the help of Table 22.12. To this may be added 5% of C_{TBM} to cover other facilities.

The factor 1.18 in Eq. (22.11) covers a contingency of 15% and a contractor fee of 3%. As with the Lang-factor method, the working capital can be estimated at 15% of the total capital investment, which is equivalent to 17.6% of the total permanent investment, or it can be estimated in detail by the method in Section 23.3.

The Guthrie method proceeds by steps as follows:

Step 1: From the process design, prepare an equipment list, giving the equipment title, label, size, material of construction, design temperature, and design pressure.

Step 2: Using the data in Step 1 with f.o.b. equipment purchase cost data, add to the equipment list the cost, C_{P_b}, and the corresponding cost index, I_b, of the cost data. In the Guthrie method, the f.o.b. purchase cost is a base cost corresponding to a near-ambient design pressure, carbon steel as the material of construction, and a base design.

Step 3: Update the cost data to the current cost index. For each piece of equipment, determine the bare-module cost—using bare-module factors, F_{BM}, from Table 22.11, being careful to determine it properly when the material of construction is not carbon steel and/or the pressure is not near ambient—as given by Eq. (22.12) and illustrated by the following example, before moving to Step 4. As discussed earlier, the bare-module cost accounts for delivery, insurance, taxes, and direct materials and labor for installation.

$$C_{BM} = C_{P_b}\left(\frac{I}{I_b}\right)[F_{BM} + (F_d F_p F_m - 1)] \quad (22.12)$$

where:

F_{MB} = bare-module factor
F_d = equipment design factor
F_p = pressure factor
F_m = material factor

Step 4: Obtain the total bare-module cost, C_{TBM}, by summing the bare-module costs of the process equipment.

Step 5: Using Eq. (22.11), estimate the total permanent investment. Add to this an estimate of the working capital to obtain the total capital investment.

EXAMPLE 22.4

The base f.o.b. purchase cost for a fabricated vertical pressure vessel, 6 ft in inside diameter and 100 ft in height (tangent-to-tangent) made of carbon steel for a design pressure of not greater than 50 psig, is given as $102,000 as of 1995 (CE index = 381). Calculate the bare-module cost for the year 2006 (CE index = 500) if the vessel is made of 316 clad stainless steel for a design pressure of 200 psig. For these conditions, Guthrie (1974) gives $F_{BM} = 4.16$, $F_d = 1$, $F_p = 1.55$, $F_m = 2.60$.

SOLUTION

Using Eq. (22.12),

$$C_{BM} = \$102{,}000\left(\frac{500}{381}\right)[4.16 + (1 \times 1.55 \times 2.60 - 1)]$$

$$= \$962{,}000$$

EXAMPLE 22.5

The total bare-module cost for a process to produce 40,000,000 lb/yr of butyl alcohols by the catalytic hydration of butylenes is $12,900,000, indexed to the year 2006. Estimate the total capital investment. The process will be an addition to an existing integrated complex and no process buildings will be required. Offsite utility plants have been estimated at $1,500,000 and the working capital has been estimated at $1,700,000.

SOLUTION

$$C_{TBM} = \$12{,}900{,}000$$

Estimates of the other terms in Eq. (22.11) are as follows:

$$C_{site} = 0.05C_{TBM} = 0.05(12{,}900{,}000) = \$645{,}000$$

$$C_{buildings} = 0.05C_{TBM} = 0.05(12{,}900{,}000) = \$645{,}000$$

$$C_{offsite\ facilities} = 1{,}500{,}000 + 0.05C_{TBM} = 1{,}500{,}000 + 0.05(12{,}900{,}000) = \$2{,}145{,}000$$

$$C_{TPI} = 12{,}900{,}000 + 645{,}000 + 645{,}000 + 2{,}145{,}000 = \$16{,}335{,}000$$

$$C_{WC} = \$1{,}700{,}000$$

$$C_{TCI} = 16{,}335{,}000 + 1{,}700{,}000 = \$18{,}035{,}000$$

For a grass-roots plant, an additional 25% of C_{TBM} is added for site development and buildings. This amounts to $3,225,000, giving a total capital investment of $21,260,000.

22.5 PURCHASE COSTS OF THE MOST WIDELY USED PROCESS EQUIPMENT

The Lang and Guthrie methods for estimating the total capital investment require f.o.b. purchase costs for all major items of process equipment. Since 1949, a number of literature articles and book chapters have presented such data. Some of the more widely used sources of equipment cost data are given in Table 22.19. Included is the cost index of the cost data. Typically, equipment cost data are presented in the form of graphs and/or equations of f.o.b. purchase cost as a function of one or more equipment size factors. Graphs show clearly the effect of the size factors on the cost and may be quickly read; however, equations are more consistent, especially compared to graphs using logarithmic coordinates. Furthermore, equations are easily incorporated into computer programs. In this section, equations and graphs are presented for f.o.b. purchase costs of the most widely used chemical processing equipment: pumps, electric motors, fans, blowers, compressors, shell-and-tube and double-pipe heat exchangers, general-purpose fired heaters (furnaces), pressure vessels and towers, trays, and packings. Then, in Section 22.6, equations alone are presented for a wide variety of other chemical processing equipment. The equipment cost equations should be used, even when one of the graphs below might apply.[1]

The form of the equations is a modification of the equation $C_P = A$ (size factor, S)^b (where A and b are constants),

[1] In developing the cost equations for the purchase cost of processing equipment presented in this chapter, available literature sources as far back as 1960 were consulted. After determining a suitable equipment size factor, all or much of the cost data were plotted. When a wide spread in the data was evident, which was not uncommon, an attempt was made to assess the validity of the data by comparison with costs of similar equipment. When the validity could not be determined, the data were averaged. In some cases, especially where available data were sparse, cost data were obtained from vendors of the equipment. It must be understood that the only accurate cost data are bids from a vendor and that bids from different vendors can sometimes differ significantly.

Table 22.19 Sources of Purchase Costs of Process Equipment

Author	Reference	Cost Index
Chilton, E.H.	*Chemical Engineering,* **56** (6), 97–106, June 1949	
Walas, S.M., and Spangler, C.D.	*Chemical Engineering,* **67** (6) 173–176, March 21, 1960	MS = 234.3
Bauman, H.C.	*Fundamentals of Cost Engineering in the Chemical Industry,* Reinhold (1964)	MS = 237.3
Mills, H.E.	*Chemical Engineering,* **71** (6), 133, March 16, 1964	MS = 238.8
Guthrie, K.M.	*Chemical Engineering,* **76** (6), 114–142, March 24, 1969	MS = 273.1
Guthrie, K.M.	*Process Plant Estimating Evaluation and Control,* Craftsman Book (1974)	MS = 303.3
Woods, D.R.	*Financial Decision Making in the Process Industry,* Prentice-Hall (1975)	MS = 300
Pikulik, A., and Diaz, H.E.	*Chemical Engineering,* **84** (21), 107–122, October 10, 1977	MS = 460
Hall, R.S., Matley, J., and McNaughton, K.J.	*Chemical Engineering,* **89** (7), 80–116, April 5, 1982	CE = 305
Walas, S.M.	*Chemical Process Equipment,* Butterworth (1988)	CE = 325
Turton, R., Bailie, R.C., Whiting, W.B., and Shaeiwitz, J.A.	*Analysis, Synthesis, and Design of Chemical Processes,* 2nd ed., Prentice Hall (2003)	CE = 397
Peters, M.S., Timmerhaus, K.D., and West, R.E.	*Plant Design and Economics for Chemical Engineers,* 5th ed., McGraw-Hill (2003)	CE = 390.4
Ulrich, G.D., and Vasudevan, P.T.	*Chemical Engineering Process Design and Economics—A Practical Guide,* 2nd ed., Process Publishing (2004)	CE = 400

obtained by taking the natural logarithm of both sides, adding additional higher-order terms as with a polynomial, and solving for C_P to obtain

$$C_P = \exp\{A_0 + A_1[\ln(S)] + A_2[\ln(S)]^2 + \cdots\}$$

The equations are usually based on the more common materials of construction, such as carbon steel. For other materials, multiplying factors are provided. Assistance in choosing the materials of construction is given in Appendix III.

As discussed by Woods (1975) and Walas (1988), when cost data are assembled from vendor quotes, they exhibit scatter due to differing qualities of equipment fabrication, design differences, market conditions, vendor profit, and other considerations. Accordingly, the accuracy of published equipment cost data, such as referenced in Table 22.19, may be no better than ±25%. More accurate estimates can be obtained from computing systems, such as the Aspen Icarus Process Evaluator (IPE) in the Aspen Engineering Suite discussed in Section 22.7 (i.e., Section 22S.1 in the file, Supplement_to_Chapter_22.pdf [in the PDF File folder, which can be downloaded from the Wiley Web site associated with this book]), which account for the details of equipment design, fabrication, and materials and labor requirements for installation, as well as related costs for site preparation, service facilities, indirect expenses, etc. Cost systems like IPE are rapidly gaining popularity in the chemical industry through the medium of personal computers. Although the equipment purchase-cost equations presented in Sections 22.5 and 22.6 may be convenient to use, the Aspen IPE system is preferred because it is more accurate, more consistent, and periodically updated. Final equipment cost estimates can only be obtained by bids from the equipment manufacturers. These require special requests that are sometimes costly to prepare and, for that reason, are often not obtained until a decision has been made to construct the plant and a final detailed capital cost estimate is needed for making an appropriation request. For some proprietary equipment systems, even a preliminary cost estimate might have to be requested from a vendor.

www.wiley.com/go/global/seider

Some chemical products, for example, home hemodialysis devices, involve small storage vessels, membrane mass exchangers, adsorption cartridges, and circulating pumps. Most often, the equations for cost estimation do not apply for these laboratory-scale vessels. In these cases, cost estimates can often be obtained from the distributors of laboratory equipment and vendors. Note that as the production level increases, volume discounts should be negotiated.

Pumps and Electric Motors

Pumps are used widely in chemical processing plants to move liquids through piping systems from one piece of equipment to another. The three most commonly used pumps are radial centrifugal, piston or plunger reciprocating, and external rotary gear, as discussed in some detail in Section 20.1. Of these three, the radial centrifugal pump (referred to here as just the centrifugal pump) is selected for industrial service approximately 90% of the time because it is:

1. Relatively inexpensive to purchase and install.
2. Operated at high speed so that it can be driven directly with an electric motor.
3. Relatively simple in construction, with no closely fitting parts that might wear, resulting in a low maintenance cost.

4. Available from a large number of vendors, many of whom comply with industry standards, such as those of American Petroleum Institute (API), American Society of Mechanical Engineers (ASME), and the International Organization for Standardization (ISO).
5. Available in a wide range of materials of construction.
6. Applicable over a wide range of volumetric flow rate and temperature.
7. Applicable by staging for the achievement of heads up to 3,200 ft.
8. Capable of pumping a liquid at a smooth flow rate with a constant discharge pressure.
9. Capable of handling slurries.
10. Easy to operate, with control of the flow rate by a valve on the discharge line.
11. Free from damage if the valve on the discharge line is inadvertently or purposely closed.

However, the centrifugal pump has the following limitations:

1. It cannot efficiently pump liquids with a kinematic viscosity greater than 100 centistokes.
2. The maximum efficiency of a particular pump is limited to a narrow range of its characteristic curve (flow rate versus head).
3. For most models, it cannot produce heads greater than 3,200 ft.
4. For most models, the volumetric flow rate must be greater than 10 gpm.
5. Most models are subject to air binding and must be primed.
6. Because of potential cavitation and the NPSH limitation, most models cannot pump liquids that are close to their bubble point.
7. Spares are normally specified because seals to prevent leakage may require attention more often than scheduled maintenance.

Pump Selection

A *centrifugal pump*, of the common radial type, should be given first consideration when the pumping requirements fall in the following ranges:

1. Volumetric flow rate from 10 gpm (0.000631 m^3/s) to 5,000 gpm (0.3155 m^3/s).
2. Head from 50 ft (15.24 m) to 3,200 ft (975.4 m).
3. Kinematic viscosity less than 100 centistokes (0.0001 m^2/s).
4. Available NPSH greater than 5 ft (1.52 m).

When one or more of these requirements is outside of the above ranges, a suitable centrifugal pump may still be available. This is particularly the case for the volumetric flow rate, where centrifugal pumps may be found for capacities up to 15,000 gpm or more. Alternatively, two or more centrifugal pumps may be placed in parallel. However, when one or more of the other three requirements cannot be met, one of the following two types of pumps should be considered for the service.

External rotary gear pumps are particularly suitable for moderate-to-very-high-kinematic viscosity liquids in the range of 100 to 500,000 centistokes for flow capacities to at least 1,500 gpm (0.252 m^3/s) and heads to at least 3,000 ft (914.4 m). They are moderately priced, but usually more expensive than radial centrifugal pumps. They cannot be used with liquids containing particles and do not give as smooth a flow rate as do radial centrifugal pumps.

Reciprocating pumps of the plunger type can achieve the highest heads, up to at least 20,000 ft and flow rates to at least 500 gpm, at a maximum Hp of 200, using 3 to 5 cylinders to reduce flow pulsations. The piston type can achieve heads to at least 5,000 ft and flow rates to at least 100 gpm, at a maximum Hp of 90, using 2 to 3 cylinders. For a given horsepower, the highest heads are only achieved with the lowest flow rates and vice versa. Reciprocating pumps can handle liquids of moderate kinematic viscosity, up to 100,000 centistokes. Reciprocating pumps are more expensive than gear pumps and deliver a pulsating, rather than a smooth, steady flow, which is reduced but not eliminated when multiple cylinders are used.

Many other specialized pumps are available for services that cannot be handled by radial centrifugal, external rotary gear, and plunger and piston reciprocating pumps. Specialized pumps are not considered here. Purchase-cost data are given next only for the three most commonly used types of pumps.

Pump and Motor Purchase Costs

Centrifugal Pumps

Much purchase-cost data for centrifugal pumps, of the most common radial type, have been published. The cost most often includes the pump, a base plate, and a direct-drive coupling. In some cases, an electric motor drive is included. There is no general agreement on the equipment size factor to be used for correlating purchase costs. Most common are (1) brake horsepower and (2) the product of the capacity and the head (or pressure increase). Here, the cost correlation method used is that of the Monsanto Company in their FLOWTRAN simulation program, which was subsequently adopted by Corripio, Chrien, and Evans (1982a). Their size factor, S, which recognizes the fact that a given centrifugal pump can operate over a range of flow rate and head combinations, is

$$S = Q(H)^{0.5} \tag{22.13}$$

where Q is the flow rate through the pump in gallons per minute and H is the pump head in feet of fluid flowing (pressure rise/liquid density). The pump purchase cost is

Table 22.20 Typical Types of Radial Centrifugal Pumps and F_T Factors

No. of Stages	Shaft rpm	Case-Split Orientation	Flow Rate Range (gpm)	Pump Head Range (ft)	Maximum Motor Hp	Type Factor [F_T in Eq. (22.15)]
1	3,600	VSC	50–900	50–400	75	1.00
1	1,800	VSC	50–3,500	50–200	200	1.50
1	3,600	HSC	100–1,500	100–450	150	1.70
1	1,800	HSC	250–5,000	50–500	250	2.00
2	3,600	HSC	50–1,100	300–1,100	250	2.70
2+	3,600	HSC	100–1,500	650–3,200	1,450	8.90

correlated with the maximum value of S that the pump can handle.

In addition to the size factor, the purchase cost of a centrifugal pump depends on its rate of rotation (usually in the range of 1,800 to 3,600 rpm); the number of impellers (usually in the range of 1 to 4) in series (called stages) to reach the desired head; the orientation of the splitting of the bolted-together pump case (HSC, horizontal split case, or VSC, vertical split case); and the material of construction. Typical ranges of flow rate and head, and maximum horsepower of the electric motor used to drive the pump, taken from Corripio et al. (1982a), are given in Table 22.20. From their cost data, indexed to 2006 (CE = 500), the f.o.b. purchase cost of a single-stage centrifugal pump with VSC construction of cast iron and operating at 3,600 rpm (referred to here as the base cost, C_B), is plotted in Figure 22.3. The cost includes the base plate and driver coupling but not the electric motor. The cost curve in Figure 22.3 is given by the following equation, which is valid from $S = 400$ to $S = 100{,}000$:

$$C_B = \exp\{9.7171 - 0.6019[\ln(S)] + 0.0519[\ln(S)]^2\} \quad (22.14)$$

For other types of centrifugal pumps and other materials of construction, the f.o.b. purchase cost is given by:

$$C_P = F_T F_M C_B \quad (22.15)$$

where F_M is a material factor given in Table 22.21 and F_T is a pump-type factor included in Table 22.20.

Electric Motors

A centrifugal pump is usually driven by an electric motor whose cost is added to the pump cost from Eq. (22.15). The size parameter for the motor is its power consumption, P_C, which is determined from the theoretical horsepower of the pump, P_T, its fractional efficiency, η_P, and the fractional efficiency of the electric motor, η_M, by the equation:

$$P_C = \frac{P_T}{\eta_P \eta_M} = \frac{P_B}{\eta_M} = \frac{QH\rho}{33{,}000\eta_P\eta_M} \quad (22.16)$$

where, as previously, Q is the flow rate through the pump in gallons per minute, H is the pump head in feet of fluid flowing, and P_B is the pump brake horsepower, with ρ equal

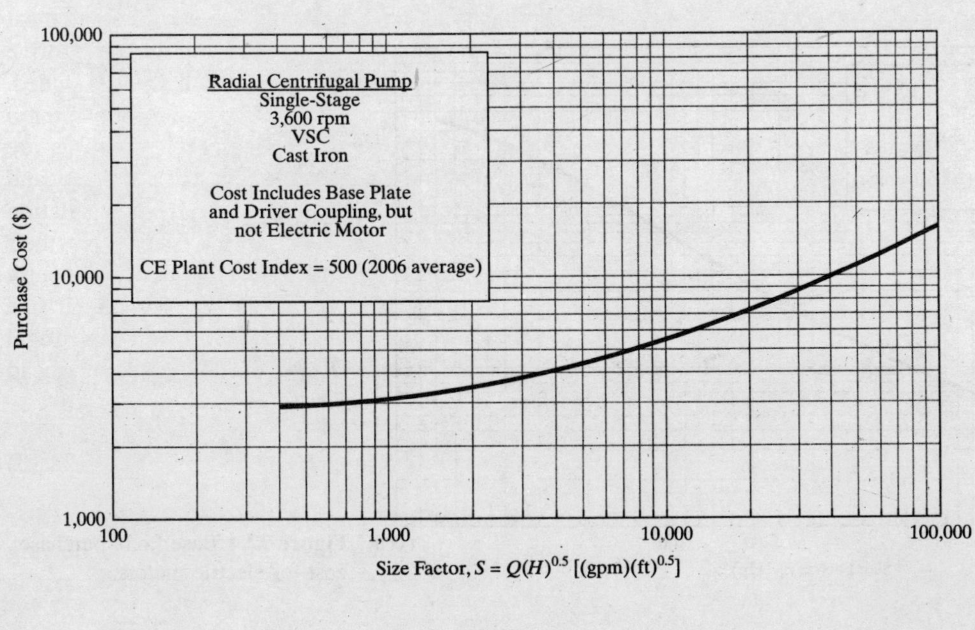

Figure 22.3 Base f.o.b. purchase cost for radial centrifugal pumps.

Table 22.21 Materials of Construction Factors, F_M, for Centrifugal Pumps

Material of Construction	Material Factor [F_M, in Eq. (22.15)]
Cast iron	1.00
Ductile iron	1.15
Cast steel	1.35
Bronze	1.90
Stainless steel	2.00
Hastelloy C	2.95
Monel	3.30
Nickel	3.50
Titanium	9.70

to the liquid density in pounds per gallon. Corripio et al. (1982a) give the following equations for estimating η_P as a function of the volumetric flow rate, and η_M as a function of pump brake horsepower:

$$\eta_P = -0.316 + 0.24015(\ln Q) - 0.01199(\ln Q)^2 \quad (22.17)$$

for Q in the range of 50 to 5,000 gpm;

$$\eta_M = 0.80 + 0.0319(\ln P_B) - 0.00182(\ln P_B)^2 \quad (22.18)$$

for P_B in the range of 1 to 1,500 Hp.

The f.o.b. purchase cost of an electric motor depends on its power consumption, P_C, the rotation rate of its shaft, and the type of motor enclosure. Cost data are given here for two common motor speeds (3,600 rpm and 1,800 rpm) and three common types of motor enclosures:

1. *Open, drip-proof enclosure*, which is designed to prevent the entrance of liquid and dirt particles, but not airborne moisture, dust, and corrosive fumes, into the internal working parts of the motor.
2. *Totally enclosed, fan-cooled* (TEFC) *enclosure*, which prevents any air from getting inside, thus protecting against moisture, dust, dirt, and corrosive vapors.
3. *Explosion-proof enclosure*, which protects the motor against explosion hazards from combustible gases, liquids, and dust by pressurizing the enclosure with a safe gas.

From the electric motor cost correlations of Corripio et al. (1982a), indexed to 2006 (CE = 500), the f.o.b. purchase cost of an electric motor operating at 3,600 rpm, with an open, drip-proof enclosure (referred to here as the base cost, C_B), is plotted in Figure 22.4 as a function of the horsepower consumption, P_C. The cost curve is given by the equation:

$$C_B = \exp\{5.8259 + 0.13141[\ln(P_C)] + 0.053255\,[\ln(P_C)]^2 + 0.028628\,[\ln(P_C)]^3 - 0.0035549[\ln(P_C)]^4\} \quad (22.19)$$

which applies over the range of 1 to 700 Hp. For other motor speeds and type enclosures, the f.o.b. purchase cost is given by:

$$C_P = F_T C_B \quad (22.20)$$

where F_T is a motor-type factor given in Table 22.22, applicable within a range of electric motor power consumption, P_C, from 1 to 1,500 Hp.

External Gear Pumps

Purchase-cost data for external gear pumps are not as widely available as they are for radial centrifugal pumps. The cost most often includes the gear pump, a base plate, and a driver

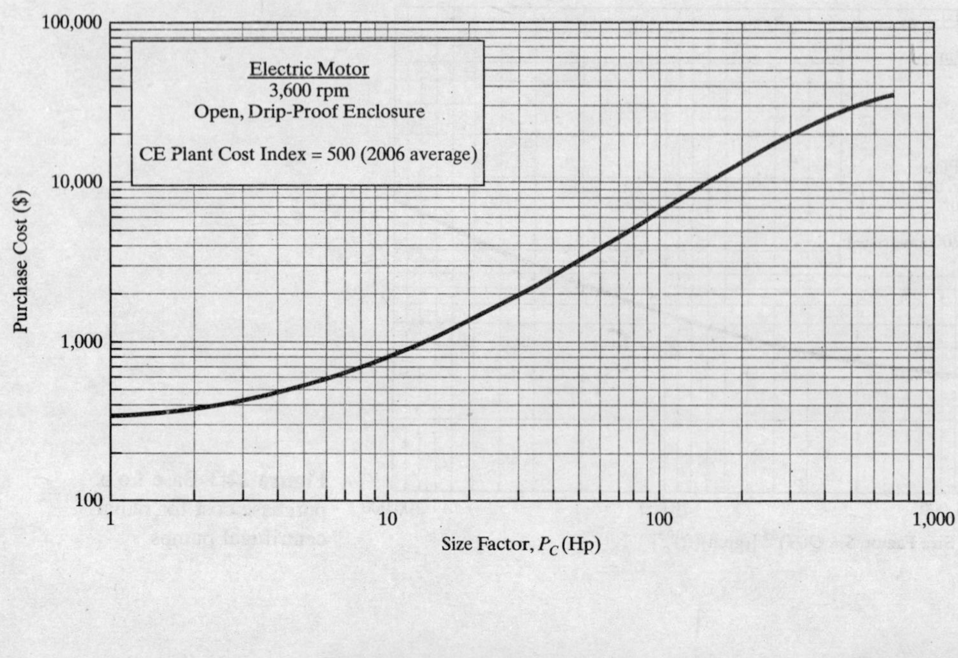

Figure 22.4 Base f.o.b. purchase cost for electric motors.

Table 22.22 F_T Factors in Eq. (22.20) and Ranges for Electric Motors

Type Motor Enclosure	3,600 rpm	1,800 rpm
Open, drip-proof enclosure, 1 to 700 Hp	1.0	0.90
Totally enclosed, fan-cooled, 1 to 250 Hp	1.4	1.3
Explosion-proof enclosure, 1 to 250 Hp	1.8	1.7

coupling. In some cases an electric motor drive is included. There is no general agreement on what equipment size factor to use for correlating purchase costs. Most common are (1) brake horsepower and (2) flow capacity. Here, the cost correlation method used is in terms of flow capacity, Q, in gallons per minute, as used by Walas (1988).

In addition to the size factor, the purchase cost of a gear pump depends on the material of construction. Although gear pumps can be designed to operate over a wide range of flow rates and discharge pressure, typical ranges are 10 to 1,500 gpm and up to 200 psia for high-viscosity fluids. Typical pump efficiencies are 80% for low-kinematic viscosity liquids (20 centistokes) and 50% for high-kinematic viscosity liquids (500 centistokes). The f.o.b. purchase cost of an external gear pump of cast iron construction and for a cost index for 2006 (CE = 500) (referred to here as the base cost, C_B) is plotted in Figure 22.5. The cost includes the base plate and driver coupling, but not the electric motor. The cost curve in Figure 22.5 is given in terms of Q by the equation:

$$C_B = \exp\{7.6964 + 0.1986[\ln(Q)] + 0.0291[\ln(Q)]^2\} \quad (22.21)$$

which is applicable over a range of 10 to 900 gpm. For other materials of construction, the f.o.b. purchase cost is given by:

$$C_P = F_M C_B \quad (22.22)$$

where F_M is a material factor given above in Table 22.21. The power requirement for the electric motor to drive the pump depends on the head, H, and the flow rate, Q, as given by Eq. (22.16).

Reciprocating Plunger Pumps

Although piston pumps are common, the plunger type is the best choice for the most demanding applications and is available for a wider range of flow rates. Purchase-cost data for reciprocating plunger pumps are not as widely available as they are for radial centrifugal pumps. The cost most often includes the pump and a driver coupling for a motor or a V-belt drive. In most cases an electric motor drive is not included. There is no general agreement on what equipment size factor to use for correlating purchase costs. Most common are (1) brake horsepower and (2) flow capacity. The cost of most models is based on the brake horsepower. By changing the plunger and cylinder diameter, a reciprocating pump of a specified horsepower can operate over a 10-fold range of flow rate and head. Here, the cost correlation method used is in terms of brake horsepower, P_B, as given by Eq. (22.16), where the pump efficiency, η_P, is typically 0.90 (90%).

In addition to the size factor, the purchase cost of a reciprocating plunger pump depends on the material of construction. The f.o.b. purchase cost of a reciprocating plunger pump of ductile iron construction and a cost index for 2006 (CE = 500) (referred to here as the base cost, C_B), is plotted in Figure 22.6. The cost includes a V-belt drive, but not the electric motor. The cost curve in Figure 22.6 is given by the equation:

$$C_B = \exp\{7.8103 + 0.26986[\ln(P_B)] + 0.06718[\ln(P_B)]^2\} \quad (22.23)$$

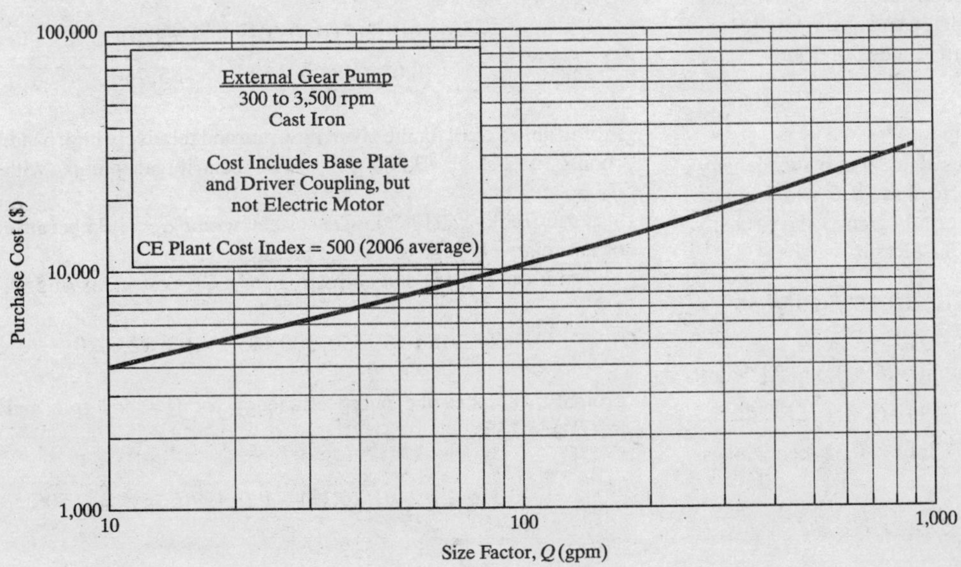

Figure 22.5 Base f.o.b. purchase cost for external gear pumps.

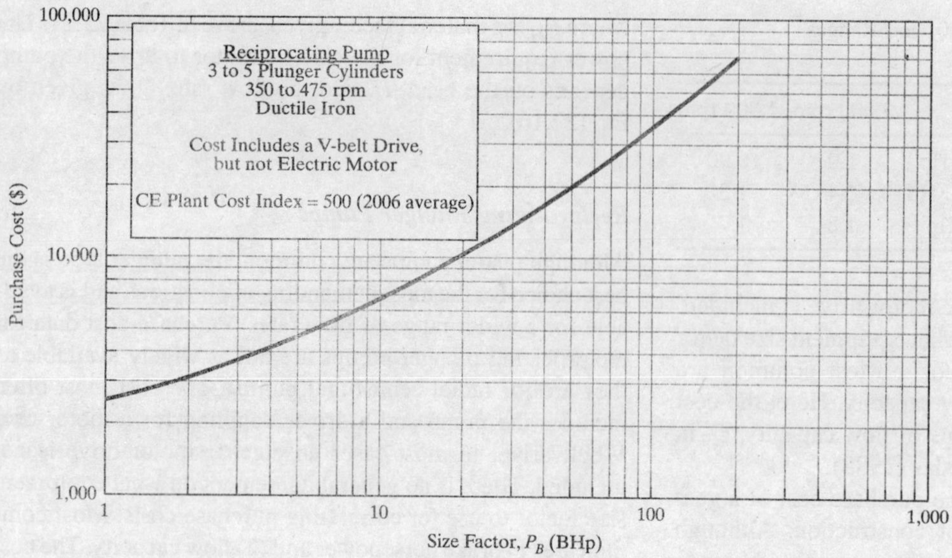

Figure 22.6 Base f.o.b. purchase cost for reciprocating plunger pumps.

which is applicable over the range of 1 to 200 BHp. For other materials of construction, the f.o.b. purchase cost is given by Eq. (22.15), where F_M is a material factor, as follows:

Ductile iron	$F_M = 1.00$
Ni–Al–Bronze	$F_M = 1.15$
Carbon steel	$F_M = 1.50$
Stainless steel	$F_M = 2.20$

EXAMPLE 22.6

In Chapter 4, a vinyl-chloride process is synthesized, with a detailed process flow diagram shown in Figure 4.19. In that process, Reactor Pump P-100 takes stream 4 (a mixture of streams 3 and 16) of 263,800 lb/hr of 1,2-dichloroethane at 90°C and 1.5 atm and delivers it to an evaporator operating at a much higher pressure of 26 atm. Select a suitable pump and electric motor and estimate the f.o.b. purchase cost at a CE index of 550.

SOLUTION

A process simulation program is used to obtain the density, viscosity, and vapor pressure of the feed at 90°C and 1.5 atm. The density is 9.54 lb/gal (71.4 lb/ft^3 or 1.14 g/cm^3), the viscosity is 0.37 cP, and the vapor pressure is 1.212 atm.

The feed volumetric flow rate is $263,800/[(60)(9.54)] = Q = 461$ gpm.

The pressure increase across the pump is $26 - 1.5 = 24.5$ atm.

The pump head is $(24.5)(14.696)(144)/71.4 = H = 726$ ft.

The kinematic viscosity is $0.37/1.14 = 0.32$ centistokes, which is quite low.

Choose a radial centrifugal pump.

However, it is necessary to first check the available NPSH, as discussed in Section 20.1.

$$\text{NPSH}_A = \frac{\text{Suction pressure} - \text{Vapor pressure}}{\text{Liquid density}} = \frac{(1.5 - 1.212)(14.696)(144)}{71.4} = 8.54 \text{ ft}$$

Assume we can purchase a radial centrifugal pump with a required NPSH > 5.

From Eq. (22.13), the centrifugal pump size parameter is

$$S = Q(H)^{0.5} = 461(726)^{0.5} = 12,420(\text{gpm})(\text{ft})^{0.5}$$

and $\ln(S) = 9.427$

From Eq. (22.14), the base pump purchase cost at a CE cost index of 500 is

$$C_B = \exp\{9.7171 - 0.6019[\ln(12,420)] + 0.0519[\ln(12,420)]^2\} = \$5,740$$

From Table 22.20, for the given flow rate and relatively high head, choose a 2-stage, 3,600-rpm, HSC centrifugal pump, with $F_T = 2.70$.

From Table 22.21, choose cast steel, with $F_M = 1.35$ because of the relatively high discharge pressure.

From Eq. (22.15) and correcting for a CE cost index of 550,

$$C_P = (2.70)(1.35)(550/500)(\$5,740) = \$23,020$$

From Eq. (22.17), the pump efficiency for $Q = 461$ gpm and $\ln(Q) = 6.13$ is

$$\eta_P = -0.316 + 0.24015(6.13) - 0.01199(6.13)^2 = 0.706$$

From Eq. (22.16), the pump brake horsepower, P_B, is

$$P_B = \frac{QH\rho}{33{,}000\eta_P} = \frac{(461)(726)(9.54)}{33{,}000(0.706)} = 137\,\text{BHp}$$

From Eq. (22.18), the motor efficiency for $\ln(P_B) = 4.92$ is

$$\eta_M = 0.80 + 0.0319(4.92) - 0.00182(4.92)^2 = 0.913$$

From Eq. (22.16), the power consumption of the motor is

$$P_C = \frac{P_B}{\eta_M} = \frac{137}{0.913} = 150\,\text{Hp}$$

From Eq. (22.19), the base cost of the motor for $\ln(P_C) = 5.01$ and a CE index of 500 is

$$C_B = \exp\{5.8259 + 0.13141(5.01) + 0.053255(5.01)^2 + 0.028628(5.01)^3 - 0.0035549(5.01)^4\} = \$9{,}710$$

Because of the possible flammability hazard of 1,2-dichloroethane, specify an explosion-proof electric motor to drive the pump. From Table 22.22, for 3,600 rpm, $F_T = 1.8$ and using Eq. (22.20), but with added updating of the CE cost index to 550,

$$C_P = F_T C_B = (1.80)(\$9{,}710)\left(\frac{550}{500}\right) = \$19{,}230$$

Total cost of centrifugal pump and motor = $\$23,020 + \$19,230 = \$42,250$. Consideration should be given to purchasing a spare pump and motor.

Fans, Blowers, and Compressors

When energy input is required to move a gas through various pipelines or ducts in a chemical processing plant, a fan, blower, or compressor is used. As discussed in Section 20.3, the power input to the gas mover increases the total head of the gas, which, ignoring a change in potential energy of the gas due to change in elevation above sea level, includes the velocity (dynamic) head and the pressure (static) head. According to Papanastasiou (1994), the gas mover is defined as: (1) a *fan*, if almost all of the energy input increases the velocity head; (2) a *blower*, if the energy input increases both the velocity head and the pressure head; and (3) a *compressor*, if almost all of the energy input increases the pressure head. However, that definition is not widely accepted. In practice, one vendor may refer to a particular gas mover as a fan, while another vendor may refer to it as a blower. The same situation applies to blowers and low-compression-ratio compressors. Here, we classify a fan as a gas mover that is generally limited to near-ambient suction pressures and pressure increases of less than 10%. Blowers can operate at any suction pressure, with compression ratios of up to 2. Thus, the main purpose of a fan is to move large quantities of gas with an increase in pressure head of up to 40 in. of H_2O head, while a blower can take a gas at 1 atm and deliver it at up to 2 atm. For larger compression ratios, a compressor is generally specified.

Fans are widely used for high-flow, low-pressure-increase applications such as heating and ventilating systems; air supply to cooling towers, low-pressure-drop dryers, and finned-tube air coolers; and removal of fumes, flue gas, and gas from a baghouse. Blowers are used for supplying combustion air to boilers and fired heaters, air to strippers, purge gas for regeneration of fixed-bed adsorbers, and air to dryers with more pressure drop than a fan can handle and for pneumatic conveying of particles. Compressors are widely used with a variety of gases and gas mixtures to increase their pressure to required levels for chemical reaction and separation.

Because a gas is compressible with a density much lower than that of a liquid, the temperature of a gas rises when compressed, and this temperature rise is a limiting factor in determining the permissible compression ratio for a single-stage gas compressor. As discussed in Section 20.3, frictional dissipation in the gas mover causes a further rise in the gas temperature. Although some specialized compressors permit a gas discharge temperature of up to 600°F, the more widely used compressors are limited to discharge temperatures in the range of 375 to 400°F. For a diatomic gas or gas mixture (e.g., air) with a specific heat ratio of 1.4, a 400°F limit corresponds to a maximum single-stage compression ratio of 3.75, after taking into account an assumed compressor isentropic efficiency of 85%. This limiting compression ratio decreases to 2.50 for a monatomic gas with a higher specific heat ratio of 1.67, but increases to 6.0 for a gas with a lower specific heat ratio of 1.30 (e.g., methane). For gases with a specific heat ratio less than 1.30, even higher compression ratios may be possible, but most compressors are limited to a compression ratio of 8.0. If a higher compression ratio is required, the compression is accomplished in stages that are separated by heat exchangers that cool the gas to about 100°F before entering the next stage. Because the cooling may cause some condensation, a vapor–liquid separator, which is usually a vertical vessel containing a demister pad to coalesce small liquid droplets and is called a *knock-out drum*, must be provided after each intercooler to remove liquid from the gas prior to its entry into the next stage of compression. Because most compressors cannot tolerate any liquid in the gas, compressor inlet and outlet phase conditions should always be checked. As discussed below in the subsection "Compressors," the maximum compression ratio for a centrifugal compressor may be limited by the maximum velocity at the blade tips rather than by the exit temperature.

Fans

Most fans are of the centrifugal or axial-flow type. *Centrifugal fans* achieve the highest discharge pressures, while *axial-flow fans* provide the highest flow rates. Although forward-curved and airfoil blade designs are available for centrifugal fans, the two most popular are the *backward-curved blade* and the *straight-radial blade*. The former is the cheapest for a given flow capacity and the most efficient, but the discharge pressure decreases rapidly from its maximum value as the

Table 22.23 Typical Operating Ranges of Fans

Fan Type	Flow Rate (ACFM)[a]	Total Head (in. H_2O)
Centrifugal backward curved	1,000–100,000	1–40
Centrifugal straight radial	1,000–20,000	1–30
Vane axial	1,000–800,000	0.02–16
Tube axial	2,000–800,000	0.00–10

[a]ACFM = actual cubic feet per minute.

Table 22.24 Head Factor, F_H, for Fans in Eq. (22.24)

Head (in. H_2O)	Centrifugal Backward Curved	Centrifugal Straight Radial	Vane Axial	Tube Axial
5–8	1.15	1.15	1.15	1.15
9–15	1.30	1.30	1.30	
16–30	1.45	1.45		
31–40	1.55			

flow rate is increased. It is only suitable for air and clean gases. The straight-radial centrifugal fan is less efficient, but is suitable for dust-laden gases and maintains the discharge pressure, up to a compression ratio of 1.2, over a wider range of flow rates. However, at this level of pressure increase, a fan may be called a blower. Axial-flow fans come in two main types: *vane axial* (compression ratio to 1.04) and *tube axial* (compression ratio to 1.025). Less efficient are propeller fans. Typical operating ranges of centrifugal and axial-flow fans are given in Table 22.23.

The equipment size factor for a fan is the actual cubic feet per minute, ACFM, entering the fan. Fans are usually driven by an electric motor with either a direct drive or a belt. Base f.o.b. purchase costs, C_B, for the four most common types of fans, averaged from several published sources listed in Table 22.19 and vendor quotes, are plotted in Figure 22.7 as a function of Q in ACFM at a cost index for 2006 (CE = 500). The base cost, which includes an electric motor drive, is for carbon-steel construction and total discharge heads to 4 in. H_2O. For other materials of construction and higher discharge heads, the f.o.b. purchase cost is given by

$$C_P = F_H F_M C_B \tag{22.24}$$

where the following values of F_M apply to other materials of construction:

Fiberglass	$F_M = 1.8$
Stainless steel	$F_M = 2.5$
Nickel alloy	$F_M = 5.0$

The head factor, F_H, for total heads greater than 4 in. H_2O is given in Table 22.24.

The base cost curves in Figure 22.7 for a CE cost index of 500 are given by the following equations, with Q in actual cubic feet per minute (ACFM) of gas entering the fan.

Centrifugal backward-curved fan (valid from $Q = 1{,}000$ to 100,000 ACFM):

$$C_B = \exp\{11.0757 - 1.12906[\ln(Q)] + 0.08860[\ln(Q)]^2\} \tag{22.25}$$

Centrifugal straight-radial fan (valid from $Q = 1{,}000$ to 20,000 ACFM):

$$C_B = \exp\{12.1678 - 1.31363[\ln(Q)] + 0.09974[\ln(Q)]^2\} \tag{22.26}$$

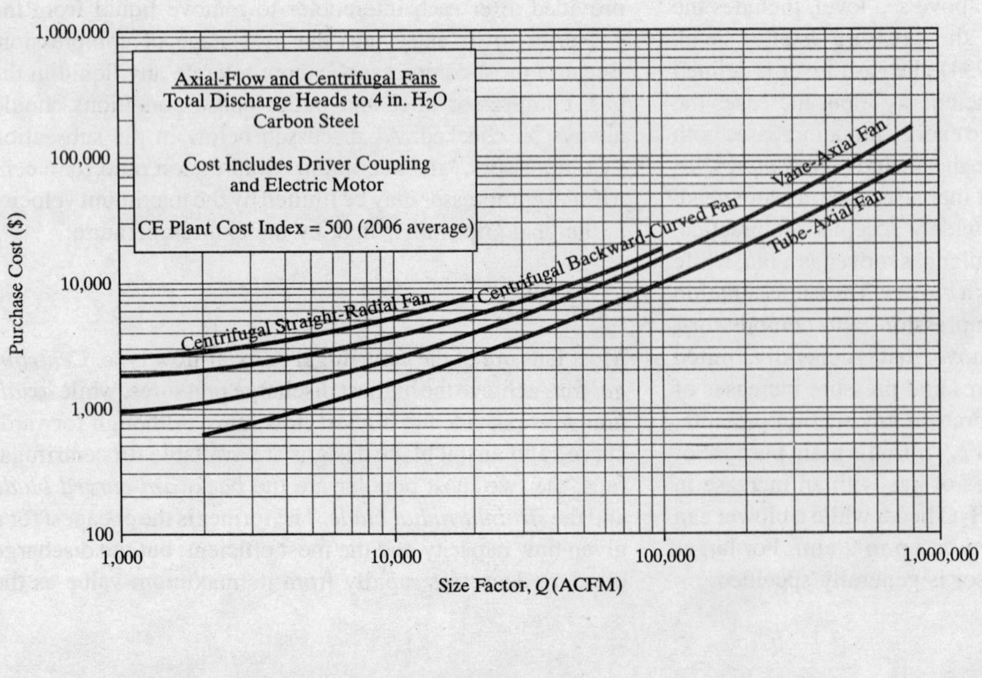

Figure 22.7 Base f.o.b. purchase costs for axial-flow and centrifugal fans.

Vane-axial fan (valid from $Q = 1,000$ to 800,000 ACFM):

$$C_B = \exp\{9.5229 - 0.97566[\ln(Q)] + 0.08532[\ln(Q)]^2\} \quad (22.27)$$

Tube-axial fan (valid from $Q = 2,000$ to 800,000 ACFM):

$$C_B = \exp\{6.12905 - 0.40254[\ln(Q)] + 0.05787[\ln(Q)]^2\} \quad (22.28)$$

The brake horsepower for a fan may be computed in any of three ways, depending on whether the total change in head is mostly dynamic, static, or a mixture of the two. The corresponding nominal fan efficiency, η_F, is 40% for mostly a dynamic change, 60% for mostly a static change, and 70% for a mixture of the two. The power consumption is given by the following equation, which is similar to Eq. (22.16) and where the electric motor efficiency, η_M, can be taken as 90%:

$$P_C = \frac{P_B}{\eta_M} = \frac{QH_t}{6,350\eta_F\eta_M} \quad (22.29)$$

where Q = gas inlet flow rate, in cubic feet per minute, and H_t = change in total head, in inches of water.

EXAMPLE 22.7

A flue gas at 200°F and 740 torr, with an average molecular weight of 31.3, is to be discharged at a rate of 12,000 standard cubic feet per minute (SCFM) at 60°F and 1 atm to a pressure of 768 torr in a duct where the velocity, V, will be 150 ft/s. Calculate the actual inlet flow rate in cubic feet per minute and the power consumption. Select a suitable fan and estimate the purchase cost for CE = 550.

SOLUTION

The actual fan inlet flow rate is $12,000\left(\frac{660}{520}\right)\left(\frac{740}{760}\right) = 14,830$ ft^3/min.

Assume the inlet velocity is zero.

The increase in dynamic head is $\frac{V^2}{2g_c} = \frac{150^2}{2(32.2)} = 349$ ft-lbf/lbm of gas.

From the ideal gas law, the average density of the gas in passing through the fan, assuming no change in temperature, is 0.0644 lb/ft^3.

The increase in pressure head $= \frac{\Delta P}{\rho} = \frac{\frac{(768-740)}{760}(14.7)(144)}{0.0644}$ = 1,211 ft-lbf/lbm of gas. The total change in head is 349 + 1,211 = 1,560 ft-lbf/lbm of gas = 19.3 in. of H_2O. Because static head is predominant, let $\eta_F = 0.60$. From Eq. (22.29),

$$P_C = \frac{14,830(19.3)}{6,350(0.60)(0.90)} = 83.5 \text{ Hp}$$

Because the head is greater than 16 in. of H_2O and the flue gas is not likely to be clean, using Table 22.23, select a centrifugal fan with a straight-radial impeller. From Table 22.24, the head factor, F_H, is 1.45. From Eqs. (22.24) and (22.26), correcting for the cost index, the purchase cost, including the motor, is

$$C_B = 1.45\left(\frac{550}{500}\right)\exp\{12.1678 - 1.31363[\ln(14,830)] + 0.09974[\ln(14,830)]^2\} = \$10,080$$

Blowers

Most blowers, with compression ratios up to 2, are of the multistage centrifugal (often called turboblower) type or the rotary positive-displacement type. Axial-flow units can also be used, but must be multistaged. The centrifugal units are similar to centrifugal fans, with the same type of blades, but operate at higher speeds and are built to withstand higher discharge pressures. The most common rotary blower is the straight-two-lobe (Roots) blower, developed by the Roots brothers in 1854, or a modification with three straight lobes. Typical operating ranges are 100 to 50,000 ICFM for centrifugal blowers and 20 to 50,000 ICFM for rotary straight-lobe blowers with two lobes, where ICFM is the cubic feet per minute at inlet conditions. Typical mechanical efficiencies, η_B, are 70–80% for centrifugal blowers and 50–70% for straight-lobe blowers. However, with straight-lobe blowers, the higher efficiency is only achieved for compression ratios from 1.2 to 1.3; from 1.3 to 2.0, the efficiency falls off rapidly. The centrifugal blower delivers a smooth flow rate, but as discussed in Section 20.2, the straight-lobe units deliver a somewhat pulsing flow. As with pumps, rotary blowers deliver a fixed volumetric flow rate with varying inlet and outlet pressures, while the volumetric throughput of centrifugal blowers varies with changes in inlet or discharge pressures. Both types of blowers have found a wide range of applications, but centrifugal blowers are more common in chemical processing plants and are widely used to supply air to strippers, dryers, and combustion devices. Rotary blowers are useful for pneumatic conveying and are well suited to other applications when a fixed volumetric flow rate is essential.

The equipment size factor for a blower is the brake horsepower, P_B, which is computed from the inlet volumetric flow rate, Q_I, in cubic feet per minute and pressures in lbf/in^2 at the inlet, P_I, and outlet, P_O, by the following equation, which assumes the ideal gas law and a constant specific heat ratio, k:

$$P_B = 0.00436\left(\frac{k}{k-1}\right)\frac{Q_I P_I}{\eta_B}\left[\left(\frac{P_O}{P_I}\right)^{\frac{k-1}{k}} - 1\right] \quad (22.30)$$

Blowers are usually driven by an electric motor with a direct drive for the centrifugal type and a belt or chain drive for the rotary type. Except for very large units, centrifugal blowers must be staged to achieve pressures greater than 40 in. of H_2O gauge. Base f.o.b. purchase costs, C_B, for the two major types

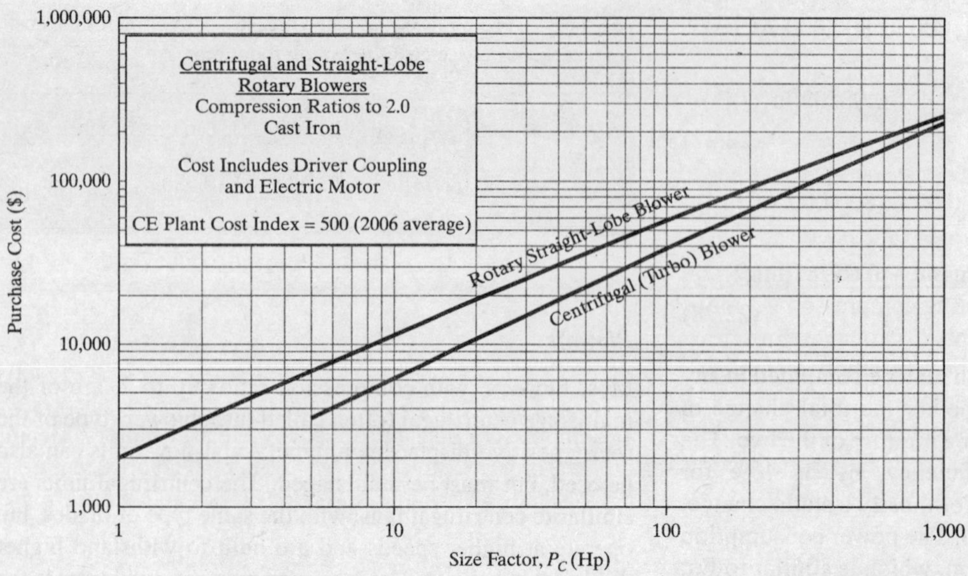

Figure 22.8 Base f.o.b. purchase costs for centrifugal and straight-lobe blowers.

of blowers, based on data from Garrett (1989) and recent data from vendors, are plotted in Figure 22.8 as a function of P_C in horsepower for a cost index in 2006 (CE = 500). The base cost, which includes an electric motor drive, is for a cast-iron construction housing, with compression ratios up to 2. The centrifugal blower uses sheet metal blades. For other materials of construction, the f.o.b. purchase cost is given by:

$$C_P = F_M C_B \tag{22.31}$$

where the metal material factors given above for fans can be used. In addition, centrifugal blowers are available with cast aluminum blades with $F_M = 0.60$.

The base blower purchase cost curves in Figure 22.8, for a CE cost index of 500, are given by the following equations, with P_C in horsepower:

Centrifugal (turbo) blower (valid from $P_C = 5$ to 1,000 Hp):

$$C_B = \exp\{6.8929 + 0.7900[\ln(P_C)]\} \tag{22.32}$$

Rotary straight-lobe blower (valid from $P_C = 1$ to 1,000 Hp):

$$C_B = \exp\{7.59176 + 0.79320[\ln(P_C)] - 0.012900[\ln(P_C)]^2\} \tag{22.33}$$

EXAMPLE 22.8

Air, available at 70°F and 14.5 psia, is to be supplied by a blower at 6,400 ACFM to a column with 15 plates to strip VOCs (volatile organic compounds) from 1,000 gpm of wastewater. The column is 6 ft in diameter. The pressure drop through the inlet line, the column, and the outlet line has been estimated to be 3 psi. The gas exiting from the column is to be sent to the next unit at a pressure of 18 psia. Select and size a blower, calculate the required power consumption, and estimate the f.o.b. purchase cost for a CE cost index of 550.

SOLUTION

The total pressure increase required across the blower is $18.0 - 14.5 + 3.0 = 6.5$ psi. At the blower inlet, the pressure is 14.5 psia, giving an air density of 0.074 lbm/ft^3. This gives an inlet pressure head of $(14.5)(144)/0.074 = 28,200$ ft-lbf/lbm. At the blower exit, the pressure is $18.0 + 3.0 = 21.0$ psia, giving an air density of approximately 0.095 lbm/ft^3 and an outlet pressure head of $(21)(144)/0.095 = 31,800$ ft-lbf/lbm. This gives a change in pressure head of $31,800 - 28,200 = 3,600$ ft-lbf/lbm. For the increase in kinetic-energy head, assume the blower inlet velocity is zero and the blower discharge air velocity in the exit line is 75 ft/s. The increase in kinetic energy is $(75)^2/[2(32.2)] = 87.3$ ft-lbf/lbm. In this example, the change in kinetic-energy head is only about 2.5% of the total increase in head. Neglecting the increase in kinetic-energy head, the blower brake horsepower from Eq. (22.30), using $k = 1.4$ and $\eta_B = 0.75$ for a centrifugal blower, is:

$$P_B = 0.00436\left(\frac{1.4}{1.4-1}\right)\frac{6,400(14.5)}{0.75}\left[\left(\frac{21}{14.5}\right)^{\frac{1.4-1}{1.4}} - 1\right]$$

$$= 211 \text{ BHp}$$

Using Eq. (22.18), the motor efficiency $= \eta_M = 0.92$. From Eq. (22.16), the consumed power for a centrifugal blower is $211/0.92 = 229$ Hp. A straight-lobe blower with $\eta_B = 0.65$ requires a consumed power of 265 Hp. For this application, either the centrifugal blower or straight-lobe blower is suitable, but with a compression ratio of $21/14.5 = 1.45$, the centrifugal blower is more efficient and is more widely used for this application. Using Eq. (22.2), with a CE cost index of 500, the estimated f.o.b. purchase cost of the centrifugal blower of iron and steel construction, which would be a multistage unit, including an electric motor with a direct drive, is

$$C_B = \frac{550}{500}\exp\{6.8929 + 0.7900[\ln(229)]\} = \$79,300$$

A centrifugal blower with aluminum blades, with $F_M = 0.60$, is also suitable and reduces the f.o.b. purchase cost to $0.60(\$79,300) = \$47,600$.

Compressors

Compressors are used widely to move gases for compression ratios greater than 2. As discussed in Section 22.3, the major types are the trunk-piston and crosshead reciprocating compressors, diaphragm compressor, centrifugal compressor, axial compressor, and the screw, sliding-vane, and liquid-ring (piston) rotary compressors. Of these, the most commonly used in chemical processing plants are the: (1) double-acting crosshead reciprocating compressor, (2) multistage centrifugal compressor, and (3) rotary twin-screw compressor. These are referred to here as simply reciprocating, centrifugal, and screw compressors.

Reciprocating compressors can handle the widest range of pressure, from vacuum to 100,000 psig, but the narrowest range of flow rates, from 5 to 7,000 ACFM, with horsepowers up to 20,000 per machine. By using many stages, centrifugal compressors can deliver pressures up to 5,000 psig for the largest flow rates, from 1,000 to 150,000 ACFM, with horsepowers to 2,000 per machine. Screw compressors have the smallest pressure range, up to 400 psig, for flow rates from 800 to 20,000 ACFM, with horsepowers to 6,000 per machine.

Because reciprocating and screw compressors are of the positive-displacement type, they are designed for a particular flow rate, with their discharge pressure set by the downstream system, provided that the power input to the compressor is sufficient. The maximum compression ratio per stage is set by a limiting temperature rise of the gas being compressed, as discussed at the beginning of this subsection. Compared to the screw compressor, the reciprocating compressor is more efficient (80–90% compared to 75–85%), more expensive, larger in size, somewhat more flexible in operation, accompanied in operation by large shaking forces that require a large foundation and more maintenance, is less noisy, and does not deliver as smooth a flow rate. Reciprocating compressors cannot tolerate the presence of liquid or solid particles in the feed gas, and consequently, must be protected by a knock-out drum.

The centrifugal compressor has become exceedingly popular in the last few decades because it is easily controlled, delivers a smooth flow rate (which, however, is dependent on the required discharge pressure), has small foundations and low maintenance, and can handle large flow rates and fairly high pressures. However, it is less efficient (70–75%) and more expensive than a screw compressor for the same application. The velocity of the blade tips sets the maximum compression ratio per machine stage. This limitation almost always translates into multiple stages in a single machine for compression ratios greater than 2, even though a limiting temperature is not achieved. Single machines may have as many as 10 stages. Further compression after reaching the temperature limit requires an intercooler followed by another machine.

Process simulation programs are preferred to compute the theoretical and brake horsepower requirements, as well as the exit temperature, of a compressor because the ideal gas law is not usually applicable for pressures above two atmospheres. However, Eq. (22.30) can be used to obtain a preliminary estimate of the brake horsepower. An estimate of the exit temperature, including the effect of compressor efficiency, η_c, can be made with the following modification of the equation for the isentropic exit temperature:

$$T_O = T_I + \frac{T_I\left[\left(\frac{P_O}{P_I}\right)^{\frac{k-1}{k}} - 1\right]}{\eta_C} \qquad (22.34)$$

Compressors may be driven by electric motors, steam turbines, or gas turbines, but the former is the most common driver and the gas turbine is the least common. All drivers are available up to at least 20,000 Hp. For applications below about 200 Hp, electric motors are used almost exclusively. Most efficient is the electric motor; least efficient is the gas turbine. Efficiencies of all three drivers increase with horsepower. At 1,000 Hp, typical efficiencies are 95%, 65%, and 35%, respectively, for the electric motor, steam turbine, and gas turbine. Therefore, unless excess steam or low-cost combustion gas is available, the electric motor is the driver of choice over the entire horsepower range. An exception is sometimes made for centrifugal compressors, where the steam turbine is an ideal driver because the speeds of the two devices can be matched. See Section 22.6 and Table 22.32 for cost equations for steam and gas turbines.

Base f.o.b. purchase costs, C_B, for the three major types of compressors, based on data from Garrett (1989) and Walas (1988) in Table 22.19, are plotted in Figure 22.9 as a function of consumed power, P_C, in horsepower for a cost index in 2006 (CE = 500). The base cost, which includes an electric motor drive, is for cast iron or carbon-steel construction. For other drives and materials of construction, the f.o.b. purchase cost is given by:

$$C_P = F_D F_M C_B \qquad (22.35)$$

where, in place of an electric motor drive, $F_D = 1.15$ for a steam turbine drive and 1.25 for a gas turbine drive, and $F_M = 2.5$ for stainless steel and 5.0 for nickel alloy.

The base compressor purchase cost curves in Figure 22.9, for a CE cost index of 500, are given by the following equations, with P_C in horsepower:

Centrifugal Compressor (valid from $P_C = 200$ to 30,000 Hp):

$$C_B = \exp\{7.5800 + 0.80[\ln(P_C)]\} \qquad (22.36)$$

Reciprocating Compressor (valid from $P_C \doteq 100$ to 20,000 Hp):

$$C_B = \exp\{7.9661 + 0.80[\ln(P_C)]\} \qquad (22.37)$$

Screw Compressor (valid from $P_C = 10$ to 750 Hp):

$$C_B = \exp\{8.1238 + 0.7243[\ln(P_C)]\} \qquad (22.38)$$

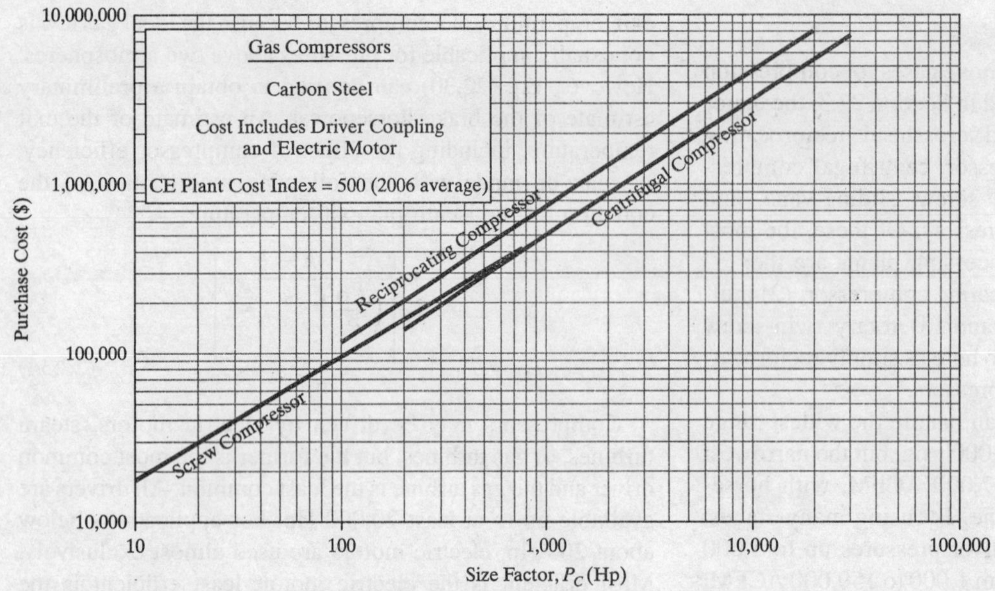

Figure 22.9 Base f.o.b. purchase costs for centrifugal, reciprocating, and screw compressors.

EXAMPLE 22.9

In an ammonia process, where hydrogen and nitrogen are combined at high temperature and high pressure in a catalytic reactor, a multistage gas compression system with intercoolers is needed to compress the feed gas to the reactor pressure. For one of the compression stages, the feed gas, at 320 K, 30 bar, and 6,815 kmol/hr, has a composition in mol% of 72.21 H_2, 27.13 N_2, 0.61 CH_4, and 0.05 Ar. It is to be compressed to 70 bar in an uncooled, adiabatic compressor. Size and select the compressor and the drive and estimate the f.o.b. purchase cost for a CE index of 550.

SOLUTION

At the high-pressure conditions, the ideal gas law does not apply and, therefore, it is preferred to size the compressor with a process simulation program. Using the SRK equation of state, the results give an inlet volumetric flow rate of 6,120 m^3/hr or 3,602 cfm. This is in the range of all three compressor types. However, the discharge pressure of 70 bar or 1,016 psi is beyond the range of the screw compressor. Select a centrifugal compressor of steel construction, with an assumed isentropic efficiency of 75%. This gives a discharge temperature of 439 K or 331°F, which is below the suggested 400°F discharge limit for a compressor. The corresponding theoretical kilowatts is 4,940 or 6,630 THp (theoretical horsepower). With a 75% compressor efficiency, the brake horsepower is 8,840. Assume a 95% efficiency for the electric motor. This gives $P_C = 9,300$ Hp. From Eq. (22.36), the f.o.b. purchase cost for CE = 550 is

$$C_B = \left(\frac{550}{500}\right)\exp\{7.5800 + 0.80[\ln(9,300)]\} = \$3,222,000$$

An alternative that might be considered is the use of two identical centrifugal compressors, each delivering 50% of the required flow, in place of the single compressor.

Heat Exchangers

As discussed in Chapter 18, a wide variety of heat exchangers is available for heating, cooling, condensing, and vaporizing process streams, particularly liquids and gases. The most important types are shell-and-tube, double-pipe, air-cooled fin-fan, and compact heat exchangers, including plate-and-frame, spiral-plate, spiral-tube, and plate-fin types. For most applications in chemical processing plants, shell-and-tube heat exchangers, which are governed by TEMA (Tubular Exchanger Manufacturers Association) standards and ASME (American Society of Mechanical Engineers) pressure-vessel code as well as other standards, are selected. However, for heat exchanger areas less than 200 ft^2, double-pipe heat exchangers are often selected, and when streams are cooled by air, fin-fan units are common. Compact heat exchangers are usually reserved for nondemanding applications. In this section, graphs and equations are presented only for shell-and-tube and double-pipe heat exchangers. Equations for air-cooled fin-fan and compact heat exchangers that are described in Section 18.2 are presented below in Section 22.6 and Table 22.32.

Shell-and-Tube Heat Exchangers

These exchangers cover a wide range of geometrical variables including tube diameter, wall thickness, length, spacing, and arrangement; baffle type and spacing; numbers of tube and shell passes; and fixed-head, floating-head, U-tube, and kettle designs. However, most published purchase-cost data are correlated in terms of heat exchange surface area (usually based on the outside surface area of the tubes) for a base-case design, with correction factors only for pressure and materials for the shell and tubes. In some cases, corrections for tube length are given. Here, the following cost correlations are based on several of the references in Table 22.19 and in Corripio et al.

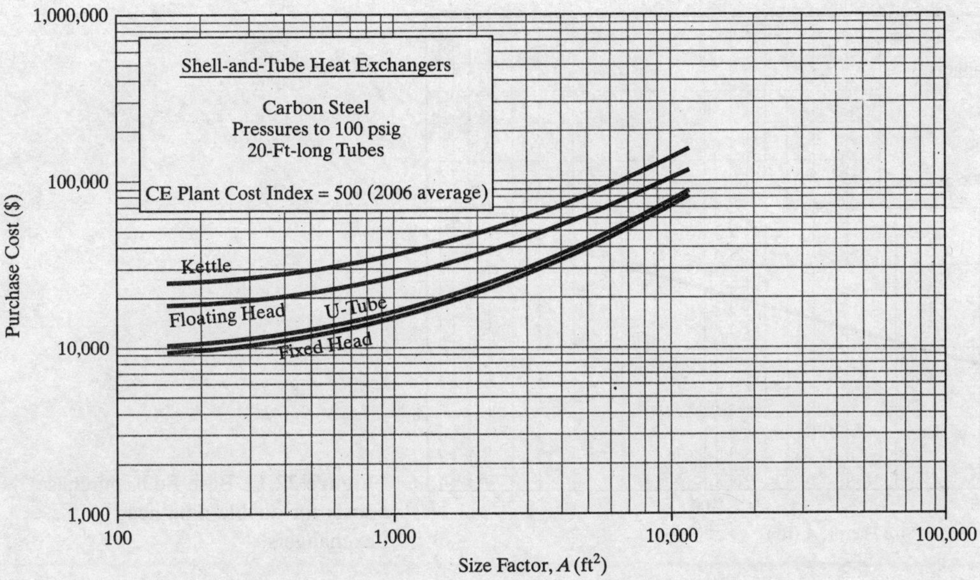

Figure 22.10 Base f.o.b. purchase costs for shell-and-tube heat exchangers.

(1982b). The base cost curves in Figure 22.10 for a CE cost index of 500 are given by the following equations, with tube outside surface area, A, in square feet, ranging from 150 to $12,000\ \text{ft}^2$. These base-case exchangers include 3/4-in. or 1-in. O.D., 16 BWG (Birmingham Wire Gage) carbon-steel tubes, 20 ft long, on square or triangular pitch in a carbon-steel shell for use with shell-side pressures up to 100 psig.

Floating head:

$$C_B = \exp\{11.667 - 0.8709[\ln(A)] + 0.09005[\ln(A)]^2\} \tag{22.39}$$

Fixed head:

$$C_B = \exp\{11.0545 - 0.9228[\ln(A)] + 0.09861[\ln(A)]^2\} \tag{22.40}$$

U-tube:

$$C_B = \exp\{11.147 - 0.9186[\ln(A)] + 0.09790[\ln(A)]^2\} \tag{22.41}$$

Kettle vaporizer:

$$C_B = \exp\{11.967 - 0.8709[\ln(A)] + 0.09005[\ln(A)]^2\} \tag{22.42}$$

The f.o.b. purchase cost for each of these four types of heat exchangers is determined from

$$C_P = F_P F_M F_L C_B \tag{22.43}$$

where F_M is a material factor for various combinations of tube and shell materials, as given in Table 22.25 as a function of the surface area, A, in square feet according to the equation:

$$F_M = a + \left(\frac{A}{100}\right)^b \tag{22.44}$$

The factor F_L is a tube-length correction as follows:

Tube Length (ft)	F_L
8	1.25
12	1.12
16	1.05
20	1.00

The pressure factor, F_P, is based on the shell-side pressure, P, in psig and is given by the following equation, which is applicable from 100 to 2,000 psig:

$$F_P = 0.9803 + 0.018\left(\frac{P}{100}\right) + 0.0017\left(\frac{P}{100}\right)^2 \tag{22.45}$$

Double-Pipe Heat Exchangers

For heat exchange surface areas of less than $200\ \text{ft}^2$ and as low as $2\ \text{ft}^2$, double-pipe heat exchangers are often selected

Table 22.25 Materials of Construction Factors, F_M, for Shell-and-Tube Heat Exchangers

Materials of Construction Shell/Tube	a in Eq. (22.44)	b in Eq. (22.44)
Carbon steel/carbon steel	0.00	0.00
Carbon steel/brass	1.08	0.05
Carbon steel/stainless steel	1.75	0.13
Carbon steel/Monel	2.1	0.13
Carbon steel/titanium	5.2	0.16
Carbon steel/Cr–Mo steel	1.55	0.05
Cr–Mo steel/Cr–Mo steel	1.70	0.07
Stainless steel/stainless steel	2.70	0.07
Monel/Monel	3.3	0.08
Titanium/titanium	9.6	0.06

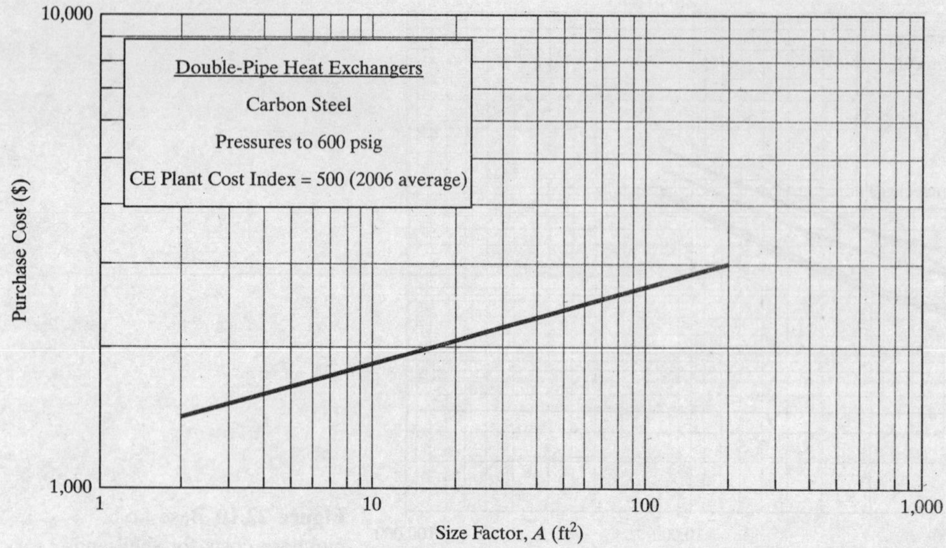

Figure 22.11 Base f.o.b. purchase costs for double-pipe heat exchangers.

over shell-and-tube heat exchangers. The area, A, is usually based on the outside surface area of the inner pipe. The cost correlation here is based on the average of several of the references in Table 22.19. The base cost curve in Figure 22.11 for a CE cost index of 500 is for carbon-steel construction for pressures to 600 psig, with the area in square feet. The correlating equation is

$$C_B = \exp\{7.1460 + 0.16[\ln(A)]\} \qquad (22.46)$$

The f.o.b. purchase cost is determined from

$$C_P = F_P F_M C_B \qquad (22.47)$$

where the material factor, F_M, is 2.0 for an outer pipe of carbon steel and an inner pipe of stainless steel. If both pipes are stainless steel, the factor is 3.0. The pressure factor, F_P, for the range of pressure, P, from 600 to 3,000 psig is given by:

$$F_P = 0.8510 + 0.1292\left(\frac{P}{600}\right) + 0.0198\left(\frac{P}{600}\right)^2 \qquad (22.48)$$

EXAMPLE 22.10

In Section 5.3, a toluene hydrodealkylation process is synthesized. Material- and energy-balance calculations on that process give a combined feed to the hydrodealkylation reactor of 5,802 lbmol/hr, containing mainly 35 vol% hydrogen, 58 vol% methane, and 7 vol% toluene, at 127.6°F and 569 psia. This stream is heated to 1,000°F in a heat exchanger by 6,010 lbmol/hr of quenched reactor effluent (which also contains a significant percentage of hydrogen), entering the exchanger at 1,150°F and 494 psia, and exiting at 364.2°F and 489 psia. The calculated heat duty, Q, is 69,360,000 Btu/hr. Estimate the area of the heat exchanger and the f.o.b. purchase cost at a CE index of 550.

SOLUTION

The combined feed enters the heat exchanger with a molar vapor fraction of 93.1% and leaves as a superheated vapor. The quenched effluent enters and exits as a superheated vapor. Thus, some vaporization occurs on the combined feed side. A zone analysis for a countercurrent heat exchanger gives a mean temperature driving force, ΔT_M, of 190.4°F, compared to end driving forces of 150°F and 236.6°F. Assuming an overall heat-transfer coefficient, U, of 50 Btu/hr-ft^2-°F, the heat exchanger area is

$$A = \frac{Q}{U(\Delta T_M)} = \frac{69{,}360{,}000}{50(190.4)} = 7{,}209 \text{ ft}^2$$

For these size- and temperature-difference conditions, select a floating-head shell-and-tube heat exchanger with 20-ft-long tubes. Pressures on both the shell and tube sides are in the range of 500 to 600 psig. Select a design pressure of 700 psig. Because temperatures are as high as 1,000 to 1,150°F, carbon steel cannot be used as the material of construction for either the shell or tubes. Because the hydrogen content of both streams is significant, Cr–Mo alloy steel, which is often used in the temperature range of this exchanger, is not suitable either, and stainless steel must be selected. From Eq. (22.39), the base purchase cost at a CE index of 500 is

$$C_B = \exp\{11.9052 - 0.8709[\ln(7{,}290)] + 0.09005[\ln(7{,}290)]^2\}$$
$$= \$79{,}500$$

From Eq. (22.44) and Table 22.25, for stainless steel construction, $F_M = 2.70 + (7{,}290/100)^{0.07} = 4.05$. For a pressure of 700 psig, using Eq. (22.45):

$$F_P = 0.9803 + 0.018\left(\frac{700}{100}\right) + 0.0017\left(\frac{700}{100}\right)^2 = 1.19$$

From Eq. (22.43), the f.o.b. purchase cost for a CE index of 550 is:

$$C_P = 1.19(4.05)(550/500)(79{,}500) = \$421{,}500$$

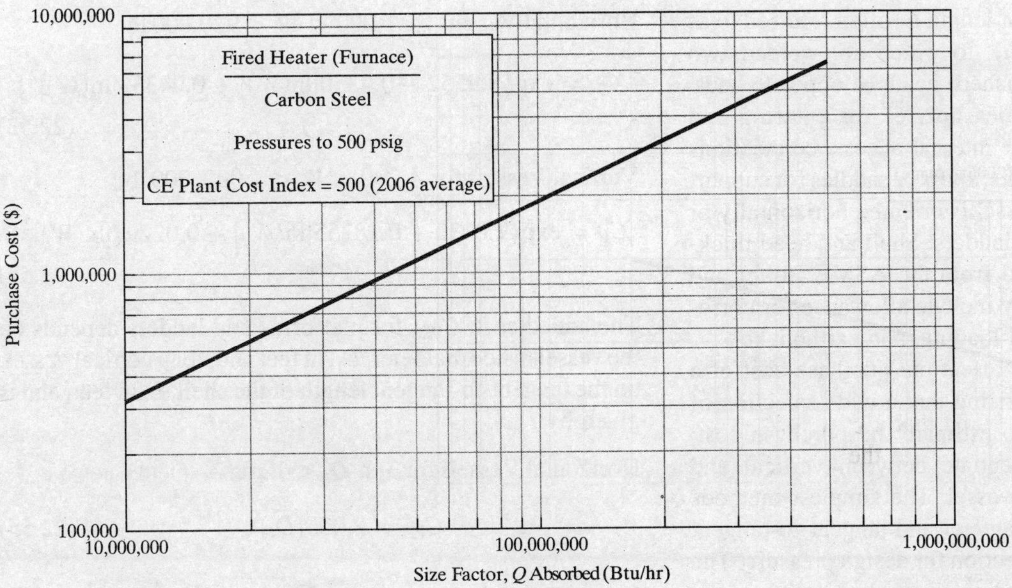

Figure 22.12 Base f.o.b. purchase costs for indirect-fired heaters of the box type.

Fired Heaters

Indirect-fired heaters of the box type, also called fired heaters, process heaters, and furnaces, are commonly used to heat and/or vaporize non-reacting process streams at elevated temperatures beyond where steam is normally employed. The fuel for combustion is either gas or fuel oil. As discussed in Section 18.2, heat duties of fired heaters are in the range of 10 to 340 million Btu/hr (3,000 to 100, 000 kJ/s or 3 to 100 MW). Typically, fired heaters are complete package units with standard horizontal tubes of carbon steel, adequate for temperatures to 1,100°F and pressures to 500 psig. For higher temperatures and/or pressures, other materials of construction may be needed and tubes must have an increased wall thickness. Thermal efficiencies range from 70 to 90%, with the higher value corresponding to units designed for energy conservation. The base cost depends on the heat duty, Q, absorbed by the process stream in Btu/hr. The cost correlation, shown in Figure 22.12, is based on the average of several of the references in Table 22.19. For CE = 500, the base cost for Q from 10 to 500 million Btu/hr is:

$$C_B = \exp\{0.32325 + 0.766[\ln(Q)]\} \qquad (22.49)$$

The f.o.b. purchase cost is determined from:

$$C_P = F_P F_M C_B \qquad (22.50)$$

where the material factor, F_M, is 1.4 for tubes of Cr–Mo alloy steel and 1.7 for stainless steel. The pressure factor, F_P, for the range of pressure, P, from 500 to 3,000 psig is given by:

$$F_P = 0.986 - 0.0035\left(\frac{P}{500}\right) + 0.0175\left(\frac{P}{500}\right)^2 \qquad (22.51)$$

Fired heaters for specific purposes are discussed below in Section 22.6.

EXAMPLE 22.11

After being heated to 1,000°F and before entering the reactor, the combined feed in Example 22.10 is heated further to 1,200°F in a fired heater. Determine the f.o.b. purchase cost of a fired heater at a CE index of 550.

SOLUTION

The calculated absorbed heat duty, Q, is 18,390,000 Btu/hr. Assume a design pressure for the tubes of 700 psig. Because of the significant hydrogen concentration in the combined feed, stainless steel tubes are required. From Eq. (22.49), the base cost is

$$C_B = \exp\{0.32325 + 0.766[\ln(18{,}390{,}000)]\} = \$507{,}100$$

From stainless steel tubes, $F_M = 1.7$. For a pressure of 700 psig, using Eq. (22.51):

$$F_P = 0.986 - 0.0035\left(\frac{700}{500}\right) + 0.0175\left(\frac{700}{500}\right)^2 = 1.015$$

From Eq. (22.50), the f.o.b. purchase cost for a CE index = 550 is

$$C_P = 1.015(1.7)(550/500)(507{,}100) = \$962{,}500$$

Pressure Vessels and Towers for Distillation, Absorption, and Stripping

Pressure vessels containing little or no internals (largely empty) are widely used in chemical processing plants. Applications include reflux drums, flash drums, knock-out drums, settlers, chemical reactors, mixing vessels, vessels for fixed-bed adsorption, and storage drums. These vessels are usually cylindrical in shape, with an inside diameter, D_i, and consist of a cylindrical shell of length L (often referred to as the *tangent-to-tangent length*), to which are welded two ellipsoidal or torispherical (dished) heads at opposite ends. In addition, the vessel includes nozzles for entering and exiting streams, manholes for internal access, connections for relief valves and instruments, skirts or saddles for support depending on whether the vessel is oriented horizontally or vertically, and platforms and ladders. Shell and head thicknesses are usually determined from the ASME Boiler and Pressure Vessel Code and may include allowance for corrosion, vacuum operation, wind loading, and earthquake.

Because many factors can affect the purchase cost of a pressure vessel, it is not surprising that a wide selection of size factors has been used to estimate the purchase cost; however, all methods differentiate between vertical and horizontal orientation of the vessel. The simplest methods base the cost on the inside diameter and tangent-to-tangent length of the shell, with a correction for design pressure. The most elaborate method is based on a complete design of the pressure vessel to obtain the vessel weight, and a sizing and count of the nozzles and manholes. Here, the method of Mulet, Corripio, and Evans (1981a, b) is employed, which is a method of intermediate complexity, based on the weight of the shell and two 2:1 elliptical heads. The f.o.b. purchase cost, which is for carbon-steel construction and includes an allowance for platforms, ladders, and a nominal number of nozzles and manholes, is given by

$$C_P = F_M C_V + C_{PL} \quad (22.52)$$

The f.o.b. purchase cost—at a CE index = 500, of the empty vessel, C_V, but including nozzles, manholes, and supports, based on the weight in pounds of the shell and the two heads, W—depends on orientation, as shown in Figure 22.13. The correlating equations are

Horizontal Vessels for $1{,}000 < W < 920{,}000$ lb:

$$C_V = \exp\{8.9552 - 0.2330[\ln(W)] + 0.04333[\ln(W)]^2\} \quad (22.53)$$

Vertical Vessels for $4{,}200 < W < 1{,}000{,}000$ lb :

$$C_V = \exp\{7.0132 + 0.18255[\ln(W)] + 0.02297[\ln(W)]^2\} \quad (22.54)$$

The added cost, C_{PL}, for platforms and ladders depends on the vessel inside diameter, D_i, in feet and, for a vertical vessel, on the tangent-to-tangent length of the shell, L, in feet, and is given by:

Horizontal Vessels for $3 < D_i < 12$ ft:

$$C_{PL} = 2{,}005(D_i)^{0.20294} \quad (22.55)$$

Vertical Vessels for $3 < D_i < 21$ ft and $12 < L < 40$ ft:

$$C_{PL} = 361.8(D_i)^{0.73960}(L)^{0.70684} \quad (22.56)$$

In Figure 22.13, it is seen that for a given shell weight, vertical vessels cost more than horizontal vessels up to 500,000 lb.

Towers are vertical pressure vessels for various separation operations, including distillation, absorption, and stripping. They contain plates and/or packing plus additional nozzles and manholes, and internals for multiple feed entries and management of bottoms liquid and its withdrawal. Figure 22.13 includes a curve for the f.o.b. purchase cost in U.S. dollars, as of a CE index = 500, of vertical towers, C_T, including nozzles, manholes, a skirt, and internals (but not plates and/or packing), based on the

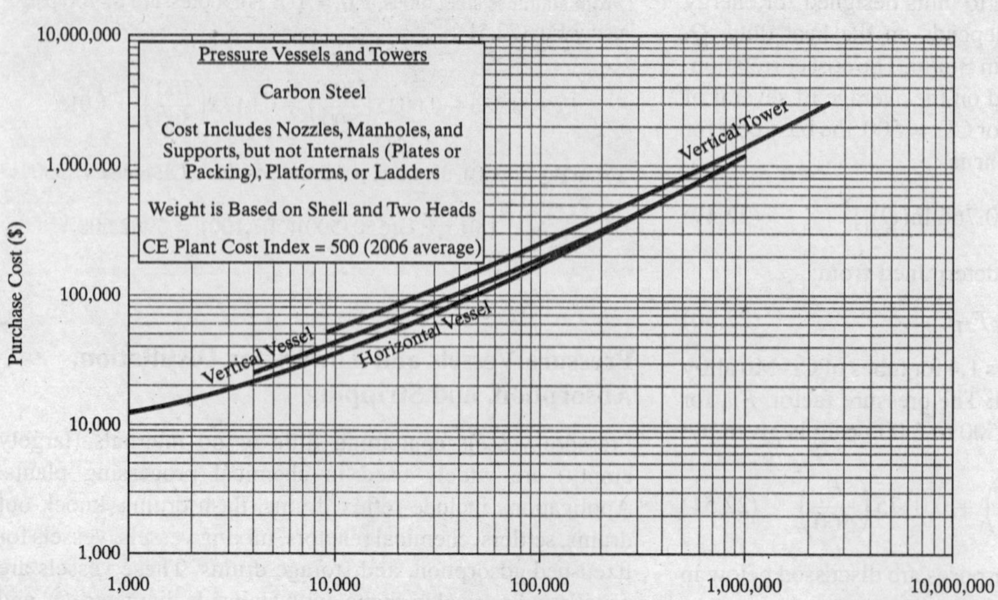

Figure 22.13 Base f.o.b. purchase costs for pressure vessels and towers.

weight in pounds of the shell and the two heads, W. The correlating equation is

Towers for $9{,}000 < W < 2{,}500{,}000$ lb:

$$C_V = \exp\{7.2756 + 0.18255[\ln(W)] + 0.02297[\ln(W)]^2\} \tag{22.57}$$

The added cost, C_{PL}, for platforms and ladders for towers depends on the tower inside diameter, D_i, in feet and on the tangent-to-tangent length of the shell, L, in feet, and is given by:

Towers for $3 < D_i < 24$ ft and $27 < L < 170$ ft:

$$C_{PL} = 300.9(D_i)^{0.63316}(L)^{0.80161} \tag{22.58}$$

The weight, W, in the cost corrrelations for a pressure vessel or tower depends on the wall thicknesses of the shell and the two heads. Although the thickness of the heads may be required to be somewhat thicker than the shell, particularly at high pressures, it is sufficient for cost-estimation purposes to assume head thicknesses equal to the shell thickness, t_S. Then, with 2:1 elliptical heads, the weight of the shell and the two heads is approximately:

$$W = \pi(D_i + t_S)(L + 0.8D_i)t_S\rho \tag{22.59}$$

where the term in L accounts for the cylinder, the term in $0.8\ D_i$ accounts for the two heads, and ρ is the density of the carbon steel, which can be taken as 490 lb/ft^3 or 0.284 lb/in.^{3}

In the absence of corrosion, wind, and earthquake considerations and for internal pressures greater than the external pressure (i.e., excluding vacuum operation), the cylindrical shell wall thickness is computed from the ASME pressure-vessel code formula:

$$t_P = \frac{P_d D_i}{2SE - 1.2P_d} \tag{22.60}$$

where t_P = wall thickness in inches to withstand the internal pressure, P_d = internal design gauge pressure in psig, D_i = inside shell diameter in inches, S = maximum allowable stress of the shell material at the design temperature in pounds per square inch, and E = fractional weld efficiency.

Sandler and Luckiewicz (1987) recommend that the design pressure, P_d in psig, be greater than the operating pressure, P_o. The following recommendations are similar to theirs. For operating pressures between 0 and 5 psig, use a design pressure of 10 psig. In the range of operating pressure from 10 psig to 1,000 psig, use the following equation:

$$P_d = \exp\{0.60608 + 0.91615[\ln(P_o)] + 0.0015655[\ln(P_o)]^2\} \tag{22.61}$$

For operating pressures greater than 1,000 psig, use a design pressure equal to 1.1 times the operating pressure. However, safety considerations may dictate an even larger difference between design pressure and operating pressure, especially when runaway reactions are possible.

The maximum allowable stress, S, in Eq. (22.60) depends on the design temperature and the material of construction. The design temperature may be taken as the operating temperature plus 50°F. However, again, safety considerations may dictate an even larger difference. At a given temperature, different steel compositions have different values for the maximum allowable stress. In the design temperature range of −20°F to 650°F, in a non-corrosive environment that is free of hydrogen, a commonly used carbon steel is SA-285, grade C, with a maximum allowable stress of 13,750 psi. In the temperature range from 650°F to 900°F, in a non-corrosive environment including the presence of hydrogen, a commonly used low-alloy (1% Cr and 0.5% Mo) steel is SA-387B. Its maximum allowable stress in its recommended temperature range is as follows:

Temperature (°F)	Maximum Allowable Stress (psi)
−20 to 650	15,000
700	15,000
750	15,000
800	14,750
850	14,200
900	13,100

The weld efficiency, E, in Eq. (22.60) accounts mainly for the integrity of the weld for the longitudinal seam. For carbon steel up to 1.25 in. in thickness, only a 10% spot X-ray check of the weld is necessary and a value of 0.85 for E should be used. For larger wall thicknesses, a 100% X-ray check is required, giving a value of 1.0 for E.

At low pressures, wall thicknesses calculated from Eq. (22.60) may be too small to give sufficient rigidity to vessels. Accordingly, the following minimum wall thicknesses should be used:

Vessel Inside Diameter (ft)	Minimum Wall Thickness (in.)
Up to 4	1/4
4–6	5/16
6–8	3/8
8–10	7/16
10–12	1/2

Equation (22.60) is suitable for calculating the thickness of a horizontal pressure vessel, but does not account for the effect of wind or an earthquake on a vertical vessel or column, and is not applicable to vessels or columns under vacuum. Mulet et al. (1981a) present a method for determining average wall thickness, t_V, of a vertical vessel or tower to withstand the internal pressure at the top of the column, and to withstand the wind load or an equivalent earthquake, in addition to the internal pressure, at the bottom of the column. The method assumes a substantial wind load based on a wind velocity of 140 miles/hr, acting uniformly over the height of the column. Their simplified equation is as follows, where t_W = the necessary thickness in inches to withstand the wind load or earthquake at the bottom of the column, D_o = the outside diameter of the vertical vessel in inches, L = the tangent-to-tangent vessel height in inches, and S = the maximum allowable stress in pounds per square inch.

$$t_W = \frac{0.22(D_o + 18)L^2}{SD_o^2} \tag{22.62}$$

where the term 18 (in inches) accounts for column cage ladders. The average vessel wall thickness, t_V, is then computed from the average of the thickness at the top, t_P, and the thickness at the bottom, $t_P + t_W$.

Equation (22.60) does not apply to vacuum vessels for which the internal pressure is less than the external pressure. Such vessels must be sufficiently thick to withstand a collapsing pressure, or they must be provided with stiffening rings around their outer periphery. For the former alternative, Mulet et. al. (1981a) present a method for computing the necessary wall thickness, t_E, based mainly on the vessel length-to-outside diameter ratio and the modulus of elasticity, E_M, of the metal wall. A simple approximation of their method, which is applicable to $t_E/D_o < 0.05$, is given by the following equation, where D_o = outside diameter:

$$t_E = 1.3D_o\left(\frac{P_d L}{E_M D_o}\right)^{0.4} \tag{22.63}$$

However, to the value of t_E the following correction, t_{EC}, must be added:

$$t_{EC} = L(0.18D_i - 2.2) \times 10^{-5} - 0.19 \tag{22.64}$$

where all variables are in inches. The total thickness for a vacuum vessel is

$$t_V = t_E + t_{EC} \tag{22.65}$$

The modulus of elasticity, E_M, depends on temperature, with the following values for carbon steel and low-alloy steel:

E_M, Modulus of Elasticity, psi (multiply values by 10^6)

Temperature (°F)	Carbon Steel	Low-Alloy Steel
−20	30.2	30.2
200	29.5	29.5
400	28.3	28.6
650	26.0	27.0
700	—	26.6
800	—	25.7
900	—	24.5

Table 22.26 Materials-of-Construction Factors, F_M, for Pressure Vessels

Material of Construction	Material Factor [F_M, in Eq. (22.52)]
Carbon steel	1.0
Low-alloy steel	1.2
Stainless steel 304	1.7
Stainless steel 316	2.1
Carpenter 20CB-3	3.2
Nickel-200	5.4
Monel-400	3.6
Inconel-600	3.9
Incoloy-825	3.7
Titanium	7.7

Even for non-corrosive conditions, a corrosion allowance, t_C, of $\frac{1}{8}$ in. should be added to t_V to give the value of t_S to be used in Eq. (22.59) for vessel weight. In addition, it is important to note that vessels are generally fabricated from metal plate, whose thicknesses can be assumed to come in the following increments:

$\frac{1}{16}$ in. increments for $\frac{3}{16}$ to $\frac{1}{2}$ in. inclusive
$\frac{1}{8}$ in. increments for $\frac{5}{8}$ to 2 in. inclusive
$\frac{1}{4}$ in. increments for $2\frac{1}{4}$ to 3 in. inclusive

The final vessel thickness is obtained by rounding up to the next increment.

The material-of-construction factor for pressure vessels and towers, F_M, is given in Table 22.26.

Before presenting cost data for plates and packings in distillation, absorption, and stripping towers, the following example is presented to estimate the purchase cost of an adiabatic, homogeneous gas-phase reactor.

EXAMPLE 22.12

An adiabatic reactor consists of a cylindrical vessel with elliptical heads, with an inside diameter of 6.5 ft (78 in.) and a tangent-to-tangent length of 40 ft (480 in.). Gas enters the reactor at a pressure of 484 psia and 800°F. Exit conditions are 482 psia and 850°F. The vessel will be oriented in a horizontal position. Estimate the vessel thickness in inches, weight in pounds, and purchase cost in dollars for a CE cost index of 550. The vessel contains no internals and the gas is non-corrosive. The barometric pressure at the plant site is 14 psia.

SOLUTION

The operating pressure, based on the higher pressure, is (484 − 14) = 470 psig. Using Eq. (22.61), the design pressure is

$$P_d = \exp\{0.60608 + 0.91615[\ln(470)] + 0.0015655[\ln(470)]^2\}$$
$$= 546 \text{ psig}$$

The higher operating temperature is 850°F. Take a design temperature of 50°F higher, or 900°F. A suitable material of construction is low-alloy steel. From a table above, its maximum allowable stress, S, is 13,100 psi. Assume that the wall thickness will be greater than 1.25 in., giving a weld efficiency, E, of 1.0. From the pressure-vessel code formula of Eq. (22.60),

$$t_P = \frac{(546)(78)}{2(13{,}100)(1.0) - 1.2(546)} = 1.667 \text{ in.}$$

which is greater than the assumed 1.25 in. and also greater than the minimum value of $\frac{3}{8}$ in. required for rigidity. Because the orientation of the vessel is horizontal, the vessel is not subject to wind load or earthquake considerations. Adding a corrosion allowance of $\frac{1}{8}$ in. gives a total thickness of 1.792 in. Since this is in the range of $\frac{5}{8}$ to 2 in., the steel plate comes in increments of $\frac{1}{8}$ in. Since 1.792 in. is greater than $1\frac{6}{8}$ in., specify a plate thickness, t_S, of $1\frac{7}{8}$ in. or 1.875 in. for use in Eq. (22.59), to give a vessel weight of:

$$W = (3.14)(78 + 1.875)[480 + 0.8(78)](1.875)(0.284)$$
$$= 72{,}500 \text{ lb}$$

The purchase cost of the vessel given by Eq. (22.53) is

$$C_V = \exp\{8.9552 - 0.2330[\ln(72{,}500)] + 0.04333[\ln(72{,}500)]^2\}$$
$$= \$129{,}900$$

From Table 22.26, the material factor is 1.20. Thus, $F_M C_V =$ \$155,900.

To this is added the cost of the platforms and ladders from Eq. (22.55):

$$C_{PL} = 2{,}005(78)^{0.20294} = \$4{,}850$$

Using Eq. (22.52), the purchase cost at a CE index of 550 is

$$C_P = \left(\frac{550}{500}\right)(155{,}900 + 4{,}850) = \$176{,}800$$

Plates

Vertical towers for absorption, distillation, and stripping utilize trays (plates) and/or packing. Mulet et al. (1981b) present a method for estimating the purchase cost of trays installed in a vertical tower. This cost is added to Eq. (22.52) to obtain the total purchase cost. The cost for the installed trays, C_T, all with downcomers, is given by

$$C_T = N_T F_{NT} F_{TT} F_{TM} C_{BT} \tag{22.66}$$

Here, the base cost, C_{BT}, is for sieve trays at a CE cost index of 500, where the inside diameter of the tower is in feet and the equation is valid for 2 to 16 ft.

$$C_{BT} = 468 \exp(0.1739\, D_i) \tag{22.67}$$

In Eq. (22.66), N_T = the number of trays. If that number is greater than 20, the factor $F_{NT} = 1$. If $N_T < 20$, the factor is greater than 1, as given by

$$F_{NT} = \frac{2.25}{1.0414^{NT}} \tag{22.68}$$

The factor F_{TT} accounts for the type of tray:

Tray Type	F_{TT}
Sieve	1.0
Valve	1.18
Bubble cap	1.87

The factor F_{TM}, which depends on column diameter in feet, corrects for the material of construction:

Material of Construction	F_{TM}
Carbon steel	1.0
303 Stainless steel	$1.189 + 0.0577\, D_i$
316 Stainless steel	$1.401 + 0.0724\, D_i$
Carpenter 20CB-3	$1.525 + 0.0788\, D_i$
Monel	$2.306 + 0.1120\, D_i$

EXAMPLE 22.13

A distillation column is to be used to separate isobutane from n-butane. The column, which is equipped with 100 sieve trays, has an inside diameter of 10 ft (120 in.) and a tangent-to-tangent length of 212 ft (2,544 in.). Operating conditions are 110 psia and 150°F at the bottom of the tower and 100 psia and 120°F at the top. The material of construction is carbon steel. The barometric pressure at the plant location is 14.5 psia. Estimate the purchase cost of the distillation column at a CE index of 550.

SOLUTION

The operating pressure, based on the higher pressure, is (110 − 14.5) = 95.5 psig. Using Eq. (22.61), the design pressure is

$$P_d = \exp\{0.60608 + 0.91615[\ln(95.5)]$$
$$+ 0.0015655[\ln(95.50)]^2\} = 123 \text{ psig}$$

The higher operating temperature is 150°F. Take a design temperature of 50°F higher, or 200°F. For carbon steel at this temperature, the maximum allowable stress, S, is 15,000 psi. Assume that the wall thickness will be less than 1.25 inches, giving a weld efficiency, E, of 0.85. From the pressure-vessel code formula, Eq. (22.60),

$$t_P = \frac{(123)(120)}{2(15{,}000)(0.85) - 1.2(123)} = 0.582 \text{ inch}$$

which is greater than the minimum value of 7/16 (or 0.4375) inch required for rigidity. Because orientation of the vessel is vertical and because it is quite tall, the tower may be subject to wind load or earthquake. From Eq. (22.62), assuming a wall thickness of 1.25 inches, which gives an outside diameter, D_o, of 122.5 inches, the additional tower wall thickness at the bottom of the tower is

$$t_W = \frac{0.22(122.5+18)[(212)(12)]^2}{15{,}000(122.5)^2} = 0.889 \text{ inch}$$

Therefore, at the bottom of the column, the required thickness to withstand internal pressure and wind load (or earthquake) is $0.582 + 0.889 = 1.471$ inches, compared to 0.582 inch at the top of the column. The average thickness $= t_V = (0.582 + 1.471)/2 = 1.027$ inch. To this, add a corrosion allowance of $\frac{1}{8}$ inch, giving a thickness of $1.027 + 0.125 = 1.152$ inches. Therefore, specify a steel plate thickness, t_S, of 1.250 inches. From Eq. (22.59), the vessel weight, W, is

$$W = (3.14)(120 + 1.250)[2{,}544 + 0.8(120)](1.250)(0.284)$$
$$= 356{,}800 \text{ lb}$$

The purchase cost at the vertical tower given by Eq. (22.57) is

$$C_V = \exp\{7.2756 + 0.18255[\ln(356{,}800)] + 0.02297[\ln(356{,}800)]^2\}$$
$$= \$636{,}700$$

From Table 22.26, the material factor is 1.00. Thus, $F_M C_V =$ \$636,700. To this is added the cost of the platforms and ladders from Eq. (22.58), which, however, is applied outside of its range of 27 to 170 ft for the tangent-to-tangent length of the shell.

$$C_{PL} = 300.9(10)^{0.63316}(212)^{0.80161} = \$94{,}700$$

The purchase cost at the CE index of 550 for just the tower, platforms, and ladders is

$$C_P = \left(\frac{550}{500}\right)(636{,}700 + 94{,}700) = \$804{,}500$$

To this must be added the cost of 100 sieve trays. From Eq. (22.67),

$$C_{BT} = 468 \exp[0.1739(10)] = \$2{,}660 \text{ per tray}$$

Using Eq. (22.66), and upgrading the cost index to 550,

$$C_T = N_T F_{NT} F_{TT} F_{TM} C_{BT} = 100(1.0)(1.0)(1.0)(2{,}660)\left(\frac{550}{500}\right)$$
$$= \$292{,}600$$

The total purchase cost of the distillation column is

$$\$804{,}500 + \$292{,}600 = \$1{,}097{,}100$$

Packings

Packings for towers are classified as dumped (random) or structured. When packings are used, the total purchase cost of the packed tower becomes

$$C_P = F_M C_V + C_{PL} + V_P C_{PK} + C_{DR} \tag{22.69}$$

where F_M for the vessel is given in Table 22.26, C_V for a vertical tower is given by Eq. (22.57), C_{PL} for the platforms and ladders is given by Eq. (22.58), V_P is the volume of the packing in cubic feet, C_{PK} is the installed cost of the packing in dollars per cubic foot, and C_{DR} is the installed cost of high-performance liquid distributors and redistributors required for obtaining satisfactory performance with packings.

Installed costs for dumped packings are given by several of the references in Table 22.19. Table 22.27, which was developed by taking averages of those costs, indexed to CE = 500, and adding some additional values from vendors, includes costs for six different dumped packings and five different materials.

Compared to dumped packings and trays, structured packings offer reduced pressure drop, higher stage efficiency in terms of reduced HETS (height equivalent to a theoretical stage), and capacity in terms of reduced diameter. However, they are more expensive in dollars per cubic foot of packing and are not normally available in carbon steel. They are most often installed when revamping existing towers to reduce pressure drop, increase capacity, and/or increase purity of products. Accordingly, installed costs of structured packings are not widely available. In the absence of a vendor quote and

Table 22.27 Installed Costs of Some Dumped Packings

	Installed Cost ($\$/ft^3$)				
Size	1 in.	1.5 in.	2 in.	3 in.	4 in.
Berl saddles					
Ceramic	48	37	28		
Raschig rings					
Carbon steel	54	40	34	26	
Stainless steel	180	139	110	63	
Ceramic	26	21	19	15	
Intalox saddles					
Ceramic	34	28	24	19	
Polypropylene	37		23	12	
Pall rings					
Carbon steel	49	37	32		
Stainless steel	168	129	110		
Polypropylene	37	26	21	16	
Cascade mini-rings					
Stainless steel	134		95	70	52
Ceramic	90		70	56	
Polypropylene	90		70	56	
Tellerettes					
Polyethylene	76				

for a very approximate estimate, the installed cost of structured packing of the corrugated-sheet type in stainless steel can be taken as $\$250/\text{ft}^3$.

Installed-cost data for high-performance liquid distributors and redistributors are also not widely available. Distributors should be placed at every feed point and, conservatively, redistributors should be used every 20 ft. In the absence of a vendor quote and for a very approximate estimate, the installed cost of a liquid distributor can be taken as $\$125/\text{ft}^2$ of column cross-sectional area.

EXAMPLE 22.14

A distillation column has two sections. The one above the feed is 14 ft in inside diameter with a 20-ft height, 15 ft of which is packed with structured packing of the corrugated-sheet type. The bottom section is 16 ft in diameter with a 70-ft height, 60 ft of which is packed with 4-in. Cascade mini-rings. The column is made of carbon steel, but both packings are of stainless steel. The column will operate under vacuum with conditions of 55 kPa and 60°C at the top and 60 kPa and 125°C at the bottom. A total of four liquid distributors or redistributors will be used. Estimate the f.o.b. purchase cost of the column, including installed packings, distributors, and redistributors, for a CE cost index of 550. The barometric pressure is 100 kPa.

SOLUTION

Take a design temperature of 50°F higher than the highest temperature of 125°C (257°F), or 307°F. For a vacuum vessel, use Eq. (22.63) to estimate the wall thickness. The maximum pressure difference between the inside and outside of the vessel is 100 − 55 = 45 kPa or 6.5 psig. Use this as the design pressure, P_d. From a table above, the modulus of elasticity for carbon steel at 307°F is 28.9×10^6 psi. For the top section, $D_i = 14$ ft or 168 in. Assume for this large a diameter that $D_i = D_o$. But for L, use the total length of (20 + 70) = 90 ft. Then the wall thickness for the top section is

$$t_E = 1.3(168)\left(\frac{6.5(90)}{28.9 \times 10^6(14)}\right)^{0.4} = 1.01 \text{ in.}$$

This is well within the limit of applicability of $t_E/D_o < 0.05$. To this must be added the correction of Eq. (22.64):

$$t_{EC} = (90)(12)[0.18(168) - 2.2] \times 10^{-5} - 0.19 = 0.11 \text{ in.}$$

The total thickness for the top section, from Eq. (22.65), is $t_V = 1.01 + 0.11 = 1.12$ in. Adding a 1/8-in. corrosion allowance gives 1.245 in. and, therefore, use a plate thickness including the next 1/8-in. increment or 1.25 in. Similar calculations for the bottom section give a wall thickness of 1.375 in.

The weights of the two sections are estimated from Eq. (22.59), but with only one head per section. However, as recommended by Mulet et al. (1981a) for a two-diameter tower, the weight of each section is based on the total length. The base cost for the two sections is then calculated from:

$$C_V = \frac{L_1 C_{V1} + L_2 C_{V2}}{L_1 + L_2} \tag{22.70}$$

For the top section,

$$W_1 = 3.14(168 + 1.25)[1{,}080 + 0.8(168)](1.25)(0.284)$$
$$= 229{,}100 \text{ lb}$$

A similar calculation for the bottom section gives $W_2 = 319{,}300$ lb. Base purchase costs for a tower section are given by Eq. (22.57). For the top section,

$$C_{V1} = \exp\{7.2756 + 0.18255[\ln(229{,}100)] + 0.02297[\ln(229{,}100)]^2\} = \$454{,}800$$

A similar calculation for the bottom section gives $C_{V2} = \$584{,}800$. Using Eq. (22.70), the purchase cost of the entire empty tower is

$$C_V = \frac{20(454{,}800) + 70(584{,}800)}{20 + 70} = \$555{,}900$$

To this is added the costs of the platforms and ladders for the tower. The concept of Eq. (22.70) is again applied after substituting C_{PL} for C_V for each section. From Eq. (22.58) for the top section, again using the total length of the tower:

$$C_{PL1} = 300.9(14)^{0.63316}(90)^{0.80161} = \$59{,}000$$

A similar calculation for the bottom section gives $C_{PL2} = \$64{,}200$. Using the form of Eq. (22.70), the total cost of the platforms and ladders is

$$C_{PL} = \frac{20(59{,}000) + 70(64{,}200)}{20 + 70} = \$63{,}100$$

The structured packing occupies a volume, V_{P1}, of $3.14(14)^2(15)/4 = 2{,}310 \text{ ft}^3$. The estimated installed cost, $C_{PK,1}$, is $\$250/\text{ft}^3$. The random packing occupies a volume, V_{P2}, of $3.14(16)^2(60)/4 = 12{,}060 \text{ ft}^3$. From Table 22.27, the installed cost of 4-in. Cascade mini-rings in stainless steel = $C_{PK,2} = \$52/\text{ft}^3$. For the four distributors, assume one has a diameter of 14 ft, while the other three have a diameter of 16 ft. This corresponds to areas of 154 ft² and 201 ft², respectively. At a cost of $\$125/\text{ft}^2$, the total installed cost of the four distributors or redistributors is

$$C_{DR} = 154(125) + 3(201)(125) = \$94{,}600$$

From Equation (22.67), after including the CE cost index ratio, the f.o.b. purchase cost of the vacuum tower, complete with packings, distributors and redistributors, platforms, and ladders, is

$$C_P = (550/500)[(1.0)(555{,}900) + 63{,}100 + [2{,}310(250) + 12{,}060(52)] + 94{,}600]$$
$$= (550/500)[555{,}900 + 63{,}100 + 1{,}205{,}000 + 94{,}600]$$
$$= \$2{,}110{,}000$$

Note that the packings are a large fraction of the total cost of the tower.

22.6 PURCHASE COSTS OF OTHER CHEMICAL PROCESSING EQUIPMENT

In this section, equations are presented for the estimation of the f.o.b. purchase cost of chemical processing equipment not covered in Section 22.5. In each equipment category, so many different designs are available that it is not possible to consider them all. Instead, an attempt has been made to include only the most widely used designs for which, in some cases, heuristics are included for estimating equipment sizes. This should be sufficient for preliminary estimates of capital cost. For final plant design, vendors of the different types of equipment should be consulted for assistance in selecting and sizing the most suitable design and to obtain more accurate estimates of purchase cost so as to achieve the most operable and economical process. The purchase-cost equations for the equipment in this section, which are based on a size factor, S, valid for a stated range and an average cost index for the year 2006 (CE = 500), are listed in Table 22.32, which appears at the end of this chapter. For the most part, the purchase-cost equations were developed from the sources of cost data given in Table 22.19 and the data at the Internet site www.matche.com/EquipCost. When the pressure and material of construction are not mentioned in Table 22.32, low-to-moderate pressures and conventional materials of construction such as carbon steel may be assumed. In lieu of data for other materials of construction, the purchase cost for another material may be estimated by multiplying the cost obtained from the equation in Table 22.32 by one of the following factors:

Material	Factor
Carbon steel	1.0
Copper	1.2
Stainless steel	2.0
Nickel	2.5
Monel	2.7
Titanium-clad	3.0
Titanium	6.0

Table 22.32 is accompanied by the following brief equipment descriptions, which include, in some cases, heuristics for estimating the size factor when design data are not readily available. More equipment descriptions and detailed methods for determining size for most of the types of equipment described below may be found in *Perry's Chemical Engineers' Handbook* (Green and Perry, 2008).

Adsorption Equipment

Adsorption from liquids is carried out in stirred vessels (*slurry adsorption*) or in fixed beds, while *fixed-bed adsorption* is used for gases. For slurry adsorption, which is usually conducted batchwise, the purchase cost includes the vessel, a motor-driven agitator, and the adsorbent particles. The size of the vessel and the amount of adsorbent required depend on the amount of solute to be adsorbed from the feed, the adsorption equilibrium, the solids content of the slurry in the vessel, and the desired time of treatment. A reasonable slurry composition is 5 vol% solids. An agitator sized at 10 Hp/1,000 gal being stirred is generally sufficient to keep the solid particles in suspension. Costs of adsorbents and motor-driven agitators are included in Table 22.32.

For fixed-bed adsorption of gases, a reasonable superficial gas velocity through the bed is 100 ft/min, while for liquids, 1 ft/min. Typical bed heights are 1 to 3 times the bed diameter. A conservative equilibrium adsorbent loading is 10 lb of adsorbate per 100 lb of adsorbent. Breakthrough times should account for mass-transfer resistance effects by adding 2 ft to the bed height calculated on the basis of equilibrium loading.

Agitators (Propellers and Turbines)

Motor-driven propellers and turbines are the most widely used devices for agitation in vessels. Propellers are small in diameter and use high rates of rotation by direct coupling, with motors running typically at 1,150 or 1,750 rpm. They are often used to agitate large liquid storage tanks by mounting several propellers sideways at locations around the periphery of the tank. Turbines are larger in diameter, typically one-third of the vessel diameter, and, by using speed reducers with electric motors, rotate at from 37 to 320 rpm. Turbines, which are available in several designs, are more versatile than propellers and are usually the preferred type of agitator for applications involving mixing of miscible and immiscible liquids in reactors, mixing of immiscible liquids in liquid-liquid extraction vessels, suspension of fine adsorbent particles in slurry adsorption, enhancement of heat transfer to or from a liquid in a jacketed tank, and dispersion of a gas into a liquid in a tank. Typical horsepower requirements for turbines, based on the fluid volume in the vessel or tank, are

Application	Hp/1,000 Gallons
Blending miscible liquids	0.5
Homogeneous liquid reaction	1.5
Reaction with heat transfer	3
Liquid-liquid reaction or extraction	5
Gas dispersion in a liquid	10
Suspension of solid particles	10

Autoclaves

An autoclave is predominantly a vertical, cylindrical stirred-tank reactor. It can be operated continuously or batchwise over a wide range of production rates, temperatures, and pressures. The stirring is achieved by internal agitators (turbines or propellers) or by forced circulation through the vessel with an external pump. However, the contents of small autoclaves are stirred by rocking, shaking, or tumbling the vessel. Most autoclaves are provided with a means of transferring heat to or from the contents of the vessel. That means may be a jacket, internal coils or tubes, an external pump–heat exchanger combination, an external reflux condenser when vapors are evolved, an electrically heated mantel, or direct firing by partial submergence of the autoclave in a furnace. In Table 22.32, equations for estimating f.o.b. purchase costs are listed for autoclaves made of steel, stainless steel, and glass-lined steel. These autoclaves are provided with turbine agitators and heat-transfer jackets.

Crystallizers

Most industrial crystallization operations are *solution crystallization* involving the crystallization of inorganic compounds from aqueous solutions. Only the inorganic compound crystallizes. However, a growing number of applications are being made for *melt crystallization*, which involves a mixture of two or more organic components whose freezing points are not far removed from each other. In that case, impure crystals (solid solutions) may be obtained that require repeated melting and freezing steps to obtain pure crystals of the component with the highest freezing temperature. Only solution crystallization is considered here.

Solution crystallization occurs from a supersaturated aqueous solution, which is achieved from the feed by cooling, evaporation, or a combination of cooling and evaporation. The application of cooling crystallization is limited because for many dissolved inorganic compounds, the decrease in solubility with decreasing temperature is not sufficient to make the method practical. Therefore, evaporative crystallizers are more common. Table 22.32 contains f.o.b. purchase costs for four types of crystallizers. The continuous jacketed scraped-wall crystallizer is based on length, which can be estimated by a heat-transfer rate using a scraped-surface cooling area of 3 ft^2 per foot of length and an overall heat-transfer coefficient of 20 Btu/hr-ft^2-°F. The heat-transfer rate is obtained by an energy balance that accounts for both sensible heat and the heat of crystallization. The purchase costs of continuous forced-circulation evaporative crystallizers or the popular continuous draft-tube baffled (DTB) crystallizers are based on the rate of production of crystals in tons (2,000 lbs) per day. The purchase cost of batch evaporative crystallizers, which usually operate under vacuum, depends on the vessel size.

Drives Other than Electric Motors

When the required shaft horsepower for power input to an item of process equipment is less than 100 Hp, an electric motor is usually the selected drive. For higher horsepower input, consideration is given to combustion gas turbines, steam turbines, and internal combustion gas engines. However, except for remote, mobile, or special situations, steam turbines are the most common alternative to electric motors. Furthermore, steam turbines are considerably more efficient, 50–80%, than gas turbines or engines, which have efficiencies of only 30–40%. Equations for f.o.b. purchase costs of steam and gas turbine drives are included in Table 22.32 as a function of shaft horsepower.

Dryers

No single drying device can handle efficiently the wide variety of feed materials, which includes granular solids, pastes, slabs, films, slurries, and liquid. Accordingly, many different types of commercial dryers have been developed for both continuous and batchwise operation. Batch dryers include tray and agitated types. Continuous dryers include tunnel, belt, tray, direct and indirect rotary, screw conveyor, fluidized-bed, spouted-bed, pneumatic-conveyor, spray, drum, infrared, dielectric, microwave, and freeze types. The selection and sizing of dryers often involves testing on a pilot-plant level. The f.o.b. purchase costs for several of the more widely used dryers are included in Table 22.32. Different size factors are used depending on the type of dryer. In the *batch compartment dryer*, the feed is placed in stacked trays over which hot air passes. Trays typically measure 30 × 30 × 3 in. Typical drying time is a few hours. The cost depends on the tray surface area.

Two types of *rotary dryers* are available for continuous drying. In the direct-heat type, longitudinal flights, which extend inward radially from the inner periphery of the slightly inclined rotating dryer cylinder, lift and shower the granular solids through hot air flowing through the cylinder. The inclination of the cylinder causes the solids to flow from the feed end to the discharge end of the cylinder. Moisture evaporation rates are generally in the range of 0.3–3 lb/hr-ft^3 of dryer volume depending on the amount of free moisture and the desired moisture content of the product. Direct-heat rotary dryers vary in size from 1-ft diameter by 4-ft long to 20-ft diameter by 150-ft long. The cost depends on the peripheral area of the shell. Typical length-to-diameter ratios vary from 4 for small dryers to 8 for large dryers. In the indirect-heat type, the material is dried by contact with the outer surface area of tubes, arranged in one or two circular rows around the inner periphery of the rotating shell. The purchase cost depends on the outside surface area of the tubes that carry condensing steam. Typical heat fluxes range from 600 to 2,000 Btu/hr-ft^2. Indirect-heat rotary dryers vary in size from 3 ft in diameter by 15 ft long to 15 ft in diameter by 80 ft long.

Drum dryers take a solution or thin slurry and spread it as a thin film over a rotating drum heated internally by condensing steam to produce a flaked product. Typical moisture evaporation rates are 3–6 lb/hr-ft^2. One or two drums (side by side) may be used. Drum dimensions range from 1 ft in diameter by 1.5 ft long to 5 ft in diameter by 12 ft long. The cost of drum dryers depends on the surface area of the drum.

Spray dryers produce small porous particles, such as powdered milk and laundry detergent, from a liquid solution by evaporation of the volatile component of the feed, with the purchase cost correlated with the evaporation rate. Size and cost data for other dryers as well as considerations in dryer selection are found in Section 12 of *Perry's Chemical Engineers' Handbook* (Green and Perry, 2008).

Dust Collectors

The removal of dust particles, typically 1 to 1,000 microns in diameter, from gas streams (also called gas cleaning) is accomplished on an industrial scale by four main types of equipment: (1) *bag filters*, (2) *cyclones* using centrifugal force, (3) *electrostatic precipitators*, and (4) *venturi scrubbers* using washing with a liquid. Reasons for dust collection include air-pollution control, elimination of safety and health hazards, recovery of a valuable product, improvement in the quality of other products, and reduction of equipment maintenance. Typical ranges of particle size that can be efficiently removed by each of the four methods are as follows:

Equipment	Dust Particle Range (microns)
Bag filters	0.1 to 50
Cyclones	10 to 1,000
Electrostatic precipitators	0.01 to 10
Venturi scrubbers	0.1 to 100

Bag filters, also referred to as fabric filters and baghouses, use natural or synthetic woven fabrics or felts. Typically, the openings in the filter fabric are larger than the particles to be removed. Therefore, an initiation period must take place during which particles build up on the fabric surface by impingement. After this occurs, the collected particles themselves act as a precoat to become the actual filter medium for subsequent dust removal. Bag filters are automatically cleaned periodically by shaking, using reverse flow of clean gas, or pulsing with clean gas. Dust collection efficiencies approach 100%.

The theory of particle removal by cyclones is well developed and equipment dimension ratios are well established, making it possible to design cyclones to remove particles of a known size range with reasonable estimates of particle-collection efficiency as a function of particle size. However, the efficiency of cyclones is greater for larger particles. Smaller particles can only be removed with cyclones of small diameter. Pressure drops in cyclones can be substantial and often limit the smallest particle size that can be removed. Multiple cyclones are sometimes used in series or parallel.

Electrostatic precipitators operate in a fashion that is just the opposite of a cyclone, removing the smallest particles at the highest collection efficiencies. They can remove the smallest dust particles as well as even smaller particles in the fume and smoke ranges down to 0.01 μm. In the precipitator, the particles are charged by passing the gas between two electrodes charged to a potential difference of tens of thousands of volts. This requires that the gas contain ionizable components such as carbon dioxide, carbon monoxide, sulfur dioxide, and water vapor. If not, water vapor can be added. The ions attach to the particles, carrying them to a collecting electrode. The particles are continuously or periodically removed from the electrode by rapping the collecting electrode.

The venturi scrubber usually consists of a venturi contactor followed by a cyclone separator. The particle-laden gas flows downward to the approach section of the venturi, where water is injected tangentially to flood the wall. After achieving intimate contact of the particles with the liquid in the throat of the venturi, the stream then turns and enters the cyclone, where clean gas leaves at the top and a slurry leaves at the bottom. Venturi scrubbers are particularly effective at high concentrations of particles.

Equations for estimating the f.o.b. purchase costs of dust collectors are given in Table 22.32, where the size factor is the actual gas flow rate. All collectors are of mild steel construction. For relatively large dust particles, cyclones are adequate and are the least expensive alternative. For gases containing a wide range of particle sizes, a cyclone followed by an electrostatic precipitator is a common combination.

Evaporators

Aqueous solutions of inorganic salts and bases are concentrated, without crystallization, in evaporators. Because the volatility of water is much higher than that of the dissolved inorganic salts and bases, only water is evaporated. Evaporators are also used with aqueous solutions of organic compounds that have little volatility. Most popular is the *long-tube vertical (rising-film) evaporator*, with tubes from 12 to 35 ft long with boiling inside the tubes. For viscous solutions, a pump is added to give a *forced-circulation evaporator*. Less efficient, but less expensive, is the *horizontal-tube evaporator*, where boiling occurs outside the tubes. For temperature-sensitive applications, a (long-tube vertical) *falling-film evaporator* is used, typically with a small overall temperature driving force of less than 15°F. Many evaporators operate under vacuum and are frequently multistaged to reduce the cost of the heating steam. The f.o.b. purchase costs for evaporators are included in Table 22.32, in terms of the heat-transfer area, for carbon-steel construction and pressures to 10 atmospheres. Typical ranges of overall heat-transfer coefficients, *U*, which increase with the boiling temperature, are as follows for a boiling-point range of 120 to 220°F:

Evaporator Type	U(Btu/hr-ft^2-°F)
Horizontal tube	80–400
Long-tube vertical	150–650
Forced circulation	450–650
Falling film	350–750

Fired Heaters for Specific Purposes

Fired heaters are available for providing heat-transfer media of the types described in Section 18.2, including hot water, steam, mineral oils, silicon oils, chlorinated diphenyls (Dowtherm A), and molten (fused) salts. Fired heaters are also used as reactors, such as reformers in petroleum refineries and for pyrolysis of organic chemicals. Purchase costs for these specific types of fired heaters, based on heat duty, are presented in Table 22.32.

Liquid-Liquid Extractors

A wide variety of commercial equipment is available for carrying out separations by liquid–liquid extraction. The most efficient are those that provide mechanical agitation of the liquid phases. When the number of equilibrium stages is small, for example, five or less, and floor space is available but headroom is at a premium, a battery of *mixer-settler vessels* may be the best choice. For preliminary cost estimates, and in lieu of mass-transfer data, the mixing vessels can be pressure vessels that have a height-to-diameter ratio of 1 and provide 5 min or less residence time, depending on the liquid viscosity and the interfacial tension. The mixers are equipped with vertical side baffles and a flat-blade turbine agitator that delivers 4 Hp/1,000 gal. The settlers can be horizontal vessels with inlet baffles and a length-to-diameter ratio of 4. The capacity of the settler can be determined based on 5 gal/min of feed/ft^2 of phase disengaging area (diameter $\times$ length), provided that the specific-gravity difference between the two liquid phases is greater than 0.10. Each mixing vessel in the battery will approximate an equilibrium stage. Purchase costs of mixers and settlers can be estimated from the costs for pressure vessels and agitators discussed above.

When headroom is available, a number of column designs, with mechanical agitation from impellers on a vertical shaft, can be used. Typical of these is the *rotating-disk contactor (RDC)*, which has a maximum diameter of 25 ft and maximum total liquid throughput of 120 ft^3 of liquid/hr-ft^2 of column cross-sectional area. Typical HETP (height equivalent to a theoretical stage) values range from 2 to 4 ft, depending on column diameter and interfacial tension. The f.o.b. purchase cost for an RDC unit is included in Table 22.32, where the size factor is the product of the column diameter raised to a 1.5 exponent and the column height.

EXAMPLE 22.15

An aqueous feed of 30,260 lb/hr of 22 wt% acetic acid is contacted with 71,100 lb/hr of a solvent of 96.5 wt% ethyl acetate at 100°F and 25 psia to extract 99.8% of the acetic acid. A process simulation program computes 6 equilibrium stages for the separation. The densities of the two entering liquid phases are 62.4 lb/ft^3 for the feed and 55.0 lb/ft^3 for the solvent. Estimate the size and f.o.b. purchase cost of an RDC liquid–liquid extraction column for a CE cost index of 550.

SOLUTION

The volumetric flow rate of the feed = 30,260/62.4 = 485 ft^3/hr.

The volumetric flow rate of the solvent = 71,100/55.0 = 1,293 ft^3/ hr.

The total volumetric flow rate through the column = 485 + 1,293 = 1,778 ft^3/hr.

For a maximum throughput of 120 ft^3/ hr-ft^2, cited above, the minimum cross-sectional area for flow = 1,778/120 = 14.8 ft^2.

Assume a throughput of 60% of the maximum value.

Actual cross-sectional area for flow = 14.8/0.6 = 24.7 ft^2.

Column diameter = $[24.7(4)/3.14]^{0.5}$ = 5.6 ft. Specify a diameter, D, of 5.5 ft.

Assume an HETP of 4 ft. This gives a total stage height of 4(6) = 24 ft. Add an additional 3 ft at the top and 3 ft at the bottom to give a total tangent-to-tangent height, H, of 24 + 3 + 3 = 30 ft.

From Table 22.32, the size factor = $S = H(D)^{1.5} = 30(5.5)^{1.5} =$ 387 ft$^{2.5}$.

For carbon steel at a CE index of 500, the f.o.b. purchase cost is

$$C_p = 317(387)^{0.84} = \$47,300$$

Because the feed contains water and acetic acid, assume stainless steel construction with a material factor of 2.0 and correct for the cost index. This gives an estimated f.o.b. purchase cost of

$$C_P = 47,300(2.0)(550/500) = \$104,000$$

Membrane Separations

Commercial membrane-separation processes include *reverse osmosis, gas permeation, dialysis, electrodialysis, pervaporation, ultrafiltration, and microfiltration.* Membranes are mainly synthetic or natural polymers in the form of sheets that are spiral wound or hollow fibers that are bundled together. Reverse osmosis, operating at a feed pressure of 1,000 psia, produces water of 99.95% purity from seawater (3.5 wt% dissolved salts) at a 45% recovery, or with a feed pressure of 250 psia from brackish water (less than 0.5 wt% dissolved salts). Bare-module costs of reverse osmosis plants based on purified water rate in gallons per day are included in Table 22.32. Other membrane-separation costs in Table 22.32 are f.o.b. purchase costs.

Gas permeation is used to separate gas mixtures, for example, hydrogen from methane. High pressures on the order of 500 psia are used to force the molecules through a dense polymer membrane, which is packaged in pressure-vessel modules, each containing up to 4,000 ft^2 of membrane surface area. Membrane modules cost approximately \$45/ft^2 of membrane surface area. Multiple modules are arranged in parallel to achieve the desired total membrane area.

Pervaporation is used to separate water–organic and organic–organic mixtures that form azeotropes and may be difficult to separate by enhanced distillation. Typical membrane modules cost \$38/ft^2 of membrane surface area.

Ultrafiltration uses a microporous polymer membrane, which allows water and molecules of less than some cut-off molecular weight to pass through, depending on the pore diameter, while retaining larger molecules. A typical membrane module may contain 30 ft^2 of membrane surface area at a cost of from \$10 to \$25/ft^2 of surface area.

Mixers for Powders, Pastes, and Doughs

A wide variety of designs is available for mixing powders or pastes, polymers, and doughs of high viscosity. Among the more widely used designs are *ribbon and tumbler mixers* for dry powders, and *kneaders and mullers* for pastes and doughs. Equations for f.o.b. purchase costs of these devices are included in Table 22.32 in terms of the volumetric size. All designs operate batchwise; some can operate continuously. Typical residence times for mixing are less than 5 min. A comprehensive treatment of this type of mixing is given in *Perry's Chemical Engineers' Handbook* (Green and Perry, 2008).

Power Recovery

Valves are often used to reduce the pressure of a gas or liquid process stream. By replacing the valve with a turbine, called an *expander, turboexpander*, or *expansion turbine* in the case of a gas and a *liquid expander* or *radial-inflow, power-recovery turbine* in the case of a liquid, power can be recovered for use elsewhere. Power recovery from gases is far more common than from liquids because for a given change in pressure and mass flow rate, far more power can be recovered from a gas than from a liquid because of the lower density of the gas. Equations for f.o.b. purchase costs of power-recovery devices are included in Table 22.32 in terms of horsepower that can be extracted. Typical efficiencies are 75–85% for gases and 50–60% for liquids. Condensation of gases in expanders up to 20% can be tolerated, but vapor evolution from liquid expansion requires a special design. Whenever more than 100 Hp for a gas and more than 150 Hp for a liquid can be extracted, a power-recovery device should be considered.

Screens

Solid particles are separated according to particle size by screening. Ideally, particles of size smaller than the opening of the screen surface, called *undersize* or *fines*, pass through, while larger particles, called *oversize* or *tails*, do not. However, a perfect separation is not possible. If the undersize is the desired product, the screen efficiency is the mass ratio of undersize obtained from the screen to the undersize in the feed. A typical efficiency might be 75%. For particles with sizes greater than 2 in., a vibrating, inclined *grizzly*, which consists of parallel bars of fixed spacing, is commonly used. The opening between adjacent parallel bars may be from 2 to 12 in. Inclined *vibrating screens* are used to separate particles smaller than 2 in. Standard screen sizes of the U.S. Sieve Series are used. The screens consist of woven wire with square apertures (openings). Screen sizes are quoted in millimeters above 8 mm and in *mesh* for 8 mm and lower. Mesh refers to the number of openings per inch. However, because wire diameter is not constant, the actual opening size cannot be easily calculated from the mesh. Instead, a table, such as found in *Perry's Chemical Engineers' Handbook* (Green and Perry, 2008), must be used to obtain the opening size. For example, a 20-mesh (No. 20) screen has square openings of 0.841 mm. Purchase costs of grizzlies and vibrating screens, included in Table 22.32, depend on the screen surface area. Screen capacities are directly proportional to screen surface area and approximately proportional to the opening between bars or the screen aperture. However, capacity for vibrating screens drops off dramatically for particle sizes below that for 140 mesh (0.105 mm). Typical capacities for grizzly screens range from 1 to 4 tons of feed/ft^2 of screen surface-hr-inch of opening between bars, with the lower value for coal and the higher value for gravel. For vibrating screens, typical capacities range from 0.2 to 0.8 ton/ft^2-hr-mm of aperture. Vibrating screens are available in single-, double-, and triple-deck machines, where the screen surface area refers to the total area in square feet of all screens in the deck. For the separation of very small particles of less than 0.1 mm in diameter, air classifiers are used, which can be costed as a cyclone separator.

Size Enlargement

Solid products are frequently produced in preferred shapes, such as tablets, rods, sheets, etc. Such shapes are produced by a variety of size enlargement or agglomeration equipment by pressure compaction, as by a *pellet mill, pug mill extruder, roll-type press, screw extruder*, or *tableting press*; or by tumbling compaction in *disk or pan agglomerators*. Equations for f.o.b. purchase costs of these devices are included in Table 22.32 in terms of the feed rate. A comprehensive treatment of size enlargement is given in *Perry's Chemical Engineers' Handbook* (Green and Perry, 2008).

Size Reduction Equipment

The size of solid particles can be reduced by one or more of the following actions: (1) impact by a single rigid force, (2) compression between two rigid forces, (3) shear, and (4) attrition between particles or a particle and a wall. The energy required to reduce the size of particles is far more than that theoretically required to increase the surface area of the particles, with the excess energy causing an increase in the temperature of the particles and the surroundings. A wide variety of equipment is available for particle size reduction, but except for special cases, most crushing and grinding tasks can be accomplished with the types of equipment listed in Table 22.28, which is taken from Walas (1988) and is organized by feed particle size. Included in the table are representative solids feed rates and power consumption. In general, primary and secondary crushers, including *jaw crushers, gyratory crushers*, and *cone crushers*, are used with feed particle sizes greater than about 2 in. They accomplish a size (diameter) reduction ratio of approximately 8. For feeds of smaller particle sizes, grinding is used with *hammer mills* and *ball mills*, but at lower capacities, especially with hammer mills. With these two mills, the size reduction ratio is much larger, 100 to 400, to achieve particle sizes in the 0.01 to

Table 22.28 Operating Ranges of Widely Used Crushing and Grinding Equipment[a]

Equipment	Feed Size (mm)	Product Size (mm)	Size Reduction Ratio	Capacity [ton/hr (1 ton = 2,000 lb)]	Power (Hp)
Gyratory crushers	200–2,000	25–250	8	100–500	135–940
Jaw crushers	100–1,000	25–100	8	10–1,000	7–270
Cone crushers	50–300	5–50	8	10–1,000	27–335
Hammer mills	5–30	0.01–0.1	400	0.1–5	1.3–135
Ball mills	1–10	0.01–0.1	100	10–300	65–6,700
Jet mills	1–10	0.003–0.05	300	0.1–2	2.7–135

[a]Reprinted with permission from Walas (1988).

0.1 mm range. For the production of even smaller particles, in the range of 3 to 50 μm, a *jet mill* (pulverizer), which uses gas jets, can be used. The f.o.b. purchase costs of the crushers and grinders in Table 22.28, including electric motor drives, are listed in Table 22.32.

Solid-Liquid Separation Equipment (Thickeners, Clarifiers, Filters, Centrifuges, and Expression)

Slurries of solid particles and liquid are separated into more concentrated slurries or wet cakes and relatively clear liquid (overflow) by means of gravity, pressure, vacuum, or centrifugal force. The solid particles and/or the clear liquid may be the product of value. Continuous *clarifiers* remove small concentrations of solid particles by settling (sedimentation) to produce a clear liquid. Continuous *thickeners* are similar in design to clarifiers, but the emphasis is on producing a more concentrated slurry that can be fed to a filter or centrifuge to produce a wet cake. With thickeners, the solid particles are usually more valuable. Clarifiers and thickeners operate continuously, most often with gravity causing the solid particles to settle to the bottom of the equipment, where a rake is used to remove concentrated slurry. Typically, the feed to a thickener is 1–30 wt% solids, while the underflow product is 10–70 wt% solids. For a clarifier, the wt% solids in the feed is usually less than 1%. The f.o.b. purchase costs for thickeners and clarifiers listed in Table 22.32 are based on a size factor of the settling area. Most thickeners and clarifiers are circular, with a diameter from 10 to 400 ft. Below 100 ft, construction is usually of carbon steel, while concrete is used for diameters greater than 100 ft. The required settling area is best determined by scaling up laboratory sedimentation experiments. In the absence of such data, very preliminary cost estimates can be made by estimating a settling area based on the equation: Settling area, $ft^2 = C_1$ (tons solids/day), where C_1 typically is in the range of 2–50, with the lower value applying to large particles of high density and the higher value applying to fine particles of low density. The settling area for a clarifier can be estimated by the equation: Settling area, $ft^2 = C_2$ (gpm of overflow liquid), where C_2 is typically 0.5 to 1.5. Power requirements for thickeners and clarifiers are relatively low because of the low rotation rate of the rake. For example, a 200-ft-diameter unit only requires 16 Hp.

Continuous wet classifiers, of the *rake* and *spiral* type, and *hydroclones* (hydrocyclones) can also produce concentrated slurries, but the overflow liquid is not clear, containing the smaller particles. A separation of particles by size and density occurs. Hydroclones are inexpensive, typically 6 in. in diameter or less, and a single unit can handle only low flow rates. For high flow rates, parallel units are employed in a *multiclone*. Table 22.32 contains f.o.b. purchase costs for wet classifiers and hydroclones. The size factor is the solids flow rate for classifiers and the liquid flow rate for hydroclones.

Wet cakes can be produced by sending concentrated slurries to filters operating under pressure or vacuum. Liquid passes through a porous barrier (filter media), which retains most of the solid particles to form a wet cake. Filters are designed to operate either continuously (e.g., the *continuous rotary-drum vacuum filter* or the *rotary pan filter*) or batchwise (e.g., the *plate-and-frame filter* or the *pressure leaf filter*). Table 22.32 contains f.o.b. purchase costs for most of the widely used filters. All costs are based on filtering area, which must be determined by laboratory experiments with a handheld vacuum or pressure leaf filter. For preliminary cost estimates, in the absence of such tests, select a continuous rotary-drum vacuum filter operating at a rotation rate of 20 rev/hr, with a filtering area estimated for fine particles (e.g., those produced by a precipitation reaction to produce relatively insoluble inorganic particles) at a filtrate rate of 1,500 lb/day-ft^2 (vacuum of 18–25 in. of Hg), and, for coarse solids (e.g., those produced by crystallization) at a filtrate rate of 6,000 lb/day-ft^2 (vacuum of 2–6 in. of Hg).

An alternative to a filter is a centrifuge, where the removal of some of the liquid in the feed is accomplished under a high centrifugal force, up to 50,000 times the gravitational force on the earth, by sedimentation with a solid bowl or by filtration with a perforated bowl, either continuously or batchwise. Table 22.32 contains f.o.b. purchase costs for many of the more widely used centrifuges in 304 stainless steel. Included are two manual *batch centrifugal filters*, two *automatic-batch centrifugal filters*, and two *continuous centrifuges*. The size factor for the latter two centrifuges is the tons per hour of solid particles, while the size factor for the other four centrifuges is the bowl diameter. With auto-batch operation, from 1 to 24 ton/hr of solids can be processed. For manual batch operation, a cycle time of 2 min for coarse particles and 30 min for fine particles is representative. From 0.15 to 5 ton/hr of solids can be processed in a 40-in.-diameter manual batch centrifuge.

Some wet solids consist of fibrous pulps or other compressible materials from which liquid cannot be removed by settling, filtering, or centrifuging. Instead *expression* is used with *screw presses* or *roll presses*. Table 22.32 contains f.o.b. purchase costs for these two presses in stainless steel.

EXAMPLE 22.16

Consider a process similar to that of Figure 8.45 for the continuous production of crystals of $MgSO_4.7H_2O$ from an aqueous solution of 10 wt% $MgSO_4$. However, replace the hydroclone-centrifugal filter combination with a rotary-drum vacuum filter. Thus, in Figure 8.45, the magma goes directly to the filter, which produces filtrate (mother liquor) that is recycled to the crystallizer and a wet cake that is sent to the dryer. The crystallizer operates adiabatically. Therefore, the filtrate is heated with an external heat exchanger before being recycled to the crystallizer. For the production of 3,504 lb/hr of $MgSO_4.7H_2O$, containing 1.5 wt% moisture, from a feed of a 10 wt% aqueous solution of $MgSO_4$ at 14.7 psia and 70°F, estimate the f.o.b. purchase costs of all major items of equipment at a CE cost index of 500, using the following results from material and energy balances, obtained with a process simulation program and preliminary equipment sizing:

Feed to the process:

$H_2O = 15,174$ lb/hr
$MgSO_4 = 1,686$ lb/hr

Evaporator Effect 1 (long-tube vertical):

Operating pressure = 7.51 psia
Evaporation rate = 6,197 lb/hr
$U = 450$ Btu/hr-ft^2-°F
$\Delta T = 30$°F
Heat duty = 7,895,000 Btu/hr
Area, $A = 585$ ft^2

Evaporator Effect 2 (long-tube vertical):

Operating pressure = 2.20 psia
Evaporation rate = 6,197 lb/hr
$U = 250$ Btu/hr-ft^2-°F
$\Delta T = 40$°F
Heat duty = 5,855,000 Btu/hr
Area, $A = 585$ ft^2

Feed to crystallizer from Evaporator Effect 2:

$H_2O = 2,780$ lb/hr
$MgSO_4 = 1,686$ lb/hr

Recycle filtrate (mother liquor) to crystallizer from external heater:

$H_2O = 7,046$ lb/hr
$MgSO_4 = 2,792$ lb/hr
Temperature after being heated = 180°F

Crystallizer (continuous adiabatic draft-tube baffled):

Operating pressure = 0.577 psia
Temperature = 85.6°F
Solubility of $MgSO_4$ at these conditions = 28.3 wt% $MgSO_4$
H_2O evaporation rate = 583 lb/hr

Magma flow rate to filter:

$H_2O = 7,783$ lb/hr
Dissolved $MgSO_4 = 3,084$ lb/hr
$MgSO_4.7H_2O$ crystals produced = 2,854 lb/hr or 34.2 ton/day

Filter (continuous rotary vacuum):

Cake, assumed to contain 73.5 wt% solids = 3,883 lb/hr
Total filtrate = 9,838 lb/hr = 236,100 lb/day
Assumed filtrate flux = 5,000 lb/day-ft^2 because crystals are fairly coarse
Filter area = 236,100/5,000 = 47 ft^2

Dryer (continuous direct-heat rotary):

Production rate of crystals with 1.5 wt% moisture = 3,504 lb/hr
Moisture evaporation rate by contact with hot air = 379 lb/hr
Moisture in crystals = 52 lb/hr
Outlet temperature of crystals = 113°F
Assumed volumetric moisture evaporation rate = 2 lb/hr/ft^3 of dryer volume because crystal moisture will be mainly free and final moisture content is not particularly low.
Dryer volume = 379/2 = 190 ft^3
Dryer dimensions: 3.5 ft diameter by 20 ft long
Peripheral area = 220 ft^2

SOLUTION

The process system consists of two long-tube vertical evaporators, a draft-tube baffled crystallizer, a rotary-drum vacuum filter, and a direct-heat rotary dryer. Also, pumps are needed to move the solution from evaporator 1 to evaporator 2, to recycle the filtrate from the filter to the crystallizer, and to move the magma from the crystallizer to the filter; and a heat exchanger is needed to heat the recycle filtrate. However, the purchase costs for the three pumps and the heat exchanger are not considered here because examples for these types of equipment are presented in Section 22.5. For the equipment considered here, assume fabrication from stainless steel, with a material factor of 2 for the ratio of stainless steel cost to carbon steel cost. For the process, using the following size factors and the equations in Table 22.32, the estimated f.o.b. equipment purchase costs at a CE index of 500 are included in the following table.

Equipment	Type	Size Factor	f.o.b. Purchase Cost, in Carbon Steel	f.o.b. Purchase Cost, in Stainless Steel
Evaporator 1	Long-tube vertical	$A = 585$ ft^2	\$190,000	\$380,000
Evaporator 2	Long-tube vertical	$A = 585$ ft^2	\$190,000	\$380,000
Crystallizer	Draft-tube baffled	$W = 34.2$ ton/day	\$260,000	\$520,000
Filter	Rotary vacuum	$A = 47$ ft^2	\$128,000	\$256,000
Dryer	Direct-heat rotary	$A = 220$ ft^2	—	\$197,000

Solids-Handling Systems

The handling of bulk solids in a chemical process is considerably more complex than the handling of liquids and gases, and requires much more attention by operators. A typical handling system may include a storage bin, hopper, or silo; a feeder; and a conveyor and/or elevator to send the bulk solids to a piece of processing equipment. The selection of equipment for handling bulk solids depends strongly on the nature of the solids. The classification scheme of the FMC Corporation, as presented in Section 21 of *Perry's Chemical Engineers' Handbook* (Green and Perry, 2008), is particularly useful. In that scheme, bulk solids are classified by size, flowability, abrasiveness, and other special characteristics. For example, fine sodium chloride is classified as being particles between 100 mesh and 1/8 in. diameter, free flowing with an angle of repose of 30 to 45°, mildly abrasive, mildly corrosive, and hygroscopic. Titanium dioxide particles, widely used as a pigment for whiteness, are particularly difficult to handle because of their small particle size (minus 325 mesh), irregular shape, and stickiness. In general, increased moisture content, increased temperature, decreased particle size, and increased time of storage at rest cause increased cohesiveness of the particles and decreased flowability.

A typical storage vessel for bulk solids, called a *bin*, consists of an upper section with vertical walls and a lower section with at least one sloping side, referred to as a hopper. The upper part of the bin is square or circular in cross section, while the hopper is frequently conical in shape. Below the hopper is a chute, gate, or a rotary star valve. The design of a bin is best accomplished by a method devised by Andrew W. Jenike in the 1960s and described in *Perry's Chemical Engineers' Handbook* (Green and Perry, 2008). This method differentiates between the more desirable uniform mass flow with all particles moving downward and the less desirable funnel flow where all particles are not in motion and solids flow downward only in a channel in the center of the vessel. The elimination of bridging and assistance in obtaining uniform flow across the cross section of the bin can often be achieved by using a vibrating hopper. Typically, bin storage is provided for 8 hrs of operation. Most bins are constructed of carbon steel with or without rubber lining, fiberglass, or stainless steel. An equation for estimating the f.o.b. purchase cost of carbon steel bins is included in Table 22.32.

Bins may discharge bulk solids directly into a piece of processing equipment. Alternatively, the solids may be dropped onto a conveyor or into a feeder. Feeders are classified as volumetric or gravimetric. Volumetric feeders, which are the most common, discharge a volume of material per unit time, while more expensive gravimetric feeders weigh the solids being discharged. If the bulk density of the solids is reasonably constant, volumetric feeders can confine mass flow rates to within a range of 5%. Volumetric feeders include *belts* (aprons), *rotary valves*, *screws*, *tables*, and *vibratory feeders*. Screw feeders are best for sticky materials, but belt and vibratory feeders also work in some cases. Gravimetric feeders work only with free-flowing solids and include weigh belts, loss-in-weight systems, and gain-in-weight systems. Purchase-cost equations for belt, screw, and vibratory feeders are included in Table 22.32.

Bulk solids are moved to other locations by conveyors (usually horizontal) or elevators (usually vertical). A wide range of conveyors is available, but the most common are the belt, screw, and vibratory conveyors. *Belt conveyors* can move coarse, corrosive, and abrasive particles, of 100 mesh to several inches in size, for distances up to 1,000 ft, with a modest degree of inclination but at temperatures normally limited to 150°F. Typical belt widths range from 14 in. to 60 in., with belt speeds ranging from 100 to 600 ft/min. Typical heights of bulk solids on the belt range from 1 in. for the narrowest belt to 6 in. for the widest belt. Typical volumetric flow capacities range from 660 ft^3/hr for a 14-in.-wide belt moving at 100 ft/min to 86,000 ft^3/hr for a 60-in.-wide belt moving at 600 ft/min.

Screw conveyors, consisting of a screw mounted in a trough, are widely used. They can move particles of any size up to a few inches, in any straight direction, for distances up to 300 ft horizontally and up to 30 ft vertically, and at temperatures up to 900°F. Screw conveyors can be fitted with special screws for sticky materials, can be sealed to keep in dust and keep out moisture, and can be jacketed for cooling and heating. The screw, which ranges from 6 to 20 in. in diameter, typically rotates at from 25 to 100 rpm. Volumetric flow capacities, when the trough is 30% full and the screw is rotating at 50 rpm, range from 75 ft^3/hr for a 6-in.-diameter screw to 3,000 ft^3/hr for a 20-in.-diameter screw.

Vibratory conveyors are limited to straight distances, usually horizontal, up to 100 ft, but are not suited for fine particles less than 100 mesh in size. Solids must be free flowing, but temperatures up to 250°F can be handled. Widths range from 1 to 36 in. with pan heights to at least 5 in. Experiments are usually necessary to properly size a vibratory conveyor. For an 18-in.-wide conveyor with a 5-in. pan height and a 20-ft length, a typical mass flow capacity is 70,000 lb/hr or 700 ft^3/hr for solid particles having a bulk density of 100 lb/ft^3.

A *bucket elevator*, consisting of an endless chain of buckets, is best for moving solids vertically. The elevator loads at one level and discharges at another. Vertical distances of more than 3,000 ft have been spanned, but commonly available elevators are limited to 150 ft. Most often, discharge is by centrifugal force from buckets moving at speeds up to 1,200 ft/min. However, for materials that are sticky or that tend to pack, discharge is by gravity by completely inverting the buckets, which travel at lower speeds, up to 400 ft/min. Typical buckets range in width from 6 to 20 in., with bucket volumes from 0.06 to 0.6 ft^3 at bucket spacings from 1 to 1.5 ft. For a bucket speed of 150 ft/min, typical volumetric capacities range from 300 to 7,500 ft^3/hr.

Equations for estimating the purchase costs for the above four conveyor and elevator systems are included in Table 22.32. Belt and vibratory conveyors are priced by the width

Table 22.29 Power Requirements of Mechanical Conveyors[a]

Conveyor Type	Power Equation[b]
Belt	$P = 0.00058\,(m)^{0.82}L$
Screw	$P = 0.0146\,(m)^{0.85}L$
Vibratory	$P = 0.0046\,(m)^{0.72}L$
Bucket	$P = 0.020\,m(L)^{0.63}$

[a]Reproduced with permission from Ulrich (1984).

[b]Units: P = Hp, m = lb/s, L = ft. For an elevation change, h, in ft, add or subtract $P = 0.00182\,mh$.

and the length of the conveyor. Screw conveyors are priced by the diameter of the screw and the length of the conveyor. Bucket elevators are priced by the width of the bucket and the elevated height. The costs do not include the electric motor or drive system. The required horsepower input of the electric motor drive, which depends on the mass flow rate, m, and the length of the conveyor, L, may be estimated from the equations given in Table 22.29, taken from Ulrich (1984). To the equations, additional power must be added for elevating the material by a height, h. The screw conveyor requires the largest amount of power, while the belt conveyor requires the least.

EXAMPLE 22.17

Flakes of phthalic anhydride with a bulk density of 30 lb/ft^3 are to be conveyed a horizontal distance of 40 ft from a bin to a packaging facility at the rate of 1,200 ft^3/hr. Size and cost a bin and conveying system as of a CE index of 500.

SOLUTION

Assume a bin storage time of 8 hr. Therefore, the bulk solids volume is $8(1,200) = 9,600$ ft^3. Assume an outage (gas space above the bulk solids) of 20%. Thus, the bin volume above the hopper $= 9,600/(1-0.20) = 12,000$ ft^3. Neglecting the volume of the hopper below the bin and assuming a cylindrical bin with a height equal to 150% of the diameter, the hopper dimensions are 22 ft in diameter by 33 ft high. From Table 22.32, the purchase cost of the bin in carbon steel is $570(12,000)^{0.46} =$ \$42,890. Because flakes may tend to mat and interlock, consideration should be given to the addition of a vibrator to the hopper.

A screw conveyor is a reasonable choice to transport the flakes and it can also act as a feeder to remove the flakes from the hopper. From the above discussion, for a trough running 30% full at 50 rpm, a 6-in. screw can transport 75 ft^3/hr while a 20-in. screw can transport 3,000 ft^3/hr. Assume that the flow rate is proportional to the screw diameter raised to the exponent x. Then, the exponent is computed as 3. Therefore, the required screw diameter is computed as 15 in. From the equation in Table 22.32, for a length of 40 ft, the purchase cost of the screw conveyor is $70.5(15)(40)^{0.59} =$ \$9,300. The cost of the motor and a belt drive must be added. From Table 22.29, for a mass flow rate, m, equal to $(1,200/3,600)30 = 10$ lb/s and with no elevation change, the electric motor Hp is $0.0146(10)^{0.85}40 = 4.13$. Assume a 5-Hp motor. From Table 22.22, select a totally enclosed, fan-cooled motor rotating at 1,800 rpm. Thus, F_T in Eq. (22.20) is 1.3. From Eq. (22.19), for $P_C = 5$ Hp, the base cost is $C_B =$ approximately \$530, giving a purchase cost of 1.3(530) = \$690. This gives a total of \$9,990 for the conveyor with motor, or a total of \$52,880 for the bin and conveyor. Add 10% to vibrate the hopper, cover the conveyor, and add a belt drive to the motor. This gives a total purchase cost for the system of \$58,170.

Bulk solids may also be conveyed pneumatically as a dilute suspension in a gas, often air, through a piping system over distances of up to several hundred feet. Materials ranging in size from fine powders to 1/4-in.-diameter pellets and in bulk densities up to 200 lb/ft^3 have been routinely conveyed in this manner. A pneumatic conveying system usually includes a blower to move the gas, a rotary air lock valve to control the rate of addition of the bulk solids to the gas, a piping system, and a cyclone to separate the solids from the gas at the discharge point. The pressure in the piping system may be below or above ambient pressure. Air velocities required to convey the solids typically range from 50 ft/s for low-bulk-density solids to 200 ft/s for high-bulk-density solids. The purchase cost of a *pneumatic conveying* system depends mainly on the bulk solids flow rate and the equivalent length (pipe plus fittings) of the piping system, but the particle size and bulk density of the solids are also factors. Table 22.32 includes an equation for estimating the purchase cost of a system to convey solids having a bulk density of 30 lb/ft^3. The power requirement, P, in horsepower depends mainly on the solids flow rate, m, in pounds per second as estimated by:

$$P = 10\,m^{0.95} \tag{22.71}$$

Storage Tanks and Vessels

Storage tanks and vessels are used to store process liquid and gas feeds and products, as well as to provide intermediate storage between sections of a plant operating continuously or between operations for a batch or semicontinuous process. For liquid storage at pressures less than approximately 3 psig, so-called atmospheric tanks are used. These tanks may be open (no roof), cone roof, or floating-roof types. *Open tanks*, which may be made of fiberglass, are commonly used only for water and some aqueous solutions because they are subject to moisture, weather, and atmospheric pollution. *Cone-roof* (or *other fixed-roof*) *tanks* require a vent system to prevent pressure changes due to changes in temperature and during changes in liquid level during filling or emptying. When the vapor pressure of the liquid over the expected range of storage temperature causes a significant rate of evaporation, a *floating-roof* (or *other variable volume*) *tank* should be used. Such tanks do not vent. Current EPA regulations dictate the use of a floating-roof tank when, at the maximum atmospheric temperature at the plant site, the vapor pressure of the liquid is greater than 3.9 psia for storage of less than 40,000 gal or greater than 0.75 psia for storage of more than 40,000 gal.

Storage of liquid feeds and products is usually provided offsite with residence times varying from one week to one month, depending on the frequency of delivery and distribution.

The capacity of atmospheric liquid storage tanks should be at least 1.5 times the size of the transportation equipment, typically 4,000 to 7,500 gal for tank trucks and up to 34,500 gal for tank cars. A shipment by barge may be as large as 420,000 gal. The maximum size for a single cone-roof or floating-roof tank is approximately 20,000,000 gal, which corresponds to a diameter of about 300 ft and a height of about 50 ft. Storage of liquid feeds, products, and intermediates may also be provided on-site in so-called surge tanks or day tanks, which provide residence times of 10 min to one day. Equations for estimating the f.o.b. purchase costs of open, cone-roof, and floating-roof tanks are included in Table 22.32.

For liquid stored at pressures greater than 3 psig or under vacuum, spherical or horizontal (or vertical) cylindrical (bullet) pressure vessels are used. Vertical vessels are not normally used for volumes greater than 1,000 gal. Horizontal pressure vessels for storage are at least as large as 350,000 gal. Spherical pressure vessels are also common, with more than 5,000 having been constructed worldwide. For liquid storage, spheres as large as 94 ft in diameter (3,000,000 gal) have been installed. The design and costing of cylindrical pressure vessels is considered in detail in Section 22.5. Purchase costs are plotted in Figure 22.13. For spherical pressure vessels, Eq. (22.60) for cylindrical pressure vessels is revised to:

$$t_P = \frac{P_d D_i}{4SE - 0.4P_d} \qquad (22.72)$$

Equations for estimating the f.o.b. purchase costs of spherical pressure vessels, based on just the vessel volume, are included in Table 22.32 for two different pressure ranges.

Pressure vessels are also used for the storage of gases at pressures greater than 3 psig. For pressures between 0 and 3 psig, a *gas holder*, similar to a floating-roof tank for liquids, is used. An equation for estimating the f.o.b. purchase cost of a gas holder is included in Table 22.32.

Vacuum Systems

In some chemical processes, operations are conducted at pressures less than ambient. To achieve such pressures, vacuum systems are required. Pressures below ambient are commonly divided into four vacuum regions:

Vacuum Region	**Pressure Range (torr)**
Rough	760 to 1
Medium	1 to 0.001
High	0.001 to 10^{-7}
Ultrahigh	10^{-7} and below

Of greatest interest to chemical processing is the rough region, which covers most polymer reactors, vacuum distillation columns, vacuum stripping columns, pervaporation membrane separations, vacuum-swing adsorbers, and vacuum crystallizers, evaporators, filters, and dryers.

Table 22.30 Lower Limits of Suction Pressure and Capacities of Vacuum Systems[a]

System Type	Lower Limit of Suction Pressure (torr)	Volumetric Flow Range at Suction Conditions (ft^3/min)
Steam-jet ejectors		10–1,000,000
One-stage	100	
Two-stage	15	
Three-stage	2	
Liquid-ring pumps		3–18,000
One-stage water sealed	50	
Two-stage water sealed	25	
Oil-sealed	10	
Dry vacuum pumps		
Three-stage rotary-lobe	1.5	60–240
Three-stage claw	0.3	60–270
Screw compressor	0.1	50–1,400

[a]Reprinted with permission from Ryans and Bays (2001).

In the rough region, the available vacuum systems include: (1) one-, two-, and three-stage *ejectors* driven with steam and with or without interstage surface or barometric (direct-contact) condensers, (2) one- or two-stage *liquid-ring pumps* using oil or water as the sealant, and (3) *dry vacuum pumps* including rotary lobe, claw, and screw compressors. Although the first two systems have been the most widely used, dry vacuum pumps are gaining attention because they are more efficient and do not require a working fluid such as steam, water, or oil, which can contribute to air pollution. Typical flow capacities and lower limits of suction pressure for process applications of these three types of vacuum systems are given in Table 22.30, taken from Ryans and Bays (2001). This table is useful in making a preliminary selection of candidate vacuum systems based on the flow rate and pressure at suction conditions.

For batch processes where a vessel is being evacuated, the flow rate to be handled by the vacuum system depends on the selected time period for evacuation and on the contents of the vessel. When the flow contains condensables, a precondenser upstream of the vacuum system should be considered so as to reduce the flow rate to the vacuum pump. For continuous processes, the flow rate to be handled by the vacuum system is usually based on an estimate of air leakage into the equipment operating under vacuum. Air leakage occurs at gasketed joints, porous welds, and cracks and fissures in vessel walls. A simple, but often adequate estimate can be made based on the equipment volume and operating pressure with the following equation, derived from recommendations of the Heat Exchange Institute:

$$W = 5 + \{0.0298 + 0.03088[\ln(P)] - 0.0005733 \times [\ln(P)]^2\}V^{0.66} \qquad (22.73)$$

where W is the air leakage rate in lb/hr, P is the absolute operating pressure in torr assuming a barometric pressure of 760 torr, and V is the vessel volume in ft^3. For many pieces of equipment operating under a vacuum, the air leakage leaving the equipment will be accompanied by volatile process components. To partially recover these components and reduce the load on the vacuum pump, the exiting gas should first pass through a precondenser before proceeding to the vacuum system. The flow rates of process components still in the gas leaving the precondenser with the air can be determined by a flash calculation as illustrated in the example below.

Note, in Table 22.30, that steam-jet ejector systems can handle a very wide range of conditions. They have no moving parts and are inexpensive to maintain, but are very inefficient because of the high usage of motive steam. The maximum compression ratio per stage is approximately 7.5. The required motive steam rate for each stage depends on the ratio of suction pressure-to-discharge pressure, steam pressure, temperature, gas properties, and ejector nozzle-to-throat ratio. A reasonably conservative range for the total motive steam requirement for all stages, when using 100-psig steam to evacuate mostly air, is 5–10 lb of steam per pound of gas being pumped. A detailed procedure for designing an ejector vacuum system is given by Sandler and Luckiewicz (1987).

Liquid-ring pumps are limited to a suction pressure of 10 torr with a moderate capacity and efficiency (25–50%). Dry vacuum pumps can achieve very low pressures at higher efficiencies, but only for low capacities. Since vacuum pumps are actually gas compressors, a tendency exists for the gas temperature to increase in an amount corresponding to the compression ratio. However, this temperature rise is greatly minimized or eliminated in ejector systems and with the liquid-ring pump because of the addition of another fluid. The temperature rise must not be overlooked with dry vacuum pumps.

The f.o.b. purchase costs for vacuum systems are included in Table 22.32. The equation for the one-stage ejector system in carbon steel is based on indexed data from Pikulik and Diaz (1977). Use the multiplying factors in Table 22.31 to add stages and interstage condensers, and change materials of construction. For the other vacuum systems, f.o.b. purchase costs in Table 22.32 were taken from Ryans and Bays (2001).

Table 22.31 Multiplying Factors for Steam-Jet Ejector Vacuum Systems

Items	Cost Multiplying Factors
1 Stage	1.0
2 Stages	1.8
3 Stages	2.1
1 Surface condenser	1.6
2 Surface condensers	2.3
1 Barometric condenser	1.7
2 Barometric condensers	1.9
Carbon steel	1.0
Stainless steel	2.0
Hastelloy	3.0

EXAMPLE 22.18

A vacuum distillation column produces an overhead vapor of 1,365 kmol/hr of ethylbenzene and 63 kmol/hr of styrene at 30 kPa. The volume of the column, vapor line, condenser, and reflux drum is 50,000 ft^3. The overhead vapor is sent to a condenser where most of the vapor is condensed. The remaining vapor at 50°C and 25 kPa is sent to a vacuum system. Determine the air leakage rate in the distillation operation and the flow rate to the vacuum system. Select an appropriate vacuum system and determine its f.o.b. purchase cost at a CE cost index of 550.

SOLUTION

The amount of air leakage, W, is estimated from Eq. (22.73). Using a pressure of 25 kPa = 188 torr:

$$W = 5 + \{0.0298 + 0.03088[\ln(188)] - 0.0005733[\ln(188)]^2\}50{,}000^{0.66} = 227 \text{ lb/hr}$$

This is equivalent to 103 kg/hr or 3.6 kmol/hr. Adding this to the overhead vapor and performing a flash calculation at 50°C and 25 kPa (188 torr) gives a vapor leaving the reflux drum and entering the vacuum system of 3.6 kmol/hr of air and 0.7 kmol/hr of ethylbenzene. The volumetric flow rate to the vacuum system is 272 ft^3/min. The flow rate in pounds per hour is 394. From Table 22.30, applicable vacuum systems are a single-stage steam-jet ejector, a single-stage liquid-ring pump, and a screw compressor. The three-stage claw unit is just out of the range of the volumetric flow rate.

From Table 22.32, the size factor for the ejector = S = 394/188 = 2.1. From the cost equation in Table 22.32, the f.o.b. purchase cost of the ejector in carbon steel and for a cost index of 550 is

$$C_P = (550/500)(1{,}690)(2.1)^{0.41} = \$2{,}520$$

The estimated 100-psig steam consumption is 10(394) = 3,940 lb/hr. Assuming a steam cost of \$5.00/1,000 lb and operation for 8,000 hr/yr, the annual steam cost is 3.94(5.00)(8,000) = \$157,600/yr, which is far more than the cost of the ejector.

Next, consider the liquid-ring pump. From Table 22.32, with a size factor of 272 ft^3/min, in stainless steel and at CE = 550, the f.o.b. purchase cost is

$$C_P = (550/500)(8{,}250)(272)^{0.35} = \$64{,}600$$

At an overall efficiency of 40% for compression to 100 kPa, the calculated horsepower input is 16.8 or 12.6 kW. Assuming an electricity cost of \$0.05 kW/hr for 8,000 hr/yr, the annual electricity cost is 12.6(0.05)(8,000) = \$5,040/yr, which is much less than for the ejector system.

These two vacuum systems can be compared on an annualized cost basis as discussed in Section 23.4, but it seems clear that the higher cost of the liquid-ring pump is more than offset by the much higher utility cost to operate the ejector system. The screw compressor is also a candidate, but its purchase cost, \$102,000, is significantly higher and the annual electricity cost, at an overall efficiency of 70%, is only about \$2,100/yr less than for the liquid-ring pump.

Table 22.32 Purchase Costs (f.o.b.) of Other Chemical Processing Equipment, CE Index = 500. Equations for pumps, compressors, motors, heat exchangers, and pressure vessels are in Section 22.5

Equipment Type	Size Factor (S)	Range of S	f.o.b. Purchase Cost Equation (\$)	Notes
Adsorbents				
Activated alumina	Bulk volume, ft^3		$C_P = 45\,S$	
Activated carbon	Bulk volume, ft^3		$C_P = 30\,S$	
Silica gel	Bulk volume, ft^3		$C_P = 115\,S$	
Molecular sieves	Bulk volume, ft^3		$C_P = 75\,S$	
Agitators				Includes motor and shaft
Propeller, open tank	Motor Hp	1–8 Hp	$C_P = 2,300\,S^{0.34}$	Direct coupling to motor
Propeller, closed vessel	Motor Hp	1–8 Hp	$C_P = 3,300\,S^{0.17}$	Direct coupling to motor, pressures to 150 psig
Turbine, open tank	Motor Hp	2–60 Hp	$C_P = 3,290\,S^{0.54}$	Includes speed reducer
Turbine, closed vessel	Motor Hp	2–60 Hp	$C_P = 3,620\,S^{0.57}$	Includes speed reducer, pressures to 150 psig
Autoclaves				Includes turbine agitator and heat-transfer jacket
Steel	Vessel volume, gal	30–8,000 gal	$C_P = 1,045\,S^{0.52}$	Pressures to 300 psig
Stainless steel	Vessel volume, gal	30–2,000 gal	$C_P = 1,980\,S^{0.58}$	Pressures to 300 psig
Glass-lined	Vessel volume, gal	30–4,000 gal	$C_P = 1,840\,S^{0.54}$	Pressures to 100 psig
Crystallizers				
Continuous cooling				
Jacketed scraped wall	Length, L, ft	15–200 ft	$C_P = 14,500\,L^{0.67}$	Stainless steel
Continuous evaporative				
Forced circulation	Tons crystals/day, W	10–1,000 ton/day	$C_P = 34,900\,W^{0.56}$	Carbon steel
Draft-tube baffled	Tons crystals/day, W	10–250 ton/day	$C_P = 28,200\,W^{0.63}$	Carbon steel
Batch evaporative	Volume, V, ft^3	50–1,000 ft^3	$C_P = 40,900\,V^{0.41}$	Stainless steel
Drives other than electric motors				
Steam turbines (noncondensing)	Shaft power, P, Hp	250–10,000 Hp	$C_P = 9,400\,P^{0.41}$	Carbon steel
Steam turbines (condensing)	Shaft power, P, Hp	250–10,000 Hp	$C_P = 25,000\,P^{0.41}$	Carbon steel
Gas turbines	Shaft power, P, Hp	100–10,000 Hp	$C_P = 2,500\,P^{0.76}$	Carbon steel
Internal combustion engines	Shaft power, P, Hp	100–4,000 Hp	$C_P = 1,400\,P^{0.75}$	Carbon steel
Dryers				
Batch tray	Tray area, A, ft^2	20–200 ft^2	$C_P = 4,400\,A^{0.38}$	Stainless steel
Direct-heat rotary	Drum peripheral area, A, ft^2	200–3,000 ft^2	$C_P = \exp\{10.396 + 0.1003[\ln(A)] + 0.04303[\ln(A)]^2\}$	Stainless steel
Indirect-heat steam-tube rotary	Heat-transfer area, A, ft^2	100–1,400 ft^2	$C_P = 1,520\,A^{0.92}$	Stainless steel

(Continued)

Table 22.32 *(Continued)*

Equipment Type	Size Factor (S)	Range of S	f.o.b. Purchase Cost Equation ($)	Notes
Dryers (continued)				
Drum	Heat-transfer area, A, ft^2	60–480 ft^2	$C_P = 32,000 A^{0.38}$	Stainless steel
Spray	Evaporation rate, W, lb/hr	30–3,000 lb/hr	$C_P = \exp\{8.2938 + 0.8526[\ln(W)] - 0.0229[\ln(W)]^2\}$	Stainless steel
Dust collectors				
Bag filters	Gas flow rate, actual ft^3/min	5,000–2,000,000	$C_P = \exp\{10.258 - 0.4381[\ln(S)] + 0.05563[\ln(S)]^2\}$	Carbon steel
Cyclones	Gas flow rate, actual ft^3/min	200–100,000	$C_P = \exp\{9,2227 - 0.7892[\ln(S)] + 0.08487[\ln(S)]^2\}$	Carbon steel
Electrostatic precipitators	Gas flow rate, actual ft^3/min	10,000–2,000,000	$C_P = \exp\{11.680 - 0.5300[\ln(S)] + 0.05454[\ln(S)]^2\}$	Carbon steel
Venturi scrubbers	Gas flow rate, actual ft^3/min	2,000–20,000	$C_P = \exp\{9.6155 - 0.3281[\ln(S)] + 0.0500[\ln(S)]^2\}$	Carbon steel
Evaporators				
Horizontal tube	Heat-transfer area, A, ft^2	100–8,000 ft^2	$C_P = 4,060 A^{0.53}$	Carbon steel
Long-tube vertical (rising film)	Heat-transfer area, A, ft^2	100–8,000 ft^2	$C_P = 5,700 A^{0.55}$	Carbon steel
Forced circulation	Heat-transfer area, A, ft^2	150–8,000 ft^2	$C_P = \exp\{8.2986 + 0.5329[\ln(A)] - 0.000196[\ln(A)]^2\}$	Carbon steel
Falling film	Heat-transfer area, A, ft^2	150–4,000 ft^2	$C_P = 13,700 A^{0.55}$	Stainless steel tubes, carbon steel shell
Fired heaters for specific purposes				
Reformer	Heat absorbed, Q, Btu/hr	10–500 million Btu/hr	$C_P = 0.859 Q^{0.81}$	Carbon steel, pressure to 10 atm
Pyrolysis	Heat absorbed, Q, Btu/hr	10–500 million Btu/hr	$C_P = 0.650 Q^{0.81}$	Carbon steel, pressure to 10 atm
Hot water	Heat absorbed, Q, Btu/hr	0.5–70 million Btu/hr	$C_P = \exp\{9.593 - 0.3769[\ln(Q)] + 0.03434[\ln(Q)]^2\}$	
Molten salt, mineral and silicon oils	Heat absorbed, Q, Btu/hr	0.5–70 million Btu/hr	$C_P = 12.32 Q^{0.64}$	
Dowtherm A	Heat absorbed, Q, Btu/hr	0.5–70 million Btu/hr	$C_P = 12.74 Q^{0.65}$	
Steam boiler	Heat absorbed, Q, Btu/hr	0.5–70 million Btu/hr	$C_P = 0.367 Q^{0.77}$	Carbon steel, pressure to 20 atm
Heat exchangers, other				
Air-cooled fin-fan	Bare-tube heat-transfer area, A, ft^2	40–150,000 ft^2	$C_P = 2,500 A^{0.40}$	Carbon steel
Compact units				
Plate-and-frame	Heat-transfer area, A, ft^2	150–15,000 ft^2	$C_P = 8,880 A^{0.42}$	Stainless steel
Spiral plate	Heat-transfer area, A, ft^2	20–2,000 ft^2	$C_P = 6,200 A^{0.42}$	Stainless steel
Spiral tube	Heat-transfer area, A, ft^2	1–500 ft^2	$C_P = \exp\{8.0757 + 0.4343[\ln(A)] + 0.03812[\ln(A)]^2\}$	Stainless steel
Liquid-liquid extractors				
Rotating-disk contactors (RDC)	S = (Height, H, ft)(Diameter, D, ft)$^{1.5}$	3–2,000 ft$^{2.5}$	$C_P = 320 S^{0.84}$	Carbon steel

Membrane separations				
Reverse osmosis, seawater	Purified water, Q, gal/day	2–50 million gal/day	$C_{BM} = \exp\{1.802[\ln(Q)] + 0.01775[\ln(Q)]^2\}$	Bare-module cost
Reverse osmosis, brackish water	Purified water, Q, gal/day	0.2–14 million gal/day	$C_{BM} = 2.7\,Q$	Bare-module cost
Gas permeation	Membrane surface area, A, ft^2	—	$C_P = 45\,A$	Membrane module
Pervaporation	Membrane surface area, A, ft^2	—	$C_P = 38\,A$	Membrane module
Ultrafiltration	Membrane surface area, A, ft^2	—	$C_P = 10\,A$ to $25\,A$	Membrane cartridge
Mixers for powders, pastes, polymers, and doughs				
Kneaders, tilting double arm	Volume, V, ft^3	10–56 ft^3	$C_P = 1,800\,V^{0.58}$	Carbon steel
Kneaders, sigma double arm	Volume, V, ft^3	20–380 ft^3	$C_P = 1,650\,V^{0.60}$	Carbon steel
Muller	Volume, V, ft^3	10–380 ft^3	$C_P = 14,000\,V^{0.56}$	Carbon steel
Ribbon	Volume, V, ft^3	25–320 ft^3	$C_P = 2,150\,V^{0.60}$	Carbon steel
Tumblers, double cone	Volume, V, ft^3	50–270 ft^3	$C_P = 3,400\,V^{0.42}$	Carbon steel
Tumblers, twin shell	Volume, V, ft^3	35–330 ft^3	$C_P = 1,500\,V^{0.60}$	Carbon steel
Power-recovery turbines				
Gas expanders (pressure discharge)	Power extracted, P, Hp	20–5,000 Hp	$C_P = 530\,P^{0.81}$	Carbon steel
Gas expanders (vacuum discharge)	Power extracted, P, Hp	200–8,000 Hp	$C_P = 1,190\,P^{0.81}$	Carbon steel
Liquid expanders	Power extracted, P, Hp	150–2,000 Hp	$C_P = 1,400\,P^{0.70}$	Carbon steel
Screens				
Vibrating grizzlies	Surface area, A, ft^2	6–40 ft^2	$C_P = 5,800\,A^{0.34}$	Carbon steel
Vibrating screens, 1 deck	Surface area, A, ft^2	32–60 ft^2	$C_p = 1,400\,A^{0.71}$	Carbon steel
Vibrating screens, 2 decks	Surface area, A, ft^2	32–192 ft^2	$C_P = 1,230\,A^{0.78}$	Carbon steel
Vibrating screens, 3 decks	Surface area, A, ft^2	48–192 ft^2	$C_P = 890\,A^{0.91}$	Carbon steel
Size enlargement				
Disk agglomerators	Feed rate, F, lb/hr	800–80,000 lb/hr	$C_P = \exp\{10.7329 - 0.4915[\ln(F)] + 0.03648[\ln(F)]^2\}$	Carbon steel
Drum agglomerators	Feed rate, F, lb/hr	800–240,000 lb/hr	$C_P = \exp\{11.4267 - 0.5981[\ln(F)] + 0.04451[\ln(F)]^2\}$	Carbon steel
Pellet mills	Feed rate, F, lb/hr	800–80,000 lb/hr	$C_P = 7,000\,F^{0.11}$	Carbon steel
Pug mill extruders	Feed rate, F, lb/hr	80–40,000 lb/hr	$C_P = \exp\{9.4868 - 0.01453[\ln(F)] + 0.01019[\ln(F)]^2\}$	Carbon steel
Screw extruders	Feed rate, F, lb/hr	8–800 lb/hr	$C_P = \exp\{10.7928 + 0.02099[\ln(F)]^2\}$	Carbon steel
Roll-type presses	Feed rate, F, lb/hr	8,000–140,000 lb/hr	$C_P = 115\,F^{0.59}$	Carbon steel
Tableting presses	Feed rate, F, lb/hr	800–8,000 lb/hr	$C_P = \exp\{9.1570 + 0.1050[\ln(F)] + 0.01885[\ln(F)]^2\}$	Carbon steel
Size reduction equipment				
Gyratory crushers	Feed rate, W, ton/hr	25–1,200 ton/hr	$C_P = 10,500\,W^{0.60}$	Includes motor and drive
Jaw crushers	Feed rate, W, ton/hr	10–200 ton/hr	$C_P = 2,300\,W^{0.89}$	Includes motor and drive
Cone crushers	Feed rate, W, ton/hr	20–300 ton/hr	$C_P = 1,800\,W^{1.05}$	Includes motor and drive

(*Continued*)

Table 22.32 *(Continued)*

Equipment Type	Size Factor (S)	Range of S	f.o.b. Purchase Cost Equation ($)	Notes
Size reduction equipment (continued)				
Hammer mills	Feed rate, W, ton/hr	2–200 ton/hr	$C_P = 3,800\,W^{0.78}$	Includes motor and drive
Ball mills	Feed rate, W, ton/hr	1–30 ton/hr	$C_P = 57,000\,W^{0.69}$	Includes motor and drive
Jet mills	Feed rate, W, ton/hr	1–5 ton/hr	$C_P = 34,000\,W^{0.39}$	Includes motor and drive
Solid-liquid separators				
Thickener, steel	Settling area, A, ft^2	80–8,000 ft^2	$C_P = 3,360\,A^{0.58}$	Carbon steel
Thickener, concrete	Settling area, A, ft^2	8,000–125,000 ft^2	$C_P = 2,400\,A^{0.58}$	Concrete
Clarifier, steel	Settling area, A, ft^2	80–8,000 ft^2	$C_P = 3,050\,A^{0.58}$	Carbon steel
Clarifier, concrete	Settling area, A, ft^2	8,000–125,000 ft^2	$C_P = 2,160\,A^{0.58}$	Concrete
Filters				
Plate-and-frame	Filtering area, A, ft^2	130–800 ft^2	$C_P = 4,800\,A^{0.52}$	Carbon steel
Pressure leaf	Filtering area, A, ft^2	30–2,500 ft^2	$C_P = 1,220\,A^{0.71}$	Carbon steel
Rotary-drum vacuum	Filtering area, A, ft^2	10–800 ft^2	$C_P = \exp\{11.670 - 0.1905[\ln(A)] + 0.0554[\ln(A)]^2\}$	Carbon steel
Rotary pan	Filtering area, A, ft^2	100–1,100 ft^2	$C_P = 24,700\,A^{0.48}$	Carbon steel
Wet classifiers (rake and spiral)	Solids feed rate, F, lb/hr	8,000–800,000 lb/hr	$C_P = 0.016\,F^{1.33}$	Carbon steel
Hydroclones	Liquid feed rate, Q, gal/min	8–1,200 gal/min	$C_P = 240\,Q^{0.50}$	Carbon steel
Centrifuges				
Batch top-drive vertical basket	Bowl diameter, D, in.	20–43 in.	$C_P = 2,000\,D^{0.95}$	Stainless steel
Batch bottom-drive vertical basket	Bowl diameter, D, in.	20–43 in.	$C_P = 860\,D^{1.00}$	Stainless steel
Vertical auto-batch	Bowl diameter, D, in.	20–70 in.	$C_P = 5,450\,D^{0.94}$	Stainless steel
Horizontal auto-batch	Bowl diameter, D, in.	20–43 in.	$C_P = 2,150\,D^{1.11}$	Stainless steel
Continuous reciprocating pusher	Tons solids/hr, S	1–20 tons solids/hr	$C_P = 150,000\,S^{0.30}$	Stainless steel
Continuous scroll solid bowl	Tons solids/hr, S	2–40 tons solids/hr	$C_P = 60,000\,S^{0.50}$	Stainless steel
Expression				
Screw presses	Wet solids flow rate, F, lb/hr	150–12,000 lb/hr	$C_P = \exp\{10.9733 - 0.3580[\ln(F)] + 0.05853[\ln(F)]^2\}$	Stainless steel
Roll presses	Wet solids flow rate, F, lb/hr	150–12,000 lb/hr	$C_P = \exp\{10.8549 - 0.4467[\ln(F)] + 0.06136[\ln(F)]^2\}$	Stainless steel
Solids-handling systems				
Bins	Volume, ft^3	10–100,000 ft^3	$C_P = 570\,S^{0.46}$	Carbon steel at atmospheric pressure
Feeders				
Belt	Volumetric flow rate, ft^3/hr	120–500 ft^3/hr	$C_P = 717\,S^{0.38}$	Includes motor and belt drive
Screw	Volumetric flow rate, ft^3/hr	400–10,000 ft^3/hr	$C_P = 965\,S^{0.22}$	
Vibratory	Volumetric flow rate, ft^3/hr	40–900 ft^3/hr	$C_P = 41.1\,S^{0.90}$	Includes motor and belt drive

Conveyors				
Belt	Width, W, in. Length, L, ft	14–60 in., up to 300 ft	$C_P = 21.5\,WL$	Does not include motor or drive
Screw	Diameter, D, in. Length, L, ft	6–20 in., up to 300 ft	$C_P = 70.5\,DL^{0.59}$	Does not include motor, drive, lid, or jacket
Vibratory	Width, W, in. Length, L, ft	12–36 in., up to 100 ft	$C_P = 81.6\,W^{0.57}L^{0.87}$	Does not include motor or drive
Bucket elevators	Bucket width, W, in. Height, L, ft	6–20 in., 15–150 ft	$C_P = 610\,W^{0.5}L^{0.57}$	Does not include motor or drive
Pneumatic conveyors	Solids flow rate, m, lb/s Equivalent length, L, feet	3–30 lb/s, 30–600 ft	$C_P = 15,200M^{0.63}L^{0.20}$	Includes blower, motor, piping, rotary valve, and cyclone
Storage tanks				
Open	Volume, V, gal	1,000–30,000 gal	$C_P = 18\,V^{0.72}$	Fiberglass
Cone roof	Volume, V, gal	10,000–1,000,000 gal	$C_P = 265\,V^{0.51}$	Carbon steel, pressure to 3 psig
Floating roof	Volume, V, gal	30,000–1,000,000 gal	$C_P = 475\,V^{0.51}$	Carbon steel, pressure to 3 psig
Spherical, 0–30 psig	Volume, V, gal	10,000–1,000,000 gal	$C_P = 60\,V^{0.72}$	Carbon steel
Spherical, 30–200 psig	Volume, V, gal	10,000–750,000 gal	$C_P = 47\,V^{0.78}$	Carbon steel
Gas holders	Volume, V, ft^3	4,000–400,000 ft^3	$C_P = 3,170\,V^{0.43}$	Carbon steel, pressure to 3 psig
Vacuum systems				
One-stage jet ejector	(lb/hr)/(suction pressure, torr)	0.1–100 lb/hr-torr	$C_P = 1,690\,S^{0.41}$	See Table 22.31 for multistage units and condensers
Liquid-ring pumps	Flow at suction, ft^3/min	50–350 ft^3/min	$C_P = 8,250\,S^{0.35}$	Stainless steel with sealant recirculation
Three-stage lobe	Flow at suction, ft^3/min	60–240 ft^3/min	$C_P = 7,120\,S^{0.41}$	Includes intercoolers
Three-stage claw	Flow at suction, ft^3/min	60–270 ft^3/min	$C_P = 8,630\,S^{0.36}$	Includes intercoolers
Screw compressors	Flow at suction, ft^3/min	50–350 ft^3/min	$C_P = 9,590\,S^{0.38}$	With protective controls
Wastewater treatment				
Primary	Wastewater rate, Q, gal/min	75–75,000 gal/min	$C_{BM} = 14,800\,Q^{0.64}$	Bare-module cost
Primary + Secondary	Wastewater rate, Q, gal/min	75–75,000 gal/min	$C_{BM} = 43,000\,Q^{0.64}$	Bare-module cost
Primary + Secondary + Tertiary	Wastewater rate, Q, gal/min	75–75,000 gal/min	$C_{BM} = 88,000\,Q^{0.64}$	Bare-module cost

Wastewater Treatment

Wastewater can contain inorganic and organic compounds in soluble, colloidal, insoluble liquid, and solid particulate forms. Before wastewater can be sent to a sewer or converted to drinking water, process water, boiler-feed water, or cooling water, it must be treated to remove certain impurities. Such treatment may consist of as many as three major treatment steps: primary, secondary, and tertiary. Primary treatment involves physical separation operations such as screening to remove large solids and sedimentation to remove smaller particulate matter, which settles to the bottom, and insoluble organic liquid, which floats to the top and is skimmed. Secondary treatment removes dissolved organic compounds by biological degradation with aerobic or anaerobic microorganisms in a recirculating activated sludge. This may produce settleable solids, which are removed by filtration. Removal of nitrogen and phosphorus nutrients, residual organic compounds, and dissolved inorganic compounds is accomplished in a tertiary treatment, which involves such operations as carbon adsorption, demineralization, and reverse osmosis. The water may also be disinfected with chlorine, ozone, or ultraviolet light. Equations for typical investment costs for wastewater treatment are included in Table 22.32. These are bare-module costs, rather than f.o.b. purchase costs.

22.7 EQUIPMENT SIZING AND CAPITAL COST ESTIMATION USING THE ASPEN ICARUS PROCESS EVALUATOR (IPE)

This is Section 22S.1 in the file Supplement_to_Chapter_22.pdf in the PDF File folder, which can be downloaded from the Wiley Web site associated with this book. It introduces Aspen IPE and shows how to estimate equipment sizes, purchase costs, installation costs, and the total permanent investment. Two examples are provided:

www.wiley.com/go/global/seider

1. ***Depropanizer distillation tower.*** This tower is presented in the multimedia modules (either HYSYS → *Separations* → *Distillation* or ASPEN PLUS → *Separations* → *Distillation*). Reference is made to the design procedure, which is carried out prior to the estimation of equipment sizes and costs. Beginning with the ASPEN PLUS file RADFRAC.bkp (in the Program and Simulation Files folder, which can be downloaded from the Wiley Web site associated with this book), Aspen IPE maps the distillation unit (that is, estimates equipment sizes for the column, condenser, reflux accumulator, condenser pump, reboiler, and reboiler pump) and estimates its purchase and installation costs.
2. ***Monochlorobenzene separation process.*** This process was introduced in Section 5.4, with simulation results using ASPEN PLUS provided in the multimedia modules (ASPEN PLUS → *Principles of Flowsheet Simulation* → *Interpretation of Input and Output* → *Sample Problem*). Beginning with the file MCB.bkp, the equipment sizes, purchase costs, and installation costs are estimated using Aspen IPE.

22.8 SUMMARY

Having completed this chapter and some of the associated exercises, the reader should

1. Be able to assess the financial condition of a company by applying financial ratio analysis to data given in its annual report.
2. Be able to estimate the purchase costs of equipment items using the provided equations together with cost indexes to update those costs.
3. Be able to estimate each of the other costs included in the capital cost of a plant and apply the concept of the bare-module cost.
4. Be able to estimate the total capital investment of a plant by three methods of increasing complexity that range from order-of-magnitude to preliminary estimates.
5. Be prepared to use the Aspen Icarus Process Evaluator (IPE) system provided by Aspen Technology, Inc., to prepare a more definitive estimate of capital cost.

REFERENCES

1. Bauman, H.C., *Fundamentals of Cost Engineering in the Chemical Industry*, Reinhold, (1964).

2. Busche, R.M., *Venture Analysis: A Framework for Venture Planning—Course Notes*, Bio-en-gene-er Associates, Wilmington, Delaware, 1995.

3. Chilton, E.H., *Chemical Engineering*, **56** (6), 97–106, June (1949).

4. Corripio, A.B., K.S. Chrien, and L.B. Evans, "Estimate Costs of Centrifugal Pumps and Electric Motors," *Chem. Eng.*, **89**, 115–118, February 22 (1982a).

5. Corripio, A.B., K.S. Chrien, and L.B. Evans, "Estimate Costs of Heat Exchangers and Storage Tanks via Correlations,," *Chem. Eng.*, **89**, 125–127, January 25 (1982b).

6. FELDMAN, R.P., "Economics of Plant Startups," *Chem. Eng.*, **76**, 87–90, November 3 (1969).

7. GARRETT, D.E., *Chemical Engineering Economics*, Van Nostrand-Rineholt, New York, 1989.

8. GREEN, D.W., and R.H. PERRY, Eds., *Perry's Chemical Engineers' Handbook*, 8th ed., McGraw-Hill, New York, 2008.

9. GUTHRIE, K.M., "Data and Techniques for Preliminary Capital Cost Estimating," *Chem. Eng.*, **76**, 114–142, March 24 (1969).

10. GUTHRIE, K.M., "Capital and Operating Costs for 54 Chemical Processes," *Chem. Eng.*, **77** (13), 140 (1970).

11. GUTHRIE, K.M., *Process Plant Estimating, Evaluation, and Control*, Craftsman, Solano Beach, California, 1974.

12. HALL, R.S., J. MATLEY, and K.J. MCNAUGHTON, *Chemical Engineering*, **89** (7), 80–116, April 5, (1982).

13. HAND, W.E., "From Flow Sheet to Cost Estimate," *Petroleum Refiner*, **37** (9), 331–334, (Sept. 1958).

14. HASELBARTH, J.E., "Updated Investment Cost for 60 Types of Chemical Plants," *Chem. Eng.*, **74** (25), 214–215, December 4 (1967).

15. HILL, R.D., "What Petrochemical Plants Cost," *Petroleum Refiner*, **35** (8), 106–110, (Aug. 1956).

16. LANG, H.J., "Engineering Approach to Preliminary Cost Estimates," *Chem. Eng.*, **54** (9), 130–133, (1947a).

17. LANG, H.J., "Cost Relationship in Preliminary Cost Estimation," *Chem. Eng.*, **54** (10), 117–121, (1947b).

18. LANG, H.J., "Simplified Approach to Preliminary Cost Estimates," *Chem. Eng.*, **55** (6), 112–113, (1948).

19. MILLS, H.E., *Chemical Engineering*, **71** (6), 133, March 16, (1964).

20. MULET, A., A.B. CORRIPIO, and L.B. EVANS, "Estimate Costs of Pressure Vessels via Correlations," *Chem. Eng.*, **88** (20), 145–150, October 5 (1981a).

21. MULET, A., A.B. CORRIPIO, and L.B. EVANS, "Estimate Costs of Distillation and Absorption Towers via Correlations," *Chem. Eng.*, **88** (26), 77–82, December 28 (1981b).

22. PAPANASTASIOU, T.C., *Applied Fluid Mechanics*, Prentice-Hall, Englewood Cliffs, New Jersey, 1994.

23. PETERS, M.S., K.D. TIMMERHAUS, and R.E. WEST, *Plant Design and Economics for Chemical Engineers*, 5th ed., McGraw-Hill, New York (2003).

24. PIKULIK, A., and H.E. DIAZ, *Chemical Engineering*, **84** (21), 107–116, October 10 (1977).

25. RYANS, J., and J. BAYS, "Run Clean with Dry Vacuum Pumps," *Chem. Eng. Prog.*, **97** (10), 32–41, (2001).

26. SANDLER, H.J., and E.T. LUCKIEWICZ, *Practical Process Engineering*, McGraw-Hill, New York (1987).

27. TURTON, R., R.C. BAILLE, W.B. WHITING, and J.A. SHAEIWITZ, *Analysis, Synthesis, and Design of Chemical Processes*, 2nd ed., Prentice Hall, New Jersey (2003).

28. ULRICH, G.D., *A Guide to Chemical Engineering Process Design and Economics*, John Wiley & Sons, Inc., New York (1984).

29. ULRICH, G.D., and P.T. VASUDEVAN, *Chemical Engineering Process Design and Economics—A Practical Guide*, 2nd ed., Process Publishing, New Hampshire (2004).

30. WALAS, S.M., and C.D. SPANGLER, *Chemical Engineering*, **67** (6), 173–176, March 21, (1960).

31. WALAS, S.M., *Chemical Process Equipment*, Butterworth, London (1988).

32. WILLIAMS, R., "Standardizing Cost Data on Process Equipment," *Chem. Eng.*, **54** (6), 102, June (1947a).

33. WILLIAMS, R., "Six-Tenths Factor Aids in Approximating Costs," *Chem. Eng.*, **54** (12), 124–125, December (1947b).

34. WOODS, D.R., *Financial Decision Making in the Process Industry*, Prentice-Hall, Englewood Cliffs, NJ (1975).

EXERCISES

22.1 On the Internet, find a recent annual report for the company Merck & Co., Inc. As of August 2002, the 1998 report was available in PDF format. Based on information in that report, determine the following:

(a) The nature of the business of Merck.
(b) The new developments by Merck.
(c) The new acquisitions or partnerships, if any.
(d) Stated concerns of the company.
(e) A financial ratio analysis, including:

1. Current ratio.
2. Acid-test ratio.
3. Equity ratio.
4. Return on total assets.
5. Return on equity.
6. Operating margin.
7. Profit margin.

(f) Your stock purchase recommendation, including reasons for buying or selling the stock.

22.2 At the beginning of the year 2006, Company XYZ had an inventory of 8,000 widgets with a unit cost of $6.00. During that year, the following purchases of widgets were made:

Month of Purchase	Number of Units	Unit Cost ($)
February	10,000	$7.00
May	5,000	$8.00
June	15,000	$9.00
August	25,000	$10.00
November	20,000	$10.50

At the end of 2006, the number of units in the inventory was 2,900. Use both the fifo and lifo methods to determine the cost of goods sold for 2006.

22.3 Based on the following data for the reactors, compressors/expanders, and distillation columns of a plant to produce 1,500 metric ton/day of methanol, with an operating factor of 0.95, estimate by Method 1 (order-of-magnitude estimate) the total capital investment. The two methanol reactors are of the shell-

Equipment	Size	Material	Pressure (kPa)	Temperature (°C)
Steam reformer	620 million Btu/hr	316 ss	2,000	350
2 Methanol reactors	Each with 4,000 1.5-in. o.d. tubes by 30-ft long on 2.25-in. triangular pitch	cs shell 316 ss tubes	6,000	320
Reformed gas centrifugal compressor	16,000 kW	cs	6,000	200
Recycle gas centrifugal compressor	5,000 kW	cs	6,000	200
Tail-gas expander	4,500 kW	cs	6,000	200
Light ends tower	3-ft diameter 60 sieve trays	ss	500	200
Finishing tower	18-ft diameter 80 sieve trays	ss	200	200

Note: ss = stainless steel, cs = carbon steel

and-tube type. The plant will be constructed outdoors and is a major addition to existing facilities. Use a CE cost index of 550.

22.4 The feed to a sieve-tray distillation column operating at 1 atm is 700 lbmol/hr of 45 mol% benzene and 55 mol% toluene at 1 atm, with a bubble-point temperature of 201°F. The distillate contains 91.6 mol% benzene and boils at 179.4°F. The bottoms product contains 94.6 mol% toluene and boils at 226.6°F. The column has 23 trays spaced 18 in. apart, and its reflux ratio is 1.25. Column pressure drop is neglected. Estimate the total bare-module cost of the column, condenser, reflux accumulator, combined reflux and distillate pump, reboiler, and reboiler pump. Also, estimate the total permanent investment using either the Lang or Guthrie methods. Compute the results using: (1) the equations in Chapter 22, and (2) ASPEN IPE (Icarus Process Evaluator). Compare the results.

Data

Overall U of total condenser = 100 Btu/hr-ft^2-°F

Cooling water from 90°F to 120°F

Reboiler heat flux to avoid film boiling = 12,000 Btu/hr-ft^2

Reflux accumulator residence time = 5 min at half full. $L/D = 2$.

Centrifugal pump pressure rise = 100 psi (for each pump). Suction pressure = 1 atm.

Use sieve trays with $A_h/A_a = 0.1$.

Calculate the flooding velocity of the column using the procedure in Example 19.3. Use 85% of the flooding velocity to determine the column diameter.

Saturated steam available at 60 psia.

Notes

The file BENTOLDIST.bkp is included in the Program and Simulation Files folder, which can be downloaded from the Wiley Web site associated with this book. It contains the simulation results using the RADFRAC subroutine in ASPEN PLUS. This file should be used to determine the physical properties, flow rates, and heat exchanger duties needed for the above calculations. Also, the file should be used to prepare the report file for Aspen IPE. Note that the simulation was carried out using 18 trays at 100% efficiency. When using Aspen IPE, set the tray efficiency to 0.8 and Aspen IPE will adjust the number of trays to 23.

Since Aspen IPE does not size and cost a bottoms pump, a centrifugal pump should be added.

Aspen IPE estimates the physical properties and heat-transfer coefficients. Do not adjust these.

In Aspen IPE, reset the temperatures of cooling water (90°F to 120°F) and add a utility for 60 psia steam. Use steam tables to estimate the physical properties.

Use a kettle reboiler with a floating head.

Aspen IPE sizes the tower using a 24-in. tray spacing as the default. After sizing (mapping) is complete, adjust the tray spacing to 18 in. Note that the height of the tower must be adjusted accordingly.

Note that Aspen IPE estimates the *Direct Material and Manpower* for each equipment item. These are also referred to as the cost of direct materials and labor, $C_{DML} = C_P + C_M + C_L$.

22.5 Figure 22.14 shows a system designed to recover argon from the purge stream in an ammonia synthesis plant. Estimate the total bare-module cost associated with the addition of this argon recovery system to an existing plant. Assuming no allocated costs for utilities and related facilities, estimate the direct permanent investment, the total depreciable capital, and the total permanent investment for the process. Include only the equipment shown in the flowsheet and specified below. Use CE = 550.

Equipment Specifications

Molecular sieve adsorbers (ignore packing):

	A1	A2
Diameter (ft)	10	10
Height (ft)	7	7
Pressure (psia)	2,000	1,000
Material	s.s.	s.s.

www.wiley.com/go/global/seider

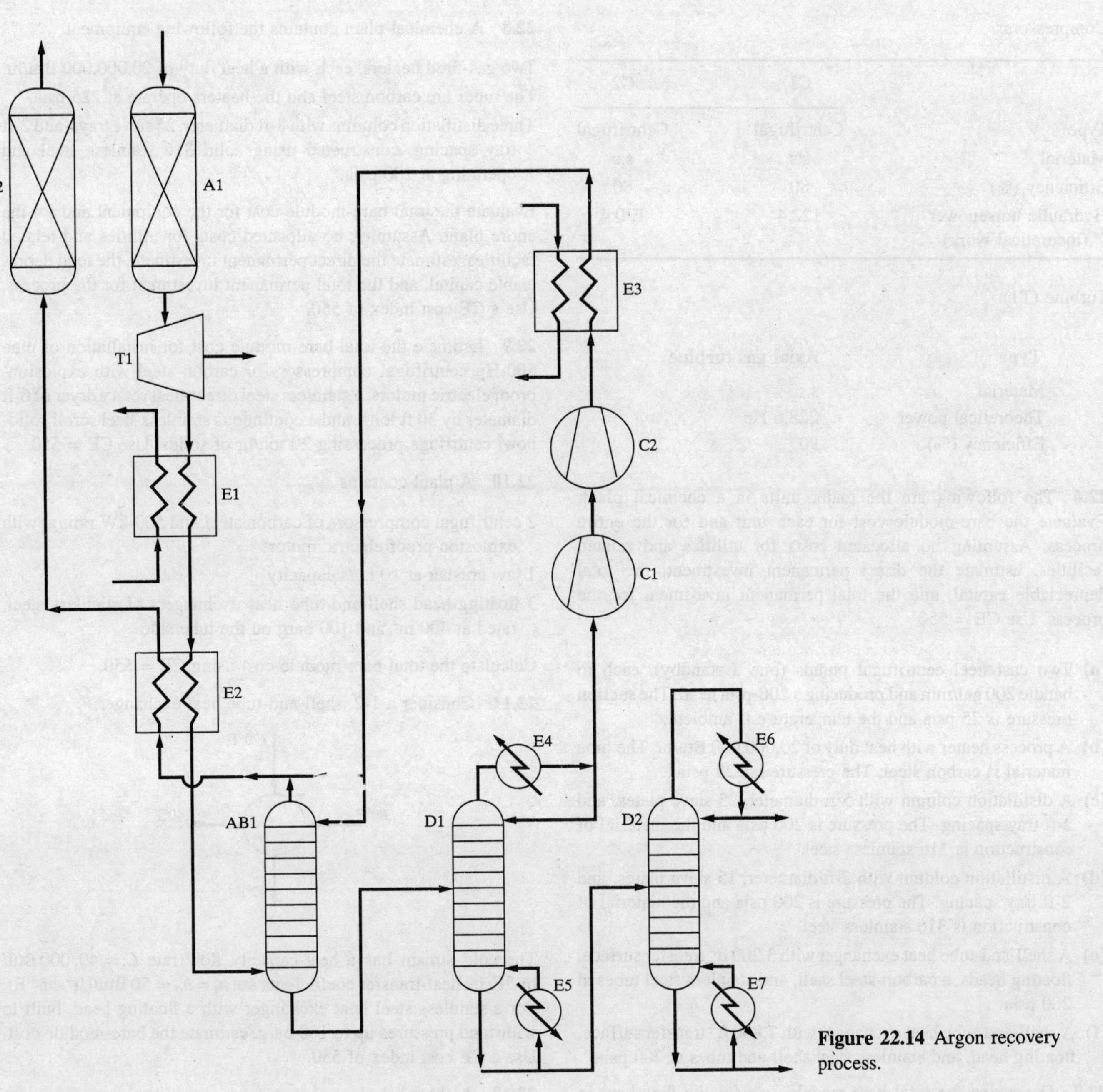

Figure 22.14 Argon recovery process.

Absorber and distillation columns:

	AB1	D1	D2
Material	s.s.	s.s.	s.s.
Diameter (ft)	2	3.5	1.7
Height (ft)	45	50	15
Pressure (psia)	1,000	70	14.7
No. of trays	40	45	12
Tray spacing (ft)	1	1	1
Tray type	Sieve	Sieve	Sieve

Heat exchangers, reboilers, and condensers:

	E1	E2	E3	E4	E5	E6	E7
Area (ft^2)	993	401	891	2,423	336	2,828	55
Pressure (psia)	1,000	1,000	1,000	70	70	14.7	14.7

E1, E2, E3, E4, and E6 are shell-and-tube, floating-head, stainless steel heat exchangers.
E5 and E7 are stainless steel kettle reboilers.

Compressors:

	C1	C2
Type	Centrifugal	Centrifugal
Material	s.s	s.s
Efficiency (%)	80	80
Hydraulic horsepower (theoretical work)	122.4	130.4

Turbine (T1):

Type	Axial gas turbine
Material	s.s
Theoretical power	228.6 Hp
Efficiency (%)	90

22.6 The following are the major units in a chemical plant. Evaluate the bare-module cost for each unit and for the entire process. Assuming no allocated costs for utilities and related facilities, estimate the direct permanent investment, the total depreciable capital, and the total permanent investment for the process. Use CE = 550.

(a) Two cast-steel centrifugal pumps (one a standby), each to handle 200 gal/min and producing a 200-psia head. The suction pressure is 25 psia and the temperature is ambient.

(b) A process heater with heat duty of 20,000,000 Btu/hr. The tube material is carbon steel. The pressure is 225 psia.

(c) A distillation column with 5-ft diameter, 25 sieve plates, and 2-ft tray spacing. The pressure is 200 psia and the material of construction is 316 stainless steel.

(d) A distillation column with 2-ft diameter, 15 sieve plates, and 2-ft tray spacing. The pressure is 200 psia and the material of construction is 316 stainless steel.

(e) A shell-and-tube heat exchanger with 3,200 ft^2 transfer surface, floating heads, a carbon-steel shell, and stainless steel tubes at 200 psia.

(f) A shell-and-tube heat exchanger with 7,800 ft^2 transfer surface, floating head, and stainless steel shell and tubes at 200 psia.

22.7 Determine the total bare-module cost for the flowsheet in Figure 22.15 at ambient temperature and pressure. Use CE = 550.

Design Specifications

Pumps: reciprocating, motor driven, 25 psia head

Heat exchangers: floating-head, $\Delta T_{LM} = 30°F$, $U = 15$ Btu/hr-ft^2-°F

Reslurry vessel and crystallizer: vertical, with $H/D = 2$

All equipment: carbon steel

22.8 A chemical plant contains the following equipment:

Two gas-fired heaters, each with a heat duty of 20,000,000 Btu/hr. The tubes are carbon steel and the heaters operate at 225 psia.

Three distillation columns with 8-ft diameter, 25 sieve trays, and 2-ft tray spacing, constructed using solid 316 stainless steel and operating at 200 psia.

Evaluate the total bare-module cost for the equipment and for the entire plant. Assuming no allocated costs for utilities and related facilities, estimate the direct permanent investment, the total depreciable capital, and the total permanent investment for the process. Use a CE cost index of 550.

22.9 Estimate the total bare-module cost for installation of nine 600-Hp centrifugal compressors of carbon steel with explosion-proof electric motors, a stainless steel direct-heat rotary dryer of 6-ft diameter by 30 ft long, and a continuous stainless steel scroll solid-bowl centrifuge processing 20 ton/hr of solids. Use CE = 550.

22.10 A plant contains

2 centrifugal compressors of carbon steel and 500-kW rating, with explosion-proof electric motors

1 jaw crusher at 10 kg/s capacity

3 floating-head shell-and-tube heat exchangers of stainless steel, rated at 400 m^2 and 100 barg on the tube side

Calculate the total bare-module cost using CE = 550.

22.11 Consider a 1-2, shell-and-tube heat exchanger:

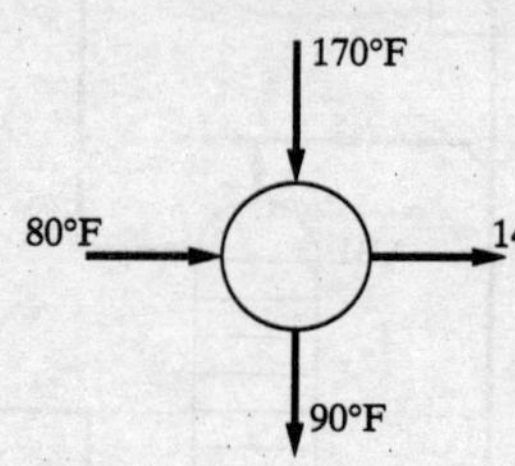

The cold stream has a heat-capacity flow rate $C = 40{,}000$ Btu/hr-°F. Its heat-transfer coefficients are $h_i = h_o = 50$ Btu/(ft^2-hr-°F). For a stainless steel heat exchanger with a floating head, built to withstand pressures up to 100 barg, estimate the bare-module cost. Use a CE cost index of 550.

22.12 A chemical plant contains

3 drum dryers of nickel alloy, each containing 540 ft^2

40 kettle reboilers, with a carbon-steel shell and copper tubes, at 1,450 psia, each containing 325 ft^2

Calculate the total bare-module cost for CE = 550.

22.13 Toluene Hydrodealkylation Process—Capital Cost Estimation. See Exercise 23.21 for a complete economic analysis, including equipment sizing, cost estimation, and calculation of the total capital investment.

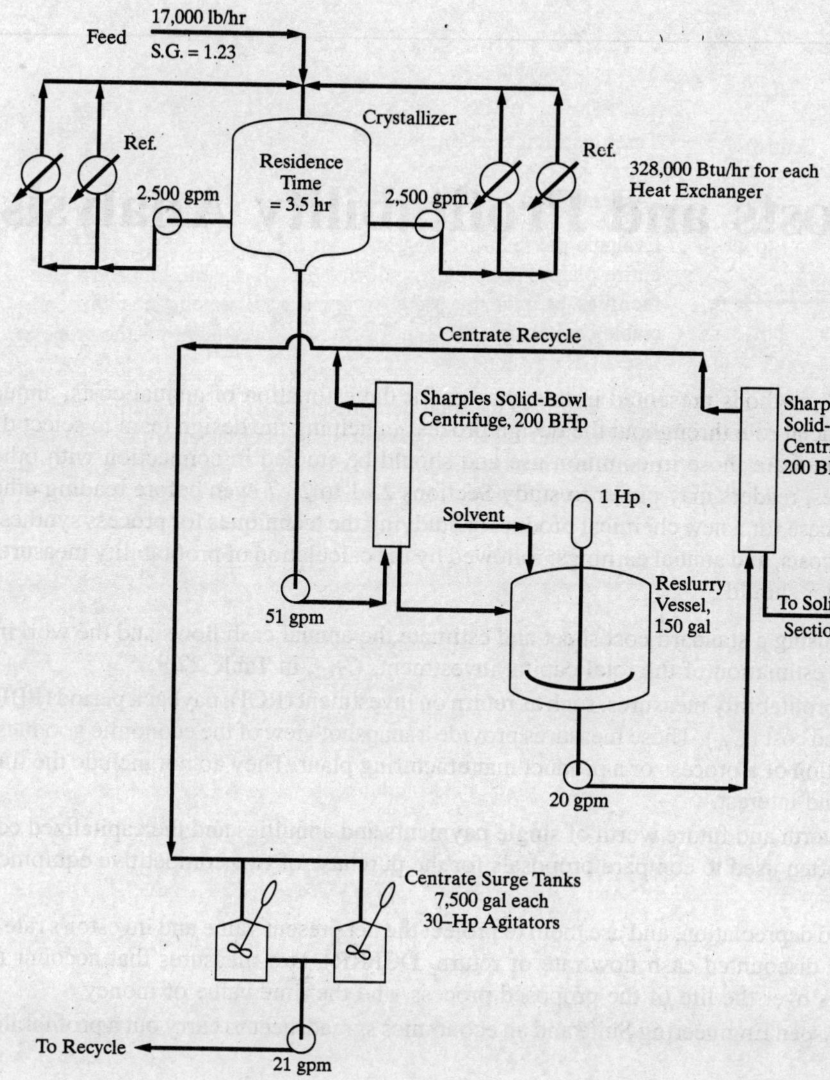

Figure 22.15 Flowsheet for Exercise 22.7.

Chapter 23

Manufacturing Costs and Profitability Analysis

23.0 OBJECTIVES

Like the capital cost estimates of Chapter 22, methods presented in this chapter for the estimation of annual costs, annual earnings, and profitability measures play a crucial role throughout the design process in helping the design team to select the best design alternatives. The methods presented are those in common use and should be studied in connection with other chapters in this book as needed. In many cases, readers may prefer to study Sections 23.1 to 23.7 even before reading other chapters, especially when creating a business case for a new chemical product or studying the techniques for process synthesis that require estimates of capital costs, annual costs, and annual earnings, followed by the calculation of profitability measures.

After studying this chapter, the reader should

1. Be able to estimate annual costs using a standard cost sheet and estimate the annual cash flows and the working capital. The latter completes the estimation of the total capital investment, C_{TCI}, in Table 22.9.
2. Be able to compute approximate profitability measures, such as return on investment (ROI), payback period (PBP), venture profit (VP), and annualized cost (C_A). These measures provide a snapshot view of the economic goodness, usually in the third year of operation of a process or a product manufacturing plant. They do not include the time value of money, that is, compound interest.
3. Be able to compute the present worth and future worth of single payments and annuities and the capitalized cost perpetuity. These measures are often used to compare proposals for the purchase of two competitive equipment items.
4. Be able to compute cash flows and depreciation, and use them to project the net present value and investor's rate of return (IRR) (also known as the discounted cash flow rate of return, DCFRR), two measures that account for projections of revenues and costs over the life of the proposed process, and the time value of money.
5. Be able to use Aspen IPE in the Aspen Engineering Suite and an economics spreadsheet to carry out a profitability analysis for a potential process.

23.1 INTRODUCTION

Having completed an estimate for the total permanent investment, C_{TPI}, in Table 22.9, of a proposed plant or of a proposed chemical product manufacturing facility, it remains to estimate the total annual sales revenue, S, the total annual production cost, C, and the annual pre-tax and after-tax earnings. This includes the development of the so-called *Cost Sheet*. Then the working capital can be estimated and added to the total permanent investment to give the total capital investment for the plant or product manufacturing facility, as shown in Table 22.9. These provide the ingredients for an approximate measure of economic goodness, called the *return on investment,* defined by

$$\text{Return on Investment} = \frac{\text{annual earnings}}{\text{capital investment}} \tag{23.1}$$

which is generally stated as a percentage per year. This definition includes a number of alternatives, depending on whether the annual earnings are before or after taxes and whether the capital investment includes land and working capital. The most common alternative for return on investment is based on the annual earnings after taxes and the total capital investment. This alternative is referred to here as ROI. A new commercial venture must compete with the *commercial interest rate* (or *cost of capital), i,* which is the annual rate at which money is returned to investors for the use of their capital, say in the purchase of high-grade bonds. The commercial interest rate is considered to be essentially without risk. Investments in chemical processing plants and product manufacturing facilities always entail *risk*. Therefore, to be attractive, an investment in a venture involving a new or revamped chemical processing plant or product manufacturing facility must have an ROI greater than *i*. The greater the risk of the venture, the greater must be the difference between a financially attractive ROI and *i*. Establishment of the degree of risk involves answers to the following questions:

1. Is a new chemical product to be produced? If so, are uses for it established and is there a sure market for it at its projected price?
2. Is an already commercially available chemical product to be produced? If so, is the new plant going to utilize new technology that is predicted to reduce investment and/or operating costs? If so, how certain is the new technology? Does the technology involve potential environmental, safety, and/or control issues? Does the technology involve uncertainties with respect to materials of construction? If the new plant is going to use established, mature technology, is future demand for the chemical predicted to be greater than the current supply?
3. For the new plant, is the availability of the feedstocks (raw materials) ensured at a known price, or are the feedstocks controlled and/or produced by the company installing the new process or product manufacturing facility?

As an example, suppose the current commercial interest rate is 10%. The proposed venture involves the manufacture of a new chemical product at conditions of high temperature and pressure using new technology. The uses of the new chemical product have been established and buyers have signed contracts to purchase the new chemical product at an agreed-upon price. The feedstocks are produced by and available from the company that will produce the new chemical product. This degree of risk might be considered moderate, requiring an (ROI – *i*) of 15% or an ROI of at least 25%. A high-risk venture might require an (ROI – *i*) of 50%.

This chapter begins with the methods for estimating the remaining elements of the approximate ROI measure, as well as other comparable measures such as the *venture profit* (VP), *payback period* (PBP), and *annualized cost*(C_A), which are utilized often to compare alternatives during the early stages of product/process design, particularly during process synthesis. Rigorous profitability measures, which involve consideration of the *time value of money* and estimates of the *cash flows* throughout the life of the proposed product/process, are applied before making a final decision on a project when a company must assess carefully how it expends its limited capital. These measures, which include the *net present value* (NPV) and the *investor's rate of return* (IRR) (also referred to as the *discounted cash flow rate of return,* DCFRR), incorporate one of a number of equipment depreciation schedules, based on U.S. tax laws, and account for the time value of money over the life of the product/process. These rigorous measures permit the design team to account for anticipated changes as well—for example, the need to replace the catalyst charge every 4 years, or the recognition that in 7 years the company patent will expire and the selling price will be reduced, and so on. To initiate the discussion of these rigorous measures, several subjects that involve the time value of money are discussed, including compound interest, annuities, and perpetuities such as capitalized costs. The effects of depreciation, inflation, depletion, and salvage value at the end of the life of a product and/or processing plant are also discussed.

The chapter concludes with a discussion of the use of the Aspen Icarus Process Evaluator (IPE) in the Aspen Engineering Suite and an economics spreadsheet to calculate profitability measures. As described in the previous chapter, Aspen IPE can be used in connection with a simulation program, such as ASPEN PLUS, CHEMCAD, HYSYS, or PRO/II, and it can be used independently.

23.2 ANNUAL SALES REVENUES, PRODUCTION COSTS, AND THE COST SHEET

Many continuing costs are associated with the operation of a chemical plant or product manufacturing facility. These are included in the cost sheet shown in Table 23.1, which is patterned after one prepared by Busche (1995) and includes representative unit costs (typical factors) that can be used for early estimates when more exact costs are not available.

Sales Revenue

Before estimating the annual costs listed on the Cost Sheet, the total annual sales revenue, S, should be estimated. If S is not greater than the costs of the feedstock(s), there is no need to consider the process further. Typically, this calculation is made early, during the *concept* stage of the Stage-Gate™ Product-Development Process (SGPDP). When a chemical process is being designed, this is the preliminary process synthesis step, as discussed in Section 4.4 and shown in Table 4.3 for the development of a vinyl-chloride process, when five different reaction paths are being considered. As the process design proceeds, this calculation needs to be repeated after conversions and yields are better established by process design. The total sales revenue is based on the unit selling

Table 23.1 Cost Sheet Outline[a]

Cost Factor	Typical Factor in American Engineering Units	Typical Factor in SI Units
Feedstocks (raw materials)		
Utilities		
Steam, 450 psig	\$6.60/1,000 lb	\$14.50/1,000 kg
Steam, 150 psig	\$4.80/1,000 lb	\$10.50/1,000 kg
Steam, 50 psig	\$3.00/1,000 lb	\$6.60/1,000 kg
Electricity	\$0.060/kW-hr	\$0.060/kW-hr
Cooling water (cw)	\$0.075/1,000 gal	\$0.020/m^3
Process water	\$0.75/1,000 gal	\$0.20/m^3
Boiler-feed water (bfw)	\$1.80/1,000 gal	\$0.50/m^3
Refrigeration, −150°F	\$3.80/ton-day	\$12.60/GJ
Refrigeration, −90°F	\$3.10/ton-day	\$10.30/GJ
Refrigeration, −30°F	\$2.40/ton-day	\$7.90/GJ
Refrigeration, 10°F	\$1.70/ton-day	\$5.50/GJ
Chilled water, 40°F	\$1.20/ton-day	\$4.00/GJ
Natural gas	\$3.20/1,000 SCF	\$0.136/SCM
Fuel oil	\$1.50/gal	\$400/m^3
Coal	\$60/ton	\$66/1,000 kg
Wastewater treatment	\$0.15/lb organic removed	\$0.33/kg organic removed
Landfill	\$0.08/dry lb	\$0.17/drykg
Operations (labor-related) (O) (See Table 23.3)		
Direct wages and benefits (DW&B)	\$35/operator-hr	\$35/operator-hr
Direct salaries and benefits	15% of DW&B	15% of DW&B
Operating supplies and services	6% of DW&B	6% of DW&B
Technical assistance to manufacturing	\$60,000/(operator/shift)-yr	\$60,000/(operator/shift)-yr
Control laboratory	\$65,000/(operator/shift)-yr	\$65,000/(operator/shift)-yr
Maintenance (M)		
Wages and benefits (MW&B)		
Fluid handling process	3.5% of C_{TDC}	3.5% of C_{TDC}
Solids–fluids handling process	4.5% of C_{TDC}	4.5% of C_{TDC}
Solids-handling process	5.0% of C_{TDC}	5.0% of C_{TDC}
Salaries and benefits	25% of MW&B	25% of MW&B
Materials and services	100% of MW&B	100% of MW&B
Maintenance overhead	5% of MW&B	5% of MW&B
Operating overhead		
General plant overhead	7.1% of M&O-SW&B	7.1% of M&O-SW&B
Mechanical department services	2.4% of M&O-SW&B	2.4% of M&O-SW&B
Employee relations department	5.9% of M&O-SW&B	5.9% of M&O-SW&B
Business services	7.4% of M&O-SW&B	7.4% of M&O-SW&B
Property taxes and insurance	2% of C_{TDC}	2% of C_{TDC}
Depreciation (see also Section 23.6)		
Direct plant	8% of ($C_{TDC} - 1.18C_{alloc}$)	8% of ($C_{TDC} - 1.18C_{alloc}$)
Allocated plant	6% of $1.18C_{alloc}$	6% of $1.18C_{alloc}$
Rental fees (Office and lab space)	(no guideline)	(no guideline)
Licensing fees	(no guideline)	(no guideline)
COST OF MANUFACTURE (COM)	Sum of above	Sum of above
General Expenses		
Selling (or transfer) expense	3% (1%) of sales	3% (1%) of sales
Direct research	4.8% of sales	4.8% of sales
Allocated research	0.5% of sales	0.5% of sales
Administrative expense	2.0% of sales	2.0% of sales
Management incentive compensation	1.25% of sales	1.25% of sales
TOTAL GENERAL EXPENSES (GE)		
TOTAL PRODUCTION COST (C)	COM + GE	COM + GE

[a]DW&B = direct wages and benefits; MW&B = maintenance wages and benefits; M&O-SW&B = maintenance and operations salary, wages, and benefits. See Table 22.9 for C_{TDC} and C_{alloc}. 1 ton of refrigeration = 12,000 Btu/hr.

Source: Busche (1995) with modifications.

price(s) and on the quantity of product(s) produced for sale. If the process produces more than one main product, such as in Table 22.8, where plants produce ethylene–propylene, ammonia–urea, and chlorine–sodium hydroxide, the total sales revenue can include both products as co-products. Otherwise, additional products can be considered byproducts, for which an annual credit can be taken toward the cost of manufacture. Other possible credits include (1) gas, liquid, or solid effluents that can be used for fuel, (2) steam produced from boiler-feed water, and (3) electrical energy produced from a gas expander (turbine). If the streams with fuel value, steam, and/or electricity are used within the process, then the credit will be automatically accounted for. Otherwise, if to be used elsewhere, a transfer cost can be assigned to determine the credit against the cost of manufacture. The quantity of product(s) is obtained from the process design material balance and the estimated plant-operating factor or annual hours of plant operation.

Feedstocks

A major consideration in determining the cost of manufacture are the costs of the feedstocks, which may be natural resources such as petroleum, commodity chemicals such as chlorine, or fine chemicals. For product designs, these can be industrial and/or configured consumer products, such as peristaltic pumps, tubing, storage tanks, and flow meters, as in the home hemodialysis product (see Section 17.3). In the production of commodity chemicals, feedstock costs can be a significant contribution to the cost of manufacture, often in the range of 40 to 60% and even higher. The required quantity of feedstock is obtained from the process design material balance and the estimated plant-operating factor. For example, in the earlier discussion in Chapter 4 of the production of vinyl chloride at the rate of 100,000 lb/hr, the process design material balance gives a chlorine feedstock flow rate of 113,400 lb/hr. If the plant-operating factor is based on 330 days/yr (operating factor $= 330/365 = 0.904$), the annual chlorine flow rate is $113{,}400(24)(365)(0.904) = 898.02$ million lb/yr. If the delivered purchase cost of the chlorine is \$0.18/lb, the annual chlorine feedstock cost is \$161,644,000/yr. Similarly, the annual flow rate of ethylene, the other feedstock, is 44,900 lb/hr or 355.56 million lb/yr. At \$0.30/lb, the annual ethylene feedstock cost is \$106,668,000/yr. The total annual feedstock cost is \$268,312,000. The process produces 100,000 lb/hr of vinyl chloride or 791.90 million lb/yr. At a selling price of \$0.35/lb, the annual sales revenue is \$277,165,000, which is greater than the total annual feedstock costs. The process also produces a byproduct of gaseous HCl, which, if it could be sold, could enhance the potential earnings of the process.

The feedstocks may be purchased from suppliers, or the company itself may produce one or more of the feedstocks or have control over a required natural resource. If a feedstock is to be purchased, availability from more than one supplier can keep the cost down. However, a long-range contract can ensure the availability of the feedstock. In the absence of price quotations from prospective suppliers of feedstocks, especially for early cost evaluations where the raw materials are commodity chemicals, the weekly newspaper *ICIS Chemical Business Americas,* formerly *Chemical Market Reporter,* can be consulted, where most of the costs are for tank-car quantities. The prices quoted are representative of prices in the United States, but they are not location specific and may require an added delivery cost. Furthermore, the prices do not reflect the discounts that usually accompany long-term contracts. Also, larger production rates can increase the supply to such an extent that the price is reduced for a given demand. For specialty chemicals such as pharmaceuticals, prices can be obtained from their manufacturers. If the company manufactures a feedstock or controls it, a *transfer price* must be assigned. The transfer price is an agreed-upon price between the company division that supplies the feedstock and the division that manufactures the products from the feedstock. The transfer price may be: (1) the market price, or (2) a price negotiated between the two divisions, recognizing that the price influences the selling division's revenue and the buying division's cost.

Utilities

Except for certain processes involving inexpensive feedstocks, such as the manufacture of oxygen, nitrogen, and argon from air or the production of hydrogen and oxygen from water, the annual cost of utilities, while much smaller than the feedstock costs, is not an insignificant contribution to the selling price of the product(s), often in the range of 5 to 10%. As listed in the Cost Sheet of Table 23.1, utilities include steam for heating at two or more pressure levels, electricity, cooling water, process water, demineralized boiler-feed water, refrigeration at different temperature levels, fuels such as natural gas, wastewater treatment, waste disposal, and landfill. Often, the largest utility cost is that of steam.

The company may purchase utilities from a public or private utility company or build its own utility plants. Credit can be taken for any utilities, for example, fuel, steam, and electricity, produced by the process. Purchased utility costs are based on consumption. For company-owned utilities, both capital costs and operating costs apply. A cogeneration unit using a fuel can supply electricity accompanied by low-, medium-, and high-pressure steam. For early estimates, the purchase of utilities can be assumed, using the unit costs in Table 23.1.

Steam

Steam has many potential uses in a process, both as a process fluid and as a utility. In the former category, it may be used as a feedstock, as an inert diluent in a reactor to absorb heat of reaction, as a direct heating agent, and as a stripping agent in absorbers and adsorbers. As a utility, it can be used in place of electricity to drive pumps and compressors, and in ejectors to produce a vacuum. Steam is used in heat exchangers to heat liquids and gases and vaporize liquids. Typical pressure levels for steam are 50, 150, and 450 psig, with high-pressure steam more costly than steam at the lower pressures. In computing steam utility requirements, credit is only taken for the latent heat of vaporization. No credit is taken for sensible heat of the steam even though the steam may arrive at the heat exchanger superheated and leave as subcooled condensate. This is illustrated in the following example.

EXAMPLE 23.1

A kettle reboiler is used to evaporate toluene at 375°F with a heat duty of 3,000,000 Btu/hr. Steam is available at 50, 150, and 450 psig. Determine the steam pressure level to use, the steam flow rate in lb/hr and lb/yr, and the estimated annual steam cost if the plant-operating factor is 0.90.

SOLUTION

Assuming a barometric pressure of 14 psia, the saturation temperature of 150-psig steam is 365°F. Since this is lower than 375°F, 450-psig steam must be used. However, to avoid film boiling, as discussed in Chapter 13, use an overall temperature-driving force of 45°F, which sets the steam condensing temperature at $375 + 45 = 420$°F, corresponding to a saturated steam pressure of 309 psia. With a valve, assumed to operate adiabatically, the pressure of the high-pressure steam is reduced from 464 psia to 309 psia, producing a superheated vapor. However, the steam is assumed to enter the heat exchanger as a saturated vapor at 420°F and 309 psia and leave as a saturated liquid at 420°F. At these conditions, the latent heat of vaporization of steam is 806 Btu/lb. By an energy balance, the hourly steam requirement is $3{,}000{,}000/806 = 3{,}722$ lb/hr or $3{,}722(24)(365)(0.9) = 29{,}350{,}000$ lb/yr. From Table 23.1, the cost of 450-psig steam is \$6.60/1,000 lb. Therefore, the annual steam cost $= 29{,}350{,}000(6.60)/1{,}000 = \$193{,}700$/yr.

Electricity

The operation of many items of processing equipment requires energy input in the form of a drive or motor. This equipment includes pumps, compressors, blowers, fans, agitators and mixers, feeders, conveyors, elevators, crushers, grinders, mills, scraped-wall crystallizers, agitated-film evaporators, agitated and centrifugal liquid–liquid extractors, centrifuges and rotary vacuum filters, rotary dryers and kilns, and spray and drum dryers. Although many types of drives are available, including air expanders, combustion gas turbines, internal combustion engines, and steam turbines, the most common drives are electric motors because they are very efficient ($>90\%$); very reliable; readily available in a wide range of wattages (Hp), shaft speeds, and designs; long lasting; and offer convenience, small footprint, favorable cost, and ease of maintenance. Electric motors are almost always used for power up to 200 Hp and are even available for applications requiring in excess of 10,000 Hp.

Alternating, rather than direct, current is used almost exclusively in electric motors. In the United States the alternating current cycles from positive to negative and back to positive 60 times a second, referred to as 60 hertz (Hz). This rate works well with electric clocks, since there are 60 seconds in a minute. In Europe, for some reason, 50 Hz is used. Alternating-current electricity originates at offsite private, public, governmental, or company facilities, generally at 18, 22, or 24 kilovolts (kV). For transmission to the plant on-site location, transformers step up the voltage to as high as 700 kV and then step it down to user voltages that are mainly in the range of 120 to 600 volts, but may be as high as 13,800 volts for motors of very high horsepower. Of the many electric motor designs, the most common is the three-phase, alternating-current, constant-speed, squirrel-cage induction motor. Induction motors of 60 Hz are capable of being operated satisfactorily on 50 Hz circuits if their voltage and horsepower ratings are reduced by a factor of 50/60. Motor enclosures may be explosion-proof (for locations near combustible fluids and dust), open drip-proof (to prevent entrance of liquid drips and dirt particles, but not vapor, dust, or fumes), or totally enclosed. For a given amount of energy transfer, the cost of electricity usually is greater than the cost of steam, as illustrated in the following example.

EXAMPLE 23.2

An electric motor is to be used to drive a compressor of 1,119 brake horsepower (BHp). The efficiency of the motor is 95%. Therefore, the electrical input to the motor must be $1{,}119(0.7457)/0.95 = 878$ kW. This is equivalent to 3,000,000 Btu/hr, which is the basis for the previous example. Calculate the kW-hr required per year for the motor if the plant-operating factor is 0.9, and calculate the cost of electricity per year.

SOLUTION

The plant will operate $(365)(24)(0.9) = 7{,}884$ hr/yr. Therefore, the motor requires $878(7{,}884) = 6{,}922{,}000$ kW-hr/yr. From Table 23.1, the cost of the electricity is $6{,}922{,}000(0.06) = \$415{,}300$/yr.

Cooling Water

Cooling water is used to cool liquids and gases and condense vapors. Typically cooling water circulates between a cooling tower and process heat exchangers by means of a pump. For preliminary design purposes, it can be assumed that the cooling water enters a heat exchanger at 90°F and exits at 120°F. In the cooling tower, direct contact of downward-flowing water with air, forced upward by a fan, causes the water temperature to approach within about 5°F of the wet-bulb temperature of the air. Approximately 80% of the reduction of the temperature of the cooling water is accomplished by evaporation of a small amount of the cooling water, with the balance caused by the transfer of heat from the cooling water to the surrounding air. In addition to the evaporation in the cooling tower, cooling water is also lost by drift (entrained water droplets in the cooling tower air discharge) and blowdown (deliberate purging of untreated cooling water to prevent buildup and subsequent precipitation) of dissolved salts in the cooling water. Typically, makeup cooling water is 1.5 to 3% of the circulating cooling water rate. Alternatives to cooling towers include spray ponds and cooling ponds. Warm water spread out over a large area of impervious ground in an open pond can cool by evaporation, convection, and radiation. The rate of cooling can be increased by recirculating the water through spray nozzles, much like a fountain.

When a plant is located near a river or large body of water, cooling water can be drawn off at one location, pumped to the heat exchangers, and discharged downstream in a river or to another location in a lake, bay, or ocean. It is customary to filter this water, but not treat it to remove salts and similar impurities.

In general, the cost of cooling water is much less than the cost of steam for a given heat exchanger duty, as illustrated in the following example.

EXAMPLE 23.3

Cooling water is used in the overhead condenser of a distillation column, with a heat duty of 3,000,000 Btu/hr. Determine the gallons per minute (gpm) of cooling water required and the annual cost of the cooling water, if the plant has an operating factor of 0.90.

SOLUTION

Assume the cooling water enters the condenser at 90°F and exits at 120°F. Water has a specific heat of 1 Btu/lb-°F and a density of 8.33 lb/gal. Therefore, by an energy balance, the condenser requires $3{,}000{,}000/[(1)(120-90)] = 100{,}000$ lb/hr or $100{,}000/[(60)(8.33)] = 200$ gpm. The total gallons per year = $200(60)(24)(365)(0.9) = 94{,}600{,}000$ gal. From Table 23.1, the cost of cooling water = ($0.075/1,000 gal. Therefore the annual cost = $0.075(94{,}600{,}000)/(1{,}000)$ = $7,100/yr.

Process Water and Boiler-Feed Water

Water is needed for many purposes in a chemical processing plant, including cooling water (discussed above), boiler-feed water, and process water, which is water that enters directly into the process, rather than being used indirectly as a heat-transfer agent. Process water must be purified to the extent necessary to avoid introduction of any undesirable chemicals into the process that could poison catalysts, foul equipment, and/or introduce impurities into products. Boiler-feed water (BFW) is used to produce steam in offsite boiler or co-generation facilities. BFW can also be used, in place of cooling water, as a cooling agent in a process when the temperature of a process stream to be cooled exceeds approximately 300°F. In the heat exchanger, the BFW is vaporized to steam, which may find use elsewhere in the process. Whether BFW is used on-site or offsite, it must be demineralized before use to avoid fouling of heat exchanger tubes. Sources of water include municipal water, well water, river water, lake water, ocean water, brackish water, treated wastewater, and condensate. The cost of process water, given in Table 23.1 as $0.75/1,000 gal, corresponds to only a moderate degree of pretreatment. The annual cost of process water, when needed, is usually very small compared to other feedstocks. When BFW is used to produce steam in a process heat exchanger, the cost of the BFW is partially offset by the value of the steam produced from it. This value is taken as a credit. In Table 23.1, the cost of BFW is given as $1.80/1,000 gal, which accounts for the credit. Extensive treatment of water containing large amounts of impurities can raise the cost to as much as $6.00/1,000 gal. Sterilized water for the manufacture of pharmaceuticals can cost as much as $550/1,000 gal. The use of BFW in a process heat exchanger is illustrated in the following example.

EXAMPLE 23.4

A process for hydrogenating benzene to cyclohexane, described in Example 9S.5, includes a well-mixed reactor, where an exothermic reaction occurs at 392°F and 315 psia. A total of 4,704,200 Btu/hr of heat must be transferred out of the reactor. Although this heat could be transferred into cooling water, the temperature in the reactor is sufficiently high to consider transferring the heat, by means of a heat exchanger, to boiler-feed water to produce steam. Determine the pressure level of the steam that could be produced, the pounds per hour of BFW required, and the annual cost of the BFW. The plant-operating factor is 0.9.

SOLUTION

To ensure nucleate boiling of the BFW, assume an overall driving force of 45°F, as discussed in Chapter 18. Thus, the BFW will be converted to steam at $(392 - 45) = 347$°F. This corresponds to a saturation pressure of 130 psia. Assume the BFW enters the heat exchanger as liquid at 90°F and exits as saturated vapor at 347°F. The change in enthalpy = 1,134 Btu/lb. Therefore, by an energy balance, the steam produced from the BFW = $4{,}704{,}200/1{,}134 = 4{,}148$ lb/hr. The annual cost of the BFW at $1.80/1,000 gal with water at a density of 8.33 lb/gal is $4{,}148(24)(365)(0.9)(1.80)/[(8.33)(1{,}000)]$ = $7,070/yr.

Refrigeration

The two most common coolants are cooling water and air. In general, cooling water from a cooling tower or pond can be used to cool a process stream to 100°F, while air, which is used in desert locations or where water is in short supply, can cool to only 120°F. To cool and/or condense process streams to temperatures below 100°F, chilled water, chilled brine, or a refrigerant is necessary, with the latter category being the most common. Several refrigerants are listed in Table 18.1. Prior to 1995, when the U.S. Clean Air Act Amendments of 1990 went into effect, two of the most popular refrigerants were CFC Freon R-12 (dichlorodifluoromethane) and HCFC Freon R-22 (chlorodifluoromethane). The chlorine atoms in these refrigerants were found to be released in the stratosphere, causing a depletion of the ozone layer. Since 1995, production of these two refrigerants and other chlorofluorocarbons has ceased or has been curtailed, as discussed in Section 3.5. A common replacement for R-12 is 1,1,1,2-tetrafluoroethane (R-134a, which is the same as HFC-134a), which according to the EPA will not propagate a flame under normal conditions in open air, shows no evidence of toxicity below 400 ppm, and does not deplete the ozone layer. This refrigerant, as well as ammonia and several light hydrocarbons, cools by transferring heat from the process stream in a heat exchanger where the refrigerant is evaporated. A typical propane refrigeration system is shown in Figure 9S.20, where the propane is circulated by a compressor through a condenser, a valve or turbine, and an evaporator where the cooling takes place. The temperature-driving force in the refrigerant evaporator is typically in the range of 2 to 10°F. Thus, with R-134a, which from Table 18.1 has an operating range of −15 to 60°F, a process stream can be cooled down to within 2°F of −15°F or −13°F. For lower temperatures, ammonia or the light hydrocarbons can be used. For example, an ethylene refrigerant can cool a stream down to as low as −148°F. The lower-limit temperature of operation of the refrigerant corresponds approximately to its normal boiling point. Lower temperatures would require an undesirable vacuum on the refrigerant side of the evaporator.

Typically, a chemical plant must provide its own refrigeration, either offsite, or more typically on-site. Petroleum refineries use light-hydrocarbon refrigerants, while other plants may consider ammonia and R-134a, unless very low temperatures (<-30°F) are required, where cascade refrigeration systems are often used, as discussed in Examples 9.19 and 9.21. These systems are often included with the equipment in the process flowsheet.

The temperature level is the key factor in determining the cost of refrigeration. For moderate temperatures, estimates of annual operating cost are based on a ton-day of refrigeration, where a ton is defined as the heat removal to freeze 1 ton (2,000 lb) per day of water at 32°F, which corresponds to 12,000 Btu/hr. Calculation of the annual refrigeration cost is illustrated in the following example. For a given energy-transfer rate, the cost of a moderate level of refrigeration compares to the cost of steam for heating.

EXAMPLE 23.5

A process stream in a petroleum refinery is to be partially condensed and cooled to 10°F, with a cooling duty of 3,000,000 Btu/hr. Select a suitable refrigerant, and calculate the tons of refrigeration required and the annual operating cost if the plant-operating factor is 0.9.

SOLUTION

From Table 18.1, a suitable refrigerant for a petroleum refinery is propane, since it has an evaporation range of −40 to 20°F. The propane evaporation temperature would be approximately 5°F. The tons of refrigeration = 3,000,000/12,000 = 250 tons. The annual refrigeration is 250(365)(0.9) = 82,130 ton-day. From Table 23.1, for this temperature level, the annual operating cost = 82,130(2.40) = \$197,100/yr. Most of this cost is the cost of electricity to drive the propane compressor.

Fuels

Various fuels may be combusted in a chemical process to provide heat or work. Besides the fuels needed for the offsite facilities such as boilers, electrical power generation, and cogeneration, fuels such as coal, natural gas, manufactured gas, and/or fuel oil may be needed for high-temperature heating in furnaces and fired heaters. Also, fuels may be used to drive pumps and compressors. Typically, the fuel, whether it be solid, liquid, or gas, is burned completely with an excess amount of air. To determine the amount of fuel required, the heating value (heat of combustion) of the fuel must be known. Two heating values are in common use, the *higher heating value* (also called the gross heating value), HHV, and the *lower heating value* (also called the net heating value), LHV. The heating value is the total heat evolved by complete combustion of a fuel with dry air when the fuel and air are at 60°F before combustion and all of the flue gas (product from combustion) is brought to 60°F after combustion. If the water vapor in the cooled flue gas is not condensed, the total heat is the LHV. If the water vapor is condensed, additional heat is evolved, giving the HHV. Some typical heating values for common fuels are given in Table 23.2. A manufactured gas is not listed; typically, it contains mainly H_2, CO, CH_4, and N_2 over wide ranges of composition. Note that heating values for solid and liquid fuels are usually quoted on a mass basis, while gaseous fuels are on a volume basis, usually standard cubic foot (SCF) at 1 atm and 60°F. For a given heat-transfer rate to a process stream being heated and/or vaporized in a fired heater, the amount of fuel required is greater than that based on its HHV because of heat losses, a flue gas temperature much greater than 60°F, and the presence of water as vapor in the flue gas. The ratio of the amount of fuel based on the HHV to the actual amount is the fired heater thermal efficiency, which may range from 50 to 80%. Typical fuel costs are included in Table 23.1. The calculation of the fuel requirement for a fired heater is illustrated in the following example.

Table 23.2 Typical Heating Values of Fuels

Fuel	HHV	LHV
Pennsylvania anthracite coal	13,500 Btu/lb	
Illinois bituminous coal	12,500 Btu/lb	
Wyoming subbituminous coal	9,500 Btu/lb	
North Dakota lignite coal	7,200 Btu/lb	
No. 2 fuel oil (33° API)	139,000 Btu/gal	131,000 Btu/gal
No. 4 fuel oil (23.2° API)	145,000 Btu/gal	137,000 Btu/gal
Low-sulfur No. 6 fuel oil (12.6° API)	153,000 Btu/gal	145,000 Btu/gal
Methyl alcohol	9,550 Btu/lb	
Ethyl alcohol	12,780 Btu/lb	
Benzene	17,986 Btu/lb	17,259 Btu/lb
Hydrogen	322 Btu/SCF	272 Btu/SCF
Carbon monoxide	321 Btu/SCF	321 Btu/SCF
Methane	1,012 Btu/SCF	907 Btu/SCF
Ethane	1,786 Btu/SCF	1,601 Btu/SCF
Propane	2,522 Btu/SCF	2,312 Btu/SCF
Natural gas (85–95 vol% methane)	1,020–1,090 Btu/SCF	920–990 Btu/SCF

EXAMPLE 23.6

A fired heater is to be used to heat and vaporize, from 1,000 to 1,200°F, the feed to a reactor. The heat duty is 3,000,000 Btu/hr. The fuel is natural gas with an HHV of 1,050 Btu/SCF. The thermal efficiency is 70%. If the plant-operating factor is 0.9, compute the SCF/hr and SCF/yr of natural gas required and the annual fuel cost.

SOLUTION

For an efficiency of 70%, the heat evolved from combustion of the fuel is $3{,}000{,}000/0.7 = 4{,}286{,}000$ Btu/hr. The natural gas must be supplied at a rate of $4{,}286{,}000/1{,}050 = 4{,}082$ SCF/hr or a rate of $4{,}082(24)(365)(0.9) = 32{,}180{,}000$ SCF/yr. From Table 23.1, the cost of natural gas is \$3.20/1,000 SCF. Therefore, the annual cost is $32{,}180{,}000(3.20)/1{,}000 = \$103{,}000$/yr.

Waste Treatment

Most chemical processes produce waste streams: gaseous (with or without particles), liquid (with or without particles, dissolved gases, and dissolved solids), solids (wet or dry), and slurries. In some cases, valuable byproducts can be removed from waste streams by additional processing. However, when this is not economical, federal regulations require that waste streams be treated to remove pollutants before being sent to the surrounding air, a sewer, a pond, a nearby river, a lake, an ocean, or a landfill.

Air-Pollution Abatement

Waste gases may contain particulates and/or gaseous pollutants, inorganic or organic. Additional equipment must be added to the process to remove these pollutants. If that equipment requires utilities, their costs must be added to the other utility costs. The removal of particles is usually accomplished with cyclone collectors, wet scrubbers, electrostatic precipitators, and fabric-filter systems. Inorganic gaseous pollutants such as ammonia; chlorine and fluorine; oxides of sulfur, carbon, and nitrogen; hydrogen sulfide, chloride, fluoride, and cyanide; and organic gaseous pollutants such as hydrocarbons and oxygenated organic compounds can be removed by absorption, adsorption, condensation, and/or combustion.

Wastewater Treatment

When water is fed into a process and/or the process produces water, wastewater is usually one of the process effluents. This wastewater must be treated for the removal of pollutants before being discharged to a sewer, pond, or body of water. In the United States, the treatment is regulated by the U.S. Clean Water Act of 1977. The treatments necessary depend on the nature of the foreign material, whether it is suspended or dissolved in the water. When private or municipal sewage treatment plants are nearby, the wastewater can be sent directly to those plants. However, pretreatment may be required to neutralize the water, remove large solids, and remove grease and oil. If the treatment facilities are located on-site or offsite, equipment for several treatments should be considered. *Primary treatment,* using gravity sedimentation or clarification, is used to remove suspended solids. *Secondary treatment* adds aerobic biological organisms (those requiring molecular oxygen for metabolism) in the form of a sludge to cause oxidation of the dissolved biodegradable organic compounds to carbon dioxide, water, sulfates, etc. Excess flocculated suspension (activated sludge) is removed from the water by clarification or air flotation. A measure of the biodegradability is the biochemical oxygen demand (BOD), which is the amount of oxygen required in parts per million of water by mass, in milligrams per liter of water. *Tertiary treatment* adds one or more additional chemical treatments to remove acids, alkalies, colloidal matter, color, odor, metals, and other pollutants not removed in earlier steps. Of most concern in chemical processing plants is the removal of dissolved organic compounds, particularly those that are carcinogenic, such as benzene. The following example illustrates the calculation of the cost of removal of dissolved organic compounds by biodegradation.

EXAMPLE 23.7

A wastewater stream of 500 gpm at 70°F contains 150 mg/L of benzene that is to be removed by biodegradation. If 99.9% of the benzene is removed, determine the amount of benzene removed per year and the operating cost of removal using a cost from Table 23.1 of \$0.15/lb benzene and a plant-operating factor of 0.9.

SOLUTION

First, determine if all of the benzene is dissolved in the wastewater. The solubility of benzene in water at 70°F is a mole fraction of 0.00040. The mole fraction of benzene (MW = 78) in the wastewater (55.5 mol/L) is $(0.150/78)/55.5 = 0.000035$. Therefore, all of the benzene is dissolved. The flow rate of 500 gpm is equivalent to $500(3.785)(60) = 113{,}600$ L/hr. The pounds of benzene removed per year = $(0.150/454)(113{,}600)(24)(365)(0.9) = 296{,}000$ lb/yr. The cost of benzene removal = $0.15(296{,}000) = \$44{,}400$/yr.

Solid Wastes

According to U.S. federal regulations, solid wastes must be classified as hazardous or nonhazardous. Hazardous wastes, due to their ignitability, corrosivity, reactivity, and/or toxicity, pose a substantial threat to human, plant, or animal life and must be treated on-site or near-site by physical, chemical, thermal, or biological means before being put in containers and removed. Nonhazardous solid wastes may be placed in containers and removed to a landfill or, in some cases, incinerated. The annual cost of solid waste treatment and disposal varies widely. Typical costs are \$0.03/lb of waste for nonhazardous dry or wet solids and \$0.10/lb for hazardous dry or wet solids.

Labor-Related Operations, O

One of the most difficult annual costs to estimate is direct wages and benefits (DW&B) for operating a chemical plant. It and the other annual costs that are proportional to it are often an important fraction of the cost of manufacture. Table 23.1 lists the labor-related charges associated with operations. These include *direct wages and benefits* (DW&B), calculated from an hourly rate for the operators of a proposed plant. To estimate all labor-related operations, it is necessary to estimate the number of operators for the plant per shift and to account for three shifts daily, except for small businesses that operate over one shift daily, five days per week. Typically, each shift operator works 40 hr per week, and, hence, for each operator required during a $7(24) = 168$-hr week, 4.2 shifts must be covered. In practice, due to illness, vacations, holidays, training, special assignments, overtime during startups, etc., it is common to provide for 5 shifts for each operator required.

Estimates of the number of plant operators needed per shift are based on the type and arrangement of the equipment, the multiplicity of units, the amount of instrumentation and control for the process, whether solids are handled, whether the process is continuous or batchwise or includes semicontinuous operations, and company policy in establishing labor requirements, particularly as it relates to operator unions. For preliminary estimates of the number of operators required per shift, the process may be divided into sections as discussed in Chapter 8 and shown in Figures 8.1 and 8.3. These sections may include: (1) feed preparation system using separation steps, (2) reactor system, (3) vapor recovery system, (4) liquid-separation system, (5) solids separation and purification system, and (6) pollution abatement system. When a process includes two or more reactor systems and/or two or more liquid-separation systems, each is counted separately. As given in Table 23.3, for a continuously operating, automatically controlled fluids-processing plant with a low-to-medium capacity of 10 to 100 ton/day of product, one operator/shift is assigned to each section. For solids–fluids processing and solids processing, the number of operators per shift is increased as noted in Table 23.3. For large capacities, for example, 1,000 ton/day of product, the number of operators/shift in Table 23.3 are doubled for each section. Batch and semicontinuous processing also require more operators than a continuous process, as indicated in Table 23.3. A process should always have at least two operators present per shift. Each shift operator is paid for 40 hr/week and 52 weeks/yr or a total of 2,080 hr/yr. The annual cost of direct wages and benefits (DW&B) is obtained from:

$$\text{DW\&B, \$/yr} = (\text{operators/shift})(5\text{ shifts}) \times (2{,}080\text{ hr/yr-operator})(\$/\text{hr}) \quad (23.2)$$

where the \$/hr covers wages and benefits, and depends on locality and whether operators are unionized. In Table 23.1, a figure of \$35/hr is typical in the United States.

To obtain the total annual labor-related operations cost, O, direct salaries and benefits for supervisory and

Table 23.3 Direct Operating Labor Requirements for Chemical Processing Plants. Basis: Plant with Automatic Controls and 10–100 Ton/Day of Product

Type of Process	Number of Operators per Process Section[a]
Continuous operation	
Fluids processing	1
Solids–fluids processing	2
Solids processing	3
Batch or semibatch operation	
Fluids processing	2
Solids–fluids processing	3
Solids processing	4

[a]Note: For large continuous-flow processes (e.g., 1,000 ton/day of product), multiply the number of operators by 2.

engineering personnel at 15% of DW&B and operating supplies and services at 6% of DW&B are added to DW&B. In addition, \$60,000/(operator/shift)-yr for technical assistance to manufacturing and \$65,000/(operator/shift)-yr for control laboratory are added. An estimate of the total annual cost of labor-related operations is illustrated in the following example.

EXAMPLE 23.8

The vinyl-chloride process discussed in Sections 4.4 and 4.5 and shown in Figures 4.8 and 4.19 produces 100,000 lb/hr of vinyl chloride or 1,200 ton/day. Estimate the annual cost of labor-related operations, O.

SOLUTION

This is a continuous fluids process of large capacity. Assume it is automatically controlled. From the block flow diagram, the process is comprised of two reactor sections and one liquid-separation section. Therefore, from Table 23.3, three operators per shift are required for a moderate-capacity plant. However, this is a large-capacity plant, requiring twice that number or 6 operators per shift and five shifts or a total of 30 shift operators. Also, a large-capacity plant requires one labor-yr each for technical assistance and control laboratory. Using Eq. (23.2), the annual costs are

$$\text{Annual DW\&B} = (30\ \text{operators})(2{,}080\ \text{hr/yr})(\$35.00/\text{hr}) = \$2{,}184{,}000$$

Using Table 23.1, the other annual labor-related operation costs are

$$\text{Direct salaries and benefits} = 0.15(\$2{,}184{,}000) = \$328{,}000$$

$$\text{Operating supplies and services} = 0.06(\$2{,}184{,}000) = \$131{,}000$$

$$\text{Technical assistance to manufacturing} = \$60{,}000(5) = \$300{,}000$$

$$\text{Control laboratory} = \$65{,}000(5) = \$325{,}000$$

The total labor-related operations annual cost, O, is

$$O = \$2{,}184{,}000 + \$327{,}600 + \$131{,}000 + \$300{,}000 + \$325{,}000 = \$3{,}268{,}100/\text{yr}$$

Maintenance, M

A second category of labor-related costs is associated with the maintenance of a proposed plant. Processing equipment must be kept in acceptable working order, with repairs and replacement of parts made as needed. Annual maintenance costs, M, are sometimes greater than the cost of labor-related operations, O. Included in Table 23.1, under annual maintenance costs, M, is the main item, maintenance wages and benefits (MW&B), which is estimated as a fraction of the total depreciable capital, C_{TDC}, depending on whether the process handles fluids, solids, or a combination of fluids and solids. The range is from a low of 3.5% for fluids to 5.0% for solids, with 4.5% for solids–fluids processing. Salaries and benefits for the engineers and supervisory personnel are estimated at 25% of MW&B. Materials and services for maintenance are estimated at 100% of MW&B, while maintenance overhead is estimated at 5% of MW&B. Thus, the total annual cost of maintenance varies from 8.05 to 11.5% of C_{TDC}. Maintenance costs can be controlled by selecting the proper materials of construction for the processing equipment, sparing pumps, avoiding high rotation speeds of shafts, restricting the highest fouling streams to the tube side of heat exchangers, selecting long-life catalysts for reactors, scheduling routine maintenance, and practicing preventative maintenance based on experience, supplier information, and record-keeping. Routine maintenance includes cleaning of heat exchanger tubing and lubrication and replacement of packing and mechanical seals in pumps, compressors, blowers, and agitators. A main goal should be to provide most of the maintenance during scheduled plant shutdowns, which might be during a two- or three-week period each year.

EXAMPLE 23.9

The total depreciable capital investment, C_{TDC}, for a plant to produce 300,000 tons per year of cumene is estimated to be \$31,000,000. The process only involves fluids processing. Estimate the annual plant maintenance cost, M.

SOLUTION

Using Table 23.1, the annual maintenance costs are

$$\text{Wages and benefits (MW\&B) at 3.5\% of } C_{TDC} = \$1{,}085{,}000$$

Salaries and benefits at 25% of MW&B = 271,000
Materials and services at 100% of MW&B = 1,085,000
Maintenance overhead at 5% of MW&B = 54,000

The total annual maintenance cost, M, is $2,495,000/yr.

Operating Overhead

To this point in Table 23.1, all costs have been directly related to plant operation. However, a company always incurs many other expenses, which while not directly related to plant operation can be estimated as a fraction of the combined salary, wages, and benefits for maintenance and labor-related operations, referred to here as M&O-SW&B. Overhead expenses include the costs of providing the following services: cafeteria; employment and personnel; fire protection, inspection, and safety; first aid and medical; industrial relations; janitorial; purchasing, receiving, and warehousing; automotive and other transportation; and recreation. In Table 23.1, overhead costs are divided into four categories: general plant overhead, provision for the services of the mechanical department and for the employee relations department, as well as business services, with the total annual operating overhead cost equal to the sum of these four categories or $(7.1 + 2.4 + 5.9 + 7.4) = 22.8\%$ of M&O-SW&B.

EXAMPLE 23.10

Estimate the annual cost of operating overhead for the cumene plant of Example 23.9, assuming that the cost of labor-related operations is the same as in Example 23.8.

SOLUTION

The previous two examples provide the following wages, salaries, and benefits per year for labor-related operations and maintenance:

Direct wages and benefits (DW&B) = $2,184,000
Direct salaries and benefits = 328,000
Maintenance wages and benefits (MW&B) = 1,085,000
Maintenance salaries and benefits = 271,000

The total annual M&O-SW&B is the sum, which equals $3,868,000/yr.
The total annual operating overhead cost is 22.8% of M&O-SW&B or

$$0.228(3,868,000) = \$882,000/\text{yr}$$

Property Taxes and Insurance

Annual property taxes are assessed by the local municipality as a percentage of the total depreciable capital, C_{TDC}, with a range from 1% for plants located in sparsely populated areas to 3% when located in heavily populated areas. Property taxes are not related to federal income taxes levied by the Internal Revenue Service and considered below. Liability insurance costs depend on the pressure and temperature levels of plant operation and on whether flammable, explosive, or toxic chemicals are involved. The annual cost of insurance is also estimated as a percentage of the total depreciable capital, C_{TDC}, with a range of 0.5 to 1.5%. In the absence of data, annual property taxes and insurance may be estimated at 2% of C_{TDC}, as given in Table 23.1. This corresponds to a process of low risk located away from a heavily populated area.

Depreciation, D

The subject of depreciation is complex and often confusing because depreciation has several definitions and applications. Most commonly, it is simply a measure of the decrease in value of an asset over time. Some companies use depreciation as a means to set aside a fund to replace a plant when it is no longer operable. In its most complex application, depreciation is an annual allowance, whose calculation is controlled by the U.S. federal government when determining federal income tax. The larger the depreciation in a given year, the smaller the federal income tax and the greater the net profit. This is considered in detail in the discussion of cash flow in Section 23.6.

For use with approximate profitability measures, as applied here to the preliminary calculation of the annual manufacturing cost, depreciation, D, is estimated as a constant percentage of the total depreciable capital, C_{TDC}. This type of depreciation is referred to as *straight-line (SL) depreciation*. Although it has been customary to take that percentage as 10% for each of 10 yr, here, in Table 23.1, the direct plant (on-site) depreciation is taken as 8% of $(C_{TDC} - 1.18C_{alloc})$ (equivalent to a plant life of about 12 yr), while the allocated plant (offsite) depreciation is taken as 6% of the contribution of the allocated costs for utilities and related facilities to the total depreciable capital, C_{TDC}, $1.18C_{alloc}$ (equivalent to a life of about 16 yr), where the 1.18 factor accounts for the share of the contingency and contractor's fee, C_{cont}.

Rental Fees

In product design, especially in the early years of startup companies, it is frequently necessary to rent office and laboratory space. These facilities are often available in research-oriented complexes and science centers that are often located in the vicinity of universities, at beltways surrounding large cities, etc. Often the laboratory facilities are available at different cleanliness ratings, with high levels usually required for the manufacture of pharmaceuticals, electronic materials, and the like.

Licensing Fees

When chemical products and processes under patent protection are utilized, an annual licensing fee may be negotiated, often based on the amount or dollar value of product sold. The amount of the annual licensing fee depends on the uniqueness of the process and the chemical being produced, with a range of 1–5% of product sales. In many cases, an initial royalty fee, as a portion of the capital investment, is included. In the absence of data, an initial royalty fee of 2% of C_{TDC} may be assumed, together with an annual licensing fee of 3% of product sales, as discussed in Section 22.3.

Cost of Manufacture, COM

The total annual cost of manufacture, COM, as shown in Table 23.1, is the sum of (1) *direct manufacturing costs:* feedstocks, utilities, labor-related operations, and maintenance; (2) *operating overhead;* and (3) *fixed costs:* property taxes, insurance, and depreciation.

Total Production Cost, C

The total annual production cost equals the sum of the cost of manufacture and general expenses,

$$C = \text{COM} + \text{general expenses} \tag{23.3}$$

General expenses refer to activities that are conducted by the central operations of a company, perhaps at the corporate headquarters, and are financed from profits made by the company from their operating plants. In Table 23.1, general expenses comprise five categories: selling (or transfer) expense, research (direct and allocated), administrative expense, and management incentive compensation. The *selling expense* covers all the costs involved in selling the products, including expenses of the sales office, advertising, traveling sales representatives, containers and shipping, commissions, and technical sales service. The research expense covers both research and development costs for new products and new manufacturing methods for existing products. Administrative expense covers those top-management and general administrative activities that are not direct manufacturing costs. In Table 23.1, all general expenses are estimated as a percentage of the total sales revenue. The total general expenses range from 9.55 to 11.55% of S. Note that the direct research allocation by pharmaceutical companies often exceeds 4.8% of sales (Pisano, 1997).

EXAMPLE 23.11

For the MCB separation process shown in Figure 5.23, estimate the annual production cost, C, where products will be used in-house, and the total annual sales, S. Basis:

Continuous plant operation	330 day/yr or 7,920 hr/yr
Feedstock	9,117 lb/hr @ $0.50/lb
MCB product	5,572 lb/hr @ $0.82/lb
Benzene co-product	3,133 lb/hr @ $0.32/lb
HCl gas co-product	412 lb/hr @ $0.06/lb
Total bare-module costs, C_{TBM}	$2,050,000
Cost of site preparation and service facilities, $C_{site}+C_{serv}$	$170,000
Allocated costs for utilities and related facilities, C_{alloc}	$90,000
Cost of contingencies @18% of C_{DPI}	3,480 lb/hr@$4.80/1,000 lb
150-psig steam	9.60 kW@$0.06/kW-hr
Electricity	182 gpm@$0.075/1,000 gal
Cooling water	two/shift
Operators	

SOLUTION

The total depreciable capital, using Table 22.9, is computed as follows:

$$C_{DPI} = \$2{,}050{,}000 + \$170{,}000 + \$90{,}000 = \$2{,}310{,}000$$
$$C_{cont} = 0.18\,C_{DPI} = 0.18(2{,}310{,}000) = \$416{,}000$$
$$C_{TDC} = C_{DPI} + C_{cont} = \$2{,}310{,}000 + \$416{,}000 = \$2{,}726{,}000$$

For this moderate-size plant, with one section, use 2 operators/shift. Using Table 23.1 with the above data, the following annual costs are computed:

Cost Factor	Annual Cost
Feedstocks (raw materials)	$36,103,000
Utilities	
Steam, 150 psig	132,300
Steam, 50 psig	29,900
Electricity	4,600
Cooling water (CW)	6,500
Total utilities	$173,300
Operations (O)	
Direct wages and benefits (DW&B)	728,000
Direct salaries and benefits	109,200
Operating supplies and services	43,700
Technical assistance to manufacturing	120,000
Control laboratory	130,000
Total labor-related operations	$1,130,900
Maintenance (M)	
Wages and benefits (MW&B)	95,400
Salaries and benefits	23,900
Materials and services	95,400
Maintenance overhead	4,800
Total maintenance	$219,500

Cost Factor	Annual Cost
Total of M&O-SW&B	956,500
Operating overhead	
General plant overhead	67,900
Mechanical department services	23,000
Employee relations department	56,400
Business services	70,800
Total operating overhead	\$218,100
Property taxes and insurance	\$54,500
Depreciation (D)	
Direct plant	209,600
Allocated plant	6,400
Total depreciation	\$216,000
COST OF MANUFACTURE (COM)	\$38,115,300
General Expenses	
Transfer expense	443,200
Direct research	2,127,500
Allocated research	221,600
Administrative expense	886,500
Management incentive compensation	554,000
TOTAL GENERAL EXPENSES (GE)	\$4,232,800
TOTAL PRODUCTION COST (C)	\$42,348,100
Sales	
Monochlorobenzene product	36,186,800
Benzene co-product	7,940,300
HCl co-product	195,800
TOTAL SALES, S	\$44,322,900

Pre-Tax (Gross) Earnings and After-Tax (Net) Earnings (Profit)

The annual *pre-tax earnings or profit,* also called the *gross earnings or profit,* is the difference between the annual sales revenue and the annual product cost:

$$\text{Gross earnings or profit} = S - C \qquad (23.4)$$

The annual *after-tax earnings or profit,* also called the *net earnings or profit,* is the gross earnings minus U.S. federal and state income taxes on the gross earnings. Since at least the year 1913, U.S. corporations have been subject to federal income tax on gross earnings. During the period from 1913 to 2007, the federal corporation income tax was as low as 1% and as high as 52%. During World War II and the Korean War, an additional excise tax brought the total tax to as high as 80%. The current schedule for U.S. corporate income tax rates, shown in Table 23.4, was established in September 2000. The smallest corporations have a rate of only 15%. Above gross earnings of \$50,000, the rate increases in steps until it reaches 35% for corporations with gross earnings that equal or exceed \$18,333,333. Corporations are also subject to state income tax, which for large corporations varies widely from as low as 3% in Illinois to as high as 9.5% in Vermont, with 5% being a reasonable assumption for preliminary

Table 23.4 Federal Income Tax Rate Schedule for Corporations

Gross Earnings Over	But Not Over	Income Tax
\$ 0	\$ 50,000	15%
50,000	75,000	\$7,500 + 25% over \$50,000
75,000	100,000	\$13,750 + 34% over \$75,000
100,000	335,000	\$22,250 + 39% over \$100,000
335,000	10,000,000	\$113,900 + 34% over \$335,000
10,000,000	15,000,000	\$3,400,000 + 35% over \$10,000,000
15,000,000	81,333,333	\$5,150,000 + 38% over \$15,000,000
18,333,333	—	\$6,416,667 + 35% over \$18,333,333 (equivalent to 35% on total gross earnings)

economic analyses. Here, we will use a combined federal and state income tax rate, t, of 35 + 5 = 40%. Thus,

$$\text{Net earnings or profit} = (1 - t)\,\text{gross earnings}$$
$$= 0.60(S - C) \qquad (23.5)$$

EXAMPLE 23.12

For the data and results of Example 23.11, calculate the annual gross earnings and annual net earnings.

SOLUTION

From Eq. (23.4),

$$\text{Annual gross earnings or profit} = S - C$$
$$= \$44{,}322{,}900 - \$42{,}348{,}100$$
$$= \$1{,}974{,}800/\text{yr}$$

From Eq. (23.5),

$$\text{Annual net earnings or profit} = 0.60(1{,}974{,}800) = \$1{,}184{,}900/\text{yr}$$

Of the costs in the cost sheet of Table 23.1, only the costs for labor-related operations, maintenance, operations overhead, property taxes and insurance, and depreciation are considered to be *fixed costs,* which do not vary with the production rate of the plant. Fixed costs are contrasted with the costs of feedstocks, utilities, and general expenses, which are referred to as *variable costs,* because they vary directly with the production rate. For a large plant, with a large total capital investment and significant economies-of-scale, the profitability can be sharply increased by substantial savings in utilities such as steam. A smaller plant, in contrast, has a larger fraction of its costs in the investment and the fixed costs of operation, and hence the same percentage decrease in the utilization of steam results in a smaller increase in its profitability. This is one of the reasons why small plants usually have difficulty competing with larger plants in the chemical industry. The reader should keep this in mind while the profitability measures in Sections 23.4 and 23.7 are studied. Before leaving Table 23.1, the reader is reminded that the prices listed there refer to the year 2006 and should be adjusted in subsequent years, possibly escalated by the rate of inflation.

23.3 WORKING CAPITAL AND TOTAL CAPITAL INVESTMENT

To complete the estimation of the total capital investment, a more accurate estimate of working capital is needed to replace the 15% of total capital investment used in conjunction with Eq. (22.10). In general, working capital is funds, in addition to fixed capital and startup funds, needed by a company to meet its obligations until payments are received from others for goods they have received. Accountants define working capital as current assets minus current liabilities, where current assets consist of cash reserves, inventories, and accounts receivable, while current liabilities include accounts payable. It is fairly standard to provide working capital for a one-month period of plant operation, because those buying the product are usually given 30 days to make their payments, while the company has 30 days to pay for raw materials. Inventories of products may be much less than 30 days. Here, 7 days is assumed. Working capital is fully recoverable and, therefore, is not depreciated. If we apply the definition of working capital to the operation of a chemical plant, working capital is

$$C_{WC} = \text{cash reserves} + \text{inventory} + \text{accounts receivable} - \text{accounts payable} \quad (23.6)$$

with the following basis for calculation, which follows general accounting practices:

1. 30 days of cash reserves for raw materials, utilities, operations, maintenance, operating overhead, property taxes, insurance, and depreciation. This amounts to 8.33% of the annual cost of manufacture, COM (assuming 30 days is $1/12$ of a year).
2. 7 days of inventories of liquid and solid (but not gas) products at their sales price, which assumes that these products are shipped out once each week, while gas products are not stored, but are pipelined. This amounts to 1.92% of the annual sales of liquid and solid products.
3. 30 days of accounts receivable for product at the sales price. This amounts to 8.33% of the annual sales of all products.
4. 30 days of accounts payable by the company for feedstocks at the purchase price. This amounts to 8.33% of the annual feedstock costs.

EXAMPLE 23.13

For the MCB plant considered in Example 23.11, estimate the working capital and compute the total capital investment if land cost and royalty costs are zero, but the startup cost is taken as 2% of C_{TDC}.

SOLUTION

The data required, obtained from Example 23.11, are

C_{TDC} = \$2,726,000

COM = \$38,115,300/yr

Sales for MCB and benzene = \$44,127,100/yr

Sales for MCB, benzene, and HCl = \$44,332,900/yr

Cost of feedstock = \$36,103,000/yr

Therefore,

Cash reserves = 0.0833(38,115,300) = \$3,175,000

Inventories = 0.0192(44,127,100) = \$847,200

Accounts receivable = 0.0833(44,332,900) = \$3,692,900

Accounts payable = 0.0833(36,103,000) = \$3,007,400

From Eq. (23.6),

Working capital, C_{WC} = \$3,175,000 + \$847,200 + \$3,692,900 − \$3,007,400 = \$4,707,700, which is much greater than the total depreciable capital

The startup cost = C_{start} = 0.02(2,726,000) = \$54,500

The total capital investment = $C_{TCI} = C_{TDC} + C_{start} + C_{WC}$ = \$2,726,000 + \$54,500 + \$4,707,700 = \$7,488,200

Note that for this case, the working capital is 63% of the total capital investment and much more than the commonly used approximate estimate of 15% of total capital investment. In this example, it appears that working capital is more a function of annual sales (perhaps 10%) than of total depreciable capital.

23.4 APPROXIMATE PROFITABILITY MEASURES

To be a worthwhile investment, a venture for the installation of a new chemical product manufacturing facility, a new chemical plant, or a revamp of an existing plant must be profitable. However, it is not sufficient that a venture make a large net profit. That profit over the life of the venture must be more than the original capital investment for the venture. The greater the excess of profits over investment, the more attractive is the venture. To compare alternative ventures that vie for capital investment, a number of profitability measures have been developed. They are all based on the estimates of capital investment and annual earnings that have been presented in Chapter 22 and the previous sections of this chapter. The simpler, approximate measures discussed in this section and summarized in Table 23.5 ignore the effect of inflation or so-called time value of money and use simple straight-line depreciation. Therefore, they are only useful in

Table 23.5 Approximate Profitability Measures
Time Value of Money Is Ignored and Straight-Line Depreciation Is Used
(Detaiis Presented in Section 23.4)

Approximate Profitability Measure	Formula[a]
Return on investment (ROI)	$\text{ROI} = \dfrac{\text{net earnings}}{\text{total capital investment}} = \dfrac{(1-t)(\text{S}-\text{C})}{C_{\text{TCI}}}$
Payback period (PBP)	$\text{PBP} = \dfrac{C_{\text{TDC}}}{(1-t)(\text{S}-\text{C})+\text{D}}$
Venture profit (VP)	$\text{VP} = (1-t)(\text{S}-\text{C}) - i_{\min}(C_{\text{TCI}})$
Annualized cost (AC)	$\text{AC} = C_{\text{A}} = \text{C} + i_{\min}(C_{\text{TCI}})$

[a] $i_{\min}$ = reasonable return on investment; t = sum of U.S. federal and state income tax rates; C = annual production cost; D = annual depreciation; S = annual sales revenues; C_{TCI} = total capital investment; C_{TDC} = total depreciable capital.

the early stages of project evaluation. The rigorous measures that account for the time value of money and faster depreciation are considered in the three subsequent sections and must be considered before a final decision is made on whether to proceed with a new venture.

Return on Investment (ROI)

This profitability measure, introduced earlier as Eq. (23.1), is also called rate of return on investment (ROROI), simple rate of return (ROR), return on original investment, engineer's method, and operator's method. ROI is the annual interest rate made by the profits on the original investment. ROI provides a snapshot view of the profitability of the plant, normally using estimates of the elements of the investment, in Table 22.9, and the pre-tax or after-tax earnings in, say, the third year of operation and assuming that they remain unchanged during the life of the process. For ROI, and all of the approximate profitability measures of this section, the production cost is computed using straight-line depreciation, and, after some startup period, the plant is assumed to operate each year at full capacity (or at some percentage of full capacity) for the same number of days per year. As was stated earlier, many definitions of ROI have been suggested and used. Here, the most common definition is applied.

$$\text{ROI} = \frac{\text{net earnings}}{\text{total capital investment}} = \frac{(1-t)(\text{S}-\text{C})}{C_{\text{TCI}}} \tag{23.7}$$

The calculation of ROI is readily made and the concept is easy to understand. However, as stated above, the definition of ROI involves many assumptions. Furthermore, ROI does not consider the size of the venture. Would a large company favor many small projects over a few large projects, when the small projects have just slightly more favorable values of ROI?

Payback Period (PBP)

The *payback period* is the time required for the annual earnings to equal the original investment. Payback period is also called payout time, payout period, payoff period, and cash recovery period. Because it is simple and even more understandable than ROI, PBP is widely used in early evaluations to compare alternatives. Like ROI, the payback period in years has several definitions, but the following is used here. This definition is not consistent with the definition of ROI in Eq. (23.7), because only the depreciable capital is used and the annual depreciation, D, is added back to the net earnings because that depreciation is retained by the company.

$$\begin{aligned}\text{PBP} &= \frac{C_{\text{TDC}}}{(1-t)(\text{S}-\text{C})+\text{D}} \\ &= \frac{C_{\text{TDC}}}{\text{net earnings} + \text{annual depreciation}} \\ &= \frac{C_{\text{TDC}}}{\text{cash flow}}\end{aligned} \tag{23.8}$$

High-risk ventures should have payback periods of less than 2 yr. In these times of rapid progress in technology, most companies will not consider a project with a PBP of more than 4 yr. PBP is especially useful for simple equipment replacement problems. For example, should an old, inefficient pump be replaced with a new, energy-efficient model? This decision is clear if the PBP is less than 1 yr. PBP should never be used for final decisions on large projects because it gives no consideration to the period of plant operation after the payback period.

EXAMPLE 23.14

A process, projected to have a total depreciable capital, C_{TDC}, of \$90 million, with no allocated costs for offsite utilities, is to be installed over a 3-yr period (2007–2009). Just prior to startup, \$40 million of working capital is required. At 90% of production capacity (projected for the third and subsequent operating years), sales revenues, S, are projected to be \$150 million/yr and the total annual production cost, excluding depreciation, is projected to be \$100 million/yr. Also, the plant is projected to operate at 0.5 of 90% and 0.75 of 90% of capacity during the first and second operating years. Thus, during those years, S = \$75 million/yr and \$113 million/yr, respectively. Take straight-line depreciation at 8%/yr. Using the third operating year as a basis, compute

(a) return on investment (ROI)

(b) payback period (PBP)

SOLUTION

$$\begin{aligned}
\text{Depreciation} &= 0.08(\$90{,}000{,}000) = \$7{,}200{,}000/\text{yr} \\
\text{Total production cost} &= \$100{,}000{,}000 + \$7{,}200{,}000 \\
&= \$107{,}200{,}000/\text{yr} \\
\text{Pre-tax earnings} &= \$150{,}000{,}000 - \$107{,}200{,}000 \\
&= \$42{,}800{,}000/\text{yr} \\
\text{Income taxes} &= 0.40(\$42{,}800{,}000) = \$17{,}100{,}000/\text{yr} \\
\text{After-tax earnings} &= \$42{,}800{,}000 - \$17{,}100{,}000 \\
&= \$25{,}700{,}000/\text{yr} \\
C_{TCI} &= \$90{,}000{,}000 + \$40{,}000{,}000 \\
&= \$130{,}000{,}000
\end{aligned}$$

(a) From Eq. (23.7),

$$\text{ROI} = \frac{\$25{,}700{,}000}{\$130{,}000{,}000} = 0.198 \text{ or } 19.8\%$$

(b) From Eq. (23.8),

$$\text{PBP} = \frac{\$90{,}000{,}000}{\$25{,}700{,}000 + \$7{,}200{,}000} = 2.74\,\text{yr}$$

In this example, values of both ROI and PBP are sufficient to merit some interest in the project, but they are not sufficient to attract a high degree of interest unless the process is of very low risk and only less-profitable ventures are under consideration.

Venture Profit (VP)

An approximate measure of the profitability of a potential process or product that does take into account the size of the project is *venture profit*. It is used often for preliminary estimates when comparing alternative flowsheets during the process synthesis stage of process design and/or the *concept* stage of the Stage-Gate^{TM} Product-Development Process (SGPDP). VP is the annual net earnings in excess of a minimum acceptable return on investment, $i_{\min}$. Thus,

$$\begin{aligned}
\text{VP} &= (1-t)(\text{S}-\text{C}) - i_{\min}C_{TCI} \\
&= \text{net earnings} - i_{\min}C_{TCI} \qquad (23.9)
\end{aligned}$$

Sometimes, for crude comparisons of flowsheets with different arrangements of process units, the total capital investment in Eq. (23.9) is estimated as the sum of the bare-module costs, or even the sum of the purchase costs; and annual production cost, C, includes only the cost of the raw materials, the utilities, and the labor-related operations. The return on investment, $i_{\min}$, is that desired by the company. Here, we take $i_{\min} = 0.20(20\%)$.

EXAMPLE 23.15

For the MCB process considered in Examples 23.11, 23.12, and 23.13, calculate

(a) return on investment (ROI)

(b) payback period (PBP)

(c) venture profit (VP)

SOLUTION

From the previous examples,

$$\begin{aligned}
C_{TCI} &= \$7{,}488{,}200 \\
C_{TDC} &= \$2{,}726{,}000 \\
\text{Net earnings} &= \$1{,}184{,}900/\text{yr} \\
\text{Depreciation} &= \$216{,}000/\text{yr}
\end{aligned}$$

(a) From Eq. (23.7),

$$\text{ROI} = \frac{\$1{,}184{,}900}{\$7{,}488{,}200} = 0.16 \text{ or } 16\%$$

(b) From Eq. (23.8),

$$\text{PBP} = \frac{\$2{,}726{,}000}{\$1{,}184{,}900 + \$216{,}000} = 1.95\,\text{yr}$$

(c) From Eq. (23.9),

$$\text{VP} = \$1{,}184{,}900 - 0.20(\$7{,}488{,}200) = -\$312{,}700/\text{yr}$$

These results are conflicting with respect to the profitability of the MCB process. The PBP is good and the process is low risk. The ROI is not outstanding unless the going interest rate is low. The VP is negative for $i_{\min} = 0.20$, eliminating the plant from further consideration.

The discrepancy among the three profitability measures is caused mainly by the large magnitude of the working capital compared to the depreciable capital.

Annualized Cost (C_A)

A measure of economic goodness, which does not involve sales revenues for products and is also used for preliminary estimates when comparing alternative flowsheets during process synthesis or alternative product concepts during the *concept* stage of the SGPDP, is the *annualized cost*. It is the sum of the production cost and a reasonable return on the original capital investment where, again, the reasonable return on investment, $i_{\min}$, is taken here as 0.2. Thus,

$$C_A = \text{C} + i_{\min}(C_{TCI}) \qquad (23.10)$$

This criterion is also useful for comparing alternative items of equipment in a process or alternative replacements for existing equipment.

EXAMPLE 23.16

Several alternative distillation sequences are being examined for the separation of a mixture of light hydrocarbons. The sequences are to be compared on the basis of annualized cost, given by Eq. (23.10). However, for the total capital investment, only the bare-module costs of the columns, trays, condensers, reboilers, and reflux accumulators will be summed. For the total annual production cost, C, only the annual utility costs for the condenser cooling water and reboiler steam will be summed. For one of the columns, design calculations have been completed and the costs have been computed, with the results given below. The column is a deisobutanizer with a saturated liquid feed of 500 lbmol/hr of isobutane and 500 lbmol/hr of *n*-butane. The distillate is 99 mol% isobutane and the bottoms is 99 mol% *n*-butane. The column shell is carbon steel, with carbon-steel sieve trays on 24-in. spacing. The trays have 0.25-in.-diameter holes with a hole area of 10%. The weir height is 2 in. The column pressure is set at 100 psia at the top so that cooling water can be used in the total condenser, while the bottoms pressure is 110 psia. Calculations give 100 trays, at a reflux ratio of 7.4. This corresponds to a condenser duty of 33,600,000 Btu/hr and a reboiler duty of 33,800,000 Btu/hr. For 24-in. tray spacing, allowing a 10-ft-high bottoms sump below the bottom tray and a 4-ft disengagement height above the top tray, the column height is 212 ft (tangent to tangent). Based on entrainment flooding, the column diameter is determined to be constant at 10 ft.

The bare-module cost of the tower vessel is estimated to be \$3,350,000 and the accompanying tray cost is \$300,000, giving a total bare-module cost for the column of \$3,650,000. The bare-module costs for the column auxiliaries are computed to be

2 Condensers in parallel	\$680,000
Reboiler	170,000
Reflux drum	200,000
Reflux pump + a spare	120,000

The total bare-module cost for the column and its auxiliaries = \$4,820,000

The annual heating steam cost for the reboiler is computed = \$2,180,000/yr

The annual cooling water cost for the two condensers = \$90,000/yr

The annual electricity cost for the reflux pump = \$48,000/yr

The total utility cost = \$2,318,000/yr

Compute the annualized cost.

SOLUTION

For purposes of comparison of alternatives, the bare-module cost of the distillation column and its auxiliary equipment replaces C_{TCI}. The annual utility cost replaces the total annual production cost.

From Eq. (23.10), $C_A = \$2,318,000 + 0.20(\$4,820,000) = \$3,282,000/\text{yr}$.

Product Selling Price for Profitability

In some cases, especially when a new chemical product is to be produced, the selling price may not be known or easily established. For basic chemical products, especially, rather than guess a selling price, a desired return on investment (say, 20%) can be assumed and Eq. (23.7) can then be used to back-calculate the selling price necessary to achieve this objective. Another useful procedure is to set the venture profit to zero and use Eq. (23.9) to back-calculate a minimum selling price. More elaborate methods for determining a selling price are implemented using the rigorous profitability measures in Section 23.7 that account for the time value of money.

For new configured consumer chemical products, such as home hemodialysis products and labs-on-a-chip for high-throughout screening, pricing strategies are dependent on the consumer market. As discussed in Section 2.8, *Product-Introduction Stage*, there are no simple recipes for setting prices.

EXAMPLE 23.17

In Example 23.15, approximate profitability measures, when applied to the MCB plant (for production of a basic chemical product), are not favorable. However, one of the chemicals produced, MCB, is given a selling price of \$0.82/lb, which is not well established by current competition.

(a) Use the ROI measure of Eq. (23.7) to estimate a selling price for a 20% return on investment.

(b) Use the VP measure of Eq. (23.9) to estimate a minimum selling price.

SOLUTION

(a) From Example 23.11, C = \$42,348,100/yr and the total annual sales of all three products is \$44,332,900/yr. This includes 44,130,200 lb/yr of MCB at \$0.82/lb. Thus, if we let x = the selling price of MCB, the total annual sales of all three products, in terms of x, becomes

$$S = \$44,332,900 - 44,130,200(0.82 - x)$$
$$= \$8,146,100 + 44,130,200x$$

From Example 23.13, $C_{TCI} = \$7,488,200$. Substitution into Eq. (23.7) gives

$$\text{ROI} = 0.20 = \frac{(1-0.40)(8,146,100 + 44,130,200x - 42,348,100)}{7,488,200}$$

Solving this equation, x = selling price of MCB = \$0.83/lb, which is slightly higher than the price of \$0.82. Clearly, the ROI is very sensitive to the selling price of MCB.

(b) Substitution into Eq. (23.9), with VP = 0, gives

$$VP = 0$$
$$= (1 - 0.40)[(8{,}146{,}100 + 44{,}130{,}200x) - 42{,}348{,}100] - 0.20(7{,}488{,}200)$$

Solving this equation, x = selling price of MCB = \$0.83/lb, which is the same result as in part (a). This is not surprising because Eqs. (23.7) and (23.9) are identical when VP is set to zero and ROI = i_{min}.

23.5 TIME VALUE OF MONEY

All of the profitability measures discussed so far give only a snapshot view at a given point in time. The total annual sales revenues, S, and the total annual production cost, C, are estimated at critical points, normally for the third operating year. Furthermore, a simple depreciation schedule, typically straight-line depreciation, is used. As mentioned earlier, company resources are often sufficiently limited so as not to justify a more careful examination of the revenues and costs over the life of a proposed product and/or plant at the early stages of consideration. However, because of the compounding effect of interest and inflation, it eventually becomes important to account for the time value of money and to charge for depreciation in accordance with the schedule required by the U.S. Internal Revenue Service since 1986 (modified in 1988). In the next section, the methods of calculating cash flows for each year in the life of a proposed product and/or plant project are presented, with which rigorous profitability measures can be computed in Section 23.7. Before doing this, however, it is necessary, in this section, to examine how interest is compounded and to discuss annuities and perpetuities. A number of useful formulas are derived and/or presented for single-payment interest in this section. They are summarized in Table 23.6

Table 23.6 Time Value of Money. Interest. Formulas—Single Payments (Details Presented in Section 23.5)

Interest Type	Formula[a]
Amount of simple interest	$I_S = F - P = niP$
Single-payment simple-amount factor	$\frac{F}{P} = 1 + ni$
Single-payment simple present-worth factor	$\frac{P}{F} = \frac{1}{1 + ni}$
Compound interest	
Amount of compound interest	$I_C = F - P = P[(1+i)^n - 1]$
Single-payment compound-amount factor	$\frac{F}{P} = (1+i)^n$
Single-payment present-worth factor	$\frac{P}{F} = \frac{1}{(1+i)^n}$
Nominal interest rate per year	$r = im$
Effective discrete compound interest	
Effective discrete annual compound interest rate	$i_{eff} = (1+i)^m - 1 = \left(1 + \frac{r}{m}\right)^m - 1$
Amount of discrete compound interest	$I_C = F - P = P[(1 + i_{eff})^{n_y} - 1]$
Discrete single-payment compound-amount factor	$\frac{F}{P} = (1 + i_{eff})^{n_y}$
Discrete single-payment present-worth factor	$\frac{P}{F} = \frac{1}{(1 + i_{eff})^{n_y}}$
Effective continuous compound interest	
Effective continuous annual compound interest rate	$i_{eff} = e^r - 1 = e^{im} - 1$
Amount of continuous compound interest	$I_C = F - P = P[(1 + i_{eff})^{n_y} - 1]$
Continuous single-payment compound-amount factor	$\frac{F}{P} = (1 + i_{eff})^{n_y} = e^{rn_y}$
Continuous single-payment present-worth factor	$\frac{P}{F} = \frac{1}{(1 + i_{eff})^{n_y}} = e^{-rn_y}$

[a] i = interest rate per period; m = number of periods per year; r = nominal interest rate per year; n_y = number of years; i_{eff} = effective annual compound interest rate; n = number of interest periods.

Compound Interest

The time value of money recognizes the fact that an amount of money at the current time, referred to as *present amount, present sum, present value,* or *present worth* and given the symbol P, may not be the same at a future date. Instead, if that money is invested at an *interest rate, i,* and the interest is added to P, the amount of money at the future date will be a *future amount, future value,* or *future worth,* here given the symbol F. The *interest,* which is the compensation for the use of the money or capital over a period of time, is the difference between F and P. The concept of interest is complicated because: (1) the *interest period* is not necessarily 1 yr, (2) interest may be simple or compound, and (3) compounding may be discrete or continuous.

Let us call the starting present worth or present sum the capital or principal, P. Simple interest over several interest time periods is calculated only on P. No interest is calculated on interest accrued in previous interest periods. Thus, the total amount of simple interest for n interest periods, where i is the simple interest rate per period, is

$$\text{Simple interest} = I_S = F - P = niP \qquad (23.11)$$

Simple interest is rarely used. It has been largely replaced by compound interest, which is calculated at each period on the principal plus the accumulated interest. The interest rate, i, is now referred to as the compound interest rate per period. The effect of compounding is shown in Table 23.7, where the future worth, F, of the principal, P, is calculated for n periods. Beginning at the start of the first period with principal (present worth) P, the interest accumulated during the first period is Pi, which when added to P gives the future worth at the end of the first period as $F = P + Pi = P(1+i)^1$. After each period, the power to which $(1+i)$ is raised increases, and consequently, after n compound-interest periods, the principal has grown to

$$F = P(1+i)^n \qquad (23.12)$$

Table 23.7 Compound Interest

No. of Periods	Capital at Start of Period	Interest Paid During Period	F = Future Worth at End of Period
1	P	Pi	$P + Pi = P(1+i)$
2	$P(1+i)$	$P(1+i)i$	$P(1+i)^2$
3	$P(1+i)^2$	$P(1+i)^2 i$	$P(1+i)^3$
.	.	.	.
.	.	.	.
.	.	.	.
n	$P(1+i)^{n-1}$	$P(1+i)^{n-1} i$	$P(1+i)^n$

With compound interest, the total amount of interest after n periods is

$$\text{Compound interest} = I_C = F - P = P[(1+i)^n - 1] \qquad (23.13)$$

The factor $(1+i)^n$ in Eqs. (23.12) and (23.13) is commonly referred to as the *discrete single-payment compound-amount factor*. As shown in Eq. (23.12), when this factor is multiplied by P, we obtain the future worth, F, after n periods with interest rate per period i. If Eq. (23.12) is solved for P, we obtain

$$P = F\left[\frac{1}{(1+i)^n}\right] \qquad (23.14)$$

where the factor $[1/(1+i)^n]$ is the discrete single-payment present-worth factor. When applied in this manner, this factor is a *discount factor* because the present worth is less than (is discounted from) the future worth.

EXAMPLE 23.18

Determine the interest rate per year required to dcuble $10,000 in 10 yr if the interest rate is

(a) simple

(b) compound

SOLUTION

$$P = \$10{,}000, \quad F = 2(10{,}000) = \$20{,}000, \quad n = 10\,\text{yr}$$
$$F - P = \$20{,}000 - \$10{,}000 = \$10{,}000$$

(a) From Eq. (23.11), $\$10{,}000 = niP = 10i(\$10{,}000)$

Solving, $i = 0.10$ or 10%

(b) From Eq. (23.13), $\$10{,}000 = P[(1+i)^n - 1] = \$10{,}000 \times [(1+i)^{10} - 1]$

Solving, $i = 0.0718$ or 7.18%

Thus, money can double in 10 yr with an interest of just over 7% compounded annually.

As seen in Example 23.18, there is a significant difference between simple interest and compound interest. Looking at this example from another perspective, if the compound interest rate for 10 yr were 10%, from Eq. (23.13), the future worth would be $25,937, compared to $20,000 for simple interest. When investing money, one should always seek compound interest so that interest is obtained on the interest.

Nominal and Effective Interest Rates

The interest period can be a day, week, month, year, etc. However, it is commonly defined in fractions of a year, for example, 1 yr, 1/2yr, ..., $1/m$ yr, where m is the number of periods per year. When the interest period is not 1 yr, it is common to use the concepts of *nominal interest rate* and *effective interest rate* for compound interest, both based on 1 yr. The use of these two concepts permits the calculations to be carried out on an annual basis.

Given the value of m, the number of times per year to calculate interest at i, and the interest rate per period of $1/m$ yr (m times per year), the *nominal interest rate* per year, r, is

$$r = im \tag{23.15}$$

If the interest rate is 3%/quarter, then with four quarters per year, the nominal interest rate, r, is 0.03(4) = 0.12 or 12%/yr. In the case of simple interest (no compounding), \$1,000 at the beginning of a year would yield 1,000(1.12) = \$1,200. But, more commonly, nominal interest rates are stated on an annual basis with a compounding period, for example, 12% compounded quarterly.

To handle compound interest when the interest period is some fraction of a year, $(1/m)$, an effective interest rate per year, i_{eff}, is defined by

$$F_{\text{end of 1 yr}} = P(1 + i_{\text{eff}}) \tag{23.16}$$

Based on i, the actual interest rate per $1/m$ yr, we can also write

$$F_{\text{end of 1 yr}} = P(1 + i)^m = P\left(1 + \frac{r}{m}\right)^m \tag{23.17}$$

Equating Eqs. (23.16) and (23.17) and solving for i_{eff} gives

$$i_{\text{eff}} = (1 + i)^m - 1 = \left(1 + \frac{r}{m}\right)^m - 1 \tag{23.18}$$

EXAMPLE 23.19

An interest rate is reported as 3% compounded quarterly. Determine the nominal and effective interest rates per year.

SOLUTION

$$i = 3\%\text{/quarter of a year, } m = 4 \text{ times per year}$$

The nominal interest rate per year, from Eq. (23.15), is

$$r = 0.03(4) = 0.12 \text{ or } 12\%\text{/yr compounded quarterly}$$

From Eq. (23.18), the effective interest rate per year is

$$i_{\text{eff}} = \left(1 + \frac{0.12}{4}\right)^4 - 1 = 0.1255 \text{ or } 12.55\%, \text{ which is larger than}$$

the nominal rate.

Continuous Compounding of Interest

In the limit, as the number of periods per year approaches infinity, that is, as $m \to \infty$, *continuous compounding* occurs and i_{eff} tends to a maximum value for a given value of i. Equation (23.18) becomes

$$i_{\text{eff,cont}} = \lim_{m\to\infty}\left(1 + \frac{r}{m}\right)^m - 1 = \lim_{m\to\infty}\left(1 + \frac{1}{(m/r)}\right)^{(m/r)} - 1$$

Since

$$\lim_{x\to\infty}\left(1 + \frac{1}{x}\right)^x = e = 2.71828\ldots$$

$$\text{and } \lim_{x\to\infty}\left(1 + \frac{1}{x}\right)^{xr} = e^r$$

therefore,

$$i_{\text{eff}} = e^r - 1 \tag{23.19}$$

where r is now the nominal annual interest rate compounded continuously, while i_{eff} is the effective annual interest rate compounded continuously. If r = 10% per year, from Eq. (23.18), $i_{\text{eff}} = \exp^{0.1} - 1 = 0.10517$ or 10.517%.

With continuous compounding, Eq. (23.12) for the future worth in terms of i_{eff} and the number of years, n_y, becomes

$$F = P(1 + i_{\text{eff}})^{n_y} = Pe^{rn_y} \tag{23.20}$$

With continuous compound interest, the total amount of interest after n_y years is

$$\begin{aligned}\text{Continuous compound interest} &= F - P \\ &= P[(1 + i_{\text{eff}})^{n_y} - 1]\end{aligned} \tag{23.21}$$

The factor $(1 + i_{\text{eff}})^{n_y}$, in Eqs. (23.12) and (23.13), which from Eq. (23.19) equals e^{rn_y}, is commonly referred to as the *continuous single-payment compound-amount factor.*

EXAMPLE 23.20

If it is assumed that \$200,000 will be needed for a 4-yr college education starting 10 yr from now, how much must be invested today at a 6% nominal annual interest rate compounded (a) continuously and (b) twice annually?

SOLUTION

F = \$200,000, r = 0.06,

(a) For continuous compounding, from Eq. (23.19), $i_{\text{eff}} = e^{0.06} - 1 = 0.06184$

From Eq. (23.20), with n_y = 10 yr,

$$P = \frac{F}{(1 + i_{\text{eff}})^{n_y}} = \frac{\$200{,}000}{(1 + 0.06184)^{10}} = \$109{,}760$$

(b) Eq. (23.17) gives the future worth for m periods per year of compounding at the end of the first year. For n_y years, that equation becomes

$$F = P\left(1+\frac{r}{m}\right)^{mn_y} \tag{23.22}$$

For compounding twice annually ($m = 2$),

$$P = \frac{F}{\left(1+\frac{r}{m}\right)^{mn_y}} = \frac{\$200{,}000}{\left(1+\frac{0.06}{2}\right)^{2(10)}} = \$110{,}740$$

Note that the minimum capital is obtained when the interest is compounded continuously, with a difference of \$980 between it and the result for semi-annual compounding.

Annuities

Early in this section, only two sums of money were considered, one at the beginning, called present worth, P, and one at the end, called future worth, F. One of these was referred to as the single payment. The two were related by equations involving the interest rate/period and the number of periods that interest was applied. The use of compound interest to determine sums earlier in time (e.g., present worth) that are equivalent to a later, larger sum (e.g., future worth) was referred to as *discounting*. Factors such as $1/(1+i)^n$ are called *discount factors*. The concepts in the previous section can be extended to a very common situation, called the *annuity,* where instead of a single payment, a series of equal payments is made at equal time intervals. Annuities also involve discounting and discount factors.

Everyday applications of annuities include house, automobile, and other loan payments (installments), where the total amount paid back over the loan period includes not only the principal (original amount of the loan), but also interest, sometimes in substantial amounts. Those saving for retirement put payments into an annuity over a period of years, with interest added to their payments. Upon retirement, retirees receive periodic payments over a specified period of years, with the unpaid amount at any period still accumulating interest. Periodic payments are also made to life insurance policies. Other kinds of annuities are created for corporations to accumulate capital, perhaps for building a new chemical processing plant.

In this section, so-called *ordinary annuities* are defined, in which the payments are made at the end of each of n interest periods and interest, i, is compounded per period. The annuity begins at the start of the first period and finishes at the end of the last period, with the duration referred to as the *annuity term.* At the close of the last period, the future worth, F, of all of the payments made is known as the *amount of the annuity.* A number of formulas are derived or presented below in this and the subsequent subsection of Section 23.5. For convenience, they are summarized in Table 23.8. Less common than ordinary annuities and not discussed here are *annuity due,* in which payments are made at the beginning of the period, and the *deferred annuity,* in which the first payment is delayed to a specified date. A *perpetuity* is another form of annuity that continues payments forever.

Discrete Compounding

To determine F when discrete uniform payments of A each are made at the end of each of the n discrete interest periods, the future worth of all the accumulated amounts, payments, and interest is summed to give the amount of the annuity. Thus, starting with the first payment at the end of the first period and finishing with the last payment at the end of the last period,

$$F = A(1+i)^{n-1} + A(1+i)^{n-2} + \ldots + A(1+i) + A \tag{23.23}$$

Table 23.8 Time Value of Money. Annuity Factors—Uniform-Series Payments—Compound Interest (Details Presented in Section 23.5)[a]

Discrete or Continuous Factor	Periodic Interest A, End of Year, Discrete Factor	Continuous Interest A, End of Year, Continuous Factor	Continuous Interest A, Continuous Factor
Uniform-series sinking-fund deposit factor	$\frac{A}{F} = \frac{i}{(1+i)^n - 1}$	$\frac{\bar{A}}{F} = \frac{e^r - 1}{e^{rn_y} - 1}$	$\frac{\bar{A}}{F} = \frac{r}{e^{rn_y} - 1}$
Uniform-series compound-amount factor	$\frac{F}{A} = \frac{(1+i)^n - 1}{i}$	$\frac{F}{\bar{A}} = \frac{e^{rn_y} - 1}{e^r - 1}$	$\frac{F}{\bar{A}} = \frac{e^{rn_y} - 1}{r}$
Uniform-series capital-recovery factor	$\frac{A}{P} = \frac{i(1+i)^n}{(1+i)^n - 1}$	$\frac{\bar{A}}{P} = \frac{e^r - 1}{1 - e^{-rn_y}}$	$\frac{\bar{A}}{P} = \frac{r}{1 - e^{-rn_y}}$
Uniform-series present-worth factor	$\frac{P}{A} = \frac{(1+i)^n - 1}{i(1+i)^n}$	$\frac{P}{\bar{A}} = \frac{1 - e^{-rn_y}}{e^r - 1}$	$\frac{P}{\bar{A}} = \frac{1 - e^{-rn_y}}{r}$

[a] i = periodic interest rate; A = payment per interest period; n = number of interest periods; $\bar{A}$ = total annual payments per year; r = nominal annual interest rate.

Note that because the first payment is made at the end of the first period, it is compounded over the remaining $(n-1)$ periods. Also, the last payment is made at the end of the last period, and consequently it is not compounded. Because as n becomes large, Eq. (23.23) becomes cumbersome to evaluate, it is useful to simplify the equation. This is accomplished by multiplying both sides of Eq. (23.23) by $(1+i)$ to give

$$F(1+i) = A(1+i)^n + A(1+i)^{n-1} + \dots + A(1+i)^2 + A(1+i) \tag{23.24}$$

Then, if Eq. (23.23) is subtracted from Eq. (23.24), we obtain

$$Fi = A(1+i)^n - A \tag{23.25}$$

which, when rearranged, gives

$$F = A\left[\frac{(1+i)^n - 1}{i}\right] \tag{23.26}$$

where the factor $[(1+i)^n - 1]/i$ is referred to as the *discrete uniform-series compound-amount factor.* If Eq. (23.26) is solved for A, we obtain

$$A = F\left[\frac{i}{(1+i)^n - 1}\right] \tag{23.27}$$

where the factor $i/[(1+i)^n - 1]$ is referred to as the *discrete uniform-series sinking-fund deposit factor.* A sinking fund consists of periodic deposits that accumulate with interest up to a maturity date. In the past, some companies have used a sinking fund as a depreciation allowance to recover an original capital investment.

Sometimes, periodic payments, A, are made two or more times per year and interest is also compounded the same number of times per year, that is, m times each year. In that case, it is convenient to express the annual total of all annuity payments by the variable $\overline{A}$. Then, the payment per period, A, is simply $\overline{A}/m$. Since $i = r/m$ and $n = mn_y$, Eq. (23.26) can be rewritten as

$$F = \frac{\overline{A}}{m}\left[\frac{\left(1+\frac{r}{m}\right)^{mn_y} - 1}{r/m}\right] = \overline{A}\left[\frac{\left(1+\frac{r}{m}\right)^{mn_y} - 1}{r}\right] \tag{23.28}$$

If, in the more general case, equal payments are made p times per year while interest is compounded m times per year, then, according to Bauman (1964), the future worth becomes

$$F = \widehat{A}\left[\frac{\left(1+\frac{r}{m}\right)^{mn_y} - 1}{\left(1+\frac{r}{m}\right)^{m/p} - 1}\right] \tag{23.29}$$

where $\widehat{A}$ is the amount of each payment and $\overline{A} = p\widehat{A}$. Equation (23.29) is not included in Table 23.8, but is considered in Example 23.21.

Continuous Compounding

For continuous compounding of interest with continuous payments, as $m \rightarrow \infty$, Eq. (23.26) can be expressed as follows, with the limit obtained, as before, from the derivation of Eq. (23.19):

$$F = \lim_{m\to\infty} \overline{A}\left[\frac{(1+r/m)^{(m/r)(rn_y)} - 1}{r}\right] = \overline{A}\left(\frac{e^{rn_y} - 1}{r}\right) \tag{23.30}$$

The factor $\left(\frac{e^{rn_y} - 1}{r}\right)$ is referred to as the *continuous uniform-series compound-amount factor.* Equation (23.30) seems hypothetical because, although interest can be credited continuously, payments cannot be made continuously.

More practical is continuous compounding of interest, but with equal discrete payments at p times per year and totaling $\overline{A}$ each year, giving the limit of Eq. (23.29) as $m \rightarrow \infty$ as

$$F = \lim_{m\to\infty} \widehat{A}\left[\frac{(1+r/m)^{(m/r)(rn_y)} - 1}{\left(1+\frac{r}{m}\right)^{(m/r)(r/p)} - 1}\right] = \widehat{A}\left(\frac{e^{rn_y} - 1}{e^{(r/p)} - 1}\right) \tag{23.31}$$

Equation (23.31) is not included in Table 23.8, but the case for just one payment per year, $p = 1$, with continuous compounding, is included.

EXAMPLE 23.21

For the college education savings plan considered in Example 23.20, which is estimated to require \$200,000 10 yr from now, calculate the total of the payments made each year to an annuity at a 6% nominal interest rate for the following conditions:

(a) Interest compounded continuously and payments continuous.

(b) Interest compounded continuously but payments quarterly.

(c) Interest compounded continuously but payments annually.

(d) Interest compounded quarterly and payments quarterly.

(e) Interest compounded semi-annually and payments semi-annually.

(f) Interest compounded quarterly and payments monthly.

For the lowest and highest payments, on an annual basis, compute the total amount of payments.

SOLUTION

For this example, $r = 0.06, F = \$200{,}000, n_y = 10$ yr.

Use the $\widehat{A}/F$ or $\overline{A}/F$ uniform-series sinking-fund deposit factors.

(a) Equation (23.30) applies:

$$\overline{A} = F\frac{r}{e^{rn_y}-1} = \$200{,}000\,\frac{0.06}{e^{0.06(10)}-1} = \$14{,}596/\text{yr}$$

(b) Equation (23.31) applies with $p = 4$ payment/yr and $A = \$/\text{payment}\ \overline{A} = p\widehat{A} = 4\widehat{A}$:

$$\widehat{A} = F\left(\frac{e^{(r/p)}-1}{e^{rn_y}-1}\right) = \$200{,}000\left(\frac{e^{0.06/4}-1}{e^{0.06(10)}-1}\right)$$

$$= \$3{,}677/\text{payment}$$

Therefore, $\overline{A} = 4(\$3{,}677) = \$14{,}706/\text{yr}$.

(c) Equation (23.31) applies with $p = 1$ payment/yr. Therefore, $\overline{A} = \widehat{A}$.

$$\overline{A} = \widehat{A} = F\left(\frac{e^{(r/p)}-1}{e^{rn_y}-1}\right) = \$200{,}000\left(\frac{e^{0.06/1}-1}{e^{0.06(10)}-1}\right)$$

$$= \$15{,}043/\text{yr}$$

(d) Equation (23.28) applies with $m = p = 4$ payment/yr.

$$\overline{A} = F\left[\frac{r}{\left(1+\frac{r}{m}\right)^{mn_y}-1}\right] = \$200{,}000\left[\frac{0.06}{\left(1+\frac{0.06}{4}\right)^{4(10)}-1}\right]$$

$$= \$14{,}742/\text{yr}$$

(e) Equation (23.28) applies with $m = p = 2$ payment/yr.

$$\overline{A} = F\left[\frac{r}{\left(1+\frac{r}{m}\right)^{mn_y}-1}\right] = \$200{,}000\left[\frac{0.06}{\left(1+\frac{0.06}{2}\right)^{2(10)}-1}\right]$$

$$= \$14{,}886/\text{yr}$$

(f) Equation (23.29) applies with $m = 4$ payment/yr and $p = 12$ payment/yr. $\overline{A} = p\widehat{A}$.

$$\widehat{A} = F\left[\frac{\left(1+\frac{r}{m}\right)^{m/p}-1}{\left(1+\frac{r}{m}\right)^{mn_y}-1}\right] = \$200{,}000\left[\frac{\left(1+\frac{0.06}{4}\right)^{4/12}-1}{\left(1+\frac{0.06}{4}\right)^{4(10)}-1}\right]$$

$$= \$1{,}222/\text{payment}$$

$$\overline{A} = p\widehat{A} = 12(\$1{,}222) = \$14{,}669/\text{yr}$$

Because the nominal interest rate is relatively low, the differences between the answers are not large, ranging from a low of \$14,596/yr for continuous compounding of interest and continuous payments to a high of \$15,043/yr for annual payments with interest compounded continuously. Thus, the total amount of payments over the 10 yr of payments ranges from \$145,960 to \$150,430. The annual payment is even higher for discrete annual payments with interest compounded annually: \$15,174/yr or a total of \$151,740 for 10 yr.

EXAMPLE 23.22

An engineer begins employment at the age of 25 and plans to invest enough money to have \$1,000,000 at a retirement age of 65. Assume that payments to the retirement fund will be made each month and that the money will receive interest at 8% compounded quarterly. Calculate the amount of each payment and the total amount of the payments made during the 40-yr savings period.

SOLUTION

Equation (23.29) applies with $F = \$1{,}000{,}000$, $r = 0.08$, $m = 4$ times/yr, and $p = 12$ times/yr.

$$\widehat{A} = F\left[\frac{\left(1+\frac{r}{m}\right)^{m/p}-1}{\left(1+\frac{r}{m}\right)^{mn_y}-1}\right]$$

$$= \$1{,}000{,}000\left[\frac{\left(1+\frac{0.08}{4}\right)^{4/12}-1}{\left(1+\frac{0.08}{4}\right)^{4(40)}-1}\right] = \$290.85/\text{month}$$

For the $12(40) = 480$ payments, the total amount of payments is only $480(\$291) = \$139{,}680$. The growth of the future worth is exponential, as shown in the following table, where an additional 10 yr of payments is added, giving a future worth of \$2,261,096.

End of Year	Future Worth (\$)	Total Payments (\$)
10	53,054	34,902
20	170,200	69,805
30	428,862	104,707
40	1,000,000	139,680
50	2,261,096	174,582

This example shows clearly the remarkable power of compound interest.

Present Worth of an Annuity

The *present worth of an annuity, P,* is the amount of money at the present time that, if invested at a compound interest rate, will yield the amount of the annuity, *F,* at a future time. This is useful for determining the periodic payments from an annuity that can be made over a specified number of years in the future.

Annuity equations relating F and the periodic payments, A, are converted to equations relating P to A by combining them with Eq. (23.12) for discrete interest or Eq. (23.20) for continuous interest. This is often referred to as discounting the amount of the annuity to determine its present worth. In Table 23.8, under periodic interest, the discrete uniform-series sinking-fund deposit factor becomes the discrete uniform-series capital-recovery factor in the following manner:

$$P = \frac{F}{(1+i)^n} = A\left[\frac{(1+i)^n - 1}{i(1+i)^n}\right] \qquad (23.32)$$

Similarly, the continuous uniform-series capital-recovery factor with payments A at the end of each year is obtained:

$$P = \frac{F}{e^{rn_y}} = \frac{A}{e^{rn_y}}\left[\frac{e^{rn_y} - 1}{e^r - 1}\right] = A\left[\frac{1 - e^{-rn_y}}{e^r - 1}\right] \qquad (23.33)$$

When comparing two annuities involving many payments into the future, it can be very helpful to discount all of the payments to their present worth. This gives the principal required at the current time, invested at the current interest rate, to enable the payments to be made at the end of each annuity period. While the annuity is making payments, interest continues to be paid on the remaining balance. At the end of the term of the annuity, the balance is zero.

EXAMPLE 23.23

Upon retirement at the age of 65, an employee has a retirement fund of \$1,000,000. If this fund is invested at 8% compounded quarterly, how much can be paid to the retiree at the end of each month if the fund is to diminish to zero at the end of 20 yr when the retiree would be 85?

$$r = 0.08, m = 4, p = 12, \text{and } P = \$1{,}000{,}000$$

SOLUTION

Since the period of compounding and the payment period are different, none of the equations in Table 23.8 apply. Instead, use the following extension of Eq. (23.14),

$$P = \frac{F}{(1+i)^n} = \frac{F}{\left(1+\frac{r}{m}\right)^{mn_y}} \qquad (23.34)$$

with Eq. (23.29), to give

$$\widehat{A} = P\left[\frac{\left(1+\frac{r}{m}\right)^{m/p} - 1}{1 - \left(1+\frac{r}{m}\right)^{-mn_y}}\right]$$

Thus,

$$\widehat{A} = \$1{,}000{,}000\left[\frac{\left(1+\frac{0.08}{4}\right)^{4/12} - 1}{1 - \left(1+\frac{0.08}{4}\right)^{-4(20)}}\right]$$

$$= \$8{,}332/\text{month or } \$99{,}979/\text{yr}$$

It is interesting to note that if the monthly payments are reduced to \$6,536, or \$19,608 quarterly, then at the end of the first quarter, 2% interest will be paid on the balance of (\$1,000,000 − \$19,608 = \$980,392), giving \$19,608, which is the same as the amount paid out during the quarter. Thus, the amount of the annuity will remain at \$1,000,000 and payments can continue forever.

Comparing Alternative Equipment Purchases

It is often desirable to compare the purchases of two or more alternative items of equipment, each having a different installed cost and estimated performance life, maintenance cost, and salvage value. The two main methods for comparison, *present worth* and *capitalized cost,* are covered in this subsection. At the outset, it is important to recognize that these measures are examined, often on an ad-hoc basis, primarily for the purchase of an equipment item after the plant has been designed. During the comparison of alternative plant designs or major retrofits, when it is important to account for sales revenues, the calculation of *cash flows* is recommended, as described in the next section, for use in computing the *net present value* (NPV) or the *investor's rate of return* (IRR or DCFRR). Note, however, that the present worth and NPV are identical when there are no revenues, for example, when comparing alternative methods for treating a waste stream. In general, when making comparisons of alternatives, it is not necessary to consider so-called *sunk costs,* which are costs that occurred in the past but have no effect on current or future decisions.

Present Worth

In the present-worth technique, all of the costs and revenues are discounted to calculate the *present worth* of each alternative. Note that it is crucial to compare the alternatives over the same time period. This approach is illustrated in Example 23.24, in which diagrams show the projected costs and the recovery of the salvage values in time.

EXAMPLE 23.24

Two alternative pumps, A (carbon steel) and B (aluminum), have different installed and maintenance costs, salvage values, and anticipated service lives, as indicated below. It is desired to select one of the pumps on the basis of present worth when the effective interest rate is 10%.

	A	B
Installed cost	$18,000	$25,000
Uniform end-of-year maintenance	$4,000	$3,000
Salvage value	$500	$1,500
Service life	2 yr	3 yr

SOLUTION

Six years is the shortest time period for which the two pumps can be compared on a common basis because six is the smallest number divisible by both two and three. Thus, pump A is replaced twice and salvaged three times over 6 yr, during which time pump B is replaced once and salvaged twice. For pump A, the costs and salvage values are shown on the following diagram, in which the installed and maintenance costs are represented by downward vectors (i.e., negative, compared to zero along the horizontal axis) and the salvage values are represented by upward vectors (i.e., positive). Notice that at the end of the second, fourth, and sixth years, the maintenance costs appear even though the pump is being salvaged and replaced. These are charges that have accumulated over the prior year.

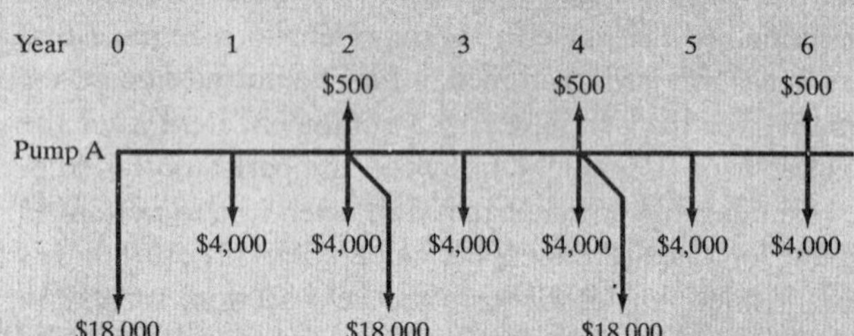

When discounting the costs and salvage values, the maintenance costs can be treated as an annuity, computed using Eq. (23.32) with $i = 0.10$ and $n = 6$, because the cost is periodic and constant at $4,000/yr. Also, the salvage value can be credited against the purchase cost, giving $18,000 – $500 = $17,500, because they both occur at the end of the second and fourth years. Thus,

$$P_A = -\$18,000 - \$4,000\left[\frac{(1+0.1)^6 - 1}{0.1(1+0.1)^6}\right] - \frac{\$17,500}{(1+0.1)^2} - \frac{\$17,500}{(1+0.1)^4} + \frac{\$500}{(1+0.1)^6} = -\$61,554$$

The corresponding diagram for pump B is

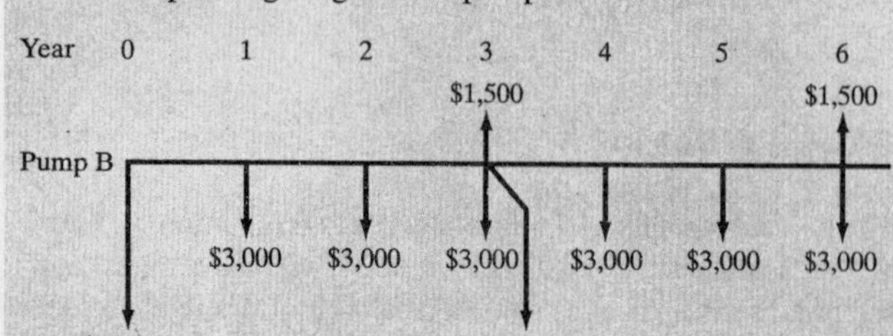

and the discounted costs and salvage values are

$$P_B = -\$25,000 - \$3,000\left[\frac{(1+0.1)^6 - 1}{0.1(1+0.1)^6}\right] - \frac{\$23,500}{(1+0.1)^3} + \frac{\$1,500}{(1+0.1)^6} = -\$54,875$$

Although pump B has the higher installation cost, it is selected because its present worth is lower than that of pump A.

Capitalized Costs and Perpetuities

Another method for comparing alternatives, which leads to conclusions identical to those of present worth, is to compute *capitalized costs*. This involves the creation of a *perpetuity*, in which periodic replacements continue indefinitely for each alternative. The capitalized cost, K, is defined as the original cost, C_I, plus the present value of the perpetuity for an infinite number of replacements made every n_R years. When a replacement is made, it is common to assign a salvage value, S_{equip}. Thus, if inflation of costs is ignored, the replacement cost is constant at $C_R = C_I - S_{\text{equip}}$. Note that better estimates for the replacement costs, taking into account inflation, likely market conditions, and similar factors, are not normally justified for comparisons involving perpetuities. To account for such factors, the cash flow analysis in the next section is preferred. Assuming a nominal interest rate r compounded m times per year, and using the general form of the discount factor in Eq. (23.34), the investment must provide a future worth, F, every n_y years, sufficient to pay for the replacement of the equipment item and replace the principal, P, so that it can be reinvested for another n_y years. Thus,

$$F_{n_y} = C_R + P = P\left(1 + \frac{r}{m}\right)^{mn_y} \tag{23.35}$$

Rearranging Eq. (23.35), the present value of the perpetuity is

$$P = \frac{C_R}{\left(1 + \frac{r}{m}\right)^{mn_y} - 1} \tag{23.36}$$

From the definition of the capitalized cost,

$$K = C_I + P = C_I + \frac{C_R}{\left(1 + \frac{r}{m}\right)^{mn_y} - 1} \tag{23.37}$$

When comparing alternatives, the equipment item having the lowest capitalized cost is selected.

EXAMPLE 23.25

Select one of the two pumps in Example 23.24 on the basis of capitalized costs.

SOLUTION

To use capitalized costs when annual operating costs are also required (in this case for maintenance), it is appropriate to discount the operating cost payments to present worth, using the annuity equation [Eq. (23.32)], and to add this to the equipment installed cost. For pump A, the initial adjusted installed cost is

$$C_{I_A} = \$18{,}000 + \$4{,}000\left[\frac{(1+0.1)^2 - 1}{0.1(1+0.1)^2}\right] = \$18{,}000 + \$6{,}940$$

$$= \$24{,}940$$

and the capitalized cost, using Eq. (23.37) with $r = 0.1$, $m = 1$, and $n_y = 2$, is

$$K_A = \$24{,}940 + \frac{\$24{,}940 - \$500}{(1+0.1)^2 - 1} = \$24{,}940 + \$116{,}380$$

$$= \$141{,}320$$

Turning to pump B,

$$C_{I_B} = \$25{,}000 + \$3{,}000\left[\frac{(1+0.1)^3 - 1}{0.1(1+0.1)^3}\right] = \$25{,}000 + \$7{,}460$$

$$= \$32{,}460$$

and

$$K_B = \$32{,}460 + \frac{\$32{,}460 - \$1{,}500}{(1+0.1)^3 - 1} = \$32{,}460 + \$93{,}540$$

$$= \$126{,}000$$

Indeed, pump B has the lower capitalized cost, a result consistent with the comparison in Example 23.24 of the two present worths. In fact, P_A/P_B in Example 23.24 equals K_A/K_B in Example 23.25.

23.6 CASH FLOW AND DEPRECIATION

The approximate profitability measures in Section 23.4—although often used to select the most promising products during the *concept* stage of the Stage-Gate™ Product-Development Process (SGPDP) and the most promising flowsheets during process synthesis (when the details of the process units are deemphasized relative to the arrangements and sequences of the units in the flowsheet)—are inadequate to enable management to make a final decision regarding the financial feasibility of a potential chemical product or process. Dr. Robert M. Busche, former President of Bio-en-gene-er Associates and a long-time engineer and venture analyst at DuPont, often reminds students that chemical companies like DuPont were awash in cash assets during the period following World War II, throughout the 1950s and 1960s. At that time, the principal challenges were to develop new products and commercialize them quickly. Approximate measures of financial goodness, like the return on investment (ROI), were used routinely to decide whether a chemical product or process was sufficiently promising to fund. In the 1970s and 1980s, however, cash assets became less plentiful and the competition among potential chemical products and processes for the limited resources of a company became much stiffer. To arrive at decisions, management required more accurate assessments of the financial aspects of potential processes, and consequently, companies began to require cash flow estimates for each year of operation of the most promising chemical products and processes. This change in perspective was signaled in a pioneering paper by Souders (1966). Note, however, that the ROI, as well as the PBP, continue to be computed, primarily because they are easy to calculate. They permit a quick comparison of investments, with relatively few calculations, and are especially useful when the costs and revenues are not anticipated to change significantly over the life of the project.

As was discussed in Chapter 22, *cash flow* for a company has become an important financial factor. Cash flow is defined as the net passage of money into or out of a company due to an investment. It may be positive (into) or negative (out of). For investment evaluation, investments are considered a negative cash flow, while after-tax profits plus depreciation are positive cash flows. During the years of plant construction or creation of a chemical product manufacturing facility, the cash flow, CF, for a particular year, is

$$\text{CF} = -fC_{\text{TDC}} - C_{\text{WC}} - C_{\text{land}} \tag{23.38}$$

where f is the fraction of the total depreciable capital, C_{TDC}, expended that year, and C_{WC} and C_{land} are the working capital and the cost of land that are expended, if any, during that year of construction.

To estimate the cash flow for a particular year of plant operation, the pre-tax earnings are computed from Eq. (23.4) and the after-tax earnings from Eq. (23.5). However, in the calculation of the production cost, a more elaborate depreciation schedule, discussed in the next section, replaces the straight-line depreciation used in Section 23.2. The cash flow from plant and product manufacturing operations is the after-tax earnings plus the depreciation:

$$\text{CF} = (1-t)(\text{S} - \text{C}) + \text{D} = 0.60(\text{S} - \text{C}) + \text{D} \tag{23.39}$$

During the first year of operation, startup costs may occur. During all years of operation, there may be royalty and additional investment costs. At the conclusion of plant operations, there may be a salvage value for used equipment, S_{equip}.

The annual cash flow, CF, for any year of the project, including the construction phase and possible salvage at the end of operation, is

$$\text{CF} = (1-t)(\text{S} - \text{C}) + \text{D} - fC_{\text{TDC}} - C_{\text{WC}} - C_{\text{land}} - C_{\text{startup}} - C_{\text{royal}} + S_{\text{equip}} \tag{23.40}$$

In Eq. (23.40), the depreciable capital is normally expended before operation of the plant or product manufacturing facility begins, and the working capital is usually expended in the year preceding the beginning of operation. The working capital is recovered during the last year of operation as a negative entry (to give a positive cash flow) in Eq. (23.40). A common convention is that all cash transactions take place at the end of the year. It is also common to project cash flows for new commodity chemical products over 10-plus years, typically 15 yr, whereas cash flows are projected for established commodity chemical products for 20 yr or more. Another convention recommended by Busche (1995) is to design commodity chemical plants on a capacity basis to operate during 330 days (7,920 hr) per year, with 35 days for shutdowns due to maintenance needs and malfunctions. This corresponds to an operating factor of 0.9041. Some companies prefer to round this figure to 8,000 hr/yr. In addition, Busche recommends that the total production cost, C (computed using the cost sheet of Table 23.1), and sales, S, be computed for production at less than 100% of capacity during the first and, perhaps, second year of operation, while the plant is being started up and any design flaws are being remedied.

For specialty chemicals, including many pharmaceuticals and many new industrial and configured consumer products, the rapid development of new technologies shortens the projected life of new chemical products. As illustrated for the home hemodialysis and lab-on-a-chip products in Sections 17.3 and 17.4, a projected life of just five years is common. Also, some small businesses prefer to operate batch processing operations during the normal workweek; that is, in just one 40-hr shift per week.

Depreciation

Depreciation is the reduction in value of an asset. Recall from the cost sheet of Table 23.1 that a company is allowed to treat depreciation as a cost of production, thereby reducing its income tax liability even though depreciation is not an actual cash flow out of the company. When calculating approximate profitability measures such as the return on investment (ROI), it is common to calculate the cost of sales using straight-line (SL) depreciation, as in Table 23.1. Since the approximate profitability measures give just a snapshot view of the economic goodness of a proposed project, usually projected for the third year of operation, no other method of accounting for the depreciation of the total depreciable capital is justified.

Two profitability measures discussed below that provide more rigor—net present value (NPV) and investor's rate of return (IRR)—involve the discounting of cash flows to present worth, as discussed in the next section. These measures increase in magnitude when a larger fraction of the total depreciation is taken in the early years of operation, when the plant or product manufacturing facility is probably operating below capacity and the discount factors are low. For these reasons, it is advantageous for a company to rapidly depreciate its capital investment early in the life of a process instead of using straight-line depreciation. Depreciation methods that favor the earlier years include the *declining-balance method* (DB), the *double declining-balance method* (DDB), and the *sum-of-the-years digits method* (SYD). More recently, in 1981, the U.S. federal income tax regulations provided an *Accelerated Cost Recovery System* (ACRS) for early depreciation. A *Modified Accelerated Cost Recovery System* (MACRS) went into effect in 1987. The ACRS and MACRS methods combine aspects of the DB or DDB methods with the SL method. These five methods are discussed next and compared to straight-line depreciation. Another depreciation method that is sometimes referred to is the sinking-fund method. However, it decelerates rather than accelerates depreciation and, therefore, is not of interest to most industries and is not considered here. It is important to note that a company may use two or more different depreciation methods, most commonly: (1) one method for *book depreciation* for internal financial accounting and (2) one method for *tax depreciation* that follows government regulations.

All of the depreciation methods to be discussed are based on the asset *book value,* which at any year is defined as the original cost of the asset (e.g., the depreciable capital investment) minus the sum of the depreciation charges made to the asset up to that year. This is in contrast to the *market value,* which is the price that could be obtained for the asset if it were placed for sale in the open market, and the *replacement value,* which is the cost of replacing the asset. The book value is the value shown on the accounting records. The book value decreases each year until it reaches a salvage value, at which time it is completely depreciated. The number of years, n, over which an asset can be depreciated is usually related to an estimate of the useful life of the asset, which is discussed below.

Declining-Balance (DB) and Double Declining-Balance (DDB) Methods

The declining-balance method is also referred to as the fixed percentage or uniform percentage method because the amount of depreciation each year is a fixed fraction, d, of the book value of the depreciable asset. Let B = the original cost of the asset, which is usually called the *basis*; t = years of service of the asset; and BV_t = book value at the end of year t. Then, the amount of annual depreciation, D_t, for the year t is given by

$$D_t = BV_{t-1} - BV_t = dBV_{t-1} \tag{23.41}$$

Consequently, after $t - 1$ yr, the book value is

$$BV_{t-1} = B(1 - d)^{t-1} \tag{23.42}$$

Combining Eqs. (23.41) and (23.42) gives

$$D_t = dB(1 - d)^{t-1} \tag{23.43}$$

Limits are placed on the value of d, allowing it to range only from $1/n$ to $2/n$, with $1.5/n$ (150% declining balance) and $2/n$ (200% or double declining balance) being common values. With the declining-balance methods, a salvage value is not used. However, the book value, which never reaches zero because it is only decreased each year by a fixed fraction, is not permitted to drop below the estimated salvage value. To force the book value to the salvage value at the end of year n, it is considered desirable to use the *combination method,* which involves switching from the declining-balance method to the straight-line method partway through the service life. Another scheme is to back-calculate the value of d that will give a book value equal to the salvage value at year $t = n$. This value of d is obtained from Eq. (23.42) by setting $(t - 1) = n$. Thus,

$$S_{\text{equip}} = \mathrm{B}(1 - d)^n \tag{23.44}$$

Solving Eq. (23.44) for d gives

$$d = 1 - \left(\frac{S_{\text{equip}}}{\mathrm{B}}\right)^{1/n} \tag{23.45}$$

However, if the computed $d > 0.2$, it is out of the accepted range and cannot be used. In that case, the only alternative declining-balance method is the combination method, as illustrated in the following example.

EXAMPLE 23.26

A new instrument is purchased for the control laboratory of a plant at a cost of \$200,000. It is estimated to have a 10-yr useful life with a salvage value of \$30,000. Estimate the amount of depreciation each year by the following methods:

(a) Straight-line depreciation over 10 yr based on \$200,000–\$30,000 = \$170,000.

(b) Declining-balance depreciation with $d = 1/n$.

(c) 150% declining-balance depreciation.

(d) Double declining-balance depreciation.

(e) Combination method of double declining-balance depreciation switching to straight-line depreciation after 5 yr.

SOLUTION

(a) The amount of depreciation each year is constant at \$170,000/10 = \$17,000.

From Eq. (23.45), for declining-balance methods,

$$d = 1 - \left(\frac{S_{\text{equip}}}{\mathrm{B}}\right)^{1/n} = 1 - \left(\frac{\$30{,}000}{\$200{,}000}\right)^{1/10} = 0.173$$

Therefore, the 100% and 150% declining-balance methods will not be attractive, since this $d > 0.10$ and 0.15. The 200% declining-balance method and the combination method will lead to good results.

(b) The depreciation each year is computed from Eq. (23.41) with B = \$200,000 and $d = 1/10 = 0.10$. See the table below.

(c) Use Eq. (23.41) with B = \$200,000 and $d = 1.5/10 = 0.15$. See the table below.

(d) Use Eq. (23.41) with B = \$200,000 and $d = 2/10 = 0.20$. See the table below.

(e) Use Eq. (23.41) with B = \$200,000 and $d = 1.5/10 = 0.15$ for the first five years and then subtract the salvage value from the book value and continue with straight-line depreciation. See the table below.

The results of the calculations, which are readily carried out on a spreadsheet, are as follows: For parts (a), (b), and (c):

	Straight-Line Depreciation		Declining Balance ($d = 0.1$)		Declining Balance ($d = 0.15$)	
End of Year	D (\$/yr)	BV (\$)	D (\$/yr)	BV (\$)	D (\$/yr)	BV (\$)
0		200,000		200,000		200,000
1	17,000	183,000	20,000	180,000	30,000	170,000
2	17,000	166,000	18,000	162,000	25,500	144,500
3	17,000	149,000	16,200	145,800	21,675	122,825
4	17,000	132,000	14,580	131,220	18,424	104,401
5	17,000	115,000	13,122	118,098	15,660	88,741
6	17,000	98,000	11,810	106,288	13,311	75,430
7	17,000	81,000	10,629	95,659	11,314	64,115
8	17,000	64,000	9,566	86,093	9,617	54,498
9	17,000	47,000	8,609	77,484	8,175	46,323
10	17,000	30,000	7,748	69,736	6,949	39,375

Note in the above table for parts (b) and (c) that the salvage value of \$30,000 is not reached by year 10. Thus, these are not good methods to apply.

For parts (d) and (e):

	Declining Balance ($d = 0.20$)		Combination Method	
End of Year	D (\$/yr)	BV (\$)	D (\$/yr)	BV (\$)
0		200,000		200,000
1	40,000	160,000	40,000	160,000
2	32,000	128,000	32,000	128,000
3	25,600	102,400	25,600	102,400
4	20,480	81,920	20,480	81,920
5	16,384	65,536	16,384	65,536
6	13,107	52,429	7,107	58,429
7	10,486	41,943	7,107	51,322
8	8,389	33,554	7,107	44,214
9	3,554	30,000	7,107	37,107
10	0	30,000	7,107	30,000

For part (d), the double declining-balance method, with $d = 0.2 > 0.173$ above, the salvage value is reached by the book value before 10 yr. As shown in the table, it is reached in year 9, so that no depreciation is taken in year 10. In part (e), the combination method switches from the double declining-balance method to the straight-line method in year 6, such that the book value becomes the salvage value in year 10. For these two methods, the depreciation is greatly accelerated over the straight-line method in the first 4 yr.

Sum-of-the-Years Digits Method (SYD)

This is a classic depreciation acceleration method that has the advantage of being able to handle a salvage value, including zero. Its disadvantage is that the depreciation acceleration is less than the double declining-balance method. Its name is derived from the use of the sum of the digits from 1 to *n*, the number of years of useful life of the asset. This sum in compact form is given by

$$\text{SUM} = \sum_{j=1}^{n} j = \frac{n(n+1)}{2} \qquad (23.46)$$

Thus, for $n = 10$ yr, $\text{SUM} = 10(10+1)/2 = 55$. The annual depreciation is

$$D_t = \frac{\text{depreciable years remaining}}{\text{SUM}}(B - S_{\text{equip}}) \qquad (23.47)$$

Thus, if $n = 10$ yr, the fraction depreciated the first year is $10/55 = 0.1818$, which is almost twice that of the straight-line method. For the next year, the fraction is $9/55 = 0.1636$. In year 6, the fraction is $5/55 = 0.0909$, which is now less than the straight-line depreciation of 0.10. If Example 23.26 is applied to the SYD method, the following results are obtained, which are compared to the DDB method:

	Sum-of-the-Years Digits		Declining Balance ($d = 0.20$)	
End of Year	D ($/yr)	BV ($)	D ($/yr)	BV ($)
0		200,000		200,000
1	30,909	169,091	40,000	160,000
2	27,818	141,273	32,000	128,000
3	24,727	116,545	25,600	102,400
4	21,636	94,909	20,480	81,920
5	18,545	76,364	16,384	65,536
6	15,455	60,909	13,107	52,429
7	12,364	48,545	10,486	41,943
8	9,273	39,273	8,389	33,554
9	6,182	33,091	3,554	30,000
10	3,091	30,000	0	30,000

In the first 3 yr, the DDB method accelerates the depreciation much more than the SYD method, although the depreciation in the first year of the SYD method, $30,909, is considerably higher than the $17,000 of the straight-line method.

ACRS and MACRS Methods for Tax Depreciation

From 1982 to 1986, the U.S. federal income tax regulations required companies to use the Accelerated Cost Recovery System (ACRS) to depreciate property when computing federal income tax. In 1987, the Modified Accelerated Cost Recovery System (MACRS) replaced that system. Both systems are based on the declining-balance method with a switch to the straight-line method when it offers a faster depreciation write-off. However, both methods assume that assets are placed in service at the midpoint of the tax year.

Table 23.9 MACRS Tax-Basis Depreciation

	Percent of total depreciable capital (C_{TDC}) for class life of:			
Year	5 Yr	7 Yr	10 Yr	15 Yr
1	20.00	14.29	10.00	5.00
2	32.00	24.49	18.00	9.50
3	19.20	17.49	14.40	8.55
4	11.52	12.49	11.52	7.70
5	11.52	8.93	9.22	6.93
6	5.76	8.92	7.37	6.23
7	100.00	8.93	6.55	5.90
8		4.46	6.55	5.90
9		100.00	6.56	5.91
10			6.55	5.90
11			3.28	5.91
12			100.00	5.90
13				5.91
14				5.90
15				5.91
16				2.95
				100.00

Table 23.10 GDS Class Life for Use with the MACRS Depreciation Method

Type of Asset	GDS Class Life (years)
Special manufacturing and handling devices, e.g., tractors	3
Autos, trucks, buses; cargo containers, computers and peripherals; copy and duplicating equipment; some manufacturing equipment	5
Railroad cars, engines, tracks; agricultural machinery; office furniture; petroleum and natural gas equipment and some other manufacturing equipment; all other business assets not listed in another class	7
Equipment for water transportation, petroleum refining, agriculture product processing, durable-goods manufacturing, and shipbuilding	10
Land improvements, docks, roads, drainage, bridges, pipelines, landscaping, nuclear-power production, and telephone distribution	15
Farm buildings, telephone switching buildings, power production equipment, municipal sewers, and water utilities	20
Residential rental property, including mobile homes	27.5
Nonresidential real property attached to the land	39

Therefore, for both methods, only 50% of the DB depreciation is allowed in the first year. Another departure occurs for the MACRS method, wherein the depreciation is continued for 1 yr beyond the life, but only 50% of the straight-line depreciation is taken in that final year. For both methods, the service life (called *class life*) is fixed by regulations for from 3 to 15 yr, and to 20 yr and even longer (in the case of some structures) for the MACRS method. The depreciation calculations are best carried out using the U.S. tables. The MACRS depreciation table is shown in Table 23.9 for class lives of 5, 7, 10, and 15 yr. The selection of class life is also regulated by the U.S. federal government, which offers two options: (1) the General Depreciation System (GDS) and (2) the Alternative Depreciation System (ADS). The GDS allows a more desirable, shorter class life and is the preferred choice. However, the ADS is sometimes used by new businesses that do not need the tax benefit of accelerated depreciation. Table 23.10 gives the GDS class life for a number of different kinds of assets. For most new chemical plant projects, a class life of 5, 7, or 10 yr is used. For these three class lives, Table 23.9 shows that depreciation begins with the double declining-balance method. For example, for a class life of 10 yr, depreciation in the first year is 50% of $2/n = 2/10 = 0.20$, which gives the 10% shown in the table. When 10% of the basis, B, is subtracted from B, the book value is 90% of the basis. In year 2, the DDB depreciation is 20% of the 90% or 18%, which is the value shown in the table. Also, for a same class life of 10 yr, the table shows a switch to straight-line depreciation of 6.55% in year 7, because the calculated DDB depreciation would be lower at 5.90%.

EXAMPLE 23.27

In Example 23.14, the total depreciable capital of a new plant is projected to be $90 million. Compute the annual depreciation by the MACRS method for class lives of 5, 7, and 10 yr and the income taxes saved because of depreciation during an 11-year period for a combined federal and state income tax rate of 37%, rather than the recommended 40%. Assume no salvage value.

SOLUTION

The basis for depreciation is $90,000,000. The amount of depreciation for each year is the product of the basis and the fractional percentage depreciation from Table 23.9. The savings in income tax each year is 37% of the amount of depreciation. The calculations are readily made with a spreadsheet, which gives the following results:

	Class Life = 5 yr		Class Life = 7 yr		Class Life = 10 yr	
Year	D ($/yr)	Taxes Saved ($/yr)	D ($/yr)	Taxes Saved ($/yr)	D ($/yr)	Taxes Saved ($/yr)
1	18,000,000	6,660,000	12,861,000	4,758,570	9,000,000	3,330,000
2	28,800,000	10,656,000	22,041,000	8,155,170	16,200,000	5,994,000
3	17,280,000	6,393,600	15,741,000	5,824,170	12,960,000	4,795,200
4	10,368,000	3,836,160	11,241,000	4,159,170	10,368,000	3,836,160
5	10,368,000	3,836,160	8,037,000	2,973,690	8,298,000	3,070,260
6	5,184,000	1,918,080	8,028,000	2,970,360	6,633,000	2,454,210
7	0	0	8,037,000	2,973,690	5,895,000	2,181,150
8	0	0	4,014,000	1,485,180	5,895,000	2,181,150
9	0	0	0	0	5,904,000	2,184,480
10	0	0	0	0	5,895,000	2,181,150
11	0	0	0	0	2,952,000	1,092,240
Total $	90,000,000	33,300,000	90,000,000	33,300,000	90,000,000	33,300,000

These results show for the three cases the same total depreciation of $90,000,000, which equals the basis, and the same total income tax savings of $33,300,000 because of depreciation. However, when the present values of the tax savings for each year are computed from Eq. (23.14) and summed for each of the three cases, the results are different, with the shorter class life favored, as shown below for a nominal interest rate of 10% compounded annually.

The class life of 5 yr is superior to 7 yr, and even more so to 10 yr.

Class Life (yr)	Present Value of Income Tax Savings
5	$25,750,000
7	$24,024,000
10	$21,783,000

Before leaving this complex topic, it is important to emphasize that depreciation does not involve a transfer of cash; it is just an accounting artifact. In the calculation of cash flows, it is needed to calculate the taxable earnings, from which income tax is computed [Income tax = 0.40(S − C)]. Then, as shown in Eq. (23.39), depreciation is added back to the after-tax earnings to obtain cash flows. From an income tax standpoint, depreciation should be taken as rapidly as the law permits.

Depletion

Whereas depreciation applies to assets that can be replaced, *depletion* applies to natural resources, which when removed for processing disappear forever, or are only renewed by nature over a period of many years. Depletion is applicable to fisheries, forests, mineral deposits, natural gas wells, oil deposits, orchards, quarries, vineyards, etc. The U.S. federal government permits those using natural resources a depletion allowance, which acts like depreciation as an expense against sales revenue. Two methods are used to calculate the annual depletion allowance: *cost* (or factor) *depletion* and *percentage depletion*.

Cost Depletion

This method is based on the usage of the resource each year, starting with an estimate of the amount of resource that can be removed (recovered) and its cost. Since it may be difficult to make an initial estimate of the amount of recoverable resource, the estimate can be revised at a later date. To use this method, a cost depletion factor, p_t, for the year t is defined

$$p_t = \frac{\text{first cost of the resource}}{\text{estimated units of recoverable resource}} \tag{23.48}$$

where the units are barrels for oil, tons for ore, standard cubic feet for gas, board feet for lumber, etc. The depletion charge for year t is the product of the cost depletion factor and the recovered number of units in year t. The total depletion charge cannot exceed the first cost of the resource.

Percentage Depletion

For the natural resources listed below, a special consideration is given. A constant percentage of the sales revenue (referred to as the gross income) from the resource may be depleted, provided that it does not exceed 50% of the taxable income (before the depletion allowance) of the company. Total depletion charges are allowed to exceed the first cost of the resource. The allowable percentage depends on the type of resource, as given in Table 23.11. However, for oil and gas, only small producers are allowed to use percentage depletion. When percentage depletion is applicable, cost depletion is also computed and the method giving the largest annual depletion charge is used.

Table 23.11 Allowable Percentages of Gross Income for Depletion of Natural Resources When Using Percentage Depletion

Natural Resource	Percentage of Gross Income
Gravel, peat, sand, and some stones	5
Coal, lignite, and sodium chloride	10
Most other minerals and metal ores	14
Copper, gold, iron ore, and silver	15
Oil and gas wells (only for small producers)	15
Lead, nickel, sulfur, uranium, and zinc	22

EXAMPLE 23.28

A mining property, containing an estimated 900,000 tons of lead and zinc ore, is purchased for \$4,500,000. In the first year of operation, 100,000 tons of the ore is mined and sold for \$20/ton. The expenses that year are \$1,200,000. Calculate the net profit and cash flow by (a) cost depletion and (b) percentage depletion. Assume a tax rate of 40%.

SOLUTION

The sales revenue (gross income) = 20(100,000) = \$2,000,000.

(a) From Eq. (23.48), p_t = \$4,500,000/900,000 = \$5/ton

Depletion charge = 5(100,000) = \$500,000

Profit before taxes = \$2,000,000 − \$1,200,000 − \$500,000 = \$300,000

Income tax = 0.40(300,000) = \$120,000

Net profit (after tax)= \$300,000 − \$120,000 = \$180,000

Cash flow = net profit + depletion charge = \$180,000 + \$500,000 = \$680,000

(b) From Table 23.11, the allowable % of gross income for depletion = 22%

Depletion allowance = 0.22(\$2,000,000) = \$440,000

Taxable income before depletion = \$2,000,000 − \$1,200,000 = \$800,000

The percentage depletion allowance of \$440,000 exceeds 50% of the taxable income before the depletion allowance. Therefore, the depletion allowance can only be 0.50 (\$800,000) = \$400,000.

Profit before taxes = \$800,000 − \$400,000 = \$400,000

Income tax = 0.40(\$400,000) = \$160,000

Net profit(after tax) = \$400,000−\$160,000 = \$240,000

Cash flow = \$240,000 + \$400,000 = \$640,000

In this example, cost depletion is better than percentage depletion.

23.7 RIGOROUS PROFITABILITY MEASURES

The two principal profitability measures that involve the time value of money in terms of discounted cash flows are the net present value or worth (NPV) and the investor's rate of return (IRR), which is also referred to as the discounted cash flow rate of return (DCFRR). These measures are anomalous in that, when used to compare alternative chemical products and processes, they often give different results. This has led to substantial disagreement within the finance community (Brealey and Myers, 1984).

When using NPV and IRR, the discounting is normally made with Eq. (23.34) using a nominal interest rate, r, that is compounded annually ($m = 1$), with n_y starting from the beginning of the first year of chemical product or plant design. It is possible to account for *investment creep* in the projection of the cash flows. This usually arises through small annual increases in the investment, of the order of 1–1.5%, due to small projects to install additional equipment as the capacity of the process or product manufacturing facility is increased or process modifications are needed. The additional investment is depreciated on the same schedule as the original investment. The calculations are more complex, but are readily made when calculating cash flows.

When carrying out a rigorous profitability analysis, some design teams adopt the convention of reducing the process yield by a small amount, such as 2%, to account for the loss of raw materials and products during startups, shutdowns, and periods when there are malfunctions. Often, raw materials are vented or flared during startup. In other cases, one part of the plant shuts down while the remainder continues to operate, with small amounts of intermediate products vented when they are nontoxic and not easily stored, until the idle portion of the plant is restarted.

Finally, when calculating the cash flow for the last year of operation, it is common to take credit for the working capital investment. Some companies also take credit for a projection of the salvage value of the plant or product manufacturing facility, assuming that it is dismantled and sold at the end of its useful life. Because salvage values are difficult to estimate and in some cases distort the NPV and IRR, many companies prefer to be conservative and assume a zero salvage value.

The NPV method is simpler to implement than the IRR method and is well defined, whereas the IRR is not defined in all situations. Because the latter involves an iterative computation of the net present value, the simpler-to-calculate NPV is discussed first.

Net Present Value (NPV)

To evaluate the *net present value* (NPV) of a proposed plant or product manufacturing facility, its cash flows are computed for each year of the projected life of the plant or product manufacturing facility, including construction and startup phases. Then, given the interest rate specified by company management (typically 15%), each cash flow is discounted to its present worth. The sum of all the discounted cash flows is the net present value. The NPV provides a quantitative measure for comparing the capital required for competing products and processes in current terms. However, the result is usually quite sensitive to the assumed interest rate, with proposed products and processes changing favor as the interest rate varies. An illustration of the calculation of NPV is given below in Example 23.29, following a discussion of the IRR method.

Investor's Rate of Return (IRR or DCFRR)

The *investor's rate of return* (IRR), also called the *discounted cash flow rate of return* (DCFRR), is the interest rate that gives a net present value of zero. Since the net present value is a complex nonlinear function of the interest rate, an iterative procedure (easily accomplished using a spreadsheet) is required to solve

$$\text{NPV}\{r\} = 0 \quad \text{for } r \tag{23.49}$$

When comparing proposed products and processes, the largest IRR is the most desirable. Note, however, that sometimes the product and process having the largest IRR has the smallest NPV. In many cases, especially when the alternatives have widely disparate investments, both the NPV and the IRR are effective measures. This is especially true when the alternatives are comparable in one measure but are very different in the other. The following example computes both the NPV and the IRR.

EXAMPLE 23.29 *(Example 23.14 Revisited)*

For the process considered in Example 23.14, but with MACRS depreciation for a 5-yr class life as determined in Example 23.27, calculate over an estimated life of 15 yr, including years 2007–2009 when the plant is being constructed, (a) the NPV for a nominal interest rate of 15% compounded annually and (b) the nominal interest rate for the IRR method (i.e., for NPV = 0). For the first 2 yr of plant operation, when at 45 and 67.5% of capacity, the cost of production, exclusive of depreciation, is \$55 million and \$78 million, respectively.

SOLUTION

(a) The cash flows are listed in the following table in millions of dollars per year. Note that the total depreciable capital of \$90 million is divided into three equal parts for the first 3 yr. The working capital of \$40 million appears in the third year. Plant startup costs in the years 2010 and 2011 are not included and no salvage is taken at the end of the project. In the year 2009, the investment in millions is $-\$30 - \$40 = -\$70$ and the discount factor is $1/(1+0.15)^2 = 0.7561$. Therefore, the PV is $0.7561(-70) = -52.93$ or $-\$52.93$ million. Instead of showing negative signs in the table, negative values are enclosed in parentheses. When

Calculation of Cash Flows (Millions of Dollars) for Example 23.29 (Nominal interest rate = 15%)

Year	Investment fC_{TDC}	Investment C_{WC}	D	$C_{Excl.\ Dep.}$	S	Net Earnings	Discounted Cash Flow	Cash Flow (PV)	Cum. PV
2007	(30.00)						(30.00)	(30.00)	(30.00)
2008	(30.00)						(30.00)	(26.09)	(56.09)
2009	(30.00)	(40.00)					(70.00)	(52.93)	(109.02)
2010			18.00	55.00	75.00	1.20	19.20	12.62	(96.39)
2011			28.80	78.00	113.00	3.72	32.52	18.59	(77.80)
2012			17.28	100.00	150.00	19.63	36.91	18.35	(59.45)
2013			10.37	100.00	150.00	23.78	34.15	14.76	(44.68)
2014			10.37	100.00	150.00	23.78	34.15	12.84	(31.85)
2015			5.18	100.00	150.00	26.89	32.07	10.48	(21.36)
2016				100.00	150.00	30.00	30.00	8.53	(12.83)
2017				100.00	150.00	30.00	30.00	7.42	(5.42)
2018				100.00	150.00	30.00	30.00	6.45	1.03
2019				100.00	150.00	30.00	30.00	5.61	6.64
2020				100.00	150.00	30.00	30.00	4.88	11.51
2021		40.00		100.00	150.00	30.00	70.00	9.89	21.41

Net earnings = $(S - C_{Excl.\ Dep.} - D) \times (1.0 - \text{income tax rate})$
Annual cash flow = C = (net earnings + D) $- fC_{TDC} - C_{WC}$

added to the −$56.09 million for the cumulative PV for 2008, a cumulative PV of −$109.02 million is obtained for 2009. In the first year of plant operation, 2010, sales revenue is $75 million, MACRS depreciation is $18 million, and production cost exclusive of depreciation is $55 million. Therefore, pre-tax earnings are 75 − 55 − 18 = $2 million. The combined federal and state income tax is 0.40(2) = $0.80 million. This gives an after-tax or net earnings of 2 − 0.80 = $1.20 million. The cash flow for year 2010 is 1.20 + the 18 depreciation allowance = $19.20 million. The discount factor for that year is $1/1.15^3 = 0.6575$. Therefore, the PV for 2010 is 0.6575(19.20) or $12.62 million. When added to the cumulative PV of −$109.02 million, a cumulative PV of −$96.39 is obtained. The calculations for the remaining years of the project are carried out in a similar manner, most conveniently with a spreadsheet. The final NPV at the end of year 2021 is $21.41 million. Notice that the cumulative PV remains negative until the year 2018. This is equivalent to a payback time of more than 8 yr from the start of plant operation. This is very different from the 2.74-yr payback period computed in Example 23.14, where both the time value of money and the first 2 yr of operation at reduced capacity were ignored.

(b) The IRR (or DCFRR) is obtained iteratively by conducting the same calculations as in part (a), but with a sequence of values for the nominal interest rate until an NPV of zero is obtained. Since an interest rate of 15% in part (a) produced a positive NPV, we know that the interest rate for a zero NPV must be greater than 15%. In fact, an IRR of 18.5% is obtained.

It is of interest to examine the annual cash flows on nondiscounted and discounted bases, as shown in the following bar graphs. Graph (a) is for the former. For discounted cash flows, graphs (b) and (c) are for an interest rate of 15%, while graph (d) is for the IRR of 18.5%. Note that discounted cash flows during the time of plant operation are much smaller than those for the nondiscounted cash flows in the first graph.

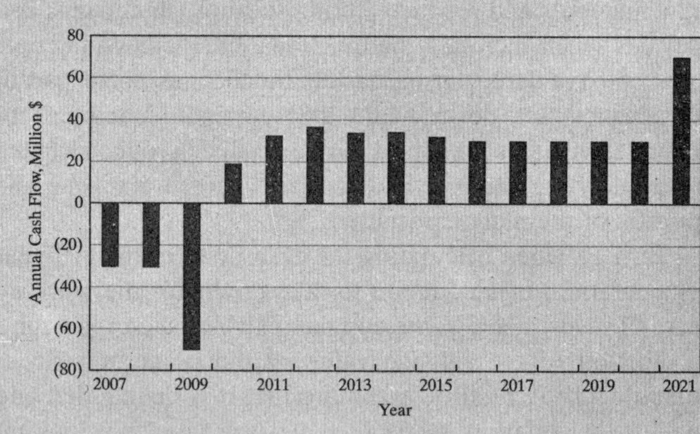

(a) Annual Cash Flows

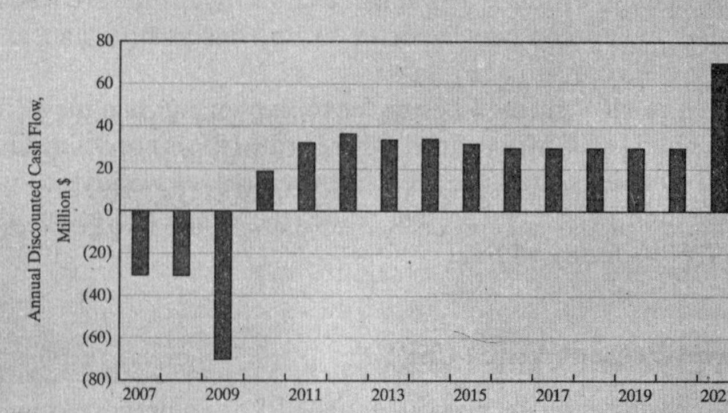

(b) Annual Discounted Cash Flows (Annual Present Values) Nominal Interest Rate = 15%

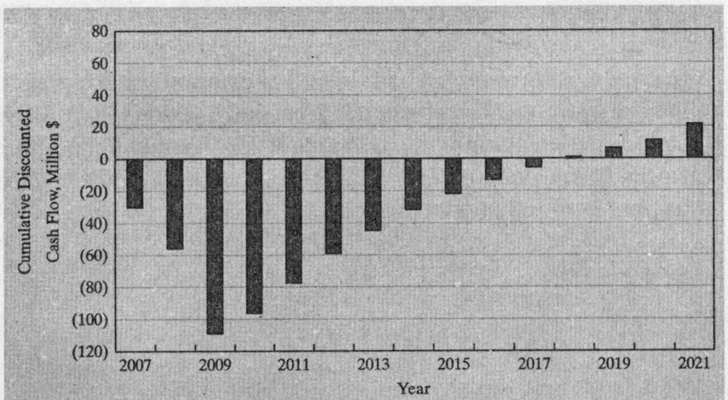

(c) Cum. Discounted Cash Flows (Cum. Present Values) Nominal Interest Rate = 15%

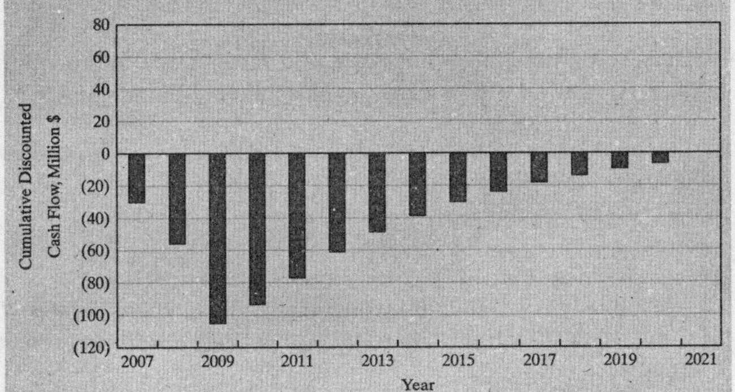

(d) Cum. Discounted Cash Flows (Cum. Present Values) Nominal Interest Rate = 18.5% (IRR or DCFRR)

Finally, when calculating annual discounted cash flows, it is not difficult to account for inflation in estimating revenues and costs, when the inflation factors are known. Inflation, considered in the following subsection, was not included in this example so as to enable the reader to trace the calculations of the cash flows more easily.

Inflation

Inflation is the change in the value of a currency over time. Most often, the change is a loss in value. The effect of inflation on a profitability analysis is difficult to include because the future inflation rate is not known and there is no general agreement on how the effect of inflation should be incorporated into present-worth calculations. Some argue that revenues and costs increase in the same proportion to the inflation rate factor, making it unnecessary to consider inflation when using a rigorous profitability measure. However, this ignores the fact that depreciation allowances are not adjusted for inflation and, therefore, if revenues and costs increase by the same percentage, the gross earnings increase, making it necessary to pay more income tax, as shown in the following example.

EXAMPLE 23.30

Consider the years 2012 and 2013 of the results in the table of Example 23.29. In 2012 and 2013, income tax in millions of dollars is, respectively,

$$0.40(150.00 - 100.00 - 17.28) = 13.09$$
$$0.40(150.00 - 100.00 - 10.37) = 15.85, \text{ with a cash flow of } 34.15$$

Assume that the results of 2012 are unchanged, but that the sales revenue and production cost, excluding depreciation, both increase by 8% due to inflation. Recompute the entries in the table in Example 23.29.

SOLUTION

Now the income tax in millions of dollars for 2012 is

$$0.40[150(1.08) - 100(1.08) - 10.37)]$$
$$= 17.45, \text{ a } 10.1\% \text{ increase in income tax.}$$

The cash flow for 2013 is now 36.55, an increase of about 7% over the 34.15 million dollars of the first case of no inflation in 2013. Based on the effect of inflation on the value of currency, the company has fallen behind somewhat. Thus, it would appear that it is important to make some correction for inflation, if the rate is high.

The historic effect of inflation on costs was seen in Chapter 22 in Table 22.6, where four cost indexes and the Consumer Price Index (CPI) were compared. From that table, the following average annual inflation rates, shown in Table 23.12, are obtained for the 10-yr periods of 1980–1990 and 1990–2000, and for the 5-year period 2000–2005. Also included in the table are the average hourly percentage rate increases in labor wages of private production workers in the United States for the same periods, taken from the Bureau of Labor Statistics. In the period 1980–1990, the average hourly labor wage increased from $6.85 to $10.20, while for 1990–2000, the increase was from $10.20 to $14.01. For 2000–2005, the increase was from $14.01 to $16.12. Thus, assuming a 40-hour workweek for 52 weeks per year, the private production worker had an average annual wage of $33,530 in 2005.

The data in Tables 22.6 and 23.12 show that attempts in the United States to curb inflation during the 25-year period of 1980–2005 met with some success, when compared to the great inflation of the 1970s. The average annual increase in the cost of building a chemical plant has been only 2.63%. Average annual inflation as measured by the Consumer Price Index has been only 3.64%, while hourly wages have increased at an average rate of 3.48%, only 5% lower than the increase in the CPI. During the 1970s, the average annual

Table 23.12 Average Annual Inflation Rates of Cost Indexes and Hourly Labor Wages as Average Percent Increase per Year

Period	CE	MS	ENR	CPI	Hourly Labor Wages
1980–1990	3.21	3.31	3.82	4.72	4.06
1990–2000	0.97	1.67	2.76	2.80	3.22
2000–2005	3.50	3.13	3.66	2.55	2.85

increase in the CPI was almost 8%, while the average increase in wages was 7%.

It is also of interest to consider the effect of inflation on the prices of commodity chemicals, such as those in Table 22.7. In the past 50 yr, the prices of the inorganic chemicals in that list have changed at the most by a factor of two. This represents an average annual inflation rate of only 1.4%. However, in the case of petrochemicals, their prices, like the price of gasoline, are tied to the price of oil, which has fluctuated greatly since the 1950s. For example, the price of ethylene was \$0.05/lb in 1963, \$0.24/lb in 1990, and as much as \$0.43/lb in 2006, making it impossible to predict the effect of inflation unless the future price of oil can be predicted. In addition, for a new chemical plant, raw-material prices and product prices are often negotiated with contracts for at least a few years.

For the purpose of comparing alternative processes with rigorous profitability measures and in the absence of future inflation rates, the following average inflation rates can be used.

Cost of raw materials and price of products	2.5%/yr
Cost of utilities	2.5%/yr
Cost of processing equipment	2.5%/yr
Cost of hourly labor	3.0%/yr

In effect, these rates correspond to approximately 2.5% for material and 3.0% for labor.

Another aspect of inflation is its effect on purchasing power, both for consumers and companies. Equation (23.12) gives the future worth of a present amount of money, if it earns compound interest.

$$F = P(1+i)^n$$

However, if inflation occurs, the purchasing power of that future sum will not be the same as that of the present sum. To account for a constant rate of inflation, Eq. (23.12) is modified to give the future worth in terms of the purchasing power of present dollars. Let the nominal interest rate compounded annually $= r = i$, the number of years $= n$, the annual inflation rate $= f$, and the future worth in present purchasing power $= F_{PP,0}$. Then,

$$F_{PP,0} = \frac{F}{(1+f)^n} = P\left[\frac{(1+i)^n}{(1+f)^n}\right] \quad (23.50)$$

EXAMPLE 23.31

A present sum of money of \$10,000 is invested for 10 yr at an interest rate of 7% compounded annually. During that 10-yr period, the inflation rate is constant at 3%. Compute the future worth, F, and the future worth in terms of purchasing power at the beginning of the investment, $F_{PP,0}$.

SOLUTION

$$P = \$10{,}000, i = 0.07, f = 0.03$$

From Eq. (23.12),

$$F = \$10{,}000(1+0.07)^{10} = \$19{,}672$$

Thus, when inflation is not taken into account, the future worth after 10 yr is almost twice as much as the present amount.

From Eq. (23.50),

$$F_{PP,0} = \frac{\$19{,}672}{(1+0.03)^{10}} = \frac{\$19{,}672}{1.3439} = \$14{,}638$$

When inflation is taken into account, the future purchasing power is only about 50% (rather than 100%) more than that of the present amount.

23.8 PROFITABILITY ANALYSIS SPREADSHEET

This is Section 23S.1 in the file, Supplement_ to_Chapter_ 23.pdf, in the PDF File folder, which can be downloaded from the Wiley Web site associated with this book. It shows how to use purchase and installation cost estimates from Aspen IPE and other sources, together with an economics spreadsheet by Holger Nickisch (2003) to estimate profitability measures for the monochlorobenzene separation process, which was introduced in Section 5.4., and for the lab-on-a-chip for high-throughput screening of kinase inhibitors presented in Sections 16.4 and 17.4. In Section 22.7, Aspen IPE was used to estimate the total permanent investment for the MCB separation process. The economics spreadsheet, Profitability Analysis–2.0.xls, is in the Program and Simulation Files folder, which can be downloaded from the Wiley Web site associated with this book.

23.9 SUMMARY

Having studied this chapter and completed several of the exercises, the reader should have learned to

1. Estimate annual costs using a standard cost sheet like that in Table 23.1.
2. Estimate annual cash flows, working capital, and the total capital investment in Table 22.9.
3. Compute approximate profitability measures, ROI, PBP, VP, and annualized cost.
4. Compute present worth and future worth of single payments and annuities.
5. Compute profitability measures that account for the time value of money, including net present value, IRR, and DCFRR.
6. Use the Aspen IPE in the Aspen Engineering Suite and an economics spreadsheet to perform a profitability analysis.

REFERENCES

1. BAUMAN, H.C., *Fundamentals of Cost Engineering in the Chemical Industry*, Reinhold Publishing Co., New York (1964).

2. BREALEY, R., and S. MYERS, *Principles of Corporate Finance*, McGraw-Hill, New York (1984).

3. BUSCHE, R.M., *Venture Analysis: A Framework for Venture Planning*, Course Notes, Bio-en-gene-er Associates, Wilmington, Delaware (1995).

4. NICKISCH, H., *Profitability Analysis Spreadsheet*, Univ. of Pennsylvania, Philadelphia (2003).

5. PISANO, G.P., *The Development Factory: Unlocking the Potential of Process Innovation*, Harvard Business School Press, Cambridge (1997).

6. SOUDERS, M., *Engineering Economy*, *Chem. Eng. Prog.*, **62** (3), 79–81 (1966).

EXERCISES

23.1 In the design of a chemical plant, the following costs and revenues (in the third year of production) are projected:

Total depreciable capital, excluding allocated power	\$10,000,000
Allocated power utility	\$ 2,000,000
Working capital	\$ 500,000
Annual sales	\$ 8,000,000/yr
Annual cost of sales excluding depreciation	\$ 1,500,000/yr

Assume the costs of land, royalties, and startup are zero.
Determine

(a) The return on investment (ROI)
(b) The payback period (PBP)

23.2 For Exercise 23.1, the return on investment desired by the chemical company is 20%. Determine the venture profit.

23.3 It is desired to have \$9,000 available 12 yr from now. If \$5,000 is available for investment at the present time, what discrete annual rate of compound interest on the investment would be necessary to yield the desired amount?

23.4 What will be the total amount available 10 yr from now if \$2,000 is deposited at the present time with nominal interest at the rate of 6% compounded semi-annually?

23.5 An original loan of \$2,000 was made at 6% simple interest per year for 4 yr. At the end of this time, no interest had been paid and the loan was extended for 6 yr more at a new effective compound interest rate of 8%/yr. What is the total amount owed at the end of the 10 yr if no intermediate payments are made?

23.6 A concern borrows \$50,000 at an annual effective compound interest rate of 10%. The concern wishes to pay off the debt in 5 yr by making equal payments at the end of each year. How much will each payment have to be?

23.7 How many years are required for money to double when invested at a nominal interest rate of 14% compounded semi-annually? Determine the shortest time in years, allowing any number of compounding periods.

23.8 A person at age 30 is planning for retirement at age 60. He projects that he will need \$100,000 a year until age 80. Determine the uniform annual contribution (by him and his company) needed to provide these funds. Assume that the effective interest rate is 8%/yr and the rate of inflation is zero.

23.9 A heat exchanger has been designed for use in a chemical process. A standard type of heat exchanger with a negligible scrap value costs \$4,000 and will have a useful life of 6 yr. Another proposed heat exchanger of equivalent design capacity costs \$6,800 but will have a useful life of 10 yr and a scrap value of \$800. Assuming an effective compound interest rate of 8%/yr, determine which heat exchanger is cheaper by comparing the capitalized costs.

23.10 Two machines, each with a service life of 5 yr, have the following cost comparison. If the effective interest rate is 10%/yr, which machine is more economical?

	A	B
First cost	\$25,000	\$15,000
Uniform end-of-year maintenance	2,000	4,000
Overhaul, end of third year		3,500
Salvage value	3,000	
Benefit from quality as a uniform end-of-year amount	500	

23.11 Two pumps are being considered for purchase:

	A	B
Initial cost	\$8,450	\$10,000
Salvage value	1,500	4,000

Determine the service life, n_y, at which the two pumps are competitive. The annual effective interest rate is 10%.

23.12 Two heat exchangers are being considered for installation in a chemical plant. It is projected that:

	A	B
Installed cost	\$70,000	\$80,000
Uniform end-of-year maintenance	?	\$4,000
Salvage value	\$7,000	\$8,000
Service life	8 yr	7 yr

For an effective interest rate of 10%, determine the uniform end-of-year maintenance for heat exchanger A at which the two are competitive.

23.13 Consider the following two alternatives for the installation of a pump:

	A	B
Installed cost	\$30,000	\$16,000
Uniform end-of-year maintenance	\$1,600	\$2,400
Salvage value	?	\$1,000
Life	5 yr	3 yr

The effective annual interest rate is 10%. Determine the salvage value for pump A when the two pumps are competitive.

23.14 You are offered two options to finance a compressor, with a nominal interest rate at 7.25% compounded monthly.

(a) You pay \$590 per month for 134 months.

(b) You pay \$545 per month for 151 months.

Compare these with the alternative of \$590 per month for 151 months (at a nominal interest rate of 8.75% compounded monthly).

Compared to the alternative, it is claimed that Option (a) saves \$10,030 and Option (b) saves \$6,795. Do you agree? Justify your response.

(*Hint*: Use present-value analysis.)

23.15 Two pumps are under consideration:

	A	B
Installed cost	\$10,000	\$18,000
Service life	1 yr	2 yr

Determine the interest rate at which the two pumps are competitive.

23.16 A chemical plant constructed in 2007 began operating in 2008. In 2010, the plant is projected to operate at 90% of capacity, with

Sales	\$10 MM
Cost of sales (Excl. Dep.)	\$ 5 MM

(a) Calculate the return on investment (ROI) in 2010 given that the total depreciable capital is \$18 MM and the working capital is \$2 MM. Assume straight-line depreciation at 8% per year.

(b) Calculate the cash flow in 2010 and discount it to present value assuming an effective interest rate of 15%. Use the MACRS depreciation schedule for a class life of 5 yr.

23.17 A proposed chemical plant has the following projected costs and revenues in millions of dollars:

	Investment	Working Capital	Cost of Sales (Excl. Dep.)	Sales
2003	(40.0)	(4.0)		
2004			4.0	10.0
2005			5.6	14.0
2006			7.0	17.5
2007			8.0	20.0
2008			9.0	22.5
2009			9.6	24.0
2010		4.0	10.0	25.0

Using an MACRS depreciation schedule having a class life of 5 yr,

(a) Compute the cash flows.

(b) At an effective interest rate of 20%, determine the net present value.

(c) Is the investor's rate of return less than or greater than 20%? Explain.

(d) Compute the investor's rate of return.

23.18 An engineer in charge of the design of a plant must choose either a batch or a continuous system. The batch system offers a lower initial outlay but, owing to high labor requirements, exhibits a higher operating cost. The cash flows relevant to this problem have been estimated as follows:

	Year 0	Year 1–10	Investor's Rate of Return	Net Present Worth at 10%
Batch system	−\$20, 000	\$5,600	25%	\$14,400
Continuous system	−\$30, 000	\$7,650	22	\$17,000

Check the values given for the investor's rate of return and net present worth. If the company requires a minimum rate of return of 10%, which system should be chosen?

23.19 An oil company is offered a lease of a group of oil wells on which the primary reserves are close to exhaustion. The major condition of the purchase is that the oil company must agree to undertake a water-flood project at the end of 5 yr to make possible secondary recovery. No immediate payment by the oil company is required. The relevant cash flows have been estimated as follows:

Year 0	Year 1–4	Year 5	Year 6–20	Investor's Rate of Return	Net Present Worth at 10%
0	\$50,000	−\$650,000	\$100,000	?	\$227,000

Should the lease-and-flood arrangement be accepted? How should this proposal be presented to the company board of directors, who understand and make it a policy to evaluate by the investor's rate of return?

23.20 *Sequencing of two distillation columns*. The demand for styrene monomer continues to increase. Other companies produce styrene by alkylating benzene with ethylene to ethylbenzene, followed by dehydrogenation to styrene. Our chemists have developed a new reaction path to styrene that involves other chemicals that appear to be available from our own supplies at a relatively low cost. These chemicals are toluene and methanol. We have been asked to prepare a preliminary design and economic evaluation for this new route to determine if it merits further consideration. If so, we will consider entering the styrene manufacturing business.

The new process can be broken down into four sections: (1) the reactor section, where toluene is partially reacted with methanol to produce styrene, water, and hydrogen, with an unfortunate side reaction that produces ethylbenzene and water; (2) an aqueous stream separation system; (3) an organic stream separation system; and (4) a vapor separation system. Fortunately, we have a potential buyer for the ethylbenzene byproduct. We are assigning you the design and economic calculations of just the organic stream separation system. Others are preparing the designs for the other three sections.

The chemical engineer working on the reactor section has already calculated the following reactor effluent:

Component	kmol/hr
Hydrogen	352
Methanol	107
Water	484
Toluene	107
Ethylbenzene	137
Styrene	352

This effluent is cooled to 38°C and enters a flash-decanter vessel at 278 kPa. Three phases leave that vessel. The vapor phase (hydrogen rich) is sent to the vapor separation system. The aqueous phase (mostly water, with some methanol) is sent to the aqueous stream separation system. The organic-rich phase is sent to the organic stream separation system, which you will design. To obtain the composition of the feed to your section, use a simulator with the UNIFAC method to perform a three-phase flash for the above conditions. If the resulting organic liquid stream contains small amounts of hydrogen and water, assume they can be completely removed at no cost before your stream enters your separation section.

Your separation system must produce the following streams with two distillation operations in series:

A methanol–toluene-rich distillate stream for recycle back to the reactor. This stream should not contain more than 5 wt% of combined ethylbenzene and styrene.

An ethylbenzene byproduct stream containing 0.8 wt% max. toluene and 3.9 wt% max. styrene.

A styrene product stream containing 300 ppm (by wt.) max. ethylbenzene.

We have a serious problem with styrene. If any stream contains more than 50 wt% styrene, the temperature of the stream must not exceed 145°C. Otherwise, the styrene will polymerize. This must be carefully considered when establishing the operation conditions for the two distillation operations. You may have to operate one or both columns under vacuum. This will require you to estimate the amount of air that leaks into the vacuum columns and select and cost vacuum systems. In designing the distillation system, you are to consider the direct sequence and the indirect sequence.

Please submit a report on the two designs and cost estimates (fixed capital and utility operating costs only). For the capital cost of each of the two alternative sequences, sum the purchase costs of the distillation columns, heat exchangers, and any vacuum equipment. Multiply that cost by the appropriate Lang factor. To annualize the capital cost, multiply by 0.333. Add to this annualized cost the annual utility cost for steam and cooling water. Call this the total annualized cost for the alternative.

Use a simulator to do as many of the calculations as possible, including the very important column pressure-drop calculations (because of the need for vacuum in one or more columns). Assume a condenser pressure drop of 5 kPa and no pressure drop across the reboiler. You may select column internals from the following list:

Sieve trays on 18-in. spacing with overall tray efficiency of 75%.

Pall rings random packing with HETP = 24 in.

Mellapak structured packing with HETP = 12 in.

Submit your results as a short report complete with an introduction to the problem, a process description, process flow diagram, discussion of equipment operating conditions and how you arrived at them, a material-balance table, cost tables, conclusions, and your recommendations. Make it clear which alternative you favor and whether it might offer some technical challenges if it is selected for final design.

23.21 *Toluene hydrodealkylation process–economic analysis using ASPEN PLUS, Aspen IPE, and the economics spreadsheet*. This assignment begins with a completed simulation of the toluene hydrodealkylation process in Figure 23.1 and involves the completion of an economic evaluation. Note that the simulation results for this process were developed in Chapter 5 and can be reproduced using the HDA.bkp file on the CD-ROM that accompanies this book. Aspen IPE will be used for equipment cost and capital cost estimation. The economics spreadsheet Profitability Analysis—2.0.xls, in the Program and Simulation Files folder, which can be downloaded from the Wiley Web site associated wtih this book, will be used for profitability analysis. This spreadsheet should enable you to complete an economic analysis using the specifications recommended for capital investment costs (Section 22.3, Tables 22.9 and 22.12) and for the cost sheet (Section 23.2, Table 23.1), and using the approximate profitability measures in Section 23.4 and those involving cash flows in Section 23.6. In most cases, only the nondefault Aspen IPE entries are provided in the problem statement. This, together with the description in Section 22.7, should enable you to understand the items on the Aspen IPE input forms.

The information you will use to complete the economic analysis is as follows.

Cost Options

The project startup date is January 2009 and the project duration is 1 yr. For simplicity, the effect of inflation is disregarded in this assignment.

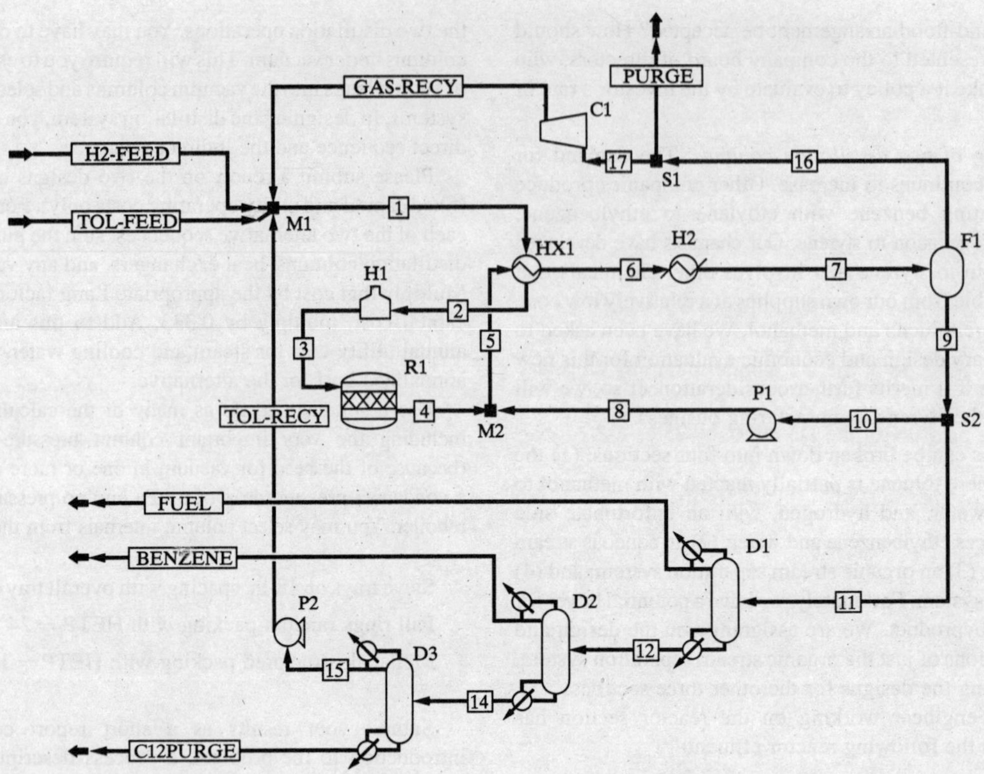

Figure 23.1 ASPEN PLUS simulation flowsheet for the toluene hydrodealkylation process.

Equipment

Unless otherwise stated, the default equipment types and materials of construction are acceptable.

Tanks

Toluene storage tank
Create a tank to store a 2-day supply of toluene.

Benzene storage tank
Create a tank to store a 2-day supply of the benzene product.

Stream 11 storage tank
Create a tank to store a 2-day supply of the intermediate stream 11.

Flash vessel F1
Size a vertical flash vessel to have a 10-minute liquid retention time.

Reactor R1
The elevated temperature (1, 268°F) and pressure (494 psia) present special complications when designing and sizing the hydrodealkylation reactor. This is a large-diameter vessel that is lined with refractory brick to insulate the steel alloy that comprises the retaining wall. Initially, use Aspen IPE to design a vessel that has a 10-ft diameter and a 60-ft tangent-to-tangent length. To estimate the thickness of the refractory brick and the temperature at the brick-steel interface, a heat balance is necessary. In practice, the brick thickness is adjusted to give an interface temperature of 900°F. When using Aspen IPE, select refractory-lined carbon steel. Using the pull-down menu, select a 9-inch layer of 90% alumina fire brick, symbol 9FB90. Also, consider the possibility of using two or three smaller-diameter vessels in parallel. Finally, compare the costs of horizontal and vertical vessels.

Before mapping the reactor model in the ASPEN PLUS simulation, it is necessary to replace the RSTOIC subroutine with the RPLUG subroutine in ASPEN PLUS. Note, however, that the kinetics of the side reaction cannot be modeled using the RPLUG subroutine. Since the conversion of this reaction is small, its kinetics can be neglected in the reactor design. Rather, it is sufficient to account for this reaction using a dummy reactor unit, R1D, modeled with the RSTOIC subroutine, which follows the reactor unit R1, modeled with the RPLUG subroutine. The dummy unit, R1D, cannot be sized by Aspen IPE.

Heat Exchangers

H2
Use cooling water.

D1, D2, and D3 condensers
Use cooling water.

D1, D2, and D3 reboilers
Use steam.

Distillation Towers

In the ASPEN PLUS simulation, for each distillation column, use the RADFRAC subroutine in place of the DISTL subroutine. As a result, the reflux ratios change somewhat to achieve the same product specifications using the same number of trays.

In Aspen IPE, set the tray efficiency of the three columns to 90%. The reflux accumulators should be horizontal vessels with a liquid holdup time of 10 min.

Pumps

Toluene pump

Create a pump to bring the toluene feed from atmospheric pressure to 569 psia.

Benzene pump

Increase the pressure of the benzene product stream by 25 psia.

D1 condenser pump

Increase the pressure of the D1 reflux by 25 psia.

D1 reboiler pump

Increase the pressure of the D1 reboiler pump by 25 psia.

D2 condenser pump

Increase the pressure of the D2 reflux by 25 psia.

D2 reboiler pump

Increase the pressure of the D2 reboiler pump by 25 psia.

D3 condenser pump

Increase the pressure of the D3 reflux by 25 psia.

D3 reboiler pump

Increase the pressure of the D3 reboiler pump by 25 psia.

Other Equipment

Compressor C1

Use a centrifugal compressor.

Use a motor drive and an electrical utility.

Fired heater H1

In the ASPEN PLUS simulation, the HEATER subroutine was used. By default, units modeled with the HEATER subroutine are mapped as heat exchangers by IPE. Instead, this unit must be mapped as a furnace. Furthermore, before estimating the cost of the furnace, the default material, carbon steel, must be replaced by a material that can withstand temperatures at 1,200°F in the furnace. Note that Incoloy 1800 is selected from among the materials available in Aspen IPE that can withstand this temperature. To replace carbon steel, in row nine of the window that displays the equipment sizes for the furnace, depress *material selection,* which opens the pull-down menu on the right. Then, select the last entry, *TUBE MATERIAL*. This produces a second pull-down menu, from which the tube material "1800" is selected. These changes must be saved before leaving this window.

Labor Costs

Use a wage rate of \$40/hr.

Materials

For the feed, product, and byproduct streams H2-FEED, TOL-FEED, BENZENE, PURGE, FUEL, and C12PURGE, the following prices are typical:

Hydrogen feed	\$1.00/lb
Toluene feed	\$1.50/gal
Benzene product	\$2.50/gal
FUEL, PURGE	\$3.00/MM Btu heating value
C12PURGE	0

The prices are for toluene and benzene at 1 atm and 75°F. You can use ASPEN PLUS to estimate the densities needed to obtain the prices on a mass basis. (You can also use ASPEN PLUS to estimate the heating value of a stream, rather than calculate it independently. To compute and display the heating values of all streams, follow the sequence: *Data→Setup→Report Options→Property Sets* and select HEAT-VAL. Note that the simulation results are arranged to report the heating values in the stream section of the report file. It is also possible to estimate the heating value of the PURGE and FUEL streams using Table 23.2.)

Utilities

For this process, cooling water, steam, electricity, and fuel are purchased.

Cooling water: \$0.06/1,000 gal. Use inlet and outlet temperatures of 90°F and 120°F.

Steam: Use prices in Table 23.1.

Electricity: \$0.05/kW-hr.

Fuel (to heater H1): Assume the heater is gas fired at a cost of \$2.60/MM Btu.

Operating Costs

The plant is anticipated to operate 330 days a year.

Average hourly wage rate is \$40/hr with two operators per shift.

Profitability Analysis

Project a 15-yr life for the plant.

Assume a 15% interest rate to calculate the net present value.

For cash flow analysis, use the 5-yr MACRS depreciation schedule. Estimate the effective tax rate to be 40% with no investment tax credit.

Assume a production schedule such that the plant operates at 50% of full scale (90% of capacity) in the first year, 75% of full scale in the second year, and 100% of full scale thereafter.

Provide for the cost of startup at 20% of the total materials and labor cost of the process units.

Provide for working capital to cover 2 days of raw material inventory, 2 days of in-process chemicals, 2 days of finished-product inventory, and 30 days of accounts receivable.

Chapter 24

Optimization for Design

24.0 OBJECTIVES

This chapter begins with a brief discussion of the fundamental principles of optimization, which are presented in much more detail by Beveridge and Schechter (1970), Edgar et al. (2001), and Ravindran et al. (2006). These principles are then applied to the use of flowsheet simulators to optimize the most promising flowsheets during process design. Two process examples are presented. The first involves maximizing the venture profit (VP) of a process for the production of ethyl chloride. In the second example, the separation afforded by a distillation column with sidestreams is maximized, typical of an optimization carried out during the development of a base-case design.

After studying this chapter and the multimedia modules, which can be downloaded from the Wiley Web site associated with this book, the reader should

www.wiley.com/go/global/seider

1. Understand the fundamentals of optimization.
2. Be able to formulate a nonlinear optimization problem (nonlinear program, NLP) to maximize or minimize an objective function by adjusting continuous decision variables in the model of the process.
3. Understand the nature of algorithms that optimize the process while simultaneously converging the recycle loops and design specifications associated with the process simulation.
4. Begin to understand the advantages and disadvantages of converting design specifications associated with a simulation model to equality constraints in the NLP.
5. Be able to use process simulators to solve the NLP.

24.1 INTRODUCTION

From a mathematical point of view, chemical engineers deal with three types of problems when solving equations. The first type, which is the subject of most undergraduate textbooks in chemical engineering and has been the main subject of this book to this point, is the *completely specified* case, where the number of equations, $N_{\text{Equations}}$, to be solved is equal to the number of variables, $N_{\text{Variables}}$, to be determined. The equations to be solved are usually at least partially nonlinear and the challenge is to find a method that will solve them. This is complicated by the fact that more than one practical solution may exist, a challenge that has been largely ignored in process simulators. The second type is common in experimental work and process operations, where $N_{\text{Variables}} < N_{\text{Equations}}$. This is the *overspecified* case, which is commonly referred to as the reconciliation (or data reconciliation and rectification) problem. An example of this case is a piece of equipment operating under steady-state conditions with all or some of the variables, for example, component flow rates, measured. Because the measurements are subject to error, they do not satisfy material-balance equations. The task with this case is to determine the most likely values for the variables. This case is not covered in this book but is discussed by Mah (1990). The third case is the subject of this chapter, *optimization*, where the problem is *underspecified* such that $N_{\text{Variables}} > N_{\text{Equations}}$. The problem is to select from the set of variables, $\underline{x}$, a subset of size

$$N_{\text{Variables}} - N_{\text{Equations}} = N_D \tag{24.1}$$

called the *decision variables*, $\underline{d}$, and to iteratively adjust the decision variables to achieve the optimal solution to a specified objective.

Optimization is normally applied to improve all designs, both product and process, at various stages in the design process. In the *concept* stage, it usually involves approximate models and objective functions, sufficient to select from among numerous alternatives. Gradually, through the *feasibility* and *development* stages, after less promising alternatives have been screened, the models and objectives become better understood and more rigorous.

In Part One of this textbook, *Product Design for Basic Chemicals*, the optimization of process designs is very important, especially for commodity chemicals (produced in large quantities at relatively low prices). After reading Chapters 4–11, and having studied the steps in process creation and solved problems involving the synthesis of alternative process flowsheets presented in previous chapters, it is appropriate to question whether a given flowsheet can and should be optimized by adjusting its key equipment parameters to, for example, increase some measure of its economic attractiveness. This optimization step, shown in connection with the detailed design stage in Figure PI.1, is applied to the most promising process flowsheets after a base case has been evaluated, where the equipment has been sized, the capital and operating costs have been estimated, and a profitability analysis has been completed. In many cases, however, optimizations are performed much earlier in process design using more approximate cost and profitability measures, as shown in Examples 24.4 and 24.5.

As introduced in Section 9.7, formal methods of optimization can be utilized to optimize a superstructure of process units with streams that can be turned on and off using binary (integer) variables. In principle, the *mixed-integer formulation* (involving both continuous and integer variables) of the optimization problem permits the optimizer to select simultaneously the best flowsheet and optimize it with respect to its continuous variables, such as pressure levels, reflux ratios, residence times, and split fractions. In practice, however, most design problems are not solved using superstructures and mixed-integer optimization algorithms. Rather, as described throughout the earlier chapters in Part One, heuristics, many of which are presented in Chapter 6, together with simulation and algorithmic methods, are utilized to build and analyze *synthesis trees*. Although substructures such as networks of heat exchangers can be optimized conveniently using mixed-integer methods, it is impractical, except for simple processes, to attempt the optimization of entire process flowsheets in this manner. Accordingly, this chapter is restricted to the case of optimization problems involving continuous variables, of either the LP (linear programming) type or the NLP (nonlinear programming) type. Thus, MILP (mixed integer-linear programs) and MINLP (mixed-integer nonlinear programs) are only briefly mentioned.

Emphasis is placed in this chapter on the usage of process simulators to carry out the optimization simultaneously while converging the recycle loops and/or decision variables. To do the optimization efficiently, simulators use one of three methods: (1) *successive linear programming* (SLP), (2) *successive quadratic programming* (SQP), or (3) *generalized reduced gradient* (GRG). Emphasis in this chapter is placed on SQP, used by ASPEN PLUS, ASPEN HYSYS, and UNISIM. GRG, which is used by CHEMCAD, is not discussed here, but is covered by Edgar et al. (2001).

24.2 GENERAL FORMULATION OF THE OPTIMIZATION PROBLEM

The formulation of the optimization problem involves:

1. A set of $N_{\text{Variables}}$ variables, $\underline{x}$.
2. The selection of a set of appropriate *decision variables*, $\underline{d}$, from the set $\underline{x}$.
3. A measure of goodness called an *objective function*, $f\{\underline{x}\}$. (24.2)
4. A set of $N_{\text{Equations}}$ equality constraints, $\underline{c}\{\underline{x}\} = 0$. (24.3)
5. A set of N_{Inequal} inequality constraints, $\underline{g}\{\underline{x}\} \leq 0$. (24.4)
6. Lower and upper bounds on some or all of the variables, $\underline{x}^L \leq \underline{x} \leq \underline{x}^U$. (24.5)

A general optimization problem is stated as follows:

Minimize (or maximize) with respect to (w.r.t.) $\underline{d}$, the design variables — $f\{\underline{x}\}$, the objective function **(NLP1)**

subject to (s. t.):
$\underline{c}\{\underline{x}\} = 0$, the equality constraints
$\underline{g}\{\underline{x}\} \leq 0$, the inequality constraints
$\underline{x}^L \leq \underline{x} \leq \underline{x}^U$, the lower and upper bounds

Objective Function and Decision Variables

Candidates for the objective function (measure of goodness), $f\{\underline{x}\}$, are often the profitability measures of Chapter 23, beginning with the approximate measures such as the return on investment (ROI), the venture profit (VP), the payback period (PBP), and the annualized cost (C_A). For more thorough analyses, the rigorous measures that involve the time value of money and cash flows are used. These include the net present value (NPV) and the investor's rate of return (IRR). Other objective functions may involve measures that are related to costs or may involve safety, control, or pollution aspects. A multiobjective function consisting of two or more measures, each with a weighting coefficient, may also be employed. In any case, the objective function is a function of all of the decision variables, $\underline{d}$, and may also be a function of some or all of the other variables of the set $\underline{x}$. Depending upon the nature of the objective function, it is minimized or maximized analytically or numerically by adjusting the values of the decision variables until the optimal solution is reached. When possible, it is best to select decision variables for which significant tradeoffs are anticipated in the objective function. It is common to begin an optimization study with an approximate measure of goodness and switch to more rigorous measures when the design continues to be promising as the model of the flowsheet is refined.

Equality Constraints

In process design, the largest fraction of equality constraints, $c\{\underset{\sim}{x}\}$, is the modeling equations (usually algebraic) associated with processing equipment. For example, a distillation column with equilibrium stages may be modeled with hundreds of material-balance, energy-balance, and phase-equilibria equations in terms of the set of variables, $\underset{\sim}{x}$. The NLP problem often involves hundreds and even thousands of equations. However, in the implementation of most process simulators, these equations are normally solved for each process unit, given equipment parameters and stream variables (usually for the feed streams), using subroutines in a program library. Hence, the equations for the process units are normally not shown explicitly in the statement of the nonlinear programming problem. Given assumed values for the decision variables, the simulators normally call upon these subroutines to solve the appropriate equations and obtain the unknowns that are needed to perform the optimization. However, certain specifications, especially those that involve more than one process unit, may require the user to formulate an equality constraint.

As discussed in Section 5.2, *Equation-Oriented Architectures*, increasingly, process simulators offer the option of using libraries of equations, that is, equality constraints, in place of subroutines. This option is available in ASPEN PLUS and ASPEN HYSYS (Aspen Technology, Inc.).

Inequality Constraints

Inequality constraints, $g\{\underset{\sim}{x}\}$, are expressions that involve any or all of the set of variables, $\underset{\sim}{x}$, and are used to bound the feasible region of operation. For example, when operating a centrifugal pump, the developed head decreases with increasing flow rate according to a pump characteristic curve. Hence, if the flow rate is varied when optimizing the process, care must be taken to make sure that the required pressure increase (head) does not exceed that available from the pump. The expression might be of the form,

$$\text{Pump head} \leq \text{a} - \text{b}\,(\text{Flow rate}) - \text{c}(\text{Flow rate})^2$$

Similar kinds of constraints involve the reflux ratio in distillation, which must exceed the minimum value for the required separation. If the distillation tower pressure is adjusted, the minimum reflux ratio will change and the actual ratio must be maintained above the minimum value. Even when optimization is not performed, the decision variable values must be selected to avoid violating the inequality constraints. In some cases, the violations can be detected when examining the simulation results. In other cases, the unit subroutines are unable to solve the equations as, for example, when the reflux ratio is adjusted to a value below the minimum value for a specified split of the key components.

Lower and Upper Bounds

Some inequality constraints simply place lower and upper bounds, $\underset{\sim}{x}^L \leq \underset{\sim}{x} \leq \underset{\sim}{x}^U$, on any or all of the variables, $\underset{\sim}{x}$. Others permit the specification of just a lower bound or just an upper bound, for example, a lower bound on the fractional recovery of a species in a product stream. Sometimes the upper and lower bounds are included with the inequality constraints, but here, they are considered separately.

24.3 CLASSIFICATION OF OPTIMIZATION PROBLEMS

The combination of the equality constraints, inequality constraints, and lower and upper bounds defines a *feasible region*. A *feasible solution* is one that satisfies the equality constraints, the inequality constraints, and the upper and lower bounds for a feasible set of decision variables. If the solution also minimizes (or maximizes) the objective function, it is a *local optimal solution*. Sometimes, other local optimal solutions exist in the feasible region, with one or more being a *global optimal solution*.

Many numerical methods have been devised for solving optimization problems. The choice of method depends upon the nature of the formulation of the problem. Therefore, it is useful to classify optimization problems with respect to certain categories.

When the objective function, equality constraints, and inequality constraints are linear with respect to the variables, $\underset{\sim}{x}$, the problem is referred to as a *linear programming* (LP) problem. If the objective function, any of the equality constraints, and/or any of the inequality constraints are nonlinear with respect to the variables, the problem is referred to as a *nonlinear programming* (NLP) problem.

The simplest optimization problems are those without equality constraints, inequality constraints, and lower and upper bounds. They are referred to as *unconstrained optimization*. Otherwise, if one or more constraints apply, the problem is one in *constrained optimization*.

For just a single decision variable (*unidimensional*), the feasible region and optimal solution(s) are conveniently displayed on a plot of $f\{x\}$ against x. This display and solution is easily carried out with a spreadsheet. The types of optimal solutions obtained depend on the nature of both the objective function and the constraints. Fig. 24.1 shows two different linear cases, where the only variable is designated x for the abscissa. An unconstrained *linear* objective function is shown in Figure 24.1a. In the absence of inequality constraints, no solution exists, unless one at $\pm\infty$ is meaningful.

In Figure 24.1b, a linear objective function is again shown, but an upper bound is placed on x. Therefore, the feasible region is to the left of this upper bound, as indicated by the arrows pointing to the left. A maximum objective function is desired. Thus, the optimal solution is at the bound. For linear objective functions, it can be shown that optimal

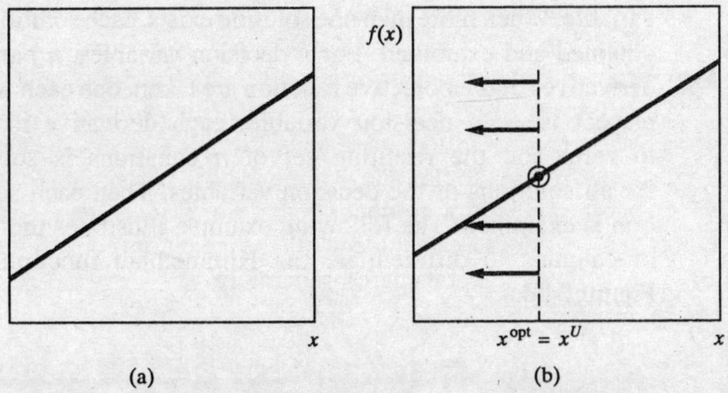

Figure 24.1 Linear objective function: (a) unconstrained; (b) subject to linear inequality constraint, $x \leq x^U$.

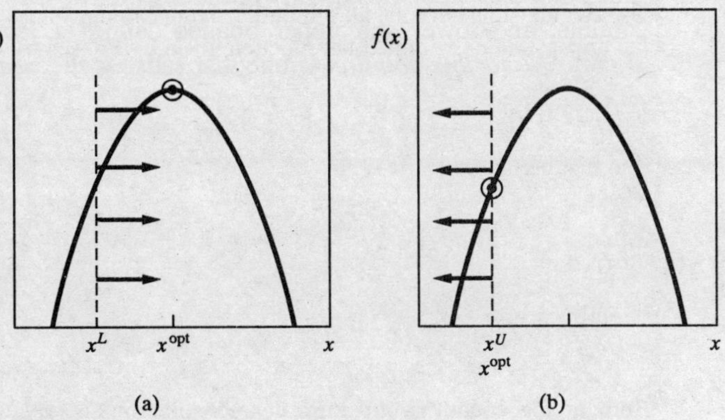

Figure 24.2 Nonlinear objective function: (a) subject to a slack constraint; (b) subject to a binding constraint.

solutions always occur at intersections between the objective function and the inequality constraints and/or lower and upper bounds.

Two cases of a nonlinear objective function that passes through a maximum are shown in Figure 24.2, with both constraints being a simple bound. The objective function is to be maximized. In Figure 24.2a, the constraint is a lower limit on x. Since the constraint is at a value of x less than the value at the maximum value of the objective function, the optimal value of x is that corresponding to the maximum value of the objective function, and the constraint is referred to as a *slack constraint.*

In Figure 24.2b, an upper bound is placed on x, and that bound is below the value at the maximum value of the objective function. Therefore, the optimal value of x is its upper bound, and the constraint is referred to as a *binding constraint.*

For two decision variables, a common display for the feasible region and optimal solution(s) is a plot of one decision variable against the other, with contours of the objective function, as shown in Figure 24.3. The optimization problem is to minimize the objective function:

$$f\{\underline{x}\} = y = 4x_1^2 + 5x_2^2$$

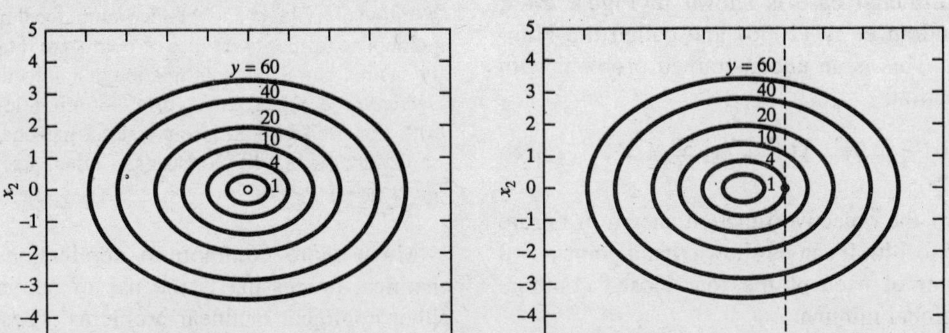

Figure 24.3 Two-variable optimization: (a) unconstrained; (b) constrained.

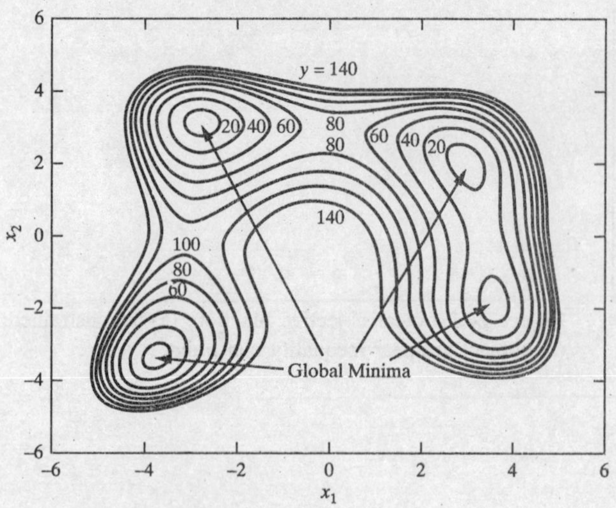

Figure 24.4 Example of a multimodal problem—Himmelblau's function.

Two solutions are shown in Figure 24.3. The first is for the unconstrained case, where the entire region is feasible. The solution, shown in Figure 24.3a, is at $x_2 = 0$ and $x_1 = 0$, where $y = 0$. The second is a constrained case with the inequality constraint:

$$g\{\underline{x}\} = 1 - x_1 \leq 0$$

This is a linear constraint that is easily converted to $x_1 \geq 1$. Now the feasible region is situated to the right of the constraint, as shown in Figure 24.3b. The optimal point for the unconstrained case is not located in the feasible region. Now the minimum value of the objective function occurs at $x_2 = 0$ and $x_1 = 1$, where $y = 4$. For more than two decision variables, a graphical representation is not usually attempted.

When one optimal solution exists in the feasible region, the objective function is *unimodal*. When two optimal solutions exist, it is bimodal; if more than two, it is *multimodal*. LP problems are unimodal unless the constraints are inconsistent such that no feasible region exists. The solutions in Figures 24.1 to 24.3 are unimodal. A two-dimensional, multimodal case is shown in Figure 24.4, taken from Ravindran et al. (2006) and called the Himmelblau problem. This is an unconstrained problem with the objective function:

$$f\{x_1, x_2\} = y = (x_1^2 + x_2 - 11)^2 + (x_1 + x_2^2 - 7)^2 \quad (24.6)$$

The contours of the objective function shown in Figure 24.4 range from 5 to 140. It can be shown that a point $y = 0$ exists at the center of each of the four closed contours identified as the global minima.

In some cases, using calculus to form derivatives, nonlinear optimization problems can be solved analytically. For a one-dimensional problem, it is only necessary to differentiate the objective function with respect to the decision variable, set that derivative to zero, and solve for the decision variable. When more than one solution exists, each solution is obtained and examined. For n decision variables, n partial derivatives of the objective function are taken, one each with respect to each decision variable; each derivative is set to zero; and the resulting set of n equations is solved for all solutions of the decision variables. Then each solution is examined. The following example illustrates the use of calculus to differentiate the Himmelblau function of Figure 24.4.

EXAMPLE 24.1 *Determining Extrema for the Himmelblau Function*

Use calculus to determine all minimum and maximum values of the unconstrained Himmelblau function given by Eq. (24.6).

SOLUTION

The two partial derivatives are

$$\frac{\partial y}{\partial x_1} = F_1 = 4x_1(2x_1^2 + x_2 - 11) + 2(x_1 + x_2^2 - 7) = 0 \quad (24.7)$$

$$\frac{\partial y}{\partial x_2} = F_2 = 2(x_1^2 + x_2 - 11) + 4x_2(x_1 + x_2^2 - 7) = 0 \quad (24.8)$$

Both of the equations are third-degree polynomials. By the Theorem of Bezout, as discussed by Morgan (1987), the maximum number of solutions to a set of polynomial equations is the product of the highest degrees of the equations, which is called the total degree of the set of functions, and is equal to nine for the set consisting of Eqs. (24.7) and (24.8). Using a solver for a set of polynomial equations, nine solutions are found, as listed in Table 24.1.

The nine solutions, referred to as *stationary points*, correspond well with the contours of constant values of the objective function plotted on Figure 24.4. The type of solution is also listed in Table 24.1. There is one maximum at an objective function, $y = 181.62$. Larger values of the objective function occur as the two decision variables are increased to infinity. Four minima occur at an objective function of zero, which is the global minimum. There are four saddle points. Edgar et al. (2001) give necessary and sufficient conditions for determining whether a stationary point is a local maximum, local minimum, or saddle point. The former two types of points are referred to as *extrema*. For a maximum, any combination of small changes in $\underline{x}$ can only decrease the objective function. For a minimum, the opposite is true. For a saddle point, from which small changes in $\underline{x}$ are made, some directions will increase and others decrease the objective function.

Much more common in applications to problems in chemical processing is the use of numerical methods for either nonlinear or linear problems. These methods, which are covered in the following sections of this chapter, are mostly *search* methods that start from an assumed solution for $\underline{d}$ and then move $\underline{d}$ in a series of iterations, by some strategy, to reduce (increase) the objective function to achieve a minimum (maximum).

Table 24.1 Solutions to Himmelblau's Function

x_1	x_2	$f\{x_1, x_2\}$	F_1	F_2	Solution Type
−0.2709	−0.9230	181.62	0.0023	−0.0004	Local maximum
−0.1279	−1.9538	178.34	−0.0023	−0.0022	Saddle point
3.5844	−1.8481	0	−0.0028	0.0006	Global minimum
3.3852	0.0739	13.31	0.0051	0.0000	Saddle point
3.0000	2.0000	0	0.0000	0.0000	Global minimum
0.0867	2.8843	67.72	−0.0001	0.0036	Saddle point
−2.8051	3.1313	0	0.0012	−0.0010	Global minimum
−3.0730	−0.0814	104.02	0.0024	0.0015	Saddle point
−3.7793	−3.2832	0	0.0016	−0.0015	Global minimum

Additional complications can be present in optimization problems. For example, the objective function and/or one or more equality constraints may be *discontinuous*. This might occur, for example, when steam is available to a process at two or three pressure levels, with a different cost at each level. When steam is used where pressure is a variable, the steam cost could change abruptly at a certain value of the pressure, causing a discontinuity in the objective function.

Another complication arises when one or more of the decision variables is an integer, rather than continuous. The most common case is when that integer is binary with just two values, 1 or 0. This gives rise to mixed-integer linear or nonlinear programming (MILP and MINLP) formulations. Although not covered in this chapter, examples of mixed-integer applications are presented in Chapter 9. MILP and MINLP arise in process optimization from the need to deal with binary as well as continuous decision variables, the former being a convenient way of representing alternative locations of a given equipment item in a flowsheet. MILP formulations are appropriate when both the objective function and the constraints are linear, such as in Example 9.8, which shows how a MILP is formulated and solved for design of a heat exchanger network (HEN) having the minimum energy requirement (MER). More commonly, both the objective function and the constraints are nonlinear, leading to a MINLP formulation. Example 9.16 shows how a MINLP is set up and solved to minimize the annual cost of a HEN. The *superstructure* for the MINLP, which incorporates all possible heat exchanger locations and flow configurations in the HEN using binary decision variables is described in Example 9.15. For comprehensive coverage of MILP and MINLP formulations in process design, the reader is referred to Floudas (1995) and Biegler et al. (1997).

24.4 LINEAR PROGRAMMING (LP)

Although LP problems are not common when optimizing product and process designs, they are common in many other applications of chemical engineering. Furthermore, a numerical solution of a NLP problem is sometimes achieved by approximating the nonlinear functions with linear functions at each step of the iterative procedure, using a method called *successive linear programming* (SLP). Therefore, it is useful to have a basic understanding of LP methods.

Some of the common applications of LP methods are for: (1) assignment, (2) blending, (3) distribution, (4) determining network flows, (5) scheduling, (6) transportation, and (7) scheduling traveling sales people. Example 9.4 demonstrates how an LP is used to determine the minimum hot and cold utilities for a HEN. For small problems that can be reduced to two decision variables, a graphical solution is instructive. The graphical solution method involves: (a) a definition of the decision variables; (b) formulation of the objective function; (c) formulation of the model; (d) reduction of the number of decision variables using equality constraints, applying Eq. (24.1); and (e) solution of the LP graphically, when the resulting number of decision variables is less than three. The following blending example, with three decision variables, although solved graphically, illustrates some of the characteristics of all LP problems.

EXAMPLE 24.2 ***Beer Supply Problem***

During the 2002 Winter Olympics in Salt Lake City, Utah, a local microbrewery received a rush order for 100 gal of beer containing 4.0 vol% alcohol. Although no 4% beer was in stock, large quantities of Beer A with 4.5% alcohol at a price of \$6.40/gal and Beer B with 3.7% alcohol priced at \$5.00/gal were available, as well as water suitable for adding to the blend at no cost. The brewery manager wanted to use at least 10 gal of Beer A. Neglecting any volume change due to mixing, determine the gallons each of Beer A, Beer B, and water that should be blended together to produce the desired order at the minimum cost.

SOLUTION

Let V_A = gallons of Beer A, V_B = gallons of Beer B, and V_W = gallons of water. The optimization problem is stated as follows

$$\begin{array}{ll}\text{Minimize} & \text{Cost, \$} = 6.40V_A + 5.00V_B + 0.00V_W \qquad (24.9) \\ \text{w.r.t} & \text{s.t.} \\ V_A, V_B, V_W & \end{array}$$

$$0.045V_A + 0.037V_B + 0.00V_W = 0.04(100) = 4.00 \quad (24.10)$$

$$V_A + V_B + V_W = 100$$
$$10 \le V_A, 0 \le V_B, 0 \le V_W \quad (24.11)$$

The problem consists of three variables, two equality constraints, and three lower bounds. The problem can be reduced to two decision variables by solving Eq. (24.11) for V_B,

$$V_B = 100 - V_A - V_W \quad (24.12)$$

and substituting it into Eqs. (24.9) and (24.10) to give the following restatement of the problem:

$$\begin{array}{lll} \text{Minimize} & \text{Cost, \$} = 1.40V_A - 5.00V_W + 500 & (24.13) \\ \text{w. r. t.} & \text{s. t.} & \\ V_A, V_W & 0.008V_A + 0.037V_W = 0.3 & (24.14) \\ & V_B = 100 - V_A - V_W & (24.15) \\ & 10 \le V_A, 0 \le V_B, 0 \le V_W & \end{array}$$

Now, the optimal volume of Beer B need only be calculated from Eq. (24.15), after the optimal volumes of Beer A and water have been determined. Since the objective function, the equality constraints, and the lower and upper bounds are all linear, this constitutes an LP problem. With just two decision variables, the problem can be shown graphically on a plot of V_A against V_W, as in Figure 24.5. The plot includes not only the lower bounds on the volumes of Beer A and water, but also the equality constraint, Eq. (24.14). The optimal solution to an LP problem occurs at a vertex of the set of constraints. Note that Eq. (24.14) can be rearranged to give

$$V_A = 4.625\,V_W + 37.5 \quad (24.16)$$

Thus, at the lower limit of V_W, $V_A = 37.5$. Thus, one vertex is at $V_A = 37.5$ and $V_W = 0$. The other vertex is at the intersection of two lower bounds ($V_A = 10$, $V_W = 0$). One might argue that another vertex exists at an upper bound on the volume of Beer A, corresponding to an upper bound on the volume of water. This occurs where the blend contains only Beer A and water ($V_A = 88.89$, $V_W = 11.11$). Now, evaluate the cost at each of these three vertices. The results are

V_A (gal)	V_B (gal)	V_W (gal)	% Alcohol	Cost ($)
37.50	62.50	0.00	4.00	552.50
10.0	90.00	0.00	3.78	514.00
88.89	0.00	11.11	4.00	568.90

The optimal solution is the first one in the table because it is the lower of the two costs that satisfy the equality constraint that fixes the percentage of alcohol at 4.0. The second result need not have been calculated because it does not satisfy that constraint, giving only a 3.78% alcohol content. The solution must lie along the equality constraint line, Eq. (24.16), which was obtained from Eq. (24.14). It is perhaps surprising that the optimal blend involves no addition of water. An acceptable blend that includes water is 5 gal water, 60.63 gal Beer A, and

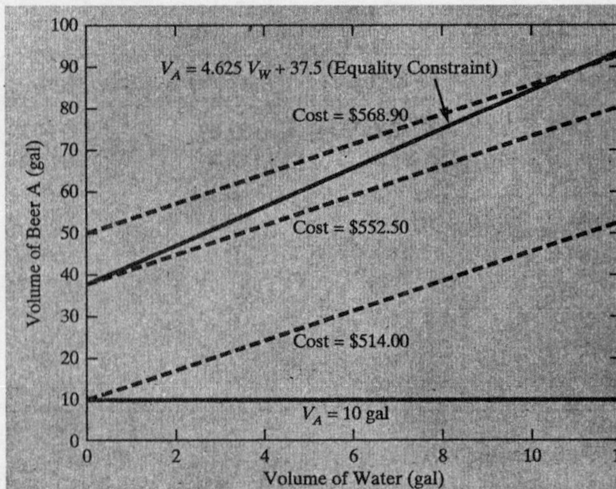

Figure 24.5 Constraints and costs for Example 24.2.

34.37 gal Beer B, but at a cost of \$559.88. Figure 24.5 includes three lines of constant cost.

Note that the two equality constraints, Eqs. (24.10) and (24.11), permit two of the decision variables to be eliminated, leaving just one decision variable, say V_A. Consequently, the objective function, Eq. (24.9), can be minimized easily using a univariable search. If the problem statement is altered to place a *lower bound* on the beer concentration of 4%, Eq. (24.10) becomes an inequality constraint:

$$0.045\,V_A + 0.037\,V_B \ge 4.00$$

Then, two decision variables result and the graphical approach in Figure 24.5 is altered, with the equality constraint replaced by the inequality constraint:

$$V_A \le 4.625\,V_W + 37.5$$

At the minimum cost, the solution remains unchanged.

For large LP problems, which may involve more than 10,000 decision variables, one of two methods is applied. The first method, developed by Dantzig (1949) in the 1940s, is referred to as the *simplex method*. It is an iterative method that begins with initial values for the design variables (iterates) that satisfy the constraints at one of the vertices, and for which the objective function is computed. The optimal solution must be at this or another vertex. Therefore, subsequent iterations generate, in a systematic procedure, a sequence of iterates that move from one vertex of the feasible region to an adjacent vertex, each time finding an improved value for the objective function, until the vertex corresponding to the optimal solution is found. As described by Edgar et al. (2001), LP problems can be solved by the simplex method with the linear model of the Solver routine of the Microsoft Excel spreadsheet. An LP solver is also available in MATLAB and GAMS.

When the number of iterations required by the simplex method is found to increase exponentially with the number of decision variables, the second method, which is also iterative and was developed in the 1980s, may be more efficient because it may require far fewer iterations, although each iteration requires more calculations. This method, introduced by Karmarkar in 1984 and described in detail by Vanderbei (1999), is called the *interior-point method.* It differs from the simplex method in that all iterates are not required to satisfy the constraints and, therefore, need not be located on vertices. This allows the iterates to be points interior to the feasible region, which on successive iterations can cut clear across the feasible region so as to locate the optimal point more quickly. Software for both LP methods is widely available from the Internet and is included in libraries of mathematical software.

24.5 NONLINEAR PROGRAMMING (NLP) WITH A SINGLE VARIABLE

Nonlinear optimization problems in just a single decision variable frequently arise in chemical engineering applications. If the objective function is unconstrained, the optimal solution(s) can often be obtained analytically using derivatives from calculus. When constrained, numerical methods are frequently necessary. Some applications that have appeared often in chemical engineering textbooks include the following, many of which involve a balance between capital and operating costs:

1. Optimal thickness of insulation for a pipe carrying steam, for which the insulation cost is balanced against the heat loss causing steam condensation.
2. Optimal reflux ratio for a distillation column, for which the capital cost of the column and heat exchangers is balanced against the utility costs for cooling water in the condenser and steam in the reboiler.
3. Optimal absorbent (stripping agent or extraction solvent) flow rate in an absorber (stripper or liquid–liquid extractor), for which the number of stages is balanced against the column diameter and absorbent (stripping agent or extraction solvent) cost.
4. Optimal pipe diameter for a flowing liquid, for which the capital cost of the pipe and the pump is balanced against the operating cost of the pump.
5. Optimal length (or height) and diameter of a cylindrical pressure vessel of a given volume to minimize the capital cost.
6. Optimal number of stages in a multieffect evaporation system, for which the capital cost is balanced against the cost of heating steam.
7. Optimal interstage pressures of a multistage gas compression system with intercoolers, for which the total power requirement is to be a minimum.
8. Optimal cooling water outlet temperature from a heat exchanger, for which the capital cost of the heat exchanger is balanced against the cost of the cooling water.
9. Optimal filter cake thickness in a batch filter, for which the rate of filtration is balanced against the cost of removing the cake.
10. Optimal number of CSTRs in series, for which the capital cost is to be minimized.

Solutions to optimization problems such as these have led to many of the heuristics presented in Chapter 6.

When the NLP problem consists of only one decision variable (or can be reduced to one) with lower and upper bounds, the optimal solution can be found readily with a spreadsheet, or by one of several structured and efficient search methods, including *region elimination, derivative based*, and *point estimation*, as described in detail by Ravindran et al. (2006). Of the search methods, the golden-section method (involving region elimination) is reasonably efficient, reliable, easily implemented, and widely used. Therefore, it is described and illustrated by example here.

Golden-Section Search

The golden-section search method determines the optimal solution to a bounded objective function that is one-dimensional and unimodal. However, the function need not be continuous in either the function or its derivative. Thus, the method can solve functions like those shown in Figure 24.6.

Referring to Figure 24.7, let a and b equal the lower and upper bounds of x, the only decision variable. It is not necessary to compute the objective function at these two bounds. The distance between the two points, $b - a = L^{(1)}$. The strategy employed in the golden-section search begins by locating two points in x that are symmetrically placed within the interval from a to b by means of a factor, τ. Thus, if the point farthest from a is located at $(a + \tau L^{(1)})$, then the other point is located at $(b - \tau L^{(1)})$, which is equal to $[a + (1 - \tau)L^{(1)}]$. The objective function is computed for each of the first two points. It is desired to eliminate one of the two points and the region between it and its closest bound. Next, a new point is tested, positioned symmetrical to the remaining point within the new interval. This enables the value of τ to be determined. Suppose a minimum in the objective function is sought, and let the point closest to a have the lowest value of the objective function. Then, because of the assumption of unimodality, the optimal value of x cannot lie to the right of the point

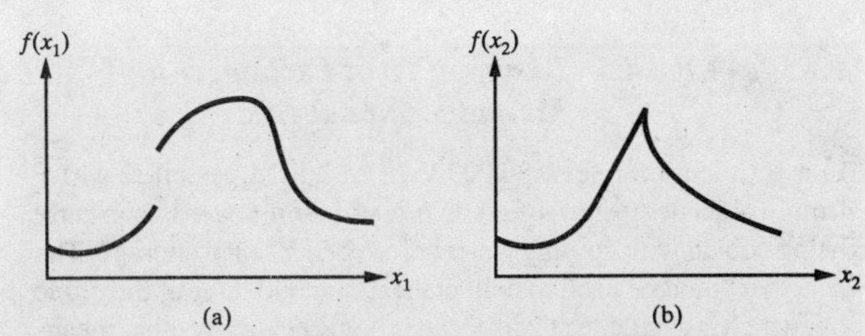

Figure 24.6 Objective function with discontinuities in: (a) the function and (b) the derivative of the function.

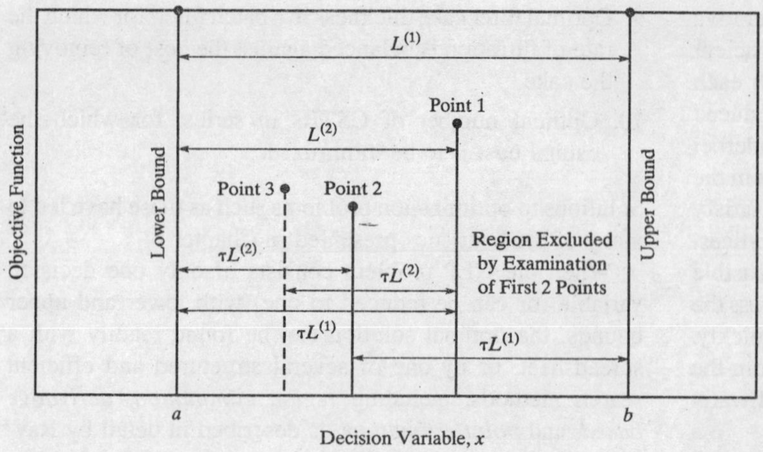

Figure 24.7 Development of the golden-section method.

closest to b. Therefore, the region between the point closest to b and the point b is eliminated from consideration, leaving a shorter search region interval of length $L^{(2)} = \tau L^{(1)}$. Where should the new point be placed? Note that the remaining point, which was located at $[a + (1-\tau)L^{(1)}]$, is now located on the new interval at $(a + \tau L^{(2)})$, which is the same as $(a + \tau^2 L^{(1)})$. Hence, $(1-\tau) = \tau^2$, whose positive solution is $\tau = 0.61803$.

Using this value of τ, subsequent steps in the golden-section method add new points symmetrical to the remaining point, calculate the objective function for the new point, and eliminate a point and the region between it and the closest bound of the remaining region. How many steps are required? Since each step reduces the search space by a factor $\tau = 0.61803$, the fraction of the search interval remaining after M steps is τ^M, requiring the computation of $M+1$ objective functions. Thus in 10 steps, the optimal solution is located in an interval that is less than 1% of the distance between the lower and upper bounds. In 20 steps, that interval is reduced to less than 0.01% of that distance.

Rudd and Watson (1968) point out that the ratio of 0.618 was known in ancient times as the "golden section." Greek temples were designed with this ratio because it was most pleasing to the eye. The ancient Badge of the Pythagoreans, a five-pointed star, consists of five isosceles triangles, whose bases are the sides of a pentagon. If the length of each side of the pentagon is 1, the length of each of the two long sides of the triangles is $1 + \tau$.

EXAMPLE 24.3 ***Design of Heat Exchanger to Minimize Annual Costs***

In a petroleum refinery, 80,000 lb/hr of a light gas oil at 440°F from a sidecut stripper of a crude distillation tower is currently being cooled with cooling water before being sent to storage. The heat lost could be used to help preheat 500,000 lb/hr of the crude oil, which is available at 240°F and is being heated by other means at a cost of \$3.00/million Btu. The plant operates 8,200 hr/yr. Based on the following data, determine what should be done, if anything, when a reasonable return on investment is $i_{min} = 0.20$. The savings in cooling water cost can be assumed negligible.

Data

Average specific heat of light gas oil = 0.50 Btu/lb-°F

Average specific heat of crude oil = 0.45 Btu/lb-°F

Use a floating-head shell-and-tube heat exchanger for areas greater than 200 ft^2.

For areas greater than 12,000 ft^2, use parallel units.

Delivered cost of a heat exchanger = 1.05 times Eq. (22.39) = $1.05 C_P\{A\}$

Bare-module factor for a heat exchanger, F_{BM}, = 3.17 from Table 22.11

Add 5% for site preparation and 18% for contingency and contractor's fee.

SOLUTION

For an objective function, use annualized cost, given by Eq. (23.10). Thus, if Q is the duty, in Btu/hr, of the light gas oil-to-crude oil heat exchanger of area A in ft^2, the annualized cost, to be minimized, is given by

$$\begin{aligned} C_A &= C + i_{min}(C_{TCI}) \\ &= -\frac{8{,}200(3.00)Q}{1{,}000{,}000} + 0.20(1.05)(1.05)(1.18)(3.17)C_P\{A\} \\ &= -0.0246Q + 0.8248 C_P\{A\} \end{aligned} \tag{24.17}$$

Thus, to be attractive, the annualized cost must be negative so that the absolute value of the savings in the annual cost of heating the crude oil (a negative quantity) is greater than the annualized cost of the heat exchanger installation. The more negative C_A is, the better.

From Table 18.5, for gas oil-to-oil, the overall heat-transfer coefficient, U, is 20 to 35 Btu/hr-°F-ft^2. Since the gas oil is a light gas oil, use $U = 35$ Btu/hr-°F-ft^2. For a mean temperature-driving force, use 0.7 times the log mean = $0.7\Delta T_{LM}$.

The equality constraints are

1. Energy balances:

$$Q = 80{,}000(0.50)(440 - T_{LGO,out}) \tag{24.18a}$$

$$Q = 500{,}000(0.45)(T_{CO,out} - 240) \tag{24.18b}$$

Table 24.2 Golden-Section Search Results for Example 24.3

Point	$T_{LGO,out}$(°F)	$T_{CO,out}$(°F)	ΔT_{LM}(°F)	$Q(\times10^6$Btu/hr)	A (ft^2)	$C_A(\times10^3\$)$
1	316.39	261.97	120.12	4.94	1,680	−100.2
2	363.61	253.58	152.87	3.06	815.8	−59.1
3	287.21	267.16	96.80	6.11	2,577	−123.7
4	269.18	270.37	79.80	6.83	3,495	−136.4
5	258.03	272.35	67.10	7.28	4,428	−142.4
6	251.15	273.57	57.44	7.55	5,368	−144.2
7	246.89	274.33	49.93	7.72	6,314	−143.4
8	253.77	273.10	61.38	7.45	4,954	−143.8
9	249.52	273.86	54.77	7.62	5,678	−144.2
10	252.15	273.39	58.99	7.51	5,199	−144.1
11	250.52	273.68	56.53	7.58	5,482	−144.2
12	250.14	273.75	55.81	7.59	5,554	−144.2
13	250.76	273.64	56.82	7.57	5,437	−144.2

2. Heat-transfer rate:

$$Q = 0.7\,UA\Delta T_{LM} = 0.7(35)A\Delta T_{LM} \tag{24.19}$$

3. Definition of log-mean temperature-driving force:

$$\Delta T_{LM} = \frac{(440 - T_{CO,out}) - (T_{LGO,out} - 240)}{\ln\left(\dfrac{440 - T_{CO,out}}{T_{LGO,out} - 240}\right)} \tag{24.20}$$

Thus, four equations relate five independent variables: $T_{LGO,out}$, $T_{CO,out}$, ΔT_{LM}, Q, and A, yielding one decision variable, which must be selected from the five variables. The best choice is the exit temperature of the light gas oil, $T_{LGO,out}$, because it is easily bounded and permits the remaining four variables to be calculated sequentially using the four equality constraints. An upper bound on its value is for no heat exchange, where $T_{LGO,out} = 440$°F and $C_A = \$0$. A lower limit assumes infinite heat exchange area, where $T_{LGO,out} = 240$°F (the inlet temperature of the crude oil), because the light gas oil has a much lower flow rate than the crude oil, and C_A is infinite.

The optimization problem is one-dimensional with a nonlinear objective function, which may be discontinuous, depending on the heat exchanger area. The single decision variable is bounded. Therefore, the golden-section search is suitable for determining the optimal solution. The calculations can be carried out conveniently in the following manner for each selection of $T_{LGO,out}$:

1. Calculate Q using Eq. (24.18a).
2. Calculate $T_{CO,out}$ using Eq. (24.18b).
3. Calculate ΔT_{LM} using Eq. (24.20).
4. Calculate A using Eq. (24.19).
5. Calculate C_A using Eq. (24.17).

To begin the steps in the golden-section search, note that the interval that bounds the decision variable is 440 − 240 = 200°F. With $\tau = 0.61803$, the first two points are located at $T_{LGO,out} =$ [240 + 0.61803(200)] = 363.606°F and [240 + (1 − 0.61803)(200)] = 316.394°F.

Table 24.2 gives the results of the golden-section search, indicating that it is attractive to install the heat exchanger. The optimal exit temperature of light gas oil from the heat exchanger is approximately 250°F, giving an annualized cost of approximately −\$144,200 or a savings of \$144,200 per year. The final interval for search is less than 1°F. More steps could reduce this interval further, but the area of the heat exchanger would change less than 2%. The crude oil outlet temperature from the heat exchanger is approximately 274°F, so it is heated up only 34°F, compared to a decrease of 190°F in temperature of the light gas oil. The optimal minimum approach temperature is approximately (250 − 240) = 10°F. The heat exchangers in the table are all within the range of 200 to 12,000 ft^2 in area so that a single shell-and-tube heat exchanger is sufficient, giving a smooth curve for the objective function. That function is plotted in Figure 24.8, where it is observed that the optimum is not particularly sharp. Consequently, other factors, such as operability and reliability, might enter into a final decision on the size of the heat exchanger.

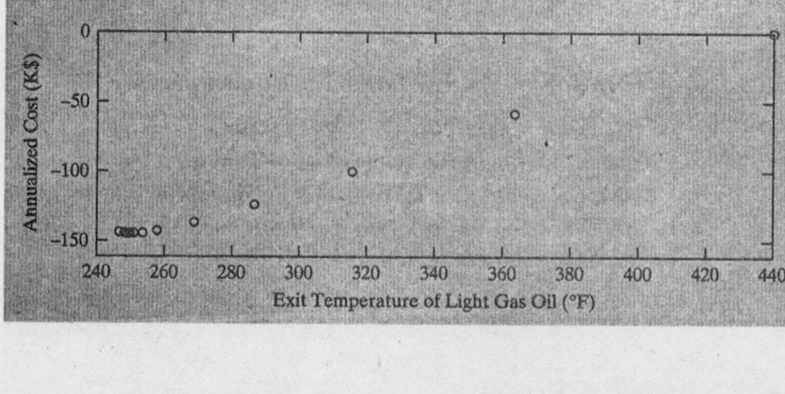

Figure 24.8 Golden-section search results for Example 24.3.

24.6 CONDITIONS FOR NONLINEAR PROGRAMMING (NLP) BY GRADIENT METHODS WITH TWO OR MORE DECISION VARIABLES

Optimization problems encountered in product and process design are often nonlinear programming (NLP) problems with two or more decision variables. Accordingly, much effort has been expended by researchers in the development of efficient search methods for finding an optimal solution. The remainder of this chapter deals with some of these methods, particularly those that are implemented in process simulators.

The formulation of the NLP for application to large process design problems begins with the steady-state simulation of the process flowsheet for a *nominal* set of specifications or decision variables. As described in Section 5.2, during the creation of the simulation model (involving the material and energy balances, kinetic equations, etc., for the process units), a degrees-of-freedom analysis is performed. For the simulation model, the number of variables, $N_{\text{Variables}}$, normally exceeds the number of equations, $N_{\text{Equations}}$, with the difference between them referred to as the number of degrees of freedom or decision variables, N_D. The best procedure is to carry out a base-case simulation, where the specifications for the decision variables are set using heuristics, such as those of Chapter 6. Then, gradually, as experience is gained by carrying out several simulations, the values of the decision variables are adjusted to better achieve the design objectives. In addition, the process units are simulated with more accurate models; the thermodynamic and transport properties are tuned, often using experimental and pilot-plant data; and profitability measures are computed. Having completed these steps, and having gained a good appreciation of the operation of the process and some indication of the key optimization tradeoffs, the engineer is well prepared to formulate a detailed NLP for solution by a process simulator.

General Formulation

The general formulation of the NLP, as applied to the optimization of product and process designs, is given above at the beginning of Section 24.2. The NLP is usually solved using gradient-based methods. By using numerical partial derivatives of the objective function with respect to the decision variables, gradient-based methods are faster than nongradient methods, with the advantage increasing with the number of decision variables. The use of these methods require conditions for optimality, referred to as stationary or stationarity conditions. These conditions involve slack variables and Lagrangian functions and are described next, used with *successive quadratic programming* (SQP), which is a widely used gradient method in simulators. Readers who prefer to skip these details should follow the text from Section 24.7.

Stationarity Conditions

To obtain the stationarity conditions, the Lagrangian is formed and differentiated, with the details of this procedure described by McMillan (1970) and by Beveridge and Schecter (1970). The development begins by converting the inequality constraints to equality constraints through the addition of *slack variables*, z_i^2, $i = 1, \ldots, N_{\text{Inequal}}$, such that constraints (24.4) become

$$\underline{g}_i\{\underline{x}\} + z_i^2 = 0 \qquad i = 1, \ldots, N_{\text{Inequal}} \tag{24.21}$$

where z_i^2 takes up the slackness when $\underline{g}_i\{\underline{x}\} < 0$. Then, the *unconstrained objective function*, or Lagrangian, is formed:

$$L\{\underline{x}, \underline{\pi}, \underline{\lambda}, \underline{z}\} = f\{\underline{x}\} + \underline{\pi}^{\mathrm{T}}\underline{c}\{\underline{x}\} + \underline{\lambda}^{\mathrm{T}}[\underline{g}\{\underline{x}\} + \underline{z}^2] = 0 \tag{24.22}$$

where $\underline{\pi}$ and $\underline{\lambda}$ are vectors of the Lagrange and Kuhn–Tucker multipliers. At the minimum,

$$\nabla L = 0, \tag{24.23}$$

which can be expanded to give the *stationarity conditions*, or the Karush–Kuhn–Tucker (KKT) conditions:

$$\nabla_x L = \nabla_x f\{\underline{x}\} + \underline{\pi}^{\mathrm{T}}\nabla_x \underline{c}\{\underline{x}\} + \underline{\lambda}^{\mathrm{T}}\nabla_x \underline{g}\{\underline{x}\} = 0 \tag{24.24}$$

$$\nabla_\pi L = \underline{c}\{\underline{x}\} = 0 \tag{24.25}$$

$$\nabla_{z_i} L = g_i \lambda_i = 0, \qquad i = 1, \ldots, m \tag{24.26}$$

$$\underline{\lambda} \geq 0 \tag{24.27}$$

Note that $\nabla_\lambda L = 0$ gives Eq. (24.21), which is the definition of the slack variables and needs not be expressed in the KKT conditions. Note also that $\nabla_{z_i} L = 2\lambda_i z_i = 0$, and, using Eq. (24.21), Eq. (24.26) results. These are the so-called *complementary slackness* equations. For constraint *i*, either the residual of the constraint is zero, $g_i = 0$, or the Kuhn–Tucker multiplier is zero, $\lambda_i = 0$, or both are zero; that is, when the constraint is *inactive* ($g_i > 0$), the Kuhn–Tucker multiplier is zero, and when the Kuhn–Tucker multiplier is greater than zero, the constraint must be *active* ($g_i = 0$). Stated differently, there is slackness in either the constraint or the Kuhn–Tucker multiplier. Finally, it is noted that $\nabla_x \underline{c}\{\underline{x}\}$ is the Jacobian matrix of the equality constraints, $\underline{\underline{J}}\{\underline{x}\}$, and $\nabla_x \underline{g}\{\underline{x}\}$ is the Jacobian matrix of the inequality constraints, $\underline{\underline{K}}\{\underline{x}\}$.

Solution of the Stationarity Equations

The KKT conditions are a set of $N_{\text{Variables}} + N_{\text{Equations}} + N_{\text{Inequal}}$ nonlinear equations (NLEs) in $N_{\text{Variables}} + N_{\text{Equations}} + N_{\text{Inequal}}$ unknowns that can be solved, in principle, using an algorithm for the solution of NLEs such as the Newton–Raphson method, which for the equation

$$\underline{F}\{\underline{X}\} = 0 \tag{24.28}$$

takes the form

$$\underline{\Delta X}^{(k)} = -\underline{\underline{J}}\{\underline{X}^{(k)}\}\underline{F}\{\underline{X}^{(k)}\} \quad (24.29)$$

$$\underline{X}^{(k+i)} = \underline{X}^{(k)} + \underline{\Delta X}^{(k)} \quad (24.30)$$

Here, $\underline{X}^{(k)}$ is the vector of guessed values at the kth iteration (the initial guesses when $k = 0$), $\underline{F}\{\underline{X}^{(k)}\}$ is the vector of residuals at $\underline{X}^{(k)}$, $\underline{\underline{J}}\{\underline{X}^{(k)}\}$ is the Jacobian at $\underline{X}^{(k)}$, $\underline{\Delta X}^{(k)}$ is the vector of corrections computed using the Newton–Raphson linearization, and $\underline{X}^{(k+1)}$ is the vector of unknowns after the kth iteration.

When the Newton–Raphson method is applied to solve the KKT Eqs. (24.24)–(24.26), $\underline{X}^T = [\underline{x}^T, \underline{\pi}^T, \underline{\lambda}^T]$, and Eqs. (24.29) and (24.30) can be rewritten in terms of these variables. This was accomplished by Jirapongphan (1980), who showed that beginning with the vector of guesses, $\underline{X}^{(k)}$, one iteration of the Newton–Raphson method is equivalent to solving the following quadratic program (QP):

$$\text{Minimize} \quad \nabla_x f\{\underline{x}^{(k)}\}^T \underline{\Delta x}^{(k+1)} + (1/2)\underline{\Delta x}^{(k+1)^T x} \times \nabla^2_{xx} \mathrm{L}\{\underline{x}^{(k)}, \underline{\pi}^{(k)}, \underline{\lambda}^{(k)}\}\underline{\Delta x}^{(k+1)}$$

$$\text{w.r.t.} \quad \underline{\Delta x}^{(k+1)}, \Delta\underline{\pi}^{(k+1)}, \underline{\Delta\lambda}^{(k+1)} \quad \textbf{(QP)}$$

s. t.

$$\underline{\underline{J}}\{\underline{x}^{(k)}\}\underline{\Delta x}^{(k+1)} + \underline{c}\{\underline{x}^{(k)}\} = 0 \quad (24.31)$$

$$\underline{\underline{K}}\{\underline{x}^{(k)}\}\underline{\Delta x}^{(k+1)} + \underline{g}\{\underline{x}^{(k)}\} = 0 \quad (24.32)$$

Algorithms for the solution of quadratic programs, such as the Wolfe (1959) algorithm, are very reliable and readily available. Hence, these have been used in preference to the implementation of the Newton–Raphson method. For each iteration, the quadratic objective function is minimized subject to linearized equality and inequality constraints. Clearly, the most computationally expensive step in carrying out an iteration is in the evaluation of the Laplacian of the Lagrangian, $\nabla^2_{xx}\mathrm{L}\{\underline{x}^{(k)}, \underline{\pi}^{(k)}, \underline{\lambda}^{(k)}\}$, which is also the Hessian matrix of the Lagrangian; that is, the matrix of second derivatives with respect to $\underline{x}^{(k)}, \underline{\pi}^{(k)}, \underline{\lambda}^{(k)}$.

To circumvent this calculation, Powell (1977) used the Broyden, Fletcher, Goldfarb, Shanno (BFGS) quasi-Newton method to approximate $\nabla^2_{xx}\mathrm{L}\{\underline{x}^{(k)}, \underline{\pi}^{(k)}, \underline{\lambda}^{(k)}\}$. This saves considerable computation time and is the basis of Powell's SQP method.

A key problem that arises in the implementation of Powell's algorithm is due to the linearization that produces a quadratic objective function and linear constraints, which often lead to infeasible solution vectors, $\underline{X}^T = [\underline{x}^T, \underline{\pi}^T, \underline{\lambda}^T]$. This problem manifests itself in solutions, $\underline{x}^{(k+1)}$, that violate the inequality constraints, as well as multipliers that are often driven to zero prematurely. Assuming that the initial guesses do not violate the inequality constraints, Han (1977) designed a unidirectional search in the direction of $\underline{\Delta X}^{(k)}$ that is designed to reduce the steps taken so as to keep the solution vector within the feasible space.

For more complete presentations of the SQP algorithm, the reader is referred to the textbooks by Biegler et al. (1997), Edgar et al. (2001), and Ravindran et al. (2006).

24.7 OPTIMIZATION ALGORITHM

The most straightforward way to improve the objective function is by *repeated simulation*. In this procedure, the designer selects values of the decision variables and completes a simulation. Then, usually using a systematic strategy, the decision variables are adjusted and the simulation is repeated—for example, using *sensitivity analysis* in the process simulators, in which simulation results are recomputed as a decision variable is adjusted using uniform increments between bounds specified by the user. However, sensitivity analyses can be very time-consuming and can generate large files of information, much of which is associated with suboptimal processes.

For process design optimization, alternatively the designer can select a formal optimization algorithm built into the process simulator to adjust the decision variables, a strategy that is usually more efficient. However, when many recycle loops are present, the simulation calculations can be very time-consuming, with 20 or more iterations required to achieve an optimum and, for each of these iterations, as many as 20 iterations to converge each of the recycle and control loops (involving design specifications). To overcome these inefficiencies, especially for integrated flowsheets, the latest strategies adjust the decision variables and the tear variables to converge the recycle loops simultaneously (Lang and Biegler, 1987; Seader et al., 1987). In these strategies, the optimization algorithm does not converge the recycle loops for each set of decision variables. Instead, it performs just one pass through the recycle loops before adjusting the decision variables, and consequently, the strategies are referred to as *infeasible path algorithms* because the solution for that path is not converged. In most cases, the infeasible path strategy is successful in ultimately converging the recycle loops to a feasible design while optimizing the process.

To clarify the infeasible path strategy, consider the simulation flowsheets in Figure 24.9. In Figure 24.9a, the ASPEN PLUS simulation flowsheet—in which a continuous-stirred-tank reactor (CSTR) is followed by a recycle loop involving another CSTR, a distillation column, a purge splitter, and a heater—is optimized. In this case, the recycle convergence unit, $OLVER01, is positioned arbitrarily so as to tear stream S9, with x^* being the vector of guessed values for the tear variables and $w\{d, x^*\}$ being the vector of tear variables after one pass through the recycle loop. In Figure 24.9b, the HYSYS PFD, which consists of a recycle loop involving a **conversion reactor**,

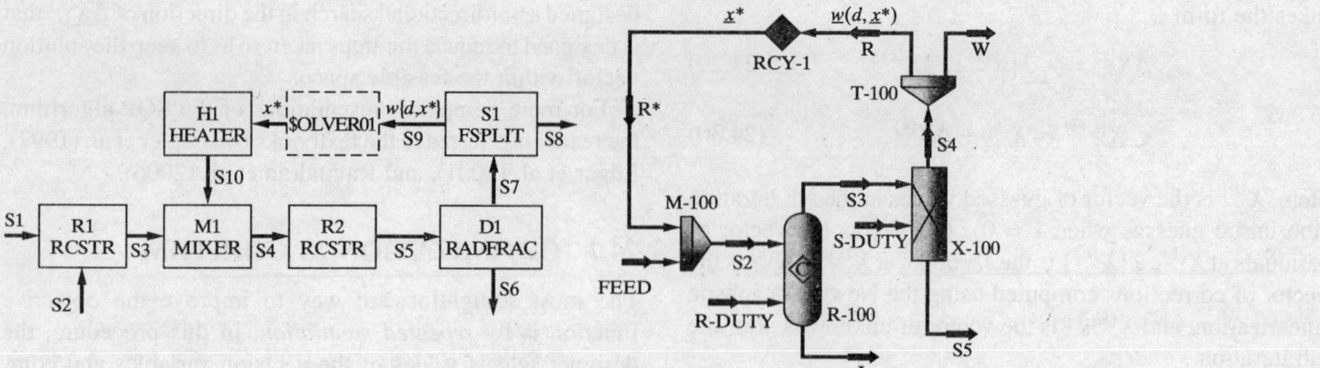

Figure 24.9 Process flowsheet with material recycle and tear stream: (a) ASPEN PLUS; (b) HYSYS.

a **component splitter**, and a **tee**, is optimized. Note that in HYSYS, a recycle convergence unit is explicitly positioned to tear stream R, with $\underline{x}^*$ being the vector of guessed values for the tear variables and $\underline{w}\{d, \underline{x}^*\}$ being the vector of tear variables after one pass through the recycle loop. Note that w is used because f is reserved for the objective function, as shown in the revised NLP that follows:

$$\begin{aligned}
&\text{Minimize} \quad f\{\underline{x}\} && \textbf{(NLP2)}\\
&\text{w.r.t. } \underline{d} &&\\
&\text{s.t.} &&\\
&\quad \underline{h}\{\underline{x}^*\} = \underline{x}^* - \underline{w}\{\underline{x}^*\} = 0 && \text{(tear equations)}\\
&\quad \underline{c}\{\underline{x}\} = 0 &&\\
&\quad \underline{g}\{\underline{x}\} \le 0 &&\\
&\quad \underline{x}^L \le \underline{x} \le \underline{x}^U &&
\end{aligned}$$

Here, the equality constraints are augmented by the tear equations $\underline{h}\{\underline{x}^*\} = 0$, which must be satisfied as well at the minimum of $f\{\underline{x}\}$. For this and similar flowsheets, the decision variables include the residence times in the reactors, the reflux ratio of the distillation tower, and the purge/recycle ratio. In one-dimensional space (i.e., with one decision variable), as $\underline{d}$ varies, the objective function can be displayed as shown in Figure 24.10a. Clearly, the optimizer seeks to locate the minimum efficiently, a task that is complicated when multiple minima exist and it is desired to locate the *global* minimum.

As the decision variables are adjusted by the optimizer, the values of the tear variables and the objective function change, and it is helpful to show these functionalities in a more explicit form of the NLP:

$$\begin{aligned}
&\text{Minimize} \quad f\{\underline{x}, \underline{d}\} && \textbf{(NLP3)}\\
&\text{w.r.t. } \underline{d} &&\\
&\text{s.t.} &&\\
&\quad \underline{h}\{\underline{x}^*, \underline{d}\} = \underline{x}^* - \underline{w}\{\underline{x}^*, \underline{d}\} = 0 &&\\
&\quad \underline{c}\{\underline{x}, \underline{d}\} = 0 &&\\
&\quad \underline{g}\{\underline{x}, \underline{d}\} \le 0 &&\\
&\quad \underline{x}^L \le \underline{x} \le \underline{x}^U &&
\end{aligned}$$

Here, $\underline{x}$ is the vector of process variables excluding the decision variables, and $\underline{x}^*$ is the vector of guessed values for the tear variables (which equal $\underline{w}$ at the minimum). As will be seen later, it helps to show the variations of the objective function, f, and the tear functions, h, in the schematic diagram of Figure 24.10b. In this diagram, level contours are displayed as a function of just one decision variable and one tear variable, and the locus of points is displayed at which the tear

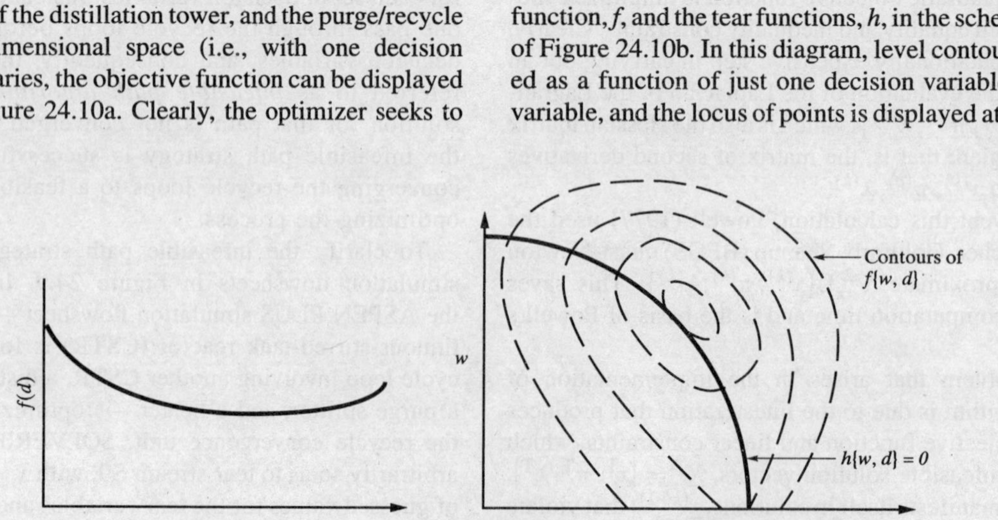

Figure 24.10 Optimization of a process with recycle: (a) objective function; (b) level contours and tear equation.

equation is satisfied as d varies along the abscissa. Clearly, the minimum of f must lie on the latter curve, with the unconstrained minimum of $f\{w, d\}$ being infeasible.

Repeated Simulation

The repeated simulation approach is illustrated in Figure 24.11a. Beginning with an initial guess for the decision and tear variables, in the small box, a simulation is completed in which the recycle loop is converged; that is, the tear equation is satisfied. Then, d is adjusted, somewhat arbitrarily, by Δ_d and the simulation is repeated using the previous solution for w as the initial guess. This strategy is repeated until convergence to the minimum is achieved.

Infeasible Path Approach

In the infeasible path approach, as illustrated in Figure 24.11b, both d and w are adjusted simultaneously by the optimizer (with $w \rightarrow x^*$ for the next iteration), usually using the SQP algorithm. This algorithm involves just one pass through the flowsheet per iteration, so the tear equations are normally not satisfied until the optimum is located. As will be seen in the examples that follow, convergence is usually achieved in a few iterations. Figure 24.11b shows a schematic of $\underline{\Delta}$, the vector of the changes in d, Δ_d, and w, Δ_w, as computed by the SQP algorithm.

Compromise Approach

In a compromise approach, which is often necessary to achieve convergence after the SQP step is taken (i.e., $\underline{\Delta}$), the tear equations, $\underline{h}\{\underline{x}^*\} = 0$, are converged as shown in Figure 24.11c. Often the tear equations are converged loosely using a convergence method, typically Wegstein's method, with a maximum number of iterations assigned, typically three or four. This compromise approach is often utilized when convergence cannot be achieved after several attempts using the infeasible path approach.

Practical Aspects of Flowsheet Optimization

In most flowsheet simulations, design specifications (or control loops) are included. The iterative calculations to achieve convergence of these specifications are often embedded within the recycle loops and are converged during each pass through the recycle loops. In other cases, these

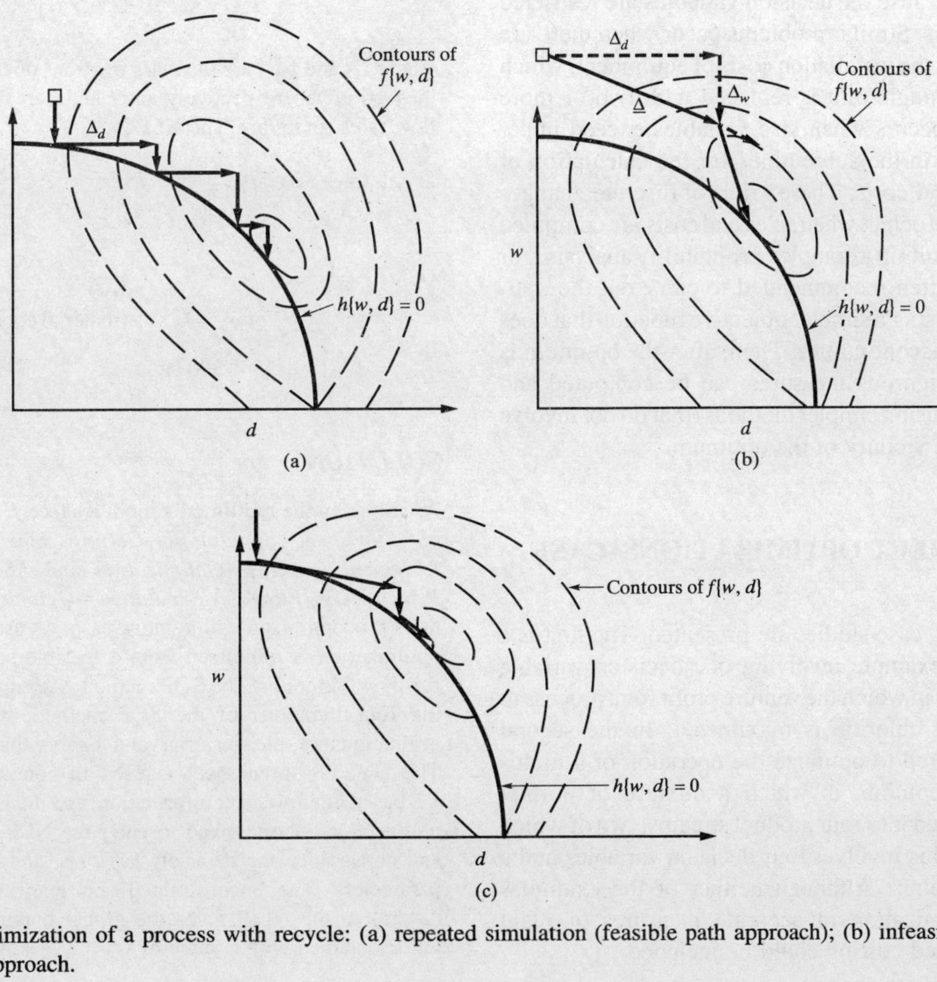

Figure 24.11 Optimization of a process with recycle: (a) repeated simulation (feasible path approach); (b) infeasible path approach; (c) compromise approach.

specifications are implemented as outer loops, with the recycle loops converged entirely during each iteration of the outer loop. Yet another alternative is to converge these loops simultaneously, usually using just one pass through the unit subroutines in the recycle and design specification loops.

When solving a NLP to optimize a flowsheet, still another alternative exists. In many cases, it is preferable to incorporate the design specifications as equality constraints, $\underline{c}\{\underline{x}, \underline{d}\} = 0$, as shown in NLP3. Then, it is necessary to remove these design specifications when adding the optimization convergence unit. The latter usually replaces the recycle convergence units in the simulation flowsheet.

Before implementing an infeasible path optimization, it is very helpful to carry out preliminary searches by varying the key decision variables somewhat randomly, to gain insights into the key tradeoffs. For these searches, it is probably best not to use optimization algorithms that require derivatives, or approximations to them, such as SQP. A common approach is to use the sensitivity analysis facilities of the process simulators referred to earlier.

As a final caution, be sure not to use a gradient-based approach, such as SQP, when selecting a discrete decision variable such as the number of trays in a distillation tower. In these cases, it is meaningless to estimate partial derivatives in expressions for the gradient of the objective function or its Hessian matrix, because the decision variables are restricted to integer quantities. Similar problems occur when there are discrete changes in the installation costs of equipment, which can arise when a single unit is replaced with two or more units. This often occurs when size variables exceed upper bounds embedded in the subroutines for the calculation of equipment sizes and costs. These kinds of discrete changes are more difficult to detect when sizes and costs are computed for many process units in a complex profitability analysis. For this reason, it is often recommended to carry out the optimization initially using a simpler objective function that does not involve such discontinuities. Then, after the optimum is computed, more rigorous measures can be computed and further optimized, using simpler methods (that do not involve derivatives), in the vicinity of the optimum.

24.8 FLOWSHEET OPTIMIZATIONS–CASE STUDIES

In this section, two case studies are presented. The first is a relatively simple example involving one decision variable and one constraint, in which the venture profit for a process to manufacture ethyl chloride is maximized. In the second example, it is desired to optimize the operation of a multidraw distillation column, in which a mixture of normal paraffins is separated into four product streams, two of which are sidestreams. This involves four decision variables and a number of constraints. Although neither of the examples involves the optimization of accurate measures of plant profitability, detailed costing could be included.

EXAMPLE 24.4 ***Maximizing the Venture Profit in Ethyl Chloride Manufacture***

In this example, the venture profit of the ethyl chloride process in Figure 24.12, introduced in the multimedia modules, which can be downloaded from the Wiley Web site associated with this text (see *HYSYS→ Tutorials→Material and Energy Balances→Ethyl Chloride Manufacture* and *ASPEN→Tutorials→Material and Energy Balances→Ethyl Chloride Manufacture*), is maximized by adjusting the purge (W) flow rate. To estimate the venture profit, the following information is supplied:

www.wiley.com/go/global/seider

Installed cost of equipment	$500\left(\dfrac{300 \times 24}{1{,}000}(F_R)\right)^{0.6}$ monetary units
Cost of ethylene	1.5×10^{-3} monetary units/kg
Cost of HCL	1.0×10^{-3} monetary units/kg
Revenue for ethyl chloride	2.5×10^{-3} monetary units/kg

where F_R is the reactor feed rate (in kg/hr). The venture profit [Eq. (23.9)] is formulated (in monetary units) assuming a 10% return on investment (ROI) and 330 operating day/yr:

$$\text{VP} = 330 \times 24 \times 10^{-3}[2.5P - (1.5x_{\text{Et}} + x_{\text{HCl}})F] - 0.1\left[500\left(\frac{330 \times 24}{1{,}000}(F_R)\right)^{0.6}\right] \tag{24.33}$$

where x_{Et} and x_{HCl} are the mass fractions of ethylene and HCl in the feed stream, respectively, and F and P are the feed and product flow rates (in kg/hr). The NLP is

$$\underset{W}{\text{Minimize}} \quad \text{VP} \tag{24.34}$$

s.t.

$$\underline{c}\{\underline{x}\} = 0 \quad \text{(material equations)} \tag{24.35}$$

$$R < 300\,\text{kg/hr} \tag{24.36}$$

SOLUTION

As shown in the multimedia modules (see *HYSYS→ Principles of Flowsheet Simulation→Getting Started in HYSYS→ Advanced features→Optimization* and *ASPEN→ Principles of Flowheet Simulation→Optimization*), the VP is optimized with relative ease. As usual, the optimization is initialized from a feasible solution, for $W = 5$ kmol/hr. The SQP method is used, requiring four iterations of the SQP method, with 13 evaluations of the material and energy balances. The HYSYS spreadsheet is used to compute the VP based on flowsheet information, and the HYSYS optimizer is then invoked to enter the NLP objective function and constraints, the decision variables, and numerical method parameters. The unconstrained solution [i.e., neglecting Eq. (24.36) in the NLP] gives the global maximum VP of 4,730 units, obtained with a value of $W = 7.27$ kmol/hr. Augmenting

www.wiley.com/go/global/seider

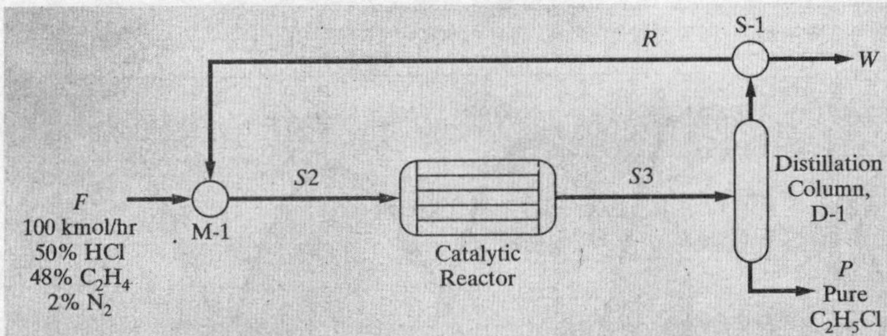

Figure 24.12 Process for the production of ethyl chloride.

the NLP with the inequality constraint in Eq. (24.36) gives VP = 4,450 units, obtained with a value of $W = 8.96$ kmol/hr, for which R is at its upper bound of 300 kg/hr. Use the file ETHYLCHLORIDE_OPT.hsc in the Program and Simulation Files folder, which can be downloaded from the Wiley Web site associated with this book, to reproduce the results presented above.

EXAMPLE 24.5 *Optimization of a Distillation Tower with Sidedraws*

In this example, the distillation tower in Figure 24.13 is optimized. A feed containing the normal paraffins from nC_5 to nC_9 is fed to the 25-stage tower (including the condenser and reboiler) on stage 15, counting upward from the partial-reboiler stage. The objective is to adjust the operating conditions so as to achieve a distillate (D) concentrated in nC_5, a sidedraw at stage 20 ($S1$) concentrated in nC_6, a second sidedraw at stage 10 ($S2$) concentrated in nC_7 and nC_8, and a bottoms product (B) concentrated in nC_9. No costing is involved. The operating conditions to be adjusted are the reflux ratio and flow rates of the distillate and two sidedraws. This is accomplished by formulating a NLP in which the feed stage and the stages of the sidedraws are held fixed during the optimization:

$$\begin{array}{ll} \text{Minimize} & D_{C5} + S1_{C6} + S2_{C7} + S2_{C8} + B_{C9} \\ \text{w.r.t.} & \\ R, D, S1, S2 & \\ \text{s.t.} & \end{array} \tag{24.37}$$

$$5 \leq R \leq 10 \tag{24.38}$$

$$0.1 \leq D/F \leq 0.7 \tag{24.39}$$

$$0.1 \leq S1/F \leq 0.7 \tag{24.40}$$

$$0.1 \leq S2/F \leq 0.7 \tag{24.41}$$

$$(D + S1 + S2)/F \leq 0.95 \tag{24.42}$$

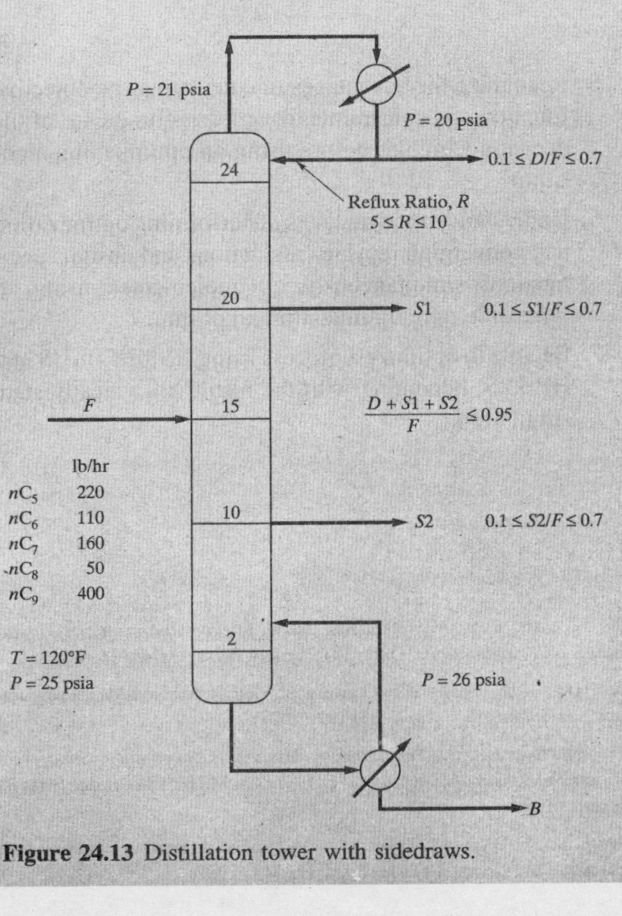

Figure 24.13 Distillation tower with sidedraws.

where R is the reflux ratio; F, D, $S1$, $S2$, and B are the molar flow rates of the feed, distillate, two sidedraws, and bottoms product streams; and the subscript denotes the molar flow rate of a specific chemical species in that stream. Note that the product streams are withdrawn as saturated liquids at the low pressures shown. The liquid and vapor phases are assumed to be ideal and at equilibrium on the stages of the tower. All of the decision variables have inequality constraints as indicated by Eqs. (24.38)–(24.42). This example does not involve recycle convergence or any user-supplied equality constraints.

SOLUTION

As shown in the multimedia modules, which can be downloaded from the Wiley Web site associated with this book (see *HYSYS→Tutorials→Process Design→Multi-draw Tower Optimization* and *ASPEN→Tutorials→Process Design→Multi-draw Tower Optimization*), the solution of this NLP requires care, since the gradient-based SQP method is sensitive to the numerical estimates of the derivatives in the Jacobian matrices $\underline{\underline{J}}\{\underline{x}\}$ and $\underline{\underline{K}}\{\underline{x}\}$. The optimization is initialized at a feasible solution: $R = 5$, $D = S1 = S2 = 2$ lbmol/hr, for which the composition profiles in the column are given in Figure 24.14a and the product recoveries summarized in Table 24.3.

Note that the peaks in the composition profiles are not at the draw stream locations, which explains the rather poor product recoveries, especially for C_8. After optimization, the decision variables are: $R = 10$, $D = 3.35$ lbmol/hr, $S1 = 1$ lbmol/hr, and $S2 = 2.16$ lbmol/hr, giving the composition profiles in

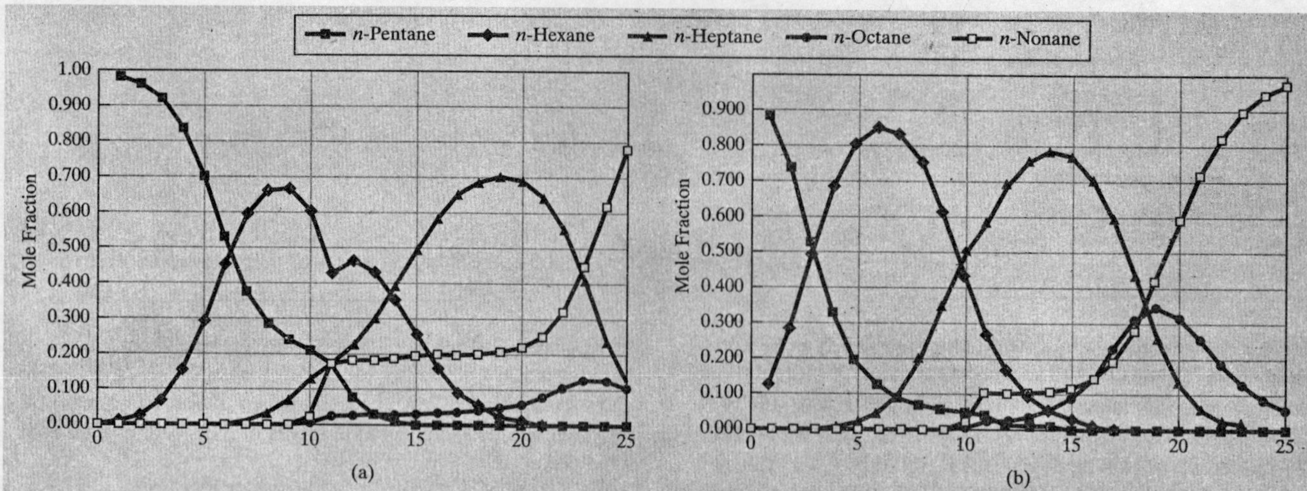

Figure 24.14 Composition profiles in multidraw column as a function of the tray number, counting from the top: (a) initial design; (b) optimized design.

Figure 24.14b, which show a marked improvement in positioning the composition peaks. This explains the significant improvement in the product recoveries shown in Table 24.3. Use the files MULTIDRAW_ OPT.hsc and MULTIDRAW_OPT.bkp in the Program and Simulation Files folder, which can be downloaded from the Wiley Web site associated with this book.

Table 24.3 Product Recoveries in the Multidraw Column

	Percentage Molar Recovery					
	D-C_5	$S1$-C_6	$S2$-C_7	$S2$-C_8	B-C_9	Objective Function
Initial design	65	72	74	18	87	6.87
Optimized design	96	65	96	68	91	8.44

24.9 SUMMARY

Having studied this chapter and the multimedia modules that accompany this book, the reader should:

1. Understand the fundamentals of optimization concerning the use of analytical or numerical methods.
2. Be able to solve LP problems in one or two decision variables.
3. Be able to create a nonlinear program (NLP) to optimize a process using equality and inequality constraints.
4. Be able to use the golden-section search to solve a constrained NLP problem in one decision variable.
5. Recognize the advantages of calculating the objective function and constraints for at least a base case of the decision variables before using an optimization algorithm.
6. Understand the advantages of performing optimization and converging recycle calculations and design specifications simultaneously, as implemented using an *infeasible path* optimization algorithm.
7. Be able to optimize a process using ASPEN PLUS and HYSYS, beginning with the results of a steady-state simulation.

REFERENCES

1. BEVERIDGE, G.S.G., and R.S. SCHECTER, *Optimization: Theory and Practice*, McGraw-Hill, New York (1970).

2. BIEGLER, L.T., I.E. GROSSMANN, and A.W. WESTERBERG, *Systematic Methods of Chemical Process Design*, Prentice-Hall, Englewood Cliffs, New Jersey (1997).

3. DANTZIG, G.B., "Programming of Independent Activities, II. Mathematical Model," *Econometrica*, **17**, 200–211 (1949).

4. EDGAR, T.F., D.M. HIMMELBLAU, and L.S. LASDON, *Optimization of Chemical Processes*, 2nd ed., McGraw-Hill, New York (2001).

5. FLOUDAS, C.A., *Nonlinear and Mixed-integer Optimization: Fundamentals and Applications*, Oxford University Press, Oxford (1995).

6. HAN, S.-P., "A Globally Convergent Method for Nonlinear Programming," *J. Optimization Appl.*, **22**, 297 (1977).

7. JIRAPONGPHAN, S., *Simultaneous Modular Convergence Concept in Process Flowsheet Optimization*, Sc.D. Thesis, M.I.T., Cambridge, Massachusetts (1980).

8. LANG, Y.-D., and L.T. BIEGLER, "A Unified Algorithm for Flowsheet Optimization," *Comput. Chem. Eng.*, **11**, 143 (1987).

9. MAH, R.S.H., *Chemical Process Structures and Information Flows*, Butterworth, Boston (1990).

10. MCMILLAN, Jr., C., *Mathematical Programming: An Introduction to Design and Application of Optimal Decision Machines*, Wiley, New York (1970).

11. MORGAN, A., *Solving Polynomial Systems Using Continuation for Engineering and Scientific Problems*, Prentice-Hall, Englewood Cliffs, New Jersey (1987).

12. POWELL, M.J.D., *A Fast Algorithm for Nonlinearly Constrained Optimization Calculations*, AERE Harwell, England (1977).

13. RAVINDRAN, A., K. M. RAGSDELL, and G.V. REKLAITIS, *Engineering Optimization—Methods and Applications*, 2nd ed., Wiley-Interscience, New York (2006).

14. RUDD, D.F., and C.C. WATSON, *Strategy of Process Engineering*, John Wiley & Sons, New York (1968).

15. SEADER, J.D., W.D. SEIDER, and A.C. PAULS, *FLOWTRAN Simulation—An Introduction*, 3rd ed., CACHE, Austin, Texas (1987).

16. VANDERBEI, R.J., *Linear Programming Foundations and Extensions*, Kluwer Academic Publishers, Norwell, Massachusetts (1999).

17. WOLFE, P., "The Simplex Method of Quadratic Programming," *Econometrica*, **27**, 382–398 (1959).

EXERCISES

24.1 *Scheduling of batch distillation.* A batch distillation facility has a bank of columns of Type 1 and another bank of Type 2. Type 1 columns are available for processing 6,000 hr/week, while Type 2 columns are available 10,000 hr/week. It is desired to use these columns to manufacture two different slates of products, A and B. Distillation time to produce 100 gal of product A is 2 hr in Type 1 columns and 1 hr in Type 2 columns. Distillation time to produce 100 gal of product B is 1 hr in Type 1 columns and 4 hr in Type 2 columns. The net profit is \$5.00/gal for product A and \$3.50/gal for product B. Use an LP with a graph to determine the production schedule that maximizes the net profit in \$/week.

24.2 *Analytical optimization.* Determine using calculus all maxima, minima, and saddle points for the following unconstrained two-dimensional objective functions:

(a) $f\{x_1, x_2\} = 2x_1^3 + 4x_1x_2^2 - 10x_1x_2 + x_2^2$

(b) $f\{x_1, x_2\} = 1{,}000x_1 + 4 \times 10^9 x_1^{-1} x_2^{-1} + 2.5 \times 10^5 x_2$

(c) $f\{x_1, x_2\} = (1 - x_1)^2 + 100(x_2 - x_1^2)^2$

24.3 *Golden-section search.* Use 10 steps of the golden-section search method to find the optimal dimensions for the cylindrical reactor vessel in Example 22.11. In that example, the dimensions of the vessel are given as the inside diameter, $D = 6.5$ ft, and tangent-to-tangent length, $L = 40$ ft. These dimensions are not critical as long as the volume is maintained. Determine the optimal diameter and length, if the permissible range of the aspect ratio, *L/D*, is 1 to 50.

24.4 *Optimization of minimum temperature approach in a heat exchanger.* As shown in the ASPEN PLUS simulation flowsheet in Figure 24.15, liquid toluene is to be heated from 100 to 350°F, while liquid styrene is to be cooled from 300 to 100°F. Auxiliary heat exchangers E2 and E3, which use steam and cooling water, respectively, are provided to meet the target temperatures when they cannot be achieved by heat exchanger E1. The process is to be optimized with respect to the minimum temperature of approach in E1, which is to be within 1°F and 50°F. The temperature of stream S3 is constrained to be less than or equal to 200°F, and the temperature of stream S4 is constrained to be less than or equal to 300°F. The annualized cost is to be minimized with the return on investment, *r*, equal to 0.5. All of the necessary data are included in Figure 24.15.

24.5 *Separation of propylene and propane by high-pressure distillation.* A process to separate propylene and propane, to produce 99 mol% propylene and 95 mol% propane, is shown in Figure 24.16. Because of the high product purities and the low relative volatility, 200 stages may be required. Assuming 100% tray efficiency and a tray spacing of 24 in., two towers are installed because a single tower would be too tall. The distillate is vapor at 280 psia and a pressure increase of 0.1 psi is assumed on each tray, with a 0.2-psi increase in the condenser. Note that the stage numbers and reflux ratio are for a nominal design. In this exercise, a suitable objective function is to be selected and the number of stages and reflux ratio are to be adjusted to find the optimum. Pay close attention to the determination of the proper feed stage to avoid pinch or near-pinch conditions.

24.6 *Petlyuk columns.* A process design for the disproportionation of toluene to benzene and the xylene isomers is being completed. Your assistance is needed on the design of the liquid-separation section. It has been established that the feed to this section is at 100°F and 50 psia with the following flow rates in lbmol/hr:

Benzene	16.3
Toluene	70.9
p-Xylene	4.0
m-Xylene	7.5
o-Xylene	3.5

From this feed, it is desired to produce 99.5 mol% benzene, 98 mol% toluene for recycle, and 99 mol% mixed xylenes, by distillation in two columns. The assigned plant operators have informed us that they prefer the direct sequence of two columns. However, because of the high percentage of toluene in the feed, a thermally coupled system, shown in Figure 24.17, known as Petlyuk towers after their Russian inventor, may be a less expensive alternative. Please prepare process designs for these two alternatives, together with estimates of capital and operating costs, and indicate whether the Petlyuk towers are attractive. The plant-operating factor is assumed to be 95%. The following instructions and data are provided for the design of the towers:

1. Determine optimal feed preheat using the bottoms product where applicable.
2. Determine optimal reflux ratios.
3. Set a pressure of 15 psia at the top of each column, assuming no pressure drop through the condenser. Determine column bottom pressure drop from tray pressure drops.

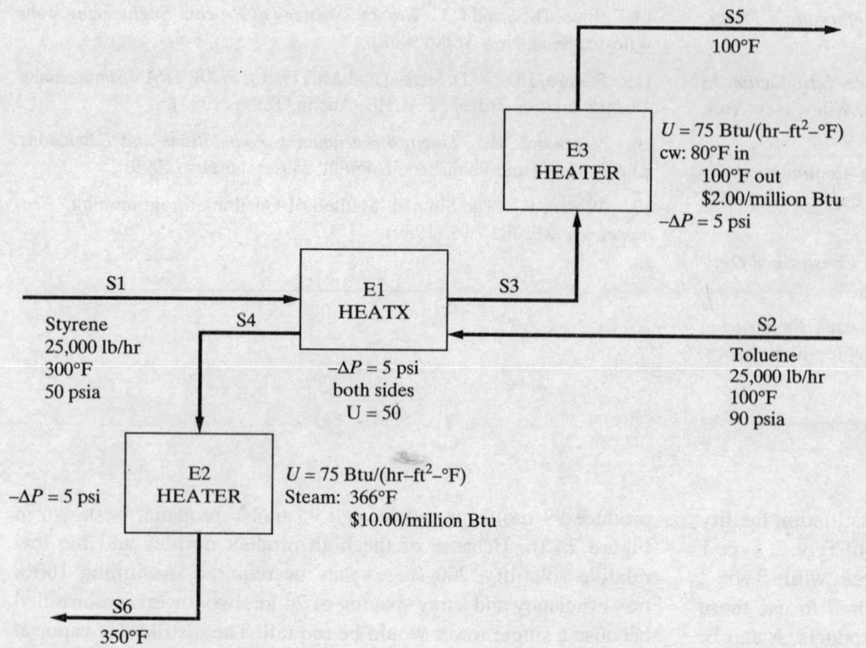

Figure 24.15 Heat exchange network.

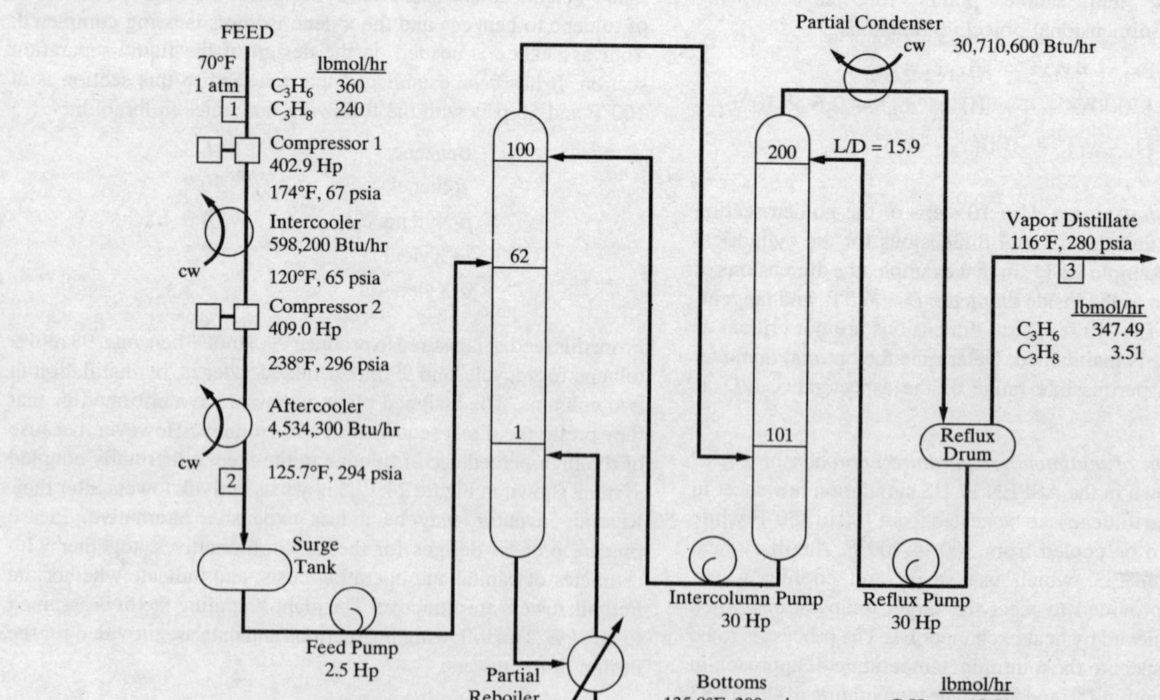

Figure 24.16 Propylene–propane distillation tower.

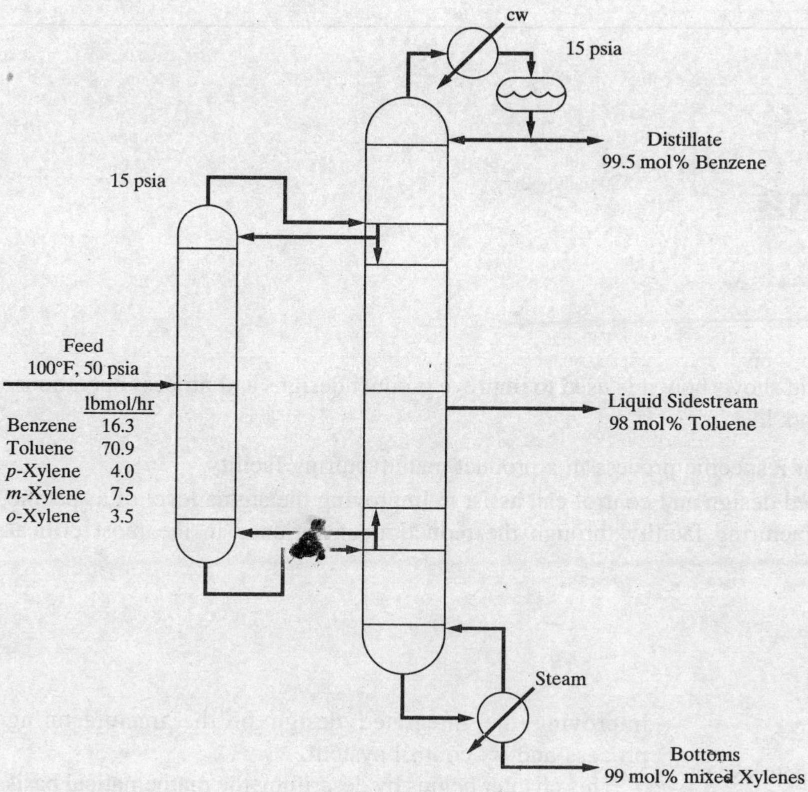

Figure 24.17 Petlyuk towers.

4. Use standard sieve trays. Overall tray efficiency is estimated to be 90% for operation at 85% of flooding.
5. Determine column diameters to the next increment of 0.5 ft for each section and swedge the columns, when section diameters differ by more than 1.0 ft.
6. Use reflux subcooled to 120°F from each condenser.
7. Standard materials of construction, for example, carbon steel, can be used.
8. Minimum shell thickness for columns and vessels is as follows

   0.25 in. for diameters less than 4 ft
   $\frac{5}{16}$ in. for diameters from 4 to 5.5 ft
   $\frac{3}{8}$ in. for diameters from 6 to 7.5 ft
   $\frac{7}{16}$ in. for diameters from 8 to 11.5 ft
   0.5 in. for diameters from 10 to 12 ft

9. Provide horizontal reflux drums that can hold liquid reflux and distillate for 5 min at half full.
10. Include all necessary liquid pumps with spares.

Chapter 25

Design for Six-Sigma

25.0 OBJECTIVES

This chapter introduces six-sigma methodology and shows how it is used to improve product designs and process operations. After studying this chapter, the reader should

1. Be able to compute the sigma level of a specific process in a product manufacturing facility.
2. Have an appreciation of how integrated design and control can assist in improving the sigma level of a specific process or the entire product manufacturing facility through the reduction of variance in the most critical manufacturing steps.

25.1 INTRODUCTION

Espresso coffee is prepared in a machine that injects water under high-pressure steam through a cake of ground coffee. In a conventional espresso machine, the user manually loads ground coffee into a metal filter cup, locks the cup under the water head, and then activates the water heater. A manufacturer of espresso machines would like to guarantee that each cup of coffee processed by the machine has a consistent quality. It is noted that the quality of each cup of espresso depends on a large number of variables, among them the grade and freshness of the coffee beans, the extent to which the beans have been ground, the operating temperature and pressure, the degree to which the ground coffee is packed into the metal filter holder, and the total amount of water used. Since many of the sources of product variability cannot be controlled by the manufacturer, the development of an improved espresso machine would be driven by the desire to either reduce the influence of these variables or eliminate as many as necessary to ensure a satisfactory product.

This chapter describes the role of integrated design and control, together with six-sigma methodology (Rath and Strong, 2000, 2002), in the manufacture of products such as espresso machines, integrated circuits, and drugs and specialty chemicals, which are either defect-free, in the case of manufactured items, or delivered on-specification, in the case of pharmaceuticals. It will be shown that these aims can be achieved by utilizing six-sigma methodology and additional statistical tools to quantify quality and, more importantly, the loss of quality and its cost. These tools assist in identifying the main sources of product variance, which are then attenuated or eliminated by improving the integrated design of the manufacturing process and its control system.

This chapter begins by describing the mathematical basis for six-sigma methodology, and its role in quantifying the cost of manufacturing defects or abnormal operation in processing steps and in guiding manufacturing to reduce product variance. Next, its role in product design is described, showing how six-sigma methodology is enhanced by incorporating integrated design and control into the product design process. The chapter concludes with examples of how the combined approach assists in improving product manufacturing and processing.

25.2 SIX-SIGMA METHODOLOGY IN PRODUCT DESIGN AND MANUFACTURING

Definitions

Six-sigma (6σ) is a structured methodology for eliminating defects and, hence, improving product quality in manufacturing and services. The methodology aims to identify and reduce the variance in product quality and involves a combination of statistical quality control, data-analysis methods, and the training of personnel.

The critical-to-quality (CTQ) variables (see Section 2.4) are monitored and used to track production to ensure that a sufficient number of measurements are within the control limits, commonly using the Shewart Chart, shown in Figure 25.1. The measurements shown in the chart could be the composition of the chemical produced from a reactor when producing a *basic chemical product*, or a characteristic attribute associated with finished items from the production

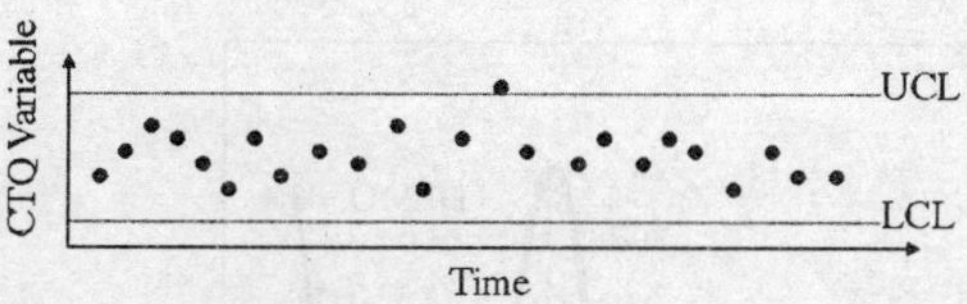

Figure 25.1 Monitoring product quality in a Shewart Chart.

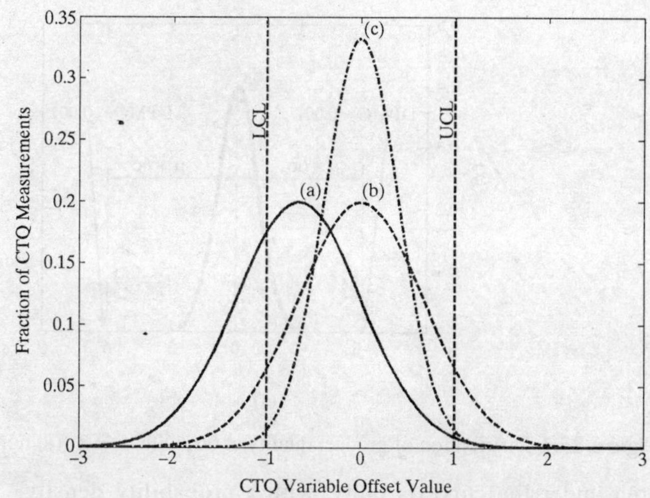

Figure 25.2 Probability distributions for CTQ variable: (a) solid line—without regulatory control; (b) dashed line—with regulatory control, but with a large variance; (c) dashed-dotted line—with regulatory control and a lower variance.

line for a *configured consumer product*. The degrees to which both of these products are satisfactory are quantified by the proportion of measurements that lie within the specification bounds, demarcated from above by the upper control limit (UCL) and from below by the lower control limit (LCL). In both cases, improved production involves reducing the number of off-specification measurements, which in six-sigma methodology is expressed as the "number of defects per million opportunities" (DPMO), with the term *six-sigma* defining a desired level of quality: 3.4 defects per million opportunities (DPMO). For example, the data in Figure 25.1 show 1 defect in 24 opportunities; that is, 1 off-specification measurement in 24.

As discussed by Lewin et al. (2007), on many production lines, open-loop, recipe-driven, feedforward control strategies are implemented. Often, the desired operating point is determined following a statistical design-of-experiment (DOE) study that locates a "stable" processing window. Subsequently, the degrees-of-freedom of the process (the manipulated variables) are fixed according to the DOE results. Feedback control, if implemented at all, is usually limited to single PID control loops, usually only the lower-level loops (e.g., for temperature control). The main disadvantages of feedforward control are well recognized: (a) Unmeasured and/or unknown disturbances are neglected, and (b) because the feedforward correction is based on an imperfect process model, generally the product is not produced consistently on target, even in the absence of unknown disturbances. Process models are usually not implemented, and when used are usually limited to empirical, polynomial-like formulations. Typically, when loss-of-control (LOC) incidents occur, the process is shut down, with associated production losses, and a new DOE study is initiated to diagnose the problem and suggest corrections.

Figure 25.2 shows typical performances using three alternative control strategies. The solid line (a) shows the expected distribution of a CTQ variable when using only feedforward control, which ignores the effects of unmeasured disturbances, leading in this case to a large fraction of CTQ measurements below the lower control limit (LCL). The dashed line (b) indicates the anticipated improvement when implementing a feedback-control strategy designed to maintain the average CTQ measurement on target. To significantly improve the product yield, however, in addition to feedback control, the CTQ variance must be reduced, corresponding to the dashed-dotted line (c).

The symbol σ (sigma) is the standard deviation of the value of a quality variable, a measure of its variance, which is assumed to have a normal distribution. Figure 25.3a shows such a distribution of measurements with $\sigma = 2$. Note that the distribution is normalized such that the total

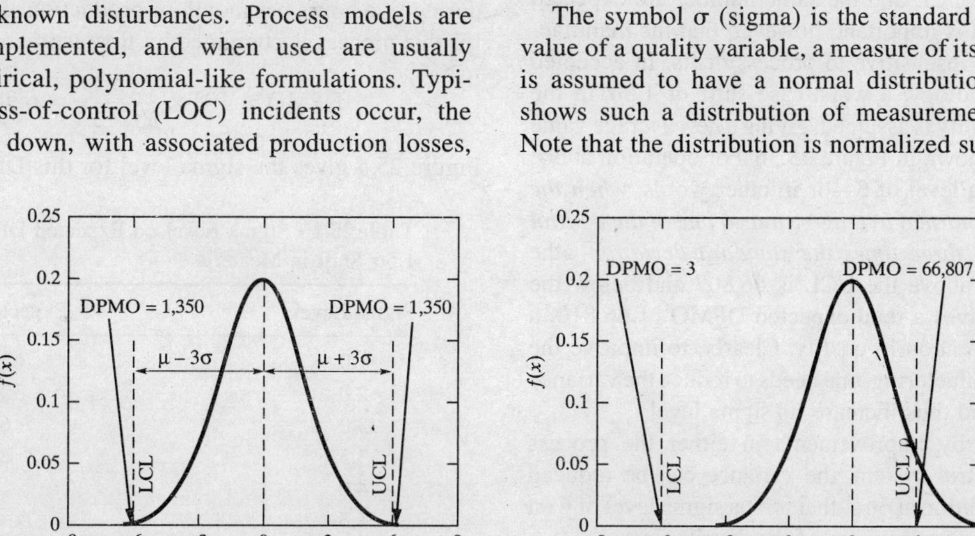

Figure 25.3 Distribution of product quality at 3σ, with $\sigma = 2$: (a) normal operation at $\mu = 0$; (b) abnormal operation shifted to $\mu + 1.5\sigma$.

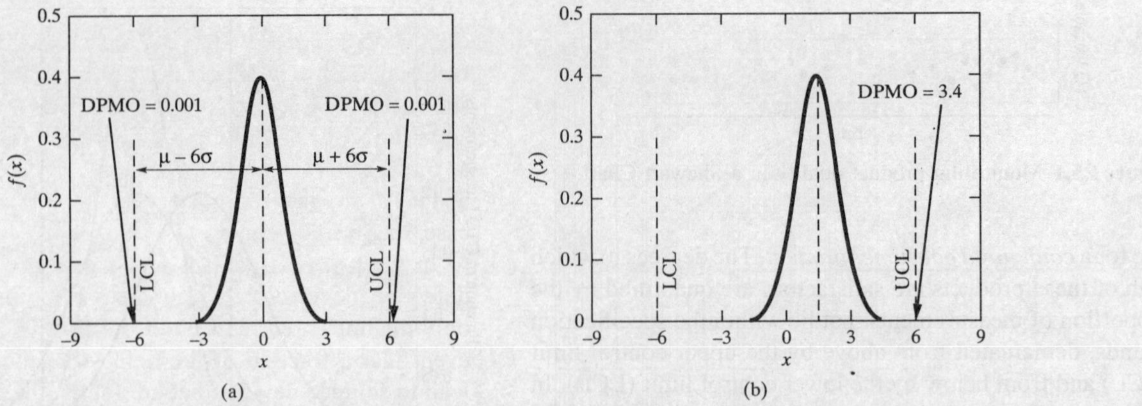

Figure 25.4 Distribution of product quality at 6σ, with $\sigma = 1$: (a) normal operation at $\mu = 0$; (b) abnormal operation shifted to $\mu + 1.5\sigma$.

area under the curve is unity, with a probability density function given by:

$$f(x) = \frac{1}{\sigma\sqrt{2\pi}}\exp\left[-\frac{1}{2}\left(\frac{x-\mu}{\sigma}\right)^2\right], \tag{25.1}$$

where $f(x)$ is the probability of the quality at a value of x, and μ is the average value of x. Assuming that operation at 3σ on either side of μ is considered normal, the UCL is at $\mu + 3\sigma$, and the LCL is at $\mu - 3\sigma$. As shown in Figure 25.3a, the number of defects per million opportunities (DPMO) above the UCL is:

$$\text{DPMO} = 10^6 \int_{\mu+3\sigma}^{\infty} f(x)dx$$

$$= \frac{1}{2}10^6\left(1 - \int_{\mu-3\sigma}^{\mu+3\sigma} f(x)dx\right) = 1{,}350 \tag{25.2}$$

This means that 1,350 DPMO are expected in a normal sample above the UCL and the same number are expected below the LCL. It is important, however, that the manufacturing process be insensitive to process drifts. In accepted six-sigma methodology, a worst-case shift of 1.5σ in the distribution of quality is assumed, giving a new average value of $\mu + 1.5\sigma$, as shown in Figure 25.3b. For operation at 3σ, that is, at a sigma level of 3—or in other words, *when the distance from the normal average value to one of the control limits is equal to three times the standard deviation*—the expected DPMO above the UCL is 66,807 and below the LCL is 3. This gives a total expected DPMO of 66,810, a significant deterioration in quality. Clearly, to improve the reliability of manufacturing, one needs to reduce the variance in the product, and thus increase its sigma level.

Suppose that, by improvements in either the process design or its control system, the variance can be reduced to $\sigma = 1$. An operation at 6σ—that is, at a sigma level of 6 on either side of the average value of the distribution $\mu = 0$—defines the UCL at $\mu + 6\sigma$ and the LCL at $\mu - 6\sigma$, as shown in Figure 25.4a. Here, there is 1 defect per billion opportunities on either side of the acceptance limits, which are insignificant defect levels. The improvement in performance is apparent when considering a shift of 1.5σ as before; for 6σ operation, or in other words, *when the distance from the normal average value to one of the control limits is equal to six times the standard deviation*, the DPMO (above the UCL) increases to only 3.4, as shown in Figure 25.4b.

Cost of Defects

Table 25.1 and Figure 25.5 present the effect of the sigma level on the DPMO, assuming a 1.5σ shift in mean, as in Figures 25.3b and 25.4b. Note that Figure 25.5 accounts for the *total* DPMO above the UCL and below the LCL. For example, for the data in Figure 25.1, with 1 defect recorded in 24 measurements, the DPMO is $(1/24) \times 10^6 = 41{,}667$, which from Figure 25.5 is equivalent to a sigma level of approximately 3.3. Although originally developed for the analysis of product manufacturing, it is relatively easy to compute the sigma level for a continuous process (Trivedi, 2002). As an example, suppose that on average, the distillate from a distillation column fails to meet its specifications during five hours per month of production. The sigma level for this process is computed by first estimating the DPMO:

$$\text{DPMO} = 10^6 \times \frac{5}{30 \times 24} = 6{,}944$$

Figure 25.5 gives the sigma level for this DPMO at 3.8.

Table 25.1 Sigma Level on Expected DPMO with 1.5σ Shift in Mean, μ

Sigma Level	Expected DPMO
1.0	697,672
2.0	308,770
3.0	66,810
3.5	22,750
4.0	6,210
4.5	1,350
5.0	233
5.5	32
6.0	3.4

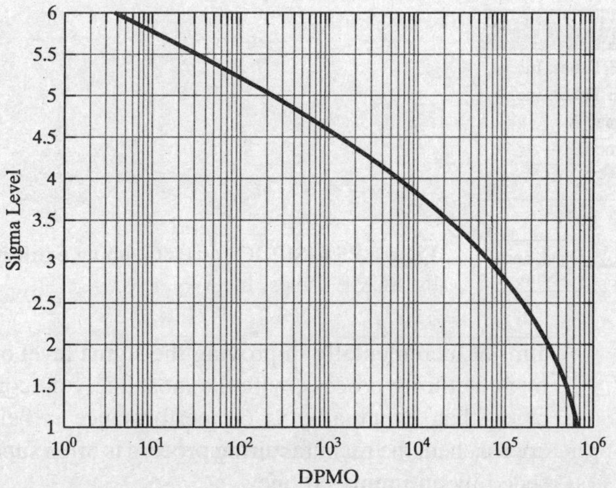

Figure 25.5 The relationship between DPMO and the sigma level.

If improved operations were to reduce the specification violations to 0.5 hr/month, the DPMO would be reduced by a factor of 10, giving an increase in the sigma level to 4.7. The increased sigma level is a consequence of the reduction in the variance in the CTQ variable. This improved operation is normally achieved by enhancements in the process and/or its control system. Evidently, lower sigma levels are achieved for processes in which abnormal operation is prevalent compared to processes in which abnormal operation seldom occurs. Thus, for example, a crude-oil distillation unit with frequent feedstock changes is expected to have a lower sigma level than one relying on a single feedstock. This is because feedstock changes cause process upsets that propagate throughout the entire unit, leading to off-specification product until corrections are made by the feedback-control system.

The expected number of defects presented in Figure 25.5 applies to a single manufacturing step. Usually, the manufacture of devices involves a number of steps. For n steps, assuming that all defective components of the device are removed from the production sequence at the step where they occur, the overall defect-free throughput yield, TY, is:

$$\mathrm{TY} = \prod_{i=1}^{n}\left(1 - \frac{\mathrm{DPMO}_i}{10^6}\right), \tag{25.3}$$

where DPMO_i is the expected number of defects per million opportunities in step i. If the DPMO is identical in each step, Eq. (25.3) reduces to:

$$\mathrm{TY} = \left(1 - \frac{\mathrm{DPMO}}{10^6}\right)^n \tag{25.4}$$

The fraction of the production capacity lost due to defects is 1 − TY. For example, consider the manufacture of a device involving 40 steps, each of which operates at 4σ. From Figure 25.5, the expected DPMO is 6,210 per step, so $\mathrm{TY} = (1 - 0.00621)^{40} = 0.779$. Thus, 22% of production capacity is lost due to defects, rendering the overall manufacturing operation a 2.3σ process (noting that Figure 25.5 shows that 22% defects, or 220,000 DPMO, corresponds to a sigma level of 2.3). In contrast, if each of the 40 steps operates at 6σ, $\mathrm{TY} = (1 - 3.4/10^6)^{40} = 0.99986$, corresponding to about 1 faulty device for every 10,000 produced, and in this case, the overall operation is a 5.2σ process.

In the preceding discussion, it has been assumed that defective devices are eliminated in production, leaving only the impact on reduced throughput yield. In the likely event that a fraction of the defects are undiscovered and lead to shipped devices that are faulty, the impact on sales resulting from customer dissatisfaction could be much greater. Noting that many manufacturing operations involve hundreds of steps (e.g., integrated-circuit chip manufacturing), it is clear that high levels of reliability, as expressed by low DPMO values, are generally required to ensure profitable manufacture. This is the driving force behind the extensive proliferation of six-sigma methodology (Wheeler, 2002).

Methods to Monitor and Reduce Variance

As described in detail by Rath and Strong (2000), an iterative five-step procedure is followed to progressively improve product quality. The five steps are: (a) Define, (b) Measure (c) Analyze, (d) Improve, and (e) Control, referred to by the acronym DMAIC:

(a) Define: First, a clear statement is made defining the intended improvement. Next, the project team is selected, and the responsibilities of each team member assigned. To assist in project management, a map is prepared showing the suppliers, inputs, process, outputs, and customers (referred to by the acronym SIPOC). A simplified block diagram usually accompanies a SIPOC, showing the principal steps in the process (usually 4–7 steps). At this stage, the main focus is on customer concerns, which are used to define critical-to-quality (CTQ) output variables, as discussed in Section 2.4. As an example, suppose the company ACME Tubes, Inc., manufactures PVC tubing by extrusion of PVC melt. A SIPOC describing its operations is presented in Figure 25.6. The quality of the PVC tubing, measured in terms of its impact strength, is considered to be the principal CTQ, and customer specifications define the LCL and UCL.

(b) Measure: The CTQ variables are monitored to check their compliance with the LCLs and UCLs. Most commonly, univariate statistical process control (SPC) techniques, such as the Shewart Chart, are utilized (see Chapter 28 in Ogunnaike and Ray, 1994). The data for the critical-to-quality variables are analyzed and used to compute the DPMO. This enables the sigma level of the process to be assessed

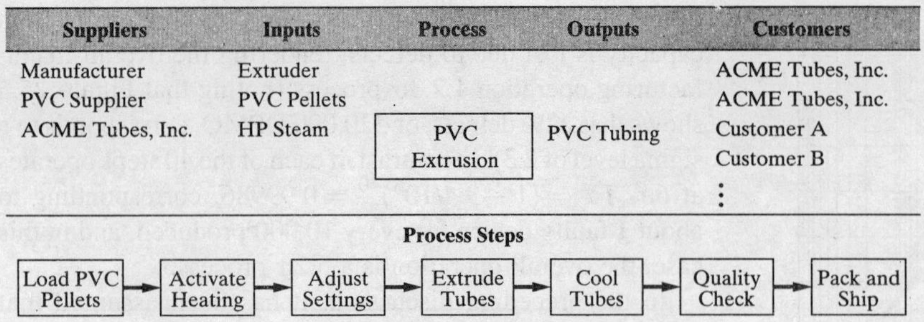

Figure 25.6 SIPOC for PVC tubing extrusion by ACME Tubes, Inc.

using Figure 25.5. As noted above, while the DPMO is relatively easy to compute for device manufacture, it is also readily applied to improve continuous processes (Trivedi, 2002; Wheeler, 2002). Continuing the PVC extrusion example, suppose this analysis indicates operation at 3σ, with a target to attain 5σ performance.

(c) Analyze: When the sigma level is below its target, steps are taken to increase it, starting by defining the most significant causes for the excessive variability. This is assisted by a systematic analysis of the sequence of steps in the manufacturing process, and the interactions between them. Using this analysis, the common root cause of the variance is identified. Continuing the PVC extrusion example, note that several factors contribute to an excessively high variance in product quality, among them the variance in the purity of the PVC pellets, the variance in the fraction of volatiles in the pellets, and the variance in the operating temperature of the steam heater. Clearly, all of these factors interact; but suppose that after analysis, it is determined that the variance in the operating temperature has the greatest impact on quality.

(d) Improve: Having identified the common root cause of variance, it is eliminated or attenuated by redesign of the manufacturing process or by employing process control. Continuing the PVC tubing example, one possible solution would be to redesign the steam heater. As will be demonstrated, systematic process redesign can improve the controllability and resiliency of a process, and hence, reduce the variance in the controlled output variables. Alternatively, a feedback controller could be installed, which manipulates the steam valve to enable tighter control of the operating temperature (through control of the steam pressure). In this way, the variance in the temperature is transferred to that of the mass flow rate of steam.

(e) Control: After implementing steps to reduce the variance in the CTQ variable, the effect of the change is quantified, analyzed, and used to drive the DMAIC procedure further. Thus, steps (b) to (e) are repeated to improve process quality in a stepwise fashion. Note that achieving 6σ performance is rarely the goal, and is seldom achieved. In fact, six-sigma methodology aims at incrementally improving the sigma level of a manufacturing process, with the most likely outcome being that, eventually, as 6σ performance is being approached, the manufacturing process is often superseded by an improved one.

Six-Sigma for Product Design

As detailed in Rath and Strong (2000), the DMAIC procedure is combined with ideas specific to product design to create a methodology that assists in applying the six-sigma approach to product design. Again, a five-step procedure is recommended:

Step 1: ***Define Project.*** In this step, the market opportunities are identified, a design team is assigned, and resources are allocated. Typically, the project time line is summarized in a Gantt chart (see Section 11.4).

Step 2: ***Identify Requirements.*** As in DMAIC, the requirements of the product are defined in terms of the needs of customers. For the design of a process such as a heat exchanger network, appropriate specifications would quantify the desired dynamic performance of the process—with the objective to reduce the occurrence of violations of the UCLs and LCLs defined for each target temperature.

Step 3: ***Select Concept.*** Innovative concepts for the new design are generated, first by brainstorming. These are evaluated, with the best selected for further development. Often for product design, a rather qualitative approach is applied to make progressive improvements. As will be seen in the first two examples of Section 25.3, this can be strengthened by adopting a quantitative, model-based methodology.

Step 4: ***Develop Design:*** Often, several teams work in parallel to develop and test competing designs, making modifications as necessary. The goal of this step is to prepare a detailed design together with a plan for its management, manufacture, and quality assurance.

Step 5: ***Implement Design:*** The detailed designs in Step 4 are critically tested. The most promising design is pilot-tested and if successful, proceeds to full-scale implementation.

25.3 EXAMPLE APPLICATIONS

This section presents three examples that show how product quality is improved by reducing the variance in the CTQ variables, as guided by six-sigma principles. In the first example, the effluent temperatures in a heat exchanger network are required to lie within control limits, with improvements made in the HEN design and the control configuration to achieve the desired sigma level. In the second example, a six-sigma approach guides improvements to a process for the manufacture of penicillin, with the main goal being to satisfy the highest possible sigma level while improving the penicillin yield. Finally, the last example, which is more qualitative, shows how similar ideas are applied in the design of a new product, that is, an improved espresso machine.

EXAMPLE 25.1 *Six-Sigma Design of a Heat Exchanger Network*

Figure 25.7 shows a heat exchanger network, introduced in Chapter 12, which is designed to cool hot stream 1 from 500 to 300°F, using two cold streams to be heated: stream 2, from 300 to 371.4°F, and stream 3, from 200 to 400°F. The network must be resilient to ±5% changes in the feed rate and ±5°F in the feed temperature of stream 1, disturbances that occur as step changes not exceeding 15 minutes. The CTQ variables are selected as the target temperatures of the three streams, with their UCLs and LCLs selected to bound the setpoints by ±3°F. Use six-sigma methodology to determine the sigma level of the existing process. If necessary, consider appropriate modifications to the network to achieve a sigma level of at least 4.

SOLUTION

Without altering the configuration of the heat exchanger network, only two degrees-of-freedom are available for control, namely, valves V-1 and V-2, which manipulate the flow rates of the two cold streams.

First, the configuration of the control system is selected using controllability and resiliency (C&R) analysis, as presented in the supplement to Chapter 12. See the file, Supplement_to_Chapter_12.pdf, in the Program and Simulation Files folder,

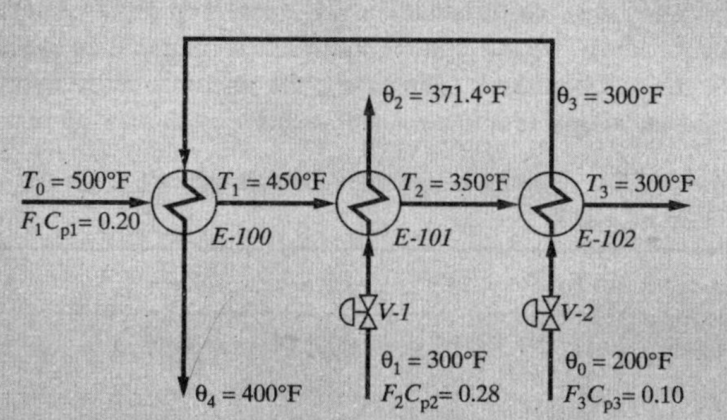

Figure 25.7 Heat exchanger network, with heat-capacity flow rates in millions (MM) of Btu/hr.

which can be downloaded from the Wiley Web site associated with this book. In the network of Figure 25.8, only two of the target temperatures, θ_2 and θ_4, are controlled by manipulation of the flow rates of the two cold streams, F_2 and F_3, leaving the third target temperature, T_3, uncontrolled. The energy balances for this system involve 15 variables: F_1, F_2, F_3, T_0, T_1, T_2, T_3, θ_0, θ_1, θ_2, θ_3, θ_4, Q_1, Q_2, and Q_3, two of which, θ_0 and θ_1, are considered to be fixed, and two of which, F_1 and T_0, are considered to be external disturbances. Two energy balances and one heat-transfer rate equation apply for each heat exchanger. For the first heat exchanger, E-100, with heat-transfer rate Q_1, they are

$$f_1\{\underline{x}\} = Q_1 - F_1C_{p1}(T_0 - T_1) = 0 \tag{25.5}$$

$$f_2\{\underline{x}\} = Q_1 - F_3C_{p3}(\theta_4 - \theta_3) = 0 \tag{25.6}$$

$$f_3\{\underline{x}\} = Q_1 - U_1A_1\frac{(T_0 - \theta_4) - (T_1 - \theta_3)}{\ln[(T_0 - \theta_4)/(T_1 - \theta_3)]} = 0 \tag{25.7}$$

For E-101, with heat-transfer rate Q_2, the equations are

$$f_4\{\underline{x}\} = Q_2 - F_1C_{p1}(T_1 - T_2) = 0 \tag{25.8}$$

$$f_5\{\underline{x}\} = Q_2 - F_2C_{p2}(\theta_2 - \theta_1) = 0 \tag{25.9}$$

$$f_6\{\underline{x}\} = Q_2 - U_2A_2\frac{(T_1 - \theta_2) - (T_2 - \theta_1)}{\ln[(T_1 - \theta_2)/(T_2 - \theta_1)]} = 0 \tag{25.10}$$

Finally, for E-102, with heat-transfer rate Q_3,

$$f_7\{\underline{x}\} = Q_3 - F_1C_{p1}(T_2 - T_3) = 0 \tag{25.11}$$

$$f_8\{\underline{x}\} = Q_3 - F_3C_{p3}(\theta_3 - \theta_0) = 0 \tag{25.12}$$

$$f_9\{\underline{x}\} = Q_3 - U_3A_3\frac{(T_2 - \theta_3) - (T_3 - \theta_0)}{\ln[(T_2 - \theta_3)/(T_3 - \theta_0)]} = 0 \tag{25.13}$$

where U_i and A_i are the heat-transfer coefficient and heat-transfer area for exchanger i, respectively, such that $U_1A_1 = 0.0811$ MMBtu/hr-°F, $U_2A_2 = 0.3162$ MMBtu/hr-°F, and $U_3A_3 = 0.1386$ MMBtu/hr-°F. The number of independent manipulated variables is $N_{\text{Manipulated}} = N_{\text{Variables}} - N_{\text{Externally Defined}} - N_{\text{Equations}} = 15 - 4 - 9 = 2$, and the pairings for control purposes can be selected using the relative gain array (RGA); see Section 12S.2. To accomplish this, a linearized model is generated using the following procedure:

1. The nonlinear state equations, $f\{\underline{x}\} = 0$, in Eqs. (25.5)–(25.13) are solved for the nominal values of the manipulated variables, $\underline{u} = [F_2, F_3]^{\text{T}}$; disturbances, $\underline{d} = [F_1, T_0]^{\text{T}}$; and constants θ_0 and θ_1, to determine 9 state variables: $\underline{x} = [T_1, T_2, T_3, \theta_2, \theta_3, \theta_4, Q_1, Q_2, Q_3]^{\text{T}}$. This is accomplished using an appropriate numerical method (e.g., the Newton–Raphson method).
2. The output vector, $\underline{y} = [\theta_2, \theta_4]^{\text{T}}$, is recomputed for small positive and negative perturbations of magnitude Δu_i to each manipulated variable u_i, one at a time, with the results stored

in the vectors $\underline{y}_{p,i}$ and $\underline{y}_{n,i}$, respectively. Then, column i of the steady-state gain matrix, $\underline{\underline{P}}(0)$, is computed: $p_{ij}(0) = \Delta u_i^{max} \cdot (y_{p,ij} - y_{n,ij})/\Delta u_i, j = 1, \ldots, 3$. Note that a factor of Δu_i^{max} scales the input variables such that $|u_i| \leq 1$.

3. The output vector is recomputed for small positive and negative perturbations of magnitude Δd_i to each disturbance variable d_i, one at a time, with the results stored in the vectors $\underline{y}_{p,i}$ and $\underline{y}_{n,i}$, respectively. Then, column i of the steady-state gain matrix, $\underline{\underline{P}}_d(0)$, is computed: $p_{dji}(0) = \Delta d_i^{max} \cdot (y_{p,ij} - y_{n,ij})/\Delta d_i, j = 1, \ldots, 3$. The disturbance gain matrix is scaled arbitrarily relative to the inputs using the scaling $\underline{\Delta d}^{max} = [5\%,\ 5°\text{F}]^{\text{T}}$.

Since the nominal values of the manipulated variables are $\underline{u} = [F_2, F_3]^{\text{T}} = [1.00, 1.00]^{\text{T}}$, the maximum perturbations are $\underline{\Delta u}^{max} = [1.00, 1.00]^{\text{T}}$. The resulting linearized model is:

$$\begin{bmatrix} \Delta\theta_2 \\ \Delta\theta_4 \\ \hline \Delta T_3 \end{bmatrix} = \underbrace{\begin{bmatrix} -58.7 & -73.3 \\ -7.14 & -112 \\ \hline -14.3 & -41.6 \end{bmatrix}}_{\begin{bmatrix} \underline{\underline{P}}_1(0) \\ \hline \underline{\underline{P}}_2(0) \end{bmatrix}} \cdot \begin{bmatrix} \Delta F_2 \\ \Delta F_3 \end{bmatrix} + \underbrace{\begin{bmatrix} 2.83 & 1.89 \\ 2.23 & 2.94 \\ \hline 4.92 & 0.883 \end{bmatrix}}_{\begin{bmatrix} \underline{\underline{P}}_{d_1}(0) \\ \hline \underline{\underline{P}}_{d_2}(0) \end{bmatrix}} \cdot \begin{bmatrix} \Delta F_1 \\ \Delta T_0 \end{bmatrix} \quad (25.14)$$

Note that the gains in Eq. (25.14) are presented as the change in °F in response to a full-scale change of each input. Thus, for example, the linear model predicts a 4.92°F increase in T_3 in response to a 5% increase in F_1. The steady-state RGA (Bristol, 1966) is computed using $\underline{\underline{P}}_1(0)$:

$$\underline{\Lambda} = \underline{\underline{P}}_1(0) \otimes \left(\underline{\underline{P}}_1^{-1}(0)\right)^{\text{T}} = \begin{bmatrix} 1.09 & -0.09 \\ -0.09 & 1.09 \end{bmatrix}, \quad (25.15)$$

where $\otimes$ is the Schur product. The RGA indicates that a control system paired diagonally, that is, $\theta_2 - F_2$ and $\theta_4 - F_3$, shown in Figure 25.8, provides responses that are almost perfectly decoupled.

Next, the resiliency of the HEN is examined by computing the linear disturbance cost (DC, Lewin (1996)) in the steady state for disturbances of ±5% in F_1 and ±5°F in T_0:

$$\begin{bmatrix} \Delta F_2(0) \\ \Delta F_3(0) \end{bmatrix} = -\underline{\underline{P}}_1^{-1}(0) \cdot \underline{\underline{P}}_{d_1}(0) \cdot \begin{bmatrix} \Delta F_1 \\ \Delta T_0 \end{bmatrix},$$
$$\text{DC} = \left\| \begin{bmatrix} \Delta F_2(0) \\ \Delta F_3(0) \end{bmatrix} \right\|_2 \quad (25.16)$$

The values of the two manipulated variables, computed to completely reject the effect of the disturbances on θ_2 and θ_4, lead to changes in T_3, computed by substituting Eq. (25.16) into Eq. (25.14):

$$\Delta T_3(0) = \left(\underline{\underline{P}}_{d_2}(0) - \underline{\underline{P}}_2(0)\underline{\underline{P}}_1^{-1}(0) \cdot \underline{\underline{P}}_{d_1}(0)\right) \begin{bmatrix} \Delta F_1 \\ \Delta T_0 \end{bmatrix} \quad (25.17)$$

Table 25.2 shows the changes in the control variables, ΔF_2 and ΔF_3 (assuming perfect control), the disturbance cost, and the resulting change in T_3, computed using Eq. (25.17) for four disturbance vectors. The results indicate that perfect disturbance rejection is achieved for θ_2 and θ_4 with negligible control effort. However, the uncontrolled temperature, T_3, is significantly perturbed, with the worst-case disturbance where ΔF_1 and ΔT_0 are in opposite directions. Variations of ±5% in F_1 and ∓5°F in T_0 lead to variations of approximately ±4°F in T_3.

To check these findings, dynamic simulations of the process, using PI controllers, are performed with HYSYS. At steady state, the hot stream of n-octane at 2,350 lbmol/hr is cooled from 500 to 300°F using n-decane as the coolant, with $F_2 = 3{,}070$ lbmol/hr and $F_3 = 1{,}200$ lbmol/hr. Note that these species and flow rates are chosen to match the heat-capacity flow rates defined by McAvoy (1983), with F_1 slightly increased to avoid temperature crossovers in the heat exchangers due to temperature variations in the heat capacities. Additional details of the HYSYS simulation are:

(a) The tubes and shells for the heat exchangers provide 2-min residence times.

(b) The feed pressures of all three streams are set at 250 psia, with nominal pressure drops of 5 psia defined for the tubes and shells. Subsequently, these pressure drops are computed based on the equipment sizes and the pressure-flow relationships.

(c) Controllers are tuned using the IMC-PI rules (see the file, Supplement_to_Chapter_12.pdf, in the Program and Simulation Files folder, which can be downloaded from the Wiley Web site associated with this book).

The regulatory response shown in Figure 25.9 indicates that, as predicted by the DC analysis, even the worst-case disturbance has little effect on the two controlled variables, whose control loops are decoupled, as indicated by the RGA analysis. Moreover, the uncontrolled output, T_3, exhibits offsets of about ±4.5°F, which compares well with the

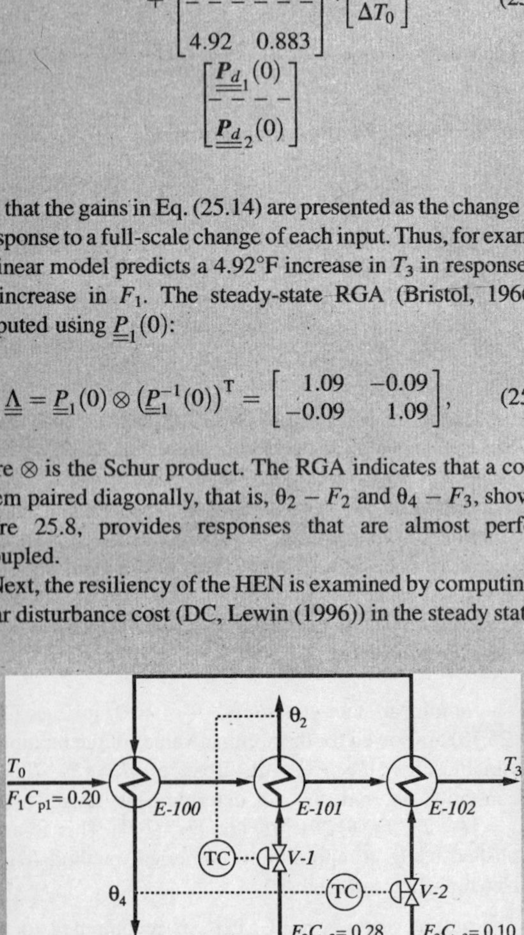

Figure 25.8 Control configuration for original HEN.

Table 25.2 Input Changes and Disturbance Cost for the Original HEN

ΔF_2	ΔT_0	ΔF_2	ΔF_3	$\text{DC} = \|\|\underline{u}\|\|_2$	ΔT_3
+5%	0	0.0253	0.0184	0.0313	3.79
+5%	+5°F	0.0246	0.0447	0.0511	3.59
0	+5°F	−0.0007	0.0264	0.0264	−0.20
−5%	+5°F	−0.0261	0.0080	0.0273	−4.00

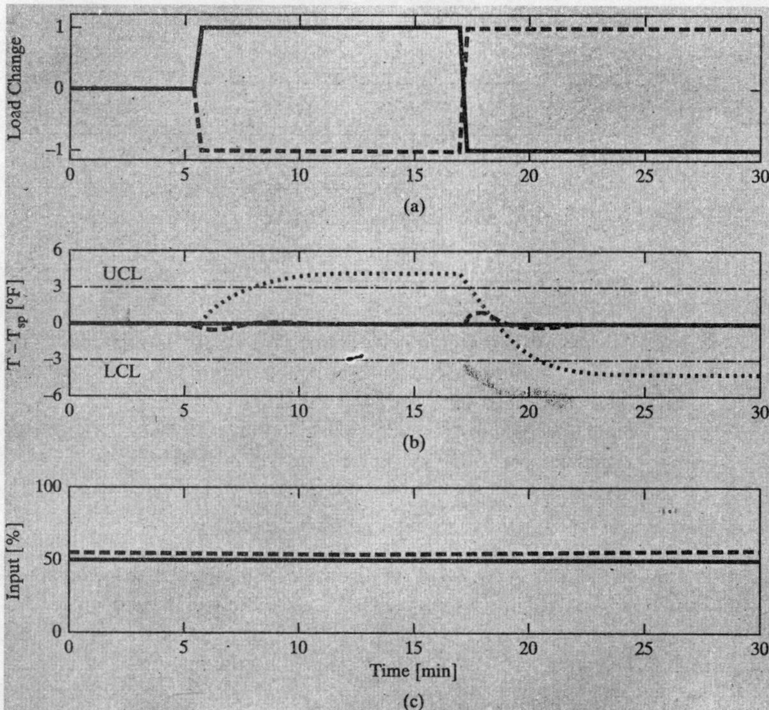

Figure 25.9 Response of HEN without bypass to the worst-case disturbances: (a) normalized changes in F_1 (solid) and T_0 (dashed); (b) tracking errors (θ_2—solid; θ_4—dashed; T_3—dotted; UCL and LCL—dot-dashed); (c) manipulated variables (F_2—solid; F_3—dashed).

value of ±4.5°F predicted by the linear DC analysis. Although both θ_2 and θ_4 are maintained within the desired operating window, the large variability in T_3 violates the control limits for this variable, with a DPMO value of 633,330, equivalent to a sigma level of 1.17. (See Figure 25.5.) Clearly, the process needs to be improved significantly.

As discussed in Chapter 12, it is often necessary to augment the process degrees-of-freedom to meet control objectives. For HENs, this is accomplished either by the addition of trim-utility exchangers or by the addition of bypasses, as illustrated in Figure 25.10. Next, resiliency analysis is used to select between these design configurations and to adjust the nominal operating conditions.

The PFD for the modified HEN, including a bypass around E-102 to eliminate the offsets in the third target temperature, is shown in Figure 25.10. Resiliency analysis is used to determine the required bypass fraction. The energy balances involve 17 variables: F_1, F_2, F_3, T_0, T_1, T_2, T_3, θ_0, θ_1, θ_2, θ_3, θ_3', θ_4, Q_1, Q_2, Q_3, and ϕ, two of which, θ_0 and θ_1, are assumed to be fixed, and two of which, F_1 and T_0, are considered to be external disturbances. The first six equations, (25.5)–(25.10), for the HEN

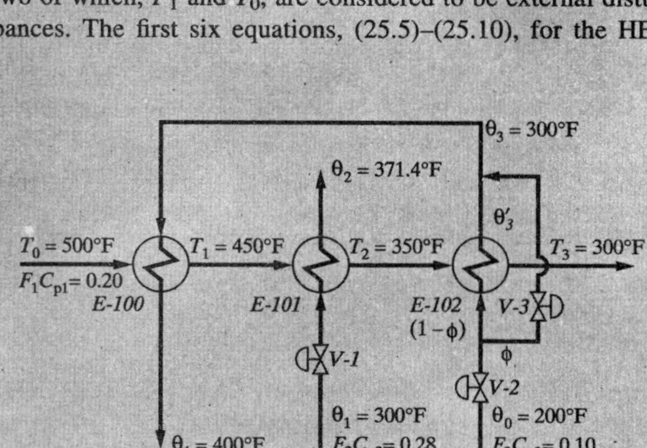

Figure 25.10 Modified heat exchanger network.

without bypasses apply. For heat exchanger E-102 and its bypass, the material and energy balances are:

$$f_7\{\underline{x}\} = Q_3 - F_1C_{p1}(T_2 - T_3) = 0 \quad (25.18)$$

$$f_8\{\underline{x}\} = Q_3 - F_3C_{p3}(\theta_3' - \theta_0) = 0 \quad (25.19)$$

$$f_9\{\underline{x}\} = Q_3 - K_3U_3A_3\frac{(T_2 - \theta_3') - (T_3 - \theta_0)}{\ln[(T_2 - \theta_3')/(T_3 - \theta_0)]} = 0 \quad (25.20)$$

$$f_{10}\{\underline{x}\} = (1 - \phi)\theta_0 + \phi\theta_3' - \theta_3 = 0 \quad (25.21)$$

In Eq. (25.20), the product U_3A_3 is identical to that for the network without bypasses (i.e., 0.1386 MM Btu/hr-°F). As the bypass fraction, ϕ, increases, K_3 increases beyond unity, corresponding to an increase in the heat-transfer area. The number of independent manipulated variables is $N_{\text{Manipulated}} = N_{\text{Variables}} - N_{\text{Externally Defined}} - N_{\text{Equations}} = 17 - 4 - 10 = 3$. This leaves F_2, F_3, and ϕ as the manipulated variables, which are paired with the controlled variables, θ_2, θ_4, and T_3.

A linearized model is generated and used to assist in the selection of an appropriate bypass fraction, ϕ. The procedure followed for the HEN without bypasses is used, parameterized by values of ϕ. Since the nominal values of the manipulated variables are $\underline{u} = [F_2,\ F_3,\ \phi]^T = [1,\ 1,\ \phi]^T$, the maximum perturbations are $\underline{\Delta u}^{max} = [1, 1, \phi]^T$. For example, for $\phi = 0.1$, the linearized model is:

$$\begin{bmatrix}\Delta\theta_2\\ \Delta\theta_4\\ \Delta T_3\end{bmatrix} = \underbrace{\begin{bmatrix}-58.7 & -72.3 & -0.068\\ -7.15 & -108 & -0.285\\ -14.3 & -44.9 & 0.237\end{bmatrix}}_{\underline{\underline{P}}(0)} \cdot \begin{bmatrix}\Delta F_2\\ \Delta F_3\\ \Delta\phi\end{bmatrix} + \underbrace{\begin{bmatrix}2.80 & 1.89\\ 2.20 & 2.94\\ 4.95 & 0.88\end{bmatrix}}_{\underline{\underline{P}}_d(0)} \cdot \begin{bmatrix}\Delta F_1\\ \Delta T_0\end{bmatrix} \quad (25.22)$$

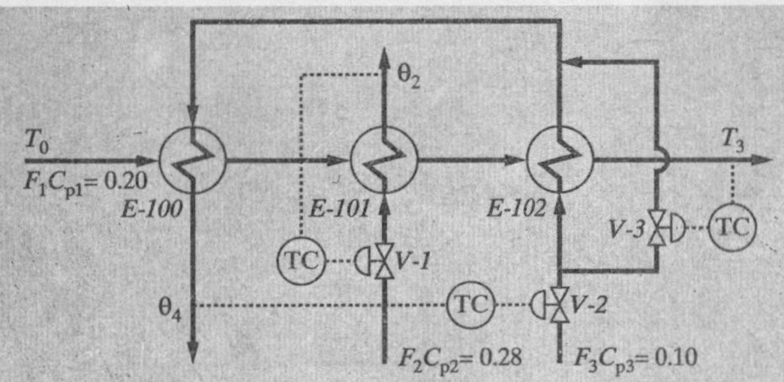

Figure 25.11 Control system for the modified heat exchanger network.

Using $\underline{\underline{P}}(0)$ in Eq. (25.22), the steady-state RGA is:

$$\underline{\underline{\Lambda}} = \underline{\underline{P}}(0) \otimes (\underline{\underline{P}}^{-1}(0))^{\mathrm{T}} = \begin{bmatrix} 1.17 & -0.22 & 0.04 \\ -0.07 & 0.84 & 0.23 \\ -0.10 & 0.38 & 0.72 \end{bmatrix} \quad (25.23)$$

Hence, the diagonal pairing is preferred, that is, $\theta_2 - F_2, \theta_4 - F_3$, and $T_3 - \phi$, with significant interactions between the second andthird loops anticipated. This configuration is shown in Figure 25.11.

The impact of the bypass fraction on the resiliency of the HEN is examined next. The manipulated variable values and the disturbance cost are computed for disturbances of ±5% in F_1 and ±5%°F in T_0. Table 25.3 shows the changes in the control variables, ΔF_2, ΔF_3, and $\Delta\phi$ (assuming perfect control), and the disturbance cost, for four disturbance vectors, $\underline{d} = [F_1 + \Delta F_1, T_0 + \Delta T_0]^{\mathrm{T}}$. Note that for the worst-case disturbance ($\Delta F_1 = -5\%$ and $\Delta T_0 = +5°\mathrm{F}$), the scaled change in the bypass fraction is $\Delta\phi = 12.3$, which far exceeds unity. To avoid this, the nominal bypass fraction is increased further to account for the expected disturbance levels, noting that heat exchanger E-102 must be resized.

With the nominal bypass fractional flow increased to $\phi = 0.25$, the linearized model is recomputed:

$$\begin{bmatrix} \Delta\theta_2 \\ \Delta\theta_4 \\ \Delta T_3 \end{bmatrix} = \underbrace{\begin{bmatrix} -58.7 & -69.8 & -0.720 \\ -7.15 & -97.1 & -3.02 \\ -14.3 & -53.7 & 2.52 \end{bmatrix}}_{\underline{\underline{P}}(0)} \cdot \begin{bmatrix} \Delta F_2 \\ \Delta F_3 \\ \Delta\phi \end{bmatrix} + \underbrace{\begin{bmatrix} 2.80 & 1.89 \\ 2.10 & 2.94 \\ 5.03 & 0.88 \end{bmatrix}}_{\underline{\underline{P_d}}(0)} \cdot \begin{bmatrix} \Delta F_1 \\ \Delta T_0 \end{bmatrix} \quad (25.24)$$

In this case, the steady-state RGA is:

$$\underline{\underline{\Lambda}} = \underline{\underline{P}}(0) \otimes (\underline{\underline{P}}^{-1}(0))^{T} = \begin{bmatrix} 1.17 & -0.21 & 0.04 \\ -0.07 & 0.75 & 0.32 \\ -0.10 & 0.46 & 0.64 \end{bmatrix} \quad (25.25)$$

Table 25.3 Input Changes and Disturbance Cost for the HEN with $\phi = 0.1$

ΔF_1	ΔT_0	ΔF_2	ΔF_3	$\Delta\phi$	$DC = \|\underline{u}\|_2$
+5%	0	−0.0010	0.051	−11.4	11.4
+5%	+5°F	−0.0003	0.075	−10.3	10.3
0	+5°F	0.0007	0.025	0.98	0.98
−5%	+5°F	0.0017	−0.026	12.3	12.3

Table 25.4 Input Changes and Disturbance Cost for the HEN with $\phi = 0.25$

ΔF_1	ΔT_0	ΔF_2	ΔF_3	$\Delta\phi$	$DC = \|\underline{u}\|_2$
+5%	0	−0.0010	0.051	−0.93	0.93
+5%	+5°F	−0.0003	0.075	0.75	0.75
0	+5°F	0.0007	0.025	0.18	0.18
−5%	+5°F	0.0017	−0.026	1.11	1.11

This RGA is similar to that obtained with $\phi = 0.1$, again indicating a diagonal pairing, as shown in Figure 25.11. Next, the resiliency is tested, with the results reported in Table 25.4. Note that when $\phi = 0.25$, the disturbance rejection is nearly acceptable, with $DC_{max} = 1.1$, only slightly above unity.

Clearly, the resiliency of the HEN increases with the nominal bypass fraction, but at the cost of increased heat-transfer area. Table 25.5 shows the tradeoff between resiliency and heat-transfer area. Note that while only 12% additional heat exchange area is required for $\phi = 0.1$, the resiliency is inadequate. In contrast, when $\phi = 0.30$, the resiliency is satisfactory (with DC significantly lower than unity), but the heat-transfer area is doubled. A good compromise is to select $\phi = 0.25$, which approximates the desired resiliency while requiring only 55% more heat-exchange area.

The C&R analysis in the steady state predicts the superior performance of the modified HEN, which allows all three target temperatures to be controlled at their setpoints in the face of disturbances in the feed flow rate and temperature of the hot stream. More specifically, the steady-state RGA indicates that a decentralized control system can be configured for the modified HEN in which $\theta_2 - F_2$, $\theta_4 - F_3$, and $T_3 - \phi$ are paired, and in which the first loop is almost perfectly decoupled, with moderate coupling between the other two loops. Finally, aided by DC analysis, the nominal bypass fraction is selected to be 0.25, providing the best tradeoff between increased plant costs and adequate resiliency.

Given the design decision to use $\phi = 0.25$ based upon the steady-state C&R analysis, verification is performed, as before, by dynamic simulations with HYSYS. The bypass valve V-3 is sized carefully, ensuring that the nominal bypass fraction is 0.25, with the nominal valve position being 50% open (selecting a linear characteristic curve).

The regulatory response of the new configuration is shown in Figure 25.12. Note that the design with $\phi = 0.25$ rejects the worst-case disturbance with no saturation, indicating that the DC analysis is slightly conservative. In addition, the first control loop $(\theta_2 - F_2)$ is perfectly decoupled, with slight interactions

Table 25.5 Tradeoff Between the Heat Exchanger Area and Bypass Fraction

ϕ	$DC = \|\underline{u}\|_2$	K_3
0.10	12.3	1.12
0.15	4.63	1.21
0.20	2.16	1.33
0.25	1.11	1.55
0.30	0.58	2.05

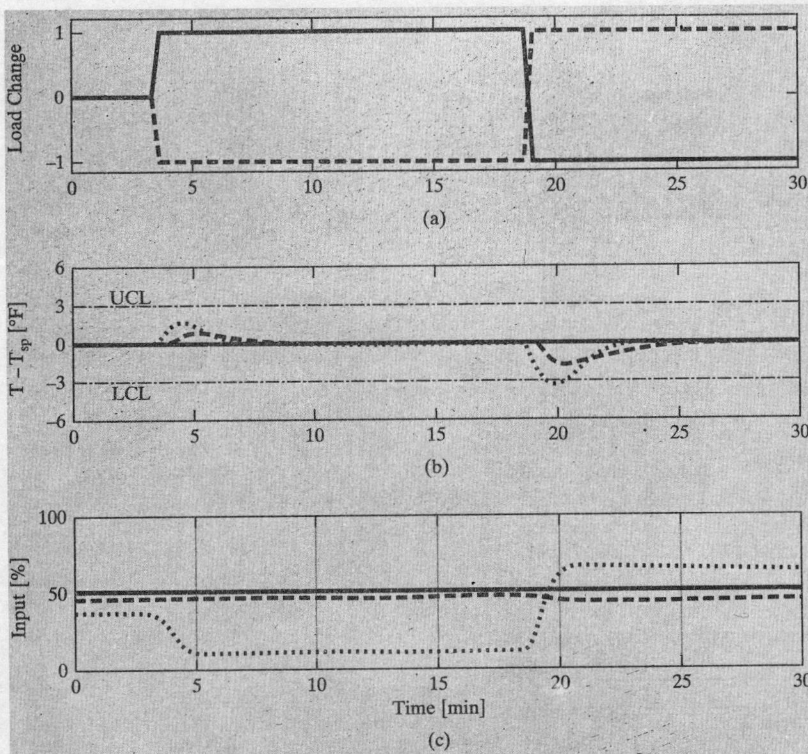

Figure 25.12 Response of HEN with bypass to the worst-case disturbance: (a) normalized changes in F_1 (solid) and T_0 (dashed); (b) tracking errors (θ_2—solid; θ_4—dashed; T_3—dotted; UCL and LCL—dot-dashed); (c) manipulated variables (F_2—solid; F_3—dashed; V_1—dotted).

seen in the other two loops, again as predicted by the static RGA analysis.

As with the original configuration, only T_3 violates its specified operating range, but here, just the LCL is violated, and for a small fraction of the operating cycle (2 samples in 1,000). Thus, DPMO = 2,000, equivalent to a sigma level of 4.38, which meets the target. The sigma level of the process can be further increased by reducing the frequency of disturbances that affect the HEN, possibly by improving the process operations. Note that increasing the nominal bypass fraction only increases the capital investment, with little or no expected reduction in process variance.

EXAMPLE 25.2 *Improving the Design and Control of Penicillin Manufacture*

The production of an active pharmaceutical ingredient (API) usually involves two principal phases: reaction/fermentation, in which the API is produced from its biosystem, and separation/purification, in which product quantity and quality specifications are satisfied. While upstream processing (i.e., in bioreactors) is important, downstream processing (i.e., in product purification operations) is often more important, because a product that fails to meet purity specifications cannot be marketed. For these reasons, a plantwide approach to the design and operation of an API process is greatly assisted by six-sigma methodology, which is the driving force for continuous improvement. Six-sigma methodology identifies the root cause or causes of low yields due to excessive variance in the desired performance of one or more processing units (Dassau et al., 2006).

A typical pharmaceutical process, illustrated in Figure 25.13 for the production of penicillin, involves batch and semi-batch operations rather than continuous processing. Although these operations are inherently transient, typically with nonlinear dynamics, they nonetheless enable the flexible production of high-value-added products in the pharmaceutical industry. These unit operations, often referred to as "unit procedures," are well known, including size reduction and classification, sterilization, mixing, filtration, evaporation and distillation, crystallization, solid–liquid extraction, drying, and bioreactors. In this example, six-sigma methodology is used for a simplified penicillin process, considering only the fermentation and the first downstream processing steps in Figure 25.14. The following analysis begins with a discussion of the process models.

SOLUTION

Modeling

The models of the fermenter and extractors, for primary recovery, and their control systems are presented in the file Supplement_to_Chapter_25.pdf, in the PDF Files folder, which can be downloaded from the Wiley Web site associated with this book. These were implemented in Matlab® and Simulink® and calibrated using nonlinear regression to compute key model parameters by minimizing the sum-of-the-square errors between the model predictions and data reported in the literature.

www.wiley.com/go/global/seider

Fermenter: The penicillin fermentation stage is simulated using the model described in Section 25S.1. Its control system, shown in Figure 25.15, manipulates the coolant flow rate, using a PI controller to regulate the fermenter temperature, and manipulates the flow rates of the acid and base streams to the fermenter using a PI controller with a nonlinear gain, to regulate the pH. The latter is intended to approximately "invert" its titration curve, as discussed in Section 25S.3. The product recipe calls for a makeup stream of substrate to be introduced when the substrate concentration is reduced to a threshold value, with the oxygen flow rate held

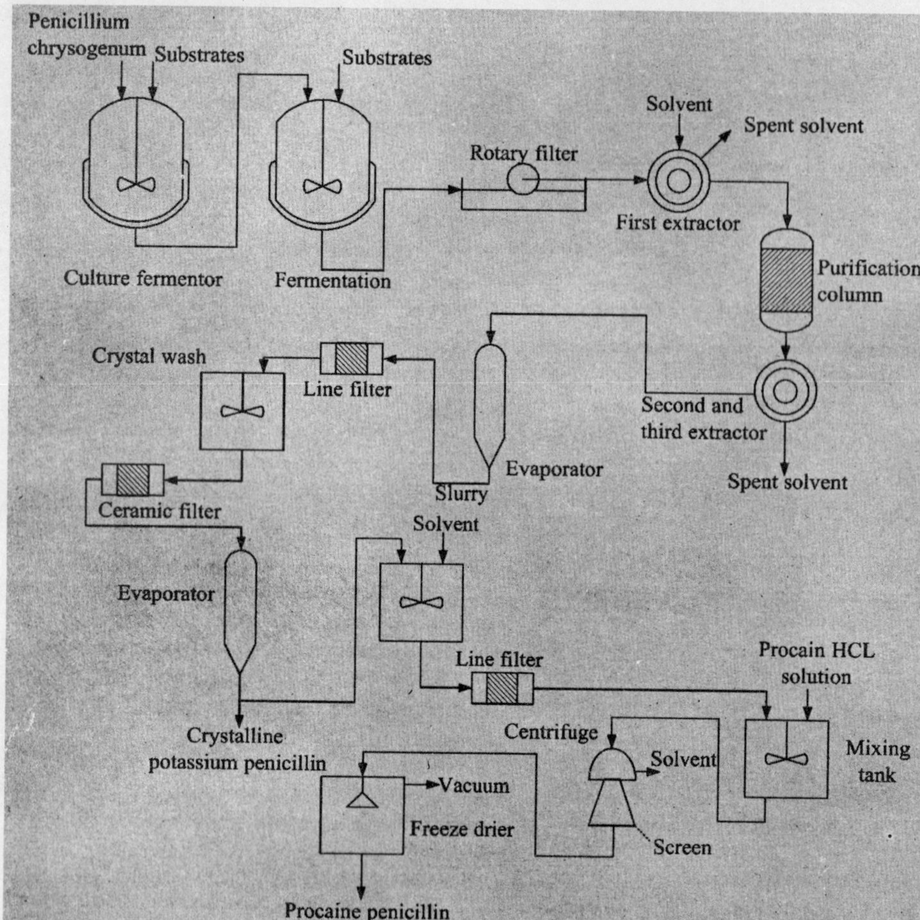

Figure 25.13 Penicillin process.

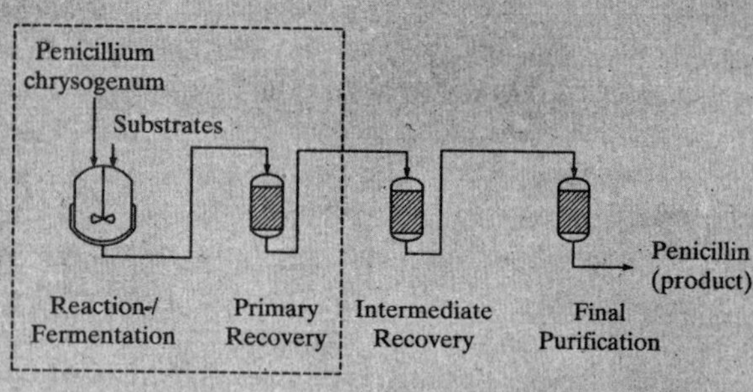

Figure 25.14 Schematic of simplified penicillin process.

constant. The simulation results, presented in Figure 25.16, are in good agreement with the published results of Bajpai et al. (1980) and Birol et al. (2002), where the temperature and pH setpoints were 25°C and 5, and the substrate threshold concentration was 0.3 g/l. About 422 simulation hours are needed to reach the maximum penicillin concentration, 1.5 g/l. Makeup substrate is introduced after 48 hours.

Primary Recovery: To recover penicillin from the fermentation broth, reactive extraction is often used, with typical organic solvents being *n*-butyr-acetate and amines such as Amberlite LA-2. Although Reschke and Schuegerl (1984, 1985, 1986) present a model describing the reactive extraction of penicillin,

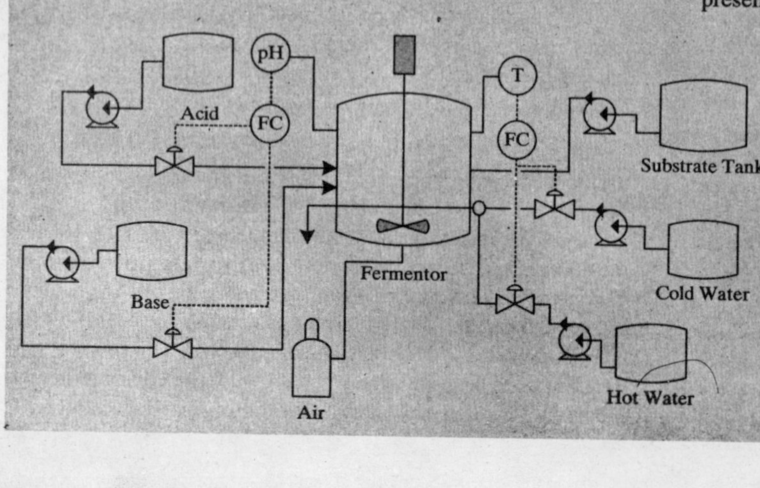

Figure 25.15 Fermenter and its control system.

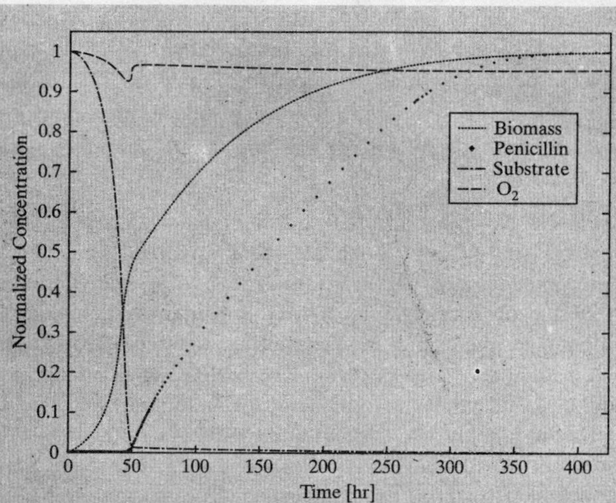

Figure 25.16 Fermentation trajectories (base case).

it lacks the degrees of freedom (DOFs) to enable the control of both the pH and the solvent and extract flow rates. To add DOFs permitting improved control, Dassau et al. (2006) developed a two-film model, as described in Section 25S.2.

Using the DMAIC Procedure for Process Refinement

For the penicillin production process described above, the DMAIC procedure is applied to define the base-case conditions in Table 25.6. As shown for the control limits on the critical-to-quality (CTQ) variables, large DPMO values are computed, accompanied by large production times and low throughput yields (TYs). Subsequently, cycles of the DMAIC procedure are implemented to improve the process iteratively, with improvements at each cycle implemented to reduce the variance of the unit exhibiting the highest DPMO value.

Cycle I: For the base-case operation, the reactive extractor has the highest DPMO, 462,456. As shown in Figure 25.17, only

Table 25.6 Summary of Control Limits, DPMO, and Throughput Yield for the Base-Case Conditions

	LCL	UCL	DPMO	Production Time (hr)
Fermenter				
pH	4.9	5.1	45,445	422
Temperature	22	28	465	
Reactive Extractor—TY = 73%				
pH	4.8	5.2	462,456	5
C_x (mole/liter)	6.75×10^{-5}			
Reactive Re-Extractor—TY = 86%				
pH	7	9	31,264	5
C_x (mole/liter)	4.2×10^{-5}			
Total Production Time (hr)				432
Total TY %				63

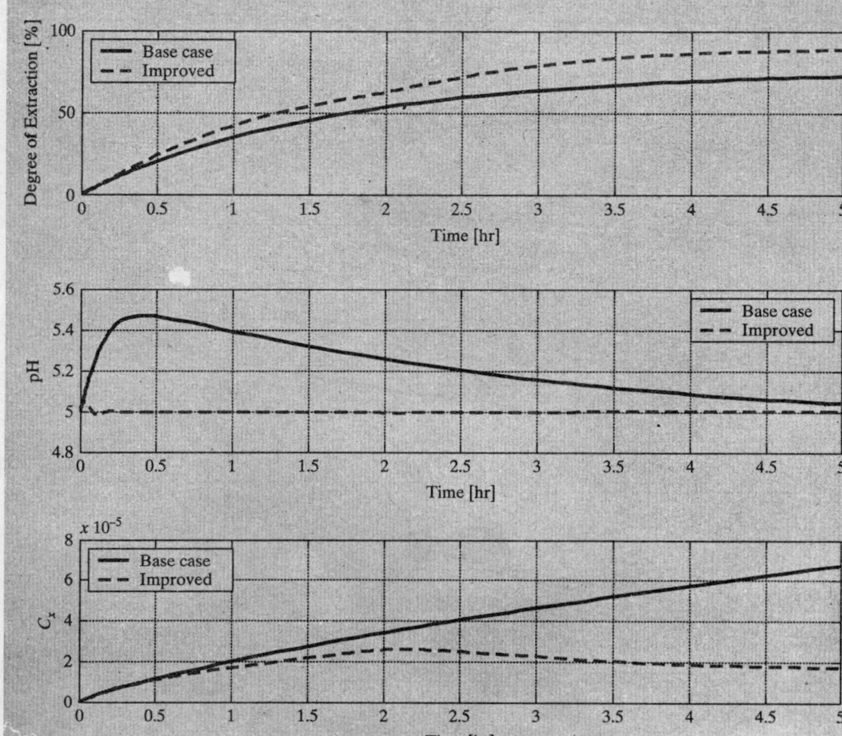

Figure 25.17 Comparison between trajectories in the reactive extractor before and after Cycle I improvements.

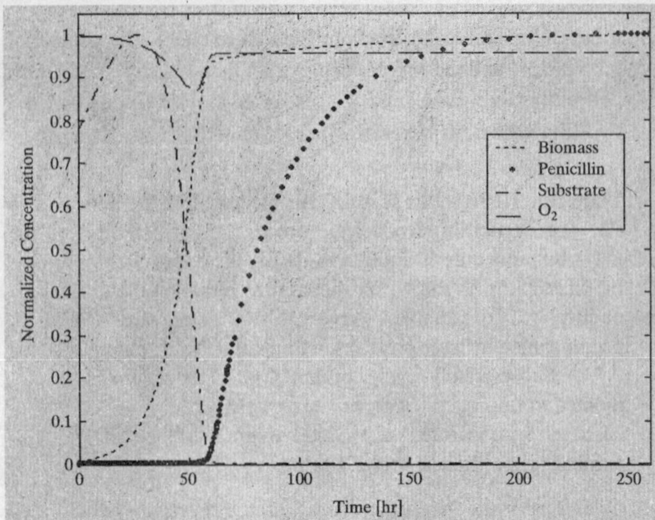

Figure 25.18 Fermentation trajectories after Cycle II improvements.

73% of the penicillin is extracted after 5 hours, and the pH value settles slowly toward its setpoint of 5. Moreover, the value of C_X, the concentration of degradation products, rises throughout the batch. The total throughput yield (TY) of the process is 63% and the production time is 432 hours. Evidently, this poor performance is due to the absence of pH control in a negative effect on the reactive re-extractor. Consequently, the process is improved by installing a control system to maintain the pH at 5, which, as shown in Figure 25.17, not only regulates the pH as required but also reduces the impurities by 74%, increasing the TY of the unit to 89%, and the overall TY to 77%.

Cycle II: Note that the pH control implemented in Cycle I improves the quality of the feed to the reactive re-extractor, thereby reducing its DPMO from 31,264 to 13,378. Moreover, the fermenter is selected for the Cycle II improvement because it dominates the overall production time and its DPMO exceeds that of the reactive re-extractor. The DMAIC cycle is repeated, with a significant decrease in the fermentation time achieved by increasing the glucose concentration at the feed outlet (i.e., the threshold value) from 0.3 g/l to above 15 g/l. This reduces the time needed to achieve a maximum penicillin concentration of 1.5 g/l, from 422 to 258 hr, as shown in Figure 25.18. But the pH and temperature distributions have significantly higher variances than in the base case, with DPMO levels of 49,628 and 15,625, both exceeding the base-case values of 45,445 and 465. These increases must be weighed against the 40% reduction in batch time with no decrease in the total TY.

Cycle III: Once again, the DMAIC procedure is repeated, this time with improvements in the reactive re-extractor unit. For the base case, without pH control, the degree of extraction reaches 86%, as shown in Figure 25.19. Note, however, that penicillin degradation is rather high. Here, also, a pH controller is introduced, decreasing the concentration of impurities in this unit by

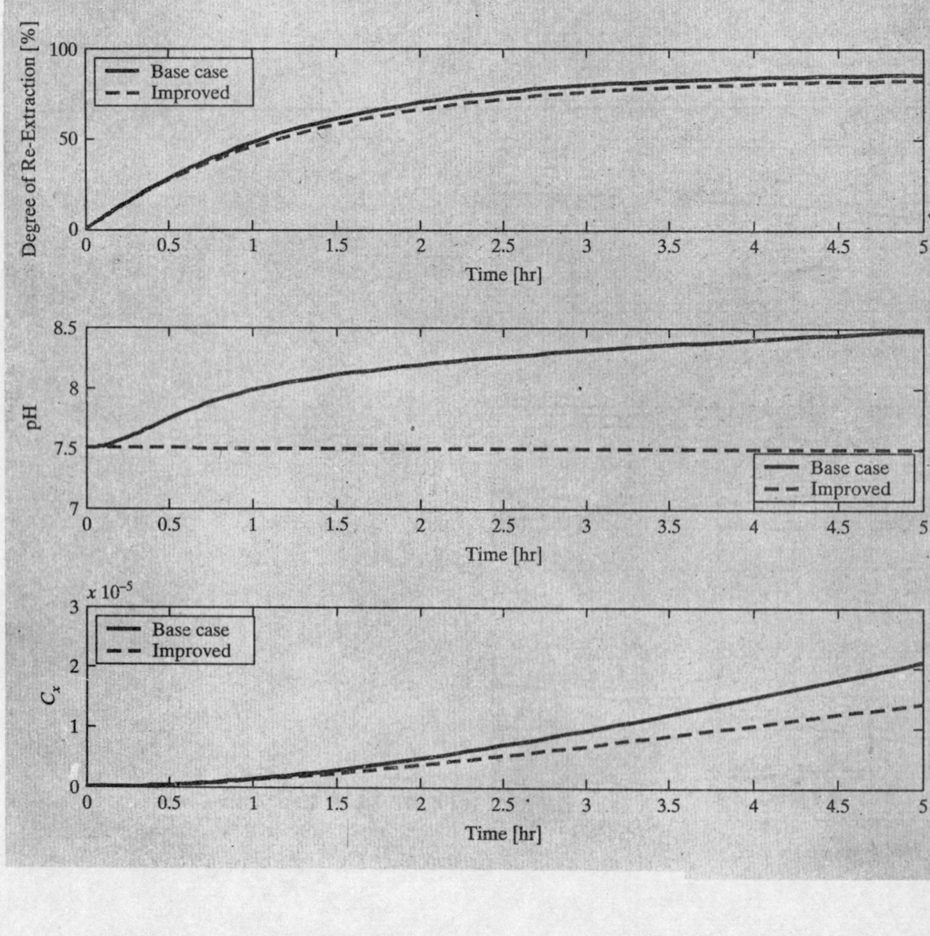

Figure 25.19 Re-extractor trajectories before and after Cycle III.

Table 25.7 Summary of Improvements Using the DMAIC Procedure

	Base Case	Cycle I	Cycle II	Cycle III
Fermenter				
DPMO—pH	45,445	45,445	49,628	49,628
DPMO—Temperature	465	465	15,625	15,625
Reactive Extractor				
DPMO—pH	462,456	<1	<1	<1
C_x (mole/liter)	6.75×10^{-5}	1.74×10^{-5}	1.74×10^{-5}	1.74×10^{-5}
Reactive Re-Extractor				
DPMO—pH	31,264	13,378	13,378	<1
C_x (mole/liter)	4.2×10^{-5}	2.1×10^{-5}	2.1×10^{-5}	1.4×10^{-5}
TY %	73	77	77	74
Production Time (hr)	432	432	268	268

33%. But there is a slight decrease in the degree of extraction from 86% to 83%, reducing the total TY to 74%. Consequently, the 33% decrease in impurity level must be weighed against a 3% decrease in penicillin yield.

Summary
As demonstrated, the DMAIC approach, involving a combination of improved process control, modified substrate feeding profiles in the fermenter, and improvements in the downstream processing section, can achieve a 40% reduction in batch time, a 17% increase in throughput yield, and a 33% reduction of impurities, as summarized in Table 25.7. These improvements arise from adopting a *plantwide* approach, noting that each improvement has its price tag, budget, and time constraints, which usually limit the total number of improvements performed. Evidently, this systematic approach can have a substantial impact in the pharmaceutical industry through improved overall process yield, quality, and return on investment.

EXAMPLE 25.3 *Designing the Ultimate Espresso Machine*

Today, the coffee industry is globally situated, employing more than 20 million people. As a commodity, coffee ranks second only to petroleum in dollars traded worldwide. Furthermore, coffee is the most popular beverage in the world, with over 400 billion cups consumed annually. Espresso, a relatively recent innovation in the preparation of coffee, originated in 1822, with the innovation of the first crude espresso machine in France, later perfected and first manufactured in Italy. Espresso has become an integral part of Italian life and culture, currently with over 200,000 espresso bars in Italy.

Espresso coffee is prepared in a machine that pumps cold water at high pressure (commonly 10–20 bar) into a hot water boiler, which displaces near-boiling hot water. This water is then forced through a cake of ground coffee, as illustrated in Figure 25.20. In a conventional machine, the user manually loads ground coffee into a metal filter housing, referred to as a *portafilter*, ensuring that the ground coffee is adequately packed, locks the portafilter under the hot water exit head, and activates the heater. A coffee cup, placed under the portafilter, is then filled with the freshly extracted espresso coffee, which is produced by the leaching action of the high-pressure hot water as it passes through the bed of ground coffee.

Suppose that a new product is envisaged as an alternative to a conventional espresso machine, with the objective being to improve the quality of the espresso obtained by a home user. The solution that follows describes a typical scenario using the DMAIC procedure.

SOLUTION

Applying the DMAIC procedure could lead to the following sequence of events:

(a) **Define:** A typical SIPOC, shown in Figure 25.21, identifies the principal steps that a home user would follow when using the machine. At this stage, the main focus would be on ensuring customer satisfaction with the quality of the espresso produced by the new machine. The quality of the coffee is considered to be the principal CTQ variable, and customer specifications define its LCL and UCL. Note that the control chart in Figure 25.22 identifies conditions under which acceptable coffee is prepared, which are constrained in a tight operating window delineated between 18–22% extraction of solubles and 1.15–1.35% solubles concentration (Sivetz and Desrosier, 1979). Note, however, that for such a food product, the issue of quality is complex, as flavor and odor attributes are difficult to quantify. As described by Andueza et al. (2002, 2003, 2007), a panel of judges typically conducts a sensory descriptive analysis. In their assessment, the appearance of foam is defined by color (clear, hazelnut, or dark),

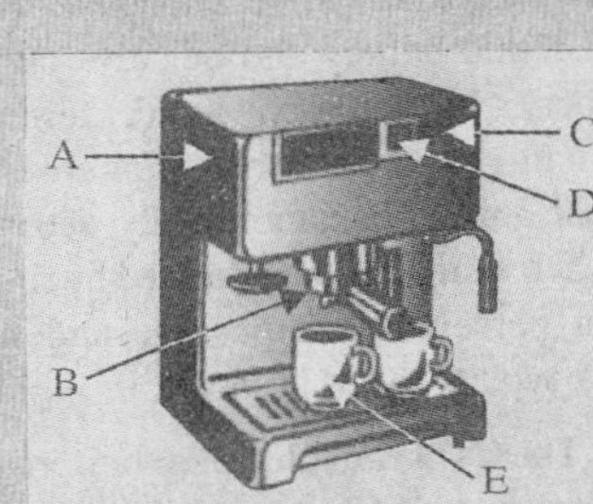

Figure 25.20 A typical espresso machine: (A) pressure vessel; (B) portafilter holding ground coffee; (C) on/off switch, with built-in pressure indicator; (D) solenoid valve for espresso coffee; (E) cup holding leached espresso coffee.

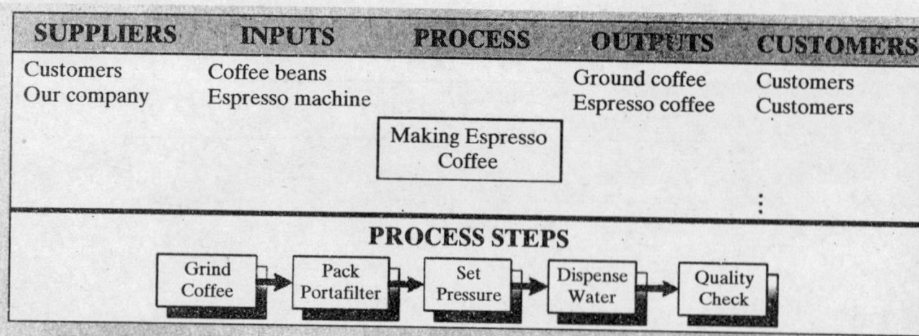

Figure 25.21 SIPOC for the preparation of espresso coffee.

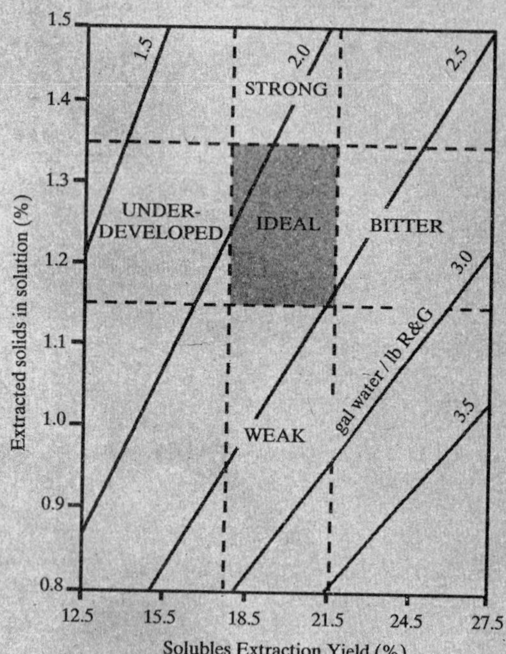

Figure 25.22 Coffee brewing control chart (developed by the Coffee Brewing Institute—Sivetz and Desrosier, 1979).

consistency (consistent or inconsistent), and persistence (with a hole in the center, evanescent or persistent), noting the percentage of judges that observe each attribute. Attributes such as odor intensity, body, acidity, bitterness, astringency, flavor intensity, and aftertaste intensity are typically assessed on a scale of "none" (0) to "very high" (10). More specifically, the judges gauge odor/flavor attributes, making a distinction between positive flavor attributes such as fruity/winey, malty/cereal, fresh, straw-like, caramel-like, chocolate-like, spicy, nutty, tobacco-like, and buttery and negative attributes such as woody/papery, burnt/roasty, acrid, fermented, earthy/musty, rancid, burnt rubbery, sulphurous, flat, grassy/green/herbal, animal-like, motor-oily, and ashy. The flavor profile of each sample is then defined by the percentage of judges that perceive each positive and negative flavor attribute. A typical result is shown in Figure 25.23.

(b) **Measure:** The CTQ variables are monitored to check their compliance with the LCLs and UCLs. Suppose this analysis indicates that using the existing machine, one cup in three, on average, has attributes outside the *ideal* operating window in Figure 25.22, indicating operation at lower than 2σ, with an immediate target to reduce this to 1 in 250, that is, to attain 4σ performance.

(c) **Analyze:** Several factors contribute to an excessively high variance in product quality (Sivetz and Desrosier, 1979; Illy, 2002; Andueza et al., 2002, 2003, 2007):

1. *Freshness of the ground coffee.* When the coffee is stale, the taste of the coffee is affected.
2. *Grade of the ground coffee.* When too coarse, leaching is insufficient, affecting the taste of the coffee. In contrast, when the coffee beans are ground too fine, the pressure drop across the packed grinds is too high, detrimentally affecting the leaching and producing harsh, bitter flavors. Serious espresso drinkers prepare their own roasted coffee beans and personally grind them fresh. However, such extreme behavior is atypical.
3. *Ground coffee packed evenly in the portafilter, to the correct degree.* Since the brew water is under high pressure, it finds the path of least resistance through the coffee. Uneven packing leads to channeling, with the coffee in and in the proximity of the channels over-extracted, and underextracted elsewhere. The resulting beverage is bitter and astringent, with many potentially good flavors remaining in the portafilter basket. In contrast, when the ground coffee is evenly and tightly packed, the water flows uniformly through all of the coffee.

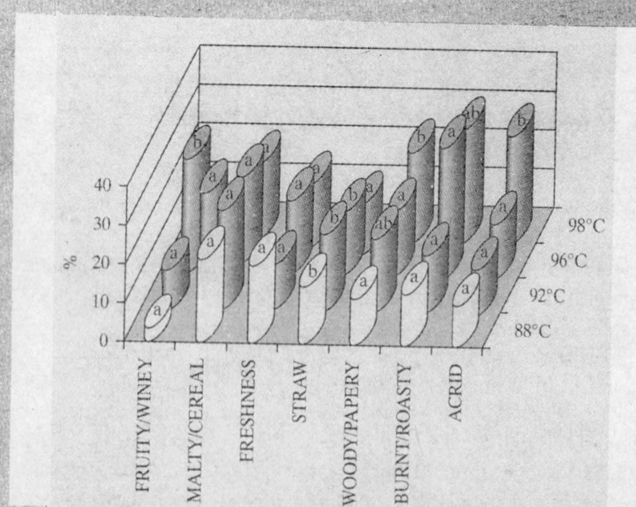

Figure 25.23 Influence of water temperature on flavor characteristics of Arabica espresso coffee samples (Andueza et al., 2003), noting the unacceptably high percentage of responders indicating undesirable flavor attributes at temperatures higher than 92°C. Reproduced with permission.

4. *Correct amount of coffee loaded.* An insufficient amount of coffee leads to overextraction and to a flat and watery drink (see Andueza et al., 2007).
5. *Sufficiently high water pressure.* This controls the temperature at which the leaching takes place (see Andueza et al., 2002).
6. *Proper amount of water passed through the ground coffee.* As indicated in Figure 25.22, the degree of extraction is critical to ensuring an acceptable product (see also Andueza et al., 2007).
7. *Quality of the water.* Since espresso coffee is 99% water, poor water quality to extract coffee (e.g., involving chlorine impurities, organic content, hardness, and alkalinity) has a detrimental effect on the quality of the product (Sivetz and Desrosier, 1979).

(d) Improve: Having identified the root cause of variance, it is eliminated or attenuated by redesign of the manufacturing process or by employing process control. For the espresso machine, there are several ways to reduce the sources of variance identified above. These include equipping the espresso machine with a water filter to reduce the variance due to item 7. In some machines, a solenoid valve is installed to dispense a precise amount of water, thus attenuating the variance due to item 6. Furthermore, by increasing the degrees-of-freedom in the design through installation of a pressure-control loop, the pressure can be maintained between its UCL and LCL, reducing the variations due to item 5. Note, however, that items 1 to 4 constitute sources of variance *not* under the control of the manufacturer of the espresso machine, as described in the introduction. To eliminate these four sources of variance, the manufacturer of a novel espresso machine can provide its users with vacuum-sealed containers of ground coffee having a built-in filter. On insertion into the machine, the container is perforated and used to prepare a single cup of coffee. Since the containers are vacuum-sealed, this ensures that the ground coffee is fresh, reducing the variance due to item 1. These containers of coffee must be manufactured by a process with a sufficiently high sigma level to ensure that variations due to items 2 and 4 do not occur. Furthermore, rather than relying on suitable packing of the ground coffee into the portafilter, a fixed flow resistance can be installed in the portafilter to ensure the correct degree of coffee extraction, reducing variations due to item 3. Moreover, when the manufacturer controls the coffee supply, the annual sales of coffee containers are likely to far exceed that of the new espresso machines.

(e) Control: After implementing steps to reduce the variance in the CTQ variables, the results are evaluated, and possible improvements are considered. Thus, steps (b) to (e) in the DMAIC procedure are repeated to improve the process quality in a stepwise fashion. For the espresso example, it is evident that a manufacturer of espresso machines would not necessarily implement all of the alternatives identified previously for the reduction of the variances in the CTQ variables. Rather, the manufacturer would first introduce either the most practical alternative, the cheapest, or those alternatives with the greatest impact.

25.4 SUMMARY

This chapter has introduced the potential advantages of using six-sigma methodology to quantify and ensure product and process quality, when integrating design and control strategies. As shown in the examples, this design methodology benefits from reducing the variance in the critical-to-quality variables by exploiting, and when necessary, increasing, the process degrees-of-freedom, using integrated design and control procedures. While product manufacturing has traditionally relied solely on statistical process control, the trend to improve profitability through increasing yields is driving many industries to embrace six-sigma methodology and advanced control strategies. For example, in integrated-circuit manufacturing, the increased reliance on advanced process control (APC), and in particular, on multivariable control, reflects the need to utilize the potential degrees-of-freedom in processes to assist in the reduction of CTQ variable variance.

REFERENCES

1. Andueza, S., L. Maeztu, B. Dean, M. Paz de Peña, J. Bello, and C. Cid, "Influence of Water Pressure on the Final Quality of Arabica Espresso Coffee. Application of Multivariate Analysis," *J. Agricul. & Food Chem.*, **50** (25), 7426–7431 (2002).

2. Andueza, S., L. Maeztu, L. Pascual, C. Ibáñez, M. Paz de Peña, and C. Cid, "Influence of Extraction Temperature on the Final Quality of Espresso Coffee," *J. Sci. Food & Agricul.*, **83** (3), 240–248 (2003).

3. Andueza, S., M.A. Vila, M. Paz de Peña, and C. Cid, "Influence of Coffee/water Ratio on the Final Quality of Espresso Coffee," *J. Sci. Food & Agricul.*, **87** (4), 586–592 (2007).

4. Bajpai, R.K., and M. Reuss, "A Mechanistic Model for Penicillin Production," *J. Chem. Tech. & Biotech.*, **30** (6), 332–344 (1980).

5. Birol, G., C. Undey, and A. Cinar, "A Modular Simulation Package for Fed-Batch Fermentation: Penicillin Production," *Comput. Chem. Eng.*, **26** (11), 1553–1565 (2002).

6. Bristol, E.H., "On a New Measure of Interactions for Multivariable Process Control," *IEEE Trans. Auto. Cont.*, AC-11, 133 (1966).

7. Dassau, E., I. Zadok, and D.R. Lewin, "Combining Six-Sigma with Integrated Design and Control for Yield Enhancement in Bioprocessing," *Ind. Eng. Chem. Res.*, **45** (25), 8299–8309 (2006).

8. Illy, E. "The Complexity of Coffee," *Scien. Amer.*, 72–77 (June, 2002).

9. Lachman-Shalem, S., B. Grosman, and D.R. Lewin, "Nonlinear Modeling and Multivariable Control of Photolithography," *IEEE Trans. of Semiconductor Manufact.*, **15** (3), 310–322 (2002).

10. Lewin, D.R. "A Simple Tool for Disturbance Resiliency Diagnosis and Feedforward Control Design," *Comput. Chem. Eng.*, **20** (1), 13–25 (1996).

11. Lewin, D.R., S. Lachman-Shalem, and B. Grosman, "The Role of Process System Engineering (PSE) in Integrated Circuit (IC) Manufacturing," *Cont. Eng. Prac.*, **15** (7), 293–309 (2007).

12. McAvoy, T.J., *Interaction Analysis*, Instrument Society of America, Research Triangle Park, North Carolina (1983).

13. Ogunnaike, B.A., and W.H. Ray, *Process Dynamics, Modeling and Control*, Oxford University Press, New York (1994).

14. Rath and Strong, *Design for Six Sigma Pocket Guide*, Rath and Strong Management Consultants/AON Management Consulting, Lexington, Massachusetts (2002).

15. Rath and Strong, *Six Sigma Pocket Guide*, Rath and Strong Management Consultants/AON Management Consulting, Lexington, Massachusetts (2000).

16. Reschke, M., and K. Schuegerl, "Continuous Reactive Extraction of Penicillin G in a Karr Column," *Chem. Eng. Jour.*, **31** (3), B19–B26 (1985).

17. Reschke, M., and K. Schugerl, "Reactive Extraction of Penicillin I, II and III," *Chem. Eng. Jour.*, **28** (1), B1–B29 (1984).

18. Reschke, M., and K. Schuegerl, "Simulation of the Continuous Reactive Extraction of Penicillin G in a Karr Column," *Chem. Eng. Jour*, **32** (1), B1–B5 (1986).

19. Sivetz, M., and N.W. Desrosier, *Coffee Technology*, AVI Publishing Co., Westport (1979).

20. Trivedi, Y.B., "Applying 6 Sigma," *Chem. Eng. Prog.*, **98** (7), 76–81 (2002).

21. Wheeler, J.M., "Getting Started: Six-Sigma Control of Chemical Operations," *Chem. Eng. Prog.*, **98** (6), 76–81 (2002).

EXERCISES

25.1

(a) Prepare a SIPOC for the manufacture of the hemodialysis device in Section 16.3.

(b) Identify all of the sources of variance in the urea concentration in treated blood after a 4-hour treatment. Suggest improvements in the design to increase the sigma level of the product.

25.2 Photolithography is an important process in integrated-circuit manufacture, in which a circuit pattern is transferred from a *mask* onto a photosensitive polymer (the PR), ultimately replicating that pattern on the surface of a silicon wafer (Lachman-Shalem et al., 2002). A typical photolithography process consists of seven steps: spin coat of the PR, prebake, chill, expose, post-exposure bake (PEB), chill, and development. The overall objective is to produce printed lines with accurate and consistent width (referred to as the *critical dimension*, CD). Table 25.8 shows the steps required, together with the sigma level of each step.

(a) Compute the sigma level of the complete process.

(b) You are required to reduce the variance in this process by "process improvements." Given the limited engineering time available, you can allocate only three instances of "process improvements," each of which will increase the sigma level of the selected step by 0.5. Allocate process improvements optimally to maximize the increase in sigma level for the overall process.

Table 25.8 Steps in Photolithography and Their Sigma Levels

Step	Subprocess	Function	Sigma Level
1	PR Coating	Coats wafer with a thick layer of PR	3.5
2	Prebake	Hardens the PR before exposure in the stepper	4.5
3	Chill 1	Cools the wafer after prebake	5
4	Stepper	Exposes the PR through a negative of the pattern to be reproduced	4
5	PEB	Fully hardens the PR after exposure	4.5
6	Chill 2	Cools the wafer after PEB	5
7	Development	Develops the image imprinted on the PR	3

Part Five

Design Report

In this part, which is comprised of Chapter 26, the contents of the written design report are presented and recommendations are made concerning its completion. Emphasis is placed on the need to document the design throughout the design process, including the project charter, the innovation map, the data used, the flowsheets considered, the material and energy balances, the detailed calculations for the process units, the cost estimates, and the profitability analysis. To accomplish these tasks, the design team needs to keep a repository of its materials. From time to time, it needs to report to its supervisor and, at universities, to its faculty advisor, industrial consultants, and fellow students. When documented properly, much of the material presented, together with the background calculations and drawings, can be used in the final design report. Hence, an objective of Part Five is to help the design team set milestones for the completion of aspects of the design as well as its design report.

In addition, the oral design presentation is discussed and suggestions are made for the organization of the presentation, the media used, methods for rehearsing, and written handouts. Also discussed are typical criteria for evaluation and the usage of videotapes.

Chapter 26

Written and Oral Reports

26.0 OBJECTIVES

At the completion of its design work, the design team is required to prepare a detailed report. If the design is in the final stages of consideration, the report should document the details of the design and how it was produced, projecting its profitability and making a recommendation as to whether or not management should make an investment in the product and/or the process. Also, a design report identifies the key assumptions in the design and their potential impact on the performance and profitability of the product and/or process. This is particularly important for designs completed by undergraduate students at universities, where facilities and time are rarely available for laboratory work or pilot-plant testing, or for the construction and testing of a product prototype.

It is these uncertainties, especially when data are lacking, that engineers encounter throughout their careers. Even when laboratories and pilot plants are available, engineering judgments are needed to determine when the investment of money and time is justified to organize an experimental program. In this respect, engineers are asked regularly to estimate the profitability of products and processes about which they have too little information. For these reasons, design teams usually expend considerable effort in trying to eliminate as much of the uncertainty as possible (by locating data in the literature, conducting process simulations, etc.). Invariably, however, uncertainties remain, and it is important that the design report identify the uncertainties and, when it recommends that a product be launched and/or a process be constructed, make recommendations as to how the uncertainties should be resolved.

A design report documents all of the steps leading up to the design of the product and/or process. This includes a discussion of the need for the new product and/or process, a summary of all the possible ways the design team generated to satisfy the need, the rationale behind the selection of the best ideas and ultimately the one selected, and the design of the new product and/or process. In industry, the report might also include the development of a pilot plant as well as the performance of a product prototype and the manufacturing tests. However, in universities, undergraduate students may not have the time to produce a prototype of the product and determine its performance. On the other hand, a product design developed by a student design team in a previous year could be given to a new design team in the current year for the development and assessment of a prototype.

The design team should view its design report as an opportunity to showcase for management its most creative engineering efforts. Wherever possible, a design report should highlight the engineering work that the team believes will lead to greater economies than are achievable using alternative or conventional technologies—emphasizing the soundness of its ideas, but the need for data to prove their validity. A product design should emphasize the superiority of the new product over available products with similar functions.

For most student design teams, the design report is the first extensive report of their professional careers. It is the culmination of a major engineering effort and, when done well, is deserving of considerable attention from other students, faculty, industrial consultants, and prospective employers. In this respect, the professional reputation of the design team depends, in part, on how well the design problem has been analyzed, how ingeniously the product has been developed and/or the process laid out, and how thoroughly the engineering calculations and design work have been done. The efforts of a design team are judged almost entirely by the quality of the engineering report provided to its supervisor that describes the work that has been accomplished. Of particular importance to management is the strength of and justification for the recommendations made in the report. Almost always, the report will be accompanied by an oral presentation by the design team, where questions can be asked by management.

There is perhaps a tendency to view the preparation of the design report and the oral presentation to management as activities reserved for the completion of a design project. Although, indeed, the level of activity in writing builds steadily toward the end, especially as the design becomes more promising, an objective of this chapter is to present the many reasons for documenting the results gradually, as the design project proceeds. In fact, for the gate reviews in Figures PI.1, PII.1, and PIII.1, as discussed in Section 1.3 and Chapter 2 and in the case studies throughout the book, up-to-date documentation is very important to the success of a design team, especially as the composition of the team changes.

After studying this chapter, the reader should

1. Understand the template that prescribes the sections in most design reports, and have a good appreciation of the materials to be included in each section.
2. Be prepared to coordinate the preparation of the design report with the other members of the design team, beginning early in the design process, and to recognize the important milestones; that is, those portions of the report that are best prepared before work begins on the next step in the design process.
3. Understand the role and format of a typical oral presentation, including the alternative media for the presentation and related topics such as the need for rehearsals and the desirability of a written handout.

26.1 CONTENTS OF THE WRITTEN REPORT

This section begins with a template of items to be included in most design reports and is followed by a discussion of several techniques found to be helpful in their preparation, as well as recommendations for the page format to permit the design report to be bound for distribution.

Sections of the Report

This listing is presented in outline form to identify, at a glance, the sections that are normally included in the sequence shown. The first six items are common to reports involving product and process designs.

1. ***Letter of Transmittal.*** This letter, on professional letterhead, is normally directed to the supervisor who requested that the design work be done. It should be signed by all members of the design team.
2. ***Title Page.*** In addition to the title, in uppercase, the authors and their affiliations are listed, as well as the publication date. The title should be short, but very descriptive.
3. ***Table of Contents.*** All sections in the report should be listed, including the page numbers on which they begin. Hence, all pages in the report, *without exception*, must be numbered. This applies to text pages, for which the word processor will probably have provided the page numbers, as well as tables, figures, and appendixes, whose pages may have to be numbered manually. Note that unnumbered pages are not readily found by the reader, who may resent the time wasted thumbing through the report to find pages that are missing or not numbered.
4. ***Abstract.*** The abstract is a brief description, in one or two paragraphs, of the design report, its key conclusions, special features, and assumptions. These include projections of any applicable economic measures of goodness (e.g., the return on investment and the net present value) and recommendations to management.
5. ***Introduction and Project Charter.*** This presents the origin of the design project and focuses on the project charter created when the design team began its work. The latter should include the specific goals, a project scope, and the deliverables and time line followed.
6. ***Innovation Map.*** This section of the report could be titled "Technology-Readiness Assessment." An innovation map should be presented showing the new technologies (materials, process/manufacturing, and product technologies) upon which the new product is based. A brief history of the technologies is helpful when discussing the innovation map, which should show how the voice of the customer for the new product is linked to the new technologies. Questions to be answered include whether all technologies are ready or can be used with minor variations.
7. ***Concept Stage***

Sections for Each Design

(a) Market and Competitive Analyses. These describe the market(s) for the new product and identify the principal competitors. When available, the production levels and annual sales of existing products are provided. Also, sales projections for the new product are included.

(b) Customer Requirements. The voice of the customer is presented in the form of customer requirements. While student design groups have limited access to customers, expressions of need should be stated based upon information from trade journals, advertisements, Web searches, and industrial consultants to design courses at universities. The customer requirements should be classified as fitness-to-standard (FTS) or new-unique-difficult (NUD).

Sections that Emphasize Product Design

These sections are prepared when the design team has focused on designing new chemical products, especially when designing *industrial* and *configured consumer* products. They document the steps followed in the *concept* stage of the SGPDP in Section 2.4. For examples, see the product designs for the halogen light bulbs (Section 17.2), LCD glass substrates (Section 15.2), washable crayons (Section 15.3), water-dispersible β-carotene (Section 13.4), and high-throughput screening of kinase inhibitors (Section 17.4). For many *basic* chemical products, where the chemical structure is well-known, these sections are skipped because the emphasis of the design team is on process design.

(c) Critical-to-Quality (CTQ) Variables—Product Requirements. This section should begin by identifying the CTQ variables, which are normally the NUD variables. Then, the translation of the customer requirements (CTQ variables) into technical requirements should be discussed. Their relationships should be presented in the rectangular section of the House of Quality (HOQ). The triangular interaction matrix, showing the synergistic technical requirements, should be added to the HOQ.

(d) Product Concepts. The alternative product concepts should be presented, preferably using the Pugh matrix. The advantages and disadvantages of each product concept should be discussed relative to a reference concept (normally the best existing product).

(e) Superior Product Concept. The superior concept should be presented, with justification for its selection.

(f) Competitive (Patent) Analysis. Having identified the superior concept for the new product, normally the competitive analysis is revisited. In this discussion, the results of more specific market analysis are presented. Also, the results of a more specific patent search are discussed in a so-called IP assessment.

(g) Other Important Considerations. In most design reports, the following considerations may deserve separate sections. Often, they are sufficiently important to warrant coverage apart from any discussion in the other parts of the report. These include those aspects of the design that address

1. Environmental problems and methods used to eliminate them.
2. Safety and health concerns, including a HAZOP (hazard and operability) study and a HAZAN (hazard analysis), as discussed in Section 1.5.

This subsection is intended to allow for a more thorough discussion of these subjects than might be appropriate elsewhere, and to enable the design team to draw attention to their importance in developing the product design.

(h) Business Case—Profitability Analysis. For most product designs, an approximate profitability analysis is carried out in the *concept* stage—to be refined with more accurate calculations in the later stages of the SGPDP. Normally, at least, an approximate profitability analysis is presented when documenting the *concept* stage.

This subsection provides estimates of the cost sheet(s) of annual costs, as discussed in Section 23.2 and shown in Table 23.1. Note that when cash flows are computed for different production rates from year to year, a separate cost sheet is required for each unique production rate.

Next, the working capital is presented, with a discussion of how it was estimated. Then the total capital investment is presented.

This subsection concludes with a presentation of the calculations used to obtain several of the profitability measures. Normally, this includes one or more of the approximate measures, such as the return on investment (ROI) and the venture profit (VP), and one or more of the rigorous methods that involve cash flows, such as the net present value (NPV) and the investor's return on investment (IRR). The latter is also referred to as the discounted cash flow rate of return (DCFRR). In all cases, it is important to indicate clearly the depreciation schedule and, for the rigorous methods, to provide a table that shows the calculation of the annual cash flows, as shown in Example 23.29. Finally, the design team should present its judgment of the profitability of the proposed plant. Where possible, the results of sensitivity analyses and optimizations are presented.

Sections that Emphasize Process Design

These sections discuss the alternatives considered during preliminary process synthesis, plus the assembly of the database and bench-scale laboratory work. Note that for product designs, where manufacturing processes are normally not synthesized in the *concept* stage of the SGPDP, these sections are skipped. For *basic* chemicals, on the other hand, where the chemical structure is normally well known, the emphasis in the *concept stage* is on process design. Examples include the manufacture of vinyl chloride and, even, tissue plasminogen activator (tPA), as discussed in Section 4.4.

(c) Preliminary Process Synthesis. The alternative process flowsheets should be presented, and possibly the synthesis tree, with a discussion of the most promising flowsheets.

(d) Assembly of Database. The principal thermophysical and transport property data should be presented, together with chemical kinetics data and toxicity data, with prices for the principal chemicals.

(e) **Bench-Scale Laboratory Work.** When laboratory data are available, they should be presented. Otherwise, the need for a laboratory program should be discussed.

8. ***Feasibility, Development, Manufacturing, and Product-Introduction Stages***

Sections that Emphasize Product Design. When items under Section 7 are *not* carried out in the *concept* stage, such as the creation of a prototype product or pilot-plant testing (which would normally be performed in a later stage such as the *feasibility* stage), these sections should be devoted to a discussion of the missing items and the stage in which they will be carried out. Also, as the business case (profitability analysis) is refined, a section should be included to present the more complete analysis.

Sections that Emphasize Process Design. The following items should be documented. Normally, they will be carried out in the *feasibility* and *development* stages.

(a) **Process Flow Diagram and Material Balances.** This is the detailed process flow diagram discussed in Section 4.5 and shown for a vinyl-chloride process in Figure 4.19. All of the streams are numbered clearly and all of the process units are labeled. At some point on the arc for each stream, the temperature and pressure should appear, or the information should be tabulated (e.g., see Table 4.6). Note that, as mentioned in Section 4.5, many software packages are available to simplify the preparation of flow diagrams, most notably those associated with the process simulators.

In addition, the drawing should contain a *material-balance block* similar to the one shown for the vinyl-chloride process in Table 4.6, that is, a table showing for each numbered stream:

1. Total flow rate
2. Flow rate of each chemical species
3. Temperature
4. Pressure

and other properties of importance (density, enthalpy, etc.). It is desirable that the flow diagram and the material-balance block appear on a single sheet for continuous reference, preferably $8\frac{1}{2}$ by 11 in., so that it can be bound easily with the remainder of the report. Most commonly, this combination is prepared by computer, using the latest software, such as Microsoft VISIO. The symbols on the drawing should follow a standard list, such as those provided in Figure 4.20 and by Peters et al. (2003), Sandler and Luckiewicz (1993), and Ulrich and Vasudevan (2004).

(b) **Process Description.** This section provides an explanation of the flow diagram. It best begins, however, with reference to a block flow diagram, similar to that in Figure 4.18, which shows just the process steps that involve chemical reactions and the separation of chemical mixtures. Then, a more detailed description is presented that refers to all steps in the process that are shown in the process flow diagram (e.g., Figure 4.19). The detailed description describes the function of each equipment item and discusses the reasons for each particular choice. Note that the details of each major equipment item are presented below in Subsection (d), on unit descriptions. To aid the reader, however, the discussion of each item in Subsection (b) should be accompanied by a reference to the page number in Subsection (d). As in the introduction, when this flow diagram has been selected from among alternatives, it is appropriate to present the alternative flow diagrams and process descriptions, and to describe the reasons for the final choice.

(c) **Energy Balance and Utility Requirements.** In describing most chemical processes, it is desirable to have a section that discusses the energy requirements of the process, and the measures adopted to improve the plant economics by energy and mass conservation, usually through the application of the methods described in Chapter 9 on heat and power integration (including second-law analysis), and Chapter 10 on mass integration. In this section, all of the heating, cooling, power, and other utility and mass-separating-agent demands should be identified (with numerical values provided), and the methods of satisfying these demands shown. A list should be provided of each demand (e.g., 500,000 Btu/hr to heat stream 5 from 80 to 200°F) and the vehicle for its satisfaction (e.g., 500,000 Btu/hr from stream 15 as it is cooled from 250 to 100°F). When power generated by a turbine is used to drive a compressor and pumps, these integrations should be listed as well. Methods used to minimize the need for solvents and other mass-separating agents, as well as to minimize wastes, should be described.

(d) **Equipment List and Unit Descriptions.** In this section, every process unit in the flow diagram is described in terms of its specifications and the design methodologies (e.g., the methods for estimating the heat-transfer coefficients, the rigorous design of a distillation tower by means of a simulation program, and the recommendations of industrial consultants) and the data employed (e.g., to characterize the reaction kinetics and vapor–liquid equilibria). The important approximations should be discussed, as well as any difficulties encountered in performing the design calculations (e.g., in converging equilibrium-stage calculations with a simulator). In addition, the materials of construction should be indicated, together with the reasons for their selection.

Each process unit described in Subsection (d) should refer to the page number in the appendix on which the design calculations appear or are described. Note that the latter calculations are either handprinted neatly or done by computer. In addition, the description of each process unit should refer to a corresponding specification sheet, discussed below, that is assembled with the other specification sheets in Subsection (e). Finally, the descriptions should refer to the estimated installed and operating costs for the process unit in cost summaries, discussed below.

The identification of each process unit (e.g., Unit No. E-154, the condenser on an ethanol still) should be very clear, so that the concerned reader is able, without confusion, not only to relate each unit description to the corresponding specification sheet, its estimated costs in the cost summaries, and its design calculations in the appendix, but also to locate that additional information readily and to check it when necessary.

The process units described in Subsection (d) should include: (1) storage facilities for the feed, product, byproduct, and intermediate chemicals, (2) spare equipment items (pumps, adsorption towers, etc.) required to avoid shutdowns in the event of operating difficulties, and (3) equipment for startup, which is often not needed during normal operation.

The descriptions are accompanied by an equipment list, which includes the unit number, unit type, brief function, material of construction, size, and operating conditions of temperature and pressure.

(e) Specification Sheets. Specification sheets are required to guide purchasing agents in locating vendors of desired equipment and to enable vendors to prepare bids. These sheets provide the design specifications for each of the process units in the process flow diagram, as referred to in the unit descriptions. A typical example is shown in Figure 26.1.

It is recommended that students at universities, before preparing specification sheets, have more experienced individuals (e.g., faculty and industrial consultants) review the specifications to identify, hopefully, impractical specifications and significant inconsistencies.

(f) Equipment Cost Summary. In this subsection a table is prepared that contains the estimated purchase price of every equipment unit in the process flow diagram, identified according to the unit number and unit type on the process flow diagram and in the equipment list. The sources of the prices should be identified (graphical or tabulated cost data, a quotation from a specific manufacturer, etc.).

(g) Fixed-Capital Investment Summary. In this subsection, the fixed-capital investment is related to the estimated purchase cost of the equipment items. If desired, the equipment list and the list of equipment purchase costs can be combined. The methods for estimating the fixed-capital investment, beginning with the purchase costs, should be clearly stated. If a factored cost estimate is used, the overall factor or individual equipment factors should be noted.

(h) Other Important Considerations. In most design reports, the following considerations may deserve separate subsections. Often, they are sufficiently important to warrant coverage apart from any discussion in the other parts of the report. These include those aspects of the design that address

1. Environmental problems and methods used to eliminate them.
2. Safety and health concerns, including a HAZOP (hazard and operability) study and a HAZAN (hazard analysis), as discussed in the supplement to Chapter 1.
3. Process controllability and instrumentation, including a piping and instrumentation diagram (P&ID).
4. Startup, including additional equipment and costs.
5. Plant layout when critical.

To the extent that these matters influence the choice of particular or additional items of equipment as well as operating strategies, at least some discussion should be included in Section 7 and in Section 8 (a–g). This section is intended to allow for a more thorough discussion of these subjects than might be appropriate elsewhere, and to enable the design team to draw attention to their importance in developing the design.

(i) Operating Cost and Economic Analysis. This is the same subsection that appeared earlier for the documentation of product designs (Subsection 7h, Business Case–Profitability Analysis). For process designs, it is normally completed and documented in the *development* stage of the SGPDP.

This subsection begins with a presentation of the estimated annual costs of operating the proposed plant, that is, the cost sheet, as discussed in

DISTILLATION COLUMN

Identification: **Item** *Distillation Column*
Item No. *T-700*
No. required 1

Date: *9 April 1997*

By: *SFG*

Function: Separate Benzoic Acid and Benzaldehyde from VCH, Styrene, and other organics.

Operation: Continuous

Materials handled:	*Feed*	*Feed 2*	*Liquid Dist.*	*Bottoms*	*Vapor Dist.*
Quantity (lb/hr):	161,527		153,022	6947	1558
Composition:					
Butadiene	4 PPB		2 PPB	trace	236 PPB
VCH	0.059		0.061	2 PPM	0.109
Styrene	0.861		0.899	0.087	0.630
Butene	10 PPB		5 PPB	trace	604 PPB
Cis-Butene	29 PPB		16 PPB	trace	2 PPM
Trans-Butene	9 PPB		5 PPB	trace	545 PPB
n-butane	3 PPB		1 PPB	trace	171 PPB
Isobutylene	7 PPB		3 PPB	trace	454 PPB
Isobutane	trace		trace	trace	9 PPB
Ethyl Benzene	0.039		0.041	96 PPM	0.041
Benzoic Acid	0.011		trace	0.244	trace
Benzaldehyde	0.028		31 PPM	0.647	10 PPM
H_2O	0.004		0.002	trace	0.205
N_2	139 PPM		2 PPM	trace	0.014
CO_2	6 PPB		1 PPB	trace	559 PPB
O_2	150 PPB		5 PPB	trace	15 PPM
Tar	902 PPM		trace	0.021	trace
Stabilizer					
Temperature (°F):	70.0		126.3	255.9	126.3

Design Data: Number of trays: 23
Pressure: 3.2 psig
Functional height: 70.5 ft
Material of construction: Carbon steel
Recommended inside diameter: 21.0 ft
Tray efficiency: 0.70
Feed stage: 13
Feed 2 stage:
Sidestream stage: 1

Molar reflux ratio: 10
Tray spacing: 3.0 ft
Skirt height: 14.5 ft

Utilities: Cooling water at 1.09 MM lb/hr and 370.52 M lb/hr 100 # stream
Controls:
Tolerances:
Comments and drawings: See Process Flow Sheet, 7 and Appendix F, 222-4.

Figure 26.1 Typical specification sheet for a process unit.

Section 23.2 and shown in Table 23.1. In addition to the total production cost on the cost sheet, it should provide an estimate of the cost per unit weight of the product (e.g., $ per lb, kg, ton, or tonne). Note that when cash flows are computed for different production rates from year to year, a separate cost sheet is required for each unique production rate. Note also that, in addition to appearing on the cost sheet, the utilities for each equipment unit and their costs should be summarized in a separate table.

Next, the working capital is presented, with a discussion of how it was estimated. Then the total capital investment is presented.

This subsection concludes with a presentation of the calculations used to obtain several of the profitability measures. Normally, this includes one or more of the approximate measures, such as the return on investment (ROI) and the venture profit (VP), and one or more of the rigorous methods that involve cash flows, such as the net present value (NPV) and the investor's return on investment (IRR). The latter is also referred to as the discounted cash flow rate of return (DCFRR). In all cases, it is important to indicate clearly the depreciation schedule and, for the rigorous methods, to provide a table that shows the calculation of the annual cash flows, as shown in Example 23.29. Finally, the design team should present its judgment of the profitability of the proposed plant. Where possible, the results of sensitivity analyses and optimizations are presented.

9. *Conclusions and Recommendations.* The principal conclusions of the design study should be presented, together with a clear statement of the recommendations, accompanied by justifications, for management. At this point, before the remaining sections of the report are discussed, it is important to emphasize that an engineering supervisor may find it necessary to check the calculations of the engineers in the design team. For this purpose, when documenting process designs, Subsections (d–g) and (i) in Section 8, as well as the associated sections of the appendix, are very important. References to the specific pages in each of these sections for every equipment item are equally important. Neither the supervisor responsible for the work of the design team, nor the faculty member who grades the design report, will regard with favor references to various sections of the report, including the appendix, that are absent or difficult to locate. The same is true of an industrial supervisor who causes such a report to be created.

10. ***Acknowledgments.*** Most design teams obtain considerable assistance and advice from industrial consultants, equipment vendors, librarians, fellow students, faculty, and the like. This section provides an opportunity to acknowledge their contributions with an expression of appreciation and thanks.

11. ***Bibliography.*** All works referred to in the design report, including the appendix, should be listed in this section. It is recommended that the references appear in the form shown in the Reference sections near the end of each chapter in this textbook.

12. ***Appendix.*** The following items are typically included in the appendix, whose pages should be numbered sequentially with the body of the report.

    (a) For each process design, the design procedures and detailed calculations for all of the equipment items in Section 8(d) must be included here. These are normally *not* typed, but must be sufficiently neat to be easily read and understood. Photocopies of legible calculation sheets, even bearing erasures or lined-out corrections, are adequate.

    (b) Computer programs developed for the design should be listed with sufficient documentation to enable the principal sections to be identified. This can normally be accomplished through the use of comment statements at the beginning of each section, including definitions of the key variables.

    (c) Relevant portions of the computer output (the variables at each stage of a distillation column, a graph showing the variables as a function of the stage number, etc.) should be included here. It is important that the output be sufficiently well annotated to permit the reader to read it intelligently. In some cases, handwritten annotations are helpful and adequate.

    (d) Pertinent printed material (e.g., materials provided by equipment vendors that describe their products) should be included here. At the risk of stating the obvious, it cannot be emphasized too strongly that the appendix is not a repository in which large quantities of computer printouts, pertinent or not, are included to increase the weight and thickness of the report. Unless the information in the appendix can readily be located by appropriate references in Sections 5–8, a responsible supervisor may doubt the results that appear in the foregoing sections. This can only adversely affect the evaluation of the report and the quality of the proposed design.

Preparation of the Written Report

Coordination of the Design Team

As mentioned in the introduction to this chapter, it is important for a design team to document its work throughout the design process. In this regard, each member is normally assigned responsibilities for a portion of the design work, as well as for its documentation. In industry, the assignments are usually coordinated by the head of the design team, who is normally appointed by the project supervisor. At a university, it is also recommended that a member of a student design team be appointed the team leader. The team leader schedules meetings to review progress of the team, plan its next steps, make assignments, and set due dates. The faculty advisor is often very helpful in advising the team as it reviews its progress and plans its next steps.

Project Notebook

When carrying out a design, the design team normally maintains a project notebook, most likely a loose-leaf binder, in which important sources of information are placed. These include articles from the literature, data from the laboratory or the literature, design calculations, and computer programs and printed outputs. This repository of information is updated regularly and is particularly helpful during the meetings of the design team, especially when visitors, such as the team's faculty advisor and industrial consultants, are present.

Milestones

Since no two design projects follow exactly the same sequence of steps, it is not possible to suggest a timetable with specific milestones to be met by all design teams. Rather, in this subsection, it should suffice to identify the milestones, with emphasis on the steps to be accomplished and the portions of the design report that can be written. It is up to the team leader to prepare the timetable so that the final completion date can be met. The following pertain to process designs. Similar steps, not given here, can be formulated for product designs.

(a) ***Complete the block flow diagram and detailed process flow diagram showing the material balances.*** Most design teams spend considerable time in the process creation steps, identifying alternative process flow diagrams and creating the synthesis tree, as discussed in Section 4.4. While these steps, and the application of the algorithmic methods for process synthesis (which are usually carried out in parallel), are very important in leading to the most profitable processes, it is crucial not to spend too much time generating alternatives. Fairly early in the design process, the team should begin to focus its attention on the base-case design, as discussed in Section 4.5. This involves the preparation of a detailed process flow diagram (see Figure 4.19) and the completion of the material balances. While this is completed, the design team should prepare a draft of Subsections (a) and (b) of Section 8 for each process design in the report. Should the base-case design be modified, the section is revised accordingly to show how the modifications improve upon the original design.

(b) ***Complete the heat integration.*** In many cases, an attempt to achieve a high degree of heat and power integration is not undertaken until after mass integration is complete and the reactor(s) and separation equipment have been designed. After heat and power integration is complete and the heat exchangers, pumps, and compressors are installed in the base-case design, it is recommended that Subsection (c) of Section 8, on the energy balance, be completed for each process design in the report.

(c) ***Complete the detailed equipment design.*** After this step is completed, Subsections (d) and (e) of Section 8, on the unit descriptions and the specification sheets, should be written for each process design in the report. Note that it helps to complete hand calculations neatly so that they can be inserted into the appendix without any additional work. Furthermore, it is recommended that the important sections of the computer outputs be removed and annotated when necessary for insertion into the appendix.

(d) ***Complete the fixed-capital investment and the profitability analysis.*** After these steps are completed, Subsections (f, g, and i) of Section 8 should be written for each process design.

For the novice design team, it is hoped that the preceding pointers will help to simplify both the preparation of the design report and the design process. Although many pointers merely follow common sense, they are included to help the design team set milestones to achieve throughout the design process.

Word Processing and Desktop Publishing

The advent of the word processor has had a major impact on the preparation of the design report. Because sections of text can be cut and pasted with ease, it is possible to write drafts of many sections, as discussed previously. As the base-case design is modified, new sections can be composed and added easily to the previously prepared sections, which can usually be included with minor modifications. For technical writing. Word, WordPerfect, PageMaker, and LaTeX are the most commonly used word processors. Except for highly mathematical manuscripts, the former two word processors are preferred.

Many design reports have on the order of 100 pages that include the sections discussed earlier. Since there are many cross-references between the sections, it can be very helpful to add headers to the pages that identify the section numbers and titles. Furthermore, in addition to the table of contents, an index can be very helpful when the reader is searching for coverage of a specific topic.

Editing

No matter how careful an author is, it is difficult to compose concise text without redundant terms and the use of words that add little, if any, meaning. Most novice designers and writers examine their manuscripts carefully for spelling errors with the help of the spelling checkers in their word processors. They also seek to confirm that their statements are technically correct. However, many are inexperienced in the art of editing.

To obtain a more tightly structured document, it is recommended that the design team read its text carefully with the objectives of improving the grammatical constructions (eliminating split infinitives, avoiding the use of long strings of adjectives, etc), avoiding the usage of redundant terms, and eliminating terms that add no meaning to the sentence. This step is important even for the most experienced writers, who can take advantage of recent versions of word processors that include grammar checkers. Checks are made and suggestions sometimes given for:

1. Incomplete sentences.
2. Use of passive voice when active voice would give more punch.
3. Improper use of *who, whom, which,* and *that.*
4. Capitalization.
5. Hyphenation.
6. Punctuation.
7. Subject and verb agreement.
8. Possessives and plurals.
9. Sentence structure.
10. Wordiness.

It must be noted, however, that grammar checkers are not always correct and suggested corrections should, therefore, not always be accepted.

Page Format

At many companies and universities, the design reports are bound for storage in technical libraries and repositories. When this is the case, to save space on the bookshelves and simplify the usage of the reports, the following guidelines are recommended for the preparation of a manuscript for binding.

(a) The pages of the report, including the appendix, should be numbered at the bottom center of each page.

(b) The pages of the report should be printed back-on-back (two-sided), with the odd page numbers appearing on the right-hand page, as shown below.

(c) All pages, including the appendix, should have left and right margins that are at least 1 in. wide, as shown below.

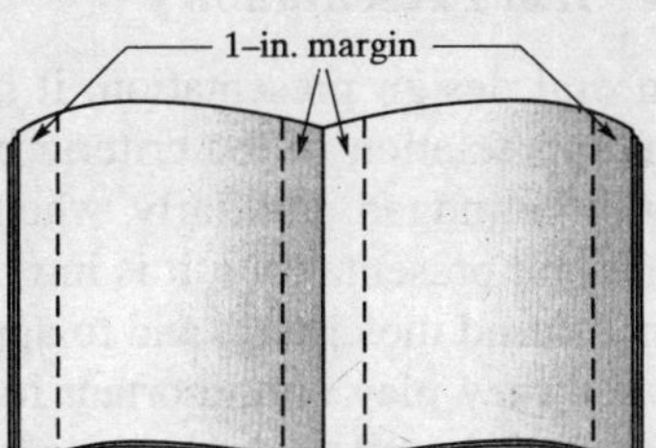

(d) Sheets that appear sideways (broadside) should be mounted so that their tops face the left margin, as shown below. Remember that, for sideways sheets, the top and bottom (which become the left and right side of the page when rotated, must have 1-in. margins.

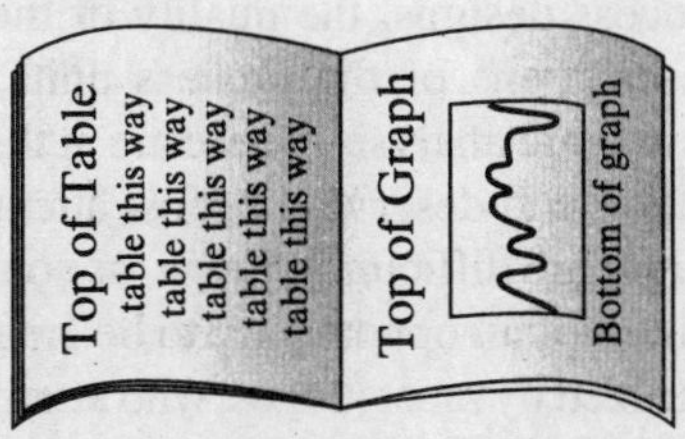

(e) Black ink should be used for the printed calculations, to ensure that the pages will photocopy adequately.

(f) When a large flow diagram is prepared by hand, as discussed in Section 8(a), it cannot be bound into the report, which is printed on $8\frac{1}{2}$-in.-by-11-in. pages. Such a flow diagram should be folded for insertion into a cover pocket, which is pasted onto the inside back cover of the report after the binding is completed.

(g) The complete manuscript should be submitted in a file folder for binding.

Sample Design Reports

Samples of design reports are available in the libraries and repositories of technical reports maintained by companies and universities. In a few cases, they are available from Interlibrary Loan, for example, the design reports prepared by students at the University of Pennsylvania since 1993. Note that titles of the problem statements that led to these reports are reproduced in Appendix II of this textbook, with full problem statements included in the file Supplement_to_Appendix_II .pdf in the PDF File folder, which can be downloaded from the Wiley Web site associated with this text book.

26.2 ORAL DESIGN PRESENTATION

It is probably most common for the oral design report in industry to be presented to the immediate supervisor of a design team, together with managers who are responsible for deciding upon the prudence of investing funds in the proposed design. Similar presentations to a somewhat different audience (including industrial consultants, faculty, and fellow students) are prepared at universities, usually to provide the students with an experience similar to what they are likely to encounter in industry. It should be noted, however, that only occasionally do young engineers have the opportunity to attend a meeting where their work and ideas are presented to the decision makers among their employers, especially to make the presentation in person.

Typical Presentation

A typical oral design presentation by a team comprised of three students at a university is scheduled for 30 min with an additional 10 min for questions and discussion. This provides sufficient time to

1. Introduce the design problem and project charter.
2. Provide an overview of the technologies involved and the customer needs; that is, introduce the innovation map.
3. For a product design, present the superior concept(s), emphasizing their strengths. For a process design, discuss the sections of the proposed process (emphasizing the strengths of the design).
4. Present the results of the economic analysis.
5. Discuss other considerations.
6. Summarize the design and make recommendations.

Normally, each student speaks for 10 min, although it is not uncommon to split the presentation into as many as six or seven segments, with each member of the design team covering those portions of the design with which he or she is most familiar.

Media for the Presentation

Overhead Projector

Until recently, the overhead projector was the most actively used vehicle for displaying the key concepts, graphs, figures, and tables that accompany a design presentation. In some cases, two overhead projectors permitted the presenters to describe concepts that benefited from the simultaneous display of two complementary figures. Although most engineers have extensive experience in the use of these projectors, it continues to be helpful to remind presenters of the importance of using sufficiently large fonts and maintaining appropriate borders to enable the entire transparency to be displayed clearly and simultaneously.

Computer Projection Software

Rapidly gaining favor with speakers in recent years, with the availability of LCD pads and computer projection devices, are computerized projection facilities. To prepare and display the images, several software packages have been developed, including PowerPoint. This software is capable of displaying animated sequences, halftones, and videos. In most cases, the quality is significantly improved relative to the use of overhead projectors. Students using this technique of presentation should be warned to carefully check out the system just prior to the presentation. Otherwise, failure of the computer system to project the presentation could result in severe criticism of the speaker(s).

Preparation of Exhibits

To avoid duplicate work effort, it is recommended that design teams prepare all of the figures and tables for their written reports in such a way that they are displayed properly by an overhead projector or computer projection software. This requires that the figures and tables be prepared with sufficiently large fonts and the information placed less compactly than when an oral presentation is not required.

Rehearsing the Presentation

One of the most difficult tasks a design team encounters is the organization of a 30-min presentation to summarize the most salient features of an extensive written report. It is especially challenging because the members of the team have usually been so involved in the details of the design calculations that they find it difficult to summarize the really important results without overemphasizing the details. For this reason, and to help the team see the forest through the trees, it is important for them to rehearse the presentation in the presence of a colleague or teacher. In the best situation, this person will have attended many design presentations in the past, and will be well positioned to recommend that certain topics be expanded upon while others be deemphasized or eliminated entirely.

In one format for the rehearsal, the team makes a complete presentation without any interruptions, with the critic sitting toward the back of the room, to check that the exhibits can be seen and that the speakers can be heard easily. In addition, the critic takes notes and records the time that each speaker begins his or her presentation. Then, when the presentation is completed, the critic reviews the timing and offers some general comments. Often the design team makes a brief pass through its exhibits to enable the critic to offer more specific criticisms. The critic often has a major impact on the organization of the report and, more specifically, in helping the design team to achieve a well-balanced presentation.

Written Handout

In some situations, design teams find it easier to make their points through the preparation of a small written handout. Often, this includes an innovation map for the new technologies associated with the product, a 3-D sketch of the product, or a detailed process flow diagram, similar to that shown in Figure 4.19.

Evaluation of the Oral Presentation

When preparing an oral design presentation, it helps the speakers to have an appreciation of the criteria by which their presentation will be judged. Similarly, when serving on a team to evaluate oral presentations, it is important for the evaluators to understand the criteria and to apply them fairly, especially when they play an important role in the preparation of a course grade and the selection of an award winner.

One possible list of items to be evaluated is shown in Figure 26.2. Included for the product designs are the quality of the product charter and the innovation map, the description of the customer and technical requirements, the quality of the superior concept, and the discussion of the economic analysis; and, for the process designs, the quality of the process description, the descriptions of the process units, and the discussion of the economic analysis. These are at the heart of the design presentation and deserve the most attention. The next item, novelty, is more difficult to judge, as some design problems provide more of an opportunity to be creative than others. This is recognized by most judges, who attempt to rate the creativity of the design work in the context of the design problem and the opportunities it provides to develop novel solutions. The next items address the organization of the presentation and its execution. Then, the quality of the exhibits and visual aids is evaluated. Finally, the overall presentation is rated, which includes a recommendation for a grade when the design report is the work of a student design team.

Videotapes and DVDs

Increasingly, oral design presentations are recorded on videotape or DVD to provide a record of the presentations, as well as to enable each design team to critique its own presentation with a view toward improving the next one. In most cases, a portable camcorder, mounted on a tripod, is adequate to capture the bulk of the presentation.

26.3 AWARD COMPETITION

At many universities, an award is presented to the design team that prepares a design judged to be the most outstanding. Normally, the criteria are a combination of those discussed for both the written and oral design reports. However, since the written reports become available using Interlibrary Loan, and the best reports are often submitted for regional competitions in which the judges select from among reports that originate from other universities, it is common to place more emphasis on the written reports.

Usually, a small awards committee, comprised of academic and industrial members, is appointed to make the judgment. It begins by reading the reports of those design teams whose oral presentations were judged to be among the best.

At many universities, the design award is presented to the design team at the commencement exercises, either for the Chemical Engineering Department, the Engineering School, or the entire university. It often involves a small stipend and a certificate or plaque.

Finally, it is important to mention the annual National Student Design Competition prepared by AIChE members from industry and academia for the AIChE Student Chapters. The design contest is timed to be completed by the end of the spring semester, after which the awardees are selected to receive their awards at the Annual Meeting of the AIChE, usually in November, and to make oral presentations at the associated Student Chapter Meeting.

Name of Presenter(s):	
Title of Presentation:	
Date of Presentation:	
Name of Examiner/Appraiser:	

Content	**Noteworthy**	**Acceptable**	**Needs Improving**
Product Design			
Project charter			
Innovation map			
Customer requirements			
Superior concept			
Economics			
Totals			
Process Design			
Process description			
Unit descriptions			
Economics			
Novelty of design			
Totals			

Figure 26.2 Oral design presentation evaluation form. (*Continued*)

Presentation—Organization	Noteworthy	Acceptable	Needs Improving
Core message			
Clear objectives			
Overall structure			
Visible logic			
Totals			

Presentation—Execution	Noteworthy	Acceptable	Needs Improving
Confident, enthusiastic, forceful, convincing			
Controlled pace/natural finish			
Voice quality (clear, calm, understandable)			
Frequent eye contact			
Totals			

Visual Aids	Noteworthy	Acceptable	Needs Improving
Interesting, relevant			
Easy to read			
Totals			

Figure 26.2 *(Continued)*

26.4 SUMMARY

In this chapter, readers have been presented with a template, associated milestones that must be completed, and guidance in the preparation of the written design report. No exercises are included because the template is intended to be used by design teams when writing their written reports.

Furthermore, readers have learned how to organize an oral design presentation. In this way, they have become familiar with the alternative media for the presentation, along with the reasons for rehearsing the presentation and the methods used to evaluate presentations.

REFERENCES

1. Peters, M.S., K.D. Timmerhaus, and R. West, *Plant Design and Economics for Chemical Engineers*, 5th ed., McGraw-Hill, New York (2003).

2. Sandler, H.J., and E.T. Luckiewicz, *Practical Process Engineering*, XIMIX, Philadelphia, Pennsylvania (1993).

3. Ulrich, G.D., and P.T. Vasudevan, *Chemical Engineering Process Design and Economics: A Practical Guide*, 2nd ed., Process (Ulrich) Publishing, www.ulrichvasudesign.com (2004).

Appendix I

Heterogeneous Residue Curves

Beginning with Eq. (8.19), which also applies for heterogeneous systems, the liquid mole fractions, x_j, are replaced by the overall liquid mole fractions, x_j^o. These are accompanied by the equations that define the vapor–liquid and liquid–liquid equilibrium constants, K_j^{VL} and K_j^{LL}, respectively, and the component mass balances, to give

$$\frac{dx_j^o}{d\hat{t}} = x_j^o - y_j, \qquad j = 1, \ldots, C \qquad \text{(A-I.1a)}$$

$$y_j = x_j^I K_j^{VL}\{T, P, \underline{x}^I, \underline{y}\} \qquad j = 1, \ldots, C \qquad \text{(A-I.1b)}$$

$$x_j^{II} = x_j^I K_j^{LL}\{T, P, \underline{x}^I, \underline{x}^{II}\} \qquad j = 1, \ldots, C \qquad \text{(A-I.1c)}$$

$$x_j^o = \alpha x_j^I + (1 - \alpha) x_j^{II} \qquad j = 1, \ldots, C \qquad \text{(A-I.1d)}$$

$$\sum_{j=1}^{C} x_j^I - \sum_{j=1}^{C} x_j^{II} = 0 \qquad \text{(A-I.1e)}$$

$$\sum_{j=1}^{C} y_j = 1 \qquad \text{(A-I.1f)}$$

where x_j^I and x_j^{II} are the mole fractions of species j in the first and second liquid phases, respectively, and α is the mole fraction of the first liquid phase in the total liquid. To trace a residue curve from some starting composition, this system of differential-algebraic equations is solved by numerical integration. Equations (A-I.1b)–(A-I.1f) are solved to determine the compositions in vapor–liquid–liquid equilibrium as $\hat{t}$ is advanced in the integration.

Appendix II*

Design Problems

A-II.0 CONTENTS AND INTRODUCTION

www.wiley.com/go/global/seider

*The complete Appendix II appears in the file Supplement_to_Appendix_II.pdf in the PDF File folder, which can be downloaded from the Wiley Web site associated with this text. Only the titles of the design problem statements are listed here.

This appendix contains the problem statements for 71 design projects, each prepared for design teams of three or four students at the University of Pennsylvania by chemical engineers in the local chemical industry and by the chemical and biomolecular engineering faculty. At Penn, each team selects its design project during the first lecture course in the fall, and spends the spring semester completing the design. In the spring, each group meets regularly with its faculty advisor

www.wiley.com/go/global/seider

and industrial consultants, including the individual who provided the problem statement, to report on its progress and gain advice.

The problem statements in the file, Supplement_to_Appendix_II.pdf, in the PDF File folder, which can be downloaded from the Wiley Web site associated with this book, are in their original forms, as they were presented to the student design teams on the date indicated. Some provide relatively little information, whereas others are fairly detailed concerning the specific problems that need to be solved to complete the design. The reader should recognize that, in nearly every case, as the design team proceeded to assess the problem statement and carry out a literature search, the specific problems it formulated were somewhat different than stated herein. Still, these problem statements should be useful to students and faculty in several respects. For students, they should help to show the broad spectrum of design problems that chemical engineers have been tackling in recent years. For the faculty, they should provide a basis for similar design projects to be created for their courses.

In formulating design problem statements, the industrial consultants and faculty strive to create product and process opportunities that lead to designs that are timely, challenging, and offer a reasonable likelihood that the final design will be attractive economically. Every effort is made to formulate problems that can be tackled by chemical engineering seniors without unduly gross assumptions and for which good sources of data exist for the reaction kinetics and thermophysical and transport properties. In this respect, this was accomplished in each of the problems included herein; furthermore, successful designs were completed by a student design team for most of these problems.

As seen in the contents, the projects have been assigned to one of the following areas, in some cases arbitrarily: Petrochemicals, Petroleum Products, Gas Manufacture, Foods, Pharmaceuticals, Biomedical, Polymers, Electronic Materials, and Environmental.

Most of the problem statements focus on process design, although in recent years, more emphasis is shifting toward product design. See especially the recent projects under Pharmaceuticals, Biomedical, and Electronic Materials. Gradually, all student design groups are creating project charters, innovation maps, and are following the Stage-Gate™ Product-Development Process, as outlined in Figures PI.1, PII.1, and PIII.1. Emphasis is placed on the *concept*, *feasibility*, and *development* stages.

Credit is given to each formulator on his problem statement. In addition, the names of the contributors are listed below with many thanks, as their contributions in preparing these design problems have been crucial to the success of the design course.

Rakesh Agrawal	Air Products and Chemicals (Currently of Purdue University)
E. Robert Becker	Environex, Wayne, PA
David D. Brengel	Air Products and Chemicals
Adam A. Brostow	Air Products and Chemicals
Robert M. Busche	Bio-en-gene-er Associates, Wilmington, DE
Leonard A. Fabiano	CDI Engineering Group (formerly ARCO Chemical and Lyondell)
Brian E. Farrell	Air Products and Chemicals
Mike Herron	Air Products and Chemicals
F. Miles Julian	DuPont
Ralph N. Miller	DuPont
Robert Nedwick	Pennsylvania State University (formerly ARCO Chemical and Lyondell)
Frank Petrocelli	Air Products and Chemicals
Mark R. Pillarella	Air Products and Chemicals
William B. Retallick	Consultant, West Chester, PA
Matthew J. Quale	Mobil Technology Company
Gary Sawyer	Lyondell Chemical Company
David G. R. Short	University of Delaware (formerly DuPont)
Peter Staffeld	Exxon/Mobil
Albert Stella	General Electric (formerly AlliedSignal)
Edward H. Steve	CDI Engineering Group
Bjorn D. Tyreus	DuPont
Kamesh G. Venugopal	Air Products and Chemicals
Bruce Vrana	DuPont
Andrew Wang	Air Products and Chemicals
Steve Webb	Air Products and Chemicals
John Wismer	Arkema, Inc. (formerly Atochem North America)
Jianguo Xu	Air Products and Chemicals

Appendix III

Materials of Construction

The selection of materials of construction, based on strength, corrosivity, and cost of fabrication, is vital to product and process design and economic evaluation. The most common materials for process equipment, which include metals, glass, plastics, and ceramics, are listed in Table A-III.1, together with typical applications. Much more extensive tables are given by M. S. Peters, K. D. Timmerhaus, and R. E. West in *Plant Design and Economics for Chemical Engineers*, fifth ed. (McGraw-Hill, New York, 2003); by G. D. Ulrich and P. T. Vasudevan in *Chemical Engineering Process Design and Economics: A Practical Guide*, second ed. (Process (Ulrich) Publishing, Durham, New Hampshire, 2004); and in Section 25 of the eigth edition of *Perry's Chemical Engineers' Handbook* (McGraw-Hill, New York, 2008). Table A-III.1 should be used only for preliminary process design and economic evaluation. For final process design, corrosion and strength data as a function of temperature are needed for the expected chemical compositions within the process. Equipment vendors can also assist in the final selection of materials. In general, carbon steel is used whenever possible because of its low cost and ease of fabrication.

Table A-III.1 Materials of Construction for Process Equipment

Material	Maximum Temperature, °C (°F)	Typical Applications
Carbon steel (e.g., SA-285C) Cast iron (not strong) Ductile iron (stronger)	400 (750)	Cooling-tower water, boiler-feed water, steam, air, hydrocarbons, glycols, mercury, molten salts, acetone
Low alloy (Cr-Mo) steel (e.g., SA-387B)	500 (930)	Same as carbon steel, hydrogen
Stainless steels	700 (1,300)	Aqueous salt solutions, aqueous nitric acid, aqueous basic solutions, food intermediates, alcohols, ethers, freons, hydrogen, hydrogen sulfide, molten salts, molten metals
Aluminum	150 (300)	Aqueous calcium hydroxide, hydrogen, oxygen
Copper and copper alloys, aluminum bronze, brass, bronze	150 (300)	Aqueous sulfate and sulfite solutions, hydrogen, nitrogen, alcohols and other organic chemicals, cooling-tower water, boiler-feed water
Nickel-based alloys (e.g., Hastelloy, Inconel, Monel, Incoloy, Carpenter 20)	400 (750)	Aqueous nitric and organic acids, flue gases, chlorine, bromine, halogenated hydrocarbons, ammonia, sulfur dioxide, sulfur trioxide, organic solvents, brackish water, seawater
Titanium-based alloys	400 (750)	Aqueous solutions, carbon dioxide, organic solvents
Conventional plastics (polyethylene, polypropylene, ABS)	50–120 (120–250)	Aqueous solutions at near-ambient temperatures

(*continued*)

Table A-III.1 (*Continued*)

Material	Maximum Temperature, °C (°F)	Typical Applications
Fluorocarbon plastics	250 (480)	Almost everything except halogens and halogenated chemicals
Rubber lining	250 (480)	Aqueous salt solutions and aqueous basic solutions at near-ambient temperatures
Glass lining	250 (480)	Aqueous sulfuric acid solutions, almost everything except fluorine and hydrogen fluoride
Ceramics	2,000 (3,630)	Almost all aqueous solutions, except hydrogen fluoride and sodium hydroxide, at near-ambient temperatures; most gases, except fluorine and hydrogen fluoride; most solvents; water
Graphite	2,000 (3,630)	Aqueous salt and base solutions; organic solvents Cl_2, HCl, H_2, H_2S, N_2, Hg; hydrocarbons; molten salts

Table of Acronyms

Acronym	Description
3M	Minnesota Mining and Manufacturing
a-Si	Amorphous Silicon
ABET	Accreditation Board for Engineering and Technology
ABS	acrylonitrile-butadiene-styrene
ACFM	Actual Cubic Feet per Minute
ACRS	Accelerated Cost Recovery System
ADP	adenosine diphosphate
AIChE	American Institute of Chemical Engineers
AM-LCD	Active Matrix Liquid Crystal Display
APC	Advanced Process Control
API	American Petroleum Institute
ASCE	American Society of Civil Engineers
ASHRAE	American Society of Heating, Refrigerating, and Air-Conditioning Engineers
ASME	American Society for Mechanical Engineers
ATP	adenosine triphosphate
ATR	Adiabatic Temperature Rise
BFD	Block Flow Diagram
bfw	Boiler Feed Water
BOD	Biochemical Oxygen Demand
BWG	Birmingham Wire Gauge
C&R	Controllability and Resiliency
CAGR	Compounded Annual Growth Rate
CCD	Charge Coupled Device
CCPS	Center for Chemical Process Safety
CD	Compact Disc
CD-ROM	Compact Disc Read Only Memory
CE	Chemical Engineering Plant Cost Index
CEO	Chief Executive Officer
CFD	Computational Fluid Dynamics

Acronym	Description
CFL	Compact Fluorescent Lamp
CHO	Chinese hamster ovary
CI	Composition Interval; Concentration Interval
CIP	Clean In Place
COG	Coke Oven Gas
COM	Cost of Manufacture
CPI	Consumer Price Index
CSB	Chemical Safety and Hazard Investigation Board
CSTR	Continuous Stirred Tank Reactor
CTE	Coefficient of Thermal Expansion
CTQ	Critical To Quality
cw	Cooling Water
DB	Declining Balance
DBEF	Dual Brightness Enhancement Film
DC	Disturbance Cost
DCFRR	Discounted Cash Flow Rate of Return
DDB	Double Declining Balance
DDT	dichloro-diphenyl trichloroethane
DIERS	Design Institute for Emergency Relief Systems
DJIA	Dow Jones Industrial Average
DMAIC	Define, Measure, Analyze, Improve, and Control
DOE	Design of Experiment
DOF	Degree of Freedom
DPI	Direct Permanent Investment
DPMO	Defects per Million Opportunities
DSC	Differential Scanning Calorimetry
DTB	Draft Tube Baffled
DTBP	di-tertiary-butyl peroxide
DVD	Digital Video Disc
DW&B	Direct Wages and Benefits
EDGE	Enhanced Data rates for Global Evolution
ENIAC	Electronic Numerical Integrator And Computer
ENR	Engineering News Record Construction Cost Index
EPA	Environmental Protection Agency
EPL	Emek Project Ltd
ESA	Energy Separating Agent

Acronym	Description
ESRD	End State Renal Disease
FC	Fully Coupled; Flow Controller
FDA	Food and Drug Administration
fg	fuel gas
FMC	Food, Machinery, and Chemical
FP	Fluorescence Polarization
FRET	Fluorescent Resonance Energy Transfer
FTS	Fitness To Standard
FUG	Fenske Underwood Gilliland
FWR	Flash With Recycle
GAC	Granular Activated Carbon
GAMS	General Algebraic Modeling System
GCC	Grand Composite Curve
GE	General Electric
GE	General Expenses
GRG	Generalized Reduced Gradient
GUI	Graphical User Interface
HAP	Hazardous Air Pollutant
HAZAN	Hazard Analysis
HAZOP	Hazard and Operability
HEN	Heat Exchanger Network
HETP	Height Equivalent of a Theoretical Plate
HETS	Height Equivalent of a Theoretical Stage
HF	Hydrofluoric acid
HHV	Higher Heating Value
HME	Heat and Mass Exchanger
HOQ	House of Quality
hps	High-Pressure Steam
HSC	Horizontal Split Case
HTML	HyperText Markup Language
HTRF	Homogeneous Time-Resolved Fluorescence
HTU	Height of a Transfer Unit
IC	Integrated Circuit
IEEE	Institute of Electrical and Electronics Engineers
IMC	Internal Model Control
IND	Investigational New Drug

Acronym	Description
IP	Intellectual Property
IPE	Icarus Process Evaluator
ips	Intermediate Pressure Steam
IR	Infrared
IRR	Investor's Rate of Return
IS	Intermediate Storage
ISO	International Organization for Standardization
ITO	Indium Tin Oxide
KE	Kinase Enzyme
KI	Kinase Inhibitor
KJ	Kawakita, Jiro
KKT	Karush–Kuhn–Tucker
LC	Level Controller
LCD	Liquid Crystal Display
LCL	Lower Control Limit
LDPE	low-density polyethylene
LED	Light Emitting Diode
LFL	Lower Flammability Limit
LHV	Lower Heating Value
LOC	Loss of Control
LP	Linear Program
lps	Low Pressure Steam
M&O-SW&B	Maintenance and Operations - Salary, Wages, & Benefits
MAC	Marginal Annualized Cost
MACRS	Modified Accelerated Cost Recovery System
MBBA	4-methoxy benzylidene-4-butylaniline
MCB	monochlorobenzene
MD	Molecular Dynamics
MEN	Mass Exchange Network
MER	Minimum Energy Requirement; Maximum Energy Recovery
MESH	Material balance, Equilibrium, Summation of mole fractions, Heat balance
MIC	methyl isocyanate
MILP	Mixed Integer Linear Program
MINLP	Mixed Integer Nonlinear Program
MIT	Massachusetts Institute of Technology
ML	Direct Materials & Labor

Acronym	Description
MO	Magneto Optical
MOC	Minimum Oxygen Concentration
MOC	Materials of Construction; Minimum Operating Cost
MOCVD	Metal Organic Chemical Vapor Deposition
mps	Medium Pressure Steam
MS	Marshall & Swift Equipment Cost Index
MSA	Mass Separating Agent
MSDS	Material Safety Data Sheet
MTBE	methyl tertiary-butyl ether
MVR	Marginal Vapor Rate
MW&B	Maintenance, Wages, and Benefits
NF	Nelson Farrar Refinery Construction Cost Index
NFPA	National Fire Protection Association
NLP	Nonlinear Program
Nortel	Northern Telecom
NPSH	Net Positive Suction Head
NPV	Net Present Value
NSF	National Science Foundation
NSPE	National Society of Professional Engineers
NTU	Number of Transfer Units
NUD	New, Unique, and Difficult
ODP	Ozone Depletion Potential
OEC	Online Ethics Center
OPEC	Organization of the Petroleum Exporting Countries
OSHA	Occupational Safety and Health Agency
OTL	Outer Tube Limit
p-Si	Polymorphous Silicon
P&G	Proctor & Gamble
P&ID	Piping and Instrumentation Diagram
PBA	Packed Bed Adsorption
PBP	Payback Period
PBS	Phosphate Buffer Solution
PDA	Personal Data Assistant
PDMS	poly-dimethylsiloxane
PDP	Plasma Display Panel
PET	polyethylene terephthalate

Acronym	Description
PFD	Process Flow Diagram
PFR	Plug Flow Reactor
PFTR	Plug Flow Tubular Reactor
PLS	Personal Laboratory System
pr	propane refrigerant
PR	Peng-Robinson
PVC	poly-vinyl chloride
QFD	Quality Functional Deployment
R&D	Research and Development
rb	refrigerated brine
RDC	Rotating Disk Contactor
RDT	RainDance Technology
RGA	Relative Gain Array
ROA	Return On Total Assets
ROE	Return On Equity
ROI	Return On Investment
ROR	Simple Rate of Return
ROROI	Rate of Return On Investment
SCC	Stress Corrosion Crack
SCFM	Standard Cubic Feet per Minute
SCN	Steam Cracked Naphtha
SGPDP	Stage-Gate Product-Development Process
SGTDP	Stage-Gate Technology-Development Process
SIP	Steam In Place
SL	Straight Line
SLP	Successive Linear Programming
SMART	Specific, Measurable, Agreed-upon, Realistic, and Time-based
SQP	Successive Quadratic Programming
SRI	Stanford Research Institute
SRK	Soave Redlich-Kwong
SS	Steady State
SYD	Sum of the Years Digits
TBA	tertiary-butyl alcohol
TBM	Total Bare Module
TC	Temperature Controller
TC	Total Capital

Acronym	Description
TCI	Total Capital Investment
TDC	Total Depreciable Capital
TDE	Thermodynamic Data Engine
TEFC	Totally Enclosed Fan Cooled
TEMA	Tubular Exchangers Manufacturers Association
TFT	Thin Film Transistor
TGA	Thermo Gravimetric Analysis
TI	Temperature Interval
TNT	trinitro-toluene
tPA	Tissue Plasminogen Activator
TPI	Total Permanent Investment
TRC	Thermodynamics Research Center
TRI	Toxic Chemical Release Inventory
TY	Throughput Yield
UCL	Upper Control Limit
UFL	Upper Flammability Limit
UIS	Unlimited Intermediate Storage
UL	Underwriters Laboratories
UV	Ultraviolet
VED	Viscous Energy Dissipation
VOC	Voice of the Customer; Volatile Organic Compound
VOM	Voice of the Market
VP	Venture Profit
VSC	Vertical Split Case
WC	Working Capital
WFI	Water For Injection
WWW	World Wide Web
ZW	Zero Wait

Author Index

Page Numbers	*Definitions*
1, 2, . . .	Page numbers in textbook
C1S.X	Section number X in Supplement_to_Chapter_1.pdf in PDF Files folder on Web Site
C3S.X	Section number X in Supplement_to_Chapter_3.pdf in PDF Files folder on Web Site
C7S.X	Section number X in Supplement_to_Chapter_7.pdf in PDF Files folder on Web Site
C9S.X	Section number X in Supplement_to_Chapter_9.pdf in PDF Files folder on Web Site
C12S.X	Section number X in Supplement_to_Chapter_12.pdf in PDF Files folder on Web Site
C16S.X	Section number X in Supplement_to_Chapter_16.pdf in PDF Files folder on Web Site
C22S.X	Section number X in Supplement_to_Chapter_22.pdf in PDF Files folder on Web Site
C23S.X	Section number X in Supplement_to_Chapter_23.pdf in PDF Files folder on Web Site
C25S.X	Section number X in Supplement_to_Chapter_25.pdf in PDF Files folder on Web Site
AIIS-X to Y	Page X to Y in Supplement_to_Appendix_II.pdf in PDF Files folder on Web Site
PDF-GAMS	In GAMS.pdf Iin PDF Files folder on Web Site
PDF-IPE	In Aspen IPE Course Notes.pdf in PDF Files folder on Web Site

Subject Index

Page Numbers	*Definitions*
1, 2, . . .	Page numbers in textbook
C1S.X	Section number X in Supplement_to_Chapter_1.pdf in PDF Files folder on Web Site
C3S.X	Section number X in Supplement_to_Chapter_3.pdf in PDF Files folder on Web Site
C7S.X	Section number X in Supplement_to_Chapter_7.pdf in PDF Files folder on Web Site
C9S.X	Section number X in Supplement_to_Chapter_9.pdf in PDF Files folder on Web Site
C12S.X	Section number X in Supplement_to_Chapter_12.pdf in PDF Files folder on Web Site
C16S.X	Section number X in Supplement_to_Chapter_16.pdf in PDF Files folder on Web Site
C22S.X	Section number X in Supplement_to_Chapter_22.pdf in PDF Files folder on Web Site
C23S.X	Section number X in Supplement_to_Chapter_23.pdf in PDF Files folder on Web Site
C25S.X	Section number X in Supplement_to_Chapter_25.pdf in PDF Files folder on Web Site
AIIS-X to Y	Page X to Y in Supplement_to_Appendix_II.pdf in PDF Files folder on Web Site
PDF-GAMS	In GAMS.pdf Iin PDF Files folder on Web Site
PDF-IPE	In Aspen IPE Course Notes.pdf in PDF Files folder on Web Site
MMM-ASPEN	Refers to the multimeda ASPEN menu on the Web Site
MMM-HYSYS	Refers to the multimeda HYSYS menu on the Web Site
PSF-ASPEN	Refers to the Aspen Eng. Suite folder in Program and Simulation Files folder on the Web Site
PSF-HYSYS	Refers to the HYSYS folder in Program and Simulation Files folder on the Web Site
PSF-MATLAB	Refers to the MATLAB folder in the Program and Simulation Files folder on the Web Site